INSECT ECOLOGY

INSECT ECOLOGY

AN ECOSYSTEM APPROACH

FOURTH EDITION

TIMOTHY D. SCHOWALTER

Entomology Department
Louisiana State University
Baton Rouge, LA, United States

AMSTERDAM • BOSTON • HEIDELBERG • LONDON
NEW YORK • OXFORD • PARIS • SAN DIEGO
SAN FRANCISCO • SINGAPORE • SYDNEY • TOKYO

Academic Press is an imprint of Elsevier

Academic Press is an imprint of Elsevier
125 London Wall, London EC2Y 5AS, United Kingdom
525 B Street, Suite 1800, San Diego, CA 92101-4495, United States
50 Hampshire Street, 5th Floor, Cambridge, MA 02139, United States
The Boulevard, Langford Lane, Kidlington, Oxford OX5 1GB, United Kingdom

Library of Congress Cataloging-in-Publication Data
A catalog record for this book is available from the Library of Congress

British Library Cataloguing-in-Publication Data
A catalogue record for this book is available from the British Library

ISBN: 978-0-12-803033-2

For information on all Academic Press publications
visit our website at https://www.elsevier.com/

  Working together
to grow libraries in
developing countries

ELSEVIER   Book Aid International

www.elsevier.com • www.bookaid.org

Publisher: Sara Tenney
Acquisition Editor: Kristi Gomez
Editorial Project Manager: Pat Gonzalez
Production Project Manager: Lucía Pérez
Designer: Matthew Limbert

Typeset by Thomson Digital

Contents

II
POPULATION ECOLOGY

5. Population Systems

6. Population Dynamics

7. Biogeography

III
COMMUNITY ECOLOGY

8. Species Interactions

9. Community Structure

IV

ECOSYSTEM LEVEL

11. Ecosystem Structure and Function

12. Herbivory

13. Pollination, Seed Predation, and Seed Dispersal

14. Decomposition and Pedogenesis

15. Insects as Regulators of Ecosystem Processes

V

APPLICATIONS AND SYNTHESIS

16. Application to Sustainability of Ecosystem Services

17. Management of Insect Populations

18. Summary and Synthesis

Preface

This fourth edition provides an updated and expanded synthesis of feedbacks and interactions between insects and their environment and adds expanded chapters on insect roles in sustainability of ecosystem services and on pest management and conservation. Some of the most exciting recent advances are (1) evaluation of fitness costs of species interactions, (2) new studies on indirect interactions among species and their effects on communities and ecosystems, (3) identification of specific genes selected by feedback between insect phenotype and environmental conditions, and (4) increased attention to insect contributions to ecosystem services on which human survival depends. Although other insect ecology textbooks continue to focus on population and community levels, this book includes seven chapters on the effects of insects on ecosystem structure and function. A new Chapter 16 expands the application of insect ecology to effects on ecosystem services (such as food and fiber supply, water yield and quality, primary production, pollination, and soil formation) on which humans depend. A new Chapter 17 expands application of insect ecology to pest management and insect conservation.

Substantial advances have been made in understanding linkages between insect populations and ecosystem conditions since the first edition of this book in 2000. Advances in mechanisms by which insects affect, and potentially regulate, ecosystem structure and function are particularly important to development of ecosystem and pest management policies and practices that protect, rather than threaten, the sustainable delivery of ecosystem services.

Ecology, especially ecosystem ecology, necessarily requires a multidisciplinary approach that involves not only biological disciplines, including molecular and evolutionary biology, but also chemists, geologists, climatologists, hydrologists, soil scientists, geographers, and mathematicians, to fully understand and manage the complexity of interactions among organisms and their environment. Despite their small size, insects have demonstrated capacity to regulate ecosystem processes that control local-to-global environment. However, much of what we know about insect ecology has been contributed by studies of economically or medically important species. We know relatively little about the 6–10 million species that do not interfere with human interests and how these species may respond to changing environmental conditions.

This text provides the data base for predicting and managing insect population dynamics and their effects on ecosystem services. As we become increasingly aware that global changes must be addressed from a global (rather than local) perspective, we need models with greater integration of ecological processes at various levels of resolution and across regional landscapes. Insect population structure, interactions with other species, and effects on ecosystem processes are integral to explaining and mitigating global changes. Furthermore, increasing recognition that insects have various short- and long-term effects on multiple ecosystem services requires a shift in approach from traditional crop "protection" to integration of compensatory benefits for sustained yield of ecosystem services. Integrated pest management (IPM) is founded on such ecological principles.

The hierarchical organization of this text highlights interactions and feedbacks among individual, population, community, and ecosystem levels of organization. This arrangement illustrates how models at the global or ecosystem levels must integrate submodels at the community level, which in turn must integrate multiple population submodels, and so on. This arrangement also contributes to integration of evolutionary and ecosystem approaches by illustrating how conditions at higher levels of resolution (eg, the community or ecosystem) contribute to the environment perceived at lower levels (eg, populations and individuals) and how responses at these lower levels contribute to conditions at higher levels of this hierarchy. An alternative organization by major topics of current concern, such as evolution, biodiversity, insect–plant interactions, omnivory, biological control, and climate change, has limited potential to integrate the fundamental feedbacks that are necessary to address these topics.

A useful textbook must balance coverage with brevity. Some overlap among sections and chapters is necessary to emphasize linkages among levels. Where possible, overlap is minimized through cross-referencing. A guiding principle for this book has been emphasis on insect responses to, and effects on, ecosystem structure and function, including how individual insect phenotypes can influence ecosystem conditions, as well as reflect evolutionary selection by ecosystem conditions. Evolution is emphasized in earlier chapters dealing with individual and population responses to environmental conditions whereas later chapters emphasize the ways in which in-

sects affect the ecosystem conditions to which they also respond.

Science advances incrementally, requiring sequential editions of this and other textbooks to incorporate new material. It also is important to note that all of us have stood on the shoulders of giants to reach our current state of knowledge. Despite the temptation to focus on the most recent literature and examples, I have included references from early work to show development, or in some cases rediscovery, of important concepts.

A number of colleagues have contributed enormously to the concepts developed in this book. I am especially grateful to N.V. Brokaw, J.T. Callahan, J.-T. Chao, S.L. Collins, R.N. Coulson, D.A. Crossley, Jr, R.F. Dame, D.A. Distler, L.R. Fox, J.F. Franklin, F.B. Golley, G. González, J.R. Gosz, V.P. Gutschick, C.M. Hammond, S.M. Heuberger, M.D. Hunter, F. Kozár, M.D. Lowman, G.L. Lovett, H.-K. Luh, J.C. Moore, M.P. North, E.P. Odum, H.T. Odum, G.T. Reagan, J.O. Schmidt, T.R. Seastedt, D.J. Shure, M.J. Stout, T. Tscharntke, P. Turchin, R.B. Waide, D.C. Weber, W.G. Whitford, R.G. Wiegert, M.R. Willig, W.-J. Wu, and J.K. Zimmerman for sharing ideas, data, and encouragement. A. Covich, L.R. Fox, T.R. Seastedt, D. Simberloff, T. Tscharntke, and M.R. Willig reviewed drafts of previous editions. I also have benefited from collaboration with colleagues at Louisiana State University and Oregon State University and associated with US Long Term Ecological Research (LTER) sites, International LTER projects in Hungary and Taiwan, the Smithsonian Tropical Research Institute, Wind River Canopy Crane Research Facility, USDA Forest Service Pacific Northwest, Pacific Southwest and Southern Research Stations, USDA Forest Service Teakettle Ecosystem Experiment, USDA Forest Service Demonstration of Ecosystem Management Options (DEMO) Project, USDA Western Regional Project on Bark Beetle–Pathogen Interactions, USDA National

Formosan Subterranean Termite Program, and the National Science Foundation. Several anonymous reviewers provided useful comments. I also am indebted to C. Schowalter for encouragement and feedback. K. Gomez, P. Gonzalez and L. Pérez at Elsevier provided valuable editorial assistance. I am, of course, solely responsible for the selection and organization of material in this book.

Timothy D. Schowalter
Entomology Department, Louisiana State
University, Baton Rouge, LA, United States

Overview

Insects are the most taxonomically diverse group of organisms on Earth, representing >50% of all described species (Fig. 1.1, E. Wilson, 1992). Beetles alone represent 25% of all known species. Insects represent the vast majority of species in terrestrial and freshwater ecosystems and are important components of near-shore marine ecosystems as well.

Approximately 1 million species of insects have been described, and estimates of the total number of species are in the range of 4–6 million (Novotný et al., 2002). This diversity of insect species represents an equivalent variety of adaptations to variable environmental conditions. In fact, insects represent the widest range of ecological strategies, spanning sessile species (such as scale insects), with life history strategies similar to plants, to social, cognitive species (such as honey bees), with life history strategies as complex as those of many vertebrates, as well as strategies encompassing both aquatic and terrestrial attributes. Many species responses are complementary, that is, some species increase while others decrease in response to any particular environmental change, ensuring continuity of representation and ecological function. Insect responses to environmental changes affect other species (including humans) and ecosystem conditions in a variety of ways. The diversity of insect species and responses represents a major challenge to insect ecologists, who have explored the responses or functions of only a very small percentage of insect species.

Insects also play critical roles in ecosystem structure and function. Their biomass often equals or exceeds that of more conspicuous vertebrates in terrestrial and freshwater ecosystems (Schowalter, 2013). They represent important food resources, predators, parasites, or disease vectors for many other organisms, including humans, and they have the capacity to alter rates and directions of energy and matter fluxes (eg, as herbivores, pollinators, detritivores, and predators) in ways that affect global conditions. In some ecosystems, insects and other arthropods represent the dominant pathways of energy and matter flow. Some species are capable of removing virtually all vegetation from a site. They affect, and are affected by, environmental issues as diverse as ecosystem health, air and water quality, food production, genetically modified crops, disease epidemiology, frequency and severity of fire and other disturbances, control of invasive exotic species, land use, water and air pollution, and climate change. The rapid change in frequencies of particular genes in response to changing environmental conditions has provided some of the best confirmation of evolutionary principles.

Adaptation and explosive population growth in response to environmental changes, especially those resulting from anthropogenic activities, have the capacity to exacerbate or

Insect Ecology. http://dx.doi.org/10.1016/B978-0-12-803033-2.00001-7

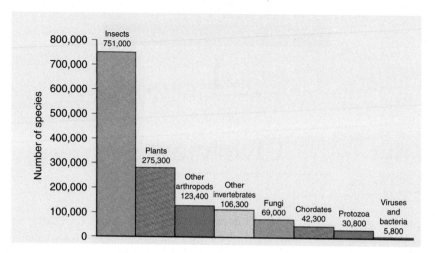

FIGURE 1.1 **Distribution of described species within major taxonomic groups.** Species numbers for insects, bacteria, and fungi likely will increase greatly as these groups become better known. *Source: Data from Wilson, E., 1992. The Diversity of Life. Harvard University Press, Cambridge, MA.*

mitigate changes in ecosystem conditions and, perhaps, global processes. Next to humans, insects have the greatest capacity to alter ecosystems over large areas, functioning as major engineers and potential regulators of ecosystem conditions. This capacity frequently brings them into conflict with human demands for ecosystem resources. On the other hand, efforts to control insects often have unintended and/or undesirable consequences for nontarget species and the sustainability of ecosystem services, such as food production and fresh water supply. Clearly, understanding insect ecology, especially the triggers for explosive population growth, is critical for effectively managing insects and ecosystem services.

A primary challenge for insect ecologists has been to place insect ecology in an ecosystem context, that represents insect effects on ecosystem structure and function, as well as the diversity of their adaptations and responses to changes in environmental conditions (Fig. 1.2). Ironically, some of the earliest entomological literature, establishing the roots of insect ecology, revealed understanding of plant chemical defenses, the role of plant diversity and predation in regulating insect populations, and the stimulation of crop production in the wake of locust outbreaks, as a result of the fertilization effect of locust feces and carcasses and reduced

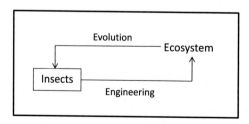

FIGURE 1.2 **Conceptual model that represents the natural selection of individuals by changing ecosystem conditions (evolution) and the effect of insects on ecosystem conditions (engineering).** Both processes are critical to understanding interactions between insects and their environment.

abundances of other crop-feeding insects (Howard, 1896; Packard, 1877; C. Riley, 1883). Insect ecologists have tended to focus on the evolutionary significance of life history strategies and interactions with other species, especially as pollinators, herbivores, and predators (Price, 1997; Price et al., 2011; Speight et al., 2008). This focus has yielded much valuable information about the ecology of individual species, species associations, and responses to environmental changes, demonstrated the function of particular genes, and provided the basis for pest management or recovery of threatened and endangered species. However, relatively little attention has been given to the important role of insects as ecosystem engineers, other than to their apparently negative effects on vegetation (especially commercial crop) or animal (especially human and livestock) production.

Ecosystem ecology has advanced rapidly during the past 60 years. Major strides have been made in understanding how species interactions and environmental conditions affect rates of energy and nutrient fluxes in different ecosystem types and how these processes regulate atmospheric chemistry and global climate and provide free ecosystem services (Chapman et al., 2003; Classen et al., 2005; Frost and Hunter, 2007; Lowman et al., 2012; Milcu et al., 2015; Schowalter, 2013; Whitham et al., 2006). Interpreting the responses of a diverse community to multiple interacting environmental factors in integrated ecosystems requires new approaches, such as multivariate statistical analysis and modeling approaches (Gutierrez, 1996; Liebhold et al., 1993; Lowman et al., 2012; Marcot et al., 2001). Such approaches often involve loss of detail, such as combination of species into phylogenetic or functional groupings. However, an ecosystem approach provides a framework for integrating insect ecology with variation in ecosystem structure and function and for applying insect ecology to the understanding of mechanisms underlying global processes, such as climate change or sustainability of ecosystem services. Unfortunately, few ecosystem studies have involved insect ecologists and, therefore, on the whole they have tended to underrepresent insect responses and contributions to ecosystem changes.

1 SCOPE OF INSECT ECOLOGY

Insect ecology is the study of interactions between insects and their environment, including other species and abiotic conditions. Ecology is necessarily multidisciplinary and integrative, requiring the contributions of biologists, chemists, geologists, climatologists, hydrologists, soil scientists, geographers, mathematicians, and others, to fully understand the complex interactions among organisms and their environment. Insect ecology requires integration of subdisciplines from the level of molecular/genomic changes in response to environmental selection to the level of global climate change in response to large-scale defoliation by insects.

Fig. 1.2 provides the conceptual framework for this book. Insect ecology integrates insect responses to changing environmental conditions and insect effects on environmental (ecosystem) conditions. These two components represent a feedback loop. The ecosystem constitutes the environment that selects for adapted phenotypes among all individuals. Such natural selection represents evolutionary feedback by the ecosystem. Despite their small size, insects have demonstrated capacity to alter ecosystem conditions (Milcu et al., 2015; Schowalter et al., 2011; Whitham et al., 2006), constituting feedback to the ecosystem by adapted individuals. This engineering effect is subject to further evolutionary feedback, as individuals, populations, and communities that interact to stabilize ecosystem conditions would be favored over those that fail to stabilize ecosystem conditions.

Insect ecology has both basic and applied goals. Basic goals are to improve our understanding and ability to model interactions and feedbacks (Price, 1997; Schowalter, 2013). Applied goals are to evaluate and manage the extent to which insect responses to environmental changes, including those resulting from anthropogenic activities, mitigate or exacerbate ecosystem change (Croft and Gutierrez, 1991; Kogan, 1998; Schowalter, 2012), especially in managed ecosystems. Some of the earliest and most valuable data on insect ecology has been contributed from studies designed to address factors affecting the population growth of "pests" (Howard, 1896; Packard, 1877; C. Riley, 1878, 1880, 1883, 1885, 1893).

Research on insects and associated arthropods (eg, spiders, mites, centipedes, millipedes, crustaceans) has been critical to development of the fundamental principles of ecology, such as evolution of social organization (Haldane, 1932; W. Hamilton, 1964; E. Wilson, 1973), population dynamics (Coulson, 1979; Morris, 1969; Nicholson, 1958; Varley and Gradwell, 1970; Varley et al., 1973; Wellington et al., 1975), competition (Park, 1948, 1954), plant–herbivore interaction (I. Baldwin and Schultz, 1983; Feeny, 1969; Fraenkel, 1953; Rosenthal and Janzen, 1979), predator–prey interaction (Nicholson and Bailey, 1935), mutualism (Batra, 1966; Bronstein, 1998; Janzen, 1966; Morgan, 1968; Rickson, 1971, 1977), island biogeography (Darlington, 1943; MacArthur and Wilson, 1967; Simberloff, 1969, 1978), metapopulation ecology (Hanski, 1989), heterotrophic succession (Howden and Vogt, 1951; Payne, 1965; Savely, 1939), and regulation of ecosystem processes, such as primary productivity, nutrient cycling, and succession (Crossley, 1966; Crossley and Howden, 1961; Mattson and Addy, 1975; J.C. Moore et al., 1988; Schowalter, 1981; Seastedt, 1984; Smalley, 1960). Insects and other arthropods are small and easily manipulated subjects. Their rapid numerical responses to environmental changes facilitate statistical discrimination of responses and make them particularly useful models for experimental study.

Insects fill a variety of important ecological (functional) roles and affect virtually all ecosystem services. Many species are key pollinators. Pollinators and plants have adapted a variety of mechanisms for ensuring transfer of pollen, especially in tropical ecosystems where sparse distributions of many plant species require a high degree of pollinator fidelity to ensure pollination among conspecific plants (Feinsinger, 1983). Virtually all fruits and vegetables produced for human consumption require pollination by insects (Klein et al., 2007). As a consequence, pollinating insects can be viewed as a critical element in the reproduction and recruitment of plants at the ecosystem scale. Other species are important agents for dispersal of plant seeds, fungal spores, bacteria, viruses, or other invertebrates (J. Moser, 1985; Nault and Ammar, 1989; Sallabanks and Courtney, 1992). Herbivorous species are particularly well-known as agricultural and forestry "pests," but their ecological roles are far more complex, often stimulating plant growth, affecting water and nutrient fluxes, and altering the rate and direction of ecological succession (Frost and Hunter, 2007; Maschinski and Whitham, 1989; Mattson and Addy, 1975; Schowalter and Lowman, 1999; Schowalter et al., 2011; Trumble et al., 1993; Whitham et al., 2006). Insects and associated arthropods are instrumental in processing of organic detritus in terrestrial and aquatic ecosystems and influence soil fertility and water quality (Coleman et al., 2004; Kitchell et al., 1979; Seastedt and Crossley, 1984). Dung beetles and termites are particularly important in removal of livestock dung and maintenance of pasture productivity (Coe, 1977; Herrick and Lal, 1996; Whitford, 1986; Whitford et al., 1982;

Yamada et al., 2007). Woody litter decomposition typically is delayed until insects penetrate the bark barrier and inoculate the wood with saprophytic fungi and other microorganisms (Ausmus, 1977; Dowding, 1984; Swift, 1977). Insects are important food resources for a variety of fish, amphibians, reptiles, birds, and mammals, as well as other invertebrate predators and parasites (J. Allan et al., 2003; Baxter et al., 2005). Humans have used insects or their products for food and for medical and industrial products (Anelli and Prischmann-Voldseth, 2009; Namba et al., 1988; Ramos-Elorduy, 2009). In addition, some insects are important vectors of plant and animal diseases, including malaria, plague, and typhus, that have affected human populations and military campaigns (Amoo et al., 1993; Diamond, 1999; Edman, 2000; Marra et al., 2004; R. Peterson, 1995; Stapp et al., 2004; Steelman, 1976; J. Zhou et al., 2002).

The significant economic and public health importance of many insect species has justified the distinct entomology programs in land-grant universities and government agencies. Damage to agricultural crops and transmission of human and livestock diseases have stimulated interest in, and support for, the study of factors influencing the abundance and effects of targeted insect species. Much of this research has focused on evolution of life history strategies, orientation to host cues, interaction with host chemistry, and predator–prey interactions as these contribute to our understanding of "pest" population dynamics, especially population regulation by biotic and abiotic factors. However, failure to understand these aspects of insect ecology within an ecosystem context undermines our ability to predict and manage insect populations and ecosystem services effectively, especially with respect to changes in land use and sustainability of pollination, water yield, and soil fertility (Kogan, 1998; Millennium Ecosystem Assessment, 2005). Suppression efforts may be counterproductive to the extent that insect outbreaks represent ecosystem-level regulation of primary production or other critical processes.

2 ECOSYSTEM ECOLOGY

The ecosystem is a fundamental unit of ecological organization, although its boundaries are not easily defined. An ecosystem generally is considered to represent the integration of a more or less discrete community of organisms and the abiotic conditions at a site (Fig. 1.3). However, research and environmental policy decisions are recognizing the importance of scale in ecosystem studies, that is, extending research or extrapolating results to landscape, regional, and even global scales (Holling, 1992; Turner, 1989; Lowman et al., 2012). Ecosystems are interconnected, just as the species within them are interconnected. Exports from one ecosystem become imports for others (Baxter et al., 2005; Dreyer et al., 2015) (Fig. 1.4). Energy, water, organic matter, and nutrients from terrestrial ecosystems are major sources of these resources for many aquatic ecosystems. Organic matter and nutrients eroded by wind from arid ecosystems are filtered from the airstream by ecosystems downwind. Some ecosystems within a landscape or watershed are the sources of colonists for recently disturbed ecosystems. Insect outbreaks often spread from one ecosystem to another over large areas. Toxic or exogenous materials introduced into some ecosystems can adversely affect remote ecosystems, for example, agricultural chemicals causing hypoxic (dead) zones in coastal waters (Howarth et al., 2011; Krug, 2007). Therefore, our perspective of the ecosystem needs to incorporate the concept of interactions among ecosystem types (patches) within the landscape or watershed and how these interactions control global conditions.

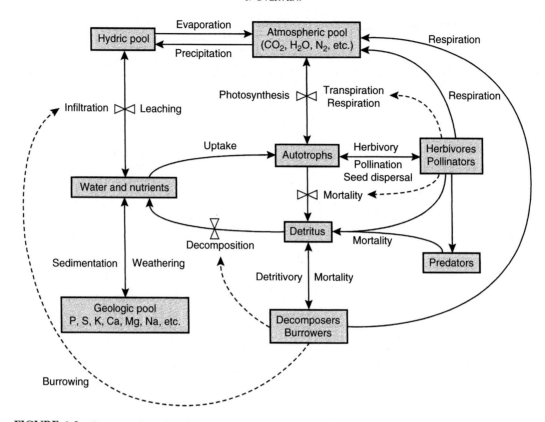

FIGURE 1.3 **Conceptual model of ecosystem structure and function.** Boxes represent storage compartments, lines represent fluxes, and hourglasses represent regulation. Solid lines are direct transfers of energy and matter, and dashed lines are informational or regulatory pathways.

Overlapping gradients in abiotic conditions establish the template that limits options for community development, but established communities can modify abiotic conditions to varying degrees. For example, minimum rates of water and nutrient supply are necessary for establishment of grasslands or forests, but once canopy cover and water and nutrient storage capacity in organic material have developed, the ecosystem is relatively buffered against changes in water and nutrient supply (J. Foley et al., 2003a; E. Odum, 1969; Webster et al., 1975). Although ecosystems typically are defined on the basis of the dominant vegetation (eg, tundra, desert, marsh, grassland, forest) or type of water body (stream, pond, lake), characteristic insect assemblages also differ among ecosystems. For example, freeze-tolerant species characterize high-latitude and high-elevation ecosystems, whereas wood-boring insects (eg, ambrosia beetles, wood wasps) are characteristic of communities in wooded ecosystems (shrub and forest ecosystems). The latter clearly could not survive in ecosystems lacking woody resources. The perspective of ecosystems represented in this text emphasizes three attributes that are fundamental to ecosystems: complexity, hierarchical organization, and self-regulation of structure and function.

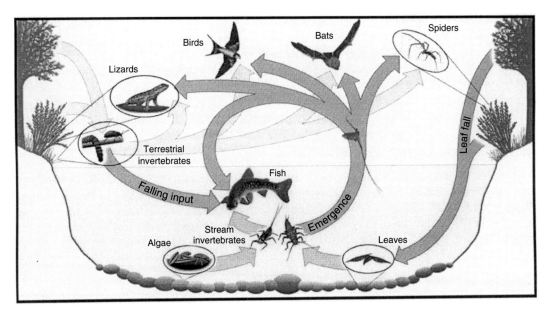

FIGURE 1.4 **Diagram of exchange of aquatic and terrestrial invertebrate prey and plant material that have direct and indirect effects in stream and riparian ecosystem food webs.** *Source: From Baxter, C.V, et al., 2005. Tangled webs: reciprocal flows of invertebrate prey link streams and riparian zones. Freshwater Biol. 50, 201–220.*

2.1 Ecosystem Complexity

Ecosystems are complex systems with *structure*, represented by abiotic resources and a diverse assemblage of component species and their products (such as organic detritus and tunnels), and *function*, represented by fluxes of energy and matter among biotic and abiotic components (Fig. 1.3). Heterogeneous distribution of environmental conditions, resources, and organisms is a fundamental ecological property (Scheiner and Willig, 2008) that controls individual foraging and dispersal strategies, patterns of population density, interactions with other species populations, and resulting patterns of energy and biogeochemical fluxes. Ecosystems can be identified at micro- and mesoscales (eg, decomposing logs or treehole pools), patch scale (area encompassing a particular community type on the landscape), landscape scale (the mosaic of patch types representing different edaphic conditions or successional stages that compose a broader ecosystem type), regional or biome scale, continental scale, and global scale. Furthermore, ecosystems change over time as populations appear or disappear, changing community and ecosystem structure and function.

Addressing taxonomic, temporal, and spatial complexity has proven a daunting challenge to ecologists, who must decide how much complexity can be ignored safely in developing predictive models (Gutierrez, 1996; Polis, 1991a,b). Evolutionary and ecosystem ecologists have taken contrasting approaches to dealing with complexity in ecological studies. The evolutionary approach emphasizes adaptive aspects of life histories, population dynamics, and species interactions. This approach restricts complexity to interactions among one or a few species and their hosts, competitors, predators, or other biotic and abiotic environmental factors and often ignores the complex direct and indirect feedbacks at the ecosystem level. On the other hand, the ecosystem approach emphasizes rates and directions of energy and matter fluxes and community control of climate and soil development. This approach restricts complexity to fluxes among broad taxonomic or functional groups and often ignores the contributions of individual species. Either approach, by itself, limits our ability to understand feedbacks among individual, population, community, and ecosystem parameters and to predict the effects of changing communities or changing global environment on these feedbacks.

2.2 The Hierarchy of Subsystems

Complex systems with feedback mechanisms can be partitioned into component subsystems, which are composed of subsubsystems. Viewing the ecosystem as a nested hierarchy of subsystems (Table 1.1), each with its particular properties and processes (Coulson and Crossley, 1987; Kogan, 1998; O'Neill et al., 1986), facilitates understanding and modeling of complexity.

TABLE 1.1   Ecological Hierarchy and the Structural and Functional Properties Characterizing Each Level

Ecological level	Structure	Function
Global	Biome distribution	Gas, water, nutrient exchange between terrestrial and marine systems
	Atmospheric condition	
	Climate	Total NPP
	Sea level	
Biome	Landscape pattern	Energy and matter fluxes
	Temperature, moisture profile	Integrated NPP of ecosystems
	Disturbance regime	Migration
Landscape	Disturbance pattern	Energy and matter fluxes
	Community distribution	Integrated NPP of ecosystems
	Metapopulation structure	Colonization and extinction
Ecosystem	Vertical and horizontal structure	Energy and matter fluxes
	Disturbance type and frequency	Succession NPP, herbivory, decomposition, pedogenesis
	Biomass	
	Functional organization	
Community	Diversity	Species interactions
	Trophic organization	Temporal and spatial changes
Population	Density	Natality
	Dispersion	Mortality
	Age structure	Dispersal
	Genetic structure	Gene flow
		Temporal and spatial changes
Individual	Anatomy	Physiology/learning/behavior
	Genome	Resource acquisition and allocation

The community is the biotic component of the ecosystem and often is composed of component communities. Component communities are more or less discrete assemblages of organisms based on particular resources within the overall community. For example, the relatively distinct soil faunas associated with fungal, bacterial, or plant root resources represent different component communities within grassland ecosystems (J.C. Moore and Hunt, 1988). The relatively distinct floras and faunas associated with canopies of different tree species represent distinct component communities within forest ecosystems (Rambo et al., 2014; Schowalter and Ganio, 2003). The distinct invertebrate assemblages associated with riffle and pool habitats represent distinct component communities within stream ecosystems (Huryn and Wallace, 1987). Communities are composed of multiple species populations, with varying strategies for acquiring and allocating resources. Species populations, in turn, are composed of individual organisms that vary in their physiology and behavior.

Each level of the hierarchy can be studied at an appropriate level of detail and its properties explained by the integration of its subsystems. For example, population responses to changing environmental conditions reflect the net phenotypic (physiological and behavioral) responses of individuals that determine their survival and reproduction. Changes in community structure reflect the dynamics and interactions of component populations. Fluxes of energy and matter through the ecosystem reflect food web structure and community organization. Landscape structure reflects ecosystem processes that affect movement of individuals or materials. Hence, the integration of structure and function at each level determines properties at higher levels.

At the same time, the conditions produced at each level establish the context, or template, for responses at lower levels. Population structure resulting from individual survival, dispersal, and reproduction determines future survival, dispersal, and reproduction of individuals. For example, crowding at high population densities reduces reproduction and encourages dispersal (Chapter 5). Ecosystem conditions resulting from community interactions affect the distribution of resources and individuals and the subsequent behavior of individual organisms, populations, and the community. Recognition of feedbacks from higher hierarchical levels has led to developing concepts of inclusive fitness (fitness accruing through feedback from benefit to a group of organisms) and ecosystem self-regulation (see Chapter 15). The hypothesis that insects function as cybernetic regulators that can stabilize ecosystem properties (M.D. Hunter, 2001b; Mattson and Addy, 1975; Schowalter, 1981) has been one of the most important and controversial concepts to emerge from insect ecology, in terms of its implications for perceptions and management of insects and ecosystems.

Spatial and temporal scales vary across levels of this hierarchy. Whereas individual physiology and behavior operate on small scales of space and time (ie, limited to the home range and life span of the individual), population dynamics span landscape and decadal scales, and ecosystem processes, such as patterns of resource turnover, recovery from disturbance, or contributions to atmospheric carbon, operate at scales from the patch to the biome and from decades to millennia.

Modeling approaches have greatly facilitated understanding of the complexity and consequences of interactions and linkages within and among these organizational levels of ecosystems. The most significant challenges to ecosystem modelers remain (1) the integration of appropriately detailed submodels at each level to improve prediction of causes and consequences of environmental changes and (2) the evaluation of contributions of various taxa (including particular insect species) or functional groups to ecosystem structure and function. Some species or structures have effects that are disproportionate to their abundance or biomass (ie, keystone species). Studies focused on the most abundant or conspicuous species or structures fail to address the substantial contributions of rare or inconspicuous components, such as many insects, that can affect ecosystem conditions to an unexpected degree.

2.3 Regulation

An important aspect of this functional hierarchy is the "emergence" of properties that are not easily predictable by simply adding the contributions of constitutive components. Emergent properties include feedback processes at each level of the hierarchy. For example, individual organisms acquire and allocate energy and biochemical resources, affecting resource availability and population structure in ways that change the environment and determine future options for acquisition and allocation of these resources. Regulation of density and resource use emerges at the population level through negative feedback, as a result of declining resource availability and increasing predation as populations grow, that functions to prevent overexploitation and/or through positive feedback that prevents extinction (see Chapter 5). Similarly, species populations acquire, transport, and exchange resources (through feeding interactions), but regulation of energy flow and biogeochemical cycling emerge at the ecosystem level. Potential regulation of atmospheric and oceanic pools of carbon and nutrients at the global level reflects integration of biogeochemical cycling and energy fluxes among the Earth's ecosystems, for example, sequestration of excess atmospheric carbon from wildfire or fossil fuel combustion in wood (in forests) or calcium carbonate (in reefs).

Information flow and feedback processes are the mechanisms of regulation. Although much research has addressed energy and material flow through food webs, relatively little research has quantified the importance of indirect interactions or information flow. Indirect interactions and feedbacks are ubiquitous features of ecosystems. For example, herbivores feeding on foliage or shoots alter the availability of resources for root-feeding organisms (Gehring and Whitham, 1991, 1995; Masters et al., 1993); early-season herbivory can affect plant suitability for later-season herbivores (Harrison and Karban, 1986; M.D. Hunter, 1987). Information can be transmitted as volatile compounds that advertise the location and physiological condition of prey, the proximity of potential mates, and the population status of predators (Bruinsma and Dicke, 2008; Kessler and Baldwin, 2001; Turlings et al., 1995). Such information exchange is critical to discovery of suitable hosts, attraction of mates, regulation of population density, and avoidance of or defense against predators by many (if not all) insects.

This ecosystem information network among the members of the community, along with resource supply/demand relationships, provides the basis for regulation of ecosystem processes. Levels of herbivory and predation are sensitive to resource availability. If environmental conditions increase resource abundance at any trophic level, communication to, and response by, the next trophic level provides negative feedback that reduces resource abundance. Negative feedback is a primary mechanism for limiting population sizes, species interactions, and process rates in ecosystems. Some interactions provide positive feedback, such as cooperation or mutualism. Although positive feedback is potentially destabilizing, it may reduce the probability of population decline to extinction. The apparent ability of many ecosystems to reduce variation in structure and function suggests that ecosystems are self-regulating, that is, they behave like cybernetic systems (Kratz et al., 1995; E. Odum, 1969; B.

Patten and Odum, 1981). Insects could be viewed as important mechanisms of regulation because their normally small biomass requires relatively little energy or matter to maintain, but their rapid and dramatic population response to environmental changes (especially host stress or overabundance) constitutes an effective and efficient means for limiting host density and primary production, thereby reducing deviation in nominal ecosystem structure and function (see Chapter 15). This developing concept of ecosystem self-regulation has major implications for ecosystem responses to anthropogenic change in environmental conditions and for our approaches to managing insects and ecosystem resources.

3 ENVIRONMENTAL CHANGE AND DISTURBANCE

Environmental changes across temporal and spatial gradients are critical components of an ecosystem approach to insect ecology. Insects are highly responsive to environmental changes, including those resulting from anthropogenic activity. Many insects have considerable capacity for long-distance dispersal, enabling them to find and colonize isolated resources as these appear. Other insects are flightless and vulnerable to environmental change or habitat fragmentation. Because of their small size, short life spans, and high reproductive rates, abundances of many species can change by several orders of magnitude on a seasonal or annual timescale, minimizing time lags between environmental changes and population adjustment to new conditions. Such changes are easily detectable and make insects more useful as indicators of environmental changes than are larger or longer-lived organisms that respond more slowly. In turn, insect responses to environmental change can affect ecosystem structure and function dramatically. Some phytophagous species are well-known for their ability, at high population levels triggered by host stress and/or abundance, to reduce host plant density and productivity greatly over large areas. Effects of other species may be more subtle but equally significant from the standpoint of long-term ecosystem structure and function.

Environmental change operates on a continuum of spatial and temporal scales. Although strict definitions of environmental change and disturbance have proven problematic, environmental change generally occurs over a longer term, whereas disturbances are acute, short-term events (Walker and Willig, 1999; P. White and Pickett, 1985). Chronic changes in temperature or precipitation patterns, such as following the last glaciation, occur on a scale of 10^3–10^5 years and may be barely detectable on human timescales. Long-term changes may be difficult to distinguish from cycles operating over decades or centuries. Acute events, such as fires or storms, are more recognizable as disturbances that have dramatic effects on time scales of seconds to hours. However, the duration at which a severe drought, for example, is considered a climate change, rather than a disturbance, has not been determined. Anthropogenic activities often function as disturbances because of the speed with which they alter ecosystem conditions. The combination of climate and geological patterns, disturbances, and environmental changes creates a constantly shifting landscape mosaic of various habitat and resource patches that determine where and how insects and other organisms find suitable conditions and resources.

Insect outbreaks traditionally have been viewed as disturbances (P. White and Pickett, 1985; Walker and Willig, 1999). P. White and Pickett (1985) defined disturbance as any relatively discrete event in time that causes measurable change in population, community, or ecosystem structure or function. This definition clearly incorporates insect outbreaks. Similarly, human activities have become increasingly prominent agents of disturbance and environmental change.

Insect outbreaks are comparable to physical disturbances in terms of severity, frequency and scale. Insects can defoliate or kill most host plants over large areas, up to 10^3–10^6 ha (Furniss and Carolin, 1977). For example, 39% of a montane forest landscape in Colorado has been affected by insect outbreaks (spruce beetle, *Dendroctonus rufipennis*) since about 1633, compared to 59% by fire and 9% by snow avalanches (Veblen et al., 1994), with an average return interval of 117 years, compared to 202 years for fire. Frequent, especially cyclic, outbreaks of herbivorous insects probably have been important in selection for plant adaptations, such as chemical defenses (see Chapter 3).

However, unlike abiotic disturbances, insect outbreaks are biotic responses to a change in environmental conditions. Recent outbreaks most commonly reflect anthropogenic redistribution of resources, especially increased density of commercially valuable (often exotic) plant species, and exotic insect species. Outbreaks typically develop in dense patches of host plants and function to reduce host density, increase vegetation diversity, increase water and nutrient availability, and prevent primary productivity from exceeding the carrying capacity (a maximum population size that can be sustained indefinitely by prevailing conditions) of the ecosystem (Cairns et al., 2008; T. Coleman et al., 2008; Schowalter and Turchin, 1993). Management responses to insect outbreaks often are more damaging to ecosystem conditions than is the insect outbreak. For example, insecticides, such as arsenicals and chlorinated hydrocarbons, had long-term, nonselective effects on nontarget organisms. Removing dead or dying host plants, and even living plants, in advance of insect colonization, has caused serious soil disturbance and erosion, as well as change in community structure. Principles of integrated pest management (IPM) improved approaches to managing insects by emphasizing adherence to ecological principles (see Chapter 17). Recognizing insects as integral components of potentially self-maintaining ecosystems could further improve our management of insects and ecosystem resources, within the context of global change.

Currently, human alteration of Earth's ecosystems is substantial and accelerating (Burney and Flannery, 2005; J. Thomas et al., 2004; Vitousek et al., 1997). Anthropogenic changes to the global environment affect insects in various ways. Combustion of fossil fuels has elevated atmospheric concentrations of CO_2 (Beedlow et al., 2004; Keeling et al., 1995), methane, ozone, nitrous oxides, and sulfur dioxide, leading to increasingly acidic precipitation and global warming. Petrochemical leaks and spills are toxic to most organisms and prevent oxygen exchange between aquatic ecosystems and the atmosphere. Insecticide pollution is pervasive, even in remote Arctic and Antarctic ecosystems (Falkowska et al., 2013). Some insect species show high mortality as a direct result of toxins in air or water, whereas other species are affected indirectly by changes in resource conditions induced by atmospheric change (see Chapter 2). However, the most immediate anthropogenic threats to ecosystems are changes in land use patterns and redistribution of exotic species, including plants, insects, and livestock. Conversion of natural ecosystems is altering and isolating natural communities at an unprecedented rate, leading to outbreaks of insect "pests" in crop monocultures and fragmented ecosystems and potentially threatening species incapable of surviving in increasingly inhospitable landscapes. Invasive species affect community and ecosystem structure and processes directly and indirectly (Kizlinski et al., 2002; Orwig, 2002; N. Sanders et al., 2003). J. Thomas et al. (2004) compared species losses of British butterflies, birds, and plants and found that loss of butterfly species has been greater than that of birds and plants. Current rates of species disappearance represent the sixth major extinction event, the last event causing the disappearance of the dinosaurs 65 million years ago. Predicting and mitigating species losses or pest outbreaks depends strongly on our understanding of insect ecology within the context of ecosystem structure and function.

4 ECOSYSTEM APPROACH TO INSECT ECOLOGY

Insect ecology can be approached using the hierarchical model described previously (Coulson and Crossley, 1987). Ecosystem conditions represent the environment, that is, the combination of physical conditions, interacting species, and availability of resources that determine survival and reproduction by individual insects, but insect activities, in turn, alter vegetation cover, soil properties, community organization, etc. (Fig. 1.2). This hierarchical approach offers a means of integrating the evolutionary and ecosystem approaches to studying insect ecology. The evolutionary approach focuses at lower levels of the hierarchy (individual, population, and community) and emphasizes individual and population adaptation to variable environmental conditions (established by higher levels of organization) through natural selection (Price et al., 2011; Speight et al., 2008). The ecosystem approach focuses at higher levels of resolution (community, ecosystem, and landscape) and emphasizes the effects of organisms on environmental (ecosystem) conditions. Natural selection can be viewed as feedback from the alteration of ecosystem conditions by coevolving organisms. Ecosystem structure and function result from the interactions among species. The evolutionary and ecosystem perspectives are most complementary at the community level, where species diversity emphasized by the evolutionary approach is the basis for functional organization emphasized by the ecosystem approach.

Although the traditional evolutionary approach has provided valuable explanations for how complex interactions have arisen, current environmental issues require an understanding of how insects affect ecosystem, landscape, and global processes, including climate. Both evolutionary and ecosystem approaches require consideration of the variety of evolutionary trade-offs that insects, and other organisms, face in selecting among and competing for resources that vary in quality, acceptability, and abundance over time and space, defending against predators, and responding to inevitable changes in abiotic conditions of their environment (ecosystem). However, whereas evolutionary ecologists have recognized insects as important components of ecosystem food webs, they have largely neglected the key roles insects play as ecosystem engineers, determining ecosystem structure and function through their effects on primary production, decomposition, and nutrient fluxes. On the other hand, ecosystem ecologists often have failed to recognize the importance of insect diversity to ecosystem function. Different species have different effects on rates and directions of ecosystem processes, for example, at different times and in different ways.

Insects can greatly alter, and potentially regulate, ecological succession, biogeochemical cycling, energy fluxes, and albedo, all of which affect regional and global climate. These roles may complement or exacerbate changes associated with human activities. Therefore, the purpose of this book is to address the fundamental issues of insect ecology as they relate to ecosystem, landscape, and global processes.

5 SCOPE OF THIS BOOK

This book is organized hierarchically, as described earlier, to emphasize feedbacks among individual, population, and community levels and the ecosystems they represent (Fig. 1.2). Five questions have been used to develop this text:

1. How do insects respond to variation in environmental conditions, especially spatial and temporal gradients in abiotic factors and resource availability (Section I)?
2. How do interactions among individuals affect the structure and function of populations (Section II)?
3. How do interactions among species affect the structure and function of communities (Section III)?
4. How do insects affect ecosystem properties and alter environmental conditions to which individuals respond (Section IV)?
5. How can this information be incorporated into management decisions and environmental policies that affect the delivery of ecosystem services (Section V)?

Most people, including many scientists, do not appreciate the challenges posed by the sheer diversity of insect species. Research at the ecosystem level frequently combines all insects into a single category, despite the wide range of ecological attributes they represent. Insect phylogeny reflects species radiation resulting from early detritivores adapting to plant resources, herbivores interacting with plant chemicals, and predators and parasites adapting to host behavior and physiology. To assist in understanding some of these patterns, the major orders of insects and their ecological importance, according to recent taxonomic changes, are listed in Table 1.2. However, readers should note that within many orders, food resources and other ecological attributes vary widely among species. Although the focus of this book clearly is on insects, related arthropods, including centipedes (class Chilopoda), millipedes (Diplopoda), and mites and spiders (Arachnida), and examples from studies of other organisms are used where appropriate to illustrate concepts and/or the wider applicability of these concepts.

Chapter and topic organization is intended to address the questions above by emphasizing key spatial and temporal patterns and processes at each level and their integration among levels. Environmental policy and management decisions (Section V) depend on evaluation of insect effects on ecosystem conditions and their responses to environmental change. The evaluation of insect effects on ecosystem conditions and their responses to environmental change (Section IV) depends on understanding of species diversity, interactions, and community organization (Section III) that, in turn, depends on understanding of population dynamics and biogeography (Section II), that reflect individual physiological and behavioral responses to environmental variation (Section I).

Three themes integrate these ecological levels. First, spatial and temporal patterns of environmental variability and disturbance determine survival and reproduction of individuals and patterns of population, community, and ecosystem structure and dynamics. Individual acquisition and allocation of resources, population distribution, colonization and extinction rates, community patterns and successional processes, and ecosystem structure and function reflect environmental conditions. Second, energy and nutrients move through individuals, populations, communities, and abiotic pools. The net foraging success and resource use by

TABLE 1.2 Orders of Insects and Related Arthropods and Their Ecological Importance

Class/order	Common names	Important as
HEXAPODA		
Collembola	Springtails	Detritivores (note that Collembola are noninsect hexapods)
INSECTA		
Exopterogota—insects with simple metamorphosis		
Ephemeroptera	Mayflies	Aquatic scrapers
Odonata	Dragonflies and damselflies	Aquatic and terrestrial predators
Orthoptera	Grasshoppers, locusts, katydids, and crickets	Herbivores and detritivores, including many crop pests
Phasmatodea	Walkingsticks and leaf insects	Herbivores
Dermaptera	Earwigs	Primarily detritivorous, but some herbivores
Plecoptera	Stoneflies	Aquatic herbivores, detritivores, and predators
Blattodea	Cockroaches and termites	Detritivores, including household pests
Mantodea	Mantises	Predators
Hemiptera	True bugs, cicadas, hoppers, aphids, and scale insects	Herbivores, including many crop pests and disease vectors, and predators
Thysanoptera	Thrips	Pollen feeders and predators
Psocoptera	Bark and book lice	Detritivores
Phthiraptera	Lice	Parasites and disease vectors
Endopterogota—insects with complex metamorphosis		
Coleoptera	Beetles	The largest order, with 40% of the known species of hexapods; herbivores, including major crop pests, detritivores, and predators
Neuroptera	Lacewings, snakeflies, alderflies, and antlions	Predators
Hymenoptera	Sawflies, bees, wasps, and ants	Herbivores, including pollinators, predators, and omnivores
Trichoptera	Caddisflies	Aquatic filter-feeders
Lepidoptera	Butterflies and moths	Herbivores, including many crop pests, and pollinators
Siphonaptera	Fleas	Parasites and disease vectors
Mecoptera	Scorpionflies	Predators and detritivores
Diptera	True flies	Herbivores, detritivores, predators, and parasites, including many disease vectors
Chilopoda	Centipedes	Predators
Diplopoda	Millipedes	Detritivores
Crustacea	Crustaceans	Aquatic or semiaquatic herbivores, detritivores, and predators
Arachnida	Spiders, mites, and scorpions	Herbivores, detritivores, predators, and parasites, including many disease vectors

individuals determine energy and nutrient fluxes at the population level. Trophic interactions among populations determine energy and nutrient fluxes at the community and ecosystem levels. Third, regulatory mechanisms at each level serve to balance resource demands with resource availability (carrying capacity) or to dampen responses to environmental changes. Regulation results from a balance between negative feedbacks that reduce population sizes or process rates and positive feedbacks that increase population sizes or process rates. Regulation of population sizes and process rates tends to stabilize ecosystem conditions within ranges favorable to most members. The capacity to regulate environmental conditions increases from individual to ecosystem level. If feedbacks within or among levels contribute to ecosystem stability, then human alteration of ecosystem structure and function could impair this function and lead to ecosystem degradation, with serious consequences for ecosystem capacity to deliver services (food, fresh water, building materials, etc.) on which human survival depends.

Section I (Chapters 2–4) addresses the physiological and behavioral ecology of individual insects. Physiology and behavior represent the phenotypic means by which organisms interact with their environment. Physiology represents "fixed" adaptations to predictable variation in environmental conditions, such as seasonal dormancy, whereas behavior represents a more flexible means of adjusting to unpredictable variation, for example, by escaping. Chapter 2 summarizes physiological and behavioral responses to variable habitat conditions, especially gradients in climate, soil, and chemical conditions. Chapter 3 describes physiological and behavioral mechanisms for acquiring energy and matter resources, and Chapter 4 compares strategies for allocating assimilated resources to various metabolic and behavioral pathways. These chapters provide a basis for understanding patterns of population distribution and movement of energy and matter through populations and communities.

Section II (Chapters 5–7) focuses at the population level. Populations of organisms integrate variation in adaptive strategies and foraging patterns among individuals. Chapter 5 describes population systems, including population structure and the processes of reproduction, mortality, and dispersal. Chapter 6 addresses processes and models of population change; Chapter 7 describes biogeography, processes and models of colonization and extinction, and metapopulation dynamics over landscapes. These population parameters determine population effects on ecological processes through time in various patches across regional landscapes.

Section III (Chapters 8–10) emphasizes community structure and dynamics. Each species interacts with other species in a variety of ways that determine patterns in community structure through time and space. Chapter 8 describes species interactions (eg, competition, predation, symbioses). Chapter 9 addresses spatial patterns in community structure. Chapter 10 addresses changes in community structure over varying temporal scales. Changes in community structure determine spatial and temporal patterns of energy and nutrient storage and flux through ecosystems.

Section IV (Chapters 11–15) focuses on ecosystems and is the major contribution of this text to graduate education in insect ecology. Chapter 11 describes general aspects of ecosystem structure and function, especially processes of energy and matter storage and flux that determine resource availability and environmental conditions, and ecosystem capacity to control climate. Chapter 12 describes patterns of herbivory and its effects on ecosystem conditions; Chapter 13 describes patterns and effects of pollination, seed predation, and seed dispersal; and Chapter 14 describes patterns of detritivory and burrowing and their effects on ecosystem conditions. Chapter 15 addresses the developing concept of ecosystem self-regulation and mechanisms, including species diversity and herbivory, that contribute to ecosystem stability.

Section V (Chapters 16–18) represents a synthesis that includes applications of insect ecology to management of ecosystem services and "pest" populations. Chapter 16 provides examples of applications to sustainability of ecosystem services. This topic is becoming increasingly important as we see widespread ecosystem deterioration in response to anthropogenic changes, threatening the delivery of services on which we depend and potentially exposing displaced human populations to epidemics of crowd diseases. Chapter 17 provides examples of applications to improved pest management and protection of endangered species. How should we decide when, and by what methods, insects should be controlled, in order to avoid undesirable consequences for nontarget species and ecosystem services? Chapter 18 summarizes and synthesizes previous chapters and suggests future directions and data necessary to improve understanding of linkages and feedbacks among hierarchical levels.

Solutions to our growing environmental problems require consideration of insect ecology at ecosystem, landscape, and global levels. Although many insect species conflict with human interests, their adaptive ecological functions also sustain the delivery of important ecosystem services. In the absence of insects, even apparently undesirable species, we would soon starve, but perhaps not before being buried in undecomposed plant debris, animal dung, and carcasses. Therefore, a broader understanding of insect ecology in an ecosystem context can lead to improved management of insects and ecosystems that sustain the delivery of ecosystem services.

ECOLOGY OF INDIVIDUAL INSECTS

The individual organism is a fundamental unit of ecology. Responses to environmental conditions determine an individual's survival and reproduction, and therefore its fitness. Individual use of habitat and food resources alters spatial and temporal patterns of habitat structure and resource distribution for other organisms. Insects have been particularly successful in adapting to environmental changes over 400 million years (Romoser and Stoffolano, 1998). Several attributes have contributed to their evolutionary and ecological success.

Small size (an attribute shared with other invertebrates and microorganisms) has permitted exploitation of habitat and food resources at a microscopic scale. Insects find protection from adverse conditions in microsites too small for larger organisms, for example, within individual leaves or soil pores. Large numbers of insects can exploit the resources represented by a single leaf, by partitioning leaf resources, with some species feeding on cell contents, others on sap in leaf veins, some on top of the leaf, others on the underside, some internally. At the same time, small size makes insects highly sensitive to changes in temperature, moisture, air or water chemistry, and other factors.

The exoskeleton (shared with other arthropods) provides protection against predation and desiccation or waterlogging (necessary for small organisms) and innumerable points of muscle attachment (for flexibility). However, the exoskeleton also limits the size attainable by arthropods. The weight of exoskeleton required to support a larger body size would limit mobility. Higher atmospheric oxygen concentrations during the Carboniferous Period facilitated activity of larger arthropods (Dudley, 1998), and faster, more flexible vertebrate predators had not appeared yet. These large-sized arthropods disappeared during the late Permian transition to more hypoxic atmospheric conditions (Chapelle and Peck, 1999; Dudley, 1998). Larger extant arthropods occur in aquatic environments, where water helps support their weight.

Metamorphosis is necessary for exoskeleton-limited growth but permits partitioning of habitats and resources among life stages, especially among insects that undergo complete (or holometabolous) metamorphosis (see Table 1.2). Immature and adult insects can differ dramatically in form and function and thereby live in different habitats and feed on different resources, reducing intraspecific competition. For example, dragonflies and mayflies live in aquatic ecosystems as immatures, but in terrestrial ecosystems as adults. Many butterflies and beetles feed on foliage as immatures and on nectar as adults. Among holometabolous insects, the quiescent, pupal stage facilitates survival during unfavorable environmental conditions. However, insects, as well as other arthropods, are particularly vulnerable to desiccation and predation during ecdysis (molting).

Finally, flight evolved first among insects and conferred a distinct advantage over other organisms. Flying organisms can travel longer distances per unit time than can flightless organisms. Flight permits rapid long-distance movement that facilitates discovery of new resources, as well as escape from predators or unfavorable conditions. The aerial acrobatics of many dipterans has inspired awe and led to development of advanced robotic spyware. Flight remains a dominant feature of insect ecology.

This section focuses on aspects of morphology, physiology, and behavior that affect insect interactions with their environment (including other organisms), specifically adaptations that favor survival and reproduction under variable environmental conditions and mechanisms for finding, exploiting, and allocating resources. Physiology and behavior are closely integrated. For example, insect movement, including dispersal, is affected by physiological perception of food resources and temperature and chemical gradients, fat storage, rapid oxygen supply, etc. Similarly, physiological processes are affected by insect selection of thermally suitable location, choice of food resources, etc. Chemical defenses against predation are based on physiological processes, but often are enhanced by behaviors that increase their effect, for example, thrashing or regurgitation to maximize predator exposure to chemical defenses. Organisms affect ecosystem processes, such as energy and nutrient fluxes, through spatial and temporal patterns of foraging and energy and nutrient allocation.

Chapter 2 describes physiological and behavioral responses to changing environmental conditions. Chapter 3 focuses on physiological and behavioral mechanisms for finding and exploiting resources. Chapter 4 describes allocation of resources to various metabolic pathways and behaviors that facilitate resource acquisition, mate selection, reproduction, and interaction with other organisms. Physiology and behavior interact to determine the conditions under which insects can survive and the means by which they acquire and use available resources. These ecological attributes affect population ecology (such as population structure, changes in population size, biogeography, etc., Section II), community attributes (such as use of, or use by, other organisms as resources, Section III), and ecosystem attributes (such as rates and directions of energy and matter flows, Section IV).

2

Responses to Abiotic Conditions

DISEASE VECTOR RESPONSE TO ANTHROPOGENIC DISTURBANCE

Human alteration of environmental conditions affects insect populations, in some cases bringing insects and humans into greater conflict. Póvoa et al. (2003) suggested that the reappearance of *Anopheles darlingi* and malaria in Belém, Brazil in 1992, after its presumed elimination in 1968, resulted from human encroachment into deforested areas that had become more favorable mosquito habitat. Vittor et al. (2006) tested this hypothesis in northeastern Peru, where malaria also had dropped dramatically during the 1960s as a result of eradication efforts and remained below 2 cases per 1000 population until the 1990s. Construction of the Iquitos–Nauta road into the region during the 1980s and 1990s initiated deforestation and allowed rapid settlement and small-scale subsistence agriculture. A sudden increase in the incidence of malaria was observed during the 1990s, reaching more than 120,000 cases (340 per 1000 population) in 1997. During 2000 Vittor et al. (2006) selected replicate sites along the Iquitos–Nauta road to represent high, medium, or low percentages of deforestation (based on satellite imagery) and human population density (within a 500-m radius around the sample site). Rates of mosquito landing on research personnel were measured at each site between 1800 and 2400 h (the period of peak mosquito activity) and compared among land use and demography treatments. Because mosquito reproduction occurred primarily in ponds and fish farms associated with cleared or naturally open areas, and adult mosquitoes did not fly far from breeding sites, biting rates reflected local populations of mosquitoes. Sites with <20% forest and >30% grass/crop cover had a 278-fold higher biting rate than sites with >70% forest and <10% grass/crop cover. Based on mean percentages of infective *A. darlingi* in the Amazon region (0.5–2.1% infective mosquitoes), Vittor et al. (2006) calculated 38 infective bites per year per km^2 in areas with >35% grass/crop cover and 8–11 infective bites per year per km^2 in areas with 2–35% grass/crop cover, compared to 0.1 infective bites per year per km^2 in areas with <2% grass/crop cover, presenting serious challenges for disease control as deforestation progresses and the human population increases. Human contact with novel zoonotic diseases is likely to increase as intrusion into previously unpopulated areas increases (K.F. Smith et al., 2007).

Insect Ecology. http://dx.doi.org/10.1016/B978-0-12-803033-2.00002-9

INTRODUCTION

Insects are a dominant group of organisms found in virtually all terrestrial, freshwater, and near-coastal marine habitats, from the Arctic to the Antarctic, including many of the harshest ecosystems on the globe (eg, deserts, hot springs, and tundra), but individual species have more restricted ranges of occurrence. One of the earliest (and still important) objectives of ecologists was explanation of the spatial and temporal patterns of species distributions (Andrewartha and Birch, 1954; Hooker, 1847, 1853, 1860; Shelford, 1918; A. Wallace, 1876, 1911). The geographical ranges of insect species generally are determined by their tolerances, or the tolerances of their food resources and predators, to variation in abiotic conditions, or to geographic barriers, such as mountain ranges or oceans.

Tolerance ranges are under genetic control (W. Bradshaw and Holzapfel, 2001; Teets and Denlinger, 2014; Teets et al., 2012a,b). Insect phenotypes (morphological, physiological, and behavioral adaptations) reflect the characteristic physical conditions of the habitats in which they occur, together with adaptations that favor their interactions with other organisms. However, variation in physical conditions requires some flexibility in physiological and behavioral traits.

The distribution of most species is determined largely by the global template of temperature and precipitation combinations, interrupted by mountain ranges, oceans, or other barriers to spread. However, all ecosystems experience variation in climate, as well as periodic disturbances, that affect the survival of organisms in the community. Warming global climate is causing evolutionary changes in adaptive traits and shifts in geographic ranges for many species (Balanyá et al., 2006; Hill et al., 2011; Kozár, 1991; Lehmann et al., 2014; Parmesan et al., 1999). Anthropogenic changes in habitat conditions also are altering the range of conditions to which organisms must respond (Gutierrez and Wilson, 2014; O'Connor et al., 2014), including increased frequency and severity of extreme weather events (Bender et al., 2010; Gleason et al., 2008; Gutschick and BassiriRad, 2010; Irwin et al., 2014; Lubchenco and Karl, 2012; W. Zhang et al., 2015) that may induce outbreaks of some species and cause the disappearance of others.

1 THE PHYSICAL TEMPLATE

1.1 Biomes

Global patterns of temperature and precipitation, reflecting the interaction between latitude, global atmospheric and oceanic circulation patterns, and topography, establish a template of physical conditions that support characteristic regional community types, *biomes* (Fig. 2.1) (Finch and Trewartha, 1949). Latitudinal gradients in temperature from Earth's equator to its poles define the tropical, subtropical, temperate, and arctic zones. Precipitation patterns overlay these temperature gradients. Warm, humid air rises in the tropics, drawing air from higher latitudes into this *equatorial convergence zone*. The rising air cools and condenses moisture, resulting in a band of high precipitation and tropical rainforests centered on the equator. The cooled, dried air flows away from the equatorial zone and warms as it descends in the "horse latitudes," centered around 30°N and S. These latitudes are dominated by arid

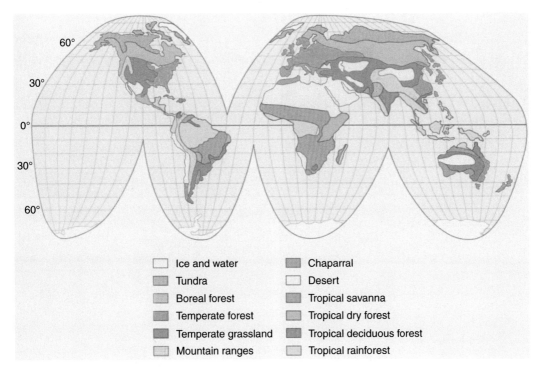

FIGURE 2.1 Global distribution of the major terrestrial biomes. The distribution of biomes is affected by latitude, global atmospheric and oceanic circulation patterns, and major mountain ranges. *Source: Modified from Finch, V.C., Trewartha, G.T., 1957. Elements of Geography: Physical and Cultural. McGraw-Hill, New York.*

grassland and desert ecosystems because of high evaporation rates in warm, dry air. Airflow at these latitudes diverges to the equatorial convergence zone and to similar convergence zones at about 60°N and S latitudes. Rising air at 60°N and S latitudes creates bands of relatively high precipitation and low temperature that support boreal forests. These latitudinal gradients in climate restrict the distribution of organisms on the basis of their tolerance ranges for these combinations of temperature and moisture. No individual species is capable of tolerating the entire range of tropical to arctic temperatures or desert to mesic moisture conditions.

The relatively distinct combinations of temperature and precipitation (MacMahon, 1981) determine the distinct assemblages of species capable of surviving and defining the characteristic community types, which are also defined by their dominant vegetation, that is, tundra, boreal coniferous forest, temperate deciduous forest, tropical rainforest, tropical dry forest, grassland, savanna, chaparral, and desert biomes. Representative terrestrial biomes and their seasonal patterns of temperature and precipitation are shown in Figs. 2.2 and 2.3.

Habitat conditions in terrestrial biomes are influenced further by topographic relief, substrate structure and chemistry, and exposure to wind. For example, topographic relief creates gradients in solar exposure and soil drainage, as well as in temperature and moisture, providing local habitats for unique communities. Mountain ranges interact with oceanic and atmospheric circulation patterns to modify latitudinal patterns of temperature and

FIGURE 2.2 Examples of ecosystem structure in representative terrestrial biomes. (A) Tundra (alpine) (western USA), (B) desert shrubland (southwestern USA), (C) grassland, supporting herds of grazing ungulates, such as American bison, *Bison bison* (northcentral USA), (D) tropical savanna, supporting herds of grazers and browsers, such as giraffe, *Giraffa camelopardalis* (South Africa), (E) boreal forest (northwestern USA), (F) temperate deciduous forest (southeastern USA), and (G) tropical rainforest (northern Panama). *Source: Photo D courtesy of C.A.S. Schowalter.*

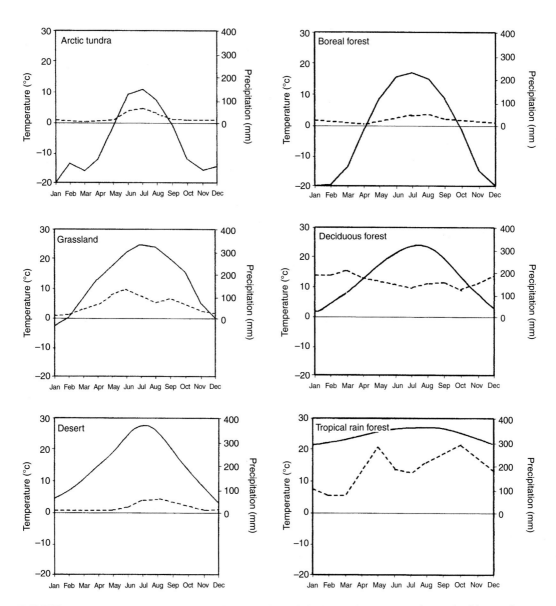

FIGURE 2.3 **Seasonal variation in temperature and precipitation at sites representing major biomes.** *Source: Data from Van Cleve, K., Martin, S., 1991. Long-term Ecological Research in the United States, sixth ed. University of Washington, Seattle, WA.*

FIGURE 2.4　**Orographic effect of mountain ranges.** Interruption of airflow and condensation of precipitation on the windward side and clear sky on the leeward side of Mt. Hood, Cascade Mountains, Oregon, USA.

precipitation. Mountains force airflow upward, causing cooling, condensation, and precipitation on the windward side (Fig. 2.4). Drier air descends on the leeward side where it gains moisture through evaporation. This orographic effect leads to development of mesic environments on the windward side and arid environments on the leeward side of mountain ranges. Mountains are characterized further by elevational gradients of temperature, moisture, and atmospheric conditions, that is, lower elevations tend to be warmer and drier, whereas higher elevations are cooler and moister. Concentrations of oxygen and other gases decline with elevation, so that species occurring at higher elevations must be capable of surviving at low gas concentrations.

The montane gradient is much shorter than the corresponding latitudinal gradient, with the same temperature change occurring in a 1000-m difference in elevation or an 880-km difference in latitude. Hence, the range of habitat conditions that occur over a wide latitudinal gradient occur on a smaller scale in montane areas, leading to development of relatively distinct *life zones* at increasing elevation ranges (Fig. 2.5). Although life zones are most distinct in western North America, where the concept was developed (Merriam, 1890) and remains widely used, gradients in community structure with elevation are a common feature of montane landscapes.

Local differences in substrate structure and chemistry may limit the ability of many species of plants and animals, characteristic of the surrounding biome, to survive but offer local opportunities for colonizing species from other areas, thereby increasing biodiversity. Some soils (eg, sandy loams) are more fertile or more conducive to excavation than others; serpentine soils and basalt flows require special adaptations for survival by plants and animals.

Aquatic biomes are formed by topographic depressions and gradients that create zones of standing or flowing water. Aquatic biomes vary in seasonality, size, depth, flow rate, and marine influence, that is, temporary pools and streams, permanent ponds, lakes, streams, rivers,

FIGURE 2.5 Example of montane life zones in southeastern Arizona, USA, distinguished by differences in color. Lowest elevations (up to 1200 m) in foreground are relatively warm and dry, supporting desert grassland and shrubland (Lower Sonoran Life Zone); higher elevations are characterized by decreasing temperature and increasing precipitation that support a dark green zone (1000-2500 m) of pinyon pine, *Pinus edulis*/juniper, *Juniperus* spp., shrubland; then a lighter green zone of oak, *Quercus* spp., woodland and chaparral (Upper Sonoran Life Zone); and finally a dark green zone (2000-3000 m) of ponderosa pine, *Pinus ponderosa*, pine forest (Transition Life Zone). In higher mountain ranges mixed conifer, *Pinus/Abies/Picea* spp., forest predominates at 2700-3500 m (Canadian and Hudsonian Life Zones), and alpine tundra (Arctic-Alpine Life Zone) occurs above treeline (>3500 m).

estuaries, and tidal marshes (Fig. 2.6). Lotic habitats often show considerable gradation in temperature and solute concentrations with depth. Because water has a high specific heat capacity, water changes temperature slowly relative to air temperature. However, because water is most dense at 4°C, changes in density as temperature changes result in seasonal stratification of water temperature. Thermal stratification develops in the summer, as the surfaces of standing bodies of water warm and trap cooler, denser water below the *thermocline* (the zone of rapid temperature change), and again in the winter, as freezing water rises to the surface, trapping warmer and denser water below the ice. During fall and spring, changing surface temperatures result in mixing of water layers and movement of oxygen and nutrients throughout the water column. Hence, deeper zones in aquatic habitats show relatively little variation in temperature, allowing aquatic insects to continue development and activity throughout the year, even in temperate regions.

Habitat conditions in aquatic biomes are influenced further by substrate structure and chemistry, amount and chemistry of regional precipitation, and the characteristics of surrounding terrestrial communities, including conditions upstream. Substrate structure and

FIGURE 2.6   **Examples of aquatic biomes.** (A) Stream (western USA), (B) beaver pond (western USA), (C) swamp (southern USA), (D) coastal saltmarsh (southeastern USA), and (E) lake (Hungary).

chemistry determine flow characteristics (including turbulence), pH, and inputs of nutrients from substrates. The amount and chemistry of regional precipitation determine the regularity of water flow and inputs of atmospheric gases and nutrients. Characteristics of surrounding communities determine the degree of exposure to sunlight and the character and condition of allocthonous inputs of organic matter, sediments, and nutrients.

1.2 Environmental variation

Physical conditions vary seasonally in most biomes (Fig. 2.3). Temperate ecosystems are characterized by obvious seasonality in temperature, with cooler winters and warmer summers, and also may show distinct seasonality in precipitation patterns, resulting from seasonal changes in the orientation of Earth's axis relative to the sun. Although tropical ecosystems experience relatively consistent temperatures, precipitation often shows pronounced seasonal variation (Fig. 2.3). Aquatic habitats show seasonal variation in water level and circulation patterns related to seasonal patterns of temperature, precipitation, and evaporation. Seasonal variation in circulation patterns can result in stratification of thermal layers and water chemistry in lotic systems. Intermittent streams and ponds may disappear during dry periods or when evapotranspiration exceeds precipitation.

Physical conditions also vary through time as a result of irregular events. Changes in global circulation patterns of air or water are particularly important. For example, the east–west gradient in surface water temperature in the southern Pacific diminishes in some years, altering oceanic and atmospheric currents globally, known as the El Niño/southern oscillation (ENSO) phenomenon (Rasmussen and Wallace, 1983; Windsor, 1990). Particularly strong El Niño years, for example, 1982–83 and 1997–98, are characterized by extreme drought conditions in some tropical ecosystems and severe storms and wetter conditions in some higher-latitude ecosystems. Seasonal patterns of precipitation can be reversed, that is, causing a drier wet season and wetter dry season. The year following an El Niño year may show a rebound, an opposite but less intense effect (La Niña). Windsor (1990) found a strong positive correlation between El Niño index and precipitation during the preceding year in Panama. Precipitation in Panama typically is lower than normal during El Niño years, in contrast to the greater precipitation accompanying El Niño in Peru and Ecuador (Windsor, 1990; J. Zhou et al., 2002).

Many insects are sensitive to the changes in temperature and precipitation that accompany such events, some with consequences for human health. Stapp et al. (2004) found that local extinction of black-tailed prairie dog, *Cynomys ludovicianus*, colonies in the western Great Plains of North America was significantly greater during El Niño years, due to flea-transmitted plague, *Yersinia pestis*, which spreads more rapidly during warmer, wetter conditions (Parmenter et al., 1999). J. Zhou et al. (2002) reported that extremely high populations of sand flies, *Lutzomyia verrucarum*, were associated with El Niño conditions in Peru, resulting in near doubling of human cases of bartonellosis, an emerging, vector-borne, highly fatal, infectious disease in the region (Fig. 2.7). Similarly, van Panhuis et al. (2015) reported that increased transmission of dengue fever across Southeast Asia coincided with elevated temperatures throughout the region in 1997–98, associated with the strongest El Niño episode of the century. Regional drying triggers outbreaks of many herbivorous species (Mattson and Haack, 1987; Stige et al., 2007; Van Bael et al., 2004).

Terrestrial and aquatic biomes differ in the type and extent of variation in physical conditions. Terrestrial habitats are sensitive to changes in air temperature, wind speed, relative humidity, and other atmospheric conditions. Aquatic habitats are relatively buffered from sudden changes in air temperature but are sensitive to changes in flow rate, depth, and chemistry, especially changes in pH and concentrations of dissolved gases, nutrients, and pollutants. Vegetation cover insulates the soil surface and reduces albedo, thereby reducing diurnal and seasonal variation in soil and near-surface temperatures (J. Foley et al., 2003a). Hence, desert biomes with sparse vegetation cover typically show the widest diurnal and seasonal variation in physical conditions. Areas with high proportions of impervious surfaces (such as roads, roofs, parking lots) greatly alter conditions of both terrestrial and aquatic systems by increasing albedo and precipitation runoff (Elvidge et al., 2004).

Physiological tolerances of organisms, including insects, generally reflect the physical conditions of the biomes in which they occur. Insects associated with the tundra biome tolerate a lower range of temperatures than do insects associated with tropical biomes. The upper threshold temperature for survival of a tundra species might be the lower threshold

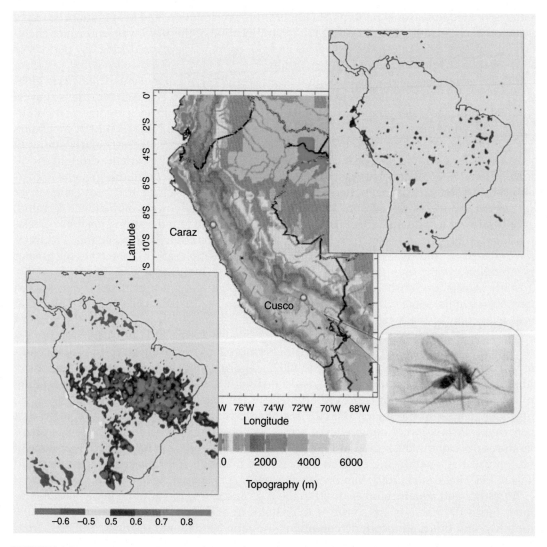

FIGURE 2.7   **Topography of Peru (center), comparison of TRMM TMI rainfall at Cuzco (lower left) and Caraz (upper right) relative to their surroundings, and the sand fly, *Lutzomyia verrucarum*, vector of bartonellosis that shows increased spread associated with higher rainfall during El Niño events.** *Source: From Zhou, J., et al., 2002. El Niño helps spread Bartonellosis epidemics in Peru. EOS Trans. Am. Geophys. Union 83, 160–161.*

temperature for survival of a tropical species (Deutsch et al., 2008; Hazell et al., 2010). Eisen et al. (2008) reported that 27 mosquito species showed distinct elevational ranges of occurrence in the Front Range of the Rocky Mountains in Colorado as a result of differential tolerance to temperature among species. Similarly, insects characterizing mesic or aquatic biomes generally have less tolerance for desiccation than do insects characterizing xeric biomes. However, species characterizing temporary streams or ponds may have adapted mechanisms for withstanding desiccation during dry periods (Batzer and Wissinger, 1996). On the other hand, concentration of prey species in reduced pool depths during dry periods may increase exposure to predation (Covich and McDowell, 1996; Covich et al., 2003, 2009). Some species show greater capacity than do others to adapt to changing environmental conditions, especially rapid changes resulting from anthropogenic activity. Many species may be predisposed to adapt to rapid changes because of evolution in frequently disturbed ecosystems.

1.3 Disturbances

Within biomes, characteristic abiotic and biotic factors interact to influence the pattern of disturbances, relatively discrete events that alter ecosystem conditions, and create a finer-scale landscape mosaic of patches with different disturbance and recovery histories (Harding et al., 1998; Schowalter et al., 2003; Willig and Walker, 1999). Disturbances, such as fire, storms, drought, flooding, anthropogenic conversion, etc. (Fig. 2.8), alter vertical and horizontal gradients in temperature, moisture, and air or water chemistry (T. Lewis, 1998; P. White and Pickett, 1985), as well as distribution of organic materials, significantly altering the abiotic and biotic conditions to which organisms are exposed (Agee, 1993; Schowalter, 2012; Schowalter and Lowman, 1999).

Disturbances can be characterized by several criteria that determine their effect on various organisms (see Schowalter, 2012; Walker and Willig, 1999; P. White and Pickett, 1985). Disturbance type, such as fire, drought, flood, storm, or volcanic eruption, determines which ecosystem components will be most affected. Above-ground versus below-ground species or terrestrial versus aquatic species are affected differently by fire versus flood. Magnitude is a measure of the extent of departure from normal ranges in abiotic conditions and can be expressed as intensity, the physical force of the event, or severity, its effect on species and the ecosystem. A fire or storm of given intensity, based on temperature or wind speed, will affect organisms differently in a grassland versus a forest. Scale or extent is the area affected by the disturbance and determines the distance between population sources and available habitat and the rate at which organisms recolonize the interior portions of the disturbed area. Frequency is the mean number of events per time period; reliability is measured as the inverse of variability in the time between successive events (recurrence interval) and determines the probability that successive generations will experience selection for adaptive characteristics.

Insects show a variety of responses to particular disturbance types (Schowalter, 2012). Some species respond positively, others negatively to particular disturbances, based on adaptive characteristics (Cleary and Grill, 2004; E. Evans, 1988; Gandhi et al., 2007; Grimbacher and Stork, 2009; Hanula and Wade, 2003; Paquin and Coderre, 1997; Schowalter, 2012; Schowalter et al., 1999; Ulyshen et al., 2010; Wikars and Schimmel, 2001). Insect outbreaks often are considered to be disturbances because they kill plants and alter distribution of biomass over large areas (Breshears et al., 2005; M. Brown et al., 2010; Kurz et al., 2008; MacLean, 2004; Willig and Walker, 1999). Insect outbreaks are comparable to physical disturbances in terms of their ability to defoliate or kill most host plants and alter ecosystem conditions over large areas, up to 10^3–10^6 ha (Furniss and Carolin, 1977). For example, 39% of a montane forest landscape in Colorado has been affected by insect outbreaks (spruce beetle, *Dendroctonus rufipennis*) since about 1633, compared to 59% by fire and 9% by snow avalanches (Veblen et al., 1994), with an average return interval of 117 years, compared to 202 years for fire.

FIGURE 2.8   Natural and anthropogenic disturbances. Natural disturbances include (A) fire, especially in grasslands and savannas (north central USA), (B) storms (north central USA), and (C) floods (northwestern USA). Anthropogenic disturbances include (D) arid land conversion to agricultural use (center-pivot irrigation, western USA), (E) forest harvest fragmentation (northwestern USA), and (F) overgrazing and desertification (right of fence, compared to natural grassland on left; southwestern USA). *Source: Photo F courtesy of D.C. Lightfoot.*

However, unlike physical disturbances, outbreaks of native insect species typically are biotic responses to environmental changes or disturbances that stress host plants and weaken their defenses or that result in abnormally high densities of host plants (see Chapters 3 and 6). Drought is a particularly important trigger for a large number of insect species (Mattson and Haack, 1987; Stige et al., 2007; Van Bael et al., 2004). Plagues of locusts most commonly result from drought conditions, but migratory swarms spread populations over vast areas (C. Riley, 1878; Stige et al., 2007; Yu et al., 2009). Defoliator outbreaks also commonly reflect abundant water-stressed plants (Mattson and Haack, 1987; Schowalter et al., 1999; Van Bael et al., 2004), perhaps reflecting changes in plant defensive chemistry, but this is not always the case (Hale et al., 2005, see Chapter 3). Bark beetle populations increasing in drought-stressed or storm-injured conifers can reach sizes capable of killing surrounding uninjured trees over large areas (Furniss and Carolin, 1977; Mattson and Haack, 1987; Schowalter and Turchin, 1993; Selikhovkin, 2005; Veblen et al., 1994). Although risk of bark beetle outbreaks in burned forests often has been used to justify salvage harvesting, Fettig et al. (2010) reported that only 4% of pine trees were killed by bark beetles following forest fires in California. Outbreaks can predispose ecosystems to subsequent abiotic or biotic disturbances, for example, through accumulated dry litter fueling subsequent fire or greater exposure to wind during storms (T. Parker et al., 2006; Taylor and MacLean, 2009). Consideration of outbreaks as disturbances, rather than changes in trophic interactions, remains a matter of perspective (Schowalter and Lowman, 1999). However, insects clearly are capable of altering ecosystem conditions on a scale second only to anthropogenic changes.

Disturbances vary in intensity and severity. A low-intensity ground fire affects primarily surface-dwelling organisms, many of which may be adapted to this level of disturbance, whereas a high-intensity crown fire can transform a large proportion of the community into ash and atmospheric gases. Plant species capable of withstanding low-to-moderate wind speeds may topple at high wind speeds (Canham et al., 2010). Hurricane winds damage large areas of forest and can virtually eliminate many arthropod species (Koptur et al., 2002; Schowalter and Ganio, 2003; Willig and Camilo, 1991).

Disturbances range in scale from local to global. Less mobile insects may require longer periods to recolonize large disturbed areas (Knight and Holt, 2005; Shure and Phillips, 1991). However, edges of disturbed areas may provide unique resources, for example, forest understory species eliminated from the burned area but able to exploit the greater light available at the edge, that favor insect herbivores (Knight and Holt, 2005).

Frequency and reliability of recurrence, with respect to generation times of characteristic organisms, of a particular disturbance type are the most important factors driving directional selection for adaptation to disturbance, for example, traits that confer tolerance (resistance) to frequent fire or flooding. Effects of disturbances may be most pronounced in ecosystems, such as mesic forests and lakes, which have the greatest capacity to modify abiotic conditions and, therefore, have the lowest exposure and species tolerances to sudden or extreme departures from nominal conditions.

Ecosystems can be subject to various combinations of disturbance type, frequency, and intensity. For example, over a 20-year period the rainforest in eastern Puerto Rico experienced two major hurricanes (Hugo in 1989; Georges in 1998) that broke or toppled trees on windward slopes over large areas, several moderate hurricanes (Luis and Marilyn in 1995, Bertha and Hortense in 1996, Erika in 1997, Jose in 1999, and Debby in 2000) that caused

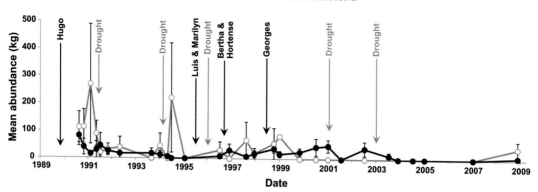

FIGURE 2.9   **Mean abundances (number of individuals ± 1 SE) per sample per kg dry weight of foliage sample for *Coccus acutissimus* on *Dacryodes excelsa*, in tropical rainforest at the Luquillo Experimental Forest in Puerto Rico as affected by hurricanes** (*black text and arrows*) **and droughts** (*gray text and arrows*) **over a 19-year period.** Data are from trees located in gaps originally created by hurricane Hugo (*open circles and gray lines*) or located under closed canopy forest following Hugo (*filled circles and black lines*). Data analysis indicated significant effects of prior disturbances on responses to subsequent disturbance. *Source: Figure courtesy of S.J. Presley; from Schowalter et al., unpubl. data.*

substantial defoliation and flooding, a number of minor hurricanes and tropical storms (Heartsill-Scalley et al., 2007), and hundreds of landslides resulting from rainstorms. A significant drought occurred during 1994–95, when precipitation was only 41% of the long-term annual average, and minor droughts occurred in 1991, 1996, 2001, and 2003 (Heartsill-Scalley et al., 2007). Biotic responses may show *legacy effects* (Fig. 2.9), with responses to each disturbance reflecting which species survived the preceding disturbance, either as surviving individuals or propagules, for example, seed banks (J. Chase, 2007; Harding et al., 1998; Schowalter et al., 2003; Summerville et al., 2009; Willig et al., 2011).

Individual insects have specific tolerance ranges to abiotic conditions that dictate their ability to survive prevailing local conditions, but may be exposed during disturbances to lethal extremes of temperature, water availability, or other factors (Buddle et al., 2006; Hirao et al., 2008; Mertl et al., 2009; Moretti and Legg, 2009). Variable ecosystem conditions typically select for wider tolerance ranges than do more stable conditions. Although abiotic conditions can affect insects directly (eg, burning, drowning, particle blocking of spiracles), they also affect insects indirectly through changes in resource quality and availability and exposure to predation or parasitism (Alstad et al., 1982; K. Miller and Wagner, 1984; Mopper et al., 2004; Schowalter, 2012; Shure and Wilson, 1993). Population size and the degree of genetic heterogeneity within populations affect the number of individuals that survive altered conditions (G. Bell and Gonzalez, 2009). As habitat conditions change, intolerant individuals disappear, leaving a higher frequency of genes for tolerance of the new conditions in any surviving portion of the population.

de Mazancourt et al. (2008) suggested that high biodiversity increases the likelihood that some species have genotypes that are adapted to the new conditions. Arrival of adapted colonists of various species from other areas augments community recovery. Even within families and genera, individual species may respond quite differently. Among Hemiptera, some scale insects increase in numbers and others decline in numbers following forest canopy

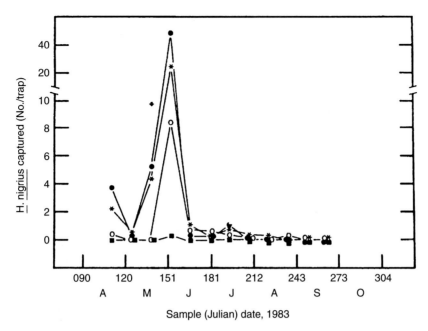

FIGURE 2.10 Changes in abundance of a root bark beetle, *Hylastes nigrinus*, in undisturbed, 12-year-old plantations *(black squares)* of Douglas fir and plantations thinned in September 1982 *(asterisks)*, January 1983 *(black circles)*, or May 1983 *(white circles)* in western Oregon. Arrow indicates time of thinning in May 1983. *Source: From Witcosky, J.J., et al., 1986. The influence of time of precommercial thinning on the colonization of Douglasfir by three species of root-colonizing insects. Can. J. Forest. Res. 16, 745–749.*

disturbance resulting from hurricane winds (Schowalter and Ganio, 2003). Root bark beetles (eg, *Hylastes nigrinus*) are attracted to chemicals that emanate from exposed stump surfaces and advertise suitable resource conditions for brood development. Consequently, these insects become more abundant following forest thinning (Fig. 2.10) (Witcosky et al., 1986), whereas related stem-feeding bark beetles (eg, *Dendroctonus* spp.) are sensitive to tree spacing and become less abundant in thinned forests (Amman et al., 1988; Sartwell and Stevens, 1975; Schowalter and Turchin, 1993).

Increasing disturbance frequency or intensity generally reduces species diversity because fewer individuals or populations are able to survive or recover (Haddad et al., 2008). Hanula and Wade (2003) found that abundances of most forest floor species (especially predators) were reduced by prescribed burning, and reduced more by annual than by biennial or quadrennial burning, but a few species (especially detritivores) increased in abundance with more frequent burning. Reice (1985) experimentally disturbed benthic invertebrate communities in a low-order stream in the eastern United States by tumbling cobbles in patches of stream bottom 0, 1, or 2 times in a 6-week period. Most insect and other invertebrate taxa decreased in abundance with increasing disturbance. Two invertebrate taxa increased in abundance following a single disturbance, but no taxa increased in abundance with increasing disturbance. However, all populations rebounded quickly following cessation of disturbance, suggesting that these taxa were adapted to this disturbance.

Timing of disturbances, relative to developmental stage, also affects insect responses. For example, Voelz et al. (1994) reported that caddisfly species surviving a period of increased water temperature (+4–5°C, from surface water release over a dam) were those with abundant terrestrial adults during this period and those with broad thermal tolerance. On the other hand, Martin-R et al. (1999) reported that experimental fires during different developmental stages of spittlebug, *Aeneolamia albofasciata*, in buffelgrass, *Cenchrus ciliaris*, grassland in Sonora, Mexico eliminated spittlebugs for at least 4 years after burning, regardless of developmental stage at the time of burning.

Following disturbance, populations may recover to their predisturbance condition at rates that reflect the magnitude of change and the extent of the disturbed area. Recovery can be as quick as a few months for rapidly reproducing species or assemblages, such as insect-dominated stream ecosystems (Reice, 1985; Gandhi et al., 2007; Matthaei and Townsend, 2000; Murdock et al., 2010), or many years to centuries for long-lived, slowly reproducing species or assemblages, such as forest ecosystems (see Chapter 10).

Disturbances that disrupt interactions among species may prolong recovery. For example, factors that delay recovery of plant communities also influence recovery of habitat conditions and delay colonization by insects. Sikes and Slowik (2010) found that the arthropod community on Kasatochi Island, Alaska, following a volcanic eruption that buried the island in ash, had shifted from a plant-based assemblage to a necromass-based assemblage that likely consisted of survivors and their offspring that could exploit nonliving resources. Elderd (2006) and Elderd and Doak (2006) reported that a common riparian plant, *Mimulus guttatus*, is favored by flooding disturbance, leading to increased abundances of grasshoppers following flood events. Overall diversity and abundance of stream invertebrate communities did not differ between young (30–40 year-old) and old-growth (500 year-old) forests in Oregon, USA, but the deciduous riparian vegetation (with seasonal litterfall and canopy opening) characterizing the young forest continued to affect stream invertebrate species composition, for example, higher densities of stoneflies, compared to streams in old-growth coniferous forest, for many years (Frady et al., 2007).

Furthermore, disturbance-induced changes in interaction strength among species (see Chapters 8 and 9) can persist longer than changes in species presence or abundance. Tylianakis et al. (2007) reported that tropical forest conversion to agricultural uses altered the evenness of interactions among cavity-nesting bees and wasps and their parasitoids in ways that focused parasitism on the most abundant bee species in the most modified agricultural systems, potentially affecting pollination and biological control services. Benstead et al. (2007) reported that recovery of aquatic insect populations in Arctic streams was delayed by a disturbance-induced shift in the dominance of primary producers from algae to bryophytes, although algal biomass recovered to predisturbance levels within 2 years.

If the disturbance-free interval is shorter than the time needed for recovery, then the altered community may persist. Conversely, anthropogenic alteration of disturbance dynamics, such as fire suppression, also changes community composition and structure, affecting habitat conditions for associated insects. Because survival and reproduction of individual insects determine population size, distribution, and effects on community and ecosystem processes, the remainder of this chapter focuses on physiological and behavioral characteristics that affect individual responses to variable abiotic conditions.

2 SURVIVING VARIABLE ABIOTIC CONDITIONS

Insects are particularly vulnerable to changes in temperature, water availability, and air or water chemistry because of their relatively large surface area-to-volume ratios. However, many insects can live within suitable microsites that buffer exposure to environmental changes. Insects in aquatic environments or deep in soil or woody habitats may be relatively protected from large changes in air temperature and relative humidity (Curry, 1994; Seastedt and Crossley, 1981a). Moist soils can mitigate heat penetration and protect soil fauna from extreme air temperatures.

Most insects are subject to environmental variability that includes periods of potentially lethal or stressful abiotic conditions (Shelford, 1918; Schowalter, 2012). Therefore, maintaining optimal body temperature, water content, and chemical processes are challenges for survival in variable environments. Insects possess a remarkable variety of physiological and behavioral mechanisms for surviving in variable environments.

Adaptive physiological responses can mitigate exposure to suboptimal conditions. For example, diapause is a general physiological mechanism for surviving seasonally adverse conditions, typically in a resistant stage, such as the pupa of holometabolous insects. Our understanding of the genetic and molecular mechanisms that govern these physiological processes has increased dramatically in the past 20 years.

Diapause induction and termination are controlled by cues such as photoperiod and *degree-day accumulation* (daily degrees above a threshold temperature × number of days) which induce chemical signals from the brain (Denlinger, 2002; Giebultowicz, 2000; Giebultowicz and Denlinger, 1986). In particular, photoreceptors that distinguish day from night trigger expression of genes that measure and accumulate information on day and/or night length and produce proteins that induce diapause (Hardie, 2001). Denlinger (2002) and Giebultowicz (2000) reported that photoperiod affects patterns of expression, whereas temperature affects the amount, of several clock mRNAs (cryptochrome, *cry*, clock, *clk*, period, *per*, and timeless, *tim*) that also regulate circadian rhythms. The relative amounts of these mRNAs show distinct trends from long, warm days to shorter, cooler days, but their precise role in triggering the onset of diapause remains unknown (Denlinger, 2002; Goto and Denlinger, 2002). Various antibiotic proteins also are produced only during diapause, apparently to prevent infection from tissue exposure to gut microorganisms while gut tissues are being reorganized (P. Dunn et al., 1994; K.Y. Lee et al., 2002). Diapause termination often requires a minimum duration of freezing temperatures, or other factors that maximize synchronization of development with seasonally suitable conditions (Ruberson et al., 1998).

Beaver et al. (2002) reported that *Drosophila* males with mutated genes that govern circadian rhythm produced fewer offspring than did wild flies, demonstrating the importance of the genes controlling periodicity. Nevertheless, exposed insects often are killed by sudden or unexpected changes in the temperature, moisture, or chemical conditions of a habitat. Even diapausing insects suffer high mortality due to a combination of temperature, disease, predation, or other factors (Ruberson et al., 1998).

Heat shock proteins (*Hsp*) are commonly expressed in response to environmental stress. These proteins were originally named for their upregulation by *Drosophila* in response to high temperature treatments. However, these proteins are upregulated by a variety of stresses, including cold shock, desiccation, anoxia, and exposure to a wide range of toxic chemicals including heavy metals, ethanol, and pollutants (Bubliy et al., 2012; Rinehart et al., 2007; Sun et al., 2016). These proteins bind to other proteins during periods of stress and prevent changes in protein conformation that would cause abnormal function. Heat shock genes are downregulated during favorable periods, because they could cause deleterious effects on development during nonstressful conditions.

Behavior represents a more flexible means of responding to environmental variation, compared to physiology, because an animal can respond actively to sensory information to avoid or mitigate lethal conditions. Mobile insects have an advantage over sessile species in avoiding or mitigating exposure to extreme temperatures, water availability, or chemical conditions. Limited mobility often is sufficient within steep environmental gradients. Many small, flightless litter species need move vertically only a few mm within the soil profile to avoid lethal temperatures and desiccation at the surface following fire or vegetation removal (Seastedt and Crossley, 1981a). Some species select protected habitats prior to entering diapause to reduce their vulnerability to potential disturbances. K. Miller and Wagner (1984) reported that pandora moth, *Coloradia pandora*, pupae in a ponderosa pine, *Pinus ponderosa*, forest were significantly more abundant on the forest floor in areas with open canopy and sparse litter than in areas with closed canopy and deeper litter. Although other factors also differ between these microhabitats, avoidance of accumulated litter may represent an adaptation to survive frequent ground fires in this ecosystem.

In addition, mobile insects may be able to escape disturbed or unfavorable patches and often can detect and colonize suitable patches within variable environments (D. Johnson, 2004; Roff, 1990). Roff (1990) concluded that flight capability is under strong selection pressure by environmental variability. Consequently, capacity for flight is best developed in insect species associated with variable environmental conditions, whereas secondary flightlessness is most often associated with stable or persistent ecosystem conditions.

Although small body size limits ability to regulate body temperature and water content, many insects are capable of at least limited homeostasis through physiological and/or behavioral mechanisms. Some insects also must deal with variability in chemical or other abiotic conditions.

2.1 Temperature

Insects, as well as other invertebrates, are heterothermic, meaning that their body temperatures are determined primarily by ambient temperature. Rates of metabolic activity (hence, energy and carbon flux) generally increase with temperature. Developmental rate and processes reflect this temperature dependence, but at least some species can regulate body temperature to some degree through physiological or behavioral responses to extreme temperatures.

Insect species show characteristic ranges in temperatures suitable for activity. Aquatic ecosystems have relatively consistent temperatures, but insects in terrestrial ecosystems can experience considerable temperature fluctuation, even on a daily basis. Tolerance limits reflect individual age or maturity, size, color, previous exposure, extent of water reserves, and other factors. For example, pigmented workers of the desert harvester termite, *Hodotermes mossambicus*, forage aboveground throughout the year, over a temperature range of about 8–45°C, whereas unpigmented workers are never seen above-ground (J. Mitchell et al., 1993). Meisel (2006) reported that the army ant, *Eciton burchellii*, is restricted to forest fragments in Costa Rica because workers survived for <3 min at 51°C (the midday temperature of surrounding pastures) and only 18 min at 43°C.

As a group, insects can survive at temperatures from well below freezing to 40–50°C (Whitford, 1992), depending on adapted tolerance ranges and acclimation (preconditioning). Typically, species from higher latitudes survive lower temperatures than do insects from lower latitudes (Deutsch et al., 2008; Hazell et al., 2010; R. Lee, 2010). Similarly, insect species from warmer environments survive higher maximum temperatures than do cold-adapted species (Cockerell et al., 2014; Deutsch et al., 2008). Chironomid larvae living in hot springs survive water temperatures of 49–51°C (R. Chapman, 1982).

In general, the developmental rate of heterotherms increases with temperature. Both terrestrial and aquatic insects respond to the accumulation of thermal units (the sum of degree-days above a threshold temperature) (Baskerville and Emin, 1969; Ward, 1992; Ward and Stanford, 1982). Degree-day accumulation can be similar under different conditions, for example, mild winter/cool summer and cold winter/hot summer, or quite different along elevational or latitudinal gradients. Anthropogenic activities can significantly alter thermal conditions, especially in aquatic habitats. Discharge of heated water, artificial mixing of thermal strata, impoundment, diversion, regulation of water level and flow, and canopy opening in riparian zones, through harvest or grazing, severely modify the thermal environment for aquatic species and favor heat-tolerant individuals and species over heat-intolerant individuals and species (M. Stone and Wallace, 1998; Ward and Stanford, 1982).

Many insects survive temperatures as low as −30°C, and some Arctic and Antarctic species survive below −50°C (N. Hadley, 1994; R. Lee, 2010; R. Lee et al., 2006; Lundheim and Zachariassen, 1993; B. Sinclair et al., 2003; Teets and Denlinger, 2014). Tolerance to freezing temperatures can be enhanced by prior exposure to nonlethal chilling for periods as short as minutes or hours, a process known as *rapid cold hardening*, or RCH (R. Lee and Denlinger, 2010; R. Lee et al., 2006; Teets et al., 2012b). Chilling to 0°C caused a 40% increase in calcium concentration in tracheal cells in the freeze-tolerant goldenrod gall fly, *Eurosta solidaginis*, and the freeze-intolerant flesh fly, *Sarcophaga bullata* (Teets et al., 2013). These conditions increased the activity of calcium/calmodulin-dependent protein kinase II (CaMKII). Inhibition of calcium entry, calmodulin activation, and CaMKII activity prevented RCH in tissue cultures, indicating that calcium signaling is required for RCH to occur and that calcium-mediated cold sensing is a general feature of insects.

Freeze-tolerant species typically undergo a variety of physiological changes that lower the freezing point of body tissues, that is, *supercooling* (R. Lee, 2010). Elimination of ice-nucleating materials is essential to avoid ice formation in tissues. Some species can survive ice formation in extracellular fluids but not ice formation in intracellular fluids (N. Hadley, 1994; Lundheim and Zachariassen, 1993). Ice-nucleating lipids and/or lipoproteins can be used to ensure that ice forms in extracellular fluids at relatively high temperatures, that is, above −10°C (Duman et al., 2010; N. Hadley, 1994; R. Lee, 2010; B. Sinclair et al., 1999). Extracellular freezing draws water osmotically from cells, thereby dehydrating cells and lowering the freezing point of intracellular fluids (N. Hadley, 1994). B. Sinclair et al. (1999) showed that temperatures at which supercooling and ice nucleation occurred differed among alpine, subalpine, and lowland species and populations of the New Zealand weta, *Hemideina* spp., in ways suggesting that freeze tolerance among species in this genus is not a specific adaptation to alpine climate.

Other species have various mechanisms for lowering their freezing or supercooling points. Voiding the gut at the onset of cold conditions may prevent food particles from serving as nuclei for ice formation. Similarly, nonfeeding stages may have lower supercooling points than do feeding stages (N. Hadley, 1994; Kim and Kim, 1997). Some insects prevent freezing to temperatures as low as −50°C by producing high concentrations (up to 25% of fresh weight) of alcohols and sugars, such as glycerol, glucose, and trehalose, as well as peptides and proteins in the hemolymph (N. Hadley, 1994; Lundheim and Zachariassen, 1993). In many cases, a multicomponent cryoprotectant system involving a number of compounds prevents accumulation of potentially toxic levels of any single component (N. Hadley, 1994). Cold tolerance varies with life stage, rate of cooling, lowest temperature, and exposure time and can be enhanced by preconditioning to sublethal temperatures (Kim and Kim, 1997; B. Sinclair et al., 1999). Rivers et al. (2000) reported that cold hardiness in a pupal parasitoid, *Nasonia vitripennis*, was enhanced by encasement within the flesh fly host, *Sarcophaga crassipalpis*, and by acquisition of host cryoprotectants, especially glycerol and alanine, during larval feeding.

The cuticle of the Oriental hornet, *Vespa orientalis*, has a surface structure that enhances absorption of solar energy by pigments in the endocuticle: melanin in the brown-colored cuticle or xanthopterin in the yellow-colored cuticle (Plotkin et al., 2010). The hornet maximizes interception of solar energy through timing of activity and body orientation to maintain body temperature during cold periods.

Many insects can reduce body temperature at high ambient temperatures, above 45°C (Casey, 1988; Heinrich, 1974, 1979, 1981, 1993). An Australian montane grasshopper, *Kosciuscola*, changes color from black at night to pale blue during the day (Key and Day, 1954), thereby regulating heat absorption.

Evaporative cooling, through secretion, regurgitation, ventilation, or other means, can lower body temperature 5–8°C below high ambient temperatures when the air is dry (N. Hadley, 1994). Prange and Pinshow (1994) reported that both sexes of a sexually dimorphic desert grasshopper, *Poekilocerus bufonius*, depress their internal temperatures through evaporative cooling. However, males lost proportionately more water through evaporation, but retained more water from food, than did the much larger females, indicating that thermoregulation by smaller insects is more constrained by water availability.

Long-term exposure to high temperatures requires high body water content or access to water, to avoid desiccation (R. Chapman, 1982; N. Hadley, 1994). N. Hadley (1994) described experiments demonstrating that males of a Sonoran Desert cicada, *Diceroprocta apache*, maintain evaporative cooling by ingesting xylem water from twigs on which they perch while singing. Although this species has high cuticular permeability, even at nonstressful temperatures, water loss ceases at death, indicating active cuticular pumping of body water. A 0.6 g cicada maintaining a temperature differential of 5°C must siphon at least 69 mg xylem fluid h^{-1}. Laboratory experiments indicated that maintaining this temperature differential resulted in a 5% increase in metabolic rate over resting levels. These cicadas probably have additional energetic costs associated with rapid extraction and transport of ingested water to the cuticle.

Thermoregulation also can be accomplished behaviorally. Some insects regulate body temperature by optimal positioning (Heinrich, 1974, 1993). Web-building spiders adjust their posture to control their exposure to solar radiation (Robinson and Robinson, 1974). Desert beetles, grasshoppers, and scorpions prevent overheating by stilting, that is, extending their legs and elevating the body above the heated soil surface and by orienting the body to minimize the surface area exposed to the sun (Heinrich, 1993).

Heinrich (1974, 1979, 1981, 1993), Casey (1988), and Bishop and Armbruster (1999) reviewed studies demonstrating that a variety of insects are capable of thermoregulation through activities that generate metabolic heat, such as fanning the wings and flexing the abdomen (Fig. 2.11). Flight can elevate body temperature 10–30°C above ambient (R. Chapman, 1982; Heinrich, 1993). A single bumble bee, *Bombus vosnesenskii*, queen can raise the temperature of the nest as much as 25°C above air temperatures as low as 2°C, even in the absence of insulating materials (Heinrich, 1979). Bees also can cool nests by fanning wings to increase ventilation (M. Yang et al., 2010).

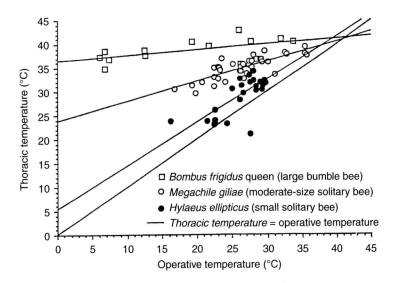

FIGURE 2.11 **Thoracic temperature of live bees relative to operative temperature, measured with a thermocouple inserted in a freshly killed bee, equilibrated to ambient temperature, to account for effects of solar radiation and convection ambient temperature, for representative small, medium, and large bee species.** Each point represents a measurement for an individual bee. Thermoregulatory capability for each bee species is indicated by respective regression lines and can be compared to the isothermal line (thoracic temperature = operative temperature). *Source: From Bishop, J.A., Armbruster, W.S., 1999. Thermoregulatory abilities of Alaskan bees: effects of size, phylogeny and ecology. Funct. Ecol. 13, 711–724.*

Some insects use or construct shelters to trap or avoid heat. Tent caterpillars, *Malacosoma* spp., build silken tents that slow dissipation of metabolic heat and increase colony temperature above ambient (Fig. 2.12) (Fitzgerald, 1995; Heinrich, 1993). L. Moore et al. (1988) reported that overwintering egg masses and tents of the western tent caterpillar, *Malacosoma californicum*, occurred significantly more often on sides of trees, or isolated trees, exposed to the sun. Tents of overwintering larvae of the arctiid moth, *Lophocampa argentata*, occur almost exclusively in the exposed upper canopy and significantly more often on the south-facing sides of host conifers in western Washington, USA (D. Shaw, 1998).

Insects can sense and often move within temperature gradients to thermally optimal habitats. Gutiérrez and Wilson (2014) reported that upward movement of the brimstone butterfly, *Gonepteryx rhamni*, during summer in montane regions of southern Spain was due to a negative response to increasing temperatures at lower elevations. Movement of individuals also was positively related to forest cover, which maintained lower temperatures than did nonforested areas. Aquatic insects move both vertically and horizontally within temperature gradients to select sites of optimal temperatures (Ward, 1992). Terrestrial insects frequently bask on exposed surfaces to absorb heat during early morning or cool periods (Fig. 2.13), and retreat to less exposed sites during warmer periods. Other species burrow to depths at which diurnal temperature fluctuation is minimal (Polis et al., 1986). Seastedt and Crossley (1981a) reported significant redistribution of soil/litter arthropods from the upper 5 cm of the soil profile to deeper levels following canopy removal and consequent soil surface exposure and warming in a forested ecosystem. Light is an important cue that often attracts insects to sources of heat or repels them to darker, cooler areas.

FIGURE 2.12   Tent caterpillars, *Malacosoma* spp., and other tent-constructing Lepidoptera, reduce airflow and variation in temperatures within their tents.

Finally, heat tolerance can be conferred by symbiotic microorganisms. J. Russell and Moran (2006) demonstrated that some bacterial symbionts confer heat tolerance on *Aphis pisum*, significantly increasing aphid survival at temperatures above 25°C. However, whereas *Serratia symbiotica* and *Hamiltonella defensa* contributed to heat tolerance, *Regiella insecticola* had the opposite effect.

2.2 Water balance

Maintenance of homeostatic water balance also is a challenge for organisms with high surface area-to-volume ratios (Edney, 1977; N. Hadley, 1994). The arthropod exoskeleton is an important mechanism for control of water loss. Larger, more heavily sclerotized arthropods

FIGURE 2.13 **Many insects, such as dragonflies, raise their body temperatures by basking. Heat absorption is enhanced by dark coloration and orientation.** *Source: Photo courtesy of S.D. Senter.*

are less susceptible to desiccation than are smaller, more delicate species (Alstad et al., 1982; Kharboutli and Mack, 1993). Nevertheless, many insects are vulnerable to dehydration. For example, Gautam and Henderson (2015) reported that Formosan subterranean termite, *Coptotermes formosanus*, workers were unable to recover when loss of body water exceeded 35%.

Arthropods in xeric environments typically are larger, have a thicker cuticle, and secrete more waxes to inhibit water loss, compared to insects in mesic environments (Crawford, 1986; Edney, 1977; N. Hadley, 1994; Kharboutli and Mack, 1993). Cuticular lipids with higher melting points might be expected to be less permeable to water loss than are lipids with lower melting points. Gibbs (2002a) evaluated cuticular permeability relative to water loss for several arthropod species and found that all species produced lipids with low melting points as well as high melting points. Furthermore, lipids with high melting points did not reduce rates of water loss (Gibbs, 2002a; Gibbs et al., 2003).

Some species in xeric environments conserve metabolic water (from oxidation of food) or acquire water from condensation on hairs or spines (R. Chapman, 1982; N. Hadley, 1994). Carbohydrate metabolism, to release bound water, increases several-fold in some insects subjected to desiccation stress (Marron et al., 2003). Others tolerate water loss of 17–89% of total body water content (Gibbs, 2002b; N. Hadley, 1994). Dehydration tolerance in *Drosophila* apparently reflects phylogeny rather than adaptation to desert environments (Gibbs and Matzkin, 2001). Some insects regulate respiratory water loss by controlling spiracular activity under dry conditions (Fielden et al., 1994; N. Hadley, 1994; Kharboutli and Mack, 1993). Water conservation is under hormonal control in some species. An antidiuretic hormone is released in desert locusts, *Schistocerca gregaria*, and other species under conditions of water loss (Delphin, 1965).

Gibbs et al. (2003) compared the three main water loss pathways among *Drosophila* species from xeric and mesic habitats. Excretory loss was <6% of the total and did not differ among species from different habitats. No consistent relationship was found between cuticular properties and water loss. Cuticular water loss rates did not appear to differ among flies from different habitats. Respiratory water loss differed significantly between xeric and mesic species. Xeric species of the same size had lower metabolic rates, were less active, and showed a cyclic pattern of CO_2 release, compared to mesic species, indicating adaptation to reduce respiratory loss.

Some insect species tolerate extreme water loss. A majority of Antarctic arthropods have highly permeable cuticles but tolerate 50–70% loss of body water (Teets and Denlinger, 2014).

Extreme dehydration may trigger the onset of *anhydrobiosis*, a physiological state characterized by an absence of free water and of measurable metabolism (N. Hadley, 1994; Whitford, 1992). Survival during anhydrobiosis requires stabilization of membranes and enzymes by compounds other than water, for example, glycerol and trehalose, whose synthesis is stimulated by dehydration (N. Hadley, 1994). Anhydrobiosis is common among plant seeds, fungi, and lower invertebrates, but among insects only some larval Diptera and adult Collembola have been shown to undergo anhydrobiosis (N. Hadley, 1994). Hinton (1960a,b) reported that a chironomid fly, *Polypedilum vanderplanki*, found in temporary pools in central Africa, withstands repeated dehydration to 8% of body water content. At 3% body water content, this midge is capable of surviving temperatures from −270 to 100°C, a range that contrasts dramatically with its tolerance range when hydrated.

Insects and other arthropods are most vulnerable to desiccation at times when a new exoskeleton is forming, that is, during eclosion from eggs, during molts, and during diapause (Crawford, 1978; Willmer et al., 1996). Tisdale and Wagner (1990) found that the percentage of sawfly, *Neodiprion fulviceps*, eggs hatched was significantly higher at relative humidities ≥50%. Yoder et al. (1996) reported that slow water loss through the integument and respiration by diapausing fly pupae was balanced by passive water vapor absorption from the air at sufficiently high humidities. The ability of adult insects to regulate water loss may decline with age (Gibbs and Markow, 2001).

Insects in diapause at subfreezing temperatures are subject to freeze-drying. Lundheim and Zachariassen (1993) reported that beetles that tolerate ice formation in extracellular fluids have lower rates of water loss than do insects that have supercooled body fluids, perhaps because the hemolymph in frozen beetles is in vapor pressure equilibrium with the surrounding ice, whereas the hemolymph in supercooled insects has a vapor pressure higher than the environment.

On the other hand, insects subjected to flooding must contend with excess water. Subterranean termites can survive short periods of inundation by entering a quiescent state; the relative abilities of species to withstand periods of flooding correspond to their utilization of above-ground or below-ground wood resources (Forschler and Henderson, 1995). Litter-dwelling ants are vulnerable to seasonal flooding in Amazonian forests (Mertl et al., 2009). Specialist predators were virtually eliminated by flooding; one *Hypoponera* species was adapted to a high degree of flooding, increasing in abundance with the frequency and duration of flooding. M. Webb and Pullin (1998) found that pupae of a wetland butterfly, *Lycaena dispar batavus*, could tolerate 28 days of submergence, but survival was negatively correlated with duration of submergence between 28 and 84 days. However, inundation also affects oxygen availability (see Section 2.3), as well as water balance.

Insects that ingest liquid food immediately excrete large amounts of water to concentrate dissolved nutrients. Elimination of excess water (and carbohydrates) in sap-feeding Hemiptera is accomplished in the midgut by rapid diffusion across a steep moisture gradient created by a filter loop (R. Chapman, 1982). The resulting concentration of sugars in honeydew excreted by phloem-feeding Hemiptera is an important resource for ants, hummingbirds, predaceous Hymenoptera, and sooty molds (Dixon, 1985; E. Edwards, 1982; N. Elliott et al., 1987; Huxley and Cutler, 1991). The abundant water excreted by xylem-feeding spittlebugs is used to create the frothy mass that hides the insect. Excretion in some species, such as the blood-feeding *Rhodnius* (Hemiptera), is controlled by a diuretic hormone (Maddrell, 1962).

Water balance also can be maintained behaviorally, to some extent, by retreating to cooler or moister areas to prevent desiccation. Burrowing provides access to more mesic subterranean environments (Polis et al., 1986). Termites construct their colonies to optimize temperature and moisture conditions. Formosan subterranean termites, *Coptotermes formosanus*, prefer nest sites with high moisture availability (Fei and Henderson, 1998). Metabolic heat generated in the core of the nest rises by convection into large upper cavities and diffuses to the sides of the nest where air is cooled and gaseous exchange occurs through the thin walls. Cooled air sinks into lower passages (Lüscher, 1961). The interior chambers of termite colonies typically have high relative humidities.

2.3 Air and water chemistry

Air and water chemistry affect insect physiology. Oxygen supply is critical to survival, and limits arthropod size (Chapelle and Peck, 1999; Dudley, 1998). Oxygen availability may become more limited under certain conditions. Airborne or dissolved chemicals can affect respiration and development. Soil or water pH can affect exoskeleton function and other physiological processes. Changes in concentrations of various chemicals, especially those affected by industrial activities, affect many organisms, including insects.

Oxygen supply can limit activity and survival of aquatic species and some terrestrial species living in enclosed habitats. Less oxygen can remain dissolved in warm water than in cold water. Stagnant water can undergo oxygen depletion as a result of algal and bacterial respiration (Ward, 1992). Some insect species living in oxygen-poor environments have more efficient oxygen delivery systems, such as increased tracheal supply, gills, or breathing tubes that extend to air supply (L. Chapman et al., 2004; R. Chapman, 1982). For example, the hemolymph of some aquatic chironomid larvae and endoparasitic fly larvae is unique among insects in containing a hemoglobin that has a higher affinity for oxygen than does mammalian hemoglobin (R. Chapman, 1982; Pinder and Morley, 1995). Oxygen supply can be enhanced by ventilatory movement, that is, movement of gills or other body parts to create currents that maintain oxygen supply and reduce the diffusion barrier (Ward, 1992). Other species must employ siphon tubes (eg, mosquito and syrphid fly larvae) or return to the surface (diving beetles) to obtain atmospheric oxygen (L. Chapman et al., 2004).

Insect species that live in riparian or other frequently flooded habitats often show adaptations to survive periods of anoxia in saturated soils. Larval tiger beetles, *Phaeoxantha klugii*, found in central Amazonian floodplains are able to tolerate anoxic conditions in flooded soils for up to 3.5 months at 29°C (Zerm and Adis, 2003). This exceptional degree of anoxia tolerance appeared to require several days of induction as water levels rose, suggesting vulnerability to more rapid inundation. Brust and Hoback (2009) found that tolerance to hypoxia among some other tiger beetle species, *Cicindela* spp., was not related to likelihood of immersion. Some wood-boring species must be able to tolerate low oxygen concentrations deep in decomposing wood, although O_2 limitation may occur only in relatively sound wood or water-soaked wood (Hicks and Harmon, 2002).

Increased atmospheric CO_2 appears to have little direct effect on insects or other arthropods. However, relatively few insect species have been studied with respect to CO_2 enrichment. Increased atmospheric CO_2 can significantly affect the quality of plant material for some herbivores (Arnone et al., 1995; Bezemer and Jones, 1998; Bezemer et al., 1998; Fajer et al., 1989; Kinney et al., 1997; Lincoln et al., 1993; Lindroth et al., 1993a,b; Roth and Lindroth, 1994) and detritivores (Grime et al., 1996; Hirschel et al., 1997), although plant response to CO_2 enrichment depends on a variety of environmental factors (Lawton, 1995; Lindroth et al., 1993a,b,c; Watt et al., 1995, see Chapter 3). In general, leaf chewers compensate for effects of elevated CO_2 by increasing consumption rates, whereas sap-suckers show reduced development times and increased population size (Bezemer and Jones, 1998; Zavala et al., 2013). Mondor et al. (2005) reported that two color morphs of pea aphids, *Acyrthosiphon pisum*, differed in response to elevated CO_2. The green genotype responded positively to CO_2 enrichment, but the pink genotype did not, altering genetic frequencies from 1:1 to 9:1 and potentially altering other aspects of aphid life history and interactions. However, plant and herbivore responses to elevated CO_2 are highly idiosyncratic, making prediction difficult. Nevertheless, at least some herbivorous species are likely to become more abundant and cause greater crop losses as a result of increased atmospheric CO_2 (Bezemer et al., 1998).

Soil and water pH affects a variety of chemical reactions, including enzymatic activity. Changes in pH resulting from acidification (such as from volcanic or anthropogenic activity) affect osmotic exchange, gill and spiracular surfaces, and digestive processes. Changes in pH often are correlated with other chemical changes, such as increased N or S, and effects of pH change may be difficult to separate from other factors. van Straalen and Verhoef (1997) found that several species of soil collembolans and oribatid mites varied in their responses to acidic or alkaline soil conditions. Pinder and Morley (1995) reported that many chironomid species are relatively tolerant of alkaline water, but few are tolerant of pH < 6.3. Other aquatic species also may be unable to survive in low-pH water (Batzer and Wissinger, 1996). Acid deposition and loss of pH buffering capacity likely will affect survival and reproduction of aquatic and soil/litter arthropods (Curry, 1994; Pinder and Morley, 1995).

2.4 Other abiotic factors

Dust and ash from volcanic eruptions or fires kill many insects, as a result of burial, ingestion of toxic ash, or abrasion of the thin epicuticular wax-lipid film that is the principal barrier to water loss, causing death by desiccation (Alstad et al., 1982; R. Cook et al., 1981; Fernández-Arhex et al., 2013; Marske et al., 2007; Masciocchi et al., 2013). Insects exposed to volcanic debris also can suffer gut epithelial stress and cell death from accumulation of heavy metals (Rodrigues et al., 2008). Ash accumulation and retention by aquatic insects following the eruption of Mount St. Helens in Washington USA was affected by exoskeletal sculpturing, armature, and pubescence (Gersich and Brusven, 1982). Substantial accumulation was noted on respiratory structures, potentially interfering with respiration. Ash-covered insects showed increased activity and orientation upstream, which successfully washed ash off within 24 hr. However, ash coating over cobbles, pebbles, and sand significantly inhibited colonization of these substrates (Brusven and Hornig, 1984). Miserendino et al. (2012) reported that the eruption of Chaitén Volcano in Argentina in 2008 significantly reduced the abundances of at least 25 aquatic macroinvertebrate taxa in five river systems; species richness and density did not recover to preeruption values for at least 2 years.

Heavy metals, such as Al, Co, Cr, Fe, Mg, Ni, Se, and Zn, also may be toxic to insects feeding on plant tissues containing these elements (Boyd, 2013; Boyd and Martens, 1994; Boyd and Moar, 1999, see Chapter 3). Meindl and Ashman (2013) reported that bumble bees, *Bombus impatiens*, feed for shorter periods on Ni-contaminated flowers and subsequently avoided Ni-contaminated flowers, but were not affected by Al contamination. Meindl et al. (2013) further reported that plants growing on serpentine soils, with high concentrations of Co, Cr, Ni, Fe, Mg, and Zn, were visited less often by pollinators and suffered less damage from flower-feeding herbivores.

Many aquatic insects are sensitive to water level and flow rate (Ward, 1992). These factors can fluctuate dramatically, especially in seasonal habitats, such as desert playas, intermittent streams, wetlands, and perched pools in treeholes and bromeliads (phytotelmata). Water level affects both temperature and water quality: temperature because smaller volumes absorb or lose heat more quickly than do larger volumes, water quality because various solutes become more concentrated as water evaporates. Insects, and other aquatic arthropods, show life history adaptations to seasonal patterns of water availability or quality, often undergoing physiological diapause as water resources disappear (Batzer and Wissinger, 1996; Ward, 1992). Although most mosquitoes oviposit in surface water, floodwater mosquitoes, *Aedes* spp. and *Psorophora* spp., oviposit in soil at the high water line. Their eggs are resistant to desiccation and can remain dormant for several years. Egg hatch is stimulated by flooding, and the number of generations at a site depends on the frequency of flooding (Schäfer and Lundström, 2006; Wiggins et al., 1980). Flow rate affects temperature and oxygenation, with cooler temperature and higher oxygen content at higher flow rates, but high flow rates can physically dislodge and remove exposed insects. McCreadie and Colbo (1993) and P. Adler and McCreadie (1997) reported that sibling species of black flies, *Simulium*, select different stream microhabitats on the basis of their adaptations to water velocity (Fig. 2.14).

Many insects are sensitive to changes in photoperiod or to ultraviolet and polarized radiation. Kelly et al. (2003) demonstrated, using experimental filters, that aquatic insects in the Little Qualicum River on Vancouver Island showed differential sensitivities to UVA and UVB exposure. Caddisflies (Trichoptera), especially *Dicosmoecus* spp. (Limnephilidae), were most sensitive to UV exposure. Final abundances under photosynthetically active radiation (PAR) alone were 15 times higher than under PAR + UVA and 40 times higher than under PAR + UVA + UVB. Stoneflies (Plecoptera) were 51% more abundant under PAR than under UV exposure. In contrast, chironomids (Diptera) were more abundant in the UV treatments. Mazza et al. (2002) and Rousseaux et al. (2004) reported that feeding by herbivorous insects was higher on foliage exposed to lower levels of UVB radiation, perhaps mediated by effects of UVB radiation on plant defensive compounds (see Chapter 3).

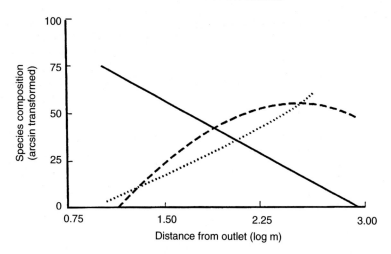

FIGURE 2.14   **Relationship between sibling species composition of black flies (*Simulium truncatum, solid line; S. verecundum, dashed line;* and *S. venustum, dotted line*) and distance from lake outlets on the Avalon Peninsula, Newfoundland, in early June.** Least squares regression equations were significant at $p < 0.01$; adjusted R^2 values were 92%, 85%, and 68% for the three species, respectively. *Source: From Adler, P.H., McCreadie, J.W., 1997. The hidden ecology of black flies: sibling species and ecological scale. Am. Entomol. 43, 153–161.*

3 DISPERSAL BEHAVIOR

Insects have a considerable capacity to escape adverse conditions and to find optimal conditions within temperature, moisture, and chemical gradients across landscapes or watersheds. Dispersal, the movement of individuals away from their source population, is an important adaptive behavior that minimizes the risk that an entire population will be destroyed by disturbance or resource depletion, maximizes the chance that some individuals will find and exploit new resources, and maximizes genetic heterogeneity (D. Johnson, 2004; Schowalter, 2012; Wellington, 1980, see Chapter 5).

Flight capacity contributes enormously to insects' ability to disperse. Adult aquatic insects can disperse from an intermittent pond or stream before the water disappears and search for other bodies of water. Dispersal may be particularly important for distributing populations and minimizing risk in ecosystems characterized by frequent disturbances (Cantrell et al., 2010; Roff, 1990). Lytle and Smith (2004) reported that 13 species of giant water bugs, including *Lethocerus medius* and *Abedus herberti*, use rainfall as a cue to abandon an aquatic habitat. In addition to allowing individuals to escape flash floods, this behavior has led to a correlation between rainfall and migration to seasonal breeding habitats. A number of ant species that characterize frequently flooded riparian ecosystems are capable of forming large rafts or floating balls of interlocked individuals that can disperse downstream en masse during floods, thereby both surviving flooding and dispersing to new habitats (C. Anderson et al., 2002; Mertl et al., 2009; Mlot et al., 2011; E. Wilson, 1986).

Nevertheless, dispersal entails considerable risk and requires considerable energy expenditure (Rankin and Burchsted, 1992, see Chapters 4 and 5). Torres (1988) documented cases of exotic insects being introduced into Puerto Rico by hurricane winds, including a swarm of desert locusts blown across the Atlantic Ocean from Africa. Few, if any, of these species became established in Puerto Rico. Many insects (and other organisms) fail to find or reach suitable habitats. Similarly, whereas aquatic invertebrates may actively enter the water column to escape nocturnal benthic predators (Schäffer et al., 2013), such drift also provides food resources for active diurnal drift-feeding predators, such as trout (Hammock et al., 2012).

4 RESPONSES TO ANTHROPOGENIC CHANGES

The current rate and extent of ecosystem fragmentation, urbanization, and other land use changes are universal, pervasive, and unprecedented. Probably no ecosystem on Earth remains unaffected by invasive species, changes in atmospheric chemistry, or climate change. Many species will be unable to adapt to these multiple changes quickly enough to survive, given that life history attributes (natality, longevity, dispersal rate, etc.) are adapted trade-offs under prevailing conditions (Bouget et al., 2015).

Human activities have increasingly introduced novel changes in ecosystems that may affect species in unpredictable ways. Insect responses to anthropogenic disturbances reflect the degree to which habitat conditions following anthropogenic disturbance are similar to those following natural disturbances, for example, forest harvest may elicit responses similar to natural canopy opening disturbances (Buddle et al., 2006; Grimbacher and Stork, 2009); vegetation conversion to crop production elicits insect responses to changes in density and apparency of early successional hosts (see Chapter 3); river impoundment may elicit responses similar to landslides that also alter drainage patterns.

4.1 Land use

Anthropogenic changes seriously affect population distribution by reducing habitat area, isolating demes (subpopulations restricted to habitat remnants), and interfering with dispersal, potentially threatening species incapable of surviving in, or escaping, increasingly inhospitable landscapes (DeBano, 2006; de Jesús-Crespo and Ramírez, 2011; R. Fox, 2013; Samways et al., 1996; Shure and Phillips, 1991; A. Suarez et al., 1998; Summerville and Crist, 2001). Ecosystem fragmentation and urbanization alter regional abiotic variables and have the most dramatic effects on survival or movement of various insects (J. Chen et al., 1995; Franklin et al., 1992; Raffa et al., 2008; Roland, 1993; Rubenstein, 1992, see Chapter 11). Braschler et al. (2009) found that many orthopterans avoided the mown matrix in a fragmented grassland, likely because of the lack of shelter (Fig. 2.15). Consequently, small populations became increasingly isolated in remnant patches of grassland, increasing their vulnerability to local extinction if large areas are mown simultaneously. Kimberling et al. (2001) reported that physical disturbance related to construction or waste disposal had relatively less effect on invertebrate communities than did conversion of shrub-steppe to agricultural use in eastern Washington, USA. Nessimian et al. (2008) found that aquatic insect assemblages in central Amazonia differed significantly between streams passing through forest versus pasture, but not among streams passing through primary forest, forest fragments, or secondary forests. Altered drainage patterns affect the temperature and chemical conditions of aquatic ecosystems and opportunities for organisms to disperse upstream or downstream (Pringle et al., 2000).

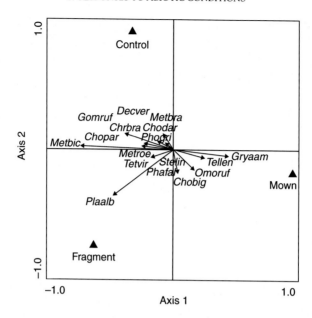

FIGURE 2.15 **Redundancy analysis (RDA) biplot showing the effect of treatment (control, mown, and fragmented grassland plots) on orthopteran species composition.** Most of the species were most abundant in the undisturbed plots, with the exception of *Gryllus campestris, Tetrix tenuicornis,* and *Omocestus rufipes,* which were more abundant in mown plots and *Platycleis albopunctata,* which was more abundant in the isolated fragments of undisturbed grassland. *Source: From Braschler, B., et al., 2009. Effects of small-scale grassland fragmentation and frequent mowing on population density and species diversity of orthopterans: a long-term study. Ecol. Entomol. 34, 321–329; with permission.*

Paving previously vegetated surfaces, especially in concentrated urban settings, has created the most extreme changes in habitat conditions for organisms sensitive to high temperature and desiccation (Akbari et al., 2009; Arnfield, 2003). Urban heat islands can be as much as 10°C hotter than surrounding vegetated ecosystems (Arnfield, 2003). Furthermore, anthropogenic changes in climate and land use appear to be increasing the frequency of extreme weather events (Bender et al., 2010; Hossain et al., 2009; Kishtawal et al., 2010; Lubchenco and Karl, 2012), which likely will affect species survival more than will changes in average conditions (Estay et al., 2014; Gutschick and BassiriRad, 2010; Jentsch et al., 2007; Kaushal et al., 2010; Reusch et al., 2005).

Human extraction of ecosystem services also changes habitat conditions. Andresen and Laurance (2007) found that hunting of forest mammals in Panama reduced mammal abundances and the availability of fecal resources for dung beetles, reducing dung beetle abundance and species composition. Diversion of streams for irrigation purposes alters water level, flow rate, and temperature, changing habitat conditions for aquatic insects and altering community structure (S. Miller et al., 2007). Bredenhand and Samways (2009) reported that a single dam on the Eerste River in the Cape Floristic Region of South Africa altered the composition and abundance of aquatic macroinvertebrates. Species diversity below the dam was only half that in the pristine section above the dam. Diversity and abundance of Ephemeroptera, Plecoptera, Trichoptera (excluding Hydropsychidae), and Coleoptera below the dam were 30%, 1%, 25%, and 6%, respectively, of their values above the dam, but Hydropsychidae were 15 times more abundant below the dam, largely reflecting changes in riparian vegetation and stream conditions below the dam. Excavation of shallow pits during brickmaking in subSaharan Africa increased predator-free habitat for, and larval abundance of, *Anopheles gambiae*, the primary vector of malaria in the region (Carlson et al., 2009). Predator diversity in these pools increased with time since disturbance and was a major factor in ultimately reducing *A. gambiae* larval populations.

4.2 Pollutants and other novel introductions

Some anthropogenic changes are unique, particularly the introduction of novel elements into the environment. Fires and other natural disturbances do not generate large numbers of stumps with exposed surfaces and in-ground root systems that offer unique opportunities for insect and microbial colonization.

Airborne and dissolved pollutants include volatile emissions of industrial origin. Fluorides, sulfur compounds, nitrogen oxides, and ozone affect many insect species directly, although the physiological mechanisms of toxicity are not well-known (Alstad et al., 1982; Heliövaara, 1986; Heliövaara and Väisänen, 1986, 1993; Pinder and Morley, 1995). Disruption of epicuticular or spiracular tissues by these reactive chemicals may be involved. Trumble and Jensen (2004) reported that concentrations of 500–1000 μg g^{-1} of hexavalent Cr (within the range of environmental contamination) in diet fed to a terrestrial dipteran detritivore, *Megaselia scalaris*, increased development time by 65–100% and reduced survival by 50–94%. Since females did not discriminate among substrates varying in Cr levels, populations are not likely to persist in contaminated areas, thereby reducing rates of decomposition and nutrient flux.

Although there is little evidence for direct effects of realistic concentrations of these major air pollutants on terrestrial herbivores, many herbivorous species respond to changes in the quality of plant resources or abundance of predators resulting from exposure to these pollutants (V.C. Brown, 1995), thereby threatening food web interactions and ecosystem processes (Butler and Trumble, 2008; Butler et al., 2009; Mogren and Trumble, 2010). Kainulainen et al. (1994) found that exposure of Scots pine, *Pinus sylvestris*, seedlings to ozone significantly reduced amounts of starch, and total amino acids at the highest ozone concentration (0.3 ppm), but did not affect other sugars or other secondary compounds. Reproduction of grey pine aphids, *Schizolachnus pineti*, was not significantly affected by ozone exposure. Jøndrup et al. (2002) reported that susceptibility of *Brassica rapa* to ozone interacted with ozone-induced biochemical changes that affected its suitability for a specialized herbivore, *Pieris brassicae*.

However, pollutants may cause sublethal effects that increase predation risk (Stark and Banks, 2003) or interfere with olfactory detection of hosts by predators or parasites. Gate et al. (1995) exposed braconid parasitoids, *Asobara tabida*, to ozone, sulfur dioxide, and nitrogen dioxide in chambers with aggregations of its host, *Drosophila subobscura*. Ozone, but not sulfur dioxide or nitrogen dioxide, significantly reduced searching efficiency and the proportion of hosts that were parasitized. Parasitoids were able to avoid patches with no hosts, but appeared to be less able to distinguish different host densities, indicating that air pollutants could reduce the effect of predation or parasitism.

Aquatic organisms historically had minimal exposure to the variety of toxins introduced into aquatic systems through agricultural and industrial production. Petrochemical contamination in streams may have little direct effect on aquatic organisms but can cause oxygen depletion and increased CO_2 concentration, leading to changes in aquatic invertebrate composition and eventual elimination of all fauna (Couceiro et al., 2007; Harrel, 1985). Oil spills and urban sewage in streams affect not only the aquatic fauna but also terrestrial fauna in seasonally flooded habitats (Couceiro et al., 2007). Many toxic metals, such as arsenic, copper, and mercury, accumulate in aquatic environments as a result of runoff or effluents from mining, smelting, and other industrial processes (Mogren and Trumble, 2010). These toxins can accumulate in aquatic insects and be transferred through aquatic and terrestrial food webs via predation, thereby affecting population and community dynamics (Mogren and Trumble, 2010; Mogren et al., 2013; Stark and Banks, 2003).

Photoperiod is an important factor affecting development, behavior, and/or distribution of many, if not most, insects. Some aquatic insects are negatively phototactic during most of their lives, but may move toward light under conditions of oxygen depletion (Ward, 1992). Algal feeders are more likely to occur in illuminated portions of streams. Moonlight affects drift rates for species that disperse in stream currents and is a synchronizing agent for emergence of a number of aquatic species, especially nocturnal feeders, with different species emerging during different lunar phases (Ward, 1992). A variety of insects are attracted to lights at night, an attribute that facilitates collection and measurement of diversity (see Chapter 9), and normal dispersal or foraging activities may be disrupted by artificial lights (K. Gaston et al., 2014). Aquatic insects often are deceived by horizontally polarized light from dark-colored reflective surfaces, such as automobiles, asphalt roads, and oil spills, that they mistake for surfaces of aquatic habitats (Horváth et al., 2009; Kriska et al., 2006).

Although insects may be less sensitive to radioactive contamination than are vertebrates, they can be affected by exposure to radioactive materials. Yoshimura and Akama (2014) reported that aquatic insects showed significantly elevated concentrations of ^{134}Cs and ^{137}Cs 1 year after the 2011 earthquake and resulting Fukushima Daiichi Nuclear Power Plant accident, apparently due to consumption of contaminated algae. The long-term effects of this radioactive contamination are not yet known. D.D. Williams et al. (2001) found morphological deformities in larvae of 9 of 13 chironomid species 4–8 years after the Chernobyl disaster in Ukraine. For species of *Chironomus*, the most abundant genus, the incidence of deformity was significantly positively related to radiation level. Møller and Mousseau (2009) reported that abundances of bumble bees, butterflies, grasshoppers, dragonflies, and spiders declined significantly with increasing radiation levels at forested sites 25 years after the Chernobyl disaster (Fig. 2.16). Møller et al. (2012) added that abundances of pollinating insects also were reduced significantly in the vicinity of Chernobyl, resulting in a significantly reduced fruit set of trees requiring pollination by insects.

Invasive species can be considered to be biotic pollution because of their dramatic effects on native species and ecosystems. Once established, they can alter abundances and interactions among other species in the same way as abiotic pollutants (Adkins and Rieske, 2013; Herms and McCullough, 2014; Twardochleb et al., 2013). For example, Twardochleb et al. (2013) reported that invasive crayfish are capable is disrupting the entire aquatic food web by feeding on macrophytes, invertebrates, and vertebrates to a much greater extent than do native crayfish. Terrestrial invaders often are favored by habitat disturbance. Red imported fire ants, *Solenopsis invicta*, are most abundant in disturbed habitats (Stiles and Jones, 1998; Zettler et al., 2004). Where abundant, they displace native ants, and negatively affect many ground nesting birds, small mammals, and herpetofauna, through aggressive foraging behavior, high reproductive rates, and lack of predators (C. Allen et al., 2004; S. Porter and Savignano, 1990; Zettler et al., 2004). A. Suarez et al. (2001) found that habitat fragmentation also favored the invasive Argentine ant, *Linepithema humile*, at the expense of native ant species.

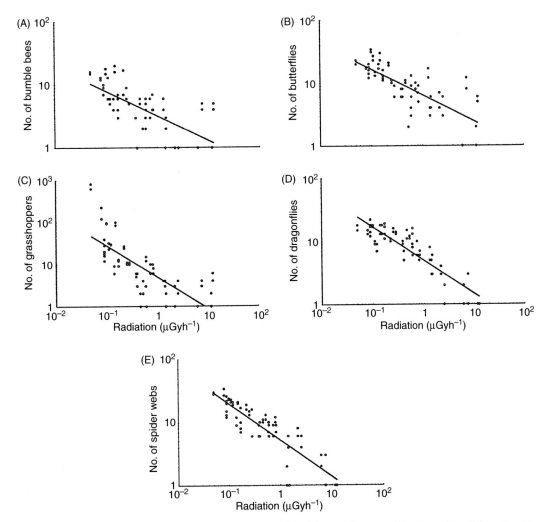

FIGURE 2.16 Abundances of bumble bees (A), butterflies (B), grasshoppers (C), dragonflies (D), and spider webs (E) along four line transects at 17 sites around Chernobyl (Ukraine and Belarus) in July 2008 in relation to background radiation (μGy hr^{-1}). *Source: From Møller, A.P., Mousseau, T.A., 2009. Reduced abundance of insects and spiders linked to radiation at Chernobyl 20 years after the accident. Biol. Lett-UK 5, 356–359.*

4.3 Climate change

Insects are particularly responsive to changes in temperature and humidity, as described in Sections 2.1 and 2.2. Some species are likely to shift geographic ranges toward higher latitudes as cold limitation is alleviated under warming scenarios (Battisti et al., 2006; de la Giroday et al., 2012; Hazell et al., 2010; Netherer and Schopf, 2010). A particularly noteworthy example is the breach of its geoclimatic barrier by the mountain pine beetle, *Dendroctonus ponderosae*, during an outbreak that coincided with warming temperatures during 2002 (de la Giroday et al., 2012; Robertson et al., 2009). For the first time, this species is now colonizing

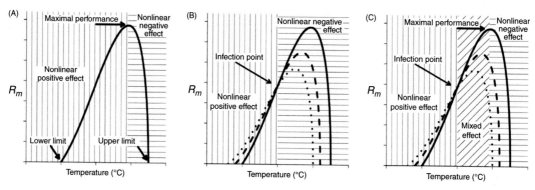

FIGURE 2.17 **Expected population growth (R_m) under three climate change scenarios.** (A) Increased average temperature without change in variance would have a positive nonlinear effect on populations occupying a range between the lower limit and the maximum performance point *(red area)*, but a negative effect between this last point and the upper limit *(blue area)*. (B) Increased thermal variability without change in average temperature would have a positive nonlinear effect on populations occupying a range between the lower limit and the inflection point *(red area)*, but a negative effect between this last point and the upper limit *(blue area)*. (C) Increased thermal variability and average temperature would have a positive nonlinear effect on populations occupying a range between the lower limit and the inflection point *(red area)*, but a negative effect between the point of maximum performance and the upper limit *(blue area)*. Between the inflection point and maximum performance point *(green area)*, the net effect depends on the relative magnitude of the increments of average temperature and variability. Solid (low), *dashed* and *dotted* (high) *lines* show a sequence of increased variability. *Source: From Estay, S.A., et al., 2014. The role of temperature variability on insect performance and population dynamics in a warming world. Oikos 123, 131–140.*

jack pine, *Pinus banksiana*, in eastern Alberta and threatening to continue eastward across Canada (Negrón and Fettig, 2014). Other species at high latitudes or elevations may disappear if heat intolerance forces range change in advance of host range shift or beyond habitable sites (Franklin et al., 1992; Klok et al., 2003; Netherer and Schopf, 2010).

However, thermal variability, as well as general warming, will affect insect ability to survive at higher latitudes (Fig. 2.17, Estay et al., 2014). The frequency and duration of extreme high temperatures have increased since 2000 (W. Zhang et al., 2015). Many insect species, and even life stages, will respond differently to extreme heat waves, increasing or decreasing survival or reproduction (W. Zhang et al., 2015). Abarca and Lill (2015) found that warmer winter and spring temperatures advance hatching time and reduce the starvation endurance of eastern tent caterpillars, *Malacosoma americanum*, thereby increasing their vulnerability to extreme spring weather. Furthermore, outbreaks of many herbivorous species are triggered by drought and are likely to increase in frequency and severity under future climate scenarios (Mattson and Haack, 1987; Stige et al., 2007; Van Bael et al., 2004).

Detection of responses to global climate change is complicated because effects of nonclimatic, especially land use, factors tend to dominate local, short-term responses (Parmesan and Yohe, 2003; Raffa et al., 2008). Warren et al. (2001) evaluated changes, over the past 30 years, in the distribution and abundances of 46 butterfly species that approach their northern range limits in Great Britain. Most species (75%) declined in distribution and abundance, indicating that negative responses to habitat loss outweighed positive responses to climate warming in the region.

Parmesan et al. (1999) found that 63% of 35 nonmigratory European butterflies have shifted their geographical range to the north by 35–240 km during the last 100 years, in response to warming temperatures; only 3% have shifted their geographic range to the south. Parmesan and Yohe (2003) conducted a global metaanalysis for 1700 species. Results showed a 6.1 km decade^{-1} range shift toward the poles (or 6 m decade^{-1} increase in elevation). Parmesan (2007) reviewed data that indicate a 2.8 day decade^{-1} advance in spring events. This diagnostic pattern was found for 279 species (including plants, insects, and vertebrates), indicating that climate change is already affecting ecosystems. S. Richardson et al. (2002) simulated temperature and nutrient changes in a subArctic dwarf shrub-heath ecosystem and found that most insect responses were related to increased grass abundance in warmed and fertilized plots. More recently, I. Chen et al. (2011) conducted a second metaanalysis and concluded that species distributions have shifted to higher latitudes at a median rate of 17 km decade^{-1} and to higher elevations at a median rate of 11 m decade^{-1}.

Balanyá et al. (2006) reported that 22 populations of a cosmopolitan fly, *Drosophila subobscura*, on three continents had experienced the equivalent of a 1° change in latitude toward the equator and showed a corresponding shift in genotypic composition equivalent to a 1° lower-latitude genotype. Bradshaw and Holzapfel (2001) and Mathias et al. (2007) found a detectable shift in the genetically controlled response of pitcher plant mosquitoes, *Wyeomyia smithii*, to changing photoperiod as a result of warming temperatures. This shift was detectable over time intervals as short as 5 years. Faster evolutionary response has occurred in northern populations where selection for response to seasonal change in photoperiod is stronger and genetic variation was greater than in southern populations. However, such genetic shifts are unlikely to mitigate negative effects on species populations (Parmesan, 2006).

A number of studies have documented insect responses to elevated temperature, increased atmospheric or aqueous concentrations of CO_2 or various pollutants, including pesticides, and habitat disturbance and fragmentation (Alstad et al., 1982; Arnone et al., 1995; Bezemer and Jones, 1998; Chung et al., 2013; Heliövaara and Väisänen, 1986, 1993; Kinney et al., 1997; Lincoln et al., 1993; Marks and Lincoln, 1996; Valkama et al., 2007; Zavala et al., 2008). Although insect herbivores respond to a variety of factors, including photoperiod, relative humidity, and host condition, that interact with effects of temperature (Bale et al., 2002), a number of studies suggest increased likelihood of herbivore outbreaks under future warming and/or drying scenarios (Breshears et al., 2005; J.A. Logan et al., 2003; Mattson and Haack, 1987; Stireman et al., 2005). Zavala et al. (2008) demonstrated that elevated atmospheric CO_2 compromised plant defense against herbivores by downregulating gene expression for defensive compounds. Chung et al. (2013) reviewed results of experimental warming in forest ecosystems and concluded that warming typically resulted in leaves emerging earlier and senescing later, increased abundance and enhanced performance of herbivorous insects and accelerated leaf litter decomposition and soil nitrogen mineralization. Stireman et al. (2005) compared predator-parasitoid responses to climate change in 15 studies and concluded that increasing temperatures will disrupt the ability of parasitoids to track prey populations.

Furthermore, paleontological data show increased herbivory during warming periods. Currano et al. (2008) reported that the frequency and diversity of leaf damage by herbivorous insects were significantly greater (mean 57%) during the Paleocene-Eocene thermal maximum (PETM) 56 million years ago than during either the preceding Paleocene (<38%) or subsequent Eocene (33%) epochs (Fig. 2.18). Both the amount and diversity of herbivore damage on angiosperm leaves were correlated positively with temperature change 55–59 million years ago. The indirect effects of these changes may be greater than the direct effects.

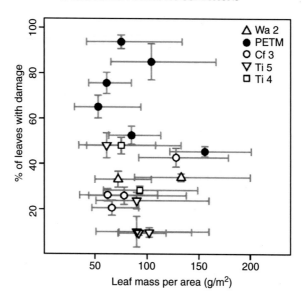

FIGURE 2.18 **Estimated leaf mass per area (LMA) and damage frequency for individual plant species from sites representing time periods before (Tiffanian 4a and 5b [Ti 4 and 5, 57.5–58.9 Mya] and Clarkforkian 3 [Cf 3, 55.9 Mya]), during (PETM, 55.8 Mya), and after (Wasatchian 2 [Wa 2, 55.2 Mya]) the Paleocene-Eocene Thermal Maximum (PETM).** LMA values are species means, and error bars represent 95% confidence intervals. Error bars for herbivory represent 1 standard deviation, based on a binomial sampling distribution. *Source: From Currano, E.D., et al., 2008. Sharply increased insect herbivory during the Paleocene-Eocene Thermal Maximum. Proc. Natl. Acad. Sci. USA 105, 1960–1964.*

Climate change may disrupt synchrony between herbivores or pollinators and their host plants or between predator and prey populations when interacting species' phenologies do not respond equally to changing temperature (Andrew and Hughes, 2007; Bale et al., 2002; Cornelissen, 2011; Hance et al., 2007; A.F. Hunter and Elkinton, 2000; Klapwijk et al., 2010; Klok et al., 2003; R. Lawrence et al., 1997; J.D. Logan et al., 2006; Nooten and Hughes, 2014; Ris et al., 2004; Singer and Parmesan, 2010; Visser and Both, 2005; Visser and Holleman, 2001; Watt and McFarlane, 2002). For example, R. Lawrence et al. (1997) and A.F. Hunter and Elkinton (2000) found that cohorts of lepidopteran larvae placed on foliage at increasing time before or after budbreak showed reduced survival and increased development times, relative to larvae placed on foliage near the time of budbreak. A.F. Hunter and Elkinton (2000) found that predation had the opposite effect, indicating that disruption of phenological synchrony of budbreak, herbivore egg hatch, and predation as a result of climate change could greatly affect herbivore population dynamics.

Ris et al. (2004) reported that parasitoid survival declined more rapidly with increasing temperature than did host survival, perhaps in part because of increased encapsulation ability of hosts at higher temperatures (Fellowes et al., 1999). However, plant emission of volatile chemicals that attract parasitoids to their herbivore hosts increases with temperature (Gouinguené and Turlings, 2002).

Mooring and Samuel (1998, 1999) reported that warming temperatures in boreal North America have led to increased exposure of moose, *Alces alces*, to a novel parasite, the winter tick, *Dermacentor albipictus*, historically associated with deer, *Odocoileus* spp. As a naïve host, moose are poorly adapted to this parasite and frequently show severe hair loss as a result of ineffective grooming behavior. Consequently, this vertebrate species is suffering greater winter mortality. Similarly, climate change is permitting European castor bean ticks, *Ixodes ricinus*, to move northward and to higher elevations, thereby maintaining longer seasonal contact with red deer, *Cervus elaphus*, which historically have escaped ticks during summer migration to higher elevations (Mysterud et al., 2016).

4.4 Interactions among factors

However, humans are changing environmental conditions in many ways simultaneously, through fossil fuel combustion, industrial effluents, water impoundment and diversion, pesticide application, and land use changes. Large areas have been planted to genetically modified crops or occupied by invasive exotic species. Global atmospheric concentrations of CO_2 and other greenhouse gases are clearly increasing, and global climate has shown a distinct warming trend (Beedlow et al., 2004; Keeling et al., 1995). Acidic precipitation has greatly reduced the pH of many aquatic ecosystems in northern temperate countries. Nitrogen subsidies resulting from increased atmospheric NO_x may provide a short-term fertilization effect in N-limited ecosystems, until the pH buffering capacity of the soil is depleted. Deforestation, desertification, urbanization and other changes in regional landscapes are fragmenting and converting habitats and altering habitat suitability for organisms around the globe, potentially undermining sustainability of ecosystem processes and services (Foley et al., 2005). Mining activities and industrial effluents add highly toxic minerals to terrestrial and aquatic ecosystems.

The interactions among environmental factors are poorly understood, but often synergistic. For example, land use changes alter surface albedo and can exacerbate regional warming and storm intensity (Foley et al., 2003b; Hossain et al., 2009; T. Lewis, 1998). Studies in the Amazon basin indicate that smoke from fires that accompany forest conversion to agricultural or urban land use reduces cloud cover (from 38% in clean air to 0% in heavy smoke), reduces droplet size, and increases the altitude at which water condenses, leading to more violent thunderstorms and hail, rather than warm rain (A. Ackerman et al., 2000; Andreae et al., 2004; Koren et al., 2004). Impounded reservoirs, along with associated land use changes (such as irrigated agriculture), alter surface fluxes of heat and moisture and potentially increase the frequency and intensity of precipitation (Hossain et al., 2009). The increased frequency of such extreme events may have greater effects on species survival than will overall trends (Bender et al., 2010; Estay et al., 2014; Gleason et al., 2008; Gutschick and BassiriRad, 2010; Irwin et al., 2014; Lubchenco and Karl, 2012; W. Zhang et al., 2015).

Few studies have measured insect responses to multiple changes in ecosystem conditions (Lindroth et al., 1993a,b,c). However, given insect sensitivity to environmental changes (as described earlier in Section 2), any change will alter insect abundance and distribution and may increase the incidence of crop pests and vectors of human and animal diseases (Dobson et al., 2006; Gottdenker et al., 2014; Keesing et al., 2006; Stapp et al., 2004; Summerville and Crist, 2001; Vittor et al., 2006, 2009; D. Williams and Liebhold, 2002; J. Zhou et al., 2002). Although climate warming might be expected to promote insect range shifts to higher latitudes, Estay et al. (2014) noted that thermal variability, as well as general warming, will affect insect survival at higher latitudes. Chapin et al. (1987) addressed plant responses to multiple stressors and concluded that multiple factors can have additive or synergistic effects that can alter plant biochemistry and vulnerability to herbivores. Song and Brown (2006) reported that the estuarine mosquito, *Aedes taeniorhynchus*, was more tolerant of pesticide exposure during periods of changing salinity and was more tolerant during decreasing salinity than during increasing salinity. Valkama et al. (2007) conducted a metaanalysis of experiments on effects of CO_2 and O_3 on insect–plant interactions. They found that elevated O_3 tended to reduce foliage quality, but also improved indices of insect performance, whereas elevated CO_2 counteracted this effect.

The effects of such changes may be difficult to predict, based on adaptations to natural disturbances, and may persist for long periods, depending on biotic mechanisms of recovery. Benstead et al. (2007) reported that effects of low levels of nutrient enrichment to freshwater streams in the Arctic depended on the duration of enrichment as mediated by bryophyte colonization and subsequent physical disturbances that removed bryophytes. Harding et al. (1998) reported that responses of aquatic invertebrates to stream restoration treatments reflected differences in community structure among stream segments with different histories of anthropogenic disturbances. Schowalter et al. (2003) found that litter arthropod responses to variable density thinning of conifer forests for restoration purposes reflected different initial community structures, resulting from previous thinning as much as 30 years earlier. Summerville et al. (2009) reported that the legacy of forest management practices could be seen in forest lepidoptera assemblages more than 60 years after timber harvest.

Long-term changes may be difficult to reverse and could have significant implications for human health. Vittor et al. (2006) found that deforestation and road development in Peru increased the abundance and biting rate of mosquitoes, *Anopheles darlingi*, that transmit malaria to humans. Pascual et al. (2006) reported that mosquito responses to increasing temperatures in East Africa during 1950–2002 have driven a resurgence of malaria in the region. Clearly, more studies are needed on insect responses to multiple natural and anthropogenic changes in order to improve prediction of the effects of environmental changes.

5 SUMMARY

Insects are affected by abiotic conditions that reflect latitudinal gradients in temperature and moisture, as modified by circulation patterns and mountain ranges. At the global scale, latitudinal patterns of temperature and precipitation produce bands of tropical rainforests along the equatorial convergence zone (where warming air rises and condenses moisture), deserts centered at 30°N and S latitudes (where cooled, dried air descends), and moist boreal forests centered at 60°N and S latitudes (where converging air masses rise and condense moisture). Mountains affect the movement of air masses across continents, forcing air to rise and condense on the windward side, and dried air to descend on the leeward side. The combination of mountain ranges and latitudinal gradients in climatic conditions creates a template of regional ecosystem types known as biomes, characterized by distinctive vegetation (eg, tundra, desert, grassland, forest). Aquatic biomes are distinguished by size, depth, flow rate, and marine influence (eg, ponds, lakes, streams, rivers, estuaries).

Environmental conditions are not static but vary seasonally and annually. In addition, environment conditions change over longer periods as a result of global processes and anthropogenic activities. Acute events (disturbances), such as storm or fire, can dramatically alter habitat conditions and resource availability for various organisms. Hence, insects must be able to avoid or adjust to changing conditions.

The inherent problems of maintaining body heat and water content and avoiding adverse chemical conditions by small, heterothermic organisms has led to an astounding variety of physiological and behavioral mechanisms by which insects adjust to and interact with environmental conditions. Recent research on genetic control of physiological processes is improving our understanding of the mechanisms of adaptation. Mechanisms for tolerating or mitigating the effects of variation in abiotic factors determine the seasonal, latitudinal, and elevational distributions of insect species.

Many insects have a largely unappreciated physiological capacity to cope with the extreme temperatures and relative humidities found in the harshest ecosystems on the planet. However, even insects in more favorable environments must cope with variation in abiotic conditions through diapause, color change, evaporative cooling, supercooling, voiding of the gut, control of respiratory water loss, etc. Many species exhibit at least limited homeostatic ability, that is, ability to regulate internal temperature and water content.

Behavior represents the active means by which animals respond to their environment. Insects are sensitive to a variety of environmental cues, and most insects are able to modify their behavior in response to environmental gradients or changes. Insects, especially those that can fly, move within gradients of temperature, moisture, chemicals, or other abiotic factors to escape adverse conditions. Many species are able to regulate body heat or water content by rapid muscle contraction, elevating the body above hot surfaces, seeking shade, or burrowing. Social insects appear to be particularly flexible in the use of colony activity and nest construction to facilitate thermoregulation. Many insects are capable of flying long distances, but dispersal entails considerable risk, and many individuals do not reach suitable habitats.

Environmental changes resulting from anthropogenic activities are occurring at an unprecedented rate. A number of studies have demonstrated significant effects of anthropogenic changes on abundances of various insects and indicate that global changes could increase the incidence of herbivore outbreaks. However, few studies have addressed the effects of multiple interacting changes on insects. More information is needed on insect responses to multiple natural and anthropogenic changes in order to improve the prediction of responses to environmental changes.

Resource Acquisition

ADVANCES IN UNDERSTANDING BIOCHEMICAL INTERACTIONS BETWEEN PLANTS AND INSECTS

Biochemical interaction between insects and plants has been a cornerstone of insect ecology, providing critical data on mechanisms of evolution and applications for pest management, and integrating species interactions with biogeochemical cycling. Important groundwork for this topic had been laid by 1900, for example, plant pharmaceutical properties were recognized at least 2500 years ago (Farnsworth, 1966; Hamilton and Baskett, 2000; Millspaugh, 1892), plant variation in susceptibility to insect herbivores by the late 1700s (C. Smith, 2005; Sorensen et al., 2008), and insecticidal properties of some plant products, such as the powdered flower heads of *Pyrethrum roseum*, by 1800 (C. Riley, 1885).

Conceptual synthesis began when R. Painter (1951) recognized that plant resistance to insects could be expressed through antibiosis, antixenosis, and tolerance. Fraenkel (1953) subsequently noted that, since plant species show relatively little variation in nutritional value, resistance to insects must reside in their "secondary metabolites." Major advances in analytical technology (eg, GC–MS equipment) during the 1960s, coupled with conceptual stimulus and serendipity, spurred progress in understanding the chemical mechanisms of insect–plant interactions.

Hairston et al. (1960) stimulated research on plant defense by suggesting that plants had little role in regulating herbivore populations. Their "green world hypothesis" was based on a perception that, since herbivores eat so little plant material overall, they must not be food-limited and, therefore, must be limited by predators. We soon learned that a variety of phytochemicals provide effective defense against herbivory (Feeny, 1969, 1970; Rosenthal and Janzen, 1979) and that herbivorous insects show specific adaptations to avoid or detoxify plant defenses (Euw et al., 1967; Xu et al., 2015), thereby limiting the number of host species that could be exploited by a particular herbivore. These data supported a view of an "evolutionary arms race" between plants and insects (Whittaker and Feeny, 1971, but see Berenbaum and Zangerl, 1988).

Serendipity played a role when Rhoades (1983) and I. Baldwin and Schultz (1983) independently found evidence, in environmental chambers, that tree seedlings damaged by herbivorous insects communicated this information to neighboring undamaged seedlings, which initiated production of antiherbivore defenses in advance of herbivory. Although their hypothesis that plants communicate the

Insect Ecology. http://dx.doi.org/10.1016/B978-0-12-803033-2.00003-0

threat of herbivory (popularly termed the "talking tree" hypothesis) was challenged because of apparent lack of an evolutionary mechanism (Fowler and Lawton, 1985), additional research confirmed that volatile elicitors induce defenses in a wide variety of plants, even species unrelated to the signaler (Dolch and Tscharntke, 2000; Farmer and Ryan, 1990; Heil and Karban, 2010; Izaguirre et al., 2013; Karban, 2001; Pearse et al., 2013; M. Stout et al., 2006) and that these plant volatiles also attract predators (Bruinsma and Dicke, 2008; Hare, 2011; Turlings et al., 1990; Wason and Hunter, 2014). However, the fitness benefits for emitter and receiver are still poorly understood (Heil and Karban, 2010).

Recent advances in molecular technology have demonstrated genetic mechanisms governing these interactions, typically involving P-450 genes (Berenbaum and Zangerl, 2008; Y. Mao et al., 2007). Insertion of genes for bacterial toxins into crop genomes has permitted measurement of rates at which herbivorous insects adapt and which insect genes are targets for selection (F. Huang et al., 2007).

However, plant production of defenses requires energy and nutrients (especially nitrogen) and affects litter quality and the return of nutrients to soil (Holopainen et al., 1995; Schweitzer et al., 2005; Whitham et al., 2006), thereby linking insect–plant interactions to energy flow, biogeochemical cycling, and disturbance regime at the ecosystem level (see Chapters 11 and 12). The contemporary view of insect–plant interaction recognizes a variety of mechanisms and trade-offs that will continue to generate new insights into evolutionary and ecosystem theory.

1 INTRODUCTION

All organisms are examples of negative entropy, in contrast to the tendency for energy to be dissipated, according to the Second Law of Thermodynamics. Organisms acquire energy in order to collect resources and synthesize the organic molecules that are the basis for life processes, particularly maintenance, growth, and reproduction. Hence, the acquisition and concentration of energy and matter are necessary goals of all organisms and largely determine individual fitness.

Insects, like other animals, are heterotrophic, that is, they must acquire their energy and material resources from other organisms (see Chapter 11). As a group, insects exploit a wide range of resources, including plant, animal, and detrital material, but individual organisms must find and acquire more specific resources to support growth, maintenance, and reproduction.

The organic resources used by insects vary widely in quality (nutritional value), acceptability (preference ranking, given choices and trade-offs), and availability (size, density, and ease of detection by insects), depending on environmental conditions. Physiological and behavioral mechanisms for evaluating and acquiring food resources, and their efficiencies under different developmental and environmental conditions, are the focus of this chapter.

2 RESOURCE QUALITY

Resource quality is the net energy and nutrient value of food resources after deducting the energetic and nutrient costs to detoxify (if necessary), digest, and assimilate the resource. The energy and nutrient value of organic molecules is a product of the number, elemental composition, and bonding energy of their constituent atoms. However, organic resources are not equally digestible into useable components. Some resources provide little nutritional value for the expense of acquiring and digesting them, others cannot be digested by common enzymes, and many organic molecules are essentially unavailable, or even toxic, to a majority of organisms. Some organic molecules are cleaved into toxic components by commonly occurring digestive enzymes. Therefore, acquiring sufficient amounts of suitable resources is a challenge for all animals.

2.1 Resource Requirements

Dietary requirements for all insects include carbohydrates, amino acids, cholesterol, B vitamins, and inorganic nutrients such as P, K, Ca, Na, etc. (Behmer, 2009; R. Chapman, 2003; Rodriguez, 1972; Sterner and Elser, 2002). With the exception of the higher termites (Bujang et al., 2014; Lo et al., 2011), insects lack the ability to produce their own cellulases to digest cellulose and rely on microorganisms (bacteria, fungi, protozoa) in their guts when cellulose digestion is necessary (M. Ayres et al., 2000; Batra, 1966; Breznak and Brune, 1994; French and Roeper, 1972; Geib et al., 2008; Morgan, 1968). The nutritional value of plant material often is limited further by deficiency in certain requirements, such as low concentrations of N (Mattson, 1980), Na (Seastedt and Crossley, 1981b; Smedley and Eisner, 1995), or linoleic acid (Fraenkel and Blewett, 1946). Resources differ in balance among essential nutrients, resulting in relative limitation of some nutrients and adversely high and potentially toxic levels of others (Behmer, 2009; Sterner and Elser, 2002). High lignin content toughens foliage and other plant tissues and limits feeding by herbivores that do not possess reinforced mandibles. Toxins or feeding deterrents in food resources increase the cost, in terms of search time, energy, and nutrients, necessary to exploit nutritional value.

For individual insects, several factors influence food requirements. The most important of these are size and maturity of the insect and the quality (nutritional balance) of food resources. Larger organisms require more food and consume more oxygen per unit time than do smaller organisms, although smaller organisms consume more food and oxygen per unit biomass (Reichle, 1968). Insects require more food, but often are able to digest a wider variety of resources, as they mature. Holometabolous species must store sufficient resources during larval feeding to support pupal diapause and adult development and, for some species, to support dispersal and reproduction by nonfeeding adult stages (Hahn and Denlinger, 2011).

2.2 Factors that Reduce Food Quality

Food resources often are defended in ways that limit their utilization by consumers. Physical defenses include spines, toughened exterior layers, and other barriers. Spines and hairs can inhibit attachment or penetration by small insects or interfere with ingestion by larger organisms. These structures often are associated with glands that augment delivery of toxins. Some plants entrap phytophagous insects in adhesives (R.W. Gibson and Pickett, 1983). Toughened exteriors include lignified epidermis of foliage and bark of woody plants and heavily armored exoskeletons of arthropods. Bark is a particularly effective barrier to penetration by most organisms (Ausmus, 1977). Ferrenberg and Mitton (2014) found that smooth bark on limber pine, *Pinus flexilis*, inhibited penetration by bark

FIGURE 3.1   **The wound response of conifers constitutes a physical-chemical defense against invasion by insects and pathogens.** The oleoresin, or pitch, flowing from severed resin ducts hinders penetration of the bark.

beetles, compared to rough bark surfaces. Lignin reduces the ability of many insects to utilize toughened foliage (Scriber and Slansky, 1981). Viscous latex and resins (pitch) can push insects out of plant tissues (Fig. 3.1). Ferrenberg et al. (2014) reported that pine trees, *Pinus contorta* and *P. flexilis*, that resisted mountain pine beetle, *Dendroctonus ponderosae*, attack had 23 and 37%, respectively, more resin ducts in their most recent 5-year growth increment than did trees killed by beetles.

Many plant and animal species are protected indirectly through associations with other organisms, especially nonhost plants, ants, or endophytic fungi (see Chapter 8). An otherwise suitable plant can be "hidden," or made nonapparent, by unsuitable plants that mask its presence (Barbosa et al., 2009; Courtney, 1985, 1986; Sholes, 2008; M. Stanton, 1983; Visser, 1986), a situation known as *associational resistance* (Kos et al., 2015; Stastny and Agrawal, 2014). Conversely, a plant surrounded by more suitable neighbors may experience *associational susceptibility* (T. Kim and Underwood, 2015; Stastny and Agrawal, 2014). A number of plant species provide food sources (extrafloral nectaries) or habitable structures (domatia) to attract or house colonies of aggressive ants or predaceous mites that provide protection against herbivores (Brenes-Arguedas et al., 2008; R. Fischer et al., 2002; Huxley and Cutler, 1991). Some plants emit chemical signals that attract parasitic wasps, especially when injured by herbivores (Kessler and Baldwin, 2001; Turlings et al., 1993, 1995). G. Carroll (1988), Clay et al.

(1993), and D. Wilson and Faeth (2001) reported that foliar infection by endophytic fungi that produce additional defensive compounds reduced herbivory by insects.

Both plants and insects produce a remarkable range of compounds that have been important sources of pharmaceutical or industrial compounds (see Chapter 16), as well as effective defenses. These compounds often function as toxins or feeding deterrents and also may increase herbivore exposure and mortality to predators and parasites (Lill and Marquis, 2001). Major categories of defenses are summarized next.

2.3 Plant Chemical Defenses

Plant chemical defenses generally are classified as nonnitrogenous, nitrogenous, and elemental. Nitrogenous and nonnitrogenous categories are represented by a wide variety of compounds, many differing only in the structure and composition of attached radicals. Elemental defenses are conferred by plant accumulation of toxic elements from the soil.

2.3.1 Nonnitrogenous Defenses

Nonnitrogenous defenses include phenolics, terpenoids, photooxidants, insect hormone, or pheromone analogs, pyrethroids, and aflatoxins (Figs. 3.1–3.4). These compounds function as feeding deterrents or toxins.

Phenolics, or flavenoids, are distributed widely among terrestrial plants and are likely among the oldest plant secondary (ie, nonmetabolic) compounds. Although phenolics are perhaps best known as defenses against herbivores and plant pathogens, they also protect plants from damage by UV radiation, provide support for vascular plants (lignins), compose pigments that determine flower color for angiosperms, and play a role in plant nutrient acquisition by affecting soil chemistry. Phenolics include the hydrolyzable tannins, derivatives of simple phenolic acids, and condensed tannins, polymers of higher molecular weight hydroxyflavenol units (Fig. 3.2). Polymerized tannins are highly resistant to decomposition, eventually composing the humic materials that largely determine soil properties. Tannins are distasteful, typically bitter and astringent, and act as feeding deterrents for many herbivores. When ingested, tannins can chelate N-bearing molecules to form indigestible complexes (Feeny, 1969), but their primary toxic effect is through oxidation in the herbivore gut to form semiquinones and other highly oxidative compounds that damage cell membranes and DNA (Barbehenn et al., 2008). Insects incapable of catabolizing tannins or preventing chelation suffer gut damage and are unable to assimilate nitrogen from their food.

Rhoades (1977) examined the effect of phenolic resins, primarily nordihydroguaiaretic acid, on the foliage surface of creosote bushes, *Larrea tridentata* from the southwestern USA and *L. cuneifolia* from Argentina. Young leaves contained about twice as much resin (26% dry wt. for *L. tridentata*, 44% for *L. cuneifolia*) as did mature leaves (10% for *L. tridentata*, 15% for *L. cuneifolia*), but amounts of nitrogen and water did not differ between leaf ages. Leaf-feeding insects that consume entire leaves all preferred mature foliage. Extraction of resins from foliage increased feeding on both young and mature leaves by a generalist grasshopper, *Cibolacris parviceps*, but reduced feeding on mature leaves by a specialist geometrid, *Semiothesia colorata*, in laboratory experiments, suggesting that low concentrations of resins may be a feeding stimulant for *S. colarata*.

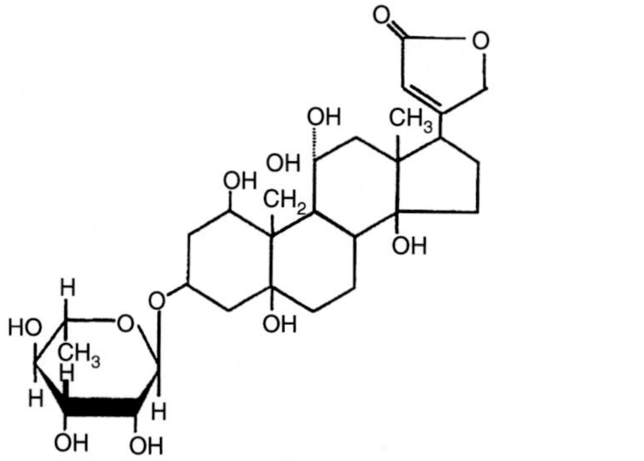

Terpenoid cardiac glycoside, ouabain, from *Acokanthera ouabaio*

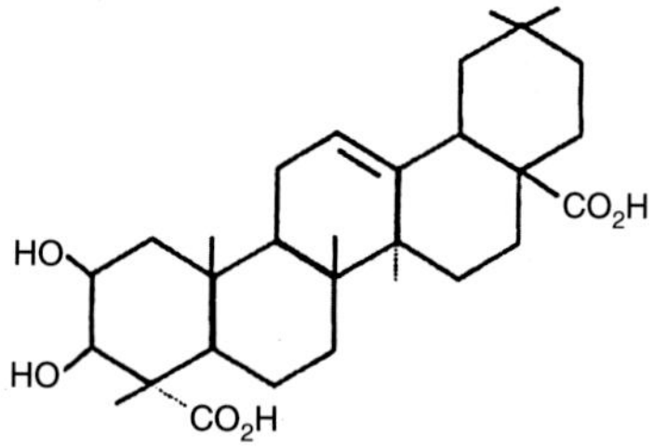

Terpenoid saponin, medicagenic acid, from *Medicago sativa*

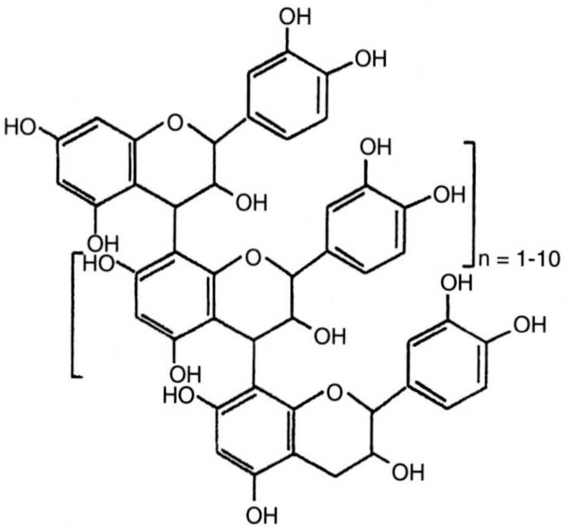

Flavonoid tannin, procyanidin, from *Quercus* spp.

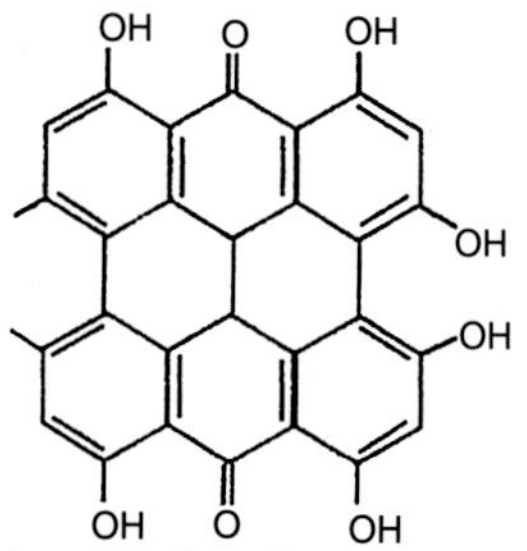

Quinone, hypericin, from *Hypericum perforatum*

FIGURE 3.2   Examples of nonnitrogenous defenses of plants. *Source: From Harborne, J.B., 1994. Introduction to Ecological Biochemistry, fourth ed. Academic Press, London.*

Terpenoids also are widely represented among plant taxa. These compounds are synthesized by linking isoprene subunits. The lower molecular weight monoterpenes and sesquiterpenes are highly volatile compounds that function as floral and foliar scents that attract pollinators and herbivores, and often associated predators and parasites, to their hosts, in addition to serving defensive functions. Some insects modify plant terpenes for use as pheromones (see Chapter 4). Terpenoids with higher molecular weights include plant resins, cardiac glycosides, saponins, and latex (Figs. 3.1 and 3.2). Terpenoids typically are distasteful or toxic to herbivores. Latex or pitch flow is a general response to wounding, but can physically push insects away, adhere to mouthparts and interfere with further feeding, and/or kill insects and associated microorganisms (Agrawal and Konno, 2009; Nebeker et al., 1993). An advantage of this mode of defense is its accumulation at the site where the latex or resin canals are severed.

Becerra (1994) reported that the tropical succulent shrub, *Bursera schlechtendalii*, stores terpenes under pressure in a network of canals in its leaves and stems. When these canals are broken during insect feeding, the terpenes are squirted up to 150 cm, bathing the herbivore

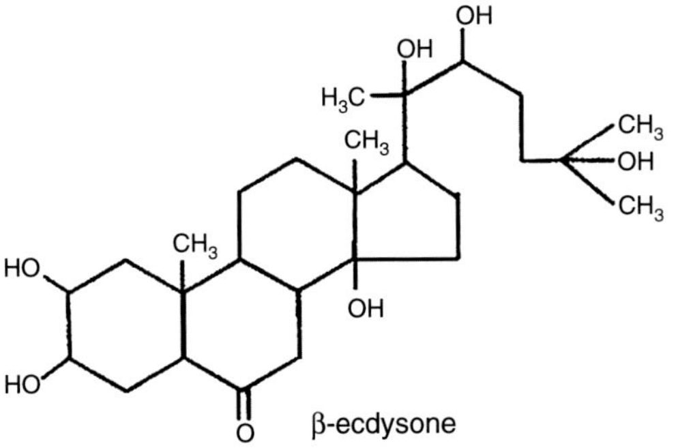

β-ecdysone

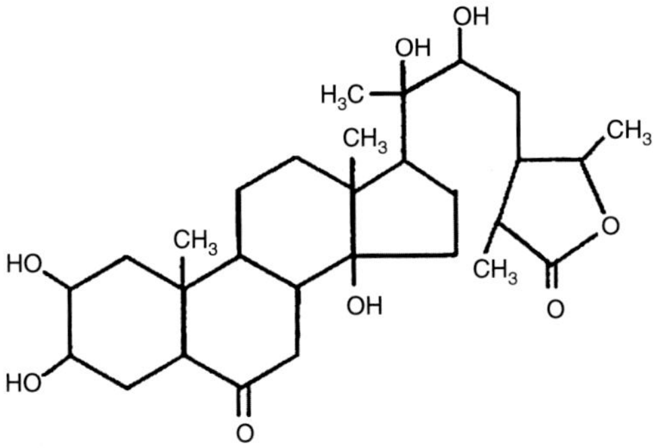

Cycasterone from *Cycas*

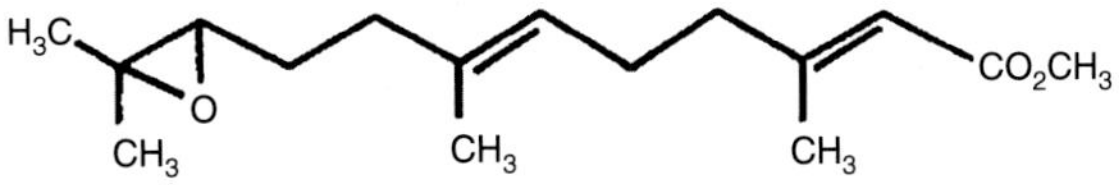

Juvenile hormone III

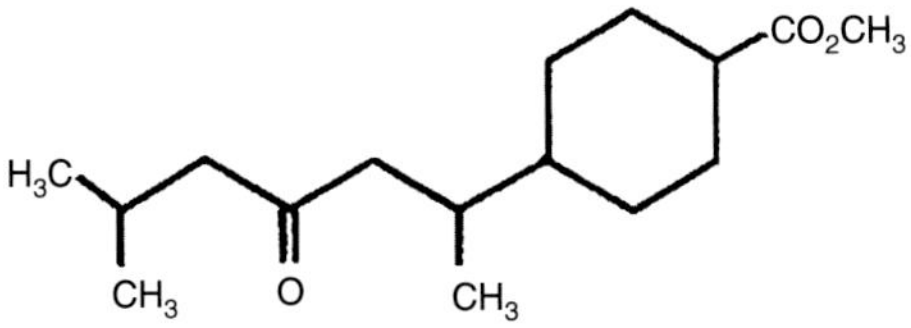

Juvabione from *Abies balsamea*

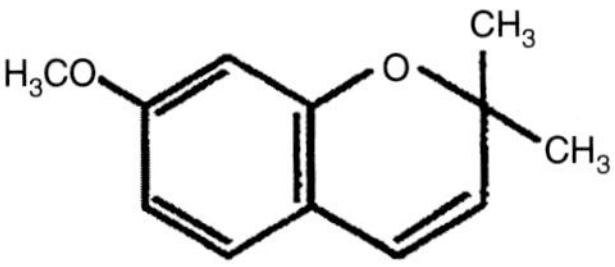

Precocene 1 from *Ageratum houstoniatum*

FIGURE 3.3 Insect developmental hormones and examples of their analogues in plants. *Source: From Harborne, J.B., 1994. Introduction to Ecological Biochemistry, fourth ed. Academic Press, London.*

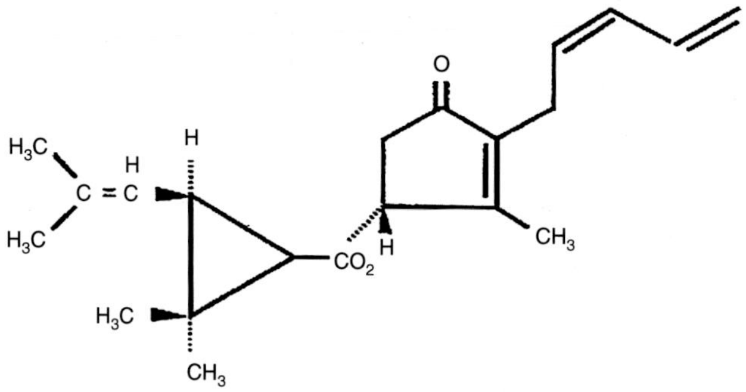

Pyrethrin I, from *Chrysanthemum cinearifolium*

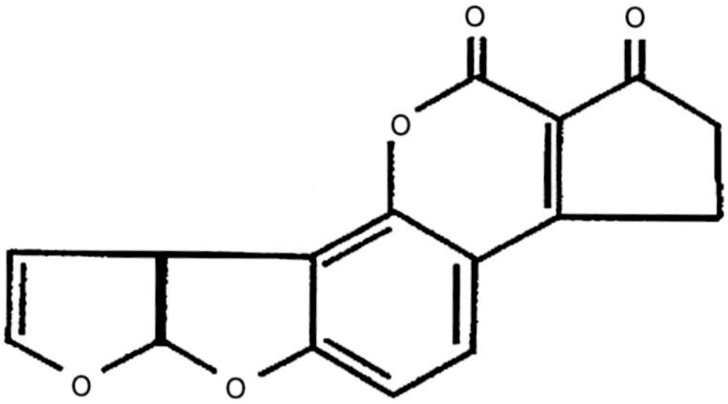

Aflatoxin B, from *Aspergillus flavus*

FIGURE 3.4 **Examples of pyrethroid and aflatoxin defenses.** *Source: From Harborne, J.B., 1994. Introduction to Ecological Biochemistry, fourth ed. Academic Press, London.*

and drenching the leaf surface. A specialized chrysomelid, *Blepharida* sp., partially avoids this defense by severing leaf veins before feeding, but nevertheless suffers high mortality and may spend more time cutting veins than feeding, thereby suffering reduced growth.

Cardiac glycosides (cardenolides) are terpenoids best known as the milkweed (Asclepiadaceae) compounds sequestered by monarch butterflies, *Danaus plexippus*. The primary cardenolide in milkweeds, ouabain (Fig. 3.2), inhibits Na$^+$, K$^+$-ATPase by binding to its α-subunit (Holzinger and Wink, 1996) and is thereby toxic to most herbivores. Monarch butterflies substitute histidine for the more widespread asparagine at position 122 in the putative ouabain binding site, making them less sensitive to ouabain than are other herbivores (Holzinger and Wink, 1996). Ingestion of butterflies containing ouabain by vertebrates induces vomiting or results in cardiac arrest. The butterflies thereby gain protection against predation by birds (L. Brower et al., 1968). However, monarch caterpillars suffer high mortality on milkweed species, for example, *Asclepias humistrata*, that have high concentrations of cardenolides (Zalucki et al., 2001).

Photooxidants, such as quinones (Fig. 3.2) and furanocoumarins, increase epidermal sensitivity to solar radiation. Assimilation of these compounds results in severe sunburn, necrosis of the skin, and other epidermal damage upon exposure to sunlight. Feeding on furanocoumarin-producing plants in daylight can cause 100% mortality to insects, whereas feeding in the dark causes only 60% mortality (Harborne, 1994). Adapted insects circumvent this defense by becoming leaf rollers or nocturnal feeders (Harborne, 1994) or by sequestering antioxidants (Blum, 1992).

Plants produce a variety of insect hormone analogues that disrupt insect development, typically preventing maturation or producing sterile adults (Harborne, 1994). Insect development is governed primarily by molting hormone (ecdysone) and juvenile hormone (Fig. 3.3). The relative concentrations of these two hormones dictate the timing of ecdysis and the subsequent stage of development. Some phytoecdysones are as much as 20-fold more active than are ecdysones produced by insects and resist inactivation by insect enzymes (Harborne, 1994). Schmelz et al. (2002) reported that spinach, *Spinacia oleracea*, produces 20-hydroxyecdysone in roots in response to root damage or root herbivory. Root feeding by a fly, *Bradysia impatiens*, increased production of 20-hydroxyecdysone by >4-fold. Fly larvae preferred diets with low concentrations of 20-hydroxyecdysone and showed significantly reduced survival when reared on diets containing higher concentrations. Plants also produce juvenile hormone analogues (primarily juvabione) and compounds that interfere with juvenile hormone activity (primarily precocene, Fig. 3.3). The antijuvenile hormones typically cause precocious development.

Some plants produce insect alarm pheromones that induce rapid departure of colonizing insects. For example, wild potato, *Solanum berthaultii*, produces (E)-β-farnesene, the major component of alarm pheromones for many aphid species (R. Gibson and Pickett, 1983). This compound is released from glandular hairs on the foliage in sufficient quantities to induce avoidance by host-seeking aphids and departure of settled colonies (R. Gibson and Pickett, 1983).

Pyrethroids (Fig. 3.4) are an important group of plant toxins and among the earliest recognized plant defenses (C. Riley, 1885). Many synthetic pyrethroids are widely used as contact insecticides, that is, absorbed through the exoskeleton, because of their rapid effect on insect pests.

Aflatoxins (Fig. 3.4) are highly toxic fungal compounds. Aflatoxins produced by mutualistic endophytic or mycorrhizal fungi augment defense by plant hosts (G. Carroll, 1988; Clay, 1990; Clay et al., 1985, 1993; Harborne, 1994; Van Bael et al., 2009). Endophytic and mycorrhizal fungi also may induce host plants to increase production of defensive compounds when injured by herbivores (Hartley and Gange, 2009).

2.3.2 Nitrogenous Defenses

Nitrogenous defenses include nonprotein amino acids, toxic proteins and proteinases, cyanogenic glucosides, glucosinolates, and alkaloids (Zenk and Juenger, 2007). These compounds typically interfere with protein function or physiological processes and can be highly toxic to herbivorous insects and vertebrates.

Nonprotein amino acids are analogues of essential amino acids (Fig. 3.5). Their substitution for essential amino acids in proteins results in improper conformation, loss of enzyme function, and inability to maintain physiological processes critical to survival. Others, such as 3,4-dihydroxyphenylalanine (L-DOPA), interfere with tyrosinase (an enzyme critical to hardening of the insect cuticle). Over 300 nonprotein amino acids are known, primarily from seeds of legumes (Harborne, 1994). Proteinase inhibitors, produced by a variety of plants, interfere with insect digestive enzymes and inhibit digestion (Kessler and Baldwin, 2002; Thaler et al., 2001).

Cyanogenic glycosides are distributed widely among plant families (Fig. 3.5). These compounds are inert in plant cells, but when crushed plant cells enter the herbivore gut, the glycoside is hydrolyzed by glucosidases into glucose and a cyanohydrin, that spontaneously

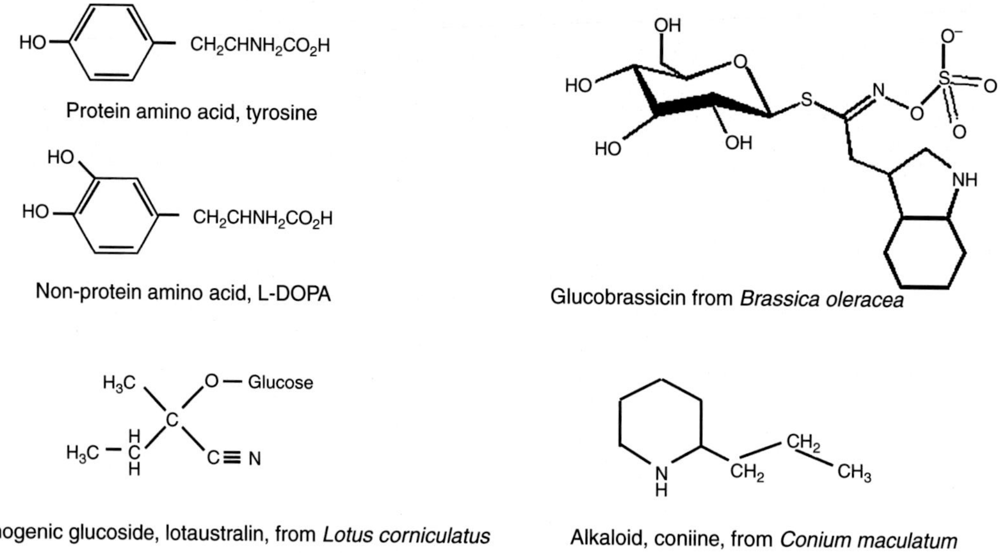

FIGURE 3.5 **Examples of nitrogenous defenses of plants.** *Source: From Harborne, J.B., 1994. Introduction to Ecological Biochemistry, fourth ed. Academic Press, London.*

decomposes into a ketone, or aldehyde, and hydrogen cyanide (Zenk and Juenger, 2007). Hydrogen cyanide is toxic to most organisms because of its inhibition of cytochromes in the electron transport system (Harborne, 1994).

Glucosinolates, characteristic of the Brassicaceae, have been shown to deter feeding and reduce growth in a variety of herbivores (Renwick, 2002; Strauss et al., 2004). Intact glucosinolates confer some resistance to herbivores, but damaged plant cells release the enzyme myrosinase that converts glucosinolates to toxic isothiocyanates, nitriles, and oxazolidinethiones that are more toxic (Hopkins et al., 2009). Rotem et al. (2003) reported that young larvae of the cabbage white butterfly, *Pieris rapae*, a specialized herbivore, showed reduced growth with increasing glucosinolate concentration in *Brassica napus* hosts, but that older larvae were relatively tolerant of glucosinolates.

Alkaloids include over 5000 known structures from about 20% of the higher plant families (Harborne, 1994), ranging in molecular size and complexity from the relatively simple coniine (Fig. 3.5) to multicyclic compounds such as solanine. These compounds are highly toxic and teratogenic, even at relatively low concentrations, because they interfere with major physiological processes, especially cardiovascular and nervous system functions, but include important pharmaceuticals. D. Jackson et al. (2002) reported that larval weights and survival of tobacco budworm, *Helicoverpa virescens*, were negatively related to pyridine alkaloid concentrations among 18 tobacco, *Nicotiana tabacum*, cultivars. Survivorship after 8 weeks declined from 60% to 0% as total alkaloid concentration increased from 0% to 2% fresh wt. Shonle and Bergelson (2000) found that generalist herbivore feeding on *Datura stramonium* was negatively correlated with hyoscyamine concentration; however, feeding by specialist flea beetles, *Epitrix* spp., was positively correlated with concentrations of scopolamine, indicating that this compound has become a phagostimulant for this adapted herbivore.

2.3.3 Elemental Defenses

Some plants accumulate and tolerate high concentrations of toxic elements, including Se, Mn, Cu, Ni, Zn, Cd, Cr, Pb, Co, Al, and As (Boyd, 2004, 2007, 2009; Trumble and Sorensen, 2008). In some cases, foliage concentrations of these metals can exceed 2% (Jhee et al., 1999). High concentrations in some plants confer protection against herbivores (Boyd, 2004, 2007; Boyd and Moar, 1999; Galeas et al., 2008; Pollard and Baker, 1997; Vickerman et al., 2002) but also deter pollinators (Meindl and Ashman, 2013).

Boyd and Martens (1994) found that larvae of the cabbage white butterfly fed *Thlaspi montanum* grown in high-Ni soil showed 100% mortality after 12 days, compared to 21% mortality for larvae fed on plants grown in low-Ni soil. Hanson et al. (2004) reported that Indian mustard, *Brassica juncea*, can accumulate Se up to 1000 mg kg^{-1} d.w., even from low-Se soils. Green peach aphids, *Myzus persicae*, avoided Se-containing leaves when offered a choice of foliage from plants grown in Se or nonSe soil. In nonchoice experiments, aphid population growth was reduced 15% at 1.5 mg Se kg^{-1} d.w. and few, if any, aphids survived at leaf concentrations > 125 mg Se kg^{-1}. Jhee et al. (1999) found that young larvae of *Pieris napi* showed no preference for high- or low-Zn leaves of *Thlaspi caerulescens*, but later-instar larvae showed highly significant avoidance of high-Zn leaves.

In a unique field study to compare arthropod diversity and intensity on Se-hyperaccumulator plants, *Astragalus bisulcatus* and *Stanleya pinnata*, and nonhyperaccumulator relatives, *Astragalus americanus*, *Camelina microcarpa*, *Descurainia pinnata*, and *Medicago sativa*, Galeas et al. (2008) found that nonhyperaccumulators hosted significantly higher (> 2X) arthropod abundance and species diversity per m^2 plant surface than did the hyperaccumulators. Arthropods on the hyperaccumulator plants contained 3–10-fold higher Se concentrations than those on nonhyperaccumulator plants, but >10-fold lower concentrations than did their hyperaccumulator hosts.

2.4 Arthropod Defenses

Arthropods also employ various defenses that reduce their availability, palatability, or nutritional value for predators and parasites. These include physical and chemical defenses, including chemicals produced by the insect, as well as those sequestered from plant sources, and cellular immunity.

2.4.1 Antipredator Defenses

Physical defenses include a hardened exoskeleton, spines, claws, and mandibles. The spines or hairs of some insects are capable of causing serious allergy or internal injury to vertebrate predators (E. Wood and Yasutake, 1956) or humans (Perlman et al., 1976). Chemical defenses are nearly as varied as those for plants. Hence, predaceous species also must be capable of evaluating and exploiting defended prey. The compounds used by arthropods generally belong to the same categories of compounds described previously for plants.

Many insect herbivores sequester plant defenses for their own defense (Blum, 1981, 1990, 1992; Boyd and Wall, 2001; L. Brower et al., 1968; Lampert et al., 2010; Opitz and Müller, 2009; Stamp et al., 1997; Tallamy et al., 1998; Traugott and Stamp, 1996). The relatively inert

FIGURE 3.6 **Defensive froth of an adult lubber grasshopper,** *Romalea guttata*. This secretion includes repellent chemicals sequestered from host plants. *Source: From Blum, M.S., 1992. Ingested allelochemicals in insect wonderland: a menu of remarkable functions. Am. Entomol. 38, 222–234.*

exoskeleton provides an ideal site for storage of toxic compounds. Toxins can be stored in scales on the wings of Lepidoptera, for example, cardiac glycosides in the wings of monarch butterflies. Sequestered compounds also can be stored in various glands for more directed defense. The eastern lubber grasshopper, *Romalea guttata*, stores a mixture of sequestered phenolics and quinones from various host plants in paired defensive glands. These defensive compounds are exuded as a defensive froth when the insect is disturbed (Fig. 3.6). Sawfly (Diprionidae) larvae store resinous defenses from host conifer foliage in diverticular pouches in the foregut and regurgitate the fluid to repel predators (Codella and Raffa, 1993). Conner et al. (2000) reported that males of an arctiid moth, *Cosmosoma myrodora*, acquire pyrrolizidine alkaloids systematically from excrescent fluids of certain plants, such as *Eupatorium capillifolium* (but not from larval food plants), and discharge alkaloid-laden filaments from abdominal pouches on the female cuticle during courtship. This topical application significantly reduced predation of females by spiders, *Nephila clavipes*, compared to virgin females and females mated with alkaloid-free males. Additional alkaloid was transmitted to the female in seminal fluid, and was partially invested in the eggs. Lampert et al. (2010) reported that sequestered plant defenses may not be effective against parasites, thereby providing hosts that also protect parasites from predation.

Reudler et al. (2015) measured costs and benefits of sequestered iridoid glycosides (catalpol) by the wood tiger moth, *Parasemia plantaginis*, an aposematic, generalist herbivore.

Although larvae excreted most of the glycosides obtained from plantain, *Plantago lanceolata*, sequestered concentrations in larvae were sufficient to deter predaceous ants and also reduced survival of a parasitoid wasp, *Cotesia villana*. However, excreting and storing plant glycosides were costly, resulting in longer development time and lower pupal mass for the moth. Furthermore, catalpol concentration in larvae was significantly negatively correlated with the size of the orange warning patch, indicating a trade-off between sequestration ability and production of warning coloration.

Accumulation of Ni from *Thlaspi montanum* by a mirid plant bug, *Melanotrichus boydi*, protected it against some predators (Boyd and Wall, 2001), but not against entomopathogens (Boyd, 2002). Boyd (2009) reported that 15 insect species have been found to have whole-body concentrations of Ni > 500 mg kg^{-1}, and one species accumulates levels up to 3500 mg kg^{-1}. Vickerman and Trumble (2003) found that a generalist predator, *Podisus maculiventris*, fed on beet armyworm, *Spodoptera exigua*, larvae reared on a Se-enhanced diet showed slower growth and higher mortality compared to predators fed larvae reared on a control diet. L. Peterson et al. (2003) reported that grasshoppers and spiders, as well as other invertebrates, had elevated Ni concentrations at sites where the Ni-accumulating plant, *Alyssum pintodasilvae*, was present, but not at sites where this plant was absent, indicating the spread of Ni through trophic interactions. Concentrations of Ni in invertebrate tissues approached levels that are toxic to birds and mammals, suggesting that using hyperaccumulating plant species for bioremediation may, instead, spread toxic metals through food chains at hazardous concentrations.

Many arthropods synthesize their own defensive compounds (Meinwald and Eisner, 1995). A number of Orthoptera, Hemiptera, and Coleoptera exude noxious, irritating, or repellent fluids or froths when disturbed (Fig. 3.6). Blister beetles (Meloidae) synthesize the terpenoid, cantharidin, and ladybird beetles (Coccinellidae) the alkaloid, coccinelline (Meinwald and Eisner, 1995). Both compounds, unique to insects, occur in the hemolymph and are exuded by reflex bleeding from leg joints to deter both invertebrate and vertebrate predators. Cantharidin is used medicinally to remove warts. Whiptail scorpions spray acetic acid from their "tail," and the millipede, *Harpaphe*, sprays cyanide (Meinwald and Eisner, 1995). The bombardier beetle, *Brachinus*, sprays a hot (100°C) cloud of benzoquinone produced by mixing, at the time of discharge, a phenolic substrate (hydroquinone), peroxide, and an enzyme catalase (Harborne, 1994). The Formosan subterranean termite, *Coptotermes formosanus*, incorporates naphthalene, a chemical with general antiseptic properties, as well as a repellent effect on many animals, into their nest material (J. Chen et al., 1998).

Several arthropod groups produce venoms, primarily peptides, including phospholipases, histamines, proteases, and esterases, for defense as well as predation (Habermann, 1972; Meinwald and Eisner, 1995; J. Schmidt, 1982). Some species of Hemiptera, Diptera, Neuroptera, and Coleoptera produce orally derived venoms that facilitate prey capture, as well as defense (J. Schmidt, 1982), whereas many Hymenoptera and some Lepidoptera are capable of injecting venom into attackers. Both neurotoxic and hemolytic venoms are represented among insects. Phospholipases are particularly well known because of their high toxicity and their strong antigen activity capable of inducing life-threatening allergy.

Venoms are most common among the Hymenoptera and consist of a variety of enzymes, biogenic amines (such as histamine and dopamine), epinephrine, norepinephrine, and

FIGURE 3.7   **Physical and chemical defensives of a limacodid (Lepidoptera) larva, the saddleback caterpillar,** *Acharia stimulea.* This is the most dangerous stinging caterpillar in North America. Its venom causes severe pain, but can cause kidney failure and cerebral hemorrhaging in susceptible victims.

acetylcholine. This combination produces severe pain and affects cardiovascular, central nervous, and endocrine systems in vertebrates (J. Schmidt, 1982). Melittin, found in bee venom, disrupts erythrocyte membranes (Habermann, 1972). Some venoms include nonpeptide components. For example, venom of red imported fire ants, *Solenopsis invicta*, contains piperidine alkaloids, with hemolytic, insecticidal, and antibiotic effects (Lai et al., 2008).

Larvae of several families of Lepidoptera, especially Saturniidae and Limacodidae (Fig. 3.7), deliver venoms passively through urticating spines, although defensive flailing behavior by many species increases the likelihood of striking an attacker. Venoms of some Lepidoptera, such as the saddleback caterpillar, *Acharia stimulea*, are capable of inducing renal failure or cerebral hemorrhaging in sensitive humans (Diaz, 2005).

Some insects apparently produce thiaminase I in their guts. This enzyme has been shown to cause acute thiamine deficiency in humans eating African silkworms, *Anaphe* spp., or variegated grasshoppers, *Zonocerus variegatus* (Ehigie et al., 2013; Nishimune et al., 2000). This thiaminase is stable at relatively high temperatures, but thorough cooking at high heat can make these insects safe for human consumption.

2.4.2 Antimicrobial Defenses

Arthropods can defend themselves against internal parasites and pathogens. Major mechanisms include ingested or synthesized antibiotics (Blum, 1992; Tallamy et al., 1998), gut modifications that prevent growth or penetration by pathogens, and cellular immunity against parasites and pathogens in the hemocoel (Vega and Kaya, 2012). Behavioral mechanisms also may be employed for protection against pathogens.

Insects produce a variety of antibiotic and anticancer proteins capable of targeting foreign microorganisms (Boman and Hultmark, 1987; Boman et al., 1991; P. Dunn et al., 1994; Hultmark et al., 1982; A. Moore et al., 1996; Morishima et al., 1995). The proteins are induced within as little as 30–60 min of injury or infection, generally bind to bacterial or fungal membranes (increasing their permeability), are effective against a wide variety of infectious organisms (Gross et al., 1996; Jarosz, 1995; A. Moore et al., 1996), and can persist for up to several days (Brey et al., 1993; Gross et al., 1996; Jarosz, 1995). *Drosophila* spp. are known to produce more than 10 different antimicrobial proteins (Cociancich et al., 1994).

Cecropin, originally isolated from cecropia moths, *Hyalophora cecropia*, is produced in particularly large amounts immediately before, and during, pupation. Similarly, hemolin (known from several moths) is produced primarily during embryonic diapause in the gypsy moth, *Lymantria dispar* (K.Y. Lee et al., 2002). Peak concentration during pupation may function to protect the insect from exposure of internal organs to entomopathogens in the gut during diapause or metamorphosis (P. Dunn et al., 1994). In mosquitoes, cecropins may protect against some blood-borne pathogenic microfiliae (Chalk et al., 1995). The entomopathogenic nematode, *Heterorhabditis bacteriophora*, produces anticecropin to permit its pathogenic bacteria to kill the host, the greater wax moth, *Galleria mellonella* (Jarosz, 1995).

Lepidoptera susceptible to the entomopathogenic bacterium, *Bacillus thuringiensis*, typically have high gut pH and large quantities of reducing substances and proteolytic enzymes, conditions that limit protein chelation by phenolics but that facilitate dissolution of the bacterial crystal protein and subsequent production of the delta-endotoxin. By contrast, resistant species have a lower gut pH and lower quantities of reducing substances and proteolytic enzymes (Vega and Kaya, 2012).

Cellular immunity is based on cell recognition of "self" and "nonself" and includes endocytosis and cellular encapsulation. Endocytosis involves infolding of the plasma membrane and enclosure of foreign substances within a phagocyte, without penetration of the plasma membrane. This process removes viruses, bacteria, fungi, protozoans, and other foreign particles from the hemolymph, although some of these pathogens then can infect the phagocytes. Cellular encapsulation occurs when the foreign particle is too large to be engulfed by phagocytes. Aggregation and adhesion by hemocytes forms a dense covering around the particle. Surface recognition may be involved because parasitoid larvae normally protected (by viral associates) from encapsulation are encapsulated when wounded or when their surfaces are altered (Vega and Kaya, 2012). Hemocytes normally encapsulate hyphae of the fungus, *Entomophthora egressa*, but do not adhere to hyphal bodies that have surface proteins protecting them from attachment of hemocytes (Vega and Kaya, 2012). S. Davis et al. (2015) reported that variation in the encapsulation response among three sympatric bumble bees, *Bombus* spp., likely determines the relative abundances of species that share a common parasitoid.

Behavioral mechanisms include grooming and isolation of infected individuals and behavioral fever. Grooming may remove ectoparasites or pathogens.

Behavioral fever, in which thermoregulatory behavior is adjusted to elevate body temperature above the optimal range, may eliminate many pathogens (Elliot et al., 2002). Elliot et al. (2002) compared survival and oviposition by desert locusts, *Schistocerca gregaria*, infected with a pathogenic fungus, *Metarhizium anisopliae*, or uninfected in an arena that permitted locusts to seek suboptimal, optimal, and supraoptimal temperature ranges. Infected locusts spent significantly less time feeding and about twice as much time near the heat source, compared to uninfected locusts. All but 4 of 234 uninfected locusts survived to the end of the 53-day study period. Average survival of infected locusts at suboptimal, optimal, and supraoptimal temperatures was 8, 15, and 21 days, respectively, and only locusts that elevated their body temperature were able to produce viable offspring, although elevated temperature did not kill the pathogen, and all but 7 of 81 infected locusts died within the study period. Infected locusts with behavioral fever produced more solitary phase offspring, compared to uninfected locusts at the same density (Elliot et al., 2003). Elliot et al. (2005) reported that reduced flight capacity and mate competition in infected locusts with behavioral fever could be viewed as a fitness cost of the fever or as a mechanism to induce the solitary phase which would restrict disease transmission.

Social insects display additional behaviors for reducing pathogen transmission. Myles (2002) reported that eastern subterranean termites, *Reticulitermes flavipes*, rapidly aggregate around, immobilize, and entomb individuals infected by the pathogenic fungus, *Metarhizium anisopliae*. Simone-Finstrom and Spivak (2010) reported that honey bees, *Apis mellifera*, increased foraging for plant resins (propolis) following experimental infection with a fungal parasite, *Ascophaera apis*, the agent of chalkbrood disease. Colonies experimentally enriched with resin showed lower infection intensities of this fungal parasite. Such behaviors protect colonies from spread of pathogens.

2.5 Variation in Resource Quality

Organisms are subject to a variety of selective factors in the environment. Intense herbivory is only one factor that affects plant fitness and expression of defenses (Bostock et al., 2001; Horvitz et al., 2005; Izaguirre et al., 2013; Koricheva, 2002). Factors that select most intensively or consistently among generations are likely to result in directional adaptation. Hence, the variety of specific biochemical defenses against herbivores among plant species is evidence of significant selection by herbivory. Nevertheless, at least some biochemical defenses have multiple functions (eg, phenolics as UV filters, pigments, and structural components, as well as defense), so their selection would be enhanced by meeting multiple plant needs. In fact, Carmona et al. (2011) conducted a metaanalysis to evaluate the effect of a variety of genetically variable plant traits on herbivory. They concluded that plant life history traits were more consistently associated with resistance to herbivory than was defensive chemistry, suggesting that defensive chemistry may have evolved because of weaker evolutionary constraints, compared to life history characteristics, rather than because they are most effective against herbivory.

Similarly, insect survival is affected by climate, disturbances, and host condition, as well as a variety of predators, in addition to plant defenses. Short generation time confers a capacity to adapt quickly to strong selective factors, such as particular plant defenses.

Defensive compounds are energetically expensive to produce, and their production competes with other metabolic pathways (I. Baldwin, 1998; Chapin et al., 1987; Herms and Mattson, 1992; Izaguirre et al., 2013; Kessler and Baldwin, 2002; Massad et al., 2012; Schwachtje et al., 2006; Stamp, 2004; Stevens et al., 2007; Strauss and Murch, 2004). Some, such as the complex phenolics and terpenoids, are highly resistant to degradation and cannot be catabolized to retrieve constituent energy or nutrients for other needs. Others, such as alkaloids and nonprotein amino acids, can be catabolized and nitrogen, in particular, retrieved for other uses, but such catabolism involves metabolic costs that reduce net gain in energy or nutrient budgets (Massad et al., 2012). Given sufficient water and nutrient availability, many plants are capable of tolerating, or even increasing, growth in response to herbivore feeding (Schwachtje et al., 2006; Stevens et al., 2007; Trumble et al., 1993, see Chapter 12).

Few studies have addressed the fitness costs of defense, in terms of plant growth or reproduction, as predicted by the *growth–differentiation balance hypothesis* (Herms and Mattson, 1992;

Loomis, 1932; Lorio, 1993; Stamp, 2004) and *carbon–nutrient balance hypothesis* (Bryant et al., 1983; Chapin et al., 1987). Karban and Maron (2002) found that defenses induced via interplant communication may or may not increase plant fitness. Glynn et al. (2007) tested the hypothesis by comparing relative growth rate, net assimilation rate, and phenylpropanoid concentrations in two willow species, *Salix eriocephala* and *S. sericea*, across five soil fertility levels. A generally negative relationship between growth and defense over the fertility gradient supported the growth-differentiation hypothesis, but indicated complex interactions between plant physiological status and soil nutrient availability (see Chapter 12). I. Baldwin (1998) evaluated seed production by plants treated or not treated with jasmonate, a phytohormone that induces plant defenses (see section 3.1.5.3 and Chapter 8). Induction of defense did not significantly increase seed production of plants that came under herbivore attack, but significantly reduced seed production of plants that were not attacked. Ballhorn et al. (2013) demonstrated that lima bean plants, *Phaseolus lunatus*, with high concentrations of cyanogenic glucosides had significantly lower above-ground biomass and seed production than did plants with low concentrations when grown in the absence of herbivores, indicating that plant defense can have significant fitness costs. However, in the presence of herbivores the performance of plants with high or low concentrations was comparable. Sthultz et al. (2009) reported that *Pinus edulis* trees that were resistant to the shoot moth, *Dioryctria albovittella*, suffered greater water stress and three times higher mortality during drought than did trees that were susceptible to the moth, indicating a trade-off between resistance to herbivory or to drought. The differential mortality caused a shift in stand structure from dominance by resistant trees to parity between the resistant and susceptible trees. Finally, Massad et al. (2012) uncovered trade-offs among photosynthesis, growth, competition, resource availability, and flavan, but not saponin, defense in seedlings of a tropical tree, *Pentaclethra macroloba*, when biomass and defenses were positively correlated (Fig. 3.8).

Given competition among metabolic pathways for limited energy and nutrients, production of defensive compounds should be sensitive to risk of herbivory or predation, as well as to environmental conditions (Chapin et al., 1987; Coley, 1986; Coley et al., 1985; Glynn et al., 2003; Hatcher et al., 2004; Herms and Mattson, 1992; M. Hunter and Schultz, 1995; Izaguirre et al., 2013; Karban and Niiho, 1995). Plants at low risk of herbivory may produce fewer chemical defenses or favor tolerance over defense (Stevens et al., 2007). L. Dyer et al. (2001) reported that *Piper cenocladum* plants hosting aggressive ant, *Pheidole bicornis*, colonies produced lower concentrations of amides to deter leaf-cutting ants and orthopterans, indicating a trade-off in costs between amides and support of ants. Nevertheless, the combination of defenses minimized losses to a diversity of herbivores. Izaguirre et al. (2013) reported that *Passiflora edulis* increased production of extrafloral nectar to attract predators in response to simulated herbivory or jasmonate application, but suppressed this response when exposed to far-red radiation, which signals the proximity of competitors.

The studies summarized in this section indicate complex relationships among plant condition, resource availability, and defenses. As a result, expression of defenses, and plant suitability for herbivores, vary spatially and temporally.

2.5.1 Variation Among Plants and Tissues

Plants and plant tissues vary in their physical and chemical attributes in space and time. Variation among plant species reflects phylogenetic differences in nutrient allocation to various metabolic pathways, including specific defenses. Some plant taxa are characterized by

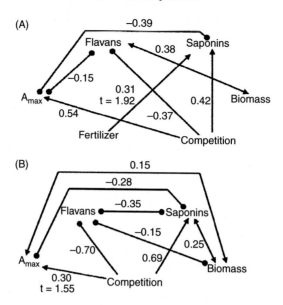

FIGURE 3.8 Interactions and trade-offs among photosynthesis (A_{max}), growth (biomass), and defense production (flavans and saponins), in *Pentaclethra macroloba* seedlings grown in the shade (A, 20% light) or sun (B, 100% light) and with low, intermediate, or high nitrogen availability. Causal relationships are indicated as single-headed arrows, and correlations as double-headed arrows. Filled circles indicate negative relationships, and arrows indicate positive relationships. Numbers are the standardized parameter estimates for relationships between variables. All relationships were significant with t-values >1.96 except where smaller t-values are indicated. *Source: From Massad et al., 2012. Costs of defense and a test of the carbon-nutrient balance and growth-differentiation balance hypotheses for two co-occurring classes of plant defense. PLoS ONE 7 (10), e47554.*

particular defensive compounds, making secondary chemistry a useful taxonomic characteristic (Waterman, 2007). For example, ferns and gymnosperms rely primarily on phenolics, terpenoids, and insect hormone analogues, whereas angiosperms more commonly produce alkaloids, phenolics, and other types of compounds. However, most plants (including aquatic species) produce compounds representing a variety of chemical classes (Fornoff and Gross, 2013; Harborne, 1994; Newman, 1990).

Each plant species can be distinguished by a unique "chemical fingerprint" conferred by these chemicals (Richards et al., 2015). Production of alkaloids and other nitrogenous defenses depends on the availability of nitrogen (Harborne, 1994), but at least four species of spruce, *Picea* spp., and seven species of pines, *Pinus* spp., are known to produce piperidine alkaloids (Stermitz et al., 1994), despite low N concentrations. Plant defenses can reduce feeding by insects substantially. Richards et al. (2015) found that increasing phytochemical diversity resulted in increased herbivore specialization and reduced overall herbivore damage. However, insects also identify potential hosts by their chemical profile.

Within plant species, individual genotypes vary widely in food quality for insects. Genung et al. (2012) reported that aphid, *Uroleucon nigrotuberculatum*, abundance was 80 times higher on the most susceptible goldenrod, *Solidago altissima*, genotype (1821 aphids plot[−1]) relative to the least susceptible genotype (24 aphids plot[−1]). Plant allocation of nitrogen to various tissues may be particularly important to associated herbivores. Low nitrogen levels in plant tissues frequently limit insect herbivore growth and reproduction (Scriber and Slansky, 1981). Ohmart et al. (1985) reported that *Eucalyptus blakelyi* subjected to different N fertilization levels significantly affected fecundity of *Paropsis atomaria*, a chrysomelid beetle. An increase in foliar N from 1.5% to 4.0% increased the number of eggs laid by 500% and the rate of egg production by 400%. However, increasing nitrogen in plant tissues does not necessarily improve herbivore performance. Zehnder and Hunter (2009) reported that *Aphis nerii* population growth on milkweed, *Asclepias syriaca*, peaked at intermediate levels of nitrogen addition (Fig. 3.9), and Cease et al. (2012) reported that *Oedaleus asiaticus*, a dominant locust of north Asian grasslands, preferred plants with low N content and artificial diets with low protein and high carbohydrate content.

Plant tissues also vary in nutritional value and concentration of defensive compounds, depending on risk of, or response to, herbivory and value to the plant (Dirzo, 1984; Feeny, 1970; Paschold et al., 2007; Strauss et al., 2004). Foliage tissues, which are the source of photosynthates and have a high risk of herbivory, typically have high concentrations of defensive compounds. Roots also produce defensive compounds (van Dam, 2009). Defensive compounds in shoots are concentrated in bark tissues, perhaps reducing risk to subcortical tissues, which have relatively low concentrations of defensive compounds (Schowalter et al., 1998). Izaguirre et al. (2013) reported that increased extrafloral nectary production by *P. edulis*, in response to simulated herbivory, was highly localized within the treated portion of the plant; untreated portions showed no response.

2.5.2 Variation Through Time

Defensive strategies change as plants or tissues mature (Dirzo, 1984; Forkner et al., 2004). A visible example is the reduced production of thorns on foliage and branches of acacia, locust, and other trees when the crown grows above the grazing height of vertebrate herbivores (S. Cooper and Owen-Smith, 1986; P. White, 1988).

The nutritional value of plant resources also changes seasonally. Filip et al. (1995) reported that the foliage of many tropical trees has higher nitrogen and water content early in the wet season than late in the wet season. R. Lawrence et al. (1997) caged several cohorts of spruce budworm, *Choristoneura fumiferana*, larvae on white spruce at different phenological stages of the host. Cohorts that began feeding 3–4 weeks before budbreak and completed larval development prior to the end of shoot elongation developed significantly faster and showed significantly greater survival rate and adult mass than did cohorts caged later (Fig. 3.10). These results indicated that the phenological window of opportunity for this insect was sharply defined by the period of shoot elongation, during which foliar nitrogen, phosphorus, potassium, copper, sugars, and water were higher than in mature needles. Concentrations of condensed tannins in oak, *Quercus* spp., leaves generally increase from low levels at budbreak to high levels at leaf maturity (Feeny, 1970; Forkner et al., 2004). Consequently, herbivores tend to be most active during periods of leaf emergence (Coley and Aide, 1991; Feeny, 1970; M. Hunter and Schultz, 1995; R. Jackson et al., 1999; Lowman, 1985, 1992). Lorio (1993) reported that production of resin ducts by loblolly pine, *Pinus taeda*, is restricted to latewood formed during summer. The rate of earlywood formation in the spring determines the likelihood that southern pine beetles, *Dendroctonus frontalis*, colonizing trees in spring will sever resin ducts and induce pitch flow. Hence, tree susceptibility to colonization by this insect increases with stem growth rate. Ruel and Whitham (2002) also found that susceptibility to stem-boring

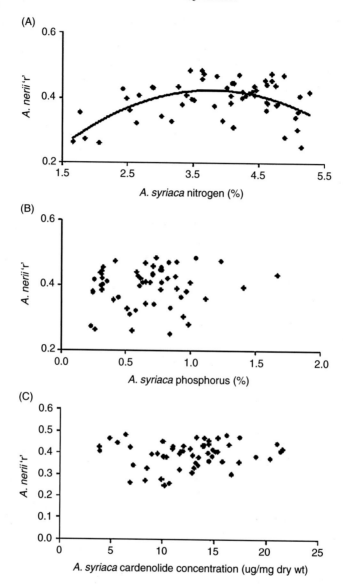

FIGURE 3.9 Relationship between per capita population growth rate (r) of the aphid, *Aphis nerii*, and (A) foliar nitrogen concentration, (B) foliar phosphorus concentration, and (C) foliar cardenolide concentration of its milkweed host, *Asclepias syriaca*. Each point represents a single aphid–host plant pair. *Source: From Zehnder, C.B., Hunter, M.D., 2009. More is not necessarily better: the impact of limiting and excessive nutrients on herbivore population growth rates. Ecol. Entomol. 34, 535–543.*

moths, *Dioryctria albovittella*, increased among pinyon pines, *P. edulis*, that grew faster as juveniles, compared to slower-growing pines.

Concentrations of various defensive chemicals are affected by environmental changes (Cronin et al., 2001; Mopper et al., 2004) and disturbance (M.D. Hunter and Forkner, 1999; Nebeker

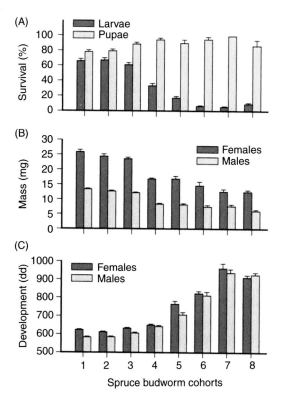

FIGURE 3.10   (A) Larval and pupal survival, (B) adult dry mass, and (C) development time from second instar through adult for eight cohorts of spruce budworm caged on white spruce in 1985. The first six cohorts were started at weekly intervals beginning on Julian date 113 (Apr. 23) for cohort 1. Cohort 7 started on Julian date 176 (Jun. 25), and cohort 8 on Julian date 204 (Jul. 23). Each cohort remained on the tree through completion of larval development, 6–7 weeks. Budbreak occurred during Julian dates 118–136, and shoot elongation during Julian dates 118–170. *Source: From Lawrence, R.K., et al., 1997. White spruce and the spruce budworm: defining the phenological window of susceptibility. Can. Entomol. 129, 291–318.*

et al., 1993). Cronin et al. (2001) monitored preferences of a stem-galling fly, *Eurosta solidaginis*, among the same 20 clones of goldenrod, *Solidago altissima*, over a 12-year period and found that preference for, and performance on, the different clones were uncorrelated among years. These data indicated that genotype × environmental interaction affected the nutritional quality of clones for this herbivore. Increased exposure to UVB reduced concentration of gallic acid and increased concentration of flavenoid aglycone in southern beech, *Nothofagus antarctica* (Rousseaux et al., 2004). Cipollini (1997) found that wind increased concentrations of peroxidase, cinnamyl alcohol-dehydrogenase, and lignin in the bean, *Phaseolus vulgaris*, and reduced oviposition and population growth of two-spotted spider mites, *Tetranychus urticae*.

2.5.3 *Variation in Response to Injury*

Plants balance the trade-off between the cost of defense and the risk of severe herbivory (Coley, 1986; Coley et al., 1985). In addition to *constitutive defenses* that are normally present in plant tissues, plants initiate production of *inducible defenses* in response to injury

(Haukioja, 1990; Kaplan et al., 2008; Karban and Baldwin, 1997; Klepzig et al., 1996; Nebeker et al., 1993; M. Stout and Bostock, 1999; Strauss et al., 2004). Constitutive defenses generally are less specific compounds that are relatively effective against a wide variety of herbivores, whereas inducible defenses are more specific compounds produced in response to particular types of injury (Hatcher et al., 2004; Uesugi et al., 2013). Herbivore feeding or regurgitants and pathogen infection are known to trigger plant wound hormones, particularly jasmonic acid, salicylic acid, and ethylene (Creelman and Mullet, 1997; Farmer and Ryan, 1990; Karban and Baldwin, 1997; Kessler and Baldwin, 2002; McCloud and Baldwin, 1997; Paschold et al., 2007; Ralph et al., 2006; Schmelz et al., 2006, 2007; Thaler, 1999a; Thaler et al., 2001). These elicitors, in turn, induce production of defenses, such as proteinase inhibitors that interfere with insect digestive enzymes (Kessler and Baldwin, 2002; Thaler et al., 2001; Zeringue, 1987). Recently, Appel and Cocroft (2014) demonstrated that vibrations caused by insect feeding also elicited production of glucosinolate and anthocyanin defenses in *Arabidopsis thaliana* that protected rosettes from subsequent feeding by the cabbage white butterfly, *Pieris rapae*. Plants also discriminated between vibrations caused by herbivory and those caused by wind or leafhopper singing.

Plants often respond to injury with a combination of induced defenses that reflect expression of specific gene sets triggered by, and targeted against, particular herbivore or pathogen species but that also confer generalized defense against associated or subsequent herbivores or pathogens (Hatcher et al., 2004; Kessler and Baldwin, 2002; T. Parsons et al., 1989; Ralph et al., 2006; D. Schmidt et al., 2005; Schwachtje and Baldwin, 2008; M. Stout and Bostock, 1999; Uesugi et al., 2013). Klepzig et al. (1996) reported that initial penetration of *Pinus resinosa* bark by bark beetles and associated pathogenic fungi was not affected by plant constitutive defenses, but elicited elevated concentrations of phenolics and monoterpenes that significantly inhibited germination of fungal spores or subsequent hyphal development. Continued insect tunneling and fungal development elicited further host reactions in healthy trees (but not stressed trees) that were usually sufficient to repel the invasion.

Plant defenses can be induced through multiple pathways that encode for different targets, such as internal specialists versus more mobile generalists, and interaction (crosstalk) among pathways may enhance or compromise defenses against associated consumers (Cusumano et al., 2015; Kessler and Baldwin, 2002; Rodriguez-Saona et al., 2005; Schultz and Appel, 2004; Schwachtje and Baldwin, 2008; M. Stout et al., 2006; Thaler, 1999a; Thaler et al., 2001; Uesugi et al., 2013). For example, Uesugi et al. (2013) reported that *Solidago altissima* responded to herbivory by one gelechiid moth, *Dichomeris leuconotella*, and two chrysomelid beetles, *Microrhopala vittata* and *Trirhabda virgata*, with distinct combinations of phenolic and diterpene acids that targeted each species in a unique way (Fig. 3.11). Schmelz et al. (2006, 2007) demonstrated that fall armyworm, *Spodoptera frugiperda*, caterpillar oral secretions contain inceptin, a small peptide resulting from proteolytic cleavage of chloroplastic ATP synthase from its cowpea, *Vigna unguiculata*, host. This unique product of herbivore digestion allows the plant to distinguish injury by herbivores from abiotic injury and triggers plant induction of antiherbivore compounds. Little et al. (2007) studied the gene expression profile for *Arabidopsis thaliana*, following oviposition by two pierid butterflies, *Pieris brassicae* and *P. rapae*. Histochemical analysis indicated that oviposition caused localized plant cell death, resulting in accumulation of callose and initiation of jasmonic acid and salicylic acid signaling pathways, indicating early perception of, and response to, incipient herbivory.

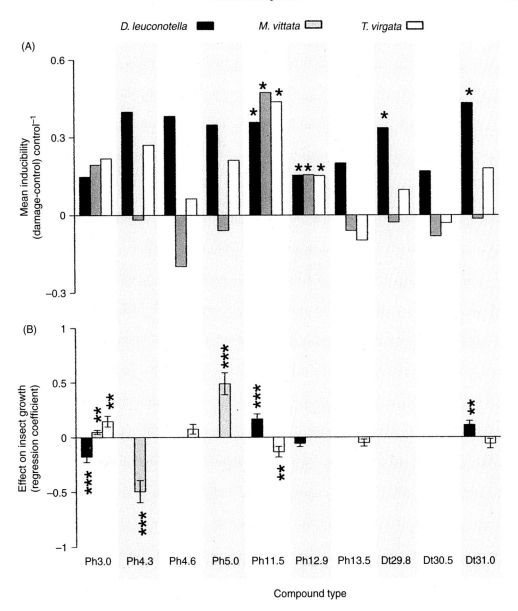

FIGURE 3.11 Phenolic *(Ph)* and diterpene acid *(Dt)* compounds that were (A) induced by *Dichomeris leuconotella*, *Microrhopala vittata*, and *Trirhabda virgata* damage, and (B) correlated with herbivore specific growth rate. Closed bars indicate values for *D. leuconotella*, gray bars for *M. vittata*, and open bars for *T. virgata*. Asterisks indicate significance: *$p < 0.05$, **$p < 0.01$, ***$p < 0.001$. Source: *From Uesugi, A., et al., 2013. A test of genotypic variation in specificity of herbivore-induced responses in Solidago altissima L. (Asteraceae). Oecologia 173, 1387–1396.*

Thaler et al. (2002) found that wild-type tomato plants capable of producing jasmonate in response to herbivory produced more defensive chemicals and attracted more predators when damaged by herbivores than did a jasmonate-deficient tomato variety. Emission of jasmonate from damaged plants can communicate injury and elicit production of induced defenses by neighboring, even unrelated, plants (Dolch and Tscharntke, 2000; Farmer and Ryan, 1990; Hudgins et al., 2004; Karban and Maron, 2002; Karban et al., 2000; Schmelz et al., 2002; M. Stout et al., 2006; Thaler et al., 2001; Tscharntke et al., 2001, see Chapter 8). Kessler et al. (2006) demonstrated that communication of injury via volatile chemicals may induce priming and accelerated defense in response to subsequent injury, rather than directly eliciting defensive chemicals, among neighboring plants. Herbivorous insects may have limited ability to detect, or learn to avoid, jasmonic acid (Daly et al., 2001). However, some insects are able to suppress jasmonate-induced defenses. Voelckel et al. (2001) demonstrated that oral secretions of the tobacco hornworm, *Manduca sexta*, suppress jasmonate-induced nicotine production in tobacco, *Nicotiana attenuata*, but instead trigger a burst of ethylene production that induces release of volatile terpenoids that, in turn, attract parasitoids (known to be sensitive to nicotine) to feeding *M. sexta* larvae.

Molecular techniques have greatly enhanced our ability to test the effects of defensive mechanisms. Kessler et al. (2004) and Paschold et al. (2007) genetically engineered tobacco plants to silence gene expression for jasmonate induction. Herbivore performance and feeding injury were significantly higher on the jasmonate-silenced plants than on untreated plants.

2.5.4 Factors Affecting Expression of Defenses

Plants show trade-offs among defense pathways, for example, between constitutive and induced defenses and between C-based and N-based defenses. Kazemi-Dinan et al. (2015) demonstrated trade-offs between organic and inorganic defenses. Healthy plants growing under optimal environmental conditions should be capable of maintaining the full array of metabolic processes and may provide greater nutritional value to insects capable of countering plant defenses. Such plants may allocate more resources to growth, relative to defenses, thereby compensating for losses to herbivores (Glynn et al., 2003; Trumble et al., 1993, see Chapter 12). By contrast, unhealthy plants or plants growing under adverse environmental conditions (such as water or nutrient limitation) may sacrifice some metabolic pathways in order to maintain those most critical to survival (Herms and Mattson, 1992; Lorio, 1993; Mattson and Haack, 1987; Mopper et al., 2004; Tuomi et al., 1984; Wang et al., 2001; R. Waring and Pitman, 1983). In particular, stressed plants often reduce production of defensive chemicals in order to maximize allocation of limited resources to maintenance pathways and thereby become relatively more vulnerable to herbivores (Fig. 3.12).

Spatial and temporal variation in plant defensive capability creates a mosaic of food quality for herbivores (L. Brower et al., 1968). In turn, herbivore employment of plant defenses affects their vulnerability to predators (L. Brower et al., 1968; Malcolm, 1992; Stamp et al., 1997; Traugott and Stamp, 1996). Herbivore feeding strategies represent a trade-off between maximizing food quality and minimizing vulnerability to predators (Murphy and Loewy, 2015; Schultz, 1983; and as discussed in section 3).

The frequent association of insect outbreaks with stressed plants (V.C. Brown, 1995; Heliövaara, 1986; Heliövaara and Väisänen, 1986, 1993; W. Smith, 1981) led T. White (1969, 1976, 1984) to propose the Plant Stress Hypothesis, that is, that stressed plants are more suitable hosts for herbivores. However, some herbivore species prefer more vigorously growing,

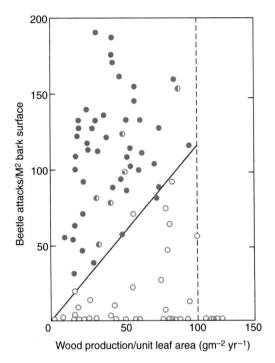

FIGURE 3.12 **The density of mountain pine beetle attacks necessary to kill lodgepole pine increases with increasing host vigor, measured as growth efficiency.** The filled portion of circles represents the degree of tree mortality. The solid line indicates the attack level predicted to kill trees of a specified growth efficiency (index of radial growth); the dotted line indicates the threshold above which beetle attacks are unlikely to cause mortality. *Source: From Waring, R.H., Pitman, G.B., 1983. Physiological stress in lodgepole pine as a precursor for mountain pine beetle attack. Z. Angew. Entomol. 96, 265–270.*

apparently nonstressed plants (G. Waring and Price, 1990), leading Price (1991) to propose the alternative Plant Vigor Hypothesis. Reviews by Koricheva et al. (1998) and G. Waring and Cobb (1992) indicated that response to plant condition varies widely among herbivore species. Schowalter et al. (1999) manipulated water supply to creosote bushes, *L. tridentata*, in New Mexico and found significant positive, negative, nonlinear, and nonsignificant responses to moisture availability among the assemblage of herbivore and predator species on this single plant species, demonstrating that both hypotheses are supported by some species, a conclusion supported by T. White (2009).

Regardless of the direction of response, water and nutrient subsidy or limitation clearly affect herbivore–plant interactions (Coley et al., 1985; M.D. Hunter and Schultz, 1995; Mattson and Haack, 1987; Van Bael et al., 2004). Therefore, resource acquisition by insects is moderated, at least in part, by ecosystem processes or environmental changes that affect the availability of water and nutrients for plants (Chapter 11).

Some plant species respond to increased atmospheric concentrations of CO_2 by allocating more carbon to defenses, such as phenolics or terpenoids, especially if other critical nutrients remain limiting (Arnone et al., 1995; Chapin et al., 1987; Grime et al., 1996; Kinney et al., 1997; Landosky and Karowe, 2014; Lindroth et al., 1993a,b; Roth and Lindroth, 1994).

However, plant responses to CO_2 enrichment vary considerably among species and as a result of other environmental conditions such as light, water, and nutrient availability or stress (Bazzaz, 1990; Dudt and Shure, 1994; P. Edwards, 1989; M. Hall et al., 2005; Landosky and Karowe, 2014; Lindroth et al., 1993a,b,c; Niesenbaum, 1992), with equally varied responses among herbivore species (Bezemer and Jones, 1998; M. Hall et al., 2005; Salt et al., 1996; Watt et al., 1995). Zavala et al. (2008) demonstrated that elevated atmospheric CO_2 resulted in downregulation of gene expression for defense-signaling compounds and, consequently, proteinase inhibitors, in soybean, *Glycine max*. Subsequently, Zavala et al. (2013) reviewed effects of elevated CO_2 on plant defenses and herbivory and concluded that, in general, elevated CO_2 downregulates constitutive and induced defenses associated with the jasmonic acid pathway and upregulates induced defenses derived from the salicylic acid pathway. Thus, plants growing under elevated CO_2 are likely to become more susceptible to chewing herbivores and less susceptible to sap-sucking herbivores (Fig. 3.13). However, changes in concentrations of particular defensive compounds in response to elevated CO_2, as well as herbivore responses, are highly idiosyncratic (Lindroth et al., 1993a,b,c; Zavala et al., 2013). Such complexity of responses precludes general prediction of the effects of CO_2 enrichment on insect–plant interactions (Bazzaz, 1990; Watt et al., 1995; Zavala et al., 2013).

Atmospheric deposition of typically limited nutrients, especially nitrogen, also affects insect–plant interactions, although the mechanisms are not clear. In general, nitrogen deposition increases growth and survival of individual insect herbivores and promotes population growth (Throop and Lerdau, 2004). N enrichment may permit plants to allocate more C to growth, and reduce production of nonnitrogenous defenses, making plants more vulnerable to herbivores, as predicted by the carbon–nutrient balance hypothesis (Holopainen et al., 1995). M. Jones et al. (2004) reported that nitrogen deposition increased bark beetle activity and pine tree mortality. Zehnder and Hunter (2008) found that experimental simulation of nitrogen deposition significantly increased milkweed, *Asclepias tuberosa*, foliar nitrogen concentration, plant biomass, and per capita aphid, *Aphis nerii*, population growth, up to a point. However, increasing dietary nitrogen beyond an optimal level did not improve insect performance (Zehnder and Hunter, 2009). Joern and Behmer (1998) reported that two grasshopper species differed in their growth and reproduction on diets varying in carbohydrate and nitrogen contents. For *Melanoplus sanguinipes*, reproductive rate showed a significant negative linear response to increasing carbohydrate and a significant quadratic response to increasing nitrogen, with a peak in egg production at 4% nitrogen. *Phoetaliotes nebrascensis*, on the other hand, showed a much weaker response to increasing nitrogen and no response to increasing carbohydrate.

Experimental fertilization has produced apparently contradictory results (Kytö et al., 1996; G. Waring and Cobb, 1992). In some cases, this inconsistency may reflect nonlinear responses of insects to increasing nitrogen in plant tissue (Joern and Behmer, 1998; Zehnder and Hunter, 2009) or different feeding strategies relative to plant allocation of subsidized nutrients (Kytö et al., 1996; Schowalter et al., 1999). In other cases, conflicting results may reflect changes in nutrient balances, that is, which nutrients were most limiting (Behmer, 2009; Elser and Urabe, 1999; Elser et al., 1996; Sterner and Elser, 2002). Furthermore, plants differ in their allocation of subsidized nutrients, for example, to increased production of N-based defenses versus increased protein content. Other associated species also may influence insect response to subsidized nutrients. Kytö et al. (1996) found that positive responses to N fertilization at the individual insect level were often associated with negative responses at the

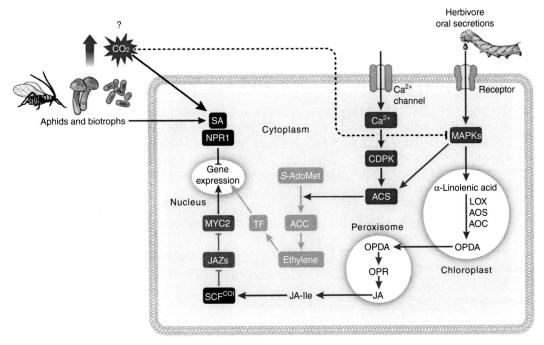

FIGURE 3.13 **Effects of elevated CO_2 on early signaling events in response to herbivory or pathogen infection.** Plant defense is controlled by cellular responses to injury or infection that activate jasmonic acid *(JA)*, ethylene, and salicylic acid *(SA)* signaling pathways. Oral secretions from herbivores trigger the activation of CA^{+2} channels and CA^{+2} fluxes in the cell. Ca^{+2} binds to calmodulins and calcium-dependent protein kinase *(CDPK)*. Mitogen-activated protein kinase *(MAPK)*, such as salicylic acid-induced protein kinase *(SIPK)* and wound-induced protein kinase *(WIPK)*, are also rapidly activated and trigger the biosynthesis of jasmonic acid and JA-Ile. JA-Ile binds to the coronatine insensitive 1 protein *(COI1)* receptor that leads to degradation of Jasmonate-Zim *(JAZ)* proteins and release of their inhibitory effect on transcription factor *(TF)* MYC2. This, in turn, induces defense genes in the nucleus. MAPK and CDPK phosphorylate 1-aminocyclopropane-1-carboxylic acid synthase *(ACS)* protein increase ethylene production, which leads to increased activity of ethylene-responsive transcription factors, inducing defense genes. Elevated CO_2 induces thioredoxins and glutathione S-transferase *(black arrow)*, leading to accumulation of SA and activation of nonexpressor of pathogenesis-related genes1 *(NPR1)*. The activated NPR1 functions as a transcription factor in the nucleus and inhibits JA-induced defense gene expression. In addition, elevated CO_2 may decrease MAPK activity and activate downstream genes of SA signaling *(dotted black line)*. The inhibition of early signaling events by elevated CO_2 decreases the accumulation of metabolites that function as a defense against herbivores. Abbreviations: *ACC*, 1-aminocyclopropane-1-carboxylic acid; *OPDA*, 12-oxo-phytodienoic acid; *OPR*, OPDA reductase; *LOX*, lipoxygenase; *AOS*, allene oxide synthase; *AOC*, allene oxide cyclase. *Source: From Zavala, J.A., et al., 2013. An emerging understanding of mechanisms governing insect herbivory under elevated CO2. Annu. Rev. Entomol. 58, 79–97.*

population level, perhaps indicating indirect effects of fertilization on attraction of predators and parasites.

2.6 Mechanisms for Exploiting Variable Resource Quality

Although plant defensive chemistry clearly affects insect performance, insects are capable of feeding on defended hosts. Feeding preferences for less-defended hosts reflect one mechanism for avoiding defenses. However, insects exhibit a variety of mechanisms for improving plant suitability and/or avoiding, circumventing, or detoxifying host defenses.

Gall-forming insects control gall formation and the chemical composition of colonized plant tissues, to the benefit of the insect (Saltzmann et al., 2008). Gall formation apparently is induced by salivary compounds, rather than mechanical injury (Sopow et al., 2003) and reflects the relationship between shoot length and the dose of gall induction stimulus (Flaherty and Quiring, 2008). Gall chemistry returns to that of surrounding tissues if the gall-former is killed (Hartley, 1998). The inner lining of galls is nutritive tissue rich in free amino acids (Price et al., 1987; Saltzmann et al., 2008), but gall tissues outside this lining often are lower in nitrogen and higher in phenolics than are ungalled tissues (Hartley, 1998). Y. Koyama et al. (2004) reported that the amount of amino acids exuding from leaves galled by the aphid, *Sorbaphis chaetosiphon*, was five times that from ungalled leaves. Furthermore, galls retained high amino acid concentrations throughout Apr., whereas amino acid concentrations declined rapidly during this period in ungalled leaves. Y. Koyama et al. (2004) also compared growth and reproduction of another aphid, *Rhopalosiphum insertum*, which can displace gall aphids or colonize ungalled leaves. Growth and reproduction by this aphid were significantly higher for colonies experimentally established in galls, compared to colonies established on ungalled leaves, indicating a positive effect of gall formation.

Some insects vector plant pathogens that induce favorable nutritional conditions or inhibit host defense (Bridges, 1983). However, not all insects that vector plant pathogens benefit from host infection (Kluth et al., 2002).

Insects that exploit nutritionally poor resources require extended periods (several years to decades) of larval feeding, or other adaptations, in order to concentrate sufficient nutrients (especially N and P) to complete development. Many have obligate associations with microorganisms that provide, or increase access to, limiting nutrients. Termites host mutualistic gut bacteria or protozoa that catabolize cellulose, fix nitrogen, and concentrate or synthesize other nutrients and vitamins needed by the insect (Breznak and Brune, 1994; Mankowski et al., 1998). Termites and some other detritivores feed on feces (coprophagy) after sufficient incubation time for microbial digestion and enhancement of nutritive quality. If coprophagy is prevented, these insects often compensate by increasing consumption of detritus (McBrayer, 1975). Aphids also may rely on endosymbiotic bacteria to provide requisite amino acids, vitamins, or proteins necessary for normal development and reproduction (Baumann et al., 1995).

Food resources typically do not have the proper proportions of nutritional components required by animals for optimal nutrition. Insects have evolved a variety of strategies that govern the extent (trade-off) to which they will overeat a limiting nutrient and undereat an overabundant nutrient for optimal nutrition. Simpson and Raubenheimer (1993) pioneered efforts to describe the fundamental variables of nutritional homeostasis, that is, rules governing trade-offs when diets have suboptimal nutrient balance. K.P. Lee et al. (2002, 2003), Raubenheimer and Simpson (1999, 2003), and Simpson et al. (2002) described nutrient balancing strategies for several grasshopper and caterpillar species varying in host range. Halvorson et al. (2016) used altered C:P ratios in microbial resources for detritivorous caddisfly larvae and used ^{14}C and ^{33}P to measure assimilation rates for these elements (see Chapter 4). They reported that caddisflies feeding on lower C:P diets assimilated P less efficiently, reflecting excess dietary P, and that species with lower body C:P must assimilate dietary C and P more efficiently to support rapid growth of P-rich tissues.

Behmer (2009) reviewed insect strategies for dealing with nutritional imbalances in food resources. According to the Nutritional Heterogeneity Hypothesis, the amount of nutritionally imbalanced food consumed should reflect the probability of encountering food that is equally and oppositely imbalanced. This probability is higher for insects with wide diet breadth, compared to insects specializing on a single food source. Therefore, insects specializing on particular resource will be unable to compensate for nutritional imbalance and should evolve to use small amounts of imbalanced food most efficiently, rather than suffer fitness costs of overeating imbalanced food (closest distance rule). In contrast, mobile generalists should eat as much as possible of imbalanced foods as they are encountered, with a high probability of achieving nutritional balance overall (fixed proportion rule).

Insects also must balance the fitness costs of ingesting harmful chemicals and the costs of regulating nutritional balance. Behmer (2009) analysis indicated that plant defenses may have little effect on insect growth and survival when nutrient balance is optimal, but become increasingly deleterious, even fatal, as protein/carbohydrate imbalance increases.

Several mechanisms are employed to avoid or circumvent host defensive chemicals. Some herbivores avoid exposure to plant defenses by moving to new resources in advance of induced responses (Paschold et al., 2007). Others sever the petiole or major leaf veins to inhibit translocation of induced defenses during feeding (Becerra, 1994; Karban and Agrawal, 2002). Sawflies (Diprionidae) sever the resin canals of their conifer hosts or feed gregariously to consume foliage before defenses can be induced (McCullough and Wagner, 1993). Species feeding on plants with photooxidant defenses often feed at night or inside rolled leaves to avoid sunlight (Berenbaum, 1987; Karban and Agrawal, 2002).

Sequestration and excretion are alternative means of avoiding the effects of host toxins that cannot be detoxified. Sequestered toxins are transported quickly to specialized storage tissues (the exoskeleton or protected pouches), whereas remaining toxins are transported to the Malphigian tubules for elimination. Boyd (2009) noted that high-Ni insects had elevated concentrations in Malphigian tubules and exuviae, indicating elimination as a strategy for feeding on hyperaccumulating plants. Sequestered toxins also may be used in the insect's own defensive strategy (Blum, 1981, 1990, 1992; Boyd and Wall, 2001; Conner et al., 2000; Holzinger and Wink, 1996; L. Peterson et al., 2003; Zalucki et al., 2001). In fact, grazing insects that ingest and sequester combinations of plant defensive compounds may benefit from increased protection from predators, compared to those that sequester single defensive compounds (C. Mason et al., 2014).

Herbivorous insects produce a variety of catalytic enzymes, in particular those associated with cytochrome P-450, to detoxify plant or prey defenses (Feyereisen, 1999; Karban and Agrawal, 2002; W. Mao et al., 2006; Y. Mao et al., 2007). Some insects produce salivary or gut enzymes that minimize the effectiveness of plant defenses. Salivary enzymes, such as glucose oxidase applied to feeding surfaces by caterpillars, inhibit expression of genes responsible for activation of induced defenses (Bede et al., 2006; Felton and Eichenseer, 1999; Musser et al., 2005, 2006). Saliva of Hemiptera gels into a sheath that separates the insect's stylet from plant cells, perhaps reducing induced plant responses (Felton and Eichenseer, 1999). In addition, gut microbial symbionts may augment the detoxification ability of herbivorous insects (Hammer and Bowers, 2015).

Digestive enzymes responsible for detoxification typically are microsomal monooxygenases, glutathione S-transferases, and carboxylesterases (Hung et al., 1990; Xu et al., 2015) that fragment defensive compounds into inert molecules. Microsomal monooxygenases provide a general-purpose detoxification system for most herbivores (Hung et al., 1990). More specific digestive enzymes also are produced by species that encounter particular defenses. Ascorbate is a primary antioxidant found in the gut fluids of foliar-feeding insects to reduce the effect of phenolic oxidation (Barbehenn et al., 2008). However, plant tissues with high concentrations of particularly reactive tannins can overwhelm this antioxidative capacity (Barbehenn et al., 2008). Exposure to plant toxins can induce production of detoxification enzymes (Karban and Agrawal, 2002). For example, caterpillars feeding on diets containing proteinase inhibitors showed reduced function of particular proteinases, but responded by producing other proteinases that were relatively insensitive to dietary proteinase inhibitors (Broadway, 1995, 1997). The compounds produced through detoxification pathways may be used to meet the insect's nutritional needs (Bernays and Woodhead, 1982), as in the case of the sawfly, *Gilpinia hercyniae*, that detoxifies and uses the phenolics from its conifer host (Schöpf et al., 1982).

The ability to detoxify plant defenses may predispose many insects to detoxify synthetic insecticides (Feyereisen, 1999; Ffrench-Constant, 2013; Plapp, 1976; Xu et al., 2015). At least 500 arthropod species are resistant to major insecticides used against them, primarily through a limited number of resistance mechanisms that confer cross-resistance to plant defenses and structurally related toxicants, and, in some cases, to chemically unrelated compounds (Hsu et al., 2004; Soderlund and Bloomquist, 1990). Le Goff et al. (2003) reported that several cytochrome P-450 genes code for detoxification of DDT, imidacloprid, and malathion. In some cases, insect adaptation reflects mutations that reduce binding to, or sensitivity of, target enzymes (Hsu et al., 2004, 2006, 2008).

Gut pH is a factor affecting the chelation of nitrogenous compounds by tannins. Some insect species are adapted to digest food at high gut pH to inhibit chelation. The insect thus is relatively unaffected by high tannin contents of its food. Examples include the gypsy moth, feeding on oak, *Quercus* spp., and chrysomelid beetles, *Paropsis atomaria*, feeding on *Eucalyptus* spp. (Feeny, 1969; Fox and Macauley, 1977). Some insects, such as monarch butterflies, are able to feed on defended plants because of amino acid substitutions that inactivate the binding sites for ingested plant toxins (Holzinger and Wink, 1996).

Many predaceous insects use their venoms primarily for subduing prey, and secondarily for defense. Venoms produced by predaceous Hemiptera, Diptera, Neuroptera, Coleoptera, and Hymenoptera function to paralyze or kill prey (J. Schmidt, 1982), thereby minimizing injury to the predator during prey capture. The carabid beetle, *Promecognathus*, a specialist predator on *Harpaphe* spp. and other polydesmid millipedes, avoids the cyanogenic secretions of its prey by quickly biting through the ventral nerve cord at the neck, inducing paralysis (G. Parsons et al., 1991). Nevertheless, host defenses increase handling time and risk of injury and mortality for the consumer (Becerra, 1994; J. Schmidt, 1982).

Diversion of limited resources to detoxification enzymes or to avoidance efforts all involve metabolic costs (Karban and Agrawal, 2002; Kessler and Baldwin, 2002). Lindroth et al. (1991) evaluated the effect of several specific nutrient deficiencies on detoxification enzyme activity in the gypsy moth. They found that larvae on a low-protein diet showed compensatory feeding behavior (though not enough to offset the reduced protein intake). Soluble esterase and carbonyl reductase activities increased in response to protein deficiency but decreased in response to vitamin deficiency. Polysubstrate monooxygenase and glutathione transferase activities showed no significant response.

Some caterpillar species are able to suppress plant induction of defenses by means of prostaglandins in their oral secretions (Schultz and Appel, 2004). Schultz and Appel (2004) reported that application of prostaglandin E_2 or oral regurgitant from the gypsy moth or forest tent caterpillar, *Malacosoma disstria*, reduced production of tannins by wounded red oak, *Quercus rubra*, leaves by 30–90%, compared to untreated controls which increased tannin production by 50–80% in response to wounding.

Although insect adaptations to exploit defended resources provide access to these resources, they also incur fitness costs. Carrière et al. (2001b) reported that pink bollworm, *Pectinophora gosypiella*, resistance to transgenic (Bt) cotton was associated with reduced percentage emergence from diapause, compared to nonresistant bollworm, indicating fitness costs of developing resistance strategies. Williams et al. (2011) added that pink bollworm larvae with cadherin resistance alleles accumulated higher concentrations of gossypol, a cotton defensive chemical that reduces larval growth and survival, than did larvae lacking the resistance alleles. Anilkumar et al. (2008) found similar fitness costs of Bt resistance in the bollworm, *Helicoverpa zea*.

3 RESOURCE ACCEPTABILITY

The variety of resources and their physical and biochemical properties, including defensive mechanisms, limits the range of resources to which any species can adapt. The particular physiological and behavioral adaptations that allow insect species to obtain sufficient nutrients and avoid toxic or indigestible materials determine which resources they can or will exploit. Insects adapted to exploit particular resources lose their ability to exploit other resources. Even species that feed on a wide variety of resource types (eg, host species) are limited in the range of resources they can exploit. For example, gypsy moths feed on a wide range of plant species (representing diverse plant families) that share primarily phenolic defenses; plants with alkaloid defenses generally are not exploited (Barbosa and Krischik, 1987). First instars are more restricted in diet breadth than are fifth instars (Martinat and Barbosa, 1987).

Insects face an evolutionary trade-off between maximizing the efficiency of exploiting a particular resource (*specialists*) or maximizing the range of resources exploited (*generalists*). Specialists maximize the efficiency of exploiting a particular host plant through specific detoxification enzymes or avoidance strategies, minimizing the effect of host constitutive and induced defenses, but sacrifice ability to feed on other plant species with different defenses (Bowers and Puttick, 1988). By contrast, generalists maximize the range of resources exploited through generalized detoxification or avoidance mechanisms, such as broad-spectrum microsomal monooxygenases, but sacrifice efficiency in exploiting any particular resource because unique biochemicals reduce digestion or survival (Bowers and Puttick, 1988). Plant compounds that provide effective defense against generalists may be largely ineffective against specialists, and may even be phagostimulants for adapted species (Shonle and Bergelson, 2000). Tallamy et al. (1997) reported that cucurbitacins (bitter triterpenes characterizing the Cucurbitaceae) deter feeding and oviposition by nonadapted mandibulate insect herbivores but stimulate feeding by haustellate insect herbivores.

Although specialists might be expected to outperform generalists on their preferred host but not on less preferred hosts, this is not necessarily the case. Friberg et al. (2015) compared two closely related butterfly species, *P. napi* and *P. rapae*, that differ in their degree of host specialization. They found that the offspring of the specialist did not outperform the offspring of the generalist on the specialist's most preferred host plant, nor did the generalist outperform the specialist on the less preferred plants.

Generalists may benefit from a mixed diet by optimizing nutrient balances or through dilution of any single host's defensive compounds (Behmer, 2009; Bernays et al., 1994) or by increasing their energetic efficiency on stressed hosts that have sacrificed production of defenses (Kessler et al., 2004). Kessler et al. (2004) demonstrated that when tobacco, *Nicotiana attenuata*, was transformed to silence its induced defense genes, it became suitable for new (nonadapted) herbivores, such as the western cucumber beetle, *Diabrotica undecimpunctata*, that fed and reproduced successfully. Generalists may be favored over specialists when host plants are rare or occur inconsistently. Wiklund and Friberg (2009) reported that fitness of a generalist pierid butterfly, *Anthocharis cardamines*, was increased by its ability to reproduce on any of a variety of host species, each of which varied widely in abundance and suitability over time.

Some generalists that occur over large geographic areas may be more specialized at the local level. Parry and Goyer (2004) demonstrated that the forest tent caterpillar is a composite of regionally specialized populations rather than an extreme generalist. In a reciprocal transplant experiment, tent caterpillars from Louisiana and Michigan, USA, and Manitoba, Canada, were reared on the variety of hosts exploited by northern and southern populations. Tent caterpillars from northern populations showed greatest growth and survival on trembling aspen, *Populus tremuloides*, and red oak, *Quercus rubra*, both northern host species, and poorest growth and survival on water tupelo, *Nyssa aquatica*, a southern host species. Tent caterpillars from southern populations showed greatest growth and survival on water tupelo and poorest growth and survival on sugar maple, *Acer saccharum*, a northern host species. Preferences among potentially suitable resources reflect resource quality, susceptibility, and acceptability.

Resource quality, as described in section 3.2, represents the net nutritional value of the resource after deducting the energy and resources needed to detoxify or avoid defenses. Some of the nutrients in any food acquired must be allocated to production of detoxification enzymes or to energy expended in searching for more suitable food. Although diversion of dietary N to production of detoxification enzymes should be reduced if N is limiting, Lindroth et al. (1991) found little change in detoxification enzyme activity in response to nutrient deficiencies in gypsy moth larvae. However, defended resources can have beneficial side effects for the consumer. M. Hunter and Schultz (1993) reported that phenolic defenses in oak leaves reduced susceptibility of gypsy moth larvae to nuclear polyhedrosis virus.

Resource susceptibility represents the physiological condition of the host. Injury or adverse environmental conditions that stress organisms can impair their ability to defend themselves. Initially, stress may prevent expression of induced defenses, an added cost, or reduce production of constitutive defenses but allow induction, as needed. Nitrogen limitation may prevent production of nitrogenous defenses but increase production of nonnitrogenous defenses. In any event, impaired defenses reduce the cost to the consumer of acquiring the resource. Therefore, specialists can allocate more energy and resources to growth and reproduction, and generalists can expand their host range as biochemical barriers are removed (Kessler et al., 2004).

Resource acceptability represents the willingness of the insect to feed on a particular resource, given the probability of finding more suitable resources or in view of other trade-offs.

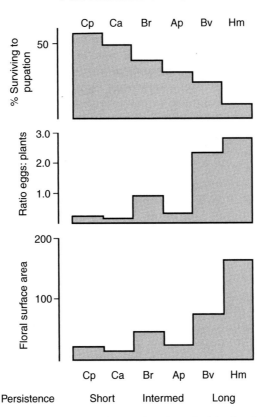

FIGURE 3.14 **Trade-off between plant susceptibility for larval survival (top) and efficiency of oviposition site selection by adult pierids, *Anthocharis cardamines*, as indicated by the ratio of eggs per host species (middle) and plant apparency, measured as floral surface area and longevity (bottom).** Searching females preferentially oviposit on the most conspicuous plants although these are not the most suitable food plants for their offspring. *Cp, Cardamine pratensis; Ca, C. amara; Br, Brassica rapa; Ap, Alliaria petiolata; Bv, Barbarea vulgaris;* and *Hm, Hesperis matronalis. Source: From Courtney, S.P., 1985. Apparency in coevolving relationships. Oikos 44, 91–98.*

Most insects have relatively limited time and energy resources to spend searching for food. Therefore, marginally suitable resources may become sufficiently acceptable when the probability of finding more suitable resources is low, as in diverse communities composed primarily of nonhosts (Courtney, 1985, 1986), or when increased risk of predation on preferred host plants maintains a wider host range (Murphy and Loewy, 2015). Courtney (1985, 1986) reported that oviposition by a pierid butterfly, *A. cardamines*, among several potential host plant species was nearly inversely related to the suitability of those plant species for larval development and survival (Fig. 3.14). The more suitable host species were relatively rare and inconspicuous compared to the less suitable host species. Given a short adult life span, butterfly fitness was maximized by laying eggs on the most conspicuous (apparent) plants, thereby ensuring reproduction, rather than by risking reproductive failure during a continued search for more suitable hosts. Nevertheless, offspring forced to feed on less suitable resources show reduced growth and survival rates (Bozer et al., 1996; Courtney, 1985, 1986). On the other hand, the ability to exploit a variety of resources, each of which may vary widely in abundance and suitability over time, spreads risk over time and space and increases the likelihood that the population will persist across the landscape (Wiklund and Friberg, 2009).

Searching insects initially identify acceptable hosts, then select particular host tissues based on nutritional value. For example, insects may target particular portions of leaves, based on gradients in the ratio among amino acids along the leaf blade (Haglund, 1980; K. Parsons and de la Cruz, 1980), and particular heights on tree boles, based on gradients in ratios among amino acids and carbohydrates (Hodges et al., 1968). Loaiza et al. (2011) reported that grasshopper densities were 67% greater in N-fertilized plots, compared to control plots, but grasshoppers showed no response to P fertilization.

Many insects feed on different resources at different stages of development. Most larval Lepidoptera feed on plant foliage, stems, or roots, but many adults feed on nectar. Some cerambycid beetles feed in wood as larvae but on pollen or nectar as adults. Most aquatic insects have terrestrial adults. Many aphids alternate generations between two host plant species (Dixon, 1985). Clearly, these changes in food resources require changes in digestive abilities between life stages. Furthermore, population survival requires the presence of all necessary resources at an appropriate landscape scale.

The primacy of resource exploitation for development and survival places strong selective pressure on insects to adapt to changing resource quality. This has led to the so-called "evolutionary arms race," in which herbivory selects for new plant defenses and the new plant defenses select for insect countermeasures. This process has driven reciprocal speciation in both plants and insects, with examples of cladograms for plants and associated insects that mirror each other (Becerra, 1997). However, Agrawal and Fishbein (2008) tested predictions of this plant defense theory with a molecular phylogeny of 38 milkweed species, *Asclepias* and *Gomphocarpus* spp., and found a pattern of phyletic decline in the three most potent defensive traits (cardenolides, latex, and trichomes) and an escalation in compensatory growth ability. Furthermore, M. Wise (2009) noted that any gain a plant might receive from increased resistance to one herbivore would be at least partially offset by increased damage from competing herbivores. These data suggest that selection ultimately may favor tolerance of herbivory over defensive ability.

4 RESOURCE AVAILABILITY

The abundance, distribution, and apparency (ease of discovery) of acceptable resources determine their availability in space and time to foraging organisms (Bozer et al., 1996; Courtney, 1985, 1986; S. Eggert and Wallace, 2003; Sholes, 2008, see also Chapters 6 and 7). Resources are most available when distributed evenly at nonlimiting concentrations or densities. Organisms living under such conditions need not move widely to locate new resources and tend to be relatively sedentary. Microorganisms suspended in a concentrated solution of organic molecules (such as in eutrophic aquatic ecosystems or in decomposing detritus) and filter feeders and scale insects that capture resources from flowing solutions of resources may enjoy relatively nonlimiting resources for many generations.

Necessary resources typically are less concentrated, are available at suboptimal ratios with other resources, or are unevenly distributed in space and time. This requires that organisms select habitats where required resources are most concentrated or in most efficient balance and seek new sources as current resources become depleted. Insects and other animals employ various physiological and behavioral mechanisms to detect, orient toward, and move to concentrations of food.

4.1 Discovering Suitable Resources

Resource quality and availability vary among plants and tissues and change seasonally and annually in both temperate and tropical ecosystems (Schultz, 1983; Whitham, 1983). The life history phenology of many species is synchronized with periods of most favorable host nutritional chemistry to improve an insect's chance of finding optimal food (Feeny, 1970; R. Lawrence et al., 1997; Varley and Gradwell, 1970). Diapause can be prolonged for a portion of a population in cases of unpredictable availability of food resources, as for insects feeding on seeds of trees that produce seed crops irregularly (masting). Turgeon et al. (1994) reported that 70 species of Diptera, Lepidoptera, and Hymenoptera that feed on conifer cones or seeds can remain in diapause for as long as 7 years. In other words, diapause can give insect populations considerable capacity to persist during long periods of unsuitable resource conditions.

Most insects also must seek food resources that are distributed unevenly in space and may be hidden from discovery by associated unsuitable resources (Courtney, 1985, 1986; Scholes, 2008; Visser, 1986). *Foraging theory* focuses on rules for optimization of diet quality, predation risk, and foraging efficiency (Behmer, 2009; Kamil et al., 1987; McArthur et al., 2014; Schultz, 1983; Stephens and Krebs, 1986; Townsend and Hughes, 1981, see also Chapter 4). Profitable resources provide a gain to the consumer, but nonnutritive or toxic resources represent a cost in terms of time, energy, or nutrient resources expended in detoxification or continued search for more suitable resources. Continued search also increases exposure to predators or other mortality agents (McArthur et al., 2014; Murphy and Loewy, 2015). Schultz (1983) developed a trade-off surface to illustrate four foraging strategies for arboreal caterpillars (Fig. 3.15). Foraging can be optimized by searching for more nutritive food and risking attention of predators, accepting less nutritive food, or defending against predation. Natural selection can favor a reduction in cost along any of the three axes, within constraints of the other two costs.

Consumers should maximize foraging efficiency by focusing on resources or patches with high profitability (ie, where hosts are concentrated or most apparent, the Resource Concentration Hypothesis), and ignore low-profitability patches, until their resource value declines below the average for the landscape matrix (W. Bell, 1990; Kareiva, 1983; Sholes, 2008). Species vary in their ability to detect and orient toward patches of concentrated resources, depending on their functional traits (Andersson et al., 2013a,b). Foraging strategy represents a trade-off between costs (in terms of reduced growth and survival) of searching, of feeding on less suitable food, and of exposure to predators (McArthur et al., 2014). Foraging efficiency can be improved by ability to detect and orient toward cues indicating suitable resources, and further enhanced by learning.

4.2 Orientation

Some insects forage randomly, with some eventually (at considerable individual risk) discovering suitable resources (Dixon, 1985; Raffa et al., 1993). However, many insects respond

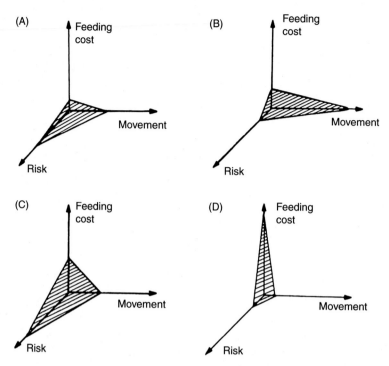

FIGURE 3.15 Trade-off planes of selected caterpillar foraging strategies. Costs of feeding (ie, metabolic costs of digestion, reduced growth, etc.), movement (metabolic costs of reduced growth), and risks (eg, probability of capture or reduced growth due to time spent hiding) increase in the direction of the arrows: (A) selective diurnal feeder, (B) selective nocturnal aposomatic feeder, (C) diurnal cryptic feeder, and (D) food mimic. *Source: From Schultz, J.C., 1983. Habitat selection and foraging tactics of caterpillars in heterogeneous trees. In: Denno, R.F., McClure, M.S. (Eds.), Variable Plants and Herbivores in Natural and Managed Systems. Academic Press, New York, pp. 61–90.*

to particular cues that indicate the availability of suitable resources. The cues to which searching insects respond may differ among stages in the search process. For example, gross cues, indicative of certain habitats, might initially guide insects to a potentially suitable location. They then respond to cues that indicate suitable patches of resources, and finally focus on cues characteristic of the necessary resources (W. Bell, 1990; Mustaparta, 1984). Insects search longer in patches where suitable resources have been detected than in patches without suitable resources, resulting in gradual increase in population density on their resources (W. Bell, 1990; S. Risch, 1980, 1981; Root, 1973; Turchin, 1988). Orientation toward cues involves the following steps.

4.2.1 Information Processing

Several types of information are processed by searching insects. Some cues are nondirectional, but alert insects to the presence of resources or initiate search behavior. A nondirectional cue may alter the threshold for response to other cues (cross-channel potentiation) or initiate behaviors that provide more precise information (W. Bell, 1990). For example, flying bark beetles typically initiate a search for their host trees only after exhausting their fat reserves. Emerging adults of parasitic wasps gather information about their host from odors emanating from host frass or food plant material associated with the emergence site (Godfray, 1994). Wasps emerging in the absence of these cues may be unable to identify potential hosts.

Directional information provides the stimulus to orient in the direction of the perceived resource. For example, detection of attractive chemicals without airflow initiates a nondirectional local search, whereas the addition of airflow stimulates orientation upwind (W. Bell, 1990). Accuracy of orientation increases with signal intensity. Signal intensity increases with density of the source and decreases with distance (Elkinton et al., 1987; M. Stanton, 1983). Concentration of attractive odors remains higher at greater distances from patches of high host density compared to patches of low host density (Fig. 3.16). Insects move upwind in circuitous fashion at low vapor concentration, but movement becomes increasingly directed as vapor concentration increases upwind (Cardé, 1996). Insects integrate visual, chemical, acoustic, and geomagnetic signals to find their resources, switching from less precise to more precise signals as these become available (W. Bell, 1990; J. Gould et al., 1978; Johnsen and Lohmann, 2005; Schiff, 1991).

4.2.2 *Responses to Cues*

Many plant chemicals, especially monoterpenes, are highly volatile and provide strong long-distance signals to pollinators and herbivores. Plants that depend on dipteran pollinators produce odors that resemble those of carrion or feces to attract these insects. Sex pheromones (see Chapter 4) often are more attractive when mixed with host volatiles (Raffa et al., 1993), indicating prior discovery and evaluation of suitable hosts. Predators can be attracted to prey pheromones or odors from damaged plants that indicate the presence of prey (Kessler and Baldwin, 2001; Stephen et al., 1993; Thaler et al., 2002; Turlings et al., 1990, 1993, 1995).

Recent studies have shown that detection of relevant odors is genetically encoded but response can be modified through learning (see section 3.4.3). Insects have relatively simple nervous systems composed of receptor neurons that detect chemical signals, interneurons that integrate and convey information, and motor neurons that elicit the behavioral response. Olfactory receptor neurons are located in various sensilla, primarily on the antennae but also on the abdomen or legs (Schmitz and Trenner, 2003). Volatile chemicals diffuse through the cuticle and bind to receptor proteins that are highly selective for biologically relevant molecules (Mustaparta, 2002). These proteins transport the odor molecule to a neuronal membrane that contains receptor proteins genetically coded for specific molecules; each receptor neuron expresses proteins specific to certain odor molecules. Therefore, the discrimination power of an organism depends on the number of different neuron types (Mustaparta, 2002). Response time for neuronal discharge is on the order of 100–300 ms (Schmitz and Trenner, 2003).

Having detected an attractive chemical, the insect begins a circuitous search pattern that involves continually turning in the direction of increasing odor concentration (Cardé, 1996, see Chapter 4). However, odor plumes are disrupted by turbulence resulting from habitat heterogeneity, such as, surface irregularities of substrate or vegetation (Mafra-Neto and Cardé, 1995; Murlis et al., 1992). For example, openings in forest canopies create sites of soil warming and convective eddies that dissipate chemical plumes (Fares et al., 1980). Elkinton et al. (1987) reported that the proportion of male gypsy moths that responded to a caged

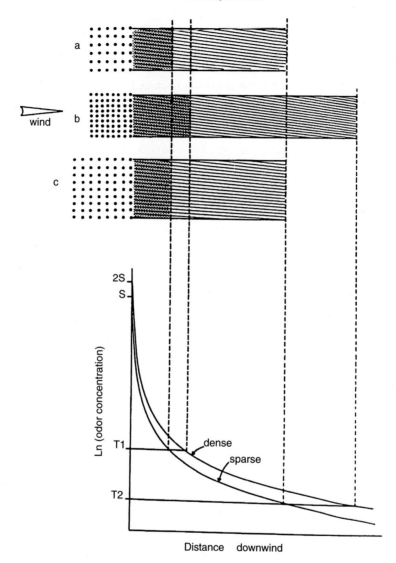

FIGURE 3.16 **Odor concentration downwind from patches of two host densities: the low-density odor curve represents patches a and c, whereas the high-density curve represents patch b.** The curves reflect an ideal situation in which diffusion is overshadowed by convection due to wind. In still air, odor concentration cannot be changed by altering host-plant density. Attractive areas shown as rectangles are actually irregular in shape. Attractive zones for low-sensitivity herbivores (threshold T1) are stippled; those for high-sensitivity herbivores (T2) are shaded. *Source: From Stanton, M.L., 1983. Spatial patterns in the plant community and their effects upon insect search. In: Ahmad, S. (Ed.), Herbivorous Insects: Host-seeking Behavior and Mechanisms. Academic Press, New York, pp. 125–157.*

female declined from 89% at 20 m distance to 65% at 120 m. Of those that responded, arrival at the female's cage declined from 45% at 20 m to 8% at 120 m (see Chapter 4).

At least some insects, as well as other animals, can orient along Earth's magnetic field (Johnsen and Lohmann, 2005; Towne and Gould, 1985). Honey bees, *A. mellifera*, contain magnetite that can amplify moderately strong external magnetic fields (MacFadden and Jones, 1985; Schiff, 1991; Towne and Gould, 1985). Towne and Gould (1985) investigated the apparent misdirection (up to 20 degree) in the honey bee waggle dance and found that misdirection disappeared when dances were oriented along the projection of the magnetic field lines onto the dance floor, regardless of the polarity of the magnetic field.

As an insect gets closer to an attractive source, other cues become important in distinguishing target resources. Visual cues include host silhouettes and radiant energy. Some species orient toward light. Many bark beetles are attracted to dark-colored silhouettes of tree boles and can be attracted to other cylindrical objects or prevented from landing on tree boles painted white (Goyer et al., 2004; Strom et al., 1999). The aquatic milfoil weevil, *Euhrychiopsis lecontei*, is attracted to host, *Myriophyllum* spp., stems under well-illuminated conditions (Reeves et al., 2009). Aphids often are attracted to young, succulent foliage and to older, senescent foliage by longer-wavelength yellow, but this cue is not a good indicator of host species (Dixon, 1985; Farnier et al., 2014). Aphids, *Pemphigus betae*, migrating in autumn may discriminate among susceptible and resistant poplar, *Populus* spp., trees on the basis of prolonged leaf retention by more susceptible hosts (N. Moran and Whitham, 1990). Some parasitic wasps detect their wood-boring hosts by means of infrared receptors on their antennae (Matthews and Matthews, 2010).

Pollinator orientation to particular floral colors or patterns is particularly well known (Chittka and Menzel, 1992; Heinrich, 1979; Spaethe et al., 2001; Wickler, 1968). Red and blue are most easily detected in open or well-lighted ecosystems, such as tropical canopies and grasslands, whereas white is more readily detected under low-light conditions, such as in forest understories. Spaethe et al. (2001) found that bumble bee, *Bombus terrestris*, foraging efficiency depended on the degree of contrast between floral color and the green foliage background when flowers were large. Search times were shortest for lemon yellow flowers and longest for red and UV-reflecting white. Search times increased for smaller flowers, and detection shifted to dependence on the green receptor signal, favoring detection of white flowers. Ultraviolet designs provide important cues for insect pollinators. Insects can detect ultraviolet "runways" or "nectar guides" leading to the nectaries (Eisner et al., 1969; Heinrich, 1979; Matthews and Matthews, 2010). Some floral designs in the orchid genus *Ophrys* resemble female bees or wasps and produce odors similar to the mating pheromones of these insects. Male bees or wasps are attracted and pollinate these flowers while attempting to copulate (Wickler, 1968).

Many wood-boring insects, such as wood wasps (Siricidae) and beetles (especially Buprestidae), are attracted to sources of smoke, infrared radiation, or volatile tree chemicals emitted from burned or injured trees over distances of up to 50 km (W. Evans, 1966; Gara et al., 1984; Matthews and Matthews, 2010; R.G. Mitchell and Martin, 1980; Raffa et al., 1993; S. Schütz et al., 1999; Wickman, 1964). H. Schmitz et al. (1997) and H. Schmitz and Trenner (2003) described the thermosensitive multipolar neurons in abdominal IR receptors of the Australian buprestid beetle *Merimna atrata*, and thoracic IR receptors of the North American buprestid, *Melanophila acuminata*.

Acoustic signals include the sounds produced by cavitating plant cells and by potential mates. Cavitation, the collapse of cell walls as turgor pressure falls, produces an audible signal that provides a valuable cue to stressed plants (Mattson and Haack, 1987). Attraction to this signal may partly explain the association of bark beetles with water-stressed trees (Mattson and Haack, 1987; Raffa et al., 1993).

Nonhosts can interfere with insect orientation (Shepherd and Sullivan, 2013; Shepherd et al., 2008; Sholes, 2008). Hambäck et al. (2003) reported that leaf-feeding beetles, *Galerucella* spp., were significantly less abundant on purple loosestrife, *Lythrum salicaria*, that were surrounded by nonhost or artificial shrubs than on hosts that were not surrounded by nonhost or artificial shrubs. Similarly, Meisner et al. (2007) reported that the presence of alfalfa aphids, *Therioaphis maculata*, reduced the foraging efficiency of the hymenopteran parasitoid, *Aphidius ervi*, for its primary host, pea aphids, *Acyrthosiphon pisum*. Nonattractive or repellent odors from nonhosts can mix with attractive odors in the airstream of more diverse ecosystems and disrupt orientation (Shepherd and Sullivan, 2013; Visser, 1986). Verbenone and 4-allylanisole, found in the resin of various nonhost angiosperm trees, repel some conifer-feeding bark beetle species and protect host trees within the repellent plume (Hayes et al., 1994; Poland et al., 1998; Q. Zhang and Schlyter, 2004).

If an insect successfully reaches the source of attractive cues, it engages in close-range gustatory, olfactory, or sound reception (Dixon, 1985; Raffa et al., 1993; Städler, 1984). Contact chemoreceptive sensilla generally are located on antennae, mouthparts, or feet (Dixon, 1985; Städler, 1984). These sensors provide information about the nutritive value and defensive chemistry of the resource (R. Chapman, 2003; Raffa et al., 1993). Some plant chemicals act as phagostimulants or as deterrents (R. Chapman, 2003; Q. Zhang and Schlyter, 2004). For example, cucurbitacins deter feeding and oviposition by nonadapted mandibulate insects, but are phagostimulants for diabroticine chrysomelid beetles (Tallamy and Halaweish, 1993). Predators also may avoid prey containing toxic or deterrent chemicals (Stamp et al., 1997; Stephens and Krebs, 1986). Many parasitic wasps avoid hosts marked chemically by wasps that oviposited previously in that host (Godfray, 1994). Because hosts support only a limited number of parasitoid offspring, often no more than one, avoidance of previously parasitized hosts reduces competition among larvae within a host.

4.2.3 Attraction of Conspecific Insects

Some insects can communicate the presence of suitable resources to conspecific insects. Cooperation facilitates acquisition of shared resources or larger prey and improves mating success (see Chapter 4).Visual signals are illustrated by the elaborate movements of the "bee dance" used by honey bees to communicate distance and direction to suitable resources to other foragers (F. Dyer, 2002; von Frisch, 1967).

Attractive and repellent chemicals produced by insects (pheromones) advertise the location of suitable resources and potential mates (Fry and Wehner, 2002; Raffa et al., 1993; Rudinsky and Ryker, 1976). Most insects produce pheromones, but those of Lepidoptera, bark beetles, and social insects have been studied most widely. Social insects mark foraging trails to guide other members of their colony to food resources and back to the colony (B. Smith and Breed, 1995; Traniello and Robson, 1995). A variety of chemical structures are used to mark trails (Fig. 3.17). A plant-derived monoterpene, geraniol, is obtained from flower scents, concentrated, and used by honey bees to mark trails and floral resources (Harborne, 1994). Trail markers can be highly effective. The trail marker produced by the leaf-cutting ant, *Atta texana*, is detectable by ants at concentrations of 3.48×10^8 molecules cm^{-1}, indicating that 0.33 mg of the pheromone would be sufficient to mark a detectable trail around the world (Harborne, 1994). Although trail markers were once thought to be species specific, more recent work has shown that multiple species may use the same compounds as trail markers,

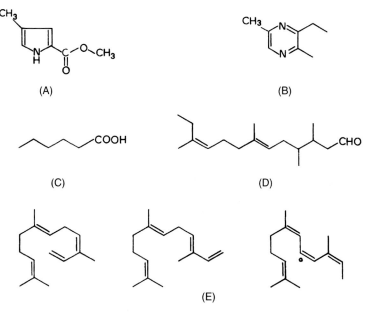

FIGURE 3.17 **Trail pheromones of myrmicine ants.** (A) *Atta texana* and *A. cephalotes*, (B) *A. sexdens rubropilosa* and *Myrmica* spp., (C) *Lasius fuliginosus*, (D) *Monomorium pharaonis*, and (E) *Solenopsis invicta*. *Source: From Bradshaw, J.W.S., Howse, P.E., 1984. Sociochemicals of ants. In: Bell, W.J., Carde, R.T. (Eds.), Chemical Ecology of Insects. Chapman and Hall, London, UK, pp. 429–473.*

with varying degrees of interspecific recognition (Traniello and Robson, 1995). Furthermore, synthetic analogues, for example, 2-phenoxyethanol, also may elicit trail-following behavior, despite little structural similarity to natural trail markers (J. Chen et al., 1988).

Acoustic signals (stridulation) from potential mates, especially when combined with attractive host cues, advertise discovery and evaluation of suitable resources. Stridulation contributes to optimal spacing and resource exploitation by colonizing bark beetles (Raffa et al., 1993; Rudinsky and Ryker, 1976).

4.3 Learning

Learning improves the efficiency of resource acquisition. A simple definition of learning is improved behavior due to experience (Papaj and Prokopy, 1989; Shettleworth, 1984). Learning is difficult to demonstrate because improved performance with experience can result from maturation of neuromuscular systems rather than from learning (Papaj and Prokopy, 1989). Although learning by insects has been appreciated less widely than has learning by vertebrates, various insect groups have demonstrated considerable learning capacity (cf., Cunningham et al., 1998; Daly et al., 2001; Drukker et al., 2000; Dukas, 2008; Gong et al., 1998; J. Gould and Towne, 1988; Gutiérrez-Ibáñez et al., 2007; A. Lewis, 1986; Li and Liu, 2004; Meller and Davis, 1996; Raubenheimer and Tucker, 1997; Schnierla, 1953; von Frisch, 1967; Wehner, 2003).

Schnierla, 1953 was among the first to report that ants can improve their ability to find food in a maze. However, ants learned more slowly and applied experience less efficiently to new situations than did rats. Learning is best developed in the social and parasitic Hymenoptera and in some other predaceous insects but also has been demonstrated in phytophagous species representing six orders (Dukas, 2008; R. Chapman and Bernays, 1989; Li and Liu, 2004; Papaj and Prokopy, 1989). Furthermore, learning during larval stages has been shown to persist through metamorphosis to adult memory (Gutiérrez-Ibáñez et al., 2007). Molecular and neurological studies have contributed to understanding the mechanisms of memory formation (R. Davis, 2005). Ishimoto et al. (2009) demonstrated that 20-hydroxyecdysone is instrumental in the formation of long-term memory in adult *Drosophila melanogaster*. Several types of learning by insects have been identified.

Habituation is loss of responsiveness to an unimportant stimulus, as a result of continued exposure. Habituation may explain parasitoid emigration from patches that are depleted of unparasitized hosts (Papaj and Prokopy, 1989). Although host odors are still present, a wasp is no longer responsive to these odors.

Imprinting is the acceptance of a particular stimulus in a situation in which the organism has an innate tendency to respond. Parasitic wasps imprint on host or plant stimuli at the site of adult emergence. Odors from host frass or the host's food plant present at the emergence site offer important information used by the emerging wasp during subsequent foraging (W. Lewis and Tumlinson, 1988). A number of studies have demonstrated that if the parasitoid is removed from its cocoon or reared on an artificial diet, it may be unable to learn the odor of its host or its host's food plant, and hence be unable to locate hosts (Godfray, 1994).

Associative learning is the linking of one stimulus with another, based on a discerned relationship between the stimuli. Most commonly, the presence of food is associated with cues consistently associated with food. Several types of associative learning have been identified.

Classical conditioning involves substitution of one stimulus for another. For example, parasitic wasps respond to empty food trays after learning to associate food trays with hosts or respond to novel odors after learning to associate them with provision of hosts (Godfray, 1994). Kelber (1996) demonstrated that hawk moths, *Macroglossum stellatarum*, could be trained to prefer floral colors for which rewards were given and to avoid floral colors that were unrewarded.

Operant conditioning, or trial-and-error learning, is demonstrated when an animal learns to associate its behavior with reward or punishment and then tends to repeat or avoid that behavior, accordingly. Association of ingested food with postingestion malaise often results in subsequent avoidance of that food (R. Chapman and Bernays, 1989; Papaj and Prokopy, 1989). Laboratory experiments by Stamp (1992) and Traugott and Stamp (1996) demonstrated that predatory wasps initially attack caterpillars that sequester plant defenses, but after a few days will reject unpalatable prey.

Pollinators provide some of the best examples of associative learning because floral structures present difficult challenges to acquiring nectar and/or pollen resources. Honey bees, trained to approach a particular flower from different directions at different times of day, will subsequently approach other flowers from the direction appropriate to the time of day at which rewards were provided during training (Fig. 3.18). Fry and Wehner (2002) and Horridge (2003) found that honey bees can distinguish pattern and landmark orientations and return to food resources even when the associated landmark orientation is altered. Similarly, experience enables traplining bumble bees, *Bombus impatiens*, to travel more rapidly among floral resources, acquire more nectar, and increase foraging performance in competition with less

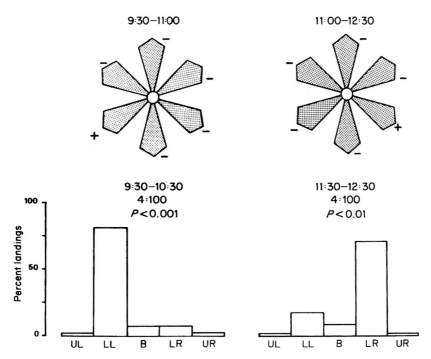

FIGURE 3.18 **Honey bees can remember how to approach specific flowers in relation to time of day.** Bees trained to land at different positions (+) of an artificial flower at different times in the morning subsequently preferred to land on the petal on which they were trained during the same part of the morning. *UL*, upper left; *LL*, lower left; *B*, bottom; *LR*, lower right; *UR*, upper right. *Source: From Gould, J.L., Towne, W.F., 1988. Honey bee learning. Adv. Insect Physiol. 20, 55–86.*

experienced individuals (Ohashi and Thomson, 2009; Ohashi et al., 2008). However, certain spatial configurations of floral resources limit optimization of the foraging route, suggesting that trapliners may select plants representing an appropriate spatial configuration (Ohashi et al., 2007). A. Lewis (1986) reported that cabbage white butterflies became more efficient at obtaining floral rewards by selectively foraging on a particular floral type, based on experience. Such floral fidelity can increase pollination efficiency (see Chapter 13). However, increased nectar foraging on larval food plants may increase the likelihood that females will use the same plant for nectar foraging and oviposition and thereby increase herbivory on pollinated plants (Cunningham et al., 1998, 1999).

Associative learning improves performance of parasitic Hymenoptera (Godfray, 1994). Information gathered during searching contributes to increased efficiency of host discovery (W. Lewis and Tumlinson, 1988). Searching wasps learn to associate host insects with plant odors, including odors induced by herbivory (Fukushima et al., 2002). Subsequently, they preferentially search similar microhabitats (Godfray, 1994; Steidle, 1998). However, exposure to new hosts or hosts in novel habitats can lead to increased responsiveness to new cues. Bjorksten and Hoffmann (1998) reported that such learned stimuli can be remembered for at least 5 days.

Insects are capable of complex associative learning. Raubenheimer and Tucker (1997) trained locust, *Locusta migratoria*, nymphs to distinguish between food containers, differing in color, with synthetic diet deficient in either protein or carbohydrate. Feeding from both containers in the arena was necessary to obtain a balanced diet. The nymphs subsequently were deprived of either protein or carbohydrate and tested for ability to acquire the deficient nutrient. Locusts most frequently selected food containers of the color previously associated with the deficient nutrient, regardless of color or whether the nutrient was protein or carbohydrate. Wäckers et al. (2002) demonstrated that parasitoid wasps, *Microplitis croceipes*, could learn multiple tasks representing feeding and reproduction. Stach et al. (2004) found that honey bees can learn multiple conditioning patterns and generalize their response to novel stimuli based on linkage among conditioned stimuli.

Foraging social insects must be able to return to the colony site. Honey bees displaced from their colony tend to follow a search pattern consisting of frequently occurring relatively short straight segments that are punctuated by infrequent longer segments that, in turn, are punctuated by rare, very long segments. This flight pattern can cover a greater area more efficiently than a random walk of the same length and maximizes the probability of locating the hive (A. Reynolds et al., 2007). Several social Hymenoptera remember geometric features along paths between their colony and food resources. Bisch-Knaden and Wehner (2003) demonstrated that desert ants, *Cataglyphis fortis*, learned to associate local foraging trail vectors with individual cylindrical landmarks during homebound runs, but not during outbound runs. However, ants returning to the nest initially reverse the outbound vector, then start a systematic search for the nest, indicating that these ants cannot learn separate inbound and outbound vectors that are not 180° reversals and that recalibration during homebound runs is dominated by the outbound vector (Wehner et al., 2002). Ants are thus able to reach the nest along the shortest route and later return to the food source by 180° vector reversal. Lent et al. (2009) found that foraging ants remembered direction to a landmark and continued a straight or curving direction of travel between intermittent sightings of landmarks (due to uneven terrain). Wystrach and Beugnon (2009) reported that ants made rotational errors in locating a target in a rectangular arena, similar to vertebrates, and suggested that ants may be guided more by a global view of large features than by individual landmarks that could be confused.

Observational learning occurs when animals gather information and modify their behavior in response to observation of other individuals. Observational learning is epitomized by social bees that communicate the location of rich floral resources to other members of the colony through the "bee dance" (F. Dyer, 2002; J. Gould and Towne, 1988; Srinivasan et al., 2000; von Frisch, 1967). Movements of this dance inform other foragers of the direction and distance to a food source. Direction relative to the sun or Earth's magnetic field is conveyed by orientation on the vertical dance floor (Johnsen and Lohmann, 2005; Towne and Gould, 1985). Distance apparently is estimated visually from the rate of change in image angle and communicated to hive members as 17.7 degree of image motion per millisecond of waggle dance (Srinivasan et al., 2000). Abbott and Dukas (2009) further demonstrated that honey bees can effectively warn naïve recruits away from dangerous flowers (treated with dead bees) and thereby focus foraging on "safe" flowers.

Cognition, characterized by awareness, memory, and judgement, is evident when information gathered during previous experiences is applied to novel situations. This basic form of thinking is widely associated with higher vertebrates. However, J. Gould (1986) demonstrated that honey bees are capable of constructing cognitive maps of their foraging area. Bees were trained to forage at either of two widely separated sites, then captured at the hive and transported in the dark to an unfamiliar site, the same distance from the hive but in a different direction, within a complex foraging area (open areas interspersed with forest). If released bees were disoriented or could not accommodate a sudden change in landmarks, they should fly in random directions. If they had only route-specific landmark memory and were familiar with a foraging route to their release point, they should be able to return to the hive and fly from there to their intended destination (the site to which they had been trained). Only if bees are capable of constructing true cognitive maps should they fly from the release point directly to their intended destination. All bees flew directly to their intended destinations. Although some studies indicate limits to large-scale cognitive mapping by bees (Dukas and Real, 1993; Menzel et al., 1998), substantial evidence indicates that honey bees construct and maintain at least local geometric representation, referenced to the time of day and to landmarks and line angles to floral resources (J. Gould, 1985, 1986; J. Gould and Towne, 1988). Wei et al. (2002) further demonstrated that honey bees intensively examine the area around a food source, through "learning flights." Bees turn back and face the direction of the food source and surrounding landmarks, then circle around, before returning to the hive. The duration of learning flights increases with the sugar concentration of food and the visual complexity of the surrounding landmarks, and is longer following initial discovery of food than during subsequent reorientation. These results indicate that bees adjust learning effort in response to the need for visual information. Such advanced learning greatly facilitates the efficiency with which resources can be acquired.

5 SUMMARY

Insects, as do all organisms, must acquire energy and material resources to synthesize the organic molecules necessary for life processes of maintenance, growth, and reproduction. Dietary requirements reflect the size and life stage of the insect and the quality of food resources. Insects exhibit a variety of physiological and behavioral strategies for finding, evaluating, and exploiting potential resources.

Defensive chemistry of plants and insects affects their quality as food and is a basis for host choice by herbivorous and entomophagous insects, respectively. Nutritional value of resources varies among host species, among tissues of a single organism, and even within tissues of a particular type. Production of defensive chemicals is expensive in terms of energy and nutrient resources and may be sacrificed during unfavorable periods (such as during water or nutrient shortages or following disturbances) to meet more immediate metabolic needs. Such hosts become more vulnerable to predation. Insect adaptations to avoid or detoxify host defenses determine host choice and the range of host species exploited. Generalists exploit a relatively broad range of host species, but exploit each host species rather inefficiently, whereas specialists are more efficient in exploiting a single or a few related hosts that produce similar chemical defenses. Major advances have occurred in the past decade on the specific plant genes that confer defense against insects and insect genes that govern tolerance or detoxification.

Chemicals also communicate the availability of food and provide powerful cues that influence insect foraging behavior. Insects are capable of detecting food resources over considerable distances. Perception of chemical cues that indicate availability of hosts is influenced by concentration gradients in air or water, environmental factors that affect downwind or downstream dispersion of the chemical, and sensitivity to particular odors. Orientation to food resources over shorter distances is affected by visual cues (such as color or pattern) and acoustic cues (such as stridulation). Once an insect finds a potential resource, it engages in tasting or other sampling behaviors that permit evaluation of resource acceptability.

Efficiency of resource acquisition may improve over time as a result of learning. Although much of insect behavior may be innate, learning has been documented for many insects. The ability to learn among insects ranges from simple habituation to continuous unimportant stimuli, to widespread associative learning among both phytophagous and predaceous species, to observational learning, and even cognitive ability. Learning represents the most flexible means of responding to environmental variation and allows many insects to adjust to changing environments during short lifetimes.

Resource Allocation

HOW ARE ENERGY AND NUTRIENT BUDGETS MEASURED?

The efficiency with which various organisms assimilate food resources and allocate the acquired energy and nutrients to growth and reproduction largely determines their fitness. High efficiency can translate into population growth, a variable of considerable interest to resource managers concerned with controlling pest populations or conserving threatened species. The energy and nutrients converted into biomass also represent an important food resource for various predators and, ultimately, detritivores.

The earliest work on energy and nutrient budgets was directed toward improved agricultural and silvicultural production. Attention to insects developed during the 1950s when ecosystem ecologists began to address pathways of energy and radioisotope fluxes through ecosystems. Using relatively crude mass balance techniques, Smalley (1960) calculated assimilation by salt marsh grasshoppers, *Orchelimum fidicinium*, as the sum of respiration (measured as replacement by water of respired CO_2 absorbed by sodalime in a flask immersed in water) and growth, or production, during the life cycle. Results indicated ingestion of 650 cal h^{-1}, assimilation of 180 cal h^{-1}, egestion of 470 cal h^{-1}, and an assimilation efficiency of 27%. Crossley (1966) used radioactive ^{137}Cs that occurred in both plants and chrysomelid beetle, *Chrysomela knabi*, larvae in a contaminated area to calculate an ingestion rate of 9.2 mg plant larva^{-1} day^{-1} that was similar to a rate of 9 mg larva^{-1} day^{-1} measured in the laboratory using conventional mass balance techniques. Early investigation of allocation by sap-sucking insects involved fine pipettes to measure mass of honeydew excreted, as an estimate of ingestion, and concentrations of carbohydrates and nutrients in sap and honeydew, to estimate assimilation of energy and nutrients (Auclair, 1958, 1959; Banks and Nixon, 1959; M. Day and Irzykiewicz, 1953; M. Day and McKinnon, 1951; Mittler, 1958; M. Watson and Nixon, 1953).

These early studies demonstrated that insects convert ingested energy and nutrients into biomass more efficiently (5–17%) than do homeothermic organisms (<2%) (Wiegert and Evans, 1967), the difference indicating the energetic cost of homeothermy. Although high production efficiency versus homeothermy represents an important ecological trade-off, the high production efficiency of insects underlies their ability to maintain population viability while providing the food base for a wide variety of predators.

Insect Ecology. http://dx.doi.org/10.1016/B978-0-12-803033-2.00004-2

1 INTRODUCTION

Insects allocate acquired resources in various ways, depending on the energy and nutrient requirements of their physiological and behavioral processes. In addition to meeting their basic requirements for metabolism, foraging, growth, and reproduction, individual organisms allocate resources to pathways that influence their interactions with other organisms and abiotic nutrient pools (Elser et al., 1996). The efficiency with which an organism assimilates acquired food resources and converts these to growth and reproduction largely determines its fitness.

Interestingly, much of the early data on energy and nutrient allocation by insects was a byproduct of studies during the 1950s and 1960s on anticipated effects of nuclear war on radioisotope movement through ecosystems (Crossley and Howden, 1961; Crossley and Witkamp, 1964). Research focused on effects of radioactive fallout on organisms that affect human health and food supply. Radiation effects on insects and other arthropods were perceived to be of special concern because of the recognized importance of these organisms to human health and crop production. Radioactive isotopes, such as ^{31}P, ^{137}Cs (assimilated and allocated as is K), and ^{85}Sr (assimilated and allocated as is Ca), became useful tools for tracking the assimilation and allocation of nutrients through organisms, food webs, and ecosystems.

2 RESOURCE BUDGET

The energy or nutrient budget of an individual can be expressed by the equation

$$I = P + R + E,$$
(4.1)

in which I = ingestion, P = production, R = respiration, E = egestion. Assimilation (A) = $I - E = P + R$. Energy is required to fuel metabolism, so only a fraction of the assimilated energy is available for growth and reproduction (Fig. 4.1). Energy used in metabolism is lost through respiration. Insects and other heterotherms require little energy to maintain thermal homeostasis. Hence, arthropods generally respire only 60–90% of assimilated energy, compared to > 97% for homeotherms (Fitzgerald, 1995; Golley, 1968; Phillipson, 1981; Schowalter et al., 1977; Wiegert and Petersen, 1983).

Resource limitation clearly affects growth, survival, and reproduction. For many holometabolous insects, resources necessary for reproduction by adults must be accumulated during larval feeding stages (Hahn and Denlinger, 2011). Boggs and Freeman (2005) demonstrated that food limitation for larvae of the butterfly, *Speyeria mormonia*, reduced adult body mass and survival (which reduced fecundity, but independently of larval treatment), compared to well-fed larvae. By contrast, food limitation for adults directly reduced fecundity (Boggs and Ross, 1993), demonstrating a survival/reproduction trade-off across life stages.

Availability of some nutrients can affect an organism's use of others, that is, acquisition and allocation pathways are based on differences in ratios among various nutrients between a resource and the needs of an organism (Behmer, 2009; Elser et al., 1996; Holopainen et al., 1995, see Chapter 3). Generalists can, to some extent, select multiple food resources that collectively achieve nutritional balance (Behmer, 2009; K.P. Lee et al., 2002, 2003, see Chapter 3). Specialists, on the other hand, must optimize the trade-off between overeating nutrients that

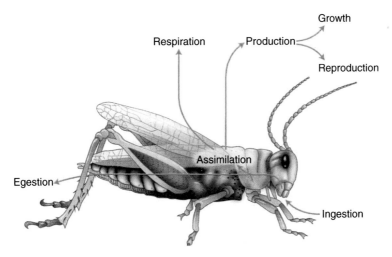

FIGURE 4.1 **Model of energy and nutrient allocation by insects and other animals.** Ingested food is only partially assimilable, depending on digestive efficiency. Unassimilated food is egested. Assimilated food used for maintenance is lost as carbon dioxide and heat energy; the remainder is used for growth and reproduction.

occur in excess and undereating nutrients that occur in insufficient amounts (Behmer, 2009; Raubenheimer and Simpson, 2003; Simpson et al., 2002). *Ecological stoichiometry* and *metabolic theory* have become useful approaches to account for mass balances among multiple nutrients and energy as they flow within and among organisms (A. Allen et al., 2002; Behmer, 2009; J. Brown et al., 2004; Elser and Urabe, 1999; Sterner and Elser, 2002, see Chapters 3 and 11).

Arthropod species vary considerably in their requirements for, and assimilation of, energy and various nutrients (Joern and Behmer, 1998). Reichle et al. (1969) and Gist and Crossley (1975) reported significant variation in cation accumulation among forest floor arthropods, and Schowalter and Crossley (1983) reported significant variation in cation accumulation among forest canopy arthropods. Millipedes accumulated the highest concentrations of Ca among litter arthropods (Gist and Crossley, 1975; Reichle et al., 1969); caterpillars and sawfly larvae accumulated the highest concentrations of K and Mg, and spiders accumulated the highest concentrations of Na among arboreal arthropods (Schowalter and Crossley, 1983). Although nitrogen is frequently limiting for herbivores, excess nitrogen may be a liability, requiring elimination of toxic ammonia. Therefore, insect performance may decline above an optimal nitrogen content in the diet (Fig. 3.9) (Behmer, 2009; Joern and Behmer, 1998; Sterner and Elser, 2002; Zehnder and Hunter, 2008).

Resources varying in nutritional quality affect assimilation efficiency (A/I). For example, S. Eggert and Wallace (2007) reported that several aquatic detritivore taxa assimilated wood epoxylon (the surface biofilm of microbes, fragmented detritus, and exoenzymes) more efficiently (26–36%) than they did leaf detritus (9–17%), although they consumed more leaf detritus (0.09–0.47 g g^{-1} day^{-1} vs 0.04–0.07 g g^{-1} day^{-1}).

Assimilation efficiency also varies among developmental stages. Schowalter et al. (1977) found that assimilation efficiency of the range caterpillar, *Hemileuca oliviae*, declined significantly from 69% for first instars to 41% for the prepupal stage (Fig. 4.2). Respiration by pupae was quite low, amounting to only a few percent of larval production. This species does

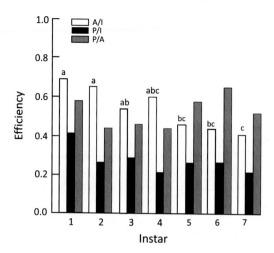

FIGURE 4.2   Assimilation efficiency, A/I, gross production efficiency, P/I, and net production efficiency, P/A, for larval stages of the saturniid moth, *Hemileuca oliviae.* Letters above A/I values denote significant differences at $p > 0.05$. P/I and P/A did not differ significantly among instars. *Source: Data from Schowalter, T.D., et al., 1977. Bioenergetics of the range caterpillar,* Hemileuca oliviae *(Ckll.). Oecologia 28, 153–161.*

not feed as an adult, so resources acquired by larvae must be sufficient for adult dispersal and reproduction.

3 ALLOCATION OF ASSIMILATED RESOURCES

Assimilated resources are allocated to various metabolic pathways. The relative amounts of resources used in these pathways depend on stage of development, quality of food resources, physiological condition, and metabolic demands for physiological processes (such as digestion and thermoregulation), activities (such as foraging and mating), and interactions with other organisms (including competitors, predators, and mutualists). For example, many immature insects are relatively inactive and expend energy primarily for feeding and defense, whereas adults expend additional energy and nutrient resources for dispersal and reproduction. Major demands for energy and nutrient resources include foraging activity, mating and reproduction, and competitive and defensive behavior.

3.1 Foraging and Dispersal Behavior

Many insects gain protected access to food (and habitat) resources through symbiotic interactions, that is, living on or in food resources (see Chapter 8). Specialized phytophagous species spend most or all of their developmental period on host food resources. A variety of myrmecophilous or termitophilous species are tolerated, or even share food with their hosts, as a result of morphological (size, shape, and coloration), physiological (chemical communication), or behavioral (imitation of ant behavior, trophallaxis) adaptations (Wickler, 1968). Resemblance to ants also may confer protection from other predators. Living within a food source minimizes energy expenditure for acquiring food.

However, many phytophagous and predaceous species must search for suitable food and habitat resources. Foraging, therefore, represents a necessary investment of energy in an individual's fitness. Dispersal activity is an extension of foraging activity and also constitutes an energy expense. Energy and nutrients invested in anatomical features that permit locomotion may conflict with other physiological or morphological needs, for example, wing development and flight that improve foraging and dispersal ability at the expense of reproductive ability (Table 4.1). Zera and Zhao (2006) demonstrated this trade-off through differential use of radiolabeled glycine by wing-dimorphic crickets, *Gryllus firmus*. The longer-winged, flight-capable morph oxidized a larger proportion of glycine and converted most into triglycerides to fuel flight, whereas the flightless morph utilized more glycine in ovarian protein for greater reproduction (S. Tanaka and Suzuki, 1998).

In at least one case, an insect may be able to augment its energy supply by directly harvesting solar energy to fuel its activity. Plotkin et al. (2010) reported that the cuticle of the Oriental hornet, *Vespa orientalis*, has a surface structure that enhances absorption of solar energy by pigments in the endocuticle: melanin in the brown-colored cuticle or xanthopterin in the yellow-colored cuticle (Plotkin et al., 2010). This hornet is most active at midday, when solar energy is maximum, and may be able to convert the harvested solar energy into chemical energy.

Foraging, and dispersal-related travel, bring an animal into contact with a wider variety of resources, compared to sedentary species. As discussed in Chapter 3, feeding on a wider variety of resources enables an insect to achieve nutritional balance (Behmer, 2009; K.P. Lee et al., 2002, 2003; Raubenheimer and Simpson, 2003; Simpson et al., 2002). However, movement also increases the risk of predation and exposure to adverse environmental conditions and, therefore, requires a trade-off between sufficient movement to acquire necessary resources and not increasing predation risk beyond the individual's defensive ability (Heads, 1986).

The nutritional value of the food resource affects the degree of energy and nutrient investment. Bumble bees, *Bombus* spp., forage on low-value resources only at high temperatures when the insects do not require large amounts of energy to maintain sufficiently high body temperature for flight (W. Bell, 1990; Heinrich, 1979, 1993). As described in Chapter 3, defended prey require production of detoxification enzymes or expenditure of energy to avoid defenses or avoid injury during prey capture. Alternatively, energy must be expended for continued search if the resource cannot be acquired successfully. However, searching

TABLE 4.1 Allocation of Assimilated Resources to Body Mass and Thoracic Muscle at Adult Eclosion and Adult Fecundity and Longevity, When Provided 16 mg Food the Day After Adult Eclosion, for Long- and Short-Winged Morphs of the Cricket, *Modicogryllus confirmatus*

Morph	Sex	Body mass (dry mg)	% Fat	Thoracic muscle mass (wet mg)	Fecundity (no. eggs)	Longevity (days)
Short-winged	M	30±5	18±5	4.6±1.0		8±2
	F	35±6	23±7	5.2±1.2	24±10	9±2
Long-winged	M	34±4	21±3	12.7±1.4		12±2
	F	39±5	25±4	13.3±1.9	4±6	13±2

Source: Modified with permisison from Tanaka, S., Suzuki, Y., 1998. Physiological trade-offs between reproduction, flight capability and longevity in a wing-dimorphic cricket, Modicogryllus confirmatus. J. Insect Physiol. 44, 121–129.

behavior also may bring the forager to the attention of predators (Schultz, 1983; Folgarait and Gilbert, 1999), and expenditure of carbohydrate resources for locomotion may result in a relative excess of nitrogen. Since most insects are short-lived, as well as energy-limited, they often maximize fitness by accepting less suitable, but more available or apparent, resources in lieu of continued searching for superior resources (Behmer, 2009; Courtney, 1985, 1986; Kogan, 1975; Sholes, 2008). van der Zee et al. (2002) offered desert locust, *Schistocerca gregaria*, nymphs food choices varying in nutritional composition (from high protein, low carbohydrate to low protein, high carbohydrate) and distance apart. They found that movement between dishes offering complementary nutritional value declined as distance between food dishes increased, demonstrating a trade-off between movement and diet optimization.

Energy expenditure varies among foraging strategies, depending on distances covered and the efficiency of orientation toward resource cues (see Chapters 2 and 3). Larger animals travel more efficiently than do smaller animals, expending less energy for a given distance traversed. Hence, larger animals often cover larger areas in search of resources. Flying against the wind requires additional energy expenditure. Orchid bees, *Euglossa imperialis*, extend their hind legs ventrally to improve stability at high wind speeds, but this position increases drag by 30%, increasing the energy expenditure of flight (Combes and Dudley, 2009). Flight is more energy-efficient and covers greater distances in less time than walking, and efficiency increases with flight speed (Heinrich, 1979), enabling flying insects to cover large areas with relatively small energy reserves. Hunting requires considerable expenditure of energy searching for prey, but yields a high return, depending on ability to detect prey from a distance. Ambushers either sit and wait or employ traps to capture prey. As examples, dragonfly larvae hide in the substrate of aquatic habitats and grasp prey coming within reach, antlion larvae excavate conical depressions in loose sandy soil that prevents the escape of ants and other insects that wander into the pit, and webspinning spiders construct sticky orb or tangled webs that trap flying or crawling insects. Movement costs are minimal for these species, but prey encounter is uncertain. Frequency of prey encounter can be increased by selecting ambush sites along prey foraging trails, near prey nest sites, etc.

Foraging efficiency can be improved greatly by detection of suitable host odors over long distances. Andersson et al. (2013b) reported that the signal strength of a resource is mediated by the size of the resource, making larger patches or more concentrated resources attractive over longer distances. Detection can be increased by orienting toward host/prey odors or plant odors indicative of prey (see Chapter 3). Accordingly, many predaceous species are attracted to mating pheromones of their prey (Stephen et al., 1993) or to volatile chemicals released by plants in response to herbivory (Thaler et al., 2001; Turlings et al., 1993). Social insects improve foraging efficiency by rapidly recruiting nestmates to discovered resources (see Chapter 3).

The actual energy costs of foraging have been measured rarely. McNab (1963) proposed that the area (home range, H) required for an individual to acquire sufficient resources is proportional to the individual's mass (M) to the ¾ power:

$$H = cM^{3/4}$$

(4.2)

However, subsequent research demonstrated that mobile animals require larger home ranges than predicted by the ¾ power equation, reflecting larger home ranges and maintenance costs necessary to defend portions that overlap with home ranges of conspecific individuals

(Jetz et al., 2004). Swenson et al. (2007) tested the relationship for sessile ant lion, *Myrmeleon* sp., and concluded that the ¾ power equation applied to the pit density of these sessile animals, despite competition for pit space.

Fewell et al. (1996) compared the ratios of benefit to cost for a canopy-foraging tropical ant, *Paraponera clavata*, and an arid grassland–seed harvesting ant, *Pogonomyrmex occidentalis* (Table 4.2). They found that the ratio ranged from 3.9 for nectar-foraging *P. clavata* and 67 for predaceous *P. clavata* to > 1000 for granivorous *P. occidentalis*. Differences were due to the quality and amount of the resource, the distance traveled, and the individual cost of transport. In general, the smaller *P. occidentalis* had a higher ratio of benefit to cost because of the higher energy return of seeds, shorter average foraging, and lower energy cost m^{-1} traveled. The results indicated that *P. clavata* colonies have similar daily rates of energy intake and expenditure, potentially limiting colony growth, whereas *P. occidentalis* colonies have a much higher daily intake rate, compared to expenditure, reducing the likelihood of short-term energy limitation.

Farji-Brener et al. (2015) calculated how branching angles in foraging trails should affect the relative costs of trail maintenance or travel distances for leaf-cutting ants, *Atta* spp., in tropical. They found that ant colonies in open sites favored acute branching angles, which best shorten travel distances but create longer new trail sections to maintain, compared to more perpendicular angles, suggesting that trail maintenance costs are smaller than travel costs. Conversely, ant colonies in forested sites showed less acute branching angles, indicating that maintenance costs are larger relative to travel costs, perhaps because large inputs of litterfall in forests increase trail maintenance costs and/or canopy cover reduces travel costs by mitigating exposure to sunlight and rain. Farji-Brener et al. (2015) concluded that branching angles of foraging trails represent a trade-off between minimizing maintenance costs and travel costs.

Insects produce a variety of biochemicals to exploit food resources. Insects that feed on chemically defended food resources often produce more-or-less specific enzymes to detoxify these defenses (see Chapter 3). On one hand, production of detoxification enzymes (typically complex, energetically- and nitrogen-expensive molecules) reduces the net energy and nutritional value of food. On the other hand, these enzymes permit exploitation of a resource and derivation of nutritional value otherwise unavailable to the insect. Some insects not only detoxify host defenses but also digest the products for use in their own metabolism and growth (Schöpf et al., 1982), thereby compensating for the expense of detoxification.

TABLE 4.2 Components of the Benefit to Cost (B/C) Ratio for Individual *Paraponera clavata* and *Pogonomyrmex occidentalis* Foragers

	Paraponera		Pogonomyrmex
	Nectar forager	**Prey forager**	
Energy cost per meter (J m^{-1})	0.042		0.007
Foraging trip distance (m)	125		12
Energy expenditure per trip (J)	5.3		0.09
Average reward per trip (J)	20.8	356	100
B/C	3.9	67	1111

Source: Reprinted with permission from Fewell, J.H., et al., 1996. Foraging energetics of the ant, Paraponera clavata. *Oecologia 105, 419–427.*

Social insects employ communication to recruit nestmates to discovered resources. Ants produce trail pheromones that provide an odor trail to guide other members of a colony to food resources and back to the colony (Fig. 3.17, see Chapter 3). Honey bees, *Apis mellifera*, have the most sophisticated communication among invertebrates. The elaborate movements of the "waggle dance" communicate distance and direction to suitable resources to other foragers (F. Dyer, 2002; von Frisch, 1967, see Chapter 3). By means of communication and recruitment of large numbers of nestmates, social insects can exploit discovered resources quickly.

Agricultural systems typically concentrate particular crop species over large areas, reducing the expense of searching for suitable food resources for adapted insects. For example, the Colorado potato beetle, *Leptinotarsa decemlineata*, probably originated in South America and subsisted on wild solanaceous hosts. Spread of buffalo burr, *Solanum rostratum*, into western North America facilitated spread of the beetle, but it was not a pest until westward movement of settlers brought it into contact with cultivated potato in the Midwest during the late 1800s (Hitchner et al., 2008; C. Riley, 1883; Stern et al., 1959), eventually allowing it to spread to Europe. Similarly, the cotton boll weevil, *Anthonomus grandis*, coevolved with scattered wild *Gossypium* spp., including *G. hirsutum*, in tropical Mesoamerica until the spread of citrus cultivation, which provided food resources that could sustain overwintering adults, in the 1890s allowed the insect to spread into subtropical cotton-growing regions of south Texas and northern Argentina (Showler, 2009). Subsequently, rapid reproduction in the spring by adults that survived a winter dormancy permitted spread throughout the US Cotton Belt (Showler, 2009).

3.2 Mating Activity

Mate attraction and courtship behavior often are highly elaborated and ritualized and can be energetically costly. Nevertheless, such behaviors that distinguish species, especially sibling species, ensure appropriate mating and reproductive success and contribute to individual fitness through improved survival of offspring from selected mates.

3.2.1 Attraction

Chemical, visual, and acoustic signaling are used to attract potential mates. Attraction of mates can be accomplished by either sex in Coleoptera, but only female Lepidoptera release sex pheromones, and only male Orthoptera stridulate.

Sex pheromones greatly improve the efficiency with which insects find potential mates over long distances in heterogeneous environments (Cardé, 1996; Law and Regnier, 1971; Mustaparta, 1984). Pheromones typically are complex blends of compounds that may or may not be attractive to potential mates when present in ratios different from that produced by the "calling" sex (McElfresh et al., 2001; D. Weber et al., 2014b). The particular blend of compounds and their enantiomers, as well as the time of calling, varies considerably among species (Allison et al., 2012). These mechanisms represent the first step in maintaining reproductive isolation (Kaae et al., 1973; McElfresh and Millar, 2001). For example, among tortricids in eastern North America, *Archips mortuanus* uses a 90:10 blend of (Z)-11- and (E)-11-tetradecenyl acetate, *A. argyrospilus* uses a 60:40 blend, and *A. cervasivoranus* uses a 30:70 blend. A related species, *Argyrotaenia velutinana*, also uses a 90:10 blend but is repelled by (Z)-9-tetradecenyl acetate that is incorporated by *A. mortuanus* (Cardé and Baker, 1984). Among three species of saturniids in South Carolina, *Callosamia promethea* calls from about

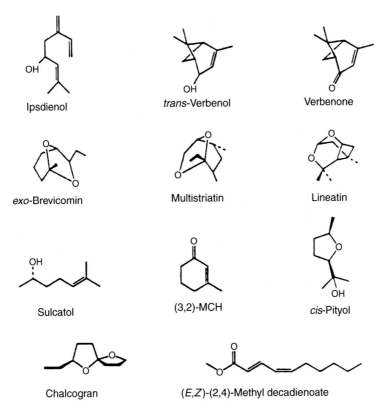

Ipsdienol

trans-Verbenol

Verbenone

exo-Brevicomin

Multistriatin

Lineatin

Sulcatol

(3,2)-MCH

cis-Pityol

Chalcogran

(*E,Z*)-(2,4)-Methyl decadienoate

FIGURE 4.3 **Representative pheromones produced by bark beetles.** Pheromones directly converted from plant compounds include ipsdienol (from mycene), trans-verbenol, and verbenone (from α-pinene). The other pheromones shown are presumed to be synthesized by the beetles. *Source: From Raffa, K.F., et al., 1993. Strategies and mechanisms of host colonization by bark beetles. In: Schowalter, T.D., Filip, G.M. (Eds.), Beetle-Pathogen Interactions in Conifer Forests. Academic Press, London, pp. 103–128.*

1000–1600, *C. securifera* from about 1600–1900, and *C. angulifera* from 1900–2400 (Cardé and Baker, 1984). Bark beetle pheromones have been studied extensively (Allison et al., 2012; Raffa et al., 1993). Representative bark beetle pheromones are shown in Fig. 4.3.

The evolution of male response to pheromone blends emitted by females apparently is constrained by a trade-off between breadth of response and sensitivity (Hemmann et al., 2008). Males responding over wide variation in pheromone composition likely fail to encounter conspecific females. Hence, selection should favor males with narrow response but high sensitivity. Hemmann et al. (2008) demonstrated, using a mutant pheromone strain of *Trichoplusia ni*, that hybrid males had the narrow breadth of wild males but the low sensitivity of mutant males, suggesting a hybrid disadvantage and a mechanism for reinforcement of male pheromone response traits. Groot et al. (2006) reported that 10 times more male *Heliothis virescens* were attracted to female *H. subflexa* with introgressed quantitative trait locus from *H. virescens* (that decreased the amount of acetate esters in pheromone glands) than to normal *H. subflexa* females. Hybrid infertility resulting from *H. virescens/H. subflexa* mating drove strong directional selection for higher acetate ester concentrations in female *H. subflexa* pheromone.

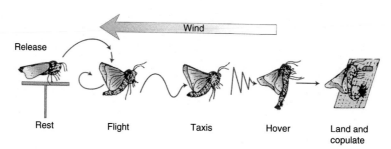

FIGURE 4.4 **Typical responses of male noctuid moths to the sex pheromone released by female moths.** *Source: From Tumlinson, J.H., Teal, P.E.A., 1987. Relationship of structure and function to biochemistry in insect pheromone systems. In: Prestwich, G.D., Blomquist, G.J. (Eds.), Pheromone Biochemistry. Academic Press, Orlando, FL, pp. 3–26.*

Sex pheromones may be released passively, as in the feces of bark beetles (Raffa et al., 1993), or actively through extrusion of scent glands and calling behavior (Cardé and Baker, 1984). The attracted sex locates the signaler by following the concentration gradient (Fig. 4.4). Early studies suggested that the odor from a point source diffuses in a cone-shaped plume that expands downwind, the shape of the plume depending on airspeed and vegetation structure (Matthews and Matthews, 2010). However, research by Cardé (1996), Mafra-Neto and Cardé (1995), Murlis et al. (1992), Roelofs (1995), and Strand et al. (2009) indicated that this plume is neither straight nor homogeneous over long distances, but is influenced by wind speed and direction and by structures that cause turbulence in the airstream, deflecting the plume and forming pockets of higher concentration or absence of the vapor (Fig. 4.5). Heterogeneity in vapor concentration is augmented by pulsed emission by many insects. An insect downwind would detect the plume as odor bursts rather than as a constant stream. Pulses in emission and reception may facilitate orientation, because the antennal receptors require intermittent stimulation to avoid saturation and sustain upwind flight (Roelofs, 1995). However, Cardé (1996) noted that the heterogeneous nature of the pheromone plume may make direct upwind orientation difficult over long distances. Pockets of little or no odor may cause the attracted insect to lose the odor trail. Detection can be inhibited further by openings in the vegetation canopy that create warmer convection zones or "chimneys" that carry the

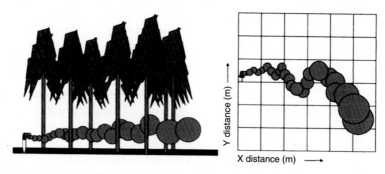

FIGURE 4.5 **Pheromone plume dispersion through a forest, as modified by variation in wind speed and direction, and deflection by vegetation or other structures.** *Source: Reprinted with permission from Strand, T., et al., 2009. A simple model for simulation of insect pheromone dispersion within forest canopies. Ecol. Model. 220, 640–656.*

pheromone through the canopy (Fares et al., 1980; Girling et al., 2013; Thistle et al., 2004). Attracted insects may increase their chances of finding the plume again by casting, that is, sweeping back and forth in an arcing pattern until the plume is contacted again (Baker and Kuenen, 1982; Cardé, 1996).

Given the small size of most insects and limited quantities of pheromones for release, mates must be able to respond to very low concentrations. Release of less than 1 μg sec^{-1} by female gypsy moth, *Lymantria dispar*, or silkworm, *Bombyx mori*, can attract males, which respond at molecular concentrations as low as 100 molecules ml^{-1} of air (Harborne, 1994). Nevertheless, the likelihood of attracted insects reaching a mate is small. Elkinton et al. (1987) reported that the proportion of male gypsy moths responding to a caged female declined from 89% at 20 m distance to 65% at 120 m. Of those males that responded, the proportion arriving at the female's cage declined from 45% at 20 m to 8% at 120 m, and the average minimum time to reach the female increased from 1.7 min at 20 m to 8.9 min at 120 m (Fig. 4.6). Therefore, the probability of successful attraction of mates is low, and exposure to predators or other mortality factors relatively high, over modest distances.

Visual signaling is exemplified by the fireflies (Coleoptera: Lampyridae) (Fu et al., 2005; Lloyd, 1983). In this group of insects, different species distinguish each other by variation in the rhythm of flashing and by the perceived "shape" of flashes produced by distinctive

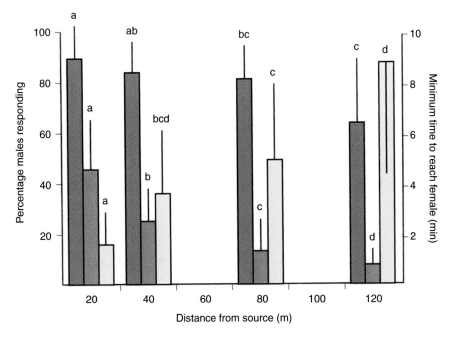

FIGURE 4.6 **Effect of distance on insect perception of and arrival at a pheromone source.** Proportion (mean ± SD) of male gypsy moths responding at 20, 40, 80, and 120 m from a pheromone source *(blue bar)*, mean proportion of those responding that reached the source within a 40-min period *(orange bar)*, and the average minimum time to reach the source *(yellow bar)*; n = 23. Values followed by the same letter do not differ significantly at $p < 0.05$. *Source: Data from Elkinton, J.S., et al., 1987. Pheromone puff trajectory and upwind flight of male gypsy moths in a forest. Physiol. Entomol. 12, 399–406.*

movements while flashing. Other insects, including glow worms (Coleoptera: Phengodidae) and several midges, also attract mates by producing luminescent signals.

Acoustic signaling is produced by stridulation, particularly in the Orthoptera, Hemiptera, and Coleoptera, or by muscular vibration of a membrane, common in the Hemiptera. Resulting sounds can be quite loud and detectable over considerable distances. For example, the acoustic signals of mole crickets, *Gryllotalpa vinae*, amplified by the double horn configuration of the cricket's burrow, are detectable by humans up to 600 m away (Matthews and Matthews, 2010).

During stridulation, one body part, the file (consisting of a series of teeth or pegs), is rubbed over an opposing body part, the scraper. Generally, these structures occur on the wings and legs (R. Chapman, 1982), but in some Hymenoptera, sound also is produced by the friction between abdominal segments as the abdomen is extended and retracted. The frictional sound produced can be modulated by various types of resonating systems. The frequency and pattern of sound pulses are species specific.

In cicadas, the sound produced by vibrating membranes (tymbals) is accomplished by contracting the tymbal muscle to produce one sound pulse and relaxing the muscle to produce another sound pulse. Muscle contraction is so rapid (170–480 contractions sec^{-1}) that the sound appears to be continuous (Matthews and Matthews, 2010). The intensity of the sound is modified by air sacs operated like a bellows and by opening and closing opercula that cover the sound organs (R. Chapman, 1982).

Such mechanisms greatly increase the probability of attracting mates. However, many predators also are attracted to, or imitate, signaling prey. For example, some firefly species that prey on other fireflies imitate the flash pattern of prey species (Lloyd, 1983).

3.2.2 Courtship Behavior

Courtship behavior is under genetic control (Demir and Dickson, 2005) and subject to strong selection to ensure appropriate reproductive partnerships. Courtship often involves an elaborate, highly ritualized sequence of stimulus and response actions that must be completed before copulation occurs (Fig. 4.7). This provides an important mechanism that identifies species and sex, thereby enhancing reproductive isolation. Color patterns, odors, and tactile stimuli are important aspects of courtship. For many species, ultraviolet patterns are revealed, close-range pheromones are emitted, or legs or mouthparts stroke the mate as necessary stimuli (L. Brower et al., 1965; Matthews and Matthews, 2010).

Another important function of courtship displays in predatory insects is appeasement, or inhibition of predatory responses, especially of females. Nuptial feeding occurs in several insect groups, particularly the Mecoptera, empidid flies, and some Hymenoptera and Hemiptera (Fig. 4.8). The male provides a food gift (such as a prey item, nectar, seed, or glandular product) that serves at least two functions (Matthews and Matthews, 2010; Price, 1997; Thornhill, 1976). Males with food may be more conspicuous to females, and feeding the female prior to oviposition may increase fecundity and fitness. Nuptial feeding has become ritualistic in some insects. Rather than prey, some flies simply offer a silk packet.

Conner et al. (2000) reported that male arctiid moths, *Cosmosoma myrodora*, acquire pyrrolizidine alkaloids from excrescent fluids of some plants, such as *Eupatorium capillifolium*. The alkaloids are incorporated into cuticular filaments that are stored in abdominal pouches

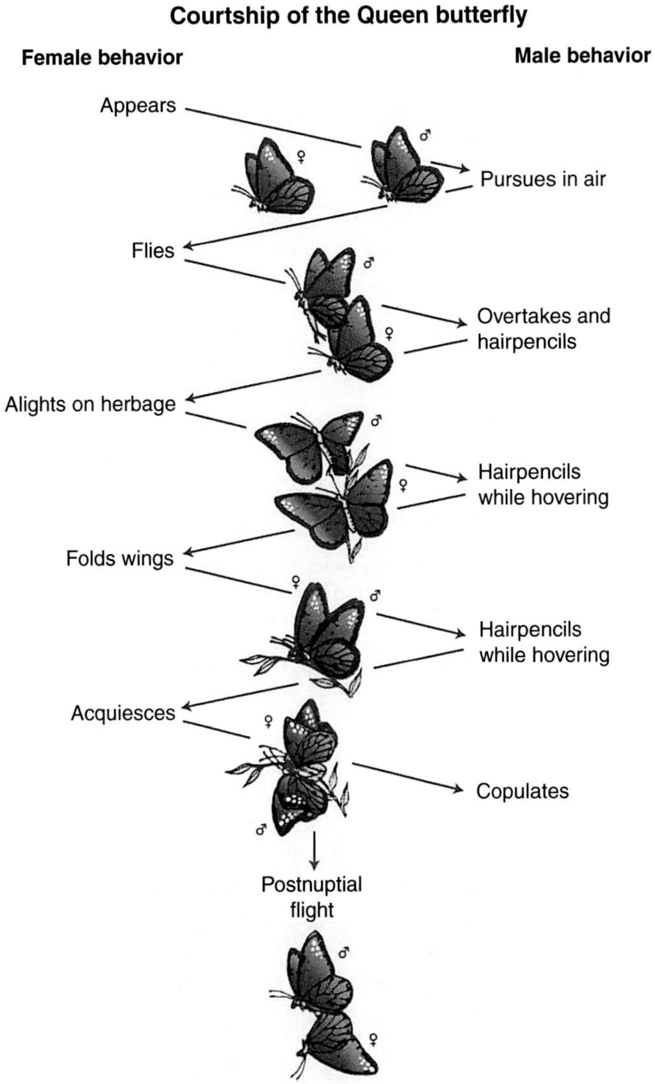

Courtship of the Queen butterfly

Female behavior   Male behavior

Appears

Pursues in air

Flies

Overtakes and
hairpencils

Alights on herbage

Hairpencils
while hovering

Folds wings

Hairpencils
while hovering

Acquiesces

Copulates

Postnuptial
flight

FIGURE 4.7   Courtship stimulus–response sequence of the Queen butterfly, *Danaus gilippus*, from top to bottom, with male behavior on the right and female behavior on the left. *Source: From Brower, L.P., et al., 1965. Courtship behavior of the queen butterfly, Danaus gilippus berenice (Cramer). Zoologica 50, 1–39.*

and discharged on the female during courtship. This topical application makes the female distasteful to spiders. Alkaloid-deprived males do not provide this protection, and females mated with such males are suitable prey for spiders.

Males of some flies, euglossine bees, Asian fireflies, and some dragonflies gather in groups, called leks, to attract and court females (Fig. 4.8). Such aggregations allow females to compare and choose among potential mates and facilitate mate selection.

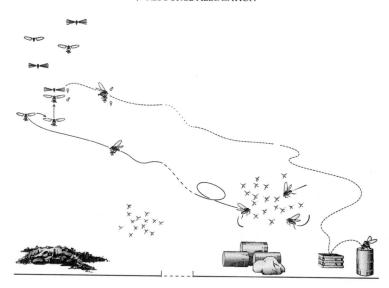

FIGURE 4.8   Example of lekking and appeasement behavior in the courtship of an empidid fly, *Rhamphomyia nigripes*. Males capture a small insect, such as a mosquito or midge, then fly to a mating swarm (lek) which attracts females. Females select their mates and obtain the food offering. The pair then leaves the swarm and completes copulation on nearby vegetation. *Source: From Downes, J.A., 1970. The feeding and mating behaviour of the specialized* Empididae (Diptera); *observations on four species of* Rhamphomyia *in the high Arctic and a general discussion. Can. Entomol. 102, 769–791.*

3.3 Reproductive and Social Behavior

Insects, like other organisms, invest much of their assimilated energy and nutrient resources in the production of offspring. Reproductive behavior includes varying degrees of parental investment in offspring that determines the survival of eggs and juveniles. Selection of suitable sites for oviposition affects the exposure of eggs to abiotic conditions suitable for hatching. The choice of oviposition site also affects the exposure of eggs and hatching immatures to predators and parasites and their proximity to suitable food resources. Nesting behavior, brood care, and sociality represent stages in a gradient of parental investment in survival of offspring.

3.3.1 Oviposition Behavior

Insects deposit their eggs in a variety of ways. Most commonly, the female is solely responsible for selection of the oviposition site(s). The behaviors leading to oviposition are as complex as those leading to mating, because successful oviposition contributes to individual fitness and is under strong selective pressure.

A diversity of stimuli affects choice of oviposition sites by female insects. Mosquitoes are attracted to water by the presence of vegetation and reflected light, but lay eggs only if salt content, pH, or other factors sensed through tarsal sensillae, are suitable (Matthews and Matthews, 2010). Grasshoppers assess the texture, salinity, and moisture of soil selected for oviposition.

Many phytophagous insects assess host suitability for development of offspring. This assessment may be on the basis of host chemistry or existing feeding pressure. Ovipositing insects tend to avoid host materials with deleterious levels of secondary chemicals. However, short adult life spans may encourage oviposition on more easily discovered resources, regardless of suitability for offspring (Courtney, 1985, 1986; Kogan, 1975). Insects also tend to avoid ovipositing on resources that are already occupied by eggs or competitors. For example, female bean weevils, *Callosobruchus maculatus*, assess each potential host bean by comparison to the previous bean and lay an egg only if the present bean is larger or has fewer eggs. The resulting pattern of oviposition nearly doubles larval survival compared to random oviposition (R. Mitchell, 1975). Many parasitic wasps mark hosts in which eggs have been deposited and avoid ovipositing in marked hosts, thereby minimizing larval competition within a host (Godfray, 1994). Parasitic wasps can minimize hyperparasitism by not ovipositing in more than one host in an aggregation. This reduces the risk that all of its offspring are found and parasitized (W. Bell, 1990). Cannibalistic species, such as *Heliconius* butterflies, may avoid laying eggs near each other, to minimize cannibalism and predation.

Selection also determines whether insects lay all their eggs during one period (semelparity) or produce eggs over more protracted periods (iteroparity). Most insects with short life cycles (eg, < 1 year) typically have relatively short adult life spans and lay all their eggs in a relatively brief period. Insects with longer life spans, especially social insects, reproduce continually for many years.

In some cases, choices of oviposition sites by adults clearly conflict with the suitability of resources for offspring. Kogan (1975) and Courtney (1985, 1986) reported that some species preferentially oviposit on the most conspicuous (apparent) host species that are relatively unsuitable for larval development. However, this behavior represents a trade-off between the prohibitive search time required to find the most suitable hosts and the reduced larval survival on the more easily discovered hosts. Albeny-Simões et al. (2014) found that the mosquito, *Aedes aegypti*, preferentially oviposited in habitats with predatory mosquito, *Toxorhynchites theobaldi*, larvae because predation increased chemical cues indicative of bacterial food resources, again reflecting a trade-off between factors affecting offspring survival.

Schäpers et al. (2016) compared oviposition patterns and larval feeding and survival among two generalist, *Polygonia c-album* and *Vanessa cardui*, and three specialist, *Aglais urticae*, *A. io* and *Vanessa atalanta*, nymphalid butterflies that vary in host preferences but all oviposit and feed on common nettle, *Urtica dioica*. They reported that species in which females discriminated among potential hosts showed poor larval performance on suboptimal hosts, whereas species in which females discriminated poorly among potential hosts showed better larval performance on suboptimal hosts, reflecting a general congruence between female preference and larval performance.

Some insects influence host suitability for their offspring. For example, female sawflies typically sever the resin ducts at the base of a conifer needle prior to laying eggs in slits cut distally to the severed ducts. This behavior prevents or reduces egg mortality resulting from resin flow into the oviposition slits (McCullough and Wagner, 1993). Parasitic Hymenoptera often inject mutualistic viruses into the host along with their eggs. The virus inhibits cellular encapsulation of the egg or larva by the host (Vega and Kaya, 2012).

3.3.2 Nesting and Brood Care

Although brood care is best known among the social insects, other insects exhibit maternal care of offspring and even maternal tailoring of habitat conditions to enhance the survival of offspring. Primitive social behavior appears as parental involvement extends further through the development of their offspring.

Several environmental factors are necessary for evolution of parental care (E. Wilson, 1975). A stable environment favors larger, longer-lived species that reproduce at intervals, rather than all at once. Establishment in new, physically stressful environments may select for protection of offspring, at least during vulnerable periods. Intense predation may favor species that guard their young to improve their chances of reaching breeding age. Finally, selection may favor species that invest in their young, which, in turn, help the parent find, exploit, or guard food resources. Cooperative brood care, involving reciprocal communication, among many adults is the basis of social organization (E. Wilson, 1975, see Chapter 5).

A variety of insect species from several orders exhibit protection of eggs by a parent (Matthews and Matthews, 2010). In most cases, the female remains near her eggs and guards them against predators. However, in some species of giant water bugs (*Belostoma* and *Abedus*), the eggs are laid on the back of the male, which carries them until they hatch. Among dung beetles (Scarabaeidae), adults of some species limit their investment in offspring to providing protected dung balls in which eggs are laid, whereas females in the genus *Copris* remain with the young until they reach adulthood (Matthews and Matthews, 2010).

Extended maternal care, including provision of food for offspring, is seen in crickets, cockroaches, some Hemiptera and Coleoptera, and nonsocial Hymenoptera. Maternal care can substantially increase the growth and survival of offspring (A. Eggert et al., 1998). For example, females of the membracid, *Umbonia crassicornis*, enhance offspring survival by brooding eggs, cutting slits in the bark of twigs to facilitate feeding by nymphs, and defending nymphs against predators (T.K. Wood, 1976). Survival of nymphs with their mother present was 80%, compared to 60% when the mother was removed 2–3 days after egg hatch, and 10% when the mother was removed prior to making bark slits. Females responded to predators or to alarm pheromones from injured offspring by fanning wings and buzzing, typically driving the predator away (T.K. Wood, 1976).

A number of arthropod species are characterized by aggregations of individuals. Groups can benefit their members in a number of ways. Large groups often are able to modify environmental conditions, such as through retention of body heat or moisture. Aggregations also increase the availability of potential mates (Matthews and Matthews, 2010) and minimize exposure of individuals to plant toxins (McCullough and Wagner, 1993; Nebeker et al., 1993) and to predators (Fitzgerald, 1995). Aggregated, cooperative feeding on plants, such as by sawflies and bark beetles, can remove plant tissues or kill the plant before induced defenses become effective (McCullough and Wagner, 1993; Nebeker et al., 1993). Groups limit predator ability to avoid detection and to separate an individual to attack from within a fluid group. Predators are more vulnerable to injury by surrounding individuals, compared to attacking isolated individuals.

Cooperative behavior is evident within groups of some spiders and communal herbivores, such as tent-building caterpillars and gregarious sawflies. Dozens of individuals of the spider, *Mallos gregalis*, cooperate in construction of a communal web and in subduing prey (Matthews and Matthews, 2010). Tent-building caterpillars cooperatively construct their web, which affords protection from predators and may facilitate feeding and retention of heat and moisture (Fitzgerald, 1995). Similarly, gregarious sawflies cooperatively defend against predators and distribute plant resin among many individuals, thereby limiting the effectiveness of the resin defense (McCullough and Wagner, 1993).

Primitive social behavior is exhibited by the wood cockroach, *Cryptocercus punctulatus*, by passalid beetles, and by many Hymenoptera. In these species, the young remain with the parents in a family nest for long periods of time, are fed by the parents, and assist in nest maintenance (Matthews and Matthews, 2010). However, these insects do not exhibit coordinated behavior or division of labor among distinct castes.

The complex eusociality characterizing termites and the social Hymenoptera has attracted considerable attention (Matthews and Matthews, 2010; E. Wilson, 1975). Eusociality is characterized by multiple adult generations, highly integrated cooperative behavior, and efficient division of labor among castes (Matthews and Matthews, 2010; Michener, 1969). Members of these insect societies cooperate in food location and acquisition, feeding of immatures, and defense of the nest. This cooperation is maintained through complex pheromonal communication, including trail and alarm pheromones (Hölldobler, 1995, see Chapter 3), and reciprocal exchange of regurgitated liquid foods (trophallaxis) between colony members. Trophallaxis facilitates recognition of nest mates by maintaining a colony-specific odor, ensures exchange of important nutritional resources and (in the case of termites) of microbial symbionts that digest cellulose, and may be critical to colony survival during periods of food limitation (Matthews and Matthews, 2010). Trophallaxis distributes material rapidly throughout a colony (M. Suarez and Thorne, 2000). E. Wilson and Eisner (1957) fed honey mixed with radioactive iodide to a single worker ant and within 1 day detected some tracer in every colony member, including the two queens. Such behavior may also facilitate spread of pathogens or toxins throughout the colony (J.K. Grace and Su, 2001; Shelton and Grace, 2003).

In some ant species, capture of slaves from other ant colonies augments or replaces various worker caste functions in the colony. However, some species of slave-making ants, for example, *Protomognathus americanus*, have little or no effect on the demographics of ant species exploited as slaves, whereas other slave-making species, for example, *Leptothorax duloticus*, significantly reduce abundances of queens, workers, and larvae of host species (Hare and Alloway, 2001).

Development of altruistic behaviors such as social cooperation can be explained largely as a consequence of kin selection and reciprocal cooperation (Axelrod and Hamilton, 1981; Haldane, 1932; W. Hamilton, 1964; Trivers, 1971; E. Wilson, 1973; Wynne-Edwards, 1963, 1965, see also Chapter 15). Self-sacrifice that increases reproduction by closely related individuals increases inclusive fitness, that is, the individual's own fitness plus the fitness accruing to the individual through its contribution to reproduction of relatives with shared genes. In the case of the eusocial Hymenoptera, because of haploid males, relatedness among siblings is greater than that between parent and offspring, making cooperation among colony members highly adaptive. The epitome of "altruism" among insects may be the development of the barbed sting in the worker honey bee that ensures its death in defense of the colony (Haldane, 1932; W. Hamilton, 1964). Termites do not share the Hymenopteran model for sibling relatedness. Genetic data for termites indicate relatively high inbreeding and relatedness within colonies and kin-biased foraging behavior for some species (Kaib et al., 1996; Vargo et al., 2003). However, Husseneder et al. (1999) reported that DNA analysis of colonies of the African termite, *Schedorhinotermes lamanianus*, did not indicate effective kin selection through inbreeding, or translocation complexes of sex-linked chromosomes that could generate higher relatedness within than between sexes. They concluded that ecological factors, such as predation and food availability, may be more important than genetics in maintaining termite eusociality, at least in this species.

Social behavior provides the greatest degree of foraging efficiency, permitting recruitment of multiple individuals to exploit large food resources and to share these resources with nest-mates. Furthermore, social behavior allows for division of labor and more efficient performance of the multiple energy-demanding activities, including foraging, reproduction, and defense, typically by means of morphologically specialized castes.

3.4 Competitive, Defensive and Mutualistic Behavior

Insects, like all animals, interact with other species in a variety of ways, as competitors, predators, prey, and mutualists. Interactions among species will be discussed in greater detail in Chapter 8. These interactions require varying degrees of energy and/or nutrient expenditure. Contests among individuals for resources occasionally involve combat. Subduing prey and defending against predators also involve strenuous activity. Mutualism requires reciprocal exchange of resources or services. Obviously, these interactions affect the energy and nutrient budgets of individual organisms.

3.4.1 Competitive Behavior

Competition occurs among individuals using the same limiting resources at the same site. Energy expended, or injury suffered, defending resources or searching for uncontested resources affects fitness. Competition often is mediated by mechanisms that determine a dominance hierarchy. Establishment of dominant and subordinate status among individuals limits the need for physical combat to determine access to resources and ensures that dominant individuals get more resources than do subordinate individuals.

Visual determination of dominance status is relatively rare among insects, largely because of their small size, the complexity of the environment which restricts visual range, and the limitations of fixed-focus compound eyes for long-distance vision (Matthews and Matthews, 2010). Dragonflies have well-developed eyes and exhibit ritualized aggressive displays that maintain spacing among individuals. For example, male *Plathemis lydia* have abdomens that are bright silvery-white above. Intrusion of a male into another male's territory initiates a sequence of pursuit and retreat, covering a distance of 8–16 m. The two dragonflies alternate roles and directions, with the abdomens raised during pursuit and lowered during retreat, until the intruder moves to another site (P. Corbet, 1962).

Mediation of competition by pheromones has been documented for several groups of insects. Adult flour beetles, *Tribolium*, switch from aggregated distribution at low densities to random distribution at intermediate densities, to uniform distribution at high densities. This spacing is mediated by secretion of quinones, repellent above a certain concentration, from thoracic and abdominal glands (Matthews and Matthews, 2010). Bark beetles employ repellent pheromones, as well as acoustic signals, to maintain minimum distances between individuals boring through the bark of colonized trees (Raffa et al., 1993; Rudinsky and Ryker, 1976). Ant colonies also maintain spacing through marking of foraging trails with chemical signals (as described in 3.3.2, see also Chapter 3).

Acoustic signals are used by many Orthoptera and some Coleoptera to deter competitors. Bark beetles stridulate to deter other colonizing beetles from the vicinity of their gallery entrances (Rudinsky and Ryker, 1976). Subsequently, excavating adults and larvae respond to the sounds of approaching excavators by mining in a different direction, thus preventing intersection of galleries.

Some male crickets and grasshoppers produce a distinctive rivalry song when approaching each other (Matthews and Matthews, 2010; Schowalter and Whitford, 1979). The winner (continued occupant) typically is the male that produces more of this aggressive stridulation.

When resources are relatively patchy, males may increase their access to females by marking and defending territories that contain resources attractive to females. Territorial behavior is less adaptive (ie, costs of defending resources exceed benefits) when resources are highly concentrated and competition is severe or when resources are uniformly distributed and female distribution is less predictable (Baker, 1972; Schowalter and Whitford, 1979).

Marking territorial boundaries takes a variety of forms among animal taxa. Male birds mark territories by calling from perches along the perimeter. Male deer rub scent glands and scrape trees with their antlers to advertise their territory. Social insects, including ants, bees, and termites, mark nest sites and foraging areas with trail pheromones that advertise their presence. These trail markers can be perceived by other insects at minute concentrations (see 3.2.1). Many orthopterans and some beetles advertise their territories by stridulating.

However, many insects advertise their presence simply to maintain spacing and do not actively defend territories. Similarly, males of many species, including insects, fight over receptive females. E. Wilson (1975) considered defense of occupied areas to be the defining criterion for territoriality. Territorial defense is best known among vertebrates, but a variety of insects representing at least eight orders defend territories against competitors (Matthews and Matthews, 2010; Price, 1997). Because territorial defense represents an energetic cost, an animal must gain more of the resource by defending it against competitors than by searching for new resources. Nonaggressive males often "cheat" by nonadvertisement and quiet interception of resources or of females attracted to the territory of the advertising male (Schowalter and Whitford, 1979).

The type of territory differs among insect taxa, but usually is associated with competition for food or mates (Matthews and Matthews, 2010; Price, 1997). Male crickets defend the area around their dens and mate with females attracted to their stridulation. Male eastern wood cockroaches, C. punctulatus, defend mating chambers in rotten wood (Ritter, 1964). Some insects that form leks defend small territories within the lek. Presumably, more females are attracted to this concentration of males, increasing mating success, than to isolated males (Price, 1997). Such mating territories apparently are not related to food or oviposition sites but may maximize attraction of females.

Two grasshopper species, *Ligurotettix coquilletti* and *Bootettix argentatus*, that feed on creosote bush, *Larrea tridentata*, in the deserts of southwestern North America are perhaps the only territorial acridoids (Otte and Joern, 1975; Schowalter and Whitford, 1979). These grasshoppers defend individual creosote bushes. The larger bushes are more likely to harbor females, and opportunities for mating are increased by defending larger shrubs, especially at low grasshopper population densities. Schowalter and Whitford (1979) reported that male movement from small shrubs was greater than movement from larger shrubs, and contests for larger shrubs occurred more frequently. However, fewer males defended territories at high population densities, apparently because interception of females by nonstridulating males and more frequent combat decreased the mating success of territorial defenders.

Males of the speckled wood butterfly, *Pararge aegeria* (Satyridae), defend sunspots on the forest floor, apparently because females are attracted to resources that occur in sunspots (Price, 1997). Only 60% of the males held such territories, but these encountered many more females than did the nonterritorial males that searched for mates in the forest canopy.

Defense of an oviposition site may be advantageous where sperm competition cannot be avoided by anatomical or physiological means, such as with mating plugs that prevent subsequent mating. A butterfly, *Inachis io*, defends territories at the approach to oviposition sites, perhaps because of selective pressure from strong competition at the oviposition sites (Baker, 1972). Other insects, especially the social Hymenoptera, defend nests, foraging trails, or food (Price, 1997).

The benefits of defending food resources or mates must be weighed against the costs of fighting, in terms of time, energy, and risk of injury. Territorial insects may abandon territorial defense at high population densities when time spent fighting detracts from feeding or mating success (Otte and Joern, 1975; Schowalter and Whitford, 1979).

3.4.2 Defensive Behavior

Insects employ a number of strategies to defend themselves against predators. Caddisfly (Trichoptera) larvae construct shelters from benthic debris, but the energy required for construction may limit survival when organic resources are limited (S. Eggert and Wallace, 2003). Mandibulate species frequently bite, and haustellate species may stab with their stylets. Kicking, wing fanning, and buzzing also are effective against some predators (Robinson, 1969; T.K. Wood, 1976). Many species eject or inject toxic or urticating chemicals, as shown in Figs. 3.6 and 3.7. Insects armed with urticating spines or setae often increase the effectiveness of this defense by thrashing body movements that increase contact of the spines or setae with an attacker. Many caterpillars and sawfly larvae rear up and strike like a snake when attacked (Fig. 4.9). This behavior may be enhanced by conspicuous eyespots that increase resemblance to predators (Janzen et al., 2010).

Insects produce a variety of defensive compounds that can deter or injure predators, as described in Chapter 3. Many of these compounds are energetically expensive to produce,

FIGURE 4.9   **Defensive posture of the black swallowtail, *Papilio polyxenes*, caterpillar.** This snake-like posture, together with emission of noxious volatiles from the orange protuberances, deters many would-be predators.

FIGURE 4.10 **Examples of aposematic coloration.** (A) Saint Andrew's cotton stainer bug, *Dysdercus andrea* and (B) some butterflies, *Heliconius* spp., sequester toxins from their host plants and advertise their distasteful or toxic attributes to would-be predators, or mimic species that do.

and may be toxic to the producer as well as to predators, requiring special mechanisms for storage or delivery. Nevertheless, their production sufficiently improves the probability of survival and reproduction to represent a net benefit to the producer (Conner et al., 2000; Sillén-Tullberg, 1985). Such species typically are conspicuously colored (*aposematic*) to facilitate avoidance learning by predators (Fig. 4.10).

Defense conferred by camouflage reduces the energy costs of active defense but may require greater efficiency in foraging or other activities that affect attraction of predators (Schultz, 1983). Insects that rely on resemblance to their background (*crypsis*) must minimize movement to avoid detection (Fig. 4.11). For example, many Hemiptera that are cryptically colored or that resemble thorns or debris are largely sedentary while siphoning plant fluids. Many aquatic insects resemble benthic debris and remain motionless as they filter suspended matter. Cryptic species typically restrict necessary movement to nighttime or acquire their food with minimal movement, especially in the presence of predators (Johansson, 1993). Such insects may escape predators by waiting until a predator is very close before flushing with a startle display, giving the predator insufficient warning to react. However, some birds use tail

FIGURE 4.11 **Examples of cryptic coloration.** (A) Creosote bush grasshopper, *Bootettix argentatus*, in creosote bush, *Larrea tridentata* (New Mexico, USA), (B) lasiocampid moth, *Gastropacha pardalis formosana*, with leaf-mimicking coloration and form (Taiwan), and (C) giant swallowtail, *Papilio cresphontes*, larvae imitating bird feces (southern USA).

fanning or other scare tactics to flush prey from a greater distance and thereby capture prey more efficiently (Galatowitsch and Mumme, 2004; Jabłoński, 1999; Mumme, 2002).

Disruptive and *deceptive coloration* involve color patterns that break up the body form, distract predators from vital body parts, or resemble other predators (Janzen et al., 2010). For example, many insects have distinct bars of color or other patterns that disrupt the outline of the body and inhibit their identification as prey by passing predators. Startle displays enhance the effect of color patterns (Robinson, 1969). The underwing moths (Noctuidae) are noted for their brightly colored hind wings that are hidden at rest by the cryptically colored front wings. When threatened, the moth suddenly exposes the hind wings and has an opportunity to escape its startled attacker. The giant silkworm moths (Saturniidae) and eyed elater, *Alaus oculatus* (Coleoptera: Elateridae), have conspicuous eyespots that make these insects look like birds (especially owls) or reptiles. The eyespots of moths typically are hidden on the hind wings during rest and can be exposed suddenly to startle would-be predators. The margins of the front wings in some saturniids are shaped and colored to resemble the heads of snakes (Fig. 4.12) (Grant and Miller, 1995). Sudden wing movement during escape may enhance the appearance of a striking snake.

Mimicry is resemblance to another, usually venomous or unpalatable, species and typically involves conspicuous, or aposematic, coloration. Mimicry can take two forms, *Batesian* and *Müllerian*. Batesian mimicry is resemblance of a palatable or innocuous species to a threatening species, whereas Müllerian mimicry is resemblance among threatening species. Both are exemplified by insects. A variety of relatively harmless insects (representing several orders) and other arthropods (especially spiders) benefit from resemblance to stinging

FIGURE 4.12   Image of a snake's head on the wing margins of *Attacus atlas*. *Source: From Grant, G.G., Miller, W.E., 1995. Larval images on lepidopteran wings – an unrecognized defense mechanism? Am. Entomol. 41, 44–48.*

FIGURE 4.13 **Batesian mimicry by two insects.** The predaceous asilid fly on the left and its prey, a cerambycid beetle, both display the black and yellow coloration typical of stinging Hymenoptera.

Hymenoptera. For example, clearwing moths (Sesiidae) and some sphingid moths, several cerambycid beetles, and many asilid and syrphid flies resemble bees or wasps (Fig. 4.13). Several harmless butterfly species mimic the toxic pipevine swallowtail, *Battus philenor*. A variety of insect and other species gain protection through adaptations that permit them to mimic ants (Blum, 1980, 1981). Müllerian mimicry is exemplified by sympatric species of Hymenoptera and heliconiid butterflies that sting, or are unpalatable, and resemble each other (A. Brower, 1996; P. Sheppard et al., 1985).

Mimicry systems can be complex, including a number of palatable and unpalatable species and variation in palatability among populations, depending on food source. For example, the resemblance of the viceroy butterfly, *Limenitis archippus*, to the monarch butterfly, *Danaus plexippus*, generally is considered to be an example of Batesian mimicry. However, monarch butterflies show a spectrum of palatability over their geographic range, depending on the quality of their milkweed, and other, hosts (L. Brower et al., 1968). Furthermore, populations of the viceroy and monarch in Florida are equally distasteful (Ritland and Brower, 1991). Therefore, this mimicry system may be Batesian in some locations and Müllerian in others. Conspicuous color patterns and widespread movement of the comodels/mimics maximize exposure to predators and reinforce predator avoidance, providing overall protection against predation.

Sillén-Tullberg (1985) compared predation by great tits, *Parus major*, between normal aposematic (red) and mutant cryptic (grey) nymphs of the seed bug, *Lygaeus equestris*. Both prey forms were equally distasteful. All prey were presented against a grey background. Survival of aposematic nymphs was 6.4-fold higher than for cryptic nymphs because the birds showed a greater initial reluctance to attack, learned avoidance more rapidly, and killed prey less frequently during an attack. The greater individual survival of aposematic nymphs indicated sufficient benefit to explain the evolution of aposematic coloration.

Some insects alert other members of the population to the presence of predators. Alarm pheromones are widespread among insects. These compounds typically are relatively simple hydrocarbons, but more complex terpenoids occur among ants. The venom glands of stinging Hymenoptera frequently include alarm pheromones. Alarm pheromones function either to scatter members of a group when threatened by a predator, or to concentrate attack on the predator, especially among the social insects. A diverse group of ground-dwelling arthropods produce compounds that mimic ant alarm pheromones. These function to scatter attacking ants, allowing the producer to escape (Blum, 1980).

Nevertheless, predation risk carries significant energetic and fitness costs. Culler et al. (2014) found that predation risk in the presence of sunfish caused reduced growth rates and thermal tolerance in damselfly, *Enallagma vesperum*, nymphs, compared to performance in the absence of sunfish. Hawlena and Schmitz (2010) demonstrated that the presence of predators generates sufficient fear among potential prey to induce physiological stress responses that alter carbohydrate metabolism (Fig. 4.14). Stressed prey must divert energy from production to meet the increased demand for maintenance energy. Thus, predators can indirectly affect population foraging and ecosystem processes (Fig. 4.14, see Chapters 8, 11, and 12).

3.4.3 Mutualistic Behavior

Insects participate in a variety of mutualistic interactions, including the well-known pollinator–plant, ant–plant, and wood borer–microorganism associations (see Chapter 8). Typi-

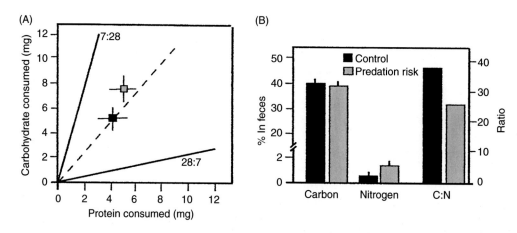

FIGURE 4.14   **Effect of predation risk on nutrient intake and excretion by grasshoppers,** *Melanoplus femurrubrum*. (A) Grasshoppers were presented with a choice of diets containing 7% protein and 28% digestible carbohydrate or 28% protein and 7% digestible carbohydrate. The solid lines represent the nutrient balances in the diets that define the boundaries for potential nutrient intake. The dashed line represents actual nutrient intake by grasshoppers feeding on the two different diets in the presence or absence of predation risk. Grasshoppers free of predation risk tended to ingest a 1:1 ratio of protein and carbohydrate, whereas grasshoppers facing risk consumed significantly greater amounts of dietary C. (B) Grasshoppers excreted similar levels of C, indicating that grasshoppers facing predation risk had higher respiration rates, resulting in excess intake of N that was released in feces and elevated fecal C:N, compared to grasshoppers free from predation risk. Values are mean ± SE. *Source: From Hawlena, D., Schmitz, O.J., 2010. Herbivore physiological response to predation risk and implications for ecosystem nutrient dynamics. Proc. Natl. Acad. Sci. USA 107, 15503–15507.*

cally, mutualism involves diversion of resources by one partner to production of rewards or inducements that maintain mutualistic interactions. Various pollinators and predators exploit resources allocated by plants to production of nectar, domatia, root exudates, etc., and thereby contribute substantially to plant fitness. At the same time, the plant limits the nectar reward in each flower to force pollinators to transport pollen among flowers. During dispersal, bark beetles secrete lipids into mycangia to nourish mutualistic microorganisms that subsequently colonize wood and improve the nutritional suitability of woody substrates for the beetles.

Obviously, the benefit gained from this association must outweigh these energetic and nutritional costs (see Chapter 8). Resources directed to support of mutualists could be allocated to growth and reproduction. These resources may be redirected if the partner is not present (Rickson, 1977), although some species maintain such allocation for long periods in the absence of partners (Janzen and Martin, 1982).

4 EFFICIENCY OF RESOURCE USE

Fitness accrues to organisms to the extent that they survive and produce more offspring than do their competitors. Hence, the efficiency with which assimilated resources are allocated to growth and reproduction determines fitness. However, except for sessile organisms, much of the assimilated energy and material must be allocated to activities pursuant to food acquisition, dispersal, mating, competition, and defense, as described in Section 3. The amount of assimilated resources allocated to these activities reduces relative growth efficiency and reproduction (Schultz, 1983; Zera and Denno, 1997; Zera and Zhao, 2006). Clearly, the diversion of resources from growth and reproduction to these other pathways must represent a net benefit to the insect.

4.1 Factors Affecting Efficiency

Efficiency is affected by a number of constraints on energy and resource allocation. Clearly, selection should favor physiological and behavioral adaptations that improve overall efficiency. However, adaptive strategies reflect the net current result of (trade-offs among) many factors that have variable and interactive effects on survival and reproduction. Hence, individual responses to current conditions vary in efficiency. Whereas physiological, and many behavioral, responses are innate (genetically based, hence relatively inflexible), the capacity to learn can improve efficiency greatly, by reducing the time and resources expended in responding to environmental variation (Cunningham et al., 1998; A. Lewis, 1986).

Hairston et al. (1960) stimulated research on the constraints of food quality on efficiency of herbivore use of resources by postulating that all plant material is equally suitable for herbivores. Just as plant chemical defenses can reduce herbivore efficiency, various animal defenses increase the resource expenditure necessary for predators to capture and assimilate prey. In addition to factors affecting the efficiency of resource acquisition, several factors affect the efficiency of resource allocation, including food quality, size, physiological condition, and learning.

4.1.1 Food Quality

Food quality affects the amount of food required to obtain sufficient nutrition for growth, reproduction, and the energy and nutrients required for detoxification and digestion (see

Chapter 3). Insects feeding on hosts with lower levels of defensive compounds invest fewer energy and nutrient resources in detoxification enzymes or continued searching behavior than do insects feeding on better-defended hosts. Herbivores process much indigestible plant material, especially cellulose, whereas predators process animal material that generally is consistent with their needs. Accordingly, we might expect higher assimilation efficiencies for predators than for herbivores (G. Turner, 1970). Although indigestible and toxic compounds in plant tissues reduce assimilation efficiency for herbivores (Scriber and Slansky, 1981), toxins sequestered or produced by prey also reduce assimilation efficiency of predators. However, few studies have addressed the effect of toxic prey on the assimilation efficiency of predators (L. Dyer, 1995; Stamp et al., 1997; Stephens and Krebs, 1986).

Insects may ingest relatively more food to obtain sufficient nutrients or energy to offset the costs of detoxification or avoidance of plant defensive chemicals. Among herbivores, species that feed on mature tree leaves have relative growth rates that are generally half the values for species that feed on forbs, because tree leaves are poor food resources compared to forbs (Scriber and Slansky, 1981). Although specialists might be expected to feed more efficiently on their hosts than do generalists, Futuyma and Wasserman (1980) reported that a specialist (the eastern tent caterpillar, *Malacosoma americanum*) had no greater assimilation or growth efficiencies than did a generalist (the forest tent caterpillar, *M. disstria*). Some wood-boring insects may require long periods (several years to decades) of larval feeding to concentrate nutrients (especially N and P) sufficient to complete development.

4.1.2 Size and Physiological Condition

Body size is a major factor affecting efficiency of energy use. Larger organisms have greater energy requirements than do smaller organisms but smaller organisms with larger surface area/volume ratios are more vulnerable to heat loss than are larger organisms. Accordingly, maintenance energy expenditure per unit body mass decreases with increasing body size (Phillipson, 1981). Larger organisms are more powerful, can move greater distances, have an advantage in competitive or predator-prey interactions and tend to use energy more efficiently during foraging and interactions with other organisms (J. Davis et al., 2015; Ernsting and van der Werf, 1988; Heinrich, 1979; Phillipson, 1981; Streams, 1994).

Physiological condition, including the general vigor of the insect as affected by parasites, also influences food requirements and assimilation efficiency. For example, hunger may induce increased effort to gain resources that would be ignored by less desperate individuals (Ernsting and van der Werf, 1988; Holling, 1965; Iwasaki, 1990, 1991; Richter, 1990; Streams, 1994). Slansky (1978) reported that cabbage white butterfly larvae parasitized by *Apanteles glomeratus* (Hymenoptera) increased food consumption, growth rate, and nitrogen assimilation efficiency. Schowalter and Crossley (1982) found that Madagascar hissing cockroaches, *Gromphadorhina portentosa*, with associated mites, *Androlaelaps schaeferi*, had a significantly more rapid egestion rate than did cockroaches with mites excluded, although assimilation efficiency did not differ significantly between mite-infested and mite-free cockroaches (Fig. 4.15).

4.1.3 Learning

Learning is a powerful tool for improving efficiency of resource use (see Chapter 3). Learning reduces the effort wasted in unsuccessful trials (Fig. 3.18). Learning to distinguish appropriate from inappropriate prey (eg, search image), to respond to cues associated with earlier

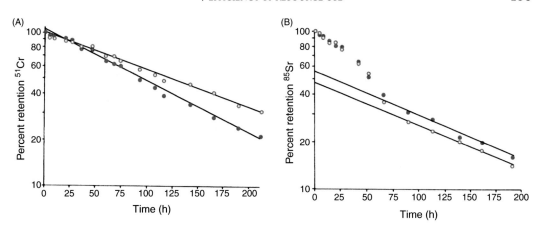

FIGURE 4.15 Bioelimination of ⁵¹Cr (A) and ⁸⁵Sr (B) by the cockroach, *Gromphadorhina portentosa* with *(solid circles)* and without *(open circles)* the associated mite, *Androlaelaps schaeferi.* ⁵¹Cr has no biological function and its elimination represents egestion; ⁸⁵Sr is an analogue of Ca and its elimination represents both egestion (regression lines similar to those for ⁵¹Cr) and excretion of the assimilated isotope (rapid initial loss). This insect appears to assimilate and begin excreting nutrients before gut passage of unassimilated nutrients is complete. *Source: From Schowalter, T.D., Crossley, Jr., D.A., 1982. Bioelimination of ⁵¹Cr and ⁸⁵Sr by cockroaches,* Gromphadorhina portentosa *(Orthoptera: Blaberidae), as affected by mites,* Gromphadorholaelaps schaeferi *(Parasitiformes: Laelapidae). Ann. Entomol. Soc. Am. 75, 158–160.*

success, and to improve foraging technique greatly increases the efficiency of energy and nutrient acquisition (Cunningham et al., 1998; Ishii and Shimada, 2010). Learning has been demonstrated by a wide variety of insects, and memory can be retained through metamorphosis in holometabolous insects (R. Davis, 2005; Dukas, 2008; Gutiérrez-Ibáñez et al., 2007; Kawecki, 2010, see Chapter 3). Honey bees represent the epitome of resource utilization efficiency among insects through their ability to communicate foraging success and location of nectar resources to nestmates (F. Dyer, 2002; J. Gould and Towne, 1988; Heinrich, 1979; von Frisch, 1967).

4.2 Trade-Offs

Allocation efficiency often is optimized by adaptations that generally tailor insect life histories, physiology or behavior to prevailing environmental conditions or resource availability. For example, synchronization of life histories with periods of suitable climatic conditions and food availability reduces the energy required for thermoregulation or search activity. Bumble bee, *Bombus* spp., anatomy optimizes heat retention during foraging in cool temperate and arctic habitats (Heinrich, 1979). Davison (1987) compared the energetics of two harvester ant species, *Chelaner rothsteini* and *C. whitei*, in Australia and found that the smaller *C. rothsteini* had lower assimilation efficiency but higher production efficiency (largely in production of offspring) than did the larger *C. whitei*. *C. rothsteini* discontinued activity during the winter, perhaps to avoid excessive metabolic heat loss, whereas *C. whitei* remained active all year. Zera and Zhao (2006) found that flight-capable morphs of the cricket, *G. firmus*, converted ingested glycine into triglycerides to fuel flight, at the expense of use in reproduction, whereas flight-incapable morphs incorporated glycine into ovarian protein and reproduction, at the expense of wing development and flight. Krams et al. (2015) demonstrated that yellow meal-worm, *Tenebrio molitor*, invested more resources in reproduction, at the expense of immune response and survival, when food resources were limited but not when food resources were unlimited.

Selection should favor individuals and species that acquire and allocate resources most efficiently. However, as discussed in Section 3, many allocation pathways compete for as-similated resources. Plant production of defenses reflects trade-offs with growth and wound repair. Similarly, insects face trade-offs in energy and nutrient allocation to various activities that interact to determine fitness. Survival of individuals and species represents the net result of various traits that often conflict (Carrière et al., 1997).

Trade-offs occur among steps in resource acquisition. Leaf fragments collected by leaf-cutting ants, *Atta columbica*, were smaller than predicted for maximum foraging efficiency (Burd and Howard, 2005). By examining leaf processing within ant nests, Burd and Howard (2005) found that the time required for below-ground processing greatly exceeded the time required to move leaf fragments from the vegetation source and that leaf fragment size strongly affected the time required for below-ground processing. Thus, selection for smaller loads by foragers may have evolved to optimize the below-ground time for processing leaf fragments.

Heinrich (1979) evaluated the trade-offs among various allocation strategies seen among bees. Some bee species begin producing queens and drones (offspring) concurrent with col-ony development (ie, production of combs and workers), whereas other bee species achieve large colony sizes before producing queens and drones. The first strategy yields immediate, but small, returns because of the competing activities of workers, and the second strategy yields no immediate returns but eventually yields much larger returns. In addition, workers must weigh the cost of foraging from particular flowers against the expected nectar returns, especially at low temperatures when nectar return must be at least sufficient to maintain high thoracic temperatures necessary for continued foraging. Because different flowers provide different amounts of nectar, bees tend to forage at flowers with high yields over a range of temperatures, but visit flowers with small nectar rewards only at high temperatures. Simi-larly, bees must weigh the benefits of foraging at various distances from the colony. Bees will fly several km, given adequate floral rewards, but respond quickly to indications of declining nectar availability, for example, they leave an inflorescence or patch after encountering empty flowers.

Even learning, which increases the efficiency of performing learned tasks, has measurable fitness costs (Kawecki, 2010). Mery and Kawecki (2005) demonstrated that *Drosophila melano-gaster* conditioned to form long-term memory through associative learning died earlier when deprived of food and water than did flies subjected to other conditioning treatments. Burger et al. (2008) found that females conditioned to form long-term memory showed a 15% shorter life span and a faster decline in fecundity with age, compared to control flies. Furthermore, flies experimentally selected for delayed aging showed poor ability to learn, compared to control populations. These studies clearly indicated significant fitness costs for formation and maintenance of long-term memory.

Several studies have demonstrated trade-offs between life stages. Female pierid butterflies oviposited preferentially on more conspicuous hosts, a more energetically efficient search strategy for the adult, but these hosts were less suitable for larval development than were less conspicuous hosts (Courtney, 1985, 1986). Similarly, females of the noctuid moth, *Autographa precationis*, preferentially oviposit on soybeans, rather than on dandelions, perhaps because the shape of dandelions is a less effective oviposition stimulus, although larvae show a marked feeding preference for dandelions (Kogan, 1975). Hoekman et al. (2007) reported that female pitcher-plant mosquitoes, *Wyeomyia smithii*, showed higher oviposition rates in pitcher plant, *Sarracenia purpurea*, leaves with low levels of experimental enrichment (0 or 2 dead ants), compared to a high rate of enrichment (20 dead ants), whereas larval development was higher at the high rate of enrichment.

Interactions with other organisms also involve trade-offs. Males that continue to defend territories when the time or energy spent on this activity interferes with mating and reproduction are less likely to contribute to the genetic composition of the next generation than are males that sacrifice territorial defense for mating opportunities under such conditions (Otte and Joern, 1975; Schowalter and Whitford, 1979). Predator avoidance behavior reduces time spent feeding (Folz et al., 2006; M.L. Johnson et al., 2006). Matsuda et al. (1993) modeled the effects of multiple predators on antipredator defenses and concluded that increasing defense against one predator comes at the expense of defenses against another. Tegelaar et al. (2013) described trade-offs made by aphids to enhance protection by ants. Initially, tended aphids showed lower colony weight and embryo size, compared to untended aphids, reflecting investment in greater honeydew excretion to attract ants. Once the tending relationship was established, the differences between tended and untended aphids disappeared. Predators learning to focus on the search image of the most frequently encountered cryptic prey can improve the detection and efficiency of foraging on more common prey, but may overlook rare prey types that do not match the search image (Ishii and Shimada, 2010). Frequent encounters with predators, especially when combined with low availability of food resources, may restrict the time an individual can spend foraging and increase the expenditure of energy to avoid predators, lowering net energy acquisition and potentially leading to inadequate energy balance for survival.

Heterogeneous habitats force many herbivores and predators to expend energy searching for scattered resources. Many individuals will be unable to maintain energy or nutrient balance under such conditions. Huberty and Denno (2006) compared differences in flight and feeding capability in two wing dimorphic–phloem feeding planthoppers to evaluate trade-offs between ingestion and dispersal capability. The more sedentary *Prokelisia dolus* allocates more resources to feeding whereas *P. marginata* allocates more resources to flight. Brachypterous forms of both species invest more energy in feeding musculature and less in flight than do macropterous forms. As a result, *P. dolus* is better adapted to increase ingestion to deal with deteriorating host quality than is *P. marginata*, which must disperse to find more suitable hosts. By contrast, abundant suitable resources reduce costs of searching for, or detoxifying, resources and facilitate maintenance of energy and nutrient budgets.

Environmental changes, especially rapid changes occurring as a result of anthropogenic activities, will change the balance among these trade-offs, affecting fitness in various ways. Papaj et al. (2007) reported that rates at which female pipevine swallowtails discovered their host plants, *Aristolochia watsoni*, declined following summer monsoon rains in Arizona, due to difficulty in finding hosts amid the flush of nonhost vegetation, but that rates at which eggs were laid on hosts increased (Fig. 4.16). These results indicated that oviposition was limited more by low host quality before the summer monsoon than by the low rate of host discovery after the monsoon, suggesting that the predicted increase in droughts for this region will

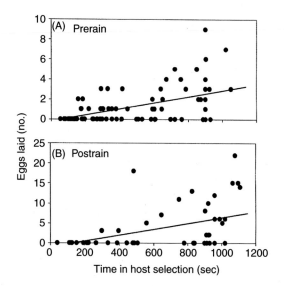

FIGURE 4.16 Relationship between total number of eggs laid in an observation period and time engaged in host selection for pipevine swallowtail females (A) prior to onset of monsoon rains (*n* = 86) and (B) after subsidence of monsoon rains (*n* = 52). Lines indicate fit to ANCOVA model. Note different scales on the *y* axes of the two plots. *Source: From Papaj, D.R., et al., 2007. Extreme weather change and the dynamics of oviposition behavior in the pipevine swallowtail,* Battus philenor. *Oecologia 152, 365–375.*

reduce reproduction by this species. Warmer global temperatures may improve energy balance for some Arctic or Antarctic species but increase respiration loss or time spent seeking shade for other species. Ecosystem fragmentation will require greater energy expenditure for sufficient foraging and dispersal, thereby impeding movement of intolerant species over inhospitable landscapes. Some species will benefit from changes that improve overall performance (eg, survival and reproduction) whereas other species will decline or disappear.

5 SUMMARY

Acquired resources are allocated to various pathways. First, they either are assimilated or egested. Assimilated resources either are allocated to production or are expended through respiration. Consumption and allocation of resources are influenced by insect size, maturity, food quality, and parasitism. Fitness accrues to the extent that assimilated resources are sufficient for growth and reproduction. However, insect allocation patterns represent trade-offs among competing requirements of growth, reproduction, and activities necessary for food acquisition, mating, reproduction, and interactions with other organisms.

Species persist to the extent that the benefits of these behaviors outweigh the costs, that is, survival and reproduction are increased by the investment of energy in particular behavior and associated biochemicals. Foraging and reproductive behaviors should provide the best return for the necessary expenditure of time and effort. Reproductive behavior should maximize survival of offspring. Among insects, selection of appropriate oviposition sites determines egg development and survival. Brood care is well represented among insects, with examples ranging from protection of young, to provision of food resources, to development of complex social systems for brood care and colony maintenance. However, efficiency of adult behaviors may be in conflict with efficiency of juvenile behaviors. For example, adults may oviposit on the most easily found hosts, whereas survival of immatures may depend on discovery of more suitable, but rare, food hosts.

Competition and defense against predators often involve considerable expenditure of resources. In many species, males engage in various forms of combat to decide which males mate successfully. Territorial behavior is characterized by both the marking of territorial boundaries and the defense of the territory against intruders. Defense of territories may maximize access to food or mates at low population densities, but becomes less advantageous and may be abandoned at high population densities. Insects defend themselves against predators physically and chemically. Behavior often enhances the effectiveness of protective coloration or toxins. For example, cryptically colored insects typically avoid movement during times when predators are active, whereas other insects may suddenly expose eyespots or brightly colored body parts to startle an attacker. Some insects imitate snakes or other predators through color patterns or movements. Such strategies minimize the energetic cost of physical defense but require greater efficiency in foraging or reproductive movements to avoid detection.

The efficiency of foraging, reproductive, competitive, and defensive behavior may be increased by use of visual, chemical, or acoustic signals that communicate information to recipients. Insects can improve foraging and mating efficiency by orienting toward chemicals produced by suitable resources or potential mates. Discovery of a potential mate initiates a courtship ritual that improves fitness by ensuring species recognition and receptivity. Competition for food or mates can be minimized by signals that deter other individuals.

Environmental changes will affect the efficiency of resource acquisition and allocation strategies. Global warming will improve energy balance for some species (eg, early-season or high-latitude pollinators) but increase respiration costs beyond the ability to acquire energy and nutrients for others. Ultimately, insect strategies for acquiring and allocating energy and nutrient resources affect community interactions, energy flow, and nutrient cycling processes.

POPULATION ECOLOGY

A population is a group of interbreeding members of a species and represents a fundamental unit of evolution and ecology. Individual fitnesses are integrated at the population level to determine species abundance and persistence or extinction. This is the level of ecological organization that is the focus of evolutionary ecology, ecological genetics, biogeography, sampling methodology, pest management, and recovery of endangered species. These disciplines all have contributed enormously to our understanding of population-level phenomena.

A number of more or less discrete subpopulations may be distributed over the geographic range of a species population. Movement of individuals among these *demes* (composing a *metapopulation*) and newly available resources compensates for local extinctions resulting from disturbances or biotic interactions (Hanski and Gilpin, 1997). Populations are characterized by structural attributes, such as density, dispersion pattern, and age, sex, and genetic composition (Chapter 5) that change through time (Chapter 6) and space (Chapter 7) as a result of responses to changing environmental conditions.

Population structure and dynamics of insects have been the subject of much ecological research. Examination of population structure and process rates can provide clues to invasiveness or conservation status in changing habitats. Population structure also can indicate future population trends. Predicting outbreaks requires understanding of relationships between population structure and changes in population size. Rapidly growing populations typically are characterized by a high production of offspring per adult. Some growing populations also show changes in geographic distribution of individuals.

Abundance of many insects can change by orders of magnitude on very short timescales, because of their small size and rapid reproductive rates. Such rapid and dramatic change in abundance in response to often subtle environmental changes facilitates statistical evaluation of population response to environmental factors and makes insects useful indicators of environmental change. The reproductive capacity of many insects enables them to colonize new habitats and exploit favorable conditions or new resources quickly. However, their small size, short life span, and dependence on chemical communication to find mates at low densities limit persistence of small or local populations during periods of adverse conditions, frequently leading to local extinction.

Population dynamics reflect the net effects of differences among individuals in their physiological and behavioral interactions with the environment (see Chapter 2). Changes in proportions of individuals that are successful in finding and exploiting resources, mating and reproducing, and avoiding mortality agents (Chapters 3 and 4) determine numbers of individuals, their spatial distribution, and genetic composition at any point in time. Population structure is a component of the environment for the members of the population and provides information that affects individual physiology and behavior, hence fitness (see Section I). For example, population density affects competition for food and oviposition sites (as well as other resources), the propensity of individuals to disperse, and the proximity of potential mates.

Population structure and dynamics also affect community structure and ecosystem processes (Sections III and IV). Each population constitutes a part of the environment for other populations in the community. Changes in abundance of any one species population affect the population(s) on which it feeds and population(s) which prey on, or compete with, it (Chapter 8). Changes in the size of any population also affect the importance of its ecological functions. A decline in pollinator abundance will reduce fertilization and seed production of host plants, thereby affecting aspects of nutrient uptake and primary productivity (Chapter 13). An increase in phytophage abundance can increase canopy "porosity", increasing light penetration, and increasing fluxes of energy, water, and nutrients to the soil (Chapter 12). A decline in predator abundance will release prey populations from regulation and contribute to increased exploitation of the prey's resources (Chapter 8). A decline in detritivore abundance can reduce decomposition rate and lead to bottlenecks in biogeochemical cycling that affect nutrient availability (Chapter 14).

Population structure across landscapes also influences source–sink relationships that determine population viability and ability to recolonize patches following disturbances. For example, the size and distribution of demes determine their ability to maintain gene flow or to diverge into separate species. Distribution of demes determines the source(s) and initial genetic composition of colonists arriving at a new habitat patch. These population attributes are critical to protection or restoration of rare or endangered species. Isolation of demes as a result of habitat fragmentation can reduce their ability to reestablish local demes and lead to permanent changes in community structure and ecosystem processes across landscapes.

5

Population Systems

SHIFT IN POPULATION STATE OF DESERT LOCUSTS

Population density often exhibits threshold levels at which a population undergoes a rapid transition to a new state, such as irruption or extinction. Small populations (solitary phase) of the desert locust, *Schistocerca gregaria*, are characterized by scattered nymphs and adults feeding and moving independently. However, when environmental conditions favor high rates of reproduction, crowding increases levels of serotonin in the thoracic ganglia, causing nymphs to undergo changes in color and form (gregarious phase) and aggregate into small groups (Anstey et al., 2009). Additional changes in gene expression accompany this phase shift (Kang et al., 2004; Pener and Simpson, 2009).

Small groups can coalesce into marching bands of hundreds to millions that cover up to several square kilometers (20–120 nymphs m^{-2}), move several kilometers per day, and consume all vegetation in their path. Adults developing from these marching nymphs have longer wings than do adults developing from solitary nymphs and form enormous swarms of millions to billions of locusts, covering thousands of hectares and moving thousands of kilometers (Gunn and Brown, 1979; Rainey, 1963; Rainey and Sayer, 1953; Uvarov, 1954), consuming all vegetation (eg, 10^8 kg, Rainey, 1963) over large areas. Various pheromones appear to cause individuals and groups to remain near each other rather than disperse (Pener and Simpson, 2009). Major outbreaks occur only when small groups remain cohesive, move into neighboring habitat, and merge with other groups to form a coordinated band. If such cohesive movement fails to occur, the nymphs disband and return to the solitary phase.

Buhl et al. (2006) studied the behavior of nymphs at various controlled densities and discovered a low incidence of alignment among nymphs at densities < 20 m^{-2} and long periods of collective motion but rapid spontaneous changes in direction at densities of 25–60 m^{-2}. At densities > 75 m^{-2} nymphs adopted a common and persistent direction, and spontaneous changes in direction no longer occurred within the 8-h observation period, demonstrating a critical density threshold underlying population transition to an irruptive state.

Entomopathogens may facilitate return to the solitary phase. Elliot et al. (2003, 2005) reported that gregarious phase locusts infected with *Metarhizium anisopliae* showed reduced flight capacity and competitiveness for mates, produced more solitary phase (rather than gregarious phase) offspring, and suffered a high rate of mortality, compared to uninfected locusts. These combined effects of infection may return the population to its solitary phase, which would limit further disease transmission.

Insect Ecology. http://dx.doi.org/10.1016/B978-0-12-803033-2.00005-4

1 INTRODUCTION

The variables that determine the abundance and distribution of a population, in time and space, constitute a population system (Berryman, 1981; Franklin, 2010; Krebs, 2008). The basic elements of this system are the individual members of the population, variables describing population size and structure, processes that affect population size and structure, and the environment. These elements of the population system largely determine the capacity of the population to increase in size and maintain itself within a shifting landscape mosaic of habitable patches and resources. This chapter summarizes these population variables and processes, their integration in life history strategies, and their contribution to change in population size (Chapter 6) and distribution (Chapter 7).

2 POPULATION STRUCTURE

Population structure reflects several variables that describe the number, age, sex and genetic composition, and spatial distribution of individuals. Population variables reflect life history and the physiological and behavioral attributes that dictate habitat preferences, home ranges, oviposition patterns, and affinity for other members of the population.

2.1 Density

Population density is the number of individuals per unit geographic area, for example, number per square meter, per hectare, or per square kilometer. This variable affects a number of other population variables. For example, mean density determines the likelihood of finding mates, hence population viability, and propensity to disperse, hence the probability of colonizing vacant habitat patches. Density also affects population dispersion pattern (see Section 2.2) and the behavior of swarming species. Buhl et al. (2006) reported that at a threshold density of 74 desert locust, *Schistocerca gregaria*, nymphs square meter, individual movements became consistently coordinated and directional (see the box). This mass marching behavior is a precursor to adult swarms of this insect.

A related measure, population intensity, is commonly used to describe insect population structure. Intensity is the number of individuals per habitat unit, such as number per leaf, per unit branch length, per square meter leaf area or bark surface, per kg foliage or wood, etc. Mean intensity indicates the degree of resource exploitation, competition for space, food, or mates, and magnitude of the effect on ecosystem processes. Intensity measures often can be converted to density measures if the density of habitat units is known (Southwood, 1978).

Densities and intensities of insect populations vary widely. Bark beetles, for example, often appear to be absent from a landscape (very low density) but, with sufficient examination, can be found at high intensities on widely scattered injured or diseased trees or in the dying tops of trees (Cairns et al., 2008). Under favorable conditions of climate and host abundance and condition, populations of these beetles can reach sizes of up to 10^5 individuals per tree over areas as large as 10^7 ha (Coulson, 1979; Furniss and Carolin, 1977). Schell and Lockwood

(1997) reported that grasshopper population densities can increase by an order of magnitude over areas of several thousand hectares within 1 year.

2.2 Dispersion

Dispersion is the spatial pattern of distribution of individuals. Dispersion is an important population characteristic that affects spatial patterns of resource use and population effect on community and ecosystem structure and function. Dispersion pattern can be regular, random, or aggregated.

A regular (uniform) dispersion pattern results from individuals spacing themselves at consistent intervals within the habitat. This dispersion pattern is typical of species that contest resource use, especially territorial species. For example, bark beetles attacking a tree show a regular dispersion pattern (Fig. 5.1). Similarly, ant lion larvae tend to space their pits, minimizing interference to ant (prey) movement among pits (Swenson et al., 2007). Such spacing reduces competition for resources. From a sampling perspective, the occurrence of one individual in a sample unit reduces the probability that other individuals will occur in the same sample unit. Variability in mean density is low, and sample densities tend to be normally distributed. Hence, regularly dispersed populations are most easily monitored because a relatively small number of samples provides estimates of mean and variance in population density that are similar to those from a larger number of samples.

In a randomly dispersed population, individuals neither space themselves apart nor are attracted to each other. The occurrence of one individual in a sample unit has no effect on the probability that other individuals will occur in the same sample unit (Fig. 5.1). Sample densities show a skewed (Poisson) distribution. Most populations are dispersed randomly.

Aggregated (or clumped) dispersion results from grouping behavior or preference for particular habitat patches. Aggregation is typical of species that occur in herds, flocks, schools, etc. (Fig. 5.1), for enhancement of resource exploitation or protection from predators (Sword et al., 2005, see Chapters 3 and 4). Gregarious sawfly larvae and tent caterpillars are examples of aggregated dispersion resulting from a tendency of individuals to form groups (Fig. 2.12). Filter feeding aquatic insects tend to be aggregated in riffles or other zones of higher flow rate within the stream continuum (Fig. 2.14), whereas predators that hide in benthic detritus, such

FIGURE 5.1 **Dispersion patterns and their frequency distributions.** (A) Regular dispersion of Douglas-fir beetle, *Dendroctonus pseudotsugae*, entrances (marked by the small piles of reddish phloem fragments) through bark on a fallen Douglas-fir tree, (B) random dispersion of aphids on an oak leaf, and (C) aggregated dispersion of forest tent caterpillars, *Malacosoma disstria*.

as dragonfly larvae or water scorpions, are aggregated in pools, due to their habitat preferences. Aphids may be aggregated as a result of rapid, parthenogenic reproduction, as well as host and habitat preferences. Massonnet et al. (2002) found that the aphid, *Macrosiphoniella tanacetaria*, a specialist on tansy, *Tanacetum vulgare*, can be aggregated at the level of individual shoots, plants, and sites.

For sampling purposes, the occurrence of an individual in a sample unit increases the probability that additional individuals occur in that sample unit. Sample densities are distributed as a negative binomial function, and variance tends to be high. Populations with this dispersion pattern require the greatest number of samples and attention to experimental design. A large number of samples is necessary to minimize the obviously high variance in numbers of individuals among sample units and to ensure adequate representation of aggregations. A stratified experimental design can facilitate adequate representation with smaller sample sizes if the distribution of aggregations among different habitat types is known.

Dispersion pattern can change during insect development, during change in population density, or across spatial scales. For example, larval stages of tent caterpillars and gregarious sawflies are aggregated at the plant branch level, but adults are randomly dispersed at this scale (Fitzgerald, 1995; McCullough and Wagner, 1993). Nam et al. (2013) reported that the ambrosia beetle, *Platypus koryoensis*, showed an aggregated distribution pattern at low population densities, when beetles were concentrated in scattered trees within stands. As beetle population densities increased, the spatial distribution of infested trees became random, perhaps due to beetle dispersal from colonized trees or avoidance of mass-attacked trees. At high population densities dispersion pattern again became aggregated, perhaps as a result of concentrated colonization in susceptible portions of the stand. Many host-specific insects are aggregated on particular hosts in diverse communities but are more regularly or randomly dispersed in more homogeneous communities dominated by hosts. Some insects, such as the western ladybird beetle, *Hippodamia convergens*, aggregate for overwintering purposes and redisperse in the spring. Aphids are randomly dispersed at low population densities, but become more aggregated as scattered colonies increase in size (Dixon, 1985). Bark beetles show a regular dispersion pattern on a tree bole, due to spacing behavior, but are aggregated on injured or diseased trees (Coulson, 1979) or at the landscape scale in patches of suitable hosts (Costa et al., 2013).

Dispersion pattern affects population interactions with other species (see Chapter 9). For example, aggregations represent more concentrated use of resources than do nonaggregations. Conversely, aggregations can become targets for concentrated predator foraging (Koenig and Liebhold, 2005), but also can increase the difficulty of focusing on an individual prey for capture (Sword et al., 2005).

2.3 Geographic Range and Metapopulation Structure

The area occupied by a species population is its geographic range. Typically the boundaries of this area are defined by species tolerances to gradients of temperature and precipitation, habitat structure, and resource conditions (Franklin, 2010; Krebs, 2008; Sheldon and Tewksbury, 2014; Waide and Willig, 2012). Range boundaries can move as a result of environmental changes, disturbances, or population growth (see Chapters 2 and 7). However, populations often become increasingly limited near their boundaries as less suitable environmental or resource conditions increase the costs of metabolic trade-offs (see Chapter 4) and vulnerability to disease, parasitism, or predation. Furthermore, suitable conditions are not distributed uniformly across landscapes. Consequently, species populations tend to be concentrated or restricted to landscape units (patches) encompassing suitable habitat and resource conditions within the broader geographic range (Franklin, 2010; Krebs, 2008; Waide and Willig, 2012).

The irregular distribution of many populations across landscapes creates a pattern of relatively distinct (often isolated) local demes (aggregations) that compose the greater metapopulation (Hanski and Gilpin, 1997). Insect species characterizing discrete habitat types often are dispersed as relatively distinct isolated demes as a result of environmental gradients or disturbances that affect the distribution of habitat types across the landscape. Obvious examples include insects associated with lotic or high elevation ecosystems. Populations of insects associated with ponds or lakes show a dispersion pattern reflecting dispersion of their habitat units. Demes of lotic species are more isolated in desert ecosystems than in mesic ecosystems. Populations of western spruce budworm, *Choristoneura occidentalis*, and fir engraver beetle, *Scolytus ventralis*, historically occurred in western North America in relatively isolated high elevation and riparian fir forests separated by more xeric patches of pine forest (Wickman, 1992). Isolated subpopulations have expanded and merged over larger areas as their host conifers invaded lower elevations in the wake of fire suppression (Schowalter, 2008). Many monophagous species show metapopulation structures associated with the discontinuous distributions of their host plants (St. Pierre et al., 2005).

Metapopulations typically are composed of demes of various sizes, reflecting the size and/or quality of habitat patches. For example, Leisnham and Jamieson (2002) found that demes of mountain stone weta, *Hemideina maori*, which shelter under rocks on isolated rock outcrops (tors) in alpine habitats in southern New Zealand, ranged in size from 0 to 6 adults on tors with 1–12 rocks to 15–40 adults on tors with 30–40 rocks. Small tors were more likely to experience extinction events (4 of 14 small tors experienced at least one extinction during the 3-year study) than were large tors (no extinction events during the study).

Population structure among suitable patches is influenced strongly by the matrix of patch types. Haynes and Cronin (2003) studied the distribution of planthoppers, *Prokelisia crocea*, among discrete patches of prairie cordgrass, *Spartina pectinata*, as affected by surrounding mudflat, native nonhost grasses, or exotic smooth brome, *Bromus inermis*. Planthoppers were released into experimental cordgrass patches constructed to be identical in size (about 24 × 24 cm), isolation (> 25 m from natural cordgrass patches), and host plant quality. Within patches, planthopper density was higher against mudflat edges (which restricted dispersal) relative to patch interior, but not against nonhost patches. Among patches, density increased with increasing proportion of the surrounding matrix being composed of mudflat. The influence of matrix composition was equal to the influence of patch size and isolation in explaining planthopper distribution.

Population distribution and degree of isolation among local demes affect gene structure and viability of the metapopulation (Pecsenye et al., 2014). If local demes become too isolated, they become inbred and may lose their ability to recolonize habitable patches following local extinction (Hedrick and Gilpin, 1997). As human activities increasingly fragment natural ecosystems, local demes become smaller and isolated more rapidly than greater dispersal (or adaptive) ability can evolve. Smaller demes tend to have lower genetic variation, produce fewer potential colonists, and greater distances between suitable habitats further limit colonization potential. In the absence of colonization of suitable habitats or adaptation to inferior habitats, species extinction becomes more likely (R. Holt and Barfield, 2010). These effects of fragmentation could be exacerbated by climate change. For example, a warming climate will push high-elevation ecosystems into smaller areas on mountaintops, and some mountaintop ecosystems will disappear (Fig. 5.2) (Franklin et al., 1992; D. Williams and Liebhold, 2002). Rubenstein (1992) showed that individual tolerances to temperature changes could affect range changes by insects under warming climate scenarios. A species with a linear response to temperature could extend its range to higher latitudes (provided that expansion is not limited by habitat fragmentation) without reducing its current habitat. Conversely, a species with a dome-shaped response to temperature could extend into higher latitudes but would be forced to retreat from lower latitudes that become too warm. If the pathway for range adjustment

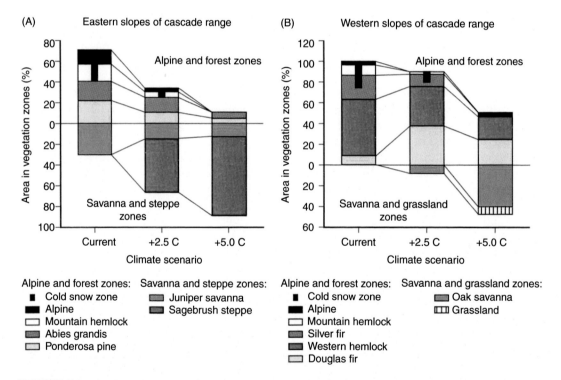

FIGURE 5.2 **Changes in the percent area in major vegetation zones on the eastern (A) and western (B) slopes of the Cascade Range in Oregon as a result of temperature increases of 2.5°C and 5°C.** Major changes are predicted in elevational boundaries and total area occupied by vegetation zones under these global climate change scenarios. Vegetation zones occupying higher elevations will decrease in area or disappear as a result of the smaller conical surface at higher elevations. Other species associated with vegetation zones also will become more or less abundant. *Source: From Franklin, J.F., et al., 1992. Effects of global climatic change on forests in northwestern North America. In: Peters, R.L., Lovejoy, T.E. (Eds.), Global Warming and Biological Diversity. Yale University Press, New Haven, CT, pp. 244–257.*

for this species were blocked by unsuitable habitat, it would face extinction. Estay et al. (2014) reached similar conclusions (Fig. 2.17). Metapopulation dynamics are discussed in more detail in Chapter 7.

2.4 Age Structure

Age structure reflects the proportions of individuals at different life stages. This variable is an important indicator of population status. Growing populations generally have larger proportions of individuals in younger age classes, whereas declining populations typically have smaller proportions of individuals in these age classes. Stable populations typically have relatively more individuals in reproductive age classes. However, populations with larger proportions of individuals in younger age classes also may reflect low survivorship in these age classes, whereas populations with smaller proportions of individuals in younger age classes may reflect high survivorship (see Section 3.2).

For most insect species, life spans are short (usually ≤ 1 year) and revolve around seasonal patterns of temperature and rainfall. Oviposition typically is timed to ensure that feeding stages coincide with the most favorable seasons and that diapausing stages occur during unfavorable seasons, for example, winter in temperate regions and the dry season in tropical and arid regions. Adults typically die after reproducing. Most temperate species have discrete, annual generations, whereas tropical species are more likely to have overlapping generations.

Periodical cicadas, *Magicicada* spp., represent a major exception. Distinct broods of 13- and 17-year periodical cicadas emerge as adults following 13- or 17-year developmental periods underground. Y. Tanaka et al. (2009) demonstrated that synchronization of prime-numbered life spans among these cicadas could be explained by the lower likelihood of hybridization with other cyclic patterns and their increased likelihood of persistence and selection under variable environmental conditions that could lead to extinction of low-density populations (see Chapter 6). Emergence densities of these insects can exceed 100 m^{-2} and represent an important resource for predators (Koenig and Liebhold, 2005; Whiles et al., 2001; Whitford and Jackson, 2007).

2.5 Sex Ratio

The proportion of females indicates the reproductive potential of a population. Sex ratio also reflects a number of life history traits, such as the importance of sexual reproduction, mating system, and ability to exploit harsh or ephemeral habitats (Pianka, 1974).

A 50:50 sex ratio generally indicates equally important roles of males and females, given that selection would minimize the less productive sex, and strong selection for sexual reproduction. The sex ratio approaches 50:50 in species where males select resources, protect or feed females, or contribute necessary genetic variability. This sex ratio maximizes the availability of males to females, and hence maximizes genetic heterogeneity. High genetic heterogeneity is particularly important for population survival in heterogeneous environments. However, when the sexes are equally abundant, only half of the population is capable of producing offspring, but all compete for resources. By contrast, a parthenogenetic population (with no males) has little or no genetic heterogeneity, but the entire population is capable of producing offspring. Parthenogenetic individuals can disperse and colonize new resources without the additional challenge of finding mates, and successful colonists can generate large population sizes rapidly, ensuring exploitation of suitable resources and large numbers of dispersants in the next generation, improving the probability of survival in ephemeral environments.

Sex ratio can be affected by environmental factors. For example, haploid males of many insect species are more sensitive to environmental variation than are diploid females, and greater mortality to haploid males may speed adaptation to changing conditions, by quickly eliminating deleterious genes (Edmunds and Alstad, 1985; J. Peterson and Merrell, 1983). Some endosymbionts also affect sex ratio. Himler et al. (2011) found that *Rickettsia*-infected whiteflies, *Bemisia tabaci*, produced more offspring, had higher survival to adulthood, developed faster, and produced a higher proportion of daughters, compared to uninfected whiteflies.

2.6 Genetic Composition

All populations show variation in genetic composition (frequencies of various alleles) among individuals and through time. The degree of genetic variability and the frequencies of various alleles depend on a number of factors, including mutation rate, environmental heterogeneity, and population size, isolation, and mobility (Hedrick and Gilpin, 1997; Mopper, 1996; Mopper and Strauss, 1998; Pecsenye et al., 2014). Genetic variation may be partitioned among isolated demes or affected by patterns of habitat use (Hirai et al., 1994). Genetic structure, in turn, affects various other population parameters, including population viability (G. Bell and Gonzalez, 2009; Hedrick and Gilpin, 1997).

Populations vary in the frequency and distribution of various alleles. Widespread species might be expected to show greater variation across their geographic range than would more restricted species. Roberds et al. (1987) measured genetic variation from local to regional scales for the southern pine beetle, *Dendroctonus frontalis*, in the southeastern United States. They reported that allelic frequencies were somewhat differentiated among populations from Arkansas, Mississippi, and North Carolina, but that a population in Texas was distinct. They found little or no variation among demes within each state, and evidence of considerable inbreeding among beetles at the individual tree level. Roberds et al. (1987) also reported that only one allele of the seven analyzed showed significant variation between demes that were growing and colonizing new trees and demes not growing or colonizing new trees.

Mock et al. (2007) examined genetic variation in mountain pine beetle, *Dendroctonus ponderosae*, populations across their range in western North America using amplified fragment length polymorphism (AFLP) and mitochondrial sequencing analysis. They concluded that genetic divergence increased with geographic distance between populations, that gene flow occurred primarily around, rather than across, the Great Basin desert, and that patterns of genetic diversity and divergence indicated a northward expansion of this species from postglacial refugia (Fig. 5.3).

The genetic variation of the founders of a new deme is relatively low, simply because of the small number of colonists and the limited proportion of the gene pool that they represent. Colonists from a population with low genetic variability initiate a deme with even lower

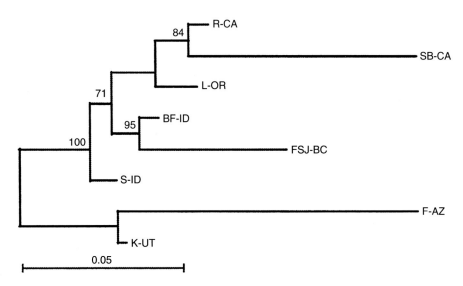

FIGURE 5.3 Unrooted neighbor-joining dendrogram of mountain pine beetle, *Dendroctonus ponderosae*, populations based on amplified fragment length polymorphism marker data (159 loci) using D_a genetic distance. The percentage of 1000 bootstrap pseudoreplicates (over loci) reproducing a particular node is provided. Population abbreviations are R-CA, Klamath, Oregon; SB-CA, San Bernadino, California; L-OR, La Grande, Oregon; BF-ID, Bonner's Ferry, Idaho; FSJ-BC, Fort St. James, British Columbia; S-ID, Stanley, Idaho; F-AZ, Flagstaff, Arizona; and K-UT, Kamas, Utah. *Source: From Mock, K.E., et al., 2007. Landscape-scale genetic variation in a forest outbreak species, the mountain pine beetle* (Dendroctonus ponderosae). *Mol. Ecol. 16, 553–568.*

genetic variability (Hedrick and Gilpin, 1997). Therefore, the size and genetic variability of the source population, as well as the number of colonists, determine genetic variability in founding populations. Genetic variability remains low during population growth unless augmented by new colonists. This is especially true for parthenogenetic species, such as aphids, for which an entire population could represent a clone derived from a founding female. Differential dispersal ability among genotypes affects heterozygosity of colonists. Florence et al. (1982) reported that the frequencies of four alleles for an esterase (esB) converged in southern pine beetles collected along a 150 m transect extending from an active infestation in east Texas. As a result, heterozygosity increased significantly with distance, approaching the theoretical maximum of 0.75 for a gene locus with four alleles. These data suggested a system that compensates for loss of genetic variability due to inbreeding by small founding populations and maximizes genetic variability in new populations coping with different selection regimes (Florence et al., 1982).

Nevertheless, dispersal among demes is critical to maintaining genetic variability (Hedrick and Gilpin, 1997; Pecsenye et al., 2014). If isolation restricts dispersal and infusion of new genetic material into local demes, inbreeding may reduce population ability to adapt to changing conditions, and recolonization following local extinction will be more difficult. G. Bell and Gonzalez (2009) demonstrated that population recovery following collapse depends on a sufficiently large precollapse population size and genetic diversity, above a threshold, to ensure survival and reproduction by adapted individuals.

Polymorphism occurs commonly among insects and underlies their rapid adaptation to environmental changes or other selective pressures, such as predation or exposure to anthropogenic chemicals (A. Brower, 1996; Sheppard et al., 1985). Polymorphism can develop in populations that use multiple habitat units or resources (Mopper, 1996; Mopper and Strauss, 1998; Via, 1990). Sturgeon and Mitton (1986) compared allelic frequencies among mountain pine beetles, *D. ponderosae*, collected from three pine hosts [ponderosa (*Pinus ponderosa*), lodgepole (*P. contorta*), and limber (*P. flexilis*)] at each of five sites in Colorado, USA. Significant variation occurred in morphological traits and allelic frequencies at five polymorphic enzyme loci among the five populations and among the three host species, suggesting that the host species is an important contributor to the genetic structure of polyphagous insect populations.

Via (1991a) compared the fitnesses (longevity, fecundity, and capacity for population increase) of *Acyrthosiphon pisum* clones from two host plants (alfalfa and red clover) on their source host or the alternate host. She reported that aphid clones had higher fitnesses on their source host, compared to the host to which they were transplanted, indicating local adaptation to factors associated with host conditions. Furthermore, significant negative correlations for fitness between source host and alternate host indicated increasing divergence between aphid genotypes associated with different hosts. In a subsequent study, Via (1991b) evaluated the relative importance of genetics and experience on aphid longevity and fecundity on source and alternate hosts. She maintained replicate lineages of the two clones (from alfalfa vs clover) on both host plants for three generations, then tested the performance of each lineage on both hosts. If genetics is the more important factor affecting aphid performance on source and alternate hosts, then aphids should have highest fitness on the host to which they were adapted, regardless of subsequent rearing on the alternate host. On the other hand, if experience is the more important factor, then aphids should have highest fitness on the host from which they were reared. Via found that three generations of experience on the alternate host did not significantly improve fitness on that host. Rather, fitness was highest on the plant from which the clone was derived originally, supporting the hypothesis that genetics is the more important factor. These data indicated that continued genetic divergence of the two subpopulations is likely, given that individuals dispersing between alternate hosts cannot improve their performance through time as a result of experience.

Biological factors that determine mate selection or mating success also affect gene frequencies, perhaps in concert with environmental conditions. In a laboratory experiment with sex-linked mutant genes in *Drosophila melanogaster* (J. Peterson and Merrell, 1983), mutant and wild male phenotypes exhibited about the same viability, but mutant males showed a significant mating disadvantage, leading to rapid elimination (ie, within a few generations) of the mutant allele. In addition, whereas the wild male phenotype tended to show a rare male advantage in mating, that is, a higher proportion of males mating at low relative abundance, mutant males showed a rare male disadvantage, that is, a lower proportion of males mating at low relative abundance, increasing their rate of elimination. Malausa et al. (2005) used a combination of genetic and stable isotope (^{13}C) techniques to identify host plant sources of 396 male and 393 female European corn borers, *Ostrinia nubilalis*, collected at multiple sites, and of 535 spermatophores carried by these females, over a 2-year period (2002–03). Moths could be differentiated unambiguously on the basis of larval host, either C_3 or C_4 plants. All but five females (three in 2002 and two in 2003) had mated with a male from the same host race, indicating > 95% assortative mating. These data indicate that nonrandom mating patterns can lead to rapid changes in gene frequencies among diverging races from different hosts.

Insect populations can adapt to environmental change more rapidly than can longer-lived, more slowly reproducing, organisms (Mopper, 1996; Mopper and Strauss, 1998). Heterogeneous environmental conditions tend to mitigate directional selection: any strong directional selection by any environmental factor during one generation can be modified in subsequent generations by a different prevailing factor. For example, Forsman et al. (2011) reported that the incidence of melanism among pygmy grasshoppers, *Tetrix subulata*, increased nearly 5-fold following fire, compared to unburned areas, in Sweden, reflecting an advantage of darker grasshoppers on a burned background. The frequency of melanism subsequently declined from about 50% of the population the first year following fire to 30% after 4 years (Fig. 5.4A),

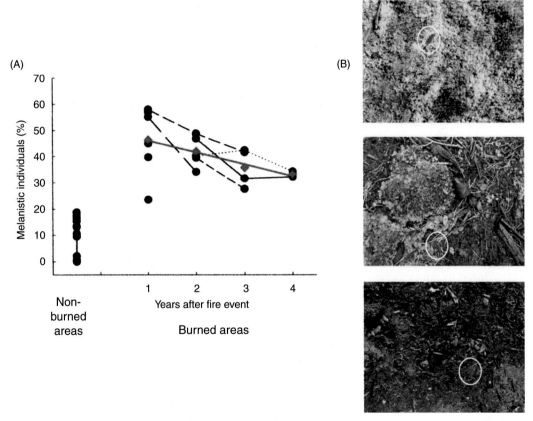

FIGURE 5.4 **Fire-induced melanism in pygmy grasshoppers, *Tetrix subulata*.** (A) Mean percentage of melanistic individuals for 11 populations in unburned areas and 9 populations in burned areas sampled at different times after fire. *Black dots* connected with lines indicate samples from successive years in the same population. *Red squares* represent mean values across populations within each year; the *red line* is the fitted linear regression of yearly means. (B) Melanistic grasshoppers presented to human viewers on samples of natural background: (top) unburned substrate; (middle) 50% burned substrate; (bottom) 100% burned substrate. *Source: Part A from Forsman, A., et al., 2011. Rapid evolution of fire melanism in replicated populations of pygmy grasshoppers. Evolution 65, 2530–2540. Part B from Karpestam, E., et al., 2012. Reduced predation risk for melanistic pygmy grasshoppers in post-fire environments. Ecol. Evol. 2, 2204–2212.*

as this advantage disappeared during vegetation recovery. Karpestam et al. (2012) subsequently presented images of black grasshoppers on backgrounds representing unburned, burned, and successional revegetation (Fig. 5.4B) to surrogate predators (humans) and found that detection time declined as the proportion of dark (burned) background decreased. These results demonstrated that visual predators are the likely drivers of selection for fire-induced melanism in these grasshoppers.

However, directional selection becomes fixed quickly under consistent selection. Among the best-known examples of population response to environmental change via polymorphism is the industrial melanism that developed in the peppered moth, *Biston betularia*, in England following the industrial revolution (Kettlewell, 1956). Selective predation by insectivorous birds was the key to the rapid shift in dominance from the white form, cryptic on light surfaces provided by lichens on tree bark, to the black form, which is more cryptic on trees blackened by industrial effluents. Birds preying on the more conspicuous morph maintained low frequencies of the black form in preindustrial England, but later they greatly reduced frequencies of the white form. Other examples of polymorphism also appear to be maintained by selective predation. In some cases, predators focusing on inferior Müllerian mimics of multiple sympatric models may select for morphs or demes that mimic different models (A. Brower, 1996; Sheppard et al., 1985).

Mondor et al. (2005) examined effects of elevated CO_2 and O_3 on a genetically polymorphic population of pea aphids, *A. pisum*, over multiple generations using a free air carbon enrichment (FACE) facility. The green genotype was positively affected by elevated CO_2 levels, but the pink genotype was not, leading to increased frequency of the green genotype in this population over time.

The shift from pesticide-susceptible to pesticide-resistant genotypes may be particularly instructive. Selective pressure imposed by insecticides causes rapid development of insecticide-resistant populations for many species, often within 5 years (Felland et al., 1990; Norris et al., 2015; Roussel and Clower, 1957). Norris et al. (2015) reported the rapid increase in frequency of insecticide resistance genes (kdr) by mosquitoes, *Anopheles coluzzii*, following widespread adoption of insecticide-treated bed nets in Mali during 2005–06, combined with a breakdown in reproductive isolation between this species and insecticide-resistant populations of *A. gambiae*.

Resistance development is facilitated by the widespread occurrence in insects, especially herbivores, of genes that encode for enzymes that detoxify plant defenses, since ingested insecticides also are susceptible to detoxification by these enzymes (see Chapter 3). Although avoidance of directional selection for resistance to any single tactic is a major objective of integrated pest management (IPM, see Chapter 17), pest management in practice still involves widespread use of the most (initially) effective tactic. Following the appearance of transgenic insect-resistant crop species in the late 1980s, genetically engineered, Bt toxin-producing, corn, cotton, soybeans, and potatoes have replaced nontransgenic varieties over large areas, raising concern that these crops might quickly select for resistance in target species (Alstad and Andow, 1995; Heuberger et al., 2008a,b; Tabashnik, 1994; Tabashnik et al., 1996).

Laboratory studies have shown that at least 16 species of Lepidoptera, Coleoptera, and Diptera are capable of developing resistance to Bt genes as a result of strong selection

(Tabashnik, 1994). However, relatively few species have shown resistance in the field. The diamondback moth, *Plutella xylostella*, has shown resistance to Bt in field populations from the USA, Philippines, Malaysia, and Thailand. Resistance in some species has been attributed to reduced binding of the toxin to membranes of the midgut epithelium. A single gene confers resistance to four Bt toxins in the diamondback moth (Tabashnik et al., 1997), and >5000-fold resistance can be achieved in a few generations (Tabashnik et al., 1996). Resistance can be reversed when exposure to Bt toxin is eliminated for several generations, probably because of fitness costs of resistance (Tabashnik, 1994), but some strains can maintain resistance in the absence of Bt for more than 20 generations (Tabashnik et al., 1996). Adding multiple Bt genes to transgenic crops has reduced the likelihood of resistance development, but depends on the ability of individual plants to maintain lethal levels of each Bt toxin throughout the growing season (Showalter et al., 2009).

2.7 Social Insects

Social insects pose some special problems for description of population structure. On the one hand, each individual requires resources and contributes to interactions with other organisms. On the other hand, colony member activity is centered on the nest, and collective foraging territory is defined by proximity to surrounding colonies. Food transfer among nestmates (trophallaxis) supports a view of colonies as sharing a collective gut. Colonies often have defined spatial structure in their parts and activities, with the queen and larvae located in the deepest chambers and progressively older workers moving upward and performing a sequence of tasks associated with vertical location, for example, brood care by young workers in the deepest levels and foraging and food storage by the oldest workers near the surface (Tschinkel, 1999). In the case of army ants, *Dorylus* spp., each colony moves as an intact entity (Schöning et al., 2005). Colony members recognize and accept other colony members, but chemosensory detection of noncolony members elicits rejection and aggression (Ozaki et al., 2005). Caste regulation within colonies may depend on colony size. L. Mao and Henderson (2010) found that increased density of Formosan subterranean termite, *Coptotermes formosanus*, workers resulted in increased concentration of juvenile hormone, responsible for transformation of workers into soldiers. However, the presence of soldiers reduced the effect of the rising juvenile hormone level, stabilizing the proportion of soldiers in colonies.

Hence, each colony appears to function as a distinct ecological unit, or *superorganism*, with colony size (number of members) determining its individual structure, physiology, and behavior. For some social insects, the number of colonies per ha may be a more useful measure of density than is number of individuals per ha. However, defining colony boundaries and distinguishing between colonies may be problematic for many species, especially those with underground nests. Molecular techniques have proven to be a valuable tool for evaluating relatedness within and among colonies in an area (Husseneder et al., 2003).

Colonies of social Hymenoptera can show varying degrees of relatedness among queens and workers (Goodisman and Hahn, 2004; Pamilo et al., 1997). Intracolony relatedness can vary among colonies and among populations. For example, Goodisman and Hahn (2004) reported that DNA microsatellite markers in the carpenter ant, *Camponotus ocreatus*, indicated that genotypes of queens, workers, and males in 15 of 16 analyzed nests were consistent with a single, once-mated queen, but that nestmate genotypes in the remaining nest were more complex, suggesting infrequent inbreeding, polygyny, and polyandry.

In other ants, such as *Solenopsis invicta* and some *Formica* species, social polymorphism can be observed, with distinct monogynous (M type, having one queen) and polygynous (P type, having multiple queens) (Pamilo et al., 1997). The two types generally show high relatedness to each other where they occur in the same area. However, gene flow is restricted in the polygynous type and between monogynous and polygynous types. Populations of polygynous colonies generally are more genetically differentiated than are those of monogynous colonies in the same area (Pamilo et al., 1997).

Polygyny may be advantageous in areas of intense competition, where the more rapid reproduction by multiple queens may confer an advantage, regardless of the relatedness of the queens. However, additional queens eventually may be eliminated, especially in ant species, with workers often favoring queens on the basis of size or condition rather than which queen is mother to most workers (Pamilo et al., 1997).

Similarly, termite colonies are cryptic and may have variable numbers of reproductive adults. Husseneder and Grace (2001b) and Husseneder et al. (1998) found DNA fingerprinting to be more reliable than aggression tests or morphometry for distinguishing termites from different colonies or sites. As expected, genetic similarity is higher among termites within collection sites than between collection sites and declines with distance between colonies (Fig. 5.5) (Husseneder and Grace, 2001a; Husseneder et al., 1998). Moderate inbreeding often is evident within termite colonies, but low levels of genetic differentiation at regional scales suggest that substantial dispersal of winged adults homogenizes population genetic

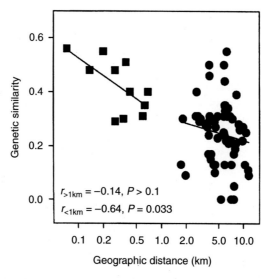

FIGURE 5.5 **Relationship between genetic similarity and spatial distance for 13 termite,** *Schedorhinotermes lamanianus,* **colonies representing 100 km² in the Shimba Hills Nature Reserve in Kenya.** *Source: From Husseneder, C., et al., 1998. Variation between and within colonies in the termite: morphology, genomic DNA, and behaviour. Mol. Ecol. 7, 983–990.*

structure (Husseneder et al., 2003). However, several species are polygynous and may show greater within-colony genetic variation, depending on the extent to which multiple reproductives are descended from a common parent (Vargo et al., 2003). Kaib et al. (1996) found that foraging termites tended to associate with close kin in polygynous and polyandrous colonies of *Schedorhinotermes lamanianus*, leading to greater genetic similarity among termites within foraging galleries than at the nest center.

Genetic studies have challenged the traditional view of the role of genetic relatedness in the evolution and maintenance of eusociality. Eusociality in the social Hymenoptera has been explained by the high degree of genetic relatedness among siblings, which share 75% of their genes as a result of their haploid father and diploid mother, compared to only 50% genes shared with their mother (W. Hamilton, 1964, see Chapter 15). However, this model does not apply to termites. Husseneder et al. (1999), Thorne (1997), and Thorne and Traniello (2003) suggested that developmental and ecological factors, such as slow development, iteroparity, overlap of generations, food-rich environment, high risk of dispersal, and group defense, may be more important than genetics in the maintenance of termite eusociality, whatever factors may have favored its original development. Myles (1999) reviewed the frequency of neoteny (reproduction by immature stages) among termite species and concluded that neoteny is a primitive element of the caste system that may have reduced the fitness cost of not dispersing, leading to further caste differentiation and early evolution of eusociality.

3 POPULATION PROCESSES

The population variables described previously change as a result of differential reproduction, movement, and death among individuals. These individual contributions to population change are integrated as three population processes: natality (birth rate), mortality (death rate), and dispersal (rate of movement of individuals into or out of the population). For example, density can increase as a result of increased birth rate and/or immigration; frequencies of various alleles change as a result of differential reproduction, survival, and dispersal. The rate of change in these processes determines the rate of population change, described in the next chapter. Therefore, these processes are fundamental to understanding population responses to changing environmental conditions.

3.1 Natality

Natality is the population birth rate, that is, the per capita production of new individuals per unit time. Realized natality is a variable that approaches potential natality, the maximum reproductive capacity of the population, only under ideal environmental conditions. Natality is affected by factors that influence production of eggs (fecundity) or production of viable offspring (fertility) by individual insects. For example, resource quality can affect the numbers of eggs produced by female insects (R. Chapman, 1982). Ohgushi (1995) reported that females of the herbivorous ladybird beetle, *Henosepilachna niponica*, feeding on the thistle, *Cirsium kagamontanum*, resorbed eggs in the ovary when leaf damage became high. Female blood-feeding mosquitoes often require a blood meal before first or subsequent oviposition can occur (R. Chapman, 1982); the ceratopogonid, *Culicoides barbosai*, produces eggs in proportion to the size of the blood meal (Linley, 1966). Hence, poor quality or insufficient food resources can reduce natality. Inadequate numbers of males can reduce fertility in sparse populations. Similarly, availability of suitable oviposition sites also affects natality.

Natality typically is higher at intermediate population densities than at low or high densities. At low densities, difficulty attracting mates may limit mating, or may limit necessary cooperation among individuals, as in the case of bark beetles that must aggregate in order to overcome host tree defenses prior to oviposition (Berryman, 1981; Coulson, 1979). At high densities, competition for food, mates, and oviposition sites reduces fecundity and fertility (Southwood, 1975, 1977). The influence of environmental conditions can be evaluated by comparing realized natality to potential natality, for example, estimated under laboratory conditions (see Section 5).

Differences among individual fitnesses are integrated in natality. Differential reproduction among genotypes in the population determines the frequency of various alleles in the filial generation. As discussed previously (Section 2.6), gene frequencies can change dramatically within a relatively short time, given strong selection and the short generation times and high reproductive capacity of insects.

3.2 Mortality

Mortality is the population death rate, that is, the per capita number of individuals dying per unit time. As with natality, we can distinguish a potential longevity or life span, resulting only from physiological senescence, from the realized longevity, resulting from the action of mortality factors. Hence, mortality can be viewed both as reducing the number of individuals in the population and as reducing survival. Both have importance consequences for population dynamics.

Organisms are vulnerable to a variety of mortality agents, including unsuitable habitat conditions (eg, extreme temperature or water conditions), toxic or unavailable food resources, competition, predation (including cannibalism), parasitism, and disease (see Chapters 2–4). These factors are a focus of studies to enhance pest management efforts (see Chapter 17). Death can result from insufficient energy or nutrient acquisition to permit detoxification of, or continued search for, suitable resources. Life stages are affected differentially by these various mortality agents (L. Fox, 1975b; Varley et al., 1973). For example, immature insects are particularly vulnerable to desiccation during molts, whereas flying insects are more vulnerable to predation by birds or bats. Many predators and parasites selectively attack certain life stages. Among parasitic Hymenoptera, species attacking the same host have different preferences for host egg, larval, or pupal stages. Predation also can be greater on hosts feeding on particular plant species, compared to other plant species, based on differential toxin sequestration, or predator attraction to plant volatiles (Stamp, 1992; Traugott and Stamp, 1996; Turlings et al., 1990, 1995).

In general, mortality due to predation tends to peak at intermediate population densities, when density is sufficient for a high rate of encounter with predators, but prior to predator satiation (Fig. 5.6) (Southwood, 1975, 1977, see Chapter 8). Mortality due to competition, cannibalism, and parasitism increases at higher population densities (Fig. 5.6) (Fox, 1975a,b;

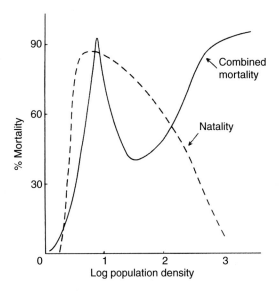

FIGURE 5.6   Relationship between population density, natality, and mortality caused by predators and parasites (peaking at lower population density) and intrapecific competition (peaking at higher population density). *Source: From Southwood, T.R.E., 1975. The dynamics of insect populations. In: Pimentel, D. (Ed.), Insects, Science, and Society. Academic Press, San Diego, CA, pp. 151–199.*

Southwood, 1975, 1977). Competition may cause mortality through starvation, cannibalism, stress, displacement of individuals from optimal habitats, disease epizootics in dense populations, and increased exposure and vulnerability to predation as a result of displacement or delayed development (see Chapter 6). For example, *Baculovirus* epizootics commonly end outbreaks of many lepidopteran species (Cooper et al., 2003; Cory and Myers, 2003). However, disease epizootics do not necessarily increase mortality above that which would occur as a result of competition for limited resources (Kistner and Belovsky, 2014) and may ensure survival of the most resistant individuals.

Survival rate represents the number of individuals still living in relation to time. These individuals continue to feed and reproduce, thereby contributing most to population size, as well as to genetic and ecological processes. Hence, survival rate is an important measure in studies of populations.

Survivorship curves reflect patterns of mortality and can be used to compare the effect of mortality in different populations. Lotka (1925) pioneered the comparison of survivorship curves among populations, by plotting the log of number or percent of living individuals against time. Pearl (1928) later identified three types of survivorship curves, based on the log of individual survival through time (Fig. 5.7). Type 1 curves represent species, including most large mammals, but also starved *Drosophila* (Price, 1997), in which mortality is concentrated near the end of the maximum life span. Type 2 curves represent species in which the probability of death is relatively constant with age, leading to a linear decline in survivorship. Many birds and reptiles approach the Type 2 curve. Type 3 curves are seen for most insects, as well as many other invertebrates and fish, which have high rates of mortality during early life stages, but relatively low mortality during

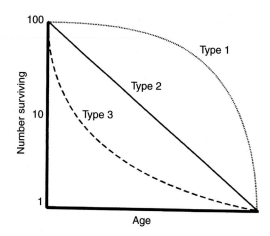

FIGURE 5.7   **Three generalized types of survivorship curves.** *Type 1* represents species with high survival rates maintained through the potential life span. *Type 2* represents species with relatively constant survivorship with age. *Type 3* represents species with low survival rates during early stages, but relatively high survival of individuals reaching more advanced ages.

later life stages (Begon and Mortimer, 1981; Pianka, 1974). Species representing Type 3 survivorship must have very high rates of natality to ensure that some offspring reach reproductive age, compared to Type 1 species which have a high probability of reaching reproductive age.

The form of the survivorship curve can change during population growth. Mason and Luck (1978) showed that survivorship curves for the Douglas-fir tussock moth, *Orgyia pseudotsugata*, changed with population growth from stable to increasing, then decreasing. Survivorship decreased less steeply during population growth and decreased more steeply during population decline, compared to stable populations.

As described for natality, mortality integrates the differential survival among various genotypes, the basis for evolution. Survivors live longer and have greater capacity to reproduce. Therefore, selective mortality can alter gene frequencies rapidly in insect populations.

3.3 Dispersal

Dispersal is the movement of individuals away from their source, and includes *spread*, the local movement of individuals, and *migration*, the cyclic mass movement of individuals among areas (L. Clark et al., 1967; Nathan et al., 2003). As discussed in Chapter 2, long-distance dispersal maximizes the probability that habitat or food resources created by environmental changes or disturbances are colonized before the source population depletes its resources or is destroyed by disturbance. Dispersal also may permit founding populations to escape predators and parasites that have concentrated in the source population. Strong selection against founders infected with pathogens, due to fitness costs associated with harboring infections, may at least initially reduce the incidence of disease among colonists (Yang et al., 2010). Dispersal also contributes to infusion of new genetic material into populations.

This contribution to genetic heterogeneity enhances population capacity to adapt to changing conditions.

Dispersal incorporates *emigration*, movement away from a source population, and *immigration*, movement of dispersing individuals into another population or vacant habitat. Immigration adds new members to the population, or founds new demes, whereas emigration reduces the number of individuals in the population. A number of factors affect the probability of successful dispersal, that is, arrival at suitable habitats, including life history strategy, crowding, nutritional status, habitat and resource conditions, and the mechanism of dispersal.

3.3.1 Life History Strategy

The degree of adaptation to disturbance affects the predisposition of individuals to disperse. Species characterizing relatively stable, infrequently disturbed, habitats tend to disperse slowly, that is, produce few offspring and move short distances (Cantrell et al., 2010; Roff, 1990). Infrequent disturbance and consistent resource availability provide little or no selection for greater dispersal ability. Many forest species (especially Lepidoptera and Coleoptera) are flightless, or at least poor fliers (Roff, 1990). Monophagous herbivore species that feed on host plants with stable distributions tend to show lower dispersal frequencies and distances than do species that feed on host plants with more variable distributions (St. Pierre et al., 2005). By contrast, species (such as aphids) that characterize temporary, frequently disturbed, habitats produce large numbers of individuals and a high proportion of dispersers. Such traits are important adaptations for species exploiting temporary, unstable conditions (Cantrell et al., 2010; Janzen, 1977; Roff, 1990).

Polymorphism in dispersal ability may be common (Bouget et al., 2015). Some dispersal-adapted species produce a specialized morph for dispersal. The dispersal form of most aphids and many scale insects is winged, whereas the feeding form typically is wingless and sedentary. Under crowded conditions, migratory locusts develop into a specialized long-winged morph (gregarious phase) capable of long-distance migration, distinct from the shorter-winged nondispersing morph (solitary phase) (Anstey et al., 2009). Some mites have dispersal stages specialized for attachment to phoretic hosts, for example, ventral suckers in the hypopus of astigmatid mites and anal pedicel in uropodid mites (Krantz, 1978).

Some species have obligatory dispersal prior to reproduction. Cronin and Strong (1999) reported that parasitoid wasps, *Anagrus sophiae*, laid >84% of their eggs in host planthoppers, *Prokelisia* spp., on cordgrass, *Spartina alterniflora*, plants isolated at 10–250 m from source populations.

3.3.2 Crowding

Dispersal increases with population size or density. Under crowded conditions, some insects spend more time eating and less time resting (R. Chapman, 1982). Crowding increases the incidence of cannibalism in many species (Fox, 1975a,b), encouraging dispersal. Crowding also may interfere with foraging or mating activity (Schowalter and Whitford, 1979), thereby encouraging individuals to seek less crowded conditions.

Crowding can induce morphological changes that promote dispersal. Uncrowded desert locusts tend to repel one another and feed quietly on clumps of vegetation, whereas crowded locusts are more active, attract one another, and march en masse, spending little time on vegetation (Anstey et al., 2009; Matthews and Matthews, 2010; Pener and Simpson, 2009). This phase shift from solitary nymphs that develop into adults with short wings and long legs to larger, gregarious nymphs that develop into long-winged, short-legged adults is accompanied by changes in color, size, wing and leg length, and diet breadth that facilitate migration and survival in new areas (Anstey et al., 2009; Maeno and Tanaka, 2008; Maeno et al., 2011; Matthews and Matthews, 2010; Pener and Simpson, 2009). Tanaka and Nishide (2012) demonstrated that this phase shift is induced in *S. gregaria* by the visual stimulus of 5–10 nearby insects. However, olfactory and tactile stimuli also may be involved (Pener and Simpson, 2009). Kang et al. (2004) identified 532 genes that were differentially expressed between the two phases, most of which were downregulated in solitary, compared to gregarious, nymphs. However, gregarious nymphs had higher numbers of genes coding for heat shock proteins and greater immunocompetence than did solitary nymphs. The solitary phase is induced by alleviation of crowding (Pener and Simpson, 2009) and, perhaps, by diseases (Elliot et al., 2003, 2005).

Because survival and fecundity are often inversely related to population density (see Section 3.1), dispersing individuals may achieve higher fitnesses than do nondispersing individuals at high population densities (Price, 1997). For example, some bark beetle species oviposit their full complement of eggs in one tree under low-density conditions, but only a portion of their eggs in one tree under high-density conditions, leaving that tree and ovipositing remaining eggs in other trees (T. Wagner et al., 1981). If all eggs were laid in the first tree under crowded conditions, the large number of offspring could deplete resources before completing development. Cronin (2003) found that emigration of planthoppers, *Prokelisia crocea*, increased linearly with density of conspecific females. Leisnham and Jamieson (2002) reported that more mountain stone weta emigrated from large tors with larger demes, but proportionately more weta emigrated from small tors, likely reflecting the greater perimeter:area ratio of small tors.

3.3.3 Nutritional Status

Nutritional status affects the endurance of dispersing insects. Populations of many insects show considerable variation in fat storage and vigor as a result of variation in food quality and quantity and maternal partitioning of nutrient resources to progeny (T. Wagner et al., 1981; Wellington, 1980). Many species exhibit obligatory flight distances that are determined by the amount of energy and nutrient reserves: dispersing individuals respond to external stimuli only after depleting these reserves to a threshold level. Hence, less vigorous individuals tend to colonize more proximal habitats, whereas more vigorous individuals fly greater distances and colonize more remote habitats. Because crowding and nutritional status have opposing effects on dispersal, the per capita accumulation of adequate energy reserves and the number of dispersing individuals should peak when resource quality and quantity are still sufficient to promote insect development and vigor.

3.3.4 Habitat and Resource Conditions

The likelihood that an insect will find a suitable patch depends strongly on patch or resource size, ease of discovery, and proximity to insect population sources. Larger or more conspicuous habitats or resources are more likely to be perceived by dispersing insects or to be intercepted by a given direction of flight (see Chapter 7). Larger habitat patches intersect a longer arc centered on a given starting point. Insects dispersing in any direction have a higher probability of contacting larger patches than they do smaller patches. Courtney (1985, 1986) reported that the pierid butterfly, *Anthocharis cardamines*, preferentially oviposited on the most conspicuous (in terms of flower size) host species, although these were less suitable for larval development than were less conspicuous hosts. This behavior by the adults represented a trade-off between the prohibitive search time required to find the most suitable hosts and the reduced larval survival on the most conspicuous hosts.

The presence of predators may affect perception of habitat quality by potential colonists (Resetarits and Binckley, 2013) and encourage emigration or a continued search (Cronin et al., 2004). Schäffer et al. (2013) reported that stream invertebrates were induced to actively enter the drift when exposed to kairomones of predaceous fish. On the other hand, Seymour et al. (2003) found that lycaenid butterflies, *Plebejus argus*, whose larvae are tended by ants, *Lasius niger*, apparently are able to orient toward patches occupied by *L. niger* colonies. Butterfly persistence in patches was influenced more strongly by ant presence than by floral resource density.

The probability of survival declines with dispersal distance, as a result of depletion of metabolic resources and protracted exposure to various mortality factors (Pope et al., 1980). Hence, more insects reach closer resources or sites. Sartwell and Stevens (1975) and Schowalter et al. (1981b) reported that, under nonoutbreak conditions, probability of bark beetle, *Dendroctonus* spp., colonization of living pine trees declined with distance from currently attacked trees. Trees more than 6 m from currently colonized trees had a negligible probability of colonization by sufficient numbers of beetles to successfully kill the tree. Under outbreak conditions, the effect of distance disappeared (Schowalter et al., 1981b). Similarly, He and Alfaro (1997) reported that, under nonoutbreak conditions, colonization of white spruce by the white pine weevil, *Pissodes strobi*, depended on host condition and distance from trees colonized the previous year, but during outbreaks most trees were sufficiently near occupied trees to be colonized.

3.3.5 *Mechanism of Dispersal*

The mechanism of dispersal strongly affects the probability that suitable resources can be found and colonized. Three general mechanisms can be identified: random, phoretic, and directed (see Matthews and Matthews, 2010).

Random dispersal direction and path is typical of most small insects with little capacity to detect or orient toward environmental cues. Such insects are at the mercy of physical barriers or wind or water currents, and their direction and path of movement are determined by obstacles and patterns of air or water movement. For example, first instar nymphs of a *Pemphigus* aphid that lives on the roots of sea aster growing in salt marshes climb the sea asters and are set adrift on the rising tide. Sea breezes enhance movement, and successful nymphs are deposited at low tide on new mud banks where they must seek hosts (Kennedy, 1975). Aquatic insect larvae often detach from substrate and disperse via streamflow, a process known as "drift" (Hammock et al., 2012; Koetsier and Bryan, 1989; Schäffer et al., 2013). Hatching gypsy moth, *Lymantria dispar*, and other tussock moth larvae (Lymantriidae), scale insect crawlers, and spiders (as well as other arthropods) disperse by launching themselves into the airstream

(Matthews and Matthews, 2010; McClure, 1990). Lymantriid females and scale insect adults have poor (if any) flight capacity. Wind-aided dispersal by larval Lepidoptera and spiders is facilitated by extrusion of silk strands, a practice known as "ballooning." Western spruce budworm, *C. occidentalis*, adults aggregate in mating swarms above the forest canopy and are carried by wind currents to new areas (Wellington, 1980).

The distance traveled by wind- or water-dispersed insects depends on several factors, including flow rate and insect size or mass. Larval settlement rates for black flies, *Simulium vittatum*, are lowest in the high stream velocity habitats preferred by the larvae, due to constraints on larval ability to control direction of movement at high flow rates (D. Fonseca and Hart, 2001). Many species can control landing, to some extent, by responding to cues indicating suitable resources. Jung and Croft (2001) measured falling speeds, relative to morphology and activity, of several wind-dispersed mite species. Heavier mites fell more rapidly than did lighter mites, as expected. However, anesthetized mites fell more rapidly than did active mites, indicating mite ability to control buoyancy and landing to some extent.

The probability that at least some insects will arrive at suitable resources depends on the number of dispersing insects and the predictability of wind or water movement in the direction of new resources. Most individuals fail to colonize suitable sites, and many become part of the aerial or aquatic plankton that eventually "falls out" and becomes deposited in remote, unsuitable locations. For example, J. Edwards and Sugg (1990) documented the fallout deposition of many aerially dispersed insect species on montane glaciers in western Washington, USA.

Phoretic dispersal is a special case in which a flightless insect or other arthropod hitches a ride on another animal (Fig. 5.8). Phoresy is particularly common among wingless

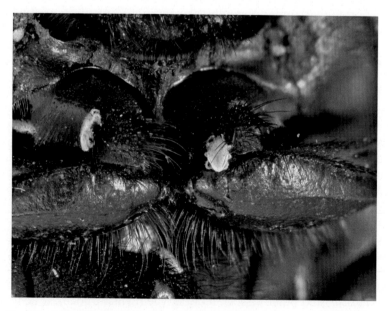

FIGURE 5.8   **Phoretic mesostigmatid mites on coxae of scarab beetle.** *Source: Photo courtesy of A. Tishechkin.*

Hymenoptera and mites. For example, scelionid wasps ride on the backs of female grass-hoppers, benefitting from both transport and the eventual opportunity to oviposit on the grasshopper's eggs. Wingless Mallophaga attach themselves to hippoboscid flies that parasitize the same bird hosts. Many species of mites attach themselves to dispersing adult insects that feed on the same dung or wood resources (Krantz and Mellott, 1972; Stephen et al., 1993). Birds and mammals provide long-distance transport for hemlock woolly adelgid, *Adelges tsugae* (McClure, 1990). The success of phoresy (as with wind- or water-aided dispersal) depends on the predictability of host dispersal. However, success is enhanced by the association of both the hitchhiker and its mobile (and perhaps cue-directed) host with the same resource.

Directed dispersal provides the highest probability of successful colonization and is observed in larger, stronger fliers capable of orienting toward suitable resources (see Chapter 3). Many wood-boring insects, such as wood wasps (Siricidae) and beetles (especially Buprestidae), are attracted to sources of smoke, infrared radiation, or volatile tree chemicals emitted from burned or injured trees over distances of up to 50 km (W. Evans, 1966; Gara et al., 1984; Matthews and Matthews, 2010; R.G. Mitchell and Martin, 1980; Raffa et al., 1993; S. Schütz et al., 1999; Wickman, 1964). Attraction to suitable hosts often is enhanced significantly by mixing of host odors with pheromones emitted by early colonists (see Chapters 3 and 4). Visual, acoustic, or magnetic cues also aid orientation (Matthews and Matthews, 2010). For example, masking the silhouette of tree boles with white paint substantially reduced numbers of attracted southern pine beetle (Strom et al., 1999), *Ips* engraver beetles, and some bark beetle predators (Goyer et al., 2004). Monarch butterflies, *Danaus plexippus*, possess organically synthesized magnetic material (probably magnetite), concentrated in the thorax, that likely facilitates orientation during migration (MacFadden and Jones, 1985).

Migration is an active mass movement of individuals that functions to displace entire populations. Migration always involves females, but not always males. Examples of migratory behavior in insects include locusts, monarch butterflies, and ladybird beetles. Locust, *S. gregaria* and *Locusta migratoria*, migration depends, at least in part, on wind patterns. Locust swarms remain compact, not because of directed flight, but because randomly oriented locusts reaching the swarm edge reorient toward the body of the swarm. Swarms are displaced downwind into equatorial areas where converging air masses rise, leading to precipitation and vegetation growth favorable to the locusts (Matthews and Matthews, 2010). In this way, migration displaces the swarm from an area of crowding and insufficient food to an area with more abundant food resources. Monarch butterfly and ladybird beetle migration occurs seasonally and displaces large numbers to and from overwintering sites, Mexico for North American monarch butterflies and sheltered sites for ladybird beetles. Merlin et al. (2009) reported that a gene-controlled circadian clock in monarch antennae provides the internal timing mechanism for a time-compensated sun compass that allows the insects to correct their flight direction relative to the position of the sun as it moves across the sky during the day.

3.3.6 Successful Colonization

Effective dispersal, the number of individuals that successfully immigrate or found new demes, is the product of source strength (the number of individuals dispersing) and the individual probability of success (Nathan et al., 2003; Price, 1997). Source strength is a function of population size, density, and life history strategy. Individual probability of successful dispersal is determined by dispersal mechanism, individual capacity for long-distance dispersal, the distance between source and sink (destination), patch size, and habitat heterogeneity, as will be described (see also Chapter 7).

The capacity of individuals for long-distance dispersal is determined by flight capacity, nutritional status, egg load, and parasitism. Winged insects disperse greater distances than wingless species (Leisnham and Jamieson, 2002). Individuals feeding on adequate resources can store sufficient energy and nutrients to live longer and travel further than can individuals feeding on marginal or inadequate resources. Although dispersal should increase as population density increases, increased competition for food may limit individual energy reserves and endurance at high densities. Isaacs and Byrne (1998) reported that egg load was negatively correlated with height above ground for dispersing female sweetpotato whiteflies, *Bemisia tabaci*, demonstrating a trade-off between dispersal and reproductive capacities for weak-flying insects. Furthermore, parasitized individuals may lose body mass more quickly during dispersal than do unparasitized individuals and consequently exhibit shorter flight distances and slower flight speeds (Bradley and Altizer, 2005). Therefore, dispersal may peak before increasing density and disease reach levels that interfere with dispersal capacity (Leonard, 1970).

Dispersing individuals become vulnerable to new mortality factors. Whereas nondispersing individuals may be relatively protected from temperature extremes and predation through selection of optimal microsites, dispersing individuals are exposed to ambient temperature and humidity, high winds, and predators as they move across the landscape or downstream. Exposure to higher temperatures increases metabolic rate and depletes energy reserves more quickly, reducing the time and distance an insect can travel (Pope et al., 1980). Actively moving insects also are more conspicuous and more likely to attract the attention of predators (Schultz, 1983; Hammock et al., 2012). Dispersal across inhospitable patches may be inhibited or ineffective (Haynes and Cronin, 2003). However, insects in patches with high abundance of predators may be induced to disperse as a result of frequent encounters with predators (Cronin et al., 2004; Schäffer et al., 2013).

The number of dispersing individuals declines with distance from the source population (Isaacs and Byrne, 1998; St. Pierre and Hendrix, 2003), with the frequency distribution of dispersal distances often described by a negative exponential or inverse power law (Fig. 5.9). However, some species show a higher proportion of long-distance dispersers than would be expected from a simple diffusion model, suggesting heterogeneity in dispersal type (Cronin et al., 2000). A general functional model of dispersal (D) can be described by the equation

$$D = \frac{c}{2\alpha \Gamma(1/c)} \exp\left(-\frac{|x|^c}{|\alpha|}\right) \tag{5.1}$$

where c and α are shape and distance parameters, respectively, and $\Gamma(1/c)$ is the gamma function (J. Clark et al., 1998; Nathan et al., 2003). The negative exponential ($c = 1$) and Gaussian ($c = 2$) are special cases of this formula. Similarly, effective dispersal declines as the probability

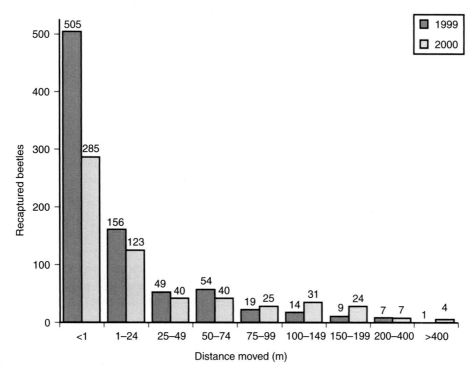

FIGURE 5.9 **Range of dispersal distances from a population source for the weevil, *Rhyssomatus lineaticollis*, in Iowa, USA.** *Source: From St. Pierre, M.J., Hendrix, S.D., 2003. Movement patterns of* Rhyssomatus lineaticollis *Say (Coleoptera: Curculionidae) within and among* Asclepias syriaca *(Asclepiadaceae) patches in a fragmented landscape. Ecol. Entomol. 28, 579–586.*

of encountering inhospitable patches increases, for example, as fragmentation reduces the occurrence of suitable patches on the landscape.

The mating status of dispersing individuals determines their value as founders when they colonize new resources. Clearly, if unmated individuals must find a mate in order to reproduce after finding a habitable patch, their value as founders may be negligible. For some species, mating occurs prior to dispersal of fertilized females (R. Mitchell, 1970). In species capable of parthenogenetic reproduction, fertilization is not required for dispersal and successful founding of populations. Some species ensure breeding at the site of colonization, such as through long-distance attraction via pheromones, for example, by bark beetles (Raffa et al., 1993), or through males accompanying females on phoretic hosts, for example, some mesostigmatid mites (Springett, 1968), or through mating swarms, for example, eastern spruce budworm, *Choristoneura fumiferana* (Greenbank, 1957).

The contribution of dispersing individuals to genetic heterogeneity in a population depends on a number of factors. The genetic heterogeneity of the source population determines the gene pool from which dispersants come. Dispersing individuals represent a proportion of the total gene pool for the population. More heterogeneous demes contribute greater genetic heterozygosity to target or founded demes than do less heterogeneous demes (Fig. 5.10) (Hedrick and Gilpin, 1997). The number or proportion of individuals that disperse affects

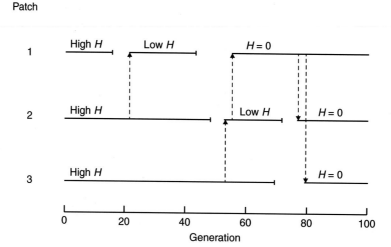

FIGURE 5.10 **Simulated population heterozygosity (H) over time in three habitat patches.** Extinction is indicated by *short vertical bars* on the right end of *horizontal lines*; recolonization is indicated by *arrows. Source: From Hedrick, P.W., Gilpin, M.E., 1997. Genetic effective size of a metapopulation. In: Hanski, I.A., Gilpin, M.E. (Eds.), Metapopulation Biology: Ecology, Genetics and Evolution. Academic Press, San Diego, CA, pp. 165–181.*

their genetic heterogeneity. If certain genotypes are more likely to disperse, then the frequencies of these genotypes in the source population may decline, unless balanced by immigration.

Distances between demes influence the degree of gene exchange through dispersal. Local demes will be influenced more by the genotypes of dispersants from neighboring demes than by more distant demes. Gene flow may be precluded for sufficiently fragmented populations. This is an increasing concern for demes restricted to isolated refugia. Populations consisting of small, isolated demes may be incapable of sufficient interaction to sustain viability. Gene flow also is affected by habitat choices made by dispersing individuals (Edelaar et al., 2008). Individuals entering an area may show particular habitat preferences based on phenotype or experience that restrict their interaction with other individuals that make different choices. Such "matching habitat choices" may limit gene flow among colonizing individuals and increase directed gene flow and speciation.

4 LIFE HISTORY CHARACTERISTICS

Life history adaptation to environmental conditions typically involves complementary selection of natality and dispersal strategies that balance expected mortality. General life history strategies appear to be related to habitat stability and are crucial to population survival and growth (Bouget et al., 2015; Huryn and Wallace, 2000; Roff, 1990).

MacArthur and Wilson (1967) distinguished two life history strategies related to habitat stability and importance of colonization and rapid population establishment. The *r strategy* generally characterizes "weedy" species adapted to colonize and dominate new or ephemeral habitats quickly (Janzen, 1977). These species are opportunists that quickly colonize new resources but are poor competitors and cannot persist when competition increases in stable

TABLE 5.1 Life History Characteristics of Species Exemplifying the r- and K-strategies

Attribute	Ecological strategy	
	r (opportunistic)	K (equilibrium)
Homeostatic ability	Limited	Extensive
Development time	Short	Long
Life span	Short	Long
Mortality rate	High	Low
Reproductive mode	Often asexual	Sexual
Age at first brood	Early	Late
Offspring/brood	Many	Few
Broods/lifetime	Usually one	Often several
Size of offspring	Small	Large
Parental care	None	Extensive
Dispersal ability	High	Limited
Numbers dispersing	Many	Few
Dispersal mode	Random	Oriented

habitats. By contrast, the *K strategy* is characterized by low rates of natality and dispersal, but high investment of resources in storage and individual offspring to ensure their survival. These species are adapted to persist under stable conditions, where competition is intense, but reproduce and disperse too slowly to be good colonizers of isolated patches. Specific characteristics of the two strategies (Table 5.1) have been the subject of debate (Boyce, 1984). For example, small size with smaller resource requirements could reflect a K strategy (Boyce, 1984), although larger organisms typically show more efficient resource use. Nevertheless, this model has been useful for understanding selection of life history attributes (Boyce, 1984).

Insects generally are considered to exemplify the r strategy because of their relatively short life spans, Type 3 survivorship, and rapid reproductive and dispersal rates. However, among insects, a wide range of r-K strategies have been identified. For example, low-order streams (characterized by narrow constrained channels and steep topographic gradients) experience wider variation in water flow and substrate movement, compared to higher-order streams (characterized by broader floodplains and shallower topographic gradients). Insects associated with lower-order streams tend to be more r selected than are insects associated with slower water and greater accumulation of detritus (Reice, 1985). Similarly, ephemeral terrestrial habitats, such as phytotelmata (water-holding plant parts), are dominated by species capable of rapidly colonizing and completing development before the resource disappears or is degraded (Yee and Willig, 2007). Many species associated with relatively stable habitats are poor dispersers, often flightless, even wingless, indicating weak selection for escape and colonization of new habitats, that is, K strategists (Roff, 1990; St. Pierre and Hendrix, 2003). Such species may be at risk if environmental change increases the frequency of disturbance.

Grime (1977) modified the r-K model by distinguishing three primary life history strategies in plants, based on their relative tolerances to disturbance, competition, and stress. Clearly, these three factors are interrelated, since disturbance can affect competition and stress, and stress can increase vulnerability to disturbance. Nevertheless, this model has proven useful for distinguishing the following strategies, characterizing harsh versus frequently disturbed and infrequently disturbed habitats.

The *ruderal* strategy generally corresponds to the r strategy and characterizes unstable habitats; the *competitive* strategy generally corresponds to the K strategy and characterizes relatively stable habitats. The *stress-adapted* strategy characterizes species adapted to persist in harsh environments. These species typically are adapted to conserve resources and minimize exposure to extreme conditions. Insects showing the stress-adapted strategy include those adapted to tolerate freezing in arctic ecosystems or to minimize water loss in desert ecosystems (Teets and Denlinger, 2014; Whitford, 2002, see Chapter 2).

Fielding and Brusven (1995) explored correlations between plant community correspondence to Grime's (1977) strategies and the species traits (abundance, habitat breadth, phenology, and diet breadth) of the associated grasshopper assemblages. They found that the three grasshopper species associated with the ruderal plant community had significantly wider habitat and diet breadths (generalists) and had higher densities than did grasshoppers associated with the competitive or stress-adapted plant communities (Fig. 5.11). Grasshopper

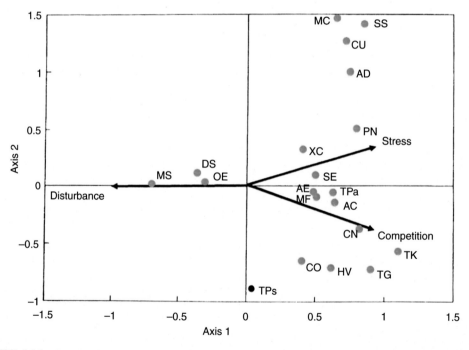

FIGURE 5.11 Constrained correspondence analysis ordination of grasshopper species in southern Idaho, using Grime's (1977) classification of life history strategies based on disturbance, competition, and stress variables *(arrows)*. Grasshoppers are denoted by the initials of their genus and species. The length of arrows is proportional to the influence of each variable on grasshopper species composition. Eigenvalues for axes 1 and 2 are 0.369 and 0.089, respectively. *Source: From Fielding, D.J., Brusven, M.A., 1995. Ecological correlates between rangeland grasshopper (Orthoptera: Acrididae) and plant communities of southern Idaho. Environ. Entomol. 24, 1432–1441.*

assemblages also could be distinguished between the competitive and stress-adapted plant communities, but these differences were only marginally significant. Nevertheless, their study suggested that insects can be classified according to Grime's (1977) model, based on their life history adaptations to disturbance, competition, or stress.

5 PARAMETER ESTIMATION

Whereas insect population structure can be measured by sampling the population, using various standard methods (Leather, 2005; Southwood, 1978), sufficiently accurate estimation of population trends for management purposes requires assessment of detection probability (Gardiner et al., 2005). Sampling techniques vary in their probability of detecting particular species, depending on ecosystem structure. For example, counting grasshoppers flushed from a transect may provide accurate density data in short or sparse vegetation but not in tall or dense vegetation (Gardiner et al., 2005). Light traps capture primarily nocturnally flying insects; branch bagging represents densities of leaf-miners and gall-formers missed by interception traps but underrepresents densities of easily alarmed or highly mobile species (see Chapter 9).

MacKenzie and Kendall (2002) described methods for addressing detection probability in assessment of population change. Estimates of natality, mortality, and dispersal require measurement of changes through time in overall rates of birth, death, and movement. A number of methods used to estimate these population processes (Southwood, 1978) are described next.

Fecundity can be estimated by measuring the numbers of eggs in dissected females or recording the numbers of eggs laid by females caged under natural conditions. Fertility can be measured if the viability of eggs can be assessed. Natality then can be estimated from these data for a large number of females. Mortality can be measured by subtracting population estimates for successive life stages, by recovering and counting dead or unhealthy individuals, or by dissection or immunoassays to identify parasitized individuals. Dispersal capacity can be measured in the laboratory using flight chambers to record duration of tethered flight. Natality, mortality, and dispersal also can be estimated from sequential recapture of marked individuals. However, these techniques require a number of assumptions about the constancy of natality, mortality, and dispersal and their net effects on the population structure of the sample, and do not measure natality, mortality, and dispersal directly.

Deevy (1947) was the first ecologist to apply the methods of actuaries, for determining life expectancy at a given age, to development of survival and reproduction budgets for animals. Life table analysis is the most reliable method to account for survival and reproduction of a population (Begon and Mortimer, 1981; Price, 1997; Southwood, 1978). The advantage of this technique over others is the accounting of survival and reproduction in a way that allows for verification and comparison. For example, a change in cohort numbers due to disappearance of individuals, at a stage when dispersal cannot occur, could signal an error that requires correction, or causal factors that merit examination.

Two types of life tables have been widely used by ecologists. The *age-specific life table* is based on the fates of individuals in a real cohort, a group of individuals born in the same time interval, whereas a *time-specific life table* is based on the fate of individuals in an imaginary cohort derived from the age structure of a stable population with overlapping generations, at a point in time. Because most insects have discrete generations and unstable populations, the age-specific life table is more applicable than the time-specific life table.

TABLE 5.2 Examples of Life Tables

x	l_x	m_x	$l_x m_x$
0	1.0	0	0
1	0.5	0	0
2	0.2	6	1.2
3	0.1	0	0
4	0	0	0
			$\overline{1.2} = R_0$
0	1.0	0	0
1	0.5	0	0
2	0.2	0	0
3	0.1	12	1.2
4	0	0	0
			$\overline{1.2} = R_0$
0	1.0	0	0
1	0.5	0	0
2	0.2	0	0
3	0.1	6	0.6
4	0	0	0
			$\overline{0.6} = R_0$

Note: in these examples, the same or different cohort replacement rates are obtained by the way in which per capita production of offspring is distributed among life stages. x, life stage; l_x, proportion surviving at x; m_x, per capita production at x; and $l_x m_x$, net production at x. The sum of $l_x m_x$ is the replacement rate, R_0.

Life table analysis begins with designation of a cohort (Table 5.2). For simplicity, the starting size of the cohort generally is corrected to a convenient number, generally 1 or 1000 females. Females are the focus of life table budgets because of their reproductive potential. Data from many cohorts representing different birth times, population densities, and environmental conditions should be analyzed and compared to gain a broad view of natality, mortality, and dispersal over a wide range of conditions.

Life tables partition the life cycle into discrete time intervals or life stages (Table 5.2). The age of females at the beginning of each period is designated by x; the proportion of females surviving at the beginning of the period, the age-specific survivorship, by l_x; and the number of daughters produced by each female surviving at age x, or age-specific reproductive rate, by m_x. Age-specific survivorship and reproduction can be compared between life stages to reveal patterns of mortality and reproduction. The products of per capita production and proportion of females surviving for each stage ($l_x \cdot m_x$) can be added to yield the net reproduction, or net *replacement rate* (R_0), of the cohort. Net replacement rate indicates population trend. A stable population has $R_0 = 1$, an increasing population has $R_0 > 1$, and a decreasing population has $R_0 < 1$. These measurements can be used to describe population dynamics, as discussed in the next chapter.

The intensive monitoring necessary to account for survival and reproduction permits identification of factors affecting survival and reproduction. Mortality factors, as well as numbers of immigrants and emigrants, are conveniently identified and evaluated. Survivorship between cohorts can be modeled as a line with a slope of $-k$. This slope variable can be partitioned among factors affecting survivorship, that is, $-k_1, -k_2, -k_3, ..., -k_i$. Such *K-factor analysis* has been used to assess the relative contributions of various factors to survival or mortality (Curry, 1994; Price, 1997; Varley et al., 1973). Factors having the greatest effect on survival and reproduction are designated *key factors* and may be useful in population management. For example, key mortality agents can be augmented for control of pest populations or mitigated for recovery of endangered species (see Chapter 17).

Measurement of insect movement and dispersal is necessary for a number of objectives (Nathan et al., 2003; Turchin, 1998). Disappearance of individuals due to emigration must be distinguished from mortality for life table analysis and assessment of effective dispersal. Movement affects the probability of contact among organisms, determining their interactions. Spatial redistribution of organisms determines population structure, colonization, and metapopulation dynamics (see Chapter 7). Several methods for measuring and modeling animal movement have been summarized by Nathan et al. (2003) and Turchin (1998). Most are labor intensive, especially for insects.

Effective dispersal can be reconstructed from biogeographic distributions, especially for island populations that must have been founded from mainland sources. This method does not reveal the number of dispersing individuals required for successful colonization.

Mark-recapture methods involve marking a large number of individuals, releasing them, and subsequently measuring their frequency in traps or observations at increasing distance from their point of release (Hagler and Jackson, 2001). Several methods can be used to mark individuals. Dye, stable isotope, and rare element incorporation through feeding or dusting provide markers that can be used to distinguish marked individuals from others in the recaptured sample. Some populations are self-marked by incorporation of rare earth elements or other markers unique to their birthplace or overwintering site (Isaacs and Byrne, 1998). Markers should persist without affecting the insect's normal behavior (Hagler and Jackson, 2001).

Large numbers must be marked in order to maximize the probability of recapture at large distances. Schneider (1999) marked c. 7,000,000 adult *Heliothis virescens* using internal dye, released moths at multiple sites over a 238 km^2 area, and trapped moths using pheromones at sites representing a 2000 km^2 area. The mean dispersal distance of male moths was c. 10 km.

Leisnham and Jamieson (2002) used mark-recapture techniques to estimate immigration and emigration rates for mountain stone weta demes among large and small tors in southern New Zealand. They found that the per capita immigration rate on large tors (0.019) slightly exceeded the emigration rate (0.017) whereas the immigration rate on small tors (0.053) was lower than the emigration rate (0.066), explaining the greater tendency for extinction of demes on small tors (4 of 14 over a 3-year study, compared to no extinctions among 4 large tors).

Sword et al. (2005) compared the survival of Mormon crickets, *Anabrus simplex*, in naturally occurring migratory bands with that of individuals transplanted from the band to nearby sites, using individuals equipped with small radio transmitters (Fig. 5.12) that permitted assessment of cricket fate with accuracy. Results indicated that transplanted crickets suffered 50–60% mortality due to predation over 2 days, compared to no mortality of crickets in migratory bands.

Wassenaar and Hobson (1998) used stable isotopes (^{2}H and ^{13}C) to identify the Midwestern USA as the source of most monarch butterflies overwintering at sites in Mexico (Fig. 5.13). Cronin et al. (2000) reported that 50% of marked checkered beetles, *Thanasimus dubius*, moved

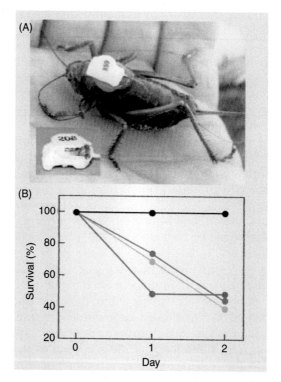

FIGURE 5.12 **Radiometric tracking of Mormon crickets, *Anabrus simplex*.** (A) Female cricket with a 0.45 g radio transmitter glued to the pronotum (inset: a recovered transmitter indicating predation by a rodent); (B) cricket survival in migratory bands (*black line*, $n = 10$ for all dates), compared to conspecifics transplanted away from the band (three replicates: *red*, $n = 10$, *blue* and *orange*, $n = 20$). An equal number of males and females were used in each treatment. Survival in migratory bands was significantly greater than for transplanted crickets. *Source: From Sword, G.A., et al., 2005. Migratory bands give crickets protection. Nature 433, 703.*

at least 1.25 km, 33% moved > 2 km, and 5% dispersed > 5 km, whereas 50% of their primary prey, the southern pine beetle, moved no more than 0.7 km and 95% moved no more than 2.25 km. St. Pierre and Hendrix (2003) demonstrated that 56% of recaptured weevils, *Rhyssomatus lineaticollis*, moved <1 m and 83% moved <50 m.

These methods can indicate the distances moved by individuals, but do not indicate the path, which requires direct observation. In some cases the pattern of dispersal has important implications for management practices. For example, McIntire et al. (2013) used a modeling approach to report that movement of the endangered Fender's blue butterfly, *Plebejus icarioides fenderi*, in Oregon, USA was best described as a random walk, rather than a nonrandom, systematic dispersal and exploration strategy. Such insight contributes to improved management of habitat fragments for species conservation.

Direct observation of movement pattern has limited value for rapidly moving individuals, although marking individuals in various ways can enhance detection at greater distances. New technology has provided for miniaturization of radio, harmonic radar, or microwave transmitters or tags that can be used with a receiver to record the location of an individual

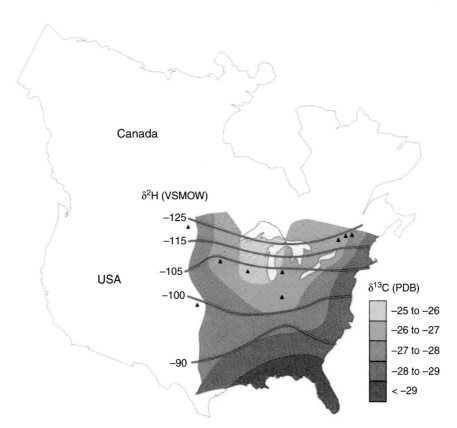

FIGURE 5.13 Geographic patterns of δ²H and δ¹³C in wings of monarch butterflies from rearing sites *(triangles)* across the breeding range in North America. *Source: Modified from Wassenaar, L.I., Hobson, K.A., 1998. Natal origins of migratory monarch butterflies at wintering colonies in Mexico: new isotopic evidence. Proc. Natl. Acad. Sci. USA 95, 15436–15439.*

continuously or at intervals (Lorch et al., 2005; Mascanzoni and Wallin, 1986; J. Riley et al., 1996; Sword et al., 2005). However, marking and electronic signaling methods could affect the behavior or predation of tagged individuals, requiring evaluation of any effects on individual behavior or predation (Lorch et al., 2005).

New genetic techniques permit identification of the source population of dispersing individuals (Nagoshi et al., 2008, 2009, 2010). However, a large number of source individuals must be genotyped in order to distinguish allelic frequencies of multiple sources. Dispersal frequency also may be measured in some cases by taking advantage of relationships between genetic differentiation and distance between demes.

A major challenge to future measurement of dispersal is the increasing homogenization of biotas by human-assisted invasion (Mack et al., 2000). Anthropogenic transport of species across geographic barriers has become a serious threat to ecosystem services globally, as well as to global economics (Roy et al., 2014). A. Suarez et al. (2001) evaluated dispersal of Argentine ants, *Linepithema humile*, at three spatial scales, local, regional, and global, based on documented rates of spread. They discovered that these ants have two discrete dispersal modes: diffusion and jump dispersal. Local diffusion occurs at a maximum rate of 150 m year^{-1}, whereas jump dispersal resulted in annual rates of spread > 160 km year^{-1}, driven largely by association with humans. Ascunce et al. (2011) provided an assessment of genetic variation in 2144 fire ant colonies from 75 sites worldwide. They concluded that at least nine separate introductions have occurred into newly invaded areas and that the southern USA is probably the source of all but one of these introductions. Nagoshi et al. (2008, 2009, 2010) discovered that most fall armyworm, *Spodoptera frugiperda*, in the USA spread from overwintering populations in Texas and Florida, which, in turn, apparently originated from endemic populations in Puerto Rico. Such genomic studies can reveal pathways of dispersal and invasion. However, as invasive species become more widespread, the source of particular populations likely will become more difficult to assess.

6 SUMMARY

Population systems can be described in terms of structural variables and processes that produce changes in population structure. These variables indicate population status and capacity for change in response to environmental heterogeneity.

Structural variables include density, dispersion pattern of individuals and demes, age structure, sex ratio, and genetic composition. Density is the number of individuals per unit area. Dispersion is a measure of how individuals are distributed geographically. Regular dispersion occurs when organisms are spaced evenly among habitat or sampling units. Aggregated dispersion occurs when individuals are found in groups, for mating, mutual defense or resource exploitation, or because of the distribution of resources. Random dispersion occurs when the locations of organisms are independent of the locations of others. Metapopulation structure describes the distribution and interaction among relatively distinct subpopulations, or demes, occurring among habitable patches over a landscape. The degree of isolation of demes influences gene flow among demes and ability to colonize or recolonize vacant patches. Age structure represents the proportion of individuals in each age class and can indicate survivorship patterns or the direction of change in population size. Sex ratio is the proportion of males in the population and indicates the importance of sexual reproduction, mating system, and capacity for reproduction. Genetic composition is described by the frequencies of various alleles in the population and reflects population capacity to adapt to environmental change. Some insect populations have been shown to change gene frequencies within relatively short times in response to strong directional selection, due to short generation times and high reproductive rates. This capacity for rapid change in gene frequencies makes insects especially capable of adapting to anthropogenic changes in environmental conditions.

Processes that produce change in population structure include natality, mortality, and dispersal. Natality is birth rate and represents the integration of individual fecundity and fertility. Natality is affected by abundance and nutritional quality of food resources, abundance and suitability of oviposition sites, availability of males, and population density. Mortality is death rate and reflects the influence of various mortality agents, including extreme weather conditions, food quality, competition, predation, and parasitism. Generally, predation has a greater effect at low to moderate densities, whereas competition and parasitism have a greater effect at high densities. Survivorship curves indicate three types of survivorship, based on whether mortality is consistent or concentrated near the beginning or end of the life span. Dispersal is the movement of individuals from a source, and is a key to genetic mixing and colonization of vacant patches. Individuals colonizing vacant patches have a considerable influence on the genetic composition and development of the deme. The probability that an insect will discover a suitable patch is a function of the tendency to disperse (as affected by life history strategy and crowding), endurance (determined by nutritional condition), patch size, distance, and the mechanism of dispersal (random, phoretic, or directed toward specific habitat cues).

Life history strategies reflect the integration of natality, mortality, and dispersal strategies selected by habitat stability. Two life history classifications have been widely used. Both reflect the importance of disturbance and environmental stress on evolution of complementary strategies for reproduction and dispersal in harsh, stable, or unstable habitats.

Whereas population structure can be described readily by sampling the population, measurement of population processes is more difficult and requires accounting for the fate of individuals. Life table analysis is the most reliable method to account for age-specific survival and reproduction by members of a cohort. The net production of offspring by the cohort is designated the replacement rate and indicates population trend. Advances in technology are creating new opportunities to explore patterns and efficiency of long-distance dispersal. Changes in these variables and processes are the basis for population dynamics. Regulatory factors and models of population change in time and space are described in the next two chapters.

Population Dynamics

RISE AND FALL OF A COTTON HERBIVORE

Insect populations are noted for dramatic fluctuations in population size and often have been targets for control efforts. However, species characterized by periodic outbreaks typically decline to very low numbers as a result of resource depletion and increased predation and parasitism. If population size falls below a critical minimum threshold, males and females may be unable to find each other to reproduce, leading to extinction.

The cotton leafworm, *Alabama argillacea*, was once the most economically important pest of cotton in the Americas and among the most abundant migratory insects in eastern North America (D.L. Wagner, 2009). This moth was a specialized herbivore on cotton, *Gossypium* spp. and close relatives, in Central and South America. It appeared in the USA in 1793, causing serious losses to cotton crops in Georgia and South Carolina, and again in 1800 (C. Riley, 1885). During the next 4 years it spread through the range of commercial cotton in the southern USA, with particularly devastating outbreaks in 1804, 1825, 1846, 1868, 1873, and 1881, and became among the earliest targets of mechanically applied arsenical insecticides (C. Riley, 1885).

This species was not a permanent resident of the US Cotton Belt. It apparently did not have a freeze-tolerant diapause or quiescent stage, so was incapable of continuous breeding beyond its natural range in the tropics and subtropics. However, it reproduced rapidly and often migrated as massive swarms into temperate regions during spring and summer, depending on environmental conditions, reproducing and causing crop damage in the Cotton Belt. The size of the migrating swarms reflected particular environmental conditions and source population size. As a result, swarms often reached the US Cotton Belt after harvest, and most records of the moth in the northern USA are from Sep.—Oct. (D.L. Wagner, 2009).

Cotton production in the USA declined in the early 1900s, as a result of devastation by the boll weevil, *Anthonomus grandis* (R.H. Smith, 2007), and agriculture in the region became much more diverse. The cotton leafworm declined in abundance during this period and apparently became extinct by the late 1900s, probably as a result of reduced cotton acreage and more intense pest management of cotton (D.L. Wagner, 2009).

Insect Ecology. http://dx.doi.org/10.1016/B978-0-12-803033-2.00006-6

1 INTRODUCTION

Changes in abundance of insect species reflect the net effects of population processes (see Chapter 5). Populations of insects can change dramatically in size over relatively short periods of time, given their short generation times and high reproductive rates. Since the beginning of the agricultural age, humans have dealt with plagues of locusts and flies that seemed to come from nowhere to destroy virtually all vegetation or torment humans and livestock over extensive areas. Such views supported early concepts of "Acts of God" and "spontaneous generation" (C. Riley, 1883).

C. Riley (1878, 1880, 1883, 1885) was among the first scientists to document the periodic irruptive behavior of insects and relate outbreaks to changes in environmental conditions. Under favorable environmental conditions, some species have the capacity to increase population size by orders of magnitude in a few years. A swarm of Rocky Mountain grasshoppers, *Melanoplus spretus*, flying over Lincoln, Nebraska on Jun. 16, 1875, a peak year, was estimated to include at least 125 billion locusts covering more than 23,000 km^2 (C. Riley, 1878). This was the largest congregation of animals ever recorded (Lockwood, 2001), although more accurate, and credible, measurement of locust swarm sizes awaited advances in aerial surveillance in the mid-20th century (Gunn et al., 1948; Rainey and Sayer, 1953; Uvarov, 1954). Under adverse conditions, populations can virtually disappear for long time periods or become extinct. Devastating outbreaks have brought insects into direct conflict with humans, but also supported focused entomological research that has expanded our understanding of factors affecting insect population dynamics and, more recently, insect effects on ecosystem conditions and services. Consequently, methods and models for describing population change are most developed for economically important species.

Predicting the effects of global change has become a major goal of research on population dynamics. Insect populations respond to changes in temperature, habitat conditions, and resource quality and abundance (Heliövaara and Väisänen, 1993; Lincoln et al., 1993; Pascual et al., 2006; Reynolds et al., 2007, see Chapters 2 and 3). Their responses to current and historic environmental changes help us to anticipate responses to future environmental changes (Currano et al., 2008). Disturbances, in particular, influence population systems abruptly, but insect responses are integrated by changes in natality, mortality, and dispersal rates. Factors that normally regulate population size, such as resource availability and predation, also are affected by disturbance. Models of population change generally do not incorporate effects of disturbances, which occur unpredictably with respect to insect generations. The population size of an insect species determines its effects on ecosystem conditions, its effects on crop production and human well-being, and its availability as food for other organisms. This chapter addresses temporal patterns of abundance, factors causing or regulating population fluctuation, and models of population dynamics.

2 POPULATION FLUCTUATION

Insect populations can fluctuate dramatically over time. Reynolds et al. (2007) reported that abundances of Lepidoptera larvae fluctuated by >20-fold over a 20-year period. Some insects have long records of fluctuation, including nearly 2000 years for the Oriental migratory locust,

Locusta migratoria manilensis, in China (Konishi and Itô, 1973; Ma, 1958; Stige et al., 2007; Tian et al., 2011), 1173 years for the larch budmoth, *Zeiraphera diniana*, in the European Alps (Esper et al., 2007), and 622 years for the Pandora moth, *Coloradia pandora*, in western North America (Speer et al., 2001). The longest record of insect fluctuation is the 1910-year compendium reported by Tian et al. (2011). Such long records aid greatly in identifying environmental factors responsible for population change. If environmental conditions change in a way that favors insect population growth, the population will increase until conditions change or regulatory factors reduce and finally reverse population growth.

Some populations can vary in density as much as 10^5-fold (Mason, 1996; Mason and Luck, 1978; Royama, 1984; Schell and Lockwood, 1997), but most populations vary less than this (Berryman, 1981; D. Strong et al., 1984). The amplitude and frequency of population fluctuations can be used to describe three general patterns. *Stable* populations fluctuate relatively little over time whereas *irruptive* and *cyclic* populations show wide fluctuations.

Irruptive populations sporadically increase to peak numbers followed by a decline. Certain combinations of life history traits may be conducive to irruptive fluctuation. Larsson et al. (1993) and Nothnagle and Schultz (1987) identified differences in life history attributes between irruptive and nonirruptive species of sawflies and Lepidoptera from European and North American forests. Irruptive species generally are controlled by only one or a few factors, especially drought (Konishi and Itô, 1973; Mattson and Haack, 1987; Preisser and Strong, 2004; Stige et al., 2007; Van Bael et al., 2004), whereas populations of nonirruptive species are controlled by many factors. In addition, irruptive Lepidoptera and sawfly species tend to be gregarious, have a single generation per year, and are sensitive to changes in the quality or availability of their particular resources, whereas nonirruptive species do not share this combination of traits.

Cyclic populations oscillate at regular intervals. Cyclic patterns of population fluctuation have generated the greatest interest among ecologists. Cyclic patterns can be seen over different timescales and may reflect a variety of interacting factors.

Strongly seasonal cycles of abundance can be seen for multivoltine species such as aphids and mosquitoes. Aphid population size is correlated with periods of active nutrient translocation by host plants (Dixon, 1985) and influenced by mid-season increases in predation (Karley et al., 2004). Hence, populations of most aphid species peak in the spring when nutrients are being translocated to new growth, followed by collapse to local extinction during midsummer. Populations of many species (especially those feeding on deciduous hosts) increase again in the fall when nutrients are being resorbed from senescing foliage. This pattern can be altered by disturbance. Schowalter et al. (1981c) reported that sustained growth of early successional vegetation following clear-cutting of a deciduous forest supported continuous growth of aphid populations during the summer (Fig. 6.1). Seven dominant mosquito species in Florida during 1998–2000 showed peak abundances at different times of the year, but the interannual pattern varied as a result of particular environmental conditions, including flooding (Zhong et al., 2003).

Longer-term cycles are apparent for many species. Several forest Lepidoptera exhibit cycles with periods of c. 10, 20, 30, or 40 years (Berryman, 1981; Esper et al., 2007; Mason and Luck, 1978; Price, 1997; Royama, 1992; Swetnam and Lynch, 1993), or combinations of cycles (Speer et al., 2001). For example, the larch budmoth has shown outbreaks every 9.3 years, on average, over a 1173-year period in the European Alps (Fig. 6.2), and spruce budworm,

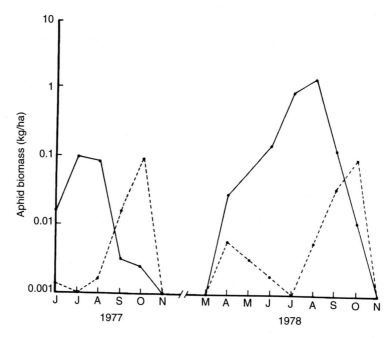

FIGURE 6.1 **Seasonal trends in aphid biomass in an undisturbed (dotted line) and an early successional (solid line) mixed-hardwood forest in North Carolina.** The early successional forest was harvested in 1976–77. Peak abundances in spring and fall in the undisturbed forest reflect nutrient translocation during periods of foliage growth and senescence; continued aphid population growth during summer in the early successional forest reflects continued production of foliage by regenerating plants. *Source: From Schowalter, T.D., 1985. Adaptations of insects to disturbance. In: Pickett, S.T.A., White, P.S. (Eds.), The Ecology of Natural Disturbance and Patch Dynamics. Academic Press, New York, pp. 235–252.*

Choristoneura fumiferana, populations have peaked at approximately 25–30-year intervals over a 250-year period in eastern North America (Royama, 1984), whereas Pandora moth populations have shown a combination of 20- and 40-year cycles over a 622-year period in western North America. Tian et al. (2011) analyzed a nearly 2000-year record of Oriental migratory locust outbreaks in China. Analysis of annual fluctuations indicated a combination of a 30-year cycle during the years 1512–1660 and around 1850 and a 10-year cycle intermittently during the entire 2000-year record. Analysis of decadal fluctuations indicated a predominant 110-year cycle during 1300–1700, with shorter cycles around 1350, 1500, and 1650.

Generally, peak abundances are maintained only for a few (2–3) years, followed by relatively precipitous declines. Changes in population size can be described by four distinct phases (Elkinton and Liebhold, 1990; Mason and Luck, 1978). The *endemic phase* (corresponding to "latency" in European terminology) is the low population level maintained between outbreaks. The beginning of an outbreak cycle is triggered by a disturbance, release from predation, or other environmental change that allows the population to increase in size above its *release threshold*. This threshold represents a population size at which reproductive momentum results in escape of at least a portion of the population from normal regulatory factors, such as predation. Despite the importance of this threshold to population outbreaks, few studies have established its size for any insect species. Schowalter et al. (1981b) reported that

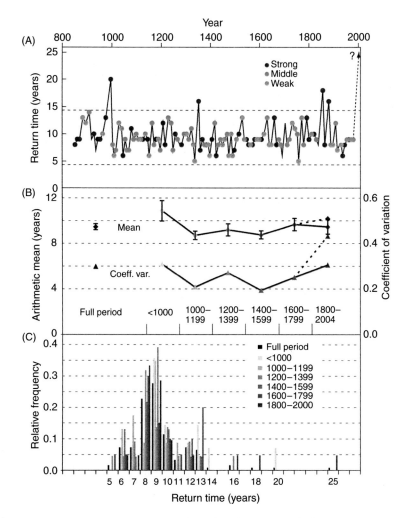

FIGURE 6.2 **Cyclic behavior of larch budmoth,** *Zeiraphera diniana*, **populations over a 1173-year period (832–2004).** (A) Return time record for 123 outbreak events, with colors indicating the severity of outbreaks, measured as reduction in wood density caused by defoliation. The dashed line at the end of the 20th century reflects the absence of outbreaks since 1981. The horizontal dashed lines represent two standard deviations. (B) Arithmetic mean and standard error (bars), and coefficient of variation computed for the full return time record and six subperiods. Dashed lines in recent centuries reflect the absence of outbreaks since 1981. (C) Relative frequency distribution of return times. Colors indicate different time periods. Single events can be represented by bars of different relative heights (eg, all bars in classes above 14 years are single events). *Source: From Esper, J., et al., 2007. 1200 years of regular outbreaks in alpine insects. Proc. Roy. Soc. B-Biol. Sci. 274, 671–679.*

local outbreaks of southern pine beetle, *Dendroctonus frontalis*, occurred when demes reached a critical size of about 100,000 beetles by early June. Above the release threshold, survival is relatively high, and population growth continues during the *release phase* ("progradation" in European terminology). During this period, emigration peaks and the population spreads to other suitable habitat patches (see Chapter 7). Resources eventually become limiting, as a result of depletion by the growing population, and predators and pathogens respond to increased prey/host density and stress. Population growth slows and abundance reaches a *peak* ("outbreak" in European terminology). Competition, predation, and pathogen epizootics initiate and accelerate population *decline* (same in European terminology). Intraspecific competition and predation rates then decline as the population reenters the endemic phase.

Outbreaks of some insect populations have become more frequent and intense in crop systems or natural monocultures where food resources are relatively unlimited or where manipulation of disturbance frequency has created favorable conditions (Heiermann and Schütz, 2008; Kareiva, 1983; Raffa et al., 2008; Wickman, 1992). In other cases, the frequency of recent outbreaks has remained within ranges for the frequencies of historic outbreaks, but the extent or severity has increased as a result of anthropogenic changes in the vegetation structure or disturbance regime (Speer et al., 2001).

Populations of many species fluctuate at amplitudes that are insufficient to cause economic damage or attract attention. Some of these species may experience more conspicuous outbreaks under changing environmental conditions, for example, climate change, introduction into new habitats, or large-scale conversion of natural ecosystems to managed ecosystems (Mattson and Haack, 1987; Van Bael et al., 2004; D. Williams and Liebhold, 2002). The Colorado potato beetle, *Leptinotarsa decemlineata*, subsisted on wild solanaceous hosts in western North America until westward movement of settlers brought it into contact with cultivated potato in the Midwest during the late 1800s (Hitchner et al., 2008; C. Riley, 1883; Stern et al., 1959), allowing it to spread eastward and, eventually, to Europe. Similarly, the cotton boll weevil, *Anthonomus grandis*, coevolved with scattered wild *Gossypium* spp., including *G. hirsutum*, in tropical Mesoamerica until citrus cultivation in the 1890s provided overwintering food resources that allowed the insect to spread into subtropical cotton-growing regions of south Texas and northern Argentina (Showler, 2009). Subsequently, rapid reproduction in the spring by overwintering adults permitted spread throughout the US Cotton Belt (Showler, 2009).

3 FACTORS AFFECTING POPULATION SIZE

Populations showing wide amplitude of fluctuation have weak intrinsic ability to regulate growth, for example, through depressed natality in response to crowding or competition, in some cases because of rapid dispersal from core areas that prevents overcrowding. Population outbreaks may represent the cumulative effects of many factors that permit population growth above successive thresholds (Fig. 6.3). These factors can influence population size in two primary ways. If the proportion of organisms affected by a factor is constant for any population density, or the effect of the factor does not depend on population density, the factor is considered to have a *density-independent* effect. Conversely, if the proportion of organisms affected varies with density, or the effect of the factor depends on population density, then the factor is considered to have a *density-dependent* effect (Begon and Mortimer, 1981; Berryman, 1981; L. Clark et al., 1967; Price, 1997; Raffa et al., 2008).

Factors affecting population size can operate over a range of time delays. For example, fire affects numbers immediately (no time lag) by killing exposed individuals, whereas predation requires some period of time (time lag) for predators to aggregate in an area of dense prey

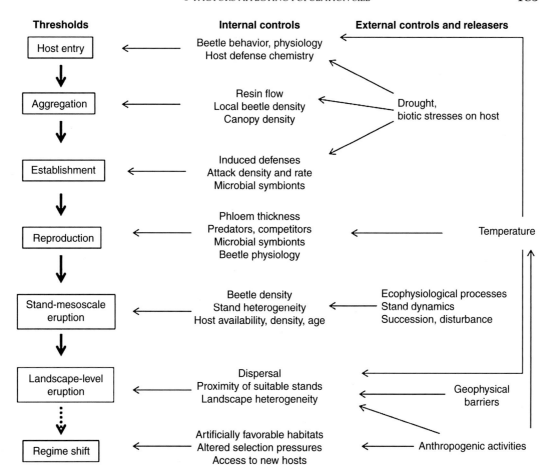

FIGURE 6.3 **Conceptual diagram of a sequence of thresholds (boxes) that must be exceeded, as a result of internal and external controlling factors, to produce a landscape-scale irruption of bark beetles.** Thresholds progress from the individual (successful host entry) to population (aggregation, establishment, reproduction) to ecosystem (stand, landscape) levels. Major direct controls that influence the likelihood of beetles surpassing a particular threshold are shown in the middle column. External controls and releasers are shown in the right column. Anthropogenic activities constitute external releasers that facilitate breach of a previously unsurpassed threshold along this continuum into an altered regime state that can explain recent changes in the irruptive behavior of many insect species. *Source: From Raffa, K.F., et al., 2008. Cross-scale drivers of natural disturbances prone to anthropogenic amplification: the dynamics of bark beetle eruptions. BioScience 58, 501–517.*

and to produce offspring. Hence, increased prey density is followed by increased predator density only after some time lag. Similarly, as prey abundance decreases, predators disperse or cease reproduction, but only after a time lag, during which predation continues to reduce prey abundance.

The distinction between density independence and density dependence is often confused, for various reasons. First, many factors may act in both density-independent and density-dependent manners, depending on circumstances. For example, climatic factors or disturbances often are thought to affect populations in a density-independent manner, because their effect does not depend on population density. However, if shelter from unfavorable conditions is limited, then the effect of the climatic factor or disturbance may appear to be related to population density. A plant defense may have a density-independent effect until herbivore densities reach a level that triggers induced defenses. Therefore, a particular factor may have a density-independent effect over one range of population densities and a density-dependent effect over another range of densities. Generally, population size is modified by abiotic factors, such as climate and disturbance, but maintained near an equilibrium level by density-dependent biotic factors (see Section 3.3).

3.1 Density-Independent Factors

Insect populations are highly sensitive to changes in abiotic conditions, such as temperature, water availability, etc., which affect insect growth and survival (see Chapter 2). Changes in population size of some insects have been related directly to changes in climate or to disturbances, especially drought (Greenbank, 1963; Kozár, 1991; Mattson and Haack, 1987; Porter and Redak, 1996; Reice, 1985; Stige et al., 2007). In some cases, climate fluctuation or disturbance affects resource values for insects. For example, loss of riparian habitat as a result of agricultural practices in western North America may have led to extinction of the historically important Rocky Mountain grasshopper, *M. spretus* (Lockwood, 2001; Lockwood and DeBrey, 1990).

Many environmental changes occur relatively slowly and cause gradual changes in insect populations as a result of subtle shifts in genetic structure and individual fitness. Other environmental changes occur more abruptly and may trigger rapid change in population size because of sudden changes in natality, mortality, or dispersal. Esper et al. (2007) found that larch budmoth, *Z. diniana*, showed regular population outbreaks at 9-year intervals for 1200 years, during warming and cooling climate periods, but has failed to increase since 1981, a period during which temperatures have increased above the historic range.

Insect populations also are likely to respond to changing global temperature, precipitation patterns, atmospheric and water pollution, and atmospheric concentrations of CO_2 and other trace gases (Alstad et al., 1982; Esper et al., 2007; Franklin et al., 1992; Heliövaara, 1986; Heliövaara and Väisänen, 1993; L. Hughes and Bazzaz, 1997; Lincoln et al., 1993; Marks and Lincoln, 1996; Reynolds et al., 2007; D. Williams and Liebhold, 2002, see Chapter 2). Many grasshopper, moth, and bark beetle populations are favored by warm and/or dry conditions (Breshears et al., 2005; Capinera, 1987; Konishi and Itô, 1973; Ma, 1958; Mattson and Haack, 1987; Reynolds et al., 2007; Stige et al., 2007) predicted by climate change models to increase in many regions. In particular, Reynolds et al. (2007) noted that Lepidoptera abundances in New Hampshire, USA, forests increased during summers that were relatively warm and/or long, because these conditions favor more rapid development and may permit a second generation each year for facultatively bivoltine species (Fig. 6.4). Contrary to expectations, mild winter temperatures and interannual variation in tree growth were not correlated with Lepidoptera abundances. D. Williams and Liebhold (2002) projected an increased outbreak area and shift northward for the southern pine beetle, *D. frontalis*, but a reduced outbreak area and shift to higher elevations for the mountain pine beetle, *D. ponderosae*, in North America as a result of increasing temperature. Vector competence of mosquitoes also is related to temperature (Watts et al., 1987). Warming temperatures are likely to increase

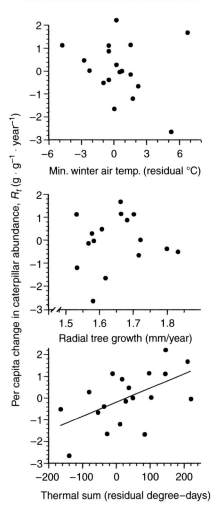

FIGURE 6.4 **Per capita change in caterpillar abundance in relation to variation in minimum winter air temperature (eg, cold tolerance), radial tree growth (eg, host condition), and thermal sum during summer (eg, development rate) in the White Mountains of New Hampshire.** *Source: From Reynolds, L.V., et al., 2007. Climatic effects on caterpillar fluctuations in northern hardwood forests. Can. J. Forest. Res. 37, 481–491.*

transmission of various arthropod-borne diseases (van Panhuis et al., 2015). Altered amplitude of variation in temperature or precipitation may be more important than changes in mean levels (Estay et al., 2014).

Interaction among multiple factors changing simultaneously may affect insects differently than predicted from responses to individual factors (Franklin et al., 1992; Marks and Lincoln, 1996). For example, insects may respond directly to warming temperatures, to increased exposure to parasitoids attracted to plant volatiles (Gouinguené and Turlings, 2002; Hance et al., 2007) or to altered host abundance or reduced mortality to parasites or predators that respond differently to temperature change (Hance et al., 2007; Klok et al., 2003; Ris et al., 2004).

Paleontological data provide additional evidence of increased herbivore populations during warmer or drier periods. Currano et al. (2008) reported that the frequency and diversity of leaf damage by herbivorous insects were significantly greater (mean 57%) during the Paleocene–Eocene thermal maximum (PETM) 56 million years ago than during either the preceding Paleocene (<38%) or subsequent Eocene (33%) epochs (Fig. 2.18). Both the amount and diversity of herbivore damage on angiosperm leaves were correlated positively with temperature change 55–59 million years ago.

Disturbances are particularly important triggers for inducing population change because of their acute disruption of population structure and of resource, substrate, and other ecosystem conditions. The disruption of population structure can alter community structure and cause changes in physical, chemical, and biological conditions of the ecosystem. Disturbances can promote or truncate population growth, depending on species tolerances to particular disturbance or postdisturbance conditions.

Drought, in particular, has been identified as an important trigger for population irruptions of many species, including locusts, moths, and bark beetles, promoted by host stress and crowding (Breshears et al., 2005; Konishi and Itô, 1973; Mattson and Haack, 1987; Preisser and Strong, 2004; Schowalter et al., 1999; Stige et al., 2007; Tian et al., 2011; Van Bael et al., 2004), but storm damage or increased precipitation also increases resources for some insects (Hanewinkel et al., 2008; Schowalter and Ganio, 2003; Schowalter et al., 2014). Interestingly, locust outbreaks appear to be triggered by either drought or flooding disturbances. A 1000-year record of locust outbreaks in China indicated that outbreaks typically originated in floodplain refuges, locations of adequate vegetation and suitable oviposition sites, during drought years and years after flooding (Fig. 6.5, Stige et al., 2007). Droughts increase the availability of suitable oviposition sites as water recedes, as well as stressed vegetation, whereas similar conditions occur in formerly flooded areas in the year after flooding.

Some species are more tolerant of particular disturbances, based on adaptation to regular recurrence. For example, plants in fire-prone ecosystems show attributes that protect meristematic tissues, whereas those in frequently flooded ecosystems can tolerate root anaerobiosis. Generally, insects do not have specific adaptations to survive disturbance, given their short generation times relative to disturbance intervals, and unprotected populations may be greatly reduced. Disturbance-adapted traits are most evident in species exposed to disturbances that occur on annual timescales [such as hypoxia tolerance in floodplains (Brust and Hoback, 2009; Zerm and Adis, 2003)] and in species with longer (2–5 year) generation times that would increase the frequency of generations experiencing a disturbance (such as orientation to smoke plumes or avoidance of litter accumulations in fire-prone ecosystems (W. Evans, 1966; K. Miller and Wagner, 1984, see Chapter 2).

Species also are affected by postdisturbance conditions. Disturbances affect insect populations both directly and indirectly, as described in Chapter 2.

Disturbances affect insects directly by creating lethal conditions for intolerant individuals (see Chapter 2). For example, fire can burn exposed insects (Porter and Redak, 1996; P. Shaw et al., 1987) or raise temperatures to lethal levels in unburned microsites. Tumbling cobbles in flooding streams can crush benthic insects (Reice, 1985). Flooding of terrestrial habitats can create anaerobic soil conditions. Drought can raise air and soil temperatures and cause desiccation (Mattson and Haack, 1987). Populations of many species can suffer severe mortality as a result of these factors, and rare species may be eliminated (P. Shaw et al., 1987;

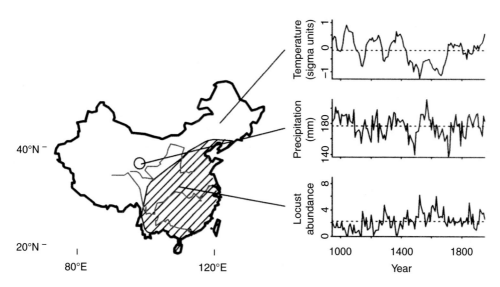

FIGURE 6.5　One thousand years of locust abundance and climate data from China. The present distribution of *Locusta migratoria manilensis* is shown by hatched lines on the map. Large rivers, blue lines, from north to south are the Yellow, Yangtze, and Pearl. Temperature data are a composite index for all of China. Precipitation data are from northeastern Qinghai (circle). Locust data are for all of China. *Source: From Stige, L.C., et al., 2007. Thousand-year-long Chinese time series reveals climatic forcing of decadal locust dynamics. Proc. Natl. Acad. Sci. USA 104, 16188–16193.*

Schowalter, 2012). Willig and Camilo (1991) reported the virtual disappearance of two species of walkingsticks, *Lamponius portoricensis* and *Agamemnon iphimedeia*, from tabonuco, *Dacryodes excelsa*, forests in Puerto Rico following Hurricane Hugo. Drought can reduce water levels in aquatic ecosystems, reducing or eliminating habitat for some aquatic insects. In contrast, storms may redistribute insects picked up by high winds. Torres (1988) documented large numbers of insects being transported to Caribbean islands by hurricane winds, including swarms of African desert locusts, *Schistocerca gregaria*.

Mortality depends on disturbance intensity and scale and species adaptation. K. Miller and Wagner (1984) reported that the pandora moth preferentially pupates on soil with sparse litter cover, under open canopy, where it is more likely to survive frequent understory fires. This habit would not protect pupae during more severe fires. Small-scale disturbances affect a smaller proportion of the population than do larger-scale disturbances. Large-scale disturbances, such as volcanic eruptions or hurricanes, could drastically reduce populations over much of the species range, making such populations vulnerable to extinction. The potential for disturbances to eliminate small populations or critical local demes of fragmented metapopulations has become a serious obstacle to restoration of endangered (or other) species (P. Foley, 1997).

Disturbances indirectly affect insect populations by altering distribution of resources or species interactions in the postdisturbance environment. Disturbance affects the abundance or physiological condition of hosts and the abundances or activity of other associated organisms (Mattson and Haack, 1987; T. Paine and Baker, 1993). Selective mortality to disturbance-intolerant plant species reduces the availability of their resources for associated herbivores.

Similarly, long disturbance-free intervals can lead to eventual replacement of ruderal plant species and their associated insects. Changes in canopy cover or plant density alter vertical and horizontal gradients in light, temperature, and moisture that influence habitat suitability for insect species, alter plant conditions, including nitrogen concentrations, and can alter vapor diffusion patterns that influence chemoorientation by insects (Cardé, 1996; Kolb et al., 1998; Mattson and Haack, 1987; J. Stone et al., 1999).

Disturbances injure or stress surviving hosts or change plant species density or apparency. The grasshopper, *Melanoplus differentialis*, prefers wilted foliage of the sunflower to turgid foliage (A. Lewis, 1979). Fire or storms can wound surviving plants and increase their susceptibility to herbivorous insects. Lightning-struck (Fig. 6.6) and windthrown or cut trees are particular targets for many bark beetles and provide refuges for these insects at low population levels (Flamm et al., 1993; T. Paine and Baker, 1993; Witcosky et al., 1986). Drought stress can cause audible cell wall cavitation that may attract insects adapted to exploit water-stressed hosts (Mattson and Haack, 1987). Stressed plants may alter their production of particular amino acids or suppress production of defensive chemicals to meet more immediate metabolic needs, thereby affecting their suitability for particular herbivores (Haglund, 1980; Lorio, 1993; R. Waring and Pitman, 1983). If drought or other dis-

FIGURE 6.6 **Lightning-strike or other injury or stress impairs tree defense systems.** Injured, diseased, or water-stressed trees typically are targets of bark beetle colonization.

turbances stress large numbers of plants surrounding these refuges, small populations can reach epidemic sizes quickly (Breshears et al., 2005; Mattson and Haack, 1987). Plant crowding, as a result of planting or long disturbance-free intervals, causes competitive stress. High densities and/or apparencies of particular plant species facilitate host colonization and population growth, frequently triggering outbreaks of herbivorous species (Mattson and Haack, 1987, see Chapter 3).

Changes in abundances of competitors, predators, and pathogens also affect postdisturbance insect populations. For example, phytopathogenic fungi establishing in, and spreading from, woody debris following fire, windthrow, or harvest can stress infected surviving trees and increase their susceptibility to bark beetles and other wood-boring insects (T. Paine and Baker, 1993). Drought or solar exposure resulting from disturbance can reduce the abundance or virulence of entomopathogenic nematodes, fungi, bacteria, or viruses (Mattson and Haack, 1987; Preisser and Strong, 2004; Roland and Kaupp, 1995). Disturbance and/or fragmentation reduce the abundances and activity of some predators and parasites (Kruess and Tscharntke, 1994; Roland and Taylor, 1997) and may induce or support outbreaks of defoliators (Roland, 1993). Alternatively, disturbance or fragmentation can interrupt spread of some insect populations by creating inhospitable barriers (Schowalter et al., 1981b).

Population responses to direct or indirect effects vary, depending on the scale of disturbance (see Chapter 7). Few natural experiments have addressed the effects of scale. Clearly, a larger-scale event should affect environmental conditions and populations within the disturbed area more than would a smaller-scale event. Shure and Phillips (1991) compared arthropod abundances in clear-cuts of different sizes in the southeastern USA (Fig. 6.7). They suggested that the greater differences in arthropod densities in larger clear-cuts reflected the steepness of environmental gradients from the clear-cut into the surrounding forest. The surrounding forest has a greater effect on environmental conditions within a small canopy opening than within a larger opening.

The capacity for insect populations to respond quickly to abrupt changes in environmental conditions (disturbances) is responsible for their responses to anthropogenic changes. Insect outbreaks have become particularly frequent and severe in landscapes that have been significantly altered by human activity (Hadley and Veblen, 1993; Huettl and Mueller-Dombois, 1993; Wickman, 1992). Anthropogenic suppression of fire, channelization and clearing of riparian areas, and conversion of natural, diverse vegetation to rapidly growing, commercially valuable crop species on a regional scale have resulted in more severe disturbances and dense monocultures of susceptible species that support widespread outbreaks of adapted insects (Schowalter, 2013; Schowalter and Lowman, 1999).

The similarity in insect population responses to natural versus anthropogenic changes in the environment depends on the degree to which anthropogenic changes create conditions similar to those created by natural changes. For example, natural disturbances typically remove less biomass from a site than do harvest or livestock grazing. This difference likely affects insects that depend on postdisturbance biomass, such as large woody debris, either as a food resource or as a refuge from exposure to altered temperature and moisture (Grove, 2002; Seastedt and Crossley, 1981a). Anthropogenic disturbances leave straighter and more distinct boundaries between disturbed and undisturbed patches (because of ownership or management boundaries), affecting the character of edges and the steepness of environmental

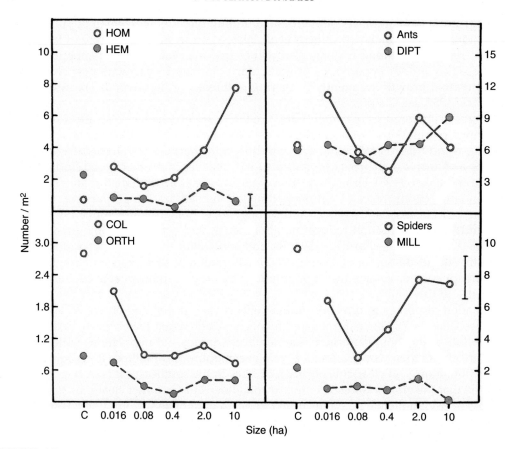

FIGURE 6.7 **Densities of arthropod groups during the first growing season in uncut forest (C) and clear-cut patches ranging in size from 0.016 to 10 ha.** For groups showing significant differences between patch sizes, vertical bars indicate the least significant difference ($p < 0.05$). HOM, Homoptera; HEM, hemiptera; COL, coleopteran; ORTH, orthoptera; DIPT, diptera, and MILL, millipedes. *Source: From Shure and Phillips (1991). Patch size of forest openings and arthropod populations. Oecologia 86, 325–334.*

gradients into undisturbed patches (J. Chen et al., 1995; Roland and Kaupp, 1995). Similarly, the scale, frequency, and intensity of prescribed fires may differ from natural fire regimes. In northern Australia, natural ignition would come from lightning during storm events at the onset of monsoon rains, whereas prescribed fires often are set during drier periods to maximize fuel reduction (Braithwaite and Estbergs, 1985). Consequently, prescribed fires burn hotter, are more homogeneous in their severity, and cover larger areas than do lower intensity, more patchy fires burning during cooler, moister periods.

Few studies have evaluated the responses of insect populations to changes in multiple abiotic factors. Habitat fragmentation, climate change, acid precipitation, multiple atmospheric pollutants, and introduction of exotic species may influence insect populations interactively in many areas (Lerdau et al., 1997; Valkama et al., 2007). For example, stepwise multiple regression indicated that persistence of native ant species in coastal scrub habitats in southern

California was best predicted by the abundance of invasive Argentine ants, *Linepithema humile*, size of habitat fragments, and time since fragment isolation (A. Suarez et al., 1998).

3.2 Density Dependent Factors

Primary density-dependent factors include intra- and interspecific competition, for limited resources, and predation. The relative importance of these factors has been the topic of much debate.

Resource limitation is a significant factor for all organisms. Malthus (1789) wrote the first theoretical treatise describing the increasing struggle for limited resources by growing populations. Effects of intraspecific competition (crowding) on natality, mortality, and dispersal have been demonstrated widely (see Section 2.2). In some species, competition may involve physical conflict among individuals attempting to gain access to limited resources. For many species, territoriality, ritual fighting, or dominance hierarchies that establish order for access limit physical conflict and injury but also limit access to resources for a portion of the population (see Chapter 4, Section 3.4). For example, foragers of the harvester ant, *Pogonomyrmex barbatus*, often encounter foragers from neighboring colonies, but relatively few encounters (about 10%) involved fighting, and fewer (21% of fights) resulted in the death of any of the combatants (Gordon and Kulig, 1996). Nevertheless, colonies were spaced at distances that indicated minimization of competition. As competition for finite resources increases, fewer individuals obtain sufficient resources to survive, reproduce, or disperse.

Similarly, a rich literature on predator–prey interactions generally, and biocontrol agents in particular, has shown the important density-dependent effects of predators, parasitoids, and parasites on prey populations (Carpenter et al., 1985; Kollberg et al., 2014; Marquis and Whelan, 1994; Parry et al., 1997; Price, 1997; Tinbergen, 1960; van den Bosch et al., 1982; Van Driesche and Bellows, 1996). Predation rates typically increase as prey abundance increases, up to a point at which predators become satiated. Predators respond both behaviorally and numerically to changes in prey density (see Chapter 8). Predators can be attracted to an area of high prey abundance, a behavioral response, and increase production of offspring as food supply increases, a numerical response. Parasites are not subject to satiation, and natural epizootics commonly terminate outbreaks (Brookes et al., 1978, see Chapter 8).

Cooperative interactions among individuals can lead to inverse density-dependence, that is, positive feedback, up to a point. Mating success (and thus natality) increases as density increases, up to a point at which male competition for females interferes with mating. Some insects show increased ability to exploit resources as density increases. Examples include bark beetles that must aggregate in order to kill trees, a necessary prelude to successful reproduction in dead or dying trees (Berryman, 1997; Coulson, 1979), and social insects that increase thermoregulation and recruitment of nestmates to harvest suitable resources as colony size increases (Heinrich, 1979; Matthews and Matthews, 2010, see Chapters 2 and 3).

3.3 Regulatory Mechanisms

When population size exceeds the number of individuals that can be supported by existing resources, competition, predation, and other factors reduce the population size until it reaches levels in balance with resource supply. This equilibrium population size, that can be sustained indefinitely by resource availability, is termed the *carrying capacity* of the environment and is designated as *K*. Carrying capacity is not constant, but depends on factors that affect both the abundance and suitability of necessary resources, including the intensity of competition with other species that also use those particular resources.

Density-independent factors modify population size, but only density-dependent factors can regulate population size, in the sense of stabilizing abundance near carrying capacity, through negative feedback processes. Regulation requires environmental feedback, such as through density-dependent mechanisms that allow population growth at low densities but reduce population growth at high densities (Isaev and Khlebopros, 1979). However, disturbances or environmental changes can change population size abruptly and either inhibit regulatory factors temporarily, and trigger population growth, or reduce population size and alleviate the induction of regulatory mechanisms (Schowalter, 2012).

Nicholson (1933, 1954a,b, 1958) first postulated that density-dependent biotic interactions are the primary factors determining population size. Andrewartha and Birch (1954) challenged this view, suggesting that density-dependent processes generally are of minor importance in determining abundance. This debate was resolved with recognition that regulation of population size requires density-dependent processes, but abundance is determined by all factors that affect the population (Begon and Mortimer, 1981; Isaev and Khlebopros, 1979). However, debate continues over the relative importance of competition and predation, the so-called *"bottom-up"* (or *resource concentration*) and *"top-down"* (or *trophic cascade*) hypotheses, for regulating population sizes (see also Chapter 8).

Bottom-up regulation is accomplished through the dependence of populations on resource supply. Suitable food is most often invoked as the limiting resource, but suitable shelter and oviposition sites also may become limiting as populations grow. Increased quality or availability of resources promotes population growth. As populations grow, resources become increasingly limited and objects of intense competition, reducing natality and increasing mortality and dispersal (Chapter 5). Eventually, resources become depleted, reducing population growth. As population size declines, resources become relatively more available and support population growth. Hence, a population should tend to fluctuate around the size (carrying capacity) that can be sustained by resource supply.

Top-down regulation is accomplished through the response of herbivores, predators, and parasites to increasing host availability. As host or prey abundance increases, predators and parasites encounter more hosts or prey. Predators respond functionally to increased abundance of a prey species by learning to acquire prey more efficiently and respond numerically by increasing population size as food supply increases (see Chapter 8). Increased intensity of predation reduces prey numbers. Reduced prey availability limits food supply for predators and reduces the intensity of predation. Hence a prey population should fluctuate around the size determined by the intensity of predation.

A number of experiments have demonstrated the dependence of insect population growth on resource availability, especially the abundance of suitable food resources, in aquatic and terrestrial ecosystems (M. Brown et al., 1987; Cappuccino, 1992; Harrison, 1994; T. Kim and Underwood, 2015; Lunderstädt, 1981; Ohgushi and Sawada, 1985; Polis and Strong, 1996; Price, 1997; Ritchie, 2000; Schowalter and Turchin, 1993; Schultz, 1988; Scriber and Slansky, 1981; Varley and Gradwell, 1970; Wallace and Webster, 1996). For example, S. Eggert and Wallace (2003) and Wallace et al. (2015) demonstrated that abundances of aquatic detritivores depended on experimental levels of detrital resources in streams. Improved food quality or

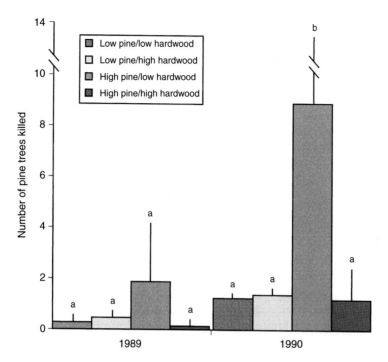

FIGURE 6.8 **Effect of host (pine) and nonhost (hardwood) densities on population growth of the southern pine beetle, measured as pine mortality in 1989 (Mississippi) and 1990 (Louisiana).** Low pine = 11–14 m² ha⁻¹ basal area; high pine = 23–29 m² ha⁻¹ basal area; low hardwood = 0–4 m² ha⁻¹ basal area; high hardwood = 9–14 m² ha⁻¹ basal area. Vertical lines indicate standard error of the mean. Bars under the same letter did not differ at an experiment-wise error rate of $p < 0.05$ for data combined for the 2 years. *Source: Data from Schowalter, T.D., Turchin, P., 1993. Southern pine beetle infestation development: interaction between pine and hardwood basal areas. Forest Sci. 39, 201–210.*

abundance as a result of environmental changes frequently results in population growth of herbivorous species. Schowalter and Turchin (1993) demonstrated that growth of southern pine beetle populations, measured as number of host trees killed, was significant only under conditions of high host density and low nonhost density (Fig. 6.8). Jactel and Brockerhoff (2007) conducted a metaanalysis of 119 studies that compared insect herbivory in monoculture and mixed-species forests. Overall, herbivory was significantly higher in monocultures than in more diverse forests. However, this pattern was more significant for specialist herbivores than for generalist herbivores and depended on the particular tree species mix, that is, the relative abundances of host versus nonhost species in diverse forests. Specialists may respond to the density of particular host species, whereas generalists may benefit from the greater total resource abundance provided by a diversity of resources (Abdala-Roberts et al., 2015). Loranger et al. (2014) reported that herbivory, primarily by grassland generalists, increased along an experimental diversity gradient that was part of the Jena Experiment in Germany, perhaps because of improved nutrient stoichiometry in more diverse plots (Abbas et al., 2014). Haase et al. (2015) found similar results in a set of forest diversity experiments.

Population regulation by herbivores, predators, or parasites has been supported by experiments demonstrating population growth following herbivore, predator, or parasite removal

(Beard et al., 2003; Carpenter and Kitchell, 1987, 1988; Dial and Roughgarden, 1995; Gruner et al., 2008; M. Johnson et al., 2010; Marquis and Whelan, 1994; Oksanen, 1983; Turchin et al., 1999) or population decline following herbivore, predator, or parasite augmentation (Preisser and Strong, 2004). Manipulative experiments have shown that increased abundance at one trophic level causes reduced abundance of the next lower trophic level and increased abundance at the second trophic level down, a *trophic cascade* (Carpenter and Kitchell, 1987, 1988; Letourneau and Dyer, 1998, see Chapter 8). However, in many cases, herbivores and predators appear simply to respond to host or prey abundance without regulating prey populations (Parry et al., 1997), and the effect of predation and parasitism often is delayed (Turchin et al., 1999) and less obvious than the effects of resource supply.

Regulation by lateral factors does not involve other trophic levels. Interference competition, territoriality, cannibalism, and density-dependent dispersal have been considered to be lateral factors that may have a primary regulatory role (Harrison and Cappuccino, 1995). For example, Fox (1975a) reviewed studies indicating that cannibalism is a predictable part of the life history of some species, acting as a population control mechanism that rapidly decreases the number of competitors, regardless of food supply. In the backswimmer, *Notonecta hoffmanni*, cannibalism of young nymphs by older nymphs occurred even when alternative prey were abundant (Fox, 1975b). In other species, any exposed or unprotected individuals are attacked (Fox, 1975a). However, competition clearly is affected by resource supply.

Harrison and Cappuccino (1995) compiled data from 60 studies in which bottom-up, top-down, or lateral density-dependent regulatory mechanisms were evaluated for populations of invertebrates, herbivorous insects, and vertebrates. They reported that bottom-up regulation was apparent in 89% of the studies, overall, compared to observation of top-down regulation in 39% and lateral regulation in 79% of the studies. Top-down regulation was observed more frequently than bottom-up regulation only for the category that included fish, amphibians, and reptiles. Bottom-up regulation may predominate in (primarily terrestrial) systems where resource suitability is more limiting than is resource density, that is, resources are defended in some way (especially through incorporation of carbohydrates into indigestible lignin and cellulose). Top-down regulation may be more important in (primarily aquatic) systems where resources are relatively undefended, or consumers are adapted to defenses, and production can compensate for consumption (D. Strong, 1992).

Gruner et al. (2008) conducted a metaanalysis of 191 factorial manipulations of herbivores and nutrient supply in freshwater ($n = 116$), marine ($n = 60$), and terrestrial ($n = 15$) ecosystems to evaluate the relative and interactive effects of fertilization and herbivory on producer biomass. They concluded that nutrient subsidy increased producer biomass in all ecosystems, but most in freshwater ecosystems, and that herbivore removal generally increased producer biomass in freshwater and marine ecosystems but had inconsistent effects in terrestrial ecosystems (Fig. 6.9). They found little evidence for interaction between bottom-up and top-down effects. Taken together, Gruner et al. (2008) and Harrison and Cappuccino (1995) indicate that bottom-up regulation is stronger and more pervasive among ecosystems and that top-down regulation provides secondary regulation that may become most effective when hosts or prey become stressed or diseased as a result of overabundance and competition for resources.

Regulating factors can have indirect, as well as direct, effects on population dynamics (see Chapter 8). For example, ants affected gall-forming psyllid, *Baccharopelma dracunculifoliae*,

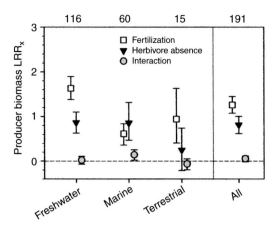

FIGURE 6.9   Effect of fertilization, herbivore removal, and their interaction on producer biomass in freshwater, marine, terrestrial, and combined ecosystems. Sample sizes are given above the panel. The effect is statistically significant when the 95% confidence intervals do not overlap the dashed line at 0 and is significantly different from other effects when the confidence intervals do not overlap. *Source: From Gruner, D.S., et al., 2008. A cross-system synthesis of consumer and nutrient resource control on producer biomass. Ecol. Lett. 11, 740–755.*

abundance directly by interfering with female oviposition behavior and indirectly by tending aphids, *Uroleucon erigeronensis*, which competed for host resources (Fagundes et al., 2005).

All populations probably are regulated simultaneously by bottom-up, top-down, and lateral factors (Denno et al., 2002, 2003; Fagundes et al., 2005; Stiling and Moon, 2005). For example, population growth of bark beetles, *Dendroctonus* spp., is known to be regulated by the availability of stressed host trees (Amman et al., 1988; Raffa et al., 2008; Sartwell and Stevens, 1975; Schowalter and Turchin, 1993; R. Waring and Pitman, 1983), but outbreaks eventually collapse as a result of a combination of depleted resources and predation (Cairns et al., 2008; Coleman et al., 2008; Schowalter et al., 1981b; Turchin et al., 1999).

Whereas density dependence acts in a regulatory (stabilizing) manner through negative feedback, that is, acting to slow or stop continued growth, inverse density dependence has been thought to act in a destabilizing manner. Allee (1931) first proposed that positive feedback creates unstable thresholds, that is, an *extinction threshold* below which a population inevitably declines to extinction, and the *release threshold* above which the population grows uncontrollably until resource depletion or epizootics decimate the population (Begon and Mortimer, 1981; Berryman, 1996, 1997; Isaev and Khlebopros, 1979). Between these thresholds, density-dependent factors should maintain stable populations near K, a property known as the *Allee effect*. However, positive feedback may ensure population persistence at low densities and is counteracted, in most species, by the effects of crowding, resource depletion, and predation at higher densities.

3.4 Causes of Outbreaks and Extinctions

Clearly, factors that cause population outbreaks or extinction are of particular interest to ecologists, as well as to resource managers. Outbreaks and extinctions affect both ecosystem services and management goals. As described in Sections 3.1–3.3, a number of factors affect population growth. Factors that generate outbreaks, in particular, have received much attention.

Host plant density and stress and disruption of predation are primary factors promoting outbreaks (Koricheva et al., 1998, 2000; Mattson and Haack, 1987; Preisser and Strong, 2004; Schowalter and Turchin, 1993, see Section 3.3). If these conditions occur over large areas, outbreaks can be extensive and affect larger areas, as insects spread to less suitable hosts and increase the probability of colonizing distant patches that might otherwise be inaccessible (see Chapters 5 and 7).

Some resources are limiting for all species, but changing environmental conditions can affect resource abundance or suitability and directly or indirectly affect higher trophic levels (Cease et al., 2012, 2015; M. Hunter and Price, 1992; Polis and Strong, 1996; Power, 1992). Environmental changes can stress vegetation over large areas and increase the suitability of a host plant without changing its abundance. For example, Cease et al. (2012, 2015) reported that reduced plant protein content promoted outbreaks of a dominant locust, *Oedaleus asiaticus*, in Eurasian grassland (Fig. 6.10). Heavy livestock grazing and consequent steppe degradation

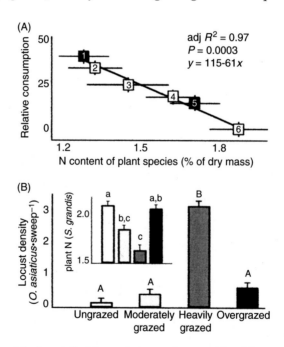

FIGURE 6.10 **Response of the locust, *Oedaleus asiaticus*, to host nutritional factors and livestock (sheep) grazing intensity in northern China.** (A) Locust consumption of the six most common host plant species, relative to their N concentration, measured over 36 h; 1 = *Stipa grandis*, 2 = *Carex korshinskyi*, 3 = *Agropyron cristatum*, 4 = *Cleistogenes squarrosa*, 5 = *Leymus chinensis*, 6 = *Achnatherum sibiricum*. Dark squares are the two most common plants, each producing 40% of above-ground biomass; the other four species each compose <5% of above-ground biomass. (B) Locust density relative to N concentration of *S. grandis* and livestock grazing intensity during 2009, an outbreak year. Grazing resulted in a shift in dominance from *L. chinensis* to *S. grandis*. The ungrazed field was fenced and protected from grazing since 1979; grazing intensity in adjacent fields has been relatively constant for >5 years. *Source: From Cease, A.J., et al., 2012. Heavy livestock grazing promotes locust outbreaks by lowering plant nitrogen content. Science 335, 467–469.*

appeared to be responsible for reduced N plant protein content via reduced N concentrations within plant species, as well as a shift in dominance from high-N species on ungrazed sites to low-N species on grazed sites (Fig. 6.10). Environmental changes also can increase the density or geographic distribution of resources. Under such circumstances, the disruption of bottom-up regulation results in population growth and, thereby, increased resource availability, and perhaps suitability (Stamp, 1992; Traugott and Stamp, 1996), for predators and parasites, resulting in increased abundance at higher trophic levels. If regulation by predation or parasitism is efficient, prey population growth may be limited quickly.

Species respond differentially to the same change in resources or predators. Ritchie (2000) reported that experimental fertilization (with nitrogen) of grassland plots resulted in increased host plant quality for, and density of, polyphagous grasshoppers but did not affect grass quality for, and reduced the density of, grass-feeding grasshoppers. Denno et al. (2002, 2003) manipulated both bottom-up (host plant biomass and nutrition) and top-down (predation by spiders) factors and concluded that bottom-up factors predominated in regulating populations of six sap-feeding insect species. The impact of predation was significant only for two species of planthoppers, *Prokelisia* spp., and was mediated by vegetation biomass and complexity. Stiling and Moon (2005) simultaneously manipulated host plant, *Borrichia frutescens*, nutritional (nitrogen) quality and density and parasitoid abundance and measured effects on the abundances of a meristem-galling fly, *Asphondylia borrichiae*, and a planthopper, *Pissonotus quadripustulatus*. Abundances of both herbivores were positively related to plant nitrogen, but not plant density at the scale of the study, and to parasitoid removal. Density-dependent competition and dispersal, as well as increased predation, eventually cause population decline to levels at which these regulatory factors become less operative.

Although increasing host availability can induce population growth, outbreaks often reflect an event that temporarily disrupts regulatory factors. Disturbances acutely stress host plants and increase their susceptibility to herbivorous insects (see Chapters 2 and 3) and typically disrupt predation (see Chapter 2). The sudden release of herbivore populations from bottom-up and/or top-down regulation leads to explosive population growth (Mattson and Haack, 1987; Raffa et al., 2008; Stige et al., 2007; Van Bael et al., 2004). Outbreaks usually end in 2–4 years, as high population density depletes host resources and/or induces epizootics of pathogens (Cairns et al., 2008; Coleman et al., 2008; Elkinton and Liebhold, 1990; Mason and Luck, 1978; K. Zimmerman et al., 2004). The increased frequency and severity of extreme weather events, predicted to result from global climate change (Bender et al., 2010; Gleason et al., 2008; Gutschick and BassiriRad, 2010; Irwin et al., 2014; Lubchenco and Karl, 2012), are likely to increase the frequency of insect population outbreaks. Population decline that approaches the extinction threshold may affect the time necessary for population recovery between outbreaks. Sparse populations may require several years to recover sufficient numbers for population growth or may continue to decline to extinction.

Some insect species are particularly prone to outbreaks under suitable circumstances. In particular, species whose populations typically are regulated by a variety of factors are less likely to experience population outbreaks than are species that have weak intrinsic mechanisms for population regulation and are regulated primarily by host suitability or factors that have a delayed effect (Larsson et al., 1993; Nothnagle and Schultz, 1987). For example, predation on pupae by small mammals is the primary factor regulating populations of gypsy moth, *Lymantria dispar*, in North America (Elkinton and Liebhold, 1990). Consequently, environmental factors that reduce abundances of small mammals undermine their regulatory effect. Furthermore, most of these small mammals are generalists, for which gypsy moth pupae are a minor food source, and which occupy or defend distinct home ranges or territories. Therefore, predation by small mammals becomes inversely density-dependent at higher prey densities (Elkinton and Liebhold, 1990).

The effect of predators and parasites often is delayed, as predators take time to aggregate and reproduce and epizootics to develop in response to increasing prey density (Turchin et al., 1999, see Chapter 8). Furthermore, populations that disperse rapidly as density increases may delay the effects of crowding and predation, by reducing the rate of density increase in the areas of most rapid population growth (see Chapter 7). Bazykin et al. (1997), Berryman et al. (1987), Turchin (1990), and Turchin et al. (1999) demonstrated that time lags weaken negative feedback and reduce the effectiveness of population regulation. Hence, populations that are controlled primarily by factors that operate through delayed negative feedback, for example, predation or parasitism, should exhibit greater amplitude of population fluctuation, whereas populations that are controlled by factors with more immediate negative feedback, for example, resource limitation or crowding, should be more stable. However, aggregation and reproduction of predators and spread of epizootics in areas of dense prey/hosts typically bring an end to outbreaks within 2–4 years (Fig. 6.11).

Cyclic outbreaks have been particularly intriguing. Explanations for regular cycles of outbreaks have included climatic cycles and changes in insect gene frequencies or behavior, food quality, or susceptibility to disease that occur during large changes in insect abundance (J. Myers, 1988). Climatic cycles may trigger insect population cycles directly through changes in mortality or indirectly through changes in host condition or susceptibility to pathogens. However, regular irruptions of *Z. diniana* have occurred for over 1000 years in the European

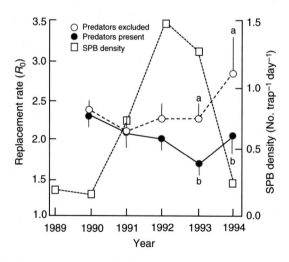

FIGURE 6.11 Survival of southern pine beetle (SPB), *Dendroctonus frontalis*, measured as the proportion of eggs surviving to become emerging adults, when protected from predation (black circles) and exposed to predation (white circles). Squares indicate the corresponding SPB population size. Different letters indicate significant difference at $p < 0.05$. *Source: Modified with permission from Turchin, P., et al., 1999. Dynamical role of predators in population cycles of a forest insect. Science 285, 1068–1971.*

Alps independently of long-term warming and cooling trends (Esper et al., 2007). Depletion of food resources during an outbreak may impose a time lag for recovery of resource levels capable of sustaining renewed population growth (W. Clark, 1979). Epizootics of entomopathogens may occur only above threshold densities. High genetic heterogeneity for disease resistance can lead to intense selection for resistance during disease-driven population collapse, followed by increasing infection risk during population growth due to the fitness cost of resistance (Cooper et al., 2003; Elderd et al., 2008).

Delayed density-dependent feedback now appears to be the primary driver of cyclic population patterns (Berryman, 1996; Klemola et al., 2014; Royama, 1992; Turchin, 1990; Turchin et al., 1999). In particular, epizootics of diseases often bring outbreaks into decline (Cooper et al., 2003; Cory and Myers, 2003; Thorne et al., 2007), with survivors showing reduced fecundity as a sublethal effect of infection (Cooper et al., 2003; Elliot et al., 2003). J. Myers (1988) and Mason (1996) concluded that delayed effects of density-dependent factors can generate outbreak cycles with an interval of about 10 years.

A critical question regarding cyclical species is the mechanism(s) synchronizing outbreaks over large areas (J. Myers and Cory, 2013). For some insects, especially Lepidoptera, population cycles are synchronized among demes up to 1200 km apart (ie, at a continental scale). Widespread triggers such as climate, sunspot, lunar, or ozone cycles have been suggested as mechanisms for synchronizing outbreaks over large areas, but these factors do not appear to be primary causes of synchrony (W. Clark, 1979; D. Johnson et al., 2005; Liebhold et al., 2004; Økland et al., 2005; Price, 1997; Royama, 1984, 1992; Speer et al., 2001). P. Moran (1953) suggested, and Royama (1992) demonstrated (using models), that synchronized cycles could result from correlations among controlling factors.

Liebhold et al. (2004) suggested that synchrony can arise from three primary mechanisms: (1) dispersal among demes transfers individuals from growing demes to smaller demes; (2) congruent dependence of multiple demes on exogenous synchronizing factors, such as temperature or precipitation (the "Moran" effect); and (3) trophic interactions with other species that integrate demes via mortality (see Chapter 9). Liebhold et al. (2006) used simulation modeling to demonstrate that geographic variation in direct and delayed density-dependence diminished synchrony resulting from stochastic forcing by geographic variation, but not synchrony resulting from dispersal processes. Hence, the cause of synchrony can be independent of the cause of the cyclic pattern of fluctuation. J. Myers and Cory (2013) considered clarification of the importance of dispersal to synchronized cyclical outbreaks to be a fundamental research need.

Raimondo et al. (2004) reported that generalist predators also could explain the observed synchronous population dynamics of multiple prey species. Synchronized emergence of periodical cicadas, *Magicicada* spp., was associated with increased abundances of 15 of 37 bird species evaluated by Koenig and Liebhold (2005). Bird populations sharing the same cicada brood showed greater intraspecific spatial synchrony than did bird populations in the ranges of different cicada broods.

Populations that decline to near their extinction threshold become vulnerable to extinction, as density reaches levels at which individuals can no longer find each other to reproduce or to provide communal protection of offspring or access to food resources. As described in the example opening this chapter, even species that can reach outbreak population sizes are vulnerable to extinction during population decline.

A large number of insect species have disappeared or are threatened with extinction (Grove, 2002; Lockwood, 2001; Samways, 1995; D.L. Wagner, 2009; M. Wilson et al., 1997). Their loss threatens the integrity of food webs and ecosystem services that support human populations (see Chapter 16). G. Bell and Gonzalez (2009) tested responses of yeast populations to normally lethal concentrations of salt and demonstrated that the ability of a population to recover from collapse reflected a threshold in initial population size, that is, sufficiently large to avoid stochastic extinction and to contain resistant genomes. Among insects, low-density populations are particularly vulnerable to the failure of potential mates to find each other (Gascoigne et al., 2009; Kramer et al., 2009; Yamanaka and Liebhold, 2009).

4 MODELS OF POPULATION CHANGE

Models are representations of complex systems and are used to understand and predict changes in those systems. Population dynamics of various organisms, especially insects, are of particular concern as population changes affect human health, ecosystem services, and the condition of terrestrial and aquatic ecosystems. Therefore, development of models to improve our ability to understand and predict changes in insect population size has a rich history.

Models take many forms. The simplest are *conceptual models* that clarify relationships between cause and effect. For example, box-and-arrow diagrams can be used to show which system components interact with each other (eg, Fig. 1.3). More complex *statistical models* represent those relationships in quantitative terms, for example, regression models that depict the relationship between population size and environmental factors (eg, Figs. 5.6 and 6.11). Such models do not necessarily represent cause and effect. Advances in computational technology have led to development of *biophysical models* that can integrate large datasets to predict responses of insect populations to a variety of interacting environmental variables. *Computerized decision support systems* integrate a user interface with component submodels that can be linked in various ways, based on user-provided keywords, to provide output to answer specific questions (C. Shaw and Eav, 1993). This section describes models for predicting insect population growth.

4.1 Exponential and Geometric Models

The simplest model of population growth describes change in numbers as the product of initial population size and the per capita rate of increase (Fig. 6.12) (Berryman, 1997; Price, 1997). This model integrates per capita natality, mortality, immigration, and emigration per unit time as the instantaneous or *intrinsic rate of increase*, designated r:

$$r = (N + I) - (M + E) \tag{6.1}$$

where N = natality, I = immigration, M = mortality, and E = emigration, all instantaneous rates.

Where cohort life table data, rather than time-specific natality, mortality, and dispersal, have been collected, r can be estimated as

$$r = \frac{\log_e R_0}{T} \tag{6.2}$$

where R_0 is replacement rate and T is generation time.

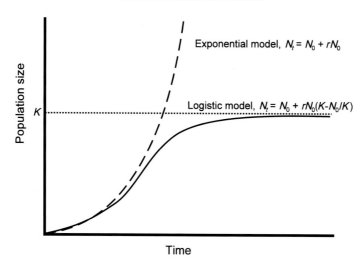

FIGURE 6.12 Exponential and logistic models of population growth. The exponential model describes an indefinitely increasing population, whereas the logistic model describes a population reaching an asymptote at the carrying capacity of the environment (K).

The rate of change for populations with overlapping generations is a function of the intrinsic (per capita) rate of increase and the current population size. The resulting model for exponential population growth is

$$N_t = N_0 + rN_0 \qquad (6.3)$$

where N_0 is the initial population size, at time 0, and N_t is the population size at a future time, t. This equation also can be written as

$$N_t = N_0 e^{rt} \qquad (6.4)$$

For insect species with nonoverlapping cohorts (generations), the replacement rate, R_0, represents the per capita rate of increase from one generation to the next. This parameter can be used in place of r for such insects. The resulting expression for geometric population growth is

$$N_t = R_0^t N_0 \qquad (6.5)$$

where N_t is the population size after t generations.

Eqs. (6.3)–(6.5) describe density-independent population growth (Fig. 6.12). However, as discussed previously, density-dependent competition, predation, and other factors interact to limit population growth.

4.2 Logistic Model

A mathematical model to account for density-dependent regulation of population growth was developed by Verhulst in 1838 and again, independently, by Pearl and Reed (1920).

This logistic model (Fig. 6.12), often called the Pearl–Verhulst equation (Berryman, 1981; Price, 1997), is expressed as

$$N_t = N_0 + rN_0 \frac{(K - N_0)}{K} \tag{6.6}$$

where K is the carrying capacity of the environment. This model describes a sigmoid (S-shaped) curve (Fig. 6.12) that reaches equilibrium at K. If $N < K$, then the population will increase up to $N = K$. If the ecosystem is disturbed in a way that $N > K$, then the population will decline to $N = K$.

4.3 Complex Models

General models such as the Pearl–Verhulst model usually do not predict the dynamics of real systems accurately. For example, the use of the logistic growth model is limited by several assumptions. First, individuals are assumed to be equal in their reproductive potential. Clearly, immature insects and males do not produce offspring, and females vary in their productivity, depending on nutrition, access to oviposition sites, etc. Second, population adjustment to changing density is assumed to be instantaneous, and effects of density-dependent factors are assumed to be a linear function of density. These assumptions ignore time lags, which may control dynamics of some populations and obscure their density dependence (Turchin, 1990). Finally, r and K are assumed to be constant. In fact, changes in factors (including K) that affect natality, mortality, and dispersal affect r. Changing environmental conditions, including elimination of resources by disturbances, affect K. Therefore, population size fluctuates with an amplitude that reflects variation in both K and the life history strategy of particular insect species. Species with the r strategy (high reproductive rates and low competitive ability) tend to undergo boom-and-bust cycles because of their tendency to overshoot K, deplete resources, and decline rapidly, often approaching their extinction threshold, whereas species with the K strategy (low reproductive rates and high competitive ability) tend to approach K more slowly and maintain relatively stable population sizes near K (Boyce, 1984). Modeling real populations of interest, then, requires development of complex models with additional parameters that correct these shortcomings, some of which are described by Eqs. 6.7–6.12.

Nonlinear density-dependent processes and delayed feedback can be addressed by allowing r to vary as

$$r = r_{max} - sN_{t-T} \tag{6.7}$$

where r_{max} is the maximum per capita rate of increase, s represents the strength of interaction between individuals in the population, and T is the time delay in the feedback response (Berryman, 1981). The sign and magnitude of s also can vary, depending on the relative dominance of competitive and cooperative interactions:

$$s = s_p - s_m N_t \tag{6.8}$$

where s_p is the maximum benefit from cooperative interactions and s_m is the competitive effect, assuming that s is a linear function of population density at time t (Berryman, 1981). The extinction threshold, E, can be incorporated by adding a term forcing population change to be negative below this threshold:

$$N_{t+1} = N_t + rN_t(K - N_t)/K)((N_t - E)/E) \tag{6.9}$$

where N_t is any point in time and N_{t+1} is an incremental time in the future. Similarly, the effects of various factors influencing natality, mortality, and dispersal can be incorporated into the model to improve representation of r.

The effect of other species interacting with a population was addressed first by Lotka (1925) and Volterra (1926). The Lotka–Volterra equation for the effect of species competing for the same resources includes a term that reflects the degree to which the competing species reduces carrying capacity:

$$N_{1(t+1)} = N_{1t} + r_1N_{1t}(K_1 - N_{1t} - \alpha N_{2t})/K_1 \tag{6.10}$$

where N_1 and N_2 are populations of two competing species, and α is a competition coefficient that measures the per capita inhibitive effect of species 2 on species 1.

Similarly, the effects of a predator on a prey population can be incorporated into the logistic model (Lotka, 1925; Volterra, 1926) as

$$N_{1(t+1)} = N_{1t} + r_1N_{1t} - \rho N_{1t}N_{2t} \tag{6.11}$$

where N_1 is prey population density, N_2 is predator population density, and ρ is a predation constant. This equation assumes random movement of prey and predator, prey capture and consumption for each encounter with a predator, and no self-limiting density effects for either population (Pianka, 1974; Price, 1997).

Pianka (1974) suggested that competition among prey could be incorporated by modifying the Lotka–Volterra competition equation as

$$N_{1(t+1)} = N_{1t} + r_1N_{1t} - r_1N_{1t}^2/K_1 - r_1N_{1t}\alpha_{12}N_{2t}/K_1 \tag{6.12}$$

where α_{12} is the per capita effect of the predator on the prey population. The prey population is density limited as carrying capacity is approached.

May (1981) and Dean (1983) modified the logistic model to include effects of mutualists on population growth. Species interaction models are discussed more fully in Chapter 8.

Gutierrez (1996) and Royama (1992) discussed additional population modeling approaches, including incorporation of age and mass structure and population refuges from predation. Clearly, the increasing complexity of these models, as more parameters are included, requires computerization for prediction of population trends.

4.4 Computerized Models

Computerized simulation models have been developed to project abundances of insect populations affecting crop and forest resources (Gutierrez, 1996; Royama, 1992; Rykiel et al., 1984). The models developed for several important forest and range insects are arguably the most sophisticated population dynamics models developed to date because they incorporate long time frames, effects of a variety of interacting factors (including climate, soils, host plant variables, competition, and predation) on insect populations, and effects of population change on ecosystem structure and processes. Often, the population dynamics model is integrated with plant growth models, impact models that address effects of population change on ecological, social, and economic variables, and management models that address effects of manipulated resource availability and insect mortality on the insect population (Colbert and Campbell, 1978; Leuschner, 1980). As more information becomes available on population responses to various factors, or effects on ecosystem processes, the model can be updated, increasing its representation of population dynamics and the accuracy of predictions.

Effects of various factors can be modeled as *deterministic* (fixed values), *stochastic* (values based on probability functions), or *chaotic* (random values) variables (Croft and Gutierrez, 1991; Hassell et al., 1991; Logan and Allen, 1992). If natality, mortality, and survival are highly correlated with temperature, these rates would be modeled as a deterministic function of temperature. On the other hand, effects of plant condition on these rates might be described best by probability functions and modeled stochastically (Fargo et al., 1982; Matis et al., 1994). A. King et al. (2004) reported that chaotic dynamics could produce near-cyclic patterns and that the particular patterns observed depend on the scale of study.

Advances in *chaos theory* are contributing to development of population models that more accurately represent the apparently erratic behavior of many insect populations (Cavalieri and Koçak, 1994, 1995a,b; Costantino et al., 1997; Cushing et al., 2003; Hassell et al., 1991; A. King et al., 2004; Logan and Allen, 1992; Lorenz, 1993). Chaos theory addresses the unpredictable ways in which initial conditions or small deviations in sensitive components of a system can affect subsequent system behavior. In other words, population trend at any instant is the result of the unique combination of population and environmental conditions at that instant. For example, changes in gene frequencies and behavior of individuals over time affect the way in which populations respond to environmental conditions. Time lags, nested cycles, and nonlinear interactions with other populations are characteristics of ecological structure that inherently destabilize mathematical models and introduce chaos (Cushing et al., 2003; Logan and Allen, 1992; Lorenz, 1993).

Chaos has been difficult to demonstrate in populations, and its importance to population dynamics is a topic of debate. Dennis et al. (2001) demonstrated that a deterministic model of flour beetle, *Tribolium castaneum*, population dynamics accounted for >92% of the variability in life stage abundances but was strongly influenced by chaotic behavior at certain values for the coefficient of adult cannibalism of pupae (Fig. 6.13).

Several recent studies suggest that insect population dynamics can undergo recurring transition between stable and chaotic phases when certain variables have values that place the system near a transition point between order and chaos (Cavalieri and Koçak, 1995a,b; Costantino et al., 1997) or when influenced by a generalist predator and specialist pathogen (Dwyer et al., 2004). Cavalieri and Koçak (1994, 1995b) found that small changes in weather-related parameters (increased mortality of pathogen-infected individuals or decreased natality of uninfected individuals) in a European corn borer, *Ostrinia nubilalis*, population dynamics model caused a regular population cycle to become erratic. When this chaotic state was reached, the population reached higher abundances than it did during stable cycles,

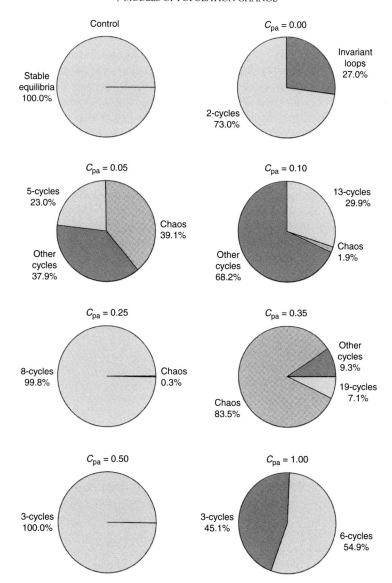

FIGURE 6.13 Frequency of predicted deterministic attractors for modeled survival probabilities of pupae in the presence of cannibalistic adults (c_{pa}) of *Tribolium castaneum* for 2000 bootstrap parameter estimates. For example, for c_{pa} = 0.35, 83.5% of estimates produced chaotic attractors, 7.1% produced stable 19-cycles, and 9.3% produced stable cycles of higher periods. *Source: From Dennis, B., et al., 2001. Estimating chaos and complex dynamics in an insect population. Ecol. Monogr. 71, 277–303.*

suggesting that small changes in population parameters resulting from biological control agents could be counterproductive. Although chaotic behavior fundamentally limits long-term prediction of insect population dynamics, improved modeling of transitions between deterministic or stochastic phases and chaotic phases may facilitate prediction of short-term dynamics (Cavalieri and Koçak, 1994; Logan and Allen, 1992).

4.5 Model Evaluation

The utility of models often is limited by a number of problems. The effects of multiple interacting factors typically must be modeled as the direct effects of individual factors, in the absence of multifactorial experiments to assess interactive effects. Effects of host condition often are particularly difficult to quantify for modeling purposes because factors affecting host biochemistry remain poorly understood for most species. Moreover, models must be initialized with adequate data on current population parameters and environmental conditions. Finally, most models are constructed from data representing relatively short time periods.

Models most accurately represent the observed dynamics of the populations from which the models were developed (Varley et al., 1973), but confidence in their utility for prediction of future population trends under a broad range of environmental conditions depends on proper validation of the model. Validation requires comparison of predicted and observed population dynamics using independent data, that is, data not used to develop the model. Such comparison using data that represent a range of environmental conditions can indicate the generality of the model and contribute to refinement of parameters subject to environmental influence, until the model predicts changes with a reasonable degree of accuracy (Hain, 1980).

Departure of predicted results from observed results can indicate several possible weaknesses in the model. First, important factors may be underrepresented in the model. For example, unmeasured changes in plant biochemistry during drought periods could significantly affect insect population dynamics. Second, model structure may be flawed. Major factors affecting populations may not be appropriately integrated in the model. Finally, the quality of data necessary to initialize the model may be inadequate. Initial values for r, N_0, or other variables must be provided or derived from historic data within the model. Clearly, inadequate data or departure of particular circumstances from tabular data will reduce the utility of model output.

Few studies have examined the consequences of using different types of data for model initialization. The importance of data quality for model initialization can be illustrated by evaluating the effect of several input options on predicted population dynamics of the southern pine beetle. The TAMBEETLE population dynamics model is a mechanistic model that integrates submodels for colonization, oviposition, and larval development with variable stand density and microclimatic functions to predict population growth and tree mortality (Fargo et al., 1982; Turnbow et al., 1982). Nine variables describing tree (diameter, infested height, and stage of beetle colonization for colonized trees), insect (density of each life stage at multiple heights on colonized trees), and environmental (landform, tree size class distribution and spatial distribution, and daily temperature and precipitation) variables are required for model initialization. Several input options were developed to satisfy these requirements. Options range in complexity from correlative information based on aerial survey or inventory records to detailed information about distribution of beetle life stages and tree characteristics that requires intensive sampling. In the absence of direct data, default values are derived from tabulated data based on intensive population monitoring studies.

Schowalter et al. (1982) compared tree mortality predicted by TAMBEETLE using four input options: all data needed for initialization (including life stage and intensity of beetles in trees), environmental data and diameter and height of each colonized tree only, environmental data and infested surface area of each colonized tree only, and environmental data and number of colonized trees only. Predicted tree mortality when all data were provided was twice the predicted mortality when only environmental and tree data were provided, and most closely resembled observed beetle population trends and tree mortality.

Insect population dynamics models typically are developed to address "pest" effects on commodity values. Few population dynamics models explicitly incorporate effects of population change on ecosystem processes or services. However, a growing number of studies are providing data on effects of insect herbivore or detritivore abundance on primary productivity, hydrology, nutrient cycling, and/or diversity and abundances of other organisms (S. Chapman et al., 2003; Christenson et al., 2002; Classen et al., 2005; Fonte and Schowalter, 2005; Frost and Hunter, 2004, 2007, 2008a,b; M.D. Hunter et al., 2003; Klock and Wickman, 1978; Leuschner, 1980; Schowalter and Sabin, 1991; Schowalter et al., 1991; Seastedt, 1984, 1985; Seastedt and Crossley, 1984; Seastedt et al., 1988; Whitham et al., 2006, see also Chapters 12–14). Colbert and Campbell (1978) documented the structure of the integrated Douglas-fir tussock moth, *Orgyia pseudotsugata*, model and the effects of simulated changes in moth density (population dynamics submodel) on density, growth rate, and timber production by tree species (stand prognosis model). Leuschner (1980) described the development of equations for evaluating direct effects of southern pine beetle population dynamics on timber, grazing and recreational values, hydrology, understory vegetation, wildlife, and likelihood of fire. Effects of southern pine beetle on these economic values and ecosystem attributes were modeled as functions of the extent of pine tree mortality resulting from changes in beetle abundance. However, for both the Douglas-fir tussock moth and southern pine beetle models, the effects of population dynamics on ecosystem processes that underlie long-term sustainability of multiple ecosystem services are based on limited data.

Modeling of insect population dynamics requires data from continuous monitoring of population size over long time periods, especially for cyclic and irruptive species, in order to evaluate the effect of changing environmental conditions on population size. Relatively few insect populations, including pest species, have been monitored for longer than a few decades, and most have been monitored only during outbreaks (Curry, 1994; Turchin, 1990). Historic records of outbreak frequency during the past 100–1000 years exist for a few species, (Fitzgerald, 1995; Greenbank, 1963; Konishi and Itô, 1973; Ma, 1958; Turchin, 1990; T. White, 1969), and, in some cases, outbreak occurrence over long time periods can be inferred from dendrochronological data in old forests (Esper et al., 2007; Royama, 1992; Speer et al., 2001; Swetnam and Lynch, 1989; Veblen et al., 1994). However, such data do not provide sufficient detail on concurrent trends in population size and environmental conditions for most modeling purposes. Data relating changes in population densities to environmental factors cover only a few decades for most species (Berryman, 1981; Mason, 1996; Price, 1997; Rácz and Bernath, 1993; Varley et al., 1973; Waloff and Thompson, 1980). For populations that irrupt infrequently, validation often must be delayed until future outbreaks occur.

Despite limitations, population dynamics models are a valuable tool for synthesizing a vast and complex body of data, for identifying critical gaps in our understanding of factors affecting populations, and for predicting or simulating responses to environmental changes. Therefore, they represent our state-of-the-art understanding of population dynamics, can be used to focus future research on key questions, and can contribute to improved efficiency of management or manipulation of important processes. Population dynamics models are the most rigorous tools available for projecting survival or recovery of endangered species and outbreaks of potential pests and their effects on ecosystem services.

5 SUMMARY

Populations of insects can fluctuate dramatically through time, with varying effects on community and ecosystem patterns and processes, as well as on the degree of crowding among members of the population. The amplitude and frequency of fluctuations distinguish irruptive populations, cyclic populations, and stable populations. Cyclic populations have stimulated the greatest interest among ecologists. The various hypotheses to explain cyclic patterns of population fluctuation all include density-dependent regulation with a time lag that generates regular oscillations.

Disturbances are particularly important to population dynamics, triggering outbreaks of some species and locally exterminating others. Disturbances can affect insect populations directly by killing intolerant individuals or indirectly by affecting abundance and suitability of resources or abundance and activity of predators and parasites. The extent to which anthropogenic changes in environmental conditions affect insect populations depends on the degree of similarity between conditions produced by natural versus anthropogenic changes.

Population growth can be regulated (stabilized) to a large extent by density-dependent factors whose probability of effect on individuals increases as density increases and declines as density decreases. Primary density-dependent factors are intra- and interspecific competition and predation. Increasing competition for food (and other) resources as density increases leads to reduced natality and increased mortality and dispersal, eventually reducing density. Similarly, predation increases as prey density increases. Although the relative importance of these two factors has been debated extensively, both clearly are critical to population regulation. Regulation by bottom-up factors (resource limitation) may be relatively more important in systems where resources are defended or vary significantly in quality, whereas regulation by top-down factors (predation) may be more important where resources are relatively abundant and show little variation in quality. Inverse density dependence results from cooperation among individuals and represents a potentially destabilizing property of populations. However, this positive feedback may prevent population decline below an extinction threshold. Populations declining below their extinction threshold may be doomed to local extinction, whereas populations increasing above a critical number of individuals (release threshold) continue to increase during an outbreak period. These thresholds represent the minimum and maximum population sizes for species targeted for special management.

Development of population dynamics models has been useful for forecasting changes in insect abundance and effects on crop, range, and forest resources. General models include the logistic (Verhulst–Pearl) equation that incorporates initial population size, per capita natality, mortality, and dispersal (instantaneous rate of population change), and carrying capacity. The logistic equation describes a sigmoid curve that reaches an asymptote at carrying capacity. This general model can be modified for particular species by adding parameters to account for nonlinear density-dependent factors, time lags, cooperation, extinction, competition, predation, etc. Chaos models have addressed the importance of initial conditions for subsequent changes in population size. More complex models incorporate specific population variables, including key factors and time lags. Models are necessarily simplifications of real systems and may represent effects of multiple interacting factors and chaotic processes poorly. Few models have been adequately validated and fewer have evaluated the effects of quality of data input on accuracy of predictions. Few population models have been developed to predict effects of insect population dynamics for ecosystem services other than commodity production. Nevertheless, models represent powerful tools for synthesizing information, identifying priorities for future research, and simulating population responses to future environmental conditions.

7

Biogeography

GYPSY MOTH SPREAD IN NORTH AMERICA

The gypsy moth, *Lymantria dispar*, has been among the most destructive invasive species in North America and has a well-documented record of spread since its accidental release in 1869 (Elkinton and Liebhold, 1990; C. Riley and Howard, 1890; C. Riley and Vasey, 1870). Adult females of North American populations are winged but do not fly (Liebhold et al., 1992). Dispersal occurs primarily as new larvae move upward, drop from foliage on silk threads, and subsequently are carried on the airstream, often for long distances. Last instar larvae move to sheltered sites, including bark crevices, leaf litter, dead trees, rocks, fence posts, firewood, machinery, and vehicles for pupation, and can be carried long distances by human movement of infested materials (Tobin et al., 2007).

The gypsy moth was initially brought to Medford, Massachusetts from France by Léopold Trouvelot as part of efforts during the 1860s to establish a silk industry in the USA (Andrews, 1868; Trouvelot, 1867). Several eggs or caterpillars apparently spilled or escaped from a jar on a windowsill (C. Riley and Howard, 1890). Trouvelot was unable to recover these and, recognizing the potential threat to North American forests, reported the introduction publicly (C. Riley and Howard, 1890; C. Riley and Vasey, 1870), but little attention was given to this at the time.

Local residents of Medford began noticing the caterpillars after about 10 years. The man who bought Trouvelot's house in 1879 noted swarms of caterpillars in a shed in the rear of the property (Forbush and Fernald, 1896). Efforts by residents to control the moth on their own properties during the 1880s were unsuccessful. In 1889 the moth reached destructive levels in an area of about 1.4 km² in Medford (Forbush and Fernald, 1896; C. Riley and Howard, 1890), leading to focused efforts by the state to contain and eradicate the population. Eradication efforts, which included burning forests and using arsenical compounds with the earliest pump spray equipment, continued until 1900, at which time three counties (4000 km²) were infested (Forbush and Fernald, 1896; Liebhold et al., 1992).

The moth subsequently spread at about 10 km year⁻¹ from 1900 to 1915, by which time much of New England was infested (about 84,000 km²), followed by a long period (1916–65) of slower spread, 3 km year⁻¹, by the end of which about 230,000 km², including eastern New York, were infested (Tobin et al., 2007). However, spread accelerated to 20 km year⁻¹ from 1965 to 1989 (580,000 km²), perhaps because the forests of Pennsylvania, Maryland, and Virginia had become older and more

Insect Ecology. http://dx.doi.org/10.1016/B978-0-12-803033-2.00007-8

continuous than they were earlier, because forests in the region had higher proportions of host species than did the forests of New England, and/or because of increased long-distance anthropogenic transportation during this period (Liebhold et al., 1992). The moth appeared across the Great Lakes in central Michigan about 1981, spread radially at about 10 km year^{-1} for the next 25 years, and was established in both Upper and Lower Peninsulas of Michigan and eastern Wisconsin (total infested area about 800,000 km^2) by 2005 (Tobin et al., 2007).

Federal quarantine of infested counties limited spread south after 1990 to only 7 km year^{-1}. However, additional population "jumps" to western states (eg, California, Oregon, Washington, British Columbia) as a result of anthropogenic transport have been treated aggressively and thus far have failed to become established (Liebhold et al., 1992).

1 INTRODUCTION

Geographic ranges of species occurrence generally reflect the tolerances of individual organisms to geographic gradients in physical conditions (see Chapter 2). However, most species do not occupy the entire area of environmental conditions potentially suitable for them. Discontinuity in geographical range reflects a number of factors, particularly distribution of resources, geographic barriers, and disturbance dynamics. By contrast, suitable habitats can be colonized over large distances from population sources, as a result of dispersal processes, often aided by anthropogenic movement. Factors determining the geographic distribution of organisms have been a particular subject of investigation for the past several centuries (Andrewartha and Birch, 1954; Price, 1997), spurred in large part by European and American exploration, and floral and faunal collections, in continental interiors during the 1700s and 1800s.

The spatial distribution of populations changes with population size. Growing populations expand over a larger area as individuals in the high-density core disperse to the fringe of the population or colonize new patches. Declining populations shrink into refuges that maintain isolated demes of a metapopulation. Spatial distribution of populations is influenced to a considerable extent by anthropogenic activities that determine landscape structure and introduce (intentionally or unintentionally) commercial and "pest" species to new regions, where they may spread and become invasive if habitat conditions are suitable. Changes in insect presence or abundance may be useful biological indicators of ecosystem conditions across landscapes or regions, depending on the degree of habitat specialization of particular species (Rykken et al., 1997). Changes in the presence and abundance of particular species affect various ecosystem properties, encouraging efforts to predict changes in distributions of insect populations.

2 GEOGRAPHIC DISTRIBUTION

Geographic distribution of species populations can be described over a range of scales. At the largest scale, some species have population distributions that span large areas of the globe, including multiple continents. At smaller scales, individual species may occur in a

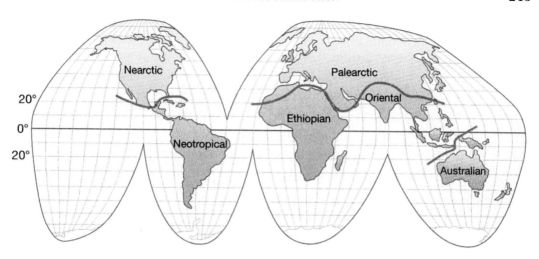

FIGURE 7.1   **Biogeographic realms identified by A. Wallace (1876).**

suitable portion of a biome or in suitable patches scattered across a biome or landscape. At the same time, species often are absent from apparently suitable habitats. The geographical distribution of individual species can change as a result of changing conditions and/or dispersal.

2.1 Global Patterns

Global patterns of distribution reflect latitudinal gradients in temperature and moisture, as well as natural barriers to dispersal (see Chapter 2). A. Wallace (1876) identified six relatively distinct faunal assemblages that largely coincide with major continental boundaries but also reflect the history of continental movement. Wallace's *biogeographic realms* (Fig. 7.1) remain a useful template for describing species distributions on a global scale. Many taxa occupy large areas within a particular biogeographic realm, for example, the unique Australian flora and fauna. Others, because of the narrow gap between the Palearctic and Nearctic realms, were able to cross this barrier and exhibit a Holarctic distribution pattern. Of course, many species occupy much smaller geographic ranges, limited by topographic barriers or other factors.

Some distribution patterns, especially of fossil species, are noticeably disjunct. Hooker (1847, 1853, 1860) was among the first to note the similarity of floras found among lands bordering the southern oceans, including Antarctica, Australia, Tasmania, New Zealand, Tierra del Fuego, the Falklands, and other islands. Many genera, and even some species, of plants were shared among these widely separated lands, suggesting a common origin.

Later in the 1800s, evidence of stratigraphic congruence of various plant and animal groups among the southern continents supported a hypothetical separation of northern and southern supercontinents. Wegener (1924) was the first to propose a geologic history of drift for all the continents, concentrated during the Cenozoic era. Wegener's *continental drift* hypothesis was criticized because this history appeared to be incompatible with nonmarine paleontology. However, a growing body of geologic and biological evidence, including stratigraphic congruence, rift valleys, uplift and subsidence zones, and distributions of both extinct and extant flora and fauna, eventually was unified into the theory of *plate tectonics.*

According to this theory, a single landmass (Pangaea) split about 200 million years ago and separated into northern (Laurasia) and southern (Gondwanaland) supercontinents that moved apart as a result of volcanic upwelling in the rift zone. About 135 million years ago India separated from Gondwanaland, moved northward, and eventually collided with Asia to form the Himalaya Mountains. Africa and South America separated about 65 million years ago, prior to the adaptive radiation of angiosperms and mammalian herbivores. South America eventually rejoined North America at the Isthmus of Panama, permitting placental mammals that evolved in North America to invade and displace marsupials (other than the generalized opossum) that had continued to dominate South America. Marsupials largely disappeared from the other continents as well, except for Australia, where they survived by virtue of continued isolation. South American flora and fauna moved northward through tropical Central America. This process of continental movement explains the similarity of fossil flora and fauna among the Gondwanaland-derived continents, and differences among biogeographic realms, for example, *Nothofagus* forests in southern continents versus *Quercus* forests in northern continents.

Continental movements result from the stresses placed on the Earth's crust by planetary motion. Fractures appear along lines of greatest stress and are the basis for volcanic and seismic activity, two powerful forces that lead to displacement of crustal masses. The mid-oceanic ridges and associated volcanism mark the original locations of the continents and preserve evidence of the direction and rate of continental movements. Rift valleys and fault lines typically provide depressions for development of aquatic ecosystems. Mountain ranges develop along lines of collision and subsidence between plates and create elevational gradients and boundaries to dispersal. Volcanic and seismic activity represents a continuing disturbance in many ecosystems.

2.2 Regional Patterns

Within biogeographic realms, a variety of biomes can be distinguished on the basis of their characteristic vegetation or aquatic characteristics (see Chapter 2). Much of the variation in environmental conditions that produce biomes at the regional scale is the result of global circulation patterns and topography. The development of similar phenotypic attributes by species in biomes belonging to different biogeographic realms is the basis for *convergent evolution.* This phenomenon reveals characteristics that are broadly adaptive in particular biomes, regardless of phylogenetic origin. Mountain ranges and large rivers may be impassable barriers that limit the distribution of many species. Furthermore, mountains show relatively distinct elevational zonation of biomes (life zones see Fig. 2.5). The area available as habitat becomes more limited at higher elevations. Mountaintops resemble oceanic islands in their degree of isolation within a matrix of lower-elevation environments and are the most vulnerable to climate changes that shift temperature and moisture combinations upward (Fig. 5.2).

Geographic ranges for most species are restricted by geographic barriers or by environmental conditions beyond their tolerance limits. Some insect species have broad geographic ranges that span multiple host ranges (eg, forest tent caterpillar, *Malacosoma disstria*, Parry and Goyer, 2004, see Chapter 3), whereas others have ranges restricted to small areas (eg, species endemic to cave ecosystems, Boecklen, 1991). Species with large geographic ranges often show considerable genetic variation among subpopulations, reflecting adaptations to regional environmental factors. For example, Istock (1981) reported that northern and southern populations of a transcontinental North American pitcher-plant mosquito, *Wyeomyia smithii*, showed distinct genetically based life history patterns. The proportion of third instars entering diapause increased with latitude, reflecting adaptation to seasonal changes in habitat or food availability. Controlled crosses between northern and southern populations yielded high proportions of diapausing progeny from northern × northern crosses, intermediate proportions from northern × southern crosses, and low proportions from southern × southern crosses, for larvae subjected to conditions simulating either northern or southern photoperiod and temperature.

2.3 Island Biogeography

Ecologists have been intrigued at least since the time of Hooker (1847, 1853, 1860) by the presence of related organisms on widely separated oceanic islands. Darwin (1859) and A. Wallace (1911) later interpreted this phenomenon as evidence of natural selection and speciation of isolated populations following separation or colonization from distant population sources. Simberloff (1969), Simberloff and Wilson (1969), and E. Wilson and Simberloff (1969) found that many arthropod species were capable of rapid colonization of experimentally defaunated islands. Although the theory of *island biogeography* originally was developed to explain patterns of equilibrium species richness among oceanic islands (MacArthur and Wilson, 1967), the same factors and processes that govern colonization of oceanic islands explain rates of species colonization and metapopulation dynamics (see Section 3.2) among isolated landscape patches (Cronin, 2003; Hanski and Simberloff, 1997; Leisnham and Jamieson, 2002; Simberloff, 1974; Soulé and Simberloff, 1986). Critics of this approach have argued that oceanic islands clearly are surrounded by habitat unsuitable for terrestrial species, whereas terrestrial patches may be surrounded by relatively more suitable patches. Some terrestrial habitat patches may be more similar to oceanic islands than others, for example, alpine tundra on mountaintops may represent substantially isolated habitats (Leisnham and Jamieson, 2002), as do isolated wetlands in a terrestrial matrix (Batzer and Wissinger, 1996), whereas disturbed patches in grassland may be less distinct (but see Cronin, 2003). A second issue concerns the extent to which the isolated populations constitute distinct species or metapopulations of a single species (Hanski and Simberloff, 1997). The resolution of this issue depends on the degree of heterogeneity and isolation among landscape patches and genetic drift among isolated populations over time.

2.4 Landscape and Stream Continuum Patterns

Within terrestrial biomes, gradients in climate and geographic factors interacting with the patch scale of disturbances across landscapes produce a shifting mosaic of habitat types that affects the distribution of populations (see Chapter 2). Local extinction of demes must be balanced by colonization of new habitats as they appear on the landscape in order for species to survive. However, colonists can arrive in terrestrial patches from various directions and distances. By contrast, distribution of aquatic species is more constrained by the linear (single dimension) pattern of water flow within a dendritic network of tributaries. Colonists are more likely to come from upstream (if movement is governed by water flow) or downstream (flying adults), with terrestrial patches between stream systems being relatively inhospitable (Fagan et al., 2010; Ganio et al., 2005). Population distributions often are relatively distinct among drainage basins (watersheds), depending on the ability of dispersants to colonize new head-waters or tributaries. New impediments to flow (such as landslides or hydroelectric dams) create more impassable barriers to movement of aquatic organisms, compared to terrestrial organisms. Species isolated in the lowest-order headwaters may be particularly vulnerable to extinction, regardless of horizontal distance, because of inhospitable stream reaches separat-ing them from a common node in the network (Fagan et al., 2010; Ganio et al., 2005). Hence, terrestrial and aquatic ecologists have developed different approaches to studying spatial dy-namics of populations, especially during the 1980s when *landscape ecology* became a paradigm for terrestrial ecologists (Turner, 1989) and *stream continuum* a paradigm for stream ecologists (Vannote et al., 1980).

Distribution of populations in terrestrial landscapes, stream continua, and oceanic islands is governed to a large extent by probabilities of extinction versus colonization in particular sites (Fig. 7.2, see Chapter 5). The dispersal ability of a species (see Chapters 2 and 5), the suit-ability of the patch, island, or stream habitat, and its size and distance from the population source determine the probability of colonization by a dispersing individual (Fig. 5.9). Island or patch size and distance from population sources influence the likelihood that an insect able to travel a given distance in a given direction will contact that island or patch.

Patch suitability reflects the abundance of resources available to colonizing insects. Clearly, suitable resources must be present for colonizing individuals to survive and reproduce. How-ever, preferences by colonizing individuals also may be important. Hanski and Singer (2001) examined the effect of two host plants, *Plantago* spp. and *Veronica* spp., that varied in their

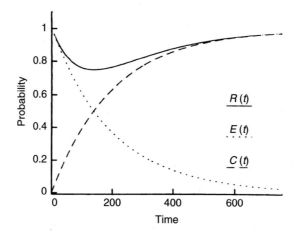

FIGURE 7.2　Probability of species presence in an ecosystem (*R*), as a function of probabilities of local ex-tinction (*E*) and colonization (*C*) over time, for specified values of v = probability of colonization over time and λ = probability of extinction over time. *Source: From Naeem, S., 1998. Species redundancy and ecosystem reliability. Conserv. Biol. 12, 39–45.*

relative abundances among patches, on colonization by the Glanville fritillary butterfly, *Melitaea cinxia*. Colonization success was strongly influenced by the correspondence between relative composition of the two host plants and the relative host use by caterpillars in the source patches, that is, colonizing butterflies strongly preferred to oviposit on the host plant they had used during larval development. The average annual colonization rate was 5% for patches dominated by the host genus less common across the connecting landscape and 15–20% for patches dominated by the host genus more common across the connecting landscape.

Individual capacity for sustained travel and for detection of cues that facilitate orientation determine colonization ability (see Chapter 5). Species that fly can travel long distances and traverse obstacles in an aquatic or terrestrial matrix better than can flightless species. Many small insects, including flightless species, catch air currents and are carried long distances at essentially no energetic cost to the insect. J. Edwards and Sugg (1990) reported that a variety of insects could be collected on montane glaciers far from the nearest potential population sources. Torres (1988) reported deposition, by hurricanes, of insect species from as far away as Africa on Caribbean islands. Clearly, many dispersing individuals are not successful colonists and instead become prey or detritus in remote locations.

Many small, flightless species have limited capacity to disperse. Any factor that increases the time to reach a suitable habitat increases the risk of mortality due to predation, extreme temperatures, desiccation, or other factors. Distances of a few meters, especially across exposed soil surfaces, can effectively preclude dispersal by many litter species sensitive to heat and desiccation or vulnerable to predation (Haynes and Cronin, 2003). On the other hand, McIntyre and Wiens (1999) found that darkling beetles, *Eleodes obsoleta*, moved faster, over longer distances, and in straighter lines through experimental landscapes with larger patch sizes, moved more slowly and spent more time in movement in heterogeneous landscapes, and showed longer residence time in intermediate-sized grass patches. Some aquatic species, for example, Ephemeroptera, have limited life spans as adults to disperse among stream systems. D. Fonseca and Hart (2001) reported that larval black flies, *Simulium vittatum*, were least able to colonize their preferred high-velocity habitats in streams because of constraints on their ability to control settlement. Courtney (1985, 1986) reported that short adult life span was a major factor influencing the common selection of less suitable larval food plants for oviposition (see Chapter 3). Clearly, the distance between an island or habitat patch and the source population is inversely related to the proportion of dispersing individuals able to reach it (Fig. 5.9).

Island or patch size and complexity also influence the probability of successful colonization. The larger the patch (or the shorter its distance from the source population), the greater the proportion of the horizon it represents and the more likely it is that a dispersing insect will contact it. Patch occupancy rate increases with patch size (Cronin, 2003). Similarly, the distribution of microsites within landscape or watershed patches affects the ability of dispersing insects to perceive and reach suitable habitats. Basset (1996) reported that the presence of arboreal insects is influenced more strongly by local factors in complex habitats, such as tropical forests, and more strongly by regional factors in less complex habitats, such as temperate forests.

The composition of surrounding patches in a landscape matrix is as important as patch size and isolation in influencing population movement and distribution. Genung et al. (2012) reported results of a common garden experiment with 1, 3, 6, or 12 randomly selected (out of

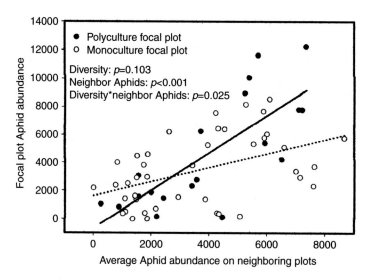

FIGURE 7.3 Aphid, *Uroleucon nigrotuberculatum*, abundance on *Solidago altissima* as a function of host-plant genotypic diversity and average aphid abundance in neighboring plots. *Filled circles* and *solid regression line* represent polyculture plots; *open circles* and *dotted regression line* represent monoculture plots. *Source: From Genung et al. (2012). Aphid and ladybird beetle abundance depend on the interaction of spatial effects and genotypic diversity. Oecologia 168, 167–174.*

21) goldenrod, *Solidago altissima*, clones per 1 m² plot. They demonstrated that aphid, *Uroleucon nigrotuberculatum*, abundance in each plot was positively correlated to aphid abundance in neighboring plots. Genotypic diversity in a given plot mediated the strength of the neighborhood effect, with stronger effects for polyculture plots and weak effects for monoculture plots (Fig. 7.3).

Haynes and Cronin (2003) manipulated the composition of the matrix (mudflat, native nonhost grasses, and exotic brome, *Bromus inermis*) surrounding small patches of prairie cordgrass, *Spartina pectinata*, that were identical in size, isolation, and host plant quality. Planthoppers, *Prokelisia crocea*, were marked and released into each host patch. Planthopper emigration rate was 1.3 times higher for patches surrounded by the two nonhost grasses compared to patches surrounded by mudflat (Fig. 7.4). Immigration rate was 5.4 times higher into patches surrounded by brome compared to patches surrounded by mudflat, and intermediate in patches surrounded by native nonhost grass. Patch occupancy and density increased with the proportion of the matrix composed of mudflat, probably reflecting the relative inhospitability of the mudflat for dispersal compared to nonhost grasses.

The increasing rate of dispersal during rapid population growth increases the number of insects moving across the landscape and the probability that some will travel sufficient distance in a given direction to discover suitable patches. Therefore, population contribution to patch colonization and genetic exchange with distant populations is maximized during population growth (see Section 3.1).

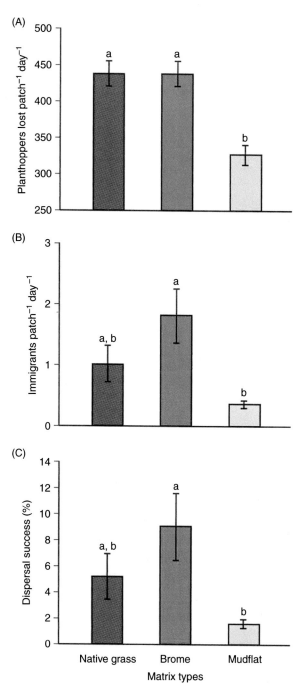

FIGURE 7.4 Effect of surrounding matrix on rate of planthopper loss from cordgrass patch in which released (A), rate of planthopper immigration into satellite patches (B), and percentage of planthoppers lost from the central release patch that successfully immigrated into any of the eight surrounding patches (C). *Vertical lines* represent 1 SE. *Bars* with different letters are significantly different at $p < 0.05$. *Source: From Haynes, K.J., Cronin, J.T., 2003. Matrix composition affects the spatial ecology of a prairie planthopper. Ecology 84, 2856–2866.*

3 SPATIAL DYNAMICS OF POPULATIONS

As populations change in size, they also change in spatial distribution of individuals. Population movement (epidemiology) across landscapes and watersheds (stream continuum) reflects integration of physiological and behavioral attributes with landscape or watershed structure. Growing populations tend to spread across the landscape as dispersal leads to colonization of new habitats, whereas declining populations tend to constrict into more or less isolated refuges. Isolated populations of irruptive or cyclic species can coalesce during outbreaks, facilitating genetic exchange.

Insect populations show considerable spatial variation in densities in response to geographic variation in habitat conditions and resource quality (Fig. 7.5). Variation can occur over relatively small scales because of the small size of insects and their sensitivity to

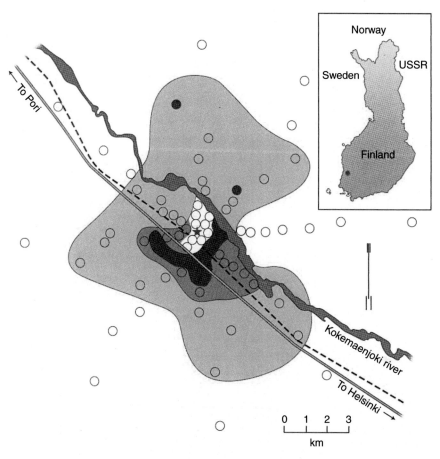

FIGURE 7.5 Gradient in pine bark bug, *Aradus cinnamomeus,* densities with distance from the industrial complex (★) at Harjavalta, Finland. *White circles* = 0–0.50 bugs 100 cm^{-2}, *yellow circles* = 0.51–1.75 bugs 100 cm^{-2}, *brown circles* = 1.76–3.50 bugs 100 cm^{-2}, and *purple circles* = 3.51–12.2 bugs 100 cm^{-2}. *Source: From Heliövaara, K., Väisänen, R., 1986. Industrial air pollution and the pine bark bug,* Aradus cinnamomeus *Panz. (Het., Aradidae). Z. Angew. Entomol. 101, 469–478.*

environmental gradients (Heliövaara and Väisänen, 1993; Lincoln et al., 1993, see Chapter 2). The spatial representation of populations can be described across a range of scales from microscopic to global (Chapter 5). Two general types of spatial variation are represented by the expansion of growing populations and by the discontinuous pattern of fragmented populations, or metapopulations.

3.1 Expanding Populations

Growing populations tend to spread geographically as density-dependent dispersal leads to colonization of nearby resources. This spread occurs in two ways. First, diffusion from the origin, as density increases, produces a gradient of decreasing density toward the fringe of the expanding population. Grilli and Gorla (1997) reported that leafhopper, *Delphacodes kuscheli*, density was highest within the epidemic area and declined toward the fringes of the population. The difference in density between pairs of sampling points increased as the distance between the sampling points increased. Second, long-distance (or jump) dispersal leads to colonization of vacant patches and "proliferation" of the population (Hanski and Simberloff, 1997; Tobin et al., 2007). Subsequent growth and expansion of these new demes can lead to population coalescence, with local "hot spots" of superabundance that eventually disappear as resources in these sites are depleted and individuals disperse.

Spread of demes from population refuges can be synchronous over landscapes, as described in Chapter 6. Alternatively, population expansion can occur as traveling waves, typified by partial synchrony with a gradient in the degree of population change (phase angle differences) as a function of distance (D. Johnson et al., 2004). For example, D Johnson et al. (2004) showed that spatial dynamics of larch budmoth, *Zeiraphera diniana*, in the European Alps from 1961 to 1998 was described as a series of traveling waves from multiple epicenters in favorable habitats and that landscape heterogeneity (gradients and connectivity of habitat suitability) alone was capable of inducing waves (Fig. 7.6). Furthermore, population spread often occurs as pulses of range expansion under favorable environmental conditions (eg, contact with patches of suitable habitat) interspersed with periods of relative stasis (D. Johnson et al., 2006).

The speed at which a population expands likely affects the efficiency of density-dependent regulatory factors. Populations that expand slowly may experience immediate density-dependent negative feedback in zones of high density, whereas induction of negative feedback may be delayed in rapidly expanding populations, as dispersal slows an increase in density. For example, rapidly expanding populations may leave diseased individuals behind, thereby escaping a major mortality factor along the expanding front (Ascunce et al., 2011). Therefore, density-dependent factors should operate with a longer time lag in populations capable of rapid dispersal during irruptive population growth.

The speed, extent, and duration of population spread are determined by abiotic and biotic factors. Insect species with annual life cycles often show incremental colonization and population expansion. Kozár (1991) reported that several insect species showed rapid range expansion northward in Europe during the 1970s, likely reflecting warming temperatures during this period. Similarly, J. Jepsen et al. (2008) reported northward expansion of two cyclic geometrid moths, *Operophtera brumata* and *Epirrita autumnata*, associated with continued warming during the past 20 years. Population expansion of spruce budworm, *Choristoneura fumiferana*,

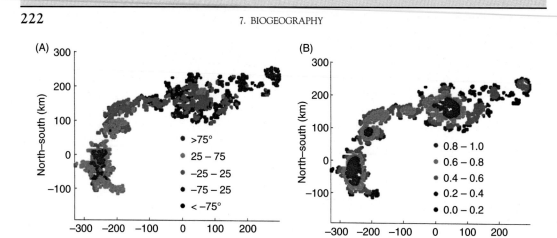

FIGURE 7.6 **(A) Time-averaged phase angles (degree of population change) for larch budmoth, *Zeiraphera diniana*, spatial dynamics in the European Alps during 1961–98.** Highest phase angles *(red)* indicate epicenter locations and lower phase angles indicate areas into which the population subsequently spread. (B) Relative connectivity measures of budmoth habitat across the Alps based on Gaussian dispersal with $\alpha = 25$. Connectivity was rescaled so that the location with highest connectivity had a value of 1. *Source: From Johnson, D.M., et al., 2004. Landscape geometry and traveling waves in the larch budmoth. Ecol. Lett. 7, 967–974.*

western harvester ant, *Pogonomyrmex occidentalis*, and grasshoppers during outbreaks is associated with warmer, drier periods (Capinera, 1987; DeMers, 1993; Greenbank, 1963). Environmental change or disturbances can terminate the spread of sensitive populations. Frequently disturbed systems, such as crop systems or streams subject to annual scouring, limit population spread to the intervals between recolonization and subsequent disturbance (Matthaei and Townsend, 2000; Reice, 1985).

Populations of species with relatively slow dispersal may expand only to the limits of a suitable patch during the favorable period. Spread beyond the patch also depends on the suitability of neighboring patches (Baum et al., 2004; Haynes and Cronin, 2003; D. Johnson et al., 2004; Liebhold and Elkinton, 1989). Populations can spread more rapidly and extensively across landscapes dominated by host species, such as agricultural and silvicultural systems, than in more heterogeneous systems in which unsuitable patches limit spread (Haynes and Cronin, 2003; D. Johnson et al., 2004; Onstad et al., 2003; Schowalter and Turchin, 1993).

Finally, the status of competitors, predators, or parasites which the expanding population escapes or contacts can facilitate or limit further expansion (Ascunce et al., 2011). Lounibos et al. (2003) examined factors responsible for the higher abundance of invasive container mosquitoes, *Aedes albopictus*, in areas where two species of native *Wyeomyia* mosquitoes are absent. Tests in experimental containers revealed that *Wyeomyia* larvae did not deter oviposition by *A. albopictus*. However, fourth instar *Wyeomyia* larvae (but not first instar larvae) significantly reduced growth and survival of *A. albopictus* larvae, indicating that interspecific larval interaction determined the geographic distribution of *A. albopictus*.

The direction of population expansion also depends on several factors. The direction of population spread is constrained by environmental gradients, by wind or water flow, and by unsuitable patches. Gradients in temperature, moisture, or chemical concentrations restrict the directions in which insect populations can spread, based on tolerance ranges to these factors (Chapter 2). Even relatively homogeneous environments, such as enclosed stored grain, are subject to gradients in internal temperatures that affect spatial change in granivore populations (Flinn et al., 1992). Furthermore, direction and flow rate of wind or water have considerable influence on insect movement. Insects with limited capability to move against air or water currents move primarily downwind or downstream, whereas insects capable of movement toward attractive cues move primarily upwind or upstream. Insects that are sensitive to stream temperature, flow rate, or chemistry may be restricted to spread along linear stretches of the stream. P. Jepson and Thacker (1990) reported that recolonization of agricultural fields by carabid beetles dispersing from population centers was delayed by extensive use of pesticides in neighboring fields.

Schowalter et al. (1981b) examined the spread of southern pine beetle, *Dendroctonus frontalis*, populations in east Texas (Fig. 7.7). They described the progressive colonization of individual trees or groups of trees through time by computing centroids of colonization activity on a daily basis (Fig. 7.8). A centroid is the center of beetle mass (numbers) calculated from the weighted abundance of beetles among the x,y coordinates of colonized trees at a given time. The distances between centroids on successive days were a measure of the rate of population movement (Fig. 7.7).

Populations moved at a rate of 0.9 m day^{-1}, primarily in the direction of the nearest group of available trees. However, since southern pine beetle populations generally were sparse during the period of this study, indicating relatively unfavorable conditions, this rate may be near the minimum necessary to sustain population growth. The probability that a tree would be colonized depended on its distance from currently occupied trees. Trees within 6 m of sources of dispersing beetles had a 14–17% probability of being colonized, compared to < 4% probability for trees further than 6 m from sources of dispersing beetles. Population spread in most cases ended at canopy gaps where no trees were available within 6 m. However, one population successfully crossed a larger gap encountered at peak abundance (Fig. 7.7), indicating that a sufficiently large number of beetles dispersed across the gap to ensure aggregation on suitable trees and sustained population spread.

Population spread in this species may be facilitated by colonization experience and cooperation between cohorts of newly emerging beetles and beetles "re-emerging" from densely colonized hosts. Many beetles re-emerge after laying some eggs, especially at high colonization densities under outbreak conditions, and seek less densely colonized trees in which to lay their remaining eggs. The success of host colonization by southern pine beetles depends on rapid attraction of sufficiently large numbers to overwhelm host defenses (see Chapter 3). For a given day, the centroid of colonization was, on average, twice as far from the centroid of new adults dispersing from brood trees as from the centroid of re-emerging beetles (Fig. 7.8). This pattern suggested that re-emerging beetles select the next available trees and provide a focus of attraction for new adults dispersing from further away.

Related research has reinforced the importance of host tree density for population spread of southern pine beetle and other bark beetles (Amman et al., 1988; M. Brown et al., 1987; R.G. Mitchell and Preisler, 1992; Sartwell and Stevens, 1975). Schowalter and Turchin (1993) demonstrated that patches of relatively dense pure pine forest are essential to the growth and spread of southern pine beetle populations from experimental refuge trees (Fig. 6.8). Experimentally established founding populations spread from initially colonized trees surrounded by dense pure pine forest, but not from trees surrounded by sparse pines or pine/hardwood mixtures.

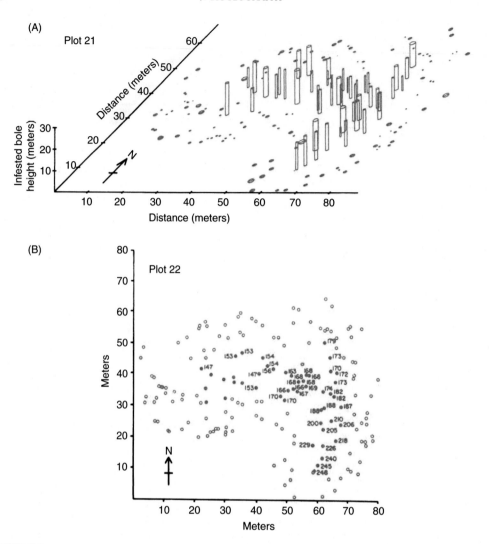

FIGURE 7.7 **Spatial and temporal pattern of spread of a southern pine beetle population in east Texas during 1977.** (A) *cylinders* are proportional in size to size of colonized trees; ellipses represent uncolonized trees within 10 m of colonized trees. (B) Julian dates of initial colonization are given for trees colonized *(green circles)* after sampling began. *Open circles* represent uncolonized trees within 10 m of colonized trees. *Source: From Schowalter, T.D., Pope, D.N., Coulson, R.N., Fargo, W.S., 1981. Patterns of southern pine beetle* (Dendroctonus frontalis Zimm.) *infestation enlargement. Forest Sci. 27, 837–849.*

Under suitable conditions, insect populations can spread rapidly. Reay-Jones et al. (2007) used pheromone-baited traps to measure the spread of invasive Mexican rice borer, *Eoreuma loftini*, through the Texas rice belt from 2000 to 2005. These data and the date of first appearance of this species by county since 1980 indicated an average rate of spread of 23 km year^{-1}. Henne et al. (2007) measured the rate of spread of a phorid fly, *Pseudacteon tricuspis*,

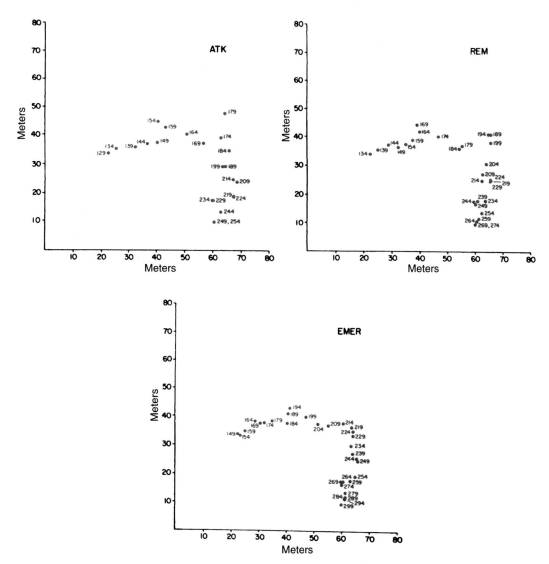

FIGURE 7.8 **Centroids of colonization *(ATK)*, reemergence *(REM)*, and emergence *(EMER)*, by Julian date, for the southern pine beetle population in Fig. 7.7**. *Source: From Schowalter, T.D., Pope, D.N., Coulson, R.N., Fargo, W.S., 1981. Patterns of southern pine beetle* (Dendroctonus frontalis *Zimm.) infestation enlargement. Forest Sci. 27, 837–849.*

introduced as a biological control agent for the red imported fire ant, *Solenopsis invicta*, in Louisiana during the first 6 years after release. Annual rate of spread was slow during the first 2 years, as the fly population became established, increased rapidly during years 3–4, and reached an asymptote of 15–25 km year^{-1} during years 5–6.

A critical aspect of population spread is the degree of continuity of hospitable resources or patches on the landscape (Schowalter et al., 1981b; Haddad, 2000; Sullivan et al., 2011). As described earlier for the southern pine beetle, unsuitable patches can interrupt population spread unless population density or growth is sufficient to maintain high dispersal rates across inhospitable patches. Similarly, Meisel (2006) reported that army ants, *Eciton burchellii*, in Costa Rica were restricted to forest fragments and consistently avoided entering pastures, where midday temperatures reached >50°C; ants survived <3 min at this temperature and only 18 min at 43°C. Heterogeneous landscapes composed of a variety of patch types force insects to expend their acquired resources detoxifying less acceptable resources or searching for more acceptable resources. Therefore, heterogeneous landscapes tend to limit population growth and spread, whereas more homogeneous landscapes, such as large areas devoted to plantation forestry, pasture grasses, or major crops, provide conditions more conducive to sustained population growth and spread. The particular composition of landscape mosaics may be as important as patch size and isolation in affecting insect movement and population distribution (Haynes and Cronin, 2003). Furthermore, herbivores and predators may respond differently to landscape structure. Herbivores were more likely to be absent from small patches than large patches, whereas predators were more likely to be absent from more isolated patches than from less isolated patches in agricultural landscapes in Germany (Zabel and Tscharntke, 1998).

Corridors or stepping stones (small intermediate patches) can facilitate population spread among suitable patches across otherwise unsuitable patches (Fig. 7.9). Haddad et al. (2003) found that 10 taxa, representing plants, insects, and mammals, consistently showed greater movement toward patches connected by corridors or stepping stones than toward unconnected patches. Populations of the western harvester ant do not expand across patches subject to frequent anthropogenic disturbance (specifically, soil disruption through agricultural activities) but are able to expand along well-drained, sheltered roadside ditches (DeMers, 1993). Roads often provide a disturbed habitat with conditions suitable for dispersal of weedy vegetation and associated insects among disturbed patches. Roadside conditions also may increase plant suitability for herbivorous insects and facilitate movement across landscapes fragmented by roads (Spencer and Port, 1988; Spencer et al., 1988). However, for some insects the effect of corridors and stepping stones may depend on the composition of the surrounding matrix. Baum et al. (2004) reported that experimental corridors and stepping stones significantly increased colonization of prairie cordgrass, *Spartina pectinata*, patches by planthoppers, *P. crocea*, in a low-resistance matrix composed of exotic, nonhost brome, *B. inermis*, that is conducive to planthopper dispersal, but not in a high-resistance matrix composed of mudflat that interferes with planthopper dispersal, relative to control matrices without corridors or stepping stones. Fried et al. (2005) found that the edge-following behavior of house flies, *Musca domestica*, results in diversion into patches with corridors.

Sullivan et al. (2011) found that the effects of corridors differed among parasite dispersal modes. Connectivity provided by corridors increased the incidence of biotically dispersed (ie, via flight) gall-forming insects on *Solidago odora*, but not of abiotically dispersed (ie, via wind or rain splash) fungal parasites on *Lespedeza hirta*, *Lespedeza stuevei*, and *S. odora*. All of these plant parasites responded to edge effects, but the direction of responses varied among species (Fig. 7.10).

An important consequence of rapid population growth and dispersal is the colonization of marginally suitable resources or patches where populations could not persist in the absence of continuous influx. Whereas small populations of herbivores, such as locusts or bark beetles, may show considerable selectivity in acceptance of potential hosts, rapidly growing

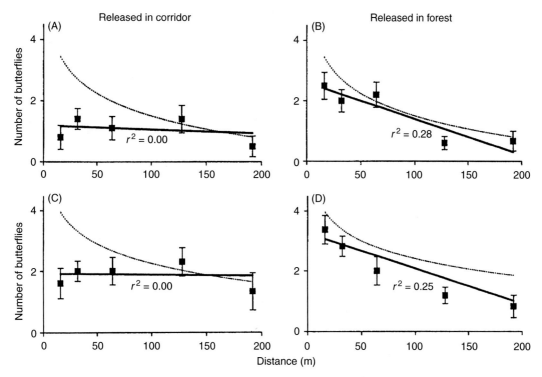

FIGURE 7.9 **Patch colonization by buckeye butterflies,** *Junonia coenia,* **an open-habitat specialist, relative to distance from release point.** (A) Number released in open-field corridors that colonized patches adjacent to their release point; (B) number released in forest that colonized patches adjacent to their release point ($p = 0.001$); (C) number released in corridors that colonized any patch; (D) number released in forest that colonized any patch ($p = 0.001$). *Dashed lines* are fit to a null model based on straight-line movement in a randomly chosen direction. *Source: From Haddad, N., 2000. Corridor length and patch colonization by a butterfly,* Junonia coenia. *Conserv. Biol. 14, 738–745.*

populations often eat all potential hosts in their path. Dense populations of the range caterpillar, *Hemileuca oliviae,* disperse away from population centers as grasses are depleted, and form an expanding ring, leaving denuded grassland in their wake. Landscapes that are conducive to population growth and spread, because of widespread homogeneity of resources, facilitate colonization of surrounding patches and more isolated resources because of the large numbers of dispersing insects. Epidemic populations of southern pine beetles, generated in the homogenous pine forests of the southern Coastal Plain during the drought years of the mid-1980s, produced sufficient numbers of dispersing insects to discover and kill most, otherwise resistant, pitch pines, *Pinus rigida,* along ridges in the southern Appalachian Mountains.

3.2 Metapopulation Dynamics

A metapopulation is a population composed of relatively isolated demes maintained by some degree of dispersal among suitable patches (Hanski and Simberloff, 1997; Harrison

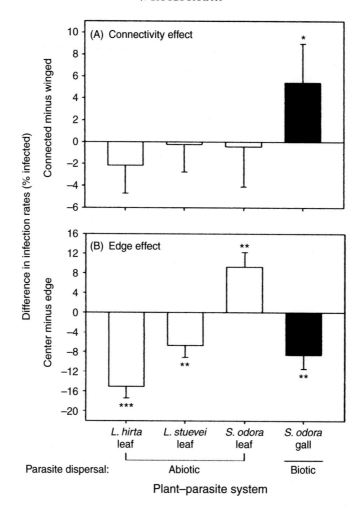

FIGURE 7.10 Effects of experimental landscape connectivity and patch edges on abiotically dispersed (wind-dispersed *Uromyces* and *Cercospora* fungi and splash-dispersed *Colletotrichum* fungi) and biotically dispersed (gall-forming tephritid and cecidomyiid flies and moths) parasites of *Lespedeza hirta*, *L. stuevei*, and *Solidago odora* leaves and stems in experimental corridors at the Savannah River site in South Carolina. (A) Connectivity effect and (B) edge effect. Data (means and SE) are the differences in infection rates between patches. *$p = 0.05$; **$p = 0.01$; ***$p = 0.0001$. *Source: From Sullivan, L.L., et al., 2011. Can dispersal mode predict corridor effects on plant parasites? Ecology 92, 1559–1564.*

and Taylor, 1997; Levins, 1970, see Chapter 5). Metapopulation structure can be identified at various scales (Massonnet et al., 2002), depending on the scale of distribution and the dispersal ability of the population (Fig. 7.11). For example, metapopulations of some sessile, host-specific insects, such as scale insects (Edmunds and Alstad, 1978), can be distinguished among host plants at a local scale, although the species occurs commonly over a wide geographic range. Local populations of black flies (Simuliidae) can be distinguished at the scale of isolated stream sections characterized by particular substrate, water velocity, temperature,

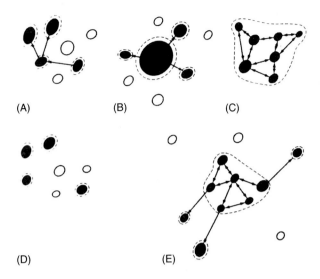

(A)   (B)   (C)

(D)   (E)

FIGURE 7.11 **Diagrammatic representation of different metapopulation models.** *Filled circles* are occupied patches; *open circles* are unoccupied patches; *dotted lines* are boundaries of local populations; *arrows* represent dispersal. (A) Classic (Levins) model of dispersal among demes, (B) island biogeography model with the mainland providing a source of colonists, (C) a network of interacting demes, (D) a nonequilibrium metapopulation with little capacity for recolonization of vacant patches, and (E) an intermediate case combining features of Parts A–D. *Source: From Harrison, S., Taylor, A.D., 1997. Empirical evidence for metapopulation dynamics. In: Hanski, I.A., Gilpin, M.E. (Eds.), Metapopulation Biology: Ecology, Genetics and Evolution. Academic Press, San Diego, CA, pp. 27–42.*

proximity to lake outlets, etc., whereas many species occur over a broad geographic area (P. Adler and McCreadie, 1997; Hirai et al., 1994). Many litter-feeding species occur throughout patches of a particular vegetation type, but that particular vegetation type and associated populations are fragmented at the landscape scale (Grove, 2002).

Metapopulation structure is most distinct where patches of suitable habitat or food resources are discontinuous and isolated due to natural environmental heterogeneity (eg, desert or montane landscapes) or to anthropogenic fragmentation. The spatial pattern of metapopulations reflects a number of interacting factors, including patch size, isolation, and quality (eg, resource availability and disturbance frequency), and insect dispersal ability (Fleishman et al., 2002; Frouz and Kindlmann, 2006; Summerville et al., 2002) and largely determines gene flow, species viability and, perhaps, evolution of life history strategies (Colegrave, 1997; Pecsenye et al., 2014; Roff, 1990). Hence, attention to spatially structured populations has increased rapidly in recent years.

Metapopulation structure can develop in a number of ways (Fig. 7.11). One is through the colonization of distant resources, and subsequent population development, that occurs during expansion of the source population (see Section 3.1). A second is through the isolation of population remnants during population decline or as a result of habitat fragmentation. A third represents a stable population structure in a heterogeneous environment, in which vacant patches are colonized as local extinction occurs in other patches.

The colonization of new patches as dispersal increases during population growth is an important mechanism for initiating new demes and facilitating population persistence on the landscape. The large number of dispersants generated during rapid population growth maximizes the probability that suitable resources will be colonized over a considerable area and that more founders will infuse the new demes with greater genetic heterogeneity (Hedrick and Gilpin, 1997). Species with ruderal life histories generally exhibit considerable dispersal capacity and often arrive at sites quite remote from their population sources (J. Edwards and Sugg, 1990). Such species quickly find and colonize disturbed sites and represent a widely occurring "weedy" fauna (Janzen, 1977). By contrast, species with competitive strategies show much slower rates of dispersal and may travel shorter distances consistent with their more stable population sizes and adaptation to more stable habitats (Roff, 1990; St. Pierre and Hendrix, 2003). Such species can be threatened by rapid changes in environmental conditions that exterminate demes more rapidly than new demes are established (Hanski, 1997; Hedrick and Gilpin, 1997).

If conditions for population growth continue, the outlying demes may grow and coalesce with the expanding source population. This process contributes to more rapid expansion of growing populations than would occur only as diffusive spread at the fringes of the source population. A well-known example of this is seen in the pattern of gypsy moth, *Lymantria dispar*, population expansion during outbreaks in eastern North America (see Box at the beginning of this chapter). New demes appear first on ridgetops in the direction of the prevailing wind, because of the wind-driven dispersal of ballooning larvae. These demes grow and spread downslope, merging in the valleys. Similarly, swarms of locusts may move great distances to initiate new demes beyond the current range of the population (Lockwood and DeBrey, 1990).

As a population retreats during decline, subpopulations typically persist in isolated refuges, establishing the postoutbreak metapopulation structure. Refuges are characterized by relatively lower population densities that escape the density-dependent decline of the surrounding population. These surviving demes may remain relatively isolated until the next episode of population growth and represent the sources of the next population expansion. The existence and distribution of refuges are extremely important to population persistence. For example, bark beetle populations typically persist as scattered demes in isolated lightning-struck, diseased, or injured trees which can be colonized by small numbers of beetles (Flamm et al., 1993). Such trees appear on the landscape with sufficient frequency and proximity to beetle refuges that endemic populations are maintained (Coulson et al., 1983). Croft and Slone (1997) and W. Strong et al. (1997) reported that predaceous mites quickly find colonies of spider mites. New leaves on expanding shoots provide important refuges for spider mite colonists by providing predator-free habitat and increasing prey distance from predators associated with source colonies, at least temporarily.

If suitable refuges are unavailable, too isolated, or of limited persistence, a population may decline to extinction (Pecsenye et al., 2014). Under these conditions, the small numbers, limited genetic variation, and low heterozygosity of dispersants generated by remnant demes are insufficient to ensure viable colonization of available habitats (Fig. 5.10) (Pecsenye et al., 2014).

For most species, life history strategies represent successful combinations of traits that balance population processes with natural rates of patch dynamics, that is, the rates of appearance and disappearance of suitable patches across the landscape. For example, Leisnham and Jamieson (2002) reported that immigration and emigration rates of the mountain stone weta, *Hemideina maori*, were equivalent (0.023 per capita). However, anthropogenic activities have dramatically altered the natural rates and landscape pattern of patch turnover and put many species at risk of extinction because their rate of adaptation is insufficient for the altered conditions (Fielding and Brusven, 1993; Lockwood and DeBrey, 1990; Vitousek et al., 1997).

Lockwood and DeBrey, 1990 suggested that loss of critical refuges as a result of anthropogenically altered landscape structure led to the extinction of a previously widespread and periodically irruptive grasshopper species. The Rocky Mountain grasshopper, *Melanoplus spretus*, occurred primarily in permanent breeding grounds in valleys of the northern Rocky Mountains but was considered to be one of the most serious agricultural pests in western North America prior to 1900 (Packard, 1877; C. Riley, 1878, 1883). Large swarms periodically migrated throughout the western United States and Canada during the mid-1800s, destroying crops over areas as large as 330,000 km^2 before declining precipitously (C. Riley, 1883). The frequency and severity of outbreaks declined during the 1880s, and the last living specimen was collected in 1902. Macroscale changes during this period (eg, climate changes, disappearance of Native Americans and bison, and introduction of livestock) do not seem adequate by themselves to explain this extinction. However, the population refuges (breeding ground) for this species during the late 1800s were riparian habitats (C. Riley, 1883) where agricultural activity, for example, tillage, irrigation, trampling by cattle, introduction of nonnative plants and birds, was concentrated. Hence, competition between humans and grasshoppers for refugia with suitable sites for oviposition and nymphal development appears to be the most likely factor leading to the extinction of *M. spretus* (Lockwood and DeBrey, 1990).

Stable metapopulation structures are maintained by a balance between source and sink habitats on the landscape. Frouz and Kindlmann (2006) described patterns of colonization and extinction for a soil-dwelling chironomid fly, *Smittia atterima*, in the Czech Republic. Larvae were most abundant in open, disturbed habitats. However, source habitats were vulnerable to desiccation and local extinction of larvae during summer. Smaller populations produced in surrounding, more densely vegetated areas tended to be more stable. Dispersing individuals from these sink habitats subsequently recolonized the source habitats, maintaining a stable population distribution.

4 HABITAT CONNECTIVITY

As described in Section 3.2, habitat homogeneity facilitates population spread over landscapes. However, habitats often are heterogeneous over landscapes, and unsuitable patches can interrupt population spread (Onstad et al., 2003). In such cases, availability of corridors connecting otherwise isolated habitat patches is critical to population growth and spread (Haddad et al., 2003). For example, roads and other disturbed corridors facilitate movement of species associated with disturbed habitats (DeMers, 1993; Haddad, 1999, 2000; Resasco et al., 2014; Spencer and Port, 1988; Spencer et al., 1988); corridors of undisturbed habitat connecting undisturbed patches are necessary to ensure adequate dispersal of species characterizing undisturbed habitats (Collinge, 2000; Várkonyi et al., 2003).

Várkonyi et al. (2003) used mark-recapture techniques to track the movement of two species of noctuid moths: *Xestia speciosa*, a habitat generalist that can be found in natural and managed spruce forests and also in pine-dominated forest throughout Finland; and *Xestia fennica*, a species more restricted to natural spruce forests in northern Finland. They found that both species preferred to move along spruce forest corridors and avoided entering the matrix of clear-cuts and regenerating forest. Movement of *X. speciosa* generally covered longer distances, whereas movement of *X. fennica* was characterized by shorter distances confined within corridors. However, *X. fennica* was capable of longer-distance dispersal across the matrix.

Haddad (1999, 2000) demonstrated that corridors between patches of open habitat embedded in pine, *Pinus* spp., forest significantly increased interpatch dispersal of buckeye, *Junonia coenia*, and variegated fritillary, *Euptoieta claudia*, butterflies. Haddad and Baum (1999) found that three butterfly species (*J. coenia*, *E. claudia*, and cloudless sulphur, *Phoebis sennae*) characterizing open habitat reached higher population densities in patches connected by corridors than in isolated patches; a fourth species, the spicebush swallowtail, *Papilio troilus*, did not show any preference for open versus pine habitat and did not differ in density between connected or isolated patches. Collinge (2000) also reported variable effects of corridors on grassland insect movement. Corridors slightly increased the probability of colonization by less vagile species, but did not affect recolonization by rare species. One of three focus species significantly preferred corridors, whereas the other two moved independently of corridors. These studies indicated that corridors may facilitate movement of organisms among patches, but their effect depends on species characteristics, landscape context, patch size, corridor length, and environmental variation.

Riparian habitats provide unique conditions for specialized terrestrial assemblages and facilitate movement of some terrestrial species through fragmented landscapes (Rykken et al., 2007a,b; Sabo et al., 2005). Riparian habitat widths of at least 30 m on either side of streams appear necessary to provide an adequate corridor effect (Rykken et al., 2007a,b). Cartron et al. (2003) and Lambeets et al. (2008) found that carabid beetle abundance and species richness were significantly higher in riparian forests subject to periodic flooding, compared to non-flooded sites, indicating the importance of flooding to maintain habitat suitability for some riparian taxa. However, the distinct habitat conditions characterizing riparian corridors may not be suitable for dispersing upland species in areas with steep elevational gradients.

Riparian corridors also may be necessary to maintain habitat conditions for populations of some stream invertebrates. Reduction in riparian canopy cover significantly increases water temperature, especially in the summer (Kiffney et al., 2003; Rykken et al., 2007b). Davies and Nelson (1994) found that mayfly (Ephemeroptera) and stonefly (Plecoptera) densities in streams were significantly and positively correlated with the width of adjacent riparian forest buffers in Tasmania, mirroring effects of buffer width on stream temperature. Changes in riparian composition, for example, deciduous versus evergreen, also influence seasonal gradients in temperature; stonefly densities were significantly higher in streams bordered by young deciduous forest, compared to streams through old-growth coniferous forest (Frady et al., 2007).

Some insect species show dramatic population displacement among habitats. Immatures of many aquatic species emerge en masse and disperse through intervening terrestrial habitats to reproduce in distant aquatic habitats (Gratton et al., 2008). Populations of monarch butterflies, *Danaus plexippus*, and other danains migrate thousands of kilometers seasonally between summer habitat in northern temperate zones and overwintering habitat in the tropics (Matthews and Matthews, 2010). Similarly, locust swarms are displaced thousands of kilometers from their source to new habitats where breeding establishes new population centers (Gunn et al., 1948; Rainey, 1963; C. Riley, 1978).

5 ANTHROPOGENIC EFFECTS ON SPATIAL DYNAMICS

The disappearance of *M. spretus* indicates the vulnerability to extinction of even cyclically abundant species when populations decline to near or below extinction thresholds as a result of habitat loss (see Chapter 6). Species persist to the extent that reproductive and dispersal capabilities are adapted to the natural frequency and scale of these changes. Species adapted to relatively unstable habitats typically have higher reproductive rates and greater dispersal capabilities than do species adapted to more stable habitats. Hence, human-dominated ecosystems are likely to favor population growth of species adapted to disturbed habitats, at the expense of species adapted to more stable habitats.

Human activities affect spatial dynamics of populations in several ways. Widespread planting of commercial crops often has facilitated spread of associated herbivores, for example, Colorado potato beetle, *Leptinotarsa decemlineata*, and cotton boll weevil, *Anthonomus grandis* (see Chapter 4). Climate changes eventually will force many species to shift their geographic ranges or face extinction as changing temperatures and humidities exceed tolerance ranges or alter energy balance in their current ranges (Franklin et al., 1992; Kozár, 1991; Rubenstein, 1992) (Fig. 5.2). Williams and Liebhold (2002) projected that southern pine beetle distribution would shift northward and expand in area with warming climate, whereas mountain pine beetle, *Dendroctonus ponderosae*, distribution would move to higher elevations with shrinking areas. However, the recent breach of the Rocky Mountain barrier by this species during a recent outbreak may permit spread of this species into jack pine, *Pinus banksiana*, forests eastward across Canada (de la Giroday et al., 2012; Negrón and Fettig, 2014; Robertson et al., 2009). Williams and Liebhold (1995) found that some climate change scenarios predicted larger areas of defoliation by gypsy moth, whereas other scenarios predicted smaller areas of defoliation (Fig. 7.12).

Fragmentation of terrestrial ecosystems, alteration and pollution of aquatic ecosystems, and redistribution of species arguably are the most serious and immediate threats to ecosystems worldwide (Samways, 1995). Patch scale and distribution and the abruptness of edges between adjoining patches have been altered as a result of habitat fragmentation (Fig. 2.8D–F). Wetlands historically occupied large portions of floodplains but have been virtually eliminated as a result of draining, filling, and stream channelization for urban and agricultural developments. Grasslands have been fragmented severely worldwide because of their suitability for agricultural uses. Forest fragmentation has made forests more vulnerable to edge-related desiccation, tree stress, treefall during storms, and invasive species establishment (Briant et al., 2010; J. Chen et al., 1995; Franklin and Forman, 1987; A. Suarez et al., 1998; Urbas et al., 2007). Reservoirs have altered drainage characteristics and reduced the distances between lake ecosystems. Industrial and agricultural pollution threatens many aquatic species. A variety of vagrant species (including various crops and "weeds," rodents, and livestock, as well as insects and pathogens) have been transported, intentionally and unintentionally, far beyond their natural ranges by human activities. These exotic species have significantly altered the structure and function of their new ecosystems.

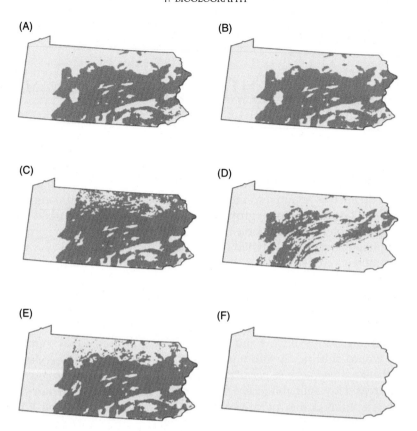

FIGURE 7.12 Potential outbreak areas of gypsy moth in Pennsylvania under climate change scenarios. (A) Current temperature and precipitation, (B) a 2°C increase, (C) a 2°C increase and 0.5 mm day^{-1} precipitation increase, (D) a 2°C increase and 0.5 mm day^{-1} precipitation decrease, (E) GISS model, and (F) GFDL model. *Source: From Williams, D.W., Liebhold, A.M., 1995. Forest defoliators and climatic change: potential changes in spatial distribution of outbreaks of western spruce budworm (Lepidoptera: Tortricidae) and gypsy moth (Lepidoptera: Lymantriidae). Environ. Entomol. 24, 1–9; Oxford University Press.*

5.1 Fragmentation

Fragmentation is the conversion of contiguous habitat into patches of different habitats or land uses. Habitat fragmentation is especially deleterious to species adapted to relatively stable ecosystems (Samways, 1995) and to rare species (Summerville and Crist, 2001). Such species typically are less adapted to rapid or long-distance dispersal and may be less able to recolonize vacant or new habitats (resulting from fragmentation) across inhospitable patches, compared to ruderal species adapted to long-distance colonization of disturbed habitats (Powell and Powell, 1987; Roff, 1990; St. Pierre and Hendrix, 2003, see Chapter 5). Furthermore, insects will not be able to colonize new habitat patches successfully until their hosts are established.

A growing number of species is becoming vulnerable to extinction as populations become more isolated in fragmented habitats (Boecklen, 1991; M. Wilson et al., 1997). Examples include a number of butterfly species, the American burying beetle, *Nicrophorus americanus*, and a number of aquatic and cave-dwelling species (Boecklen, 1991; Hanski and Simberloff, 1997; C. Thomas and Hanski, 1997; M. Wilson et al., 1997). Similarly, local extinction of insectivorous bird species and reduced species richness in isolated fragments of Amazonian rainforest reduce the effect of predation in regulating insect populations (Stratford and Stouffer, 1999).

Old-growth (500–1000 year-old) conifer forests in Pacific Northwestern North America were substantially fragmented by clear-cut harvesting over a 50-year period (1940–90). The forest landscape changed from about 75% old-growth to about 75% stands < 50 years old. A significant proportion of species associated with old-growth forest now exist as relatively small, isolated, and declining populations in a matrix of apparently inhospitable young forest (Christensen et al., 2000). Schowalter (1995) found that 70% of arboreal arthropod species in old-growth (500-year-old) conifer forests in western Oregon were not present in adjacent young (20-year-old) conifer plantations. Predators and detritivores were particularly affected. Similarly, Powell and Powell (1987) found that flower visitation by male euglossine bees declined following forest fragmentation, even in the largest, 100-ha, fragment size, and was proportional to fragment size, indicating that very large areas of forest are necessary to maintain viable population sizes for some species.

The degree of fragment isolation affects colonization and genetic variability. Steffan-Dewenter and Tscharntke (1999) demonstrated that abundance of pollinating bees and seed production declined with increasing isolation (distance) of experimental mustard, *Sinapis arvensis*, and radish, *Raphanus sativus*, plants from intact grassland in Germany. Steffan-Dewenter et al. (2002) further demonstrated that abundance and diversity of bees increased as the proportion of surrounding habitat in seminatural condition increased. Krawchuk and Taylor (2003) studied patterns of abundance of three dipterans, *W. smithii* (Culicidae), *Metriocnemus knabi* (Chironomidae), and *Fletcherimyia fletcheri* (Sarcophagidae), inhabiting pitcher plants, *Sarracenia purpurea*, in western Newfoundland, Canada. For all three insect species, habitat configuration (patch size and isolation) was more important than the total area of habitat, but the relative importance of patch size versus isolation changed with spatial scale. Patch size was more important at the scale of movement and survival of individuals, whereas patch isolation was more important at the scale of matrix configuration and metapopulation dynamics.

Effects of fragmentation will likely be exacerbated by disturbances or extreme climate events. Piessens et al. (2009) analyzed the effect of an extreme summer heat wave in Belgium in 2003 on oviposition, survival, and population size of the blue butterfly, *Cupido minimus*, a specialist herbivore of *Anthyllis vulneraria*, in calcareous grasslands. Populations of the host plant, as well as population size and oviposition rate of the butterfly, declined in size following the heat wave. By 2006 several butterfly populations had disappeared, but most plant populations had recovered, with only one population going extinct. The probability of local extinction for *C. minimus* was significantly related to initial population size, with smaller populations having a higher risk of extinction than larger populations.

Edges between patches are particularly pronounced in anthropogenic landscapes and affect dispersal of many species (Ries et al., 2004). Natural gradients of climate and geology interacting with disturbances produce relatively large patches, with broad transition zones (ecotones) between patches that dampen interference by one patch on environmental conditions in another. By contrast, human land use practices tend to produce smaller patches with abrupt edges, for example, distinct agricultural monocultures within fenced boundaries, plowed edges against grasslands, harvested and regenerating plantations against mature forests (Fig. 2.8D–F), and greater edge density, that is, edge perimeter length per ha (Radeloff et al., 2000). These distinct edges substantially influence environmental conditions of the adjacent patches. For example, an edge of tall trees along an abrupt boundary with an adjacent plantation of short trees is exposed to much greater insolation and airflow, depending on edge orientation, leading to higher temperatures, lower humidities, and greater vulnerability to windthrow than prevail when trees along the edge are buffered by surrounding trees (Franklin and Forman, 1987). J. Chen et al. (1995) discovered that microclimatic gradients (higher temperature, lower humidity) extended 180–480 m into old-growth Douglas-fir, *Pseudotsuga menziesii*, forests from clear-cut edges, affecting habitat conditions for associated organisms. They concluded that forest patches <64 ha would be completely compromised by external environmental conditions, that is, would be characterized entirely as edge habitat rather than as interior forest habitat. Similarly, grasslands overgrazed by livestock within fenced boundaries expose soil to desiccation, leading to the death of surrounding vegetation and an increasing area of desertification (Schlesinger et al., 1990; J. Foley et al., 2003a, see Chapter 11 and Fig. 2.8F). However, different effects on horizontal gradients of light, air temperature, relative humidity, soil moisture, and wind speed were found for edges created by narrow, linear openings for highways, powerlines, and streams, indicating that the nature of the edge, as well as its width, determine the magnitude of the effect on microclimatic conditions (Pohlman et al., 2007).

Insects, as well as other organisms, are sensitive to these edge effects (Ries et al., 2004). Haynes and Cronin (2003) found that planthoppers, *P. crocea*, accumulated along edges, compared to the interior, of prairie cordgrass patches adjacent to mudflat but not patches adjacent to nonhost grasses, reflecting lower rates of dispersal across inhospitable mudflats (Fig. 7.4). Ries and Fagan (2003) reported that egg case (oothecae) densities for a mantid, *Stagmomantis limbata*, were three times higher in two types of edge habitat than in interior habitats; bird predation on oothecae also was about three times higher in one type of edge habitat (and tended to be higher in the other), compared to interior habitats. This situation represents an ecological trap, in which an organism actively selects a poor habitat over superior habitat.

Roland and Kaupp (1995) found that transmission of nuclear polyhedrosis virus was reduced along forest edges, prolonging outbreaks of the forest tent caterpillar, *Malacosoma disstria*. Similar results were found for understory insectivorous birds in tropical forest, suggesting that outbreaks of some insects could be more likely in fragments from which predators have disappeared (Şekercioğlu et al., 2002). Remnant patches of natural habitat also are highly vulnerable to influx of nonindigenous species, from neighboring patches, that may compete with, or prey upon, indigenous species (Punttila et al., 1994). Ewers and Didham (2006) developed a statistical model to quantify the magnitude and extent of edge effects on four insect taxa. They found that the magnitude of edge effect and the extent of effect from the edge were not correlated but provided different information about edge effects.

Effects of edge density on the landscape can change during the course of population growth and decline. Radeloff et al. (2000) found that correlations between landscape patterns and jack pine budworm, *Choristoneura pinus*, population size varied over time, with proportion of jack pine, *P. banksiana*, and edge density (sum of perimeter length for land use classes per ha) positively correlated up to the peak of the outbreak, but edge density negatively correlated during population decline. These results probably reflected the more suitable resources represented by pollen cones that were more abundant on edge trees and the greater abundance of avian predators and the primary wasp parasitoid, *Itoplectis conquisitor*, along edges.

As reported for natural disturbances (see Chapter 2), fragmentation does not affect all species equally, nor all negatively. Tscharntke (1992) reviewed studies that examined responses of several insect species to differences in reed, *Phragmites australis*, quality in fragmented (agricultural) and unfragmented (nature reserve) wetlands. Reeds in small patches had thinner shoots but more leaves than did reeds in large patches. Two chloropid flies, *Lipara* spp., that depend on thin shoots survived only in small patches or in the unmown edges of large patches. However, the stem-boring noctuid moth, *Archanara germinipuncta*, that depends on thick shoots persisted only in large patches. Shoot damage caused by this moth created necessary habitat for >20 other herbivores, saprovores, and parasitoids. For example, the gall midge, *Lasioptera arundinis*, survived only in the side shoots induced by *A. germinipucta* damage, making this midge equally dependent on large patches. Tscharntke (1992) calculated that survival of local populations of *A. germinipuncta* requires at least 180,000 individuals or at least a 2 ha area.

Fragmentation of natural ecosystems typically is associated with homogenization of vegetation patterns. Widespread planting of commercial crops and suppression of natural disturbances have eliminated much of the diversity of vegetation patches characterizing natural landscapes. In a diverse landscape, outbreaks of particular demes most often would be confined to patches of susceptible vegetation. Agricultural and managed forest landscapes are more conducive to expansion and regionwide outbreaks of adapted species (Schowalter and Turchin, 1993).

5.2 Disturbances to Aquatic Ecosystems

Stream channelization and impoundment have reduced heterogeneity in channel morphology and flow characteristics. Channelization constrains channel morphology, removes obstacles to flow, and shortens stream length. These modifications eliminate habitats in overflow areas (such as wetlands and side channels) and in logs and other impediments, and speed drainage in the channeled sections. Impoundments replace a sequence of turbulent sections and pools behind logs and other obstacles (characterized by rocky substrates and high oxygen content) with deep reservoirs (characterized by silty substrates and stratification of oxygen content and temperature). These changes in stream conditions eliminate habitat for some species (such as species associated with high flow rates and oxygen concentrations) and increase habitat availability for others (such as species associated with lotic conditions and low oxygen concentrations).

The linear configuration of stream systems (ie, the stream continuum concept, Vannote et al., 1980) makes them particularly vulnerable to disturbances that occur upstream. For example, heavy precipitation in the watershed is concentrated in the stream channel, scouring the channel and redistributing materials and organisms downstream. Fire or harvest of riparian vegetation exposes streams or wetlands to increased sunlight, raising temperatures and increasing primary production, altering habitat and resource conditions downstream, often for long time periods (Batzer et al., 2000a; Haggerty et al., 2004; Harding et al., 1998; Kiffney et al., 2003). Industrial effluents, runoff of agricultural materials (eg, fertilizers), or accidental inputs of toxic materials (eg, oil spills, leachates from mine spoils, and pesticides) affect habitat suitability downstream until sufficient dilution has occurred (S. Smith et al., 1983; Southwick et al., 1995). Eutrophication, resulting from addition of limiting nutrients, substantially alters the biological and chemical conditions of aquatic systems.

Lake Balaton (Europe's largest lake) in Hungary has experienced incremental eutrophication since the early 1960s, when lake chemistry was relatively uniform (Somlyódy and van Straten, 1986). Since that time, phosphorus inputs from agricultural runoff and urban development have increased, starting at the west end where the Zala River enters the lake. The division of Lake Balaton into four relatively distinct basins draining distinct subwatersheds facilitated documentation of the progression of eutrophication from west to east (Somlyódy and van Straten, 1986). Dévai and Moldován (1983) and Ponyi et al. (1983) found that the abundance and species composition of chironomid larvae were correlated with this longitudinal gradient in water quality. The original species characterizing oligo-mesotrophic conditions have been replaced by species characterizing eutrophic conditions in a west to east direction. Similarly, sedimentation resulting from erosion of croplands or clear-cut forests or from trampling of streambanks by livestock alters substrate conditions and habitat suitability for organisms downstream.

Pringle (1997) reported that disturbances and anthropogenic modification of downstream areas (eg, urbanization, channelization, impoundment) also affect conditions for organisms upstream. Degraded downstream areas may be more vulnerable to the establishment of exotic species that are tolerant of stream degradation. These species subsequently invade upstream habitats. Degradation of downstream areas may restrict movement of upstream species within the watershed, thereby isolating headwater populations and limiting gene flow between watersheds. Finally, degradation of downstream zones often prevents movement of anadromous or catadromous species.

Disturbances to adjacent terrestrial ecosystems affect aquatic species. Davies and Nelson (1994) compared aquatic invertebrate responses to forest harvest within 10 m of streams, 10–30 m of streams, 30–50 m of streams, or no harvest in Tasmania. Densities of aquatic invertebrates were measured at a site upstream of the treatment and at a second site immediately downstream from the treatment. Differences in mayfly (Ephemeroptera) and stonefly (Plecoptera) densities between upstream and downstream sites were significantly, positively correlated with the width of the riparian forest buffer. Overall, mayfly density declined 62% and stonefly density 34% at sites with a buffer of < 30 m, demonstrating the importance of riparian forest buffers to these aquatic detritivores. In a similar study in western Canada, Kiffney et al. (2003) compared periphyton and chironomid midge grazer abundances in streams with no riparian buffer (clear-cut to stream edge), a 10 m or 30 m buffer, and an uncut control. Periphyton mass and chironomid abundance generally increased with increasing solar exposure, that is, with decreasing buffer width, demonstrating a different response to riparian buffer width by grazing insects.

5.3 Anthropogenic Transport Across Barriers

Human transportation of exotic species across natural barriers to their dispersal has altered dramatically the structure and function of natural ecosystems across the globe (Samways, 1995; A. Suarez et al., 1998; Wallner, 1996). Examples include the devastation of island vegetation by pigs and goats introduced intentionally by explorers, destruction of grasslands globally by domesticated, often introduced livestock, disruption of aquatic communities by introduced species (eg, water hyacinth and zebra mussel in North America), and disruption of grassland and forest communities by introduced plants (eg, spotted knapweed in North America), mammals (eg, rabbits in Australia), reptiles and amphibians (brown tree snake in Oceania, African clawed frog in North America), insects (eg, gypsy moth in North America, the European wood wasp, *Sirex noctilio*, in Australia), and pathogens (eg, chestnut blight and white pine blister rust in North America, Dutch elm disease in North America and Europe, pinewood nematode in Japan). Exotic species, especially of insects, can be found in virtually all "natural" ecosystems on all continents. Many herbivorous insects and mites have arrived on agricultural or forestry products and become plant pests in agroecosystems or forests. Some herbivorous and predaceous arthropods have been introduced intentionally for biological control of exotic weeds or plant pests (Croft, 1990; Kogan, 1998; McEvoy et al., 1991). Despite evaluation efforts, these biological control agents, especially arthropod predators, compete with native species and have the potential to colonize native hosts related to the exotic host and develop new biotypes (Louda et al., 2003). Indigenous herbivore species also can colonize exotic hosts and develop new biotypes (D. Strong et al., 1984), with unknown consequences for long-term population dynamics and community structure. Samways et al. (1996) found that different invertebrate assemblages were found on exotic vegetation, compared to indigenous vegetation, in South Africa.

Urban areas represent increasingly large and interconnected patches on regional landscapes and are particularly important ports for the spread of exotic species into surrounding ecosystems. Urban centers are the origin or destination for commercial transport of a wide variety of materials, including forest and agricultural products, and often have highly modified climatic conditions, for example, heat islands (Akbari et al., 2009; Arnfield, 2003, see Chapter 11). Urban areas are characterized by a wide variety of exotic species, especially ornamental plants and their associated exotic insects and pathogens. Exotic or native ornamental species typically are stressed by soil compaction, air and water pollutants, and hot, dry conditions (Arnfield, 2003). Arriving exotics, especially ants, termites, and cockroaches, often have little difficulty finding suitable resources, such as construction debris or food scraps, and becoming established in urban centers and subsequently spreading into surrounding ecosystems. Some native insect species have adapted to higher urban temperatures and to finding isolated hosts among urban structures (San Martin y Gomez and Van Dyck, 2012).

Road networks that connect urban centers and penetrate natural ecosystems represent unique ecosystems that can interrupt population movement or provide corridors that facilitate spread of native or exotic species (Lugo and Gucinski, 2000; Vasconcelos et al., 2006). Roadsides typically are highly disturbed by road maintenance, other human activities, and air pollution from vehicles, and provide suitable habitat for a variety of species (Fig. 7.13). Gypsy moth is particularly capable of spreading via human transportation (of pupae or egg masses attached to vehicles, outdoor equipment, or commercial products) between urban centers. Stiles and Jones (1998) demonstrated that population distribution of the red imported fire ant, *S. invicta*, was significantly related to width and disturbance frequency of road and power line corridors through forests in the southeastern USA. Mound densities were significantly highest along dirt roads not covered by forest canopy and lowest along roads covered by forest canopy. Power line and graveled or paved roads not covered by forest canopy

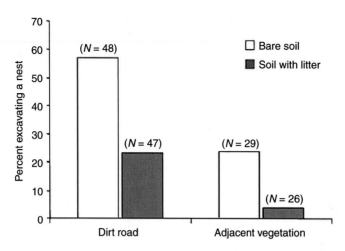

FIGURE 7.13 **Percentage of *Atta laevigata* queens excavating nests in bare soil or soil with litter in roadsides or adjacent cerrado vegetation in Brazil.** *Source: From Vasconcelos, H.L., et al., 2006. Roads alter the colonization dynamics of a keystone herbivore in Neotropical savannas. Biotropica 38, 661–665.*

supported intermediate densities of mounds. These trends suggest that canopy openings of intermediate width and high disturbance frequency are most conducive to fire ant colonization, facilitating spread among disturbed patches. Resasco et al. (2014) reported similar promotion of fire ant spread via open-canopied corridors through these forests.

6 MODELS OF SPATIAL DYNAMICS

One of the most significant advances in population dynamics research in recent years has been the development of spatially explicit models of population dynamics. A number of approaches have been used to model spatial dynamics. As with temporal dynamics, spatial dynamics can be modeled using deterministic, stochastic, or chaotic functions (Hassell et al., 1991; Matis et al., 1994; Sherratt and Jepson, 1993). Different spatial dynamics result from using these different types of functions.

The earliest attempts to model spatial dynamics either applied diffusion models to describe insect dispersal and population spread from population centers (Rudd and Gandour, 1985; Skellam, 1951; Turchin, 1998) or modeled population dynamics independently among individual landscape patches, based on local conditions within each patch, and linked patches by dispersal processes (W. Clark, 1979). Diffusion models assume that the environment is homogeneous and that individuals disperse independently and with equal probability in any direction. The diffusion approach is useful for modeling spatial dynamics of insects in stored grain or relatively homogenous crop systems but less useful in most natural landscapes where patchiness interrupts diffusion.

Advances in spatial modeling have been facilitated by the development of powerful computers that can store and manipulate large data-sets. Concurrent development of geographic positioning systems (GPS) and geographic information system (GIS) and geostatistical

6 polygons   33–170 polygons

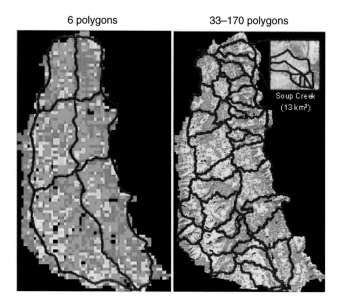

FIGURE 7.14   Examples of geostatistical representation. The Seeley–Swan watershed in Montana, USA, is depicted with 1 km² raster cells on the left, by 30 × 30 m raster cells on the right, and with vector polygons defined from topographic analysis in the inset. The more heterogeneous the landscape, the greater the number of cells required to provide an accurate assessment of spatial variation. *Source: From R. Waring, R.H., Running, S.W., 1998. Forest Ecosystems: Analysis at Multiple Scales. Academic Press, San Diego, CA.*

software has been a key to describing insect movement (Turchin, 1998) and population epidemiology (Liebhold et al., 1993) across landscapes.

A GIS is an integrated set of programs that facilitate collection, storage, manipulation, and analysis of geographically referenced data, such as topography, vegetation type and density, and insect population densities. Data for a particular set of coordinates can be represented as a value for a cell, and each cell in the matrix is given a value (Fig. 7.14). This method is called the *Raster method*. A second method, that requires less storage space, is the *vector method*, in which only data representing the vertices of polygons containing data must be stored (Fig. 7.14). Various matrices representing different map layers can be superimposed to analyze interactions. For example, a map layer representing insect population distribution can be superimposed on map layers representing the distribution of host plants, predator abundances, climatic conditions, disturbances, or topography to evaluate the effects of patchiness or gradients in these factors on the spatial dynamics of the insect population.

Geostatistics are a means of interpolating the most probable population densities between sample points, in order to improve the representation of spatial distribution over landscapes. Early attempts to characterize spatial patterns were based on modifications of s^2/x, Taylor's Power Law, Lloyd's Patchiness Index, and Iwao's patchiness regression coefficients (Liebhold et al., 1993). These indices focus on frequency distributions of samples, and are useful for identifying dispersion patterns (see Chapter 5), but ignore the spatial locations of samples. Modeling spatial dynamics across landscapes requires information on the location of sampling points, as well as population density data. The locations of population aggregations affect densities in

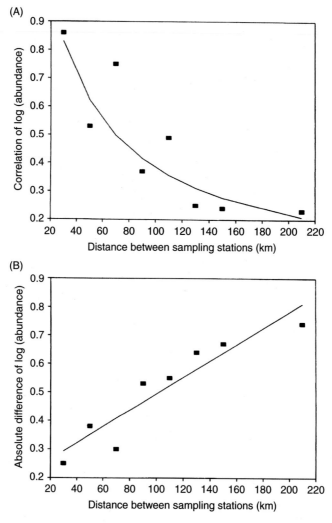

FIGURE 7.15   Relationships between the temporal correlation of *Delphacodes kuscheli* density and the distance between sampling stations (A) and between the mean absolute difference between densities of pairs of sampling stations and the distance between sampling stations (B) in Argentina. *Source: From Grilli, M.P., Gorla, D.E., 1997. The spatiotemporal pattern of* Delphacodes kuscheli *(Homoptera: Delphacidae) abundance in central Argentina. B. Entomol. Res. 87, 45–53.*

adjacent cells (Coulson et al., 1996; Liebhold and Elkinton, 1989). Development of geographic positioning systems has facilitated the incorporation of precise sample locations in GIS databases.

An underlying assumption of geostatistics is that the degree of similarity between sample points is correlated with their proximity (Fig. 7.15) (Coulson et al., 1996; Gilbert and Grégoire, 2003; Grilli and Gorla, 1997; Liebhold et al., 1993; M. Smith et al., 2004). Population structure in a given cell is influenced by the population structures in neighboring cells more than by distant cells. An autocorrelation matrix can be developed from data for different distance classes, that is, x and y coordinates differing by a given distance (Liebhold and Elkinton, 1989). This spatial autocorrelation can be used to interpolate values for un-sampled locations by taking a weighted linear average of the available samples, a technique known as *kriging* (Gilbert and Grégoire, 2003; Gribko et al., 1995; Grilli and Gorla, 1997; Hohn et al., 1993; Liebhold et al., 1993). Kriging represents an advance over traditional methods of interpolation in several ways, but its most important provision is the incorporation of several forms of information simultaneously. The joint spatial dependence of population density and factors such as climate, soil conditions, vegetation, etc. can be integrated to provide more ac-curate estimates than would be possible with any single variable.

Gilbert and Grégoire (2003) used these methods to evaluate factors affecting the spatial structure of the European bark beetle, *Dendroctonus micans*, in a French spruce forest (Fig. 7.16). They demonstrated that the *D. micans* population had a strong spatial structure, significantly related to tree density, average slope within a 250 m radius, and the abundance of a specialist predator, *Rhizophagus grandis*, released within a 300 m radius > 6 years previously. D. Wil-liams and Liebhold (1995) used these techniques to predict the spatial distribution of insect population densities under potential future climates (Fig. 7.12). K. Zimmerman et al. (2004) applied a Markov chain model to digitized, spatially explicit grasshopper abundance data in Wyoming (representing 45 years at 1-km^2 resolution) to calculate probabilities that grasshop-per populations would increase or decrease over time. They concluded that the average dura-tion of outbreaks (average density $\geq$ 9.6 grasshoppers m^{-2}) is < 2 years, and large portions of the state experience outbreaks that do not last more than 1 year. On the other hand, periods of uninfested conditions generally do not last long either: $\leq$ 10 years on 36% of suitable land.

Modeling of spatial dynamics in stream networks or montane topography with branched topology presents special challenges. In such networks, the distance between two points may not be represented adequately by Euclidean distance, because of limitations to the movement of aquatic organisms across land (Fagan et al., 2010; Ganio et al., 2005). Rather, the shortest

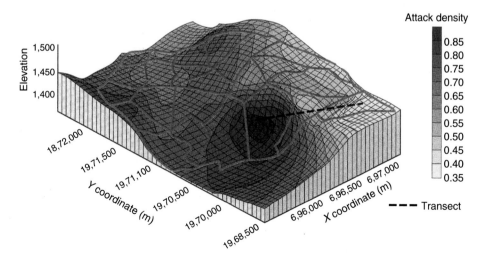

FIGURE 7.16 **Spatial structure of proportion of trees attacked by** *Dendroctonus micans*, **based on two-dimensional omnidirectional kriging, in a 600 ha spruce stand in France.** *Source: From Gilbert, M., Grégoire, J.-C., 2003. Site condition and predation influence a bark beetle's success: a spatially realistic approach. Agr. Forest Entomol. 5, 87–96.*

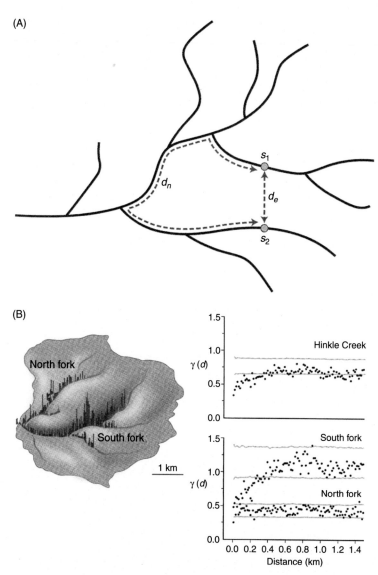

FIGURE 7.17   (A) Distance between points S_1 and S_2 in a network can be measured either as Euclidean distance (d_e) or as distance along the network pathway (d_n); (B) spatial distribution and empirical variograms of coastal cutthroat trout counts in Hinkle Creek in western Oregon. Variograms show semivariance as a function of network distance (d_n) with 2.5th and 97.5th percentiles from 5000 permutations (*red lines*) for the entire watershed and for the North and South forks separately. *Source: From Ganio, L.M., et al., 2005. A geostatistical approach for describing spatial pattern in stream networks. Front. Ecol. Environ. 3, 138–144.*

distance from the perspective of aquatic organisms is along the stream channel (Fig. 7.17A). Ganio et al. (2005) described the use of an empirical variogram, based on shortest distances along the network pathway between sample points, to evaluate spatial patterns and differences in spatial structure along stream networks in western Oregon, USA (Fig. 7.17B). Such new tools will contribute to modeling of spatial structure in aquatic populations.

7 SUMMARY

Factors affecting the geographic distributions of populations have intrigued ecologists for at least the past two centuries. Distributions can be described at different geographic scales. Six distinctive floral and faunal associations (biogeographic realms) can be identified, conforming roughly to continental boundaries, but also reflecting the history of continental movement (plate tectonics). Topography also creates gradients in environmental conditions on mountains and temperature stratification with depth in aquatic ecosystems.

The distribution of species among islands intrigued early ecologists. The ability of populations to colonize oceanic islands was found to reflect the dispersal capacity of the species, the size of the island, and its distance from the population source. Although controversial, principles of island biogeography have been applied to colonization of terrestrial habitat islands, for example, mountaintops and patches of unique habitat in otherwise inhospitable landscapes.

At more local scales, the spatial distribution of populations changes with population size. Growing populations expand over a larger area as individuals move from high-density patches to the fringe of the population. Rapidly expanding populations generate large numbers of dispersing individuals that maximize the colonization of new patches. Under favorable conditions, these satellite demes expand and coalesce with the main population, affecting ecosystem processes over large areas. Declining populations shrink into isolated refuges that maintain distinct demes of a metapopulation. The extent of dispersal among these demes determines genetic heterogeneity and ability to recolonize patches following local extinctions.

All populations are vulnerable to local extinctions due to changing environmental conditions and disturbances. Populations survive to the extent that their dispersal strategies facilitate recolonization and population movement over landscapes. Anthropogenic activities alter spatial distribution in several ways. Climate changes affect the geographic distribution of suitable habitats. However, the most serious anthropogenic effects on spatial patterns are habitat fragmentation, alteration and pollution of aquatic ecosystems, and redistribution (intentionally or unintentionally) of various species. Fragmentation increases isolation of demes and places many species at risk of extinction. At the same time, predators and parasites appear to be most vulnerable to fragmentation and habitat disturbances, often increasing opportunities for population growth by prey species. Humans also are responsible for the introduction of a large and growing number of plant and animal species to new regions as a result of transportation of commercial species and forest and agricultural products. Urban areas represent centers of commercial introductions and provide opportunities for exotic ornamental and associated species to become established and move into surrounding ecosystems. These species affect various ecosystem properties, often dramatically altering vegetation structure and competing with, or preying on, native species.

Modeling of spatial distribution patterns has been facilitated by recent development of geographic information systems (GIS) and geostatistical techniques. Early models represented population expansion as a simple diffusion process. Application of GIS techniques to the patch dynamics of metapopulations permits integration of data on population dynamics with data on other spatially varying factors across landscapes. Geostatistical techniques, such as kriging, permit interpolation of density data between sampling stations to improve mapping and projecting of population distributions. These techniques are improving our ability to evaluate population contributions to ecosystem conditions across landscapes.

COMMUNITY ECOLOGY

Species co-occurring at a site interact to various degrees, both directly and indirectly, in ways that have intrigued ecologists since the earliest times. These interactions regulate population dynamics, thereby affecting community structure, and also control rates of energy and matter fluxes among organisms, thereby affecting ecosystem function. Some organisms engage in close, direct interactions, as consumers and resources, whereas others interact indirectly in ways that affect activity, such as foraging behavior and reproduction. For example, predation on mimics depends on the presence of their models, and herbivores can be affected by their host's chemical responses to other herbivores. Direct interactions, that is, competition, predation, and symbioses, have been the focus of research on factors controlling community structure and dynamics, but recent studies have demonstrated that indirect interactions also control community organization. Species interactions are the focus of Chapter 8.

A community is composed of the plant, animal, and microbial species occupying a site. Some of these organisms are integral and characteristic components of the community and help define the community type, whereas others occur by chance as a result of dispersal, foraging, or other movement across a landscape or through a watershed. Particular combinations of species distinguish desert, grassland, or forest communities. Different species assemblages are found in turbulent stream versus lake or in eutrophic versus oligotrophic systems. The number of species and their relative abundances define species diversity, an aspect of community structure that is the focus of a number of ecological issues. Chapter 9 addresses the various approaches to describing community structure and factors determining geographic patterns of community structure.

Community structure and interactions change through time as populations respond differentially to changing environmental conditions, especially to disturbances and changes in resource availability and predation. Anthropogenic changes have become a dominant factor affecting community dynamics. Just as population dynamics reflect the net effects of individual reproduction, death, and dispersal interacting with the environment, community dynamics reflect the net effects of species population dynamics interacting with the environment. Severe disturbance or environmental changes can lead to drastic changes in community structure. Recent studies are clarifying changes in prehistoric community structure and interactions. Changes in community structure through time are the subject of Chapter 10.

Community structure largely determines the biotic environment affecting individuals (Section I) and populations (Section II). The interactions among members of the community also modify the abiotic conditions of a site (Narr and Frost, 2015). Vegetation cover reduces albedo (reflectance of solar energy), modifies temperature and humidity within the boundary layer, reduces soil erosion, and alters energy and biogeochemical fluxes, compared to nonvegetated sites, thereby determining ecosystem conditions. Species interactions, including those involving insects, modify vegetation cover and affect these processes, as discussed in Section IV. Different community structures affect these processes in different ways.

8

Species Interactions

SPECIES INTERACTIONS SPEED DISEASE SPREAD

The rapid spread of West Nile Virus (WNV) across the USA following its introduction in 1999 reflected complex interactions among the diversity of bird hosts and mosquito vectors of the virus. WNV originated in central Africa perhaps 1000 years ago, as a zoonotic disease of birds, but was not recognized until 1937 when a human case was diagnosed in the West Nile District of Uganda. It appeared in New York City in 1999, perhaps through introduction of an infected bird or mosquito vector, and caused 62 human cases of encephalitis (7 deaths), concurrent with extensive mortality of crows, *Corvus* spp., and exotic birds at a zoological park (Lanciotti et al., 1999). By 2001 the disease was epidemic in New York and Florida and increasing in incidence in other East and Gulf Coast states. In 2002, 4156 human cases and 284 deaths were recorded, primarily in the Great Lakes states and Louisiana. The following year the disease caused 264 deaths among nearly 10,000 human cases in the Midwest. The disease reached epidemic proportions on the West Coast in 2004 and the Pacific Northwest in 2006 before declining in incidence among humans through 2009.

Infection of horses and humans typically followed the spread of infected birds, the primary (reservoir) hosts of the virus. Some avian species suffered 100% mortality of infected individuals, whereas others showed virtually no mortality (LaDeau et al., 2008). Of 20 species of potential avian hosts across the USA, seven species in four families showed significant declines in abundance as a result of WNV, with American crows, *Corvus brachyrhynchos*, declining 45% following WNV introduction (LaDeau et al., 2007). Only two of the seven species had recovered to pre-WNV abundances by 2005.

The virus is vectored by mosquitoes. Mosquito species vary in their ability to transmit the virus. Among 25 mosquito species known to carry WNV in nature, the Asian tiger mosquito, *Aedes albopictus*, common house mosquito, *Culex pipiens*, southern house mosquito, *C. quinquefasciatus*, white-dotted mosquito, *C. restuans*, salt marsh mosquito, *C. salinarius*, western encephalitis mosquito, *C. tarsalis*, and Asian rock pool mosquito, *Ochlerotatus japonicus*, were highly efficient vectors, whereas the western treehole mosquito, *O. sierrensis*, black salt marsh mosquito, *O. taeniorhynchus*, and white-footed woods mosquito, *Psorophora ferox*, had little or no ability to transmit the virus (Turrell et al., 2005). Furthermore, *Culex* species feeding primarily on birds were the principal enzootic vectors, but opportunistic species, such as *A. albopictus* and *O. japonicus*, were important bridge vectors that transmitted the virus from birds to mammals (Turrell et al., 2005).

Insect Ecology. http://dx.doi.org/10.1016/B978-0-12-803033-2.00008-X

The incidence of human cases was related to the diversity of interactions among reservoir and nonreservoir hosts and competent and noncompetent vectors. Allan et al. (2009) tested hypotheses that WNV transmission is related primarily to bird species diversity, to mosquito vector abundance, or to human population density. They found that the prevalence of WNV infection in mosquitoes and humans increased with decreasing bird species diversity and increasing proportion of primary reservoir hosts. Interestingly, the primary reservoir hosts were avian species best adapted to human-dominated, for example, urban and agricultural, landscapes: crows, blue jays, *Cyanocitta cristata*, and cardinals, *Cardinalis cardinalis*. Thus, augmenting the diversity of nonreservoir bird species, for example, by increasing the availability of seminatural habitat in urban or agricultural settings, could reduce transmission and human cases of this disease.

1 INTRODUCTION

Just as individuals interact in ways that affect population structure and dynamics, species populations in a community interact in ways that affect community structure and dynamics. All species influence each other to varying degrees through feeding interactions, foraging activities, alteration of vegetation or animal abundances, provision of burrows, or other habitats or resources, etc. Species interactions vary considerably in their form, strength, and effect, and can create quite complex interaction networks among species.

One species can influence the behavior or abundance of another species directly or indirectly. Direct effects generally require some form of contact, for example, a predator feeding on its prey, whereas indirect effects do not involve contact, but rather result from the presence or activities of another species, for example, predators attracted to herbivore-induced plant chemicals. The web of interactions, direct and indirect and having positive or negative feedback, determines the structure and dynamics of the community (see Chapters 9 and 10) and controls rates of energy and matter fluxes through ecosystems (see Chapter 11).

Insects have provided some of the most interesting and complex examples of species interactions. Direct interactions among plants and associated insect herbivores and pollinators and between parasites and their hosts, including paleontological evidence of their evolutionary development (Boucot and Poinar, 2010; Labandeira, 1998, 2002; Poinar and Poinar, 2004b, 2005), have been among the most widely studied. Our understanding of plant–herbivore, predator–prey, and various symbiotic interactions is derived largely from models involving insects.

However, the complexity of indirect effects has become a major topic of investigation (Clay et al., 1993; M.L. Johnson et al., 2006; Karban and Baldwin, 1997; Kessler et al., 2006; Ohgushi, 2005, 2008; Poveda et al., 2007; Van Zandt and Agrawal, 2004). Indirect interactions often may affect population dynamics and community organization at least as much as more obvious direct effects (Buchanan and Underwood, 2013; Hawlena et al., 2012; Preisser et al., 2005) but are more difficult to measure. This chapter describes the major classes of interactions, factors that affect these interactions, and consequences of interactions for community organization.

2 DIRECT INTERACTIONS

Species interact directly with other species in various ways and with varying degrees of intimacy (interaction strength). Individuals compete with, prey on, or are prey for, various associated species, and may be involved in stronger interactions with particular species (ie, symbiosis). Categories of interactions generally have been distinguished on the basis of the direction of effect, that is, positive, neutral, or negative effects on growth or mortality of each species. Interactions can have multiple effects on the species involved, depending on abundance and condition of the partners, and can change through time, depending on environmental conditions or adaptations, requiring consideration of the net effects of an interaction in order to understand its origin and consequences (Økland et al., 2009).

2.1 Competition

Competition is the struggle for use of shared, limiting resources (Fig. 8.1). Resources can be limiting at various amounts and for various reasons. Any resource can be an object of

FIGURE 8.1 **Competition: evidence of interference between southern pine beetle,** *Dendroctonus frontalis,* **larvae (small mines) and co-occurring cerambycid,** *Monochamus titillator,* **larvae (larger mines) preserved in bark from a dead pine tree.** The larger cerambycid larvae often remove phloem resources in advance of bark beetle larvae, consume bark beetle larvae in their path, or both.

interspecific competition, for example, food resources, shelter, basking or oviposition sites. Water or nutrient resources may be largely unavailable and support only small populations or a few species in certain habitats (eg, desert and oligotrophic lakes) but abundant and support larger populations or more species in other habitats (eg, rainforest and eutrophic lakes). Newly available resources may be relatively unlimited until sufficient colonization has occurred to reduce per capita availability.

Although competition for limited resources has been a major foundation for evolutionary theory (Malthus, 1789; Darwin, 1859), its role in natural communities has been controversial (Connell, 1983; Lawton, 1982; Lawton and Strong, 1981; Schoener, 1982; Strong et al., 1984). Denno et al. (1995) and Price (1997) attributed the controversy to three major criticisms that arose during the 1980s. First, early studies were primarily laboratory experiments or field observations. Few experimental field studies were conducted prior to the 1970s (Istock, 1973). Second, Hairston et al. (1960) argued that food must rarely be limiting to herbivores because so little plant material is consumed under normal circumstances (see also Chapters 3 and 11), leading to experimental focus on effects of predators, parasites, and pathogens on herbivore populations during the 1960s through early 1980s. Third, many species that were assumed to compete for the same resource(s) co-occur and appear not to be resource limited. In addition, many communities apparently were unsaturated, that is, many niches were apparently vacant (Kozár, 1992b; Strong et al., 1984). These arguments led many ecologists to believe that competition was not a major factor in natural communities.

Fortunately, this controversy stimulated experimental approaches to studying competition. Some (but not all) experiments in which one presumed competitor was removed have demonstrated increased abundance or resource use by the remaining competitor(s), thereby providing evidence for competition (T. Davis and Hofstetter, 2009; Denno et al., 1995; Istock, 1973, 1977; Pianka, 1981; Tack et al., 2009). However, many factors affect interspecific competition (Colegrave, 1997). Denno et al. (1995), Moser and Obrycki (2009), Pianka (1981), and Tack et al. (2009) suggested that competition may operate over a gradient of intensities, depending on the relative densities of the competing species (Fig. 8.2) and the degree to which potentially competing species specialize on a portion of (partition) shared resources.

Denno et al. (1995) reviewed studies involving 193 pairs of phytophagous insect species. They found that 76% of these interactions demonstrated competition, whereas only 18% indicated no competition, although they acknowledged that published studies might be biased in favor of species expected to compete. The strength and frequency of competitive interactions varied considerably. Generally, interspecific competition was more prevalent, frequent, and symmetric among haustellate (sap-sucking) species than among mandibulate (chewing) species, or between sap-sucking and chewing species. Competition was more prevalent among species feeding internally, for example, miners and seed-, stem-, and wood-borers (Fig. 8.1), than among species feeding externally. Competition was observed least often among free-living, chewing species, that is, the most obvious potential competitors selected for earlier studies that challenged the importance of competition.

Most competitive interactions (84%) were asymmetric, that is, one species was a superior competitor and suppressed the other (Denno et al., 1995). Root feeders were consistently outcompeted by folivores, although such competitive interactions can be mediated by host plant factors (see Chapter 3 and Section 3.1 below). Istock (1973) demonstrated experimentally that

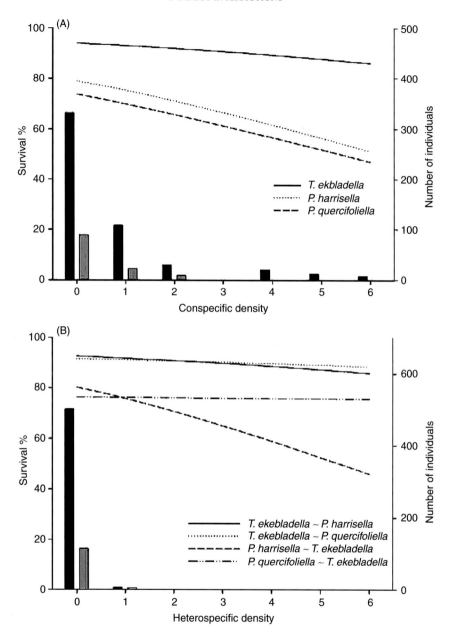

FIGURE 8.2 **Density-dependence of interspecific competitive effects among leaf miners on the same leaf in laboratory experiments (lines) and the frequency of co-occurrence in the field (histograms) on** *Quercus robur* **in Finland.** Effect of (A) conspecific larval density at the leaf levelo on survival for three species and (B) interspecific larval density on survival of the focal species of competing pairs. Survival of the focal species is shown as a function of the density, on the same leaf, of the second species in each pair. The histogram shows the number of individuals in the field that encounter competing larvae of (A) conspecific or (B) a second species on the same leaf. Black bars show numbers of *Tischeria ekebladella* and gray bars numbers of *Phyllonorycter* in each category. *Source: From Tack, A.J.M.,et al., 2009. Competition as a structuring force in leaf miner communities. Oikos 118, 809–818.*

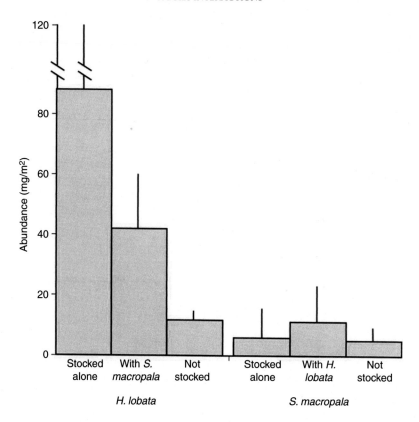

FIGURE 8.3 **Results of competition between two waterboatmen species,** ***Hesperocorixa lobata*** **and** ***Sigara macropala*,** **in 1.46 m² enclosures in a 1.2 ha pond.** Enclosures were stocked in Jun. with adult *H. lobata* and/or *S. macropala* and final abundance measured after 2 months. Waterboatmen in unstocked enclosures provided a measure of colonization. Vertical bars represent 1 SD. N = 4–8. *Source: Data from Istock, C.A., 1973. Population characteristics of a species ensemble of water-boatmen (Corixidae). Ecology 54, 535–544.*

competition between two waterboatmen species was asymmetrical (Fig. 8.3). Population size of *Hesperocorixa lobata* was significantly reduced when *Sigara macropala* was present, but population size of *S. macropala* was not significantly affected by the presence of *H. lobata*.

Competition generally is assumed to have only negative effects on both (all) competing species. As discussed in Chapter 6, competition among individuals in a population represents a major negative feedback mechanism that regulates population size. Similarly, competition among species regulates the total abundance of multiple species populations (Siepielski et al., 2010). As the total density of all individuals of competing species increases, each individual has access to a decreasing per capita share of the resource(s). If the competition is asymmetric, the superior species may competitively suppress other species, leading over sufficient time to *competitive exclusion* (Denno et al., 1995; Park, 1948; Strong et al., 1984). However, Denno et al. (1995) found evidence of competitive exclusion in <10% of the interactions they reviewed. Competitive exclusion normally may be prevented by factors that limit complete preemption of resources by any species. For example, predation often limits population growth of the superior species and reduces its ability to exclude other species (R. Paine, 1966, 1969a,b, see Section 3.3).

Interspecific competition takes two general forms and possible outcomes. *Exploitation competition* occurs when all individuals of the competing species have equal access to the resource. Species that can find or exploit a resource more quickly, develop or reproduce more rapidly, or increase resource use efficiency are favored under such circumstances. *Interference competition* involves preemptive use, and often defense, of a resource, allowing more aggressive species to increase access to, and share of, the resource, to the detriment of other species. Foraging ants often attack other predators and preempt prey resources. Halaj et al. (1997) reported that exclusion of foraging ants in young conifer plantations increased abundances of arboreal spiders > 1.5-fold.

Branson and Haferkamp (2014) described competitive interactions between grasshoppers, during a natural outbreak in which densities reached 130 m^{-2}, and livestock on rangelands in Montana during 2000 and 2001. Competition for food was stronger in plots with higher grasshopper densities and repeated or late grazing by livestock, leading to reduced grasshopper survival and reproduction. These density-dependent effects led to reduced grasshopper densities in 2001. Competition was more intense at high grasshopper densities than when livestock were present, suggesting that competition from vertebrates may be relatively less important for herbivorous insects during outbreaks.

Many species avoid resources that have been exploited previously, thereby losing access. Parasitic wasps frequently avoid hosts that are already parasitized by other wasps. Van Baaren et al. (2009) studied the host-attacking behavior of three *Aphidius* species that share the same aphid host, *Sitobion avenae*. Females of *A. rhopalosiphi* and *A. avenae* each discriminated between unparasitized hosts and hosts parasitized by the other species, but recognition of parasitized hosts occurred only after ovipositor insertion, indicating perception of an internal marker. Females of *A. rhopalosiphi* and *A. ervi* also discriminated between unparasitized hosts and hosts parasitized by the other species, but recognition in this case occurred through antennal perception, prior to ovipositor insertion, indicating an external marker.

Because competition can be expensive, in terms of lost resources, time, and/or energy spent in defending resources (see Chapter 4), evolution should favor strategies that reduce competition. Hence, competing species might be expected to minimize use of the contested portion of a resource and maximize use of noncontested portions. This results in partitioning of resource use, a strategy referred to as *niche partitioning*. Over evolutionary time, sufficiently consistent partitioning might become fixed as part of the species' adaptive strategies, and species would no longer respond to changes in abundance of their former competitor(s). In such cases, competition is not evident, although niche partitioning may provide evidence of competition in the past (Connell, 1980). However, niche partitioning among congeners may reflect speciation and divergence into unexploited niches, rather than competition (Fox and Morrow, 1981). Niche partitioning is observed commonly in natural communities. Species competing for habitat, food resources, or oviposition sites, tend to partition thermal gradients, time of day, host species, host size classes, feeding mode, etc. Several examples are noteworthy.

Granivorous ants and rodents frequently partition available seed resources, ants specializing on smaller seeds and rodents on larger seeds when the two compete. J. Brown et al. (1979) reported that both ants and rodents increased in abundance in the short term when the other was removed experimentally. However, Davidson et al. (1984) found that ant populations in rodent-removal plots declined gradually but significantly after about 2 years. Rodent populations did not decline over time in ant-removal plots. These results reflected a gradual displacement of small-seeded plants (on which ants specialize) by large-seeded plants (on which rodents specialize) in the absence of rodents, an indirect effect of rodents on seed resources for ants. Ant removal led to higher densities of small-seeded species but these species could not displace large-seeded plants.

Most bark beetle species can colonize extensive portions of dead or dying trees when other species are absent. However, given the relative scarcity of dead or dying trees and the narrow window of opportunity for colonization (the first year after tree death), these insects are adapted to finding such trees rapidly (see Chapter 3) and typically several species co-occur in colonized trees (T. Davis and Hofstetter, 2009). Under these circumstances, subcortical resources are partitioned on the basis of beetle size, because each species shows highest survival rates in phloem that is thick enough to accommodate growing larvae, and because larger species are capable of repulsing smaller species (Flamm et al., 1993). Therefore, the largest species typically occur near the base of the tree, and progressively smaller species occupy successively higher portions of the bole, with the smallest species colonizing the upper bole and branches. However, other competitors, such as wood-boring cerambycids and buprestids, often chew through bark beetle mines, feeding on bark beetle larvae and reducing survival (Fig. 8.1) (Coulson et al., 1980; Dodds et al., 2001).

Many species partition resource use in time. Partitioning can be by time of day, for example, nocturnal versus diurnal Lepidoptera (Schultz, 1983) and nocturnal bat and amphibian versus diurnal bird and lizard predators (Reagan et al., 1996), or by season, for example, asynchronous occurrence of 12 species of waterboatmen (Hemiptera) which breed at different times (Istock, 1973). However, temporal partitioning does not preclude indirect competition through preemptive use of resources or induced host defenses (see sections 3.1 and 3.2).

In addition to niche partitioning, other factors also may obscure or prevent competition. Resource turnover in frequently disturbed ecosystems may prevent species saturation on available resources and prevent competition. Similarly, spatial patchiness in resource availability may hinder resource discovery and prevent species from reaching abundances at which they would compete. Finally, indirect interactions, such as host defenses or predation, can maintain populations below sizes at which competition would occur (R. Paine, 1966, 1969a,b; Tack et al., 2009, see section 3.3).

Competition has been relatively easy to model (see Chapter 6). The Lotka–Volterra equation generalized for n competitors is

$$N_{i(t+1)} = N_{it} + r_i N_{it}(K - N_{it} - \sum_{j>1}^{n}\alpha_{\tau j}N_{jt})/K \qquad (8.1)$$

where N_i and N_j are species abundances, and α_{ij} represents the per capita effect of N_j on the growth of N_i and varies for different species, for example, if competition is asymmetric, species j could have a greater negative effect (larger α_{ij}) on species i than species i has on species j.

Istock (1977) evaluated the validity of the Lotka–Volterra equations for co-occurring species of waterboatmen, *H. lobata* (species 1) and *S. macropala* (species 2), in experimental exclosures (Fig. 8.3). He calculated the competition coefficients, α_{12} and α_{21}, as

$$\alpha_{12} = (K_1 - N_1)N_2 = 3.67 \quad \text{and} \quad \alpha_{21} = (K_2 - N_2)N_1 = -0.16 \qquad (8.2)$$

The intercepts of the zero isocline ($dN/dt = 0$) for *H. lobata* were $K_1 = 88$ and $K_1/\alpha_{12} = 24$; the intercepts for *S. macropala* were $K_2 = 6$ and $K_2/\alpha_{21} = -38$. The negative K_2/α_{21} and position of the zero isocline for *S. macropala* indicated that the competition is asymmetric, consistent with the observation that *S. macropala* population growth was not affected significantly by the interaction (Fig. 8.3). Although niche partitioning by these two species was not clearly identified, the equations correctly predicted the observed coexistence.

2.2 Predation

Predation has been defined either as a general process of feeding on other (prey) organisms (May, 1981) or as a more specific process of killing and consuming multiple prey (Price, 1997). Parasitism (and the related parasitoidism), the consumption of tissues in a living host, may or may not be included (Price, 1997). Both predation and parasitism have positive effects for the predator or parasite, but negative effects for the prey. In this section, predation is treated as the relatively opportunistic capture of multiple prey during a predator's lifetime. Section 2.3 will address the more specific parasite–host interactions.

Although typically considered in the sense of an animal killing and eating other animals (Fig. 8.4), predation applies equally well to carnivorous plants that kill and consume insect prey and to herbivores that kill and consume plant prey, especially those that feed on seeds and seedlings. Predator–prey and herbivore–plant interactions represent similar foraging strategies and are affected by similar factors (host or prey density, defensive strategy and physiological condition, herbivore or predator ability to detect and orient toward host or prey cues, etc.; see Chapter 3).

Insects and other arthropods represent major predators in terrestrial and aquatic ecosystems. The importance of many arthropods as predators of insects has been demonstrated widely through biological control programs and experimental studies (Price, 1997; Strong

FIGURE 8.4 **Predation: golden silk orb-weaver spider,** *Nephila clavipes,* **with a captured June beetle,** *Phyllophaga* **sp.** *Source: From Schowalter, T.D., 2013. Insects and Sustainability of Ecosystem Services. CRC Press/Taylor and Francis Group, Boca Raton, FL.*

et al., 1984; van den Bosch et al., 1982; Van Driesche and Bellows, 1996). However, many arthropods prey on vertebrates, as well. Dragonfly larvae, water bugs, and aquatic beetles prey on fish and amphibians, as well as other invertebrates. Terrestrial ants, spiders, and centipedes prey on amphibians, reptiles, and immature birds (C. Allen et al., 2004; Reagan et al., 1996).

Insects also represent important predators of plants or seeds. Some bark beetles might be considered to be predators to the extent that they kill trees. Seed bugs, weevils, and ants are effective seed predators, often kill seedlings, and may be capable of preventing plant reproduction under some conditions (Davidson et al., 1984; Turgeon et al., 1994, see Chapter 13).

Insects are important prey for a variety of other organisms. Carnivorous plants generally are associated with nitrogen-poor habitats and depend on insect prey for adequate nitrogen (Juniper et al., 1989; Krafft and Handel, 1991). A variety of mechanisms for entrapment of insects has evolved among carnivorous plants, including water-filled pitchers (pitcher plants), triggered changes in turgor pressure that close capture organs (flytraps and bladderworts), and sticky hairs (eg, sundews). Some carnivorous plants show conspicuous ultraviolet patterns that attract insect prey (Joel et al., 1985), similar to the floral attraction of some pollinators (see Chapters 3 and 13).

Insects also are prey for other arthropods (eg, predaceous insects, spiders, mites) and vertebrates. Many fish, amphibian, reptile, bird, and mammal taxa feed largely or exclusively on insects (J. Allan et al., 2003; Baxter et al., 2005; Dial and Roughgarden, 1995; Gardner and Thompson, 1998; C.G. Jones et al., 1998; Kawaguchi and Nakano, 2001; Tinbergen, 1960). Even apex predators, such as the American alligator, *Alligator mississippiensis*, consume a variety of aquatic and terrestrial insects and other arthropods (Rosenblatt et al., 2015).

Aquatic and terrestrial insects support major freshwater fisheries, including salmonids (Baxter et al., 2005; Cloe and Garman, 1996; Wipfli, 1997). Reduced availability of insect prey reduces the productivity and abundance of these fish (Baxter et al., 2007). Stewart and Woolbright (1996) calculated, from gut contents, that tree frog (*Eleutherodactylus coqui*) adults, at densities of about 3300 ha^{-1} in Puerto Rican rainforest, consumed 10,000 insects ha^{-1} per night; 17,000 preadult frogs ha^{-1} ate an additional 100,000 insects ha^{-1} per night. Gut contents consisted primarily of ants, crickets, and cockroaches, three of the most abundant canopy taxa on foliage at this site. A number of bird and bat species feed primarily or exclusively on insects.

Predation has been widely viewed as a primary regulator of prey population density (see Chapters 5 and 6). Appreciation for this lies at the heart of predator control policies designed to increase abundances of commercial or game species by alleviating population control by predators. However, mass starvation and declining genetic quality of populations protected from nonhuman predators have demonstrated the benefits of predation to maintenance of prey population vigor, or genetic structure, through selective predation on old, injured, diseased, or poorly defended individuals. As a result of these changing perceptions, predator reintroduction programs are being implemented in some regions. At the same time, recognition of the important role of entomophagous species in controlling populations of insect pests has justified augmentation of predator abundances, often through introduction of exotic species, for biological control purposes (van den Bosch et al., 1982; Van Driesche and Bellows, 1996, see Chapter 17). As discussed in Chapter 6, the relative importance of predation to population regulation, compared to other regulatory factors, has been a topic of considerable discussion.

Just as coevolution between competing species has favored niche partitioning for more efficient resource use, coevolution between predator and prey has produced a variety of defensive strategies balanced against predator foraging strategies. Selection favors prey that can avoid or defend against predators and favors predators that can efficiently acquire suitable prey. Prey defenses include speed, predator detection and alarm mechanisms, spines or horns, chemical defenses, cryptic, aposematic, disruptive, or deceptive coloration, and behaviors (such as aggregation or warning displays) that enhance these defenses (Conner et al., 2000; Jabłoński, 1999; Sillén-Tullberg, 1985, see Chapters 3 and 4). Prey attributes that increase the energy cost of capture or utilization restrict the number of predators able to exploit that prey.

Venoms of social Hymenoptera may have evolved as a defense against much larger vertebrate predators that exploit concentrations of prey in large colonies. Schmidt (2014) noted a strong correlation between the degree of social structure among hymenopteran species and the pain and lethality indices of their venoms. The extremely painful and hemolytic venoms of species that develop large colonies would not be necessary for defense against arthropod predators but are very effective against primates and other large vertebrate predators. By contrast, solitary species typically produce venoms that inflict less pain and have little or no hemolytic or toxic activity, requiring only the warning pain to effect escape.

Predator attributes that increase their efficiency in capturing and immobilizing prey include larger size, greater speed, detection of cues that indicate vulnerable prey, claws or sharp mouthparts, venoms, and behaviors (such as ambush, flushing, or attacking the most vulnerable body parts) or other attributes (such as bioluminescent attraction of prey) that compensate for or circumvent prey defenses and reduce the effort necessary to capture the prey (Jabłoński, 1999; Galatowitsch and Mumme, 2004; Mumme, 2002; Sivinski, 1982). For example, a carabid beetle, *Promecognathus laevissimus*, straddles polydesmid millipedes, quickly moves toward the head, pierces the neck with its mandibles, and severs the ventral nerve cord, thereby paralyzing its prey and circumventing its cyanide spray defense (Parsons et al., 1991).

A unique strategy among predaceous insects is tool use by a tropical reduviid bug, *Salyavata variegata*, to facilitate acquisition of termite, *Nasutitermes* spp., prey (McMahan, 1982, 1983). Adult and juvenile bugs are attracted quickly to breaches in termite nests where they can attack exposed workers. The bugs camouflage themselves with bits of carton material, to avoid detection by termite soldiers, grasp workers as they appear at the rim to investigate, then retreat a short distance to feed. Furthermore, 3rd, 4th, and 5th instars employed a baiting strategy to acquire additional prey. After feeding completely on the first worker, the bug holds the termite carcass in front of its head with its forelegs, moves back to the breach as it is being repaired, and pushes the carcass into the opening, jiggling it slightly. When one or more termites seize the carcass and attempt to pull it into the hole, the bug pulls the carcass and grasping termite backward slowly until the worker's head is in an accessible position, at which point the bug releases the carcass and grasps the new worker, then moves away from the breach to feed. One nymph fed on 31 successive termite workers over a 3-h period (McMahan, 1982).

Predators are relatively opportunistic with respect to prey taxa, compared to parasites, although prey frequently are selected on the basis of factors determining foraging or capture efficiency. For example, chemical defenses of prey affect attractiveness to nonadapted predators (Bowers and Puttick, 1988; Stamp et al., 1997; Traugott and Stamp, 1996). Prey size affects the resource gained per foraging effort expended. Predators generally select prey within a size range that provides sufficient energy and nutrient rewards to balance the cost of capture (Ernsting and van der Werf, 1988; Iwasaki, 1990, 1991; Richter, 1990; Streams, 1994; Tinbergen, 1960). Within these constraints, foraging predators should attack suitable prey species in proportion to their probability of encounter, that is, more abundant prey types are encountered more frequently than are less abundant prey types (Tinbergen, 1960).

Predators exhibit both *functional* (behavioral) and *numeric* responses to prey density. Functional responses reflect predator hunger level, handling time required for individual prey, ability to discover prey, handling efficiency resulting from learning, etc. (Holling, 1959, 1965; Tinbergen, 1960). For some invertebrate predators, the rate of prey capture is a linear function of prey density, up to a point of *predator satiation*, that is, the predator is incapable of capturing or eating more prey (type 1 response). For many invertebrate predators, prey capture increases at a gradually declining rate as prey density increases, eventually reaching an asymptote as predators become satiated (type 2 response). Vertebrates, and some invertebrates, are capable of increasing their efficiency of prey discovery, for example, through development of a search image that enhances recognition of appropriate prey (Tinbergen, 1960) and prey processing time through learning, up to a point. The percentage of prey captured increases rapidly as the predator learns to find and handle prey more quickly but eventually approaches a peak and subsequently declines as discovery and handling time reach maximum efficiency (type 3 response). Social ants are capable of a type 3 response because of their colonial foraging and rapid communication of prey availability to nestmates. Whitford and Jackson (2007) reported that harvester ants, *Pogonomyrmex rugosus*, responded quickly to a pulse in prey availability (grass cicada emergence), taking > 5 cicadas min^{-1} into their nests in areas where cicada density was > 3 m^{-2}; ants were inactive where cicada density was < 1 m^{-2}. All three responses are limited by predator satiation at high prey densities.

Various factors affect functional response and the resulting predation rate. The rate of prey capture tends to decline as a result of learned avoidance of distasteful prey. The maximum rate of prey capture depends on how quickly predators become satiated and on relative abundances of palatable and unpalatable prey (Holling, 1965). Some insect species, such as the periodical cicadas, *Magicicada* spp., apparently exploit the functional responses of their major predators by appearing en masse for brief periods following long periods (13 or 17 years) of inaccessibility. Predator satiation maximizes the success of such mass emergence and mating aggregations (K. Williams and Simon, 1995). Palatable species experience greater predation when associated with less palatable species than when associated with equally or more palatable species (Holling, 1965).

Numeric responses reflect predator orientation toward, and longer residence in, areas of high prey density, and subsequent reproduction in response to increased food availability. However, increased predator density also may increase competition, conflict, and predation among predators (Cardinale et al., 2006; J.M. Schmidt et al., 2014). The combination of a type 3 functional response and numeric response (total response) make predators effective at cropping abundant prey and maintaining relatively stable populations of various prey species. However, the tendency to become satiated, to reproduce more slowly than prey populations, and to show reduced per capita predation rates at high predator densities (as a result of interguild competition and predation) limits the ability of predators to regulate irruptive prey populations released from other controlling factors.

The importance of predator–prey interactions to population and community dynamics has generated considerable interest in modeling this interaction. The effect of a predator on a prey population was first incorporated into the logistic model by Lotka (1925) and Volterra (1926). As described in Eq. (6.11), their model for prey population growth was

$$N_{1(t+1)} = N_{1t} + r_1 N_{1t} - \rho_1 N_{1t} N_{2t}$$

where N_2 is the population density of the predator, and ρ_1 is a predation constant. Lotka and Volterra modeled the corresponding predator population as

$$N_{2(t+1)} = N_{2t} + \rho_2 N_{1t} N_{2t} - d_2 N_{2t} \tag{8.3}$$

where ρ_2 is a predation constant and d_2 is per capita mortality of the predator population. The Lotka–Volterra equations describe prey and predator populations oscillating cyclically and out of phase over time. Small changes in parameter values lead to extinction of one or both populations after several oscillations of increasing amplitude.

Pianka (1974) modified the Lotka–Volterra competition and predator–prey models to incorporate competition among prey and among predators for prey. Eq. (6.12) represents the prey population

$$N_{1(t+1)} = N_{1t} + r_1 N_{1t} - r_1 N_{1t}^2 / K_1 - r_1 N_{1t} \alpha_{12} N_{2t} / K_1$$

where α_{12} is the per capita effect of the predator on the prey population. The corresponding model for the predator population is

$$N_{2(t+1)} = N_{2t} + \alpha_{21} N_{1t} N_{2t} - \beta_2 N_{2t}^2 / N_{1t} \tag{8.4}$$

where α_{21} is the negative effect of predation on the prey population and β_2 incorporates the predator's carrying capacity as a function of prey density (Pianka, 1974). This refinement provided for competitive inhibition of predator population growth as a function of the relative densities of predator and prey. The predator–prey equations have been modified further to account for variable predator and prey densities (Berlow et al., 1999), predator and prey distributions (Begon and Mortimer, 1981), and functional responses and competition among predators for individual prey (Holling, 1959, 1966). Other models have been developed primarily for parasitoid–prey interactions (see section 2.3).

Current modeling approaches have focused on paired predator and prey. Real communities are composed of multiple predator species exploiting multiple prey species, resulting in complex interactions (Fig. 8.5), including *facilitation* or *intraguild predation*. Facilitation results in greater prey suppression as predators are added, at least up to a point (Long and Finke, 2014). Intraguild predation is the preying by one predator upon another (J.M. Schmidt et al., 2014). This is also a form of interference competition. Intraguild predation can reduce predation effects on prey species by reducing the number of predators (J.M. Schmidt et al., 2014).

2.3 Symbiosis

Symbiosis involves an intimate association between two unrelated species, in which one (symbiont) benefits by living in or on the other (host). Three types of interactions are considered symbiotic, although the term often has been used as a synonym for only one of these, mutualism. *Parasitism* describes interactions in which the symbiont derives a benefit at the

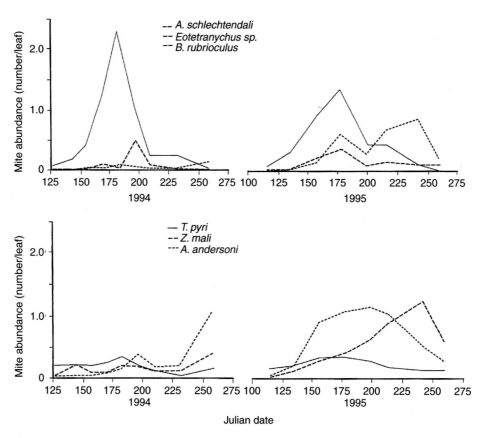

FIGURE 8.5 Densities of three phytophagous mites, *Aculus schlechtendali*, *Bryobia rubrioculus*, and *Eotetranychus sp.* (prey), and three predaceous mites, *Amblyseius andersoni*, *Typhlodromus pyri*, and *Zetzellia mali*, in untreated apple plots (*N* = 2) during 1994 and 1995. *Source: Data from Croft, B.A., Slone, D.H., 1997. Equilibrium densities of European red mite (Acari: Tetranychidae) after exposure to three levels of predaceous mite diversity on apple. Environ. Entomol. 26, 391–399.*

expense of the host, as in predation. *Commensalism* occurs when the symbiont derives a benefit without significantly affecting its partner. *Mutualism* results in benefit to both partners in the interaction. Insects have provided some of the oldest and most interesting examples of symbiosis (Boucot and Poinar, 2010; Poinar and Poinar, 2007).

2.3.1 Parasitism

Parasitism affects the host population in ways that are similar to predation and can be described using predation models. However, whereas predation involves multiple prey killed and consumed during a predator's lifetime, parasites exploit living hosts. *Parasitoidism* is unique to insects, especially flies and wasps, and combines attributes of both predation and parasitism. The adult parasitoid typically deposits eggs or larvae on, in, or near multiple hosts, and the larvae subsequently feed on their living host and eventually kill it (Fig. 8.6). Parasites must be adapted to survive long periods of exposure to the defenses of a living host (see Chapter 3). Therefore, parasitic interactions tend to be relatively specific associations

FIGURE 8.6 Parasitism: a parasitoid (sarcophagid fly) ovipositing on a host caterpillar at Nanjinshan Long Term Ecological Research Site, Taiwan.

between coevolved parasite species and their particular host species and may involve modification of host morphology, physiology, or behavior to benefit parasite development or transmission. Because of this specificity, parasites and parasitoids tend to be more effective than predators in responding to, and controlling, population irruptions of their hosts and, therefore, have been primary agents in biological control programs (Hochberg, 1989). In fact, release from parasites, coupled with slow response by generalist predators, may largely explain the rapid spread of invasive plants and animals introduced into naïve communities (Torchin and Mitchell, 2004).

Parasitic interactions can be quite diverse and complex (van den Bosch et al., 1982). Ectoparasites feed externally, by inserting mouthparts into the host (eg, lice, fleas, mosquitoes, ticks), and endoparasites feed internally, within the host's body (eg, bacteria, nematodes, bot flies, and parasitic wasps). Primary parasites develop on or in a nonparasitic host, whereas hyperparasites develop on or in another parasite. Some parasites parasitize other members of the same species (autoparasitism or adelphoparasitism), as seen in the hymenopteran, *Coccophagus scutellaris*. The female of this species parasitizes scale insects and the male is an obligate hyperparasite of the female (van den Bosch et al., 1982). Superparasitism occurs when more individuals occupy a host than can develop to maturity. Multiple parasitism occurs when more than one parasitoid species is present in the host simultaneously. In most cases of superparasitism and multiple parasitism, one dominant individual competitively suppresses others and develops to maturity. In a special case of multiple parasitism, some parasites preferentially attack hosts parasitized by other species (cleptoparasitism). The cleptoparasite is not a hyperparasite, but typically kills and consumes the original parasite as well as the host.

Slave-making ants represent a specialized form, social parasitism. These ants raid colonies of neighboring ant species and capture brood, many of which are eaten, but most are reared and augment the slave-makers' workforce. Hare and Alloway (2001) compared the effect of two slave-making species, *Protomognathus americanus* and *Leptothorax duloticus*, on fitness of their host ant species, *L. longispinosus*. They found that *L. longispinosus* colonies enclosed with *P. americanus* colonies showed no change in demographic variables, but *L. longispinosus* colonies enclosed with *L. duloticus* colonies showed significant reduction in numbers of dealate queens, workers, and larvae, compared to control colonies enclosed without slave-makers. Hare and Alloway (2001) concluded that the higher abundance of *P. americanus* compared to *L. duloticus* suggests that social parasites may evolve to minimize fitness cost to their hosts.

Some parasites alter the physiology or behavior of their hosts in ways that enhance parasite development or transmission. For example, parasitic nematodes often destroy the host's genital organs, sterilizing the host (Vega and Kaya, 2012). Parasitized insects frequently show prolonged larval development (Vega and Kaya, 2012). Flies, grasshoppers, ants, and other insects infected with insect or fungal parasites often seek exposed sites before death, facilitating dispersal of emerging parasites or transmission of windblown spores (Henne and Johnson, 2007; Vega and Kaya, 2012) (Fig. 8.7).

Insects have evolved various defenses against parasites (see Chapter 3). *Solenopsis* ants stop foraging and retreat to nests when phorid flies, *Pseudacteon* spp., appear (Feener, 1981; Feener and Brown, 1992; Folgarait and Gilbert, 1999; Mehdiabadi and Gilbert, 2002; Mottern et al., 2004; Orr et al., 2003). Hard integument, hairs and spines, defensive flailing, and antibiotics secreted by metapleural glands prevent attachment or penetration by some parasites (Hajek and St. Leger, 1994; Peakall et al., 1987). Ingested or synthesized antibiotics or gut modifications prevent penetration by some ingested parasites (Tallamy et al., 1998; Vega and Kaya, 2012). Endocytosis and cellular encapsulation are physiological mechanisms for destroying internal parasites (Vega and Kaya, 2012, see Chapter 3). However, some parasitic wasps inoculate hosts with a virus that inhibits encapsulation of their eggs or larvae (Edson et al., 1981; Godfray, 1994).

FIGURE 8.7 Parasitism: stink bug infected and killed by a parasitic fungus in Louisiana, USA.

Insects are parasitized by a number of organisms, including viruses, bacteria, fungi, protozoa, nematodes, flatworms, mites, and other insects (D. Cooper et al., 2003; Cory and Myers, 2003; Hajek and St. Leger, 1994; Vega and Kaya, 2012; C. Thorne et al., 2007; Tzean et al., 1997). Some parasites cause sufficient mortality to have been exploited as agents of biological control (van den Bosch et al., 1982; C. Thorne et al., 2007, see Chapter 17). Epizootics of parasites often are responsible for termination of host outbreaks (Cory and Myers, 2003; Hajek and St. Leger, 1994; Hochberg, 1989; C. Thorne et al., 2007). Parasites also have complex sublethal effects that make their hosts more vulnerable to other mortality factors. Bradley and Altizer (2005) reported that monarch butterflies, *Danaus plexippus*, parasitized by the protozoan, *Ophryocystis elektroscirrha*, lost 50% more body mass per kilometer flown and exhibited 10% slower flight velocity, 14% shorter flight duration, and 19% shorter flight distance, compared to uninfected butterflies. These data, together with much higher infection rates among nonmigrating monarchs (Altizer et al., 2000), suggest that long-distance migration of this species may eliminate infected individuals and reduce rates of parasitism.

Many insects and other arthropods are parasitic on other species. Although parasitism generally is associated with animal hosts, most insect herbivores can be viewed as parasites of living plants, since they feed on, but rarely kill, their plant hosts. Some herbivores, such as sap-suckers, leaf miners, and gall-formers, are analogous to blood-feeding or internal parasites of animals. The majority of insect species that are parasites of animals are wasps, flies, fleas, and lice, but some beetle species also are parasites (Price, 1997). Parasitic wasps are a highly diverse group that differentially parasitize the eggs, juveniles, pupae, or adults of various arthropods. Spider wasps, for example, tarantula hawks, provision burrows with paralyzed spiders for their parasitic larvae. Flies parasitize a wider variety of invertebrate and vertebrate hosts. Mosquitoes and other biting flies are important blood-sucking ectoparasites of vertebrates, including humans. Oestrid and tachinid flies are important endoparasites of vertebrates and insects. Fleas and lice are ectoparasites of vertebrates, including humans. Mites, chiggers, and ticks parasitize a wide variety of hosts.

Parasitoids generally attack only other arthropods, but a sarcophagid fly, *Anolisomyia rufianalis*, is a parasitoid of *Anolis* lizards in Puerto Rico. Dial and Roughgarden (1996) found a slightly higher rate of parasitism of *Anolis evermanni*, compared to *A. stratulus*. They suggested that this difference may be due to black spots on the lateral abdomen of *A. stratulus* that resemble the small holes made by emerging parasites. Host-seeking flies apparently avoid lizards showing signs of prior parasitism.

Insect parasites can significantly reduce growth, survival, reproduction, and movement of their hosts (J. Day et al., 2000; Mooring and Samuel, 1999; Norval et al., 1988; Steelman, 1976). Mehdiabadi and Gilbert (2002) reported that densities as low as a single phorid, *Pseudacteon tricuspis*, female per 200 foraging fire ants, *Solenopsis invicta*, reduced colony intake of protein by > 50% and significantly reduced numbers of large workers 50 days later in laboratory trials. Biting flies have been reported to reduce growth and survival of wildlife species through irritation and/or blood loss (J. Day et al., 2000). DeRouen et al. (2003) reported that a 14% reduction in hornfly, *Haematobia irritans*, numbers on treated cattle resulted in a significant 14% increase in cattle weight, but no effect on reproductive rate. However, Sanson et al. (2003) found reduced hornfly abundance resulted in significantly increased weight of cattle in only one of three years of study. Norval et al. (1988)

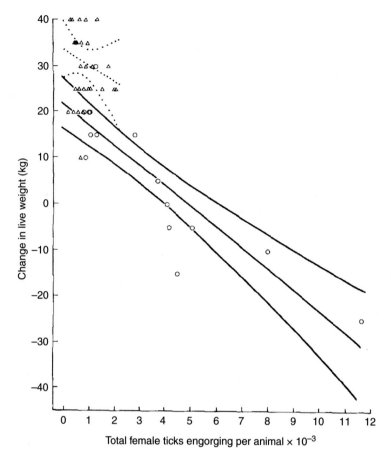

FIGURE 8.8 **Relationship between numbers of engorging female ticks, *Rhipicephalus appendiculatus*, and change in live weight of tick-naïve cattle in Zimbabwe.** *Triangles* and *dotted lines* represent change in weight of Nkoni steers (Sanga breed), and *circles* and *solid lines* represent Friesian × Hereford steers. Regression lines are mean and 95% confidence intervals. *Source: Reprinted from Norval et al. (1988). The effect of the brown ear-tick* Rhipicephalus appendiculatus *on the growth of sanga and European breed cattle. Vet. Parasitol. 30, 149–164.*

reported that brown ear-ticks, *Rhipicephalus appendiculatus*, caused sufficient blood loss in African cattle to reduce weight gain by 4 g per engorging female tick. Tick abundances of 3000–11,000 per animal (Kaiser et al., 1982; Norval et al., 1988; Pegram et al., 1986) clearly cause serious losses in weight gain (Fig. 8.8).

Parasites can weaken their hosts sufficiently to make them vulnerable to predation or other mortality factors. Grooming behavior by moose, *Alces alces*, to dislodge winter ticks, *Dermacentor albipictus*, can remove virtually all outer hair and increase vulnerability to winter mortality (Mooring and Samuel, 1999). This condition appears to be increasing in frequency as warming temperatures in North America bring moose into greater contact with these ticks that historically have been associated with deer, *Odocoileus* spp., at lower latitudes. Parasitized predators may be unable to capture prey (Wilmers et al., 2006).

Many arthropod parasites also vector vertebrate pathogens, including agents of malaria, *Plasmodium malariae,* bubonic plague, *Yersinia pestis,* typhus, *Rickettsia prowazekii,* Lyme disease, *Borrelia burgdorferi,* Texas cattle fever, *Babesia bovis,* and yellow fever, dengue, bluetongue, and encephalitis viruses (Edman, 2000; Piesman and Eisen, 2008). Some of these diseases cause substantial mortality in human, livestock, and wildlife populations, especially when contacted by nonadapted (naïve) hosts (Amoo et al., 1993; Marra et al., 2004; Stapp et al., 2004; Steelman, 1976; J. Zhou et al., 2002). Human population dynamics, including invasive military campaigns, have been substantially shaped by insect-vectored diseases (Diamond, 1999; R. Peterson, 1995, see Chapter 16).

In many cases, pathogen development in arthropod vectors requires complex interaction between the parasite and arthropod immune responses. In some cases, additional mutualistic symbionts within the arthropod gut or hemocoel interfere with the infectivity of the pathogen (Paris et al., 2008; Rio et al., 2006; Wang et al., 2009). Consequently, not all pathogens acquired by arthropods are able to exit the gut and reach the salivary glands or other organs as required for transmission to other organisms.

Nicholson and Bailey (1935) proposed a model of parasitoid–host interactions that assumed that hosts are dispersed regularly in a homogeneous environment, that parasitoids search randomly within a constant area of discovery, and the ease of host discovery and parasitoid oviposition do not vary with host density. The number of hosts in the next generation (u_s) was calculated as

$$pa = \log_e(u_i/u_s) \tag{8.5}$$

where p = parasitoid population density; a = area of discovery; and u_i = host density in the current generation.

Hassell and Varley (1969) showed that the area of discovery (a) is not constant for real parasitoids. Rather $\log a$ is linearly related to parasitoid density (p) as

$$\log a = \log Q - (m \log p) \tag{8.6}$$

where Q is a quest constant and m is a mutual interference constant. Hassell and Varley (1969) modified the Nicholson–Bailey model to incorporate density limitation (Q/p^m). By substitution,

$$pa = \log_e(u_i/u_s) = Qp^{1-m} \tag{8.7}$$

As m approaches Q, model predictions approach those of the Nicholson–Bailey model.

2.3.2 Commensalism

Commensalism is a relatively rare type of interaction, because few hosts are completely unaffected by their symbionts. Epiphytes, plants that use their hosts for aerial support but gain their resources from the atmosphere, and cattle egrets, that eat insects flushed by grazing cattle, are well-known examples of commensalism. However, epiphytes may capture and provide nutrients to the host (a benefit) and increase the likelihood that heavy branches will break during high winds (a detriment). Examples of commensalism often may be seen as other interaction types as additional information becomes available. Some interactions involving insects may be largely commensal.

Phoretic or vector interactions (Fig. 5.8) benefit the hitchhiker or pathogen, especially when both partners have the same destination, and may have little or no effect on the host, at least up to a point where hosts become overburdened, inhibiting dispersal, resource acquisition, or escape. In some cases, phoretic partners may be mutualists, with predaceous hitchhikers reducing competition or parasitism for their host at their destination (Kinn, 1980).

A number of insect and other arthropod species are commensal in ant or termite nests. Such species are called *myrmecophiles* or *termitophiles*, respectively. These symbionts gain shelter, and often detrital food, from their host colonies with little, if any, effect on their hosts. This relationship is distinct from interactions involving species that intercept host food (through trophallaxis) and, therefore, function as colony parasites. Some vertebrate species also are commensals of termite castles that may reach several meters in height and diameter and provide critical shelter for reptile, bird, and mammal species in tropical savannas (see Chapter 14).

Bark beetle galleries provide habitat and resources for a variety of invertebrate and microbial commensals, most of which have little or no effect on the bark beetles (Stephen et al., 1993). Many of the invertebrate species are fungivores or detritivores that depend on penetration of the bark by bark beetles in order to exploit resources provided by the microbial decay of wood.

2.3.3 *Mutualism*

Mutualistic interactions often involve modification of host morphology, physiology, or behavior to provide habitat or food resources for the symbiont. In return, the symbiont provides necessary services, resources, or protection from competitors or predators. Although classic examples of mutualism often involve mutually dependent (obligate) partners, that is, the disappearance of one leads to the demise of the other, some mutualists are less dependent on each other. Mutualism can be viewed as mutual exploitation or manipulation. Pollination is a byproduct of insect attraction to nectar resources; some nectar-robbing species circumvent pollinia to acquire nectar. The anatomical modification and resources that maintain interaction represent costs to the organisms involved. Provision of resources for ants by ant-protected plants requires energy and nutrients that could be allocated to growth and reproduction. Ants may provide nitrogen or other nutrients, as well as defense, for their hosts (Fischer et al., 2003), but plants may lose ant-related traits when ants are absent (Rickson, 1977).

Mutualisms have received considerable attention, and much research has focused on examples such as pollination (see Chapter 13), ant–plant, mycorrhizae–plant interactions, and other conspicuous mutualisms. Nevertheless, Price (1997) argued that ecologists have failed to appreciate mutualism as equal in importance to predation and competition, at least in temperate communities, reflecting a perception, based on early models, that mutualism is less stable than competition or predation (Goh, 1979; May, 1981; M. Williamson, 1972). However, as Goh (1979) noted, such models did not appear to reflect the widespread occurrence of mutualism in ecosystems. As a cooperative relationship, mutualism can contribute greatly to the presence and ecological function of the partners, but the extent to which such positive feedback stabilizes or destabilizes interacting species populations remains a topic of discussion.

Among the best-known mutualisms are those involving pollinator associations with plants (Chittka and Raine, 2006; Feinsinger, 1983; Huxley and Cutler, 1991; Jolivet, 1996; Stoepler et al., 2012). The variety of obligate relationships between pollinators and their floral hosts in the tropics has supported a perception that mutualism is more widespread and important in the tropics. As described in Chapter 13, the prevalence of obligate mutualisms between plants and pollinators in the tropics, compared to temperate regions, largely reflects the high diversity of plant species that precludes wind pollination between nearest conspecific neighbors. Sparsely distributed or understory plants in temperate regions also tend to have mutualistic association with pollinators. Some mutualistic associations (eg, insect–microbial association, as discussed later) may be more prominent in temperate than in tropical regions.

Seed-feeding insects, especially ants, often benefit plants by dispersing nonconsumed seeds (Ohkawara et al., 1996). This mutualism is exemplified by myrmecochorous plants that provide a nutritive body (elaiosome) attached to the seed to attract ants. The elaiosome typically is rich in lipids (Gorb and Gorb, 2003; Jolivet, 1996). The likelihood that a seed will be discarded in or near an ant nest following removal of the elaiosome increases with elaiosome size, perhaps reflecting increasing use by seed-disperser, rather than seed-predator, species with increasing elaiosome size (Gorb and Gorb, 2003; Mark and Olesen, 1996; Westoby et al., 1991).

Many plants provide nest sites or shelters (domatia), for example, hollow stems or pilose vein axils, for ants or predaceous mites that protect the plant from herbivores (R. Fischer et al., 2002; O'Dowd and Willson, 1991). Cecropia trees, *Cecropia* spp., in the tropics are protected from herbivores by aggressive ants, *Azteca* spp., housed in their hollow stems (Rickson, 1977). Central American acacias, *Acacia* spp., also are defended by colonies of aggressive ants, *Pseudomyrmex* spp., housed in swollen thorns (Janzen, 1966). Other plant species provide extrafloral nectaries rich in amino acids and lipids that attract ants (Dreisig, 1988; Jolivet, 1996; Oliveira and Brandão, 1991; Rickson, 1971; Schupp and Feener, 1991; Tilman, 1978) that, in turn, protect the plant from herbivores. In addition to defense, plants also may acquire nitrogen or other nutrients from the ants (R. Fischer et al., 2003).

Clarke and Kitching (1995) discovered an unusual mutualistic interaction between an ant and a carnivorous pitcher plant in Borneo. The ant, *Camponotus* sp., nests in hollow tendrils of the plant, *Nepenthes bicalcarata*, and is capable of swimming in pitcher-plant fluid, where it feeds on large prey items caught in the pitcher. Through ant-removal experiments, Clarke and Kitching found that accumulation of large prey (but not small prey) in ant-free pitchers led to putrefaction of the pitcher contents and disruption of prey digestion by the plant. By removing large prey, the ants prevented putrefaction and accumulation of ammonia.

Insects engage in a variety of mutualistic interactions with microorganisms. Parasitoid wasps inoculate their host with a virus that prevents cellular encapsulation of the parasitoid larva (Edson et al., 1981; Godfray, 1994, see Chapter 3). Intestinal bacteria may synthesize some of the pheromones used by bark beetles to attract mates (Byers and Wood, 1981). Most aphids harbor mutualistic bacteria or yeasts in specialized organs (bacteriomes or mycetomes) that appear to provide amino acids, vitamins, and/or proteins necessary for aphid development and reproduction (Baumann et al., 1995). Experimental elimination of the microbes results in aphid sterility, reduced weight, and reduced survival. Furthermore, some aphid endosymbionts enhance host tolerance to heat shock (Russell and Moran, 2006). Many hemipterans vector plant pathogens and may benefit from changes in host condition induced by infection (Kluth et al., 2002). Leaf-cutting ants, *Atta* spp. and *Acromyrmex* spp., cultivate fungus gardens that provide food for the ants (C. Currie, 2001; N. Weber, 1966).

Virtually all wood-feeding species interact mutualistically with cellulose-digesting micro-organisms. Ambrosia beetles are the only means of transport for ambrosia (mold) fungi, carrying hyphae in specialized invaginations of the cuticle (mycangia) that secrete lipids for fungal nourishment, and require the nutrition provided by the fungus. The adult beetles carefully cultivate fungal gardens in their galleries, removing competing fungi. Their offspring feed exclusively on the fungus, which derives its resources from the wood surrounding the gallery, and collect and transport fungal hyphae when they disperse (Batra, 1966; French and Roeper, 1972). Similarly, siricid wasps are the only means of dispersal for associated *Amylostereum* (decay) fungi, and larvae die in the absence of the fungus (Morgan, 1968). The adult female wasp collects fungal hyphae from its gallery prior to exiting, stores and nourishes the fungus in a mycangium at the base of the ovipositor, then introduces the fungus during oviposition in the wood. The wasp larva feeds on the fungal mycelium, destroying it in the gut, and passes decayed wood fragments around the body to combine posteriorly with its frass. Phloem-feeding bark beetles transport mycangial fungi and bacteria as well as opportunistic fungi (Therrien et al., 2015). M. Ayres et al. (2000) reported that mycangial fungi significantly increased nitrogen concentrations in phloem surrounding southern pine beetle, *Dendroctonus frontalis*, larvae, compared to uncolonized phloem. Termites similarly depend on mutualistic bacteria or protozoa in their guts for digestion of cellulose (Breznak and Brune, 1994).

Many mutualistic interactions pair insects with other arthropods. Homoptera, especially aphids, excrete much of the carbohydrate solution (honeydew) that composes plant sap in order to concentrate sufficient nutrients (see Chapter 3). Honeydew attracts ants that provide protection from predators and parasites (Fig. 8.9, Bristow, 1991; Dixon, 1985; Dreisig, 1988). This mutualism involves about 25% of aphid species and varies in interaction strength and benefits, perhaps reflecting plant chemical influences or the relative costs of defending aphid colonies (Bristow, 1991). Ant species show different preferences among aphid

FIGURE 8.9   Mutualism: ant tending honeydew-producing aphids in Georgia, USA. *Source: Photo courtesy of S.D. Senter.*

species, and the efficiency of protection often varies inversely with aphid and ant densities (Bristow, 1991; Cushman and Addicott, 1991; Dreisig, 1988). Dung beetles and bark beetles often have mutualistic associations with phoretic, predaceous mites. The beetles are the only means of long-distance transport for the mites, and the mites feed on the competitors or parasites of their hosts (Kinn, 1980; Krantz and Mellott, 1972).

Models of mutualistic interactions have lagged behind models for competitive or predator–prey interactions, largely because of the difficulty of simultaneously incorporating negative (density-limitation) and positive (density-increasing) feedback. The Lotka–Volterra equations may be inadequate for extension to mutualism, because they lead to unbounded exponential growth of both populations (May, 1981, but see Goh, 1979). May (1981) asserted that minimally realistic models for mutualists must allow for saturation in the magnitude of at least one of the reciprocal benefits, leading to a stable equilibrium point, with one (most often both) of the two equilibrium populations being larger than that sustained in the absence of the interaction. However, recovery from perturbations to this equilibrium may take longer than in the absence of interaction, leading to instability (May, 1981). May (1981) presented a simple model for two mutualists

$$N_{1(t+1)} = N_{1t} + r_1 N_{1t}[1-(N_{1t}+\alpha N_2)/K_1] \tag{8.8}$$

$$N_{2(t+1)} = N_{2t} + r_2 N_{2t}[1-(N_{2t}+\beta N_1)/K_2] \tag{8.9}$$

in which the carrying capacity of each population is increased by the presence of the other, with α and β representing the beneficial effect of the partner, $K_1 \to K_1 + \alpha N_2$, $K_2 \to K_2 + \beta N_1$, and $\alpha\beta < 1$ to limit uncontrolled growth of the two populations. The larger the product, $\alpha\beta$, the more tightly coupled the mutualists. For obligate mutualists, a threshold effect must be incorporated to represent the demise of either partner if the other becomes rare or absent. May (1981) concluded that mutualisms are stable when both populations are relatively large, and increasingly unstable at lower population sizes, with a minimum point for persistence.

Dean (1983) proposed an alternative model that incorporates density dependence as the means by which two mutualists can reach a stable equilibrium. As a basis for this model, Dean described the relationship between population carrying capacity (k_y) and an environmental variable (M) that limits k_y

$$dk_y/dM = a(K_y - k_y)/K_y \tag{8.10}$$

where K_y is the maximum value of k_y and the constant a is reduced by a linear function of k_y. This equation can be integrated as

$$k_y = K_y(1 - e^{(-aM+C_y)/K_y}) \tag{8.11}$$

where C_y is the integration constant. Eq. (8.11) describes the isocline where $dY/dt = 0$.

For species Y exploiting a replenishable resource provided by species X, Eq. (8.11) can be rewritten as

$$k_y = K_y(1 - e^{(-aN_x+C_y)/K_y}) \tag{8.12}$$

where N_x is the number of species X. The carrying capacity of species X depends on the value of Y and can be described as

$$k_x = K_x(1 - e^{(-bN_y + C_x)/K_x})$$ (8.13)

where N_y is the number of species Y. Mutualism will be stable when the number of one mutualist (N_y) maintained by a certain number of the other mutualist (N_x) is greater than the N_y necessary to maintain N_x. When this condition is met, both populations grow until density effects limit the population growth of X and Y, so that isoclines defined by Eqs. (8.12) and (8.13) inevitably intersect at a point of stable equilibrium. Mutualism cannot occur when the isoclines do not intersect and is unstable when the isoclines are tangential. This condition is satisfied when any value of N_x or N_y can be found to satisfy either of the following equations:

$$K_y(1 - e^{(-aN_x + C_y)/K_y}) > -(C_x + K_x[\ln(K_x - N_x) - \ln K_x])/b$$ (8.14)

$$K_x(1 - e^{(-bN_y + C_x)/K_x}) > -(C_y + K_y[\ln(K_y - N_y) - \ln K_y])/a$$ (8.15)

The values of the constants, C_x and C_y, in Eqs. (8.14) and (8.15) indicate the strength of mutualistic interaction. When C_x and $C_y > 0$, the interacting species are facultative mutualists; when C_x and $C_y = 0$, both species are obligate mutualists; when C_x and $C_y < 0$, both species are obligate mutualists and their persistence is determined by threshold densities.

The growth rates of the two mutualists can be described by modified logistic equations as

$$N_{y(t+1)} = N_{y(t)} + (r_y N_{y(t)}[k_y - N_{y(t)}])/k_y$$ (8.16)

$$N_{x(t+1)} = N_{x(t)} + (r_x N_{x(t)}[k_x - N_{x(t)}])/k_x$$ (8.17)

where r_y and r_x are the intrinsic rates of increase for species Y and X, respectively. However, k_y and k_x are not constants but are determined by Eqs. (8.12) and (8.13).

More recently, Holland and DeAngelis (2009) demonstrated that mutualism could be modeled using extensions of the Lotka–Volterra equations for species interaction, but varying parameter values for interaction strength (α_{ij}) and resource supply by one species to the other (β_{ij}) (Fig. 8.10). Their model allowed for shifts in mutualistic interaction from stable coexistence to overexploitation by one or the other species, depending on environmental conditions.

3 INDIRECT EFFECTS

Research historically has focused on pairs of species that interact directly, that is, through energy or material transfers, as described previously. Indirect interactions have received less attention but may be at least as important as direct effects in organizing communities. In fact, Aschehaug and Callaway (2015) found that competition among multiple species in field experiments was less than indicated by pairwise experiments and that indirect effects reduced direct competition, thereby contributing to species coexistence and higher species diversity than predicted. Indirect interactions occur when the presence of a species affects other species without direct contact. For example, pollinators can augment plant reproduction sufficiently to compensate for herbivory, thereby indirectly affecting plant–herbivore interaction

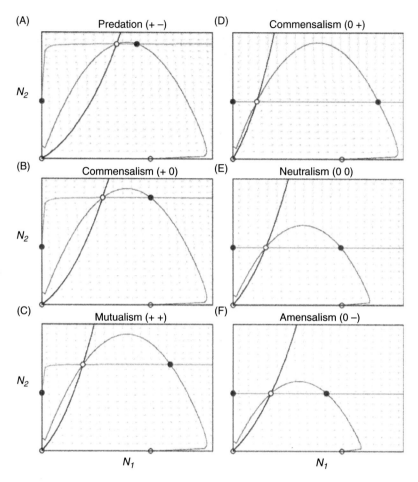

FIGURE 8.10 **Phase-plane diagrams for populations of one-way consumer–resource interactions between two species with population densities N_1 and N_2.** The sequence of panels shows how changes in interaction strengths (α_{ij} = per capita interaction strength of species j on species i) and resource supply (β_i = saturation level of resources exploited by species i) lead to dynamic transitions between (A) predation ($r_1 = 0.7$, $r_2 = 0.5$, $\alpha_{12} = 0.4$, $\alpha_{21} = 0.5$, $\beta_1 = 0.3$), (B) commensalism (species 1 benefits, $r_1 = 0.7$, $r_2 = 0.5$, $\alpha_{12} = 0.4$, $\alpha_{21} = 0.435$, $\beta_1 = 0.3$), (C) mutualism ($r_1 = 0.7$, $r_2 = 0.5$, $\alpha_{12} = 0.4$, $\alpha_{21} = 0.25$, $\beta_1 = 0.3$), (D) reverse commensalism (species 2 benefits, $r_1 = 0.7$, $r_2 = 0.5$, $\alpha_{12} = 0.4$, $\alpha_{21} = 0$, $\beta_1 = 0.3$), (E) neutralism ($r_1 = 0.7$, $r_2 = 0.5$, $\alpha_{12} = 0.21$, $\alpha_{21} = 0$, $\beta_1 = 0.3$), and (F) amensalism ($r_1 = 0.7$, $r_2 = 0.5$, $\alpha_{12} = 0.15$, $\alpha_{21} = 0$, $\beta_1 = 0.3$). The *red* and *green lines* are zero-growth isoclines for N_1 and N_2, respectively. *Grey arrows* designate vector fields in phase-plane space and denote direction and speed (size/length of arrow) of population trajectories for particular points throughout phase-plane space. Stable and unstable nodes are identified by *filled* and *open circles*, respectively. Saddle points have a black line passing through them to the origin, subdividing the phase-plane space into different basins of attraction that correspond with a particular node. Starting with the origin and moving clockwise, the equilibria for each panel are unstable node, stable node, saddle point, stable node, and saddle point. *Source: From Holland, J.N., DeAngelis, D.L., 2009. Consumer-resource theory predicts dynamic transitions between outcomes of interspecific interactions. Ecol. Lett. 12, 1357–1366.*

(L. Adler et al., 2001; Strauss and Murch, 2004). Batzer et al. (2000b) reported that indirect effects of predaceous fish on invertebrate predators and competitors of midge prey had a greater effect on midge abundance than did direct predation on midges. Among the best known indirect effects are those that involve indirect competition between herbivores feeding on different plant parts or at different times, effects of plants on predation, effects of predators on nonprey individuals and plants, effects of microorganisms on species interactions, and herbivore effects on vegetation structure.

3.1 Indirect Competition Among Herbivores

Herbivores feeding on above-ground plant resources frequently deplete root resources, through compensatory translocation, or induce production of defensive compounds that negatively affect root-feeding herbivores; root-feeding herbivores also can reduce above-ground herbivores (Masters et al., 1993; Rodgers et al., 1995; Salt et al., 1996; M. Wang et al., 2014). Conversely, root herbivores and pathogens can increase plant susceptibility to above-ground herbivores. Aukema et al. (2010) reported that feeding by root weevils, *Hylobius* spp., increased the likelihood that red pine, *Pinus resinosa*, would be colonized and killed by bark beetles, *Ips* spp.

Much research also has addressed the indirect effects of plant defenses induced by early-season herbivores on later colonists (Fig. 8.11) (Harrison and Karban, 1986; M.D. Hunter, 1987; Kogan and Paxton, 1983; N. Moran and Whitham, 1990; E. de Oliveira et al., 2016; Sticher et al., 1997; Van Zandt and Agrawal, 2004; Wold and Marquis, 1997) and on decomposers (Grime et al., 1996). Li et al. (2014) found that early feeding on cabbage, *Brassica oleracea*, by the cabbage aphid, *Brevicoryne brassicae*, enhanced the performance of the specialist diamondback moth, *Plutella xylostella*,

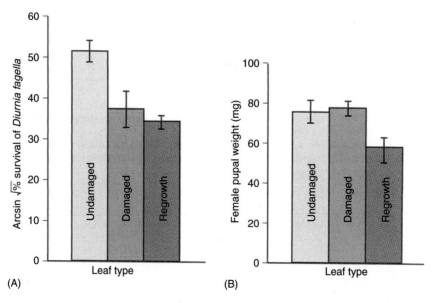

FIGURE 8.11 Differential survival to pupation (A) and mean female pupal weight (B) of *Diurnea fagella* on foliage that was undamaged, naturally damaged by folivores, and produced following damage. *Vertical lines* represent standard errors of the mean. *D. fagella* larvae feeding on regrowth foliage show both reduced survival to pupation and reduced pupal weight. *Source: From Hunter, M.D., 1987. Opposing effects of spring defoliation on late season oak caterpillars. Ecol. Entomol. 12, 373–382.*

and its larval endoparasitoid wasp, *Diadegma semiclausum*, but did not affect that of the generalist cabbage moth, *Mamestra brassicae*, nor its larval endoparasitoid wasp, *Microplitis mediator*. These data indicated that changes in plant quality induced by the cabbage aphid had differential effects on the development of associated herbivores. Branson (2010) reported that indirect effects of feeding by early-season grasshoppers, primarily *Eritettix simplex* and *Psoloessa delicatula*, on a later-season grasshopper, *Ageneotettix deorum*, depended on grasshopper density. Survival of *A. deorum* nymphs was significantly reduced at very high densities of *E. simplex* and *P. delicatula* that reduced grass biomass by 80% but not at lower levels of herbivory.

As for direct interactions, indirect interactions can be asymmetric. E. de Oliveira et al. (2016) and Godinho et al. (2016) reported that one spider mite, *Tetranychus urticae*, induces plant production of proteinase inhibitors and thereby indirectly reduces abundances of two other spider mites, *T. evani* and *T. ludeni*, whereas *T. evansi* and *T. ludeni* inhibit production of plant defenses, thereby indirectly benefiting *T. urticae*.

Such indirect competition also occurs between vertebrate and invertebrate herbivores. Hrabar and du Toit (2014) evaluated the effect of browsing by elephants, *Loxodonta africana*, on oviposition by mopane moths, *Imbrasia belina*, in replicate transects through mopane, *Colophospermum mopane*, woodland with high or low levels of browsing. After controlling for variation in canopy volume, egg mass density in woodlands recovering from severe elephant browsing was only half the density in less browsed woodlands. Muiruri et al. (2015) reported that browsing by moose, *A. alces*, altered the usual effect of tree diversity on insect herbivory on birch, *Betula pendula*. Whereas unbrowsed trees showed lower rates of herbivory in diverse stands, browsed trees showed higher rates of herbivory in diverse stands.

K. Anderson et al. (2009) extended the Lotka–Volterra competition model to describe plant-mediated interactions between two herbivore species. Their model for induction of multiple plant traits with negative or positive effects on a second herbivore is

$$H_{1(t+1)} = H_{1t} + r_1 H_1((K_1 - H_1 - f_1 I_2 + g_1 I_1)/K_1) \tag{8.18}$$

$$H_{2(t+1)} = H_{2t} + r_2 H_2((K_2 - H_2 - f_2 I_1 + g_2 I_2)/K_2) \tag{8.19}$$

$$I_{1(t+1)} = I_{1t} + \rho_1(I_1, H_1) - \delta_1 I_1 \tag{8.20}$$

$$I_{2(t+1)} = I_{2t} + \rho_2(I_2, H_2) - \delta_2 I_2 \tag{8.21}$$

where H_1 and H_2 are herbivores 1 and 2, respectively, I_1 and I_2 are induced responses of the plant with effect strength f and g, respectively, and $\delta_1 I_1$ and $\delta_2 I_2$ represent decay in induction over time.

3.2 Indirect Effects of Plants on Herbivory

Baldwin and Schultz (1983) and Rhoades (1983) independently found evidence that damage by herbivores is communicated chemically among plants, leading to induction of defense in neighboring plants in advance of herbivory (see Chapter 3). Although their hypothesis that plants communicate herbivore threat chemically with each other was challenged widely because of its apparent incongruency with natural selection theory (Fowler and Lawton, 1985), numerous studies have confirmed the induction of chemical defenses via applied volatile elicitors, particularly jasmonic acid (Fig. 8.12), salicylic acid, and ethylene (Farmer and Ryan, 1990; McCloud and Baldwin, 1997; Pearse et al., 2013; Schmelz et al., 2002;

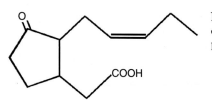

FIGURE 8.12 Structure of jasmonatic acid, a volatile plant chemical that communicates plant damage and induces defensive chemical production in neighboring plants.

Sticher et al., 1997; Thaler, 1999a; Thaler et al., 2001, see Chapter 3). Jasmonate induces production of proteinase inhibitors and other defenses against multiple insects when applied at low concentrations to a variety of plant species (Fig. 8.13) (Chamberlain et al., 2001; Hudgins et al., 2003, 2004; Thaler et al., 2001). Interplant communication via jasmonate induces production of defenses among neighboring plants (Dolch and Tscharntke, 2000; Hudgens et al., 2004; M. Stout et al., 2006; Tscharntke et al., 2001), including unrelated plant species (Fig. 8.14) (Farmer and Ryan, 1990; Karban, 2001; Karban and Maron, 2002; Karban et al., 2000; Schmelz

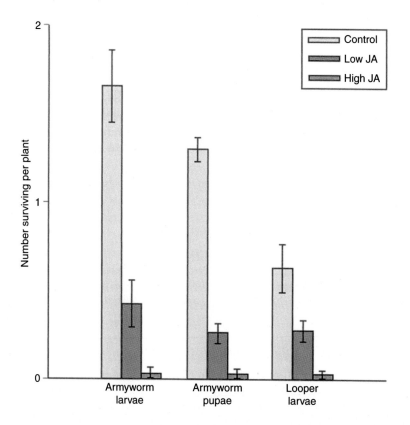

FIGURE 8.13 Survival of beet armyworm, *Spodoptera exigua*, larvae and pupae and cabbage looper, *Trichoplusia ni*, larvae on field-grown tomatoes sprayed with low (0.5 mM) or high (1.5 mM) doses of jasmonic acid (JA), or unsprayed (control). *Vertical lines* represent 1 SE. *Source: From Thaler, J.S., et al., 2001. Jasmonate-mediated induced plant resistance affects a community of herbivores. Ecol. Entomol. 26, 312–324.*

FIGURE 8.14 **Maximum proportion of leaves that were damaged by grasshoppers on tobacco plants that were near sagebrush plants that were artificially clipped or unclipped (mean + SE).** Effects of clipping were significant all 5 years ($p < 0.0001$). *Source: Reprinted from Karban, R., 2001. Communication between sagebrush and wild tobacco in the field. Biochem. Syst. Ecol. 29, 995–1005.*

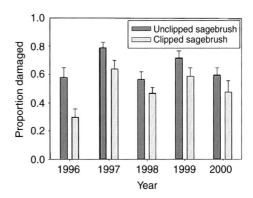

et al., 2002; Thaler et al., 2001), although the fitness consequences of interspecific communication are not clear (Hare, 2011; Karban and Maron, 2002).

In addition, volatile defenses induced by defoliators often attract predators and parasites (Chamberlain et al., 2001; Kessler and Baldwin, 2001; Price, 1986; Price et al., 1980; Thaler, 1999b; Turlings et al., 1990, 1993, 1995; Wason and Hunter, 2014). Thaler (1999b) demonstrated that tomato, *Lycopersicon esculentum*, defenses induced by jasmonate treatment doubled the rate of parasitism of armyworm, *Spodoptera exigua*, by the wasp, *Hyposoter exiguae*. However, some induced proteinase inhibitors may reduce pupal weight and survival of attracted parasitoids (Rodriguez-Saona et al., 2005). Zeng et al. (2009) found that herbivore production of P-450 detoxification enzymes in response to plant signaling chemicals, an adaptive response to induced plant defense, increased the toxicity of fungal aflatoxins ingested with plant material.

Plant domatia, floral or extrafloral nectaries, as well as aphid honeydew, commonly attract ants that, in turn, affect herbivore–plant interactions (Cushman and Addicott, 1991; Fritz, 1983; Jolivet, 1996; Oliveira and Brandão, 1991; Tilman, 1978). The strength of this interaction varies inversely with distance from ant nests. Tilman (1978) reported that ant, *Formica obscuripes*, visits to extrafloral nectaries declined with the distance between cherry trees and ant nests. The associated predation on tent caterpillars, *Malacosoma americanum*, by nectar-foraging ants also declined with distance from the ant nest.

3.3 Indirect Effects of Predators

Predators can impose a variety of indirect effects on other species. For example, the presence of a nearby predator can discourage foraging by potential prey or even nonprey. Bucher et al. (2015) found that cues left on plants by nursery web spiders, *Pisaura mirabilis*, following removal of spiders from experimental plants, reduced herbivory by 50%. On the other hand, an aphid-tending ant, *Myrmica rubra*, increased recruitment of workers to plants bearing spider cues, perhaps to protect aphids, *Microlophium carnosum*. Preisser et al. (2005) conducted a metaanalysis of 274 studies reporting consumptive and nonconsumptive effects of predators and concluded that intimidation had at least as strong an effect on prey (63% of total predator effect) as did direct consumption (51% of the total predator effect). Trophic cascades affect the entire network of feeding interactions in the community.

Herbivore and detritivore behavior can be affected by the presence of predators to a greater extent than would occur through actual predation (Belovsky et al., 2011; M.L. Johnson et al., 2006; Lagrue et al., 2015). Cronin et al. (2004) found that web-building spiders, at high densities, were more likely to affect planthoppers, *Prokelisia crocea*, through induced emigration than through direct mortality. Johansson (1993) reported that immature damselflies, *Coenagrion hastulatum*, increased avoidance behavior and reduced foraging behavior when immature dragonfly, *Aeshna juncea*, predators were introduced into experimental aquaria. Lagrue et al. (2015) also found that aquatic detritivores reduced foraging and, in some cases, moved to predator-free sites. Belovsky et al. (2011) found that grasshoppers responded to the threat of predation by reducing reproduction. Hawlena and Schmitz (2010) and Hawlena et al. (2012) found that the perceived risk of predation (fear) could cause sufficient changes in grasshopper physiology to alter carbon and nitrogen metabolism, litter quality, and decomposition rate (see also Chapters 11 and 12). Furthermore, Finke (2012) and Long and Finke (2015) reported that increased predator-avoidance movement of the bird cherry-oat aphid, *Rhopalosiphum padi*, in response to increased predation risk resulted in reduced transmission of the barley yellow dwarf virus.

Intraguild predation or competition among multiple predators affects predator foraging activity and can reduce predation on preferred prey (Finke and Denno, 2002, 2006; J.M. Schmidt et al., 2014; O. Schmitz, 2007). Alternatively, predators can be distracted by the presence of nonprey species in the community. Meisner et al. (2007) evaluated the effect of the spotted alfalfa aphid, *Therioaphis maculata*, on two parasitoids, the native *Praon pequodorum* and introduced *Aphidius ervi*, of the pea aphid, *Acyrthosiphum pisum*. The spotted alfalfa aphid had a greater distraction effect on the more common *A. ervi*, thereby contributing to persistence of *P. pequodorum* in this system.

A predator that preys indiscriminately on competing prey species, as they are encountered, will prevent the most abundant prey species from competitively suppressing others. R. Paine (1966, 1969a,b) introduced the term *keystone species* for top predators that maintain balanced populations of competing prey species. However, this term has become used more broadly to include any species with a disproportionate effect on community and ecosystem structure or function, based on its abundance (Bond, 1993; Power et al., 1996). Herbivorous insects that selectively reduce the density of abundant host species, and thereby balance abundances of host and nonhost plants (Louda et al., 1990a; Schowalter and Lowman, 1999) and their associated species, function in a keystone capacity.

Trophic cascades appear where predation, or fear induced by presence of predators, reduces herbivory and thereby increases plant growth. Conversely, reduced predation can increase herbivory, thereby reducing plant growth. Trophic cascades appear to be more common and pronounced in aquatic ecosystems (Batzer et al., 2000b; S. Carpenter and Kitchell, 1984, 1987, 1988; Vanni and Layne, 1997) than in terrestrial ecosystems (Shurin et al., 2002), but this may be related to the productivity of the system or the ease of experimental manipulation.

L. Dyer and Letourneau (1999a,b) and Letourneau and Dyer (1998) described a trophic cascade in a neotropical rainforest community. Clerid beetle, *Tarsobaenus letourneauae*, predation on ants, especially *Pheidole bicornis*, reduced ant abundance and increased herbivore abundance and herbivory on *Piper cenocladum* ant-plants. Where this beetle was absent and spiders were a less effective top predator, ant abundance was higher and reduced herbivore abundance (Fig. 8.15). Manipulation of top-down and bottom-up effects indicated that increased resources (light and nutrients) directly increased plant biomass but had no indirect effect on predators or top predators, but ant exclusion indirectly affected plant biomass by increasing herbivory (L. Dyer and Letourneau, 1999a).

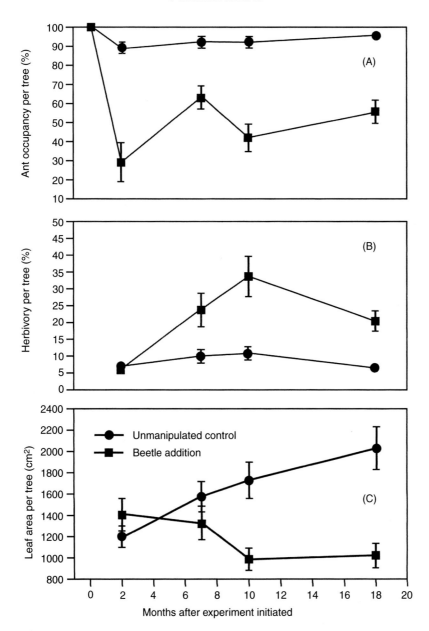

FIGURE 8.15 (A) Mean ant, *Pheidole bicornis*, abundance (percentage of occupied petiole chambers per plant), (B) mean folivory (leaf area eaten per *Piper cenocladum* shrub), and (C) mean leaf area per shrub before (0–2 months) and after (7–18 months) addition of a top predator, *Tarsobaenus letourneauae* beetles, to half the shrubs. *Vertical bars* represent 1 SE. *Source: From Dyer, L.A., Letourneau, D.K., 1999a. Relative strengths of top-down and bottom-up forces in a tropical forest community. Oecologia 119, 265–274.*

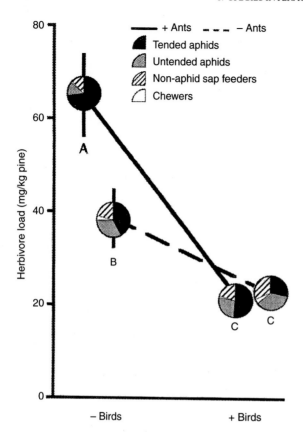

FIGURE 8.16 Effects of bird and/or ant exclusion on total herbivore density and composition (pie charts) in ponderosa pine canopies. Herbivore densities (means + SE) are for 6 post-treatment months. Means with different letters differ significantly ($p < 0.05$). The interaction between bird and ant effects was significant ($p = 0.03$). *Source: From Mooney, K.A., 2007. Tritrophic effects of birds and ants on a canopy food web, tree growth, and phytochemistry. Ecology 88, 2005–2014.*

Mooney (2007) excluded birds and/or ants, primarily *Formica podzolica*, from experimental mature ponderosa pine, *Pinus ponderosa*, trees in Colorado, USA (Fig. 8.16). She found that birds and ants reduced abundances of folivorous and predaceous arthropods in an additive manner, with the effect of ants stronger than that of birds. Abundance of ants doubled and abundance of tended aphids, *Cinara* spp., tripled when birds were excluded. However, ants only increased the abundance of tended aphids in the absence of birds, whereas birds only reduced their abundance in the presence of ants, apparently because of bird disruption of the aphid–ant interaction. Predation by birds, but not ants, resulted in increased wood and foliage growth. Similarly, Marquis and Whelan (1994) reported that insectivorous birds significantly increased above ground production of white oak, *Quercus alba*, by about 20%, compared to bird exclusion treatment. Schwenk et al. (2010) excluded birds from striped maple, *Acer pensylvanicum*, saplings and reported that birds significantly reduced arthropod biomass by 20%, compared to control saplings, but reduced herbivory in exclosures was not significantly different from controls, perhaps due to low levels of herbivory (<5%).

Beard et al. (2003) excluded frogs, *Eleutherodactylus coqui*, in tropical rainforest plots and observed reduced invertebrate abundance and herbivory and increased foliage production.

Stiling and Moon (2005) reported that parasitoid exclusion increased herbivore abundance and reduced plant growth. M.D. Hunter et al. (2003) reported that exclusion of litter predators in litterbags increased Collembola abundances and litter decomposition rate.

Predators can initiate trophic cascades even in the absence of direct consumption of prey (Z. Costa and Vonesh, 2013; Peckarsky et al., 1993; Preisser et al., 2005; H. Schmitz et al., 1997). Z. Costa and Vonesh (2013) tested the effect of presence of a caged predator, larval Amazon darner dragonfly, *Anax amazili*, on herbivory by two tadpole species, *Agalychnis callidryas* and *Dendropsophus ebraccatus*. Although predator presence affected the two herbivores to different extents, the predator treatment significantly increased perphyton production, as measured by chlorophyll a (Fig. 8.17).

FIGURE 8.17 Periphyton production, measured as μg chlorophyll (chl a) after (A) 13 and (B) 28 days of presence (filled circles) or absence (open circles) of an experimental predator, *Anax amazili*, for each herbivore (tadpole) treatment: *Agalychnis callidryas* + *Dendropsophus ebraccatus* (Ag + De), *A. callidryas* alone (Ag), *D. ebraccatus* alone (De), or no tadpoles (No Tads). Pred = predator effect; Tads = herbivore effect; Time = temporal effect using repeated measure mixed models. Significance level: * $p < 0.05$; ** $p < 0.01$; *** $p < 0.001$. *Source: From Costa, Z.J., Vonesh, J.R., 2013. Prey subsidy or predator cue? Direct and indirect effects of caged predators on aquatic consumers and resources. Oecologia 173, 1481–1490.*

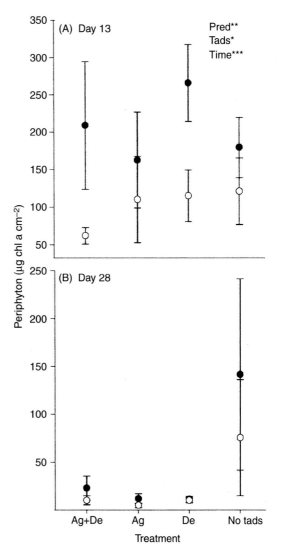

Hermann and Thaler (2014) tested the nonconsumptive effects of predatory spined soldier bugs, *Podisus maculiventris*, on feeding by Colorado potato beetle, *Leptinotarsa decemlineata*, larvae. Larvae exposed to bugs that were surgically manipulated to hunt but not kill reduced feeding by 29%, compared to controls. Larvae on leaves preexposed to predators reduced feeding by 24%, compared to leaves never exposed to predators. Larvae exposed to volatile odor cues from predators reduced feeding by 10%, compared to controls. These data indicated that tactile and visual cues are not necessary to elicit prey response but that olfactory cues play a role.

Ability to induce trophic cascades apparently differs among community types and predator species (Z. Costa and Vonesh, 2013; C. Huang et al., 2007; Leroux and Loreau, 2008). Several factors appear to determine cascade strength. Ecosystems that have high productivity or receive large amounts of allochthonous subsidies (eg, aquatic ecosystems) should have the strongest trophic cascades, especially if the subsidies are disproportionately available to plants and predators (J. Chase et al., 2000; Leroux and Loreau, 2008). Predator diversity and habitat complexity determine predator foraging efficiency and degree of intraguild predation (Finke and Denno, 2006; Letourneau et al., 2009; Straub and Snyder, 2008). Finally, variation in herbivore–plant interaction, for example, plant defenses and compensatory growth ability (Shurin et al., 2002), or other species interactions (Mooney, 2006) may buffer trophic levels from cascading effects.

3.4 Indirect Effects of Microorganisms

Microorganisms affect interactions involving insects in various ways. They can alter competitive interactions among plants in ways that affect herbivores, affect plant signaling pathways, or increase prey susceptibility to predators (C. de Oliveira et al., 2014; Janssens and Stoks, 2014; Nachappa et al., 2013; Pan et al., 2013).

Endophytic or mycorrhizal fungi affect interactions between other organisms (E. Allen and Allen, 1990; G. Carroll, 1988; Clay, 1990; Kaur et al., 2015, see Chapter 3). G. Carroll (1988) and Clay et al. (1985) reported that mycotoxins produced by mutualistic endophytic fungi complement host defenses in deterring insect herbivores. Clay et al. (1993) documented complex effects of insect herbivores and endophytic fungi on the competitive interactions among grass species. Tall fescue, *Festuca arundinacea*, competed poorly with orchard grass, *Dactylis glomerata*, when herbivores were absent, but fescue infected with its fungal endophyte, *Acremonium* spp., competed better than either orchard grass or uninfected fescue when herbivores were present. Pangesti et al. (2015) demonstrated that colonization of *Arabidopsis thaliana* roots by the rhizobacterium, *Pseudomonas fluorescens*, increased the attraction of the parasitoid, *Microplitis mediator*, to its host, caterpillars of the cabbage moth, *Mamestra brassicae*. However, Kaur et al. (2015) found that negative effects of herbivore feeding on plants infected with endophytic fungi extended to parasitoids.

Mauck et al. (2010) tested the effect of cucumber mosaic virus (CMV) on the quality and attractiveness of squash plants, *Cucurbita pepo*, for two aphid vectors, *Myzus persicae* and *Aphis gossypii*. Although aphids performed poorly on, and emigrated rapidly from, infected plants, CMV increased the attractiveness of infected plants by inducing elevated emissions of plant volatiles. Such deceptive attraction of vectors to infected plants from which they then disperse rapidly is highly conducive to virus transmission.

FIGURE 8.18 Indirect effects of associated species. The light-colored foliage at the ends of shoots is new grand fir, *Abies grandis*, foliage produced during 1994, a dry year, in western Washington; the blackened 1993 foliage was colonized by sooty mold during a wet year; normal foliage prior to 1993 was produced during extended drought. Sooty mold exploits moist conditions, especially honeydew accumulations and, in turn, may affect foliage quality for folivores.

Gange et al. (1999) and Goverde et al. (2000) experimentally inoculated plants with arbuscular mycorrhizal fungi and evaluated effects on aphids, *M. persicae*, and butterfly, *Polyommatus icarus*, larvae, respectively. In both studies, mycorrhizal inoculation increased insect growth and survival, apparently related to increased P concentrations in foliage of mycorrhizal plants. Goverde et al. (2000) further reported that herbivore performance was related to the mycorrhizal species colonizing the host plant. Sooty molds growing on foliage also may affect palatability for herbivores (Fig. 8.18).

Kersch-Becker and Thaler (2014) tested the effect of three strains of potato virus Y on the abundance and fecundity of its aphid vector, *Macrosiphum euphorbiae*, and two nonvector species, the cabbage looper, *Trichoplusia ni*, and the Colorado potato beetle. All three virus strains induced the salicylate resistance pathway and increased aphid fecundity and density. Both nonvector herbivores showed positive responses related to the strength of salicylate induction. Similarly, Mason et al. (2014) reported that bacteria associated with foliar surfaces help degrade plant toxins upon ingestion by gypsy moth, *Lymantria dispar*, larvae.

C. Currie (2001) and C. Currie et al. (1999a,b) reported complex interactions between fungus-growing ants, especially leaf-cutting species of *Atta* and *Acromyrmex*, their mutualistic fungi, *Leucocoprinus* spp. and *Leucoagaricus* spp., and associated microorganisms. The ants provide live or dead vegetable material for fungal decomposition, tend the gardens by weeding alien microbes, and feed on the fungus. Foundress queens carry fungus inoculum to establish new colonies. However, fungus gardens often host a virulent fungal pathogen, *Escovopsis*, capable of destroying the fungus garden and the dependent ant colony. The ants have an additional mutualistic association with an actinomycete bacterium that produces specialized antibiotics with potent inhibitory activity against *Escovopsis*.

Complex interactions among a community of invertebrates and fungi affect bark beetle interactions with host trees. The southern pine beetle–blue stain fungus interaction was thought at one time to be mutualistic, with beetles providing transport and the fungus contributing to tree death and beetle reproduction. However, more recent studies demonstrated that this beetle can colonize trees in the absence of the fungus (Bridges et al., 1985), that the blue stain fungus is detrimental to beetle development and is avoided by the mining beetles (Barras, 1970; Bridges, 1983; Bridges and Perry, 1985), and that other mycangial fungi are necessary for optimal beetle development (M. Ayres et al., 2000; Bridges and Perry, 1985). Blue stain fungus is associated only indirectly with the beetle. Spores are collected by phoretic tarsonemid mites in specialized sporothecae (Bridges and Moser, 1983; Hofstetter and Moser, 2014; J. Moser, 1985). Beetles carrying these mites transport the blue stain fungus significantly more often than do mite-free beetles (Bridges and Moser, 1986; Hoffstetter and Moser, 2014). The beetle–tree interaction is affected further by phoretic predaceous mites that prey on nematode parasites of the beetle (Kinn, 1980). Finally, defoliation by folivorous insects can increase tree susceptibility to colonization by bark beetles (Wallin and Raffa, 2001; Wright et al., 1986).

Termite use of plant resources is affected by interaction with mutualistic gut symbionts, that are affected by wood chemistry and associated wood-colonizing fungi. Mankowski et al. (1998) found that termite preferences among combinations of wood and fungal species generally reflected the suitability of the wood species for termite gut fauna, as indicated by changes in gut faunal densities when termites were forced to feed on wood–fungus combinations.

3.5 Indirect Effects of Herbivores on Vegetation Structure

Insect herbivores affect competition among host and nonhost plant species (see Chapter 12). Suppression of host plants provides opportunity for increased growth of nonhosts (Ritchie et al., 1998; Wickman, 1980). Callaway et al. (1999) reported that the tortricid moth, *Agapeta zoegana*, introduced to the western USA for biological control of spotted knapweed, *Centaurea maculosa*, increased the negative effect of its host on native grass, *Festuca idahoensis*. Reproductive output of grass was lower when neighboring knapweed had been defoliated by the moth, compared to grass surrounded by nondefoliated neighbors. Callaway et al. (1999) suggested that defenses induced by the moth also had allelopathic effects on neighboring plants or altered root exudates that affected competition via soil microbes.

On the other hand, shared herbivores or seed predators also affect plant competition and coexistence (Garzon-Lopez et al., 2015), in the same way that keystone predators maintain higher diversity of prey species (R. Paine, 1966, 1969a,b, see section 3.3). Garzon-Lopez et al. (2015) concluded that the probability of a seed being consumed by a generalist seed predator was a complex function of density, phenology, and spatial distribution of conspecific as well as heterospecific seeds on Barro Colorado Island, Panama. Shared seed predators were capable of either promoting or reducing coexistence among plant species.

Resulting changes in vegetation structure affect virtually all other community members indirectly via changes in habitat conditions, resource distribution, water and nutrient fluxes, and local climate (see Chapter 12). Examples include effects of termite mounds, canopy opening by African elephants, and altered fire frequency on maintenance of open woodlands in

Africa (Joseph et al., 2014; Mapaure and Moe, 2009; Shannon et al., 2011); effects of southern pine beetle and mountain pine beetle, *Dendroctonus ponderosae*, on maintenance of open pine woodlands and low-intensity fire in southern and western North America (Schowalter, 2008; Schowalter et al., 1981a); effects of spruce budworm, *Choristoneura fumiferana*, on maintenance of greater diversity of tree species in boreal forests of North America (Belle-Isle and Kneeshaw, 2007); and effects of burrowing mammals and ants, *Aphaenogaster cockerelli*, on maintenance of desert vegetation (Eldridge et al., 2009).

4 FACTORS AFFECTING INTERACTIONS

Multispecies interactions are highly complex (M. Wise, 2009). Species can simultaneously compete for space and enhance each other's food acquisition (mutualism), as described by Cardinale et al. (2002) for three caddisfly species that in combination increase substrate surface heterogeneity and near-surface velocity and turbulent flow that control food delivery. The strength, and even type, of interaction, can vary over time and space depending on biotic and abiotic conditions (B. Inouye and Stinchcombe, 2001; Økland et al., 2009; Tilman, 1978). Interactions can change during life history development or differ between sexes. For example, immature butterflies (caterpillars) are herbivores, but adult butterflies are pollinators. Insects with aquatic immatures are terrestrial as adults. Immature males of the strepsipteran family Myrmecolacidae parasitize ants, whereas immature females parasitize grasshoppers (de Carvalho and Kogan, 1991). Herbivore–plant interactions may be largely mutualistic at low herbivore population densities, with the plant providing food and the herbivore providing limited pruning, but become increasingly parasitic, or even predatory, at high herbivore densities (see Chapter 12). Holland and DeAngelis (2009) demonstrated that all possible outcomes of species interactions emerged simply from changes in the parameters of consumer–resource relationships (interaction strength and direction of exploitation), indicating that changes in abiotic or biotic conditions could alter outcomes of species interactions (Fig. 8.10).

The strength of interaction depends on the proximity of the two species, their ability to perceive each other, their relative densities, and their motivation to interact. These factors in turn are affected by abiotic conditions, resource availability, and indirect effects of other species. Modeling interaction strength for prediction of community dynamics has taken a variety of approaches that may be subject to unrecognized biases or to nonlinear or indirect effects (Abrams, 2001; Berlow et al., 1999).

4.1 Abiotic Conditions

Relatively few studies have addressed the effects of abiotic conditions on species interactions. Abiotic conditions that affect host growth or defensive capability influence predation or parasitism. Increased exposure to sunlight can increase plant production of defensive compounds and reduce herbivory (Dudt and Shure, 1994; Niesenbaum, 1992). Stamp et al. (1997) found that defensive chemicals sequestered by caterpillars had greater negative effects on a predator at higher temperatures. Light availability to plants may affect their relative investment in toxic compounds versus extrafloral nectaries and domatia to facilitate defense by ants (Davidson and Fisher, 1991). Fox et al. (1999) reported that drought stress in the UK did not affect growth of

St. John's wort, *Hypericum perforatum*, directly but increased plant vulnerability to herbivores. Le Lann et al. (2014) reported that increased temperature resulted in convergence of habitat use among four coexisting parasitoids of *Drosophila* spp., increasing competitive interactions.

Altered atmospheric chemistry, for example, CO_2 enrichment or pollutants, affects interactions (Alstad et al., 1982; Arnone et al., 1995; V.C. Brown, 1995; Heliövaara and Väisänen, 1986, 1993; Kinney et al., 1997; Roth and Lindroth, 1994; Salt et al., 1996). Hughes and Bazzaz (1997) reported that elevated CO_2 significantly increased C to N ratio and decreased percentage of nitrogen in milkweed, *Asclepias syriaca*, tissues, resulting in lower densities but greater per capita leaf damage by western flower thrips, *Frankliniella occidentalis*. However, increased plant growth at elevated CO_2 levels more than compensated for the leaf damage. Mondor et al. (2004) found that the aphid, *Chaitophorus stevensis*, exhibited reduced predator-escape behavior in an enriched CO_2 atmosphere but greater escape behavior in an enriched O_3 atmosphere, compared to ambient atmospheric conditions. Elevated CO_2 can affect litter quality and alter interactions among litter flora and fauna (Coûteaux et al., 1991). Ozone, but not nitrogen dioxide or sulfur dioxide, interfered with searching behavior and host discovery by a braconid parasitoid, *Asobara tabida* (Gate et al., 1995).

J. Chase (1996) experimentally manipulated temperature and solar radiation in grassland plots containing grasshoppers and wolf spiders. When temperature and solar radiation were reduced by shading, grasshopper activity was reduced, but spider activity was unaffected, and spiders reduced grasshopper density. In contrast, grasshoppers remained active in unshaded plots, and spiders did not reduce grasshopper density. Stamp and Bowers (1990) found that temperature affects interactions between plants, herbivores, and predators.

Fitzpatrick et al. (2014) found that three of the four ant species that defend cactus, *Ferocactus wislizeni*, in the Sonoran Desert abandoned plants during much of the day, between 0100 and 1700 hr, when plant surface temperature exceeded 40°C, returning when the surface temperature decreased in late afternoon. The fourth species occupied plants in the absence of the other three but was the least effective defender of cactus plants.

Preisser and Strong (2004) reported that neither a root-feeding lepidopteran herbivore, *Hepialus californicus*, nor its lupine host, *Lupinus arboreus*, was directly affected by variation in soil moisture, but outbreaks typically coincide with drought conditions. Experimental increase in soil moisture during a dry year demonstrated that high soil moisture favored parasitic nematodes, *Heterorhabditis marelatus*, that suppressed the herbivore outbreak and protected the plant host via trophic cascade.

Hart (1992) studied the effect of water flow on the relationship between crayfish, caddisfly prey, and algal food base in a stream ecosystem. He found that crayfish foraging activity was impaired at high flow rates, limiting predation on caddisfly grazers and altering the algae–herbivore interaction. Conversely, Covich and McDowell (1996) and Covich et al. (2003, 2009) found that reduced pool depth and volume in streams during dry conditions increase prey exposure to predation. Kelly et al. (2003) reported that exposure of stream communities to UV radiation reduced aquatic grazing and led to increased algal biomass.

Disturbances affect species interactions in several ways. First, disturbances reduce abundances of intolerant species, thereby affecting their interactions with other species. Second, disturbances contribute to landscape heterogeneity, thereby providing potential refuges from predation (Denslow, 1985; Kruess and Tscharntke, 1994; Schowalter and Ganio, 1999) but also decoupling positive interactions.

4.2 Resource Availability and Distribution

Resource availability affects competition and predation. If suitable resources (plants or animal prey) become more abundant, resource discovery becomes easier, and consumer populations grow. Increased probability of close contact and competition among consumers leads to densities at which superior competitor(s) suppress or exclude inferior competitors. As a result, the intensity of interspecific competition may peak at intermediate levels of resource availability, although the rate of resource use may continue to rise with increasing resource availability (depending on functional and numerical responses). Population outbreaks reduce resource availability and also reduce populations of competing species.

Interactions are affected by landscape heterogeneity. Sparse resources in heterogeneous habitats tend to maintain small, low-density populations of associated species. The energetic and nutrient costs of detoxifying current resources or searching for more suitable resources limit growth, survival, and reproduction (see Chapters 3 and 4). Under these conditions, potentially interacting species are decoupled in time and space, co-occurring infrequently among landscape patches (Covich et al., 2009; Tack et al., 2009). Hence, competition is minimized, and predator-free space is maximized, in patchy environments. In contrast, more homogeneous environments facilitate population spread of associated species and maximize the probability of co-occurrence.

Palmer (2003) explored the effect of termite-generated heterogeneity in resource availability on the competitive interactions of four ant species that reside on acacia, *Acacia drepanolobium*, in East Africa. Only one ant species occupied an individual tree at any given time, and violent interspecific competition for host trees by adjacent colonies was common. Acacia shoot production and densities of litter invertebrates increased with proximity to termite mounds. The competitively dominant ant, *Crematogaster sjostedti*, displaced other acacia ants, *C. mimosae, C. nigriceps*, and *Tetraponera penzigi*, near termite mounds, whereas the probability of subordinate species displacing *C. sjostedti* increased with distance from termite mounds. This variation in the outcome of competition for acacia hosts appeared to result from differential responses among the ant species to resource heterogeneity on the landscape.

Species interactions also affect habitat heterogeneity and/or resource availability. Cardinale et al. (2002) manipulated composition of three suspension-feeding caddisfly species at the same total density in experimental stream mesocosms. They reported that total consumption of suspended particulate food was 66% higher in mixtures, compared to single-species treatments. Facilitation of food capture by these potentially competing species in mixture resulted from increased stream bed complexity (reflecting variation in silk catchnet size), which in turn increased eddy turbulence and near-bed velocity, factors controlling the rate of food delivery.

5 CONSEQUENCES OF INTERACTIONS

Each species interacts with many other species in a variety of ways (competing for various food, habitat, and other resources, preying, or being preyed, on, and cooperating with mutualists, directly and indirectly) with varying degrees of positive and negative feedback on abundance. Therefore, the population status of each species represents the net effects of these feedbacks.

Modeling and predicting effects of multiple direct and indirect interactions are difficult. Even tritrophic interactions represent highly simplified models of communities (Gutierrez, 1986; C.G. Jones et al., 1998). In most communities species potentially interact directly or indirectly with hundreds of other species to alter environmental conditions for all (see Chapters 9 and 10).

Bezemer et al. (2005) reported that manipulation of soil nematodes and microorganisms significantly altered plant amino acid and phenolic content and aphid and parasitoid performance. The abundance of tick vectors of Lyme disease is related to the abundance of small mammal reservoirs, which reflect acorn production that, in turn, is affected by gypsy moth defoliation (C.G. Jones et al., 1998). The tendency for multiple interactions to stabilize or destabilize species populations and community structure has been debated (Goh, 1979; May, 1973, 1983; Price, 1997). May (1973) proposed that community stability depends on predator–prey interactions (negative feedback) being more common than mutualistic interactions (positive feedback). Because multispecies interactions control rates of energy and nutrient fluxes through ecosystems (Narr and Frost, 2015), resolution of the extent to which indirect interactions reduce variation in community structure will contribute significantly to our understanding of ecosystem stability.

5.1 Population Regulation

As discussed in Chapter 6, competition and predation have been recognized as two primary mechanisms, along with resource quality and quantity, for limiting population growth of a given species (May, 1983). A particular species typically interacts with at least 2–5 other species as prey (see Chapter 9) and with additional species as a competitor. It interacts with virtually all other species in the community as it forages or responds to the presence of potential predators. The presence of keystone species can maintain balanced populations of prey species and thereby maintain relatively constant vegetation (Mooney, 2007; Preisser et al., 2005). Life table analysis often is used to identify key factors that contribute most to population change, but a combination of interactions provides for "redundant" control of population growth.

As previously noted in section 2.3, mutualistic interactions may reduce the probability that either species will decline to extinction. Mutualistic species often are closely associated, especially in obligate relationships, and enhance each other's resource acquisition, energy and nutrient balance, or reproduction. Although mutualism is likely to become unstable at low population densities of either partner, depending on the degree of obligation (May, 1983), mutualism could maintain the populations above extinction thresholds (Dean, 1983).

Species interactions involve trade-offs among costs and benefits of multiple interactions. For example, plants supporting ants as defensive mutualists are subject to selection based on the relative benefit of protection from herbivores versus the cost of ant predation on pollinators (Ohm and Miller, 2014). Furthermore, the relative benefits and costs of various interactions are likely to change as environmental conditions or species abundances change. Therefore, the combination of various interactions involving a particular species should maintain its population levels within a narrower range than would occur with fewer interactions. Croft and Slone (1997) found that three predaceous mite species maintained populations of the European red mite, *Panonychus ulmi*, at lower equilibrium levels than did fewer predator species. However, few studies have documented the importance of species diversity or food web structure to the stability of population levels.

5.2 Community Regulation

The extent to which the network of regulatory interactions maintains stable community structure (see Chapters 9 and 10) has been a topic of considerable debate. Although some irruptive species show wide amplitude in population size over time and space, such irruptions often reflect disruption of normal interactions as a result of extreme weather, disturbances, anthropogenic habitat alteration, or introduction of invasive species into new habitats (see Chapters 6 and 7). The amplitude of population deviation may be narrower, and the duration shorter, in natural ecosystems with intact regulatory interactions.

The capacity for the network of interactions to stabilize species populations may be enhanced by compensatory interactions and changes in the nature or strength of interaction as environmental conditions change. For example, the plant species at a site can, at the same time, compete for resources, share nutrients via mycorrhizae, be growth-limited by herbivores, and limit herbivore populations through the mingling of induced defenses, attractive host odors, and repellent (or unattractive) nonhost odors (E. Allen and Allen, 1990; Dolch and Tscharntke, 2000; A. Hunter and Arssen, 1988; Visser, 1986). The net result of negative and positive feedbacks may be balanced coexistence (W. Carson and Root, 2000). Ants maximize energy gain by preying on aphids when the value of honeydew rewards is low (eg, scattered individuals or individuals dispersing from dense colonies) and by tending aphids when the value of honeydew rewards is high (Bristow, 1991; Cushman and Addicott, 1991). A competitive interaction could become mutualistic if a third, more competitive, species can be excluded (Pianka, 1981). Predation can be neutral or beneficial at low rates and increasingly negative at higher rates (see Chapters 12 and 15), thereby stabilizing host populations at intermediate size. Such flexibility in species interactions may facilitate regulation in a variable environment. Stabilization of community structure has substantial implications for the stability of ecosystem processes (see Chapter 15).

Interactions strongly affect energy or nutrient balances, survival, and reproduction of the associated species and therefore represent major selective factors. Strongly negative interactions should select for adaptive responses that minimize the negative effect, for example, niche partitioning among competitors, prey defenses, etc. Therefore, negative interactions should evolve toward more neutral or mutualistic interactions (G. Carroll, 1988; Price, 1997).

6 SUMMARY

Species interact in a variety of direct and indirect ways with the other species that co-occur at a site. These interactions produce combinations of positive, neutral, or negative effects for species pairs. Some species alter the nature or strength of pairwise interactions among other species, for example, predators can reduce the intensity of competition among prey species by maintaining their populations below levels that induce competition.

Many, if not all, species compete for shared resources, with the result that the per capita share of the resource is reduced. This interaction has negative effects on all competing species. Competition can be by exploitation, when all individuals have equal access to the resource, or interference, when individuals of one species preempt use of, or defend, the resource. In cases of asymmetric competition, the superior competitor can exclude inferior competitors over a period of time (competitive exclusion), unless the inferior competitor can escape through dispersal or survival in refuges where superior competitors are absent.

Predator–prey interactions involve a predator killing and eating prey and, therefore, have a positive effect on the predator but a negative effect on the prey. Predators and parasites affect prey populations similarly, but predators generally are opportunistic with respect to prey taxa and kill multiple prey per individual, whereas parasites generally are more specialized for association with particular host species and may or may not kill the host. Predators show preferences for prey size or defensive capability that maximize capture and utilization efficiency.

Symbiosis involves an intimate association between a symbiont and its host species, which have often coevolved to maximize the probability of association and to mitigate any host defense against the symbiont. Symbiosis includes parasitism, commensalism, and mutualism. Parasitism is beneficial to the parasite but detrimental to its host. Parasitoidism is unique to insects and involves an adult female ovipositing on or in a living host, with her offspring feeding on and eventually killing the host. Commensalism benefits the symbiont but has neutral effects for the host. Typically the symbiont uses the host or its products as habitat or as a means of transport with negligible effects on the host. Mutualism benefits both partners and is exemplified by pollinator–plant, ant–plant, ant–aphid, and detritivore–fungus interactions.

Indirect interactions potentially occur among all species in the community as a result of foraging or other movement through the community, perception of potential predators, and mixing of various plant and animal volatiles, both attractants and repellents. Predators can affect other species indirectly by reducing their foraging or feeding activities. Trophic cascades result when predation, or the fear of predation, reduces herbivore feeding, thereby increasing plant production, an indirect effect of predation.

A variety of factors influence the nature and intensity of interaction. Abiotic factors that affect the activity or condition of individuals of a species may alter their competitive, predatory, or defensive ability. Resource availability, particularly the quality and patchiness of resources, may mitigate or exacerbate competition or predation by limiting the likelihood that competitors, or predators and their prey, co-occur in time and space. Other species can influence pairwise interactions indirectly. Species whose presence significantly affects diversity or community structure have been considered keystone species. A number of insect species function as keystone species.

Competition and predation/parasitism have been recognized as important mechanisms of population regulation and have been amenable to mathematical modeling. Mutualism has been viewed largely as a curiosity, rather than an important regulatory interaction, and modeling efforts have been more limited. However, mutualism may promote both populations and reduce their risk of decline to unstable levels. The network of interactions affecting a particular species may maintain population size within a narrower range with less frequent irruptions than occurs when populations are released from their regulatory network. The extent of mutual regulation (stabilization) of populations through this network of interactions has been widely debated, but has significant implications for the stability of community structure and ecosystem processes governed by these interactions, as described in the following chapters.

Community Structure

STREAMS AS A NETWORK OF HABITAT TYPES

Streams represent a relatively simple linear continuum of habitat types (reaches) progressing from high-gradient headwater sections with narrow channels, high (and perhaps intermittent) flow rates, rocky beds, and abundant coarse woody debris to low-gradient floodplains with wide channels, low flow rates, beds composed of fine sediments, and fine suspended organic matter (Vannote et al., 1980). Habitat types are distinguished by their substrate, flow rate, water temperature, and surrounding terrestrial vegetation. Different types support distinct aquatic communities. Headwater reaches with steep gradients, rocky substrate, and abundant woody structure support primarily shredders and scrapers (particularly Trichoptera, Plecoptera, and Ephemeroptera) that fragment largely intact detritus entering the stream from overtopping vegetation; pools and lower-elevation reaches with shallow gradients and fine-textured substrates support more filter-feeders (particularly Diptera) that collect fine suspended or deposited organic matter (Huryn and Wallace, 1987, see Chapter 14).

The aquatic communities characterizing the continuum of habitat types are interconnected with each other and with adjoining riparian patches by the direction of water flow, detritus, and amphibious organisms (Baxter et al., 2005). Dislodged aquatic invertebrates drift downstream and colonize downstream habitats or become prey for fish or other aquatic predators. Allochthonous invertebrates and detritus fall or wash in from adjoining riparian patches and fuel the stream ecosystem (J. Allan et al., 2003; Baxter et al., 2005; S. Eggert and Wallace, 2003; Wipfli and Musslewhite, 2004). Aquatic insect adults emerge and move upstream and over intervening terrestrial landscapes to colonize new pools and riffles (Baxter et al., 2005). Anadromous species move nutrients back upstream to spawning sites (Kikkert et al., 2009; J. Walter et al., 2006). Terrestrial predators and floods transport material from aquatic to terrestrial habitats.

The configuration and structure of the network change over time. Storms raise water level and flow rate, altering channel morphology, washing in terrestrial debris, tumbling stones, moving boulders and logs, dislodging aquatic invertebrates, and depositing fine sediments and nutrients in the floodplain. Terrestrial disturbances change riparian vegetation structure, altering insolation, soil water retention, sedimentation, and abundance and composition of terrestrial inputs to the stream (J. Allan et al., 2003). Severe disturbances (earthquakes, landslides, and extreme storms) that alter stream configuration, geomorphology, and/or detrital structure alter the distribution of

Insect Ecology. http://dx.doi.org/10.1016/B978-0-12-803033-2.00009-1

substrate and resources on which aquatic communities can develop. The characteristic sequence of invertebrate communities may redevelop on the new continuum of habitat types or remain in flux, depending on the frequency of disturbance (Reice, 1985). Artificial dams may disrupt processes that link upstream and downstream communities and ensure adequate dispersal and recolonization (Kikkert et al., 2009; Pringle, 1997; Pringle et al., 2000), thereby threatening the integrity of aquatic communities.

1 INTRODUCTION

A community is composed of all the organisms occupying a site. The extent to which these organisms are coevolved to form a consistent and recurring integrated community or represent ad hoc assemblages of loosely interacting species remains a topic of much discussion. Considerable research has been directed toward identifying spatial and temporal patterns in community structure and evaluating factors that determine community composition and development. Such efforts have become increasingly important to conservation efforts, with recognition that many species depend on the presence of associated species with which they interact. However, comparison of community structures within, or among, broadly distributed community types that share few, if any, species requires approaches that are independent of the taxonomic composition of the community.

Ecologists have developed a variety of nontaxonomic metrics for describing community structure, providing different types of information to meet different objectives. The diversity of approaches has hindered comparison of communities described in different terms. Nevertheless, distinct geographic patterns can be seen in community structure, and some community types characterize particular habitat conditions. A number of factors determine community composition, distribution, and dynamics. This chapter focuses on approaches to describing community structure, and on biogeographic patterns and underlying factors contributing to community structure. Temporal patterns in community structure are the focus of the next chapter.

2 APPROACHES TO DESCRIBING COMMUNITIES

Although the community is understood to include all organisms at a site, few studies have attempted to describe entire communities. For example, Basset et al. (2007) described an intensive project in Panama that surveyed arthropods in 12 tropical rainforest plots from the ground to the upper canopy using a variety of canopy access techniques and 14 sampling methods for focal groups in soil, litter, understory, and mid- and upper-canopy habitats. Approximately 500,000 arthropods belonging to about 5500 species have been collected from 9400 samples and 315 plant species.

As a result of such diversity, most studies have been limited to subsets of the community, especially plants and vertebrates, but these subsets often are described as the "plant community," "arthropod community," "bird community," or communities associated with different plant species, tritrophic interactions, etc. These subsets are more appropriately termed "assemblages" or component communities. Insects have been addressed to varying degrees in studies of communities, although insects represent the vast majority of species in terrestrial and freshwater communities (Table 9.1) and clearly are integral to community structure and dynamics, for example, as pollinators, herbivores, detritivores, resources for vertebrate predators, etc. (see Chapters 11–14).

Three general approaches to describing community structure can be identified: species diversity, species interactions, and functional organization. Although the "ideal" approach is a topic of intense ecological debate (Polis, 1991a), each approach provides useful information, and the choice largely reflects objectives and practical considerations. Where possible, combining approaches can yield a broader description of the community.

2.1 Species Diversity

Species diversity is a central theme in ecology (Hutchinson, 1959). An enormous amount of research has addressed how diversity develops under different environmental conditions, how anthropogenic changes are affecting diversity, and how diversity affects the stability of natural communities and ecosystems (see Chapters 10 and 15). Global diversity of arthropods currently is estimated at 4–6 million species, with most species in the tropics (Novotný et al., 2002). In most ecosystems where diversity of insect or arthropod species has been inventoried, along with plants and vertebrates, arthropods account for 70–90% of the total number of recorded species (Table 9.1), roughly the same proportion as for the total number of described species of organisms. Given that plant and vertebrate inventories are relatively complete, whereas currently described insect species represent only a fraction of the estimated total number of species (May, 1988; Sharkey, 2001; E. Wilson, 1992), the proportional representation of invertebrates likely will increase. Clearly, the measurement of insect diversity is fundamental to understanding patterns of diversity among ecosystems, how diversity is affected by environmental changes, and how diversity affects community and ecosystem development and stability.

Species diversity can be represented in various ways (Magurran, 2004). The simplest representation is a catalog, or checklist, of the total number of species (*richness*), a measure that indicates the variety of species in a community (*α diversity*). Rarely can all species be detected and documented. Typically, the number of species recorded increases with the number of samples collected. The total number of species can be estimated by extrapolating from a species abundance curve that plots cumulative species against cumulative sample number or area. Alternatively, several parametric and nonparametric estimators can be used (Magurran, 2004). Species richness can be standardized for various ecosystems by measuring the number of species per unit area or per 1000 individuals. This measure of diversity accounts for the typical increase in number of species with increasing sample area or number of sampled individuals.

Species richness for many plant and animal groups increases from high latitudes to lower latitudes and from smaller, more isolated islands to larger islands near continents (R. Dunn et al., 2009; MacArthur and Wilson, 1967; Magurran, 2004; Stiling, 1996). Richness also increases from harsh or frequently disturbed ecosystems that restrict richness to more productive ecosystems that provide a greater number of niches, but typically declines again in very productive ecosystems (Tilman and Pacala, 1993). Species diversity appears generally to peak at intermediate

TABLE 9.1 Numbers of Species of Vascular Plants, Vertebrates, and Arthropods in Desert, Grassland, Forest, and Marshland Ecosystems

Ecosystem	Vascular plants	Vertebrates	Arthropods	% Arthropods
Desert				
USSR[a]	125	198	>1360	75
Southwestern US I[a]	174	145	>1100	77
Southwestern US 2[a]	>600	201	>2640	77
Southwestern US 3[b]	>1000	576	4600	74
Grassland/Savanna				
Hungary[c]	1311	347	8496	93
Hungary[d]	1762	289	7095	78
Central US[e]	521	355	>1750	67
Forest				
Conifer/wetland[f]	536	380	>6500	88
Conifer, western US[g]	600	88	>3500	84
Deciduous, eastern US[h]	2816	450	>4300	57
Tropical, Puerto Rico[i]	470	78	>1500	73
Tropical, Costa Rica[j]	7000	1000	100,000	93
Marsh				
Hungary[k]	804	118	5332	85
Stream				
Tropical, Puerto Rico[l]	0	7	50	88
Lake				
Balaton, Hungary[m]	>9	51	>1200	95

[a] Data from Pollis (1991b)
[b] Data from Van Pelt (2002) and Wauer (1980)
[c] Data from Mahunka (1986, 1987) and Szujko-Lacza and Kovacs (1993)
[d] Data from Mahunka (1981, 1983) and Szujko-Lacza (1982)
[e] Data from Hazlett (1998) and Lavigne et al. (1991); arthropod data are for insects only
[f] Data from Kaiser (2005), Mannville (1942), Procter (1946), and Rand and Redfield (1894)
[g] Data from Parsons et al. (1991)
[h] Data from Sharkey (2001)
[i] Data from Garrison and Willig (1996), Lawrence (1996), and Reagan et al. (1996)
[j] Data from Janzen (2000, 2004)
[k] Data from Mahunka (1991)
[l] Data from Covich and McDowell (1996); no vascular plants represented in this headwater stream
[m] Data from Benedek (1988)

levels of disturbance (the *Intermediate Disturbance Hypothesis*) due to a combination of sufficient resources and insufficient time for competitive exclusion (Connell, 1978; Huston, 1979; Lubchenco, 1978; Pickett and White, 1985; Sousa, 1979). Insect diversity may reflect primarily the diversity of plants, which affects diversity of host resources and habitat structure (Curry, 1994; Lewinsohn and Roslin, 2008; Magurran, 2004; Novotný et al., 2006; Stiling, 1996).

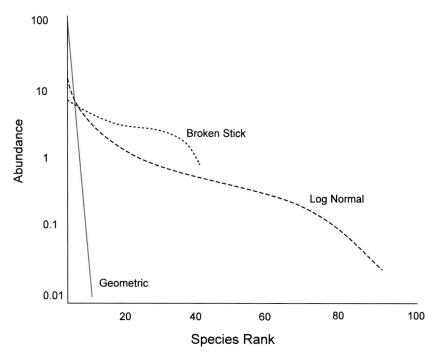

FIGURE 9.1 **Typical shapes of three rank-abundance models.** Species are ranked from most to least abundant. *Source: Modified from Whittaker, R.H., 1970. Communities and Ecosystems. Macmillan, London.*

The various species in a community are not equally abundant. Typically, relatively few species are abundant and many species are represented by only one or a few individuals. The distribution of numbers of individuals among species (*evenness*) is one measure of each species' importance. *Rank-abundance curves* are a commonly used method for presenting species abundance data (Magurran, 2004). Three rank-abundance patterns are most commonly used for comparison among different communities (Fig. 9.1). The *geometric model* (or *niche-preemption hypothesis*) describes a community in which successively less abundant species use the same proportion of resources available after preemption by the more abundant species. This situation is predicted to occur when species arrive in an unsaturated community at regular time intervals and exploit a fraction of the remaining resources. The *log normal model* has been shown to be widely applicable, because this distribution results mathematically from random variation among a large number of factors producing a normal distribution. In natural communities, the large number of environmental factors affecting species abundances fulfills this condition. This condition also can be met by increasing numbers of species randomly partitioning available niches (see Chapter 8). The *broken stick model* reflects relatively uniform use of resources among species in the community.

Generally, as richness and evenness increase, the rank-abundance pattern shifts from a geometric pattern to a log pattern and finally to a broken stick pattern. Disturbances and other environmental changes can alter rank-abundance patterns (Fig. 9.2) (Bazzaz, 1975; Kempton, 1979). For example, continuous fertilization of permanent pasture at Rothamsted,

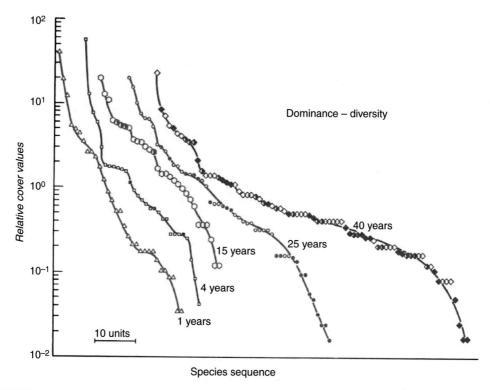

FIGURE 9.2 **Rank-abundance curves for old fields representing five postabandonment ages in southern Illinois.** Open symbols are herbs, half-open symbols are shrubs, and closed symbols are trees. *Source: From Bazzaz (1975) with permission of the Ecological Society of America.*

UK since 1856 resulted in changes in the species rank-abundance pattern from a log normal curve in 1856 to progressively more geometric curves by 1949 (Kempton, 1979).

Richness and evenness have been combined mathematically in various ways to calculate diversity indices based on proportional abundances of species (Magurran, 2004; Stiling, 1996). Two indices have been used widely, the *Shannon-Wiener*, or Shannon (often incorrectly referred to as the Shannon-Weaver), index and *Simpson's* index. The two indices differ in their emphasis on species richness (Shannon-Wiener) or abundance (Simpson's).

The Shannon-Wiener index assumes that individuals are randomly sampled from an effectively infinite population and that all species are represented in the sample. Diversity (H') is calculated as

$$H' = -\sum_{i=1}^{n} p_i \ln p_i \tag{9.1}$$

where p_i is the proportion of individuals found in the ith species. Values generally fall in the range 1.5–3.5, rarely surpassing 4.5. If the rank-abundance pattern follows a log normal model, 10^5 species are necessary to produce a value of $H' > 5$. If the index is calculated for a number of samples, the indices will be normally distributed and amenable to the use of parametric statistics, including ANOVA, to compare diversities among sets of samples (Magurran, 2004), for example, to evaluate the effects of ecosystem change (Fukami et al., 2001). If all species were equally abundant, a maximum diversity (H_{max}) can be calculated as ln S, where S is the total number of species. The ratio of observed to maximum diversity is a measure of evenness.

When randomness cannot be assured (eg, data from light trapping, with species representation based on differential attraction to light), the *Brillouin* index is a more appropriate measure of diversity (Magurran, 2004). This index (HB) is calculated as

$$HB = (\ln N! - \sum \ln n_i!)/N \tag{9.2}$$

where N is the total number of individuals, and n_i is the number of individuals in the ith species. Values of this index rarely exceed 4.5 and generally are correlated with, but lower than, Shannon indices for the same data.

Simpson's index differs from the Shannon-Wiener and Brillouin indices in being weighted toward the abundances of the most common species, rather than species richness (Magurran, 2004). This index (D) is calculated as

$$D = \sum_{i=1}^{n} (n_i(n_i - 1))/(N(N - 1)) \tag{9.3}$$

where n_i is the number of individuals in the ith species, and N is the total number of individuals. As diversity increases, D is decreased so Simpson's index generally is expressed as $1 - D$ or $1/D$. Once the number of species exceeds 10, the underlying rank-abundance pattern is important in determining the value of D.

Diversity indices have been a tool for comparing taxonomically distinct communities, based on their rank-abundance patterns. However, important information is lost when species diversities are reduced to an index (Magurran, 2004). For example, a larger diversity index can reflect the influence of increased abundances of invasive or exotic species without conveying important information about the change in community integrity or function. Very different community structures can produce the same diversity index. Furthermore, ecologically unique communities are not necessarily diverse and would be lost if conservation decisions were made on the basis of diversity alone (Magurran, 2004).

The large number of species represented by single individuals ("singletons") poses a dilemma (Novotný and Basset, 2000). Should these be included in diversity calculations, or not? Their presence may be accidental or reflect inadequate or biased sampling. Novotný and Basset (2000) found that singletons consistently represented 45% of herbivores sampled among plant species. Some singletons represented species that were more common on unsampled plant species, whereas others represented species that were relatively rare on numerous host plants. Novotný and Basset (2000) concluded that singletons are an important component of communities and should not be excluded from community studies as an artifact or a group of negligible importance.

Diversity also can be measured as variation in species composition among communities. *β diversity* reflects the extent of species sharing or replacement over landscapes or regions, whereas *γ diversity* is the total diversity of a landscape or region (Magurran, 2004; Mori et al., 2015). A landscape with distinct assemblages of species, with no overlap in species representation, will have higher β diversity than a landscape with the same total number of species but with many species shared among assemblages. Several techniques have been developed to compare communities, based on their species compositions and rank-abundance patterns, across environmental gradients or between areas (Magurran, 2004).

The simplest of these similarity measures are indices based on species presence or absence in the communities being compared. The *Jaccard* index (C_J) is calculated as

$$C_J = j/(a + b - j) \tag{9.4}$$

and the *Sorenson* index (C_S) as

$$C_S = 2j/(a + b) \tag{9.5}$$

where j is the number of species found in both sites, a is the number of species in the first site, and b is the number of species in the second site. Neither of these indices accounts for species abundances.

Three quantitative similarity indices have been used widely. A modified version of the Sorenson index (C_N) is calculated as

$$C_N = 2jN/(N_a + N_b) \tag{9.6}$$

where $2jN$ is the sum of the lower of the two abundances for each species found in both sites, N_a is the total number of individuals in the first site, and N_b is the total number of individuals in the second site. Most quantitative similarity indices are influenced strongly by species richness and sample size. The *Morisita–Horn* index (C_{mH}) is influenced less by species richness and sample size but is sensitive to the abundance of the dominant species. Nevertheless, it generally may be a satisfactory similarity index (Magurran, 2004). This index is calculated as

$$C_{mH} = 2\Sigma(an_i bn_i)/(da + db)aN \cdot bN \tag{9.7}$$

where aN is the total number of individuals in the first site, an_i is the number of individuals of the ith species in the first site, and $da = \Sigma an_i^2 / aN^2$. The *Bray–Curtis Similarity* index also has been shown to be effective and robust (Minchin, 1987). This index is calculated as

$$C_B = 1 - \sum_{i=1}^{N}|X_{ij} - X_{ik}| \Big/ \sum_{i=1}^{n}(X_{ij} - X_{ik}) \tag{9.8}$$

where n is the number of species, and X_{ij} and X_{ik} are the number of individuals of the ith species at sites j and k, respectively (Cartron et al., 2003).

More recently, multivariate statistical techniques have been applied to comparison of communities. *Cluster analysis* can be performed using either presence-absence or quantitative data. Each pair of sites is evaluated on the degree of similarity, then combined sequentially into clusters to form a dendrogram with the branching point representing the measure of similarity (Figs. 9.3 and 9.4). *Ordination* compares sites on their degree of similarity, then plots them in Euclidian space, with the distance between points representing their degree of similarity (Figs. 9.5, 9.6). Ordination techniques include principal components analysis (PCA), detrended correspondence analyses (DCA), and nonmetric multidimensional scaling (NMS).

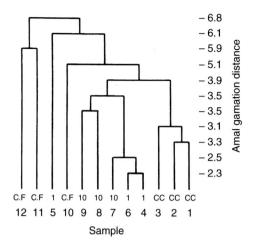

FIGURE 9.3 Dendrogram of similarity for dung beetles in clear-cuts (CC), 1 ha (1) and 10 ha (10) forest fragments, and contiguous forest (CF). *Source: From Klein, B.C., 1989. Effects of forest fragmentation on dung and carrion beetle communities in central Amazonia. Ecology 70, 1715–1725.*

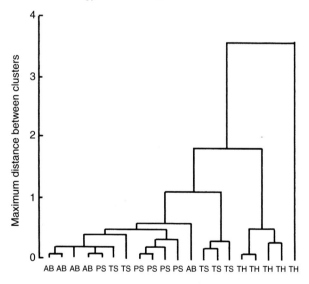

FIGURE 9.4 Dendrogram of arthropod community similarity in canopies of four old-growth conifer species at the Wind River Canopy Crane Research Facility in southwestern Washington, USA. *AB, Abies grandis* (grand fir); *PS, Pseudotsuga menziesii* (Douglas-fir); *TS, Tsuga heterophylla* (western hemlock); and *TH, Thuja plicata* (western redcedar). *Source: Data from Schowalter, T.D., Ganio, L.M., 1998. Vertical and seasonal variation in canopy arthropod communities in an old-growth conifer forest in southwestern Washington, USA. B. Entomol. Res. 88, 633–640.*

Minchin (1987) evaluated several commonly used ordination techniques for sensitivity to sampling pattern, data distribution, and geometric distortion. PCA and principal coordinates analysis both suffered from curvilinear distortion, and DCA lacked robustness to variation in sampling pattern and response model. NMS was shown to be the most robust ordination method and has become widely used in ecological studies.

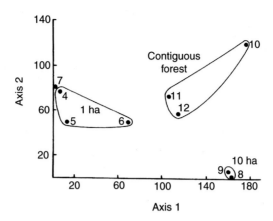

FIGURE 9.5 Detrended Correspondence Analysis ordination of dung beetle assemblages in clear-cuts, 1 ha and 10 ha forest fragments, and contiguous forest. *Source: From Klein, B.C., 1989. Effects of forest fragmentation on dung and carrion beetle communities in central Amazonia. Ecology 70, 1715–1725.*

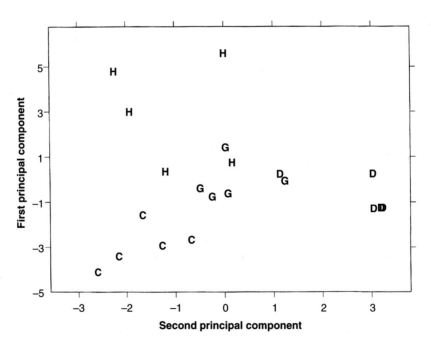

FIGURE 9.6 Principal Components Analysis ordination of arthropod communities in canopies of four old-growth conifer species at the Wind River Canopy Crane Research Facility in southwestern Washington, USA. *G*, grand fir (*Abies grandis*); *D*, Douglas-fir (*Pseudotsuga menziesii*); *H*, western hemlock (*Tsuga heterophylla*); and *C*, western redcedar (*Thuja plicata*). *Source: Data from Schowalter, T.D., Ganio, L.M., 1998. Vertical and seasonal variation in canopy arthropod communities in an old-growth conifer forest in southwestern Washington, USA. B. Entomol. Res. 88, 633–640.*

Both cluster and ordination techniques can indicate which species or environmental factors contribute most to the discrimination of groupings. *Indicator species analysis* (Dufrêne and Legendre, 1997) is another method that can be used to identify species or groups of species that characterize groups of sites, based on ecological gradients or treatments, by combining the frequency of a species occurrence in a particular site category and its degree of restriction to that site category. Dufrêne and Legendre (1997) compared this method with clustering and ordination techniques to identify carabid beetle species characterizing combinations of soil moisture and alkalinity represented by 69 sites in Belgium.

The significance of differences among groups of points representing sites, treatments, etc., can be analyzed using *multiple response permutation procedures* (MRPP) (Biondini et al., 1988). This method measures the separation among weighted means of points in *a priori* groups and tests the probability of occurrence of this mean relative to other possible separations with the same size structure that could have occurred for these points (Biondini et al., 1988).

2.2 Species Interactions

Communities can be characterized in terms of the relationships among species, most commonly trophic (feeding) interactions, that is, *food webs*. Clearly, the most complete description of the community would include all possible interactions (including indirect interactions) among the total number of species (Dyer et al., 2010; Polis, 1991a). In practice, this is difficult to accomplish, even in relatively species-poor communities (Camilo and Willig, 1995; Polis, 1991a, 1991b; Reagan et al., 1996), because of the largely unmanageable number of arthropod species (Table 9.1) and lack of complete information on their interactions. More commonly, research focuses on subsets or simplified representations of the community (eg, Figs. 9.7 and 9.8).

The simplest approach to community description emphasizes interactions between only a few species, for example, plant–herbivore or predator–prey interactions. In particular, many studies have addressed the relatively distinct assemblages of arthropods based on individual plant species (Richerson and Boldt, 1995; Schowalter and Ganio, 1998) or soil/litter resources (Moore and Hunt, 1988; Seastedt et al., 1989). This approach maximizes description of interactions among a manageable number of relatively resource-specific herbivores or detritivores and their associated predators and parasites. Detailed descriptions at this level have been useful for identifying and comparing factors affecting these trophic interactions (eg, chemical defenses, see Chapters 3 and 8), for evaluating the coevolutionary patterns of speciation between insects and their hosts (Becerra, 1997), and for comparing trophic interactions among community types, for example, comparing phenological responses of insect herbivores to leaf emergence in tropical and temperate forests (Coley and Aide, 1991). However, this approach emphasizes relatively linear trophic relationships (ie, *food chains*) and does not address linkages among members of different component communities (see section 2.3).

Broader subcommunities can be identified. For example, Hunt et al. (1987) described the trophic interactions among arthropod and microbial species composing the litter subcommunity of a grassland ecosystem. Moore and Hunt (1988) subsequently noted that relatively discrete component communities supported by particular resource bases (bacteria, fungi, or plant roots) could be distinguished within this broader subcommunity (Table 9.2). Similarly, individual

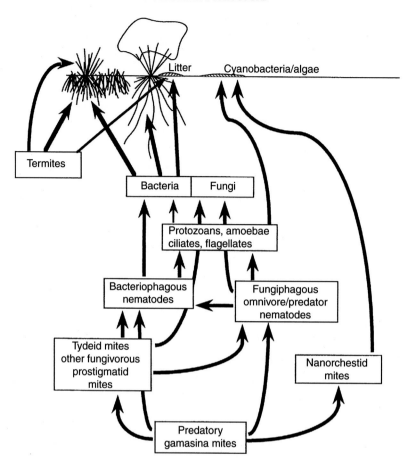

FIGURE 9.7 **A desert soil food web.** Note that each box includes many interacting species and that this food web interacts with the above-ground food web. *Source: From Whitford, W.G., 2002. Ecology of Desert Systems. Elsevier/ Academic Press, San Diego, CA.*

plant species represent resource bases for relatively discrete component communities of associated arthropods and other organisms in the above-ground subcommunity (Curry, 1994). Resource-based component communities are linked to each other by generalist herbivores and predators. The canopy and soil/litter subcommunities are linked by species that feed aboveground but pupate in the soil or that feed on litter resources but disperse and bask on foliage, and by predators and detritivores that move among substrates in search of resources.

The most inclusive approach to community description is represented by interaction webs, in which all species are connected by arrows indicating interactions. Relatively few communities are composed of sufficiently few species to depict all interactions conveniently. Hot springs and other communities subject to extreme abiotic conditions typically are composed of a few tolerant algal and invertebrate species (N.C. Collins et al., 1976). Communities composed of relatively few invertebrate and vertebrate species characterize some aquatic ecosystems, for example, vernal pools, riffles, etc. However, even the desert communities described

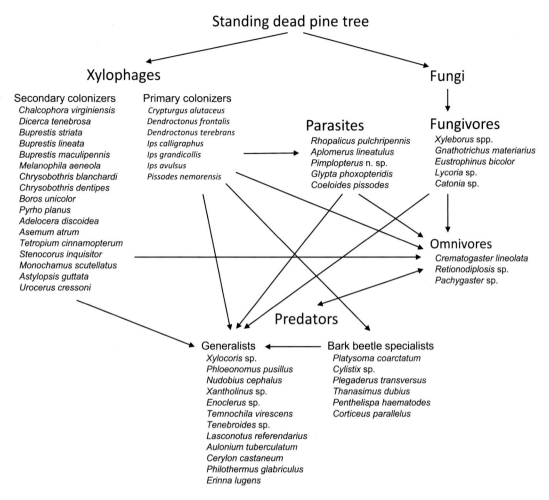

FIGURE 9.8 **Food web characteristic of the early stages of decomposition of standing dead pine,** *Pinus virginiana,* **trees.** *Source: Modified from Howden, H.F., Vogt, G.B., 1951. Insect communities of standing dead pine (Pinus virginiana Mill.). Ann. Entomol. Soc. Am. 44, 581–595.*

by Polis (1991a) were composed of >1000 arthropod species, most of which had not been studied sufficiently to provide complete information on interactions. A number of studies have addressed trophic interactions, that is, food webs, although even trophic interactions are poorly known for many species, especially insects.

A number of techniques have been used to identify trophic relationships. Early studies of food web structure tracked radioisotopes through trophic exchanges (Crossley and Howden, 1961). Stable isotopes or other tracers also can be tracked through feeding exchange (Christenson et al., 2002; D. Wise et al., 2006). Furthermore, animal tissues reflect the stable isotope ratios of their diet, with slight enrichment of ^{15}N with increasing trophic level (Blüthgen et al., 2003; Ponsard and Arditi, 2000; Scheu and Falca, 2000; Tayasu et al., 1997). However, interpretation of trophic interactions depends on the isotopic homogeneity of the

TABLE 9.2 The Proportion of Energy and Nitrogen Derived from the Bacteria, Fungal, and Root (Including Mycorrhizal Fungi) Resource Channels by Different Faunal Groups in the North American Shortgrass Steppe

Faunal group	Resource channel		
	Bacteria	Fungi	Roots
Protozoa			
Flagellates	100	0	0
Amoebae	100	0	0
Ciliates	100	0	0
Nematodes			
Bacteriovores	100	0	0
Fungivores	0	90	10
Root-feeders	0	0	100
Omnivores	100	0	0
Predators	69	3	28
Microarthropods			
Mycophagous Collembola	0	90	10
Mycophagous oribatid mites	0	90	10
Mycophagous prostigmatid mites	0	90	10
Nematophagous mites	67	4	30
Predaceous mites	40	39	21

From Moore, J.C., Hunt, H.W., 1988. Resource compartmentation and the stability of real ecosystems. Nature 333, 261–263.

diet (Gannes et al., 1997). Selective feeding on particular substrates or inefficient assimilation can affect ^{13}C or ^{15}N enrichment in animals (Dodds et al., 2014; Šantrůčková et al., 2000). T. Adams and Sterner (2000) reported that ^{15}N enrichment was linearly related to dietary C:N ratio, which could vary sufficiently to indicate as much as a 2-trophic level separation, potentially leading to misidentification of trophic level for particular species. More recently, Steffan et al. (2013) reported that using ^{15}N-labeled amino acids provided unambiguous identification of trophic level, up to five levels, an unprecedented degree of discrimination.

Advances in molecular techniques have provided new tools for identifying interactions among species in communities. Enzyme-linked immunosorbent assay (ELISA) techniques involve development of antibodies against enzymes from potential food sources. These antibodies can be used to precipitate enzymes in gut samples containing the target food source. Irby and Apperson (1988) and Savage et al. (1993) used ELISA to identify associations between various mosquito species and their particular amphibian, reptile, bird, and mammal hosts. Agustí et al. (1999a) demonstrated the utility of this technique for detecting prey, *Helicoverpa armigera*, in hemipteran, *Dicyphus tamaninii*, *Macrolophus caliginosus*, and *Orius majusculus*, predator gut contents. More recently, polymerase chain reaction (PCR) and DNA amplification techniques have been used to illuminate feeding relationships. Schoeller et al. (2012) used gut DNA analysis to demonstrate that 26 (10%) of 271 field-collected pine sawyer,

Monochamus titillator, larvae tested positive for DNA of (ie, had eaten) members of the southern pine beetle guild; 25 larvae were positive for *Ips grandicollis* and one was positive for *I. calligraphus.* Broderick et al. (2004) used this methodology to describe the microbial community in gypsy moth, *Lymantria dispar,* midgut and to demonstrate that bacterial composition was influenced by the plant species composition of the diet. PCR and DNA amplification can be used to identify prey species in gut contents for up to 28 hr after predator feeding, depending on predator hunger level and prior feeding (Agustí et al., 1999b; Y. Chen et al., 2000; Hoogendoorn and Heimpel, 2001; Schoeller et al., 2012; D. Weber and Lundgren, 2009; Zaidi et al., 1999). Although these techniques can help identify feeding relationships, developing the sequence library to distinguish all potential prey in the field presents a challenge.

More recently, S. Hall (2009) described a stoichiometrically explicit approach to food web modeling. This approach recognizes the importance of imbalances between nutrient composition required by each member of the food web and the nutrient composition available in food resources (see Chapter 3). These imbalances shape the effect of consumers on their own resources through nutrient cycling (see Chapters 11–15) and link food web structure to ecosystem function.

Several properties have appeared to characterize food webs (see Briand and Cohen, 1984; Cohen and Palka, 1990; Cohen et al., 1990; N. Martinez, 1992; May, 1983; Pimm, 1980, 1982; Pimm and Kitching, 1987; Pimm and Lawton, 1977, 1980; Pimm and Rice, 1987; Pimm et al., 1991; Polis, 1991b; Reagan et al., 1996). However, food web analysis typically has been based on combination of all insects (often all arthropods) into a single category, in contrast to resolution at the individual species level for plants and vertebrates. Nevertheless, even relatively simple arthropod communities, such as the subcortical community in dead trees (Fig. 9.8) (Howden and Vogt, 1951), pollinator–plant communities (Burkle et al., 2013), or the early successional posteruption community on Mt. St. Helens (Fagan, 1997), contain dozens of insect and other arthropod species representing multiple trophic levels interacting relatively independently of vertebrate predators.

Polis (1991b) and Reagan et al. (1996) increased the resolution of arthropod diversity to individual "kinds," based on taxonomy and similar phylogeny or trophic relationships, for evaluation of food web structure in desert and tropical rainforest communities, respectively. They found that the structure of their food webs differed from that of food webs in which arthropods were combined. Goldwasser and Roughgarden (1997) analyzed the effect of taxonomic resolution on food web structure and found that food web properties reflected the degree of taxonomic resolution. The following proposed properties of food webs, based on analyses with insects or arthropods as a single category, are evaluated with respect to challenges based on greater resolution of arthropod diversity.

2.2.1 Food Chain Length

Early analyses indicated that the length of food chains within food webs should be relatively short, at most 3–5 links (May, 1983; Pimm and Kitching, 1987; Pimm and Lawton, 1977), because the laws of thermodynamics predict energy limitation at higher trophic levels. Therefore, energy gain should be maximized by feeding lower on the food chain. At the same time, competition for prey is most severe at lower levels, perhaps restricting energy gains. Consequently, the trophic level selected by predators represents a trade-off between maximizing energy availability and minimizing competition. However, Polis (1991b) and Reagan et al. (1996) found chain lengths of

6–19 links, using food webs with greater resolution of arthropod taxonomy. Reagan et al. (1996) reported a mean chain length of 8.6, double the length of chains found when arthropods are combined into a single category. This likely reflects greater efficiency of secondary production among insects, compared to homeothermic organisms (see Chapter 4).

2.2.2 Trophic Loops and Intraguild Predation

Loops, or reciprocal predation, in which two species feed on each other or a third species feeds on one and is eaten by the other, should be rare or absent because the size range of prey is constrained by physical limits and because loops potentially reduce population recovery following disturbance (Pimm, 1982; Pimm and Rice, 1987). Intraguild predation involves predation among members of the same trophic level (Müller and Brodeur, 2002). Cannibalism is considered a "self-loop" (see Fox, 1975a).

Polis (1991b) and Reagan et al. (1996) reported the occurrence of a substantial number of loops, especially involving arthropods. In most cases, each species in the loop preys on juveniles of the other species. For example, in a tropical forest in Puerto Rico, adult centipedes prey on young frogs, whereas adult frogs prey on young centipedes. Polis (1991b) reported that several species of desert ants regularly prey on each other. Other predators constituted 9% of the overall diet of the aquatic hemipteran, *Notonecta hoffmanni*, studied by Fox (1975b). Longer loops involving up to four species have been observed (Reagan et al., 1996). Reagan et al. (1996) found that 35% of 19,800 observed chains (corrected to exclude loops) included at least one species involved in at least one loop.

Intraguild predation appears to be pervasive within arthropod food webs, with frequencies of 58–87% within trophic levels (Arim and Marquert, 2004; Gardiner et al., 2011; M.D. Hunter, 2009; Müller and Brodeur, 2002) complicating measurement of food chain length and explaining irruptions of prey species when multiple predator species are present. However, the extent of intraguild predation is influenced by prey and predator densities, habitat complexity, and the particular predator species involved (Finke and Denno, 2006; S. Moser and Obrycki, 2009; D. Sanders et al., 2011; O. Schmitz, 2007). Furthermore, habitat complexity can reduce intraguild predation by providing refuges for multiple predators (Finke and Denno, 2002, 2006).

A number of studies have demonstrated significant reduction in predator abundances and ability to regulate herbivore populations as a result of intraguild predation (Denno et al., 2004; Erbilgin et al., 2004; Finke and Denno, 2002, 2006; Mooney, 2007; Pérez-Lachaud et al., 2004; Rosenheim, 2005; D. Sanders and Platner, 2007; D. Sanders et al., 2011). Rosenheim (2005) used manipulative experiments to show that abundance of the anthocorid bug, *Orius tristicolor*, was significantly reduced as a result of intraguild predation by big-eyed bugs, *Geocoris* spp., and lacewing, *Chrysoperla* sp., larvae in cotton fields in California, USA, interfering with top-down control of spider mite, *Tetranychus* spp., prey.

Spiders and ants commonly compete for prey and also prey on each other (Halaj et al., 1997; D. Sanders and Platner, 2007; D. Sanders et al., 2011). D. Sanders and Platner (2007) used a factorial experimental design to exclude ants and/or spiders from grassland plots and measured changes in abundances of ants, spiders, and other arthropods. Exclusion of either ants or spiders resulted in increased abundance of the other. Ant exclusion increased densities of Lepidoptera larvae but reduced densities of scale insects. Spider exclusion increased densities of Collembola. Increased ^{15}N enrichment of adult spiders, relative to juveniles, indicated greater intraguild predation by adults.

2.2.3 Food Web Connectance

Community connectance, the proportion of potential feeding relationships that actually occur in the community (Pimm, 1982), should increase with increasing species richness as

$$L = 0.14S^2 \tag{9.9}$$

where L is the number of links and S is the number of species (N. Martinez, 1992). This *constant connectivity hypothesis* predicts that, on average, each species will be involved in predator–prey interactions with 14% of the other species in the community. Havens (1992) analyzed 50 pelagic food webs with species richness ranging from 10 to 74 and found that the number of links per species increased 4-fold over this range. Reagan et al. (1996) reported that the food web in a tropical forest in Puerto Rico supported constant connectance at low taxonomic resolution but that connectance dropped quickly as taxonomic resolution was increased. Polis (1991b) and Reagan et al. (1996) also found that the prediction that each species interacts with only 2–5 other species greatly underestimates the actual number of linkages per species and concluded that these properties are sensitive to taxonomic resolution.

2.2.4 Food Web Compartmentalization

Pimm and Lawton (1980) proposed that food webs should be compartmentalized between, but not within, habitats. Whereas the relatively distinct communities representing disturbed versus undisturbed patches within an ecosystem represent compartmentalization, the communities within habitat patches should not be compartmentalized. This property largely follows from the constant connectivity hypothesis, that is, compartmentalization is inconsistent with equal linkage among species.

The vague definition of habitat complicates assessment of compartmentalization. For example, does soil/litter constitute a habitat or a subunit of the site habitat? Soil/litter subcommunities tend to be relatively distinct from plant-based above-ground subcommunities but are, nevertheless, connected by predators that feed on both soil-based and plant-based prey (Wolkovich et al., 2014).

Nevertheless, compartmentalization can be identified within recognized habitats. J.C. Moore and Hunt (1988), Polis (1991b), and Reagan et al. (1996) found distinct compartmentalization within the community of a single patch when arthropod species or "kinds" were distinguished (Table 9.2). Distinct compartmentalization of arthropod assemblages has been shown among plant species (eg, Fig. 9.6) and even between trees and saplings of the same species (Basset, 2001). Fetcher et al. (2015) used ^{13}C and ^{15}N to demonstrate food web compartmentalization between C_3 and C_4 grasses in a restored grassland at a site contaminated by heavy metals from a mining operation. Compartmentalization reflects the development of component communities composed of specialists feeding on particular resources and the resulting channels of energy and material transfer. Host specificity appears to occur more frequently and at a finer spatial scale among herbivorous and detritivorous arthropods (compared to vertebrates), based on their small size, short life spans, and intricate biochemical interactions (see Chapter 3) that facilitate rapid adaptation for utilization of particular resources, even within individual leaves (Mopper and Strauss, 1998; K. Parsons and de la Cruz, 1980). Many parasitoids also are host specific, but predators feeding on prey from multiple channels may mask compartmentalization at higher trophic levels (Fetcher et al., 2015). J.C. Moore and Hunt (1988) found that compartmentalized models of food webs were more stable than noncompartmentalized webs.

Generalists at all trophic levels connect compartments and link food webs among habitats. For example, emergent stream insects enter terrestrial food webs, and riparian insects falling into streams enter the aquatic food web (Baxter et al., 2005; Burdon and Harding, 2008, see Fig. 1.4). McCann et al. (2005) employed spatially implicit food web models to demonstrate that highly mobile predators connect food webs with important effects on food web stability. They argued that mobile predators can stabilize food webs in variable and expansive spatial structures, but destabilize food webs when space is confined and predators more strongly couple local habitats.

2.2.5 Omnivory

Omnivores (defined as species feeding on more than one trophic level) should be rare (Pimm, 1982; Pimm and Rice, 1987). Pimm and Rice (1987) concluded that omnivory should reduce the stability of food web interactions. However, as noted previously (section 3.2), a number of studies have demonstrated intraguild predation (Arim and Marquert, 2004; Denno et al., 2004; Erbilgin et al., 2004; M.D. Hunter, 2009; Pérez-Lachaud et al., 2004; Rosenheim, 2005). Some herbivores or detritivores consume competitors when encountered, and some predators feed on plant materials and other predators, as well as on herbivores (M.D. Hunter, 2009). Polis (1991b) and Reagan et al. (1996) reported that omnivory is common in food webs when arthropods are resolved to species or "kinds." In fact, they found that most species fed at more than one trophic level, often from nonadjacent trophic levels, in desert and tropical rainforest communities.

Fagan (1997) tested the effect of omnivory on stability of community structure by manipulating the degree of omnivory (excluding either a specialist predator, the nabid bug, *Nabis alternatus*, or an omnivorous predator, wolf spiders, *Pardosa* spp.) in replicated plots, then disturbing the community by applying aphicide to disrupt prey, *Macrosiphum valerianae*, abundance. Plots with high levels of omnivory showed significantly reduced responses to disturbance for seven of 14 species, compared to plots with low levels of omnivory; no species showed significantly increased responses to disturbance. These data indicated that omnivory increased the stability of food web interactions.

2.2.6 Ratio of Basal to Top Species

Finally, ratios of species and links from basal to intermediate to top trophic levels (where basal species are prey only, intermediate species are prey and predators, and top predators have no predators) are expected to be constant (Briand and Cohen, 1984). This implies a large proportion of top predators. Top predators are expected to comprise 29% of all species in a given community, and prey to predator ratios should be < 1.0 (Briand and Cohen, 1984).

As shown for the properties discussed earlier, this property reflects poor resolution of arthropod diversity. Top predators appear to be common because they are easily distinguished vertebrate species, whereas poor taxonomic resolution underrepresents diversity at basal and intermediate levels. Reagan et al. (1996) reported that, in a rainforest food web that distinguished "kinds" of arthropods, representation of basal and intermediate species was 30 and 70% of all species, respectively, and the proportion of top predators was < 1%. Polis (1991b) also reported that top predators were rare or absent in desert communities. Both Polis (1991b) and Reagan et al. (1996) reported that ratios of prey species to predator species are much greater than 1.0 when the true diversity of lower trophic levels is represented.

Although the properties of food webs identified by early theorists may be flawed to the extent that arthropod diversity was not resolved adequately, they represent hypotheses that have stimulated considerable research into community organization. Future advances in food web theory will reflect efforts to address arthropods at the same level of taxonomic resolution as other taxa.

2.3 Functional Organization

A third approach to community description is based on the guild, trait or functional group, concept (Cummins, 1973; C. Hawkins and MacMahon, 1989; Körner, 1993; Root, 1967; Simberloff and Dayan, 1991). The guild concept was originally proposed by Root (1967), who defined a guild as a group of species, regardless of taxonomic affiliation, that exploit the same class of environmental resources in a similar way. This term has been useful for studying potentially coevolved species that compete for, and partition use of, a common resource. The largely equivalent trait or functional group was proposed by Cummins (1973) to refer to a group of species having a similar ecological function. Insects, as well as other organisms, have been combined into guilds or functional groups based on similarity of response to environmental conditions (Coulson et al., 1986; Fielding and Brusven, 1993; Grime, 1977; Root, 1973) or of effects on resources or ecosystem processes (Romoser and Stoffolano, 1998; Schowalter et al., 1981c; Siepel and de Ruiter-Dijkman, 1993). This method of grouping is one basis for pooling "kinds" of organisms, as discussed earlier.

Pooling species in this way has been attractive for a number of reasons (Root, 1967; Simberloff and Dayan, 1991). First, it reflects the compartmentalization of natural communities (see section 2.2.4) and focuses attention on sympatric species that share an ecological relationship, for example, competing for a resource or affecting a particular ecological process, regardless of taxonomic relationship. Second, it helps resolve multiple usage of the term "niche" to refer both to the functional role of a species and to the set of conditions that determines its presence in the community. Use of guild or functional groupings to refer to species' ecological role(s) permits limitation of the term niche to refer to the conditions that determine species presence. Third, this concept facilitates comparative studies of communities which may share no taxa but do share functional groups, for example, herbivores, pollinators, detritivores, etc. Guild or functional groupings permit focus on a particular group, with specific functional relationships, among community types. Hence, researchers avoid the necessity of cataloging and studying all species represented in the community, a nearly impossible task, before comparison is possible. Functional groupings are particularly useful for simplifying ecosystem models to emphasize effects of functional groups with particular patterns of carbon and nutrient use on fluxes of energy and matter. Finally, recent studies have shown that functional or trait groups may be related to ecosystem processes better than are species diversity indices (Gagic et al., 2015).

The designation of guild or functional groupings is largely a matter of convenience and depends on research objectives (C. Hawkins and MacMahon, 1989; Körner, 1993; Simberloff and Dayan, 1991). For example, defining "same class of resources" or "in a similar manner" is ambiguous. Each species represents a unique combination of abilities to respond to environmental conditions and to affect ecosystem processes, that is, species within functional groups are similar only on the basis of the particular criteria used to distinguish the groups. Characterization of functional groups based on response to climate change, response to a disturbance gradient, effect on carbon flux, or effect on biogeochemical cycling would involve different combinations of species.

Insects are particularly difficult to categorize because functional roles can change seasonally (wasps switching between predation and pollination) or during maturation (eg, sedentary herbivorous larvae becoming mobile pollinating adults, aquatic larvae becoming terrestrial adults), and many species are too poorly known to assign functional roles. Nearly all Lepidoptera can be assigned to a plant-feeding functional group, but various species would be assigned to different functional groups on the basis of plant part(s) affected (eg, foliage, shoots, or roots) or response to drought or other disturbances (Schowalter et al., 1999). Clearly, guilds or functional groups can be subdivided to represent a diversity of responses to different gradients or subtle differences in ecological effects. For example, a stress-adapted "functional group" could be divided into subgroups that tolerate desiccation, physiologically prevent desiccation, or avoid desiccation by feeding on plant fluids. Similarly, a foliage-feeder guild can be divided into subgroups that fragment foliage, mine foliage, suck cellular fluids, feed on different plant species, etc., each subgroup affecting energy and matter fluxes in a different manner. Luh and Croft (1999) developed a computer algorithm to classify predaceous phytoseiid mite species into functional groups (specialist vs generalist predators). The computer-generated classification confirmed the importance of the combination of life history traits used previously to distinguish functional groups.

Species included in a particular functional group should not be considered redundant (Beare et al., 1995; Lawton and Brown, 1993), but rather complementary, in terms of response to environmental change or ensuring ecological functions. Schowalter et al. (1999) reported that each functional group defined on the basis of feeding type included species that responded positively, negatively, or nonlinearly to moisture availability. Species replacement within functional groups maintained functional organization over an experimental moisture gradient, but changes in species would result in differences in pathways, phenology, and rates of energy and matter fluxes (see Chapter 4).

Changes in the relative abundance or biomass of functional groups can signal changes in the rate and direction of ecological processes. For example, changes in the relative proportions of filter-feeder versus shredder functional groups in aquatic ecosystems affect the ways in which detrital resources are processed within the stream community and their contribution to downstream communities. Similarly, changes in the relative proportions of folivores versus sap-suckers affect the flux of nutrients as solid materials versus liquid (eg, honeydew) and their effect on phylloplane (leaf surface) and detrital communities (Schowalter and Lowman, 1999; Stadler and Müller, 1996; Stadler et al., 1998).

The functional group concept permits a convenient compromise in dealing with diversity, that is, sufficient grouping to simplify taxonomic diversity while retaining an ecologically relevant level of functional diversity. Therefore, the functional group approach has become widely used in ecosystem ecology.

3 PATTERNS OF COMMUNITY STRUCTURE

A central theme of community ecology has been identification of patterns in community structure across environmental gradients in space and time (see also Chapter 10). The diversity of community types at landscape and regional scales (γ diversity) has been a largely neglected aspect of biodiversity, but is important to the maintenance of regional species pools and metapopulation dynamics for many species. In addition, the mosaic of community types on a landscape may confer conditional stability to the broader ecosystem, in terms of relatively consistent proportions of community types over space and time (see Chapters 10 and 15).

Identification of patterns in community organization has become increasingly important to population and ecosystem management goals. Introduction of exotic insects to combat noxious pests (weeds or other insects) requires attention to the ability of the biocontrol agent to establish itself within the community and to its potential effects on nontarget components of the community. Efforts to conserve or restore threatened species require consideration and maintenance of the underlying community organization, as this determines resource availability and patterns of competition, predation, or mutualism.

Depending on the descriptive approach taken (see section 2), patterns have been sought in terms of species diversity, food web structure, or guild or functional group composition. Unfortunately, comparison of data among communities has been hampered by the different approaches used to describe communities, compounded by the variety of sampling techniques, with their distinct biases, used to collect community data (Supp and Ernest, 2014). For example, sweep netting, light trapping, interception trapping, pitfall trapping, soil coring, canopy fumigation, and branch bagging are among the techniques commonly used to sample terrestrial arthropods (Leather, 2005). These techniques differ in their representation of nocturnal versus diurnal flying insects, arboreal versus soil/litter species, sessile versus mobile species, etc. (Blanton, 1990; Leather, 2005; Majer and Recher, 1988; Southwood, 1978). Variation in mesh size of sampling nets affects the representation of aquatic species (Storey and Pinder, 1985). Relatively few studies have used the same, or similar techniques, to provide comparative data among community types or locations. Some proposed patterns have been challenged, as subsequent studies provided more directly comparable data or increased resolution of arthropod taxonomy (C. Hawkins and MacMahon, 1989; Polis, 1991b; Reagan et al., 1996). Disturbance history, or stage of postdisturbance recovery, also affects community structure (Harding et al., 1998; Schowalter et al., 2003; Summerville et al., 2009; E. Wilson, 1969, see Chapter 10). However, the history of disturbance at sampled sites often is unknown, potentially confounding interpretation of differences in community structure. Nevertheless, apparent patterns identified at a variety of spatial scales may serve as useful hypotheses to guide future studies.

3.1 Global Patterns

Communities can be distinguished on a taxonomic basis at a global scale because of the distinct phylogenetic origin of faunas among biogeographic realms (Wallace, 1876, see Chapter 7). Similar community types on different continents often are dominated by unrelated species with similar attributes, termed *ecological equivalence*, often resulting from *convergence* of phenotypic attributes that are favored by similar environmental conditions. Consequently, grassland communities on every continent show similar food web structure and functional group organization, reflecting similar environmental conditions, regardless of taxonomic representation, for example, migratory antelope and zebra grazers in Africa, bison and pronghorn in North America, and pampa deer and guanaco in South America. A number of studies have indicated global patterns in community structure related to latitudinal gradients in temperature and moisture and to the ecological history of adaptive radiation of particular taxa.

Latitudinal gradients in temperature and precipitation establish a global template of habitat suitability (eg, biomes), as discussed in Chapters 2 and 7. Equatorial areas, characterized by high sun angle and generally high precipitation, provide favorable conditions of light, temperature, and moisture, although seasonal patterns of precipitation in some tropical areas create periods of adverse conditions for many organisms. The strongly seasonal climate of temperate zones requires specific adaptations for survival during unfavorable cold periods, thereby limiting species diversity. The harsh conditions of temperate deserts and high-latitude zones generally restrict the number of species that can be supported or that can adapt to these conditions.

Species richness generally decreases with latitude for a wide variety of taxa (R. Dunn et al., 2009; K. Gaston, 2000; Price, 1997; J. Stout and Vandermeer, 1975; Wiens et al., 2006; Willig and Lyons, 1998). Latitudinal gradients are especially pronounced for insects, with some studies suggesting that the tropics support several million undescribed arthropod species (Erwin, 1995; May, 1988; E. Wilson, 1992), depending on scale-dependent estimates of specialization of herbivorous groups among plant species (Gering et al., 2007). Latitudinal trends may not be reflected by all taxa (eg, aphids, Dixon, 1985) or component communities (Boyero et al., 2011, 2012; Vinson and Hawkins, 1998). Although L. Dyer et al. (2007) reported that larval diets of tropical Lepidoptera were more specialized than those of temperate forest caterpillars, contributing to higher diversity of this group in tropical forests, Novotný et al. (2006) found similar levels of specialization between tropical and temperate Lepidoptera and concluded that the greater diversity of this group in the tropics reflected the greater diversity of plant species. Lewinsohn and Roslin (2008) conducted a metaanalysis of studies comparing temperate and tropical herbivore diversity and found that correlation between plant and herbivore diversity explained 60% of the variation in insect species richness. They also concluded that higher insect diversity in the tropics reflects greater diversity of host plants. Vinson and Hawkins (1998) reviewed literature for stream communities and concluded that species richness is highly variable, and no strong latitudinal trends are apparent. Furthermore, Willig and Lyons (1998) showed that latitudinal gradients can result from chance. Nevertheless, a number of hypotheses have been proposed to explain latitudinal gradients in species richness.

Terborgh (1973) showed that the apparent trend in species richness with latitude can reflect increasing land area toward the equator. He noted that climate is relatively constant across a wide belt between 20°N and S latitudes but shows a distinct latitudinal gradient above 20°N and S latitudes. Combining climate and surface area gradients yielded a latitudinal gradient in habitat area available within each climate class, with a preponderance of global surface area in tropical habitat. These data suggest that gradients in species richness reflect habitat area available for within-habitat speciation (see section 4.1).

Latitudinal gradients in species richness also may reflect greater primary productivity in the tropics (Rosenzweig and Abramsky, 1993; Tilman and Pacala, 1993; Waide et al., 1999; see section 4.3). Hutchinson (1959) proposed that animal diversity is related to the energy available in ecosystem primary production. D. Currie (1991) subsequently demonstrated that North American patterns of plant and vertebrate diversity were related to environmentally available energy. Venarsky et al. (2014) showed that terrestrial plant detritus provided all, or nearly all, available energy to support macroinvertebrate production in a cave stream ecosystem.

More recently, A. Allen et al. (2002) and J. Brown et al. (2004) proposed a *metabolic theory of ecology* that explained latitudinal gradients in diversity as a result of relationships between temperature, body size, and metabolic rate that determine the maximum population sizes attainable, given the energy and nutrients available in the ecosystem. All other factors being equal, available energy will support more small organisms than larger organisms, and warm environments will support more organisms than will cold environments. Algar et al. (2007) and B. Hawkins et al. (2007) tested the prediction of metabolic theory that the natural logarithm of species richness (of a variety of data-sets for trees, blister beetles, tiger beetles, butterflies, amphibians, and reptiles) is a linear function of temperature. All taxa tested showed a curvilinear relationship to temperature, rather than the predicted linear relationship, indicating that energy availability alone is not a sufficient explanation for latitudinal gradients in species richness.

Finally, evolutionary time may explain latitudinal gradients for some taxa. Wiens et al. (2006) found that tree frog diversity was strongly correlated with time since colonization of a region, but not with latitude per se. However, since these frogs originated in tropical South America and spread to temperate regions relatively recently, there has been more time for speciation in the tropics than in temperate regions. R. Dunn et al. (2009) evaluated the asymmetry in ant species richness between the northern and southern hemispheres (Fig. 9.9) and

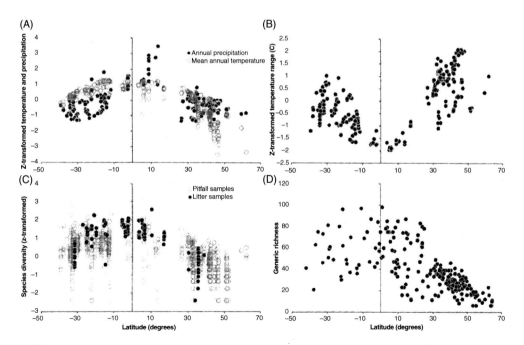

FIGURE 9.9 Latitudinal trends in (A) mean annual precipitation and temperature, (B) temperature range, (C) local species richness of ants, and (D) regional richness of ant genera. Negative latitudes are for the southern hemisphere. Generic richness is derived from lists of species and genera from countries and smaller political regions and presented for comparison. *Source: From Dunn, R.R., et al., 2009. Climatic drivers of hemispheric asymmetry in global patterns of ant species richness. Ecol. Lett. 12, 324–333.*

concluded that the greater climate change since the Eocene in the northern hemisphere resulted in more extinctions and reduced species richness relative to the southern hemisphere.

Superimposed on the latitudinal gradients are the relatively distinct biogeographic realms identified by Wallace (1876). These biogeographic realms reflect the history of continental breakup, with southern floras and faunas largely distinct from northern floras and faunas (see Chapter 7). However, the southern continents show a varied history of reconnection with the northern continents that has resulted in invasion primarily by northern species. The proximity of North America and Eurasia has facilitated movement of species between these landmasses, leading to development of a Holarctic species component, especially within the arctic and boreal biomes. Whereas many genera, and even some species, occur throughout the Holarctic realm, the flora and fauna of Australia have remained relatively distinct as a result of continued isolation.

Species richness also may be related to geological time. E. Wilson (1969) suggested that coevolution should improve the efficiency of total resource exploitation and lead to further increase in coexisting species over time. In other words, a habitat or resource that has persisted for a longer period of time would acquire more species than a more recently derived habitat or resource. Birks (1980) found that the residence time of tree species in Britain was strongly correlated with the diversity of associated insect species. Tree species that had a longer history of occurrence in Britain hosted a larger number of species than did tree species with shorter residence times. Again, because residence time is correlated with area of occurrence (habitat area), the effects of these two factors cannot be distinguished easily (Price, 1997, see sections 4.1 and 4.3).

3.2 Biome and Landscape Patterns

Patterns in species richness, food web structure, and functional organization have been observed among biomes and across landscapes. To some extent, patterns may reflect variation in occurrence or dominance of certain taxa in different biomes. Regional species pools may obscure effects of local habitat conditions on species richness (Kozár, 1992a), especially in temperate ecosystems (Basset, 1996), but few ecologists have addressed the extent to which the regional species pool may influence local species richness. Gering et al. (2003), Kitching et al. (1993), and Progar and Schowalter (2002) distinguished arthropod assemblages among sites within biomes that reflected regional gradients in environmental conditions.

Various hypotheses have been proposed to account for apparent *metacommunity* patterns at the biome and landscape levels (Cottonie, 2005; Cottonie and de Meester, 2004; Kneitel and Miller, 2003; Leibold and Mikkelson, 2002; Leibold et al., 2004). Metacommunities, like metapopulations, represent a recurring community structure or food web among similar habitat units across a landscape, interconnected by differential dispersal patterns among species (Kneitel and Miller, 2003; LeCraw et al., 2014; Leibold et al., 2004). Patterns include nested subsets (Summerville et al., 2002), checkerboards, and various types of gradients (Leibold and Mikkelson, 2002). Leibold and Mikkelson (2002) proposed a set of criteria to distinguish which pattern characterizes a given landscape. *Coherence* is the degree to which a pattern can be represented by a single dimension, *species turnover* is the number of species replacements along this dimension, and *boundary clumping* describes how edges of species ranges are distributed along this dimension. Presley et al. (2010) described additional patterns and demonstrated that combinations of patterns at fine spatial scales can aggregate to form different patterns at larger scales.

General functional groups are common to all terrestrial and aquatic biomes, for example, grazing herbivores (depending on the amount of autochthonous primary production in streams), predators, parasites, and detritivores, whereas other functional groups depend on particular resources being present, for example, sap-suckers require vascular plants and wood borers require wood resources. Proportions of the fauna representing different functional groups vary among biomes. Low-order streams have primarily detrital-based resources, and their communities are dominated by detritivores and associated predators and parasites. Other communities represent various proportions of autotroph functional groups (eg, chemoautotrophs, ruderal, competitive, and stress-tolerant vascular vs nonvascular plants) and heterotroph functional groups (herbivores, predators, detritivores) (see Chapter 11).

Different species compose these functional groups in different biomes. For example, the insect grazer functional group is composed primarily of moths, beetles, and tree crickets in broadleaved forests, by moths and sawflies in coniferous forests (Schowalter, 1995; Schowalter and Ganio, 1999; Schowalter et al., 1981c), by grasshoppers in grasslands and shrublands (Curry, 1994), and by caddisflies and flies in aquatic communities (Hart, 1992). The predator functional group in terrestrial arthropod communities is dominated by a variety of arachnids, beetles, flies, and wasps, whereas in aquatic arthropod communities this functional group is dominated by dragonflies, true bugs, and beetles.

Functional group composition has not shown consistent differences among biomes (C. Hawkins and MacMahon, 1989; Stork, 1987). Detritivores represent a relatively greater proportion of the community in boreal forests, headwater streams, and other biomes characterized by accumulated organic material, and a lower proportion in tropical forests, deserts, and other biomes with little organic matter accumulation (Haggerty et al., 2002; Seastedt, 1984). Wood borers occur only in forest or shrub ecosystems with abundant wood resources. Pollinators are more diverse in tropical forests and deserts where plant diversity and isolation have led to greater reliance on insect and vertebrate pollinators, compared to temperate grassland, temperate and boreal forest, and arctic tundra biomes. Proportional representation of species and individuals among functional groups varies widely among canopy arthropod communities in temperate and tropical forests, depending on tree species composition (Fig. 9.10) (V. Moran and Southwood, 1982; Schowalter and Ganio, 1998, 1999; Stork, 1987).

At the landscape or stream network scale, patterns in species richness and functional group organization can be related to local variation in physical conditions. The history and geographic pattern of disturbance may be particularly important factors affecting variation in community structure (Harding et al., 1998; Schowalter et al., 2003; Summerville et al., 2009). Polis et al. (1997a) concluded that the exchange of organisms and resources among the interconnected community types composing a landscape can contribute to the organization of the broader landscape community by subsidizing more resource-limited local communities. However, Basset (1996) found that diversity in tropical rainforest trees was related to five factors: numbers of young leaves available throughout the year, ant abundance, leaf palatability, leaf water content, and altitudinal range. These data suggested that local factors may be more important determinants of local species diversity and community structure in complex ecosystems, such as tropical forests, than in less complex ecosystems, such as temperate forests.

Diversity of stream insects varies among riffle and pool habitats and substrate conditions (Ward, 1992). Diversity generally is higher in running water with cobble substrates, with high oxygen supply and heterogeneous structure, than in standing water with mud, sand, or gravel

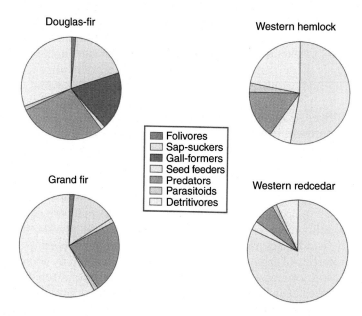

FIGURE 9.10 Functional group organization of arthropod communities in canopies of four old-growth conifer species at the Wind River Canopy Crane Research Facility in southwestern Washington, USA. *Source: Data from Schowalter, T.D., Ganio, L.M., 1998. Vertical and seasonal variation in canopy arthropod communities in an old-growth conifer forest in southwestern Washington, USA. B. Entomol. Res. 88, 633–640.*

substrates. Vinson and Hawkins (1998) reviewed six studies that compared the species richness of stream insects over drainage basins. Species diversity varied with elevation, which covaried with a number of important factors, such as stream morphology, flow rate and volume, riparian cover, and agricultural or urban land use. In one study Carter et al. (1996) used multivariate analysis (TWINSPAN) to compare species composition among 60 sites representing first-order (characterized by a narrow V-shaped channel, steep gradient, nearly complete canopy cover) to sixth-order (characterized by a wide channel, low gradient, little canopy cover) streams over a 15,540 km² drainage basin. They identified five communities distinguished largely by elevation. The highest species richness occurred in mid-order, mid-elevation streams that included species groups characterizing both higher- and lower-order streams.

Transition zones (*ecotones*) between community types typically have higher species richness because they represent habitat variables and include species from each of the neighboring communities (Muff et al., 2009). Zhong et al. (2003) reported that adult mosquito species diversity was higher at sites surrounded by freshwater and salt marsh than at sites surrounded by either freshwater or salt marsh alone. However, Sabo et al. (2005) reported that riparian zones represent unique habitats that support species not represented in the neighboring aquatic or terrestrial communities. Ecotones can move across the landscape as environmental conditions change. For example, the northern edge of Scots pine, *Pinus sylvestris*, forest in Scotland moved rapidly 70–80 km northward about 4000 years BP then retreated southward again about 400 years later (Gear and Huntley, 1991). Sharp edges between community types, such as those resulting from land use practices, reduce the value of this ecotone as a transition zone.

Patches representing different stages of postdisturbance recovery show distinct patterns of species richness, food web structure, and functional group organization (see Chapter 10). Species richness typically increases during community development up to an equilibrium, perhaps declining somewhat prior to reaching equilibrium (MacArthur and Wilson, 1967; E. Wilson, 1969). As the number of species increases, the number of species interactions increases. Food chains that characterize simpler communities develop into more complex food webs (E. Wilson, 1969). Schowalter (1995), Schowalter and Ganio (1999), and Schowalter et al. (1981c) found that patches of recently disturbed temperate and tropical forests were characterized by higher sap-sucker/folivore ratios than were patches of undisturbed forests, even when data were reported as biomass.

Shure and Phillips (1991) found that species richness and functional group composition were modified by manipulated patch size (Fig. 6.7). Species richness was lowest in mid-sized canopy openings (0.08–0.4 ha). Herbivore guilds generally had lowest biomass in mid-sized canopy openings; omnivore biomass peaked in the smallest openings (0.016 ha) and then declined as opening size increased; predator biomass was highest in the control forest and smallest openings and lowest in the mid-sized openings; and detritivore biomass was similar among most openings but much lower in the largest openings (10 ha). This pattern may indicate the scale that distinguishes communities characterizing closed-canopy and open-canopy forest. Smaller openings were influenced by surrounding forest, whereas larger openings favored species that were tolerant of solar exposure and altered plant conditions, for example, early successional species and higher phenolic concentrations (Dudt and Shure, 1994; Shure and Wilson, 1993). Intermediate-sized openings may be too exposed for forest species, but insufficiently exposed for earlier successional species. However, species richness generally increases with habitat area (M.P. Johnson and Simberloff, 1974; MacArthur and Wilson, 1967), for reasons discussed in section 4.1.

4 DETERMINANTS OF COMMUNITY STRUCTURE

A number of factors affect community structure, that is, the composition, relative abundance, population size, and distribution of species (Price, 1997). Factors associated with habitat area, habitat or resource conditions, and species interactions generally have the greatest influence on community structure.

4.1 Habitat Area and Complexity

The relationship between number of species and sampling effort, in time or space, has been widely recognized and supported (He and Legendre, 2002). The increase in number of species with increasing number of samples reflects the greater representation of the community. Similarly, a larger habitat area will "sample" a larger proportion of a regional species pool (MacArthur and Wilson, 1967; Summerville and Crist, 2004). Summerville et al. (2002) found that larger patches hosted more butterfly species because habitat generalists tended to colonize all patch sizes, whereas habitat specialists avoided the smallest patches in favor of larger patches. Increasing habitat area also tends to represent increasing heterogeneity of habitat conditions (M.P. Johnson and Simberloff, 1974; MacArthur and Wilson, 1967; D. Strong et al., 1984), providing an increasing number of niches. Larger bodies of water have higher diversities of aquatic insects (Paradise, 2004).

In developing the *Theory of Island Biogeography*, MacArthur and Wilson (1967) emphasized the relationship between species richness (*S*) and island area (*a*), expressed as

$$S = Ca^z \tag{9.10}$$

where C depends on the taxon and biogeographic region, and z is a parameter that varies little among taxa or biogeographic regions, generally falling in the range 0.20–0.35 (Fig. 9.11). The value of z increases with habitat heterogeneity and proximity to the mainland. For non-isolated sample areas within islands or within continental areas, the relationship between species number and sample area is similar, but z is smaller, generally 0.12–0.17 (MacArthur and Wilson, 1967).

Habitat area has continued to be viewed as a primary factor affecting species richness and other aspects of community structure, likely producing apparent gradients in species richness with latitude and host residence time (Birks, 1980; Price, 1997; Terborgh, 1973), as discussed earlier (see section 3.1). However, habitat area also is a surrogate for habitat heterogeneity. Larger islands are more likely than smaller islands to represent a wider range in elevation, soil types, aspects, disturbance histories, etc. Similarly, larger continental areas are more likely than smaller areas to represent a diversity of habitat conditions. Because relatively distinct component communities develop on particular resources, such as plant or microbial species (J.C. Moore and Hunt, 1988), species richness increases exponentially as representation of habitat and resource diversity increases. Furthermore, habitat heterogeneity provides for refuges from competition and/or predation, that is, local patches of competition- or

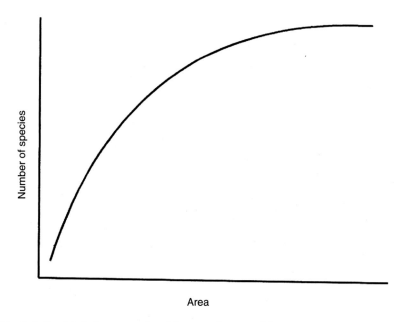

FIGURE 9.11   Relationship between species richness and geographic area.

predator-free space (Covich et al., 2009; Finke and Denno, 2006), although this effect depends on patch scale relative to the foraging scale of predators (McCarthy and White, 2016). The architectural complexity of individual plants or vegetation cover also can affect the diversity of associated fauna (Lawton, 1983). For example, Petermann et al. (2015) reported that the appearance of dragonfly larvae in bromeliad tanks of sufficient size to support these predators altered the detritivore-based community relative to smaller, predator-free tanks.

Fragmentation of habitat types often alters species richness and other measures of diversity. Larger fragments retain a greater proportion of species richness than do smaller fragments (Fig. 9.12) (Collinge, 2000; Kruess and Tscharntke, 2000; Summerville and Crist, 2004; Summerville et al., 2002). Species characteristic of the fragmented habitat often are replaced by species characterizing the surrounding matrix as environmental conditions change within fragments (J. Chen et al., 1995; Summerville and Crist, 2004). Some guilds may be more sensitive to fragmentation than are others. Golden and Crist (1999) reported that sap-sucking herbivores and parasitoids were significantly reduced by fragmentation of a goldenrod community, but chewing herbivores and predators were largely unaffected. Overall insect species richness was reduced by fragmentation, primarily through loss of rare species. However, some insect species also may be favored by fragmentation that reduces abundance and diversity of insectivorous birds (Arriaga-Weiss et al., 2008; Ferraz et al., 2007; Şekercioğlu et al., 2002; Stouffer et al., 2009, 2011).

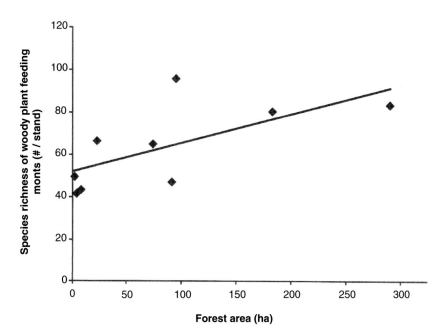

FIGURE 9.12 **Relationship between the size of forest fragments and number of woody-plant-feeding moth species in the western Allegheny Plateau of eastern North America.** The regression is significant ($p < 0.05$, $R^2 = 0.61$). *Source: From Summerville, K.S., Crist, T.O., 2004. Contrasting effects of habitat quantity and quality on moth communities in fragmented landscapes. Ecography 27, 3–12.*

4.2 Habitat Stability

Habitat stability determines the length of time available for community development (see Chapter 10). E. Wilson (1969) proposed four stages in community development. The *noninteractive stage* occurs on newly available habitat or immediately following a disturbance, when numbers of species and population sizes are low. As species number rises during the *interactive stage*, competition and predation influence community structure, with some species disappearing and new species arriving. The *assortative stage* is characterized by persistence of species that can coexist and utilize resources most efficiently, facilitating niche partitioning and species packing. Finally, the *evolutionary stage* is characterized by coevolution that increases the efficiency of overall utilization and species packing. Recurring disturbances restrict diversity to species tolerant of disturbance or altered postdisturbance conditions (Cole et al., 2008, Fig. 9.13, see also Chapter 2). Community development in frequently disturbed habitats cannot progress beyond earlier stages, whereas more stable habitats permit advanced community development and increased species richness. However, the most stable habitats also allow the most adapted species to preempt resources from other species, leading to a decline in species richness through competitive exclusion (see Chapter 8). This trend has led to the development of the *Intermediate Disturbance Hypothesis*, which predicts that species richness peaks at intermediate levels of disturbance (Connell, 1978; Sousa, 1985, but see Reice, 1985). Community recovery from disturbance is described more fully in Chapter 10.

4.3 Habitat or Resource Conditions

Habitat conditions establish the template that determines the ability of particular species to survive and persist. Physical conditions include temperature, water availability, chemistry, etc. (see Chapter 2). S. Miller et al. (2007) reported that withdrawal of water from streams for irrigation purposes altered habitat conditions for aquatic invertebrate communities. Changes in water depth, flow rate, and wetted habitat were proportional to the amount of

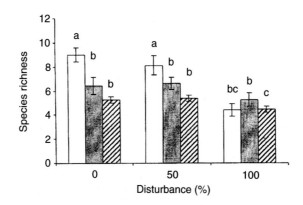

FIGURE 9.13 **Response of oribatid (*open bars*) and mesostigmatid (*shaded bars*) mites and Collembola (*hatched bars*) to soil disturbance treatment (proportion of plot with soil disrupted to simulate livestock trampling) in 2001.** Vertical lines represent standard errors, and bars with the same letter do not differ at $p < 0.05$. *Reprinted with permission from Cole, L., et al., 2008. Influence of disturbance and nitrogen addition on plant and soil animal diversity in grassland. Soil Biol. Biochem. 40, 505–514; Figure 3, p. 510.*

water removed, whereas changes in conductivity and temperature were observed only when water removal exceeded 85% of ambient levels. Water removal > 90% and temperatures above 30°C resulted in a shift in community structure from dominance by collector-gatherer and filterer Ephemeroptera, Plecoptera, and Trichoptera to dominance by scraping elmid beetles, predators, and noninsect taxa. Similarly, Stubbington et al. (2009) found the lowest species richness in streams drying during drought, but also reported that water exchange between the hyporheic zone and groundwater influenced the abundance and behavior of benthic invertebrates. Whiles and Wallace (1997) documented changes in stream invertebrate taxonomic composition resulting from conversion of adjacent deciduous forest to pine forest with accompanying changes in the quality of litter inputs to the stream community.

As discussed earlier, the availability of particular resources determines the presence of associated species. If a limiting resource (host) becomes more abundant, then associated species also become more abundant, until some other factor(s) become limiting. For example, Siemann (1998) and Cole et al. (2008) reported that experimental fertilization of grassland plots increased arthropod species richness and abundance. Doi (2009) reviewed studies of spatial patterns in allochthonous and autochthonous aquatic resources and concluded that connectivity of food webs based on these different resources along the stream continuum is critical to the sustainability of top predators that require a diversity of prey.

Limiting resources may preclude any single adaptive strategy from becoming dominant and thereby maintain high species richness. Rosenzweig and Abramsky (1993), Tilman and Pacala (1993), and Waide et al. (1999) concluded that species richness is not always linearly related to productivity. Intermediate levels of productivity often support the highest diversity because higher productivity favors dominance by the most competitive species. Mittelbach et al. (2001) compiled 171 published studies relating species richness and productivity for aquatic and terrestrial plants and animals. Hump-shaped relationships were most common, indicating that productivity generally was higher at intermediate levels of species richness.

Diversity at each trophic level is affected by the diversity of resources available (M.T. Johnson and Agrawal, 2005; M.T. Johnson et al., 2006; Wilby and Orwin, 2013). As described earlier (eg, Fig. 9.6), each plant species hosts a relatively unique assemblage of associated herbivores, each of which in turn hosts a relatively unique assemblage of parasites and, to a lesser extent, predators. Therefore, the diversity of arthropod assemblages is a function of the diversity of plant assemblages (Curry, 1994; Lewinsohn and Roslin, 2008; Magurran, 2004; Novotný et al., 2006; Stiling, 1996). Even within a given plant species, genetic diversity of the plant population influences the diversity of associated arthropods (Bangert et al., 2006; Crutsinger et al., 2006; M.T. Johnson and Agrawal, 2005; Wimp et al., 2007). Genetically diverse plant populations represent a more complex habitat than do genetically similar populations. Complexity of plant architecture also influences the diversity of associated arthropods (Jeffries et al., 2006).

Unique communities develop on particular substrates embedded within the background matrix. For example, specialized species assemblages characterize lava flows, serpentine soils, and wetlands. Sabo et al. (2005) reported that riparian zones support species that do not occur in surrounding habitats, thereby enriching regional species diversity.

A number of studies have compared species richness between relatively homogeneous and heterogeneous environments (Cromartie, 1975; Risch, 1980, 1981; Root, 1973; D. Strong et al., 1984; Tahvanainen and Root, 1972). Because organisms have greater difficulty maintaining energy and nutrient balance when resources are scattered (see Chapter 4), the abundance of individual species generally decreases with increasing resource heterogeneity, precluding exclusive use of the niche and permitting species richness to increase. By contrast, homogeneous resources facilitate rise to competitive dominance by the best-adapted species, leading to reduced species richness. Extensive planting of agricultural or silvicultural monocultures establishes the conditions necessary for some species to reach epidemic population levels across landscapes (see Chapter 7), reducing availability of resources shared with other species but providing prey resources for predators (Polis et al., 1997a, see also Chapter 8).

4.4 Species Interactions

Species interactions can enhance or preclude persistence of some species, as discussed in Chapter 8. As noted in section 4.3, species populations cannot persist where their host species are absent. However, the presence of competitors, predators, and mutualists also affects persistence of associated species, both directly and indirectly. In the past, species interactions often were viewed as evidence of coevolution. However, species colonizing new areas can occupy vacant niches through preadaptation, that is, similarity between their native and new habitats (Diamond, 1990; Janzen, 1985; Janzen and Martin, 1982). S. Gould and Vrba (1982) proposed the term "exaptation" to describe characters evolved for other purposes but that "preadapt" an organism for current conditions.

Some species can have particularly profound effects on community structure. Their presence in a community leads to a different community structure than occurs in their absence. A top predator that preferentially preys on the most abundant of several competing prey species can prevent the best-adapted species from competitively suppressing others, thereby maintaining higher diversity than would occur in the absence of the predator (see Chapter 8). R. Paine (1966, 1969a,b) considered such species to be *keystone species*. Bond (1993) and Power et al. (1996) applied this term to any species that have effects on ecosystem structure or function that are disproportionate to their abundance or biomass.

Some insect species could be considered keystone species to the extent that their abundance greatly alters diversity, productivity, rates of energy or nutrient flux, etc. Many herbivorous insects increase the diversity of plant species by selectively reducing the density of abundant host species and providing space and resources for nonhost plants (T. Coleman et al., 2008; Lawton and Brown, 1993; Schowalter and Lowman, 1999). The southern pine beetle, *Dendroctonus frontalis*, is capable, at high population densities, of killing pine trees and increasing the availability of woody resources that maintain populations of other xylophagous species (Flamm et al., 1993). Naiads of the large dragonfly, *Tramea lacerata*, prey on other dragonflies as well as on various other taxa that also are prey of other dragonfly and damselfly naiads. Wissinger and McGrady (1993) found that addition of *T. lacerata* to wetland communities had a direct negative effect on damselfly prey, but also an indirect positive effect through reduced numbers of other predaceous dragonflies. Termites and ants affect soil structure and fertility in ways that determine vegetation development (Brody et al., 2010; Fox-Dobbs et al., 2010, see Chapter 14).

Céréghino et al. (2010) described the influence of two ant species, *Pachycondyla goeldii* and *Camponotus femoratus*, on diversity of invertebrates associated with a bromeliad, *Aechmea mertensii*, an ant-garden mutualistic epiphyte. These ants incorporate fruits and seeds of the bromeliad into their arboreal carton nests, providing for dispersal and protection from herbivores, and the bromeliad subsequently covers the nest, anchoring and stabilizing the nest and providing food rewards to the ants. The ants control the size and shape of bromeliads by determining their location, from exposed to partially shaded habitats, and thereby create a gradient of aquatic habitat sizes and resources that increases the diversity of associated aquatic invertebrates.

As discussed in Chapter 6, the combination of bottom-up (resource supply) and top-down (trophic cascade) factors tends to stabilize population levels. Changes in abundance of any trophic level, however, affect abundances at other trophic levels, for example, trophic cascades (see Chapter 8). Generally, increased abundance at one trophic level increases resources available to the next higher trophic level, increasing abundance at that level, but reducing abundance at the next lower level. Reduced abundance at the lower trophic level reduces its control over the second lower trophic level, which increases in abundance and reduces abundance at the third lower trophic level.

Linkages between communities can affect trophic cascades. Terrestrial arthropod inputs to aquatic systems can represent up to half of the annual diet of many fish (J. Allan et al., 2003; Baxter et al., 2005; Kawaguchi and Nakano, 2001; Nakano and Murakami, 2001; Pray et al., 2009) and influence predator–prey interactions. Nakano et al. (1999) experimentally manipulated terrestrial arthropod inputs and predatory fish presence in forest headwater streams in northern Japan. When terrestrial arthropod inputs were reduced, predatory fish switched from terrestrial to aquatic arthropod prey, reducing aquatic arthropod abundance and increasing periphyton biomass. By contrast, removal of predatory fish did not significantly alter the food web, indicating that the trophic cascade was controlled by terrestrial arthropod inputs. On the other hand, Knight et al. (2005a) found that terrestrial–aquatic cascades worked in the opposite direction as well. Fish reduced dragonfly larvae in ponds, leading to fewer dragonfly adults and reduced predation on pollinators in adjacent terrestrial habitats. As a result, plants near ponds with fish had more pollinator visits and were less pollen limited than were plants near ponds without fish (Fig. 9.14). Such linkages control fluxes of energy and nutrients between ecosystems.

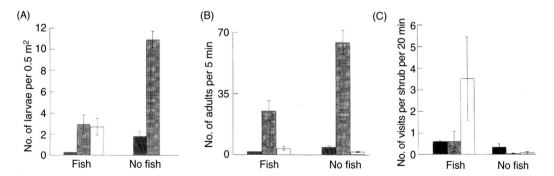

FIGURE 9.14 Trophic cascade across aquatic–terrestrial ecosystems. (A) Significantly fewer ($p < 0.01$) medium (*gray bars*) and large (*black bars*), but significantly more ($p < 0.002$) small (*white bar*), larval dragonflies were found in ponds with fish than in ponds without fish; (B) significantly fewer ($p < 0.02$) adult dragonflies near ponds with fish than near ponds without fish; and (C) significantly more pollinator visits (black bars = Diptera, gray bars = Lepidoptera, white bars = Hymenoptera) to *Hypericum fasciculatum* near ponds with fish than near ponds without fish ($p < 0.02$). Data are mean ± SEM. *Reprinted with permission from Knight, T.M., et al., 2005. Trophic cascades across ecosystems. Nature 437, 880–883.*

5 SUMMARY

Communities are composed of the species occupying a site. Identification of patterns in community structure has been a major goal of ecological research. However, no standard approach for delimiting a site and describing or comparing community structure has been adopted. Indices of species diversity, food web structure, and functional group organization are three methods used to facilitate comparison among communities.

Species diversity has two components: richness and evenness. Richness is the number of species in the community, whereas evenness is a measure of relative abundances. These two components can be represented by rank-abundance curves and by diversity indices. Geometric rank-abundance curves characterize harsh or disturbed habitats with a limited number of adapted species and strong dominance hierarchy, whereas log normal and broken stick models characterize more stable habitats with higher species accumulation and greater evenness in abundance among species. A number of diversity indices and similarity indices have been developed to integrate richness and evenness in a variable that can be compared among community types.

Food web structure represents the network of pairwise interactions among the species in the community. Unfortunately, little is known about most interactions among the thousands of arthropod species in terrestrial and freshwater communities. A number of food web attributes have been proposed, based on limited taxonomic resolution of insects and other arthropods. Our perspective of food web dynamics has been refined as greater resolution of arthropod taxonomy reveals networks of interactions within this diverse group.

Functional group organization reflects combination of species on the basis of their effects on ecological processes, regardless of taxonomic affiliation. This approach has become popular because it simplifies species diversity in an ecologically meaningful way. However, the allocation of species to functional groups is based on particular objectives and is, therefore, arbitrary to the extent that each species represents a unique combination of functional responses or effects.

The noncomparable descriptions of communities based on these three approaches, compounded by the variety of arthropod sampling techniques, each with its unique biases, have hindered comparison of community structure among habitat types. Many taxa show latitudinal gradients in abundance, with species richness increasing toward the equator. However, the climate gradient thought to underlie this trend is correlated with latitudinal gradients in habitat area and productivity. Some taxonomic groups are more diverse within biogeographic realms of origin or where resources have been available over longer time periods. Some functional groups are more abundant in certain biomes, for example, pollinators in diverse tropical habitats, detritivores and wood borers in habitats with greater organic matter or wood accumulation.

Habitat area and stability, resource availability, and species interactions are major factors that affect community structure. Habitat area affects the pool of species available and the heterogeneity of habitat conditions and resources. Habitat stability determines the length of time available for species accumulation, assortment, and species packing. Species richness generally increases with resource availability, up to a point at which the most adapted species competitively suppress other species. Species interactions often affect persistence in a particular habitat. Colonists cannot survive unless their resources are available.

Competition, predation, and mutualism also affect species directly and indirectly. Keystone species have effects on community structure or ecosystem processes that are disproportionate to their numbers or biomass. Keystone species include predators that focus on the most abundant prey species, thereby reducing competition among prey species and maintaining more species than would coexist in the absence of the predator. Some herbivorous insects function as keystone species by selectively reducing abundance of dominant host species and facilitating persistence of nonhosts. Trophic cascades reflect top-down effects of predators reducing prey abundance, thereby increasing abundance of the trophic level supporting the prey.

TEMPORAL CHANGE IN COMMUNITY STRUCTURE ON CARIBBEAN ISLANDS

Changes in community structure occur over short to long timescales as individual species respond, according to their tolerance ranges or other attributes, to variation in environmental conditions that occur seasonally, annually, and over evolutionary time. Rapid changes occur as a result of sudden disruption in habitat or resource conditions by disturbances or anthropogenic influences on global climate.

Schowalter and Ganio (2003) and Schowalter et al. (unpubl. data) sampled invertebrates seasonally over a 19-year period (1991–2009) in tropical rainforest canopies in Puerto Rico. Annual variation in community structure during this period was pronounced and reflected invertebrate responses to two major hurricanes (Hugo in 1989 and Georges in 1998), multiple minor hurricanes and tropical storms (including Luis and Marilyn, 1995, Bertha and Hortense in 1996, Erika in 1997, Jose in 1999, and Debby in 2000), and droughts of varying intensity (including a severe drought during 1994–95, when precipitation was only 41% of the long-term annual average) (Heartsill-Scalley et al., 2007). Five of seven species (three Coccoidea and one Mirid) and three of four functional groups (sap-suckers, predators, and detritivores) that were sufficiently abundant to analyze statistically showed significant responses to year × tree species interaction. Coccoidea generally were most abundant during years following major hurricanes, although this was not the case for all species on all host trees, whereas detritivores were more abundant during drought years, and predators were more abundant during drought and posthurricane years.

Disturbances altered abundances of some species, and successional trends were apparent during years between disturbances. In general, canopy folivores and detritivores increased in abundance and sap-suckers decreased over a 3–4-year period after hurricanes, during which the forest canopy closed and ecological functions recovered (Beard et al., 2005; Drew et al., 2009). However, long-term analysis indicated that responses to Georges and droughts reflected legacies of Hugo, that is, community responses to a major disturbance constrain responses to subsequent disturbances.

Longer-term changes over evolutionary time can be inferred for this forest. Dominican amber provides a wealth of information on the community structure (including insects) of tropical forests on the nearby island of Hispaniola > 25 million years ago (Poinar and Poinar, 1999). Extant forests

Insect Ecology. http://dx.doi.org/10.1016/B978-0-12-803033-2.00010-8

on Puerto Rico and Hispaniola are similar and could be expected to show similar histories of community change. Schowalter and Ganio (2003) and Schowalter (unpubl. data) found that the frequency distribution of insect families represented in canopy and sticky trap samples from Puerto Rican rainforest is similar to that found in Dominican amber samples (Poinar and Poinar, 1999), suggesting that the taxonomic and functional structure of insect assemblages in these forests has changed relatively little over at least 25 million years.

1 INTRODUCTION

Community structure changes through time as species abundances change, altering the network of interactions, with changes cascading through the community. Short-term (eg, seasonal or annual) changes in community structure represent responses to environmental changes that favor some species or affect interaction strength (see Chapter 8). Longer-term (eg, successional) changes in community structure reflect changes in habitat conditions and species turnover during community development on newly available or disturbed sites. Finally, changes in community structure over evolutionary time reflect species responses to long-term trends in environmental conditions.

Among the major environmental issues facing governments worldwide is the effect of anthropogenic activities (eg, altered atmospheric or aquatic chemistry, land use, species redistribution) on the composition of natural communities and the ecosystem services they provide to humans (see Chapter 16). Both natural and anthropogenic changes in community structure affect the delivery of food, fresh water, building materials, and the epidemiology of human diseases (Chapter 16). How stable is community structure in the face of human pressures, and how sensitive are communities and ecosystems to changes in species abundances or composition? Our perception of communities as self-organizing entities, with internal regulatory mechanisms, or as random assemblages has significant implications for policies to reduce extinctions and for our approach to managing ecosystem resources.

As described in Chapter 6 for population dynamics, study of changes in community structure requires long periods of observation. Few studies have continued over sufficiently long time periods to evaluate many of the factors presumed to affect community structure. However, paleoecological evidence and studies of community recovery following disturbances have provided useful data. Research on factors affecting community structure over a range of temporal scales can enhance understanding of the degree of stability in community structure and anticipation of responses to environmental changes.

2 SHORT-TERM CHANGE IN COMMUNITY STRUCTURE

Community structure changes over relatively short time periods. Short-term variation in community structure reflects interactions among species responding differentially to fluctuating abiotic conditions and species interactions. Relatively few studies have measured effects of seasonal or annual changes in arthropod communities over extended periods. Several studies represent annual to decadal dynamics in arthropod communities.

spiders may facilitate development of cyanobacterial crusts and early successional vegetation. Sikes and Slowik (2010) reported that the majority of species present 1 year after a volcanic eruption on Kasatochi Island, Alaska in 2008 were insects, either surviving detritivores or their offspring feeding on rotting kelp and bird carcasses. No phytophagous or fungivorous species were found. Thus, the initial community was insect-dominated rather than plant-dominated.

Disturbances to existing communities affect each species differently, depending on its particular tolerances to disturbance or postdisturbance conditions (J. Chase, 2007; Schowalter and Ganio, 2003; Schowalter et al., 1999, 2014; Shure and Phillips, 1991, see Chapter 2). Often, legacies from the predisturbance community (such as buried rhizomes, seed banks, woody litter, and plants and animals surviving in protected stages or microsites) remain following disturbance and influence the trajectory of community recovery (Keeton and Franklin, 2005; S. Yang et al., 2008). Disturbance-induced changes in some species can facilitate or inhibit recovery by others. Benstead et al. (2007) reported that algal and fish populations returned to predisturbance levels within 2 years following cessation of nutrient enrichment in Arctic freshwater streams. However, a dramatic and persistent increase in bryophyte mass resulting from nutrient enrichment appeared to delay recovery of insect species that had been positively or negatively affected by the disturbance-induced shift in dominant primary producer.

Although the succession of species and communities on newly exposed or disturbed sites is one of the best-documented phenomena in ecology, the nature of the community and mechanisms driving species replacement have been debated intensely from the beginning. Gleason (1917, 1926, 1927) argued that succession is not directed by autogenic processes, but reflects population dynamics of individual species based on their adaptations to changing environmental conditions. Egler (1954) further argued that succession could proceed along many potential pathways, depending on initial conditions and initial species pools. E. Odum (1969) integrated the Clementsian model of succession with ecosystem processes by proposing that a number of ecosystem properties, including species diversity, primary productivity, biomass, and efficiency of energy and nutrient use, increase during succession. Drury and Nisbet (1973) viewed succession as a temporal gradient in community structure, similar to the spatial gradients discussed in Chapter 9, and argued that species physiological tolerances to environmental conditions were sufficient to explain species turnover. More recently, the importance of disturbances and heterotroph activity in determining successional processes and preventing ascension to the climatic climax has been recognized (Davidson, 1993; MacMahon, 1981; Ostfeld et al., 1997; Pickett and White, 1985; Schowalter, 1981, 2012; Willig and Walker, 1999).

The concept of succession as goal-oriented toward a climax has succumbed to various challenges, especially recognition that succession can progress along various pathways to nonclimatic climaxes under different environmental conditions (Whittaker, 1953). Furthermore, the mechanism of species turnover is not necessarily facilitation by the replaced community (Botkin, 1981; Connell and Slatyer, 1977; H. Horn, 1981; McIntosh, 1981; Peet and Christensen, 1980; Whittaker, 1953, 1970). Nevertheless, debate continues over the integrity of the community, the importance of autogenic factors that influence the process, and the degree of convergence toward particular community composition (Bazzaz, 1990; Peet and Christensen, 1980; Glenn-Lewin et al., 1992; West et al., 1981).

3.1 Patterns of Succession

Two types of succession can be recognized. *Primary succession* occurs on newly exposed substrates, for example, lava flows, glacial moraine, uplifted marine deposits, dunes, newly deposited beaches, etc. Primary succession typically involves a long initial period of soil formation and colonization by species requiring little substrate modification. *Secondary succession* occurs on sites where the previous community was disturbed and is influenced by remnant substrate and surviving individuals or propagules. Although most studies of succession have dealt with trends in vegetation, heterotrophic succession, including succession dominated by insects or other arthropods, has contributed greatly to perspectives on the process. Insects and other arthropods dominate the development of freshwater communities and litter (especially woody litter and carrion) communities, and succession in these habitats occurs over shorter timescales than does succession involving longer-lived plant species. The smaller spatial and temporal scales of heterotrophic succession have facilitated study of mechanisms driving species turnover.

Succession varies in duration from weeks for communities with little biomass (eg, carrion feeders) to centuries for communities with abundant biomass (eg, forests). Shorter successions are amenable to study by individual researchers. However, forest or desert succession spans decades to centuries and has not been studied adequately throughout its duration (Fig. 10.3). Rather, forest succession typically has been studied by selecting plots of different age since disturbance or abandonment of management to represent various seres (ie, the *chronosequence approach*). Although this approach has proven convenient for comparing and contrasting various seres, it fails to account for effects of differences in initial conditions on subsequent species colonization and turnover processes, that is, the prior history of chronosequence plots is unknown but may affect current conditions (Egler, 1954; Harding et al., 1998; Progar et al., 2000; Schowalter et al., 2003; Summerville et al., 2009). Even Clements (1916) noted that comparison of the successional stages is less informative than is evaluation of the factors controlling transitions between stages. However, this approach requires establishment of long-term plots protected from confounding activities and a commitment by research institutions to continue studies beyond the usual confines of individual careers. Characterization of succession is a major goal of the network of US and International Long Term Ecological Research (LTER and ILTER) Sites (Peters et al., 2013; Van Cleve and Martin, 1991). Long-term and comparative studies will improve understanding of successional trajectories and their underlying mechanisms in different ecosystems.

3.1.1 Vegetation Succession

A number of trends have been associated with vegetation succession. Generalists or r-strategists generally dominate early successional stages, whereas specialists or K-strategists dominate later successional stages (Fig. 10.3) (Boyce, 1984; V.K. Brown, 1984, 1986; V.K. Brown and Hyman, 1986; V.K. Brown and Southwood, 1983; Grime, 1977; Janzen, 1977; D. Strong et al., 1984, see Chapter 5). Species richness typically increases during early–mid succession but reaches a plateau or declines during late succession (Peet and Christensen, 1980; Whittaker, 1970).

E. Wilson (1969), based in part on data from Simberloff and Wilson (1969), suggested that community organization progresses through four stages: noninteractive, interactive, assortative, and evolutionary (see Chapter 9). The noninteractive stage occurs early during succession when species richness and population densities are too low to induce density-dependent competition, predation, or parasitism. As species colonization and densities increase, interaction strength increases and produces a temporary decline or equilibrium in species number, as some species are excluded by competition or predation. The assortative stage occurs over long disturbance-free time periods, as a result of species persistence in the community on the basis of efficient resource use and coexistence. Niche partitioning may allow more species to colonize and persist. On the other hand, T. Miller et al. (2014) found that protozoan communities evolved during succession in pitcher-plant tanks to modify competitive interactions. Strong competitors evolved to become weaker, whereas weak competitors evolved to become stronger, contrary to predictions, thereby promoting coexistence and higher species diversity. Finally, coevolution over very long time periods increases the efficiency of interaction and permits further increase in species number. However, most communities are disturbed before reaching the assortative stage. The Intermediate Disturbance Hypothesis predicts that species richness is maximized at intermediate levels of disturbance that maintain a combination of early and late successional species (Connell, 1978; Sousa, 1985).

Arthropod communities also change during vegetative succession (V.K. Brown, 1984; Shelford, 1907; Weygoldt, 1969). Torres (1992) reported that a sequence of Lepidoptera species appeared and reached outbreak levels on a corresponding sequence of early successional host plant species during the first 6 months following Hurricane Hugo (1989) in Puerto Rico, but disappeared after depleting their resources. Schowalter (unpubl. data) observed this process repeated following Hurricane Georges (1998). Schowalter (1995), Schowalter and Ganio (2003), and Schowalter et al. (1981c) reported that sap-sucking insects (primarily Hemiptera) and ants dominated early successional temperate and tropical forests, whereas folivores, predators, and detritivores dominated later successional forests. This trend likely reflects the abundance of young, succulent tissues with high translocation rates that favor sap-suckers and tending ants during early regrowth. V.K. Brown and Southwood (1983) reported a similar trend toward increased representation of predators, scavengers, and fungivores in later successional stages. They noted, in addition, that species richness of herbivorous insects and plants were highly correlated during the earliest successional stages, but not later successional stages, whereas numbers of insects and host plants were highly correlated at later stages, but not the earliest successional stages. V.K. Brown and Southwood (1983) suggested that early colonization by herbivorous insects depends on plant species composition, but that population increases during later stages depend on the abundance of host plants (see also Chapters 6 and 7).

In many cases, particular successional habitats are necessary for species persistence. Punttila et al. (1994) reported that the diversity of ant species declined during forest succession in Finland. Most ant species were found in early successional stages, but only the three species of shade-tolerant ants were common in old (> 140-year-old) forests. They noted that forest fragmentation favored species that require open habitat by reducing the number of forest patches with sufficient interior habitat for more shade-tolerant species. Evans (1988) found that grasshopper assemblages showed predictable changes following fire in a grassland in Kansas, USA. The relative abundance of grass-feeding species initially increased following fire, reflecting increased grass growth, and subsequently declined, as the abundance of forbs increased. Similarly, Helbing et al. (2014) reported that each successional stage of pine

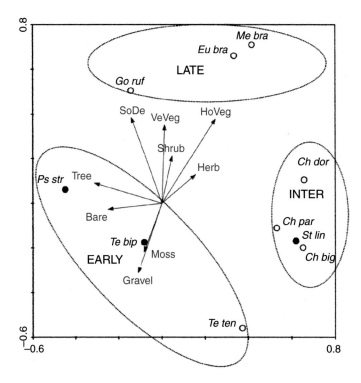

FIGURE 10.4 **Principal coordinates analysis (PCA) of densities of the most frequent Orthoptera species (constancy >6%) and associated environmental parameters.** "Early," "Inter," and "Late" refer to successional status of the sampled habitat. Filled circles = threatened species; open circles = nonthreatened species. *Source: From Helbing, F., Blaeser, T.P., Löffler, F., Fartmann, T., 2014. Response of Orthoptera communities to succession in alluvial pine woodlands. J. Insect Conserv. 18, 215–224.*

woodland in the northern Alps supported a distinct assemblage of grasshopper species, with highest diversity at the intermediate stage and lowest at the late stage, and threatened species disproportionately represented in intermediate and late successional stages (Fig. 10.4). Competition and predation likely limit diversity in the early stage, and adverse microclimate resulting from high canopy coverage limits diversity in the late stage. Schirmel and Fartmann (2014) found that butterfly diversity and abundance were highest in intermediate stages of succession, when shrubs had become established on heathland, and declined in later stages. Two threatened species, *Hipparchia semele* and *Plebeius argus*, decreased in abundance as vegetation density increased and were absent from the later successional stages dominated by shrubs and birch forest.

In other cases, late successional stages are necessary for species persistence. Starzyk and Witkowski (1981) examined the relationship between bark- and wood-feeding insect assemblages and stages of oak–hornbeam forest succession. They found the highest species richness in older forest (> 70 years old) with abundant dead wood, and in recent clear-cuts with freshly cut stumps. Densities of mining larvae also were highest in the older forest and intermediate in the recent clear-cut. Intermediate stages of forest succession supported fewer species and lower densities of bark- and wood-feeding insects. These trends reflected the decomposition of woody residues remaining during early stages and the accumulation of woody debris again during later stages. Niemelä and Spence (1994) and Niemelä et al. (1992) reported that a number of carabid beetle species were restricted to old, undisturbed forests. Similarly, Schowalter (1995) reported that several arthropod species characterizing old-growth conifer forest canopies were rare or absent in younger stages.

Patterns of insect succession mirror vegetation succession but may also direct vegetation succession, especially during outbreaks. Davidson (1993), Schowalter (1981), and Schowalter and Lowman (1999) suggested that insect outbreaks and other animal activity can advance, retard, or reverse succession by affecting replacement of host species by nonhost species (see Section 3.2 and Chapter 12).

3.1.2 *Heterotrophic Succession*

Heterotrophic successions also occur during plant development and in decomposing wood, animal carcasses, and aquatic ecosystems. These processes can be divided into distinct stages characterized by relatively discrete heterotrophic communities.

A succession of insect assemblages occurs on long-lived plants as a result of ontogenetic changes in plant physiology and morphology during growth and development. Fonseca et al. (2006) described the ontogenetic succession of gall-forming insects on a canopy laurel, *Cryptocarya aschersoniana*, in Brazil. Gall-former composition changed from dominance by an unidentified hymenopteran on small plants to dominance by an unidentified hemipteran on canopy trees. Hymenopteran gall density decreased 50-fold, whereas hemipteran density increased 10-fold, from small to large trees.

Succession occurs over relatively short times in ephemeral aquatic systems, such as low-order streams, seasonal ponds, and phytotelmata (water-filled plant structures). Murdock et al. (2010) reported that the timing and diversity of colonizing consumers determined patterns of succession in streams following drought disturbance. Murrell et al. (2014) described the sequence of dipteran assemblages in water-filled containers initiated at three times during spring and summer in deciduous forest. Habitat age was a more important driver of dipteran succession than was time of container initiation during the season.

B. Richardson and Hull (2000) and B. Richardson et al. (2000b) observed distinct sequences of arrival of dipteran filter-feeders and gatherers during community development in phytotelmata in Puerto Rico. The earliest colonizer, of barely opened *Heliconia* bracts, was a small unidentified ceratopogonid, followed by an unidentified psychodid, cf. *Pericoma*. Subsequently, phytotelmata were colonized by two syrphids, *Quichuana* sp. and *Copestylum* sp. Older bracts with accumulated detritus and low oxygen concentration supported mosquitoes, *Culex antillummagnorum*, and finally tipulids, *Limonia* sp., in the oldest bracts.

Similarly, T. Miller and terHorst (2012) and T. Miller et al. (2014) described successional changes in pitcher-plant tank communities, based on decomposition of trapped insects, over a 60-week period (Fig. 10.5). Ants were the principal prey and were captured primarily by young leaves, peaking at 3 weeks of age and providing high nutrient inputs to the tank community during early succession. Bacterial density closely tracked resource input into the leaf, peaking at 5 weeks. Total bacterivore density (protozoa and rotifers) peaked at intermediate leaf ages, but different species showed different patterns of abundance with leaf age. Larval

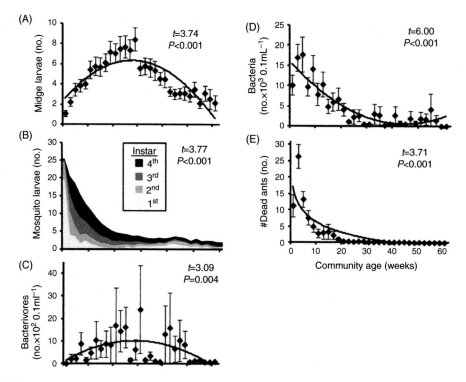

FIGURE 10.5 **Changes in abundances of species in pitcher-plant, *Sarracenia purpurea*, tank communities over time from 0 to 60 weeks: (A) specialist detritivorous midge, *Metriocnemus knabi*, larvae, (B) predaceous mosquito, *Wyeomyia smithii*, larvae, (1st–4th instar), (C) bacterivorous protozoans and rotifers, (D) bacteria, and (E) dead ants, the food base for the community.** Data are means ± SE. Nonlinear parameters (t values) were significant and fit the data better than linear models in each case. *Source: From Miller, T.E., terHorst, C.P., 2012. Testing successional hypotheses of stability, heterogeneity, and diversity in pitcher-plant inquiline communities. Oecologia 170, 243–251.*

mosquito, *Wyeomyia smithii*, abundance was highest in young leaves and limited bacteriovore population growth during this period.

Slavik et al. (2004) experimentally added phosphorus to an Arctic stream in Alaska over a 16-year period. After about 8 years, bryophytes replaced epilithic algae as the dominant primary producer in the treated section, compared to the untreated section upstream. About the same time, immature chironomids and mayflies, *Ephemerella* sp., replaced filter-feeding black flies, *Stegopterna mutata* and *Prosimulium martini*, mayflies, *Baetis* spp., and caddisflies *Brachycentrus americanus*, as the dominant invertebrates.

In general, succession in decaying wood occurs over decadal time scales and is initiated by the penetration of the bark barrier by bark and ambrosia beetles (Scolytinae and Platypodinae) at, or shortly after, tree death (Ausmus, 1977; Dowding, 1984; Howden and Vogt, 1951; Savely, 1939; Swift, 1977; Ulyshen and Hanula, 2010; Zhong and Schowalter, 1989). These beetles inoculate galleries in fresh wood (decay class I, bark still intact) with a variety of symbiotic microorganisms (Stephen et al., 1993, see Chapter 8), and provide access to interior substrates for a diverse assemblage of saprotrophs and their predators. The bark and ambrosia beetles remain only for the first year but are instrumental in penetrating bark, separating bark from wood, and facilitating drying of subcortical tissues (initiating decay class II, bark fragmented and falling off). These insects are followed by wood-boring beetles, wood wasps, and their associated saprophytic microorganisms that typically colonize wood with, or slightly later than, bark beetles and dominate decaying wood for 2–10 years (Chapter 8). Powderpost beetles, carpenter ants, *Camponotus* spp., and termites dominate later stages of wood decomposition (decay classes III–IV, extensive tunneling and decay in sapwood and heartwood, loss of structural integrity), that may persist for 5–100 years, depending on wood conditions (especially moisture content) and proximity to population sources. These insects and associated bacteria and fungi complete the decomposition of wood and incorporation of recalcitrant humic materials into the forest floor (decay class V). Wood becomes increasingly soft and porous, becoming an important source of water, as decay progresses.

Ulyshen and Hanula (2010) described distinct beetle assemblages associated with early, mid, and late stages of decay of cut loblolly pine, *Pinus taeda*, boles. Which species colonize first may determine subsequent patterns of species colonization. Weslien et al. (2011) reported that two initial colonists, a bark beetle, *Hylurgops palliatus*, and a wood borer, *Monochamus sutor*, influenced subsequent colonization, and the pattern of succession depended on which of these two species became dominant. These two species particularly affected (*H. palliatus* positively, *M. sutor* negatively) the emergence of a rare beetle, *Peltis grossa*, that may feed on brown-rotting bracket fungi, *Fomitopsis pinicola*. The effects of the two beetles on *P. grossa* were largely mediated by *F. pinicola* (Fig. 10.6).

Insect species composition follows characteristic successional patterns in decaying carrion (Figs. 10.7 and 10.8), with distinct assemblages of species defining fresh, bloated, decay, dry, and remains stages (Horenstein and Linhares, 2011; Payne, 1965; Shi et al., 2009; Tantawi et al., 1996; Tullis and Goff, 1987; E. Watson and Carlton, 2003). For small animals, several carrion beetle species initiate the successional process by burying the carcass prior to oviposition.

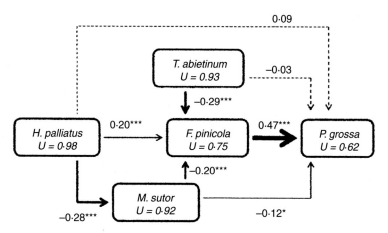

FIGURE 10.6 Alternate pathways for heterotrophic succession in decaying wood, from colonization by *Hylurgops palliatus* to emergence of *Peltis grossa*. Thickness of line indicates strength of effect. Coefficients indicate direct effects. U = unexplained variation (1–R²). * $p < 0.05$; *** $p < 0.001$. *Source: From Weslien, J., et al., 2011. Long-term priority effects among insects and fungi colonizing decaying wood. J. Anim. Ecol. 80, 1155–1162.*

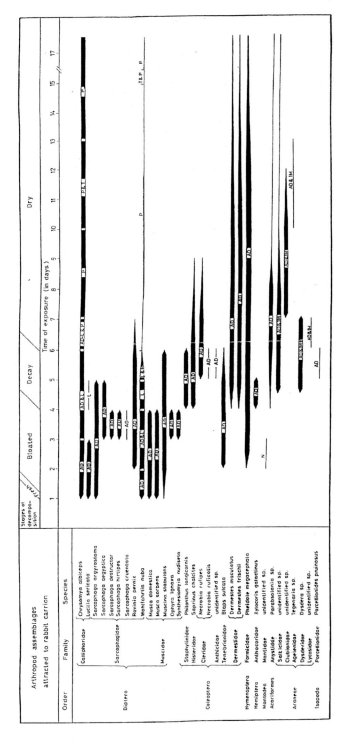

FIGURE 10.7 Succession of arthropods on rabbit carrion during summer in Egypt. *Source: From Tantawi, T.I., et al., 1996. Arthropod succession on exposed rabbit carrion in Alexandria, Egypt. J. Med. Entomol. 33, 566–580.*

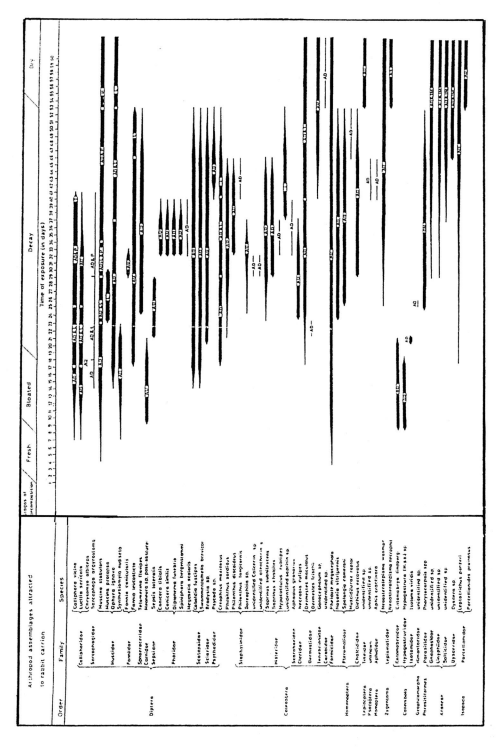

FIGURE 10.8 Succession of arthropods on rabbit carrion during winter in Egypt. *Source: From Tantawi, T.I., et al., 1996. Arthropod succession on exposed rabbit carrion in Alexandria, Egypt. J. Med. Entomol. 33, 566–580.*

Distinct assemblages of insects characterize mammalian versus reptilian carcasses (Watson and Carlton, 2003). For all animal carcasses, the fresh, bloated, and decay stages are dominated by various Diptera, especially calliphorids, whereas later stages are dominated by Coleoptera, especially dermestids.

The duration of each stage depends on environmental conditions that affect the rate of decay (compare Figs. 10.7 and 10.8). Some species occur only at certain times of the year (Shi et al., 2009). Horenstein and Linhares (2011) reported that staphylinid beetle colonization of pig carcasses in the shade was delayed, relative to the timing of colonization of carcasses in full sun. Carrion feeder abundances and duration of decay also are affected by predators, especially ants (Tullis and Goff, 1987; Wells and Greenberg, 1994) and staphylinid beetles (Horenstein and Linhares, 2011). The distinct sequence of necrophagous insect community types, as modified by local environmental factors, has been applied by forensic entomologists to determine time since death in criminal cases.

3.2 Factors Affecting Succession

Succession generally progresses toward the community type characteristic of the biome within which it occurs, for example, toward deciduous forest within the deciduous forest biome, or toward chaparral within the chaparral biome (Whittaker, 1953, 1970). However, succession can progress along various alternative pathways and reach alternative endpoints, depending on a variety of local abiotic and biotic factors (Norden et al., 2015). Substrate conditions represent an abiotic factor that selects a distinct subset of the regional species pool determined by climate. Farwig et al. (2014) reported that lower rates of carrion decomposition at lower temperatures reflected primarily the absence of larger beetle species at low temperatures, rather than the reduced total number of beetles. Distinct initial communities reflecting disturbance conditions, or unique conditions of local or regional populations, can affect the success of subsequent colonists. These initial conditions, and subsequent changes, guide succession into alternative pathways leading to distinct self-perpetuating endpoints (Egler, 1954; Norden et al., 2015; Whittaker, 1953). Herbivory and granivory also can guide succession along alternative pathways (Blatt et al., 2001; Davidson, 1993; Quesada et al., 2009).

Substrate conditions affect the ability of organisms to settle, become established, and derive necessary resources. Some substrates restrict species representation, for example, serpentine soils, gypsum dunes, and lava flows. Relatively few species can tolerate such unique substrate conditions, or the exposure resulting from limited vegetative cover. In fact, distinct subspecies often characterize the communities on these and the surrounding substrates. Contrasting communities characterize cobbled or sandy sections of streams, because of different exposure to water flow and filtration of plant or detrital resources. Finally, sites with a high water table support communities that are distinct from the surrounding communities, for example, marsh or swamp communities embedded within grassland or forested landscapes.

Successional pathways are affected by the composition of initial colonists and survivors from the previous community. The initial colonists of a site represent regional species pools, and their composition can vary depending on proximity to population sources. A site is more likely to be colonized by abundant species than by rare species. Rapidly growing and expanding populations are more likely to colonize even marginally suitable sites than are declining populations (see Chapter 7). For example, trees dying during a period of minimal bark beetle abundance would undergo a delay in initiation of heterotrophic succession, dominated by a different assemblage of insect species associated with different microorganisms (Stephen et al., 1993). Wood initially colonized by decay fungi, such as inoculated by wood-boring beetles, wasps, and termites, decays more rapidly, thereby affecting subsequent colonization, than does wood initially colonized by mold fungi, such as inoculated by bark and ambrosia beetles (Käärik, 1974; Progar et al., 2000).

Some individuals survive disturbance, depending on their tolerance to (or protection from) disturbance, and affect subsequent succession (Egler, 1954). Disturbance scale also affects the rate of colonization. Succession initiated primarily by ruderal colonists will differ from succession initiated by a combination of ruderal colonists and surviving individuals and propagules, for example, seed banks. Such legacies from the previous community contribute to the early appearance and advanced development of later successional species. These may preclude establishment of some ruderal species that would lead along a different successional pathway. Large-scale disturbances promote ruderal species that can colonize a large area rapidly, whereas small-scale disturbances may expose too little area for shade-intolerant ruderal species and be colonized instead by later successional species expanding from the edge (Brokaw, 1985; Denslow, 1985; Shure and Phillips, 1991). Fastie (1995) identified distance from each study site to the nearest seed source of Sitka spruce, *Picea sitchensis*, at the time of deglaciation as the major factor explaining among-site variance in spruce recruitment at Glacier Bay, Alaska.

The sequence of disturbances during succession determines the composition of successive species assemblages. For example, fire followed by drought would filter the community through a fire-tolerance sieve, then a drought-tolerance sieve, whereas flooding followed by fire would produce a different sequence of communities (J. Chase, 2007; Wiescher et al., 2012). Schowalter et al. (unpubl. data) found that arthropod community structure following hurricanes in Puerto Rico depended on prior drought and vice versa (Fig. 2.9). Arthropod communities reflected the influence of Hurricane Hugo for at least 19 years. In other words, community structure at any point in time depends on the prior history of disturbances and other environmental changes. Harding et al. (1998), Schowalter et al. (2003), and Summerville et al. (2009) demonstrated that arthropod communities in stream, forest litter, and forest canopies, respectively, showed responses to experimental disturbances that reflected distinct community structures among blocks with different disturbance histories.

Disturbances also can truncate community development. Grasslands and pine forests often dominate sites with climatic conditions that could support mesic forest, but succession is arrested by topographic or seasonal factors that increase the incidence of lightning-ignited fires and preclude persistence of mesic trees (Schowalter et al., 1981a).

Longer-term environmental changes (including anthropogenic suppression of disturbances) also affect the direction of community development. J. T. Chase (2007) found that species assembly in experimental ponds varied considerably among replicates and appeared to be largely stochastic. However, ponds subjected to experimental drought (by slowing draining during late summer and refilling during winter/spring) became more similar, reflecting the filtering out of species incapable of tolerating this disturbance. Ironically, fire suppression to "protect" natural communities often results in successional replacement of fire-dominated communities, such as pine forests and grasslands. The replacing communities may be more vulnerable to different disturbances. For example, fire suppression in the intermountain region of western North America has caused a shift in community structure from relatively open, pine woodland maintained by frequent, low-intensity ground fires to closed-canopy pine/fir forest that has become increasingly vulnerable to drought and stand-replacing crown fires, as well as to defoliator and bark beetle outbreaks in stressed trees (Agee, 1993; Raffa et al., 2008; Schowalter and Lowman, 1999; Wickman, 1992).

The importance of animal activity to successional transitions has become more widely recognized, given the obvious effects of many herbivores and granivores on plant species composition (Bishop, 2002; Cairns et al., 2008; T. Coleman et al., 2008; Davidson, 1993; Goodale et al., 2014; Karlsen et al., 2013; Louda et al., 1990a; Maloney and Rizzo, 2002; Norghauer and Newbery, 2014; Quesada et al., 2009; Ritchie et al., 1998; Torres, 1992; Wickman, 1980, see Chapter 12). Animals often are the drivers of successional transitions (Davidson, 1993; Goodale et al., 2014; MacMahon, 1981; Schowalter and Lowman, 1999; Willig and McGinley, 1999). Goodale et al. (2014) planted seedlings of six pioneer tree species across a gradient of light exposure from the center of canopy gaps into interior tropical forest in Sri Lanka and found that canopy light environment (highest in the centers of gaps) and herbivory (highest in the forest understory) were the primary factors determining the distribution of surviving seedlings after 550 days.

Blatt et al. (2001) showed that incorporation of herbivory into an old-field successional model helped to explain the multiple successional pathways that could be observed. Herbivorous species can delay colonization by host species (Goodale et al., 2014; Tyler, 1995; D. Wood and Andersen, 1990) and can suppress or kill host species and facilitate their replacement by nonhosts over areas as large as 10^6 ha during outbreaks (Bishop, 2002; Cairns et al., 2008; Coleman et al., 2008; Schowalter and Lowman, 1999). Bishop (2002) reported that insect herbivores limited the persistence and spread of early successional lupines, *Lupinus lepidus*, during primary succession on Mount St. Helens following the 1980 eruption. Bullock (1991) reported that the scale of disturbance can affect animal activity, thereby influencing colonization and succession. Generally, herbivory and granivory during early seres halt or advance succession (V.K. Brown, 1984; Schowalter, 1981; Torres, 1992), whereas herbivory during later seres halts or reverses succession (Davidson, 1993; Schowalter and Lowman, 1999). Similarly, Tullis and Goff (1987) and Wells and Greenberg (1994) reported that predaceous ants affected colonization and activity of necrophages, thereby affecting succession of the carrion community.

Granivores tend to feed on the largest seeds available, which most often represent later successional plant species, and thereby inhibit succession (Davidson, 1993). Herbivores and granivores can interact competitively to affect local patterns of plant species survival and succession. For example, Ostfeld et al. (1997) reported that voles dominated interior portions of old fields, fed preferentially on hardwood seedlings over white pine, *Pinus strobus*, seedlings, and competitively displaced mice, which fed preferentially on white pine seeds over hardwood seeds near the forest edge. This interaction favored growth of hardwood seedlings in the ecotone and favored growth of white pine seedlings in the old-field interior. Ants affect succession through dispersal of plant seeds and manipulation of soil and vegetation around nest sites (Q. Guo, 1998; R. Inouye et al., 1980; Jonkman, 1978, see Chapters 13 and 14).

Animals that construct burrows or mounds or that wallow or compact soils can kill all vegetation in small (several m^2) patches and/or provide suitable germination habitat and other resources for ruderal plant species (D. Andersen and MacMahon, 1985; MacMahon, 1981, see also Chapter 14), thereby reversing succession. Several studies have demonstrated that ant and termite nests create unique habitats, typically with elevated nutrient concentrations, that support distinct vegetation when the colony is active and facilitate succession following colony abandonment (Brenner and Silva, 1995; Garrettson et al., 1998; Q. Guo, 1998; T. King, 1977a,b; Lesica and Kannowski, 1998; Mahaney et al., 1999, see Chapter 14). Jonkman (1978) reported that the collapse of leaf-cutter ant, *Atta vollenweideri*, nests following colony abandonment provided small pools of water that facilitated plant colonization and accelerated development of woodlands in South American grasslands.

Relatively few studies have tested community development hypotheses experimentally. Patterns of arthropod colonization of new habitats represent a relatively short-term succession amenable to analysis. D. Strong et al. (1984) considered the unwitting movement of plants around the world by humans to represent a natural experiment for testing hypotheses about development of phytophage assemblages on a new resource. They noted that relatively few arthropod colonists on exotic plants were associated with the plant in its native habitat. Most arthropods associated with exotic plants are new recruits derived from the native fauna of the new habitat. Most of the insects that colonize introduced plants are generalists that feed on a wide range of hosts, often unrelated to the introduced plant species, and most are external folivores and sap-suckers (Kogan, 1981; D. Strong et al., 1984). Miners and gall-formers represent higher proportions of the associated fauna in the region of plant origin, likely because of the higher degree of specialization required for feeding internally. For example, endophages represented 10–30% of the phytophages associated with two species of thistles in native European communities, but represented only 1–5% of phytophages associated with these thistles in southern California where they were introduced (D. Strong et al., 1984). These results indicate that generalists are better colonists than are specialists, but adaptation over ecological time increases exploitation efficiency (Kogan, 1981; D. Strong et al., 1984).

In one of the most ambitious studies of community development, Simberloff and Wilson (Simberloff, 1969; Simberloff and Wilson, 1969; E. Wilson and Simberloff, 1969) used methyl bromide fumigation to defaunate six small mangrove islands formed by *Rhizophora mangle* in Florida Bay. They then monitored the reestablishment of the arthropod community during the following year. Simberloff and Wilson (1969) reported that by 250 days after defaunation, all but the most distant island had species richness and composition similar to those of untreated islands, but densities were lower on treated islands. Initial colonists included both strong and weak fliers, but weak fliers, especially psocopterans, showed most rapid population growth. Ants, which dominated the mangrove fauna, were among the later colonists but showed the highest consistency in colonization among islands. Simberloff and Wilson (1969) found that colonization rates for ant species were related to island size and distance from population sources (see Chapters 7 and 9). The ability of an ant species to colonize increasingly smaller islands was similar to its ability to colonize increasingly distant islands. Species richness initially increased, declined gradually as densities and interactions increased, then reached a dynamic equilibrium with species colonization balancing extinction (see also E. Wilson, 1969). Calculated species turnover rates were > 0.67 species per day (Simberloff and Wilson, 1969), consistent with the model of MacArthur and Wilson (1967).

These studies explain why early successional stages are dominated by r-selected species with wide tolerances (generalists) and rapid reproductive rates, whereas later stages are dominated by K-selected species with narrower tolerances, necessary for coexistence among more specialized species (see Chapter 5). The first arthropods to appear on newly exposed or denuded sites (also glaciated sites) typically are generalized detritivores and predators that exploit residual or exogenous dead organic material (Hodkinson et al., 2001, 2002), and dying colonists unable to survive. These arthropods feed on less toxic material than do herbivores, or on material in which the defensive compounds have decayed. Herbivores can reappear only after their host plants become established, and specialized predators similarly appear after sufficient prey appear.

3.3 Models of Succession

Clements (1916) noted that comparison of successional stages is less useful than is understanding processes affecting the transitions from one sere to another. Nevertheless, few studies have continued over sufficient periods to evaluate the mechanism(s) producing successional transitions. Rather, a number of nonmutually exclusive models, all of which may affect particular transitions to varying degrees, have been proposed and debated widely (Connell and Slatyer, 1977; Horn, 1981; McIntosh, 1981; Peet and Christensen, 1980). The debate involves competing views of succession as reflecting two distinct contrasts: (1) resulting from either population dynamics or emergent ecosystem processes and (2) resulting from stochastic assembly or converging on equilibrial community structure (Horn, 1981; McIntosh, 1981).

The *facilitation model* was proposed by Clements (1916), who viewed communities as an entity that showed progressive (facilitated) development similar to the ontogeny of individual organisms. According to this model, also called *relay floristics* (Egler, 1954), successive stages cause progressive changes in environmental conditions that facilitate their replacement by the subsequent stage, and later successional species cannot appear until sufficient environmental modification by earlier stages has occurred. For example, soil development or increased plant density during early stages makes the environment less suitable for recruitment of additional early, r-selected species but more suitable for recruitment of later, K-selected species. Fire-dominated ecosystems (in which nitrogen is volatilized during fire) typically are colonized following fire by symbiotic nitrogen fixers such as lupines, *Lupinus* spp., black locust, *Robinia pseudoacacia*, alders, *Alnus* spp., ceanothus, *Ceanothus* spp., or cherries, *Prunus* spp., depending on biome. These species are relatively shade intolerant, and increasing density eventually suppresses their photosynthesis and nitrogen-fixation, facilitating replacement by shade-tolerant species growing in the understory and exploiting the replenished organic nitrogen in the soil (Boring et al., 1988). The increasing porosity and altered nutrient content of decomposing wood, resulting from heterotroph activity, preclude further recruitment of early successional species, for example, bark beetles and anaerobic or microaerophilic microorganisms, and facilitate replacement by later successional wood borers and more aerobic microorganisms (Edmonds and Eglitis, 1989; Zhong and Schowalter, 1989).

This model was challenged early. Gleason (1917, 1926, 1927), Whittaker (1953, 1970) and, more recently, Drury and Nisbet (1973), argued that species colonization and turnover were based on life history attributes and population dynamics. Connell and Slatyer (1977), Horn (1981), and MacMahon (1981) proposed that succession could reflect multiple pathways and mechanisms.

Egler (1954) argued that secondary succession often may reflect differential longevity of colonizing species. Most of the eventual dominants colonize relatively early while competition is still low. Failure of species to become established at this early stage reduces the probability of future dominance. Juveniles of later species grow to maturity over a longer period, tolerating the early dominance of ruderal species, and eventually exclude the early successional species (eg, through shading, preemptive use of water, etc.). Connell and Slatyer (1977) referred to this model as the *tolerance model*. This model is represented best in ecosystems dominated by species that sprout from roots or stumps, germinate from seed banks, or colonize rapidly from adjacent sources. These attributes ensure early appearance along with ruderal species. However, many large-seeded trees, flightless arthropods, and other animals characterizing later successional stages of forest ecosystems require a long period of colonization and establishment and achieve dominance only during late succession, especially in large areas of disturbed habitat (Shure and Phillips, 1991).

A third model, proposed by Connell and Slatyer (1977) to explain at least some successional transitions, is the antithesis of facilitation. According to their *inhibition model*, the initial colonists preempt use of resources and exclude, suppress, or inhibit subsequent colonists for as long as these initial colonists persist. Succession can proceed only as individuals are damaged or killed, and thereby release resources (including growing space) for other species. Examples of inhibition are successional stages dominated by allelopathic species, such as shrubs that increase soil salinity or acidity, by species that preempt space, such as many perennial sod-forming grasses whose network of rhizomes restrict establishment by other plants, by species whose life spans coincide with the average interval between disturbances, and by species that create a positive feedback between disturbance and regeneration, such as eucalypts, *Eucalyptus* spp. (Shugart et al., 1981). The inhibition model has been supported by patterns of succession in decomposing wood, where the sequence of colonization by various insects determines initial fungal association, and early colonization by mold fungi and some wood borers can inhibit subsequent establishment by decay fungi or other wood borers (Käärik, 1974; Weslien et al., 2011), delaying further succession. Environmental fluctuation, disturbances, or animal activity (such as gopher mounds, bison wallows, trampling, and insect outbreaks) often are necessary to disrupt this bottleneck in succession (MacMahon, 1981; Schowalter et al., 1981a; Schowalter and Lowman, 1999). However, Agee (1993), Schowalter (1985), and Schowalter et al. (1981a) noted that bark beetle outbreaks also increase fuel accumulation and the probability of fire, perhaps ensuring the continuity of pine forest (Fig. 10.9).

Horn (1981) developed a model of forest succession as a *tree-by-tree replacement* process using the number of saplings of various species growing under each canopy species (ignoring species for which this is not a reasonable predictor of replacement) and correcting for expected longevity. This model assumes that knowing what species occupies a given position narrows the statistical range of expected future occupants, and that the probability of replacement depends only on the species occupying that position and does not change with time unless the occupant of that position changes. The model is not directly applicable to communities in which recurrent large-scale disturbances are the primary factor affecting vegetation dynamics. Interestingly, Horn (1981) found that successive iterations by a given replacement matrix invariably converged (ie, became significantly more similar than predicted by a random model) on a particular community composition, regardless of the starting composition. These results indicate that convergence is not necessarily a reflection of biotic processes (Horn, 1981) and warrant increased attention to the rate of convergence and transition states producing convergence. E. Evans (1988) reported that grasshopper assemblage structure in replicate plots in a grassland ecosystem converged during recovery from fire.

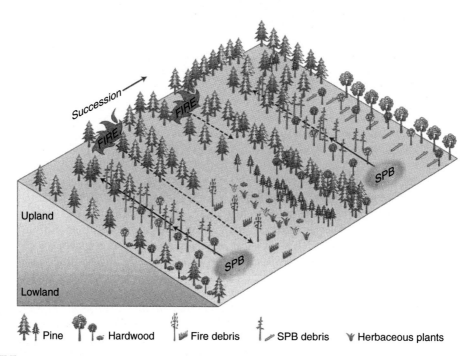

FIGURE 10.9 Diagrammatic representation of interactions between southern pine beetle and fire in the southeastern coniferous forest. Successional transitions extend from left to right; dotted arrows indicate direction of movement. Fire is a regular feature of the generally dry uplands, but moves into generally moist lowlands where drought or southern pine beetle creates favorable conditions for combustion. Southern pine beetle (SPB) is a regular feature of both forests, but is most abundant where pines occur at high density and stress levels. Fire is necessary for regeneration of pines, especially following succession to hardwoods if fire return is delayed. *Source: From Schowalter, T.D., et al., 1981. Role of southern pine beetle and fire in maintenance of structure and function of the southeastern coniferous forest. Environ. Entomol. 10, 821–825.*

Many ecologists consider vegetation changes over time to be no more than expressions of species' life history characteristics. Species distributions in time reflect their physiological tolerances to changing environmental conditions, parallel to distributions in space (Botkin, 1981; Drury and Nisbet, 1973), as well as to average life span. Several major simulation models of forest gap succession are based on species-specific growth rates and longevities as affected by stochastic mortality (T. Doyle, 1981; Shugart et al., 1981; Solomon et al., 1981). Platt and Connell (2003) explored effects of relationships between early colonists and later colonists on species replacement following catastrophic versus noncatastrophic disturbances as explanation for variable successional trajectories, depending on disturbance severity and relative survival of early and late successional species. However, Blatt et al. (2001) presented the only model that currently addresses the contribution of animals (insects) to the successional process. Norden et al. (2015) developed a model for Neotropical forest succession that integrated both deterministic and stochastic components to address dynamic interdependencies in successional pathways that varied widely among study plots, despite accounting for prior land use, initial conditions, and environmental variation. The variety of successional pathways determined by unique combinations of interacting initial and subsequent conditions may favor models that apply chaos theory (see Chapter 6).

4 PALEOECOLOGY

Paleoecology provides a context for understanding the development of extant interactions and community structure. Although most paleoecological study has focused on biogeographical patterns (Price, 1997), fossils also reveal much about prehistoric species interactions and community structure (Boucot, 1990; Boucot and Poinar, 2010; Labandeira, 1998; Labandeira and Sepkoski, 1993; Poinar and Poinar, 1999) and even the consequences of prehistoric changes in climate (Currano et al., 2008; Wilf and Labandeira, 1999; Wilf et al., 2001) or other disturbances (Labandeira et al., 2002). Similar morphological features of fossil and extant organisms imply similar functions and associated behaviors (Boucot, 1990; Boucot and Poinar, 2010; Poinar, 1993; Scott and Taylor, 1983), helping to explain fossil records as well as to understand long-term patterns of community change.

The fossil record contains abundant evidence of functions and behaviors similar to those observed currently. For example, haustellate mouthparts of proto-Hemiptera suggest the early appearance of feeding on plant sap (Labandeira and Sepkoski, 1993; Scott and Taylor, 1983). A fossil termite bug, *Termitaradus protera*, in Mexican amber has the same morphological modifications as its extant congeners for surviving in termite colonies and therefore can be assumed to have had similar interactions with termites (Poinar, 1993). Dental structure of Upper Carboniferous amphibians suggests that most were predaceous, many insectivorous (Scott and Taylor, 1983). Poinar et al. (2007) documented an example of chemical defense by a soldier beetle preserved in Burmese amber, at least 100 million years old.

Evidence of consistent species roles suggests that host selection behaviors and other species associations within communities have been conserved over time, the *Behavioral Fixity Hypothesis* (Boucot, 1990; Poinar, 1993; Poinar and Poinar, 1999). Association of potentially interacting taxa in the same deposits and anatomical evidence of interaction are common. For example, evidence of wood boring, perhaps by ancestral beetles, can be found as early as the Upper Carboniferous (Scott and Taylor, 1983). Bark beetle galleries and termite nests, complete with fecal pellets, in fossil conifers from the early- to mid-Tertiary demonstrate a long evolutionary history of association between these insects and conifers (Boucot, 1990; Labandeira et al., 2001). Some vertebrate coprolites from the Upper Carboniferous contain arthropod fragments (Scott and Taylor, 1983). The presence of fig wasps (Agaonidae) in Dominican amber suggests co-occurrence of fig trees (Poinar, 1993). Many fossil leaves from as early as the Upper Carboniferous show evidence of herbivory similar to that produced by modern insects (Boucot, 1990; Currano et al., 2008; Labandeira, 1998, 2002; Scott and Taylor, 1983). Gut contents from arthropods in Upper Carboniferous coal deposits indicate herbivorous, fungivorous, or detritivorous diets for most early arthropods (Labandeira, 1998; Scott and Taylor, 1983). Fossil dinosaur dung contains evidence of use by dung beetles (Poinar and Poinar, 2007). Dinosaur bones often show evidence of feeding by necrophilous species (Poinar and Poinar, 2007).

Demonstrated interaction between pairs or groups of particular species is uncommon (Boucot and Poinar, 2010) but provides the most convincing evidence of behavioral constancy (Fig. 10.10). Evidence of competition is particularly difficult to identify in fossil remains. Poinar and Poinar (2007) suggested that herbivorous insects likely competed with herbivorous dinosaurs for plant resources, especially during prehistoric insect outbreaks.

Co-occurrence of species that interacted directly is most likely to be preserved. Boucot (1990) reported a unique example of an extant insect species associated with extant genera in an Upper Miocene deposit in Iceland. The hickory aphid, *Longistigma caryae*, occurred in the same deposit with fossil leaves of *Carya* (or *Juglans*), *Fagus*, *Platanus*, and *Acer*. This aphid species survives on the same tree genera in eastern North America, providing strong evidence for long-term association between this insect and its hosts.

Mermithid nematodes that parasitize chironomid midges, typically castrating males and causing diagnostic changes in antennal morphology, are relatively common in Baltic and Dominican amber. Embedded male chironomids show both the altered antennal morphology and the nematode emerging at the time of host death (Boucot, 1990; Poinar, 1993). Parasitic mites frequently are found attached to their hosts in amber. Phoretic mites associated with their beetle or fly hosts are relatively rare (Boucot and Poinar, 2010), but have been found in Dominican amber (Poinar, 1993). Similarly, staphylinid beetles commensal in termite nests have been found with their termite hosts in Dominican amber (Poinar, 1993).

Microbial pathogens are more difficult to detect in fossil material. Thézé et al. (2011) used phylogenetic and Baysian analyses to infer that baculoviruses that infect insect larvae evolved during the same time period as the earliest holometabolous insects, about 310 mybp during the Carboniferous Period. Poinar and Poinar (2005) reported inclusion of cytoplasmic polyhedrosis virus and trypanosomatids in an adult ceratopogonid biting fly and nuclear polyhedrosis virus in an adult phlebotomid sand fly from early Cretaceous Burmese amber, as well as several types of fungal thalli on an adult mosquito and fungal growth on a fungus gnat in Dominican amber. These fossils represent the earliest evidence of microbial infections in arthropods. Apparently, insect polyhedrosis viruses were present 100 million years ago, and those infecting biting insects may have evolved into related insect-vectored vertebrate pathogens (Poinar and Poinar, 2005).

A few examples of demonstrated mutualistic interactions are preserved in the fossil record (Boucot and Poinar, 2010; Labandeira, 1998, 2002). Scott and Taylor (1983) noted that spores of Upper Carboniferous plants had a resistant sporoderm capable of surviving passage through animal guts, suggesting that herbivores may have served as agents of spore dispersal. An Upper Carboniferous arthropod, *Arthropleura armata*, was found with pollen grains of a medullosan seed fern attached along its posterior edge at the base of its legs. This species could have been an early pollinator of these seed ferns, whose pollen was too large for wind transport. Furthermore, some Upper Carboniferous plants produced glandular hairs that might have been an early type of nectary to attract pollinators (Scott and Taylor, 1983). An Early Cretaceous termite, *Kalotermes* sp., adult in Burmese amber and a Miocene termite, *Mastotermes electrodominicus*, in Dominican amber contained a variety of gut microbes, including specialized wood-digesting protists (Trichomonada, Hypermastigida, and Oxymonada) that have mutualistic associations with modern lower termites (Boucot and Poinar, 2010; Wier et al., 2002). Two Early Cretaceous cockroaches in Burmese amber contained protists related to mutualistic flagellates found in extant *Cryptocercus* cockroaches and lower termites (Fig. 10.10C, Boucot and Poinar, 2010;

FIGURE 10.10 **Fossil evidence of early interactions involving insects.** (A) Competitive or predaceous interaction between an ant, *Azteca alpha*, and a pseudoscorpion in Dominican amber, the ant clutching a pincher of the pseudoscorpion in its mandibles. (B) A nematode, *Heydenius formicinus*, emerging from an ant, *Prenolepis henschei*, in Baltic amber; arrow points to oak trichome, indicating habitat association similar to that of extant ants of this genus. (C) A protist, *Burmanymphus cretacea*, from the gut of a cockroach in Cretaceous Burmese amber; arrow indicates possible wood particle. Line drawing on right to clarify flagella and other structures. This protist is related to mutualistic gut flagellates occurring in extant *Cryptocercus* cockroaches. *Source: (A) from Poinar, Jr., G.O., 2001. Dominican amber. In: Briggs, D.E., Crowther, P.R. (Eds.), Palaeobiology II. Blackwell, Malden, MA; (B) from Poinar, Jr., G., 2002. First fossil record of nematode parasitism of ants: a 40 million year tale. Parasitology 125, 457–445; (C) from Poinar, Jr., G., 2009. Early Cretaceous protest flagellates (Parabasilia: Hypermastigia: Oxymonada) of cockroaches (Insecta: Blattaria) in Burmese amber. Cretaceous Res. 30, 1066–1072.*

(A)

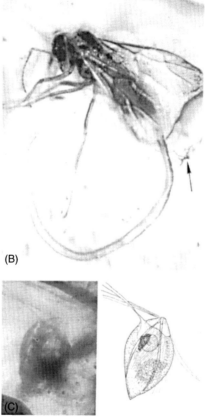
(B)

(C)

Poinar, 2009). These examples represent the earliest known records of mutualism between protists and terrestrial animals. A winged queen *Brachymyrmex* ant in Dominican amber was carrying a scale insect in its mandibles (Poinar and Poinar, 1994), transporting a honeydew source to its new nest site, a mutualism that persists today.

Complex multispecies interactions are indicated by the preservation of insects containing vectored pathogens and vertebrate blood (Poinar, 2005, 2015; Poinar and Poinar, 2004b;

Poinar and Telford, 2005). Fossils in Burmese amber from the early Cretaceous include a female ceratopogonid midge with malarial parasites in the abdominal cavity (Poinar and Telford, 2005) and a female phlebotomine sand fly containing nucleated reptilian blood cells infected with leishmanial trypanosomatids, as well as leishmanial trypanasomatids in the proboscis and midgut (Fig. 10.11) (Poinar and Poinar, 2004a,b). Insects apparently vectored major reptilian diseases as early as 100 million years ago, suggesting that novel insect-vectored diseases in naïve populations of dinosaurs could have contributed to dinosaur decline and vulnerability to eventual extinction (Poinar and Poinar, 2007).

Fossils in Dominican amber indicate transmission of avian malaria by mosquitoes as early as the mid-Tertiary and support suggestions that some forms of primate malaria evolved in the Americas. A fossil flea, *Atopopsyllus cionus*, in Dominican amber contained trypanosomes and bacteria that resemble extant pathogenic microorganisms that are vectored to vertebrate hosts (Poinar, 2015).

Fossil data permit limited comparison of diversity and species interactions between taxonomically distinct fossil and extant communities (see also Chapter 9). Insect diversity has increased at a rate of about 1.5 families per million years since the Devonian; the rise of angiosperms during the Cretaceous contributed to diversification within families but did not increase the rate of diversification at the family level (Labandeira and Sepkoski, 1993). Arthropod diversity was high in the communities recorded in Upper Carboniferous coal deposits and in Dominican and Mexican ambers (Poinar, 1993; Poinar and Poinar, 1999; Scott and Taylor, 1983). Similar associations, as discussed earlier, indicate that virtually all types of interactions represented by extant communities (eg, herbivore–plant, arthropod–fungus, predator–prey, pollinator, wood borer, detritivore, etc.) were established as early as the Upper Carboniferous. Note, however, that during the past 500 million years, there have been arguably five mass extinction events, during which the rate of species loss was more than twice the background rate, and following which speciation largely replaced lost biodiversity, albeit over long periods of time (Alroy, 2008; Barnosky et al., 2011; Foster and Twitchett, 2014).

The behavioral fixity hypothesis permits reconstruction of prehistoric communities, to the extent that organisms associated in sediments, coal, amber, or other deposits represent prehistoric communities (eg, Fig. 10.12) (Poinar, 1993; Poinar and Poinar, 1999). The Upper Carboniferous coal deposits represent a diverse, treefern-dominated, swamp ecosystem. The fossils in Dominican amber represent a tropical, evergreen angiosperm rainforest. Some insect specimens indicate the presence of large buttress-based host trees, whereas other specimens indicate the presence of palms in forest openings (Poinar, 1993; Poinar and Poinar, 1999). The presence of fig wasps indicates that fig trees were present. Baltic amber contains a combination of warm temperate and subtropical groups, suggesting a number of possible community structures. The temperate elements could have originated at a higher elevation, or Baltic amber may have formed during a climate change from subtropical to temperate conditions (Poinar, 1993). Diversity, food web structure, and functional group organization were similar between these extinct communities and extant communities (Poinar, 1993; Scott and Taylor, 1983), suggesting that broad patterns of community structure are conserved through time, even as species composition changes (Poinar and Poinar, 1999).

The fossil record can document changes in community structure at a site through time. The degree to which particular community types occur in continuous strata at a site indicates

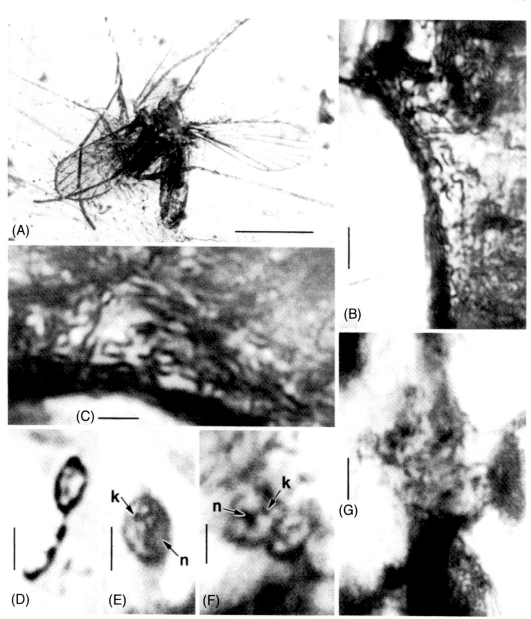

FIGURE 10.11 **Stages in the development of *Paleoleishmania proterus* n. gen., n. sp.** (A) Burmese amber sand fly containing stages of *Paleoleishmania proterus* n. gen., n. sp. Bar = 540 μm. (B) Promastigotes in the abdominal midgut of the fossil sand fly. Bar = 10 μm. (C) Detail of promastigotes in the abdominal midgut of the fossil sand fly. Bar = 5 μm. (D) Pear-shaped promastigote with nucleus and kinetoplast. Bar = 6.3 μm. (E) A single amastigote in the proboscis of the fossil sand fly. n = nucleus, k = kinetoplast. Bar = 2.2 μm. (F) Two amastigotes in the proboscis of the fossil sand fly. n = nucleus, k = kinetoplast. Bar = 2 μm. (G) Group of amastigotes in the proboscis of the fossil sand fly. Bar = 5 μm. *Source: From Poinar, Jr., G., Poinar, R., 2004a. Palaeoleishmania proterus n. gen., n. sp., (Trypanosomatidae: Kinetoplastida) from Cretaceous Burmese amber. Protist 155, 305–310.*

FIGURE 10.12   **Fossil evidence of insect species associations.** (A) Tipulid and sciarid flies (Diptera) and a beetle in Eocene shale (Green River Formation, Utah). (B) Sciarid and phorid flies (Diptera) and spider from a Columbian amber sample containing > 12 species of insects (4 orders) and spiders.

consistency of environmental conditions and community structure (Boucot, 1990; Labandeira et al., 2002). Boucot (1990) noted that, although a particular fossilized community (taxonomic association) rarely persists long in a local stratigraphic section, communities typically recur over larger areas for 10^6–10^7 years, indicating a high degree of stability within environmental constraints, similar to current landscape distributions of particular community types. Labandeira et al. (2002) compiled data for insect–plant associations spanning the Cretaceous–Tertiary

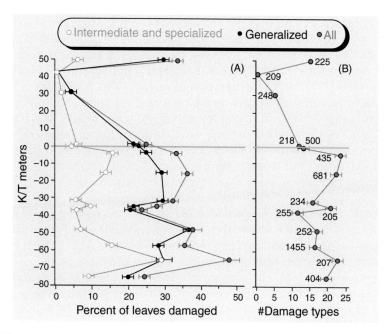

FIGURE 10.13 (A) Frequency analysis (percentage) of insect damage for 14 stratigraphic horizons (with at least 200 specimens of identified dicot leaves) across the Cretaceous/Tertiary (K/T) boundary (orange bar) from the Williston Basin of southwestern North Dakota, USA. The horizontal scale is the percentage of leaves bearing insect damage (+ 1 SD). The green line represents combined damage types; the black line is generalized damage types only; the purple line is intermediate and specialized damage types. Because some individual leaves contain more than one damage type, the total percentage (green) is usually less than the sum of the two other data series. (B) Diversity analysis of insect damage, with raw data bootstrapped to 5000 replicates. Vertical scale as in (A). The data labels show the number of leaves in each sample. Poor preservation is probably responsible for the lack of recovered insect damage around the 30–40-m interval. *Source: From Labandeira, C.C., 2002. The history of associations between plants and animals. In: Herrera, C.M., Pellmyr, O. (Eds.), Plant-Animal Interactions: an Evolutionary Approach. Blackwell, London, pp. 26–74.*

boundary. They found that specialized (monophagous) associations almost disappeared at the boundary and have not recovered to Cretaceous levels, whereas generalized (polyphagous) associations regained their Cretaceous abundances (Fig. 10.13).

Prehistoric changes in Earth's climate have been documented using changes in fossil community structure, including changes in abundances and composition of plants, pollen, and associated animals (Lindbladh et al., 2013). Arthropod assemblages should respond strongly to changes in prevailing temperature (see Chapter 2). For example, the Paleocene-Eocene Thermal Maximum (PETM) 56 million years BP, during which average temperatures were about 5°C higher than at present, can be inferred from changes in species composition and abundances. Wilf and Labandeira (1999) and Currano et al. (2008) reported that insect herbivore diversity and intensity of herbivory, as indicated by chewed plant parts, increased sharply during the global warming interval from the late Paleocene to early Eocene (Fig. 2.18). Irvine et al. (2012) used pollen and midge assemblage in sediment cores from Trout Lake, Alaska, to reconstruct environmental changes over the past 15,000 years. They found sharp transitions in plant and midge composition indicating a major warming period 14,400–12,800 years BP, a cooling period 12,200–11,200 years BP, followed by a second warming period, during which average temperature was about 2.2°C higher than at present (Fig. 10.14).

Gear and Huntley (1991) reported that dating of fossilized Scots pine, *Pinus sylvestris*, stumps in northern Scotland indicated that pine forest expanded rapidly northward 70–80 km about 4000 years BP and persisted for about 400 years before retreating southward again, suggesting a 400-year period of warmer climate and community change. However, they noted that even this remarkably rapid rate of species movement would be insufficient (by an order of magnitude) to accomplish the range change necessary for survival under future climate change scenarios, especially if population spread were impeded by landscape fragmentation.

Data from these long-term studies provide the basis for anticipating population and community responses to future environmental changes. The consistency of community responses to climate changes in the past indicates major changes resulting from current climate warming. Unfortunately, the current rate of warming is unprecedented in the past (Fig. 10.15). Past changes comparable to the current rate of change occurred over thousands of years. As Ice Ages ended over the past 1 million years, global temperature rose a total of 4–7°C over about 5000 years, whereas global temperature has increased 0.7°C during the past century, about 10-fold faster than the average rate of warming during recovery from the Ice Age (Jouzel et al., 2007; Mann et al., 2008). Many species, including dominant vegetation in some ecosystems, will be unable to adapt at this rate of change, potentially threatening the structure of their communities (Beedlow et al., 2013; Bellard et al., 2012). However, herbivore outbreaks are likely to become more frequent and intense under warmer conditions (Currano et al., 2008; Yu et al., 2009, see Figs. 2.18 and 10.1).

5 DIVERSITY VERSUS STABILITY

The relationship between community diversity and stability remains a controversial issue (de Ruiter et al., 1995; Gonzalez and Loreau, 2009; Grime, 1997; Hooper and Vitousek, 1997; Hooper et al., 2005; Schulze and Mooney, 1993; Tilman et al., 1997, see Chapter 15). An early assumption that diversity conferred stability on communities and ecosystems was challenged, beginning in the 1970s, by modeling efforts that indicated increasing vulnerability to perturbation with system complexity (May, 1973, 1981; Yodzis, 1980). However, more recent studies have addressed the importance of diversity for maintenance of ecosystem processes (de Ruiter et al., 1995; Fukami et al., 2001; Gonzalez and Loreau, 2009; Tilman and Downing, 1994; Tilman et al., 1997). Among these are studies of "pest" dynamics and their effects on community structure in diverse communities versus simple communities (Jactel and Brockerhoff, 2007).

Fundamental to our understanding of this relationship are definitions and measurements of diversity and stability (O'Neill, 2001, see Chapter 15). As noted in Chapter 9, the variety of methods for measuring diversity has complicated comparison of communities, including assessment of community change. Should diversity be measured as species richness,

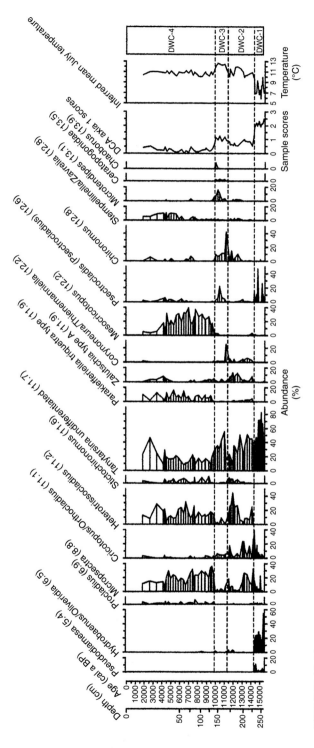

FIGURE 10.14 Midge abundances in a deep-water core from Trout Lake, Alaska. DCA axis 1 scores and midge-inferred temperatures of the fossil samples plotted by calibrated radiocarbon years. Numbers in parentheses represent the temperature optima of extant counterparts. *Source: From Irvine, F., et al., 2012. Midge-inferred temperature reconstructions and vegetation change over the last ~15,000 years from Trout Lake, northern Yukon Territory, eastern Beringia. J. Paleolimnol. 48, 133–146.*

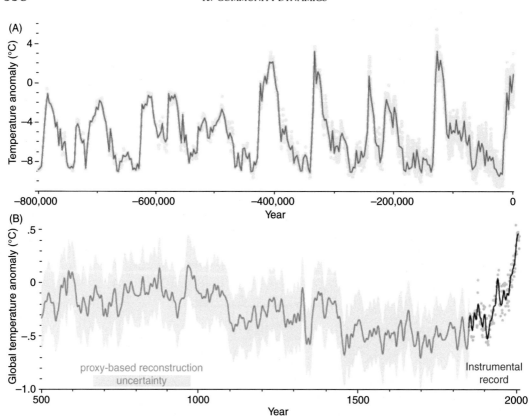

FIGURE 10.15　Temperature change over the past 800,000 years (A) and detail from last 1500 years (B). As the Earth moved out of Ice Ages over the past 1 million years, global temperature rose a total of 4–7°C over about 5000 years, whereas global temperature has increased 0.7°C during the past century, about 10 times faster than the average rate of warming during recovery from the Ice Age. Courtesy of NASA, earthobservatory.nasa.gov/Features/ GlobalWarming/page3.php. *Source: from Jouzel, J., et al., 2007. Orbital and millennial Antarctic climate variability over the ast 800,000 years. Science 317, 793–796 and Mann, M.E., et al., 2008. Proxy-based reconstructions of hemispheric and global surface temperature variations over the past two millennia. Proc. Natl. Acad. Sci. USA 105, 13252–13257.*

functional group richness, or some diversity index using species or functional groups (de Ruiter et al., 1995; Grime, 1997; Hooper and Vitousek, 1997; Tilman and Downing, 1994; Tilman et al., 1997)? Stability can be defined as reduced variability in system behavior. However, ecologists have disagreed over which variables are best to measure stability. Stability has been shown to have multiple components, one representing capacity to resist change (*resistance*), and the other representing ability to recover following a change (*resilience*, eg, through succession), which indicate different degrees of stability for a given ecosystem (see Section 5.1).

Traditionally, stability was measured by population and community ecologists as the constancy of species composition and community structure over time (Grime, 1997; May, 1973, 1983). Ecosystem ecologists have emphasized ecosystem processes such as primary productivity, energy flux, and biogeochemical cycling, especially as variability changes during succession (de Ruiter et al., 1995; Kratz et al., 1995; E. Odum, 1969; Tilman and Downing, 1994). Species diversity may stabilize some variables but not others, or at one spatiotemporal scale but not another, leading to different conclusions. The extent to which diversity contributes to ecosystem integrity will be addressed in Chapter 15.

5.1 Components of Stability

Holling (1973) originally defined stability as the ability of a community to withstand disturbance with little change in structure, whereas resilience was the capacity of the community to recover following perturbation. Webster et al. (1975) subsequently refined the definition of stability to incorporate both resistance to change and resilience following perturbation. Succession is an expression of resilience. However, the criteria for measuring stability remain elusive. What degree of change can be accommodated before resistance is considered to have been breached? Does resilience require the recovery of the particular predisturbance community composition or of ecosystem functions that support a particular community type, and over what scale of space or time?

Webster et al. (1975) developed a functional model to evaluate the relative stability of ecosystems based on the lowest turnover rates, that is, the longest time constraint, and damping factors, that is, factors that reduce amplitude of fluctuation, in the system. The system has not fully recovered from displacement until the slowest component of the response has disappeared. They concluded that ecosystems with greater structure and amounts of resource storage were more resistant to disturbance, whereas ecosystems with greater turnover (eg, via consumption and succession) were more resilient. From a community standpoint, resistance depends on the level of tolerance of the dominant species to characteristic disturbances or other environmental changes, for example, through protected meristems or propagules, or resource storage; resilience is conferred by species with rapid recolonization and growth rates. Overall, temperate forests, with high biotic and abiotic storage and slow turnover, appear to be most resistant, but least resilient, to disturbance, and stream systems, with low biotic and abiotic storage and high turnover, appear to be least resistant, but most resilient. Resistance and resilience were found to be related inversely, with their relative contributions to stability in a given ecosystem determined by the proportions of K and r specialists (see Chapter 5). Succession appears to represent a trend from more resilient to more resistant communities.

Resistance and resilience are affected by regional species abundances and distribution and by resource availability. For example, Herbert et al. (1999) found that resistance and recovery of Hawaiian forests following Hurricane Iniki (1992) depended to a large extent on the availability of soil phosphorus (P) in plots that had been established 1 year prior to the hurricane. For example, mean stem growth of trees in the +P treatment showed less departure and statistically faster recovery, compared to –P plots (Fig. 10.16). Their results supported the hypotheses that ecosystem resistance and resilience are inversely related; resistance decreased and resilience increased as the availability of a limiting resource (P) increased.

Resistance can be compromised by fragmentation, which increases community exposure to external factors (Piessens et al., 2009). For example, trees in interior forest communities typically are buffered from high temperatures and high wind speeds by surrounding trees, and typically have less buttressing than open-grown trees. Fragmentation increases the proportion of trees exposed to high temperatures and wind speeds and thereby vulnerable

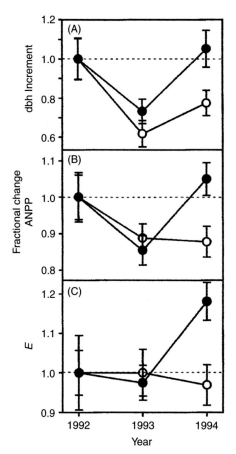

FIGURE 10.16 **Fractional change in three forest parameters on Kauai, Hawaii during 2 years following Hurricane Iniki (1992).** (A) stem diameter growth (dbh increment), (B) above-ground net primary production (ANPP), and (C) E (ANPP/leaf area). Values are means $\pm$ 1 SE. Closed circles are +P treatments; open circles are –P treatments. *Source: From Herbert, D.A., et al., 1999. Hurricane damage to a Hawaiian forest: nutrient supply rate affects resistance and resilience. Ecology 80, 908–920.*

to moisture stress or toppling (J. Chen et al., 1995; Franklin et al., 1992). Fragmentation also interferes with the adapted abilities of species in the regional pool to recolonize disturbed sites. Species are adapted to levels of dispersal and colonization sufficient to maintain populations within the characteristic habitat matrix of the landscape. If the rate of patch turnover is increased through fragmentation, the colonization capacity for many species may be insufficient to provide the necessary level of resilience for community recovery. Such changes in landscape condition may bias evaluation of community stability.

5.2 Stability of Community Variables

A number of community variables can be examined from the standpoint of their variability with respect to diversity. Among these are species composition and food web structure. Simpler communities, in terms of species composition and food web structure, often appear to be more stable than complex communities, with respect to recovery from disturbance (May, 1973, 1983). Boucot (1990) noted that simple marine communities in the fossil record continue across sedimentary discontinuities more often than do complex marine communities. Boucot (1990) noted that particular taxonomic associations typically recur over larger areas for 10^6–10^7 years, indicating a high degree of stability within environmental constraints. However, the appearance of greater stability for simpler systems may reflect their limited capacity for variation in recovery from disturbance. The variety of successional pathways leading to multiple endpoints (Horn, 1981; Whittaker, 1953) indicates that complex communities may recover most of their predisturbance composition or food web structure, but not the relative abundances of various species. Furthermore, some mechanisms lead to positive feedback between disturbance and community organization (Schowalter, 2008; Schowalter et al., 1981a; Shugart et al., 1981).

Modeling approaches have led to contrasting conclusions. May (1973, 1983) and Yodzis (1980) reported that more complex communities were more vulnerable to disruption by perturbations in any particular species population because of their cascading effects through the network of interactions involving that species. However, de Ruiter et al. (1995) incorporated the patterning of interaction strengths in real communities and found that simultaneous occurrence of strong top-down regulation of lower trophic levels and strong bottom-up regulation of higher trophic levels imposed stabilizing patterns on interaction strengths. E. Evans (1988) found that grasshopper assemblages converged toward significantly greater similarity in structure following fire in a grassland ecosystem than predicted by a random model. Fukami et al. (2001) modeled compartmentalized communities and demonstrated that increasing diversity increased similarity in composition among local communities and that greater similarity improved reliability of community structure and function. Thébault and Fontaine (2010) applied a modeling approach to 34 empirical pollination networks and 23 empirical herbivory networks to demonstrate that a highly connected and nested architecture promotes community stability in mutualistic networks, whereas a compartmented and weakly connected architecture promotes the stability of trophic networks.

A number of studies, especially in aquatic and grassland systems, have demonstrated that higher diversity permits compensatory responses in species composition (eg, replacement of intolerant species by more tolerant species) that maintain ecosystem productivity, which underlies ecosystem structural and functional attributes (Gonzalez and Loreau, 2009). Although Houlahan et al. (2007) reported that most studies of natural systems show positive covariances among species, rather than negative covariances predicted by earlier compensatory dynamics models, Loreau and de Mazancourt (2008) and Gonzalez and Loreau (2009) argued that positive covariances could be consistent with compensatory dynamics when various populations in the community are inherently synchronized by strong environmental forcing and/or fluctuating abundance of a dominant species.

Diversity may dampen the spread of insects or pathogens that could threaten some species, hence disrupt community structure. For example, the diversity of pines and hardwoods in the southern USA reduces growth and spread of southern pine beetle, *Dendroctonus frontalis*, populations (Schowalter and Turchin, 1993). Ostfeld and Keesing (2000) found that the number of human cases of Lyme disease, caused by the tick-vectored spirochaete, *Borrelia burgdorferi*, declined with species richness of small mammals and lizards, but increased with species richness of ground-dwelling birds (Fig. 10.17). Similarly, B. Allan et al. (2009) reported

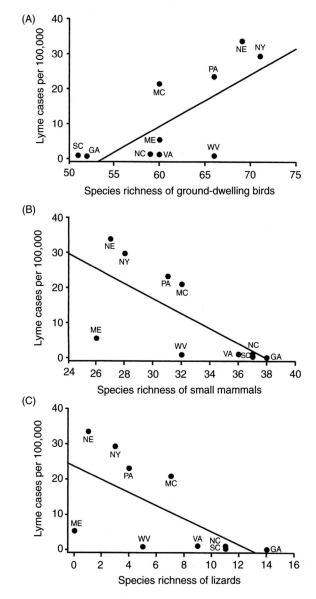

FIGURE 10.17 **Relationship between reported cases of human Lyme disease in 1996 and species richness of ground-dwelling birds (A), small mammals (B), and lizards (C) in the eastern USA.** *GA*, Georgia; *MC*, Mid-Atlantic states; *ME*, Maine; *NC*, North Carolina; *NE*, New England states; *NY*, New York; *PA*, Pennsylvania; *SC*, South Carolina; *VA*, Virginia; and *WV*, West Virginia. *Source: From Ostfeld, R.S., Keesing, F., 2000. Biodiversity and disease risk: the case of Lyme disease. Conserv. Biol. 14, 722–728.*

that the incidence of human infection with West Nile virus was negatively correlated with bird diversity, but not with human population density. These data indicated that disease epidemiology may depend on the diversity of reservoir hosts, but disease incidence generally should decline with increasing dilution of reservoir hosts by nonhosts.

Alternatively, insects could be viewed as accelerating compensatory dynamics or providing the negative feedback that prevents unsustainable production by any particular plant species (see Chapter 15). By preferentially targeting stressed, especially dense, hosts, herbivorous insects would accelerate replacement of intolerant species by more tolerant species and increase overall diversity, thereby functioning in the capacity of keystone species.

To some extent, lack of a clear correlation between diversity and stability of community variables may be an artifact of the duration of succession or the number of intermediate stages that can generate alternative pathways. More frequently disturbed communities may appear to be more stable than infrequently disturbed communities because a small and consistent group of species are selected by disturbance (J. Chase, 2007; Wiescher et al., 2012) or because the ecological attributes of ruderal species favor rapid recovery, whereas longer time periods and more intervening factors affect recovery of tree species composition in disturbed forests. Furthermore, if maximum species diversity occurs at intermediate levels of disturbance (the intermediate disturbance hypothesis), then the lower species diversity of earlier and later successional communities is associated with both high and low stability, in terms of frequency and amplitude of departure from a particular community structure.

A major source of diversity is the variety of community types and the regional species pool maintained in a shifting landscape mosaic of patch types that vary in substrate conditions and disturbance histories. Although the community of any particular site may appear unstable because of multiple factors interacting to affect its response to perturbation, the metacommunity represented by the landscape pattern of local communities (see Chapter 9) minimizes the distance between population sources and sinks and ensures proximity of colonists for species packing and assortment during site recovery (Cottonie, 2005; Cottonie and de Meester, 2004; Kneitel and Miller, 2003; LeCraw et al., 2014; Leibold et al., 2004; Supp and Ernest, 2014). Even if the community in one patch does not recover to the same predisturbance endpoint, that predisturbance community structure likely occurs in other patches.

6 SUMMARY

Community structure changes over a range of timescales, from annual to decadal to millennial time periods. Temporal patterns of community organization and their sensitivity to environmental changes can indicate their stability in response to anthropogenic changes.

Community structure changes over short time periods as population sizes respond to seasonal and annual variation in environmental conditions. Changes in resource quality, competition, and predation lead to population irruptions of some species and local extinction of others, thereby affecting their interactions with other species and leading to changes in community structure.

Ecological succession, the sequential stages of community development on newly exposed or disturbed sites, is one of the best-documented ecological phenomena and has provided a unifying concept that integrates species' life history strategies, population behavior, community dynamics, and ecosystem processes. Early successional communities typically are dominated by relatively generalized ruderal species with high mobility and rapid reproductive rates. Later successional stages are increasingly dominated by species that are more specialized, less mobile, and have lower reproductive capacities. Although most studies of succession have focused on plants, insects show successional patterns associated with changes in vegetation, and the relatively rapid heterotrophic succession in decomposing wood and animal carcasses has contributed much to successional theory.

A number of factors influence successional pathways. Local substrate conditions can restrict initial colonists to those from the surrounding species pool that can become established on distinct substrates, such as serpentine, volcanic, or water-saturated soils. The composition of the initial community, including survivors of the previous disturbance and colonists, can affect the success of subsequent colonists. Subsequent disturbances and animal activity can affect successional pathways. Animals, including insects, create germination sites for colonists and suppress some host species, thereby facilitating, inhibiting, or reversing succession. In fact, animal activity often may account for vegetation changes that have been attributed to plant senescence.

Several models of succession have augmented the early model of succession as a process of facilitated community development, in which earlier stages create conditions more conducive to successive stages. In some cases, all the eventual dominants are present in the initial community, and succession reflects differential development time and longevity among species, that is, the tolerance model. Some successional stages are able to competitively exclude later colonists, the inhibition model. Succession may advance beyond such stages as a result of plant injury or death from subsequent disturbances or animal activity.

Paleoecological research indicates that species interactions and community structures have been relatively consistent over evolutionary time. However, the communities occupying particular sites have changed over these time periods as the environmental conditions of the site have changed. Recent research has provided important evidence for development of species interactions, including insect vectoring of vertebrate diseases.

The relationship between species or functional diversity and community or ecosystem stability has been highly controversial. Much of the discussion reflects different definitions of diversity and stability. Stability can be seen to have two major components: resistance to change, and resilience following perturbation. Succession is an expression of resilience. Although much evidence indicates that a particular community composition or structure may not be replaced at a site, indicating instability at the local level, the structure and diversity of natural communities at a landscape scale may ensure that declining species are replaced by more tolerant species (compensatory dynamics) and that component communities are maintained within a shifting landscape mosaic, indicating stability at the landscape or regional level.

ECOSYSTEM LEVEL

The ecosystem level of organization integrates species interactions and community structure with their responses to, and effects on, the abiotic environment (Fig. 1.2). Interactions among organisms are the mechanism governing energy and nutrient fluxes through ecosystems. The rates and spatial patterns in which individual organisms and populations acquire and allocate energy and nutrients determine the rate and direction of these fluxes (see Chapters 4 and 8).

Communities vary in their ability to modify their abiotic environment. The relative abundance of various nutrient resources affects the efficiency with which they are cycled and retained within the ecosystem. Increasing biomass confers increased storage capacity and buffering against changes in resource availability. Community structure also can modify climatic conditions, by controlling albedo and hydric fluxes, buffering individuals against changing environmental conditions.

A major issue at the ecosystem level is the extent to which communities are organized to maintain optimal conditions for the persistence of the community. Species interactions and community structures may represent adaptive attributes at the supra-organismal level that stabilize ecosystem properties near optimal levels for the various species. If so, anthropogenic interference with community organization (eg, species redistribution, pest control, overgrazing, deforestation) may disrupt stabilizing mechanisms and contribute to ecosystem degradation.

Insects affect virtually all ecosystem properties, especially through their effects on vegetation, detritus, and soils, earning them recognition as ecosystem engineers (eg, Hölker et al., 2015). Insects clearly affect primary productivity, hence the capture and flux of energy and nutrients. In fact, insects are the dominant pathway for energy and nutrient flow in many aquatic and terrestrial ecosystems. They affect plant growth and reproduction and vegetation density and porosity, hence albedo and the penetration of light, wind, and precipitation, as well as the composition of plant species with varying resource demands. They affect accumulation and decomposition of litter, and mixing and porosity of soil and litter, thereby affecting soil fertility and moisture. Insects often influence disturbance frequency, succession, and associated changes in efficiency of ecosystem processes over time. Their small size and rapid and dramatic responses to environmental changes are ideal attributes for regulators of ecosystem processes, through positive and negative feedback mechanisms. Ironically, effects of detritivores on decomposition (largely ignored by insect ecologists) have been emphasized by ecosystem ecologists, whereas effects of herbivorous insects on primary production (the focus of insect ecologists) have been all but ignored by ecosystem ecologists until recently.

Chapter 11 summarizes key aspects of ecosystem structure and function, including energy flow, biogeochemical cycling, and climate modification. Chapters 12–14 cover the variety of ways in which insects affect ecosystem structure and function. The varied effects of herbivores are addressed in Chapter 12. Although not often viewed from an ecosystem perspective, pollination and seed predation affect patterns of plant recruitment and primary production as described in Chapter 13. The important effects of detritivores on organic matter turnover and soil development are the focus of Chapter 14. Finally, the potential roles of these organisms as regulators of ecosystem processes are explored in Chapter 15.

11

Ecosystem Structure and Function

CONNECTING ECOSYSTEM STRUCTURE AND FUNCTION WITH GLOBAL CHANGES

Lindeman (1942) launched ecosystem research by demonstrating that the biotic and abiotic components of an aquatic ecosystem were inseparably connected by the exchange of energy and matter. Subsequent ecosystem research was largely site-specific, that is, ecosystems were treated as ecological units with relatively discrete boundaries, often the ridges separating distinct watersheds. Later, starting in the latter decades of the 20th century, ecosystems were compared in terms of their efficiency in retaining energy and nutrients, ability to modify local abiotic conditions, and to resist or recover from disturbances.

The advent of landscape ecology and stream continuum concepts during the 1980s broadened the perspective of ecosystems to a mosaic of interconnected patches that shared a common regional species pool and exchanged individuals, energy, and matter. Changes in the pattern of patch types could affect the distribution and exchange of energy and matter and alter local and regional climatic conditions. With growing public concern about effects of land use change, climate change, and invasive species, networks of long-term research sites, for example, the US Long Term Ecological Research (LTER) Program and National Environmental Observatory Network (NEON) and global networks of International Long Term Ecological Research (ILTER) and carbon flux towers (FLUXNET), established platforms on which ecosystem responses to, and effects on, environmental changes and disturbances could be measured (Peters et al., 2013; Van Cleve and Martin, 1991). Many of these sites have towers or construction cranes that permit access to the tops of forest canopies to facilitate measurement of interactions between the forest and atmosphere (Lowman et al., 2012). New technologies, such as high-resolution spectrophotometers, gas-exchange analyzers, and instruments for measuring eddy-covariance, permit precise measurements of carbon and nutrient fluxes among biotic and abiotic components of ecosystems (Irvine et al., 2005; Porder et al., 2005; Treuhaft et al., 2004), while improved isotopic detection methods permit identification of sources of atmospheric gases (Aranibar et al., 2006).

The application of this technology to remote sensing techniques has revolutionized our ability to monitor changes in ecosystem conditions on a global scale. Researchers now can detect changes in local, regional, and global net primary productivity (R. Waring and Running, 1998; Yuan et al., 2007),

Insect Ecology. http://dx.doi.org/10.1016/B978-0-12-803033-2.00011-X

net ecosystem productivity (Irvine et al., 2005; Misson et al., 2007; Treuhaft et al., 2004; D. Turner et al., 2005; Xiao et al., 2008), evapotranspiration rate (J. Zhang et al., 2009), foliage area, biomass and chemistry, especially water, lignin, nitrogen and, indirectly, phosphorus (Chambers et al., 2007; Porder et al., 2005), and drought or insect stress (Carter and Knapp, 2001; Nansen et al., 2009, 2010) using hyperspectral reflectance signatures, such as from airborne visible and infrared imaging spectrometry (AVIRIS). Data permit testing of cause-and-effect relationships between these ecosystem variables and global climate (J. Foley et al., 2003b; Janssen et al., 2008; Juang et al., 2007).

We currently have an unprecedented ability to detect changes in ecosystem structure and function at local, regional, and global scales. Results of large-scale studies are demonstrating ecosystem responses to, and effects on, changing climate, land use, invasive species, etc. For example, removal of overstory trees from a site greatly increases albedo and reduces the ability of the ecosystem to modify local precipitation and ameliorate the impact of torrential rain (Trenberth, 1999), leading to dramatic increases in soil temperature, erosion, and sedimentation of streams (J. Foley et al., 2003a). Removal of all trees or vegetation from a site exacerbates these effects (J. Foley et al., 2003b; Janssen et al., 2008). Although small-scale harvest and/or conversion of forests may appear to be relatively innocuous, cumulative deforestation or reduction in average height of forests not only leads to regional warming and drying (Janssen et al., 2008; Juang et al., 2007; Meher-Homji, 1991) but has effects far downstream in terms of flooding of human communities and infilling of reservoirs (H. Guo et al., 2008). Furthermore, atmospheric turbulence generated by changes in vegetation cover can increase the intensity of storms (Hossain et al., 2009; Kishtawal et al., 2010). Increasing frequency of extreme weather events is likely to have greater effects on species survival and ecosystem function, including loss of ecosystem ability to buffer climatic conditions, than will changes in average conditions (Breshears et al., 2005; Gutschick and BassiriRad, 2010; Jentsch et al., 2007; Lubchenco and Karl, 2012).

Insects have demonstrated capacity to modify environmental conditions. J.W. Moore (2006) described ways in which aquatic insects modify substrate and resource conditions. A number of herbivorous species are capable of dramatic alteration of vegetation structure and composition, changing habitat conditions for associated organisms (C.G. Jones et al., 1994), often in response to anthropogenic changes in environmental conditions (Raffa et al., 2008; Schowalter, 2008, see Chapter 2). This capacity of insects to alter anthropogenic designs for ecosystems requires that their responses to, and effects on, ecosystem conditions be addressed in any environmental policies or management decisions.

1 INTRODUCTION

Tansley (1935) coined the term "ecosystem" to recognize the integration of the biotic community and its physical environment as a fundamental unit of ecology, within a hierarchy of physical systems that span the range from atom to universe. Shortly thereafter, Lindeman's (1942) study of energy flow through an aquatic ecosystem introduced the modern concept of an ecosystem by demonstrating that exchange of energy and matter between biotic and abiotic pools makes a community inseparable from its environment. More recently, during the 1950s–70s, concern over the fate of radioactive isotopes from nuclear fallout generated

FIGURE 11.1   The community of aquatic organisms, including microflora and invertebrates, that develops in water-holding structures of plants, such as *Heliconia* flowers, represents a small-scale ecosystem with measurable inputs of energy and matter, species interactions that determine fluxes and cycling of energy and matter, and outputs of energy and matter.

considerable research on biological control of elemental movement through ecosystems (Golley, 1993). From the beginning, insects have been recognized as important distributors of energy and matter and engineers of ecosystem conditions (Crossley and Howden, 1961; Crossley and Witkamp, 1964; Smalley, 1960; Teal, 1962; Witkamp and Crossley, 1966). Recognition of anthropogenic effects on atmospheric conditions, especially contributions of greenhouse gas and pollutant concentrations to global warming (J. Foley et al., 2003a; Janssen et al., 2008; Juang et al., 2007), has stimulated broader interest in how natural and altered communities control fluxes of energy and matter and modify abiotic conditions.

Delineation of ecosystem boundaries can be problematic. Ecosystems can be described at various scales. At one extreme, the diverse flora and fauna living on the backs of rainforest beetles (Gressitt et al., 1965, 1968) or the aquatic communities in water-holding plant structures (phytotelmata, Fig. 11.1) (B. Richardson et al., 2000a,b) constitute an ecosystem. At the other extreme, the interconnected terrestrial and marine ecosystems constitute a global ecosystem that generated and controls Earth's soil and atmospheric conditions (Golley, 1993; J. Lovelock, 1988; Tansley, 1935). Generally, ecosystems have been described at the level of the landscape patch or stream reach composed of a relatively distinct community type. However, concern about global climate change has increased attention to the interconnections among patches that compose a broader landscape-level or watershed-level ecosystem (Baxter et al., 2005; O'Neill, 2001; Polis et al., 1997a; Vannote et al., 1980).

Ecosystems are characterized by their structure and function. Structure reflects the way in which the ecosystem is organized, for example, physical features, species composition, distribution of energy and matter, and trophic or functional organization in space. Function reflects exchanges of energy and matter among individuals and between the community and abiotic pools and biological modification of abiotic conditions, including modification of soil and climate. This chapter describes the major structural and functional parameters of ecosystems in order to provide the basis for description of insect effects on these parameters in Chapters 12–14. The extent to which communities regulate ecosystem processes to maintain optimal conditions for members of the community is addressed in Chapter 15.

2 ECOSYSTEM STRUCTURE

Ecosystem structure represents the various pools (both sources and sinks) of energy and matter and their relationships to each other, that is, directions of matter or information flow (eg, Fig. 1.3). The size of these pools (ie, storage capacity) determines the buffering capacity of the system. Ecosystems can be compared on the basis of the sizes and relationships of various biotic and abiotic compartments for storage of energy and matter and on their trophic or functional group structure, biomass distribution, or spatial and temporal variability in structure. Structure generally is described in terms of spatial variability in physical and trophic structure.

2.1 Physical Structure

Physical structure refers to the size and distribution of physical features of the ecosystem. These variables determine direct and indirect interactions and pathways of energy and nutrient fluxes.

The structure of some ecosystems, especially aquatic, tundra, and desert systems, is dominated by abiotic materials, for example, rocks, sediment, water, or ice, that restrict the distribution and development of biotic material. The sparse biota is almost entirely exposed to changes in abiotic conditions and has relatively little capacity to modify environmental conditions.

At the other extreme, tropical and temperate rainforests are characterized by massive trees that are capable of considerable regulation of abiotic conditions, through buffering of variation in temperature, precipitation, and wind speed, and extensive control of energy and nutrient exchange among biotic and abiotic pools. Forest canopies shade the ground and decrease albedo, reducing diurnal and seasonal variation in temperature. Evapotranspiration directs moisture into the air, facilitating cooling, condensation, and local recycling of precipitation (Trenberth, 1999; Juang et al., 2007). Forest capacity to abate wind depends on tree structure, canopy density, and wind speed. Tree structure (eg, height, taper, rooting depth or other buttressing, wood density, and branching pattern) affects sway frequency (oscillations per minute) and damping ratio (ability to return to resting position) when exposed to wind (J.R. Moore and Maguire, 2005). Branch and foliage density affect wind resistance. Much ecosystem energy and nutrient capital is stored in biotic pools, such as wood and litter, that are connected via food webs to buffer the ecosystem from changes in supply from abiotic pools.

2.2 Trophic Structure

Trophic structure represents the various feeding levels in the community. Organisms generally can be classified as *autotrophs* (or primary producers), which synthesize organic compounds from abiotic pools, and *heterotrophs* (or secondary producers), including insects, which ultimately derive their energy and resources from autotrophs (Fig. 11.2).

Autotrophs are those organisms capable of "fixing" (acquiring and storing) inorganic resources in organic molecules. Photosynthetic plants, responsible for fixation of abiotic carbon into carbohydrates, are the primary sources of organic molecules. This chemical synthesis is powered by solar energy. Free-living and symbiotic N-fixing bacteria and cyanobacteria are an important means of converting inorganic N_2 into ammonia. Nitrifying bacteria oxidize ammonia into nitrite or nitrate (the form of nitrogen available to most green plants). Other chemoautotrophic bacteria oxidize inorganic sulfur into organic compounds. Production of autotrophic tissues must be sufficient to compensate for the amounts consumed by heterotrophs.

Heterotrophs can be divided into several trophic levels depending on their source of food. *Primary consumers* (eg, herbivores) eat plant or other autotroph tissues. *Secondary consumers* eat primary consumers, *tertiary consumers* eat secondary consumers, etc. *Omnivores* feed on more than one trophic level. Finally, *reducers* (including detritivores and decomposers) feed on dead plant and animal matter (Whittaker, 1970). Detritivores (primarily arthropods and

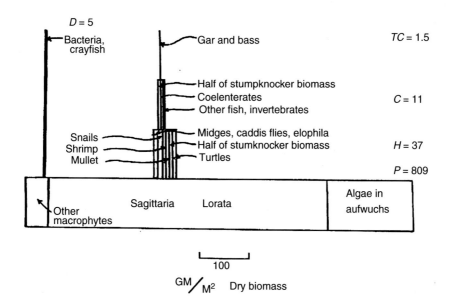

FIGURE 11.2   Biomass pyramid for the Silver Springs ecosystem. *P,* primary producers; *H,* herbivores; *C,* predators; *TC,* top predators; *D,* decomposers. *Source: From Odum, H.T., 1957. Trophic structure and productivity of Silver Springs, Florida. Ecol. Monogr. 27, 55–112.*

worms) fragment organic material and facilitate colonization by decomposers (primarily bacteria and fungi), which catabolize the organic compounds.

Each trophic level can be subdivided into functional groups, based on the way in which organisms gain or use resources (see Chapter 9). For example, autotrophs can be subdivided into photosynthetic, nitrogen-fixing, nitrifying, and other functional groups. The photosynthetic functional group can be subdivided further into ruderal, competitive, and stress-tolerant functional groups (Grime, 1977), or into C-3 and C-4, nitrogen-accumulating, calcium-accumulating, high-lignin, or low-lignin functional groups, etc., to represent their different strategies for resource use and growth. Similarly, primary consumers can be subdivided into migratory grazers (eg, many ungulates and grasshoppers), sedentary grazers (various leaf-chewing insects and some mammals), leaf miners, gall-formers, sap-suckers, root feeders, parasitic plants, plant pathogens, etc., to reflect different modes for acquiring and affecting their plant resources.

The distribution of biomass in an ecosystem is an important indicator of storage capacity, a characteristic that influences ecosystem stability (Webster et al., 1975, see Chapter 15), especially the ability to modify climate and buffer against extreme events (J. Foley et al., 2003a; Juang et al., 2007; Jentsch et al., 2007). Harsh ecosystems, such as tundra and desert, restrict autotrophs to relatively few, small plants with relatively little biomass to store energy and matter. Dominant species are adapted to retain water, but water storage capacity is limited. By contrast, wetter and more fertile ecosystems permit development of large producers with greater storage capacity in branch and root systems. Accumulated detritus represents an additional pool of stored organic matter that buffers the ecosystem from changes in resource availability. Tropical and other warm, humid ecosystems generally have relatively low detrital biomass because of rapid decomposition and turnover. Stream and tidal ecosystems lose detrital material as a result of export in flowing water. Detritus is most likely to accumulate in cool, moist ecosystems, especially boreal forest and deep lakes, in which detritus decomposes slowly. Biomass of heterotrophs is relatively small in most terrestrial ecosystems, but may be larger than primary producer biomass in some aquatic ecosystems, as a result of high production and turnover by a small biomass of algae (Whittaker, 1970).

Trophic structure can be represented by numbers, mass (biomass), or energy content of organisms in each trophic level (Fig. 11.2). Such representations are called numbers pyramids, biomass pyramids, or energy pyramids (see Elton, 1939) because the numbers, mass, or energy content of organisms generally decline(s) at successively higher trophic levels. However, the form of these pyramids differs among ecosystems. Terrestrial ecosystems typically have large numbers or biomass of primary producers that support progressively smaller numbers or biomass of consumers. Many stream ecosystems are supported primarily by allochthonous material (detritus or prey entering from the adjacent terrestrial ecosystem) and have few primary producers (Cloe and Garman, 1996; Oertli, 1993; J.B. Wallace et al., 1997; Wipfli, 1997). Numbers pyramids for terrestrial ecosystems may be inverted because individual plants can support numerous invertebrate consumers. Biomass pyramids for some aquatic ecosystems are inverted because a small biomass of plankton with a high rate of reproduction and turnover can support a larger biomass of organisms with low rates of turnover at higher trophic levels (Whittaker, 1970).

Insects have been largely ignored in many terrestrial ecosystem studies, because their small size and cryptic habits have made them seem unimportant compared to larger and

TABLE 11.1 Biomass of Plants, Insects, and Vertebrates in Ecosystems for Which These Have All Been Measured

	Plants[a] (g m^{-2})	Insects (g m^{-2})	Vertebrates (g m^{-2})
Tropical lowland forest, Peru[b]	39,000	5.4	0.15
Tropical rainforest, Puerto Rico[c]	26,000	4.0	0.69
Tropical grassland, Serengeti[d]	3000	0.76	2.3
Temperate deciduous forest, various[e]	20,000	5.0	0.11
Temperate coniferous forest, various[f]	30,000	2.4	0.08
Temperate grassland, Colorado, USA[g]	2300	0.62	1.1
Cropland, Poland[h]	1260	5.8	0.20
Spring, Florida, USA[i]	809	22	31
Stream, Arizona, USA[j]	350	3.0	50

[a] *Plant biomasses are dry mass, whereas vertebrate biomasses are fresh mass; insects are dry mass, except where noted*
[b] *DeWalt and Chave (2004), Endo et al. (2010), Lavelle and Pashanasi (1989); insects mass is fresh mass*
[c] *H. Odum (1970)*
[d] *Lamotte and Bourliére (1983); data include native ungulates*
[e] *Monk and Day (1988), Perry et al. (2008), Turcek (1971)*
[f] *Perry et al. (2008), Turcek (1971)*
[g] *Coupland and Van Dyne (1979); data for vertebrates include stocked cattle (1.09 g m^{-2})*
[h] *Ryszkowski (1979)*
[i] *H. Odum (1957)*
[j] *Fisher (1986)*
From Schowalter, T.D., 2013. Insects and Sustainability of Ecosystem Services. CRC Press/Taylor and Francis Group, Boca Raton, FL.

more conspicuous components of ecosystems. However, invertebrate biomass in many terrestrial ecosystems may exceed that of more conspicuous vertebrates (Table 11.1) and, during outbreaks, can substantially alter the biomass of plants and vertebrates.

2.3 Spatial Variability

At one time, the ecosystem was considered to be the interacting community and abiotic conditions of a site. This view gradually has expanded to incorporate the spatial pattern of interacting component communities at a landscape or watershed level (see Chapter 9). Patches within a landscape or watershed are integrated by disturbance dynamics and interact through the movement of organisms, energy, and matter (see Chapter 7). For example, the *Stream Continuum Concept* (Vannote et al., 1980) integrates the various stream sections that mutually influence each other. Downstream reaches are influenced by inputs from upstream, but upstream reaches are influenced by organisms returning materials from downstream (Pringle, 1997; Wipfli et al., 2007). The structure of stream segments determines connectivity and may provide predator-free refugia for some species (Covich et al., 2009). Riparian zones (floodplains) connect terrestrial and aquatic ecosystems. Periodic flooding and emerging arthropods move sediments and nutrients from the aquatic system to the terrestrial system; runoff and falling litter and terrestrial arthropods move sediments and nutrients from the terrestrial to the aquatic system (Baxter et al., 2005; Cloe and Garman, 1996;

J. Davis et al., 2011; Dreyer et al., 2015; Gustafsson et al., 2014; Kraus and Vonesh, 2012; Wipfli, 1997). The structure of riparian and upslope vegetation influences the interception and flow of precipitation (rain and snow) and sediment moving downhill into streams (H. Guo et al., 2008; Post and Jones, 2001).

The structure of ecosystems at a stream continuum or landscape scale may have important consequences for recovery from disturbances, by affecting proximity of population sources and sinks. Patches representing various stages of recovery from disturbance provide the sources of energy and matter (including colonists) for succession in disturbed patches. Important members of some trophic levels, especially migratory herbivores, birds, and anadromous fish, often are concentrated seasonally at particular locations along migratory routes. Social insects may forage long distances from their colonies, integrating patches through pollination, seed dispersal, predation, or other interactions. Such aggregations add spatial complexity to trophic structure and are instrumental in redistributing resources.

3 ENERGY FLOW

Life represents a balance between the tendency to increase entropy (Second Law of Thermodynamics) and the decreased entropy, through continuous energy inputs, necessary to concentrate resources for growth and reproduction. Most energy for life on Earth ultimately comes from solar radiation which powers the chemical storage of energy through photosynthesis; some comes from chemical conversion by chemoautotrophs. Given the First and Second Laws of Thermodynamics, the energy flowing through ecosystems, including resources harvested for human use, can be no greater, and typically is much less, than the amount of energy stored in carbohydrates.

Organisms have been compared to thermodynamic machines powered by the energy of carbohydrates to generate maximum power output, in terms of work and progeny (Lotka, 1925; H. Odum and Pinkerton, 1955; Wiegert, 1968). Just as organisms can be studied in terms of their energy acquisition, allocation, and energetic efficiency (Chapter 4), so ecosystems can be studied in terms of their energy acquisition, allocation, and energetic efficiency (E. Odum, 1969; H. Odum and Pinkerton, 1955). Energy acquired from the sun powers the chemical synthesis of carbohydrates, which represent storage of potential energy that is then channeled through various trophic pathways, each with its own power output, and eventually is dissipated completely as heat through the combined respiration of the community (Lindeman, 1942; E. Odum, 1969; H. Odum and Pinkerton, 1955).

The study of ecosystem energetics was pioneered by Lindeman (1942), whose model of energy flow in a lacustrine ecosystem ushered in the modern concept of the ecosystem as a thermodynamic machine. Lindeman noted that the distinction between the community of living organisms and the nonliving environment is obscured by the gradual death of living organisms and conversion of their tissues into abiotic nutrients that are reincorporated into living tissues.

The rate at which available energy is transformed into organic matter is called productivity. This energy transformation at each trophic level (as well as by each organism) represents the storage of potential energy that fuels metabolic processes and power output at each trophic level. Energy flow reflects the transfer of energy for productivity by all trophic levels.

3.1 Primary Productivity

Primary productivity is the rate of conversion of solar energy into plant matter. The rate of solar energy conversion into carbohydrates (total photosynthesis) is *gross primary productivity* (GPP). However, a portion of gross primary productivity must be expended by the plant through metabolic processes necessary for maintenance, growth, and reproduction, and is lost as heat through respiration. The net rate at which energy is converted to carbohydrates is *net primary productivity*. The energy stored as organic matter (net primary production or NPP) becomes available to heterotrophs.

Photosynthetic rates and NPP are sensitive to environmental conditions. Photosynthetic rate and NPP increase with precipitation up to a point, after which they decline, due to low light associated with cloudiness and reduced nutrient availability associated with saturated soils (Schuur et al., 2001). These rates also increase with temperature, up to a point at which water loss causes stomatal closure (Whittaker, 1970).

Photosynthetically active radiation (PAR) occurs within the range 400–700 nm. The energy content of NPP divided by the supply of short-wave radiation, on an annual basis, provides a measure of photosynthetic efficiency (W. Webb et al., 1983). Photosynthetic efficiency generally is low, ranging from 0.065% to 1.4% for ecosystems with low to high productivities, respectively (Sims and Singh, 1978; Whittaker, 1970).

Photosynthetically active radiation can be limited as a result of latitude, topography, cloud cover, or dense vegetation, which restrict penetration of short-wave radiation. Terborgh (1985) discussed the significance of differences in tree geometries among forest biomes. Boreal tree crowns are tall and narrow to maximize interception of lateral exposure to sunlight filtered through a greater thickness of atmosphere, whereas tropical tree crowns are umbrella-shaped to maximize interception of sunlight filtered through a thinner layer of atmosphere directly overhead. Solar penetration through tropical tree canopies, but not boreal tree canopies, is sufficient for development of multiple layers of understory plants.

The relationship between precipitation and potential evapotranspiration (PET) is an important factor affecting photosynthesis. Water limitation can result from insufficient soil water and/or precipitation, relative to evapotranspiration. Plants respond to water deficits by closing stomata, thereby reducing O_2 and CO_2 exchange with the atmosphere. Plants subject to frequent water deficits must solve the problem of acquiring CO_2, when stomatal opening facilitates water loss. Many desert and tropical epiphyte species are able to take up and store CO_2 as malate at night (when water loss is minimal) through crassulacean acid metabolism (CAM), then carboxylate the malate (to pyruvate) and refix the CO_2 through normal photosynthesis during the day with stomata closed (K. Winter and Smith, 1996; Woolhouse, 1981). Although CAM plants require high light levels to provide the energy for fixing CO_2 twice (Woolhouse, 1981), desert plants often have high photosynthetic efficiencies relative to foliage biomass (Webb et al., 1983).

Air circulation is necessary to replenish CO_2 within the uptake zone neighboring the leaf surface. Although atmospheric concentrations of CO_2 may appear adequate, high rates of photosynthesis, especially in still air, can deplete CO_2 in the boundary area around the leaf, reducing photosynthetic efficiency.

NPP, turnover (loss and replacement), and standing crop biomass (biomass at the time of measurement) are governed by a number of factors that differ among successional stages

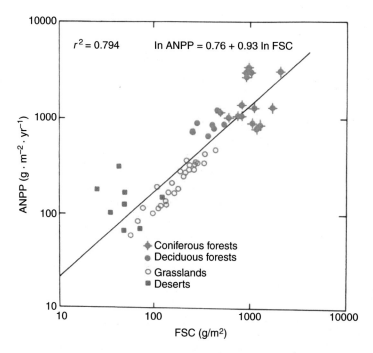

FIGURE 11.3 **Relationship between above-ground net primary production (ANPP) and peak foliar standing crop (FSC) for forest, grassland, and desert ecosystems.** *Source: From Webb, W.L., et al., 1983. Primary production and abiotic controls in forests, grasslands, and desert ecosystems in the United States. Ecology 64, 134–151.*

and between terrestrial and aquatic ecosystems. Ruderal plants in terrestrial ecosystems and phytoplankton in aquatic ecosystems typically have high turnover rates (due to short life spans) and high rates of net primary production per gram biomass, because resources are relatively nonlimiting and the plants are composed primarily of photosynthetic tissues. Net primary production by all vegetation is low, however, because of the small biomass available for photosynthesis. By contrast, later successional plant species have low turnover rates (long life spans) and lower rates of net primary production per gram, because shading reduces photosynthetic efficiency and large portions of biomass necessary for support and access to sunlight are nonphotosynthetic but still respire, for example, wood and roots (Gutschick, 1999). Furthermore, NPP is correlated with foliar standing crop biomass (Fig. 11.3). Hence, reduction of foliar standing crop biomass by herbivores can reduce NPP.

Often, only above-ground NPP is measured, although below-ground production typically exceeds above-ground production in grassland and desert ecosystems (W. Webb et al., 1983). Among major terrestrial biomes, total (above-ground + below-ground) net primary productivity ranges from < 200 g m^{-2} year^{-1} in tundra and deserts to 2000 g m^{-2} year^{-1} in tropical forests, swamps and marshes, and estuaries (Fig. 11.4) (Brown and Lugo, 1982; Waide et al., 1999; W. Webb et al., 1983; Whittaker, 1970).

Typically, the NPP that is consumed by herbivores on an annual basis is low, an observation that prompted Hairston et al. (1960) to conclude that herbivores are not resource limited

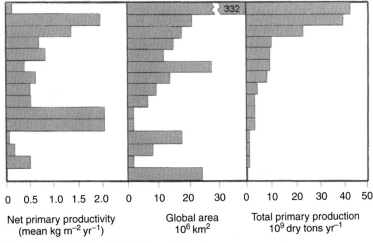

Open ocean
Tropical forest
Temperate forest
Savanna
Boreal forest
Continental shelf
Agricultural land
Temperate grassland
Woodland and shrubland
Swamp and marsh
Algal beds and estuaries
Desert scrub
Tundra and alpine
Lake and stream
Extreme desert, rock, and ice

Net primary productivity
(mean kg m^{-2} yr^{-1})

Global area
10^6 km^2

Total primary production
10^9 dry tons yr^{-1}

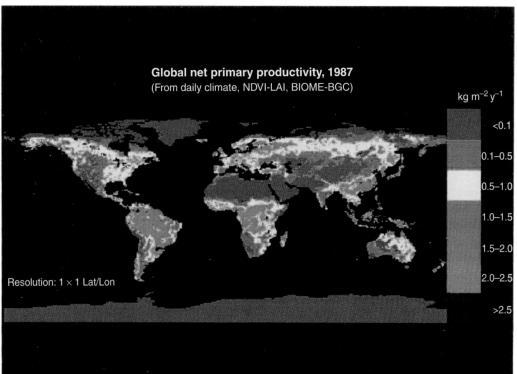

FIGURE 11.4 Net primary production, total area, and contribution to global net primary production of the major biomes (top, data from Whittaker, 1970); global calculation of total NPP using the Light Use Efficiency Model and biweekly time-integrated Normalized Difference Vegetation Index (NDVI) values for 1987. *Source: From Waring, R.H., Running, S.W., 1998. Forest Ecosystems: Analysis at Multiple Scales. Academic Press, San Diego, CA.*

and must be controlled by predators. However, early studies of energy content of plant material involved measurement of change in enthalpy (heat of combustion) rather than free, or available, energy (Wiegert, 1968). We now know that the energy initially stored as carbohydrates is incorporated, through a number of metabolic pathways, into a variety of compounds varying widely in their digestibility by herbivores. The energy stored in plant compounds often costs more to digest than the free energy it provides (see Chapters 3 and 4). Many of these herbivore-deterring compounds require energy expenditure by the plant, reducing the free energy available for growth and reproduction (Coley, 1986). The methods used to measure herbivory often overestimate consumption but underestimate the turnover of NPP (Risley and Crossley, 1993; Schowalter and Lowman, 1999, see Chapter 12).

3.2 Secondary Productivity

Net primary production provides the energy for all heterotrophic activity (see Chapter 4). Consumers capture the energy stored within the organic molecules of their food sources. Therefore, each trophic level acquires the energy represented by the biomass consumed from the lower trophic level. The rate of conversion of NPP into heterotroph tissues is *secondary productivity*. As with primary productivity, we can distinguish the total rate of energy consumption by secondary producers (gross secondary productivity) from the energy incorporated into consumer tissues (net secondary productivity) after expenditure of energy through respiration. Secondary productivity is limited by the amount of net primary production, because only the net energy stored in plant matter is available for consumers, secondary producers cannot consume more matter than is available, and energy is lost during each transfer between trophic levels.

Not all food energy removed by consumers is ingested. Consumer feeding often is wasteful. Scraps of food are dropped, or damaged plant parts are abscised (Faeth et al., 1981; Risley and Crossley, 1993), making this material available directly to reducers (detritivores and decomposers). Of the energy contained in ingested material, some is not assimilable and is egested, also becoming available to reducers. A portion of assimilated energy must be used to support metabolic work, for example, for maintenance, searching for food, mates, or oviposition sites, and various other activities, and is lost through respiration (see Chapter 4). The remainder is available for growth and reproduction (secondary production).

Secondary production varies widely among heterotrophs and ecosystems. Herbivores generally have lower efficiencies of food conversion (ingestion/GPP $\leq$ 10%) than do predators ($\leq$ 15%) because the chemical composition of animal food is more digestible than is plant food (Whittaker, 1970). Heterotherms have higher efficiencies than do homeotherms because of the greater respiratory losses associated with maintaining constant body temperature (Golley, 1968; G. Turner, 1970, see also Chapter 4). Therefore, ecosystems dominated by invertebrates or heterothermic vertebrates (eg, most freshwater ecosystems dominated by arthropods and fish) will have higher rates of secondary production, relative to net primary production, than will ecosystems with greater representation of homeothermic vertebrates. Annual secondary production by aquatic macroinvertebrates in streams averages 1–1000 g dry mass m^{-2}, with the highest rates in streams dominated by filter feeders (Huryn and Wallace, 2000).

Eventually, all plant and animal matter enters the detrital pool as organisms die. The energy in detritus then becomes available to reducers. Detritivores fragment detritus and inoculate homogenized detritus with microbial decomposers during gut passage. Terrestrial detrital material consists primarily of lignin and cellulose, but detritivores often improve their efficiency of energy assimilation by association with gut microorganisms or by reingestion of feces (coprophagy) following microbial decay of cellulose and lignin and concentration of nitrogen and other nutrients (Breznak and Brune, 1994).

3.3 Energy Budgets

Energy budgets can be developed from measurements of available solar energy, primary productivity, secondary productivity, decomposition, and respiration. Comparison of budgets and conversion efficiencies among ecosystems can indicate factors affecting energy flow and contributions to global energy budget. Development of energy budgets for agricultural ecosystems can be used to evaluate the efficiency of human resource production.

Lindeman (1942) was the first to demonstrate that ecosystem function can be represented by energy flow through a trophic pyramid or food web. He accounted for the energy stored in each trophic level, transferred between each pair of trophic levels, and lost through respiration. H. Odum (1957) and Teal (1957, 1962) calculated energy storage and rates of energy flow among trophic levels in several aquatic and wetland ecosystems (Fig. 11.5). E. Odum and Smalley (1959) and Smalley (1960) calculated energy flow through consumer populations. During the 1960s, the International Biological Programme (IBP) focused attention on the energy budgets of various ecosystems (Bormann and Likens, 1979; Misra, 1968; E. Odum, 1969; Petrusewicz, 1967; Sims and Singh, 1978), including energy flow through insect populations (Kaczmarek and Wasilewski, 1977; McNeill and Lawton, 1970; Reichle and Crossley, 1967).

More recently, the energy budgets of agricultural ecosystems have been evaluated from the standpoint of energetic efficiency and sustainability. Whereas the energy available to natural communities comes from the sun, additional energy inputs are necessary to maintain agricultural productivity. These include energy from fossil fuels (used to produce fertilizers and pesticides and to power machinery) and from human and animal labor (Bayliss-Smith, 1990; Schroll, 1994). These additional inputs of energy have been difficult to quantify (Bayliss-Smith, 1990). Although the amount and value of food production are well known, the efficiency of food production (energy content of food produced per unit of energy input) is poorly known but critical to sustainability and economic development (Patnaik and Ramakrishnan, 1989; Schowalter, 2013). Promotion of predaceous insects to control pests, as an alternative to energy-expensive pesticides, and of soil organisms (including insects) to reduce loss of soil organic matter, as an alternative to fertilizers, have been proposed as means to increase the efficiency of agricultural production (E. Elliott et al., 1984; Ostrom et al., 1997, see Chapter 17).

Costanza et al. (1997), Daily (1997), N. Myers (1996) and H. Odum (1996) attempted to account for all energy used to produce and maintain the ecosystem services that support human culture (see Chapter 16). In addition to the market and energy value of current ecosystem resources, energy was expended in the past to produce those resources. The energy inputs, over time, that must be included in the energy budget of the system. When forests are harvested, the energy or resources derived from the timber can be replaced only by cumulative inputs of solar energy to replace the harvested biomass. Additional energy is expended for transportation of resources to population centers and development of societal infrastructures. Solar

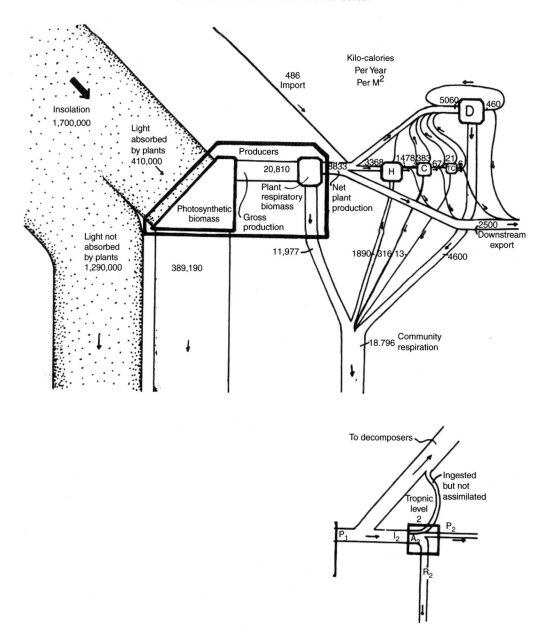

FIGURE 11.5　Energy flow (kcal m⁻² year⁻¹) in the Silver Springs ecosystem. *H*, herbivores; *C*, predators; *TC*, top predators; *D*, decomposers. *Source: From Odum, H.T., 1957. Trophic structure and productivity of Silver Springs, Florida. Ecol. Monogr. 27, 55–112.*

energy also generates tides and evaporates water necessary for maintenance of intertidal and terrestrial ecosystems and their resources.

H. Odum (1996) proposed the term, *emergy*, to denote the total amount of energy used to produce resources and cultural infrastructures. Costanza et al. (1997), Daily (1997), H. Odum (1996), and Schowalter (2013) noted that ecosystems provide a variety of "free" services (see Chapter 16), such as food, pharmaceutical and industrial products, filtration of air and water, and fertilization of floodplains, with energy derived from the sun and from topographic gradients, that must be replaced at the cost of fossil fuel expenditure when these services are lost as a result of environmental degradation (eg, desertification, channelization and impoundment of streams). Sustainability of cultural systems based on ecosystem resources thus depends on the energy derived from the ecosystem relative to the total emergy required to produce the resources. Consequently, many small-scale, subsistence agricultural systems are far more efficient and sustainable than are larger-scale, industrial agricultural systems that could not be sustained without massive inputs from nonrenewable energy sources. Unfortunately, these more sustainable agroecosystems may not provide sufficient production to feed the growing human population.

4 BIOGEOCHEMICAL CYCLING

Organisms use the energy available to them as currency to acquire, concentrate, and organize chemical resources for growth and reproduction (Sterner and Elser, 2002, see Chapter 4). Even sedentary organisms living in or on their material resources must expend energy to acquire resources against chemical gradients or to make these resources useable (eg, through oxidation and reduction reactions necessary for detoxification, digestion, and assimilation). Energy gains must be greater than energy expenditures, or resource acquisition, growth, and reproduction cannot be maintained.

Energy and matter are transferred from one trophic level to the next through consumption, but whereas energy is dissipated ultimately as heat, matter is conserved and reused. Conservation and reuse of nutrients within the ecosystem buffer organisms against resource limitation and contribute to ecosystem stability. The efficiency with which limiting elements are recycled varies among ecosystems. Biogeochemical cycling results from fluxes among biotic and abiotic storage pools.

Biogeochemical cycling occurs over a range of spatial and temporal scales. Cycling occurs within ecosystems as a result of trophic transfers and recycling of biotic materials made available through decomposition. Rapid cycling by microbial components is coupled with slower cycling by larger, longer-lived organisms within ecosystems. Nutrients exported from one ecosystem become inputs for another. Detritus washed into streams during storms is the primary source of nutrients for many stream ecosystems. Nutrients moving downstream are major sources for estuarine and marine ecosystems. Nutrients lost to marine sediments are returned to terrestrial pools through geologic uplifting. Materials stored in these long-term abiotic pools become available for extant ecosystems through weathering and erosion. The pathways and rates of nutrient movement can be described by ecological stoichiometry (Sterner and Elser, 2002).

4.1 Abiotic and Biotic Pools

The sources of all elemental nutrients necessary for life are abiotic pools, the atmosphere, oceans, and sediments. The atmosphere is the primary source of nitrogen, carbon (as carbon dioxide), and water for terrestrial ecosystems. Sediments are a major pool of carbon (as calcium carbonate), as well as the primary source of mineral elements, for example, phosphorus, sulfur, and cations such as sodium, potassium, calcium, and magnesium released through chemical weathering. The ocean is the primary source of water, but also is a major source of carbon (from carbonates) for marine organisms and of cations that enter the atmosphere when winds lift water and dissolved minerals from the ocean surface (McDowell et al., 1990).

Resources from abiotic pools are not available to all organisms, but must be transformed (fixed) into biologically useful compounds by autotrophic organisms. Photosynthetic plants acquire water and atmospheric or dissolved carbon dioxide to synthesize carbohydrates, which then are stored in biomass (see section 3.1). Nitrogen-fixing bacteria and cyanobacteria acquire atmospheric or dissolved N_2 and convert it into ammonia, which they and some plants can incorporate directly into amino acids and nucleic acids. Nitrifying bacteria oxidize ammonia into nitrite and nitrate, the form of nitrogen available to most plants. These autotrophs also acquire other essential nutrients in dissolved form. The living and dead biomass of these organisms represents the pool of energy and nutrients available to heterotrophs.

The size of biotic pools represents storage capacity that buffers the organisms representing these pools against reduced availability of nutrients from abiotic sources. Larger organisms have a greater capacity to store energy and nutrients for use during periods of limited resource availability than do smaller organisms. Many plants can mobilize stored nutrients from tubers, rhizomes, or woody tissues to maintain metabolic activity during unfavorable periods. Similarly, larger animals can store more energy, such as in the fat body of insects, and can retrieve nutrients from muscle or other tissues during periods of inadequate resource acquisition. Detritus represents a major pool of organic compounds and mineral nutrients. The nutrients from detritus become available to organisms through decomposition (Christenson et al., 2002; T.E. Wood et al., 2009). Ecosystems with greater nutrient storage in living or dead biomass tend to be more resistant to certain environmental changes than are ecosystems with more limited storage capacity (Herbert et al., 1999; Webster et al., 1975).

4.2 Major Cycles

In the absence of biota, the rate and direction of chemical fluxes would be controlled solely by the physical and chemical factors determining exchanges between abiotic pools. Chemicals would be retained at a site only to the extent that chelation or concentration gradients restricted leaching or diffusion. Exposed nutrients would continue to move with wind or water (erosion). Biotic uptake and storage of chemical resources creates a biotic pool that alters rates of exchange among abiotic pools and restricts movement of nutrients across chemical and topographic gradients. For example, biotic uptake of various sedimentary nutrients retards their transport from higher elevations back to marine sediments. The uptake and storage of atmospheric CO_2 by plants (including long-term storage in fossil biomass, ie, coal, oil, and gas) and the uptake and storage of calcium carbonate by marine animals (and deposition in marine sediments) control concentration gradients of CO_2 available for exchange between the atmosphere and ocean (Keeling et al., 1995; Sarmiento and Le Quéré, 1996). Conversely, fossil fuel combustion, deforestation and desertification, and destruction of coral reefs are reducing CO_2 uptake by biota and releasing CO_2 from biotic storage, thereby increasing global CO_2 available for exchange between the atmosphere and ocean.

Consumers, including insects, affect the rate at which nutrients are acquired and stored (see Chapters 12–14). In fact, consumers can significantly redirect or alter rates of nutrient fluxes (eg, Hölker et al., 2015). Consumption reduces the biomass of the lower trophic level, thereby affecting nutrient uptake and storage at that trophic level, and moves nutrients from consumed biomass into biomass at the higher trophic level (through secondary production) or into detritus (through secretion and excretion) where nutrients become available to detritivores, soil microorganisms, or are exported via water flow to aquatic food webs. Insects themselves can constitute significant pools of nutrients, and their dispersal can represent significant redistribution of nutrients (Dreyer et al., 2015; Whiles et al., 2001). Carlton and Goldman (1984) and Menninger et al. (2008) found that large numbers of ants and emergent periodical cicadas, respectively, falling into aquatic ecosystems provided sufficient pulses of carbon and nitrogen to stimulate aquatic productivity and respiration. Nutrients are recycled through decomposition of dead plant and animal biomass, which releases simple organic compounds or elements into solution for reacquisition by autotrophs. Conversely, Dreyer et al. (2015) reported that aquatic midge emergence during a peak year increased N and P inputs to riparian ecosystems within 50 m of the stream 3–5-fold above background deposition rates.

Some nutrients are lost during trophic transfers. Carbon is lost (exported) from ecosystems as CO_2 during respiration. Gaseous or dissolved CO_2 remains available to organisms in the atmosphere and oceanic pools. Organic biomass can be blown or washed away. Soluble nutrients (eg, dissolved organic carbon or DOC) are exported as water percolates through the ecosystem and enters streams. The efficiency with which nutrients are retained within an ecosystem reflects their relative availability. Nutrients such as nitrogen and phosphorus often are limiting and tend to be cycled and retained in biomass more efficiently than are nutrients that are more consistently available, such as potassium and calcium. The following four examples exemplify the processes involved in biogeochemical cycling.

4.2.1 Hydric Cycle

Water availability, as discussed in Chapters 2 and 9, is one of the most important factors affecting the distribution of terrestrial organisms. Many organisms are modified to optimize their water balances in arid ecosystems, for example, through their adaptations for acquiring and retaining water (Chapter 2). Water available to plants is a primary factor affecting photosynthesis and ecosystem energetics (see Section 3.1). Water absorbs solar energy, with little change in temperature, thereby buffering aquatic and humid terrestrial ecosystems against large changes in temperature. At the same time, water use by organisms significantly affects its passage through terrestrial ecosystems.

The primary source of water for terrestrial ecosystems is water vapor from evaporation over the oceans (Fig. 11.6). Additional water is evaporated from terrestrial ecosystems and redistributed. The availability of water to terrestrial ecosystems is controlled by a variety of factors, including the rate of evaporation from the ocean or upwind ecosystems, the direction of prevailing winds, atmospheric and topographic factors that affect convection and precipitation, temperature, relative humidity, and soil texture. Water enters terrestrial ecosystems as precipitation and condensation, and as subsurface flow and groundwater derived from

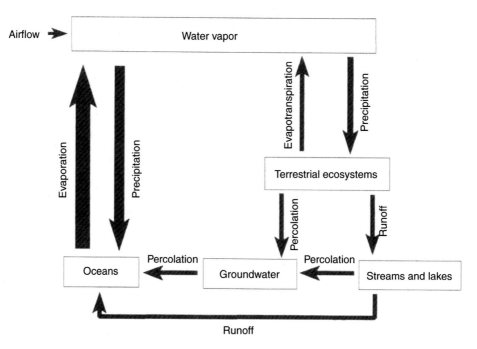

FIGURE 11.6   **The hydric cycle.** Net evaporation over the oceans is the source of water vapor carried inland by air currents. Water precipitated into terrestrial ecosystems eventually is returned to the ocean.

precipitation or condensation at higher elevations. Condensation may be a major avenue for water input to arid ecosystems. Many plants in arid regions are adapted to acquire water through condensation. Some desert insects also acquire water through condensation on specialized hairs or body parts (R. Chapman, 1982). Vegetation intercepts up to 50% of precipitation, depending on crown structure and plant surface area (G. Parker, 1983). Most intercepted water evaporates. The remainder penetrates the vegetation as throughfall (water dripping from foliage) and stemflow (water funneled to stems).

Water is necessary for photosynthesis and is actively acquired by plants. Terrestrial vegetation takes up water primarily from the soil, although most epiphytes collect precipitation directly. Vascular plants conduct water upward and transpire much of it through the stomata. Evapotranspiration is the major mechanism for maintaining the upward capillary flow of water from the soil to the canopy and is controlled primarily by atmospheric vapor pressure deficit and solar radiation (J. Wallace and McJannet, 2010). These factors determine the maximum vegetation height that can be attained in various ecosystems. With adequate sustained soil moisture, tree height could reach 120–130 m (Koch et al., 2004). This active evaporative process greatly increases the amount of water moving back into the atmosphere, rather than flowing downslope, and can significantly increase atmospheric moisture and condensation for precipitation at a particular site, as discussed in Section 5.

Vegetation stores large amounts of water intra- and extracellularly and controls the flux of water through the soil and into the atmosphere. Accumulation of organic material increases soil water storage capacity and further reduces downslope flow. Soil water storage mediates plant acquisition of other nutrients in dissolved form. Food passage through arthropods and earthworms, together with materials secreted by soil microflora, bind soil particles together, forming soil aggregates (Hendrix et al., 1990; Setälä et al., 1996). These aggregates increase water and nutrient storage capacity and reduce erodibility. Burrowing organisms increase the porosity and water storage capacity of soil and decomposing wood, for example, earthworms and wood borers (Eldridge, 1994, see also Chapter 14). Macropore flow increases the rate and depth of water infiltration.

Some organisms also control water movement in streams. Trees falling into stream channels impede water flow. Similarly, beaver dams impede water flow and store water in ponds. Swamp and marsh vegetation restricts water flow in low-gradient ecosystems. However, water eventually evaporates or reaches the ocean, completing the cycle.

4.2.2 Carbon Cycle

The carbon cycle (Fig. 11.7) is particularly important because of its intimate association with energy flow through ecosystems, via the transfer of chemical energy in carbohydrates.

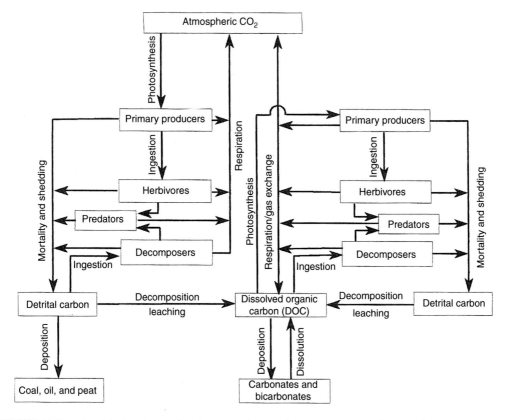

FIGURE 11.7 **The global carbon cycle.** The atmosphere is the primary source of carbon for terrestrial ecosystems (left), whereas dissolved carbonates and bicarbonates are the primary source of carbon for marine ecosystems (right). Exchange of carbon between atmosphere, hydrosphere, and geosphere is regulated largely by biotic uptake and deposition.

Carbon is stored globally both as atmospheric carbon dioxide and as sedimentary and dissolved carbonates (principally calcium carbonate). The atmosphere and ocean mediate the global cycling of carbon among terrestrial and aquatic ecosystems. The exchange of carbon between atmosphere and dissolved or precipitated carbonates is controlled by temperature, carbonate concentration, salinity, and biological uptake that affect concentration gradients (Keeling et al., 1995; Sarmiento and Le Quéré, 1996).

Carbon enters terrestrial ecosystems primarily as a result of photosynthetic fixation of CO_2 in carbohydrates. The chemical energy stored in carbohydrates is used to synthesize all the organic molecules used by plants and animals. Carbon enters many aquatic ecosystems, especially those with limited photosynthesis, primarily as allochthonous inputs of exported terrestrial materials (eg, terrestrial organisms captured by aquatic animals, detritus, and dissolved organic material entering with runoff or leachate). Carbon is transferred among trophic levels through consumption, converted into an astounding diversity of compounds for a variety of uses, and eventually is returned to the atmosphere as CO_2 from respiration, especially during decomposition of dead organic material, completing the cycle. However, loss of carbon from an ecosystem is minimized by rapid acquisition and immobilization of soluble and fine particulate carbon by soil organisms and aquatic filter feeders, from which carbon becomes available for transfer within soil and aquatic food webs (de Ruiter et al., 1995; Hölker et al., 2015; J.B. Wallace and Hutchens, 2000).

However, some carbon compounds (especially complex polyphenols, eg, lignin) decompose very slowly, if at all, and are stored for long periods as soil organic matter, peat, coal, or oil. Humic compounds are phenolic polymers that are resistant to chemical decomposition and constitute long-term carbon storage in terrestrial soils. These compounds contribute to soil water and nutrient holding capacities because of their large surface area and numerous binding sites. Plants produce organic acids that are secreted into the soil through roots. These acids facilitate extraction of mineral nutrients from soil exchange sites, maintain ionic balance (with mineral cations), reduce soil pH, and often inhibit decomposition of organic matter. Similarly, peat accumulates in bogs where low pH inhibits decomposition and eventually may be buried, contributing to formation of coal or oil. Coal and oil represent long-term storage of accumulated organic matter that decomposed incompletely as a result of burial, anaerobic conditions, and high pressure. The carbon removed from the atmosphere and stored by these fossil plants is now reentering the atmosphere rapidly, as a result of fossil fuel combustion, leading to increased atmospheric concentrations of CO_2.

4.2.3 Nitrogen Cycle

The atmosphere is the reservoir of elemental nitrogen, making nitrogen an example of a nutrient with an atmospheric cycle (Fig. 11.8). Nitrogen is a critical element for synthesis of nucleic acids and proteins and is available in limited amounts in most ecosystems. Most organisms cannot use gaseous nitrogen and many other nitrogen compounds. In fact, some common nitrogen compounds are toxic in small amounts to most organisms (eg, ammonia). Nitrogen cycling is mediated by several groups of microorganisms that transform toxic or unavailable forms of nitrogen into biologically useful compounds.

Gaseous N_2 from the atmosphere becomes available to organisms through fixation in ammonia, primarily by nitrogen-fixing bacteria and cyanobacteria. These organisms are key components of most ecosystems, but are particularly important in ecosystems subject to periodic massive losses of nitrogen, such as through fire. Ammonium compounds also are

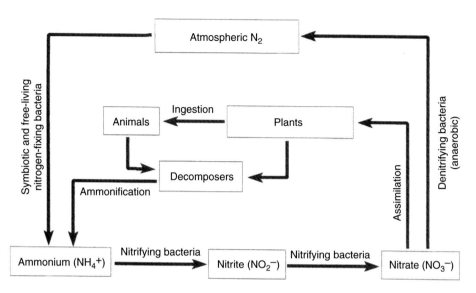

FIGURE 11.8 **The nitrogen cycle.** Bacteria are the primary organisms responsible for transforming elemental nitrogen into forms available for assimilation by plants. Note that the return of nitrogen to the atmospheric pool occurs almost exclusively under anaerobic conditions.

produced by lightning and volcanic eruptions. Many early successional plants, especially in fire-dominated ecosystems, have symbiotic association with nitrogen-fixing bacteria in root nodules. These plants can use the ammonia produced by the associated bacteria, but most plants require nitrate (NO_3) as their source of nitrogen.

Nitrifying bacteria oxidize ammonia to nitrite (NO_2) and nitrate, which then is available to plants, for synthesis of nucleic acids and amino acids, and transferred to higher trophic levels through consumption. The nitrogen compounds in dead organic matter are decomposed to ammonium by ammonifying bacteria. T.E. Wood et al. (2009) demonstrated that experimental addition of leaf litter to tropical forest floor increased leaf litter production and litter nitrogen content by 92% and 156%, respectively, within 4–5 months.

Organic nitrogen enters aquatic ecosystems as exported terrestrial organisms, detritus, or runoff and leachate solutions. Nitrogen in freshwater ecosystems similarly is transferred among trophic levels through consumption, eventually reaching marine ecosystems.

Under anaerobic conditions, which occur naturally (eg, in stagnant water and waterlogged soil) and as a result of anthropogenic eutrophication or soil compaction, anaerobic denitrifying bacteria convert nitrate to gaseous nitrogen, which is lost to the atmosphere, thereby completing the cycle. However, nitrogen loss is minimized by soil organisms that aerate the soil through excavation and by the rapid acquisition and immobilization of soluble nitrogen by soil microorganisms and aquatic filter feeders, from which nitrogen becomes available to plants and to soil and aquatic food webs.

4.2.4 Sedimentary Cycles

Many nutrients, including phosphorus and mineral cations, are available only from sedimentary sources. These nutrients are cycled in similar ways, as exemplified by phosphorus

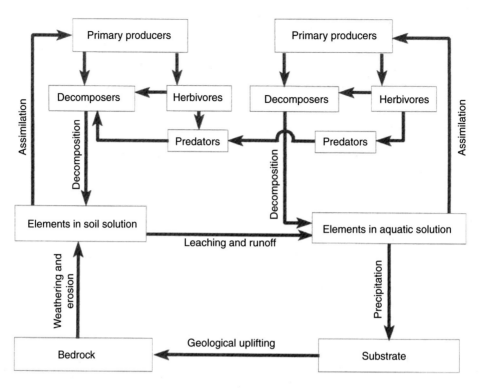

FIGURE 11.9 Sedimentary cycle. Phosphorus and other nongaseous nutrients precipitate from solution and are stored largely in sediments of marine origin. These nutrients become available to terrestrial ecosystems primarily through chemical weathering of uplifted sediments.

(Fig. 11.9). Phosphorus is biologically important in molecules that mediate energy exchange during metabolic processes (ATP and ADP) and in phospholipids. Like nitrogen, it is available to organisms only in certain forms and is in limiting supply in most ecosystems. Phosphorus and mineral cations become available to terrestrial ecosystems as a result of chemical weathering or erosion of geologically uplifted, phosphate-bearing sediments.

Phosphate enters an ecosystem from weathered bedrock and moves among terrestrial ecosystems through materials washed downslope or dust filtered from the air. Phosphorus is highly reactive, but available to plants only as phosphate, that often is bound to soil particles. Plants extract phosphorus (and mineral cations) from cation exchange and sorption sites on soil particles and from soil solution. Phosphorus then is synthesized into biological molecules and transferred to higher trophic levels through consumption; it eventually is returned to the soil as dead organic matter and is decomposed. Phosphorus enters aquatic ecosystems largely in particulate forms exported from terrestrial ecosystems. It is transferred between aquatic trophic levels through consumption, eventually being deposited in deep ocean sediments, completing the cycle. Phosphorus loss is minimized by soil organisms and aquatic filter feeders that rapidly acquire and immobilize soluble phosphorus and make it available for plant uptake and exchange among soil and aquatic organisms.

4.3 Factors Influencing Cycling Processes

A number of factors alter the rates and pathways of biogeochemical fluxes. Variation in fluxes reflects the chemical properties and source of the nutrient, interactions with other cycles, and the composition of the community, especially the presence of specialized organisms that control particular fluxes. Hence, changes in community composition resulting from disturbance and recovery alter the rates and pathways of chemical fluxes.

The chemical properties of various elements and compounds, especially their solubility in water and susceptibility to pH changes, as well as biological uses, affect cycling behavior. Some elements, such as Na and K, form compounds that are readily soluble over normal ranges of pH. These elements generally have high rates of input to ecosystems via precipitation but also high rates of export via runoff and leaching. Other elements, such as Ca and Mg, form compounds that are not as soluble over usual ranges of pH and have lower rates of input and export. Elements such as nitrogen and phosphorus are necessary for all organisms, relatively limiting, and generally conserved within organisms and ecosystems. For example, deciduous trees typically resorb nitrogen from senescing foliage prior to leaf fall (Gutschick, 1999; Marschner, 1995). Sodium has no known function in plants, and is not retained in plant tissues, but is required by animals for osmotic balance and for muscle and nerve function. Consequently, it is conserved tightly by these organisms. In fact, animals often seek mineral sources of sodium (Seastedt and Crossley, 1981b). Many decay fungi accumulate sodium (Cromack et al., 1975; Schowalter et al., 1998), despite absence of any apparent use in fungal metabolism, but perhaps to attract animal vectors of fungal spores.

Biogeochemical cycles interact with each other in complex ways (Daufresne and Loreau, 2001; Elser and Urabe, 1999; Rastetter et al., 1997; Sterner and Elser, 2002). For example, precipitation affects decomposition and carbon storage in soils (Schuur et al., 2001). Some plants respond to increased atmospheric CO_2 by reducing stomatal opening, thereby acquiring sufficient CO_2 while reducing water loss. Hence, increased size of the atmospheric pool of CO_2 may alter transpiration, permitting some plant species to colonize more arid habitats. Nitrogen subsidies, either from anthropogenic atmospheric deposition or as pulses of biogenic inputs (Carlton and Goldman, 1984), can stimulate photosynthesis and primary production but depress decomposition and mineralization (Throop et al., 2004; Treseder, 2008; G. Waring and Cobb, 1992), with varying effects on herbivory (Kytö et al., 1996; G. Waring and Cobb, 1992, see Chapter 3). Similarly, the calcium cycle interacts with cycles of several other elements. Calcium carbonate generally accumulates in arid soils as soil water evaporates. Acidic precipitation, such as resulting from industrial emission of nitrous oxides and sulfur dioxide into the atmosphere, dissolves and leaches calcium carbonate from soils and sediments. Soils with high content of calcium carbonate are relatively buffered against pH change, whereas soils depleted of calcium carbonate become acidic, increasing export (through leaching) of other cations as well.

Some biogeochemical fluxes are controlled by particular organisms. The nitrogen cycle depends on several groups of microorganisms that control the transformation of nitrogen among various forms that are available or unavailable to other organisms (see Section 4.2.3). Soil biota secrete substances that bind soil particles into aggregates that facilitate retention of soil water and nutrients. Some plants (eg, western redcedar, *Thuja plicata*, and dogwoods, *Cornus* spp.) accumulate calcium in their tissues (Kiilsgaard et al., 1987) and generally increase the pH and buffering capacity of surrounding soils. Their presence or absence thereby affects retention of other nutrients as well. Tropical and temperate trees, including oaks, *Quercus* spp., and spruces, *Picea* spp., emit large amounts of carbon as volatile compounds, especially isoprene, that affect the oxidation potential of the atmosphere (Lerdau et al., 1997). Isoprene emission by forests appears to be in balance with atmospheric oxidative capacity and may function to maintain atmospheric conditions conducive to forest production, but deforestation and conversion to agricultural or urban uses are likely to disrupt this balance (Lelieveld et al., 2008). Outbreaks of insects significantly increase fluxes of nutrients and water (Carlton and Goldman, 1984; Fonte and Schowalter, 2005; Frost and Hunter, 2004, 2007; Kurz et al., 2008; Menninger et al., 2008; Schowalter et al., 1991, 2011).

Changes in community composition following disturbance or during succession affect rates and pathways of biogeochemical fluxes. Early successional communities frequently are inefficient, because of limited competition for resources by the small biomass, and early successional species have little selective pressure to retain nutrients. For example, the early successional tropical tree, *Cecropia* spp., has large, thin leaves that transpire water more rapidly than the smaller, more sclerotized leaves of later successional species. Although later successional communities are not always efficient, declining resource supply relative to growing biomass promotes efficiency of nutrient retention within the ecosystem (E. Odum, 1969; Schowalter, 1981).

Agricultural and silvicultural systems are relatively inefficient, largely because communities composed of a single, or few, plant species cannot acquire or retain all available forms of matter effectively. Furthermore, the diversity of organisms in natural systems may increase per capita resource acquisition or provide overall resistance to herbivores and pathogens (Cardinale et al., 2002; A. Hunter and Arssen, 1988). Nitrogen fixation often is controlled by noncommercial species, such as symbiotic nitrogen-fixing lichens, herbs, and shrubs, or structures, such as large decomposing woody litter, that are suppressed or eliminated by management activities. Necessary nitrogen then must be supplied anthropogenically, often in excess amounts that leach into groundwater and streams, resulting in eutrophication and hypoxia of aquatic ecosystems (Howarth et al., 2011). Exotic species also can alter nutrient cycling processes. Liu and Zou (2002) reported that invasion of tropical pastures and wet forest in Puerto Rico by exotic earthworms significantly increased decomposition and leaching rates.

5 CLIMATE MODIFICATION

Although most previous ecology texts have emphasized the effect of climate on survival, population growth, and distribution of organisms (see Chapters 2, 6, and 7), some communities are capable of significant modification of local and regional climatic conditions, perhaps influencing global climatic gradients (T. Chase et al., 1996; J. Foley et al., 2003a; Juang et al., 2007; G. Parker, 1995; Pielke and Vidale, 1995). Climate modification largely reflects the capacity of vegetation to shade and protect the soil surface, abate airflow, and control water fluxes (Fig. 11.10). Isoprene emission by some tree species apparently increases leaf tolerance to high temperatures and also affects the oxidation potential of the atmosphere in ways that favor forest growth (Lerdau et al., 1997). Biomes and successional stages vary widely in ability to modify climate.

When vegetation development is limited or moisture is limited, as in deserts, the soil surface is exposed fully to sunlight and contains insufficient water to restrict temperature change

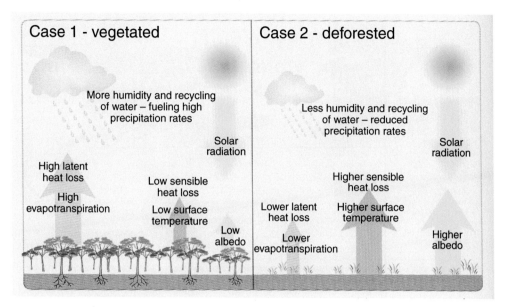

FIGURE 11.10 Diagrammatic representation of the effects of vegetation on climate and atmospheric variables. The capacity of vegetation to modify climate depends on vegetation density and vertical height and complexity. *Source: From Foley, J.A., et al., 2003). Green surprise? How terrestrial ecosystems could affect earth's climate. Front. Ecol. Environ. 1, 38–44.*

(T. Lewis, 1998). The reflectivity of the soil surface (albedo) determines absorption of solar energy and heat. Soils with high organic content have lower albedo (0.10) than does desert sand (0.30) (Monteith, 1973). Albedo also declines with increasing soil water content. In the absence of vegetation cover, surface temperatures can reach 60–70°C during the day, a lethal level for most organisms (Seastedt and Crossley, 1981a), but fall rapidly at night as a result of long-wavelength (infrared) radiation from the surface. Exposure to high wind speeds dries soil and moves soil particles into the atmosphere. Soil desiccation reduces infiltration of precipitation, leading to greater runoff and erosion (Fig. 11.11). These altered soil characteristics increase albedo, surface heating, and advective flux of moist air, leading to increased surface warming and drying (J. Foley et al., 2003a).

Vegetation modifies local and regional climate in several ways. Even thin (3 mm) biological crusts, composed of cyanobacteria, green algae, lichens, and mosses, on the surface of soils in arid and semiarid regions are capable of modifying surface conditions and reducing erosion (Belnap and Gillette, 1998). During the day, vegetation shades the surface of the ground, reducing surface temperature (T. Lewis, 1998). Vegetation also absorbs solar radiation to drive photosynthesis and evapotranspiration (G. Parker, 1995), further cooling the near-surface boundary zone. At night, vegetation absorbs reradiated infrared energy from the ground, maintaining warmer nocturnal temperatures, compared to nonvegetated areas. As a result, vegetation reduces variation in diurnal and annual temperature ranges. Vegetation also intercepts precipitation and reduces the impact of raindrops on the soil surface, although this effect depends on rainfall volume and droplet size (Calder, 2001), and impedes the downslope movement of water, thereby reducing erosion and loss of soil (Ehigiator and

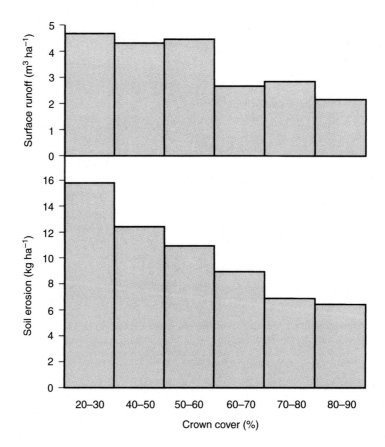

FIGURE 11.11 **Effect of canopy cover on average runoff and soil erosion, based on 41 runoff-producing storms totaling 1128 mm in northern Thailand.** *Source: From Ruangpanit, N., 1985. Percent crown cover related to water and soil losses in mountainous forest in Thailand. In: El-Swaify, S.A., Moldenhauer, W.C., Lo, A. (Eds.), Soil Erosion and Conservation. Soil Conservation Society of America, Ankeny, IA, pp. 462–471.*

Anyata, 2011). Soil organic matter retains water, increasing soil moisture-holding capacity, reducing temperature change. Resistance to airflow by vegetation reduces wind speeds and increases turbulence, contributing to deposition of airborne particles and aerosols and generating convection that increases local precipitation. Exposure of individual organisms to damaging or lethal wind speeds is reduced as a result of buffering by surrounding individuals.

The degree of climate modification depends on vegetation height, density, and "roughness" (the degree of unevenness of canopy topography). Albedo is inversely related to vegetation height and roughness, declining from 0.25 for vegetation < 1.0 m in height to 0.10 for vegetation > 30 m in height, and generally reaches lowest values in vegetation with an uneven canopy surface, for example, tropical forest, and highest values in vegetation with a smooth canopy surface, for example, agricultural crops (Monteith, 1973). Canopy surface roughness creates turbulence in air flow, contributing to surface cooling by wind (sensible heat loss) and evapotranspiration (latent heat loss) (J. Foley et al., 2003a; Juang et al., 2007).

Sparse vegetation has less capacity to modify temperature, water flux, and wind speed than does dense vegetation. Shorter vegetation traps less radiation between multiple layers of leaves

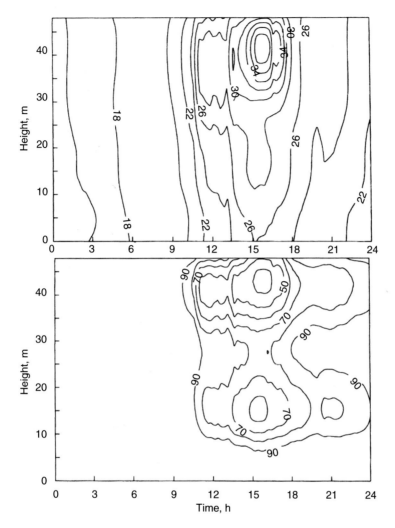

FIGURE 11.12 Height–time profiles of air temperature and relative humidity in mixed-hardwood forest in Maryland. Temperature contours are 2°C; relative humidity contours are 10% units. Nocturnal temperature gradients are weak, but a hot spot develops in the upper canopy in mid-afternoon. Humidity declined in the upper canopy in mid-afternoon, coincident with peak temperatures, and was near saturation (> 95%) outside the marked contours. *Source: From G. Parker (1995).*

and stems and modifies climatic conditions within a shorter column of air, compared to taller vegetation. Tall, multicanopied forests have the greatest capacity to modify local and regional climate because the stratified layers of foliage and denser understory successively trap filtered sunlight, intercept precipitation and throughfall, contribute to evapotranspiration, and impede airflow in the deepest column of air. G. Parker (1995) demonstrated that rising temperatures during midday had the greatest effect in upper canopy levels in a temperate forest (Fig. 11.12). Temperature between 40 and 50 m height ranged from 16°C at night to 38°C during mid-afternoon (a diurnal fluctuation of 22°C); relative humidity in this canopy zone declined from > 95% at

FIGURE 11.13 Deforestation in Panama. Removal of tropical rainforest cover has exposed soil to solar heating and severe erosion, leading to continued ecosystem deterioration and, potentially, to altered regional temperature and precipitation patterns.

night to 50% during mid-afternoon (G. Parker, 1995). Below 10 m, temperature fluctuation was only 10°C, and relative humidity was constant at > 95%. Windsor (1990) and Madigosky (2004) reported similar gradients in canopy environment in a lowland tropical forest.

Vegetation can control local and regional precipitation patterns to a significant extent through evapotranspiration. Surface cooling by vegetation lowers the altitude at which moisture condenses, while vegetation-generated evapotranspiration, turbulence, and latent heat flux combine to elevate moist air to the height of condensation, increasing local precipitation (Janssen et al., 2008; Juang et al., 2007; Trenberth, 1999). Higher rates of local recycling (>20%) occur where rates of evapotranspiration and convective flux are high and advective moisture flux is low (Trenberth, 1999). As much as 30% of precipitation in tropical rainforests in the Amazon basin is generated locally by evapotranspiration (Salati, 1987; Trenberth, 1999).

Insects and other organisms (including humans) alter vegetation and soil structure (Fig. 11.13) and thereby affect biotic control of local and regional climate (M. Brown et al., 2010; Kurz et al., 2008, see Chapters 12–14). Forest fragmentation increases wind fetch and penetration of air from surrounding crop or pasture zones into forest fragments (J. Chen et al., 1995). Belnap and Gillette (1998) found that trampling of the brittle biological crusts on desert soils by livestock greatly increased the effect of wind on soil loss. Increased deposition of airborne particulates reduces exposure to photosynthetically active radiation.

Deforestation and desertification, resulting from anthropogenic land use, expose soils to desiccation and reduce evapotranspirative cooling, thereby reversing the effect of vegetation

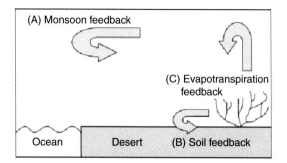

FIGURE 11.14   Three feedbacks by which semiarid vegetation contributes to positive feedback on growth conditions in sub-Saharan Africa. (A) The monsoon feedback results from reduced surface albedo that modifies monsoon circulation, (B) the soil feedback reflects increased soil water availability associated with vegetation, and (C) the evapotranspiration feedback results from increased evapotranspiration in vegetated areas, which increases atmospheric humidity and precipitation of the area. *Source: From Janssen, R.H.H., et al., 2008. Microscale vegetation-soil feedback boosts hysteresis in a regional vegetation-climate system. Glob. Change Biol. 14, 1104–1112.*

and ultimately increasing surface temperatures and reducing precipitation and relative humidity (J. Foley et al., 2003a,b; Janssen et al., 2008; Juang et al., 2007; T. Lewis, 1998; Salati, 1987; Trenberth, 1999). M Costa and Foley (2000) calculated a net warming of 1–2°C in tropical regions as a result of deforestation, an effect that would exacerbate the warming due to increased atmospheric CO_2. Deforestation and desertification initiate positive feedback between climate and vegetation change, leading to further vegetation mortality and soil exposure (J. Foley et al., 2003a; Janssen et al., 2008; Juang et al., 2007). Holocene warming led to northward advance of the boreal forest, which lowered albedo and contributed to continued warming of the ecotone (J. Foley et al., 1994). Schlesinger et al. (1990) reported that desertification in southwestern North America resulted in a destabilizing positive feedback, whereby initial vegetation removal caused surface warming and drying that stressed and killed adjacent vegetation, leading to an advancing arc of desertified land (see Fig. 2.8F). Similar processes contribute to the desertification of sub-Saharan Africa (Fig. 11.14, J. Foley et al., 2003b; Janssen et al., 2008).

The effects of similar, large-scale vegetation changes resulting from insect outbreaks on regional climatic conditions have not been evaluated, although Classen et al. (2005) reported that increased soil temperature and moisture caused by manipulated levels of herbivory were of sufficient magnitude to drive changes in ecosystem processes. K. Clark et al. (2010) and Kurz et al. (2008) reported that outbreaks significantly reduced net ecosystem productivity and can transform forests from carbon sinks to carbon sources, potentially contributing to further climate change.

6 URBAN ECOSYSTEMS

Humans have altered ecosystems globally. Although ecologists may be attracted to "natural" ecosystems in order to study interactions unaffected by anthropogenic changes, in fact no ecosystem on Earth is unaffected by altered atmospheric chemistry, climate change, altered disturbance regime, or invasive species. Consequently, the greater challenge and contribution of insect ecologists will be to document the effects of anthropogenic changes on insects and consequences for ecosystem services (see Chapters 2 and 12–16).

Conversion of natural ecosystems to urban ecosystems or other paved surfaces is the most extreme change in ecosystem conditions and results in urban conditions similar to those in deforested or desertified ecosystems described in the previous section. Loss of forest cover, in particular, represents substantial carbon loss and increased albedo (Seto et al., 2012).

Currently, urban or other built-up areas account for more than 5% of the land surface area of the United States and 2% of the land surface area of the Earth (Grimm et al., 2000; McKinney, 2002). About 60% of urban and suburban surfaces are pavement and roofs (Akbari et al., 2009; McKinney, 2002). Overall, 1.4% of the land surface area of the United States is paved surfaces (Elvidge et al., 2004). Global urban area is expected to triple by 2030 (Seto et al., 2012).

Extreme three-dimensional structure and low albedo surfaces in urban and suburban areas increase local temperature, air turbulence, and vertical shear (Akbari et al., 2009; Arnfield, 2003; Baik et al., 2001; Duckworth and Sandberg, 1954; Kishtawal et al., 2010; Savva et al., 2010). Urban heat islands can be as much as 12°C warmer than surrounding ecosystems (Arnfield, 2003; Dale and Frank, 2014; Diamond et al., 2014; Hawkins et al., 2004; Kalnay and Cai, 2003; Savva et al., 2010). A. Williams et al. (2015) reported that the urban heat island effect can increase the altitude of condensation and cloud base height, thereby reducing fog frequency and precipitation. Heat-generated convective turbulence over urban areas can increase the severity of storms over and around these areas (Changnon et al., 1991; Kishtawal et al., 2010). Industrial effluents negatively affect urban dwellers, as well as ecosystems downwind and downstream.

Exotic plant species dominate the sparse urban vegetation; consumers are primarily domestic pets, various exotic species that generally are considered pests, and a small subset of native species that are able to coexist with humans and exploit urban habitats and resources (Frankie and Ehler, 1978; McKinney, 2002). For example, Blair (2001) found only seven species of summer resident bird species in a city center in California, compared to 21 species in a natural preserve outside the city limits. However, urban parks, green roofs, and other seminatural areas embedded in urban landscapes can provide habitat for additional species (Madre et al., 2014).

Urban vegetation stressed by elevated temperatures, pollutants, fertilizer application, soil compaction, impervious surfaces, and altered drainage conditions becomes vulnerable to a variety of herbivorous insects (Cregg and Dix, 2001; Dale and Frank, 2014; Frankie and Ehler, 1978; Raupp et al., 2010). Meineke et al. (2013) and Dale and Frank (2014) reported that water-stressed urban trees become foci for elevated scale insect, *Parthenolecanium quercifex* and *Melanaspis tenebricosa*, populations, respectively, which exacerbate stress and threaten urban ecosystem services. Some native insect species have adapted to the higher temperatures and greater distances between potential resources in the urban matrix (San Martin y Gomez and Van Dyck, 2012).

Many species of insects are capable of exploiting urban habitats, including parasites supported by abundant human and pet hosts (such as bedbugs, lice, and fleas); aquatic insects supported by standing water in gutters, fountains, garden containers, and drainage ditches (particularly mosquitoes and phorid flies); household insects supported by accessible food and detrital resources (such as cockroaches, termites, and ants); and pollinators supported by the diversity of exotic garden flowers (Frankie and Ehler, 1978; Hennig and Ghazoul, 2012; M.R. Hunter and Hunter, 2008; Lowenstein et al., 2015; Matteson et al., 2013; Raupp et al., 2010).

The diversity of urban resources and habitats can promote high diversity of arthropod species adapted to urban conditions (Sattler et al., 2011). Sattler et al. (2011) compared arthropod communities in Swiss forests, agricultural, and urban ecosystems. The three ecosystem types harbored distinct arthropod assemblages. Whereas forest arthropod assemblages were distinct from the other two, agricultural assemblages shared more than a third of their characteristic species with urban arthropod assemblages. As found for nonurban ecosystems, habitat connectivity through urban gardens and rooftop vegetation greatly influences urban arthropod communities (Braaker et al., 2014).

Despite differences, urban ecosystems exhibit the same structural and functional properties described for natural ecosystems, including physical and trophic structure, species diversity and interactions, energy and biogeochemical fluxes among biotic and abiotic components, and exchange of energy and materials with surrounding ecosystems, albeit with significant differences in sources of inputs and destinations of outputs (Fig. 11.15) (de Jesús-Crespo and Ramírez, 2011; Grimm et al., 2000; Kaye et al., 2006; Nowak et al., 2013; Pataki et al., 2006; Ramírez et al., 2012). For example, most functional groups of insects are represented in urban environments (Sattler et al., 2011). The greater diversity of floral resources and structures that offer nesting habitats, compared to nearby nonurban ecosystems, may increase pollinator

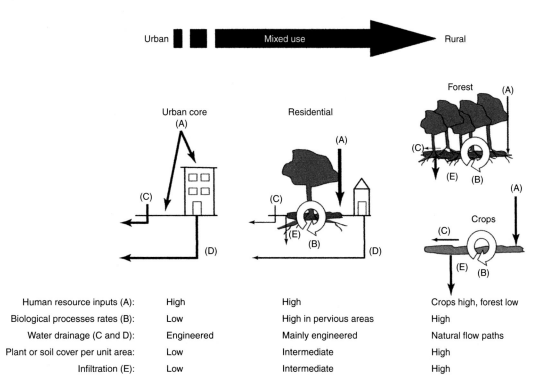

FIGURE 11.15 **Anthropogenic and natural ecosystem characteristics along a continuum from urban (almost entirely anthropogenic) to rural (least human modification) ecosystem types.** At the urban core, engineered paths for surface water (C) and wastewater (D) fluxes disconnect material inputs (A) and flows from infiltration (E) and natural processing (B) that occur in rural agricultural and forest ecosystems. In residential areas, connectivity can be relatively high, depending on the spatial arrangement of pervious and impervious surfaces. Managed and environmental inputs in residential areas can be high per unit of pervious area (A), but these areas can have high cycling rates (B), depending on site history, soil type, and concentration of flows. *Source: Reprinted from Kaye, J.P., et al., 2006. A distinct urban biogeochemistry? Trends Ecol. Evol. 21, 192–199.*

diversity and contribute to pollination in urban gardens, although floral specialists typically are absent (Lowenstein et al., 2015; Matteson et al., 2013).

Continued expansion of human activity at urban fringes propagates the urban environment into remnant "natural" ecosystems. Urban warming will affect surrounding ecosystems. As the proportion of landscapes occupied by urban, suburban, and industrial patches and connecting paved roads increases, their influence on remnant ecosystems will increase, compromising ecosystem processes and services in the surrounding landscape.

7 ECOSYSTEM MODELING

Modeling has become a useful tool for testing hypotheses concerning behavior and self-regulation of complex systems (Camilo and Willig, 1995; B. Patten, 1995; Ulanowicz, 1995) and for predicting ecosystem responses to environmental changes, as well as ecosystem contributions to environmental change, especially carbon flux (Rastetter et al., 1991; Sarmiento and Le Quéré, 1996). The logistical difficulty of measuring and manipulating all ecosystem components and processes for experimental purposes has placed greater emphasis on modeling to simulate experimental conditions and to identify critical components and processes for further study. However, despite limitations, models represent the most powerful tools for synthesizing available information and predicting effects of changing environmental conditions on ecosystem conditions and services, results of which become testable hypotheses.

Modeling at the ecosystem level necessarily starts with conceptual models of linkages among components and reflects the perception of individual modelers of the importance of particular components and interactions (eg, Figs. 1.3, 11.6–11.9). Models differ in their representation of complexity, for example, the degree to which species are distinguished in individual submodels or combined into functional group submodels (Astrup et al., 2008; de Ruiter et al., 1995; Kimmins et al., 2008; Naeem, 1998; Polis, 1991b; Reagan et al., 1996) and to which light, water, and nutrient availability are integrated simultaneously with changes in ecosystem structure and composition (R. Waring and Running, 1998). Obviously, conceptualizing the integration of the many thousands of species and other components in a given ecosystem is virtually impossible. On the other hand, some global-scale models distinguish the biota only at the community level, if at all.

Astrup et al. (2008) compared predictions of forest dynamics over a range of model complexities and found that both the most and least complex models had poor predictive capability, compared to models with intermediate complexity. The degree to which individual species are distinguished influences the representation of the variety of interactions and feedbacks that influence ecosystem parameters (Naeem, 1998; Polis, 1991b; Reagan et al., 1996). Models that attempt to incorporate all species and interactions suffer from insufficient data for many species and interactions, especially indirect interactions. Similarly, models based on a limited set of variables to predict a single type of output (eg, carbon flux) may fail to account for effects of other variables (eg, effects of limiting nutrients, such as nitrogen, on carbon flux) (R. Waring and Running, 1998). More general models require simplifying assumptions to expand their application and may lose accuracy as a consequence. Kimmins et al. (2008) concluded that ecosystem models should be as simple as possible but as complex as necessary.

After the conceptual organization of the model has been determined, interaction strengths are quantified (Figs. 11.16, 11.17), based on available data, or subjected to sensitivity analysis

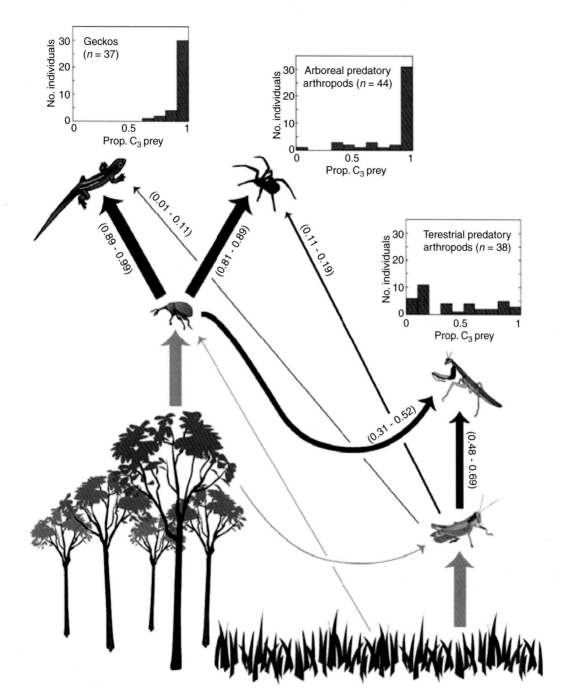

FIGURE 11.16 **Diagram of interaction strengths linking canopy and understory food webs in a savanna ecosystem in central Kenya.** Values in parentheses represent the range of three mean dietary proportions of C_3-feeding prey for each link based on $\delta^{13}C$ trophic discrimination factors (on $\Delta^{13}C$). Links between predators and prey (*black lines*) are proportional in width to the mean proportion of diet constituted by that link. Histograms illustrate the proportion of C_3-feeding prey in the diets of individual predators. The widths of the links between prey and plants (*grey lines*) reflect estimated, not calculated, diet proportions. *Source: From Pringle, R.M., Fox-Dobbs, K., 2008. Coupling of canopy and understory food webs by ground-dwelling predators. Ecol. Lett. 11, 1328–1337.*

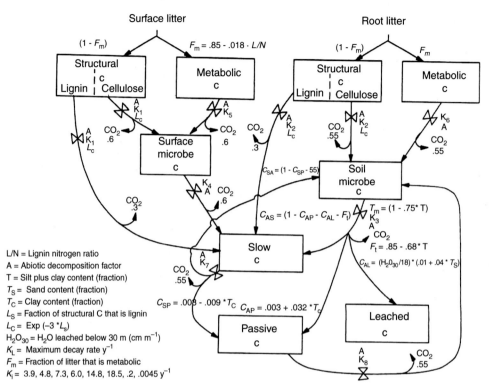

FIGURE 11.17 **Detail of carbon fluxes in the soil organic carbon submodel of the Century ecosystem model.** This model can be coupled to the nitrogen submodel. *Source: From Parton, W.J., et al., 1993. Observations and modeling of biomass and soil organic matter dynamics for the grassland biome worldwide. Global Biogeochem. Cy. 7, 785–809.*

to identify the range of values that represent observed interaction (Benke and Wallace, 1997; Dambacher et al., 2002; de Ruiter et al., 1995; T. Miller et al., 2014; Parton et al., 1993; Pringle and Fox-Dobbs, 2008; Rastetter et al., 1991, 1997; Running and Gower, 1991). Direct and indirect interactions can be represented in transition matrix form, for example,

$$
\begin{array}{cccccc}
N_1 & N_2 & N_3 & N_4 & \cdot\cdot & N_i \\
N_1 \quad \alpha_{11} & \alpha_{21} & \alpha_{31} & \alpha_{41} & \cdot\cdot & \alpha_{i1} \\
N_2 \quad \alpha_{12} & \alpha_{22} & \alpha_{32} & \alpha_{42} & \cdot\cdot & \alpha_{i2} \\
N_3 \quad \alpha_{13} & \alpha_{23} & \alpha_{33} & \alpha_{43} & \cdot\cdot & \alpha_{i3} \\
N_4 \quad \alpha_{14} & \alpha_{24} & \alpha_{34} & \alpha_{44} & \cdot\cdot & \alpha_{i4} \\
\cdot & \cdot & \cdot & \cdot & \cdot\cdot\cdot & \cdot \\
\cdot & \cdot & \cdot & \cdot & \cdot\cdot\cdot & \cdot \\
N_j \quad \alpha_{1j} & \alpha_{2j} & \alpha_{3j} & \alpha_{4j} & \cdot\cdot & \alpha_{ij}
\end{array}
$$

where N_j is the jth ecosystem component, and α_{ij} is the relative effect (direct + indirect) of N_j on N_i. When $N_i = N_j$, α_{ij} represents intrinsic (intraspecific) effects on numbers or mass. Differential equations of the general form

$$N_{i(t+1)} = N_{it} + \Sigma(\alpha_{ij}N_{jt}) \tag{11.1}$$

are used to calculate the transitional states of each component as input conditions change. Note the application of this inclusive equation to equations for growth of individual populations and interacting species in Chapters 6 and 8. Components must be linked so that changes in the number, mass, or energy or nutrient content of one component have appropriate effects on the numbers, masses, or energy or nutrient contents of other components. Models focused on species emphasize fluxes of energy or matter through food webs. Models focused on energy or matter pools emphasize fluxes of energy and matter among pools, but may include important species that affect flux rates.

Ecosystem models are sensitive to effects of indirect interactions. Nutrient availability and directions of fluxes indirectly affect all organisms. For example, a direct predator–prey interaction reduces prey abundance and directs energy and nutrients through that predator, thereby indirectly affecting resources available for other organisms, as well as interactions between that prey and its competitors, hosts, and other predators (see Chapter 8). Even the perceived risk of predation can alter herbivore allocation of nutrients sufficiently to affect ecosystem processes (Fig. 11.18) (Hawlena and Schmitz, 2010; Hawlena et al., 2012). Ultimately, indirect effects of this interaction can affect primary production, canopy cover, and resource availability in ways that determine climate, substrate, and resource conditions for the entire ecosystem.

Nontrophic interactions are difficult to recognize and measure (Dambacher et al., 1999, 2002; O'Neill, 2001); quantitative data are available for relatively few potential indirect interactions. Accordingly, the complexity of indirect, as well as direct, interactions is difficult to model, but has important implications for how ecosystems respond to environmental changes (see Chapter 15).

A number of models have been developed to predict fluxes of energy or key elements, especially carbon or nitrogen, through ecosystems. However, as noted earlier, interactions among various cycles (eg, nitrogen and carbon cycles integrated through biomolecules, carbon and calcium cycles integrated in carbonates, or nitrogen and calcium cycles integrated through soil pH change) may confound predictions based on individual resources. Furthermore, as described for population modeling (Chapter 6), ecosystem models are subject to the unpredictable ways in which initial conditions of a system can affect subsequent system behavior and to the so-called "butterfly effect," in which small changes in model parameters can cause large changes in the resulting prediction (Lorenz, 1993).

Comprehensive ecosystem models that integrate energy, carbon, water, and nutrient fluxes include FOREST-BGC/BIOME-BGC (Running and Gower, 1991) and CENTURY (eg, Fig. 11.17, Parton et al., 1993; Throop et al., 2004), which have been modified to represent a variety of ecosystem types. These models are useful for predicting global biogeochemical processes because they integrate common ecosystem processes in a logical framework, have minimum requirements for detail of inputs for ecosystem characteristics, and account for the mass balances of multiple nutrients moving through interacting plants, detritus, decomposers, and abiotic pools. This ecological stoichiometry (Daufresne and Loreau, 2001; Sterner and Elser, 2002) provides a tool for evaluating consequences of changes in mass balances among multiple elements as a result of changes in environmental conditions or community interactions.

The effects of insects and other invertebrates rarely have been incorporated in these, or other, existing ecosystem models. De Vries et al. (2013) used a hierarchical modeling approach

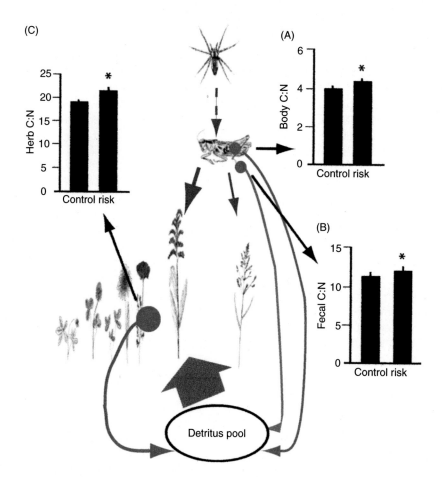

FIGURE 11.18 **Pathways through which changes in herbivore allocation of C and N in response to predation risk can lead to shifts in C:N content of organic matter and decomposition.** (A) C:N content of herbivore tissues; (B) C:N content of herbivore feces; and (C) C:N content of vegetation as mediated by herbivore promotion of *Solidago* dominance in the face of predation risk. *Solid red arrows* indicate direct trophic interactions; *dashed red arrows* indicate a nontrophic fear effect. Arrow thickness indicates food preference under risk conditions. *Green arrows* indicate the source and fate of tissue C and N in the ecosystem. Values are mean $\pm$ SE; * indicates significant difference ($p < 0.05$). *Source: From Hawlena, D., Schmitz, O.J., 2010. Herbivore physiological response to predation risk and implications for ecosystem nutrient dynamics. Proc. Natl. Acad. Sci. USA 107, 15503–15507.*

to demonstrate the importance of soil food web interactions (including earthworms, enchytraeid worms, mites, and collembolans) to the prediction of carbon and nutrient fluxes in several land use systems in Europe. Throop et al. (2004) used the CENTURY model to predict effects of atmospheric nitrogen deposition and herbivory on C and N fluxes and found that herbivory depressed plant and soil C storage and N mineralization. Most often, insects are combined as "insects" or "arthropods," thereby losing valuable information about this diverse group, species of which can respond dramatically and differentially to environmental change and have major effects on ecosystem properties (Chapters 2, 6, 12–14).

8 SUMMARY

An ecosystem represents the integration of the biotic community and the abiotic environment. The capacity of the community to modify its environment depends on its structure and the degree to which it controls energy flow, biogeochemical cycling, and climatic conditions.

Ecosystem structure reflects the organization of various abiotic and biotic pools that exchange energy and matter. Abiotic pools are the atmosphere, oceans, and sediments that represent the sources of energy and matter for biotic use. Biotic pools are the various organisms (individuals, species populations, functional groups, or trophic levels) in the community. Autotrophs (or primary producers) are those organisms that can acquire resources from abiotic pools. Heterotrophs (or secondary producers) are those organisms that must acquire their resources from other organisms. Energy and matter storage in these pools can be represented as pyramids of productivity, numbers, or biomass.

Energy available to ecosystems comes primarily from solar radiation, captured and stored in carbohydrates by primary producers (autotrophs) through the process of photosynthesis. The total rate at which energy is captured (gross primary productivity) depends on exposure to sunlight, availability of water, and biomass. Some of the energy from gross primary production is expended through plant respiration. The remaining net primary production is stored as plant biomass and is the source of energy and matter for heterotrophs. Primary heterotrophs (herbivores) feed on autotrophs, whereas secondary heterotrophs (predators) feed on other heterotrophs. Consumption transfers the energy stored in consumed biomass to the higher trophic level, with some lost as egestion and consumer respiration. Generally, < 10% of the energy available at each trophic level is converted into biomass at the next higher trophic level, although predators generally have a higher efficiency of conversion than do herbivores. Energy remaining in organisms at the time of death becomes available to decomposers that release the remaining energy through respiration.

Energy is the currency with which organisms acquire and concentrate material resources necessary for growth and reproduction. Material resources often are available in limited supply, favoring mechanisms that facilitate retention and reuse within the ecosystem. Biogeochemical cycling represents the processes whereby material resources, including water, carbon, nitrogen, and mineral elements, are acquired from abiotic pools and exchanged indefinitely among trophic levels, with eventual return to abiotic pools. The efficiency with which these materials are recycled and conserved, rather than lost to abiotic pools, buffers an ecosystem against resource depletion and reduced productivity. Hence, ecosystems become organized in ways that maximize the capture and storage of resources among organisms. Resources egested or excreted during trophic transfers, as well as dead organisms, become available to decomposers that rapidly acquire and store the nutrients from organic matter. Nutrients released by decomposers become available for exchange among soil and aquatic organisms and for plant uptake. Microorganisms are particularly instrumental in making nitrogen available for plant uptake, with different specialists fixing atmospheric nitrogen as ammonia, converting ammonia to nitrate, and organic nitrogen to ammonia. Volatilization by fire and denitrification by anaerobic bacteria complete the cycle by returning elemental nitrogen to the atmosphere.

Ecosystems also modify local and regional climatic conditions. The degree to which vegetation reduces soil warming, evaporation, erosion, and wind speed depends on density and vertical architecture. Insects and other organisms affect vegetation structure, hence canopy–atmosphere interactions. Tall, multicanopied forests are most effective at modifying surface temperatures, relative humidities, and wind speed, thereby ameliorating local and regional fluctuations in temperature, wind speed, and precipitation.

Models have become important tools for synthesizing complex, and often incomplete, data for prediction of ecosystem responses to, and effects on, global environmental changes. Ecosystem models differ in their structure and degree of simplification. Effects of insects on a variety of ecosystem parameters have been largely ignored in ecosystem models.

Herbivory

THE COMPLEXITY OF HERBIVORE EFFECTS ON ECOSYSTEM STRUCTURE AND FUNCTION

Although herbivore effects on primary production are obvious and well-known, herbivores have complex direct and indirect effects on community structure, biogeochemical processes, and climate that may affect long-term ecosystem productivity and other functions. Among the best illustrations of this complexity is a series of studies in northern Arizona, including long-term manipulation of a stem-boring moth, *Dioryctria albovittella*, and piñon needle scale, *Matsucoccus acalyptus*, on insect-resistant and susceptible pinyon pines, *Pinus edulis*, in a semiarid woodland (J. Brown et al., 2001).

In two of the earliest studies to demonstrate a range of plant responses to herbivory, Paige and Whitham (1987) and Maschinski and Whitham (1989) found that effects of grazing depended on the interaction of timing and intensity of grazing and availability of water or nutrients. Under conditions of adequate water or nutrient supply, naturally growing plants were capable of substantially overcompensating for herbivory, that is, increasing primary production in excess of the amounts necessary to replace lost plant tissues. However, when water or nutrients were insufficient for compensation, herbivory reduced plant production.

Subsequent studies addressed herbivore effects on community dynamics. K. Christensen and Whitham (1991) found that moth feeding on stems and cones reduced cone production and reduced seed dispersal by birds. Birds avoided entire stands of trees with reduced cone production, even though individual insect-resistant trees produced substantial numbers of cones. These data indicated the importance of masting to ensure sufficient cone production to attract seed dispersers. Concurrently, Gehring and Whitham (1991, 1995) showed that folivory significantly reduced mycorrhizal colonization and growth on stressed trees, but not on unstressed trees, demonstrating the importance of plant stress to folivore effects on below-ground processes. Whitham et al. (2003, 2006) and Shuster et al. (2006) developed a model of "extended phenotype" that showed how heritable traits that control interactions among organisms and their effects on community and ecosystem dynamics could lead to evolution at community and ecosystem levels.

Effects on biogeochemical processes were demonstrated by S. Chapman et al. (2003, 2006) and Classen et al. (2007a,b). Herbivory by both moth and scale insects was shown to affect litter quality,

Insect Ecology. http://dx.doi.org/10.1016/B978-0-12-803033-2.00012-1

mineralization rate, and soil microbial abundance and activity. Herbivory increased nitrogen concentration in pinyon pine litterfall but reduced microbial biomass. The direction of nutrient release changed with time, that is, net nitrogen immobilization one year and net mineralization a second year, perhaps reflecting alternating periods of nitrogen mineralization and leaching. These effects were strongly mediated by seasonal rainfall. However, Schweitzer et al. (2005) found that herbivory on *Populus* decelerated decomposition rate, but this effect depended on plant genotype.

Herbivore effects on local climate were demonstrated by Classen et al. (2005). Scale insects reduced the leaf area index of susceptible trees by 39% and increased soil temperature (by 26%) and moisture (by 35%), whereas moths had no effect on these variables. However, both insects reduced canopy interception by 51% (scale insect) and 29% (moth). The magnitude of scale insect-driven change in soil temperature and moisture was similar to global change scenarios and sufficient to drive ecosystem processes. The larger size of moth-susceptible trees may minimize herbivore effects on microclimate, despite changes in crown architecture. This study showed that herbivores are capable of influencing local climate, potentially modifying the ability of vegetation to control climate and ecosystem processes.

1 INTRODUCTION

Herbivory is the rate of consumption by animals of any plant parts, including foliage, stems, roots, flowers, fruits, or seeds. Herbivory is a key ecosystem process that reduces biomass and density of plants or plant materials, transfers mass and nutrients to the soil or water column, and affects habitat and resource conditions for other organisms. Insects are the primary herbivores in many ecosystems, and their effect on primary production can equal or exceed that of more conspicuous vertebrate grazers (A. Andersen and Lonsdale, 1990; Gandar, 1982; Sinclair, 1975; Weisser and Siemann, 2004; Wiegert and Evans, 1967).

Loss of plant material through herbivory generally is negligible, or at least inconspicuous, but periodic outbreaks of herbivores have a well-known capacity to reduce growth and survival of host species by as much as 100% and to alter vegetation structure over large areas (Schowalter and Lowman, 1999). A key aspect of herbivory is its variation in intensity among plant species, reflecting biochemical interactions between the herbivore and the various host and nonhost species that comprise the vegetation (see Chapter 3).

Effects of herbivory on ecosystem variables depend on the type of herbivore and pattern of consumption, as well as its intensity and the scale at which it is measured (B. Brown and Allen, 1989; Mauricio et al., 1993). Measurement and comparison of herbivory and its effects among ecosystems and environmental conditions remain problematic due to the lack of standardized techniques for measuring or manipulating intensity. Relatively few studies have assessed the effects of herbivory on ecosystem processes other than primary production. Nevertheless, accumulating evidence indicates that effects of herbivory on ecosystem processes, including primary production, are complex, and long-term compensatory effects may at least partially offset short-term effects. As a result, ecosystem management practices that exacerbate or suppress herbivory may be counterproductive.

This chapter focuses on the effects of herbivory on plant vegetative tissues. Effects of insects on reproductive tissues are addressed in Chapter 13.

2 TYPES AND PATTERNS OF HERBIVORY

2.1 Herbivore Functional Groups

Herbivorous insects that have similar means of exploiting plant parts for food can be classified into feeding guilds or functional groups. Groups of plant-feeders include *grazers* that chew foliage, stems, flowers, pollen, seeds, and roots; *miners* and *borers* that feed between plant surfaces; *gall-formers* that reside and feed within the plant and induce the production of abnormal growth reactions by plant tissues; *sap-suckers* that siphon plant fluids; and *seed predators* and *frugivores* that consume the reproductive parts of plants (Romoser and Stoffolano, 1998). Some species, such as seed predators, seedling-eaters, and tree-killing bark beetles, are true plant predators, but most herbivores function as plant parasites because they feed on living host plants (Price, 1980). These different modes of consumption affect plants and ecosystem conditions in different ways. For example, *folivores* (species that chew foliage) directly reduce the area of photosynthetic tissue and redirect nutrients to litter as solid materials, whereas sap-sucking insects affect the flow of fluids and nutrients within the plant and redirect nutrients to litter as solutes, and *root feeders* reduce plant capacity to acquire nutrients or remain upright and redirect nutrients to soil/litter in situ.

Folivory is the best-studied aspect of herbivory. In fact, the term herbivory often is used when folivory alone is measured because loss of foliage is the most obvious and easily quantified aspect of herbivory. The loss of leaf area can be used to indicate the effect of herbivory. In contrast, other herbivores such as sap-suckers or root borers cause less conspicuous losses that are more difficult to measure. Nonetheless, Schowalter et al. (1981c) reported that calculated loss of photosynthates to sap-suckers greatly exceeded measured foliage loss to folivores in an early successional deciduous forest. Sap-suckers and root feeders also may have long-term effects, for example, through transmission of plant diseases or altered rates of nutrient acquisition or growth (J.P. Smith and Schowalter, 2001).

2.2 Measurement of Herbivory

Effects of herbivory on plants or ecosystem conditions are determined by temporal and spatial variability in the magnitude of consumption. Clearly, evaluating the effects of herbivory requires robust methods for measuring herbivory, as well as primary production and other ecosystem conditions. Measurement of herbivory can be difficult, especially for underground plant parts and forest canopies, and has not been standardized. Several methods commonly used to measure herbivory have been compared by Filip et al. (1995), Landsberg (1989), and Lowman (1984).

The simplest and most widely used technique is the measurement of feeding rate by individual herbivores and extrapolation to feeding rate by a population. This technique provides relatively accurate rates of consumption and can be used to estimate the per capita feeding rate for sap-suckers as well as folivores (Gandar, 1982; Maino and Kearney, 2015; Schowalter et al., 1981c; B. Stadler and Müller, 1996). Insect folivores typically consume 50–150% of their dry body mass per day (Blumer and Diemer, 1996; Reichle and Crossley, 1967; Reichle et al., 1973; Schowalter et al., 1981c).

Rates of sap and root consumption are difficult to measure, but a few studies have provided limited information. For example, honeydew production by individual sap-sucking insects can be used as an estimate of their consumption rates. Stadler and Müller (1996) and Stadler et al. (1998) reported that individual spruce aphids, *Cinara* spp., produced from 0.1 mg honeydew day^{-1} for 1st instars to 1 mg day^{-1} for adults, depending on aphid species, season, and nutritional status of the host. Schowalter et al. (1981c) compiled consumption data from studies of eight herb- and tree-feeding aphids (Auclair, 1958, 1959, 1965; Banks and Macaulay, 1964; Banks and Nixon, 1959; M. Day and Irzykiewicz, 1953; M. Llewellyn, 1972; Mittler, 1958, 1970; Mittler and Sylvester, 1961; Van Hook et al., 1980; M. Watson and Nixon, 1953), a leafhopper (M. Day and McKinnon, 1951), and a spittlebug (Wiegert, 1964) that yielded an average consumption rate of 2.5 mg dry sap mg^{-1} dry insect day^{-1}.

Several factors affect the rate of sap consumption. P. Andersen et al. (1992) found that leafhopper feeding rate was related to xylem chemistry and fluid tension. Feeding rates generally increased with amino acid concentrations and decreased with xylem tension, ceasing above tensions of 2.1 Mpa when plants were water stressed. Stadler and Müller (1996) reported that aphids feeding on poor-quality hosts with yellowing needles produced twice the amount of honeydew as did aphids feeding on high-quality hosts during shoot expansion, but this difference disappeared by the end of shoot expansion. Banks and Nixon (1958) reported that aphids tended by ants approximately doubled their rates of ingestion and egestion, compared to untended aphids.

Measurement of individual consumption rate has limited utility for extrapolation to effects on plant growth because more plant material typically is removed, or not produced, than actually consumed, as a consequence of wasteful feeding or mortality to meristems (Blumer and Diemer, 1996; Gandar, 1982). For example, Schowalter (1989) reported that feeding on Douglas-fir, *Pseudotsuga menziesii*, buds by a budmoth, *Zeiraphera hesperiana*, caused an overall loss of < 1% of foliage standing crop, but the resulting bud mortality caused a 13% reduction in production of shoots and new foliage.

Herbivory can be estimated as the amount of frass collected per unit time (Fig. 12.1), adjusted for assimilation efficiency (Chapter 4). This measure is sensitive to conditions that affect frass collection, such as precipitation, which causes disintegration of frass pellets. Hence, frass generally must be collected prior to rainfall events. Mizutani and Hijii (2001) measured the effect of precipitation on frass collection in conifer and deciduous broadleaved forests in central Japan and calculated correction factors for frass disintegration due to precipitation. Such methods enhance the use of frass collection for estimation of herbivory.

Herz et al. (2007) described a method for estimating foliage harvest by leaf-cutting ants, *Atta columbica*. These insects harvest foliage for maintenance of fungal gardens in underground chambers and discard exhausted substrate in refuse piles outside the nest. Herz et al. (2007) found that the number of refuse particles deposited per day was tightly correlated with the number of harvested foliage fragments for nests of different sizes. The number of particles adjusted for average foliage fragment area or mass provided an estimate of annual rates of harvest.

Percentage leaf area missing can be measured at discrete times throughout the growing season. This percentage can be estimated visually but is sensitive to observer bias (Landsberg, 1989). Alternatively, leaf area of foliage samples is measured, then remeasured after holes and missing edges have been reconstructed (Filip et al., 1995; H. Odum and Ruíz-Reyes, 1970; Reichle et al., 1973; Schowalter et al., 1981c). Reconstruction originally was accomplished using tape or paper cutouts. More recently, computer software has become available to

FIGURE 12.1   Insect herbivore feces collected on understory vegetation in cypress-tupelo swamp in southern Louisiana, USA.

reconstruct leaf outlines and fill in missing portions (Hargrove, 1988). Neither method accounts for expansion of holes as leaves expand, for compensatory growth (to replace lost tissues), for completely consumed or prematurely abscissed foliage, for foliage loss due to high winds, nor for herbivory by sap-suckers (Faeth et al., 1981; Hargrove, 1988; Lowman, 1984; Reichle et al., 1973; Risley and Crossley, 1993; Stiling et al., 1991).

The most accurate method for measuring loss to folivores is detailed life table analysis of marked leaves at different stages of growth (Aide, 1993; Filip et al., 1995; Hargrove, 1988; Lowman, 1984). Continual monitoring permits accounting for consumption at different stages of growth or plant development, with consequent differences in degree of hole expansion, compensatory growth, and complete consumption or loss of damaged leaves (Lowman, 1984; Risley and Crossley, 1993). Estimates of herbivory based on long-term monitoring often are 3–5-fold higher than estimates based on discrete measurement of leaf area loss (Lowman, 1984, 1995). Filip et al. (1995) compared continual and discrete measurements of herbivory for 12 tree species in a tropical deciduous forest in Mexico. Continual measurement provided estimates 1–5 times higher than those based on discrete sampling. On average, measurements by the two techniques differed by a factor of 2. Broadleaved plants are more amenable to this technique than are needle-leaved plants.

Several methods also have been used to measure effects of herbivory on plants or ecosystem processes. A vast literature is available on the effects of herbivory on growth of individual plants or plant populations (Crawley, 1983; Huntly, 1991). However, most studies have focused on effects of above-ground herbivores on above-ground plant parts. Few studies have addressed root-feeding insects or root responses to herbivory (M.D. Hunter, 2001a; Morón-Ríos et al., 1997b; J.P. Smith and Schowalter, 2001; D. Strong et al., 1995). J.P. Smith and Schowalter (2001) and

D. Strong et al. (1995) found that roots can take at least a year to recover from herbivory, indicating that short-term experiments may be inadequate to estimate herbivore effects on roots.

Remote sensing techniques are being developed to measure effects of herbivory, as well as various plant stressors, on a variety of plant species from grasses to conifers and deciduous trees (Carter and Knapp, 2001). Carter and Knapp (2001) reported that stress was consistently expressed as increased reflectance at wavelengths near 700 nm. This optical response can be explained by a general tendency for stress to reduce chlorophyll concentrations in foliage. Nansen et al. (2009) reported that experimental infestation by wheat stem sawfly, *Cephus cinctus*, significantly affected reflectance at 553 nm and 725 nm but not at 452 nm or 760 nm in one data-set. Normalized difference vegetation index (NDVI) and photochemical reflectance index (PRI) decreased in response to sawfly infestation, whereas stress index (SI) increased. In a second data-set, sawfly infestation was detectable at 760 nm 3 weeks after infestation. Nansen et al. (2010) evaluated effects of severe, moderate, or no drought stress or spider mite, *Tetranychus urticae*, infestation in cereal crops. They found a particularly strong response to drought (but not spider mite) stress at 706 nm and a significant response to spider mites, as well as drought stress, at 440 nm.

At the ecosystem level, a number of studies have compared ecosystem processes between sites naturally infested or not infested, during population irruptions. Such comparison confounds herbivore effects with background environmental gradients that may be responsible for the discontinuous pattern of herbivory (see Chapter 7). Hurlbert (1984) discussed the importance of independent, geographically intermixed replicate treatment plots for robust evaluation of treatment effects. This requires manipulation of herbivore abundances in independent, replicate plots to evaluate effects on ecosystem parameters. However, comparison of naturally infested and uninfested sites may be valid in cases where sufficient preinfestation data are available to demonstrate substantial similarity among infested and uninfested sites prior to infestation (Hurlbert, 1984).

Experimental manipulation of herbivore numbers has been accomplished in some studies, especially on short vegetation (Kimmins, 1972; McNaughton, 1979; Morón-Ríos et al., 1997a; Schowalter et al., 1991; Seastedt, 1985; Seastedt et al., 1983; S. Williamson et al., 1989), but clearly is difficult in mature forests or over large areas. The most common method for manipulation is chemical suppression (V.K. Brown et al., 1987, 1988; D. Gibson et al., 1990; Louda and Rodman, 1996; Seastedt et al., 1983). However, insecticides can provide a source of limiting nutrients that may affect plant growth and confound evaluation of herbivore effects. Carbaryl, for example, contains nitrogen, frequently limiting and likely to stimulate plant growth. Furthermore, measuring differences in herbivore abundance between treatments in dense or tall vegetation may be difficult. Herbivore abundance also can be manipulated using enclosures or exclosures (Schowalter et al., 1991; S. Williamson et al., 1989), but augmenting herbivore abundance often is difficult (I. Baldwin, 1990; Crawley, 1983; Schowalter et al., 1991) and may require rearing facilities to produce sufficient herbivore numbers. Cages constructed of fencing or mesh screening can exclude or contain experimental densities of herbivores (Fonte and Schowalter, 2005; McNaughton, 1985; Palmisano and Fox, 1997). Mesh screening should be installed in a manner that does not restrict air movement or precipitation and thereby alter growing conditions within the cage.

An alternative option has been to simulate herbivory by clipping or pruning plants or by punching holes in leaves (Honkanen et al., 1994). This method avoids the problems of manipulating herbivore abundance but may fail to represent important indirect effects of herbivory (I. Baldwin, 1990; Crawley, 1983; Lyytikäinen-Saarenmaa, 1999). For example, herbivore saliva may stimulate growth of some plant species (M. Dyer et al., 1995), and natural patterns of consumption and frass deposition affect litter condition, decomposition, and nutrient supply (Christenson et al., 2002; Frost and Hunter, 2004, 2007, 2008b; Hik and Jefferies, 1990; Lovett and Ruesink, 1995; B. Stadler et al., 1998; Zlotin and Khodashova, 1980, see Section 3.3). Lyytikäinen-Saarenmaa (1999) reported that artificial defoliation of Scots pine, *Pinus sylvestris*, saplings caused greater growth reduction than did comparable herbivory by sawflies, *Diprion pini* and *Neodiprion sertifer*, in May–Jun., whereas the opposite trend was seen for trees subjected to treatments in Jul.–Aug.

The choice of technique for measuring herbivory and its effects depends on several considerations. The method of measurement must be accurate, efficient, and consistent with objectives. Percentage leaf area missing at a point in time is an appropriate measure of the effect of herbivory on canopy porosity, photosynthetic capacity, and canopy–soil or canopy–atmosphere interactions but does not represent the rate of consumption or removal of plant material. Access to some plant parts is difficult, precluding continuous monitoring. Hence, limited data are available for herbivory on roots or in forest canopies. Simulating herbivory by removing plant parts or punching holes in leaves fails to represent some important effects of herbivory, such as salivary toxins or stimulants or flux of canopy material to litter as feces, but does overcome the difficulty of manipulating abundances of herbivore species. However, as noted earlier, only manipulation of herbivore abundance can provide results for net direct and indirect effects of herbivores.

Similarly, the choice of response variables depends on objectives. Most studies have examined only effects of herbivory on above-ground primary production, consistent with the emphasis on foliage and fruit production. However, herbivores feeding above-ground affect root production and rhizosphere processes, as well (Gehring and Whitham, 1991, 1995; Holland et al., 1996; Rodgers et al., 1995; J.P. Smith and Schowalter, 2001). Effects on some fluxes, such as dissolved organic carbon in honeydew, are difficult to measure (B. Stadler et al., 1998). Some effects, such as compensatory growth and altered community structure, may not become apparent for long time periods following herbivore outbreaks (Alfaro and Shepherd, 1991; Wickman, 1980), requiring long-term measurement. Note, again, that only measurement of the suite of direct and indirect effects will provide a complete perspective on herbivore effects.

2.3 Spatial and Temporal Patterns of Herbivory

All plant species support characteristic assemblages of insect herbivores, although some plants host a greater diversity of herbivores and exhibit higher levels of herbivory than do others (Coley and Aide, 1991; de la Cruz and Dirzo, 1987). Some plants tolerate continuous high levels of herbivory, whereas other species show negligible loss of plant material (S. Carpenter and Kitchell, 1984; Lowman and Heatwole, 1992; McNaughton, 1979; Schowalter and Ganio, 2003), and some plant species suffer growth loss or mortality at lower levels of herbivory than do others. Herbivory typically is concentrated on the most nutritious or least defended plants and plant parts (Chapter 3, Aide and Zimmerman, 1990).

The consequences of herbivory vary substantially, not just among plant–herbivore interactions, but also as a result of different spatial and temporal factors (Huntly, 1991; Maschinski and Whitham, 1989). For example, water or nutrient limitation and ecosystem fragmentation can affect the ability of the host plant to respond to herbivory (Chapin et al., 1987; Kolb et al., 1999; Maschinski and Whitham, 1989; W. Webb, 1978). The timing of herbivory with respect to plant development and the time intervals between episodes of elevated herbivory also have important effects on ecosystem processes (Hik and Jefferies, 1990).

Herbivory usually is expressed as daily or annual rates of consumption and ranges from negligible to several times the standing crop biomass of foliage (Table 12.1), depending on ecosystem type, environmental conditions, and regrowth capacity of the vegetation (Lowman, 1995; Schowalter and Lowman, 1999). Herbivory for particular plant species can be integrated at the ecosystem level by weighting rates for each plant species by its biomass or leaf area. When the preferred hosts are dominant plant species, loss of plant parts can be dramatic and conspicuous, especially if these species are slow to replace lost parts (B. Brown and Ewel, 1987). For example, defoliation of evergreen forests may be visible for months, whereas deciduous forests and grasslands are adapted for periodic replacement of foliage and typically replace lost foliage quickly. Eucalypt forests are characterized by chronically high rates of herbivory (Fox and Morrow, 1992). Some species lose more than 300% of their foliage standing crop annually, based on life table studies of marked leaves (Lowman and Heatwole, 1992).

Comparison of herbivory among ecosystem types (Table 12.1) indicates considerable variation. The studies in Table 12.1 reflect the range of measurement techniques described previously. Most are short-term snapshots of folivory, often for only a few plant species, do not provide information on herbivory by sap-suckers or root feeders, and do not address any deviation in environmental conditions, plant chemistry, or herbivore densities from long-term means during the period of study. Long-term studies using standardized techniques are necessary for meaningful comparison of herbivory rates.

Cebrián and Duarte (1994) compiled data from a number of aquatic and terrestrial ecosystems and found a significant relationship between percentage of plant material consumed by herbivores and the rate of primary production. Herbivory ranged from negligible to >50% of photosynthetic biomass removed daily, with the highest rates in some phytoplankton communities where herbivores consumed all production daily and lowest rates in some forests where herbivores removed <1% of production. Insects are the primary herbivores in forest ecosystems (Janzen, 1981; Wiegert and Evans, 1967) and account for 11–73% of total herbivory in grasslands, where native vertebrate herbivores remove an additional 15–33% of production (Detling, 1987; Gandar, 1982; Sinclair, 1975). Temperate deciduous forests and tropical evergreen forests show similar annual losses of 3–20%, based on discrete sampling of leaf area loss (Coley and Aide, 1991; Landsberg and Ohmart, 1989; H. Odum and Ruíz-Reyes, 1970; Schowalter and Ganio, 1999; Schowalter and Lowman, 1999; Van Bael et al., 2004). Aquatic ecosystems, tropical evergreen forests, and grasslands, which replace lost photosynthetic tissue continuously, often lose several times their standing crop biomass to herbivores annually, based on loss of marked foliage or on herbivore exclusion (S. Carpenter and Kitchell, 1984; Cebrián and Duarte, 1994; Crawley, 1983; Landsberg, 1989; Lowman and Heatwole, 1992; McNaughton, 1979).

In addition to the conspicuous loss of photosynthetic tissues, terrestrial plants lose additional material to sap-suckers and root feeders. Schowalter et al. (1981c) compiled data on rates of sap consumption to estimate turnover of 5–23% of primary production through sap-sucking

TABLE 12.1   Herbivory Measured in Temperate and Tropical Ecosystems (Including Understory)

Location	Ecosystem type	Level of grazing	Technique[a]	Source
Tropical				
Costa Rica	Tropical forest	7.5% (new leaves)	1	N. Stanton (1975)
	Tropical evergreen forest	30% (old)	1	N. Stanton (1975)
Panama	Tropical evergreen forest	13%	1	Wint (1983)
Panama (BCI)	Tropical evergreen forest	8% (6% insect; 1–2% vertebrates)	1, 2	Leigh and Windsor (1982)
		15%	1, 2	Leigh and Smythe (1978)
	Understory only	21% (but up to 190%)	3	Coley (1983)
Puerto Rico	Tropical evergreen forest	7.8%	1	Odum and Ruíz-Reyes (1970)
		5.5–16.1%	1	Benedict (1976)
		2–13%	1	Schowalter and Ganio (1999)
Mexico	Tropical deciduous forest	7–9%	1	Filip et al. (1995)
	Tropical deciduous forest	17%	3	Filip et al. (1995)
Venezuela	Understory only	0.1–2.2%	1	Golley (1977)
New Guinea	Tropical evergreen forest	9%–12%	1	Wint (1983)
Australia	Montane or cloud forest	26%	3	Lowman (1984)
	Warm temperate forest	22%	3	Lowman (1984)
	Subtropical forest	14.6%	3	Lowman (1984)
Cameroon	Tropical evergreen forest	8–12%	3	Lowman et al. (1993)
Tanzania	Tropical grassland	14–38% (4–8% insect; 8–34% vertebrates)	4	Sinclair (1975)
South Africa	Tropical savanna	38% (14% insect; 24% vertebrates)	4	Gandar (1982)
Temperate				
North America	Deciduous forest	2–10%	1	Reichle et al. (1973)
		1–5%	1	Schowalter et al. (1981c)
	Herbaceous sere	3%	4	Crossley and Howden (1961)
	Coniferous forest	<1%	1	Schowalter (1989)
		1–6%	1	Schowalter (1995)
	Grassland	5–15%	1	Detling (1987)
Australia	Evergreen forest	15–300%	3	Lowman and Heatwole (1992)
	Dry forest	5–44%	1	Fox and Morrow (1983)
		3–6%	2	Ohmart et al. (1983)
Europe	Deciduous forest	7–10%	1	Nielsen (1978)
	Alpine grassland	19–30%	1	Blumer and Diemer (1996)

[a] 1, Leaf area missing; 2, litter or frass collection; 3, turnover of marked foliage; 4, individual consumption rates.
Source: Expanded from Lowman, M.D., 1995. Herbivory as a canopy process in rain forest trees. In: Lowman, M.D., Nadkarni, N.M. (Eds.), Forest Canopies. Academic Press, San Diego, CA, pp. 431–455.

herbivores, in addition to 1–2% turnover through folivores in a temperate deciduous forest. J.P. Smith and Schowalter (2001) found that shoot-feeding aphids, *Cinara pseudotsugae*, significantly reduced Douglas-fir root tissue density and growth and that at least 1 year was required for recovery after feeding ceased. V.K. Brown and Gange (1991) and Morón-Ríos et al. (1997a) reported that root-feeding insects can reduce primary production of grasses by 30–50%.

Factors that affect herbivore abundance, for example, host abundance and condition, predation, and environmental changes, also affect herbivory (see Chapters 3, 6 and 8), although herbivory also may vary as a result of herbivore activity. Urbas et al. (2007) found that leaf-cutting ant, *Atta cephalotes*, colonies at forest edges removed nearly twice the amount of foliage from their foraging area as did colonies in interior forest (14 vs 8%). The higher level of herbivory reflected significantly smaller foraging areas for edge, compared to interior, colonies (0.9 vs 2.3 ha colony^{-1} year^{-1}) and perhaps greater productivity of pioneer tree species at edges.

Proportional losses of foliage to herbivores generally are higher in less diverse ecosystems, compared to more diverse ecosystems (Jactel and Brockerhoff, 2007; Kareiva, 1983, see Chapter 6), but the intensity of herbivory also depends on the particular herbivore species and the species composition of the vegetation (Loranger et al., 2014; R. Moore and Francis, 1991; R. Moore et al., 1991; Stastny and Agrawal, 2014). B. Brown and Ewel (1987) demonstrated that ecosystem-level foliage losses per unit ground area were similar among four tropical ecosystems that varied in vegetation diversity, but the proportional loss of foliage standing crop was highest in the less diverse ecosystems. Nevertheless, rare plant species in diverse ecosystems can suffer intense herbivory, especially under conditions that increase their apparency or acceptability (B. Brown and Ewel, 1987; Schowalter and Ganio, 1999), (eg, associational susceptibility; Stastny and Agrawal, 2014). Whereas specialist herbivores may increase herbivory as host abundance increases, generalists may show the opposite response. Loranger et al. (2014) reported that herbivory, primarily by grassland generalists, increased along an experimental diversity gradient that was part of the Jena Experiment in Germany, perhaps because of improved nutrient stoichiometry in more diverse plots (Abbas et al., 2014). C. Fonseca (1994) reported that an Amazonian myrmecophytic canopy tree showed 10-fold greater foliage losses when ants were experimentally removed than when ants were present.

Seasonal and annual changes in herbivore abundance affect patterns and rates of herbivory, but the relationship may not be linear, depending on variation in per capita rates of consumption or wasteful feeding with increasing population density (Crawley, 1983; B. Stadler et al., 1998). Herbivory in temperate forests typically is concentrated in the spring, during leaf expansion (Feeny, 1970; M.D. Hunter, 1987). M.D. Hunter (1992) reported that over 95% of total defoliation on *Quercus robur* in Europe occurs between budburst in Apr. and the beginning of Jun. Although some herbivorous insects prefer mature foliage (Cates, 1980; Sandlin and Willig, 1993; Volney et al., 1983), most defoliation events are associated with young foliage (Coley, 1980; M.D. Hunter, 1992; R. Jackson et al., 1999; Lowman, 1985). Herbivory also is highly seasonal in tropical ecosystems. Tropical plants produce new foliage over a more protracted period than do temperate plants, but many produce new foliage in response to seasonal variation in precipitation (Aide, 1992; Coley and Aide, 1991; Lowman, 1992; Ribeiro et al., 1994). Young foliage may be grazed more extensively than older foliage in tropical rainforests (Coley and Aide, 1991; Lowman, 1984, 1992). Schowalter and Ganio (1999) reported significantly greater rates of leaf area loss during the "wet" season than during the "dry" season in a tropical rainforest in Puerto Rico (Fig. 12.2).

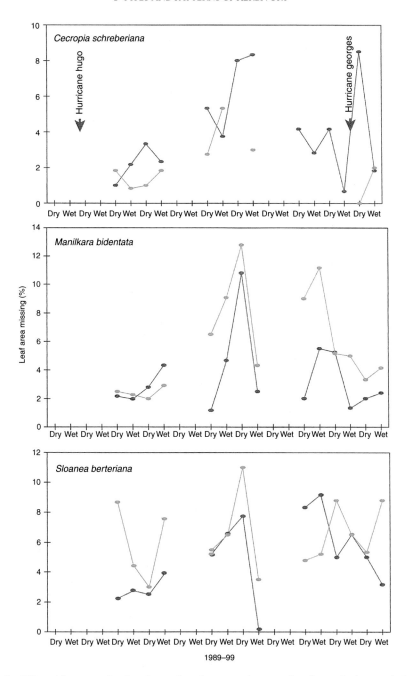

FIGURE 12.2 **Effects of tree species, hurricane disturbance, and seasonal cycles on leaf area missing in a tropical rainforest in Puerto Rico, as affected by two hurricanes (1989 and 1998) and a drought (1994–95).** *Cecropia* is an early successional tree; *Manilkara* and *Sloanea* are late successional trees. *Solid lines* represent intact forest (lightly disturbed); *dashed lines* represent treefall gaps.

Few studies have addressed long-term changes in herbivory as a result of environmental changes. However, environmental changes and disturbances often induce elevated rates of herbivory at a site, primarily through changes in host plant condition or abundance (see Chapters 2, 3, 6, and 10). Droughts and storms are well-known triggers for herbivore population outbreaks and elevated herbivory (Mattson and Haack, 1987; Prather, 2014; Schowalter and Ganio, 2003; Stige et al., 2007; Van Bael et al., 2004, see Chapters 2 and 6). Forest harvest also can affect rates and patterns of herbivory (Schowalter, 1995; Schowalter et al., 1981c).

Changes in vegetation associated with disturbance or recovery affect temporal patterns of herbivory. Bach (1990) reported that intensity of herbivory declined during succession in dune vegetation in Michigan (Fig. 12.3). Coley (1980, 1982, 1983), Coley and Aide (1991), and Lowman and Box (1983) found that rapidly growing early successional tree species showed higher rates of herbivory than did slow-growing late successional trees. Schowalter (1995), Schowalter and Ganio (1999, 2003), and Schowalter and Crossley (1988) compared canopy herbivore abundances and folivory in replicated disturbed (harvest or hurricane) and undisturbed patches of temperate deciduous, temperate coniferous, and tropical evergreen forests. In all three forest types, disturbance resulted in greatly increased abundances of sap-suckers and somewhat increased abundances of folivores on abundant, rapidly growing early successional plant species. The shift in biomass dominance from folivores to sap-suckers following disturbance resulted in an elevated flux of primary production as soluble photosynthates, relative to fragmented foliage and feces. Schowalter et al. (1981c) calculated that loss of photosynthate to sap-suckers increased from 5% of foliage standing crop in undisturbed forest to 20–23% of foliage standing crop during the first 2 years following clear-cutting, compared to relatively consistent losses of 1–2% to folivores. Continued measurement of herbivory over long time periods will be necessary to relate changes in the intensity of herbivory to environmental changes and to effects on ecosystem processes (Chapter 10).

3 EFFECTS OF HERBIVORY

Herbivory affects a variety of ecosystem properties, primarily through differential changes in survival, productivity, and growth form among plant species. Herbivory is not evenly distributed among plant species or over time. Rather, some species are subject to greater herbivory than are others, and relative herbivory among plant species varies with environmental conditions (Coley, 1980; Coley and Aide, 1991; Crawley, 1983; Prather, 2014; Schowalter and Ganio, 1999, see Chapter 3). These differential effects on host conditions alter vegetation structure, energy flow, and biogeochemical cycling, and often predispose ecosystems to characteristic disturbances.

The observed severity of herbivore effects in agroecosystems and some native ecosystems has led to a widespread perception of herbivory as a disturbance (see Chapter 2). This perception raises a number of issues. Can a normal trophic process also be a disturbance? At what level does herbivory become a disturbance, for example, do the normally low levels of 5–20% loss of NPP constitute disturbance? Although debate may continue over whether or not herbivory is a disturbance (Veblen et al., 1994; P. White and Pickett, 1985), rather than simply an ecosystem regulatory process (Coleman et al., 2008; Schowalter, 2012; Schowalter

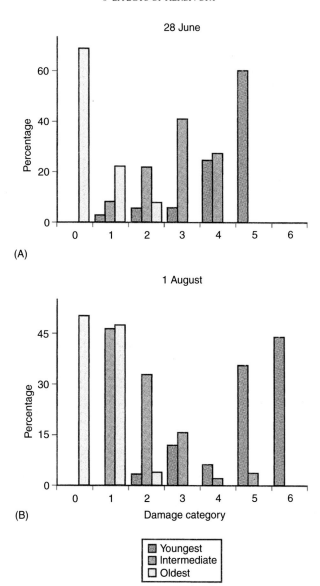

FIGURE 12.3 **Herbivore damage to plants in young, intermediate, and old successional sites in sand dune vegetation in Michigan in June (A) and August (B) 1988.** Percentages are averages for leaves on upper and lower canopy branches by damage category: 0 = 0% damage, 1 = 1–5%, 2 = 6–25%, 3 = 26–50%, 4 = 51–75%, 5 = 76–100%, and 6 = no leaves remaining. *Source: From Bach, C.E., 1990. Plant successional stage and insect herbivory: flea beetles on sand-dune willow. Ecology 71, 598–609.*

and Lowman, 1999; Willig and McGinley, 1999, see Chapter 15), herbivory can dramatically alter ecosystem structure and function over large areas.

3.1 Plant Productivity, Survival, and Growth Form

Traditionally, herbivory has been viewed solely as a process that reduces primary production. As described earlier, herbivory can remove several times the standing crop of foliage, alter plant growth form, or kill all plants of selected species over large areas during severe outbreaks. However, more recent studies indicate complex indirect effects of herbivory on plant growth, survival, and primary production (Massad et al., 2013). The way in which herbivory affects plant survival, productivity, and growth form depends on the plant part(s) affected, plant condition, the stage of plant development, the intensity of herbivory, nutrient availability, and changing environmental conditions (Massad et al., 2013).

Different herbivore species and functional groups, for example, folivores, sap-suckers, shoot borers, and root feeders (see Section 2.1), determine which plant parts are affected, but all affect plant ability to acquire and allocate carbohydrates and nutrients. Folivores and leaf miners reduce foliage surface area and photosynthetic capacity, thereby limiting plant ability to produce and accumulate photosynthates for growth and maintenance (R. Doyle et al., 2002). In addition to direct consumption of foliage, much nonconsumed foliage is lost due to wasteful feeding by folivores (Risley and Crossley, 1993) and induction of leaf abscission by leaf miners (Faeth et al., 1981; Stiling et al., 1991). Sap-suckers and gall-formers siphon photosynthates from the plant's vascular system and reduce plant ability to accumulate photosynthates for growth and maintenance. Shoot borers and bud feeders damage meristems and growing shoots, altering plant growth rate and form (A. Martínez et al., 2009). Root feeders reduce plant ability to acquire water and nutrients and weaken plant support (Blossey and Hunt-Joshi, 2003).

The direct effects of herbivory on plant performance are in proportion to the amount of plant tissue loss or damage. However, particular herbivore species can affect plants in different ways, depending on plant condition and temporal dynamics. For example, Speer et al. (2010) reported that periodical cicadas, *Magicicada* spp., had no detectable effect on growth of five tree species during the long period of root parasitism by nymphs, but that growth of three tree species was reduced the year following cicada emergence as a result of oviposition damage to branches. Three of the five tree species subsequently showed elevated growth after 5 years, consistent with the nutrient pulse hypothesis.

Herbivory also affects plant growth and reproduction indirectly. Reduced accumulation of energy often reduces flowering, pollination, or seed production (Pearse et al., 2015; Quesada et al., 1995; Simard and Payette, 2005), often completely precluding reproduction (V.K. Brown et al., 1987; Crawley, 1989). For example, M. Parker (1985) and Wisdom et al. (1989) reported that flower production by composite shrubs, *Gutierrezia microcephala*, was reduced as much as 80% as a consequence of grazing by the grasshopper, *Hesperotettix viridis*.

Herbivores may cause additional indirect injury to plants. Some herbivores inject salivary or other toxins into host plants. The European wood wasp, *Sirex noctilio*, injects conifer hosts with a glycopeptide venom, noctilisin, that facilitates subsequent lethal infection by a symbiotic fungus, *Amylostereum areolatum* (Bordeaux et al., 2014). In its native range, this insect–fungus association causes little mortality to trees, but in its invasive ranges in Australia and

South America, it causes significant needle wilting and tree mortality. Many sap-sucking and shoot- and root-feeding insects also transmit or facilitate growth of plant pathogens, including viruses, bacteria, fungi, and nematodes (C.G. Jones, 1984). Alternatively, folivory may induce production of defenses that confer resistance to subsequent infection by plant pathogens (Hatcher et al., 1995).

Plant condition is affected by developmental stage, growth rate, resource availability, and environmental conditions, including prior herbivory or parasitism, and determines herbivore population dynamics (Gruner et al., 2008; Massad, 2013; McNutt et al., 2012, see Chapters 3 and 6) and plant capacity to compensate for herbivory. The ability of aquatic vegetation to replace primary production lost to herbivores is particularly well-known (S. Carpenter and Kitchell, 1984, 1987, 1988; S. Carpenter et al., 1985; J. Wallace and O'Hop, 1985). However, low or moderate levels of herbivory in terrestrial ecosystems also can stimulate photosynthesis and primary production (E. Ayres et al., 2007; Belovsky and Slade, 2000; C. Carroll and Hoffman, 1980; Detling, 1987, 1988; Dungan et al., 2007; M. Dyer et al., 1993; Kolb et al., 1999; Lowman, 1982; McNaughton, 1979, 1993a; Pedigo et al., 1986; Trumble et al., 1993; S. Williamson et al., 1989), whereas severe herbivory usually results in mortality or decreased fitness (Detling, 1987, 1988; Marquis, 1984; S. Williamson et al., 1989). J. Wallace and O'Hop (1985) reported that new leaves of water lilies, *Nuphar luteum*, disappeared within 3 weeks as a result of grazing by the leaf beetle, *Pyrrhalta nymphaeae*. A high rate of leaf production was necessary to maintain macrophyte biomass. R. Doyle et al. (2002) concluded that 10–30% foliage removal reduced maximum daily photosynthesis of *Hydrilla verticillata* by 30–40%, relative to controls, to levels just adequate to balance respiration; 70–90% foliage removal reduced maximum daily photosynthesis by 60%, to levels less than are necessary to balance respiration and making survival of these plants unlikely. D. Inouye (1982) reported that herbivory by several insect and mammalian herbivores had a variety of positive and negative effects on fitness of a thistle, *Jurinea mollis*.

Pearson et al. (2003) evaluated factors that influenced growth and mortality of six pioneer tree species in forest gaps of different sizes in Panama. They found that herbivory varied from 2% to 10% overall, with *Croton billbergianus* showing levels of 5–30%. Most species showed a trend of increasing leaf area loss with increasing gap size, but the fastest-growing species did not have the highest levels of herbivory. Variation in growth rate and mortality of these plant species could not be explained by foliage losses to herbivores, but was strongly influenced by a trade-off between maximum growth in the wet season and ability to survive seasonal drought, particularly in small gaps.

Healthy plants with adequate supplies of light, water, and nutrients can replace lost foliage, resulting in higher annual primary production, although standing crop biomass of plants typically is reduced. McNutt et al. (2012) manipulated density and herbivory on goldenrod, *Solanum carolinense*, clones that varied in tolerance to herbivory and measured tolerance in terms of change in biomass. The herbivory treatment involved two folivores, a specialist chrysomelid beetle, *Leptinotarsa juncta*, and a generalist noctuid moth, *Spodoptera exigua*. Plants growing at high density showed increased biomass, especially of roots, when exposed to herbivory, compared to biomass loss of plants growing at low density. Only plants growing at low density benefited from tolerance in the absence of herbivory, perhaps reflecting inherently high growth rates of more tolerant genotypes. These results suggest that competition facilitates the evolution of tolerance in *S. carolinense*, while variation in density maintains genetic variation in tolerance.

Kolb et al. (1999) experimentally evaluated a number of factors that potentially influence the effect of western spruce budworm, *Choristoneura occidentalis,* defoliation on potted Douglas-fir seedling physiology and growth. They demonstrated that seedling biomass decreased, but photosynthetic rate, stomatal conductance, foliar concentrations of N, Ca, and Mg, and soil water potential increased with increasing intensity of herbivory. Welker and Menke (1990) found similar results for blue oak, *Quercus douglasii,* seedlings. Only oak seedlings that experienced gradual water stress and severe defoliation survived to produce foliage in the second year. Increased photosynthesis and reduced water stress may improve tree survival in environments where water stress has a more serious negative effect on survival than does defoliation.

Trumble et al. (1993) reviewed literature demonstrating that compensatory growth (replacement of consumed tissues) following low-to-moderate levels of herbivory is a widespread response by terrestrial plants. Increased productivity of grazed grasses, compared to ungrazed grasses, has been demonstrated experimentally in a variety of grassland ecosystems (Belovsky and Slade, 2000; Detling, 1987, 1988; McNaughton, 1979, 1986, 1993a; Seastedt, 1985; S. Williamson et al., 1989), but not all (G. Zhang et al., 2011). Growth enhancement may depend on adequate water and soil fertility, including the presence of herbivore feces (Allsup and Paige, 2016; I. Baldwin, 1990; Christenson et al., 2002; Frost and Hunter, 2007; Hik and Jefferies, 1990) or other herbivore products (I. Baldwin, 1990). M. Dyer et al. (1995) demonstrated that crop and midgut extracts present in grasshopper regurgitants during feeding stimulate coleoptile growth in grasses, but saliva may not stimulate growth of all plant species (Detling et al., 1980). Wickman (1980) and Alfaro and Shepherd (1991) reported that short-term growth losses by defoliated conifers were followed by several years, or even decades, of growth rates that exceeded predefoliation rates (Fig. 12.4). Romme et al. (1986) found that annual wood production in some pine forests in western North America reached

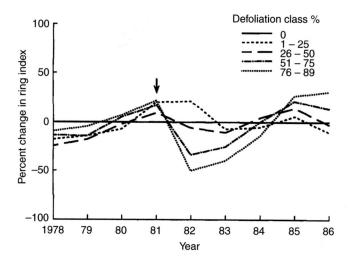

FIGURE 12.4 **Changes in ring width indices for Douglas-fir defoliated at different intensities by the Douglas-fir tussock,** *Orgyia pseudotsugata,* **moth in 1981 (arrow).** The horizontal line at 0% represents ring width index for nondefoliated trees. *Source: From Alfaro, R.I., Shepherd, R.F., 1991. Tree-ring growth of interior Douglas-fir after one year's defoliation by Douglas-fir tussock moth. Forest Sci. 37, 959–964.*

or exceeded preattack levels within 10–15 years following mountain pine beetle, *Dendroctonus ponderosae*, outbreaks.

Detling (1987, 1988), M. Dyer et al. (1993, 1995), McNaughton (1979, 1986, 1993a), and Paige and Whitham (1987) have argued that herbivory may benefit some plants, to the extent that species adapted to replace consumed tissues often disappear in the absence of grazing. Net primary productivity of some grasslands declines when grazing is precluded, due to smothering of shoots as standing dead material accumulates (Kinyamario and Imbamba, 1992; Knapp and Seastedt, 1986; McNaughton, 1979).

These observations generated the *herbivore optimization hypothesis* (Fig. 12.5), or *overcompensation hypothesis*, that primary production is maximized at low-to-moderate levels of herbivory (S. Carpenter and Kitchell, 1984; Mattson and Addy, 1975; McNaughton, 1979; Pedigo et al., 1986). This hypothesis is widely recognized among aquatic ecologists as the basis for inverted biomass pyramids (S. Carpenter and Kitchell, 1984, 1987, 1988; Carpenter et al., 1985). Its application to terrestrial systems has been challenged as contradictory to evolutionary principles (Belsky, 1986; E. Painter and Belsky, 1993; D. Patten, 1993) but also is supported by experimental tests for both insect and vertebrate herbivores in grassland (Belovsky and Slade, 2000; Detling, 1987; M. Dyer et al., 1993; McNaughton, 1979, 1993b; Seastedt, 1985), salt marsh (Hik and Jefferies, 1990), forest (Alfaro and Shepherd, 1991; Dungan et al., 2007; Feeley and Terborgh, 2005; Lovett and Tobiessen, 1993), and even agricultural (Pedigo et al., 1986) ecosystems. Resolution

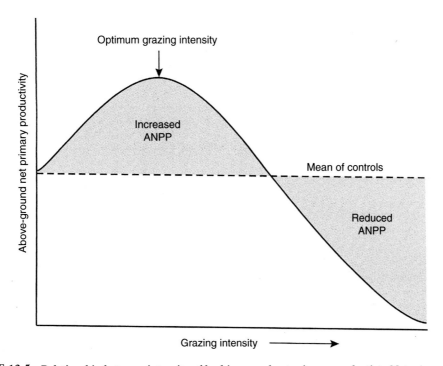

FIGURE 12.5 **Relationship between intensity of herbivory and net primary production.** Net primary production often peaks at low to moderate intensities of herbivory, supporting the grazing optimization hypothesis. *Source: From Williamson, S.C., et al., 1989. Experimental evaluation of the grazing optimization hypothesis. J. Range Manage. 42, 149–152.*

of evolutionary challenges likely rests on demonstrated benefits to individuals accruing from effects on ecosystem processes that favor individuals (Fig. 15.2, see Chapter 15).

Compensatory growth depends on environmental conditions, availability and balances of limiting nutrients, timing of herbivory, and plant adaptation to herbivory (Allsup and Paige, 2016; de Mazancourt et al., 1998; Loreau, 1995; Massad, 2013; Trlica and Rittenhouse, 1993; S. Williamson et al., 1989). C. Lovelock et al. (1999) demonstrated that CO_2 enrichment did not enhance compensation by a tropical legume, *Copaifera aromatica*, compared to compensation under ambient atmospheric CO_2, following artificial defoliation in Panama. Rastetter et al. (1997) used a multielement model to demonstrate that plant response to CO_2 enrichment could be constrained by nitrogen limitation.

De Mazancourt et al. (1998) and Loreau (1995) used a theoretical model to study conditions under which grazing optimization could occur. They found that grazing optimization required that low–moderate herbivory decreased nutrient losses from the system. They concluded that grazing optimization is most likely to occur in ecosystems with large losses of limiting nutrients during decomposition or where herbivores import nutrients from outside the ecosystem.

Plants often are able to compensate for herbivory in the spring when conditions favor plant productivity, but become less able to compensate later in the season (Akiyama et al., 1984; Hik and Jefferies, 1990; Thompson and Gardner, 1996). Grasshoppers, *Aulocara elliotti*, did not significantly reduce standing crop of blue grama grass, *Bouteloua gracilis*, when feeding occurred early in the growing season but significantly reduced standing crop when feeding occurred later in southwestern New Mexico, USA (Thompson and Gardner, 1996).

M. Dyer et al. (1991) reported that grazing-adapted and nongrazing-adapted clones of an African C_4 grass, *Panicum coloratum*, differed significantly in their responses to herbivory by grasshoppers. After 12 weeks of grazing, the grazing-adapted plants showed a 39% greater photosynthetic rate and 26% greater biomass, compared to the nongrazing-adapted plants. Lovett and Tobiessen (1993) found that experimental defoliation resulted in elevated photosynthetic rates of red oak, *Quercus rubra*, seedlings grown under conditions of low and high nitrogen availability, but that seedlings in the high-nitrogen treatment were able to maintain high photosynthetic rates for a longer time (Fig. 12.6). Vanni and Layne (1997) reported that consumer-mediated nutrient cycling strongly affected phytoplankton production and community dynamics in lakes.

Honkanen et al. (1994) artificially damaged needles or buds of Scots pine. Damage to buds increased shoot growth. Damage to needles stimulated or suppressed shoot growth, depending on the degree and timing of damage and the position of the shoot relative to damaged shoots. Growth was significantly reduced by loss of 100%, but not 50%, of needles and was significantly reduced on shoots located above damaged shoots, especially late in the season. Shoots located below damaged shoots showed increased growth. Honkanen et al. (1994) suggested that these different effects of injury indicated an important effect of physiological status of the damaged part, that is, whether it was a sink (bud) or source (needle) for resources.

Morón-Ríos et al. (1997a) reported that below-ground herbivory by root-feeding scarab beetle larvae, *Phyllophaga* sp., prevented compensatory growth in response to above-ground grazing. Furthermore, salivary toxins or plant pathogens injected into plants by some sap-sucking species can cause necrosis of plant tissues (C.G. Jones, 1984; Miles, 1972; Raven, 1983; Skarmoutsos and Millar, 1982), honeydew accumulation on foliage can promote growth of pathogenic or other phylloplane fungi and limit photosynthesis (Dik and van Pelt, 1992), and some leaf

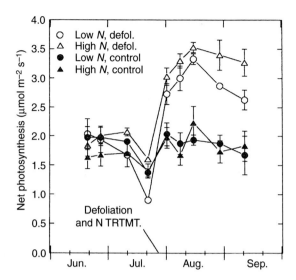

**FIGURE 12.6 Mean net photosynthetic rate in old leaves from plants subjected to four combinations of nitro-
gen fertilization and defoliation intensity.** Defoliation and fertilization treatments began Jul. 26. *Source: From Lovett,
G., Tobiessen, P., 1993. Carbon and nitrogen assimilation in red oaks (Quercus rubra L.) subject to defoliation and nitrogen
stress. Tree Physiol. 12, 259–269.*

miners induce premature abscission (Chabot and Hicks, 1982; Faeth et al., 1981; Pritchard and
James, 1984a,b; Stiling et al., 1991), thereby exacerbating the direct effects of herbivory. However, foliage injury can induce resistance to subsequent herbivory or infection by plant pathogens (Hatcher et al., 1995; M.D. Hunter, 1987; Karban and Baldwin, 1997, see Chapters 3 and 8).

Although primary productivity may be increased by low-to-moderate intensities of grazing, some plant tissues may be sacrificed by plant allocation of resources to replace lost foliage. Morrow and LaMarche (1978) and Fox and Morrow (1992) reported that incremental
growth of *Eucalyptus* stems treated with insecticide was 2–3 times greater than that of unsprayed stems.

Root growth and starch reserves are affected significantly by above-ground, as well as
below-ground, herbivory. Morón-Ríos et al. (1997a) noted that root feeders reduced root-to-
shoot ratios by 40% and the live-to-dead above-ground biomass ratio by 45% through tiller
mortality, apparently reducing plant capacity to acquire sufficient nutrients for shoot production. Rodgers et al. (1995) observed that starch concentrations in roots were related inversely to
the level of mechanical damage to shoots of a tropical tree, *Cedrela odorata* (Fig. 12.7). Gehring
and Whitham (1991, 1995) reported that folivory on pinyon pine adversely affected mycorrhizal
fungi, perhaps through reduced carbohydrate supply to roots. However, Holland et al. (1996)
reported that grasshopper grazing on maize increased carbon allocation to roots. McNaughton
(1979, 1993a) and van der Maarel and Titlyanova (1989) concluded that sufficient shoot biomass
to maintain root function is critical to plant ability to compensate for losses to herbivores.

Levels of herbivory that exceed plant ability to compensate lead to growth reduction,
stress, and mortality. Seedlings are particularly vulnerable to herbivores because of their
limited resource storage capacity and may be unable to replace tissues lost to herbivores

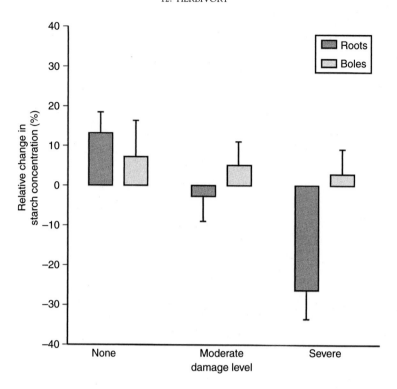

FIGURE 12.7   **Effect of intensity of artificial herbivory (to simulate terminal shoot damage by a lepidopteran,** *Hypsipyla grandella*) **on mean relative change (+ standard error) in starch concentrations (percent of initial level) in roots and lower boles of a Neotropical hardwood,** *Cedrela odorata*, **in Costa Rica.** In the moderate treatment, 0.2–0.3 cm of terminal shoot was excised; in the severe treatment, 0.5–0.6 cm of terminal shoot was excised. Data represent five sampling dates over a 12-day period beginning 18 days after treatment. *Source: From Rodgers, H.L., et al., 1995. Shoot damage effects on starch reserves of Cedrela odorata. Biotropica 27, 71–77.*

(P. Hulme, 1994; Wisdom et al., 1989). D. Clark and Clark (1985) reported that survival of tropical tree seedlings was highly correlated with the percentage of original leaf area present 1 month after germination and with the number of leaves present at 7 months of age. Continued grazing during periods of reduced plant productivity generally exacerbates stress. Resource-limited plants are more likely to succumb to herbivores than are plants with optimal resources (Belovsky and Slade, 2000; Lovett and Tobiessen, 1993). Plant species most stressed by adverse conditions suffer severe mortality to herbivores (Crawley, 1983; Painter and Belsky, 1993; Schowalter and Lowman, 1999). Wright et al. (1986) found that Douglas-fir beetles, *Dendroctonus pseudotsugae*, and fir engraver beetles, *Scolytus ventralis*, preferentially colonized Douglas-fir trees that had lost > 90% of foliage to Douglas-fir tussock moths, *Orgyia pseudotsugata*, although larval survival was greater in nondefoliated than in defoliated trees. However, Kolb et al. (1999) demonstrated that intense defoliation also could reduce moisture stress during dry periods.

Herbivory by exotic species may cause more severe or more frequent reduction in productivity and survival, in part because plant defenses may be less effective against newly associated herbivores. The most serious effects of herbivory in grasslands result from artificially high intensities of grazing by livestock or game (Oesterheld et al., 1992; D. Patten, 1993). Whereas grazing by native herbivores typically is seasonal, and grasses have sufficient time to replace lost tissues before grazing resumes, grazing by exotic species may be more continuous, allowing insufficient time for recovery (McNaughton, 1993a; Oesterheld and McNaughton, 1988, 1991; Oesterheld et al., 1992).

Herbivory also can alter plant architecture, potentially influencing future growth and susceptibility to herbivores. Gall-formers deform expanding foliage and shoots. Repeated piercing during feeding-site selection by sap-sucking species can cause deformation of foliage and shoots (Miles, 1972; Raven, 1983). Shoot borers and bud-feeders kill developing shoots and induce growth of lateral shoots (D. Clark and Clark, 1985; Nielsen, 1978; Reichle et al., 1973; Zlotin and Khodashova, 1980). Severe or repeated herbivory of this type often slows or truncates vertical growth and promotes proliferation of lateral branches (A. Martínez et al., 2009). Gange and Brown (1989) reported that herbivory increased variation in plant size. Morón-Ríos et al. (1997a) found that both above-ground and below-ground herbivory alter shoot-to-root ratios. Suppression of height or root growth restricts plant ability to acquire resources and often leads to plant death.

3.2 Community Dynamics

Differential herbivory among plants and plant species in an ecosystem affects both the distribution of individuals of a particular plant species and the opportunities for growth of other plant species resistant to or tolerant of herbivory (Zhang et al., 2011). The intensity of herbivory and the composition of host and nonhost plant species determine effects on plant communities. Low-to-moderate intensities that prevail most of the time generally ensure a slow turnover of plant parts or individual plants. High intensities during outbreaks can dramatically reduce the abundance of preferred host species and rapidly alter vegetation structure and composition.

Herbivory by exotic insect species (but rarely native species) is capable of eliminating plant species that are unable to compensate (Herms and McCullough, 2014; McClure, 1991; Orwig et al., 2002). Overgrazing by domestic livestock has initiated desertification of arid grasslands (by reducing vegetation cover, causing soil desiccation) in many parts of the globe (Schlesinger et al., 1990; Janssen et al., 2008, see Chapter 11).

Patterns of herbivory often explain observed geographic or habitat distributions of plant species (Bishop, 2002; Crawley, 1983, 1989; Fine et al., 2004; Goodale et al., 2014; Huntly, 1991; Karlsen et al., 2013; Louda et al., 1990a; Norghauer and Newbery, 2014; Schowalter and Lowman, 1999; Zhang et al., 2011), and often are the drivers of successional transitions (Cairns et al., 2008; Coleman et al., 2008; Davidson, 1993; Goodale et al., 2014; MacMahon, 1981; Schowalter, 1981; Willig and McGinley, 1999; Winter et al., 2015, see also Chapter 10). Herbivory can prevent successful establishment or continued growth of colonizing plant species, especially during the vulnerable seedling stage (Bishop, 2002; D. Clark and Clark, 1985; Goodale et al., 2014; P. Hulme, 1994; Wisdom et al., 1989). Louda et al. (1990a) reported that patterns of herbivory on two species of goldenbushes, *Haplopappus* spp., explained the significant difference between expected and observed distributions of these species across an environmental gradient from maritime to interior ecosystems in southern California (Fig. 12.8). Louda

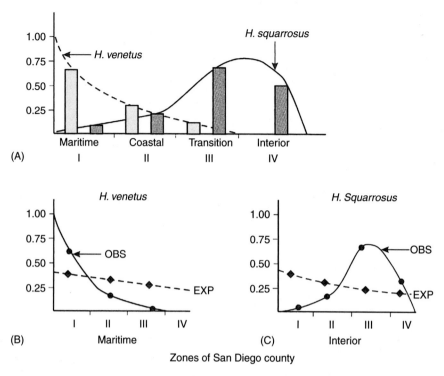

FIGURE 12.8 **Herbivore effects on plant species distribution.** (A) Gradients in observed frequencies of two goldenbushes, *Haplopappus venetus* (yellow bars, dashed line) and *H. squarrosus* (orange bars, solid line), from maritime to interior montane sites in San Diego County, California, USA. (B) and (C) Observed frequency accounting for observed herbivore effects *(solid lines)* compared to potential distribution in the absence of herbivory (dashed line), based on several measures of performance of control plants when insects were excluded. *Source: From Louda, et al., 1990a. Herbivore influences on plant performance and competitive interactions. In: Grace, J.B., Tilman, D. (Eds.), Perspectives on Plant Competition. Academic Press, San Diego, CA, pp. 413–444.*

and Rodman (1996) found that chronic herbivory by insects was concentrated on bittercress, *Cardamine cordifolia*, growing in sunny habitats and largely explained the observed restriction of this plant species to shaded habitats.

Fine et al. (2004) used herbivore exclosures and reciprocal transplants of clay and white sand specialist forest plant species in a lowland Amazonian site in Peru to evaluate the effect of herbivores on plant survival in each habitat type. They found that clay specialists grew significantly faster than white sand specialists on both soil types when protected from herbivores. However, when unprotected from herbivores, clay specialists dominated clay forests and white sand specialists dominated white sand forests, demonstrating an important role of herbivores in maintaining plant distribution.

Herbivory on dominant plant species can promote recruitment and persistence of associated plant species, a keystone species effect. Zhang et al. (2011) reported that increasing herbivory by grasshoppers, *Oedaleus asiaticus*, in Inner Mongolian steppe reduced plant biomass up to 4-fold and shifted the plant community toward dominance by nonhost plant species.

Sousa et al. (2003) found that predation by a scolytine beetle, *Coccotrypes rhizophorae*, on seedlings of the mangrove, *Rhizophora mangle*, prevented establishment of *R. mangle* in lightning-generated gaps and permitted a shade-intolerant species, *Laguncularia racemosa*, to codominate the mangrove community on the Caribbean coast of Panama. Norghauer and Newbery (2014) found that insect herbivores contributed to the coexistence of two dominant rainforest tree species in Cameroon, Africa by strongly suppressing the maximum height of the faster-growing species, *Microberlinia bisulcata*, while only weakly affecting the slower-growing species, *Tetraberlinia bifoliolata*.

Belle-Isle and Kneeshaw (2007) compared effects of spruce budworm, *Choristoneura fumiferana*, outbreaks, clear-cutting with protection of advance regeneration and soil, and precommercial thinning on boreal forest dynamics. They found that budworm-generated canopy openings had greater diversity of saplings and trees and larger perimeter/area ratios than did harvested openings, suggesting that the rates of stand recovery and influence by the surrounding forest should be greater in budworm-generated openings. Yorks et al. (2003) reported that eastern hemlock, *Tsuga canadensis*, mortality designed to simulate hemlock woolly adelgid, *Adelges tsugae*, infestation doubled the cover percentage of understory plant species within 3 years. Souza et al. (2016) concluded that whereas nutrient availability determined net primary production, herbivory by insects promoted spatial–temporal β diversity.

Herbivory often facilitates successional transitions (see Chapter 10). Selective herbivory among plant species suppresses those on which herbivory is focused and provides space and other resources to replacing species, resulting in altered plant community composition (Cairns et al., 2008; Coleman et al., 2008; Davidson, 1993; McEvoy et al., 1991; Norghauer and Newbery, 2014; Schowalter, 1981; Schowalter and Lowman, 1999; Winter et al., 2015). V.K. Brown and Gange (1989), V.K. Brown et al. (1988), and D. Gibson et al. (1990) reported that chemically reduced above-ground herbivory resulted in lower plant species richness after 2 years, whereas V.K. Brown and Gange (1989) found that reduced below-ground herbivory resulted in higher plant species richness, largely reflecting differential intensities of herbivory among various grass and forb species. V. Anderson and Briske (1995) simulated herbivory by livestock in a transplant garden containing midseral and late-seral grass species to test alternative hypotheses, that (1) midseral species have greater tolerance to herbivory, or (2) herbivory is focused on late-seral species, to explain species replacement in intensively grazed grasslands in the southern USA. They found that late-seral species had greater competitive ability and equivalent or higher tolerance to herbivory, indicating that selective herbivory on the late-successional species is the primary mechanism for reversal of succession, that is, return to dominance by midseral species under intense grazing pressure. Conversely, Bach (1990), Coley (1980, 1982, 1983), Coley and Aide (1991), and Lowman and Box (1983) reported that intensities of herbivory by insects were higher in earlier successional stages than in later successional stages. Cairns et al. (2008), Coleman et al. (2008), and Schowalter et al. (1981a) suggested that the southern pine beetle, *Dendroctonus frontalis*, is instrumental in advancing succession in the absence of fire by selectively killing early successional pines, thereby favoring their replacement by later successional hardwoods (Fig. 10.9).

Davidson (1993) compiled data indicating that herbivores may retard or reverse succession during early seres but advance succession during later seres. She suggested that herbivory is concentrated on the relatively less defended, but grazing-tolerant, midsuccessional grasses, forbs, and pioneer trees (Bach, 1990). Environmental conditions may affect this trend. For example, succession from pioneer pine forest to late successional fir forest in western North America can be retarded or advanced, depending primarily on moisture availability and condition of the dominant vegetation. Under conditions of adequate moisture (riparian corridors and high elevations), mountain pine beetle advances succession by facilitating the replacement of host pines by the more shade-tolerant, fire-intolerant, understory firs. However, limited moisture and short fire return intervals at lower elevations favor pine dominance. In the absence of fire during drought periods, herbivory by several defoliators and bark beetles is concentrated on moisture-stressed understory firs, truncating (or reversing) succession. Fire fueled by fir mortality also leads to eventual regeneration of pine forest. Similarly, each plant species that became dominant during succession following Hurricane Hugo in Puerto Rico induced elevated herbivory that facilitated its demise and replacement (Torres, 1992). The direction of succession then depends on which plant species are present and their responses to environmental conditions.

Herbivore-induced changes in plant condition, community composition, and structure affect habitat and food for other animals and microorganisms. Changes in nutritional quality or abundance of particular foliage, fruit, or seed resources affect abundances of animals that use those resources. Animals that require or prefer nesting cavities in dead trees may be promoted by tree mortality resulting from herbivore outbreaks.

Grazing on above-ground plant parts can affect litter and rhizosphere communities in a variety of ways (Bardgett et al., 1998). Reduced foliar quality resulting from induced defenses or replacement of palatable by less palatable plant species can reduce the quality of detrital

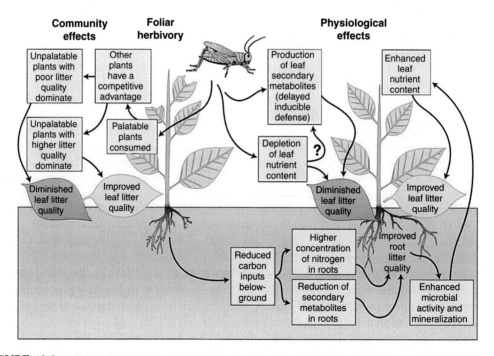

FIGURE 12.9 Effects of herbivory on host nutrient allocation and trophic interactions. *Source: Modified from Bardgett, R.D., et al., 1998. Linking above-ground and below-ground interactions: how plant responses to foliar herbivory influence soil organisms. Soil Biol. Biochem. 30, 1867–1878.*

material (Fig. 12.9). Seastedt et al. (1988) reported that simulated herbivory affected litter arthropod communities. Schowalter and Sabin (1991) found that three taxa of litter arthropods were significantly more abundant under experimentally defoliated (≤20% foliage eaten) Douglas-fir saplings, compared to nondefoliated saplings. B. Reynolds et al. (2003) found that herbivore-induced changes in forest litter increased abundances of Collembola and fungal- and bacterial-feeding nematodes; exclusion of herbivore-induced litterfall reduced abundances of oribatid and prostigmatid mites. Gehring and Whitham (1991, 1995) documented significantly reduced mycorrhizal activity on roots of pinyon pines subject to defoliation by insects, compared to nondefoliated pines.

Insect herbivores or their products constitute highly nutritious resources for insectivores and other organisms. Caterpillars concentrate essential nutrients several orders of magnitude over concentrations in foliage tissues (Schowalter and Crossley, 1983). Abundances of insectivorous birds and mammals often increase in patches experiencing insect herbivore outbreaks (Barbosa and Wagner, 1989; Koenig and Liebhold, 2005). Arthropod tissues also represent concentrations of nutrients for decomposers (Schowalter and Crossley, 1983; Seastedt and Tate, 1981).

A variety of organisms utilize honeydew accumulation from aphids, scales, and other plant-feeding Hemiptera. Ants, honey bees, *Apis mellifera*, hummingbirds, and other animals forage on the carbohydrate-rich honeydew (E. Edwards, 1982). Stadler and Müller (1996) and Stadler et al. (1998) reported that the presence of honeydew significantly increased the growth of a variety of epiphytic bacteria, yeasts, and filamentous fungi on the surface of conifer needles, potentially affecting the photosynthetic efficiency of underlying foliage.

3.3 Water and Nutrient Fluxes

Relatively few studies have addressed effects of insect herbivores on biogeochemical cycling processes, despite herbivore effects on plant chemistry and the importance of vegetation and litter structure, and turnover of material between these pools, to biogeochemical cycling. Crossley and Howden (1961) pioneered the study of nutrient fluxes from vegetation through arthropod communities and demonstrated that insect herbivores concentrate and accelerate the cycling of some nutrients. Subsequent research has demonstrated that insect herbivores affect biogeochemical cycling in a number of ways, including altered vegetation composition and structure, direct transfer of material from plants to litter, and effects on litter quality and litter communities.

Herbivory affects water fluxes by altering water uptake, sap flow, and transpiration by host plants. S. Cunningham et al. (2009) reported that sap-sucking herbivory by a psyllid, *Cardiaspina albitextura*, reduced foliage density of *Eucalyptus blakelyi* in an Australian woodland by 20% and halved sap flow velocity. Reduced metabolic demands by pruned, defoliated, or killed plants reduce water and nutrient uptake (Ford and Vose, 2007; W. Webb, 1978) and potentially contribute to plant survival during drought periods (Kolb et al., 1999; Welker and Menke, 1990).

Reduced water use by plants results in greater movement of water into streams draining sites of elevated herbivory. Schowalter et al. (1991) demonstrated that a 20% reduction in Douglas-fir foliage by the silver-spotted tussock moth, *Lophocampa argentata*, doubled the amount of precipitation that reached the forest floor as a result of reduced interception and increased throughfall (precipitation percolating through vegetation) (Fig. 12.10). Watson et al. (2001)

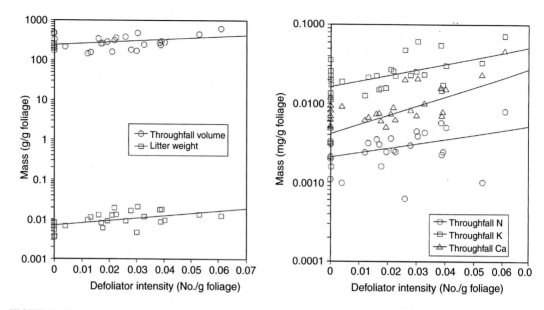

FIGURE 12.10 Folivore, *Lophocampa argentata*, effects on throughfall, litterfall, and fluxes of N, K, and Ca from young Douglas-fir during the feeding period, Apr.–Jun., in western Oregon. *Source: Reprinted from Schowalter, T.D., et al., 1991. Phytophage effects on primary production, nutrient turnover, and litter decomposition of young Douglas-fir in western Oregon. Forest Ecol. Manag. 42, 229–243.*

reported sharply increased volumes of water in several Australian streams during years of elevated psyllid densities in mountain ash, *Eucalyptus regnans*, watersheds (Fig. 12.11). Ford and Vose (2007) estimated that mortality of eastern hemlock to hemlock woolly adelgid decreased annual transpiration by 10% and winter–spring transpiration by 30%, leading to increased streamwater discharge.

Insects affect carbon flux in several ways. First, reduced carbon uptake and respiration result from loss of foliage area at high intensities of herbivory (see Section 12.3.1). Second, plants respond to herbivore feeding by altering carbon allocation patterns (see Chapter 3). Third, altered vegetation structure or composition changes patterns of carbon uptake and loss by vegetation (Amiro et al., 2010).

W. Webb and Karchesy (1977) reported that defoliation by the Douglas-fir tussock moth reduced host starch content proportional to defoliation intensity. On the other hand, Frost and Hunter (2008a) found that experimental herbivory by the white-marked tussock moth, *Orgyia leucostigma*, that removed about 20% of the foliage of red oak, *Quercus rubra*, seedlings significantly reduced ^{13}C allocation to below-ground tissues by 63%, with a corresponding increase in allocation to foliage, and reduced ^{15}N allocation to fine roots by 39%, in favor of storage in taproots and stems, compared to herbivore-free seedlings (Fig. 12.12).

Joo et al. (2010) reported that infestation of European beech, *Fagus sylvatica*, by woolly beech aphid, *Phyllaphis fagi*, altered the emission spectrum of volatile organic compounds into the atmosphere. Whereas aphid-free trees emitted primarily a complex of monoterpenes, the emission spectrum of trees with aphids shifted to α-farnesene, linalool, (E)-β-ocimene, and (E)-4,8-dimethyl-1,3,7-nonatriene, with α-farnesene and linalool becoming the second and third most important compounds emitted.

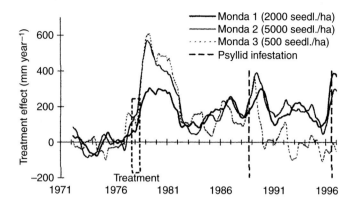

FIGURE 12.11 **Effects of 75–80% clear-cutting and replanting on monthly streamflow in experimental plots in the Maroondah study area, Victoria, Australia from 1971–97.** Harvest and replanting are indicated by the treatment box. Infestations of psyllids, *Hyalinaspis semispherula* in 1988 and *Cardiaspina bilobata* in 1996, were associated with increased streamflow (probably due to canopy dieback) that did not occur in control watersheds. *Source: Reprinted from Watson, F., et al., 2001. Improved methods to assess water yield changes from pairedcatchment studies: application to the Maroondah catchments. Forest Ecol. Manag. 143, 189–204.*

Kurz et al. (2008) reported that widespread pine mortality during outbreaks of mountain pine beetle reduced carbon uptake and increased carbon emission from decaying trees. K. Clark et al. (2010) found similar results for defoliation by gypsy moth, *Lymantria dispar*. In both studies, the change in net carbon flux converted the forest from a carbon sink to a carbon source. However, D. Moore et al. (2013) reported that reduced carbon uptake, rather than increased carbon emissions, resulted in net flux of carbon out of ecosystems experiencing mountain pine beetle outbreaks. M. Brown et al. (2010) noted that forests recovering from mortality to mountain pine beetle remained growing-season carbon sinks as a result of increased photosynthesis by remaining healthy trees and understory vegetation, whereas nearby harvested stands remained carbon sources for at least 10 years after harvest. J. Song et al. (2015) reported that simulated locust outbreak (using vegetation clipping and grasshopper carcass subsidy treatments) affected net ecosystem fluxes of carbon but in ways that suggested compensation and rapid recovery of long-term flux rates.

Amiro et al. (2010) compiled carbon flux data for 180 studies of disturbed ecosystems (including three reflecting insect outbreaks) and found that all ecosystems initially showed net carbon losses but became net carbon sinks by 20 years, and most had become net carbon sinks by 10 years. Insect outbreaks and silvicultural thinning resulted in less change in net carbon flux than did stand-replacing events, and decreases in the year of disturbance were followed by rapid recovery.

As described for carbon fluxes, herbivory affects nutrient fluxes via reduction in uptake and use of various nutrients (see Section 12.3.1), alteration of allocation patterns within plants, and changes in the composition of vegetation among plant species with different nutrient use patterns. In addition, herbivores affect nutrient fluxes indirectly via leaching of nutrients from damaged plants and stimulation of decomposition by throughfall and feces.

Griffin et al. (2011) found that the canopy N pool was reduced by 58% in recent mountain pine beetle outbreaks and was still 48% lower 30 years postoutbreak, compared to

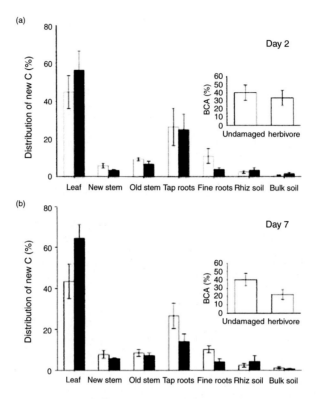

FIGURE 12.12 Distribution of recovered ¹³C among leaf, new stem, old stem, tap root, and fine root tissues in red oak, *Quercus rubra*, seedlings that were free of herbivores (light bars) or treated to herbivory by white-marked tussock moth, *Orgyia leucostigma* (dark bars) on (A) day 2 and (B) day 7 following introduction of herbivores. Inset graphs depict below-ground allocation of C (sum of roots + soil) as a percentage of total C allocation for each day. Bars are means ± SE). *Source: From Frost, C.J., Hunter, M.D., 2008. Herbivore-induced shifts in carbon and nitrogen allocation in red oak seedlings. New Phytol. 178, 835–845.*

nonoutbreak plots. This reduction in outbreak plots was balanced by higher needle litter N concentration and more rapid N mineralization rates, compared to nonoutbreak plots.

Frost and Hunter (2008a) used stable isotopes (¹³C and ¹⁵N) to explore effects of white-marked tussock moth on red oak seedlings. They found that herbivory reduced carbon allocation to fine roots by 63%, with a corresponding increase in new carbon allocation to shoots, and reduced nitrogen allocation to fine roots by 39%, with increased storage in taproot and stem tissues. Recovery of ¹³C in soil pools was similar between treatments, indicating that exudation of carbon remains an actively regulated process. Changes in nutrient concentrations among plant tissues affect their rate of turnover.

Altered carbon storage in roots and carbohydrate exudation into the soil (Filip et al., 1995; Holland et al., 1996) affect resources available for below-ground food webs that facilitate nutrient acquisition by plants (Fig. 12.13) (Coleman et al., 2004). Crutsinger et al. (2008) reported that galled goldenrod, *Solidago altissima*, had lower litter mass and N mineralization than did ungalled plants. Schweitzer et al. (2005) also found that gall aphids, *Pemphigus betae*, reduced

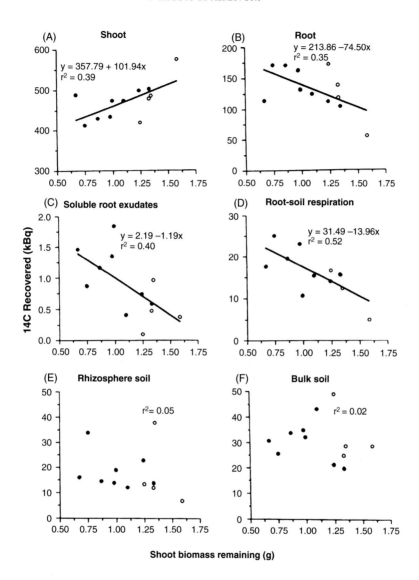

FIGURE 12.13 Carbon allocation as a function of intensity of herbivory (measured as shoot biomass remaining) in (A) shoots, (B) roots, (C) soluble root exudates, (D) respiration from roots and soil, (E) rhizosphere soil, and (F) bulk soil. Data were normalized for differences in $^{14}CO_2$ uptake; 1 kBq = 1000 disintegrations s^{-1}. Shoot biomass was inversely related to leaf area removed by herbivores. Regression lines are shown where significant at $p < 0.05$. Open circles represent ungrazed plants, solid circles grazed plants. *Source: From Holland, J.N., et al., 1996. Herbivore-induced changes in plant carbon allocation: assessment of below-ground C fluxes using carbon-14. Oecologia 107, 87–94.*

litter quality of *Populus* spp., and galled litter decomposed 34–40% slower than nongalled litter. Fonte and Schowalter (2004) demonstrated that fresh foliage of four tropical tree species in Puerto Rico had higher nitrogen concentration and decomposed significantly more rapidly than did senescent foliage from the same tree species.

Ritchie et al. (1998) reported that herbivory generally reduced the abundance of plant species with N-rich tissues, leading to replacement by plant species with lower N concentrations, in an oak savanna in the northcentral USA. Zhang et al. (2011) found that an experimental gradient in grasshopper density in Inner Mongolia, China, increased litter biomass 30% and increased C:N and C:P ratios for host plant species but decreased these ratios in nonhost plant species, suggesting either a direct grasshopper-induced flux of nutrients from host to nonhost plants or a release of nonhost plants from nutrient competition with host plants. Similarly, insects (such as bark beetles) that affect the relative composition of Douglas-fir and western redcedar, *Thuja plicata*, in the northwestern USA affect calcium dynamics and soil pH, that is, calcium accumulation and higher pH under western redcedar compared to Douglas-fir (Kiilsgaard et al., 1987). Cobb (2010) reported that replacement of eastern hemlock by black birch, *Betula lenta*, in riparian zones of eastern North America as a result of hemlock woolly adelgid infestation has resulted in dramatic changes in litter composition, decomposition rate, and N cycling rates.

Herbivory affects biogeochemical cycling directly by changing the timing, amount, and form of nutrients transferred from plants to litter or soil. In the absence of herbivory, litter accumulation may be highly seasonal (ie, concentrated at the onset of cold or dry conditions) and have low nutrient concentrations, especially of nitrogen or other nutrients that are resorbed from senescing foliage (Gutschick, 1999; Marschner, 1995). Herbivory generally increases the amount and nutrient content of litter inputs during the growing season, by transferring nutrients in fragmented plant material, insect tissues, and insect feces, as well as nutrient-enhanced throughfall (Fonte and Schowalter, 2005; Frost and Hunter, 2004, 2007, 2008b; Hollinger, 1986; le Mellec et al., 2009; Nitschke et al., 2015; Schowalter et al., 1991, 2011; Seastedt et al., 1983; Stadler et al., 2006).

Insect feces and tissues have higher concentrations of nutrients, especially N, that control litter decomposition than does senescent leaf litter (Effler et al., 2006; Frost and Hunter, 2004, 2007, 2008b; M.D. Hunter et al., 2003; Schowalter and Crossley, 1983; Speer et al., 2010; L. Yang, 2004, 2013). Consequently, herbivore outbreaks typically provide a pulse in nutrients from vegetation to litter that often stimulates decomposition and mineralization (Speer et al., 2010; L. Yang, 2004, 2013). Zlotin and Khodashova (1980) reported that herbivore feces decomposed more rapidly than did raw plant material. M.D. Hunter et al. (2003) found that deposition of folivore feces explained 62% of the variation in soil nitrate availability. Hollinger (1986) reported that during an outbreak of the California oak moth, *Phryganidia californica*, fluxes of nitrogen and phosphorus from trees to the ground more than doubled, and feces and insect remains accounted for 60–70% of the total nitrogen and phosphorus fluxes. J.R. Grace (1986) found a similar increase in nitrogen flux to the forest floor from 31 kg N ha^{-1} in nondefoliated forest to 52 kg N ha^{-1} in forest defoliated by gypsy moth in Pennsylvania, USA. L. Yang (2013) reported that addition of 17-year periodical cicada, *Magicicada septendecim*, carcasses to experimental plots of American bellflower, *Campanulastrum americanum*, increased mean plant biomass by 61%, and cicada-supplemented plants showed 20% greater foliage nitrogen concentrations, compared to nonsupplemented plants. L. Yang (2008) found that mammalian herbivory was greater on cicada-supplemented bellflowers than on control plants.

The contribution of honeydew to nutrient cycling has been a subject of considerable interest. Stadler and Müller (1996) and Stadler et al. (1998) documented significant amounts of dissolved organic carbon in aphid honeydew. Most of the honeydew in their studies was immobilized quickly by phylloplane microorganisms before reaching the ground. Owen (1978) and Owen and Wiegert (1976) suggested that the trisaccharide, melezitose, in aphid honeydew provides a rich, labile carbohydrate resource for free-living, nitrogen-fixing soil bacteria. Petelle (1980) subsequently demonstrated that fructose, also abundant in aphid honeydew, increased nitrogen fixation 9-fold more than did melezitose. However, Grier and Vogt (1990) found that chemical removal of aphids increased available soil nitrogen, nitrogen mineralization rates, net primary production, and nitrogen uptake by red alder, *Alnus rubra*. These data, together with those of Lovett and Ruesink (1995) and Stadler et al. (2001), indicate that nutrients mobilized by folivores and sap-suckers may be immobilized rapidly by soil microorganisms.

Folivory also increases the flux of nutrients in throughfall. Throughfall nutrient fluxes from canopy to litter are controlled strongly by foliage area, exposed surfaces resulting from herbivory, and amount of precipitation (Lovett et al., 1996). M.D. Hunter et al. (2003), Kimmins (1972), Nitschke et al. (2015), Schowalter et al. (1991), Seastedt et al. (1983), and Stachurski and Zimka (1984) have shown that herbivory greatly increases leaching of nutrients from chewed foliage (Fig. 12.10). Stadler et al. (2006) added that hemlock woolly adelgid infestation increased dissolved organic carbon by 25%, dissolved organic nitrogen by 29%, and potassium by 39%, but decreased dissolved inorganic nitrogen by 40%, in throughfall from eastern hemlock compared to uninfested trees. However, in ecosystems with high annual precipitation, herbivore-induced nutrient turnover may be masked by nutrient inputs via precipitation (Schowalter et al., 1991).

Several studies have experimentally addressed the effect of herbivore-derived inputs on decomposition, soil nutrient fluxes, and nutrient uptake by plants. Throughfall, senescent foliage, fresh foliage fragments lost via herbivory (greenfall), and herbivore feces differ in the amount and form of nitrogen and carbon compounds, as well as in the degree of microbial preconditioning. Hillstrom et al. (2010) reported that large inputs of herbivore frass and greenfall to experimental microcosms nearly doubled respiratory carbon loss and nitrate immobilization, compared to microcosms receiving no herbivore inputs. Schowalter et al. (2011) compared the separate effects of herbivore feces, greenfall, and throughfall enhancement in 0.1 m² plots in a tropical rainforest in Puerto Rico. Feces and greenfall treatments were increased 10-fold higher than measured background rates; throughfall N and P (but not water) treatments were double the background rates. These treatments represented the increased fluxes expected from a low–moderate herbivore outbreak. Addition of herbivore feces significantly increased NO_3 and NH_4 fluxes, and feces and throughfall additions significantly reduced decay rate, compared to controls, suggesting that the N flux was sufficient to inhibit microbial decomposition (see Chapter 14) and that low–moderate herbivory in this forest could increase N and P fluxes >30%.

Changes in nutrient content and defenses induced by herbivory affect the quality of litter for decomposers in complex ways (S. Chapman et al., 2003, 2006; Coley and Barone, 1996; M.D. Hunter et al., 2003; Kurokawa and Nakashizuka, 2008; le Mellec et al., 2009). S. Chapman et al. (2006) proposed that herbivory on evergreen trees commonly results in premature leaf abscission, higher litter quality, and accelerated decomposition, whereas herbivory on deciduous trees is more likely to result in production of induced defenses and decelerated decomposition. Although induced defenses may inhibit decomposition (R. Stout, 1989), rapid leaching in wet ecosystems may minimize their effect on decomposition (Ardón and Pringle, 2008). Schweitzer et al. (2005) found that herbivory reduced litter decomposition rate. The amount of nitrogen flux relative to background levels also may determine the direction of the herbivore effect on decomposition. Treseder (2008) conducted a metaanalysis of nitrogen subsidy studies and found that increased nitrogen flux does not necessarily increase decomposition rate. Results depend on the responses of soil microbes to increased nitrogen supply. Microbial activity may increase in response to nitrogen subsidy when background levels of nitrogen are low, but decrease in response to subsidy when background levels are high.

Lovett and Ruesink (1995) reported that gypsy moth feces contained much labile carbon and nitrogen but that microbial growth, stimulated by labile carbon, was sufficient to immobilize all the available nitrogen. In a subsequent experiment, Christenson et al. (2002) added ^{15}N-labeled leaf litter or gypsy moth feces to experimental plots, in which a red oak seedling had been planted, to evaluate pathways of nitrogen flux (Fig. 12.14). They found that gypsy moth feces significantly increased the concentration of ^{15}N in total and mineralizable nitrogen pools in surface and subsurface soils, with 40% of the recovered ^{15}N incorporated in soil. The

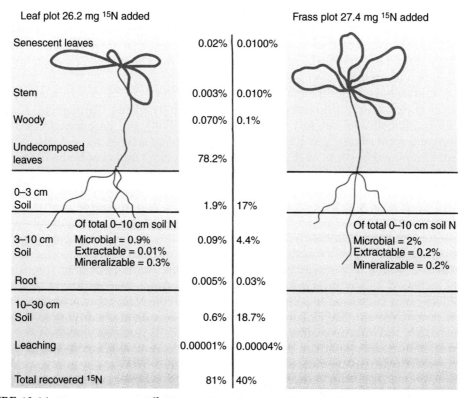

FIGURE 12.14 **Percent recovery of ^{15}N from various fractions, relative to the total amount added to plots as insect feces in Jun. or as leaves at the time of leaf senescence in Nov.** *Source: From Christenson, L.M., et al., 2002. The fate of nitrogen in gypsy moth frass deposited to an oak forest floor. Oecologia 131, 444–452.*

red oak seedlings in plots with feces addition had significantly higher ^{15}N concentrations in green leaves, stems, and roots. By comparison, 80% of the ^{15}N in plots with added leaf litter remained in undecomposed leaves. Differences in the amounts of ^{15}N recovered between the two treatments might reflect unmeasured gas fluxes or leaching of dissolved organic nitrogen. Alternatively, increased C/N ratio in frass (le Mellec et al., 2009) may stimulate microbial immobilization of N. Frost and Hunter (2007) demonstrated, using ^{15}N, that herbivore feces are rapidly decomposed, and some of the feces-N became available for plant uptake, and even incorporation into subsequent defoliators.

Belovsky and Slade (2000) reported that grasshoppers, *Melanoplus sanguinipes*, accelerated nitrogen cycling by increasing the abundance and decomposition rate of grass litter. Cobb et al. (2006) found that hemlock woolly adelgid significantly increased the decomposition rate of hemlock litter in infested stands, compared to uninfested stands, largely as a result of increased soil moisture in infested stands. However, Stadler et al. (2006) found that adelgids increased foliar N, compared to uninfested foliage, which likely affected the decomposition rate.

S. Chapman et al. (2003) used a unique long-term herbivore exclusion experiment in Arizona, USA, to evaluate the effects of folivores and sap-suckers on foliage chemistry and decomposition. The shoot moth, *Dioryctria albovittella*, feeds on shoots and cones of mature pinyon pines, *Pinus edulis*, whereas the scale insect, *Matsucoccus acalyptus*, feeds only on needles of juvenile trees. Exclusion of both herbivores from experimental trees permitted comparison of susceptible, infested and uninfested, trees and resistant trees. Foliage litter chemistry was similar for resistant and susceptible uninfested trees, but both treatments had significantly lower litter nitrogen concentration than did litter of susceptible infested trees for both insect species, demonstrating that the insects were responsible for the increased nitrogen in the litter. Furthermore, phosphorus concentration was significantly higher (by 50%) in litter from trees infested with scale insects, compared to resistant and susceptible uninfested trees. The scale insect, but not the moth, significantly increased the rate of litterfall from infested trees, relative to resistant and susceptible uninfested trees. When litter from resistant and susceptible infested trees was placed under each treatment type (Fig. 12.15), litter from susceptible, infested trees for both insects decomposed significantly more rapidly than did litter from resistant trees. For the scale treatments, both resistant and susceptible litter decomposed more rapidly under resistant and susceptible infested trees, compared to susceptible uninfested trees, perhaps because of higher throughfall, soil moisture, and soil temperature that were indirect effects of herbivory by scales. Litter from infested trees decomposed more rapidly under infested trees than did resistant litter under resistant trees. For moth treatments, both litter types decomposed more rapidly under resistant trees than under susceptible infested or uninfested trees; litter from infested trees did not decompose more rapidly under infested trees than did resistant litter under resistant trees, for unknown reasons. However, because litter from moth-infested trees was higher quality and decomposed more rapidly across all treatments, compared to resistant litter, higher rates of decomposition might be observed over longer time periods. Classen et al. (2007a,b) subsequently found that differences in decomposition and mineralization of litter from susceptible and resistant trees reflected increased litter N and increased soil moisture resulting from herbivory, as mediated by seasonal changes in soil/litter microbial abundance and activity. Madritch et al. (2007) found similar results for decomposition of litter from aspen, *Populus tremuloides*, genotypes versus gypsy moth and forest tent caterpillar, *Malacosoma disstria*, frass.

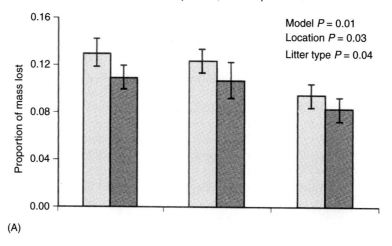

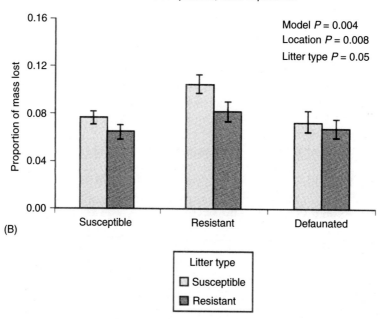

FIGURE 12.15 **Increased decomposition resulting from herbivore-induced changes in litter quality.** Litter from pinyon pine trees that were susceptible or resistant to (A) scale insects, *Matsucoccus acalyptus*, or to (B) stem-boring moths, *Dioryctria albovittella*, was placed under susceptible, resistant, or experimentally defaunated trees. N = 20 for all treatments; vertical lines are standard errors. P values are given for main effects by 2-way ANOVA. The two litter types did not differ within any location, but did differ across all three locations. *Source: From Chapman, S.K., et al., 2003. Insect herbivory increases litter quality and decomposition: an extension of the acceleration hypothesis. Ecology 84, 2867–2876.*

The shift in biomass dominance from folivores to sap-suckers following disturbance in temperate and tropical forests (Schowalter, 1995; Schowalter and Ganio, 1999; Schowalter et al., 1981c) indicates an accompanying shift in the predominant flux of nutrients. Dominance by folivores in undisturbed forests should transfer nutrients primarily as throughfall (enhanced by leaching from chewed foliage), feces, and fragmented plant parts, whereas dominance by sap-suckers in early successional communities should transfer nutrients predominantly as a labile carbohydrate solution that stimulates microbial growth and immobilizes accompanying nutrients. Holland (1995) reported that soil microbial biomass peaked at intermediate levels of herbivory in no-tillage agricultural systems, perhaps because moderate intensities of herbivory increased root exudates that fuel microbial production (Fig. 12.13) (Holland et al., 1996).

Herbivory also mediates fluxes of nutrients among plants. Carbon and nitrogen exuded from roots to support mycorrhizal and soil microbial activity often become available to neighboring plants. E. Ayres et al. (2007) used ^{15}N to demonstrate that herbivory by above-ground (but not below-ground) herbivores increased below-ground flux of N from white clover, *Trifolium repens*, to perennial ryegrass, *Lolium perenne*, 5-fold in laboratory microcosms. Growth of *L. perenne* was not affected in any of the treatments. However, given that primary production in terrestrial ecosystems is frequently N-limited, they suggested that such transfer of below-ground N among plants could alter competitive interactions and plant community structure.

Herbivory has the capacity to alter nutrient cycling processes at the landscape or watershed level. Although most studies have addressed herbivore effects during outbreaks, non-outbreak levels of herbivory can significantly affect fluxes (M.D. Hunter et al., 2003). MacDonald et al. (1992) attributed elevated nitrate concentrations in soil solution in hardwood forests in Michigan, USA, to defoliation by the forest tent caterpillar. Swank et al. (1981) reported that defoliation of hardwood forests by the fall cankerworm, *Alsophila pometaria*, in the southern Appalachian Mountains in North Carolina, USA, increased nitrate export via soil leaching and streamflow. Eshleman et al. (1998) and J. Webb et al. (1995) documented increased nitrate export in streams draining hardwood forests defoliated by the gypsy moth in the mid-Appalachian Mountains in Pennsylvania, Virginia, and West Virginia, USA. However, Bormann and Likens (1979) reported no increase in nitrate export resulting from saddled prominent caterpillar, *Heterocampa guttivitta*, in the northern Appalachian Mountains in New Hampshire, USA. Differences in results between these studies might reflect elevated nitrogen inputs throughout the northern Appalachians during the 1970s or different retention mechanisms among northern and southern study sites (Treseder, 2008). Although most of the nitrogen consumed by folivores is retained by the ecosystem (Lovett and Ruesink, 1995; Frost and Hunter, 2007, 2008b), even small increases in stream export of nitrate can alter water quality and leaching of nutrient cations such as Ca or toxic cations such as Al (Lovett et al., 2002).

3.4 Effects on Climate and Disturbance Regime

Herbivore-induced changes in vegetation structure affect soil temperature, relative humidity, erosion, soil moisture, and soil fertility (see Chapter 11). Reduced foliage surface area reduces interception of precipitation and evapotranspiration (G. Parker, 1983; Schowalter et al., 1991). Changes in litter accumulation can affect albedo and soil warming (Classen

FIGURE 12.16 **Increased canopy porosity resulting from herbivory.** Holes chewed by folivorous insects in the large leaves of *Cecropia* reduce interception of light, water, and airflow by the canopy and increase penetration to lower strata and the forest floor.

et al., 2005). However, relatively few studies have demonstrated herbivore effects on these variables.

Herbivory increases vegetation porosity (Fig. 12.16) and penetration of light, precipitation, and wind to the understory and soil surface. Canopy opening greatly affects abiotic conditions in the understory (Chazdon and Fetcher, 1984; Denslow, 1985; Fernandez and Fetcher, 1991). Increased soil warming due to penetration of sunlight may be offset to some extent by increased penetration of precipitation to the ground. Schowalter et al. (1991) reported that 20% loss of Douglas-fir foliage by a folivore, *L. argentata*, doubled the amount of water reaching the soil surface. Classen et al. (2005) reported that herbivory by scale insects increased soil moisture 35% at the primary rooting depth of 15–30 cm and increased maximum soil temperature by an average 5°C at 5 cm depth under pinyon pines with scale insects, compared to resistant trees or trees with scales removed (Fig. 12.17). Classen et al. (2007a) concluded that the mechanism by which herbivory affected litter mass loss and nutrient dynamics in this woodland system was through change in canopy cover and consequent change in litter microclimate. Increased accumulation of litter resulting from herbivory in forest ecosystems may contribute to soil water retention.

Canopy opening over large areas by herbivores could affect regional climate. Although most studies of effects of canopy opening on climate have focused on anthropogenic canopy removal (see Chapter 11), herbivory may have similar effects, for example, increased soil surface temperature, reduced evapotranspiration, and consequent regional warming and drying (Classen et al., 2005; J. Foley et al., 2003a; Salati, 1987).

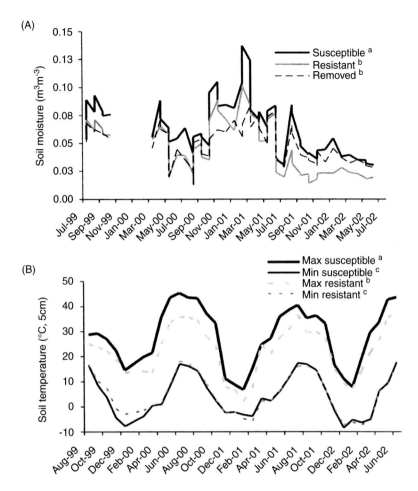

FIGURE 12.17 **Effect of herbivory by scale insects, *Matsucoccus acalyptus*, on soil moisture (A) and tempera-ture (B) in a pinyon-juniper woodland in northern Arizona during Aug. 1999–Jun. 2002. Treatments included trees that were susceptible or resistant to the scale or susceptible but with scales removed.** Different letters indicate significant differences, using contrasts. *Source: From Classen, A.T., et al., 2005. Insect infestations linked to changes in microclimate: important climate change implications. Soil Sci. Soc. Am. J. 69, 2049–2057.*

Herbivory can affect the likelihood or severity of future disturbances. Herbivory in grass-lands reduces the amount of standing dead material (Knapp and Seastedt, 1986), potentially reducing the severity of fire but increasing soil exposure to desiccation and exacerbating effects of drought. However, reduced foliage surface area may reduce water demand and mitigate the effects of drought (Kolb et al., 1999). Herbivory in forests increases fuel accumulation in the form of fine and coarse litter material, potentially increasing the likelihood and severity of fire, especially in arid forests where litter decomposes slowly and lightning strikes are frequent (McCullough et al., 1998; Schowalter, 2012). Bebi et al. (2003) concluded that spruce, *Picea engelmannii*, mortality to the spruce beetle, *Dendroctonus rufipennis*, in Colorado, USA, did not increase the occurrence of subsequent fires. Péch (1993) found similar results for balsam fir, *Abies balsamea*, mortality to the spruce budworm, *C. fumiferana*, in eastern Canada. The absence of subsequent fire in these cases may have reflected the rapid decomposition of spruce and fir litter (Bebi et al., 2003; Péch, 1993). Alternatively, outbreaks that increase the amount and distribution of fine litter materials may increase the frequency of low-intensity fire and reduce the probability of catastrophic fire.

Jenkins et al. (2008) concluded that the effect of bark beetles on fire is complex. Fuel accumulation alone had less effect on fire than did changes in stand structure. Although we might expect tree mortality to fuel catastrophic crown fire, decreased continuity of crown fuels could slow fire spread through the canopy. Simard et al. (2011) compared effects of mountain pine beetle outbreaks across a chronosequence of lodgepole pine, *Pinus contorta*, stands from undisturbed to 36 years postoutbreak. They found that stands thinned by the beetles were less likely to experience crown fire than were undisturbed stands and that these results were relatively unaffected by wind speed, based on fire model simulations. However, Harvey et al. (2014a,b) reported that several metrics of fire severity were positively correlated with the severity of prefire mountain pine beetle outbreaks under moderate burning conditions. The effect of outbreak severity disappeared under extreme burning conditions.

As with fire, few experimental studies have addressed effects of insect outbreaks on storm disturbance. Vegetation structure, especially height, density, and "roughness," affects wind speed (Gardiner et al., 2005). Vertical height and density determine the degree to which vegetation impedes near-surface winds. Greater height (especially of woody vegetation) and density confer greater wind resistance. Insect outbreaks that open the canopy should increase penetration of high wind speeds and the probability of treefall, but also should reduce wind resistance of defoliated trees. J. Moore and Maguire (2005) found that pruning at least 80% of the canopy was necessary to reduce wind stress significantly. However, trees with greater exposure to wind typically have greater root development and wind-firmness (Tamasi et al., 2005). Taylor and MacLean (2009) reported that wind-related tree mortality following spruce budworm defoliation in eastern Canada was related to outbreak severity. Tree mortality during storms peaked 11–15 years after outbreak, due to greater exposure of surviving trees to wind.

4 SUMMARY

Herbivory, the feeding on living plant parts by animals, is a key ecosystem process that has widely recognized effects on primary production and on vegetation structure and composition. The effect of herbivory depends on herbivore feeding type and intensity. Different types of herbivory affect different tissues and the production, translocation, and accumulation of photosynthates to varying degrees.

A number of methods have been used to measure the intensity and effects of herbivory. The most common method for measuring intensity has been estimation of consumption rates by individual herbivores and extrapolation to population size. This method can be used to measure consumption by sap-sucking herbivores as well as folivores. A second method is measurement, by various means, of missing plant biomass. This method does not account for completely consumed (and unobserved) parts or for compensatory growth. Measurement of turnover of marked plant parts is the most accurate, but labor-intensive, method for estimating herbivory. Estimates of herbivory can differ by 2–5 times among methods, making standardization a key to comparison among ecosystems. Evaluating the effect of herbivory requires measurement of a variety of plant and ecosystem responses, not simply plant growth or productivity.

The intensity of herbivory varies widely, but a trend is apparent among ecosystem types. Herbivory generally is lowest (<2% reduction in primary production) in some forests and highest (most primary production consumed daily) in aquatic ecosystems. Insects are the primary herbivores in forest ecosystems and may account for the bulk of herbivory in grasslands, even though vertebrate grazers are more conspicuous.

Herbivory has well-known effects on survival, productivity, and growth form of individual plants. However, the traditional view of herbivory as a negative effect on plants is being replaced by a view that recognizes more complex effects of variable intensity and timing. Moderate intensities of herbivory often stimulate production, through compensatory growth, and flowering, thereby increasing fitness. A given intensity of herbivory can have different effects at different times during the growing season or under different environmental conditions. Herbivory can affect the growth form of plants by terminating shoot growth and initiating branching and by affecting shoot-to-root ratios. Changes in survival, productivity, and growth of individual plant species affect vegetation structure and community dynamics. Herbivores often determine the geographic or habitat patterns of occurrence of plant species and facilitate successional transitions.

An increasing number of studies have addressed effects of insect herbivores on biogeochemical cycling or other abiotic conditions. Herbivores affect, often dramatically, the turnover of plant nutrients to litter as plant fragments, feces animal tissues and nutrients leached from chewed surfaces. Folivory alters seasonal patterns of nutrient fluxes by transferring material prior to plant resorption of nutrients from senescing parts. Sap-sucking insects transfer copious amounts of labile carbohydrates (as honeydew) that stimulate growth and nutrient uptake by microbes. Herbivory also can affect climate and the likelihood and intensity of future disturbances. Reducing vegetation cover greatly affects the penetration of light, precipitation, and wind to the understory and soil, affecting soil warming and water content, relative humidity, erosion, transpiration, etc. Reduced vegetation biomass or litter accumulation affects abundance of fuel to support fire and affects water-holding capacity and vegetation demand for water during drought. Therefore, herbivory can influence ecosystem stability substantially (Chapter 15).

13

Pollination, Seed Predation, and Seed Dispersal

CONSEQUENCES OF POLLINATOR DECLINE

Pollination is among the most visible and important ecosystem services (see Chapter 16), necessary for reproduction of about 80% of all plant species and 35% of global crop production (A. Klein et al., 2007; Kremen et al., 2007; Losey and Vaughan, 2006; Ollerton et al., 2011). Pollination of agricultural crops by honey bees, *Apis mellifera*, is worth US$117 billion per year globally (Costanza et al., 1997) and $8 billion per year in the USA (Isaacs et al., 2009). The value of pollination by all insects is estimated at US$175 billion, or nearly 10% of total agricultural production for human consumption in 2005 (Gallai et al., 2009). This economic importance of insect pollinators has led to widespread transport and introduction of honey bees to most regions of the globe. Recently, however, dramatic declines in pollinator abundances have generated concern for the maintenance of pollination services (Biesmeijer et al., 2006; Cox-Foster et al., 2007; Genersch, 2010). In the absence of pollinators, agricultural crop production would decline 3–8% globally (Aizen et al., 2009). The threat is 3-fold.

First, of major concern in North America and northern Europe, is the mysterious honey bee colony collapse disorder (CCD) that has resulted in the loss of 50–90% of hives in the USA (Cox-Foster et al., 2007). The disorder is indicated by the unexplained absence of live or dead adult workers in or near hives, despite abundant brood, honey, and pollen (Cox-Foster et al., 2007; R. Johnson et al., 2009). Explanations have included increased pesticide exposure, especially sublethal effects of modern systemic neonicotinoids, pathogens, parasitic mites, undetected diseases, and disorientation or other stresses during transportation of hives among agricultural regions (Ahn et al., 2012; Blacquière et al., 2012; Dainat et al., 2012; Di Prisco et al., 2013; Dively et al., 2015, Henry et al., 2012; Pettis et al., 2013; Ramirez-Romero et al., 2008; Sanchez-Bayo and Goka, 2014). Claudianos et al. (2006) and Weinstock et al. (2006) reported that the honey bee genome contains substantially fewer genes for immunity or xenobiotic detoxifying enzymes, compared to other known insect genomes, explaining the greater honey bee sensitivity to pesticides. R. Johnson et al. (2009) conducted microarray analysis of CCD and non-CCD bees and found no elevated expression of pesticide-response genes and inconclusive patterns of expression for immune genes. However, an abundance of unusual ribosomal RNA fragments, consistent with infection of picorna-like viruses, suggested that reduced protein synthesis

Insect Ecology. http://dx.doi.org/10.1016/B978-0-12-803033-2.00013-3

resulting from ribosomal damage would leave bees more vulnerable to multiple additional stresses, such as pesticides or pathogens. The parasitic mite, *Varroa destructor*, is known to suppress the honey bee's immune system and may be acting in concert with sublethal effects of insecticides and emerging diseases to cause CCD (Dainat et al., 2012; Di Prisco et al., 2013, Dively et al., 2015; Genersch, 2010; Sanchez-Bayo and Goka, 2014). Furthermore, antibiotics and acaricides used in maintenance of managed hives can, themselves, adversely affect bees (Hawthorne and Dively, 2011).

Second, abundance of native pollinators has declined in many parts of the globe, especially in agricultural regions. Many plant species, including crop species in some areas, depend primarily on native specialist pollinators, which are more efficient pollinators than the more generalist honey bee. Native pollinators, where protected, are capable of providing full pollination service and of compensating for decline or absence of honey bees (Kremen et al., 2002; Ricketts, 2004). In fact, native species can substantially improve pollination and crop yields above those with honey bees alone (Garibaldi et al., 2014). However, abundances of native pollinators are threatened by a combination of habitat destruction (Holzschuh et al., 2007; Kennedy et al., 2013; A. Klein et al., 2003; Kremen, 2005; Potts et al., 2010; Ricketts, 2004; Taki et al., 2007; N. Williams and Kremen, 2007), pesticides (Rundlöf et al., 2015), and competition from widely introduced honey bee populations (Aizen and Feinsinger, 1994).

Finally, agricultural intensification and habitat fragmentation have isolated pollinator habitats from the crops that must be pollinated. Pollination services by native bees are positively related to the amount of natural habitat in the vicinity of agricultural crops (Kennedy et al., 2013; Kremen et al., 2004; Ricketts, 2004; N. Williams and Kremen, 2007; Winfree et al., 2009, 2011). Balvanera et al. (2005) found that increasingly intensive conventional agriculture resulted in loss of 60% of native bee pollinators, reduced abundances of the most functionally important pollinators, 60–80% loss of pollination function, and reduced evenness in functional structure, compared to organic agriculture within a conserved forest matrix. Colonies of wild honey bees also typically inhabit forests and suffer from forest fragmentation (Crane, 1999).

The combination of these interacting anthropogenic changes will likely continue to undermine pollinator networks and pollination of crop and wild plant species (Biesmeijer et al., 2006; Vanbergen and the Insect Pollinators Initiative, 2013). Climate change will likely disrupt synchronization of pollinator and plant phenologies, leading to species-impoverished pollinator assemblages dominated by generalist pollinators that provide less effective pollination (Burkle et al., 2013). Declining pollination services clearly threaten sustainable crop production for an increasingly demanding human population. However, the survival of many other plant species that require insect pollinators for outcrossing also is threatened (Biesmeijer et al., 2006).

1 INTRODUCTION

Insects affect plant reproduction and associated processes in a variety of ways. Direct and indirect effects of herbivores, including florivores, on plant production and allocation of resources to reproduction were described in Chapter 12. This chapter focuses on the important roles played by pollinators and seed predators in seed production, seedling recruitment, and plant demography.

Pollinators control fertilization and reproductive rates for many plant species, especially in desert and tropical ecosystems characterized by low densities of individual plant species.

Many plant species depend on pollinators for successful reproduction and may disappear if their pollinators become rare or extinct (Powell and Powell, 1987; Steffan-Dewenter and Tscharntke, 1999). Most fruits and vegetables consumed by humans require pollination by insects (A. Klein et al., 2007; Kremen et al., 2007; Losey and Vaughan, 2006).

Seed predators consume seeds and thereby reduce plant reproductive efficiency but often move seeds to new locations and thereby contribute to plant dispersal. Many plant species depend on seed dispersers for successful movement of seeds to new habitats and may become endangered if their dispersers disappear (Caughlin et al., 2015; Janzen and Martin, 1982; R. Johnson et al., 2009; O'Dowd and Hay, 1980; Schupp, 1988; Temple, 1977).

Overall, about 88% of 352,000 plant species globally depend on pollination by animals, primarily insects, birds, and bats. Insects are the major agents of pollination, seed predation, or seed dispersal in many ecosystems (Bawa, 1990; Degen and Roubik, 2004; Sallabanks and Courtney, 1992). Given their importance to plant reproduction, pollination and seed dispersal are among the most intricate mutualisms between animals and plants (see Chapter 8) and have been studied widely from the perspective of coevolution. Unfortunately, few studies have evaluated the effects of pollinators, seed predators, and seed dispersers on ecosystem processes. Nevertheless, pollinators, seed predators, and seed dispersers clearly control plant recruitment and its effects on ecosystem structure and function (see Chapter 11).

2 TYPES AND PATTERNS OF POLLINATION

Plants exhibit a variety of reproductive mechanisms. Many, if not most, can reproduce vegetatively, but this mechanism largely limits population growth to local clonal recruitment of offspring. Similarly, many plant species are capable of self-fertilization, but colonization ability and adaptation to changing environmental conditions require outcrossing. Consequently, a large percentage of plant species (a vast majority in some ecosystems) are self-incompatible, and many are dioecious (eg, 20–30% of tropical tree species), with male and female floral structures separated among individual plants, to preclude inbreeding (Bawa, 1990; Momose et al., 1998a). Mechanisms for transporting pollen between individuals become increasingly critical for reproduction with increasing distance between male and female structures and increasing isolation of individual plants (Ghazoul and McLeish, 2001; Jakobsson and Ågren, 2014; Regal, 1982; Ricketts et al., 2008; Steffan-Dewenter and Tscharntke, 1999).

Several mechanisms have developed to move pollen among flowering individuals. Pollen can be transferred between plants through abiotic or biotic mechanisms (Regal, 1982). Wind is the primary abiotic mechanism for pollen transport. Biotic transport involves insects, birds, and bats (Fig. 13.1). The proportion of animal-pollinated plant species increases toward the tropics, from 0–20% in polar communities to 78% in temperate-zone communities to 94% in tropical communities (Bawa, 1990; Ollerton et al., 2011; Regal, 1982). Insects are the principal pollinators in many ecosystems. For example, Momose et al. (1998b) noted that social bees were the primary pollinators for 44% of 270 plant species in a lowland dipterocarp forest in Sarawak, Malaysia, beetles for 24%, solitary bees for 19%, and birds and bats for 6%. These mechanisms provide varying degrees of fertilization efficiency, depending on ecosystem conditions.

FIGURE 13.1   **Examples of pollinators.** (A) Honey bee, *Apis mellifera scutellata*, Hluhluwe National Park, South Africa. Note the pollen basket (corbicula) filled with red pollen. (B) Scarab beetle, Fushan Experimental Forest, Taiwan.

2.1 Pollinator Functional Groups

Functional groups of pollinators may be more or less restricted to groups of plants based on floral or habitat characteristics (Bawa, 1990). A large number of pollinators are *generalists* with respect to plant species. This functional group includes honey bees and many beetles, flies, thrips, etc. that forage on any floral resources available. *Specialist* pollinators exploit particular floral characteristics that may exclude other pollinators. For example, nocturnally flowering plants with large flowers attract primarily bats, whereas plants with small flowers attract primarily moths and midges. Long, bright red flowers attract birds but are largely unattractive to insects (S. Johnson and Bond, 1994). Such flowers often are narrow to hinder entry by bees and other insect pollinators (Heinrich, 1979), but may nonetheless be pollinated by some insects (Roubik, 1989). *Pollen feeders* feed primarily on pollen (eg, beetles and thrips) and are likely to transport pollen acquired during feeding, whereas others are primarily *nectar feeders* (eg, butterflies, moths, and flies) and transport pollen more coincidentally. In fact, many nectar feeders avoid the reproductive organs, some by perforating the base of the flower to reach the nectar, for example, nectar thieves (Dedej and Delaplane, 2004), or, in the case of ants, may reduce pollen viability (Peakall et al., 1987). *Bees*, especially *Apis* spp., feed on pollen and nectar. Functional groupings also reflect sensitivity and attraction to floral odors (Chittka and Raine, 2006). For example, dung-, fungus-, and carrion-feeding flies and beetles are the primary pollinators of plants that emit dung or carrion odors (Appanah, 1990; Norman and Clayton, 1986; Norman et al., 1992).

Ants frequently exploit floral resources but have little importance as pollinators. Peakall et al. (1987) suggested that antibiotic secretions produced by most ants, to inhibit infection by entomophagous fungi in a subterranean habitat, also inhibit germination of pollen. Ants lacking these secretions are known to function as pollinators. Ants attracted to nectar resources also may interfere with pollinator visitation (Ness, 2006). However, Holland et al. (2011) reported that senita cactus, *Pachycereus schottii*, distract ants from floral resources by providing extrafloral nectaries, thereby retaining the beneficial role of ants in protecting plants while minimizing ant interference with pollination by the mutualistic senita moth, *Upiga virescens*. In fact, ant visitation to extrafloral nectaries increased pollination and oviposition by the moth and reduced parasitism of the fruit- and seed-eating moth larvae by the wasp, *Temelucha* sp., thereby maintaining the moth–cactus mutualism (Holland et al., 2011).

Pollinator functional groups also have been distinguished on the basis of habitat preferences, such as vegetation stratum (Fig. 13.2). Appanah (1990) distinguished four groups of plant–pollinator associations in a tropical lowland dipterocarp forest in Malaysia. The *forest floor stratum* was characterized by low visibility and limited airflow. Floral rewards were small, reflecting low productivity of light-limited plants and low energy requirements of associated pollinators, and flowering times were extended, increasing the probability of pollination by infrequent visitors. The plant–pollinator association of this stratum was dominated largely by nonselective, low-energetic beetles, midges, and other flies. These pollinators were attracted over short distances by strong olfactory cues, often resembling dung or carrion, which have limited effective range. The *understory stratum* shared many of the environmental features of the forest floor. Plants in this stratum also offered limited visual cues and floral rewards and were pollinated by nonspecific trapliners, that is, species that revisit particular plants along an established circuit (eg, trigonid bees, solitary wasps, and butterflies). The *overstory stratum* was characterized by brightly colored flowers, held above the canopy to attract pollinators over a wide area, and brief, highly synchronized flowering within plant species. Dominant pollinators were *Apis dorsata* and trapliners such as carpenter bees, birds, and bats. Dipterocarps in the genera *Shorea*, *Hopea*, and *Dipterocarpus* formed a separate association based on tiny flowers with limited nectar rewards and nocturnal flowering. Thrips and other tiny flower-feeding insects were the primary pollinators. By contrast, Sakai et al. (1999) observed that beetles (chrysomelids and curculionids), rather than thrips, were the primary pollinators of these tree species in Sarawak. Finally, some plant species representing various canopy positions were *cauliflorous*, that is, they produced flowers along the trunk or main branches. These flowers typically were large, or small and clumped, pale colored, odiferous,

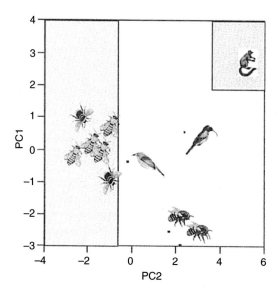

FIGURE 13.2   Principal components analysis for pollinator partitioning by 11 species of Coleeae (Bignoniaceae) trees in Madagascar, based on variation in flower size (PC1, 54% of variation among species) and phenology (PC2, 26% of variation). Timing was strongly negatively correlated with flowering height on trees. Icons represent pollinator groups. Shading indicates that, with the exception of a potentially lemur-pollinated species, trees pollinated by understory pollinators (small and medium-sized bees) are on the left, and trees pollinated by canopy pollinators are on the right. *Source: From Zjhra, M.L., 2008. Facilitating sympatric species coexistence via pollinator partitioning in endemic tropical trees of Madagascar. Plant Syst. Evol. 271, 157–176.*

and produced during a brief, highly synchronized period. Pollinators included understory and overstory insects, birds, and bats. Momose et al. (1998b) noted that long-distance pollinators tended to be less common in Malaysian forests than in Neotropical forests.

Roubik (1993) manipulated vertical location of floral resources from different canopy strata in tropical forests in Panama. Results indicated that the apparent fidelity of pollinator species to particular canopy strata reflected pollinator preferences for particular floral resources. Most pollinator species were attracted to their preferred floral resources regardless of their location in the canopy.

In deciduous forests, understory flowering may be concentrated prior to completion of canopy closure, when solar exposure and photosynthesis are maximal (Kudo et al., 2008). However, flowering time may vary widely, depending on the end of adverse seasonal conditions. Understory plants flowering after canopy closure show reduced flowering and fruiting rates as a result of declining light level. Ulyshen et al. (2010) reported that pollinator assemblages in deciduous forest canopy and understory were statistically distinct.

2.2   Measurement of Pollination

Several factors govern the transport of pollen between conspecific flowers, including the mechanism of pollen transport, proximity of conspecific plants, pollinator attraction to floral structures, adaptations for carrying pollen, fidelity, and thermodynamic constraints. Clearly, the importance of this process to plant reproduction should favor mechanisms that increase pollination efficiency, that is, the likelihood that a pollen grain will be deposited in a conspecific flower.

A variety of methods have been used to measure pollinator activity, pollen transport, and pollination efficiency. Observations of the type and frequency of floral visitors can provide a measure of pollinator activity (Aizen and Feinsinger, 1994; Ghazoul and McLeish, 2001; Holt et al., 2014; Sakai et al., 1999; Steffan-Dewenter and Tscharntke, 1999; Steffan-Dewenter et al., 2001). Interception traps also can be used to collect insects visiting flowers (S. Johnson et al., 2004). Roubik (2002) used fine mesh bags to exclude pollinators from coffee, *Coffea arabica*, in order to measure difference in pollination and seed set when pollinators had access or not. Number of fertilized seeds per flower provides a measure of pollination for self-incompatible species (Steffan-Dewenter et al., 2001; Holt et al., 2014; S. Johnson et al., 2004). Kohn and Casper (1992) used electrophoresis to identify seeds containing alleles that did not occur in neighboring plants. G. White et al. (2002) used DNA marker techniques to measure pollen transfer among trees, *Swietenia humilis*, in isolated fragments of tropical forest in Honduras. Brosi et al. (2009) used stable isotopic ratios to evaluate foraging patterns of stingless bees in a fragmented tropical forest landscape. Although this method was sufficient to distinguish bees foraging in pastures versus forest, and demonstrated significant relationships between foraging pattern and landscape structure, it could not indicate the proportion of floral resources obtained in different habitats.

Wind pollination is highly inefficient, with respect to the probability of individual pollen grains reaching a conspecific stigma and fertilizing an ovule. The probability of successful pollen transfer among conspecific flowers by wind decreases as the cube of distance between plants (Moldenke, 1976; Wragg and Johnson, 2011). However, for dominant plant species in windy habitats, the investment in individual pollen grains is small compared to the probability of reaching neighboring plants over short distances, so large numbers of pollen grains can be produced, ensuring that some will land on conspecific reproductive structures.

Directed transport of pollen by animal pollinators increases efficiency to the extent that the pollinator visits a conspecific flower before the pollen is lost or contaminated with pollen from other plant species. For example, Roubik (2002) reported that insect pollinators accounted for 36% of coffee seed production, and insect-pollinated coffee beans were 7–50% heavier than were self-pollinated beans. Heavier seeds typically have a higher probability of germination and early survival. Accordingly, animal-pollinated plant species benefit by investing energy and nutrients in adaptations to improve the attraction and fidelity of pollinators. These adaptations include nectar rewards to attract pollinators; floral and aromatic advertisements; floral structures that restrict the diversity of pollinators visiting the flowers; synchronized flowering among conspecific individuals; and divergence in time of flowering among plant species to reduce pollen contamination (Chittka and Raine, 2006; Heinrich, 1979; Liu and Huang, 2013; Wragg and Johnson, 2011).

The transition from wind pollination to animal pollination has been a rare event and requires evolution of floral and pollen traits to attract insects and facilitate pollen transfer (Wragg and Johnson, 2011). Wragg and Johnson (2011) compared flower and pollen traits of intermingled wind- and insect-pollinated species of sedge, *Cyperus* spp., at nine grassland sites in southeastern Africa. They found that flowers of insect-pollinated species produced a greater diversity and higher emission rate of volatile chemicals known to attract insects,

FIGURE 13.3 **Mean (± SE) seed set in three sedge species, as affected by wind-borne or insect-borne pollen transport.** Exclusion of insects, but not wind, reduced seed set in *Cyperus obtusiflorus* (B) and *C. sphaerocephalus* (C) but not in *Pycreus oakfortensis* (A). Additional exclusion of wind reduced seed set in *P. oakfortensis* but not in *C. obtusiflorus* or *C. sphaerocephalus*. Sample sizes in parentheses are florets/spikelets/inflorescences. Insect visitors pictured are (A) a leaf beetle *Monolepta cruciata*, (B) honey bee, *Apis mellifera*, and (C) monkey beetle, *Eriesthis fallax*. White bars provide scale, 5 mm. *Source: From Wragg, P.D., Johnson, S.D., 2011. Transition from wind pollination to insect pollination in sedges: experimental evidence and functional traits. New Phytol. 191, 1128–1140.*

but their pollen could be transported only short distances by wind, compared to flowers of wind-pollinated species. Experimental exclusion of insects resulted in 56–89% reduction in seed set in insect-pollinated species, but not in intermingled wind-pollinated species (Fig. 13.3).

Liu and Huang (2013) used a combination of visitation frequency and manipulated pollination timing to conclude that three sympatric bellflower, *Adenophora*, species in southwestern China partitioned pollinators across a diurnal to nocturnal gradient via flowering time. *Adenophora jasionifolia* showed peak flowering at midday, whereas *A. capillaris* flowered during a brief period at dusk, and *A. khasiana* showed a more protracted flowering period that peaked in late afternoon and extended through midnight (Fig. 13.4). Large bees were the primary diurnal pollinators, whereas moths were the primary nocturnal pollinators. Flowers

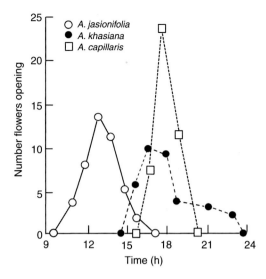

FIGURE 13.4 Partitioning of anthesis timing in three species of *Adenophora*, based on 40 plants of each species at Shangri-La Alpine Botanical Garden in southwest China. *Source: Modified from Liu, C.-Q., Huang, S.-Q., 2013. Floral divergence, pollinator partitioning and the spatiotemporal pattern of plant–pollinator interactions in three sympatric Adenophora species. Oecologia 173, 1411–1423.*

experimentally exposed during periods when their primary pollinators were active produced more seeds than did flowers that were not exposed at appropriate times (Fig. 13.5). However, Liu and Huang (2013) noted that many pollinator species were shared among the three plant species, especially with *A. khasiana* which overlapped the other two species, suggesting that floral traits did not completely filter visitors.

Nectar or pollen rewards must be sufficient to compensate the pollinator for the foraging effort. For example, Petanidou and Vokou (1990) reported that pollen of wind-pollinated plants provided lower energy content than did pollen of insect-pollinated plants in Mediterranean ecosystems. Greater nectar return is necessary to attract bees during cooler periods, when energy allocation to thermoregulation is high, compared to warmer periods (Heinrich, 1979). Heinrich (1979) noted that pollinator fidelity reflects offsetting adaptations. Plants invest the minimum amount of energy necessary to reward pollinators, but pollinators quickly learn to concentrate on flowers offering the greatest rewards (Chittka et al., 1997). Individual plants in aggregations could attract bees and be pollinated even if they produced no nectar, provided that their neighbors produced nectar. The nonproducers should be able to invest more energy in growth and seed production. However, if these "cheaters" became too common, pollinators might switch to competing plant species that offered greater food rewards (Feinsinger, 1983). Theis and Raguso (2005) reported that some plant species dramatically reduce floral scent emission rates following pollination, thereby communicating prior visitation, and honey bees were three times more likely to visit unpollinated than pollinated flowers. Some Australian orchids apparently produce chemical signals similar to ant and wasp mating pheromones and exploit the male ants and wasps that pollinate these orchids while attempting to copulate with the orchid (Peakall and Beattie, 1996).

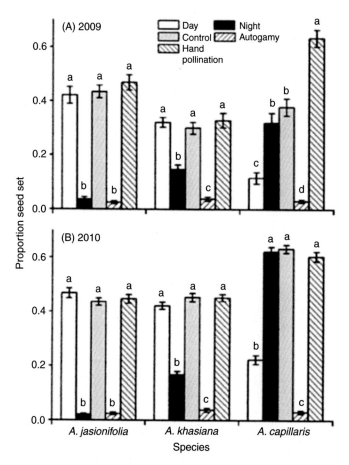

FIGURE 13.5 **Proportion seed set by flowers under different pollination treatments in three species of Adenophora during 2009 (A) and 2010 (B) at the Shangri-La Alpine Botanical Garden in southwest China.** Vertical bars represent 1 SE. Different letters above bars indicate significant differences among treatments within species. *Source: Modified from Liu, C.-Q., Huang, S.-Q., 2013. Floral divergence, pollinator partitioning and the spatiotemporal pattern of plant–pollinator interactions in three sympatric Adenophora species. Oecologia 173, 1411–1423.*

A. Lewis (1993) suggested that floral characteristics may reflect advantages accruing to the plant when pollinators must make a substantial investment in learning to handle a flower, thereby becoming facultative specialists. Plant investment in attractants and rewards for pollinators represents an evolutionary trade-off between growth and reproduction (Heinrich, 1979) and may affect the ability of light- or resource-limited species to attract pollinators. On the other hand, pollinator choice among floral resources that vary in quality represents a trade-off between investment in recognition of high-quality resources and accuracy of discrimination (Chittka and Raine, 2006).

Despite adaptations to attract pollinators, plants often are subject to pollen limitation, due to competition for pollinators or other factors, that reduces potential seed production

(Ashman et al., 2004; Knight et al., 2005b; R.J. Mitchell et al., 2009; Quesada et al., 2009). For example, Fishman and Willis (2008) reported that supplemental pollination of yellow monkeyflowers, *Mimulus guttatus*, increased seed production by 37%, compared to open-pollinated flowers.

Effects of pollination on plant seedling recruitment and ecosystem processes have been measured less frequently. Effects on seed production can be measured as the number of seeds produced when pollinators have access to or are excluded from flowers (Jakobsson and Ågren, 2014; S. Johnson et al., 2004; Norman and Clayton, 1986; Norman et al., 1992; Roubik, 2002; Steffan-Dewenter and Tscharntke, 1999; Steffan-Dewenter et al., 2001). Pollinator effects on ecosystem processes should reflect their direct influence on plant reproduction and indirect influence on vegetation dynamics.

2.3 Spatial and Temporal Patterns of Pollination

Pollination by animals varies among ecosystems. Pollinators are more common in angiosperm-dominated ecosystems than in gymnosperm-dominated ecosystems, but pollination by wind is energetically efficient for dominant species in grasslands and temperate forests.

The regularity with which conspecific plants occur in close proximity to each other largely determines their pollination mechanism. Long-lived species that dominate relatively simple ecosystems, that is, grasslands and temperate forests, are pollinated primarily by wind. These plant species do not require efficient pollination or frequent reproduction to ensure population survival. Energetically inexpensive transport of pollen by wind provides sufficient pollination (and successful reproduction) so that energy need not be diverted to production of expensive nectar rewards and floral displays to advertise availability.

Directed transport of pollen by animals is critical to reproduction of plant species that are short-lived, sparsely distributed, or occur in habitats with restricted airflow (Appanah, 1990; Moldenke, 1979; Regal, 1982; Somanathan et al., 2004). In contrast to long-lived plants, short-lived plants have limited opportunities for future reproduction and, therefore, tend to depend on more efficient pollination to ensure seed production. Sparsely distributed plants and plants in habitats with limited airflow cannot rely on inefficient transport of pollen by wind between distant or inaccessible individuals. Such species include early successional plants dominating ephemeral communities, widely spaced plants in harsh environments (eg, deserts), scattered forbs in grasslands, subdominant trees, shrubs, and herbs in temperate forests, and all, or most, plant species in tropical forests (S. Johnson et al., 2004; Momose et al., 1998b; Regal, 1982). Regal (1982) reported that fewer than 6% of desert shrub species are wind pollinated. All of the 270 plant species in a lowland dipterocarp forest in Sarawak, Malaysia, were animal pollinated, 90% by insects (Momose et al., 1998b). Bawa (1990) reviewed studies that demonstrated long-distance pollen flow and outcrossing for tropical canopy trees, but a high degree of inbreeding for many tropical herbs and shrubs.

Insects and other animal pollinators can transport pollen over considerable distances. Kohn and Casper (1992) documented gene flow among bee-pollinated buffalo gourds, *Cucurbita foetidissima*, over distances up to 0.7 km in New Mexico, USA. Somanathan et al. (2004) reported that carpenter bees, *Xylocopa tenuiscapa*, pollinated a Neotropical tree, *Heterophragma quadriloculare*, isolated from pollen sources by as much as 330 m, permitting reproduction by spatially isolated trees. G. White et al. (2002) identified sources of pollen reaching isolated *Swietenia humilis* trees and forest fragments in Honduras. Substantial proportions of pollen were transported over distances of 1.5 km to more than 4.5 km between fragments. By contrast, a Neotropical shrub, *Lasiosiphon eriocephalus*, pollinated by a weakly flying nitidulid beetle, may be particularly vulnerable to isolation or fragmentation (Somanathan et al., 2004). Duncan et al. (2004) reported that outcrossed, conspecific pollen receipt on flowers declined significantly with increasing plant isolation over distances of 50 m, even though a constant rate of heterospecific pollen deposition indicated that visitation by pollinators was not reduced. These results likely reflect increased pollinator switching among floral species as distance increases (Chittka et al., 1997).

Pollinator species vary in their efficiency of pollen transport. Specialists often have higher pollination efficiency because of specific adaptations between coevolved plant and insect that ensure proper positioning of pollen, often as a pollinium in orchids and milkweeds, on the insect, such that proper contact with a conspecific stigma is ensured (Stoepler et al., 2012). For example, Stoepler et al. (2012) reported that two bee species, *Apis mellifera* and *Bombus* sp., and a skipper butterfly, *Epargyreus clarus*, were the primary pollinators of two milkweed species, *Asclepias exaltata* and *A. syriaca*, and their hybrids. However, only *Bombus* sp. properly removed and inserted pollinia for both parent species as well as hybrids and it was, therefore, the most effective pollinator. Similarly, Woodcock et al. (2013) found that solitary bees, *Andrena*, *Osmia*, and *Lasioglossum* spp., and bumble bees had greater frequency of contact with the stigma of oilseed rape, *Brassica napus*, flowers than did honey bees. The interaction between probabilities of free pollen on the body and contact with the stigma indicated that only 34% of honey bee visits were likely to result in pollen transfer, compared to 35% for bumble bees and 71% for solitary bees. However, the greater number of honey bees from managed hives in fields resulted in their being the primary pollinators.

Primary pollinator composition varies spatially. Bees are the most important group globally (Klein et al., 2007), but tend to be most diverse and abundant in warm, dry mid-latitudes (Michener, 2007). Diptera are the second most important group (Klein et al., 2007) and are more important pollinators at higher latitudes (Eberling and Olesen, 1999). Beetles, butterflies, and moths, as well as birds and bats, are more diverse and abundant in the tropics (Scoble, 1992). Furthermore, many pollinators show species-specific spatial and temporal variation in flower visitation traits and within-flower behavior (Hoehn et al., 2008), such that different pollinator species can complement each other, and their combination provides more effective pollination (Brittain et al., 2013; Hoehn et al., 2008).

Roubik (1989) reviewed studies that distinguished seasonal patterns of pollinator activity. Primary pollinators typically were most active during periods of peak flowering. Heithaus (1979) reported that megachilid and anthophorid bees were most active during the dry season in Costa Rica, halictid bees during both wet and dry seasons, and andrenid and colletid bees during the wet season or during both seasons. Social pollinators (eg, apid bees) require a sequence of floral resources throughout the year to support long-lived colonies, and visit a succession of flowering plant species, whereas more ephemeral, solitary species with short life spans can be relatively more specialized on seasonal floral resources (S. Corbet, 1997; Roubik, 1989).

Pollination efficiency by insects is strongly affected by flower size or attractiveness and plant spacing (Brys et al., 2008; Dauber et al., 2010). Mustajärvi et al. (2001) reported that bumble bees, *Bombus* spp., preferred larger populations of *Lychnis viscaria*, but visitation rates were higher in sparse populations, where the large flowers of this plant were more visible. Momose et al. (1998a) found that pollination by thrips and consequent fruit and seed development of a small (< 8 m height) tree species, *Popowia pisocarpa*, in Sarawak declined dramatically when distances between trees exceeded 5 m. Brys et al. (2008) demonstrated that pollination efficiency, pollen transport, and fruit set for 13 populations of a European orchid, *Listera ovata*, increased with local population size up to 30–40 flowering plants, above which pollination efficiency and plant reproduction declined again.

Pollinator diversity is critical for adequate pollination of many plant species. Klein et al. (2003) reported that fruit set of highland coffee depended on the diversity, not abundance, of flower-visiting bee species and ranged from 60% when three bee species were present to 90% when 20 bee species were present. Kremen et al. (2002) reported similar dependence of watermelon production on the diversity of local bee species. In fact, Brittain et al. (2013) found that a sand bee, *Andrena cerasifolii*, and the blue orchard bee, *Osmia lignaria*, pollinated almond trees, *Prunus dulcis*, in California more efficiently per floral visit than did honey bees. Garibaldi et al. (2013) reported that comparison of pollination efficiency by honey bees and native pollinators in 600 fields representing 41 crop systems globally demonstrated significant increases in fruit set by native pollinators for all 41 crop systems, compared to only 14% of crop systems by honey bees.

Environmental changes that increase the distance between conspecific plants, such as habitat fragmentation, may threaten their survival (Bawa, 1990; Biesmeijer et al., 2006; Brosi and Briggs, 2013; Burkle et al., 2013; Didham et al., 1996; Goulson et al., 2008; Klein et al., 2003; Potts et al., 2010). Brys et al. (2008) reported decline in pollination over short distances of separation among conspecific flowering plants (Fig. 13.6). Ricketts et al. (2008) conducted a

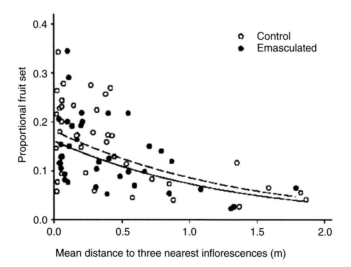

FIGURE 13.6 **Effect of local plant density (mean distance to the three nearest flowering plants) on proportional fruit set in emasculated plants** *(filled circle, solid regression line)* **and control plants** *(open circle, dashed regression line)* **within a large** *Listera ovata* **population (N = 1524).** The emasculation treatment indicated the effect of distance on cross-fertilization. *Source: From Brys, R., et al., 2008. Pollination efficiency and reproductive patterns in relation to local plant density, population size, and floral display in the rewarding Listera ovata (Orchidaceae). Bot. J. Linn. Soc. 157, 713–721.*

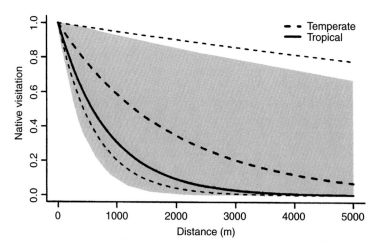

FIGURE 13.7 **Relationship between plant distance from population sources and the rate of native pollinator visitation in temperate and tropical ecosystems.** Shaded area represents 90% confidence interval for tropical studies (*n* = 11); *light dashed lines* represent 90% confidence interval for temperate studies (*n* = 11). Distances at which native pollinator visitation drops to 50% of maximum are 590 m for tropical systems and 1300 m for temperate systems. *Source: From Ricketts, T.H., et al., 2008. Landscape effects on crop pollinator services: are there general patterns? Ecol. Lett. 11, 499–515.*

metaanalysis of isolation studies and concluded that the distance at which native pollinator visitation rate declines to 50% of maximum is 590 m in tropical ecosystems and 1300 m in temperate ecosystems (Fig. 13.7).

Pauw and Hawkins (2011) used herbarium specimens of orchids, *Pterygodium catholicum*, collected from an isolated urban park and nearby rural sites in South Africa to reconstruct historical pollination rates from 1880 to 2000, a period during which the urban site became increasingly isolated from rural populations. They found that disappearance of the primary pollinator, the oil-collecting bee, *Rediviva peringueyi*, from the urban site resulted in collapse of pollination (Fig. 13.8), disappearance of a nonclonal, pollinator-dependent species, *Disperis circumflexa*, and a shift in the orchid assemblage toward greater dominance by clonally reproducing species, including *P. catholicum*, which also depends on the oil-collecting bee for pollination but can reproduce vegetatively. This shift did not occur in the rural areas. Bodbyl Roels and Kelly (2011) studied responses of yellow monkeyflowers, a predominantly outcrossing species, when pollinating bumble bees, *Bombus* spp., were experimentally excluded. The populations without pollinators showed greatly reduced fitness in early generations, compared to populations with pollinators, but by the fifth generation had developed an ability to self-fertilize and to produce seed at a rate approaching that of pollinated populations (Fig. 13.9). Similarly, Ushimaru et al. (2014) reported that a Japanese *Commelina* species displayed traits that permit self-pollination only in urban sites.

Ecosystem fragmentation and agricultural intensification are two primary factors that isolate plants and interfere with pollination. Aizen and Feinsinger (1994) compared pollinator visitation among replicated blocks containing continuous forest and large (> 2.2 ha) and small (< 1 ha) fragments in subtropical dry forest in northwestern Argentina. The diversity and visitation frequency of native pollinators decreased significantly, and the visitation

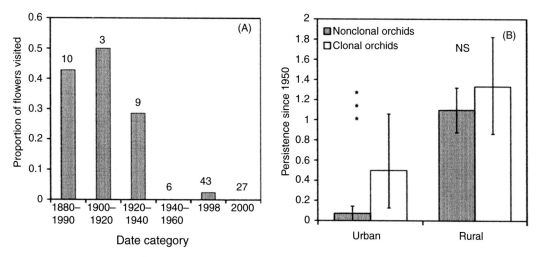

FIGURE 13.8 Historical pollination rates for *Pterygodium catholicum* from Signal Hill, Cape Town, South Africa, reconstructed from herbarium specimens (A). The orchid assemblage shifted toward greater representation by clonal species in the Cape Town area following pollination collapse during 1940–60, but not in rural areas where pollination remained intact (B). A persistence level of 1 indicates an equal number of herbarium records before and after 1950. *Source: From Pauw, A., Hawkins, J.A., 2011. Reconstruction of historical pollination rates reveals linked declines of pollinators and plants. Oikos 120, 344–349.*

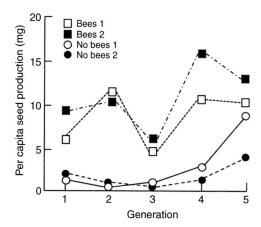

FIGURE 13.9 **Changes in population fitness, measured as per capita seed production (mg), of monkeyflowers, *Mimulus guttatus*, over five generations with bumble bee, *Bombus* spp., pollinators present or excluded.** Increased seed set in the exclusion treatment after four generations indicates evolution toward improved ability to self-fertilize flowers. *Source: Modified from Bodbyl Roels, S.A., Kelly, J.K., 2011. Rapid evolution caused by pollinator loss in Mimulus guttatus. Evolution 65, 2541–2552.*

frequency of exotic honey bees, *A. mellifera*, increased significantly, with decreasing fragment size (Fig. 13.10). Fragments supported fewer bee species than did continuous forests. Although honey bees from the surrounding agricultural matrix replaced most of the lost visitation by native pollinators, some plant species could be threatened by loss or reduced specificity of pollinators.

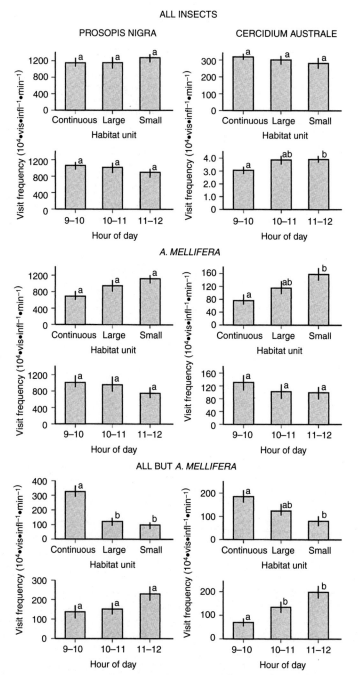

FIGURE 13.10 **Rates of visitation by all pollinating insects, exotic honey bees (*Apis mellifera*) alone, and native pollinators alone on flowers of two plant species by treatment (continuous forest, and large (2.2 ha) and small (1 ha) fragments) and by time of day in Argentina.** *Vertical lines* represent standard errors; bars under the same letter do not differ at $p < 0.05$. *Source: From Aizen, M.A., Feinsinger, P., 1994. Habitat fragmentation, native insect pollinators, and feral honey bees in Argentine 'Chaco Serano'. Ecol. Appl. 4, 378–392.*

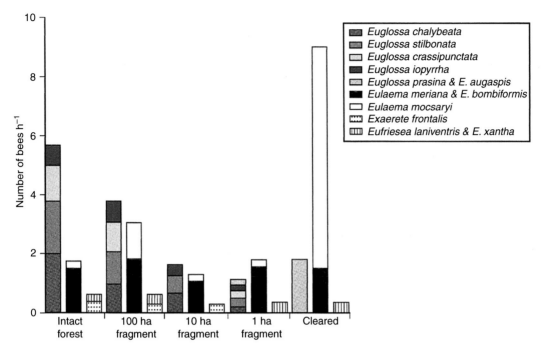

FIGURE 13.11 **Rates of visitation by male euglossine bees at chemical baits in intact forest, forest fragments of varying size (100 ha, 10 ha, and 1 ha), and recently deforested (500 ha).** *Source: Modified from Powell, A.H., Powell, G.V.N., 1987. Population dynamics of male euglossine bees in Amazonian forest fragments. Biotropica 19, 176–179.*

Powell and Powell (1987) compared attraction of male euglossine bees to floral chemical baits in forest fragments in Brazil. Abundance and species composition did not differ among sites prior to fragmentation. However, after fragmentation, visitation rates for most species were correlated to fragment size, and the bee species trapped in clearings differed from the species trapped in forests (Fig. 13.11). Powell and Powell (1987) concluded that the reduced abundance and activity of particular pollinators in fragmented forests threatened the viability of their orchid hosts.

Steffan-Dewenter and Tscharntke (1999) examined the effects of plant isolation on pollination and seed production in replicate grasslands surrounded by intensively managed farmland. They established small experimental patches of two grassland species, *Sinapis arvensis* and *Raphanus sativus*, at increasing distances from the grassland boundaries and found that the number and diversity of bees visiting flowers, and seed production, declined with increasing isolation. Number of seeds per plant was reduced by 50% at 260 m from the nearest grassland for *R. sativus* and at 1000 m for *S. arvensis*.

Agricultural intensification has led to fragmentation and isolation of native pollinator habitats within agricultural landscapes (Carvalheiro et al., 2011; Goulson et al., 2008; Klein et al., 2007; Kremen et al., 2002, 2004, 2007). However, patches of native habitat and associated floral resources within agricultural landscapes can provide more consistent floral resources, thereby increasing pollinator diversity and pollinator efficiency (Carvalheiro et al., 2011; Kennedy et al., 2013; Kremen et al., 2004, 2007; Rundlöf et al., 2014; N. Williams and Kremen, 2007). Morandin and Winston (2005) compared pollinator diversity and pollination in organic, conventional, and herbicide-resistant transgenic canola, *Brassica napus* and *B. rapa*, fields in Alberta,

FIGURE 13.12   Honey bee, *Apis mellifera*, hive in tree cavity in a remnant woodlot within an agricultural landscape.

Canada. Abundance of native bees was greatest in organic fields, followed by conventional fields. Bee abundance was lowest in transgenic fields. Bee abundance was significantly and positively correlated with pollination rate. Landscape structures that maintain pollinator habitat (Fig. 13.12) and more consistent floral resources, in time and space, will benefit as a result of greater pollination of agricultural crops, as a result of "spillover" of native pollinators from suitable to less suitable habitats (Kovács-Hostyánski et al., 2013; Pereira-Peixoto et al., 2014).

3 EFFECTS OF POLLINATION

Although much research has demonstrated the critical role of adequate pollination for reproduction by many plant species, as described previously, virtually no research has addressed pollinator effects on plant community structure and the dependent web of direct and indirect interactions. Differential reproductive success among plant species affects vegetation dynamics.

Pollination contributes to genetic recombination and survival of plant species in heterogeneous environments. Many plants can reproduce vegetatively or by self-fertilization, but these mechanisms are not conducive to long-distance colonization or genetic recombination. Species' survival and adaptation to changing environmental conditions require outcrossing and environmental selection among diverse genotypes. Some long-lived perennials may endure adverse conditions and persist by vegetative reproduction until conditions favor outcrossing and seedling recruitment. Such windows of opportunity are unpredictable, promoting annual investment in flower and seed production (Archer and Pyke, 1991).

Pollinator-facilitated reproduction is a key factor maintaining populations of ephemeral or sparsely distributed plant species. Obligate outcrossing plant species that depend on insect or vertebrate pollinators for pollination are vulnerable to loss of these mutualists. Maintenance of rare plant species or restoration of declining species depends to a large extent on protection or enhancement of associated pollinators (Archer and Pyke, 1991; Corbet, 1997). Norman and Clayton (1986) and Norman et al. (1992) found that pawpaws, *Asimina* spp., in Florida, USA, depended on beetle and fly pollinators attracted to yeasty floral odors for production of viable seeds. Self-pollinated flowers occasionally produced fruits, but only seeds from cross-pollinated flowers germinated.

Pollinator choices among floral resources that vary in nutritional quality likely affect vegetation dynamics and community structure (Heystek and Pauw, 2014; Quesada et al., 2009). For example, rewardless orchids might appear to have a disadvantage in attracting pollinators. Galizia et al. (2005) compared visual and olfactory floral cues of a rewardless orchid, *Orchis israelitica*, and its lily model, *Bellevalia flexuosa*. The orchid displayed visual, but not olfactory, mimicry. Nevertheless, bees readily switch between the two plant species, indicating dominance of visual stimuli in short-range floral choice. However, Chittka et al. (1997) reported that foraging bumble bees, *Bombus* spp., were more likely to switch to a different floral species after a short visit indicative of low nectar reward than after a long visit or when the floral species was rare. Overall, bees remained faithful to a particular species as long as rewards were sufficient and plants were close together, but switched when rewards were low or plants were not encountered at close range. Gumbert et al. (1999) found that rare plant species provided more distinct floral signals, which, combined with greater reward, would compensate pollinators for greater travel distances (Heinrich and Raven, 1972). Burkle and Irwin (2009, 2010) experimentally increased nitrogen availability to subalpine plant assemblages to evaluate effects on pollinator visitation rates. They found that pollinator visitation rate was unaffected by this bottom-up enrichment. Pollinators distributed their visitation evenly among floral resources, regardless of nitrogen treatment.

Although these studies have indicated considerable switching among floral resources by generalist pollinators, many plants compete for pollinators and may suffer from pollen limitation (Knight et al., 2005b; R.J. Mitchell et al., 2009; Quesada et al., 2009). Pollen limitation is significantly positively related to plant species richness, making the most diverse communities most vulnerable to species decline and replacement as a result of reproductive failure (Vamosi et al., 2006; Quesada et al., 2009). Changes in relative abundances of specialist and generalist pollinators would influence vegetation structure and composition. Reduced abundances of specialist pollinators, in particular, likely would threaten survival of some plant species and result in shifts in plant community composition. Brosi and Briggs (2013) manipulated abundance of the most abundant *Bombus* species in field plots and found that loss of a single pollinator species reduced floral fidelity and proportions of conspecific pollen deposition by the remaining pollinators and reduced seed production by an abundant larkspur, *Delphinium barbeyi*. This experiment demonstrated impaired plant reproduction, even when abundances of other potentially effective pollinators were unchanged.

Heystek and Pauw (2014) addressed the importance of differential pollination among congeneric dominant *Erica* spp. in structuring plant communities in South Africa. They assigned the 680 species of *Erica* to three categories based on their pollination syndrome: 80% insect pollinated, 15% bird pollinated, and 5% wind pollinated. They found that *Erica* communities were assembled randomly with respect to pollination in six regions, but that communities in the remaining three regions showed significantly lower co-occurrence of pollinator syndromes than expected, consistent with their hypothesis that communities are structured by interspecific competition for pollinators. They tested this hypothesis further by transplanting species that shared pollination by the orange-breasted sunbird, *Anthobaphes violacea*. Pollination rate generally declined for both the resident and the introduced species as density of the resident species increased, supporting a conclusion that competition for pollinators contributes to plant community structure but is not the primary factor structuring *Erica* communities.

Pollination also contributes to production of fruits and seeds that support associated food webs. Many animal species depend on fruit and seed production, at least seasonally (see Section 4). Hence, pollination of fruiting plants has consequences not only for plant reproduction, but also for the survival of frugivores and seed predators (Bawa, 1990).

Pollinators can affect ecosystem energy and nutrient fluxes through transport of energy- and nutrient-rich nectar and pollen. Roubik (1989) calculated the effects of social bees on energy and nitrogen budgets of tropical forests in Central America. He estimated that 600 colonies km^{-2} harvested 1.4 x 10^7 kJ year^{-1} and disposed of an equivalent energy value represented by dead bees scattered on the ground within a few dozen meters of each nest. This value exceeded estimates of energy fixed annually by primary producers, indicating that the energetics of flowering are greatly underestimated (Roubik, 1989). The 600 colonies also distributed about 1800 kg trash (pupal exuviae and feces) ha^{-1} year^{-1}. At 4% nitrogen content, this represents a flux of 72 kg ha^{-1} year^{-1} or about 1% of above-ground nitrogen in biomass. Although this flux represents only a small proportion of above-ground nitrogen in biomass, its concentration near nests would enrich litter and stimulate decomposition locally (see Chapter 14). These effects of pollinators on community structure and ecosystem processes warrant further study.

4 TYPES AND PATTERNS OF SEED PREDATION AND DISPERSAL

The fate of seeds is critical to plant reproduction and demography. A variety of animals feeds exclusively or facultatively on fruits or seeds, limiting potential germination and seedling recruitment but often facilitating seed dispersal. Dispersal of seeds is necessary for colonization of new habitats and for escape from high mortality near parent plants, although relatively few studies have measured the advantages of seed dispersal to plant fitness (Howe and Smallwood, 1982). In contrast to pollination, effective seed dispersal, that results in successful germination, relies less on disperser specialization than on movement to suitable habitat

(Wheelwright and Orians, 1982). Seed dispersal mechanisms confer varying degrees of seed dispersal efficiency and advantages for seedling growth, depending on ecosystem conditions.

4.1 Seed Predator and Disperser Functional Groups

Seed dispersal can be accomplished through both abiotic and biotic mechanisms. Abiotic dispersal involves wind and water; biotic dispersal involves autogenic mechanisms, such as explosive fruits, and various animal agents, including insects, fish, reptiles, birds, and mammals. Dispersal by animals typically is a consequence of frugivory or seed predation, but some species acquire seeds or spores through external attachment by various kinds of clinging devices, for example, adhesive material or barbed spines. Seeds of a majority of plant species are dispersed by animals in many ecosystems (Howe and Smallwood, 1982).

Fruits and seeds are highly nutritive food resources, as a consequence of plant provision for germination and, often, attraction of dispersal agents. Many vertebrate species are obligate fruit- or seed-feeders (eg, many birds and rodents), whereas others feed primarily on other resources but exploit fruits and/or seeds when available (eg, many ungulates and primates) (Bravo, 2008; Thorsen et al., 2011). A large number of insect species are specialized to feed on or in seeds. For example, Turgeon et al. (1994) reported that more than 400 species of insects, representing seven orders, feed on conifer cones and/or seeds.

Seed predator and seed disperser functional groups can be distinguished on the basis of consumption of fruits or seeds versus transport of seeds. *Frugivores* feed on fleshy fruits and may terminate fruit or seed development (Sallabanks and Courtney, 1992), but many vertebrate frugivores (including fish, reptiles, birds, and mammals) consume entire fruits and disperse seeds that are adapted to survive passage through the digestive tract (Bravo, 2008; Correa et al., 2007; Crawley, 1989; de Souza-Stevaux et al., 1994;, M. Horn, 1997; Sallabanks and Courtney, 1992; Temple, 1977). *Seed predators* include a number of insect, bird, and rodent species that consume seeds where found. Some seed predators eat the entire seed, for example, vertebrates and ants, but others penetrate the seed coat and consume only the endosperm, for example, seed bugs (Lygaeidae and Coreidae) and weevils (Curculionidae), or develop and feed within the seed, for example, seed wasps (Torymidae) and seed maggots (Anthomyiidae) (J. Brown et al., 1979; Crawley, 1989; Honek et al., 2009; Louda et al., 1990b; Schowalter, 1993; Turgeon et al., 1994). *Seed cachers* eat some seeds and move others from their original location to storage locations. Although ants and rodents are best known for caching seeds (J. Brown et al., 1979), at least one carabid beetle, *Synuchus impunctatus*, caches seeds of *Melampyrum* spp. in hiding places after consuming the caruncle at the end of the seed (Manley, 1971). *Seed vectors* include primarily vertebrates that passively carry seeds adapted to adhere to fur or feathers or to pass through the digestive tract. Insects generally are too small to transport seeds in this way but often transmit spores of microorganisms adapted to adhere to insect exoskeletons or to pass through insect digestive systems.

These functional groups can be subdivided on the basis of pre- or postdispersal seed predation, seed size, etc. Predispersal frugivores and seed predators feed on the concentrated fruits and seeds developing on the parent plant, whereas postdispersal frugivores and seed predators must locate scattered fruits and seeds that have fallen to the ground. Fruit size is related to primary frugivores/seed dispersers. Tree species that produce large fruits or seeds typically attract large ungulates and primates that can consume these fruits and seeds.

Such trees become vulnerable to extinction when disappearance of large mammals (often as a result of overhunting by humans) restricts seed dispersal (Caughlin et al., 2015; Janzen and Martin, 1982; C. Johnson, 2009). Smaller frugivores or seed predators often cannot move such fruits or seeds adequate distances. Rodents and birds typically exploit larger seeds than do insects, and species within taxonomic groups also partition seeds on the basis of size (J. Brown et al., 1979; Davidson et al., 1984; Whitford, 1978).

Vertebrates are more likely to disperse seeds from consumed fruits than are insects, which (due to their small size) typically feed on portions of fruits and on or in seeds. However, dung beetles and ants may be important secondary dispersers, redistributing seeds from the dung of larger primary dispersers (Andresen, 2002; Martínez-Mota et al., 2004). Insects, especially ants, are more likely to disperse small seeds, particularly of plant species adapted for dispersal by ants (*myrmecochory*) (Leal et al., 2014).

4.2 Measurement of Seed Predation and Dispersal

Several methods have been used to measure seed predation and dispersal. Predispersal seed predation can be measured by marking fruits and/or seeds on the plant and observing their fate, using a life table approach (see Chapter 5). Mature fruits and seeds can be collected for emergence of seed predators (Steffan-Dewenter et al., 2001) or dissected or radiographed for identity and number of internal seed predators or evidence of endosperm digestion by hemipterans (Schowalter, 1993). Seed-piercing Hemiptera may leave detectable pectinases or stylet sheaths on the seed coat of consumed seeds (Campbell and Shea, 1990). Postdispersal seed predation can be measured by placing marked seeds on the ground and measuring rate of disappearance (C. Chapman and Chapman, 1996; Côté et al., 2005; Heithaus, 1981; Honek et al., 2009; O'Dowd and Hay, 1980; Schupp, 1988). Marked seeds (eg, with fluorescent dyes or stable isotopes) can be identified in caches or fecal material for assessment of seed dispersal rate (O'Dowd and Hay, 1980).

Insects generally are more important predispersal seed predators than are vertebrates, but vertebrates are more important postdispersal seed predators (Crawley, 1989; Davidson et al., 1984; Louda et al., 1990b; Schupp, 1988). Predispersal seed predators greatly reduce seed production efficiency and reduce the number of seeds available for postdispersal seed predators and dispersal. At the same time, frugivores and postdispersal seed predators consume colonized seeds and can significantly reduce populations of predispersal seed predators (Bravo, 2008; Coe and Coe, 1987; Herrera, 1989).

Seed dispersal is an important mechanism for plant colonization of new sites. However, dispersal also may increase seed and seedling survival. Schupp (1988) reported that vertebrate seed predators limited seed survival under the parent tree to 15% of marked seeds but that dispersal distances of only 5 m significantly increased seed survival to nearly 40% over a 7-month period (Fig. 13.13). C. Chapman and Chapman (1996) compared fruit and seed disappearance and survival of seeds remaining under the parent canopy for six tree species in a tropical forest in Uganda. Three of the six species showed higher rates of seed removal at locations away from the parent canopy, compared to locations under the parent canopy, whereas the other three species showed no difference in seed removal between locations. However, for two of the latter species, survival of transplanted seedlings was much higher under conspecific canopies than at locations away from conspecifics, but

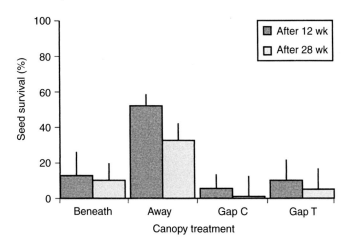

FIGURE 13.13 **Survival of *Faramea occidentalis* seeds beneath fruiting parent trees (Beneath), away from parent trees (Away; 5 m from crown perimeter of nearest fruiting adult), and within the canopy (Gap C) and trunk (Gap T) zones of treefall gaps on Barro Colorado, Panama.** Survival of seed was significantly ($p < 0.05$) higher 5 m from parent trees than beneath parent trees or in treefall gaps. *Source: Data from Schupp, E.W., 1988. Seed and early seedling predation in the forest understory and in treefall gaps. Oikos 51, 71–78.*

subsequent herbivory tended to be higher on seedlings under conspecific trees. Fruits not harvested by dispersers typically rot on the ground, destroying the seeds within (Janzen and Martin, 1982; Asquith et al., 1999). Ants often play a critical role in seedling survival and germination by foraging on fruit, cleaning seeds, and dispersing seeds to ant nests (Oliveira et al., 1995; Passos and Oliveira, 2003). Seeds not cleaned by ants succumb to decay. These results indicate that seed dispersal represents various trade-offs. Nevertheless, the efficiency with which seeds reach favorable sites is critical to plant population dynamics.

Seeds transported by wind or water often have low dispersal efficiency, for which plants must compensate by producing large numbers of seeds. Animals are presumed to be more efficient dispersal agents, but this may not always be accurate. Seeds drop from animal vectors with no more likelihood of landing on suitable germination sites than do seeds deposited by wind or water, unless animal dung, dens, or habitats provide more suitable germination sites. However, the direction of animal movement is more variable than that of wind or water. Birds, in particular, quickly cover large areas, but local seed redistribution by ants also can significantly affect plant demographies (Gorb and Gorb, 2003; O'Dowd and Hay, 1980).

A number of plant species are specifically adapted for seed dispersal by animals. Myrmecochorous species produce a lipid-rich *elaiosome* to attract ants, which move seeds variable distances, depending on whether the elaiosome is removed prior to or during transport or at the nest (Fig. 13.14, Gorb and Gorb, 2003; Leal et al., 2014). As described for plant production of rewards to attract pollinators (Section 2.2), variation in elaiosome rewards can result in differential attraction of efficient seed dispersers to seed species (Leal et al., 2014). Some species with large seeds or thick seed coats may show reduced dispersal or germination ability where

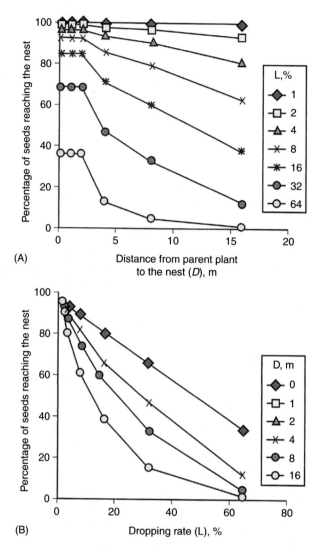

FIGURE 13.14 Relationship between seed number transported to ant nests and distance from the parent plant to the nest, for given diaspore dropping rates (A) and relationship between seed number transported to nests and dropping rate of diaspores, for given distances from the nest (B). *Source: From Gorb, E., Gorb, S., 2003. Seed Dispersal by Ants in a Deciduous Forest Ecosystem: Mechanisms, Strategies, Adaptations. Kluwer Academic Publishers, Dordrecht, The Netherlands.*

movement by animals or seed scarification is prevented (Culver and Beattie, 1980; Oberrath and Böhning-Gaese, 2002; Temple, 1977). However, many seeds are dispersed more passively by various animals, including secondary dispersers such as dung beetles that redistribute frugivore dung (Fig. 13.15).

Seed storage underground by ants and rodents may move seeds to sites of better soil conditions or reduce vulnerability to further predation. A number of studies have demonstrated

FIGURE 13.15   Dung beetles represent secondary dispersers of seeds in vertebrate dung.

that seedlings germinating in ant nests are larger and have higher survival rates than do seedlings emerging elsewhere (A. Andersen, 1988; Bennett and Krebs, 1987; Culver and Beattie, 1980; Rissing, 1986; D. Wagner, 1997). Ant nests may or may not enrich surrounding soils (Horvitz and Schemske, 1986; Westoby et al., 1991, see Chapter 14). Soil from ant nests often has significantly higher concentrations of nitrate, ammonium, phosphorus, and water and higher nitrogen mineralization rates than does soil away from nests (A. Andersen, 1988; Culver and Beattie, 1983; Herzog et al., 1976; Lesica and Kannowski, 1998; D. Wagner, 1997; D. Wagner et al., 1997). On the other hand, Rice and Westoby (1986), L. Hughes (1990), and Gorb and Gorb (2003) found that myrmecochorous plants do not necessarily show distribution patterns associated with soil fertility or with ant nests. Gorb and Gorb (2003) found that foraging *Formica polyctena* transported myrmecochorous seeds to territorial borders after removing the elaiosome, thereby distributing seeds widely, but nonmyrmecochorous seeds were transported to nests, where they remained, leading to increased competition between plants that grew on the mound.

Plants may benefit from seed deposition at suitable depths for germination or protected from intense predation by vertebrates (Cowling et al., 1994). Shea et al. (1979) found that germination of serotinous seeds of several legume species in Western Australia was enhanced by seed redistribution by ants to depths that were heated sufficiently, but protected from higher surface temperatures during high-intensity autumn fires. O'Dowd and Hay (1980) reported that transport of diaspores of *Datura discolor* by ants, to nests averaging only 2.3 m from the nearest plant, reduced seed predation by desert rodents from 25–43% of seeds in dishes under parent plants to < 1% of seeds in dishes near ant nests. Heithaus (1981) found that when seed dispersal by ants was experimentally prevented, rodents removed 70–84% of *Asarum canadense* and *Sanguinaria canadensis* seeds, compared to 13–43% of seeds lost when ants were present. Furthermore, laboratory experiments demonstrated that rodents located buried seeds less frequently than seeds on the surface and consumed buried seeds less often when elaiosomes were removed, as done by ants. L. Hughes (1990) reported that changes in nest structure, indicated by relocation of nest entrances, may provide refuges for seeds remaining in abandoned portions of nests and reduce seedling competition by preventing long-term concentration of seeds in localized sections of nests.

As found for pollination (Section 2.2), plants also may be subject to dispersal limitation (Pinto et al., 2014). For example, Pinto et al. (2014) added seeds of rare plant species and removed competitors and seed predators (mice) from experimental plots in a factorial design to evaluate the relative importance of seed limitation, competition, or seed predation on seedling recruitment and plant community structure. Added seeds germinated regardless of competition or seed predation treatment and increased species richness in seed addition plots, indicating that dispersal limitation was the primary factor affecting recruitment of rare species and overall plant community structure in their study. However, other studies have indicated significant effects of rodent removal on relative abundances of seedling species (J. Brown and Heske, 1990; Heske et al., 1993; Howe and Brown, 2001; R. Inouye et al., 1980).

4.3 Spatial and Temporal Patterns of Seed Predation and Dispersal

A number of factors influence rates of seed predation and dispersal. The extent of seed mortality, mechanism of seed transport, distance moved from the parent plant, attraction of particular dispersal agents, and thermodynamic constraints determine the probability that seeds will survive and be moved to suitable or distant locations.

Pollinators and seed predators can have opposing effects on seed production (Sallabanks and Courtney, 1992). Steffan-Dewenter et al. (2001) reported that pollinator activity decreased, but seed predation increased, on experimental *Centaurea jacea* plants, with distance from seminatural habitats in an agricultural landscape in Germany. Similarly, pre- and postdispersal seed predators can have opposing effects on seed dispersal. Bravo (2008) reported that black howler monkeys, *Alouatta caraya*, in Argentina did not avoid eating tree seeds that were heavily colonized by insects. Seeds containing early instars were defecated with the larvae killed, and the seeds were able to germinate. Seeds in which larvae had reached an advanced stage of development did not germinate. Depending on tree species and timing, ingestion of infested fruits by howler monkeys could result in destruction of infested seeds, killing of larvae in pulp or seeds and dispersal of healthy seeds, or spread of advanced larvae via defecation. On the other hand, K. Christensen and Whitham (1991) reported that seed-dispersing birds avoided foraging in pinyon pine trees in which the stem- and cone-boring moth, *Dioryctria albovittella*, had inhibited cone development and increased cone mortality. Perea et al. (2012) reported that rodents discriminated between sound acorns and acorns with weevil, *Curculio* sp., larvae inside. Acorns containing weevil larvae had a lower germination rate and were moved less often and dispersed shorter distances, compared to sound seed, thereby limiting their ability to colonize new environments.

Few studies have compared seed predation and dispersal among ecosystems. Different agents dominate these processes in different ecosystems (Moll and McKenzie, 1994). For example, dominant plant species in temperate ecosystems frequently have wind-dispersed seed, whereas plant species on oceanic islands often are water-dispersed (Howe and Smallwood, 1982). Howe and Smallwood (1982) concluded that consistently windy ecosystems promote wind-driven dispersal, whereas more mesic conditions promote animal-driven dispersal. Old World deserts have relatively few (< 5%) animal-dispersed plant species (Howe and Smallwood, 1982). More than 60% of temperate and tropical forest plant species are dispersed by animals (Howe and Smallwood, 1982). A variety of large vertebrate herbivores are important frugivores and seed dispersers in temperate and tropical ecosystems (Janzen and Martin, 1982; C. Johnson, 2009). Fruits and seeds in seasonally flooded tropical forests often are dispersed by fish during periods of inundation (de Souza-Stevaux et al., 1994; M. Horn, 1997; Howe and Smallwood, 1982). Bats and primates are more important frugivores and seed dispersers in tropical forests than in temperate ecosystems. Insects are ubiquitous frugivores and seed predators, but may be more important dispersers in grassland and desert ecosystems, where transport to ant nests may be critical to protection of seeds from vertebrate seed predators, from competition, and from fire (Louda et al., 1990b; Rice and Westoby, 1986).

Rice and Westoby (1986), Rissing (1986), and Westoby et al. (1991) discussed a number of potential factors affecting differences in the incidence of ant-dispersed seeds among biogeographic regions. Myrmecochory appears to be more prevalent in Australia and South Africa than in other regions. One hypothesis is that smaller plants (characteristic of arid biomes) generally are more likely to be ant-dispersed than are larger plants. A second hypothesis is that the relatively infertile soils of Australia and South Africa preclude nutrient allocation to fruit production, forcing plants to adapt to seed dispersal by ants rather than vertebrates. Finally, Australia and South Africa lack the large harvester ants, for example, *Pogonomyrmex* spp., *Messor* spp., and *Veromessor* spp., that are common in arid regions of North America and Eurasia. These ants consume relatively large seeds, limiting the value of an elaiosome as a food reward for seed dispersal.

Within biomes, disturbance affects patterns of seed predation and dispersal. Côté et al. (2005) reported that invertebrate predation on black spruce, *Picea mariana*, seeds was highest in recently burned sites, compared to undisturbed sites. On the other hand, Farwig et al. (2006) found that vertebrate frugivore abundances and dispersal of *Prunus africana* seeds were 1.5 times higher in disturbed areas than in forest in western Kenya.

Seed predators are capable of consuming or destroying virtually the entire production of viable seed of a given plant species in some years (Coe and Coe, 1987; Ehrlén, 1996; Honek et al., 2009; Horvitz et al., 2005; Robertson et al., 1990; Schowalter, 1993; Turgeon et al., 1994; Żywiec et al., 2013). The intensity of seed predation depends to a large extent on seed availability. Seed predators focus on the largest or most concentrated seed resources (Ehrlén, 1996). During years of poor seed production, most or all seeds may be consumed, whereas during years of abundant seed production, predator satiation enables many seeds to survive (Schowalter, 1993; Turgeon et al., 1994; Żywiec et al., 2013). Long-lived plant species need produce few offspring over time to balance mortality. Hence, many tree species produce abundant seed only once every several years, a strategy known as masting. Poor seed production during intervening years reduces seed predator populations and limits seed predation during mast years (Fig. 13.16). However, extended diapause by some frugivores and seed predators reduces the effect of masting by ensuring that portions of previous years' cohorts emerge during years of good seed availability (Schowalter, 1993; Żywiec et al., 2013).

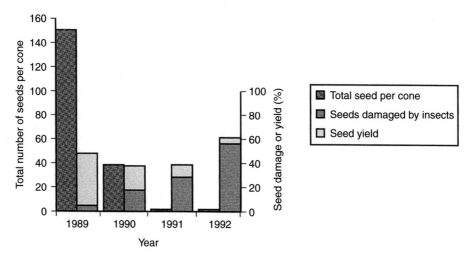

FIGURE 13.16   Relationship between total seed produced, seed loss to insects, and seed yield in a Douglas-fir seed orchard in western Oregon, USA. *Source: Data from Schowalter, T.D., 1993. Cone and seed insect phenology in a Douglas-fir seed orchard during three years in western Oregon. J. Econ. Entomol. 87, 758–765.*

5 EFFECTS OF SEED PREDATION AND DISPERSAL

Seed predators and dispersers influence plant population dynamics and community structure by affecting both seed survival and seedling recruitment (Effiom et al., 2014; Garzon-Lopez et al., 2015). Robertson et al. (1990) reported that predispersal seed predation rates varied widely among mangrove species at study sites in northeastern Australia. Three species (*Ceriops australis, C. tagal,* and *Rhizophora apiculata*) had fewer than 10% of seeds damaged by insects, whereas six species (*Avicennia marina, Bruguiera gymnorrhiza, B. parviflora, Heritiera littoralis, Xylocarpus australasicus,* and *X. granatum*) consistently had > 40% of seeds damaged. These mangrove species also showed variation in survival and growth rates (height and diameter) of seedlings from insect-damaged seeds. Schowalter (1993) and Żywiec et al. (2013) reported substantial variation in seed predation rates among years, resulting in improved survival of seeds produced by masting individuals during mast years. Ehrlén (1996) reported a significant positive correlation between the change in population growth rate and the reproductive value of seeds, as reduced by seed predation, indicating that survival of seeds and seedlings is the most important aspect of seed predator effects on plant population growth.

Postdispersal seed predators similarly affect the survival and growth of seeds and seedlings. Garzon-Lopez et al. (2015) reported that shared seed predators could either promote or reduce coexistence among plant species in tropical forests, depending on complex indirect effects on density, phenology, and spatial distribution of conspecific and heterospecific seeds. Côté et al. (2005) suggested that higher predation on black spruce seeds at recently burned sites may delay spruce regeneration on these sites.

Seeds selected for storage in ant nests or refuse piles often show increased survival and seedling growth, relative to seeds in control sites (A. Andersen, 1988; Culver and Beattie, 1980; L. Hughes, 1990; Rissing, 1986). Enhanced seedling growth on ant nests may reflect the higher

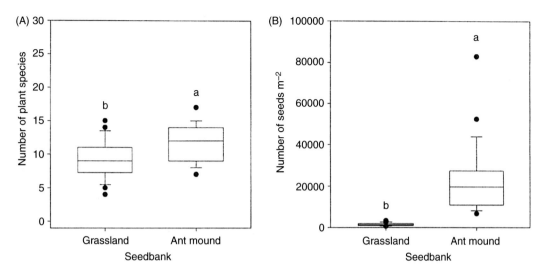

FIGURE 13.17 Seed ling (A) and bank (B) composition in samples from *Formica exsecta* mounds and surrounding alpine grassland in Switzerland. Box plots indicate median, 25th, and 75th percentiles, black lines are 10th and 90th percentiles, and filled circles are all values below 10th and above 90th percentiles. Different letters indicate significant differences ($p < 0.05$). *Source: From Schütz, M., Kretz, C., Dekoninck, L., Iravani, M., Risch, A.C., 2008. Impact of Formica exsecta Nyl. on seed bank and vegetation patterns in a subalpine grassland ecosystem. J. Appl. Entomol. 132, 295–305.*

nutrient concentrations (A. Andersen, 1988; Culver and Beattie, 1983; Herzog et al., 1976; D. Wagner, 1997; D. Wagner et al., 1997, but see Rice and Westoby, 1986) or greater water-holding capacity (Jonkman, 1978; D. Wagner, 1997) of ant nests.

The composition and seed preferences of the frugivore/granivore community affect plant community development (Garzon-Lopez et al., 2015; O. Lewis and Gripenberg, 2008; M. Schütz et al., 2008). M. Schütz et al. (2008) found that ants, *Formica exsecta*, altered vegetation structure around nests in alpine grasslands. Mound seed banks contained 15 times more seeds, 8 times more seeds of myrmecochorous plant species, and higher species richness than did samples from surrounding grassland. However, graminoids dominated edges of nest mounds, whereas legumes, shrubs, and myrmecochorous species dominated transect points at increasing distance from mound edges (Fig. 13.17). Similarly, G. Brown et al. (2012) found that nests of harvester ants, *Messor ebeninus*, provided hot spots for vegetation development in deserts in Kuwait, with nest plots showing 4-fold higher mean vegetation cover and 2-fold higher mean biomass than nonnest plots.

R. Inouye et al. (1980) reported that exclusion of granivorous rodents or ants altered densities and community composition of annual plant species in a desert site in New Mexico, USA (Table 13.1). Rodents preyed selectively on large-seeded species, for example, *Erodium* spp. and *Lotus humistratus*. In plots from which rodents were excluded, these species increased to dominate vegetative biomass and replace small-seeded plant species, especially *Euphorbia polycarpa*. Ants preyed most intensively on the most abundant plant species, *Filago californica*. When ants were excluded, this small-seeded composite became numerically dominant and reduced species diversity. J. Brown and Heske (1990) and Heske et al. (1993) reported that continued rodent exclusion over a 12-year period gradually transformed this desert site into grassland.

TABLE 13.1 Effects of Removal of Ants, Rodents, or Both on Densities of Certain Annual Plant Species, All Plants, Plant Biomass, and Two Measures of Species Diversity

	+ Rodents		− Rodents		Effects of removal of	
	+ Ants	− Ants	+ Ants	− Ants	Rodents	Ants
Initial Census Jan. 29, 1977						
1. Large plants	1.00 (35.8)	0.98	2.08	2.35	Increase **	NS
2. Small plants	1.00 (292.5)	3.30	3.32	3.17	NS	Increase **
Final Census Apr. 2, 1977						
3. *Erodium cicutarium* (seed mass = 1.6 mg)	1.00 (1.8)	1.83	7.03	16.11	Increase **	NS
4. *E. texanum* (seed mass =1.6 mg)	1.00 (0.6)	0.88	2.07	0.78	Increase *	NS
5. *Euphorbia polycarpa* (seed mass = 0.2 mg)	1.00 (0.6)	2.00	0.14	0.29	Decrease *	NS
6. *Filago californica* (seed mass = 0.04 mg)	1.00 (142.1)	1.90	1.43	2.59	NS	Increase *
7. *Lotus humistratus* (seed mass = 1.5 mg)	1.00 (11.4)	1.14	2.43	5.22	Increase **	NS
8. All plants	1.00 (209.6)	1.35	1.34	1.94	Increase *	Increase **
9. Dry mass (all species)	1.00 (5.8)	1.07	2.09	2.17	Increase **	NS
10. Species diversity (H′)	1.00 (2.78)	0.73	0.99	0.89	NS	Decrease *
11. Species evenness (E)	1.00 (0.53)	0.77	1.99	1.04	NS	Decrease *

Values given are ratios of treatment to control (+ Rodents + Ants) means; numbers in parentheses are mean values for unthinned plots except for plant biomass and the two measures of diversity, which are for control plots. Statistical analysis was by ANOVA; NS = not significant; *, significant at $p < 0.05$; **, significant at $p < 0.01$.

Reproduced from Inouye, R.S., et al., 1980. Effects of predation and competition on survivorship, fecundity, and community structure of desert annuals. Ecology 61, 1344–1351.

Many plant species have become dependent on animal mutualists for seed dispersal. Seed and seedling survival for some species depends on distance from parent plants, under which seed predation may be concentrated (O'Dowd and Hay, 1980; Schupp, 1988). As found by Powell and Powell (1987) and Steffan-Dewenter and Tscharntke (1999) for pollinators (see 2.2), decline in abundance of seed dispersal agents, especially as a result of overhunting or habitat loss, may threaten persistence of some plant species (Caughlin et al., 2015; Janzen and Martin, 1982; C. Johnson, 2009).

Plant species adapted for dispersal by vertebrates often have hardened seed coats to survive gut passage and may require scarification during passage through the digestive systems before germination is possible. Temple (1977) noted the coincidence between the age (300–400 years) of the last naturally regenerated tambalacoque trees, *Sideroxylon sessiliflorum* (= *Calvaria major*), and the disappearance of the dodo in 1680 on the South Pacific island of Mauritius. When *S. sessiliflorum* seeds were force-fed to turkeys (approximately the size of the dodo), the seed coats were sufficiently abraded during gut passage to permit germination, demonstrating a potential role of the dodo in dispersal and survival of this once-dominant tree. Although the primacy of the dodo's role in *S. sessiliflorum* survival has been challenged (Witmer and Cheke, 1991), it appears that *S. sessiliflorum* and other plant species have suffered from disappearance of seed-dispersing animals from Mauritius.

Janzen and Martin (1982) suggested that a number of tropical plants may show reduced seed dispersal as a result of the Pleistocene extinction of the large mammalian megafauna that likely fed on their fruits and dispersed seeds. Many large-fruited species experience high seed mortality in fruits rotting under trees in the absence of effective dispersal by extant (smaller) mammals (Asquith et al., 1999; Caughlin et al., 2015; Janzen and Martin, 1982; C. Johnson, 2009; Oliveira et al., 1995).

Effiom et al. (2014) compared effects of hunting on seedling composition of tropical forests in Nigeria. Seedling composition was significantly related to the presence of mammalian seed dispersers. In forests protected from hunting, seedling composition was dominated by primate-dispersed species, whereas in forests open to hunting, seedling composition was shifted toward dominance by abiotically dispersed species. They found no evidence that non-hunted animals compensated for loss of primary seed dispersers.

Disappearance of native ant seed dispersers as a result of habitat fragmentation or competition from invasive ant species (A. Suarez et al., 1998) similarly may threaten the survival of ant-dispersed plant species. On the other hand, seed dispersers have been shown to facilitate the spread of some exotic plant species (J.M.B. Smith, 1989).

The effects of seed predation and dispersal on nutrient cycling or other ecosystem processes have not been studied. However, these organisms affect nutrient cycling through the movement of nutrient-rich fruits and seeds. By dispersing fruits and seeds, frugivores in particular remove the large energy and nutrient pools in fruits from under parent trees and distribute these over a large area. Furthermore, as described for herbivores, seed predators and dispersers affect the spatial distribution of various plant species that differentially control nutrient fluxes.

6 SUMMARY

Insects are the major agents of pollination, seed predation, or seed dispersal in many ecosystems. Although few studies have evaluated the effects of pollinators, seed predators, and seed dispersers on ecosystem processes, these organisms often are critical to seedling recruitment and vegetation dynamics that affect other ecosystem processes.

Pollination is an important means of increasing genetic heterogeneity and improving plant fitness. Pollination can be accomplished by abiotic (wind) or biotic (insects, birds, and bats) agents. Wind pollination is inefficient but sufficiently effective for species that dominate temperate ecosystems. However, animal agents increase pollination efficiency for more isolated plants and are critical to survival of many plant species that typically occur as widely scattered individuals, especially in deserts and tropical forests. Pollinator functional groups can be distinguished on the basis of their degree of specialization on particular floral resources.

Seed predators often consume the entire reproductive effort of host plants. Predispersal seed predators typically focus on concentrated seed resources on the parent plants whereas postdispersal seed predators must locate more scattered seed resources on the ground. Insects are more important predispersal seed predators, but vertebrates are more important postdispersal seed predators in most ecosystems.

Seed dispersal is critical to plant species survival both because new habitats can be colonized and because seed relocation often improves seed and seedling survival. Seeds can be dispersed by abiotic (wind and water) or biotic (insect and vertebrate) agents. Animals can increase dispersal efficiency by moving seeds to more suitable germination sites, especially if seeds are buried. Ants, in particular, can increase seed survival and seedling growth by relocating seeds to nests, where seeds are protected from further predation, from suboptimal surface conditions, and from competition with parent plants. Ant nests also may provide more suitable soil conditions for germination and growth. Some seeds require scarification of hard seed coats and must pass through vertebrate digestive systems before germination can occur.

Both pollination and seed dispersal affect plant population and community dynamics. Differential pollination, seed predation, and seed dispersal efficiencies among plant species affect seedling recruitment and growth. Survival of some plant species depends on sufficient abundance of pollinators and/or seed dispersers. However, research should address the extent to which pollinators, seed predators, and seed dispersers affect ecosystem processes.

Decomposition and Pedogenesis

DETRITIVORE EFFECTS ON LITTER DECOMPOSITION

Detritivores have the best-demonstrated effects on ecosystem processes, due largely to their accessibility and ease of manipulation (Coleman et al., 2004). Given that all energy fixed in primary production eventually becomes available to detritivores and decomposers, either as producer or consumer biomass (Chapter 11), the critical role of detritivores and decomposers in the turnover of detritus for re-use of nutrients by autotrophs was recognized early by ecosystem ecologists (Engelmann, 1961, 1968).

Although some vertebrates function as scavengers of vertebrate carcasses, virtually all detritivores are arthropods, and most are small. The largest is the elephant beetle, *Megasoma elephas*, of Central and South America, larvae of which feed in coarse woody debris. Although individually small, the biomass of detritivorous arthropods typically approaches, or exceeds, biomass of above-ground consumers (Table 11.1). Detritivores as a group are responsible for the fragmentation of plant and animal debris and inoculation with saprophagous microorganisms that complete the decomposition and mineralization that make nutrients from detritus available for plant uptake (Figs. 1.3, 11.8, and 11.9).

Detritivores are responsible for up to 80% of total decay rate (Coleman et al., 2004; Seastedt, 1984; Vossbrinck et al., 1979). In the absence of detritivores, biotic detritus would accumulate, creating a bottleneck in nutrient cycling and inhibiting further primary production (Setälä and Huhta, 1991; T. Wood et al., 2009). Unfortunately, some important detritivores, for example, cockroaches, termites, mosquitoes, dung and carrion flies, and xylophagous beetles, are considered nuisances and subject to widespread control efforts, potentially interfering with their important roles in ecosystems. Such trade-offs require consideration of the importance of these insects to ecosystem services (Chapter 16) and integrated pest management approaches that minimize their threats to human interests without jeopardizing their ecological functions (Chapter 17).

1 INTRODUCTION

Decomposition is the breakdown of dead organic matter that eventually results in release of CO_2, other organic trace gases, water, mineral nutrients, and energy, reversing the process by which energy and matter were fixed in primary production. Pedogenesis (soil development)

Insect Ecology. http://dx.doi.org/10.1016/B978-0-12-803033-2.00014-5

largely reflects the activities of animals, including those involved in decomposition, that mix organic matter with mineral soil. These two processes contribute greatly to the capacity of a site to support primary productions (T. Wood et al., 2009). Accumulated organic litter represents a major pool of energy, water, and nutrients in many ecosystems. Carbon and other nutrients released through decomposition can be acquired by plants or microbes or returned to abiotic pools (see Chapter 11). Incorporation of nutrients and decay-resistant organic matter into soil increases fertility, aeration, and water-holding capacity. Release of CO_2, CH_4, and other trace gases affects atmospheric conditions and global climate.

Decomposition can be categorized into four component processes: *photo-oxidation*, the abiotic catabolism resulting from exposure to solar radiation; *leaching*, the loss of soluble materials as a result of percolation of water through material; *comminution*, the fragmentation of organic litter, largely as a result of detritivory; and *mineralization,* the catabolism of organic molecules by microorganisms. Vossbrinck et al. (1979) found that when arthropods and microbes were excluded, grassland detritus lost only 5% mass, due entirely to leaching and/or photo-oxidation. A variety of arthropods are the primary detritivores in most ecosystems. The feeding and burrowing activities of many animals, including ants, termites, and other arthropods, redistribute and mix soil and organic material. Burrowing also increases soil porosity, thereby increasing aeration and water-holding capacity.

The effects of arthropod detritivores and burrowers on decomposition and soil development have been the most widely studied effects of arthropods on ecosystem processes (Ausmus, 1977; N. Clay et al., 2013; Coleman et al., 2004; Crossley, 1977; Eldridge, 1994; Eldridge et al., 2009; Graça and Poquet, 2014; Jamali et al., 2011; Seastedt, 1984; Swift, 1977; Swift et al., 1979; J. Wallace and Webster, 1996; Whitford, 2000; Whitford and Eldridge, 2013; Wotton et al., 1998; Yamada et al., 2007). Arthropod detritivores and burrowers are relatively accessible and can be manipulated easily for experimental purposes. Their key contributions to decomposition and mineralization of litter and pedogenesis have been demonstrated in virtually all ecosystems. Indeed, some aquatic and glacial ecosystems consist entirely of arthropod detritivores and associated microorganisms feeding on allochthonous detritus (J. Edwards and Sugg, 1990; Oertli, 1993; J. Wallace et al., 1992). Effects of detritivorous and fossorial species on decomposition and soil mixing depend on the size of the organism, its food source, type and rate of detritivory, volume of displaced litter or soil, and type of saprophytic microorganisms inoculated into litter. Although most studies have addressed the effects of detritivores and burrowers on soil processes, some have documented effects on primary production, as well.

2 TYPES AND PATTERNS OF DETRITIVORY AND BURROWING

2.1 Detritivore and Burrower Functional Groups

Functional groups of detritivorous and fossorial arthropods have been distinguished on the basis of principal food source, mode of feeding, and microhabitat preferences (J.C. Moore et al., 1988; J. Wallace et al., 1992). For example, functional groups can be distinguished on the basis of seasonal occurrence, habitats, and substrates (eg, terrestrial vs aquatic, animal vs plant detritus, foliage vs wood, arboreal vs fossorial), or particular stages in the decomposition process (N. Anderson et al., 1984; C. Hawkins and MacMahon, 1989; Schowalter and Sabin, 1991; Schowalter et al., 1998; Seastedt, 1984; Siepel and de Ruiter-Dijkman, 1993; Tantawi et al., 1996; Tullis and Goff, 1987; J. Wallace et al., 1992; Winchester, 1997; Zhong and Schowalter, 1989).

General functional groupings for detritivores are based on their effect on decomposition processes. *Coarse* and *fine comminuters* are instrumental in the fragmentation of litter material. Major taxa in terrestrial ecosystems include earthworms, millipedes, termites, and beetles (coarse) and mites, collembolans, and various other small arthropods (fine). Many species are primarily *fungivores* or *bacteriovores* that fragment substrates while feeding on the surface microflora. Many fungivores and bacteriovores, including nematodes and protozoa, as well as arthropods, feed exclusively on microflora and affect the abundance and distribution of these decomposers (Santos et al., 1981). A number of species, including dung beetles, millipedes, and termites, are *coprophages,* either feeding on feces of larger species or reingesting their own feces following microbial decay and enrichment (Cambefort, 1991; Coe, 1977; Dangerfield, 1994; Holter, 1979; Kohlmann, 1991; McBrayer, 1975; Whitford, 1986; Whitford et al., 1982; Yamada et al., 2007).

In aquatic ecosystems *scrapers* (including mayflies, caddisflies, chironomid midges, and elmid beetles), that graze or scrape microflora from mineral and organic substrates, and *shredders* (including stoneflies, caddisflies, crane fly larvae, crayfish, and shrimp), that chew or gouge large pieces of decomposing material, represent coarse comminuters; *gatherers* (including stoneflies, mayflies, and copepods), that feed on fine particles of decomposing organic material deposited in streams, and *filterers* (mayflies, caddisflies, and black flies), that have specialized structures for sieving fine suspended organic material, represent fine comminuters (Cummins, 1973; J. Wallace and Webster, 1996; J. Wallace et al., 1992).

Xylophages are a diverse group of detritivores specialized to excavate and fragment woody litter. Aquatic xylophages include crane flies and elmid beetles. Terrestrial taxa include curculionid, buprestid, cerambycid, anobiid, bostrichid, and lyctid beetles, siricid wasps, carpenter ants, *Camponotus* spp., and termites (Fig. 14.1), with different species often specialized on particular wood species, sizes, or stages of decay (see Chapter 10). Most of these species either feed on fungal-colonized wood or support mutualistic, internal, or external fungi or bacteria that digest cellulose and enhance the nutritional quality of wood (Breznak and Brune, 1994; Mankowski et al., 1998; Siepel and de Ruiter-Dijkman, 1993, see Chapter 8).

Necrophages, or *carrion feeders*, represent another specialized group that breaks down animal carcasses. Major taxa include calliphorid, muscid, and sarcophagid flies, staphylinid, silphid, scarabaeid, and dermestid beetles, and various ants. Different species typically specialize on particular stages of decay (Figs. 10.7,10.8) and on particular animal groups, for example, reptiles versus mammals (Tantawi et al., 1996; E. Watson and Carlton, 2003, see Chapter 10).

An important consequence of litter fragmentation by arthropods is increased surface area for microbial colonization and decomposition. Microbes also are redistributed, either passively through transport of microbes acquired during feeding or dispersal or actively through inoculation of mutualistic associates, to fresh surfaces during feeding (see Chapter 8).

Many detritivores redistribute large amounts of soil or detritus during foraging or feeding activities (Kohlmann, 1991). However, nondetritivores also contribute to mixing of soil and organic matter. Fossorial functional groups can be distinguished on the basis of their

FIGURE 14.1 *Melanophila* **sp. (Coleoptera: Buprestidae) larva in mine in phloem of recently killed Douglas-fir tree in western Oregon, USA.** The entire phloem volume of this tree has been fragmented and converted to frass packed behind mining larvae of this species, demonstrating detritivore capacity to reduce detrital biomass.

food source and mechanism and volume of soil/detrital mixing. *Subterranean nesters* burrow primarily for shelter. Vertebrates, for example, ground squirrels, woodrats, badgers, and foxes, and many invertebrates, including crickets and solitary wasps, excavate tunnels of various sizes, typically depositing soil on the surface and introducing some organic detritus into nests. *Gatherers*, primarily social insects, actively concentrate organic materials in colonies excavated in the substrate. Ants and termites redistribute large amounts of soil and organic matter during construction of extensive subterranean, surficial, or arboreal nests (J. Anderson, 1988; Haines, 1978). Subterranean species concentrate organic matter in nests excavated in soil, but many species bring fine soil particles to the surface and mix soil with organic matter in arboreal nests or foraging tunnels. These insects can affect large volumes of substrate (up to 1000 m^3), especially as a result of restructuring and lateral movement of the colony (L. Hughes, 1990; J. Moser, 1963, 2006; Tschinkel, 1999, 2004, 2005; Whitford et al., 1976). *Fossorial feeders*, such as gophers, moles, earthworms, mole crickets (Gryllotalpidae), and benthic invertebrates feed on subsurface resources (plant, animal, or detrital substrates) as they burrow, constantly mixing mineral substrate and organic material in their wake.

2.2 Measurement of Detritivory, Burrowing, and Decomposition Rates

Evaluation of the effects of detritivory and burrowing on decomposition and soil mixing requires appropriate methods for measuring rates of these processes. Several methods have been used to measure rates of leaf litter and carrion decomposition and soil mixing (Coleman et al., 2004; Tantawi et al., 1996). Measurement of wood decomposition presents special problems, including the long time frame of wood decomposition and the logistical difficulties of experimental placement and manipulation of large, heavy material. Decomposition of large woody debris represents one of the longest ecological processes, often spanning centuries (Harmon et al., 1986). This process traditionally was studied by comparing mass of wood of an estimated age to the mass expected for the estimated original volume, based on particular tree species. However, decomposition of some wood components begins only after lag times of up to several years, decomposition of standing tree boles is much slower than that of fallen boles, and differences in chemistry and volume between bark and wood components affect overall decay rates (Harmon et al., 1986; Schowalter et al., 1998).

Detritivory can be measured by providing experimental substrates and measuring colonization and consumption rates. K. Johnson and Whitford (1975) measured the rate of termite feeding on an artificial carbohydrate source and natural substrates in a desert ecosystem. Edmonds and Eglitis (1989) and Zhong and Schowalter (1989) measured the rate of woodborer colonization and excavation in freshly cut tree boles. Dissection of wood samples may be necessary for measurement of excavated volume for small insects. Radiography can be used to measure larger volumes, for example, termite galleries.

Detritivory often has been estimated by multiplying the per capita feeding rate for each functional group by its abundance (N. Anderson et al., 1984; Cárcamo et al., 2000; Crossley et al., 1995; Dangerfield, 1994). Cárcamo et al. (2000) estimated consumption of conifer needle litter by the millipede, *Harpaphe haydeniana*, at about 90 mg g^{-1} animal biomass per day, a rate that could account for the processing of 36% of annual litterfall. Laboratory conditions, however, might not represent the choices of substrates available under field conditions. For example, Dangerfield (1994) noted that laboratory studies might encourage coprophagy by millipedes by restricting the variety of available substrates, thereby overrepresenting this aspect of consumption. Mankowski et al. (1998) used both forced-feeding and choice tests to measure wood consumption by termites, *Zootermopsis angusticollis*, when a variety of substrate types was available or restricted.

Radioisotope movement from litter provided early data on decomposition rate (Witkamp, 1971). Stable isotopes (eg, ^{13}C, ^{14}C, and ^{15}N) are becoming widely used to measure fluxes of particular organic fractions (Ågren et al., 1996; Andreux et al., 1990; Horwath et al., 1996; Mayer et al., 1995; Šantrůčková et al., 2000; Spain and Le Feuvre, 1997; Wedin et al., 1995). The most widely used techniques for measuring decomposition rates in terrestrial and aquatic ecosystems involve measurement of respiration rate, comparison of litterfall and litter standing crop, and measurement of mass loss (J. Anderson and Swift, 1983; Bernhard-Reversat, 1982; Seastedt, 1984; Witkamp, 1971; Woods and Raison, 1982). These techniques tend to oversimplify representation of the decomposition process and, consequently, yield biased estimates of decay rate.

Respiration from litter or soil represents the entire heterotrophic community, as well as living roots. Most commonly, a chamber containing soda lime or a solution of NaOH is sealed over litter for a 24-h period, and CO_2 efflux is measured as the weight gain of soda lime or volume of acid neutralized by NaOH (N. Edwards, 1982). Comparison of respiration rates between plots with litter present and plots with litter removed provides a more accurate estimate of respiration rates from decomposing litter, but separation of litter from soil is difficult and often arbitrary (J. Anderson and Swift, 1983; Woods and Raison, 1982). More recently, gas chromatography and infrared gas analysis (IRGA) have been used to measure CO_2 efflux (Nakadai et al., 1993; Parkinson, 1981; Raich et al., 1990).

The ratio of litterfall mass to litter standing crop provides an estimate of the decay constant, k, when litter standing crop is constant (Olson, 1963). Decay rate can be calculated if the rate of change in litter standing crop is known (Woods and Raison, 1982). This technique also is limited by the difficulty of separating litter from underlying soil for mass measurement (J. Anderson and Swift, 1983; Spain and Le Feuvre, 1987; Woods and Raison, 1982).

Weight loss of fine litter has been measured using tethered litter, litterbags, and litter boxes. Tethering allows litter to take a natural position in the litterbed and does not restrict detritivore activity or alter microclimate, but is subject to loss of fragmented material and to difficulty in separating litter in late stages of decay from surrounding litter and soil (N. Anderson et al., 1984; Birk, 1979; Witkamp and Olson, 1963; Woods and Raison, 1982).

Litterbags provide a convenient means for studying litter decomposition in terrestrial or aquatic ecosystems (Crossley and Hoglund, 1962; C. Edwards and Heath, 1963; Graça and Poquet, 2014). Litterbags retain selected litter material, and mesh size can be used to selectively restrict entry by larger detritivore functional groups (C. Edwards and Heath, 1963; D. Wise and Schaefer, 1994). However, litterbags may alter litter microclimate and restrict detritivore activity, depending on litter conformation and mesh size. Moisture retention between flattened leaves apparently is independent of mesh size. Exclusion of larger detritivores by small mesh sizes has little effect, at least until litter has been preconditioned by microbial colonization (J. Anderson and Swift, 1983; Macauley, 1975; O'Connell and Menagé, 1983; Spain and Le Feuvre, 1987; Woods and Raison, 1982). Exclusion of predators by small mesh sizes can significantly affect detritivore abundances and decomposition processes (M.D. Hunter et al., 2003). Large woody litter (eg, tree boles) also can be enclosed in mesh cages for experimental restriction of colonization by xylophagous insects (Harmon et al., 1994; Ulyshen, 2014; Zhong and Schowalter, 1989). The potential interference with decomposition by small mesh sizes has been addressed in some studies by minimizing leaf overlap (and prolonged moisture retention) in larger litterbags, using small mesh on the bottom to retain litter fragments and large mesh on the top to maximize exchange of moisture and detritivores, and measuring decomposition over several years to account for differences due to changing environmental conditions (J. Anderson et al., 1983; Cromack and Monk, 1975; Woods and Raison, 1982, 1983). Despite limitations, litterbags have been the simplest and most widely used method for measuring decomposition rates and probably provide reasonably accurate estimates (Seastedt, 1984; Spain and Le Feuvre, 1987; Woods and Raison, 1982).

More recently, litter boxes have been designed to solve problems associated with litterbags. Litter boxes can be inserted into the litter, with the open top providing unrestricted exchange of moisture and detritivores (Seastedt and Crossley, 1983), or used as laboratory microcosms to study effects of decomposers (Haimi and Huhta, 1990; Huhta et al., 1991). Similar constructions can be incorporated into streams for assessment of detrital decomposition (March et al., 2001).

Abundances of detritivore functional groups can be manipulated directly to some extent by use of microcosms (Setälä and Huhta, 1991; Setälä et al., 1996), selective biocides, or other exclusion techniques (Crossley and Witkamp, 1964; C. Edwards and Heath, 1963; G. González and Seastedt, 2001; E. Ingham, 1985; E. Ingham et al., 1986; Macauley, 1975; Pringle et al., 1999; Santos and Whitford, 1981; Seastedt and Crossley, 1983; J. Wallace and Webster, 1996; J. Wallace et al., 1991; Zhong and Schowalter, 1989) or by adding or simulating detritivores in new substrates (Barker, 2008; G. González and Seastedt, 2001; Progar et al., 2000). Naphthalene and chlordane in terrestrial studies (Crossley and Witkamp, 1964; Santos and Whitford, 1981; Seastedt and Crossley, 1983; Whitford, 1986) and methoxychlor or electric fields in aquatic studies (Pringle et al., 1999; J. Wallace et al., 1991) have been used to exclude arthropods. However, E. Ingham (1985) reviewed the use of selective biocides and concluded that none had effects limited to a particular target group, limiting their utility for evaluating effects of individual functional groups. Furthermore, Seastedt (1984) noted that biocides provide a carbon and, in some cases, nitrogen source that may alter the activity or composition of microflora. Mesh sizes of litterbags can be manipulated to exclude detritivores larger than particular sizes, but this technique often alters the litter environment and may reduce fragmentation, regardless of faunal presence (Seastedt, 1984). In lieu of manipulating xylophage colonization of woody litter, effects have been simulated by drilling holes into phloem or sapwood to represent bark beetles and ambrosia beetles, respectively (Progar et al., 2000).

Few experimental studies have compared effects of manipulated abundances of xylophagous insects on wood decomposition (Edmonds and Eglitis, 1989; Progar et al., 2000; Zhong and Schowalter, 1989). Some studies have compared species or functional group abundances in wood of an estimated age or decay class, but such comparison ignores the effect of initial conditions on subsequent community development and decomposition rate. Prevailing weather conditions, the physical and chemical condition of the wood at the time of plant death, and prior colonization determine the species pools and establishment of potential colonists. Penetration of the bark and transmission by wood-boring insects generally facilitate microbial colonization of subcortical tissues (Ausmus, 1977; Dowding, 1984; Swift, 1977). Käärik (1974) reported that wood previously colonized by mold fungi (Ascomyctina and Fungi imperfecti) was less suitable for establishment by decay fungi (Basidiomycotina) than was uncolonized wood. Mankowski et al. (1998) reported that wood consumption by termites was affected by wood species and fungal preconditioning. Hence, experiments should be designed to evaluate effects of species or functional groups on decomposition over long time periods using wood of standard size, composition, and condition (Progar et al., 2000).

Assessing rates of burrowing and mixing of soil and litter is even more problematic. Nest structure of social insects has been revealed by pouring liquid latex, plaster, or metal into nest entrances and excavating after solidification (Tschinkel, 2004, 2005). Larger or deeper nests require excavation by construction equipment (J. Moser, 1963, 2006; Tschinkel, 1999; Whitford et al., 1976). Excavation can be extremely labor-intensive. Leaf-cutting ant, *Atta texana*, colonies can be 20–30 m in diameter, with as many as 170 chambers and vertical tunnels extending perhaps as deep as 32 m (Fig. 14.2, J. Moser, 2006). The difficulty of separating litter from soil, except where fungal gardens are in discrete chambers, limits measurement of mixing. Tunneling through woody litter presents similar problems. Zhong and Schowalter (1989) dissected decomposing tree boles to assess volume of wood excavated and/or mixed among bark, wood, and fecal substrates.

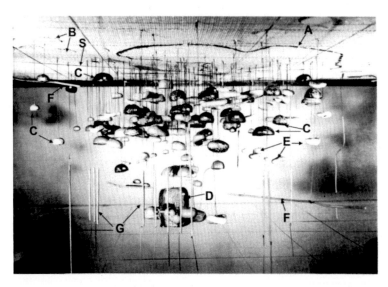

FIGURE 14.2   **Model of a Texas leaf-cutter ant, *Atta texana*, colony excavated in northern Louisiana, showing the funnel-shaped structure of the underground nest.** (A) Edge of the excavated surface subsoil, (B) feeder holes on the surface, (C) fungus garden cavities, showing typical inverted hemispherical shape (black cavities are largely full of fungus), (D) central cavity at bottom of nest, (E) irregularly shaped detritus cavities, (F) horizontal tunnels, (G) vertical tunnels, and (S) south end of N–S line across ground surface of nest. *Source: From Moser, J.C., 2006. Complete excavation and mapping of a Texas leafcutting ant nest. Ann. Entomol. Soc. Am. 99, 891–897.*

2.3 Spatial and Temporal Patterns in Processing of Detritus and Soil

Decomposition rate varies considerably, depending on substrate chemistry and environmental conditions. Some materials are decomposed more readily than are others, some processes release carbon primarily as methane, rather than carbon dioxide, and some enter long-term storage as humus, peat, coal, or oil. Litter quality (especially lignin, nitrogen and, perhaps, sodium content), moisture, oxygen supply, and the composition of the detritivore/decomposer community are the primary factors controlling decomposition rate (Aerts, 1997; Birk, 1979; N. Clay et al., 2014; Cotrufo et al., 1998; Fogel and Cromack, 1977; Fonte and Schowalter, 2004; G. González and Seastedt, 2001; Jabiol et al., 2014; Kaspari et al., 2009; Kurokawa and Nakashizuka, 2008; Meentemeyer, 1978; Patrick and Fernandez, 2013; Progar et al., 2000; Seastedt, 1984; Tian et al., 1995; Treseder, 2008; Vos et al., 2013; Whiles et al., 1993; Whitford et al., 1981).

Decomposition rate generally increases with litter nitrogen content and decreases with lignin content, but can be inhibited at low C/N ratios (Hobbie, 2008; Knorr et al., 2005; Treseder, 2008). Animal carrion, except for bones, is readily digestible by many organisms and decomposes rapidly (Payne, 1965), whereas plant materials, especially those composed largely of lignin and cellulose, can be decomposed only by relatively few species of fungi, bacteria, or protozoa and may require long time periods for complete decomposition (Harmon et al., 1986). Conifer litter tends to decompose more slowly than does angiosperm litter because of low nitrogen content and high lignin content (Harmon et al., 1986). Low soil or litter pH inhibits decomposition (Kappes et al., 2007). Decomposition rate generally increases with moisture content, but saturation with water inhibits decomposition, because of limited oxygen availability (Progar et al., 2000; Ulyshen, 2014). Submerged litter is degraded primarily by aquatic gougers and scrapers that slowly fragment and digest consumed organic matter from the surface inward (N. Anderson et al., 1984).

Nutrient-rich animal wastes can stimulate the decomposition rate of plant litter (see Chapter 12). N. Clay et al. (2013) reported that arboreal nests of canopy ants, *Azteca trigona*, provided a constant stream of excreta and other refuse to the forest floor of a tropical rainforest in Panama. Ant refuse had a 3-fold higher concentration of N, 6-fold higher concentration of P, and 20-fold higher concentration of K, compared to leaf litter. Leaf litter decomposed 20% faster below *A. trigona* nests, and supported higher abundances of invertebrate detritivores and predators, compared to controls 10 m away. Fonte and Schowalter (2005) manipulated herbivory by walkingsticks, *Lamponius portoricensis*, on *Piper glabrescens* in a tropical rainforest in Puerto Rico and found that increased input of feces stimulated litter decomposition rate. However, Schowalter et al. (2011) reported that an experimental 10-fold increase in input of herbivore feces alone in this same forest reduced litter decomposition rate, perhaps because this level of N flux was sufficient to inhibit microbial decomposition activity at this site (see Lodge et al., 2008; Treseder, 2008). Kaspari et al. (2009) and N. Clay et al. (2014, 2015) found that addition of Na, to simulate patchy urine deposition, to litter attracted invertebrate detritivores, especially termites, and increased litter decomposition rate by 43% in a Na-limited inland tropical forest, but not in a nonlimited coastal tropical forest.

Decomposition processes differ among ecosystem types. Physical factors may predominate in xeric ecosystems where decomposition reflects catabolic effects of ultraviolet light on exposed litter. Biological factors are favored by warm, moist conditions. Decomposition is most rapid in wet tropical ecosystems, where litter disappears quickly, and slowest in desert, tundra, and boreal ecosystems because of dry or cold conditions. González and Seastedt (2001) and Heneghan et al. (1999) compared decomposition of a common litter species between tropical and temperate ecosystems and demonstrated that decomposition was consistently higher in the tropical wet forests. Nevertheless, decomposition may continue underground, or under snow in tundra and boreal regions, where temperature and moisture are adequate (Santos et al., 1981). Decomposition rates may be lower in aquatic ecosystems, or very wet terrestrial ecosystems, as a result of litter saturation and limited oxygen supply. Low decomposition rates generally result in the accumulation of large standing crops of woody and fine litter.

Different groups of detritivores and decomposers dominate different ecosystems. For example, shredders and gatherers were more abundant in pools and headwater streams, characterized by substantial inputs of largely unfragmented organic matter, whereas filter-feeders were more abundant in high-gradient sections or higher-order streams (the Little Tennessee River), characterized by highly fragmented, suspended organic matter (Fig. 14.3). Fungi and associated fungivores, for example, oribatid mites and Collembola, are more prevalent in forests, whereas bacteria, bacteriovores, especially prostigmatid mites and Collembola, and earthworms are more prevalent in grasslands (Seastedt, 2000). Termites are the most important detritivores in arid and semiarid ecosystems and may largely control decomposition processes in forest and grassland ecosystems (K.E. Lee and Butler, 1977; Whitford, 1986). J. Jones (1989, 1990) reported that termites in dry tropical ecosystems in Africa so thoroughly decomposed organic matter that little or no carbon was incorporated into the soil. Wood-boring insects occur only in ecosystems with woody litter accumulation and are vulnerable to loss of this resource in managed forests (Grove, 2002). Dung feeders are important in ecosystems where vertebrate herbivores are abundant (Coe, 1977; Holter, 1979). Some patches may become "hot spots" for decomposition and nutrient fluxes. For example, animal carcasses are nutrient-rich resources whose decomposition

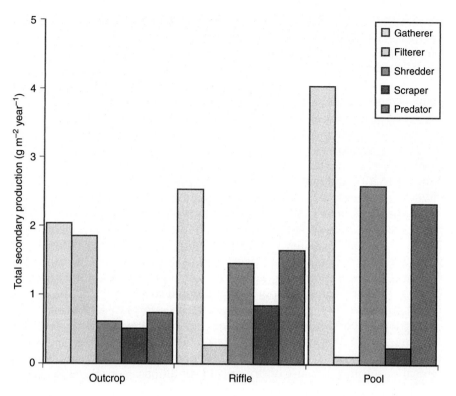

FIGURE 14.3   **Annual secondary production for aquatic functional groups in bedrock outcrop, riffle, and pool habitats of upper Ball Creek, North Carolina, during Jul. 1983–Jun. 1984.** *Source: Data from Huryn, A.D., Wallace, J.B., 1987. Local geomorphology as a determinant of macrofaunal production in a mountain stream. Ecology 68, 1932–1942.*

and mineralization provide pulses of nutrients to the surrounding area (eg, Menninger et al., 2008; Pray et al., 2009; Schowalter et al., 1998; Towne, 2000; Yang 2004, 2013).

Distinct temporal patterns in decomposition rates often reflect either the preconditioning requirements for further degradation or the inhibition or facilitation of new colonizers by established groups. For example, leaching of toxic defensive chemicals may be necessary before many groups are able to colonize litter (Barz and Weltring, 1985). M. Hulme and Shields (1970) and Käärik (1974) reported that wood decay is inhibited by competition for labile carbohydrates between colonizing decay fungi and established nondecay fungi. On the other hand, Blanchette and Shaw (1978) found that decay fungus growth in wood with bacteria and yeasts was twice that in wood without bacteria and yeasts, presumably because bacteria and yeasts provide fixed nitrogen, vitamins, and other nutrients while exploiting carbohydrates from lignocellulose degradation. Microbes typically require bark penetration, and often inoculation, by insects in order to colonize woody litter. Many saprophagic arthropods require some preconditioning of litter by bacteria, fungi, or other arthropods prior to feeding. Small comminuters typically feed on fragments or feces left by larger comminuters (O'Connell and Menagé, 1983). Inkley et al. (2008) reported that arthropod shredders in temporary wetlands responded to fungal and bacterial biomass that was higher on detritus that was consistently wet more than it was on detritus subjected to frequent wetting and drying. In turn, shredders convert coarse particulate organic matter (CPOM) to fine particulate organic matter (FPOM) that can be acquired by filterers (Patrick and Fernandez, 2013; J. Wallace and Webster, 1996; J. Wallace et al., 1991). Santos and Whitford (1981) reported that a consistent succession of microarthropods was related to the percentage of organic matter lost.

Decomposition often begins long before detritus reaches the ground. Considerable detrital accumulation occurs in forest canopies, as litter collects on large branches or crotches (Coxson and Nadkarni, 1995; Lindo and Winchester, 2007; Paoletti et al., 1991). Processes of decomposition and pedogenesis in these suspended sediments are poorly known. Paoletti et al. (1991) reported that suspended soils associated with bromeliads in a Venezuelan cloud forest had higher concentrations of organic matter, nitrogen, calcium, and magnesium and higher densities (based on bulk density of soil) of macro- and microinvertebrates than did forest floor soils. Lindo and Winchester (2007) found that suspended soils in old-growth western redcedar, *Thuja plicata*, canopies in western Canada had higher concentrations of nitrogen, phosphorus, and potassium, but lower moisture content and oribatid mite abundance and diversity, compared to forest floor soils. In both studies, rates of litter decomposition, as measured in litterbags, were similar between the canopy and forest floor. Oribatid mites and Collembola are the most abundant detritivores in temperate and tropical forest canopies (Paoletti et al., 1991; Schowalter and Ganio, 1998; D. Walter and O'Dowd, 1995; Winchester, 1997) and many are canopy specialists that do not occur on the forest floor (Lindo and Winchester, 2007; Winchester et al., 1999).

Decomposition is an easily modeled process. Typically, an initial period of leaching or microbial oxidation of simple organic molecules results in a short-term, rapid loss of mass, followed by a longer-term, slower decay of recalcitrant compounds. Decomposition of foliage litter has been expressed as a single- or double-component negative exponential model (Olson, 1963),

$$N_t = S_0 e^{-kt} + L_0 e^{-kt} \tag{14.1}$$

where N_t is mass at time t, S_0 and L_0 are masses in short- and long-term components, respectively, and ks are the respective decay constants. The short-term rate of decay reflects labile organic molecules, and the long-term rate of decay reflects lignin content and actual evapotranspiration (AET) rate, based on temperature and moisture conditions (Meentemeyer, 1978; Seastedt, 1984). Long-term decay rates for foliage litter range from -0.14 year^{-1} to -1.4 year^{-1}, depending on nutritional value for decomposers (Table 14.1) (Laskowski et al., 1995; Seastedt, 1984; Schowalter et al., 1991). Decay rates for wood range from -0.004 year^{-1} to -0.5 year^{-1} (Harmon et al., 1986). Schowalter et al. (1998) monitored decomposition of freshly cut oak, *Quercus* spp., logs over a 5-year period and found that a 3-component exponential model was necessary to account for differential decay rates among bark and wood tissues. An initial decay rate of -0.12 year^{-1} during the first year reflected primarily the rapid loss of relatively nutrient-rich inner bark (phloem), which largely disappeared by the end of the second year as a result of rapid exploitation by insects and fungi. An intermediate decay rate of -0.06 year^{-1} for years 2–5 reflected the slower decay rate for sapwood and outer bark, and a long-term decay rate of -0.012 year^{-1} was predicted, based on the slow decomposition of heartwood.

Decomposition often is not constant but shows seasonal peaks and annual variation that reflect periods of suitable temperature and moisture for decomposers. Patterns of nutrient

TABLE 14.1   Percent of Leaf Litter Decomposition Attributable to Invertebrate Detritivores

Leaf species	k_T	k_{NA}	k_F	Percent due to fauna
Dogwood	−0.00248	−0.00089	−0.00159[a]	64[b]
Sweetgum	−0.00175	−0.00050	−0.00125	71
Tulip poplar	−0.00229	−0.00113	−0.00116	51
Red maple	−0.00125	−0.00069	−0.00056	45
Water oak	−0.00174	−0.00037	−0.00137	79
White oak	−0.00216	−0.00076	−0.00140	65

[a] *Loss rate attributable to detritivores (k_F) was calculated as total decay rate (k_T) minus rate with naphthalene added (k_{NA}).*
[b] *Percent due to fauna was calculated as k_F/k_T.*
Modified with permission from Coleman, D.C., Crossley, Jr., D.A., Hendrix, P.F., 2004. Fundamentals of Soil Ecology, second ed. Elsevier, Amsterdam; Table 5.4, p. 205.

mineralization from litter reflect periods of storage and loss, depending on activities of various functional groups. For example, Schowalter and Sabin (1991) reported that nitrogen and calcium content of decomposing Douglas-fir, *Pseudotsuga menziesii*, needle litter, in litterbags, in western Oregon, USA, peaked in spring each year, when microarthropod abundances were lowest, and declined during winter, when microarthropod abundances were highest. High rates of comminution by microarthropods and decay by microorganisms during the wet winters likely contributed to release of nutrients from litter, whereas reduced comminution and decay during dry springs and summers led to nutrient immobilization in microbial biomass. Similarly, fluctuating concentrations of nutrients in decomposing oak wood over time probably reflect patterns of colonization and nutrient mobilization (Schowalter et al., 1998).

The relative contributions of physical and biological factors to pedogenesis vary among ecosystems. Erosion and earth movements (eg, soil creep and landslides) mix soil and litter in ecosystems with steep topography or high wind or raindrop impact on surface material. Burrowing animals are common in ecosystems with loose substrates suitable for excavation. Grasslands and forests on sandy or loamy soils support the highest diversity and abundances of burrowers. Ants can excavate nests through rocky, or other, substrates that would preclude burrowing by larger or softer-bodied animals and are the dominant burrowers in many ecosystems.

3 EFFECTS OF DETRITIVORY AND BURROWING

Arthropod detritivores and burrowers directly and indirectly control decomposition, carbon flux, biogeochemical cycling, pedogenesis, and primary production. The best known effects are on decomposition and mineralization (Seastedt, 1984; Coleman et al., 2004). However, some detritivorous and fossorial arthropods are capable of significantly affecting ecosystem structure, including the spatial distribution of carbon and nutrients, as well as ecosystem capacity to store and release nutrients and pollutants, making them true ecosystem engineers (Jouquet et al., 2006).

3.1 Decomposition and Mineralization

An extensive literature has addressed the effects of detritivores on decomposition and mineralization rates (Coleman et al., 2004; Hättenschwiler and Gasser, 2005). Generally, the effect of arthropods on the decay rate of litter can be calculated by subtracting the decay rate when arthropods are excluded from the decay rate when arthropods are present (Table 14.1). Detritivores affect decomposition and mineralization processes, including fluxes of carbon as CO_2 or CH_4, by fragmenting litter and by affecting rates of microbial catabolism of organic molecules. The magnitude of these effects depends on the degree to which feeding increases the surface area of litter and inoculates or reduces microbial biomass.

3.1.1 Comminution

Large comminuters are responsible for the fragmentation of large detrital materials into finer particles that can be processed by fine comminuters and saprophytic microorganisms. Cuffney et al. (1990) and J. Wallace et al. (1991) reported that 70% reduction in abundance of shredders from a small headwater stream in North Carolina, USA, reduced leaf litter decay rates by 25–28% and the export of fine particulate organic matter by 56%. As a result, unprocessed leaf litter accumulated (J. Wallace et al., 1995). D. Wise and Schaefer (1994) found that excluding macroarthropods and earthworms from leaf litter of selected plant species in a beech forest reduced decay rates 36–50% for all litter types except fresh beech litter. When all detritivores were excluded, the comparable reduction in decay rate was 36–93%, indicating the prominent role of large comminuters in decomposition. Tian et al. (1995) manipulated abundances of millipedes and earthworms in tropical agricultural ecosystems. They found that millipedes alone significantly accounted for 10–65% of total decay over a 10-week period. Earthworms did not affect decay significantly by themselves, but earthworms and millipedes combined significantly accounted for 11–72% of total decay. Haimi and Huhta (1990) demonstrated that earthworms significantly increased mass loss of litter by 13–41%. N. Anderson et al. (1984) noted that aquatic xylophagous tipulid larvae fragmented > 90% of decayed red alder, *Alnus rubra*, wood in a 1-year period.

Termites have received considerable attention because of their substantial ecological and economic importance in forest, grassland, and desert ecosystems. Based on laboratory feeding rates, K.E. Lee and Butler (1977) estimated wood consumption by termites in dry sclerophyll forest in South Australia. They reported that wood consumption by termites was equivalent to about 25% of annual woody litter increment and 5% of total annual litterfall. Based on termite exclusion plots, Whitford et al. (1982) reported that termites consumed up to 40% of surficial leaf litter in a warm desert ecosystem in the southwestern USA (Fig. 14.4). Overall, termites in this ecosystem consumed at least 50% of estimated annual litterfall (K. Johnson and Whitford, 1975; Silva et al., 1985). N.M. Collins (1981) reported that termites in tropical savannas in West Africa consumed 60% of annual wood fall and 3% of annual leaf fall (24% of total litter production), but fire removed 0.2% of annual wood fall and 49% of annual leaf fall (31% of total litter production). In that study, fungus-feeding Macrotermitinae were responsible for 95% of the litter removed by termites. Termites apparently consume virtually all litter in tropical savannas in East Africa (J. Jones, 1989, 1990). Termites consume a lower proportion of annual litter inputs in more mesic ecosystems. N.M. Collins (1983) reported that termites consumed about 16% of annual litter production in a Malaysian rainforest receiving 2000 mm

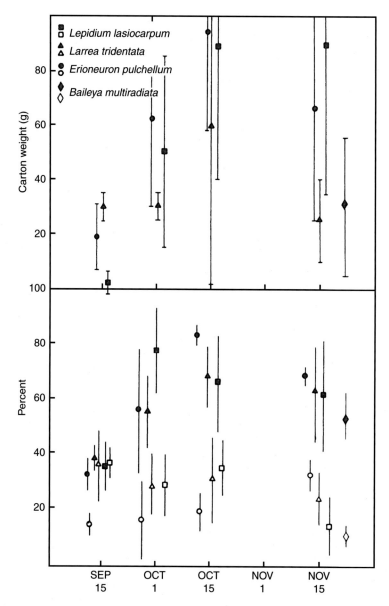

FIGURE 14.4 Rate of gallery carton deposition (top) and mass loss (bottom) of pepperweed, *Lepidium lasiocarpum*, creosote bush, *Larrea tridentata*, fluff grass, *Erioneuron pulchellum*, and desert marigold, *Baileya multiradiata* foliage when subterranean termites were present *(black symbols)* or absent *(white symbols)* in experimental plots in southern New Mexico, USA. Litter (10 g) was placed in aluminum mesh cylinders on the soil surface on Aug. 15, 1979. Vertical lines represent standard errors. *Source: Modified from Whitford, W.G., et al., 1982. Contributions of subterranean termites to the "economy" of Chihuahuan Desert ecosystems. Oecologia 55, 298–302.*

precipitation year^{-1} and 1–3% of annual litter production in a Malaysian rainforest receiving 5000 mm precipitation year^{-1}.

Dung beetles and earthworms also are important consumers of dung in many tropical and subtropical ecosystems (Coe, 1977; Holter, 1979; Kohlmann, 1991). Accumulation of dung from domestic mammalian grazers has become a serious problem in many arid and semiarid ecosystems. Termites can remove as much as 100% of cattle dung in 3 months (Coe, 1977; Herrick and Lal, 1996; Whitford et al., 1982). On average, termites in the tropics remove 33% of dung in a particular habitat within 1 month of deposition (Freymann et al., 2008). In the absence of termites, dung would require 25–30 years to disappear (Whitford, 1986).

Relatively few studies have provided estimates of wood consumption by bark- and wood-boring insects, despite their recognized importance to wood decomposition. Zhong and Schowalter (1989) used exclusion techniques to show that bark beetles consumed 0.1–7.6% of inner bark, and wood-boring beetles consumed an additional 0.05–2.3% during the first year of decomposition, depending on the conifer tree species. Ambrosia beetles consumed 0–0.2% of the sapwood during the first year. Schowalter et al. (1998) found that virtually the entire inner bark of oak logs was consumed by beetles during the first 2 years of decomposition, facilitating separation of the outer bark and exposing the sapwood surface to generalized saprophytic microorganisms. Edmonds and Eglitis (1989) used exclusion techniques to demonstrate that, over a 10-year period, bark beetles and wood borers increased decay rates of large Douglas-fir logs (42 cm diameter at breast height) by 12% and of small logs (26 cm diameter at breast height) by 70%. Ulyshen (2014) reported that insects were responsible for 21% and 14% mass loss in *Pinus taeda* coarse woody debris after 31 months in flooded and unflooded forests, respectively.

Payne (1965) explored the effects of carrion feeders on carrion decay during the summer in South Carolina, USA. He placed baby pig carcasses under replicated treatment cages, open at the bottom, that either permitted or restricted access to insects. Carcasses were weighed at intervals. Carcasses exposed to insects lost 90% of their mass in 6 days, whereas carcasses protected from insects lost only 30% of their mass in this period, followed by a gradual loss of mass, with 20% mass remaining in mummified pigs after 100 days.

Not all studies indicate significant effects of litter fragmentation by macroarthropods. Setälä et al. (1996) reported that manipulation of micro-, meso-, and macroarthropods in litter baskets resulted in slower decay rates in the presence of macroarthropods. Most litter in baskets with macroarthropods (millipedes and earthworms) was converted into large fecal pellets that decayed slowly.

A number of studies have demonstrated that microarthropods are responsible for up to 80% of the total foliage decay rate, depending on litter quality and ecosystem (Table 14.1, Fig. 14.5) (Coleman et al., 2004; González and Seastedt, 2001; Hättenschwiler and Gasser, 2005; Heneghan et al., 1999; Seastedt, 1984; Vossbrinck et al., 1979). Seastedt (1984) suggested that an apparent, but nonsignificant, inverse relationship between decay rate due to microarthropods and total decay rate indicated a greater contribution of arthropods to decomposition of recalcitrant litter fractions compared to more labile fractions. Tian et al. (1995) subsequently reported that millipedes and earthworms contributed more to the decomposition of plant residues with high C/N, lignin, and polyphenol contents than to high-quality plant residues.

In addition to accelerating litter decomposition, detritivores also synergize the effect of plant species diversity. Hättenschwiler and Gasser (2005) found that slowly decomposing

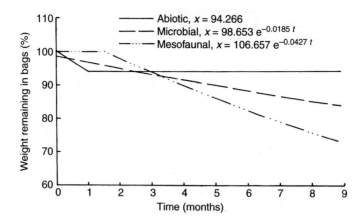

FIGURE 14.5   **Decomposition rate of blue grama grass in litterbags treated to permit decomposition by abiotic factors alone, abiotic factors + microbes, and abiotic factors + microbes + mesofauna (microarthropods).** Decomposition in the abiotic treatment was insignificant after the first month; decomposition showed a 2-month time lag in the treatment including mesofauna. *Source: From Vossbrinck, C.R., et al., 1979. Abiotic and biotic factors in litter decomposition in a semiarid grassland. Ecology 60, 265–271.*

oak, *Quercus petraea*, litter disappeared 26% faster in combination with other litter species than in single-species litter when millipedes, *Glomeris marginata*, were present, but showed no difference in decay rate among diversity treatments when millipedes were excluded, even when earthworms, *Aporrectodea longa*, were present. By contrast, faster-decomposing *Prunus avium* and *Tilia platyphyllos* decayed more slowly with increasing litter species diversity in the presence of earthworms. Variable results were found for other litter species, but this study demonstrated that detritivores interact with litter species diversity to accelerate or decelerate decomposition.

3.1.2 Microbial Respiration

Microbial decomposers are responsible for about 95% of total heterotrophic respiration in litter and soil. Several studies have documented increased microbial respiration as a result of increased arthropod access to detrital substrate and stimulation of microbial production. Arthropods generally increase microbial respiration rates and carbon flux but may reduce respiration rates if they overgraze microbial resources (Huhta et al., 1991; Seastedt, 1984).

Many arthropods directly transport and inoculate saprophytic microorganisms into organic residues. Although bark beetles and wood-boring insects are the best known vectors of saprophytic fungi (Bridges and Perry, 1985; French and Roeper, 1972; Morgan, 1968), Behan and Hill (1978) documented transmission of fungal spores by oribatid mites. Some inoculated fungi colonize wood in advance of insect excavation and degrade cellulose into labile carbohydrates that subsequently are used by the insects (Bridges and Perry, 1985; French and Roeper, 1972; Morgan, 1968). Other nutrients may be acquired incidentally by insects during feeding or movement through colonized material.

Mixing of organic material and microbes during passage through detritivore guts ensures infusion of consumed litter with decomposers and may alter litter quality in ways that stimulate microbial production (Maraun and Scheu, 1996). Gut mixing is especially important for species such as termites and other wood borers that require microbial digestion of cellulose and lignin into labile carbohydrates (Breznak and Brune, 1994) that may fuel nitrogen fixation by microbes in xylophage guts (Nardi et al., 2002).

Litter fragmentation by detritivores greatly increases oxygen penetration, as well as the surface area exposed for microbial colonization. Zhong and Schowalter (1989) reported that ambrosia beetle densities averaged 300 m^{-2} bark surface in Douglas-fir and western hemlock, *Tsuga heterophylla*, coarse woody debris, and their galleries extended 9–14 cm in 4–9 cm thick sapwood, indicating that considerable sapwood volume was made accessible to microbes colonizing gallery walls. The entire sapwood volume was colonized by various fungi within the first year after the coarse woody debris was accessible.

Fungivorous and bacteriophagous arthropods stimulate microbial activity by maximizing microbial production. As discussed for herbivore effects on plants in Chapter 12, low-to-moderate levels of grazing often stimulate productivity of the microflora by alleviating competition, altering microbial species composition, and gouging new detrital surfaces for microbial colonization. Microarthropods also can stimulate microbial respiration by preying on bacteriophagous and mycophagous nematodes (Seastedt, 1984; Setälä et al., 1996). Higher levels of grazing may depress microbial biomass and reduce respiration rates (Huhta et al., 1991; Seastedt, 1984).

Seastedt (1984) suggested a way to evaluate the importance of three pathways of microbial enhancement by arthropods, based on the tendency of microbes to immobilize nitrogen in detritus until the C:N ratio approaches 10–20:1. Where arthropods affect decomposition primarily through comminution, nitrogen content of litter should be similar with or without fauna. Alternatively, where arthropods stimulate microbial growth and respiration rates, the C:N ratio of litter with fauna should be less than the ratio without fauna. Finally, where arthropods graze microbial tissues as fast as they are produced, the C:N ratio of litter should be constant, and mass should decrease.

Seasonal variation in arthropod effects on microbial production and biomass may explain variable results and conclusions from earlier studies. Maraun and Scheu (1996) reported that fragmentation and digestion of beech leaf litter by the millipede, *G. marginata*, increased microbial biomass and respiration in Feb. and May but reduced microbial biomass and respiration in Aug. and Nov. They concluded that millipede feeding generally increased nutrient (nitrogen and phosphorus) availability, but that these nutrients were only used for microbial growth when carbon resources were adequate, as occurred early in the year. Depletion of carbon resources relative to nutrient availability in detritus limited microbial growth later in the year.

Although CO_2 is the major product of litter decomposition, incomplete oxidation of organic compounds occurs in some ecosystems, resulting in release of other trace gases, especially methane (Khalil et al., 1990). P. Zimmerman et al. (1982) first suggested that termites could contribute up to 35% of global emissions of methane. A number of arthropod species, including most tropical representatives of millipedes, cockroaches, termites, and scarab beetles, are important hosts for methanogenic bacteria and are relatively important sources of biogenic global methane emissions (Hackstein and Stumm, 1994).

Termites have received the greatest attention as sources of methane because their relatively sealed colonies are warm and humid, with low oxygen concentrations that favor fermentation processes and emission of methane or acetate (Brauman et al., 1992; Wheeler et al., 1996). Thirty of 36 temperate and tropical termite species assayed by Brauman et al. (1992), Hackstein and Stumm (1994), and Wheeler et al. (1996) produced methane and/or acetate. Generally, acetogenic bacteria outproduce methanogenic bacteria in wood- and grass-feeding termites, but methanogenic bacteria are much more important in fungus-growing and soil-feeding termites (Brauman et al., 1992).

P. Zimmerman et al. (1982) suggested that tropical deforestation and conversion to pasture and agricultural land could increase the biomass and methane emissions of fungus-growing and soil-feeding termites, but Martius et al. (1996) concluded that methane emissions from termites in deforested areas in Amazonia would not contribute significantly to global methane fluxes. Jamali et al. (2011) reported that CO_2 and CH_4 emissions from termite colonies are highly seasonal (higher during the wet season) in Australia and depend on termite abundance and activity. Khalil et al. (1990), Martius et al. (1993), and Sanderson (1996) calculated CO_2 and methane fluxes based on global distribution of termite biomass, and concluded that termites contribute c.2% of the total global flux of CO_2 (3500 tg year^{-1}) and 4–5% of the global flux of methane (≤ 20 tg year^{-1}) (Fig. 14.6). However, emissions of CO_2 by termites are 25–50% of annual emissions from fossil fuel combustion (Khalil et al., 1990). Furthermore, Cattânio et al. (2002) reported unexpected emissions of methane following experimental drought treatment in the eastern Amazon. Since dry soils should not provide the anaerobic conditions necessary for methane emission, they suggested that the most plausible explanation was that root mortality supported increased termite activity in the dried soil. Contributions to atmospheric composition by this ancient insect group may have been more substantial prior to anthropogenic production of CO_2, methane, and other trace gases.

3.1.3 Mineralization

Changes in elemental concentrations in decomposing litter represent net mineralization rates. Net mineralization includes loss of elements due to mineralization and accumulation by microflora of elements entering as microparticulates, precipitation, and leachate or transferred (eg, via hyphae) from other organic material (Harmon et al., 1994; Lodge et al., 2008; Schowalter et al., 1998; Seastedt, 1984). Although microbial biomass typically is a negligible component of litter mass, microbes often represent a large proportion of the total nutrient content of decomposing detritus and significantly affect the nutrient content of the litter-microbial complex (Lodge et al., 2008; Seastedt, 1984).

Arthropods affect net mineralization in two measurable ways: through litter consumption and assimilation of consumed nutrients and through effects on nutrient content of the litter-microbe system. Seastedt (1984) proposed the following equation to indicate the relative effect of arthropods on mineralization:

$$Y = (\% \, \mathrm{mass}_i / \% \, \mathrm{mass}_x) \times (\mathrm{concentration}_i / \mathrm{concentration}_x) \qquad (14.2)$$

where Y is the relative arthropod effect, % mass$_i$ is the percentage of initial mass remaining that has been accessible to arthropods, % mass$_x$ is the percentage of initial mass remaining that has been unavailable to arthropods, and concentration$_i$ and concentration$_x$ are the respective concentrations of a given element. Net immobilization of an element is indicated by $Y > 1$, and net loss is indicated by $Y < 1$. Temporal changes in nutrient content depend on the structural position of the element within organic molecules, microbial use of the element, and the form and amounts of the element entering the detritus from other sources.

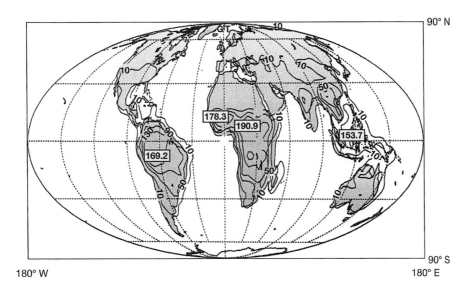

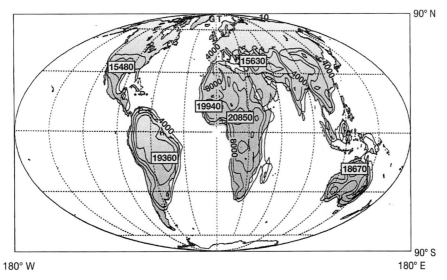

FIGURE 14.6   **Geographic distribution of emissions of methane (top) and carbon dioxide (bottom) by termites.**
Units are 10^6 kg year^{-1}. *Source: From Sanderson, M.G., 1996. Biomass of termites and their emissions of methane and carbon dioxide: a global database. Global Biogeochem. Cy. 10, 543–557.*

Nitrogen generally is considered to be the element most likely to limit growth of plants and animals, and its release from decomposing litter often is correlated with plant productivity (Scriber and Slansky, 1981; Vitousek, 1982; T. Wood et al., 2009). Saprophytic microbes typically immobilize nitrogen until sufficient carbon has been respired to make carbon or some other element more limiting than nitrogen (Maraun and Scheu, 1996; Schowalter et al., 1998; Seastedt, 1984). Thereafter, the amount of nitrogen released should equal the amount of carbon oxidized. Microbes have considerable capacity to absorb nitrogen from precipitation, canopy leachate, and animal excrement (Fig. 12.14) (Lovett and Ruesink, 1995; Seastedt and Crossley, 1983; Stadler and Müller, 1996), and fungi can translocate limited N and other nutrients from hyphae in nutrient-rich substrates to hyphae in nutrient-poor substrates (Lodge et al., 2008), permitting nitrogen mineralization and immobilization even at high C:N ratios. Generally, exclusion of microarthropods decreases the concentration of nitrogen in litter, but the absolute amounts of nitrogen in litter are decreased or unaffected by microarthropod feeding activities (Seastedt, 1984).

Yokoyama et al. (1991) compared nitrogen transformations among cattle dung (balls) colonized by dung beetles, *Onthophagus lenzii*, uncolonized dung, and residual dung remaining after beetle departure. They reported that dung beetles reduced ammonia volatilization from dung 50% by reducing pH and ammonium concentration in dung (through mixing of dung and soil). However, dung beetles increased denitrification 2–3-fold by increasing the rate of nitrate formation. Dung beetles also increased nitrogen fixation 2–10-fold, perhaps by reducing inorganic nitrogen concentrations in a substrate of easily decomposable organic matter.

Nardi et al. (2002) found that nitrogen-fixing microbes are abundant in the hindguts of a wide variety of detritivores. Based on measured rates of nitrogen fixation by gut microbes, especially among termites, they calculated rates in ecosystems as high as 10–40 kg ha^{-1} year^{-1}.

Phosphorus concentrations often show initial decline due to leaching, but subsequently reach an asymptote determined by microbial biomass (Schowalter and Sabin, 1991; Schowalter et al., 1998; Seastedt, 1984). Microarthropods can increase or decrease rates of phosphorus mineralization, presumably as a result of their effect on microbial biomass (Seastedt, 1984).

Calcium dynamics are highly variable. This element often is bound in organic acids (eg, calcium oxalate), as well as in elemental and inorganic forms in detritus. Some fungi accumulate high concentrations of this element (Cromack et al., 1975, 1977; Schowalter et al., 1998), and some litter arthropods, especially millipedes and oribatid mites, have highly calcified exoskeletons (Norton and Behan-Pelletier, 1991; Reichle et al., 1969). Nevertheless, calcium content in arthropod tissues is low compared to annual inputs in litter. No consistent arthropod effects on calcium mineralization have been apparent (Seastedt, 1984).

Potassium and sodium are highly soluble elements, and their initial losses (via leaching) from decomposing litter invariably exceed mass losses (Schowalter and Sabin, 1991; Schowalter et al., 1998; Seastedt, 1984). Amounts of these elements entering the litter in precipitation or throughfall approach or exceed the amounts entering as litterfall. In addition, these elements are not bound in organic molecules, so their supply in elemental form is adequate to meet the needs of microflora. Arthropods have been shown to affect mineralization of ^{134}Cs or ^{137}Cs, used as analogs of potassium (Crossley and Witkamp, 1964; Witkamp and Crossley, 1966), but not mineralization of potassium (Seastedt, 1984). Sodium content often increases in decomposing litter, especially decomposing wood (Cromack et al., 1977; Schowalter et al., 1998). Sollins et al. (1987) suggested that this increase represented accumulation of arthropod tissues and products, which typically contain relatively high concentrations of sodium (Reichle et al., 1969). However, Schowalter et al. (1998) reported increased concentrations of sodium during early stages of wood decomposition, prior to sufficient accumulation of arthropod tissues. They suggested that increased sodium concentrations in wood reflected accumulation by decay fungi, which contained high concentrations of sodium in fruiting structures. Fungi and bacteria have no known physiological requirement for sodium (Cromack et al., 1977). Accumulation of sodium, and other limiting nutrients, in decomposing wood may represent a mechanism for attracting sodium-limited animals that transport fungi to new wood resources.

Sulfur accumulation in decomposing wood or forest and grassland soils (Schowalter et al., 1998; Stanko-Golden et al., 1994; Strickland and Fitzgerald, 1986) reflects both physical adsorption of sulfate and biogenic formation of sulfonates by bacteria (Autry and Fitzgerald, 1993). Although arthropods have no demonstrated role in these processes, arthropod feeding on bacterial groups responsible for sulfur mobilization or immobilization should influence sulfur dynamics. Because sulfur flux plays a major role in soil acidification and cation leaching, factors affecting sulfur immobilization require further investigation.

The generally insignificant effects of arthropods on net mineralization rates, compared to their substantial effects on mass loss, can be attributed to the compensatory effects of arthropods on microbial biomass. The stimulation by arthropods of microbial respiration and immobilization of nutrients results in loss of litter mass, especially carbon flux through respiration, but not of the standing crops of other elements within litter (Seastedt, 1984). Other aspects of fragmentation also may contribute to nutrient retention, rather than loss. Aquatic comminuters generally fragment detritus into finer particles more amenable to downstream transport (J. Wallace and Webster, 1996). However, some filter-feeders concentrate fine detrital material into larger fecal pellets that are more likely to remain in the aquatic ecosystem (Wotton et al., 1998). Some shredders deposit feces in burrows, thereby incorporating the nutrients into the substrate (R. Wagner, 1991). Furthermore, Seastedt (2000) noted that most studies of terrestrial detritivore effects have been relatively short term. Accumulating data (Setälä et al., 1996) suggest that mixing of recalcitrant organic matter and mineral soil in the guts of some arthropods may produce stable soil aggregates that reduce the decay rate of organic material.

3.2 Soil Structure, Fertility, and Infiltration

Fossorial arthropods alter soil structure by redistributing soil and organic material and increasing soil porosity (J. Anderson, 1988). Porosity determines the depth to which air and water penetrate the substrate. A variety of substrate-nesting vertebrates, social arthropods, and detritivorous arthropods and earthworms affect spatial and temporal patterns of substrate structure, organic matter content, and infiltration in terrestrial and aquatic systems.

Defecation by a larval caddisfly, *Sericostoma personatum*, increases subsurface organic content in a stream ecosystem by 75–185% (R. Wagner, 1991). Larvae of this caddisfly feed on detritus on the surface of the stream bed at night and burrow into the stream bed during the day, trapping organic matter in burrows. Frouz et al. (2004) found higher survival of chironomid larvae in benthic substrates composed primarily of sand and/or accumulated chironomid fecal pellet aggregates (particle diameter > 0.25 mm), compared to substrates composed of fine organic sediment (< 0.25 mm). Accumulated fecal pellets increased mean substrate particle size, leading to more extensive larval tunneling and higher dissolved oxygen levels also seen in sand substrates.

Ants and termites are particularly important soil engineers (Dangerfield et al., 1998; Eldridge et al., 2009; James et al., 2008; Jouquet et al., 2006; MacMahon et al., 2000; Whitford and Eldridge, 2013). Colonies of these insects often occur at high densities and introduce cavities into large volumes of substrate. Eldridge (1993) reported that densities of funnel ant, *Aphaenogaster barbigula*, nest entrances could reach 37 m^{-2}, equivalent to 9% of the surface area over portions of the eastern Australian landscape. Nests of leaf-cutting ants, *Atta vollenweideri*, reach depths of > 3 m in pastures in western Paraguay (Jonkman, 1978). J. Moser (2006) excavated a leaf-cutting ant, *Atta texana*, nest in northern Louisiana, USA, and found 97 fungus-garden chambers, 27 dormancy chambers, 45 detritus chambers (for disposal of depleted foliage substrate), and a central cavity at 4 m depth in which the ants and fungus overwintered (Fig. 14.2). The nest extended over an area of 12 x 17 m on the surface and at least 4 m deep. The bottom of the colony could not be reached, but vertical tunnels extended to at least 7.5 m and might have extended to the water table at 32 m. Whitford et al. (1976) excavated nests of desert harvester ants, *Pogonomyrmex* spp., in New Mexico, USA, and mapped their 3-dimensional structure (Fig. 14.7). Colony densities were 21–23 ha^{-1} at four sites, and each colony consisted of 12–15 interconnected galleries (each about 0.035 m^3) within a 1.1 m^3 volume (1.5 m diameter $\times$ 2 m deep) of soil, equivalent to about 10 m^3 ha^{-1} of cavity space (Fig. 14.7). These colonies frequently penetrated the calcified hardpan (caliche) layer 1.7–1.8 m below the surface. E. King and Spink (1969) excavated a colony of Formosan subterranean termites, *Coptotermes formosanus*, in southwestern Louisiana that covered an area of about 0.6 ha, with galleries up to 1.2 m deep extending > 60 m from the source tree.

The infusion of large soil volumes with galleries and tunnels greatly alters soil structure and chemistry. Termite and ant nests typically represent sites of concentrated organic matter and nutrients (Ackerman et al., 2007; J. Anderson, 1988; Culver and Beattie, 1983; Herzog et al., 1976; Holdo and McDowell, 2004; James et al., 2008; J. Jones, 1990; Jurgensen et al., 2008; Lesica and Kannowski, 1998; MacMahon et al., 2000; Mahaney et al., 1999; A. Risch et al., 2005; Salick et al., 1983; D. Wagner, 1997; D. Wagner and Jones, 2004; D. Wagner et al., 1997; Whitford and Eldridge, 2013). Nests may have concentrations of macronutrients 2–3 times higher than surrounding soil (Fig. 14.8). J. Jones (1990) and Salick et al. (1983) noted that soils outside termite nest zones become relatively depleted of organic matter and nutrients. L. Parker et al. (1982) reported that experimental exclusion of termites for 4 years increased soil nitrogen concentration by 11%. Ant nests also have been found to have higher rates of microbial activity and carbon and nitrogen mineralization than do surrounding soils (Dauber and Wolters, 2000; Lenoir et al., 2001; D. Wagner and Jones, 2004) and represent sites of concentrated CO_2 efflux (Domisch et al., 2006; Jurgensen et al., 2008; A. Risch et al., 2005).

Nest pH often differs from surrounding soil. Mahaney et al. (1999) found significantly higher pH in termite mounds than in surrounding soils. Jonkman (1978) noted that soil within leaf-cutter ant, *Atta* spp., nests tended to have higher pH than did soil outside the nest. However, D. Wagner et al. (1997) measured significantly lower pH (6.1) in nests of harvester ants, *Pogonomyrmex barbatus*, than in reference soil (6.4). Lenoir et al. (2001) reported that *Formica rufa* nests had higher pH than did surrounding soil at one site and lower pH than did surrounding soil at a second site in Sweden. Ant mounds in Germany did not differ from surrounding soils (Dauber and Wolters, 2000).

Termites and ants also transport large amounts of soil from lower horizons to the surface and above for construction of nests (Fig. 14.9), gallery tunnels, and "carton," the soil deposited around litter material by termites for protection and to retain moisture during feeding above ground (Fig. 14.10) (Whitford, 1986; Whitford and Eldridge, 2013). Whitford et al. (1982) reported that termites brought 10–27 g m^{-2} of fine-textured soil material (35% coarse

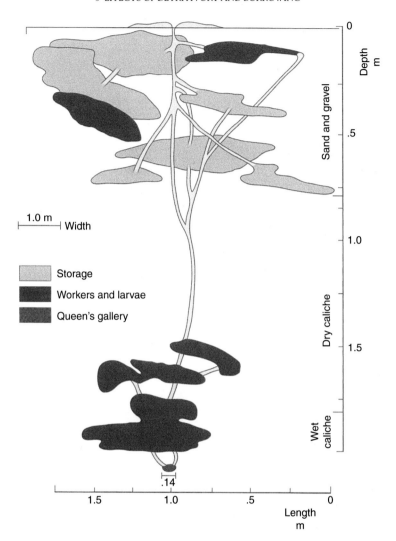

FIGURE 14.7 **Vertical structure of a harvester ant, *Pogonomyrmex rugosus*, nest in southern New Mexico, USA.** *Source: From Whitford, W.G., et al., 1976. Comparative ecology of the harvester ants* Pogonomyrmex barbatus *(F. Smith) and* Pogonomyrmex rugosus *(Emery). Insect. Soc. 23, 117–132.*

sand, 45% medium fine sand, and 21% very fine sand, clay, and silt) to the surface and de-posited 6–20 g of soil carton per gram of litter removed (Fig. 14.4). Herrick and Lal (1996) found that termites deposited an average of 2.0 g of soil at the surface for every gram of dung removed. Ulyshen et al. (2014) found that about 20% of the final dry weight of *Pinus taeda* coarse woody debris that was exposed to subterranean termites, *Reticulitermes* spp., consisted of termite-imported soil. Mahaney et al. (1999) reported that the termite mound soil con-tained significantly more (20%) clay than did surrounding soils.

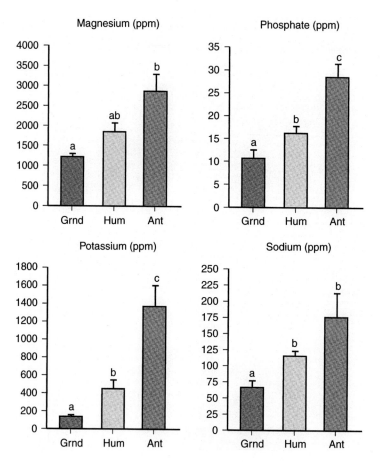

FIGURE 14.8 **Concentrations of major nutrients from bog soil (Grnd), hummocks (Hum), and *Formica* nests (Ant) in bogs in Montana, USA.** Vertical bars represent 1 SE. Means with different letters are significantly different at $p < 0.05$. *Source: From Lesica, P., Kannowski, P.B., 1998. Ants create hummocks and alter structure and vegetation of a Montana fen. Am. Midl. Nat. 139, 58–68.*

A variety of vertebrate species in Africa have been observed to selectively ingest termite mound soil (Mahaney et al., 1999). Mahaney et al. (1999) suggested that the higher clay content of termite mounds, along with higher pH and nutrient concentrations, could mitigate gastrointestinal ailments and explain termite soil consumption by chimpanzees. Termite mound soils, as well as surrounding soils, had high concentrations of metahalloysite, used pharmaceutically, and other clay minerals that showed mean binding capacities of 74–95% for four tested alkaloids. Chimpanzees could bind most of the dietary toxins present in 1–10 g of leaves by eating 100 mg of termite mound soil. Humans also ingest termite mound soil, which has been shown to meet dietary requirements for various minerals (J. Hunter, 1993).

A number of studies have demonstrated effects of soil animals on soil moisture (Fig. 14.11). Experimental reduction or removal of litter from the soil surface increases soil temperature and evaporation, and reduces infiltration of water. Burrowing and redistribution of soil and litter by animals increase soil porosity, water infiltration, and stability of soil aggregates that

FIGURE 14.9   **Termite castle in northern Australian woodland.** Dimensions are approximately 3 m height and 1.5 m diameter.

control water- and nutrient-holding capacity. Conversely, the dense pavement over mound-building termite nests restricts water infiltration but increases moisture in the runoff zone surrounding the mound (I. Ackerman et al., 2007; Eldridge, 1993, 1994).

Ant and termite nests have particularly important effects on soil moisture because of the large substrate surface areas and volumes affected (MacMahon et al., 2000). D. Wagner (1997) reported that soil near ant nests had higher moisture content than did more distant soil. Elkins et al. (1986) compared runoff and water infiltration in plots with termites present or excluded during the previous 4 years in New Mexico, USA. Plots with < 10% plant cover had higher infiltration rates when termites were present (88 mm h^{-1}) than when termites were absent (51 mm h^{-1}); runoff volumes were twice as high in the termite-free plots with low plant cover (40 mm) as in untreated plots (20 mm). Infiltration and runoff volumes did not differ between shrub-dominated plots (higher vegetation cover) with or without termites.

Eldridge (1993, 1994) measured effects of funnel ants, *A. barbigula*, and subterranean harvester termites, *Drepanotermes* spp., on infiltration of water in semiarid eastern Australia. He found that infiltration rates in soils with ant nest entrances were 4–10-fold higher (1030–1380 mm h^{-1}) than in soils without nest entrances (120–340 mm h^{-1}). Infiltration rate was correlated positively with nest entrance diameter. However, infiltration rate on

FIGURE 14.10 **Termite gallery carton on stems of dead creosote bush.** Soil particles are cemented together to provide protection and moisture control during termite feeding on detrital material.

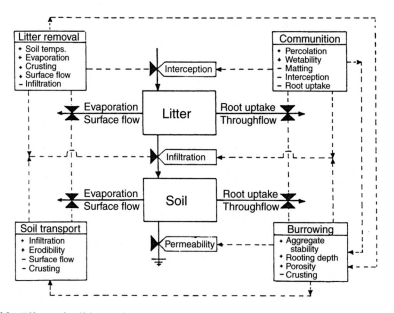

FIGURE 14.11 **Effects of soil invertebrates on soil water balance.** *Source: Reprinted from Anderson, J.M., 1988. Invertebrate-mediated transport processes in soils. Agr. Ecosyst. Environ. 24, 5–19.*

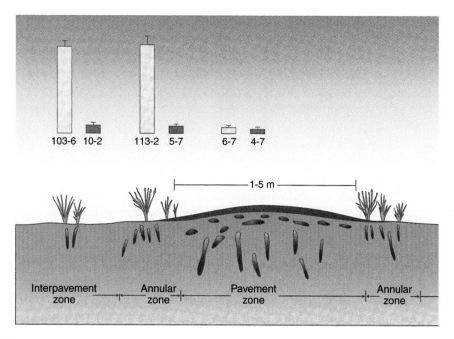

FIGURE 14.12 **Effect of termite colony structure on infiltration of water under ponded conditions (yellow) and under tension (brown).** Vertical lines indicate 1 SE of the mean. *Source: Reprinted from Eldridge, D.J., 1994. Nests of ants and termites influence infiltration in a semi-arid woodland. Pedobiologia 38, 481–492.*

the subcircular pavements covering the surface over termite nests was an order of magnitude lower than in the annular zone surrounding the pavement or in interpavement soils (Fig. 14.12). The cemented surface of the pavement redistributed water and nutrients from the pavement to the surrounding annular zone. Ant and termite control of infiltration creates wetter microsites in moisture-limited environments.

3.3 Primary Production and Vegetation Dynamics

Through control of decomposition, mineralization, and pedogenesis, detritivorous and fossorial arthropods have the capacity to control nutrient availability for, and perhaps uptake by, plants (Crossley, 1977; Setälä and Huhta, 1991). In particular, release of nitrogen and phosphorus from decaying organic matter often is correlated with plant productivity (Vitousek, 1982; T. Wood et al., 2009). Some species, especially ants, directly alter vegetation around nests by concentrating harvested seeds in nests and/or by clipping plants surrounding nests (MacMahon et al., 2000, see Chapter 13). However, relatively few studies have measured the effect of detritivores and burrowers on plant growth or vegetation dynamics.

C. Edwards and Lofty (1978) compared seedling emergence and shoot and root growth of barley between pots of intact, sterilized soil (from fields in which seed had been either drilled into the soil or planted during ploughing) with microarthropods or earthworms absent or reintroduced. Percent seedling emergence, plant height, and root weight were higher in ploughed soil and direct-drilled soil with animals, compared to sterile direct-drilled soil, suggesting important effects of soil animals on mineralization, soil porosity, and infiltration that support primary production.

R. Ingham et al. (1985) inoculated microcosms of blue grama grass, *Bouteloua gracilis*, in sandy loam soil, low in inorganic nitrogen, with bacteria or fungi. Half of each microflora treatment was inoculated with microbivorous nematodes. Plants growing in soil with bacteria and bacteriophagous nematodes grew faster and acquired more nitrogen initially than did plants in soil with bacteria only. Addition of mycophagous nematodes did not increase plant growth. These differences in plant growth resulted from greater nitrogen mineralization by bacteria (compared to fungi), excretion of NH_4^+-N by bacteriophagous (but not mycophagous) nematodes, and rapid uptake of available nitrogen by plants. Mycophagous nematodes did not increase plant growth or nitrogen uptake over fungi alone, because these nematodes excreted less NH_4^+-N, and the fungus alone mineralized sufficient nitrogen for plant growth.

In a unique, definitive study, Setälä and Huhta (1991) created laboratory microcosms with birch seedlings, *Betula pendula*, planted in partially sterilized soil reinoculated with soil microorganisms only or with soil microorganisms and a diverse soil fauna. During two growing periods the presence of soil fauna increased birch leaf, stem, and root biomass by 70, 53, and 38%, respectively, and increased foliar nitrogen and phosphorus contents 3-fold and 1.5-fold, respectively, compared to controls with microorganisms only (Fig. 14.13). Laakso and Setälä (1999) found that experimental removal of microbe- or detritus-feeding soil fauna, especially the microbi-detritivorous enchytraeid worm, *Cognettia sphagnetorum*, reduced plant biomass and uptake of N.

Yamada et al. (2007) manipulated cattle dung and numbers of dung beetles in experimental plots in pastures in Japan. They reported that dung beetles significantly increased cattle dung decomposition and nutrient uptake and herbage production (Fig. 14.14).

Soil arthropods could influence plant growth by inoculating roots with mycorrhizal fungi. Rabatin and Stinner (1988) reported that 28–97% of soil animals contained mycorrhizal spores or hyphae in their guts. Conversely, fungivore grazing on mycorrhizae could inhibit plant growth by interfering with nutrient uptake.

Soil animals also influence community dynamics. Although ant and termite nests may represent relatively minor components of total soil carbon and nutrient pools, they substantially increase spatial heterogeneity of soil water and nutrient availability (Domisch et al., 2006; Jurgensen et al., 2008; MacMahon et al., 2000; Risch et al., 2005), thereby influencing patterns of community development. Zaragoza et al. (2007) reported distinct protozoan communities associated with ant nest mounds, compared to reference soils 5 m away from ant nests. Several studies have demonstrated that ant and termite mounds typically support distinct plant communities, compared to surrounding soil (Figs. 14.15 and 14.16, Brody et al., 2010; Garrettson et al., 1998; Q. Guo, 1998; Holdo and McDowell, 2004; Joseph et al., 2014; T. King, 1977a; M. Schütz et al., 2008; Støen et al., 2013), but the effect on vegetation development may differ between active and abandoned mounds (Lesica and Kannowski, 1998; F. Smith and Yeaton, 1998).

Lesica and Kannowski (1998) reported that wood ants, *Formica podzolica*, were responsible for mound formation in peat bogs in Montana, USA. Mounds provided elevated habitat that

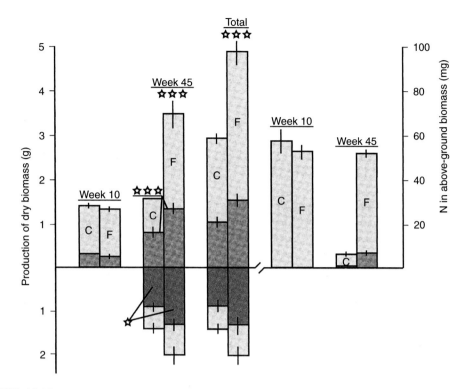

FIGURE 14.13 Biomass production (left of break in horizontal axis) and nitrogen accumulation (right of break in horizontal axis) of birch, *Betula pendula*, seedlings. Bars above the horizontal axis are stems (orange) and leaves (yellow); bars below the horizontal axis are roots in humus (blue) and roots in mineral soil (yellow). C = fauna removed; F = refaunated. Vertical lines represent 1 SD for all data (except nitrogen at week 45, where vertical lines represent minimum and maximum values). For C versus F, * = $p < 0.05$; *** = $p < 0.001$. Stem nitrogen was not measured week 10. *Source: From Setälä, H., Huhta, V., 1991. Soil fauna increase* Betula pendula *growth: laboratory experiments with coniferous forest floor. Ecology 72, 665–671.*

was warmer, better aerated, and had higher nutrient content than did surrounding peat surfaces (Fig. 14.8). Although active mounds supported only a few species of grasses, abandoned nests supported shrubs, as well as plant species that could not grow in the saturated peat surface. The ants foraged primarily on honeydew from aphids tended on shrubs, indicating a positive feedback relationship.

Abandoned nests of leaf-cutter ants, *Atta vollenweideri*, serve as sites of accelerated succession in Paraguayan pastures (Jonkman, 1978). Collapse of the nest chamber forms a depression that holds water and facilitates development of woody vegetation. At high nest densities, these oases coalesce, greatly increasing forest area. Brener and Silva (1995) found that active nests of *Atta laevigata* were associated more frequently with groves of trees, and the size of nests increased with grove size and the abundance of forest tree species in Venezuelan savanna, suggesting that active nests both facilitated and were facilitated by formation of groves.

L. Parker et al. (1982) demonstrated that termite exclusion significantly reduced biomass of four annual plant species and significantly increased biomass of one annual plant species.

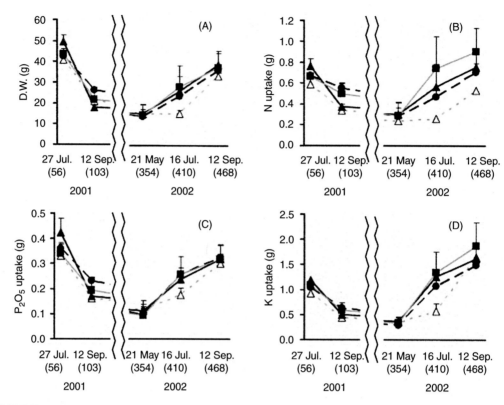

FIGURE 14.14 Changes in forage yield (A) and uptake of nutrients, nitrogen (B), phosphorus (C), and potassium (D), in experimental pasture in Japan during 2001–02. Vertical bars indicate 1 SE. Open triangles are plots with no cattle dung, filled triangles are plots with dung but no dung beetles, filled circles are plots with dung and 40 dung beetles introduced, and filled squares are plots with dung and 120 dung beetles introduced. Figures in parentheses indicate days after the beginning of the experiment. *Source: From Yamada, D., et al., 2007. Effect of tunneler dung beetles on cattle dung decomposition, soil nutrients and herbage growth. Grassland Sci. 53, 121–129.*

They observed an overall trend toward increased biomass of annual plants in plots with termites excluded. These results likely reflected increased nitrogen availability in termite exclusion plots, compared to plots with unmanipulated termite abundance. I. Ackerman et al. (2007) found that active termite mounds in abandoned farmland in central Amazonia inhibited water retention and restricted secondary succession. By contrast, Bloesch (2008) reported that *Macrotermes* spp. mounds provided spatially distinct elevated sites for development of wooded thickets in seasonally flooded savanna landscapes in East Africa. Furthermore, Støen et al. (2013) found that termites were primary determinants of plant diversity and composition on mounds, whereas ungulate grazing tended to regulate abundances of the most common plant species.

The high nutrient concentrations of termite and ant nests are incorporated by plants growing on or near nests and become available to higher trophic levels. Holdo and McDowell (2004) reported that trees growing on termite mounds had higher concentrations of all nutrients tested, except sodium and crude protein, than did trees from the surrounding

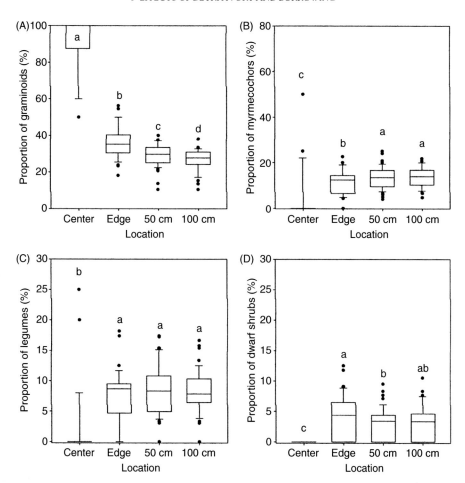

FIGURE 14.15 Vegetation composition along 46 transects from *Formica exigua* mound centers into surrounding alpine grassland in Switzerland by proportions of (A) graminoids, (B) myrmecochors, (C) legumes, and (D) dwarf shrubs. Box plots indicate median, 25th, and 75th percentiles, black lines are 10th and 90th percentiles, and filled circles are all values below 10th and above 90th percentiles. Different letters indicate significant differences ($p < 0.05$); note different scales for y axes among figures. *Source: From Schütz, M., et al., 2008. Impact of Formica exsecta Nyl. on seed bank and vegetation patterns in a subalpine grassland ecosystem. J. Appl. Entomol. 132, 295–305.*

woodland matrix in Zimbabwe. Trees on mounds also were subjected to more intense feeding by elephants, *Loxodonta africana*. Termite and ant nests thereby affect food availability and feeding patterns for herbivores, providing indirect positive feedback for herbivore effects on litter quality and availability for detritivores. Brody et al. (2010) found that *Acacia drepanolobium* trees growing at the edge of termite, *Odontotermes* spp., mounds in Kenya showed twice the foliar N content and seed production of trees growing away from mounds, due to enhanced soil N and P in mounds. Fox-Dobbs et al. (2010) further reported that *A. drepanolobium* trees off mounds acquired a higher percentage of N via fixation (55–80%) than did trees near mounds (40–50%), reflecting the higher availability and use

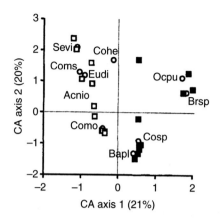

FIGURE 14.16 **Correspondence analysis of species × plot classification based on the 10 most common shrub species** *(open circles)* **found in a vegetation survey on termite mounds** *(open squares)* **and woodland plots** *(solid squares)* **in western Zimbabwe.** Termite plots and woodland plots were clearly distinguished by axis 1. Species associated with termite mounds were *Securinega virosa* (Sevi), *Combretum mossambicense* (Coms), *Euclea divinorum* (Eudi), *Acacia nigrescens* (Acni), and *Colophospermum mopane* (Como); species associated with the surrounding woodland matrix were *Combretum collinum* and *C. apiculatum* (Cosp), *Baikiaea plurijuga* (Bapl), *Ochna pulchra* (Ocpu), and *Brachystegia spiciformis* (Brsp). *Combretum heroroense* (Coho) was not clearly associated with either vegetation type. *Source: From Holdo, R.M., McDowell, L.R., 2004. Termite mounds as nutrient-rich food patches for elephants. Biotropica 36, 231–239.*

of soil-based N near termite mounds. Exclusion of vertebrate herbivores, which preferentially used mound vegetation, as evidenced by patterns of dung deposition (Fig. 14.17), did not affect these results, demonstrating that soil enrichment by termites, rather than dung and urine deposition by vertebrates, was responsible for the vegetation responses (Brody et al., 2010).

4 SUMMARY

Decomposition and pedogenesis are major ecosystem processes that affect biogeochemical cycling, CO_2 and trace gas fluxes, soil fertility, and primary production. Decomposition of organic matter involves four component processes: photo-oxidation, leaching, comminution, and mineralization. Arthropods are key factors influencing comminution and mineralization.

Functional groups involved in decomposition include coarse comminuters that fragment large materials and fine comminuters that fragment smaller materials, often those produced by large comminuters. In aquatic ecosystems, scrapers and shredders represent coarse comminuters, whereas gatherers and filterers represent fine comminuters. Xylophages represent a specialized group of comminuters that fragment woody litter in terrestrial and aquatic ecosystems. Necrophages reduce carcasses, and coprophages feed on animal excrement. Fungivores and bacteriovores fragment detrital material while grazing on microflora. Fossorial functional groups include subterranean nesters that excavate simple burrows, gatherers that return detrital or other organic materials to nesting areas, and fossorial feeders that consume organic material and/or soil and mix biotic and abiotic materials in their wake.

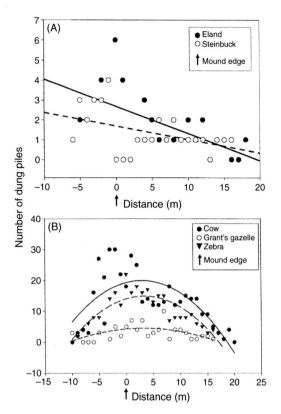

FIGURE 14.17 **Vertebrate dung density at increasing distance from termite, *Odontotermes* spp., mound centers into surrounding savanna in north-central Kenya.** (A) For eland and steinbuck, a linear regression explained 40% and 19% of variance, respectively. (B) For Grant's gazelle, zebra, and cow, a second-order polynomial explained 26%, 73%, and 54% of variance, respectively. In all cases, the first term was not significant, but the second term was highly significant ($p < 0.01$). *Source: From Brody, A.K., et al., 2010. Termites, vertebrate herbivores, and the fruiting success of Acacia drepanolobium. Ecology 91, 399–407.*

Evaluation of detritivore and burrower effects on decomposition and pedogenesis requires appropriate methods for measuring animal abundances and process rates. Abundances of detritivores or burrowers can be manipulated using exclusion and microcosm techniques, and detritivory can be measured as the product of detritivore abundance and individual consumption rate or as comparison of the rate of substrate disappearance with detritivores present or absent. Decomposition most commonly is measured as respiration rate, as the ratio of litter input to litter standing crop, or as the rate of litter disappearance. Isotopic tracers also provide data on decomposition rate.

Decomposition rate typically is higher in mesic than in arid ecosystems. Different functional groups dominate different ecosystems, depending on availability and quality of detrital resources. For example, shredders dominate headwater streams where coarse detrital inputs are the primary resource, whereas filterers dominate larger streams with greater availability of suspended fine organic material. Xylophages occur only in ecosystems with woody residues. Decomposition generally can be modeled as a multiple negative exponential decay function over time, with decay constants proportional to the quality of litter components. Typically, an initial large decay constant represents rapid loss of labile materials, and successively smaller decay constants represent slower losses of recalcitrant materials, for example, lignin and cellulose. Most studies have been relatively short term. Recent long-term studies suggest that mixing of recalcitrant materials and soil in arthropod guts may create stable aggregates that decay very slowly.

Detritivores affect decomposition in three ways: through comminution, effects on microbial biomass, and effects on mineralization. Comminution increases detrital surface area and facilitates colonization and decay by microflora. Low-to-moderate levels of grazing on microflora stimulate microbial productivity and biomass, maximizing microbial decomposition and respiration. High levels of grazing may reduce microbial biomass and decomposition. Grazers also disperse fungi and bacteria to new substrates. Not all organic material is converted to CO_2. The low oxygen concentrations characterizing warm, humid termite colonies favor incomplete reduction of organic molecules to methane and other trace gases. Arthropod detritivores affect mineralization in different ways, depending on the chemical characteristics and biological use of the element. Detritivores often increase mineralization of nitrogen, but nitrogen released from detritus may be immobilized quickly by microorganisms.

Burrowers affect soil development by redistributing soil and organic matter. Ants and termites, in particular, excavate large volumes of soil and accumulate organic material in their centralized nests, mixing soil with organic material and influencing the distribution of soil nutrients and organic matter. Surrounding soils may become depleted in soil carbon and nutrients.

Detritivore and burrower effects on mineralization and soil composition can affect primary production and vegetation dynamics. Elevated moisture and nutrient concentrations in, or adjacent to, ant and termite nests support distinct vegetation and may facilitate succession following colony abandonment. A few studies have demonstrated increased plant growth, altered vegetation structure, and increased herbivory resulting from detritivore-induced nutrient mobilization.

Insects as Regulators
of Ecosystem Processes

A NEW PERSPECTIVE OF INSECT HERBIVORES

Could insects regulate ecosystem processes, in terms of providing stabilizing feedback? This hypothesis is one of the most revolutionary ideas to emerge from integration of insect ecology and ecosystem ecology, because it requires a change in our traditional perspective of insects as crop or forest pests, disease vectors, or just nuisances, and because regulation at supra-organismal levels appears contradictory to evolutionary promotion of selfish self-interest.

Several lines of evidence indicate that native insects function as regulators of ecosystem processes. First, native herbivore populations are strongly regulated by host chemistry and availability and by predation (see Chapter 6). In turn, insects themselves are top-down mechanisms for regulating host or prey populations (Marra et al., 2004; Poinar and Poinar, 2004b, 2007, see Chapter 8). Small populations of herbivores stimulate plant production, whereas outbreaks reduce production by stressed and/or abundant plant species, thereby pruning and thinning host plants and reducing plant competition (Trumble et al., 1993; S. Williamson et al., 1989, see Chapter 12). Second, outbreaks of native species typically promote growth of less-stressed host or nonhost species with lower resource requirements, resulting in "healthier" and more diverse vegetation (Cairns et al., 2008; Coleman et al., 2008; Ritchie et al., 1998). Insects also control litter decomposition and nutrient availability for plants (Christenson et al., 2002; Frost and Hunter, 2007; Seastedt, 1984, see Chapter 14). These roles affect surface albedo, carbon and water fluxes, and global climate (Chapter 12). As a result, ecosystem conditions are affected substantially by insects and may be maintained within a narrower range of conditions than would result from an absence of insects.

Although accumulating evidence suggests that insects regulate ecosystem processes, definitive experiments are difficult (because of the difficulty of manipulating insect abundances over sufficiently large areas) and results are subject to interpretation. For example, does extensive tree mortality resulting from native bark beetle outbreaks represent a disturbance that increases variability in ecosystem structure and function (Romme et al., 1986) or a reversal of anthropogenic changes and return to a lower density of site-adapted species that prevailed prior to forest management

Insect Ecology. http://dx.doi.org/10.1016/B978-0-12-803033-2.00015-7

(North et al., 2007; Schowalter, 2008)? The forest structure reflecting outbreaks of bark beetles and defoliators, as well as historic low-intensity fires, in the central Sierras of California is the current recommendation for sustainable management of these forests (North et al., 2007). Although arguing against regulation by insects, Romme et al. (1986) noted that ecosystem production in some forests returned to preoutbreak levels within 10 years. For long-lived ecosystems, outbreaks by native insect herbivores might reduce variability in ecosystem conditions, compared to more extensive mortality and longer recovery that could result from unmitigated tree competition and stress in the absence of insect outbreaks.

1 INTRODUCTION

Insects, and other organisms, inevitably affect their environment through spatial and temporal patterns of resource acquisition and redistribution (Milcu et al., 2015; Whitham et al., 2006). Insects respond to environmental changes in ways that dramatically alter ecosystem conditions, as discussed in Chapters 12–14. These effects do not necessarily provide stabilizing regulation of ecosystem processes. However, the combination of bottom-up and top-down regulation of populations is pervasive in ecosystems and could be expected to maintain the community, and accompanying ecosystem processes, within narrower ranges of variation than would occur in the absence of such mutual regulation. The inclusion of insects in virtually all pathways through food webs and their ability to respond rapidly and dramatically to changes in host condition or abundance, in ways that often reverse these changes, appear to be ideal properties for regulation of ecosystem processes. The hypothesis that insects stabilize ecosystem properties through feedback regulation is one of the most important and revolutionary concepts to emerge from integration of insect ecology and ecosystem ecology and should be considered in making pest management decisions in natural ecosystems.

The concept of self-regulation is a key aspect of ecosystem ecology. Vegetation has a documented role in ameliorating variation in climate and biogeochemical cycling (Chapter 11), and vegetative succession facilitates recovery of ecosystem functions following disturbances. However, the concept of self-regulating ecosystems has seemed to be inconsistent with evolutionary theory (emphasizing selection of "selfish" attributes) (Pianka, 1974), with variable successional trends following disturbance (Horn, 1981), and with the lack of obvious mechanisms for maintaining homeostasis (Engelberg and Boyarsky, 1979).

Similar arguments defined earlier debates over evolutionary explanations for density-dependent feedback regulation of population size (Chapter 6) and chemical communication of herbivory among neighboring plants (Chapter 8). Both of these concepts are now well established (Harrison and Cappuccino, 1995; Heil and Karban, 2010) and provide useful examples of how science develops, with accumulating data often forcing changes in perspectives. The outcome of the debate over insect roles in ecosystem self-regulation has significant consequences for how we manage ecosystems and their biotic resources, including management of native "pests." Although controversial, this concept is an important aspect of insect ecology, and its major premises are the subject of this chapter.

2 DEVELOPMENT OF THE CONCEPT

The intellectual roots of ecosystem self-regulation lie in Darwin's (1859) recognition that some adaptations apparently benefit a group of organisms more than the individual, for example, reduced reproduction under crowded conditions (Chapter 6), leading to selection for population stability. Despite its apparent inconsistency with the selfish-gene view of evolution, natural selection at supra-organismal levels has remained an important concept (Axelrod and Hamilton, 1981; Schowalter, 1981; D.S. Wilson, 1976, 1997).

Behavioral ecologists have been challenged to explain the evolution of apparently altruistic behaviors that are fundamental to social organization. Even sexual reproduction could be considered a form of self-restraint, because each individual contributes only half of its genotype to progeny through sexual reproduction, compared to the entire genotype to progeny through asexual reproduction (Pianka, 1974). Cooperative interactions, such as mutualism and communication of herbivory among plants, that benefit receivers rather than senders, have been more difficult to explain in terms of individual selection.

Haldane (1932) proposed an early model in which altruism would have a selective advantage if the starting gene frequency were high enough and the benefits to the group outweighed individual disadvantage. This model raised obvious questions about the origin of altruist genes and the relative advantages and disadvantages that would be necessary for increased frequency of altruist genes.

Group selection theory was advanced during the early 1960s by Wynne-Edwards (1963, 1965), who proposed that social behavior arose as individuals evolved to curtail their own individual fitnesses to enhance survival of the group. According to this model, populations that do not restrain combat among their members or that overexploit their resources have a higher probability of extinction than do populations that regulate combat or resource use. Selection thus should favor demes with traits to regulate their densities, that is, maintain homeostasis in group size. Behaviors such as territoriality, restraint in conflict, and suppressed reproduction by subordinate individuals (including workers in social insect colonies) thereby reflect selection (feedback) for individual traits that prevent destructive interactions or oscillations in group size.

This hypothesis was challenged for lack of explicit evolutionary models or experimental tests that could explain the progressive evolution of homeostasis at the group level, that is, demonstration of an individual advantage to altruistic individuals over selfish individuals. Furthermore, Wynne-Edwards' proposed devices by which individuals curtail their individual fitnesses, and communicate their density and the degree to which each individual should decrease its individual fitness, were either inconsistent with available evidence or could be explained better by models of individual fitness (E. Wilson, 1973). Nevertheless, the concept of group selection was recognized as an important aspect of social evolution (E. Wilson, 1973).

W. Hamilton (1964) and J.M. Smith (1964) developed an evolutionary model, based on *kin selection*, whereby individual fitness is increased by behaviors that favor survival of relatives with similar genotypes. They introduced a new term, *inclusive fitness*, to describe the contributions of both personal reproduction and reproduction by near kin to individual fitness. For example, care for offspring of one's siblings increases an individual's fitness to the extent that it contributes to the survival of related genotypes. Failure to provide sufficient care for offspring of siblings reduces the survival of family members. This concept explained the evolution of many apparently altruistic behaviors, such as maternal care, shared rearing of offspring among related individuals, and alarm calls (that may draw the attention of predators to the caller).

In the social Hymenoptera, voluntary suppression of reproduction and suicidal defense of the colony by workers can be seen to benefit close relatives. W. Hamilton (1964) noted that males are produced from unfertilized eggs and have unpaired chromosomes. Accordingly, all the daughters in the colony inherit only one type of gamete from their father and thereby share 50% of their genes through this source. In addition, they share another 25%, on average, of their genes in common from their mother. Overall, the daughters share 75% of their genes with each other, compared to only 50% of their genes with their mother. Therefore, workers maximize their fitness by helping to rear siblings, rather than by having their own offspring.

This model does not apply to termites, which do not share the hymenopteran mating system but, nevertheless, show behaviors that benefit the colony rather than individuals, for example, suicidal defense and regulation of soldier:worker ratios (Dong et al., 2008; L. Mao and Henderson, 2010; Šobotník et al., 2012). Husseneder et al. (1999) and Thorne (1997) suggested that developmental and ecological factors, such as slow development, iteroparity, overlap of generations, food-rich environment, high risk of dispersal, and group defense, may be more important than genetics in the maintenance of termite eusociality, whatever factors may have favored its original development.

Levins (1970) and Boorman and Levitt (1972) proposed *interdemic selection* models to account for differential extinction rates among demes of metapopulations that differ in altruistic traits. In the Levins model, colonists from small populations found other small populations in habitable sites. Increasing frequency of altruist genes decreases the probability of extinction of these small populations, that is, cooperation elevates and maintains each deme above the extinction threshold (see Chapters 6 and 7). In the Boorman–Levitt model, colonists from a large, stable population found small, marginal populations in satellite habitats. Altruist genes do not influence extinction rates until marginal populations reach demographic carrying capacity, that is, altruism prevents destructive population increase above carrying capacity (see Chapters 6 and 7). Both models require restrictive conditions for evolution of altruist genes. Matthews and Matthews (2010) noted that group selection requires that an allele become established by selection at the individual level. Thereafter, selection could favor demes with altruist genes that reduce extinction rates, relative to demes without these genes. Interdemic selection has become a central theme in developing concepts of metapopulation dynamics (Chapter 7).

Meanwhile, the concept of group selection was implicit in early models of ecological succession and community development (see Chapter 10). The facilitation model of succession proposed by Clements (1916) and elaborated by E. Odum (1953, 1969) emphasized the apparently progressive development of a stable, "climax," ecosystem through succession. Each successional stage altered conditions in ways that benefitted the replacing species more than itself. However, such facilitation contradicted the individual self-interest that was fundamental to the theory of natural selection. Furthermore, identification of alternative models of succession, including the inhibition model (Chapter 10), made succession appear to be more consistent with evolutionary theory.

D.S. Wilson (1976, 1997) developed a model that specifically applied the concept of group selection to the community level. Wilson recognized that individuals and species affect their own fitness through effects on their environment, including the fitness of other individuals. For example, earthworm effects on soil development stimulate plant growth, herbivory, and litter production (see Chapters 12 and 14) and thereby increase the detrital resources exploited by the worms, an indirect positive feedback. Furthermore, spatial heterogeneity, from large geographic to microsite scales, in population distribution results in intra- and interdemic variation in effects of organisms on their community. Given sufficient iterations of Wilson's model, every effect of a species on its community eventually affects that species, positively or negatively, through all possible feedback pathways, resulting in a net inclusive fitness value. Intra- and interdemic variation in effects on the environment is subject to selection for adaptive traits of individuals.

These models help explain the increased frequency of altruist genes, but not the selective factors that maintain altruist genes in the face of evolutionary pressure to "cheat" among nonrelated individuals. Trivers (1971) and Axelrod and Hamilton (1981) developed a model of *reciprocal altruism* based on the Prisoner's Dilemma, in which each of two players (prisoners) can cooperate or defect. Each player can choose to cooperate or defect if the other player chooses to cooperate or defect. If the first player acts cooperatively, the benefit/cost for cooperation by the second player (reward for mutual cooperation) is less than that for defection (temptation for the first player to defect in the future); if the first player defects, the benefit/cost for cooperation by the second player (suckers' payoff) is less than that for defection (punishment for mutual defection). Therefore, if an interaction occurs only once, defection (noncooperation) is always the optimal strategy, despite both individuals doing worse than they would if they both cooperate. However, Axelrod and Hamilton (1981) recognized the probability of repeated interaction between pairs of unrelated individuals and addressed the initial viability (as well as final stability) of cooperative strategies in environments dominated by noncooperating individuals or more heterogeneous environments composed of other individuals using a variety of strategies. After numerous computer simulations with a variety of strategies, they concluded that the most robust strategy in an environment of multiple strategies also was the simplest, *Tit-for-Tat*. This strategy involves cooperation based on reciprocity and a memory extending only one move back, that is, never being the first to defect but retaliating after a defection by the other and forgiving after just one act of retaliation. They also found that once Tit-for-Tat was established, it resisted invasion by possible mutant strategies as long as the interacting individuals had a sufficiently large probability of meeting again.

Axelrod and Hamilton emphasized that Tit-for-Tat is not the only strategy that can be evolutionarily stable. The *Always Defect Strategy* also is evolutionarily stable, no matter what the probability of future interaction. They postulated that altruism could appear between close relatives, when each individual has part interest in the partner's gain (ie, rewards in terms of inclusive fitness), whether or not the partner cooperated. Once the altruist gene exists, selection would favor strategies that base cooperative behavior on recognition of cues, such as relatedness or previous reciprocal cooperation. Therefore, individuals in relatively stable environments are more likely to experience repeated interaction and selection for reciprocal cooperation than are individuals in unstable environments that provide low probabilities of future interaction.

S. Frank (1995) developed a mathematical model that could explain evolution of self-restraint and suppression of competition within groups as a result of the mutual benefit derived

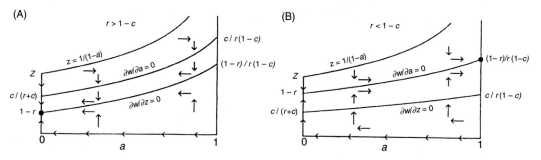

FIGURE 15.1 **A minimal model for evolutionary dynamics of competitiveness, z, and mutual policing, a.** (A) High relatedness, r, relative to benefits of mutual policing, $r > 1 - c$ (where c = cost), leads toward self-restraint, with $z^* = 1 - r$ and no mutual policing, $a^* = 0$. (B) Low relatedness relative to benefits, $r < 1 - c$, leads toward full investment in mutual policing and complete repression of competition, $a^* = 1$. In (A), $c = r = 0.7$; in (B), $c = 0.2$ and $r = 0.4$. The gradient of fitness with respect to mutual policing, $\delta w/\delta a$, is calculated as the partial derivative of w_{ij} with respect to a_{ij}, evaluated at fixed trait values for all individuals, $a_{ij} = a$ and $z_{ij} = z$. A similar approach is used for $\delta w/\delta z$. Although assumptions about costs, benefits, and genetics complicate the analysis, this minimal model demonstrates powerful selective pressure on self-restraint and mutual policing that favor increasingly complex evolutionary units. *Source: From Frank, S.A., 1995. Mutual policing and repression of competition in the evolution of cooperative groups. Nature 377, 520–522.*

by preventing overexploitation of limiting resources (Fig. 15.1). This model demonstrated that kin selection alone could not explain evolution of complex regulatory mechanisms among unrelated units of selection. Rather, individuals invest resources in mutual suppression of competition but nevertheless strive for reproductive gain within the more equitable distribution of resources they help create.

Sanctions against cheaters have been demonstrated in recent studies. A. Smith et al. (2009) found that reproductive cheaters in ant, *Aphaenogaster cockerelli*, colonies can be reliably identified by cuticular hydrocarbons, particularly pentacosane, that are associated only with fertile individuals. Experimental application of pentacosane to workers elicited aggression by nestmates in colonies with queens but not in colonies without queens, in which workers had begun to reproduce. Although cheaters might benefit from suppressing their hydrocarbon profile, they are prevented from doing so by the reproductive physiology of hydrocarbon biosynthesis.

Jandér and Herre (2010) experimentally excluded pollen from a number of fig wasp species associated with actively and passively pollinated fig species and found that actively pollinated (but not passively pollinated) fig species reduced fitness of nonpollinating fig wasps through a combination of increased abortion of nonpollinated figs that had wasp eggs and reduced production of wasp offspring in figs that were not aborted. The relative proportion of unpollinated figs that matured and the relative number of wasp offspring that matured in unaborted figs were both significantly lower in actively pollinated figs, compared to passively pollinated figs. All fig species aborted figs that received neither pollen nor wasp eggs. The strength of sanctions against nonpollinating wasps (measured as 1-wasp relative fitness) varied from 0.33 to 1.0 for actively pollinated fig species (Fig. 15.2), compared to 0 for passively pollinated species. Pollen-free wasps occurred only among actively pollinating species, and their prevalence was < 5% and negatively correlated with sanction strength. These data

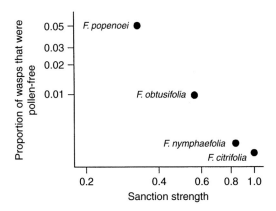

FIGURE 15.2 **The proportion of naturally occurring pollen-free fig wasps as a function of sanction strength among actively pollinated fig, Ficus, species.** *Source: From Jandér, K.C., Herre, E.A., 2010. Host sanctions and pollinator cheating in the fig tree-fig wasp mutualism. Proc. Roy. Soc. B-Biol. Sci. 277, 1481–1488.*

demonstrated substantial selection against cheaters as a means of stabilizing obligate mutualisms when providing a benefit to a host is costly, in terms of wasp time and energy.

These studies demonstrate that selection at supra-organismal levels must be viewed as contributing to the inclusive fitness (of individuals) derived from contribution to stabilizing interactions. Cooperating individuals have demonstrated greater ability in finding or exploiting uncommon or aggregated resources, defending shared resources, and mutual protection (W. Hamilton, 1964). Cooperating predators, for example, wolves and ants, have higher capture efficiency and can acquire larger prey, compared to solitary predators. The mass attack behavior of bark beetles is critical to successful colonization of living trees. Coexisting caddisfly larvae can modify substrate conditions and near-surface water velocity, thereby enhancing food delivery (Cardinale et al., 2002). Animals in groups are more difficult for predators to attack. Mutualists can impose sanctions that reduce fitness of cheaters to maintain benefits (Jandér and Herre, 2010).

Reciprocal cooperation and sanctions against cheaters reflect selection via feedback from individual effects on their environment. The strength of individual effects on the environment is greatest among directly interacting individuals and declines from the population to community levels (Ballhorn et al., 2013; Lewinsohn and Price, 1996). Reciprocal cooperation can explain the evolution of sexual reproduction and social behavior as the net result of trade-offs between maximizing the contribution of an individual's genes to its own progeny and maximizing the contribution of its genes to progeny of its relatives. Similarly, species interactions represent trade-offs among positive and negative effects (see Chapter 8).

Population size and distribution in time and space (ie, metapopulation dynamics, see Chapter 7) are major factors affecting interaction strengths. Individuals dispersed in a regular pattern (Chapter 5) over an area will affect a large proportion of the total habitat and interact widely with co-occurring populations, whereas the same total number of individuals dispersed in an aggregated pattern will affect a smaller proportion of the total habitat, but may have a higher frequency of interactions with co-occurring populations in areas of local

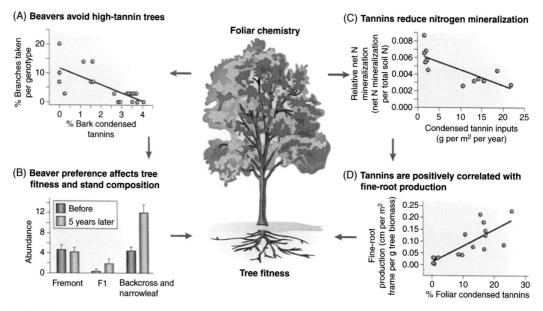

FIGURE 15.3 **Selection for species phenotype affects interactions with other species, that in turn feed back to affect fitness of the individuals that produced that phenotype.** In this example, condensed tannin phenotype in poplar affects herbivore foraging, nutrient turnover, and nutrient acquisition. (A) and (B) Beavers, *Castor canadensis*, select trees with low condensed tannin concentrations, leading to increased abundance of trees with higher condensed tannin concentrations and reduced abundance of trees with lower concentrations, affecting other species that depend on this tree for survival. (C) and (D) Increased foliar tannin concentration inhibits nitrogen mineralization from litter and soil, requiring the tree to produce more fine roots to acquire more limited soil nitrogen. *Source: Reprinted from Whitham, T.G., et al. 2006. A framework for community and ecosystem genetics: from genes to ecosystems. Nat. Rev. Genet. 7, 510–523; (A) data from Bailey, J.K., et al., 2004. Beavers as molecular geneticists: a genetic basis to the foraging of an ecosystem engineer. Ecology 85, 603–608; (C) data from Schweitzer, J.A., et al., 2004. Genetically based trait in a dominant tree affects ecosystem processes. Ecol. Lett. 7, 127–134; (D) data from Fischer, D.G., et al., 2006. Do high-tannin leaves require more roots? Oecologia 149, 668–675.*

abundance. Consistency of population dispersion through time affects the long-term frequency of interactions and reinforcement of selection from generation to generation. Metapopulation dynamics interacting with disturbance dynamics provide the template for selection of species assemblages best adapted to local environmental variation.

Finally, negative feedback among all levels of the ecosystem hierarchy can maintain system stability (Dambacher et al., 2003). Feedback at the genetic level is the basis for selection of stabilizing interactions (Fig. 15.3). Accumulating evidence demonstrates genetic variation, subject to selection, in biotic effects on ecosystem properties (Bailey et al., 2004; Classen et al., 2007a; D. Fischer et al., 2006; Schweitzer et al., 2004, 2005; Shuster et al., 2006; Whitham et al., 2003, 2006; Wimp et al., 2007), as well as responses to changing environmental conditions (Balanyá et al., 2006; W. Bradshaw and Holzapfel, 2001; Edelaar et al., 2008; J. Hsu et al., 2006; Ralph et al., 2006). For example, D. Fischer et al. (2006) found that high tannin production in foliage requires increased root growth to compensate for tannin inhibition of soil N cycling. Thus, reduced N availability at the ecosystem level provides feedback on plant fitness, along with other factors, including herbivory, that favor increased tannin production. Similarly, G. Zhang et al. (2011) reported that grasshopper herbivory maintains plant diversity and transfers nutrients from host plants to nonhost plants, perhaps maintaining nutrient stoichiometry in grasslands.

3 ECOSYSTEMS AS CYBERNETIC SYSTEMS

The cybernetic nature of ecosystems, from patch to global scales, has been a central theme of ecosystem ecology. J. Lovelock (1988) suggested that autotroph–heterotroph interactions have been responsible for the development and regulation of atmospheric composition and climate that are suitable for the persistence of life. The ability of ecosystems to minimize variability in climate and rates of energy and nutrient fluxes also would affect their regulatory responses to anthropogenic changes in global conditions, potentially delaying many effects of anthropogenic activities.

3.1 Properties of Cybernetic Systems

Cybernetic systems generally are characterized by (1) information systems that integrate system components, (2) low-energy feedback regulators that have high-energy effects, and (3) goal-directed stabilization of high-energy processes. Mechanisms that sense deviation (perturbation) in system condition communicate with mechanisms that function to reduce the amplitude and period of deviation. Negative feedback is the most commonly recognized method for stabilizing outputs. A thermostat represents a simple example of a negative feedback mechanism. The thermostat senses a departure in room temperature from a set level and communicates with a temperature control system that interacts with the thermostat to readjust the temperature to the set level. The room system is maintained at temperatures within a narrow equilibrial range. Cruise control provides similar regulation of vehicle speed. However, disturbances to the system, such as power outages, can disrupt this regulation and require repair and reestablishment of regulation.

Organisms are recognized as cybernetic systems with neurological networks for communicating physiological conditions and sensing departures from homeostasis and with various feedback loops for maintaining homeostasis of biological functions. Cybernetic function is perhaps best developed among homeotherms. These organisms are capable of self-regulating internal body temperature through physiological mechanisms that sense change in temperature and trigger changes in metabolic rate, blood flow, and sweat that increase or decrease temperature, as necessary. However, energy demand is high for such regulation (see Chapter 4), and many homeotherms hibernate or estivate to conserve energy during cold or hot seasons, respectively. Heterotherms also have physiological and behavioral mechanisms for adjusting body temperature within a somewhat wider range but with lower energy demand (see Chapters 2 and 4). Furthermore, heterotherms have other mechanisms for maintaining homeostasis in body conditions. Regardless of mechanism, the result is sufficient stability of metabolic processes for survival within normal ranges of environmental variation.

Although self-adjusting mechanical or electronic systems and organisms are the best-recognized examples of cybernetic systems, the properties of self-regulating systems have analogues at supra-organismal levels (W. Hamilton, 1964; B. Patten and Odum, 1981; Schowalter, 2000). Human families and societies express goals in terms of survival, economic growth, improved living conditions, etc., and accomplish these goals culturally through governing bodies, communication networks, and balances between reciprocal cooperation (eg, trade agreements, treaties) and negative feedback (eg, economic regulations, warfare). Social insects similarly accomplish colony survival goals through communication networks and adjustment of reproduction, caste proportions, and foraging activity, as necessary, to maintain sufficient resource acquisition, defense, etc. (K.L. Christensen et al., 2010; Dong et al., 2008; Halloy et al., 2007; L. Mao and Henderson, 2010; Šobotník et al., 2012; B. Taylor et al., 2010).

3.2 Ecosystem Homeostasis

E. Odum (1969) presented a number of testable hypotheses concerning ecosystem capacity to develop and maintain homeostasis, in terms of energy flow and biogeochemical cycling, during succession. Although subsequent research has shown that many of the predicted trends are not observed, at least in some ecosystems, Odum's hypotheses focused debate on ecosystems as cybernetic systems. Engelberg and Boyarsky (1979) argued that ecosystems do not possess the critical goal-directed communication and low-cost/large-effect feedback systems required of cybernetic systems. Although ecosystems can be shown to possess these properties of cybernetic ecosystems, as described later in this section, this debate cannot be resolved until ecosystem ecologists reach consensus on a definition and measurable criteria of stability and demonstrate that potential homeostatic mechanisms, such as biodiversity and herbivorous insects (see sections 3.4 and 3.5), function to reduce variability in ecosystem conditions.

Although discussion of ecosystem goals appears to be teleological, nonteleological goals can be identified, for example, maximizing distance from thermodynamic ground (see B. Patten, 1995), a requisite for all life, or maximizing persistence of the community. Stabilizing ecosystem conditions obviously would reduce exposure of all individuals and populations to extreme, and potentially lethal, departures from normal values of resource supply and other environmental variables and thereby contribute to community persistence. Furthermore, stable population sizes for any species would prevent extreme fluctuations in abundances that would jeopardize populations of interacting species and the ecosystem processes they control. Hence, over long periods of time, environmental heterogeneity might select for individual traits that contribute to stability of ecosystem conditions.

The argument that ecosystems do not possess centralized mechanisms for communicating departure in system condition and initiating responses (Engelberg and Boyarsky, 1979) predated recognition of the pervasive chemical communication network in ecosystems, that includes communication of herbivory among unrelated plant species and of the presence of predators (Hawlena and Schmitz, 2010; Hawlena et al., 2012; Heil and Karban, 2010, see Chapters 3 and 8). The importance of volatile chemicals for communicating resource conditions among species has been recognized relatively recently (I. Baldwin and Schultz, 1983; Dolch and Tscharntke, 2000; Farmer and Ryan, 1990; Heil and Karban, 2010; Karban, 2001; Rhoades, 1983; Sticher et al., 1997; M. Stout et al., 2006; Thaler et al., 2001; Turlings et al., 1990;

Zeng et al., 2009; Zeringue, 1987). The airstream carries a blend of volatile chemicals, produced by the various members of the community, that advertises the abundance, distribution, and condition of various organisms within the community and provides attractive or repellent cues (Raguso et al., 2015). Changes in the chemical composition of the local atmosphere indicate changes in the relative abundance and suitability of hosts or the presence and proximity of competitors and predators. Sensitivity among organisms to the chemical composition of the atmosphere or water column provides a global information network that communicates conditions for a variety of populations and initiates feedback responses through a network of indirect effects. For example, fear of predation can induce changes in herbivore physiology and foraging activity that can alter primary production and decomposition, among other potential ecosystem processes (Finke, 2012; Hawlena and Schmitz, 2010; Hawlena et al., 2012; Long and Finke, 2015, see Chapter 8).

Feedback loops are the primary mechanisms for maintaining ecosystem stability, regulating abundances and interaction strengths (W. Carson and Root, 2000; Dambacher et al., 2003; de Ruiter et al., 1995; B. Patten and Odum, 1981; Polis et al., 1997a,b, 1998). The combination of direct and indirect bottom-up (resource availability), top-down (predation), and lateral (competitive) interactions (see Chapters 6 and 8) generally represent negative feedback, stabilizing food webs by reducing the probability that populations increase to levels that threaten their resources (and, thereby, other species requiring those resources). G. Zhang et al. (2011) suggested that grasshopper-induced fluxes of nutrients from host plant species to nonhost species may maintain nutrient balance in grassland ecosystems. At the next trophic level, Hawlena and Schmitz (2010) and Hawlena et al. (2012) reported that fear induced by predation risk altered grasshopper physiology in ways that cascaded through carbon and nitrogen cycles, potentially regulating nutrient fluxes, as well as prey densities. Mutualistic interactions and other positive feedbacks reduce the probability of population decline to extinction thresholds. Although positive feedback often is viewed as destabilizing, such feedback may be most important when populations are small and should be limited by negative feedbacks as populations grow beyond threshold sizes (Ulanowicz, 1995).

Such compensatory interactions may maintain ecosystem properties within relatively narrow ranges, despite spatial and temporal variation in abiotic conditions (Kratz et al., 1995; Ulanowicz, 1995). Herbivores typically maintain plant species diversity (Cairns et al., 2008; Coleman et al., 2008; G. Zhang et al., 2011), which provides insurance against the consequences of reduced vegetation cover (Tilman and Downing, 1994). Omnivory increases ecosystem stability, perhaps by increasing the number of linkages subject to feedback (Fagan, 1997). Ecological succession represents one mechanism for recovery of ecosystem properties following disturbance-induced departures from nominal conditions (see Chapters 10 and 11).

The concept of self-regulation does not require efficient feedback by all ecosystems or ecosystem components. Just as some organisms (recognized as cybernetic systems) have greater homeostatic ability than do others (eg, homeotherms vs heterotherms), some ecosystems demonstrate greater homeostatic ability than do others (J. Webster et al., 1975). Frequently disturbed ecosystems may be reestablished by relatively random assemblages of opportunistic colonists; frequent disturbance would select genes for rapid exploitation and dispersal. Frequent disturbance provides little opportunity for repeated interaction that could lead to stabilizing cooperation (see Axelrod and Hamilton, 1981). Some species increase variability or promote disturbance, for example, brittle or flammable species (eg, easily toppled Cecropia and flammable *Eucalyptus*), and epitomize dominance of "selfish" species. Insect outbreaks increase variation in some ecosystem parameters (Romme et al., 1986), largely in the short term and often in ways that promote diversity and regeneration of resources (Cairns et al., 2008; Coleman et al., 2008; Schowalter et al., 1981a; G. Zhang et al., 2011). On the other hand, relatively stable and/or diverse ecosystems, such as tropical rainforests, also might not select for stabilizing interactions (Connell, 1980). However, stable environmental conditions should favor consistent species interactions and the evolution of reciprocal cooperation, such as demonstrated by a diversity of mutualistic interactions in tropical forests. Selection for stabilizing interactions should be greatest in ecosystems characterized by intermediate levels of environmental variation. Interactions that reduce such variation would contribute to individual inclusive fitnesses.

3.3 Definition of Stability

B. Patten and Odum (1981) proposed that a number of time-invariant or regularly oscillating ecosystem parameters represent potential goals for stabilization. These included total system production (P) and respiration (R), P:R ratio, total chlorophyll, total biomass, nutrient pool sizes, species diversity, population sizes, etc. However, the degree of spatial and temporal variability of these parameters remains poorly known for most, even intensively studied, ecosystems (Kratz et al., 1995), although long-term studies are clarifying these parameters (Peters et al., 2013).

Kratz et al. (1995) compiled data on the variability of climatic, edaphic, plant, and animal variables from 12 Long Term Ecological Research (LTER) Sites, representing forest, grassland, desert, lotic, and lacustrine ecosystems, in the USA. Unfortunately, given the common long-term goals of these projects, comparison was limited because different variables and measurement techniques were represented among these sites. Nevertheless, Kratz et al. offered several important conclusions concerning variability.

First, the level of species combination (eg, species, family, guild, total plants or animals) had a greater effect on observed variability in community structure than did spatial or temporal extent of data (see Chapter 9). For plant parameters, species- and guild-level data were more variable than were data for total plants; for animal parameters, species-level data were more variable than were guild-level data, and both were more variable than were total animal data. As discussed in Chapter 9, the tendency to ignore diversity, especially of insects (albeit for logistic reasons), clearly affects our perception of variability. Detection of long-term trends or spatial patterns depends on data collection for parameters sufficiently sensitive to show significant differences but not so sensitive that their variability hinders detection of differences.

Second, spatial variability exceeded temporal variability. This result indicates that individual sites are inadequate to describe the range of variation among ecosystems within a landscape. Variability must be examined over larger spatial scales. Edaphic data were more variable than were climatic data, indicating high spatial variation in substrate or other site properties, whereas common weather across landscapes homogenizes microclimatic conditions. This result also could be explained as the result of greater biotic modification of climatic variables, compared to substrate variables (see Chapter 11).

Third, biotic data were more variable than were climatic or edaphic data. Organisms can exhibit exponential responses to incremental changes in abiotic conditions (see Chapter 6). The ability of animals to move and alter their spatial distribution quickly in response to environmental changes is reflected in greater variation in animal data, compared to plant data. However, animals also have greater ability to hide or escape sampling devices.

Finally, two sites, a desert and a lake, provided a sufficiently complete array of biotic and abiotic variables to permit comparison. These two ecosystem types represent contrasting properties. Deserts are exposed to highly variable and harsh abiotic conditions, but are interconnected within landscapes, whereas lakes exhibit relatively constant abiotic conditions (buffered from thermal change by mass and latent heat capacity of water, from pH change by bicarbonates, and from biological invasions by their isolation), but are isolated by terrestrial barriers. Comparison of variability between these contrasting ecosystems supported the hypothesis that deserts are more variable than lakes among years, but lakes are more variable than deserts among sites.

Kratz et al. (1995) provided important data on variation in a number of ecosystem parameters among ecosystem types. However, important questions remain. Which parameters are most important for stability? How much deviation can be tolerated? What temporal and spatial scales are relevant to ecosystem stability?

Among the parameters that could be stabilized as a result of species interactions, net primary production and biomass structure (living and dead) may be particularly important, because many ecosystem variables, including energy, water and nutrient fluxes, trophic interactions, species diversity, population sizes, climate modification, and soil development, are directly or indirectly determined by net primary production or biomass structure (Boulton et al., 1992; Classen et al., 2005, see Chapter 11). In particular, the ability of ecosystems to modify climate, protect and modify soils, and provide stable resource bases for primary and secondary producers depends on NPP and biomass structure. Therefore, natural selection over long periods of coevolution should favor individuals whose interactions stabilize these ecosystem parameters. NPP may be stabilized over long time periods as a result of compensatory community dynamics and biological interactions, such as those resulting from biodiversity and herbivory (see Chapters 12–14 and Sections 3.4 and 3.5).

No studies have addressed the limits of deviation, for any parameter, within which ecosystems can be regarded as conditionally stable. Traditional views of stability have emphasized consistent species composition, at the local scale, but shifts in species composition may be a mechanism for maintaining stability in other ecosystem parameters at the landscape or watershed scale. This obviously is an important issue for evaluating stability and predicting effects of global environmental changes. However, given the variety of ecosystem parameters and their integration at the global scale, within the context of anthropogenic change, this issue will be difficult to resolve.

The range of parameter values within which ecosystems are conditionally stable may be related to characteristic fluctuations in environmental conditions or nutrient fluxes. For example, biomass accumulation increases ecosystem storage capacity and ability to resist variation in resource availability (J. Webster et al., 1975), but also increases ecosystem vulnerability to some disturbances, including fire and storms. Complex ecosystems with high storage capacity (ie, forests) are the most buffered ecosystems, in terms of regulation of climate, soil conditions, and resource supply, but also fuel the most catastrophic fires under drought conditions and suffer the greatest damage during cyclonic storms. Hence, ecosystems with lower biomass, but rapid turnover of matter or nutrients, may be more stable under some environmental conditions. Species interactions that periodically increase rates of nutrient fluxes and reduce biomass (eg, herbivore outbreaks) traditionally have been viewed as evidence of instability, but may contribute to the stability of ecosystems in which biomass accumulation or rates of nutrient turnover from detritus are destabilizing (de Mazancourt et al., 1998; Loreau, 1995), such as resulting from anthropogenic changes.

No studies have addressed the appropriate temporal and spatial scales over which stability should be evaluated, or whether these scales should be the same for all ecosystems. Most studies of ecosystem processes represent periods of less than 10 years, although some ecosystem studies now span 50–100 years (Peters et al., 2013). The long timescales representing processes such as succession exceed the scale of human lifetimes and have required substitution of temporal variation by spatial variation (eg, chronosequences within a landscape). Data from such studies have limited utility, because individual patches have unique conditions, reflecting substrate and disturbance history (Boulton et al., 1992; Harding et al., 1998; Schowalter et al., 2003; Summerville et al., 2009), and are influenced by conditions in surrounding patches (Haynes and Cronin, 2003; Kratz et al., 1995; Woodwell, 1993; J. Zimmerman et al., 2010). Therefore, temporal changes at the patch scale often follow different successional trajectories (see Chapter 10).

Boulton et al. (1992) compared rates and directions of benthic aquatic invertebrate succession following flash floods of varying magnitude among seasons in a desert stream in Arizona, USA, over a 3-year period. Several flash floods occurred each year, but the interval between floods was long relative to the life spans of the dominant fauna. Invertebrate assemblage structure changed seasonally but was highly resistant and resilient to flooding disturbance, that is, displacements resulting from flooding were less than were seasonal changes. By summer, robust algal mats supported dense invertebrate assemblages that were resistant to flooding disturbance. By autumn, algal mat disruption made the associated invertebrate community more vulnerable to flooding disturbance. Assemblages generally returned to preflood structure, although trajectories varied widely. Long-term community structure was relatively consistent, despite unpredictable short-term changes.

Van Langevelde et al. (2003) proposed a model of African savanna dynamics in which alternate vegetation states cycle over time as a result of the interactive effects of fire and herbivory. Positive feedback between grass biomass and fire intensity is disrupted by grazing, which reduces fuel load, fire intensity, and tree seedling mortality. Increased woody vegetation causes a change in state from grass dominance to tree dominance. Browsers respond to increased tree abundance, reducing woody biomass and stimulating grass growth, causing the cycle to repeat. Joseph et al. (2014), Mapaure and Moe (2009), and Shannon et al. (2011) demonstrated that termites, *Macrotermes* spp., and elephants, *Loxodonta africana*, control these transitions. Such a system may be relatively stable over long time periods but appear unstable during short transition periods.

Although individual patches may change dramatically over time, or recover to variable endpoints, the dynamic mosaic of ecosystem types (eg, successional stages or meta-community types) at the landscape or watershed scale may stabilize the proportional area represented by each ecosystem type and maintain the regional species pool required for recolonization during recovery from disturbances (Cottonie, 2005; Cottonie and de Meester, 2004; Leibold and Mikkelson, 2002; Presley et al., 2010; Zimmerman et al., 2010, see Chapter 10). Changing land use practices have disrupted this conditionally stable heterogeneity of patch types at the landscape scale and resulted in isolated community fragments, vulnerable to elimination by subsequent disturbance.

Finally, the time frame of stability must be considered within the context of the ecosystem. For example, forests appear to be less stable than grasslands because of the long time period required for recovery of forests to predisturbance conditions, compared to rapid refoliation of grasses from surviving underground rhizomes. However, forests typically are disturbed less frequently. Net primary production may recover to predisturbance levels within 2–3 years, although biomass requires longer periods to reach predisturbance levels (Boring et al., 1988; Scatena et al., 1996; J. Zimmerman et al., 1996).

3.4 Regulation of NPP by Biodiversity

The extent to which biodiversity contributes to ecosystem stability has been highly controversial (Hooper et al., 2005, see Chapter 10). Different species have been shown to control different aspects of ecosystem function, for example, primary production, decomposition, and nutrient fluxes, demonstrating that biodiversity in its broadest sense determines ecosystem function (Beare et al., 1995; Hättenschwiler and Gasser, 2005; Spehn et al., 2005; Vitousek and Hooper, 1993; Waide et al., 1999; Woodwell, 1993). The presence or absence of individual species, especially keystone species, affects biotic, atmospheric, hydrospheric, and substrate conditions (Downing and Leibold, 2002).

E. Allan et al. (2013) evaluated 418 separate measures of 38 ecosystem processes and found that, overall, 45% of these processes were significantly affected by plant species richness and that the effect of biodiversity became stronger over time. However, relatively few species have been studied sufficiently, under different conditions, to evaluate their effects on ecosystem functions. The debate depends, to a large extent, on definitions and measures of stability (as discussed earlier) and diversity (see Chapter 9).

Vitousek and Hooper (1993) suggested that the relationship between biodiversity and ecosystem function could take several forms. Their Type 1 relationship implies that each species has the same effect on ecosystem function. Therefore, the effect of adding species to the ecosystem is incremental, producing a line with constant slope. The Type 2 relationship represents a decreasing and eventually disappearing effect of additional species, producing a curve that approaches an asymptote. The Type 3 relationship indicates no further effect of additional species.

As demonstrated in Chapters 8–10, communities are not random assemblages of species but, rather, functionally linked groups of species (Bastolla et al., 2009). Therefore, the Type 2 relationship probably represents most ecosystems, with additional species contributing incrementally to ecosystem function and stability until all functional groups are represented (Gagic et al., 2015; Vitousek and Hooper, 1993). Further additions have progressively smaller effects, as species packing within functional groups simply redistributes the overall contribution among species. Hence, ecosystem function is not linearly related to diversity (Waide et al., 1999), but increased diversity within functional groups provides insurance against loss of individual species and their functional roles (Long and Finke, 2014; Reich et al., 2012; Roscher et al., 2011).

Functional diversity may be more important than species diversity for maintaining ecosystem processes (Gagic et al., 2015; Griffiths et al., 2015; Peralta et al., 2014). Within-group diversity could affect the persistence or sustainability of a given function, albeit with variation among species in its rate or regulation, and thereby increase the reliability of that function over space and time. According to the **insurance hypothesis** (Yachi and Loreau, 1999), species **redundancy** (species using the same resources) and **complementarity** (species with different resource-use strategies and functional traits that permit specialization on different resources or on the same resource but in different locations or times) within functional groups

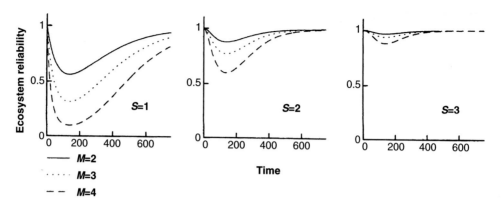

FIGURE 15.4 **Ecosystem reliability over time as a function of the number of functional groups (M) and num-ber of species per functional group (S), for a probability of species colonization over time of 0.005 and a prob-ability of species presence over time of 0.005.** *Source: From Naeem, S., 1998. Species redundancy and ecosystem reliability. Conserv. Biol. 12, 39–45.*

can minimize variation in ecosystem processes by maintaining functions when some species decline in abundance as a result of environmental changes (Fig. 15.4) (E. Allan et al., 2013; Griffiths et al., 2015; Naeem, 1998; Naeem and Li, 1997; Peralta et al., 2014; Reich et al., 2012; Rogers et al., 2014; Roscher et al., 2011; Spehn et al., 2005). However, Yachi and Loreau (2007) demonstrated that competitive balance also plays a role in maintaining ecosystem function.

Food web structure also may be important to stability of ecosystem processes. Thébault and Fontaine (2010) applied a modeling approach to 34 empirical pollination networks and 23 empirical herbivory networks to demonstrate that a highly connected and nested architecture promotes community stability in mutualistic networks, whereas a compartmented and weakly connected architecture promotes the stability of trophic networks.

Tilman et al. (1997) reported that both plant species diversity and functional diversity significantly influenced six ecosystem response variables, including primary productivity and nitrogen pools in plants and soil, when analyzed in separate univariate regressions, but that only functional diversity significantly affected these variables in a multiple regression. Hooper and Vitousek (1997) also found that variability in ecosystem parameters was significantly related to the composition of functional groups, rather than the number of functional groups, further supporting the concept of complementarity among species or functional groups. Fukami et al. (2001) investigated the mathematical relationship between such compartmentalized biodiversity and ecosystem stability. They concluded that biodiversity loss reduces similarity in species composition among local communities and thereby reduces the reliability (stability) of continued ecosystem processes.

Dominant organisms in all ecosystems are adapted to survive environmental changes or disturbances that recur regularly with respect to generation time. Therefore, adaptation to prevailing conditions (evolution) constitutes a feedback that reduces ecosystem deviation from nominal conditions. For example, many grassland and pine forest species are adapted to survive low-intensity fires and drought (eg, underground rhizomes and insulating bark, respectively) that characterize these ecosystems, thereby stabilizing vegetation structure and primary production. Nevertheless, dominant species are not necessarily most important for maintenance of ecosystem function or stability (Hooper et al., 2005). Rare species, such as top predators, may function as keystone species to maintain ecosystem diversity and stability.

For example, Støen et al. (2013) found that termites were primary determinants of plant diversity and composition on mounds, whereas ungulate grazing tended to regulate abundances of the most common plant species on the landscape. As a result of this combined effect, plant species richness, evenness, and α and β diversity (and, presumably, primary production) were more consistent than would occur in the absence of either or both animal groups.

Diverse communities may be more resistant to spread of host-specific insects or pathogens, or to invasive species (Hooper et al., 2005, see Chapter 6). However, spread of generalists may increase with diversity, where diversity ensures a greater proportion of hosts that provide balanced nutritional requirements (Ostfeld and Keesing, 2000).

All ecosystems are subject to periodic catastrophic disturbances and subsequent community recovery through species turnover (succession). Ecosystem diversity at large spatial or temporal scales may buffer the ecosystem from disturbance or provide the species pool for reestablishment of key species, from neighboring patches or seed banks (Hooper et al., 2005; Reusch et al., 2005). The rapid development of early successional communities limits loss of ecosystem assets, especially soil and limiting nutrients. Hence, succession represents a mechanism for reducing deviation in ecosystem parameters, but some early- or mid-successional stages are capable of inhibiting further succession. Herbivores may be instrumental in facilitating the replacement of inhibitive successional stages under suitable conditions (Chapter 10).

Few studies have measured the effect of biodiversity on stability of ecosystem parameters. Most are based on selection of plots that differ in plant species diversity and, therefore, potentially are confounded by other factors that could have produced differences in diversity among plots.

McNaughton (1985, 1993b) studied the effects of plant species diversity on the persistence and productivity of biomass in grazed grasslands in the Serengeti Plain in East Africa. Portions of areas differing in plant diversity were fenced to exclude ungulate grazers. Stability was measured as both resistance (change in productivity resulting from grazing) and resilience (recovery to fenced control condition following cessation of grazing). Grazing reduced diversity 27% in more diverse communities but had no effect on less diverse communities. The percentage of biomass eaten was 67% and 76% in the more and less diverse communities, respectively, a nonsignificant difference. By 4 weeks after cessation of grazing, the more diverse communities had recovered to 89% of control productivity, but the less diverse communities recovered to only 31% of control productivity, a significant difference.

McNaughton (1977, 1993b) also compared resistance of adjacent grasslands of differing diversities to environmental fluctuation. Stability, measured as resistance to deviation in photosynthetic biomass, increased with diversity, as a result of compensation between species with rapid growth following rain but rapid drying between showers and species with slower growth after showers but slower drying between showers. Eight of ten tests demonstrated a positive relationship between diversity and stability (McNaughton, 1993b).

D. Frank and McNaughton (1991) similarly compared effects of drought on plant species composition among communities of differing diversities in Yellowstone National Park in the western USA. Stability of species composition to this environmental change was strongly correlated to diversity.

Ewel (1986) and Ewel et al. (1991) evaluated effects of experimental manipulation of plant diversity on biogeochemical processes in a tropical rainforest in Costa Rica. This study included five treatments: a diverse natural succession, a modified succession with the same number and growth form of successional species but no species in common with natural succession, an enriched species diversity with species added to a natural succession, a crop monoculture (replicates of three different crop species), and bare ground (vegetation-free). After 5 years this design yielded plots with no plants (vegetation-free), a single species (monoculture), > 100 species (natural and modified succession), and 25% more species (enriched succession). Nutrient pool sizes always were significantly larger in the more diverse plots, reflecting a greater variety of mechanisms for retention of nutrients and maintenance of soil processes favorable for plant production. The results suggested a Type 2 relationship between biodiversity and stability (Vitousek and Hooper, 1993), with most change occurring at low species diversity. However, the absence of intermediate levels of diversity, between the monoculture and > 100 species treatments, limited interpolation of results.

Tilman and Downing (1994) established replicated plots, in 1982, in which the number of plant species was altered through different nitrogen fertilization treatments. These plots subsequently (1987–88) were subjected to a record drought. During the drought, plots with > 9 species averaged about half of their predrought biomass, but plots with < 5 species averaged only about 12% of their predrought biomass (Fig. 15.5). Hence, the more diverse plots were better buffered against this disturbance because they were more likely to include drought-tolerant species, compared to less diverse plots. More diverse plots also recovered biomass more quickly following the drought. When biomass was remeasured in 1992, plots with ≥ 6

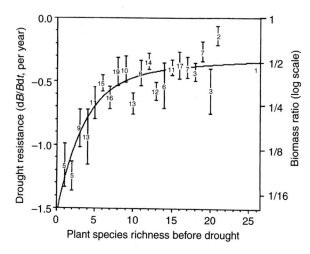

FIGURE 15.5   **Relationship between plant species diversity prior to drought and drought resistance in experimental grassland plots planted with different species diversities.** Mean, standard error, and number of plots with given species richness are shown. 1 dB/Bdt (per year) = 0.5 ln(1988 biomass/1986 biomass), where 1988 was the peak drought year and 1986 was the year preceding drought. The biomass 1988/1986 ratio (right-hand scale) indicates the proportional decrease in plant biomass associated with dB/Bdt values. *Source: From Tilman, D., Downing, J.A., 1994. Biodiversity and stability in grasslands. Nature 367, 363–365.*

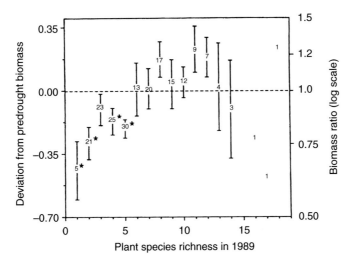

FIGURE 15.6   **Relationship between plant species diversity and deviation in 1992 biomass (following drought) from mean (1982–86) predrought biomass in experimental grassland plots planted with different species diversities.** Mean, standard error, and number of plots with given species richness are shown. Negative values indicate 1992 biomass lower than predrought mean. Biomass ratio is biomass 1992/predrought. Plots with 1, 2, 4, or 5 species (but not plots with > 5 species) differed significantly from predrought means. *Source: From Tilman, D., Downing, J.A., 1994. Biodiversity and stability in grasslands. Nature 367, 363–365.*

species had biomass equivalent to predrought levels, but plots with ≤ 5 species had significantly lower biomass, with deviations of 8–40% (Fig. 15.6). Tilman and Downing (1994) and Tilman et al. (1997) concluded that more diverse ecosystems represented a greater variety of ecological strategies that confer both greater resistance and greater resilience to environmental variation. Jentsch et al. (2011) found similar compensatory responses that stabilized processes at the ecosystem level in response to recurrent experimental drought treatment in central Europe. However, the contribution of diversity to ecosystem stability may be related to environmental heterogeneity, that is, diversity does not necessarily increase stability in more homogeneous environments.

Spehn et al. (2005) manipulated plant diversity in multiple European grassland ecosystems. They found that more diverse communities were more productive and utilized resources more completely, through greater occupation of available space and uptake and retention of nitrogen, than did less diverse communities. Reusch et al. (2005) demonstrated that experimentally increased genotypic diversity of a common seagrass, *Zostera marina*, enhanced biomass production, plant density, and faunal abundances, thereby buffering the coastal community against extreme temperatures. Vogel et al. (2012) concluded that low mowing frequency was more important for sustained primary production during drought than was species richness, but that species richness increased primary production in all management treatments under both ambient and drought conditions.

A number of studies have demonstrated that ecosystem resistance to elevated herbivory or plant disease is positively correlated to vegetation diversity (Garrett and Mundt, 1999; M.T. Johnson et al., 2006; Knops et al., 1999; Y. Liu et al., 2015; McNaughton, 1985; Schowalter and

Lowman, 1999; Schowalter and Turchin, 1993, Chapters 6 and 7). As vegetation diversity increases, relative to the host range of any particular herbivore, the ability of herbivores to find and exploit their hosts decreases (Jactel and Brockerhoff, 2007), leading to increasing stability of herbivore–plant interactions. Even genetic variation in a dominant plant species can affect herbivory (Hochwender and Fritz, 2004; M.T. Johnson et al., 2006; Wimp et al., 2007). On the other hand, overall herbivory may increase as plant diversity increases, as a result of a greater diversity of herbivores (Loranger et al., 2014).

Experimental studies relating ecosystem stability to diversity generally have been limited to manipulation of plant species diversity (Hooper et al., 2005; Spehn et al., 2005). However, insects represent the bulk of diversity and a substantial portion of consumer biomass in virtually all ecosystems (eg, Tables 9.1 and 11.1) and are capable of controlling a variety of ecosystem conditions (Chapters 12–14). A few studies have addressed the significance of diversity at higher trophic levels to ecosystem processes, but not to ecosystem stability (Downing and Leibold, 2002; Laakso and Setälä, 1999; Lewinsohn and Price, 1996; Long and Finke, 2014).

B. Klein (1989) found that diversity of dung beetles (Scarabaeidae) and the rate of dung decomposition were positively correlated to the size of forest fragments in central Amazonia. However, abiotic conditions that also affect decomposition likely differed among fragment sizes, as well. Larsen et al. (2005) discovered that dung burial rate increased with the diversity of dung beetle species retained among islands created by construction of a hydroelectric dam in Venezuela. Furthermore, absence of some species caused a greater reduction in dung burial rate than did absence of other species, indicating that species composition is as important as richness.

Coûteaux et al. (1991) manipulated diversity of decomposer communities in microcosms with ambient or elevated concentrations of CO_2. They found that decomposition and respiration rates were significantly related to decomposer diversity, as affected by species shifts following CO_2 treatment. This study demonstrated an effect of biodiversity on rates of a key ecosystem process, but did not address the long-term stability of this process.

Downing and Leibold (2002) evaluated the effects of manipulated species composition nested within multitrophic diversity treatments in pond mesocosms. The effect of species composition on productivity, respiration, and decomposition was equivalent to, or greater than, the effect of diversity per se. Productivity was highest in the highest diversity treatments.

Although herbivores or predators have been augmented or excluded en masse, for evaluation of effects on lower trophic levels, including trophic cascades (see Chapter 8), herbivore and predator diversities have not been experimentally manipulated in terrestrial ecosystems to evaluate the effect of diversity at these levels on processes at lower trophic levels. Manipulations of herbivore or predator diversities generally have addressed questions concerning biological control of pests. These manipulations may not represent interactions in natural ecosystems.

For example, McEvoy et al. (1993) manipulated the abundances of two insect species with complementary feeding strategies (cinnabar moth, *Tyria jacobaeae*, a foliage and inflorescence feeder, and ragwort flea beetle, *Longitarsus jacobaeae*, a root feeder) introduced to control the exotic ragwort, *Senecio jacobaea*, in coastal Oregon, USA. Their results indicated that increasing diversity (from no herbivores to one herbivore to both herbivores) decreased local stability of the herbivore–plant interaction, as increasing herbivory drove the host to local extinction, at the plot scale. However, this plant species persisted at low densities over the landscape, suggesting that the interaction was stable at larger spatial scales. Croft and Slone (1997) reported that European red mite, *Panonychus ulmi*, abundances in apple orchards were maintained at lower, equilibrial, levels by three predaceous mite species in combination than by any single predaceous species. Letourneau et al. (2009) conducted a metaanalysis of studies on effects of predator diversity on herbivores and concluded that increased predator diversity generally increased top-down regulation of herbivore populations. However, the particular composition of predator communities and the degree of redundancy or complementarity may be more important than is diversity per se (Long and Finke, 2014; Peralta et al., 2014).

Ultimately, the capacity of ecosystems to endure or modify the potential range of environmental conditions is the primary measure of stability (Hooper et al., 2005) (Fig. 15.4). In this regard, Boucot (1990) noted that the fossil record demonstrates that characteristic species assemblages (hence, ecosystems) often have persisted for many thousands of years over large areas. Furthermore, studies of current ecosystems must account for their degree of departure from preanthropogenic conditions, including adjustments for altered climate, disturbance regime, invasive species, and loss of native species, that likely affect their ability to maintain stability.

3.5 Regulation of NPP by Insects

Mattson and Addy (1975) introduced the hypothesis that phytophagous insects regulate primary production, based on observations that herbivory is typically focused on dense and/or stressed host plants and functions to reduce primary production in a density-dependent manner similar to predation. Schowalter (1981) proposed that herbivore outbreaks, triggered by host stress and density as resources become limiting, advance succession from communities with high demands for resources to communities with lower demands for resources. Davidson (1993) and Schowalter and Lowman (1999) refined this hypothesis by noting that herbivores and granivores can advance, retard, or reverse succession, depending on environmental conditions. Belovsky and Slade (2000) demonstrated that grasshoppers can accelerate nitrogen cycling and increase primary productivity, especially by plants that are better competitors when nitrogen is more available, at intermediate levels of herbivory. At low levels of herbivory, grasshoppers had too little influence on nitrogen cycling to affect primary production, and at high levels grasshoppers depressed plant growth and survival more than could be offset by increased nitrogen cycling and plant productivity.

Despite the obvious influence of animals on key ecosystem processes, their potential regulatory role has remained controversial and largely untested. Herbivorous insects possess the characteristics of cybernetic regulators (ie, low maintenance cost and rapidly amplified effects, sensitivity to deviation in ecosystem parameters, and capacity to dramatically alter primary production through positive and negative feedback) and appear, in many cases, to stabilize net primary production. For example, inconsequential biomass of phytophagous insects, even at outbreak densities, is capable of removing virtually all foliage from host plants and altering plant species composition (see Chapter 12). Similarly, small biomass of termites accounts for substantial decomposition, soil redistribution, and gas fluxes that could affect global climate (see Chapter 14). The following model for insect effect on ecosystem stability focuses on herbivores, but detritivores, pollinators, and seed dispersers also are capable

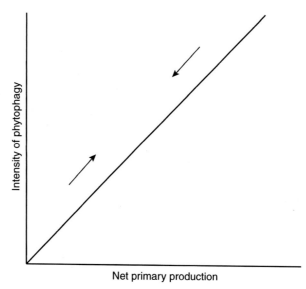

FIGURE 15.7 **Stimulation of primary production at NPP < K and suppression of primary production at NPP > K by phytophages could stabilize primary production.** *Source: Redrawn from Schowalter, T.D., 2000. Insects as regulators of ecosystem development. In: Coleman, D.C., Hendrix, P. (Eds.), Invertebrates as Webmasters in Ecosystems. CAB International, Wallingford, UK, pp. 99–114.*

of modifying ecosystem conditions in ways that might promote stability, for example, decomposer enhancement of nutrient availability, plant growth, and herbivory (Holdo and McDowell, 2004) as feedback on herbivore effects on litter quality and availability (S. Chapman et al., 2003; Whitham et al., 2006).

Primary production often peaks at low-to-moderate intensities of pruning and thinning (Fig. 12.5), which often are employed by crop managers to increase commodity production, supporting the grazing optimization hypothesis (Belovsky and Slade, 2000; S. Williamson et al., 1989). Herbivores can stimulate primary production at low levels of herbivory, when host density is low or condition good, and reduce host production at high levels, when host density is high or condition poor (Fig. 15.7), potentially stabilizing primary production at intermediate levels (Alfaro and Shepherd, 1991; Belovsky and Slade, 2000; Romme et al., 1986; S. Williamson et al., 1989). Elevated primary productivity following outbreaks suggests alleviation of stressful conditions that triggered the outbreak and could lead to instability, but also reflects greater species turnover as stressed hosts are replaced by plants or species better able to tolerate prevailing conditions (J. Chase et al., 2000; Belovsky and Slade, 2000; Ritchie et al., 1998). By stabilizing primary production, herbivores also stabilize processes controlled by primary production, including climate and soil conditions, biogeochemical fluxes, etc., that affect survival and reproduction of associated organisms. Romme et al. (1986) reported that mountain pine beetle, *Dendroctonus ponderosae*, outbreaks appeared to increase variation (destabilization) of some ecosystem properties, but some stands recovered preoutbreak NPP within 10 years following the outbreak. Significantly, these outbreaks represented a biotic response to an anthropogenic deviation in primary production, that is, increased tree density resulting from fire suppression. As outlined in the opening of this chapter, insect outbreaks often reverse anthropogenic increases in host density and return the ecosystem to conditions that prevailed prior to anthropogenic change (Schowalter, 2008).

Outbreaks of phytophagous insects are most likely to occur under two interrelated conditions, both of which represent departure from nominal ecosystem conditions that often reflect anthropogenic change (Schowalter, 2012; Schowalter and Lowman, 1999). First, adverse environmental conditions, such as inadequate water or nutrient availability, changing climate, and atmospheric pollution, stress plants and cause changes in physiological conditions that increase suitability for herbivores (see Chapters 2 and 3). High intensities of herbivory under these conditions generally reduce biomass and improve water or nutrient balance (Kolb et al., 1999; Welker and Menke, 1990; S. Wright et al., 1989) or, in extreme cases, reduce biomass of the most stressed plants, regardless of their abundance, and promote replacement by better-adapted plants (J. Chase et al., 2000; Ritchie et al., 1998; Schowalter and Lowman, 1999). Second, high densities of particular plant species, as a result of rapid colonization by pioneer species or artificial planting, enhance host availability for associated herbivores. High intensities of herbivory represent a major mechanism for reversing site dominance by such plant species (Schowalter, 2008; Torres, 1992), facilitating their replacement and increasing plant diversity (Cairns et al., 2008; Coleman et al., 2008).

If communities evolve to minimize environmental variation, then herbivore interactions with disturbances are particularly important. Although outbreaks of herbivores traditionally have been viewed as disturbances (together with events such as fire, storm damage, and drought), their response to host density or stress often appears to reduce the severity of abiotic disturbances. Herbivore outbreaks commonly are induced by drought conditions (Mattson and Haack, 1987; Stige et al., 2007; Tian et al., 2011; T. White, 1969, 1976, 1984; Yu et al., 2009), suggesting that plant moisture stress may be a particularly important trigger for feedback responses that reduce transpiration and improve water balance (Kolb et al., 1999; W. Webb, 1978; Welker and Menke, 1990). Fuel accumulation, as a result of herbivore-induced fluxes of material from living to dead biomass, often predisposes ecosystems to fire in arid environments. Whether such predisposition is stabilizing or destabilizing depends on the degree to which outbreaks modify the severity and temporal or spatial scale of such disturbances. Schowalter et al. (1981a) suggested that herbivore-induced disturbances might occur more regularly with respect to host generation times or stages of ecosystem development, as a result of specific plant–herbivore interactions, and thereby facilitate rapid adaptation to disturbance or postdisturbance conditions. Although such induction of disturbance would seem to increase variation in the short term, accelerated adaptation would contribute to stability over longer time periods. Furthermore, increased likelihood of disturbance during particular seres should maintain that sere on the landscape, contributing to stability over larger spatial scales. The following example demonstrates the potential stabilization of ecosystem properties over the large spatial scales of western North America.

Conifer forests dominate much of western North America. The large, contiguous, lower elevation zone is characterized by relatively arid conditions and frequent droughts that historically maintained a sparse woodland dominated by drought- and fire-tolerant (but shade-intolerant) pine trees and a ground cover of grasses and shrubs, with little understory (Fig. 15.8). Low-intensity ground fires occurred frequently, at intervals averaging 15–25 years, and covered large areas (Agee, 1993), minimizing drought-intolerant vegetation and litter accumulation. The relatively isolated higher elevation and riparian zones were more mesic and

FIGURE 15.8 (A) The relatively arid interior forest region of North America historically was characterized by open-canopied forests dominated by widely spaced drought- and fire-tolerant pines, and by sparse, heterogeneous understories. (B) Fire suppression beginning in the late 1800s has transformed forests into dense, multistoried ecosystems stressed by competition for water and nutrients. Note that remnant pines are surrounded by dense firs that act as fire ladders, increasing the likelihood of catastrophic fire. *Source: From Goyer, R.A., et al., 1998. Current and proposed technologies for bark beetle management. J. Forest. 96 (12), 29–33.*

supported shade-tolerant (but fire- and drought-intolerant) fir and spruce forests. Fire was less frequent (every 150–1000 years), but more catastrophic, at higher elevations as a result of the greater tree densities and understory development that facilitated fire access to tree canopies (Agee, 1993; Veblen et al., 1994).

As a result of fire suppression during the past 120 years, much of the lower elevation zone has undergone succession from pine forest to later successional fir forest (Fig. 15.8), with greatly increased plant biomass, a conspicuous deviation from historic conditions and unsustainable under prevailing environmental conditions. Outbreaks of a variety of folivore and bark beetle species have become more frequent in these altered forests. During mesic periods and in more mesic locations, for example, riparian corridors and higher elevations, the mountain pine beetle has advanced succession by facilitating the replacement of competitively stressed pines by more competitive firs. However, during inevitable drought periods, such as occurred during the 1980s and 2010s, moisture limitation increases the vulnerability of stressed firs to herbivores (Fig. 15.9). Insect-induced mortality of the firs reversed succession by favoring the remaining drought- and fire-tolerant pines. Tree mortality can increase

FIGURE 15.9 **Herbivore modification of succession in central Sierran mixed conifer ecosystems during 1998.** Understory white fir (*Abies concolor*), the late successional dominant, is increasingly stressed by competition for water in this arid forest type. An outbreak of the Douglas-fir tussock moth, *Orgyia pseudotsugata*, has completely defoliated the white fir (brown, defoliated trees), restoring the ecosystem to a more stable, and sustainable, condition dominated by earlier successional, drought- and fire-tolerant, sequoias and pines (green, foliated trees).

the severity and scale of catastrophic fires, that historically were rare in these forests, unless litter decomposition reduces fuel accumulation before fire occurs. However, this altered fire regime likely will be mitigated in ecological time by eventual reestablishment of the pine sere following catastrophic fire. A similar situation has been inferred from insect demography in pine–hardwood forests of the southern USA (Fig. 10.9). Van Langevelde et al. (2003) and Joseph et al. (2014) also suggested a cycle of alternating vegetation states maintained by interaction of fire, termites, and herbivores in African savanna.

To what extent do insects contribute to the stability and "health" of various ecosystems? Until recently, insect outbreaks and disturbances have been viewed as destructive forces. The increased productivity of ecosystems in the absence of fire and insect outbreaks supported a view that resource production could be freed from limitations imposed by these regulators. However, fire now is recognized as an important tool for restoring sustainable (stable) ecosystem conditions and characteristic communities. The outbreak described in Figs. 15.8 and 15.9 caused devastating losses in short-term timber production and required emergency salvage harvest to prevent premarket wood degradation. However, the insect outbreak effectively reversed anthropogenic change in this forest, producing a more sustainable forest structure that is now the recommended management goal for Sierran forests (North et al., 2007). From this perspective, the outbreak contributed to the stability of this forest.

Regulation of primary production by herbivorous insects could stabilize other ecosystem variables, as well. Clearly, experimental studies should address the long-term effects of herbivorous insects on variability of ecosystem parameters. Our management of ecosystem resources, and in particular our approach to managing herbivorous insects, requires that we understand the extent to which herbivores contribute to ecosystem stability. To the extent that insect outbreaks reflect stabilizing feedback, suppression could be counterproductive to ecosystem stability and to the sustainable delivery of ecosystem services (Chapter 16).

4 SUMMARY

The hypothesis that herbivorous insects regulate ecosystem processes is one of the most important and controversial concepts to emerge from research on insect ecology. The extent to which ecosystems are random assemblages of species that simply affect ecosystem processes or are tightly coevolved groups of species that stabilize ecosystem function has important implications for management of ecosystem resources and "pests." Concepts of group selection have developed from and contributed to this hypothesis.

Debate on the concept of group selection has solidified consensus on the dominance of direct selection for individual attributes. However, individual attributes affect other organisms and environmental conditions that generate feedback on individual fitness. Such feedback selection contributes to the inclusive fitness of an individual. The intensity of this feedback is proportional to the relatedness of interacting individuals. The greatest feedback selection is between individuals sharing genes (kin selection). The frequency of interaction and the intensity of feedback selection decline as interacting individuals become less related. However, frequent interspecific interaction can lead to negative feedback (eg, competition and predation) and reciprocal cooperation (mutualism), based on the trade-off between gain or loss to each individual from such interaction.

Homeostasis at supra-organismal levels depends only in part on selection for attributes that benefit assemblages of organisms (ie, group selection). The critical issue is the trade-off required to balance individual sacrifice, if any, and inclusive fitness accruing from traits that benefit the group (eg, ecosystem). Stabilization of environmental conditions through species interactions favors survival and reproduction of the constituent individuals. Therefore, feedback selection over evolutionary timescales should select for species interactions that contribute to ecosystem stability and mutually assured survival.

Major challenges for ecologists include defining stability, that is, which ecosystem properties are stabilized, what range of deviation is tolerated, and what temporal and spatial scales are appropriate levels for measurement of stability, and evaluating the effect of mechanisms, such as biodiversity and herbivory, that contribute to stability. Traditionally, stability has been viewed as constancy or recovery of species composition over narrow ranges of time and space. Alternative views include reliability of net primary production and biomass structure that affect the stability of climate and soil conditions, and biogeochemical pools and fluxes over larger ranges of time and space. Stability may be achieved, not at the patch scale, but rather at the landscape scale where conditional stability is achieved through relatively constant proportions of various ecosystem types (eg, metacommunities).

The relationship of stability to diversity has been a major topic of debate. Some species are known to control ecosystem properties, and their loss or gain can severely affect ecosystem structure or function. Furthermore, effects of different species often are complementary, such that more diverse assemblages should be buffered better against changes in ecosystem properties in heterogeneous environments. A few experimental manipulations of plant species diversity have shown that more diverse communities have lower variability in primary production than do less diverse communities.

Herbivorous insects have been identified as potentially important regulators of primary production, hence of ecosystem variables controlled by primary production. Herbivorous insects possess the key criteria of cybernetic regulators, that is, small biomass, rapid amplification of effect at the ecosystem level, sensitivity to airborne or waterborne cues indicating ecosystem conditions, and stabilizing feedback on primary production and other processes. Low intensity of herbivory, under conditions of low densities or optimal condition of hosts, tends to stimulate primary production, whereas higher intensities, under conditions of high host density or stress, tend to reduce primary production. Clearly, this aspect of insect ecology has significant implications for our approaches to managing ecosystem resources and "pests." Anthropogenic activities that disrupt ecosystem stability or regulatory mechanisms threaten the sustainable delivery of ecosystem services on which we depend (Chapter 16).

APPLICATIONS AND SYNTHESIS

The previous sections have addressed insect ecology at the individual, population, community, and ecosystem levels of organization. Resource acquisition and allocation by individuals (Section I) depend on population (Section II), community (Section III), and ecosystem conditions that also are influenced by individuals, as described in Section IV. Insects are involved in a particularly rich variety of direct and indirect interactions and feedbacks between individual, population, community, and ecosystem levels as a consequence of their dominance and diversity in terrestrial and freshwater ecosystems and their sensitivity and dramatic responses to environmental changes. The hypothesis that insects are major regulatory mechanisms in homeostatic ecosystems has important ecological and management implications and warrants critical testing.

The importance of temporal and spatial scales is evident at each level of the ecological hierarchy. Individuals have a life span and range of occurrence, populations show temporal dynamics and dispersion patterns, communities and ecosystems are represented over larger temporal and spatial scales. In particular, ecosystem stability and its effect on component individuals traditionally has been evaluated at relatively small scales, in time and space, but larger scales are more appropriate. The dynamic mosaic of ecosystem types at the landscape or biome level is conditionally stable in its proportional representation of ecosystem types.

Insect ecology has major applications to management of ecosystem services and insects, as they affect these services positively and negatively. This concluding section emphasizes application of insect ecology to management of ecosystem services (Chapter 16) and of insects (Chapter 17), then summarizes and synthesizes the study of insect ecology (Chapter 18).

Application to Sustainability of Ecosystem Services

DUNG BEETLES AND DUNG MANAGEMENT IN AUSTRALIA

Detritivores tend to be small and concealed in soil and litter and, therefore, often overlooked. However, they play a critical role in the decomposition of plant litter, carrion, and dung and, thereby, contribute to the sustainability of numerous ecosystem services. Their importance to the sustainability of pasture and livestock production, and to biological control of nuisance flies, is demonstrated by Australian efforts to manage dung accumulation and biting fly populations following introduction of domestic livestock.

Dung beetles are instrumental in burying vertebrate dung within a few days and contributing to its decomposition. Individual species are relatively specific with regard to substrate conditions and host range of colonized dung (A. Davis, 1996). When cattle were introduced to Australia, their dung accumulated in pastures, without specialized dung beetles to consume it. Although Australia is home to 437 species of native dung beetles (Doube and Marshall, 2014), these species prefer the relatively dry, pelleted feces of marsupials and were effective in shredding and burying cattle dung for only a few weeks in spring and autumn (Tyndale-Biscoe, 1994). Ferrar (1975) reported that experimentally marked cattle dung survived at least 3 months and sometimes more than a year. Dung accumulation smothered pasture vegetation and increased reproductive habitat for two hematophagous flies, the buffalo fly, *Haematobia irritans exigua*, and the bush fly, *Musca vetustissima*, that became serious pests of cattle and humans (Ferrar, 1975; Tyndale-Biscoe and Vogt, 1996).

Beginning in 1967, a number of African dung beetles were evaluated and introduced into Australia to accelerate dung disintegration and nutrient turnover and to manage fly populations (R. Hughes et al., 1978). Initial introductions resulted in substantially increased dung disintegration and burial, from < 7% week^{-1} at sites with only one exotic species to 30% at sites with five exotic species, but fluctuated from 0% to 70% depending on beetle abundance (Tyndale-Biscoe, 1994). Suppression of fly reproduction occurred primarily through dung disturbance (R. Hughes et al., 1978), but the first exotic species were most active during the warm monsoon season and relatively ineffective earlier when bush flies first appear (A. Davis, 1996; Tyndale-Biscoe and Vogt, 1996). Subsequent research identified additional dung beetle species that could be active earlier (Ridsdill-Smith and

Kirk, 1985) and demonstrated the importance of phoretic mites, for example, native *Macrocheles glaber* and exotic *M. peregrinus*, that prey on fly larvae in dung pads (J. Roth et al., 1988). Although we do not know all the consequences of these introductions, mean abundances of native dung beetles have remained similar to their preintroduction abundances (Tyndale-Biscoe and Vogt, 1996). This experience demonstrates that dung beetles are capable of dramatically reducing accumulation of cattle dung and abundances of biting flies thereby contributing to more sustainable delivery of important ecosystem services.

1 INTRODUCTION

Natural ecosystems provide a variety of services on which humans, and other organisms, depend for survival and well-being. These ecosystem services can be categorized as provisioning (production of food, fiber, water, and other resources), cultural (nonmaterial benefits, such as recreation, spiritual and other aesthetic values), supporting (primary production, pollination, decomposition, and soil formation necessary for resource production), and regulating (biological control and other feedback mechanisms that maintain relatively consistent delivery of services) (Millennium Ecosystem Assessment, 2005). These services are provided at no cost, but their value is difficult to assess because only provisioning and some cultural and supporting services have market values (Dasgupta et al., 2000). For example, the value of riparian forests for flood control and coastal wetlands for mitigation of storm surges was largely unappreciated until their removal resulted in substantial damage to human infrastructures (J.P. Schmidt et al., 2014). Global value of ecosystem services has been estimated at US$33 trillion annually (Costanza et al., 1997).

Managing insects and ecosystems in ways that ensure sustainability of ecosystem services is not optional, but rather is critical to human survival. The consequences of undermining the sustainable delivery of ecosystem services include famine, water shortages, threats to human health, and economic disruption as supply and demand become decoupled (Acuña-Soto et al., 2002; Bora et al., 2010; R. Bray, 1996; Diamond, 1999; Hsiang et al., 2011; C. Riley, 1878; R.H. Smith, 2007; Whitmee et al., 2015; D. Zhang et al., 2007). Whitmee et al. (2015) concluded that human health has improved in the last several decades through unsustainable exploitation of ecosystem services, but continued overexploitation risks degradation of human health in the future. The collapse of Mesoamerican and Mesopotamian civilizations has been linked to overexploitation of ecosystem services and resulting population displacement (Briant et al., 2010; B. Cook et al., 2012; Lentz and Hockaday, 2009; Xue et al., 1990; Zheng and Eltahir, 1998). Social unrest and population displacement increase the likelihood of epidemics of crowd diseases, often vectored by insects (Acuña-Soto et al., 2002; R. Bray, 1996; Brouqui, 2011; Diamond, 1999; Therrell et al., 2004).

Insects affect ecosystem services in a variety of ways, positively and negatively (Losey and Vaughan, 2006; Prather et al., 2013; Schowalter, 2013). Although insect outbreaks generally have been viewed as devastating to ecosystem resources, outbreaks do not necessarily threaten ecosystem services. In many cases, outbreaks are responses to, and often function to reverse, anthropogenic changes that, in fact, undermine the sustainability of ecosystem processes and services. Accordingly, any management decision should be based on consideration of the effects of the insect and of available management tactic(s) on multiple ecosystem services. Such information is necessary to determine when, and how, insects should be controlled, with the primary goal of ensuring sustainable delivery of services. This chapter summarizes ecosystem services and how these are affected by insects. Management strategies are the topic of the next chapter.

2 PROVISIONING SERVICES

Ecosystems are the source of food, water, fiber, biofuels, and medical and industrial resources for humans (Blancas et al., 2010). Many plants produce edible fruits, seeds or tubers, wood, fiber, or other tissues that have become the basis for their cultivation as crop plants. Livestock are grazed on pastures and grasslands. Wildlife and fish are important food sources worldwide, and many of these animals feed primarily or exclusively on insects (Losey and Vaughan, 2006). Macadam and Stockan (2015) listed a number of predatory fish species, harvested by humans, that feed primarily on aquatic insects. Terrestrial insects falling into streams also contribute to fish production, comprising 30–80% of the diets of young salmon (Fig. 16.1, J. Allan et al., 2003; Baxter et al., 2005; Kawaguchi and Nakano, 2001). In fact, insects or insect-inspired lures are among the most widely used tools for catching fish (Parrella, 2013). Woody materials are used for housing, furnishings, and fences, but also widely used for firewood. Phytochemicals provide important pharmaceutical compounds (Zenk and Juenger, 2007), for example, salicylic acid, morphine, quinine, epinephrine, and taxol. Synthesis of complex

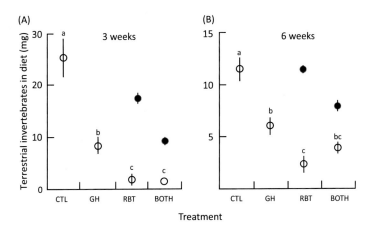

FIGURE 16.1 Mean biomass of terrestrial invertebrates in diets of individual native Dolly Varden (*Salvelinus malma, open circles*) and introduced rainbow trout (*Oncorhynchus mykiss, filled circles*) in treated reaches of Horonai Stream in Hokkaido, Japan after (A) 3 weeks and (B) 6 weeks during 2002. *CTL*, control (native Dolly Varden only); *GH*, greenhouse cover (to reduce subsidy of terrestrial invertebrates to stream); *RBT*, rainbow trout (nonnative rainbow trout added); and *BOTH*, greenhouse cover and trout added. Values are means ± 1 SE (*n* = 4 except *n* = 3 for RBT treatment). Different letters indicate significant differences (*p* < 0.05) among treatments within periods. *Source: Redrawn with permission from Baxter, C.V., et al., 2007. Invading rainbow trout usurp a terrestrial prey subsidy from native charr and reduce their growth and abundance. Oecologia 153, 461–470.*

compounds often is difficult or expensive, and exploration continues for new pharmaceutical chemicals (Helson et al., 2009). Many plant and animal products, including insects (see Sections 2.1, 2.2, and 2.3), are widely used in traditional remedies. Some animals also use plant or other materials to remedy or prevent ailments (Mahaney et al., 1999). Plant-derived tannins, resins, and other compounds are used in various industrial applications.

Short-term losses in production of plant or livestock resources are obvious (see Chapters 8 and 12), and their economic value is easily measured and used to develop economic thresholds for pest control (Pedigo et al., 1986). Furthermore, arthropods that parasitize livestock have been shown to induce production of antibodies that can cause serious allergic reactions in humans who eat red meat (Commins et al., 2011; Hamsten et al., 2013). On the other hand, insect effects on fish and wildlife production may be largely positive, and insects themselves provide important products for human use (Section 2.1).

2.1 Food and Cosmetics

Insects or their products are valuable food resources in many cultures. Honey has been among the most important commercial trade products for millennia. An early cave painting at Altamira in Spain (from about 15,500 years ago) apparently represents honey hunting (Crane, 1999). Mesolithic rock art from Spain, South Africa, and India more clearly records honey hunting, often depicting the use of ladders to climb trees or rock faces and using smoke to drive out bees (Clausen, 1954; Crane, 1999; Dams, 1978). Egyptians developed artificial hives of clay at least 4400 years ago during the First Dynasty (Clausen, 1954; Crane, 1999).

Honey production often is a major use of forests (Bradbear, 2009), and ownership of forests with bee trees has included rights to revenue from honey harvest (Crane, 1999). Honey or beeswax has been preferred or acceptable payment for taxes, tolls, tithes, and trade goods. Commercial honey production amounts to about US$2 billion globally, with China accounting for 20% of the total production and 25% of world honey exports (J. Parker, 2003). Honey has substantial antimicrobial activity against human pathogens (R. Cooper and Molan, 1999; R. Cooper et al., 1999). Beeswax is used for candles, adhesives, seals, and molds for casting (eg, lost-wax method), and during some periods was a more important resource than honey (Crane, 1999).

Grasshoppers, cicadas, caterpillars, beetles, and other insects are important sources of protein and revenue in many parts of the world (Cerritos and Cano-Santana, 2008; Clausen, 1954; DeFoliart, 1999; Leather, 2015; Mbata et al., 2002; Ramos-Elorduy, 2009; Yen, 2009). About 1500 edible insect species are consumed by 3000 ethnic groups in 113 countries (MacEvilly, 2000). In some cultures, maintenance of edible insect populations represents a primary ecosystem management goal. Mbata et al. (2002) described the process for managing harvest of edible caterpillars (primarily two saturniids, *Gynanisa maja* and *Gonimbrasia zambesina*) in Zambian forests. In one unique study, Cerritos and Cano-Santana (2008) reported that harvesting grasshoppers for food during an outbreak in Mexico substantially reduced grasshopper abundance and reproduction (Fig. 16.2) and provided US$3000 revenue per family, compared to US$150 cost per family for insecticide treatment had control tactics been implemented. In China, yartsa gunbo, a ghost moth, *Thitarodes* sp., caterpillar infected with *Ophiocordyceps sinensis* fungus, is worth US$50,000 per pound, making it among the most valued food resources globally and a growing conservation concern (Shrestha and Bawa, 2013; Weckerle et al., 2010).

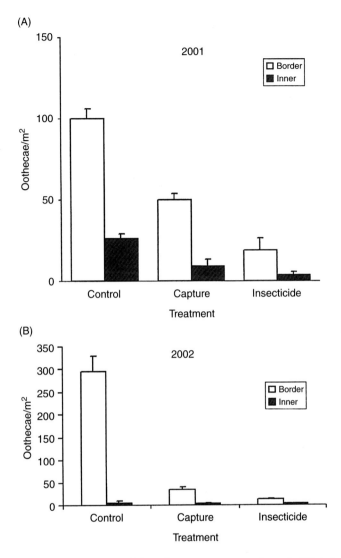

FIGURE 16.2 Mean density of grasshopper, *Sphenarium purpurascens*, oothecae (± SE), in inner and border zones of alfalfa fields, by treatment (control, hand capture for human consumption, and insecticide) in the Puebla-Tlaxcala Valley of Mexico in 2001 (A) and 2002 (B). *Source: Reprinted from Cerritos, R., Cano-Santana, Z., 2008. Harvesting grasshoppers* Sphenarium purpurascens *in Mexico for human consumption: a comparison with insecticidal control for managing pest outbreaks. Crop Prot. 27, 473–480.*

Insects produce protein much more efficiently than do homeothermic vertebrates. In fact, crickets convert food into protein 2-fold more efficiently than chickens, 4-fold more efficiently than pigs, and 12-fold more efficiently than cattle (van Huis, 2013). In large part, this is due to (1) insects investing less energy in maintenance of body temperature than do homeotherms (see Chapter 4), and (2) insects providing 80% edible mass, compared to < 55% edible mass for chickens, pigs, and cattle.

TABLE 16.1 Minerals Obtained from Geophagy by Pregnant Women in Five Countries in Southern Africa, Compared to Recommended Daily Allowances (RDA)

| Mineral | RDA | Percentage of RDA for pregnant females | | | | | |
| | | Consuming 40 g/day | | | Consuming 100 g/day | | |
		White clay	Brown clay	Termite clay	White clay	Brown clay	Termite clay
Potassium	1600–2000 mg	0.7	5.6	1.2	1.6	1.4	3.1
Calcium	800–1200 mg	0.8	0.4	33	1.9	1.1	82
Phosphorus	800–1200 mg	–	–	1.3	–	–	3.1
Magnesium	280 mg	0.2	0.1	12	0.5	3.2	29
Iron	15 mg	130	49	210	320	120	540
Zinc	12 mg	1.4	1.4	4.8	3.5	3.4	12
Manganese	2–5 mg	17	31	2300	43	75	5700
Copper	2–3 mg	28	10	86	70	26	210

Percentages represent lowest RDA values. A "–" indicates < 0.1%.
Data from Hunter, J.M., 1993. Macroterme geophagy and pregnancy clays in southern Africa. J. Cult. Geogr. 14, 69–92.

Insects can enhance food value for humans in other ways, as well. Insects are responsible for the economic value of "Oriental Beauty" oolong tea in Taiwan. This unique tea is made from wilted, yellow leaves infested by the tea green leafhopper, *Jacobiasca formosana*, and has a unique aroma resembling ripe fruits and honey. Cho et al. (2007) evaluated gene expression and chemical profiles of infested and uninfested leaves and concluded that feeding by the leafhopper elevates production of plant defensive compounds that impart the desired flavors and aromas. Humans and other primates in Africa selectively consume kaolin-rich clays from termite mounds, both to obtain essential nutrients (Table 16.1) (J. Hunter, 1993) and to bind (and deactivate) dietary toxins (Mahaney et al., 1999).

A number of insect products have been used in cosmetics. Beeswax is commonly used in lip balms, face masks, and hand and body moisturizers (Crane, 1999). Carmine (cochineal) dye, made from the crushed bodies of scale insects, particularly species of *Kermes, Porphyrophora*, and *Dactylopius*, has been used historically in lipstick and other cosmetics (Chávez-Moreno et al., 2009; Vigueras G and Portillo, 2001).

2.2 Fresh Water

Ecosystems are valued sources of fresh water, and adequate water supply often is the primary management goal for municipal watersheds. Insects falling into oligotrophic lakes and headwater streams add substantial amounts of carbon, nitrogen, and phosphorus to these nutrient-poor ecosystems and affect aquatic community dynamics (Carlton and Goldman, 1984; Mehner et al., 2005; Menninger et al., 2008; Nowlin et al., 2007; Pray et al., 2009). Herbivorous insects reduce canopy cover and increase the volume of precipitation reaching the ground and flowing into streams (Chapter 12). Soil and litter arthropods affect soil porosity and decomposition rate, factors that affect the rate of water movement through the substrate (Chapter 14). Changes in water yield and quality resulting from insect activity may or may not be desirable, depending on the needs of downstream users.

2.3 Medical and Industrial Products

A number of insect species provide widely used medical products (Leather, 2015; Singh and Jayasomu, 2002). Honey has demonstrated antimicrobial activity against human pathogens (Cooper and Molan, 1999; Cooper et al., 1999). Bee and wasp venom has been used since ancient times to treat arthritis and rheumatism (Crane, 1999). Pharmaceutical compounds from insects include cantharidin, a defensive alkaloid produced by blister beetles (Meloidae) that is used commercially to remove warts (Epstein and Kligman, 1958; Namba et al., 1988; Pemberton, 1999).

Use of ant mandibles to stitch wounds or surgical incisions was documented in India as early as 1000 BC and continued in some areas at least into the early 1900s (Gottrup and Leaper, 2004; Gudger, 1925). After live ants are induced to bite the pinched sides of a wound or incision, the ant bodies are removed, and the mandibles remain fastened until the wound heals (Gottrup and Leaper, 2004; Gudger, 1925). Silk also has been used to stitch wounds (Clausen, 1954).

Blow fly, *Lucilia* spp., maggots were used for wound healing by Maya Indians, Australian Aborigines, and ancient Burmese over thousands of years (Greenberg, 1973; Sherman and Pechter, 1988; Whitaker et al., 2007). Their medical value lies in their selective feeding on necrotic tissue, leaving clean tissue when they depart (Baer, 1931; Sherman et al., 2007; Whitaker et al., 2007). Maggots were credited with saving the lives of many wounded soldiers (Baer, 1931; Sherman et al., 2000; Whitaker et al., 2007). Modern use of medical maggots dates to World War I, when Dr. William Baer, an orthopedic surgeon, recognized that wounds of soldiers left on the battlefield for several days before receiving medical treatment healed faster when they had been colonized by blow fly maggots. In 1929, Dr. Baer experimentally introduced maggots into 21 patients with untreatable chronic osteomyelitis and observed rapid debridement, reduction in the number of pathogenic organisms, reduced odor levels, alkalinization of wound beds, and ideal rates of healing (Baer, 1931). All 21 patients healed completely after 2 months of maggot therapy. Subsequently, maggot therapy became acceptable for wound treatment, particularly in the United States (McKeever, 1933). More than 300 American hospitals employed maggot therapy through the 1940s (Whitaker et al., 2007). Maggot therapy declined after World War II, following the discovery of penicillin, but is receiving renewed attention for wound debridement, as surgical treatments increasingly risk infection by antibiotic-resistant pathogens (Kerridge et al., 2005; Sherman et al., 2000, 2007).

Insects are a source of important industrial products. Silkworms, *Bombyx mori*, and a few other moth species remain the only source of commercial silk. Silk production has been practiced at least since 2000–3000 BC in China and is among the most widely traded commercial products. The historic importance of silk is demonstrated by the construction of the Silk Road that connected Europe, the Middle East, and China for at least 500 years, and by the introduction of the gypsy moth, *Lymantria dispar*, and other silk-producing species into North America and Europe as part of unsuccessful efforts to establish silk industries in the West (Andrews, 1868; Anelli and Prischmann-Voldseth, 2009; Forbush and Fernald, 1896; C. Riley and Howard, 1890).

Carmine (cochineal) dye is one of the oldest organic pigments still in use. The dye is produced from the crushed bodies of scale insects, especially *Kermes vermilio* from the Near East and southern

Europe and *Dactylopius coccus* from Mesoamerica. Carmine dye was used by ancient Egyptians, Greeks, and Persians and by preColumbian civilizations in the New World as early as 2000 years ago (Chávez-Moreno et al., 2009; Greenfield, 2005). Cochineal was once the most valuable export from colonial Mexico, next to silver (Anelli and Prischmann-Voldseth, 2009; Chávez-Moreno et al., 2009; Donkin, 1977; Greenfield, 2005), resulting in transport of cactus and cochineal insects to other colonized territories, with disastrous results (eg, the introduced cacti became invasive in Australia, Clausen, 1954). Because this method of providing carmine dye is labor intensive, the scale insects were replaced in the 20th century with synthetic dyes, but some cochineal production continues, primarily for use by indigenous cultures (Chávez-Moreno et al., 2009).

Scale insects, *Laccifer lacca*, have been the principal source of commercial lac products, including shoe polishes, electrical insulation, sealing waxes, glazes, phonographic records, and shellac varnishes for wood finishing (Clausen, 1954). These insects remain a source of some commercial shellac.

Beeswax has been among the most valuable trade commodities in the past, often more highly valued than honey (Crane, 1999). Beeswax was the basis for the lost-wax method of casting metals as early as 5500 years ago (Crane, 1999). A beeswax model was coated with clay to form a mold that was allowed to dry. The mold was then heated to melt the wax, which was drained out (lost) through one or more holes. The molten metal was poured into the mold through the hole(s) and allowed to solidify, after which the mold was removed. Beeswax was used to make the earliest candles, invented some 3500 years ago (Crane, 1999). Candles represented an improvement over the earlier oil lamps, because the wick could be inserted into the wax candle and a pottery vessel was not required. Beeswax also was used to make models of divinities, humans, and animals that were placed within wrappings of Egyptian mummies to accompany the deceased (Crane, 1999). Later, the Romans used beeswax for seals on deeds and legal documents. Beeswax has been used as a surface finish or polish for stonework, pottery, wood, and leather for at least 5500 years. Beeswax also was used as an ingredient in paints to provide luster.

More recently, insects have provided inspiration for technological advancements (Leather, 2015). During the 1930s, Joe Cox revolutionized the timber harvest industry in the northwestern USA with a new saw chain design inspired by the alternating cuts made by the curved mandibles of the cerambycid beetle, *Ergates spiculatus*; while one mandible is cutting, the other acts as a depth gauge (Fore, 1970). The "C"-shaped cutters alternating sides along the length of the saw chain improved woodcutting efficiency and reduced maintenance requirements (Fore, 1970).

Insects also have led to advances in walking or flying robots (Aktakka et al., 2011; Delcomyn, 2004; K. Ma et al., 2013). Insects are uniquely capable of moving efficiently and stably through complex terrain (Ritzmann et al., 2004). Incorporating insect body flexion, leg articulation, and righting ability, as well as sensory feedback mechanisms to avoid collisions, has improved the use of robots in volcanic research and exploration of other planets (Frantsevich, 2004; Ritzmann et al., 2004; B. Webb et al., 2004). Aktakka et al. (2011) and K. Ma et al. (2013) demonstrated electrical generation by the wing motion of flying beetles and flies as a model for developing flying robots that could explore hostile environments more quickly.

Gut microorganisms that facilitate digestion of lignin and cellulose by wood-feeding insects currently are being explored as sources of novel and inexpensive enzymes for wood degradation. Such biocatalysts may permit economically feasible biofuel production from wood (D.M. Cook and Doran-Peterson, 2010).

3 CULTURAL SERVICES

Ecosystems provide various religious, artistic, recreational, and other cultural services, including hunting and fishing, wildlife observation, hiking and camping, educational and scientific pursuits, and spiritual renewal (Leather, 2015). For example, most remnant forests in Ghana are sacred sites set aside by indigenous groups centuries ago and protected by religious sanctions and taboos (Bossart et al., 2006).

3.1 Religious, Spiritual, and Artistic Icons

Insects have been used as religious symbols or cultural icons for millennia. They have been depicted in various art media as bringing fortune or enjoyment (eg, Fig. 16.3).

Scarab beetles were the ancient Egyptian symbol of eternal life and held sacred from prehistoric times, as demonstrated by preserved beetles and images carved in jade, malachite, and even emeralds (Clausen, 1954). Their image symbolized the sun god, Khepera, creator and highest god to the Egyptians. Scarabs symbolized resurrection when used in burial ceremonies and were often inserted in place of the removed heart of the deceased (Clausen, 1954). These Egyptian customs were later carried to Rome, and Roman soldiers customarily wore images of the sacred scarab as a talisman.

Humans have used insects to forecast weather for millennia, although most of this prediction has been based on superstition (Clausen, 1954). The Zuni Indians of the southwestern USA believe that the early appearance of butterflies is an indication of fair weather. White

FIGURE 16.3 Detail of grasshopper and cricket carved on jadeite cabbage, Ming Dynasty, 19th century.
Source: From Schowalter, T.D., 2013. Insects and Sustainability of Ecosystem Services. CRC Press/Taylor and Francis Group, Boca Raton, FL.

FIGURE 16.4 Silkworm cocoon rattles worn by a Zulu dancer in South Africa. Insects provide valued food and other resources in many cultures.

butterflies signify the onset of summer, and white butterflies flying from the southwest forecast rain. Local winter weather was predicted in the eastern United States by the width of the reddish-brown band of the common Arctiid species, *Isia isabella* (woolly bear). The narrower the band, the colder and longer the winter. Similarly, unusually large stores of honey are used by some to forecast severe winter (Clausen, 1954).

Dragonflies are symbols of victory in Japan and depicted during ceremonies to commemorate past successes (Clausen, 1954). Some giant silk moth species spin large, tough cocoons that are used ceremonially by Native Americans in the western USA and Mexico and Zulus in South Africa (Fig. 16.4). The cocoons are split open, the pupae are removed (often eaten, see Section 2.1), a few pebbles are placed in each cocoon, and the cocoons are stitched closed and tied singly or in groups to make rattles and musical instruments.

Cicadas symbolized resurrection to the Oraibi Indians of North America and the ancient Chinese, as a result of the sudden appearance of adults out of the ground. In China, a carved jade amulet in the image of a cicada was placed on the tongue of the deceased before burial, the cicada representing resurrection and jade symbolizing the triumph of good over evil (Clausen, 1954). The Chinese continue to hold cicadas in high esteem for their songs and keep cicadas in special cages as pets.

The Chinese also value crickets for their singing and keep males in bamboo cages as pets. In addition, cricket fighting is popular in China, with large sums of money wagered on combats and high values for champions (Clausen, 1954). Special, often highly ornate, cages are used to house these special insects.

Children in equatorial Africa tie a string to the leg of a goliath beetle, *Goliathus goliatus*, and use it as a toy, letting the beetle fly noisily (Clausen, 1954). During the early 20th century, butterfly collecting became a popular social pastime, with amateur societies established to promote interest (Comstock, 1911). Some tropical butterflies, including morphos (*Morpho* spp.) and birdwings (especially *Ornithoptera* spp.), have been collected so intensively that they are now endangered, and their trade is prohibited by international law (Clausen, 1954). "Jewel beetles" and dragonflies also represent colorful groups enjoyed and traded by avid amateur societies (Cave, 2001). These insect orders are the focus of many popular field guides (Ferro et al., 2010; Pyle, 1981; R. White, 1983).

3.2 Recreation

Insects often are the objects of tourist destinations, for example, overwintering aggregations of monarch butterflies, *Danaus plexippus*, in Mexico and other danaines in tropical Asia, and the glow worm, *Arachnocampa* spp., caves in Australia and New Zealand (C. Hall, 2013; Lockwood, 2013). Insect festivals, such as the Festa del grillo (Festival of Crickets) in Florence, Italy and Dragonfly Citizen Summit in Japan draw thousands of attendees (Hvenegaard et al., 2013). These events provide substantial economic benefits to local communities. Butterfly houses also are popular in many areas (Veltman, 2013), offering tourist opportunities to enjoy representative butterflies from around the world (Fig. 16.5).

FIGURE 16.5 **Brightly colored morpho butterflies, *Morpho* spp., delight visitors to a commercial butterfly house.** Butterfly houses have become popular tourist destinations in many cities. *Source: From Schowalter (2013) with permission from CRC Press.*

On the other hand, trees killed by insects can create safety hazards that may reduce recreational use or undermine spiritual value of parks or conservation areas (Michalson, 1975), unless the ecological feedbacks can be used as educational opportunities (Müller and Job, 2009). Defoliation reduces shade and may be perceived as unsightly. Frass and tissues falling on people or eating surfaces are considered a nuisance by many and may be allergenic to some (Perlman et al., 1976). These effects can cause reduced visitation to recreational sites experiencing outbreaks.

4 SUPPORTING SERVICES

Supporting services include ecosystem processes necessary for delivery of provisioning or cultural services, for example, primary production, pollination, seed dispersal, decomposition, soil formation, and water filtration. Insects affect these processes in a number of ways, as described in Chapters 12–14. Although disruption of these processes during outbreaks can impair primary production and threaten short-term delivery of some ecosystem services, long-term compensatory growth may offset these losses (Chapter 12), and insect contributions to pollination, seed dispersal, decomposition, soil formation, and water filtration may be instrumental in sustaining delivery of ecosystem services (Chapters 13 and 14).

4.1 Primary Production

Despite traditional perspectives of herbivorous insects as destructive to plants, normal rates of herbivory by native species are controlled by a combination of bottom-up and top-down feedbacks that tend to sustain, or even stimulate, primary production (Fig. 16.6, see Chapters 3, 6, and 12). In fact, herbivory by native species should be viewed as one of the feedbacks that regulates primary production, in the same way that predators regulate prey populations (Fig. 15.7). From this perspective, herbivory by insects functions to support primary production, removing weakened, inefficient, or less defended plants or plant tissues and stimulating compensatory production (Chapter 12, see also Section 4.1).

4.2 Pollination and Seed Dispersal

Pollination services are among the most widely recognized roles of insects. Bees and other pollinators are necessary for pollination for 60–90% of plant species and are responsible for 35% of global crop production (Kremen et al., 2007). For example, pollination service is required for efficient production of many fruits and vegetables. Bommarco et al. (2012) reported that insect pollinators (primarily honey bees, *Apis mellifera*, but including hover flies, bumble bees, and other insects) increased oilseed rape seed weight by 18% and market value by 20% per plant. Consequently, pollination services have recognized monetary value (see Chapter 13 and Section 6). However, adequate pollination depends on complex networks of ecological interactions that maintain a diverse pollinator guild (Aslan et al., 2013; Bascompte, 2009; Forup et al., 2008; R.H. Gibson et al., 2006; Vamosi et al., 2006, see Chapters 8 and 13) and on nutrient cycling processes that maintain adequate nutrient availability (see Chapters 11–14) for maximum fruit and vegetable development (Bos et al., 2007; Hoehn et al., 2008), values of which are difficult to calculate. Honey bees generally are credited with providing pollination

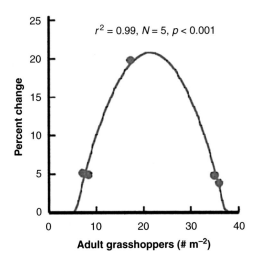

FIGURE 16.6 **Relationship between grasshopper,** *Melanoplus sanguinipes*, **density and primary production in experimental field microcosms at the National Bison Range in Montana.** Peak natural densities at the site were 4–36 adults m^{-2}. *Source: From Belovsky and Slade (2000) with permission from the National Academy of Sciences, USA.*

service, and colonies are moved widely to provide this service, but native pollinators often provide more efficient pollination. Garibaldi et al. (2013) conducted a metaanalysis of pollination efficiency for 41 crop systems representing all continents except Antarctica and reported that native pollinators increased fruit set in all crop systems, whereas honey bees increased fruit set in only 14% of the systems surveyed. Declining pollinator abundance in many areas (Biesmeijer et al., 2006; Burkle et al., 2013) has prompted efforts to protect or restore native habitats necessary to sustain pollinator diversity and abundance (Aslan et al., 2013; Bascompte, 2009; Carvalheiro et al., 2011, 2012; Carvell et al., 2007; Forup et al., 2008).

The direct reduction of total crop production in the absence of insect pollinators would be 3–8% (Aizen et al., 2009). Sustainable reproduction of many noncrop plants (eg, alfalfa) that support native vertebrates and livestock and/or affect ecosystem conditions for other species also depends on a diversity of insect pollinators (R.H. Gibson et al., 2006; Vamosi et al., 2006, see Chapter 13). Threats to honey bee pollination services by invasive varroa mites, *Varroa destructor*, and other parasites have generated widespread concern about the sustainability of many vegetable and fruit crops (see Chapter 13). Honey bees, an exotic species in much of the world, also have been blamed for losses of many native species that may be required for pollination of their mutualistic hosts (Aizen and Feinsinger, 1994; G. Yang, 2005, see Chapter 13).

4.3 Decomposition and Soil Formation

Arthropods are instrumental in the development of soil structure and fertility (see Chapter 14). Decomposition of plant and animal detritus is necessary for release of nutrients that become available for new plant growth (T. Wood et al., 2009). In the absence of detritivores, detritus would accumulate and create a bottleneck in nutrient cycling that would prevent further primary production (Frost and Hunter, 2007; Setälä and Huhta, 1991; T. Wood et al., 2009, see Chapter 14). Termites and dung beetles, in particular, provide a major agricultural service by removing and burying livestock dung, thereby preventing fouling of pasture forage by dung accumulation, reducing nitrogen loss via volatilization, and reducing livestock losses due to hematophagous flies (see Box introducing this chapter, also Chapter 14). The infusion of organic matter into soil also contributes to soil texture and water-holding capacity (Brody et al., 2010; Eldridge, 1993, 1994; Eldridge et al., 2009).

5 REGULATING SERVICES

Regulating services include the various feedbacks that minimize variation in primary production, biogeochemical fluxes, climate, etc. These feedback mechanisms are critical to stabilizing population sizes, species interactions, and nutrient fluxes that support community and ecosystem processes. Undermining these feedbacks, through overexploitation of resources or introduction of pollutants or invasive species, has led to ecosystem degradation, desertification, and loss of ecosystem services (J. Foley et al., 2003a; Janssen et al., 2008).

5.1 Herbivory

As described in Chapters 12 and 15 and Section 4.1, populations of native herbivores regulate primary production in a density-dependent manner, stimulating compensatory primary production at low-to-moderate levels of herbivory (when resource availability is adequate to support plant growth and defense) and reducing primary production at high levels of herbivory (when high plant density and/or stress favor herbivore population growth, see Fig. 15.7). As a result, primary production in unmanipulated ecosystems may be maintained near carrying capacity (Chapter 15). Outbreaks are most likely when environmental changes, especially anthropogenic alteration of plant condition (eg, through breeding programs that sacrifice defense for improved palatability for humans) or density (eg, field crops or plantation forestry), create favorable conditions for herbivore population growth (see Chapter 6). Outbreaks typically remove stressed plants, reduce density of host plants, increase vegetation diversity, and stimulate nutrient cycling, all of which combine to improve long-term primary production following outbreaks (Fig. 15.9, Chapters 12 and 15). From this perspective, herbivory functions as a negative feedback mechanism to regulate primary productivity in the same manner that predators regulate prey populations. Stabilization of primary production via herbivory maintains delivery of other ecosystem services.

5.2 Biological Control

Biological control is an essential regulatory service that typically is undervalued until lost. The consequences of predator removal programs (often intended to increase populations of game animals) were so severe (overabundances of starving game species) that predator reintroduction programs have been required to restore the role of predators in regulation of healthy, stable prey populations (Peterson, 1999; Wilmers et al., 2006). Diversity of insectivorous birds and mammals provides important regulation of insect prey, including many pest species (B. Allan et al., 2009; M.D. Johnson, 2000; M.D. Johnson et al., 2009, 2010; C.G. Jones et al., 1998; Kellermann et al., 2008, see the following section and Chapters 8 and 17). Similarly, insects frequently have been instruments of biological control for other insects or invasive weeds (see Chapter 17). L. Thomson

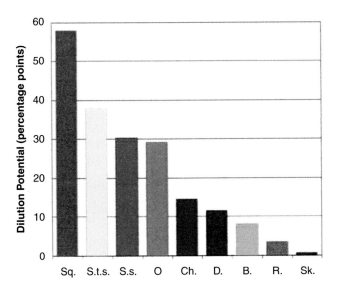

FIGURE 16.7   **The ability of associated small mammal species to reduce the effect of white-footed mice (the most competent reservoir) on tick nymphal infection prevalence (NIP) and human risk of Lyme disease.** NIP represents the probability of being exposed to Lyme bacteria if bitten by a nymphal tick and is a function of the distribution of larval meals among the community of vertebrates. Dilution potential is the difference (as percentage) between expected NIP in a two-host community consisting of mice plus the focal species and a community in which mice are the only possible host. *Sq.*, squirrel; *S.t.s.*, short-tailed shrew; *S.s.*, Sorex shrew; *O*, opossum; *Ch.*, chipmunk; *D.*, deer; *B.*, birds; *R.*, raccoon; *Sk.*, skunk. *Source: From LoGiudice et al. (2003), copyright (2003) National Academy of Sciences, USA, with permission from the National Academy of Sciences, USA.*

and Hoffman (2010) reported that biological control of insect pests was enhanced when vinyards were surrounded by natural remnant forest or planted woody vegetation, compared to no woody vegetation.

5.3 Disease Cycles

Unmanipulated ecosystems regulate abundances of pathogen and vector species through various food web interactions and habitat conditions. For example, the complex interactions that limit tick abundance and spread of Lyme disease depend on the diversity of small mammal species (Fig. 16.7) (Dobson et al., 2006; C.G. Jones et al., 1998; LoGiudice et al., 2003; Turney et al., 2014). Similarly, bird species diversity limits the abundance of reservoir hosts for West Nile Virus and reduces the incidence of human cases (B. Allan et al., 2009). Bryan and Kandulu (2009) reported that removal of cattle dung by dung beetles is among the mechanisms that can mitigate the risk of *Cryptosporidium* infection from drinking water in Australia. Vittor et al. (2006, 2009) reported that deforestation is the primary factor contributing to increased abundance of *Anopheles darlingi* and resurgence of malaria in Peru. de Castro et al. (2006) and Steiger et al. (2012) similarly reported that conversion of rainforests to grasslands increases the likelihood that mosquito species in both habitats will acquire and transmit diseases through mixing of the vector community at rainforest edges.

Keesing et al. (2006) and Gottdenker et al. (2014) reviewed studies of the relationship between biodiversity and risk of human diseases. Both groups concluded that biodiversity is critical to maintaining low risk of human diseases, primarily through a *dilution effect* of disease transmission into poor hosts. Anthropogenic changes that reduce biodiversity, especially habitat fragmentation, are likely to increase the incidence of zoonotic human diseases. More than half (57%) of 300 studies reviewed by Gottdenker et al. (2014) documented increased incidence of human disease following anthropogenic land use change, compared to 10% of studies that documented decreased incidence of human disease. Climate change also is expected to alter the distribution and epidemiology of insect-vectored diseases, although the effect of warming temperatures may be overshadowed by other factors (Lafferty, 2009).

6 VALUATION OF ECOSYSTEM SERVICES

Traditional value systems have emphasized provisioning services, for example, food, fiber, wood, fresh water, medicinal and industrial compounds, and some cultural and recreational services. Only recently has ecological research demonstrated the critical benefit of supporting and regulating services to the sustainability of extractive and cultural services, as well as to human health and well-being (J.P. Schmidt et al., 2014; Steiger et al., 2012; Vittor et al., 2006, 2009). Although the variety of medical products provided by bioprospecting (Helson et al., 2009; Kursar et al., 1999), increased crop yields provided by pollinator diversity (Garibaldi et al., 2013; Hoehn et al., 2008; A. Klein et al., 2003; Kremen et al., 2007; Winfree et al., 2007), and improved pest and disease management provided by predator and parasitoid diversity (B. Allan et al., 2009; Isaacs et al., 2009; Landis et al., 2000, 2005; Tscharntke et al., 2005, 2007; Vittor et al., 2006, 2009) demonstrate the value of biodiversity conservation, the economic value of biodiversity per se has been difficult to calculate. The diversity of species represents an equivalent diversity of functional attributes, interactions, and feedbacks that support ecosystem services (Duffy, 2009; Hoehn et al., 2008). The effects of species loss are virtually unknown until such loss threatens essential services (Biesmeijer et al., 2006; Cox-Foster et al., 2007; Garibaldi et al., 2013; Genersch, 2010). Recognition of the value of ecosystem services has promoted policies in the USA and European Union to offset biodiversity loss in agricultural landscapes through agri-environmental programs that subsidize farmers for conservation or restoration practices, for example, protection or restoration of patches or strips of native grassland, that mitigate the effects of agricultural intensification (Donald and Evans, 2006; Scheper et al., 2015).

6.1 Valuation of Insect Effects on Provisioning Services

Most economic analyses of insect effects on provisioning services have focused on losses of plant products due to insect herbivores. However, insects have largely positive effects on fish and wildlife production, including production of important fisheries (J. Allan et al., 2003; Baxter et al., 2005; Kawaguchi and Nakano, 2001). Insect effects on fresh water resources may be positive or negative, depending on the needs of downstream users.

Harvesting insects for human consumption has the potential to improve diet and reduce crop losses and pesticide use in many countries. The current rate of grasshopper, *Sphenarium purpurascens*, harvest for human consumption in Mexico is 200 tons annually (Cerritos and Cano-Santana, 2008). Cerritos Flores et al. (2015) calculated that harvesting this grasshopper from agricultural fields in Mexico has the potential to yield 350,000 tons of animal biomass annually, an amount for a single grasshopper species that represents the equivalent of traditional livestock (cattle and poultry) production in Mexico. The projected harvest of grasshopper protein would improve diet for millions of people, yield economic profits of more than US$1.8 million, and reduce crop losses and costs for pesticide use.

Plant growth loss and mortality may remain undesirable in ecosystems managed for plant products, but costs of insect outbreaks may be tolerable in public ecosystems managed for multiple uses (Torell et al., 1989). For example, the Douglas-fir tussock moth, *Orgyia pseudotsugata*, outbreak depicted in Fig. 15.9 caused serious losses in timber supply and increased the risk of wildfire in the short term. However, reduced fir density returned forest structure to historic conditions that prevailed prior to fire suppression in Sierran forests and that are the current recommended management goal for these forests (North et al., 2007). From this perspective, the outbreak improved the sustainability of ecosystem services from this forest. Furthermore, if expensive salvage harvest can be delayed, compensatory growth over the long term may largely replace lost products (see Chapter 12). To the extent that outbreaks of native insects function in a regulating capacity, suppression may be unnecessary for, or even counterproductive to, maintenance of ecosystem services.

6.2 Valuation of Insect Effects on Cultural Services

The global value of recreational services (which can be calculated in part from usage fees) has been estimated at $815 billion by Costanza et al. (1997). Wildlife observation in the USA alone is valued at US$34 billion annually, including an estimated $20 billion for observation of insect-dependent wildlife (Losey and Vaughan, 2006). Few studies have addressed values of direct insect effects on cultural services. However, insects influence these services through public perceptions or recreational use of ecosystems.

K. Downing and Williams (1978) reported that defoliation by a Douglas-fir tussock moth outbreak in Oregon did not significantly affect recreational use. Most visitors were aware of the outbreak, but few chose to avoid the outbreak area. The only negative effect was avoidance of salvage logging operations that were considered unappealing or hazardous for recreational use. On the other hand, extensive plant mortality may be viewed as unattractive or hazardous (Michalson, 1975). Several studies have evaluated preferences (measured as a visual-quality rating) among photographs showing different amounts and stages of defoliation or tree mortality due to insects (Hollenhorst et al., 1993; Leuschner, 1980; Schowalter, 2013; S. Sheppard and Picard, 2006).

Generally, visual-quality rating is unaffected by participants' backgrounds but shows relatively low thresholds (eg, at 10–20% defoliation or mortality) above which visual-quality rating declines more steeply than at lower levels (S. Sheppard and Picard, 2006). Some, but not all, studies showed that visual preference was affected by the subject's awareness of the cause (see Müller and Job, 2009). Because visitors' perceptions of changes in recreational values resulting from insect outbreaks determine willingness to stay and spend money on recreational activities or in nearby towns, negative perceptions can reduce the economic value of cultural services, regardless of longer-term costs or benefits of outbreaks. However, even when tourist perceptions of insect damage do not warrant control efforts (Müller and Job, 2009), recreational values versus provisioning values represent a trade-off. The net value of insect effects on cultural services has not been calculated.

Some insects provide lucrative tourism destinations. Huntly et al. (2005) reported that 95% of tourists to Hluhluwe-Imfolozi Park in South Africa expressed interest in information on invertebrates. Tours can be arranged to view monarch butterfly migrations, morpho butterflies, dragonflies, or jewel beetles (Huntly et al., 2005). Butterfly farmers rear a variety of temperate and tropical butterflies for butterfly houses (Fig. 16.5) and for release at weddings and other special events.

Insects also can be objects for entertainment. Butterfly gardens, ant farms, and pet tarantulas, millipedes, and scorpions are available from a variety of biological supply and pet sources. Insects have inspired much lucrative artwork and many documentaries, as well as science fiction, literature and film. The economic values of these contributions have not been calculated.

6.3 Valuation of Insect Effects on Supporting Services

The economic effects of insects on primary production may be positive or negative, depending on plant conditions, timing, and intensity of herbivory (see Chapter 12 and Section 4.1). However, pollination by insects provides recognized economic benefits. Pollination of agricultural crops by honey bees is worth US$117 billion per year globally (Costanza et al., 1997) and $8 billion per year in the USA (Isaacs et al., 2009). Native bee species also contribute substantially to pollination services and can, in many cases, provide full pollination service in the absence of honey bees (Fig. 16.8) (Kremen et al., 2002). In fact, fruit or seed set has been shown to increase with visitation by native pollinators to a greater extent than by honey bees (Fig. 16.9) (Garibaldi et al., 2013; Klein et al., 2003). In Kenya, where > 99% of crop production is attributed to pollination by feral bees, the value of bee pollination represented nearly 40% of the annual US$3.2 million in major crop production (Kasina et al., 2009). Losey and Vaughan (2006) attributed US$12–16 billion per year in crop production in the USA to honey bees and US$3.1 billion per year to native bees.

The annual value of dung burial by dung beetles in the USA is estimated at $380 million (Losey and Vaughan, 2006). Beynon et al. (2015) calculated that dung beetles in the UK currently save the cattle industry US$560 million per year via increased nutrient cycling, reduced pasture fouling, and reduced biting fly and gastrointestinal parasite abundance. In addition, protecting dung beetles under proposed agro-environment systems could save an additional US$61 million per year, whereas protecting dung beetles under organic farming systems could save US$578,000 per year. Furthermore, reduced need for treatment of adult cattle with anthelmintics could save an additional US$9.5 million per year plus the cost of the anthelmintics. These benefits of dung beetles are evident in the costs for research, foreign exploration, and quarantine operations to introduce livestock-adapted dung beetles into Australia, required for continued livestock production and biting fly remediation (Doube and Marshall, 2014; Tyndale-Biscoe, 1994; Tyndale-Biscoe and Vogt, 1996).

6.4 Valuation of Insect Effects on Regulating Services

The value of natural regulatory effects is difficult to calculate. However, these services are fundamental to sustainable delivery of other ecosystem services. The value of biological control

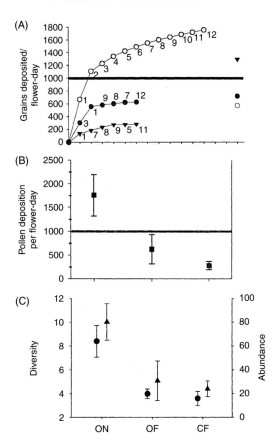

FIGURE 16.8 **Native bee contribution to pollination of watermelons in organic farms near seminatural oak woodland and chaparral habitat (ON), organic farms isolated from seminatural habitat (OF), and conventional farms isolated from seminatural habitat (CF).** (A) Estimated mean pollen deposition per flower-day (± SE) for native bees (numbered 1–12) and honey bees, *A. mellifera*. ON, Open circles; *OF*, filled circles; *CF*, filled triangles. (B) Total estimated pollen deposition (± SE) by native bees in 2001. The horizontal line represents the pollen deposition threshold for production of marketable fruit. (C) Native bee diversity *(circles)* and abundance *(triangles)* in 2001 (± SE). *Source: From Kremen et al. (2002) with permission from the National Academy of Sciences, USA. Copyright (2002) National Academy of Sciences, USA.*

of native crop pests by native or introduced insects, a single regulatory mechanism among many, has been estimated at US$5.4 billion per year in the USA (Losey and Vaughan, 2006). However, this estimate did not include the value of controlling disease vectors, invasive crop pests, or weeds. Landis et al. (2008) estimated the value of native predators and parasites in controlling a single invasive pest, the soybean aphid, *Aphis glycines*, in four US states at > US$239 million per year.

Clearly, although current economic systems encourage maximum exploitation of ecosystem services, a more consistent supply of services would prevent catastrophic oscillations that can destabilize society and lead to population displacement and epidemics of disease, as seen during famines of the past (Acuña-Soto et al., 2002; Bora et al., 2010; R. Bray, 1996;

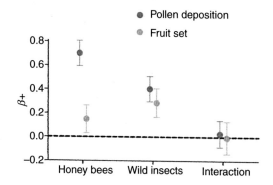

FIGURE 16.9   Effect of wild insect visitation to crop flowers compared to honey bee visitation for 41 crop systems on six continents. Visitation by wild insects enhanced reproduction in all crops examined (regression coefficient $\beta_i > 0$), whereas honey bee visitation had weaker effects overall, despite higher pollen deposition. Data are overall partial regression coefficients ($\beta_+ + 95\%$ CI) for the direct and interacting effects of visitation by wild insects and honey bees on pollen deposition or fruit set. *Source: From Garibaldi, L.A., et al., 2013. Wild pollinators enhance fruit set of crops regardless of honey bee abundance. Science 339, 1608–1611.*

Diamond, 1999; Hsiang et al., 2011; Perry et al., 2008; C. Riley, 1878; K. Smith et al., 2007; D. Zhang et al., 2007; Zorrilla-Miras et al., 2014). For example, Zorrilla-Miras et al. (2014) reported that current land-use changes reflect trade-offs among ecosystem services, with marketable provisioning services, such as cash crops and fiber, being promoted at the expense of regulating services, such as flood buffering and native biological control, and specific provisioning and cultural services traditionally used by local inhabitants.

Human history is filled with examples of social unrest, population displacement, and war resulting from food or water shortages, often caused by overexploitation of resources, as well as by insects or other natural causes (Acuña-Soto et al., 2002; Bora et al., 2010; R. Bray, 1996; Diamond, 1999; Hsiang et al., 2011; Perry et al., 2008; Riley, 1878; R.H. Smith, 2007; D. Zhang et al., 2007). In the wake of social unrest or population displacement come epidemics of crowd diseases, often vectored by insects (Acuña-Soto et al., 2002; R. Bray, 1996; Brouqui, 2011; Diamond, 1999; Therrell et al., 2004).

Recent studies have shown that Mesopotamian and Maya civilizations, even with relatively primitive technologies, were capable of sufficient deforestation to threaten their survival (Abrams and Rue, 1988; B. Cook et al., 2012; Janssen et al., 2008; Lentz and Hockaday, 2009; B. Turner and Sabloff, 2012; Xue et al., 1990; Zheng and Eltahir, 1998). Overexploitation of forest resources by the Late Classical Maya, after careful forest stewardship for at least 1200 years, likely reflected unsustainable urban population growth in the region (Abrams and Rue, 1988; B. Cook et al., 2012; Lentz and Hockaday, 2009; B. Turner and Sabloff, 2012). The collapse of both Mesopotamian and Maya civilizations, in what were originally fertile regions, resulted from long-term drought that apparently was exacerbated by deforestation and overgrazing (B. Cook et al., 2012; Lentz and Hockaday, 2009). B. Cook et al. (2012) used climate models to demonstrate that deforestation during the Maya period may have increased the severity of drought effects by 250%.

Thus, regulatory services that stabilize rates of delivery of other ecosystem services should not be undervalued. Once the value of natural regulation of ecosystem services is established, the value of regulation by insects and other ecosystem components could be calculated as the difference in long-term resource production, cultural use, and futures market values accruing or discounted from changes in ecosystem conditions resulting from the presence or absence of insect activity.

7 THREATS TO ECOSYSTEM SERVICES

The most significant threats to ecosystem services are anthropogenic activities that undermine the ecosystem processes that produce services. Ecosystem conversion, pollution, and invasive species represent the most immediate threats to ecosystem services, whereas climate change is likely to alter the distribution of services. For example, L. Thomson and Hoffman (2010) reported that biological control of insect pests was enhanced when vinyards were surrounded by natural remnant forest or planted woody vegetation, compared to no woody vegetation.

As an example, Allred et al. (2015) estimated that the land area occupied by oil well pads, roads, and storage facilities built from 2000 to 2012 in the central USA and Canada represented about 3 million ha, equivalent to three Yellowstone National Parks. The vegetation removal required for this oil and gas development reduced NPP by about 4.5 Tg of carbon or 10 Tg of dry biomass. Vegetation loss in rangelands was equivalent to loss of 420,000 head of livestock per year. Vegetation lost in croplands was equivalent to about 120 million bushels of wheat, that is, 6% of the wheat produced in 2013 within the region and 13% of the wheat exported by the USA, representing a significant trade-off between petroleum extraction and ecosystem services.

8 INSECTS AS INDICATORS OF ENVIRONMENTAL CHANGE

Given the growing need to anticipate environmental changes, indicators of environmental change have become an important ecosystem service. Insects have proven to be useful indicators of environmental changes (Dufrêne and Legendre, 1997; Maleque et al., 2009; van Straalen, 1998). Because of their short life spans, rapid reproductive rates, and sensitivity to temperature and to biochemical changes in their resources, insects provide early warning of changes that are not yet visible in the condition or abundance of larger, longer-lived plants or vertebrates, usually favored as bioindicators (Balanyá et al., 2006; Menéndez, 2007).

Aquatic insects have been particularly useful indicators of water quality (Bonada et al., 2006; C. Hawkins et al., 2000). For example, replacement of chironomid species that characterize oligo- to meso-trophic conditions by species characterizing eutrophic conditions provided early indication of pollution in Lake Balaton, Hungary (Dévai and Moldován, 1983; Ponyi et al., 1983).

Ant associations have been used as indicators of ecosystem integrity and the status of restoration efforts in Australia (A. Andersen and Majer, 2004). van Straalen (1998) and van Straalen and Verhoef (1997) reported use of soil arthropods as bioindicators of toxic residues in soil or changes in soil pH resulting from acid precipitation. Similarly, grasshoppers

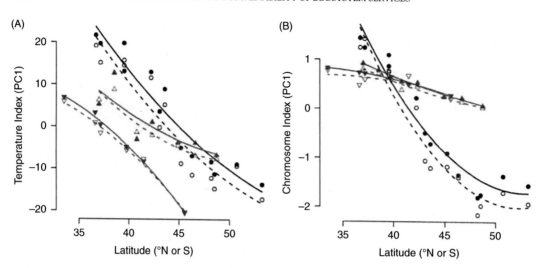

FIGURE 16.10 Temporal shifts in temperature and chromosome inversion frequencies for *Drosophila subobscura* at 26 sites at different latitudes on three continents (*black*, European sites; *red*, North American sites; *blue*, South American sites). (A) A climate temperature index is inversely correlated with latitude and has increased from historical (mean 24 years previous, open symbols, dashed regression lines) to recent (1997–2004, filled symbols, solid regression lines) samples. (B) A chromosome index is inversely correlated with latitude and also has increased from historical to contemporary samples. *Source: From Balanyá, J., et al., 2006. Global genetic change tracks global climate warming in* Drosophila subobscura. *Science 313, 1773–1775.*

(Fig. 5.11), dung beetles (Fig. 9.5), ground beetles, and xylophagous beetles can be used to assess ecosystem integrity and recovery status (Fielding and Brusven, 1995; Grove, 2002; B. Klein, 1989; Maleque et al., 2009; Niemelä and Spence, 1994; Niemelä et al., 1992). Insect herbivores could indicate changes in plant biochemistry before visible wilt, chlorosis, or other symptoms of stress become apparent.

Insect responses to changing climate can be rapid and evident before other indicators appear. Responses to elevated global temperatures include measureable shifts in the geographic ranges of species toward higher elevations and latitudes, earlier activity in spring as degree-day requirements are met earlier in the year, and measureable change in frequencies of genes conferring temperature tolerance (I. Chen et al., 2011, see Chapter 2). Menéndez (2007) analyzed distribution data for 1700 species of plants, insects, and vertebrates and found a significant range shift, averaging 6.1 km decade[-1] toward the poles (or m decade[-1] upward in elevation), and significant advancement of spring events by 2–3 day decade[-1]. Balanyá et al. (2006) compared genetic change and climate change for 26 populations of *Drosophila subobscura*, a cosmopolitan species, for which genetic composition has been known for, on average, 24 years (Fig. 16.10). Over this period, 22 of these populations experienced measureable warming, and 21 showed a shift toward the lower-latitude (warm-adapted) genotype. Both temperature and genetic shifts were equivalent to a 1 degree shift in latitude toward the equator.

9 SUMMARY

Ecosystems are the source of a variety of services, including food, fresh water, building materials, and medical and industrial products, as well as cultural and recreational services and the supporting and regulatory services that support other services. Although many people are unaware of, or ignore, their dependence on ecosystem services, human survival depends on sustainable delivery of these services.

Insect ecology provides a basis for managing the complex, often complementary, effects of insects on delivery of ecosystem services. In addition to affecting production of plant and animal resources, in both positive and negative ways, insects themselves provide valuable food, medical, and industrial products. Insects affect cultural and recreational values, as well, and contribute to supporting and regulating services that underlie resource production and cultural values. Although herbivores traditionally have been viewed as having negative effects on production of plant resources, herbivory should be viewed as regulating primary production in the same way that predation is recognized as maintaining stable, healthy populations of prey. Pollinators and detritivores have become widely recognized for their roles in supporting ecosystem services, including mitigation of human diseases. Biological control is an important regulating service, including mitigation of human diseases. Given that human survival depends on sustainable, not maximum and unsustainable, delivery of ecosystem services, pest management decisions should be based on the net benefits and costs of various options for multiple ecosystem services, as described in the next chapter.

17

Management of Insect Populations

ECOLOGICAL PRINCIPLES SOLVE THE BOLL WEEVIL PROBLEM

The cotton boll weevil, *Anthonomus grandis*, entered the USA from Mexico in 1892 (W. Hunter and Hinds, 1904) and by 1922 had spread across the entire cotton-growing region of the southern USA (Showler, 2009). At that time cotton was the primary crop in a largely agrarian region, but cotton production plummeted following appearance of the weevil. For example, weevils destroyed the entire cotton crop of southwestern Alabama in 1913 (R.H. Smith, 2007). The first year the boll weevil was found statewide in Alabama (1917), cotton production fell 70% to 500,000 bales, from a high of 1.7 million bales in 1914 (R.H. Smith, 2007).

Early efforts to control the weevil included picking and burning infested cotton squares, drowning weevils in kerosene, and a variety of home remedies, none of which improved cotton yields. Destroying all cotton residue following harvest in the fall and volunteer cotton in early spring did reduce weevil damage and was one of the earliest recommended control tactics (Howard, 1896). Failed cotton crops caused land values to plunge, leading to bankruptcy and mass migration out of the region (Hardee and Harris, 2003). However, cotton failure also led to diversification of crop production in the South and to a shift in cotton production to northern (and later western) regions where the weevil was unable to survive colder winters.

Passage by Congress of the Smith–Lever Act of 1914 created the Cooperative Extension Service, a state–federal cooperative that supported hiring of entomologists to advise cotton growers. Calcium arsenate dust became available for weevil control in 1918. Initially, this was applied by hand as a dust, with no protection for applicators. Following World War I aeronautical technology combined with veteran pilots led to aerial application methods that could cover larger areas with lower exposure to applicators (creating the term "crop duster").

Calcium arsenate application continued until the 1950s, when more effective chlorinated hydrocarbons became available after World War II. Initially, dichlorodiphenyltrichlorethane (DDT) was highly effective against boll weevil. However, weevil populations in Louisiana were resistant to DDT by 1955 (Roussel and Clower, 1957), and resistance was soon discovered in other areas. In addition, environmental research showed that these chemicals persisted in the environment and become more concentrated at higher trophic levels, leading to serious environmental concerns (Carson, 1962).

Insect Ecology. http://dx.doi.org/10.1016/B978-0-12-803033-2.00017-0

During this period, organophosphates, such as methyl parathion and malathion (developed as nerve gases during World War II), were introduced as liquid insecticides. These compounds were highly toxic to both weevils and applicators (as well as other vertebrates) but were short-lived. During the 1970s, pyrethroids and carbamates were introduced. These chemicals also were acutely toxic to insects, but had lower toxicity for vertebrates and low persistence. Frequent application (12–16 applications per season) of these compounds was required for control. In addition, growers became more dependent on insecticides to control boll weevils and relied less on cultural and other nonchemical options.

Increasing concern about environmental costs and fear that boll weevils would become resistant to newer insecticides, as they had to chlorinated hydrocarbons, led to consideration of other options. Advances in understanding of weevil "diapause" (Brazzel and Newsom, 1959) and pheromone communication (Tumlinson et al., 1969) suggested that targeting these key attributes, using a combination of tactics, could eliminate this pest. Starting in the mid-1970s, intensive crop monitoring with weevil pheromone bait trapped many weevils and indicated where weevils were most abundant, typically demes isolated by crop diversification. These more restricted sites were targeted for application of malathion, the most effective insecticide. These tactics were combined with mandated crop destruction soon after harvest to deprive weevils of late-season resources and overwintering sites.

By the mid-1990s, no yield losses to boll weevil were reported in Alabama for the first time since 1910 (R.H. Smith, 2007). Furthermore, insecticide use was reduced from an average of 10–14 applications per season to 1–4 (R.H. Smith, 2007). The last boll weevil in Alabama was reported in 2003 (R.H. Smith, 2007). The boll weevil was largely eliminated from most other states in the southern USA by 2010.

This century-long experience with the boll weevil demonstrated three major points. First, boll weevil spread was facilitated by extensive cultivation of a primary host. Second, crop diversification in the South established a more stable economy and limited spread of this and other insects. A monument to the boll weevil was dedicated in Enterprise, Alabama in 1919 to recognize the weevil's contribution to crop diversification and improved economy in Coffee County. Third, the successful removal of boll weevil as a factor affecting cotton production was achieved ultimately through application of ecological research to target critical aspects of weevil life history, with insecticides used as a last resort on isolated demes.

1 INTRODUCTION

Understanding of interactions between insects and our common environment, as described in Chapters 1–15, provides the basis for managing our interactions with insects in the broader context of ecosystem and global processes. Much of our knowledge of insect ecology, especially population dynamics and insect–plant and predator–prey interactions, has been gained from studies with more applied goals. Although insect outbreaks occur in natural ecosystems when conditions are favorable, anthropogenic activities or resource values have elevated a number of species to "pest" status (Stern et al., 1959). In particular, anthropogenic changes in ecosystem conditions have promoted population growth of many species (Raffa et al., 2008; Cease et al., 2012), and transport of others has permitted spread into novel habitats (Adkins and Rieske, 2013; Boettner et al., 2000; Evans, 2004). These changes often can be reversed or mitigated with adequate ecological information.

On the other hand, many insect species with critical roles in providing ecosystem services (see Chapter 16) are threatened by land use changes, climate change, or nontarget effects of pesticides or other pollutants (see Chapter 2). Pollinators, in particular, are in jeopardy as a result of agricultural intensification, habitat loss, and climate change (see Chapters 13 and 16). Significant declines in many species with unknown roles in ecosystem processes also may affect the delivery of ecosystem services (Warren et al., 2001).

Any management decision should be based on consideration of trade-offs between costs and benefits of alternative strategies and the various, often complementary, effects of insects on multiple ecosystem services. In all cases, the goal of ecosystem and pest management programs should be the sustainability of ecosystem services, that is, avoiding tactics that would undermine natural regulatory processes that contribute to sustained delivery of services (see Chapter 16). It is important to note that insect species achieve "pest" status as a result of their ecological attributes bringing them into conflict with human interests and that these interests can change (Stern et al., 1959). Such information is necessary to determine when suppression of outbreaks may be warranted to sustain, not maximize, delivery of ecosystem services, by avoiding undesirable nontarget effects of management tactics. The principles of integrated pest management (IPM) were developed largely to facilitate pest management decisions that contribute to sustainability of ecosystem services.

2 INTEGRATED PEST MANAGEMENT

The concept of integrated pest management (IPM) is perhaps the most important application of insect ecology in terms of contributions to reduced use of pesticides and consistency with sustainability of ecosystem services (see Chapter 16). Most, if not all, species fill important ecological roles in their native ecosystems (see Chapters 12–15). However, under suitable conditions, some species can interfere with essential ecosystem services or threaten human health or structures. Insects become viewed as pests when their activities conflict with human well-being. It is important to distinguish a species' role as a pest in a managed ecosystem from its role(s) in natural ecosystems. For example, termites often pose serious threats to the integrity of human cellulose-based building materials, but eliminating termites beyond human-dominated habitats would threaten the integrity of decomposition processes required for sustainability of ecosystem productivity (see Chapters 14 and 16).

Insects most often become pests as a result of management practices that favor insect population growth. Native insects can become pests when previously scattered native host species, or a related exotic species, become widely planted at high density (eg, cotton boll weevil, Colorado potato beetle, *Leptinotarsa decemlineata*, see Chapter 4) or when altered disturbance frequency creates an artificially high density of host plants (eg, Douglas-fir tussock moth, *Orgyia pseudotsugata*, see Chapter 15). Crop species often are bred to reduce bitter (defensive) flavors, facilitating growth and reproduction by herbivorous species (Michaud and Grant, 2009, see Chapters 3, 6, and 7).

Humans have transported many species across natural barriers, many intentionally (such as crop, livestock, and ornamental species), but others unintentionally (crop-feeding insects and vectors of human pathogens). Some of these species have become serious invasive pests in new habitats as a result of release from bottom-up and top-down regulatory factors characterizing their native habitat. Their intrusion into natural ecosystems threatens native species through competition, herbivory, predation, and parasitism and alters ecosystem structure and function in ways that threaten ecosystem services. More than 11,000 alien species in Europe cost at least US\$13 billion annually in lost ecosystem services, but because environmental and economic impacts are known for only about 10% of these invasive species, their cost is likely much higher (P. Hulme et al., 2009). Efforts to control such species are expensive and not always effective (Pimentel et al., 1992).

Recent advances in understanding the complex effects of insects and their interactions with other organisms on ecosystem services have influenced evaluation of the need for insect management. Management goals for natural ecosystems, especially forests and grasslands, have become more complex in many regions, as societal needs have changed from a focus on extractive uses, for example, fiber, timber, or livestock production, to broader goals that include protection of water yield and quality, fisheries, recreational values, biodiversity, and ecosystem integrity. In many cases, the net effects of native species on ecosystem services are neutral or positive, and/or the cost of control is not warranted by marginal benefits (see Chapter 16). The Integrated Pest Management (IPM) and Integrated Forest Protection (IFP) concepts provide a framework for deciding when insects pose a sufficient threat to ecosystem services to warrant suppression.

2.1 Development of the IPM Concept

Before synthetic chemical insecticides became available in the 20th century, entomologists combined knowledge of population biology, plant resistance, and biological control agents into multitactic control strategies (T. Harris, 1841; Kogan, 1998; R. Painter, 1936, 1951; C. Riley, 1883, 1885, 1893; C. Smith, 2005). One of the earliest insecticides was pyrethrum, prepared from the powdered flower heads of *Pyrethrum roseum*, used at least as early as 1800 to control insects in Asian countries south of the Caucasus Mountains (C. Riley, 1885). Mary Beebe reported that this insecticide, as Persian Insect Powder, was indispensable for reducing irritation from biting insects during a trip with William Beebe in Mexico (Beebe, 1905). However, pyrethrum was effective only in direct contact with target insects and became ineffective within an hour of application. As a result, pest management in the late 1800s and first half of the 20th century became dominated by broad-spectrum toxins that persisted longer.

Forbush and Fernald (1896) noted high tolerance of gypsy moth larvae to arsenical compounds (high survival at tissue concentrations 12× the lethal dose for humans), and Melander (1923) reported that San Jose scale, *Quadraspidiotus perniciosus*, in Washington, USA, was becoming progressively more tolerant of insecticides, particularly sulfur compounds, the first documentation of insecticide resistance. Nontarget effects threaten ecosystem function and stability (D. Baldwin et al., 2009; Claudianos et al., 2006; S. Smith et al., 1983). Although insecticide development during the past 50 years has emphasized shorter residual times and reduced nontarget effects, continued problems with pesticide resistance have led to increased attention to alternative strategies (Hsu et al., 2006, 2008; Ottea and Leonard, 2006). The IPM concept developed fully during the 1960s when various tactics were combined with consideration of economic thresholds and multiple pests and nonpests (Kogan, 1998; Nicholls and Altieri, 2007; Stern et al., 1959; van den Bosch and Stern, 1962).

The last legal use of DDT in the USA, in 1974 against a Douglas-fir tussock moth outbreak in Oregon and Washington, reflected growing public concern about pesticide use and demonstrated the value of alternative control methods. Because the US Environmental Protection Agency (EPA) had cancelled use of DDT in the USA in 1972, its use in 1974 required emergency authorization, based on an apparent lack of practical alternatives. As part of this authorization, the EPA mandated that research on alternative methods of control be intensified and that experimental plots be established to demonstrate the efficacy of DDT (Brookes et al., 1978). In the mandated experiment, tussock moth populations declined in all plots, regardless of DDT treatment, leading to the discovery that nuclear polyhedrosis virus (NPV), *Baculovirus* spp., naturally ends tussock moth outbreaks in 3–4 years. NPV had been recognized as a natural mortality agent (Brookes et al., 1978), but its importance had been masked by prior control programs that prevented development of epizootics. In the wake of this outbreak, aerial application of technical-grade NPV became the preferred means of control for incipient outbreaks.

The current concept of IPM represents an intersection between pest management and insect ecology that emphasizes integration of natural ecological processes with various management methods and addresses interactions among the diversity of target and nontarget species, to maintain pest population sizes below levels that would interfere with maintenance of ecosystem services (see Kogan and Jepson, 2007). Control methods for any target species must be consistent with the complex of other insects and pathogens to avoid outbreaks of nontarget species (Luck and Dahlsten, 1975). The following four IPM principles were designed to guide pest management decision-making.

First, pest management strategies should be designed to sustain, not maximize, use of ecosystem services, in order to avoid undermining continued delivery of those services (Schowalter, 2013). Killing insects does not necessarily improve crop yield, and insecticides can disrupt other ecosystem services, especially pollination (Cayuela et al., 2011; Claudianos et al., 2006) biological control (Hallmann et al., 2014) and fisheries (S. Smith et al., 1983). Furthermore, to the extent that native insects represent regulatory mechanisms in natural ecosystems (Chapter 15), insect control may be unnecessary and even counterproductive to long-term maintenance of ecosystem services.

Second, the goal of pest management is not to control a pest population, but rather to maintain population size below a threshold (injury level) that would cause losses to ecosystem services exceeding costs of control (Stern et al., 1959). Implementation of control measures should be based on predictive models that indicate when an insect population is expected to exceed the injury level (Rabb et al., 1984). Such prediction assists in establishing an action threshold at which planned action should be initiated to prevent the population from exceeding the injury level. For example, Ragsdale et al. (2011) concluded that soybean aphids, *Aphis glycines*, cause little economic loss to soybean yields below 675 aphids per plant, but control should be initiated at 250 aphids per plant to prevent growing populations from reaching the economic injury level. Controlling aphids at lower densities is an unnecessary expense and undermines natural regulatory processes.

Economic thresholds typically are based on economic losses in production of agricultural or forest commodities, although thresholds also can be established to prevent other undesirable effects on ecosystem services, such as change in vegetation composition or structure in nature preserves or reduced carbon sequestration. Unfortunately, the general public tends to be intolerant of blemished produce or increased food prices and thereby forces producers to control insects at population sizes that cause little loss in food or product value (Stern et al., 1959).

Acceptance of control costs by producers and the general public varies with the degree of visual damage or food shortage (S. Sheppard and Picard, 2006; Torell et al., 1989). When control costs are subsidized by the government, crop producers or resource managers are inclined to control insects at lower densities than would be acceptable to taxpayers. For example, an individual rancher's 50% share of the cost for control at an economic injury level of 18 grasshoppers ha^{-1} is equivalent to a rancher's economic injury level of < 3 grasshoppers ha^{-1} (Table 17.1, Torell et al., 1989). Gatto et al. (2009) conducted an economic analysis of pest management for the processionary moth, *Thaumetopoea pityocampa*, in Portugal and concluded that pest management costs outweighed market revenues for maritime pine, *Pinus pinaster*, plantations, at least in the short term, making control undesirable for private landowners (Table 17.2). Taxpayer support also would be unwise, based on provisioning service values alone, but could be justified by potential benefits to the public through other types of ecosystem services, such as improved carbon sequestration, recreation, and public health, that is, nonprovisioning ecosystem services.

Third, management strategies must be consistent with ecological principles. Any action that does not contribute to enhanced sustainability of ecosystem services is economically and environmentally unsound, regardless of how many insects can be killed. Understanding the environmental factors that promote unacceptable population growth and augmenting bottom-up and top-down regulation (see Chapter 6) should prevent most undesirable population growth of target species while minimizing unintended consequences for food web interactions and ecosystem services (Branson et al., 2006; K. Zimmerman et al., 2004). In particular, anthropogenic changes, such as resource concentration in planted crops or forests or vegetation changes resulting from overgrazing or fire suppression, often are responsible for insect outbreaks (eg, Cease et al., 2012, 2015; Jactel and Brockerhoff, 2007; Schowalter, 2008; Schowalter and Turchin, 1993). Fortunately, most economically important insect species have been studied sufficiently to provide necessary information on factors that trigger outbreaks.

Finally, when pest suppression is necessary, using multiple remedial tactics, including cultural practices, pheromone disruption, biological control, and insecticide application, can effectively reduce pest population levels and prevent or delay genetic adaptation to any particular tactic (Barbosa, 1998; Huffaker and Messenger, 1976; Kogan, 1998; Lowrance et al., 1984; Rabb et al., 1984; Reay-Jones et al., 2003; Rickson and Rickson, 1998; R.H. Smith, 2007). For example, application of entomopathogenic nematodes increases the fitness cost of Bt resistance in pink bollworm, *Pectinophora gossypiella*, thereby delaying development of Bt resistance and maintaining its effectiveness in transgenic cotton (Gassmann et al., 2008, 2009). In the boll weevil example described in the Box introducing this chapter, successful control ultimately depended on a combination of crop diversification, mating disruption with pheromones, and destruction of overwintering habitat, as well as limited use of insecticides on local residual populations.

In practice, IPM requires a stepwise decision-making process, beginning with definition of management goals or desired future conditions. Defined goals or desired future conditions determine which available management options are most appropriate in a given situation. For example, sustained timber production in the Douglas-fir tussock moth example in Chapter 15 requires reduced fir density and frequent low-intensity fire, conditions that

TABLE 17.1   Economic Injury Levels for Grasshopper Control Programs, Based on Varying Levels of Treatment Cost, Forage Value, Discount Rate, Grasshopper Population Growth Rate, and Treatment Efficacy

| | Economic threshold (no. grasshoppers m^{-2}) | |
Model assumption	Fourth instar	Adult
Average conditions[a]	18	10
Treatment cost, \$ ha^{-1}		
3.09	<4	<3
4.94	11	7
6.18	18	10
7.41	28	14
8.65	36	17
Forage value, \$ AUM^{-1}		
4	68	26
6	33	16
8	18	10
10	10	6
12	5	4
14	<4	<3
Discount rate, %		
4	17	8
7	18	10
10	19	10
13	20	11
16	22	12
Population growth rate for untreated grasshoppers		
Decrease by 0.5 year^{-1} ($r_1 = -0.695$)	49	21
Remain constant ($r_1 = 0$)	18	10
Double each year ($r_1 = 0.695$)	<4	<3
Population growth rate for treated grasshoppers		
Remain constant ($r_2 = 0$)	7	5
Double each year ($r_2 = 0.695$)	18	10
Triple each year ($r_2 = 1.10$)	32	15
Treatment efficacy rate, %		
99	5	4
95	8	6
90	18	10

(Continued)

TABLE 17.1 Economic Injury Levels for Grasshopper Control Programs, Based on Varying Levels of Treatment Cost, Forage Value, Discount Rate, Grasshopper Population Growth Rate, and Treatment Efficacy (*cont.*)

Model assumption	Economic threshold (no. grasshoppers m^{-2})	
	Fourth instar	Adult
85	29	14
80	42	19
75	65	25
Grasshopper life stage		
Before fourth instar	18	10
Before fifth instar	20	11
Before adult	24	13

a Average conditions: treatment cost = $6.18 ha^{-1}; forage value = $8 animal unit month (AUM)$^{-1}$; discount rate = 7%; untreated grasshopper density remains constant; treated grasshopper density doubles each year; treatment efficacy = 90%; and treatment occurs at the beginning of the fourth instar. These conditions were altered to compute the economic injury level for alternative model assumptions.
Source: From Torell, L.A., et al., 1989. Economic injury levels for interseasonal control of rangeland insects. J. Econ. Entomol. 82, 1289–1294.

TABLE 17.2 Benefit/cost Analysis of Processionary Moth (*Thaumetopoea pityocampa*) Effects on Ecosystem Services in Portuguese Pine Forests. Net Present Value (NPV) in thousand euros.

Total benefits (+) and costs (−)	Scenario		Net benefit (with—without)
	Management	No management	
Financial analysis			
Timber revenue (+)	27,562	22,785	+4,777
Forest management cost (−)	22,266	22,266	0
Pest management cost (−)	6,044	0	−6,044
Cumulative NPV (2%)	−748	519	−1,267
Conventional economic analysis			
Estate value (+)	33,079	28,049	+5,030
Cumulative NPV (2%)	32,331	28,568	+3,763
Extended economic analysis			
Carbon sequestration (+)	9,878	7,902	+1,976
Risk of dermatitis (−)	0	4,058	−4,058
Cumulative NPV (2%)	42,209	32,412	+9,797
Extended economic analysis			
Recreation (+)	11,160	8,928	+2,232
Cumulative NPV (2%)	53,369	41,340	+12,029

Source: Reprinted from Gatto, P., et al., 2009. Economic assessment of managing processionary moth in pine forests: a case study in Portugal. J. Environ. Manage. 90, 683–691.

minimize the likelihood of insect outbreaks (North et al., 2007; Schowalter, 2008). In all cases, the primary goal should be enhanced sustainability, not maximum use, of ecosystem services. Second, adequate estimation of population size (see Chapter 5) and predicted time to reach the action threshold establish the timeframe for initiating necessary action (Schowalter et al., 1982). Third, evaluation and comparison of available management options (including no action) on the basis of expected accomplishment of management goals, effectiveness within the established timeframe, cost, environmental and social consequences, etc. permit ranking of options in terms of their net benefits relative to costs. In many cases, insect control may not be the best option (eg, if compensatory growth prior to resource harvest would replace lost production) and often does not improve yields (Cayuela et al., 2011; Ragsdale et al., 2011). Computerized decision-support models have been developed to aid evaluation of management options. Such models integrate a user interface with submodels that predict population growth, effects of management actions, and calculated benefit/cost to provide benefit/cost analysis under various user-defined scenarios (C. Shaw and Eav, 1993). Finally, monitoring of management activity and evaluation of the benefits and costs of the completed project are essential to improved future (adaptive) management. Tactics available for crop and forest pests, disease vectors, and urban pests are described in Sections 2.2–2.4.

2.2 Ecological Tactics for Managing Crop and Forest "Pests"

Management of crop and forest "pests" has been a major application of insect ecology for at least the past two centuries (C. Riley, 1878, 1883, 1893; Kogan and Jepson, 2007). For most of the 1900s, the prevailing method for pest control was application of insecticides. However, the widespread use of insecticides often has been unnecessary and has not protected ecosystem services (Branson et al., 2006; Cayuela et al., 2011; Gatto et al., 2009; Torell et al., 1989). In fact, insecticides often affect nontarget species in ways that threaten ecosystem services (D. Baldwin et al., 2009; Claudianos et al., 2006; Hallmann et al., 2014; Sánchez-Zapata et al., 2007). Furthermore, frequent exposure to insecticides has induced resistance in many target species, often within 5–10 years, (Felland et al., 1990; Feyereisen, 1999; Ffrench-Constant, 2013; Mutunga et al., 2015; Soderlund and Bloomquist, 1990). For many target pests, insecticides have become obsolete faster than new chemicals, with new modes of action, can be developed.

Forests present special challenges for pest management. Access to tall canopies for assessment of injury levels is limited. Aerial surveys can provide data on severity and extent of injury, but ground checking is necessary to confirm the identity of the pest. This is particularly difficult in montane forests. Furthermore, forests are subject to cumulative and interactive effects of insects and pathogens, as well as abiotic conditions, that predispose forests to other factors for long periods before injury becomes apparent. This requires a degree of attention to changes in forest conditions that often is not supported within operating budgets. Forests in many regions are managed for multiple uses that may be affected by insects in complementary ways (see Chapter 16). Finally, controlling insects in forests often is not cost-effective, given marginal profits and losses due to pest control or unscheduled salvage harvests.

Cultural methods for preventing resource losses were the subject of investigation as early as the late 1700s (C. Smith, 2005; R. Painter, 1951) and remain the most sustainable option for minimizing crop damage by both insects and pathogens. In managed forests, promotion of site-adapted tree species and adequate tree spacing generally are sufficient to minimize losses

FIGURE 17.1   **Examples of multicropping to hinder spread of insect species over agricultural landscapes in northeastern China.** (A) Embedded intercropping within rows; (B) multiple crop species arranged in strips.

to insects or pathogens (Nealis et al., 2009; Schowalter and Turchin, 1993). In agricultural crops, planting resistant cultivars, mixing different crops (Fig. 17.1), including agro-forestry, across the landscape, and crop rotation among host and nonhost species have a long history of use (Barbosa et al., 2009; Gliessman, 2007; Nicholls and Altieri, 2007; S. Risch, 1980, 1981; Yao et al., 2012). Reay-Jones et al. (2003) estimated that planting resistant varieties of sugarcane could reduce crop losses to the Mexican rice borer, *Eoreuma loftini*, by 24%, and irrigation with 30 cm of water would reduce losses by 29%, substantially reducing the need for pest control. Mixing different crop species also can reduce pest populations and crop losses (S. Risch, 1980, 1981; H. Zhou et al., 2009). Even mixing resistant and susceptible varieties of the same crop in the same field can provide significant benefit for pest management (Garrett and Mundt, 1999).

Many crop varieties show reduced resistance to insects and pathogens, compared to wild progenitors, as a result of breeding for rapid growth in lieu of defense (J.K. Cronin et al., 2007; Lindig-Cisneros et al., 1997; Michaud and Grant, 2009; Rodriguez-Saona et al., 2011). Consequently, backcrossing with wild progenitors may improve insect resistance for some crops (J.K. Cronin et al., 2007).

Newer cultural practices emphasize protection of soil moisture, fertility, and nutrient cycling processes that minimize crop stress and retain a diversity of predators and parasites to control herbivore populations (Altieri, 2002; Denys and Tscharntke, 2002; Gliessman, 2007; Isaacs et al., 2009; Landis et al., 2000, 2005; Nicholls and Altieri, 2007; Pretty et al., 2003; Thies et al., 2003; Tscharntke et al., 2005, 2007; Vincent et al., 2003; H. Zhou et al., 2009). Examples include nontillage, intercropping or cover crops of N-fixing legumes, mulch application, and conservation of noncrop patches and hedgerows within the agricultural landscape. Another recent addition, trap cropping, involves planting a border strip of an attractive host variety around the crop field to concentrate pests in a small area, either to keep them out of the crop or to focus control measures more efficiently (Hokkanen, 1991; Kogan and Turnipseed, 1987). Pretty et al. (2003) surveyed 208 farming projects (representing 9 million farms covering 29 million ha in 52 developing countries in Africa, Asia, and Latin America) in which farmers have adopted environmentally sensitive practices and technologies. Of 89 farming projects with reliable data, 93% showed an increase in food production per hectare, resulting from increased water use efficiency, improved soil health and fertility, and minimal or zero pesticide use. Shifts to more sustainable agricultural practices have substantial benefits for the rural poor in developing countries.

A number of remedial options are available when suppression of pest populations becomes necessary. Insecticides remain an option when rapid reduction in pest populations is necessary. Modern insecticides includes a variety of insect growth regulators (IGRs, such as methoprene), lipid synthesis inhibitors (LSIs, such as spirotetramat), and chitin synthesis inhibitors (CSIs, such as noviflumuron) that represent novel modes of action, that minimize cross-resistance with other insecticides, and have shorter residual times and fewer non-target effects (Brück et al., 2009; Merzendorfer, 2013; Mohapatra et al., 2012). Insecticide, use can be minimized by restricting application to trap crop borders around crop fields (Buteler et al., 2009) and by employing a variety of ecologically based alternatives. Use of the options described next still requires consideration of insect adaptation and consequences for environmental conditions.

Physical control methods involve use of manual or mechanical tools or barriers for killing target insects (C. Riley, 1878, 1883; Vincent et al., 2003). Hand-picking insects from crop plants has been a basic control tool for centuries (C. Riley, 1883, 1885). Collected insects can be killed or used commercially, for human or livestock food (Cerritos and Cano-Santana, 2008; Ramos-Elorduy, 2009). Trenches or other barriers have a long history of interrupting movement of walking insects, especially marching locust nymphs (C. Riley, 1878, 1883). Horse-drawn hopper-dozers were used in the USA in the early 1900s to scrape grasshoppers into oil- or kerosene-filled trenches (C.R. Jones, 1917; R.C. Smith, 1954) or to bag them as poultry feed (Fig. 17.2). Collected insects also could provide valued protein for humans (Cerritos and Cano-Santana, 2008; see Chapter 16). Sexton and Schowalter (1991) described the cost and effectiveness of several tree bole barriers to prevent a flightless cone-feeding weevil, *Lepesoma lecontei*, from moving into seed orchard trees. Prescribed fire also may be used to control populations of ground-based insects (C. Riley, 1880; K. Miller and Wagner, 1984). Other physical control methods include cold water sprays in agricultural fields to disrupt phenological

FIGURE 17.2   **Horse-drawn "hopper dozer", near St. Ignatius, Montana, in 1917, illustrates an early nonchemical technique for grasshopper control.** Approximately 363 kg (800 lbs) of grasshoppers were caught, bagged, and dried for use as winter poultry food in 2 h 15 min. Such machines reduced grasshopper populations and provided poultry feed, while avoiding risks to livestock from the use of highly poisonous, arsenic-based baits. *Source: US Department of Agriculture photo archive.*

synchrony between insects and plants and hot water baths to kill insects in postharvest produce (Vincent et al., 2003).

Many entomopathogens or their products, including viruses (eg, *Baculovirus*, NPV), bacteria (eg, *Bacillus thuringiensis*, Bt, and spinosad), fungi (eg, *Beauveria bassiana* and *Metarhizium anisopliae*), and nematodes (*Steinernema riobrave* and *Heterorhabditis bacteriophora*), can be applied in the same manner as chemical controls and often are confused with chemical insecticides (Vega and Kaya, 2012). The advantage of live microbial control agents is their ability to replicate within their hosts and spread epizootically through the target population.

The effectiveness of many microbial pathogens has led to genetic modification of some crops to express toxic bacterial proteins, such as Bt endotoxins Cry1A, Cry1Ac, and Cry2Ab. Although controversial in many countries, transgenic corn, wheat, cotton, and other crops have reduced the use of more toxic insecticides substantially (Cattaneo et al., 2006) and show fewer effects on nontarget species than do insecticides (Marvier et al., 2007). Nevertheless, pollen and plant detritus containing Bt toxins may affect soil organisms and decomposition and are exported to other ecosystems, such as streams, where they could affect nontarget organisms (Hansen Jesse and Obrycki, 2000; Losey et al., 1999; O'Callaghan et al., 2005; Rosi-Marshall et al., 2007; Sears et al., 2001; Zangerl et al., 2001).

Widespread planting of transgenic crops in some regions threatens to undermine their long-term effectiveness, given insect ability to adapt quickly to any strong selection factor (F. Gould, 1998; Kruger et al., 2008; Tabashnik et al., 2008). Therefore, a high-dose-with-refuge strategy is recommended to prevent survival of pests on the Bt crop and maintain a large, nonadapted population in nonBt refuges (Alstad and Andow, 1995; Carrière et al., 2003; Ives and Andow, 2002; Kruger et al., 2008; Tabashnik et al., 2008), which are controlled with insecticides but in a more limited area.

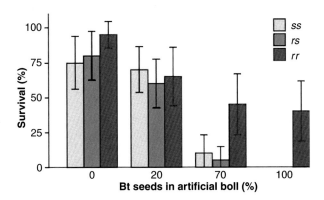

FIGURE 17.3 **Percentage survival and 95% confidence intervals for pink bollworm,** *Pectinophora gossypiella,* **larvae of susceptible** *(ss),* **resistant** *(rr),* **and heterozygous** *(rs)* **genotypes on artificial cotton bolls containing 0–100% Bt seeds.** *Source: From Heuberger, S., et al., 2008. Effects of refuge contamination by transgenes on Bt resistance in pink bollworm (Lepidoptera: Gelichiidae). J. Econ. Entomol. 101, 504–514.*

Incipient resistance to Bt toxins in some target species (F. Huang et al., 2007; Ottea and Leonard, 2006) has led to insertion of multiple Bt-toxin genes into some crop plants to delay insect adaptation, to the extent that cross-resistance to the different genes is minimized. Nevertheless, long-term effectiveness could be undermined if any or all Bt toxins are not produced at concentrations sufficient to kill all herbivores during the entire season (Brévault et al., 2013; Kranthi et al., 2005; Mahon and Olsen, 2009; Showalter et al., 2009), or if pollen or seed contamination of nontransgenic refuges or native vegetation leads to variable Bt concentrations across the landscape (Fig. 17.3, Chilcutt and Tabashnik, 2004; Heuberger et al., 2008a,b, 2010; D. Llewellyn et al., 2007; Romeis et al., 2008; Zangerl et al., 2001). This requires attention to soil moisture, nutrient availability, or other conditions that stress crop plants and influence production of Bt toxins, and to the landscape structure of Bt and nonBt crops (especially for insects with broad host ranges that might include multiple transgenic crops), as well as cooperation among scientists, growers, and government agencies (Carrière et al., 2001a).

Other genetic tools to control pest populations are being explored but also may have unintended consequences for communities and ecosystems. RNA interference (RNAi) is a transgenic method that can inhibit target insect expression of P-450 genes for detoxification of plant defenses. RNAi is accomplished by incorporating double-stranded RNA (dsRNA) specific to the detoxification gene. Y. Mao et al. (2007) demonstrated that adding dsRNA to cotton bollworm, *Helicoverpa armigera,* diet decreased the expression of CYP6AE14, the bollworm's primary detoxification gene, in the midgut and reduced larval growth. This technology could permit focused control of a specific target species in specific situations with minimal nontarget effects (Burt, 2003). Homing endonuclease technology (Burt, 2003) involves engineering a homing endonuclease gene (HEG) that encodes for an enzyme that recognizes and cleaves a specific 20–30 bp sequence on chromosomes that do not contain the HEG. The intact HEG+ homologue is used by the cell's repair system as the template, so after repair both chromosomes contain the HEG. The gene would spread quickly through the population from a relatively small number of introduced individuals, providing for rapid manipulation of population characteristics. Although such options seem attractive because of their apparent focus on

FIGURE 17.4 Natural biological control represented by a predaceous pentatomid nymph preying on a forest tent caterpillar, *Malacosoma disstria*. *Source: From Schowalter, T.D., 2013. Insects and Sustainability of Ecosystem Services. CRC Press/Taylor and Francis Group, Boca Raton, FL.*

the target species, they fail to address potential effects of the engineered population on the network of direct and indirect interactions in ecosystems. Such technology may be most useful and appropriate for control of invasive pests.

A promising new tool is application of chemical elicitors, such as jasmonic acid (see Chapters 3 and 8), to induce production of natural defenses by crop plants in advance of herbivory (Senthil-Nathan et al., 2009; M. Stout et al., 2002; Thaler, 1999b; Thaler et al., 2001). This approach would be most effective when infestations can be reliably anticipated and action thresholds are high, because of delay in expression of induced defenses. However, expression of natural or transgenic defenses by plants depends on adequate resources for growth and defense (see Chapter 3).

Biological control involves the augmentation or introduction of predators or parasites to reduce pest populations (Fig. 17.4; see also Figs. 6.11, 8.4–8.7). Biological control was practiced in China as early as the 3rd century AD, for example, ants were sold for control of citrus insect pests (Konishi and Itô, 1973). C.V. Riley (1893) initiated the practice of importing natural enemies from a pest's region of origin for control purposes. He was responsible for the first widely successful biological control program, the importation of the coccinellid, *Rodolia (Vedalia) cardinalis*, to control the cottony cushion scale, *Icerya purchasi*, that was devastating California's citrus crops (Hagen and Franz, 1973). As a result, Riley is often called the "father of biological control."

This success launched an increasing number of introductions of parasitic or predaceous organisms from a pest's region of origin for biological control purposes (Van Driesche and

Bellows, 1996). In some cases, herbivorous insects have been introduced to control invasive plant species (R. Doyle et al., 2002; Louda et al., 2003), but biological control agents often have targeted insect pests (Louda et al., 2003). Introducing biological control agents from the pest's region of origin requires consideration of the agent's ability to become established in the new community and its effects on nontarget species (Delfosse, 2005; Louda et al., 2003; McCoy and Frank, 2010; Symondson et al., 2002; van Lenteren et al., 2006). Early failures in biological control often were due to inaccurate identification of the most effective predator or parasite species. For example, sibling species may not be equally effective in regulating a host population; superparasites or cleptoparasites reduce the effectiveness of the primary biocontrol agent; endemic predators may interfere with the efficacy of introduced parasitoids (Erbilgin et al., 2004). Furthermore, ecosystem integrity can be disrupted by introduction of nonspecific biocontrol agents that attack native, as well as invasive, hosts (Boettner et al., 2000; E. Evans, 2004; Louda et al., 2003). Refinement of quarantine and testing procedures for potential biocontrol agents and selection for more specialized agents have minimized these problems, but it is difficult to anticipate all consequences of introductions (Delfosse, 2005; Hokkanen et al., 2007; Louda et al., 2003; McCoy and Frank, 2010).

An alternative to introduction of biological control agents is conservation or augmentation of native predator, parasitoid, and parasite diversity within agricultural or forest landscapes (Isaacs et al., 2009; Landis et al., 2000, 2005; C. Thorne et al., 2007; Tscharntke et al., 2005, 2007). This requires attention to noncrop features that are necessary to sustain populations of these regulatory agents over areas greater than the crop patches. In particular, conservation of hedgerows and patches of natural ecosystems maintains populations of alternate hosts and floral resources necessary to sustain predators and parasitoids during periods when their hosts on crop plants may be absent (Hassall et al., 1992; Landis et al., 2000; Marino and Landis, 1996; Thies and Tscharntke, 1999). Such practices require cooperation among multiple landowners on a landscape scale. Forest defoliators have proven to be amenable to control with native microbial agents. For example, Douglas-fir tussock moth was among the first species to be managed with biological control agents. Following discovery in the 1970s that a *Baculovirus* sp. (NPV) typically ended outbreaks of its host, this virus has been produced in laboratory colonies of its host for application as needed (C. Thorne et al., 2007; C. Thompson, 1978). *Baculovirus* spp. and *Bacillus thuringiensis* are now widely used for control of native defoliators, as well as gypsy moth, *Lymantria dispar*. However, native predators and parasites often do not recognize invasive species as potential prey or hosts (J. Adams et al., 2009).

Pheromones have become highly valued tools for pest management because of their specificity in attracting target species (Groot et al., 2006; Hemmann et al., 2008; see Chapter 4). Identification of the particular blend of volatile compounds used by a species (see Chapter 4) is necessary for best effect (L. Gaston et al., 1967, 1977; Tumlinson et al., 1969). Modern chemical techniques have produced extensive pheromone libraries for many groups, facilitating identification of pheromone blends for new species (Khrimian et al., 2014; Rodstein et al., 2009).

Pheromones can be used in a number of ways for management purposes. Pheromone-baited traps can be placed in strategic locations to detect and monitor distribution and abundance of a target species, such as a potential invasive species (Boddum et al., 2009; Gries et al., 2009; Weber et al., 2014a). Pheromones in controlled-release containers placed in the field or infused in light, plastic filaments applied aerially can be used to disrupt normal attraction of mates, that is, mating disruption (C. Alfaro et al., 2009; L. Gaston et al., 1967, 1977; Niwa et al., 1988; Vacas et al., 2009), or to repel insects from the treated site (Gillette et al., 2009). Baited traps also can be used to reduce target populations locally (D. Bray et al., 2009; Vargas et al., 2009; Weber et al., 2014a). These applications will be most effective for controlling small or relatively isolated populations, for which saturation of the populated area with pheromone is feasible and attraction of additional target insects from outside the patch is limited, for example, orchards and islands (Yamanaka and Liebhold, 2009). L. Gaston et al. (1977) first demonstrated that use of pheromones for mating disruption could significantly reduce pest abundance and insecticide use.

Sterile insect release (SIR) is an option in some circumstances (J. Myers et al., 1998; Yamanaka and Liebhold, 2009). This technique requires mass rearing, sterilization, and inundative release of sterile insects to saturate the target population with nonreproductive individuals. A critical requirement for success is a single-mating population. If females can mate more than once, unsuccessful mating with a sterile mate does not necessarily prevent reproduction. Eradication of the screwworm fly, *Cochliomyia hominivorax*, from the USA during the 1960s, and subsequently from Mexico by 1990, remains a classic example of successful use of this technique, requiring release of millions of sterile males at a total cost of US$750 million (J. Myers et al., 1998; Yamanaka and Liebhold, 2009). Other sterile insect release programs have been most successful on islands (J. Koyama et al., 2004; Yamanaka and Liebhold, 2009).

Advances in remote sensing and geographic information systems (GIS) have improved the precision of pest management activities (Bongiovanni and Lowenberg-Deboer, 2004; Eisen and Eisen, 2011). Within-field patterns of pest density can be superimposed on crop production gradients to identify areas of greatest potential increase in yield under different pest management scenarios (Willers et al., 2005). Similarly, patterns of disease vector habitat and human habitation can be used to identify target sites for prevention of transmission to humans (Eisen and Eisen, 2011). In place of traditional broadcast methods, pest management can be focused at locations where it will be most economically or ecologically beneficial. Furthermore, fertilizer and pesticide application rates can be tailored to habitat conditions to avoid overapplication and minimize contamination of groundwater or streams (Bongiovanni and Lowenberg-Deboer, 2004).

Entomophagy represents an alternative to killing insect pests that can also provide necessary dietary protein (van Huis, 2013; Lundy and Parrella, 2015; see Chapter 16 and Figs 16.2 and 17.2). Cerritos and Cano-Santana (2008) reported that harvesting grasshoppers for food, in Mexico, reduced grasshopper density nearly as well as insecticide application but provided a net benefit per family of US$3000.

2.3 Ecological Tactics for Managing Medical and Veterinary "Pests"

Insects and insect-vectored diseases have serious effects on human and livestock health (B. Allan et al., 2009; R. Bray, 1996; Lockwood, 2008; Merritt et al., 2005; Perlman et al., 1976; R. Peterson, 2009; Vittor et al., 2006; see Chapter 8). Many human diseases, particularly malaria, typhus, yellow fever, bubonic plague, and equine encephalitis, are vectored by arthropods among humans and other animal species. Rodents are reservoirs for several important

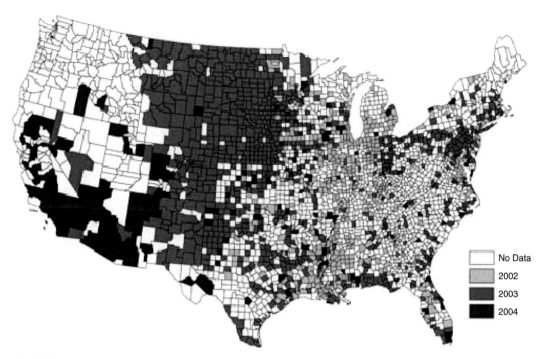

FIGURE 17.5 **Spread of West Nile Virus across the USA 2002–04 based on peak human incidence by county.** *Source: From Allan, B.F., et al., 2009. Ecological correlates of risk and incidence of West Nile virus in the United States. Oecologia 158, 699–708.*

human diseases, but livestock, wildlife, and birds also are sources of disease inoculum (Diamond, 1999). The rapid spread of West Nile Virus across North America between 1999 and 2004 reflected a combination of transmission among multiple hosts by multiple mosquito species and rapid spread across the continent by infected birds (Fig. 17.5, B. Allan et al., 2009; Marra et al., 2004; Turrell et al., 2005; see Chapter 8).

In many cases, the life-threatening nature of arthropod-borne diseases warrants a zero-tolerance approach to vector control. Nevertheless, disease prevention or control must address interactions among disease reservoirs, insect vectors, and other organisms. Mosquito control, for example, benefits hosts of vectored diseases but may undermine food web interactions critical to sustainability of ecosystem services (see Chapter 16). Many insectivorous fish and songbirds that support biological control or recreational activities feed primarily on mosquitoes. While other insect species might replace mosquitoes in some of their functions, differences in phenology or other life history attributes might make them less available or less desirable resources. C.G. Jones et al. (1998) reported that human risk of contracting Lyme disease in oak forests reflected complex interactions among tick vectors, deer and rodent reservoirs that feed on acorns, gypsy moth outbreaks that affect acorn availability, and rodent predation on gypsy moth pupae. Manipulation of acorn abundance demonstrated that gypsy moth abundance increased at low acorn abundances, due to reduced rodent abundance and

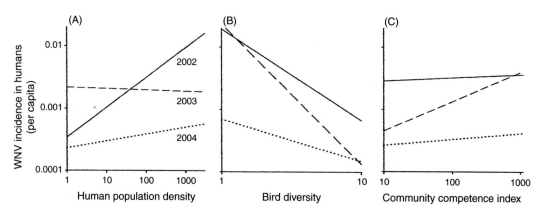

FIGURE 17.6   Incidence of human cases of West Nile Virus (WNV) relative to human population density (A), bird species diversity (B), and community competence index (C). Only the effect of bird species diversity was significant all three years, after controlling for spatial autocorrelation of data. *Source: Data from Allan, B.F., et al., 2009. Ecological correlates of risk and incidence of West Nile virus in the United States. Oecologia 158, 699–708.*

predation, and incidence of Lyme disease increased at high acorn abundances, due to increased reservoir density. Furthermore, B. Allan et al. (2009) found that bird diversity was inversely related to the competence of the West Nile Virus reservoir, indicating that less diverse bird assemblages tended to be dominated by more competent reservoirs and suggesting that protection of bird diversity could reduce human incidence of this disease (Fig. 17.6).

The importance of arthropod-vectored diseases to human population dynamics, including the success of military campaigns (R. Bray, 1996; Lockwood, 2008; R. Peterson, 1995, 2009), underscores the importance of understanding human roles in ecological interactions. Increasing human intrusion into previously unoccupied ecosystems has exposed humans to novel animal diseases that may involve insect vectors (R. Peterson, 2009; Vittor et al., 2006, 2009). Human transport of disease agents, their vectors, and/or livestock reservoirs has increased the globalization of many diseases; pathogens that require multiple reservoir species for different portions of their life cycle tend to remain more restricted in distribution (K. Smith et al., 2007). Ecosystem disturbance promotes population growth of at least some disease vectors (Merritt et al., 2005; Vittor et al., 2006, 2009). Transmission frequency increases with density and movement of human, reservoir, and/or vector populations. Best management practices must involve a combination of approaches that augment natural controls (eg, protect insectivorous fish, amphibians, birds, and mammals, as well as invertebrate predators) and reduce exotic breeding habitat for vectors (eg, water collected in discarded tires, flower pots, or roadside ditches) and/or reservoir hosts, as well as prevent inoculation of humans who may be exposed.

Control of mosquitoes and other vectors has been accomplished largely by broad-scale application of insecticides, including (historically) DDT and (currently) pyrethroids. Insect vectors are capable of adapting quickly to insecticides (Djogbénou et al., 2011; Mutunga et al., 2015; Norris et al., 2015; Ranson et al., 2002; Reimer et al., 2008). Insecticide resistance undermines the effectiveness of more targeted insecticide-treated nets and indoor residual spraying as alternatives to broadcast application (Djogbénou et al., 2011; Norris et al., 2015).

Furthermore, application of insecticides over bodies of water, where mosquito breeding is concentrated, can be counterproductive for ecosystem services, especially for aquaculture yields or biological control (see D. Baldwin et al., 2009; Okoniewski et al., 2006; S. Smith et al., 1983).

Application of *Bacillus thuringiensis* var. *israelensis* (Bti) and *Bacillus sphaericus* to aquatic habitats has controlled mosquito larvae effectively under some conditions, but the bacterial protein must be ingested and is only active for 2–3 days (T. Russell and Kay, 2008; Skovmand et al., 2009). Mosquitoes also can become resistant to these bacteria (Singh and Prakash, 2009). Several entomopathogenic fungi, including *Beauveria bassiana* and *Metarhizium anisopliae*, show promise for adult vector control (Scholte et al., 2005). Unlike bacteria or viruses, fungal entomopathogens can infect and kill insects through contact, without being ingested.

Several predators also show promise as biological control agents. Culler and Lamp (2009) reported that two dytiscid beetle species, *Agabus disintegratus* and *A. punctatus*, selectively prey on mosquito larvae, with mosquito larvae representing 76% of all consumed prey. Dhanker et al. (2013) and Murugan et al. (2011) reported that the predatory copepod, *Mesocyclops asperi-cornis*, effectively reduced larval populations of *Aedes aegypti*, especially during early instars, and could be combined successfully with neem extract to increase mosquito control.

Various chemicals, including sulfur, permethrin, and *N,N*-diethyl-meta-toluamide (DEET), have been used to repel vectors from individuals or kill vectors only on contact with treated hosts. However, these repellents can have negative effects on human health and environmental conditions. Variation in attractiveness to vectors among individual humans provides additional opportunities for protection. J.G. Logan et al. (2009) identified a number of human-emitted compounds that differed in aerosol concentration between subjects who were attractive or nonattractive to biting midges, *Culicoides impunctatus*. Two of these compounds, 6-methyl-5-hepten-2-one and geranylacetone, significantly reduced attraction of midges to baited traps in the field. D. Bray et al. (2009) reported positive results for attraction of a leishmaniasis vector, *Lutzomyia longipalpis*, to pheromone traps in the field. Future use of such repellents and attractants in human-dominated sites may avoid potential problems for human health or environmental conditions.

A new "feed-through" technique targets insect vectors during the portion of their life cycle when they depend on nonhuman hosts (Mascari and Foil, 2010; Mascari et al., 2007a,b). For example, phlebotomine sand flies vector trypanosomes, the causal agent of leishmaniasis, between rodent and human hosts. Adult sand flies are hematophagous and transmit the disease while feeding on host blood. Sand fly larvae typically live in rodent burrows and feed on rodent feces, making them difficult targets for control. Broadcast application of rodenticides negatively affects predator populations and ecosystem integrity. The feed-through technique involves provision of rodent food infused with an insecticide that is nontoxic to rodents. Several insect growth regulators and chitin synthesis inhibitors can pass through the rodent digestive system and remain sufficiently toxic to kill coprophagous fly larvae (Mascari and Foil, 2010; Mascari et al., 2007a,b). Sufficient insecticide is assimilated by the rodent to kill adult flies during bloodfeeding (Mascari et al., 2007a,b). Incorporating markers into the baits permits tracking of flies that feed on the baits (Fig. 17.7, Mascari and Foil, 2009). The advantage of this technique is that, because sand flies have relatively short dispersal distances, baiting rodents within buffer zones around human habitations could provide adequate control of vectors with minimal nontarget effect (Mascari and Foil, 2009).

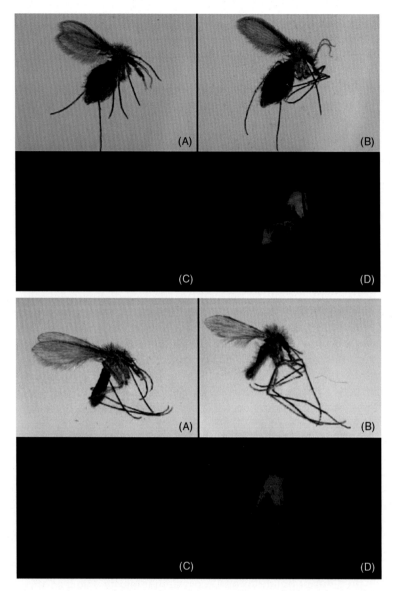

FIGURE 17.7 (Top set) Female sand fly A and C took bloodmeal from hamster fed control diet; female B and D took bloodmeal from hamster fed diet containing rhodamine-B dye. (Bottom set) Female sand fly A and C fed as larva on hamster fed control diet; female B and D fed as larva on hamster fed on diet containing rhodamine-B dye. *Source: From Mascari, T.M., Foil, L.D., 2009. Evaluation of rhodamine B as an orally delivered biomarker for rodents and a feed-through transtadial biomarker for phlebotamine sand flies (Diptera: Psychodidae). J. Med. Entomol. 46, 1131–1137.*

Pridgeon et al. (2008) reported that RNAi could be used to kill mosquitoes by causing massive cell death. Mosquitoes have been engineered to make them ineffective as disease vectors (V. Carter and Hurd, 2010; James, 2002; Riehle et al., 2007). Although such options seem attractive because of their apparent focus on the target vector (Fang, 2010), they fail to address potential effects of the engineered population on the network of direct and indirect interactions in ecosystems, for example, the variety of vertebrates that interact with mosquitoes and the potential losses to ecosystem services while other species adjust.

2.4 Ecological Tactics for Managing Urban "Pests"

The urban community is dominated by humans, pets, lawns, and exotic ornamental and weedy species (see Chapter 11). Human structures provide ideal habitats for many insect species, including ants, termites, cockroaches, and flies. Stable temperature and moisture, unsealed food, human or animal wastes, and "detrital" resources (eg, structural wood) provide attractive resources. Discarded food containers, tires, flowerpots, and other debris collect water and provide habitat for mosquitoes amid numerous vertebrate hosts (Unlu et al., 2013, 2014).

Termites, carpenter ants, and wood-boring beetles often threaten wooden structures through their excavation of structural wood. Considerable investment has been made in research to reduce damage, particularly in historically important structures (Guillot et al., 2010). Optimal management of these insects requires multiple approaches, including elimination of conducive conditions, such as construction debris, leaking roofs, mulch pushed against foundations, etc.; chemical barriers to make buildings less attractive; removal or treatment of infested building material or nearby trees; pheromone disruption of foraging behavior; non-repellent termiticides that can be transferred in lethal doses to other colony members through trophallaxis; and microbial toxins to inhibit gut flora and fauna (J.K. Grace and Su, 2001; Husseneder et al., 2003).

Other urban "pests" include nuisances and health hazards, such as cockroaches, bedbugs, exotic ants, biting or swarming flies, and even winter aggregations of ladybird beetles, that may be promoted by unsanitary conditions or proximity of lawns, gardens, and ornamental pools. Fumigation or heating of reused furniture before bringing it inside human habitations will minimize introduction of some pests, but many species are highly resistant to most household insecticides because of frequent exposure. Relatively simple approaches such as sanitation, maintenance of window and door screens, avoidance of transporting insects from infested to uninfested areas, and sealing of cracks and crevices can prevent entry or survival of many species, such as cockroaches and bedbugs. Pheromones may be useful in controlling some household pests (Liang et al., 1998).

The complexity of urban structures, including shared walls, rooftop or balcony gardens, and multiple sources of moisture and access by small insects, makes treatment and monitoring difficult. Treatment of large areas is necessary to reduce overall population sizes and dispersal into treated areas. In some cases, government-subsidized programs have been established to supplement individual property-owner efforts to control structural pests, such as the invasive Formosan subterranean termite, *Coptotermes formosanus* (Guillot et al., 2010).

Urban vegetation stressed by elevated temperatures, pollutants, fertilizer application, soil compaction, impervious surfaces, and altered drainage conditions is vulnerable to herbivorous insects (Cregg and Dix, 2001; Frankie and Ehler, 1978; Raupp et al., 2010; see Chapter 11). Although herbivorous insects can be managed through a combination of tactics (see 2.2), urban dwellers typically rely on chemical insecticides, thereby contributing to resistance development among target species. Frequent pesticide application also reduces the abundance of desirable insects, such as butterflies, dragonflies, and biological control agents. Educating urban residents about the ecological factors that promote or suppress these insects in urban settings will improve management strategies.

2.5 Invasive Species

Invasive species are among the most serious threats to ecosystem services and restoration success. Invasive species frequently displace native species and alter food web interactions in ways that threaten ecosystem services (Adkins and Rieske, 2013; Boettner et al., 2000; E. Evans, 2004; Rohr et al., 2009; Silvey et al., 2015; see Chapter 16). For example, invasive Chinese privet, *Ligustrum sinense*, has significantly reduced diversity and abundances of native plants, bees, butterflies, and ground-dwelling beetles in deciduous forests in the southeastern USA, as demonstrated by increased abundances of these taxa in privet-removal plots (Hanula and Horn, 2011; Hudson et al., 2013, 2014; Ulyshen et al., 2010). Such changes in food web structure can cascade through ecosystem processes and services. Orwig (2002), Ford and Vose (2007), and Siderhurst et al. (2010) described changes in soil and riparian conditions resulting from eastern hemlock, *Tsuga canadensis*, mortality caused by the invasive hemlock woolly adelgid, *Adelges tsugae*. However, Siderhurst et al. (2010) predicted that canopy opening and replacement of hemlock by deciduous trees would have little long-term effect on brook trout, *Salvelinus fontinalis*.

Costs of controlling damage by invasive species have been estimated at more than US$100 billion in the USA (Pimentel et al., 2000). D. Johnson et al. (2006) reported that more than US$194 million was spent on monitoring and control of gypsy moth alone in the USA during 1985–2004. D.C. Cook et al. (2007) calculated that continued exclusion of the varroa mite, *Varroa destructor*, from Australia for the next 30 years would prevent losses of US$16–39 million per year in terms of reduced pollination service. Furthermore, desperate attempts to control invasive species with pesticides seriously interfere with more sustainable natural controls.

Invasive species may be introduced passively, via storm winds or by hitchhiking on travelers or commercial shipments (Torres, 1988), or actively, as ornamental or commercial products (C. Riley and Howard, 1890). The gypsy moth and Cynthia silkworm, *Samia cynthia*, originally were introduced into the USA as efforts to establish a silk industry (Andrews, 1868; Forbush and Fernald, 1896; C. Riley and Howard, 1890). Honey bees, *Apis mellifera*, have been transported widely by humans for honey production and pollination services, but their presence in many regions has led to declines in native pollinator populations and to ineffective pollination of native plants that require more specialized pollinators (Aizen and Feinsinger, 1994; Garibaldi et al., 2013). In the absence of predators and parasites that regulate their populations in their native habitats, invasive species can spread rapidly, that is, the *Enemy Release Hypothesis* (J. Adams et al., 2009; D. Carpenter and Cappuccino, 2005; C. Yang et al., 2010).

Insects can be among the solutions or may contribute to invasibility of other species, as well as being among the most serious invaders. For example, invasive plants often require introduction of insect herbivores as biological control agents (D. Carpenter and Cappuccino, 2005; R. Doyle et al., 2002; Louda et al., 2003; McEvoy et al., 1991, 1993), although native insect herbivores also can inhibit the spread of invasive plants (N. Sanders et al., 2007). Insects may provide resources for invasive predators. Native charr in Japan derive 50% of their energy budget from terrestrial insects falling into streams (Baxter et al., 2007). An invasive trout reduced charr abundance by reducing the availability of allochthonous insects by 75% (Baxter et al., 2007).

Invasive ants often are particularly disruptive to ecosystem structure and function because of their complex interactions with a variety of other organisms (Neville et al., 2008; N. Sanders et al., 2003; see Chapters 8,13,14). Red imported fire ants, *Solenopsis invicta*, negatively affect populations of ground-nesting birds, small mammals, and reptiles, and can discourage larger animals from entering infested areas (C. Allen et al., 2004; S. Porter and Savignano, 1990). LeVan et al. (2014) found that replacement of a native ant, *Crematogaster californica*, by the invasive Argentine ant, *Linepithema humile*, disrupted pollination and fruit set by coast barrel cacti, *Ferocactus viridescens*, in California (Fig. 17.8). Similarly, Hansen and Müller (2009) reported that an invasive ant, *Technomyrmex albipes*, interfered with pollination and seed dispersal of an endangered plant, *Roussea simplex*, by endemic blue-tailed day-geckoes, *Phelsuma cepediana*, on the island of Mauritius. In areas of Kenya where the invasive big-headed ant, *Pheidole megacephala*, has displaced native *Crematogaster* spp., which defend *Acacia drepanolobium* trees from herbivory, the number of trees catastrophically damaged by elephants, *Loxodonta africana*, has increased 5–7-fold, compared to uninvaded areas (Riginos et al., 2015).

Invasive vectors also can disrupt food web interactions. The Asian tiger mosquito, *Aedes albopictus*, introduced into North America with a shipment of used tires, vectors several diseases of birds and mammals (Meyerson and Reaser, 2003; Turrell et al., 2005). Insect-vectored plague and West Nile Virus, both exotic in North America, have decimated wildlife populations (Marra et al., 2004; Stapp et al., 2004), requiring control of vector or pathogen to ensure successful conservation or restoration of vulnerable species.

Too often, efforts to control invasive species are delayed until populations are established and growing exponentially in the absence of regulatory mechanisms. Low-density populations are vulnerable to extinction (the Allee effect, Chapter 6), particularly through failure to find potential mates (Contarini et al., 2009; Gascoigne et al., 2009; Kramer et al., 2009), making the preestablishment period a window of opportunity to facilitate extinction (Yamanaka and Liebhold, 2009).

Mating success can be reduced through sterile male release, mass trapping, or mating disruption techniques. Early detection of preestablished populations is critical to maximizing the time available to produce sterile males or develop pheromones necessary for mass trapping or mating disruption. Yamanaka and Liebhold (2009) used computer models to predict eradication success for two insect "types": (1) an insect that has a long reproductive period and is capable of multiple matings but has a low daily reproductive rate, such as many Coleoptera, and (2) an insect that has a short reproductive period and is capable of mating only once per generation but has a high reproductive rate, such as many Lepidoptera. They found that sterile male release methods would be most effective for type 1 insects, and that mass-trapping would be most effective for type 2 insects, provided a sufficiently large number of female-baited traps can be installed. Mating disruption also would be effective for type 2 insects when sex attractants or pheromone-masking compounds can be identified. Yamanaka

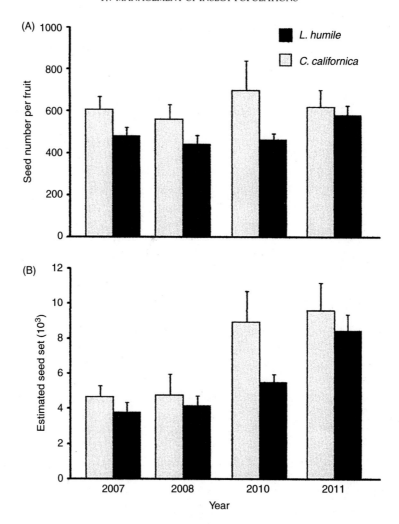

FIGURE 17.8 (A) Mean (± SE) number of seeds per fruit for cacti, *Ferocactus viridescens*, occupied by native ants, *Crematogaster californica (black bars)*, and invasive ants, *Linepithema humile (white bars)*. (B) Mean (± SE) total seed set (number of seed per fruit × number of fruit) for cacti occupied by native and invasive ants. *Source: From LeVan, K.E., et al., 2014. Floral visitation by the Argentine ant reduces pollinator visitation and seed set in the coast barrel cactus,* Ferocactus viridescens. Oecologia 174, 163–171.

and Liebhold (2009) discovered that insect pests are more likely to be eradicated than had been predicted previously by models that did not account for mating failure.

D. Johnson et al. (2006) found that invasive species often may spread through pulsed range expansion, that is, regularly punctuated range expansions interspersed with periods of static distribution. This pattern could be explained by interaction between Allee effects and stratified diffusion, a process whereby most individuals disperse locally, but a few individuals move long distances to initiate new colonies (see Chapter 7). These results suggested that invasion could be slowed considerably by suppressing population peaks along range borders.

However, the effects of invasive species should be evaluated carefully to avoid compounding problems. Invasive species are not necessarily detrimental to restoration efforts and may, in some cases, contribute to restoration success (Ewel and Putz, 2004; Lugo, 2004). For example, invasive plant species may have unique abilities to restore soil structure and fertility, canopy cover, and biodiversity on degraded sites in the absence of native species to accomplish these processes (Ewel and Putz, 2004; Lugo, 2004). Introduction of biological control agents in such cases may have undesirable consequences for restoration of ecosystem services.

3 CONSERVATION/RESTORATION ECOLOGY

Relatively few studies have addressed insects as part of ecosystem conservation or restoration projects (Boecklen, 1991; Samways, 1995; Samways et al., 1996). Some colorful insects, such as the Fender's blue butterfly, *Icaricia icarioides fenderi*, and American burying beetle, *Nicrophorus americanus*, have become targets for conservation or restoration efforts (M. Wilson et al., 1997). Some species are critical to plant reproduction (M. Schütz et al., 2008; Steffan-Dewenter et al., 2002; see Chapter 13), and may affect the success of conservation or restoration projects focused on plant species or integrated communities (Molano-Flores, 2009). Therefore, insect roles as herbivores, pollinators, seed dispersers, or prey should be considered as part of conservation/restoration planning. Furthermore, restoration of species or ecosystems requires consideration of mechanisms and processes that maintain key interactions necessary for survival of target taxa or communities (Lake et al., 2007).

Loss of key insect species or functional groups would deprive dependent species of food or other resources and lead to ecosystem degradation (Nichols et al., 2009; Tylianakis et al., 2008). Many vertebrates of conservation concern feed primarily or exclusively on insects (see Chapter 8). Stewart and Woolbright (1996) calculated that adult tree frogs, *Eleutherodactylus coqui*, at densities of about 3300 ha^{-1}, consumed 10,000 insects ha^{-1} per night in Puerto Rican rainforest; 17,000 preadult frogs ha^{-1} ate an additional 100,000 insects ha^{-1} per night. During dry periods when the availability of insect prey declined, many frogs were emaciated and had empty guts, suggesting vulnerability of insectivores to reduced abundance of insects. Similarly, restoration of aquatic systems for protection of fisheries requires attention to the composition and abundance of both aquatic insects (Lake et al., 2007) and riparian insect inputs (Baxter et al., 2005; Wipfli, 1997; Wipfli and Musslewhite, 2004) that may be equally important sources of prey for fish.

Fragmentation and/or conversion of natural ecosystems remains the most serious threat to conservation and restoration efforts regionally and globally (R. Fox, 2013). Fragmentation reduces and isolates populations of sensitive species, as described in Chapter 7, and increases the availability of degraded habitat and other resources for early successional, highly vagile species most likely to become pests or public health concerns (Póvoa et al., 2003; Vittor et al., 2006, 2009). Conversion of streamside vegetation dramatically alters water temperature, oxygen and chemical concentrations, and detrital and other resource availability, thereby threatening persistence of aquatic macroinvertebrate, fish, and other species (van Biervliet et al., 2009). Furthermore, populations of vagile species growing and spreading in extensive degraded habitats increases the likelihood that they will reach, and become invasive in, distant habitats, threatening species and services in those ecosystems.

Agricultural intensification has been identified as a primary threat to many native species, including critical pollinators (see Chapter 16). Pleasants and Oberhauser (2013) estimated that milkweed, *Asclepias syriaca*, abundance declined by 58% in the Midwestern USA from 1999 to 2010 as a result of increased herbicide use in agricultural landscapes. Concurrently, monarch butterfly, *Danaus plexippus*, abundance declined by 81% during this period, perhaps explaining the decline in overwintering populations of this iconic species in Mexico. Protection or restoration of patches of natural flora and fauna and connectivity and continuity among native ecosystems and resources are critical to sustainability of populations and interactions necessary to maintain ecosystem services (Duffy, 2009; R.H. Gibson et al., 2006; Greathouse et al., 2006; Landis et al., 2000; Schellhorn et al., 2015; Scheper et al., 2015; Steffan-Dewenter and Tscharntke, 1999). For example, Scheper et al. (2015) found that planting or retaining strips of floral resources in agricultural landscapes could protect pollinator diversity.

Xylophages may be particularly threatened as a result of deforestation, forest fragmentation, and conversion of landscapes dominated by old forests, with abundant woody litter, to landscapes dominated by young forests, with little woody litter accumulation (Grove, 2002; Seibold et al., 2015; Similä et al., 2002). Numerous wood-boring species became extinct as a result of deforestation of Europe during the past 5000 years (Grove, 2002). Loss or isolation of key species at the landscape level could threaten decomposition of woody residues in remnant natural areas. Similarly, loss of key necrophages, such as the American burying beetle, would threaten turnover of carrion and potentially create a public health hazard.

Loss of specialized pollinators or seed dispersers, as a result of habitat fragmentation or competition from introduced honey bees, threatens the reproduction and survival of their plant mutualists (Aizen and Feinsinger, 1994; Bascompte, 2009; Biesmeijer et al., 2006; R.H. Gibson et al., 2006; Hansen and Müller, 2009; Powell and Powell, 1987; Somanathan et al., 2004; Steffan-Dewenter and Tscharntke, 1999; Taki et al., 2007; Vamosi et al., 2006; C. Winter et al., 2008; see Chapter 13). Native bees are critical to the production of many plants, including agricultural crops, and often are capable of providing full pollination service in the absence of honey bees (Kremen et al., 2002), but their continued pollination service depends on proximity to, or corridors that provide access from, natural or seminatural ecosystems that provide necessary habitat requirements (Figs. 16.8 and 16.9) (Holzschuh et al., 2007; A. Klein et al., 2003; Kremen, 2005; Kremen et al., 2002; Ricketts, 2004; Taki et al., 2007; N. Williams and Kremen, 2007). Ants and ground beetles (Carabidae) are important predators in many ecosystems but are sensitive to changes in ecosystem conditions, especially resulting from agricultural intensification, potentially undermining their role as regulators of prey populations (A. Andersen and Majer, 2004; Hassall et al., 1992; Landis et al., 2000; Marino and Landis, 1996; Niemelä and Spence, 1994; Niemelä et al., 1992; Thies and Tscharntke, 1999). Such groups should be targeted in conservation or restoration efforts.

Maintenance or recovery of endangered species requires attention to the size and distribution of nature reserves for remnant populations. The *theory of island biogeography* dominated conservation biology during the 1970s and 1980s and continues to shape perspectives of nature reserves as habitat islands (Diamond and May, 1981; L. Harris, 1984). An early application of this theory was the development of rules for refuge design, such as the SLOSS (*single large or several small*) rule, based on the likelihood of colonization and persistence of large versus small islands or patches. Diamond and May (1981) noted that the value of various options for species viability depended on the habitat area required by a species and its dispersal capability. Small organisms such as insects could persist longer in small reserves than could larger organisms. In fact, insects often can persist undetected on rare hosts in relatively small, isolated patches, as was the case for Fender's blue butterfly, last seen in 1936 before being rediscovered in 1989 in small remnant patches of its host lupine, *Lupinus sulphureus kincaidii*, in western Oregon, USA (M. Wilson et al., 1997).

Rösch et al. (2015) compared conservation of a range of taxonomic groups (plants, leafhoppers, true bugs, and snails) in 14 small (0.1–0.6 ha) and 14 large (1.2–8.8 ha) fragments of calcareous grassland in Germany. They found that species richness was not significantly different between large and small fragments. However, species-area accumulation curves showed that both total and specialist species richness were higher on multiple small fragments of calcareous grassland than on fewer large fragments with comparable total area. The total area of small fragments (4.6 ha) contained 85% of the overall species richness, compared to only 37% of overall species richness contained in the total area (15.1 ha) of the two largest fragments (Fig. 17.9). This difference could be explained by the greater geographic range represented by the 14 small fragments (264 km²), compared to a few large fragments of similar total area (56 km²). In addition, community composition differed significantly between large and small fragments, with some of the rarest specialist species apparently confined to large fragments. Clearly, conservation of species and communities requires accommodation of species requiring large, as well as small, habitat areas (Rösch et al., 2015). However, species remain vulnerable to extinction as habitat area continues to decline (Lockwood and DeBrey, 1990).

Metapopulation models are based on the landscape pattern of demes and gene flow among demes across a landscape (Hanski and Simberloff, 1997; Harrison and Taylor, 1997). Small demes are most vulnerable to local extinction due to disturbances or land use change, but their presence may be critical to recolonization of vacant patches or gene exchange with nearby demes. Dispersal among patches is critical to maintaining small demes and preventing or delaying local extinction. Clearly, population recovery for such species depends on restoration or replacement of habitats.

Principles of metapopulation dynamics may be particularly important for conservation and restoration of populations of entomophagous predators and parasites in agricultural landscapes. Predators and parasitoids are recognized as important natural agents of crop pest regulation, but as a group are particularly vulnerable to habitat fragmentation (Kruess and Tscharntke, 1994; Schowalter, 1995) and pesticide application (Sherratt and Jepson, 1993). Hassell et al. (1991) and Sherratt and Jepson (1993) suggested that predator and parasite persistence in agroecosystems depends on the metapopulation dynamics of their prey, as well as on the frequency and distribution of pesticide use, and that connectivity between patches characterized by locally unstable predator–prey interactions could allow mutual persistence. M. Thomas et al. (1992) found that creation of islands of grassland habitats in agricultural landscapes increased the abundances of several groups of entomophagous arthropods.

Restoration goals must address the appropriate site conditions. For example, clear-cut harvest, burning, and planting of ponderosa pine, *Pinus ponderosa*, or Douglas-fir, *Pseudotsuga menziesii*, in western North America reflected an early perception of fire as a stand-replacing disturbance that created a mineral soil seed bed necessary for establishment of even-aged forest. The resulting even-aged monocultures have supported nearly continuous insect

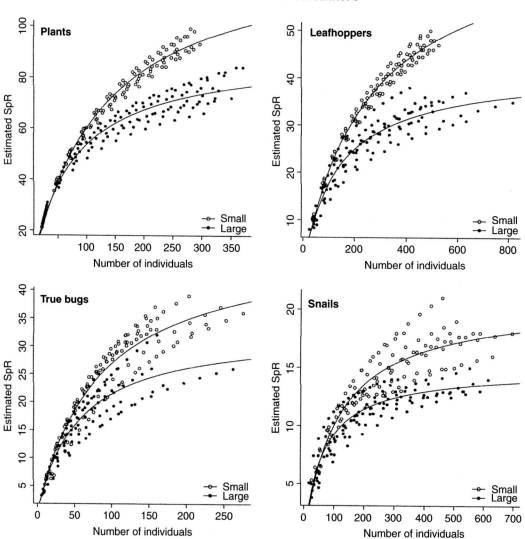

FIGURE 17.9 Rarefaction curves showing estimated species richness (SpR) relative to the number of individuals per plot for 12 randomly selected plots from 12 randomly selected small fragments (10 replications of randomly selected fragments and plots; open symbols) versus 12 randomly selected plots from each of 2 randomly selected large fragments *(filled symbols)*, for plants, leafhoppers, true bugs, and snails. For plants, the analysis was based on presence–absence data. Lines represent the average for the 10 randomly selected curves per fragment-size category. *Source: From Rösch, V., et al., 2015. Biodiversity conservation across taxa and landscapes requires many small as well as single large habitat fragments. Oecologia 179, 209–222.*

outbreaks as these forests age. Research following natural fires demonstrated more complex effects of fire, with patches of surviving trees intermingled with patches burned to mineral soil, resulting in heterogeneous, uneven-aged forest structure as regeneration spread from refuges. Harvest alone, to create patchy landscapes, does not satisfy habitat requirements for many species (Gandhi et al., 2004). Consequently, restoration efforts currently focus on thinning and prescribed fire to produce uneven-aged forest structure, with wider tree spacing, often aided by insects (North et al., 2007; J. Stone et al., 1999). Insect outbreaks may increase the likelihood of fire, at least under some circumstances (Lynch et al., 2006), or function in the absence of fire to create open-canopied patches required by some threatened species (Radeloff et al., 2000).

Planted seedlings may be insufficient, by themselves, for forest restoration on harsh sites. Amaranthus and Perry (1987) demonstrated that transfer of biologically active soil (containing invertebrates and microorganisms) from established conifer forests or plantations significantly increased the survival and growth of seedlings on clear-cut sites by up to 50%, compared to seedlings planted directly into clear-cut soils, from which soil biota had disappeared as a result of exposure to heat and desiccation.

Similarly, flooding a depression is not sufficient for wetland restoration. Attention to temporal patterns of surface and subsurface water flux and substrate conditions may be necessary for reestablishment of wetland communities (Batzer and Wissinger, 1996). For example, S.C. Brown et al. (1997) found that transplanting wetland soil resulted in significantly faster and more prolific plant growth and macroinvertebrate colonization in artificial wetlands. Wetlands are particularly vulnerable to sedimentation and concentration of agricultural chemicals draining from terrestrial habitats (Batzer and Wissinger, 1996). Insects are useful indicators of wetland conditions and restoration success (Batzer and Wissinger, 1996).

Maintaining adequate corridors among critical habitats is a key issue (see Chapter 7). Riparian corridors may be equally important for the integrity of aquatic and terrestrial ecosystems. Riparian corridors filter terrestrial runoff that can alter conditions for aquatic communities. A number of studies have demonstrated that riparian zones at least 30 m wide may be necessary to prevent changes in water temperature and other conditions that alter the aquatic ecosystem (Davies and Nelson, 1994; Kiffney et al., 2003). Riparian vegetation represents an important source of terrestrially derived invertebrates that compose at least 33% of the diet of young salmon (J. Allan et al., 2003; Baxter et al., 2005; Kawaguchi and Nakano, 2001). Furthermore, riparian corridors facilitate spread of terrestrial populations among suitable landscape patches. However, riparian corridors also represent unique, frequently flooded habitats on the terrestrial landscape. Sabo et al. (2005) conducted a metaanalysis of the diversity of riparian plant and animal taxa on seven continents and discovered that riparian habitats do not just harbor larger numbers of species, but rather support significantly different species pools, increasing landscape diversity by 50% globally.

Restoration also must address the network of interactions among species that may be at least as important to restoration success as is attention to key species (Gratton and Denno, 2006). For example, many plant species targeted for restoration cannot survive without associated pollinators and seed dispersers (Bascompte, 2009; Biesmeijer et al., 2006), but these mutualists, in turn, depend on a diversity of plant species to maintain their populations (Bascompte, 2009; Forup et al., 2008; R.H. Gibson et al., 2006; A. Klein et al., 2003; Kremen, 2005; Kremen et al., 2007). Plants in diverse communities may be more vulnerable to declines in pollinator diversity and abundance, because of greater pollen limitation due to competition for remaining pollinators, compared to plants in less diverse communities (Vamosi et al., 2006).

With increasing urban development on a global scale, urban environments will have a growing effect on surrounding ecosystems, both those adjacent to urban centers that are subject to conversion and human intrusion and those further away that are influenced by pollution, heat exchange, and other products of the urban environment. However, urban parks and other open space, roadside corridors, stormwater drainage systems, and backyard and rooftop gardens could be modified to serve some conservation purposes, for example, favoring native plants over exotic or invasive plants for landscaping (M.R. Hunter and Hunter, 2008; Madre et al., 2014). Insects generally are not popular with the urban public, given the association of some groups with filth and spread of disease. However, butterflies, dragonflies, ladybird beetles, and tiger beetles are widely appreciated, offering opportunities to use them as examples for educational programs that promote the importance of insects for maintenance of ecosystem services and that encourage stewardship of biodiversity and natural resources. For example, urban landscapes and private gardens that provide native host plants and floral resources for larval and adult butterflies could partially offset loss of natural habitats. Mosquito control in or near aquatic habitats can be modified (eg, using bacterial controls described previously) to minimize effects on aquatic predators and enhance natural regulation of vector populations.

Unfortunately, rapidly changing conditions at both local and regional levels may limit options for conservation and restoration. Changing public vision and environmental policy will be necessary to ensure ecosystem integrity and the sustainability of ecosystem services globally.

4 SUMMARY

Application of insect ecology to management of insects perceived to be pests has been a driving factor of scientific advances. Development of the integrated pest management (IPM) approach has emphasized protection or augmentation of natural regulatory mechanisms to manage crop, forest, medical/veterinary, and urban pests and to minimize insect adaptation to particular control tactics. Whereas native species in natural ecosystems may represent important feedback regulation mechanisms, warranting little interference, invasive species in agroecosystems require intensive management efforts. A variety of cultural methods to modify host and habitat conditions can reduce the likelihood of pest outbreaks. Transgenic crop varieties, biological control, pheromones, and precision application techniques are available to reduce target insect population levels below resource injury thresholds and minimize the need for insecticides. Pest management requires evaluation of the benefit to ecosystem services and to effects on nontarget species that may be critical to long-term sustainability of ecosystem services.

Few insects have been targeted for conservation efforts, but insects often affect the success of conservation and restoration programs focused on other taxa. Attention should be given to availability of key pollinators and seed dispersers that may be necessary to the reproduction and recruitment of targeted plants. Insects also are critical food resources for insectivorous vertebrates. Retention of important insect functional groups, such as xylophages or necrophages, in fragmented landscapes may be necessary to maintain ecosystem functions. Increasing urbanization of human populations affects surrounding ecosystems and the quality of urban life, but urban parks, roadside corridors, and private gardens could serve conservation purposes by offsetting habitat losses for threatened native species and communities.

Insect ecology has demonstrated its value for managing insect populations, ecosystem conditions, and ecosystem services. Understanding of ecological factors affecting boll weevil diapause and reproduction was critical to the elimination of boll weevil as a serious threat to cotton production in the southern USA. Similarly, ecological knowledge has altered the way in which many forest insects, such as the Douglas-fir tussock moth, are managed. The study of insect ecology will continue to provide the information necessary to manage insect populations, ecosystem conditions, and ecosystem services in a changing global environment.

18

Summary and Synthesis

Insect ecology addresses both insect adaptations to their environment and their effects on environmental conditions, although their adaptations have been the traditional focus of insect ecologists. Insects represent the full scope of heterotrophic adaptive life history strategies, from sessile species whose ecological strategies resemble those of plants to social insects whose range of behavioral attributes resembles that of advanced vertebrates. Insect ability to survive in some of the harshest environments on Earth, their capacity for dramatic population growth, and the variety of their direct and indirect interactions with other species span the range of ecological complexity. Their diversity brings them to the attention of biologists, natural resource managers, artists, and amateur collectors. The first three sections in this book emphasize this traditional approach to the study of insect ecology.

Insects also have considerable capacity to alter ecosystem/environmental conditions. Insects influence their environment in complex, often dramatic, ways, earning many species the designation of ecosystem engineers. Foraging patterns affect distribution of resources and interactions with other organisms. Population outbreaks of some herbivorous insects alter vegetation structure, biogeochemical cycles, and climate. Natural selection represents feedback between ecosystem conditions and individual phenotypes that affect ecosystem variables. Other feedback mechanisms between individuals, populations, and communities can stabilize or destabilize ecosystem, landscape, and global processes. Understanding these feedbacks is critical to prediction of ecosystem (and insect) responses to environmental changes. For example, herbivores dramatically alter vegetation structure across landscapes and potentially affect global climate and biogeochemistry (Chapter 12). Termites account for substantial portions of global carbon flux (Chapter 14). Major advances have been made in recent decades on the potential ability of insects to stabilize primary production and other ecosystem services. This concluding chapter summarizes key ecological issues, synthesizes key integrating variables, and identifies critical issues for future study.

1 SUMMARY

The conceptual organization of this book (Fig. 1.2) emphasizes linkages and feedbacks among levels of ecological organization that integrate evolutionary adaptations with effects on ecosystem processes. Linkages and feedbacks are strongest between neighboring levels

Insect Ecology. http://dx.doi.org/10.1016/B978-0-12-803033-2.00018-2

but are significant even between individual and ecosystem levels of the hierarchy. Physiological and behavioral responses to environmental variation are under genetic control and determine individual fitness, but also affect the rate and geographic pattern of resource acquisition and allocation that control energy and biogeochemical fluxes at the ecosystem level and climate at the global level. These feedbacks are an important and largely neglected aspect of insect ecology that affect ecosystem stability and global processes.

The geographic distribution of individual species generally reflects the environmental template established by the history of continental movement, latitude, mountain ranges, and global atmospheric and oceanic circulation patterns. The great diversity of insects reflects their rapid adaptation, conferred by small size, short life spans, and rapid reproductive rates, to environmental variation. These attributes have facilitated speciation at multiple scales: among geographic regions, habitats, resources, and at microscales on or within resources (eg, individual leaves). However, within the potential geographic range of a species, spatial and temporal patterns of abundance reflect disturbance dynamics, resource distribution, and interactions with other species that affect individual fitnesses and enhance or limit colonization and population growth.

Energy and resource budgets (Fig. 4.1) are key aspects of individual fitness, population persistence, and community interactions. All organisms require energy to accumulate resources, necessary for growth and reproduction, against resource concentration gradients, and thereby maintain the thermodynamic disequilibrium characteristic of life. Where resources are more concentrated, relative to individual needs, less energy is required for acquisition. Interactions among organisms often may be controlled by mass balances of multiple nutrients. Resource use requires adaptations to acquire necessary limiting nutrients, such as nitrogen, while avoiding or circumventing toxic or defensive chemicals, as well as overabundant nutrients.

Much research has addressed plant defenses against feeding by insects and other herbivores. Advances in molecular analyses have permitted identification of genetic mechanisms underlying biochemical interactions between plants and herbivorous insects, including induction of specific chemical defenses and their communication to other plants. Insect herbivores have evolved a variety of mechanisms for avoiding, detoxifying, or inhibiting expression of plant defenses in order to exploit plant resources. All insect species have mobile stages adapted to find new resources before current resources are depleted or destroyed. The early evolution of flight among insects greatly facilitated foraging, escape from unsuitable environmental or resource conditions, and discovery of more optimal conditions. Individuals or populations that fail to acquire sufficient energy and nutrients to grow and reproduce do not survive.

Adaptations for detecting and acquiring resources are highly developed among insects. Many insects can detect the presence and location of resources from chemical cues carried at low concentrations on wind or water currents. The diversity of strategies among insect species for acquiring resources has perhaps drawn the most ecological attention. These strategies range from ambush to active foraging, often demonstrate considerable learning ability (especially among social insects), and involve insects in all types of interactions with other organisms, including competition (eg, for food, shelter, and oviposition site resources), predation and parasitism (on plant, invertebrate, and vertebrate prey or hosts and as prey or hosts), and mutualism (eg, for protection, pollination, and seed dispersal).

Spatial and temporal variation in population and community structure reflects net effects of environmental conditions. Changes in population and community structure also constrain survival and reproduction of associated species. Population density and competitive, predatory, and mutualistic interactions affect foraging behavior and energy and nutrient balances of individuals. Individuals forced to move constantly to avoid intra- or interspecific competition or predation may be unable to forage sufficiently for energy and nutrient resources. On the other hand, energy and nutrient balances can be improved through mutualistic interactions that enhance the efficiency of resource acquisition. The relative contributions of intra- and interspecific interactions to individual survival and reproduction remain a central theme of ecology, but their study has been poorly integrated with ecosystem conditions. Debate over the importance of bottom-up versus top-down controls of populations perhaps reflects variation in the contributions of these factors among species, as well as spatial and temporal variation in their effects.

Ecosystems represent the level at which complex feedbacks among abiotic and biotic processes are integrated. Ecosystems can be viewed as dynamic energy- and nutrient-processing engines that modify global energy and nutrient fluxes. Cycling and storage processes controlled by organisms reduce variation in abiotic conditions and resource availability. Although ecosystem properties are largely determined by vegetation structure and composition, insects and other animals modify ecosystem conditions, often dramatically, through effects on primary production, decomposition and mineralization, and pedogenesis. Insect herbivore effects on vegetation structure affect albedo, evapotranspiration, and wind abatement. Changes in decomposition processes affect fluxes of carbon and trace gases, as well as soil structure and fertility. Insect roles as ecosystem engineers mitigate or exacerbate environmental changes resulting from anthropogenic activities. Resolution of environmental issues requires attention to these roles of insects, as well as to their responses to environmental changes.

An ecosystem approach to insect ecology can aid in applications to management of ecosystem services. Agricultural and forest ecosystems have been manipulated for millennia to increase commodity production, but the density and extent of planted crops and breeding for desirable traits have promoted herbivore population growth and required intensive "pest" management efforts. Insects also provide pollination services and affect fish and wildlife, water yield, cultural and other ecosystem services. Recent approaches to managing insects that interfere with ecosystem services emphasize identification and manipulation of factors affecting ecological interactions among primary production, herbivorous insects, and various predators and parasites. Target insects quickly develop genetic resistance to insecticides or other management tactics to which they are exposed frequently and widely.

Many insects may be critical to ecosystem conservation or restoration efforts. Conservation and restoration programs have advanced beyond simple efforts to protect individual species to recognition of the key contributions of associated plants, herbivores, pollinators, seed dispersers, and decomposers to the sustainability of any conservation or restoration project. Finally, many insects have become valued tools for evaluating ecosystem condition, based on their sensitivity to changing ecosystem conditions that result in dramatic and measurable changes in abundance. Key species or functional groups appear to function as regulators of ecosystem processes that underlie the sustainability of ecosystem services.

2 SYNTHESIS

Insect ecology addresses an astounding variety of interactions between insects and their environment. A unifying theme is the diversity of positive and negative direct and indirect feedbacks that govern insect responses to environmental changes and their potential regulation of primary production, energy and nutrient fluxes, and climate. Feedback integration among species and hierarchical levels occurs primarily through responses to variation in environmental conditions and functions to stabilize ecosystem conditions. Insect behavioral and physiological attributes that affect their interactions with all aspects of their environment are under genetic control. Evolution represents feedback on individual attributes that affect higher levels of organization.

The importance of environmental change and disturbance as a central theme in insect ecology has been recognized only recently. Disturbance, in particular, provides a context for understanding and predicting individual adaptations, population strategies, organization and succession of community types, and rates and regulation of ecosystem processes. Environmental changes or disturbances kill individuals or affect their postdisturbance activity and reproduction. Some populations are reduced to local extinction, but others exploit the altered conditions. Population strategies and interactions with other species affect ecosystem properties in ways that increase the probability of disturbance (or other changes) or that mitigate environmental changes and favor persistence of species less tolerant to change. Insects contribute greatly to feedback between ecosystem properties and environmental variation. This aspect of insect ecology has important consequences for ecosystem responses to global changes resulting from anthropogenic activities.

Energy and biogeochemical fluxes integrate individuals, populations, and communities with their abiotic environment. Energy flow and biogeochemical cycling processes determine rates and spatial patterns of resource availability. Many, perhaps most, species attributes can be shown to represent trade-offs between maximizing resource acquisition and optimizing resource allocation among metabolic pathways, for example, foraging activity, defensive strategies, growth, and reproduction. The patterns of energy and nutrient acquisition and allocation by individuals determine the patterns of storage and fluxes among populations, fluxes among species at the community level, and storage and flux at the ecosystem level that, in turn, determine resource availability for individuals, populations, and communities. Resource availability is fundamental to ecosystem productivity and diversity. Resource limitation, including reduced availability resulting from inhibition of water and nutrient fluxes, is a key factor affecting species interactions. Herbivore and predator populations grow when increasing numbers of hosts or prey are available and/or incapable of escape or defense because of insufficient resource acquisition or poor food quality.

Regulatory mechanisms emerge at all levels of the ecological hierarchy. Negative feedback and reciprocal cooperation are apparent at population, community, and ecosystem levels. Cooperation benefits individuals by improving ability to acquire limiting resources. This positive feedback balances the negative feedbacks that limit population density, growth, and ecological processes. At the population level, positive and negative feedbacks maintain density within narrower ranges than occur when populations are released from regulatory mechanisms. The responsiveness of insect herbivores to changes in plant density and condition, especially resulting from crop management, introduction into new habitats, and land use changes, brings some species into conflict with human interests. However, insect outbreaks in natural ecosystems appear to be restricted in time and space and function to (1) maintain net primary production within relatively narrow ranges imposed by the carrying capacity of the ecosystem and (2) increase vegetation diversity and facilitate replacement of plant species that are poorly adapted to current conditions by species that are better adapted to these conditions. Regulatory capacity appears to reflect selection for recognition of cues that signal changes in host density or condition that affect the long-term carrying capacity of the ecosystem.

The issue of ecosystem self-regulation is a key concept that significantly broadens the scope of insect ecology. Although this concept remains controversial, accumulating evidence supports a view that outbreaks of native insect species reduce long-term deviation in net primary production, at least in some ecosystems. Although outbreaks appear to increase short-term variation in some ecosystem parameters, in many cases outbreaks reverse unsustainable increases in net primary production due to management practices (Fig. 15.9), and thereby reduce long-term variation in ecosystem conditions.

Models of group selection predict that stabilizing interactions are most likely in ecosystems where pairs or groups of organisms interact consistently. Hence, selection for stabilizing selection might be least likely in ecosystems where such interactions are inconsistent, such as in harsh or frequently disturbed environments. However, selection for stabilizing interactions also might be less direct in productive, highly diverse, ecosystems with little variation in abiotic conditions or resource availability, such as tropical rainforests. Stabilizing interactions are most likely in ecosystems where selection would favor coevolution of interactions to reduce moderate levels of variation in abiotic conditions or resource availability. Research on stabilizing interactions should focus on these ecosystems.

3 CRITICAL ISSUES

Agrawal et al. (2007) proposed three core areas for future research in ecology: (1) evaluating and predicting strength and context dependence of species interactions (both direct and indirect) across multiple scales, (2) identifying and measuring feedbacks between individual and multiple interactions and ecosystem dynamics, and (3) linking spatial and temporal patterns and processes to understand species evolution and coexistence. Advances in understanding of factors controlling patterns of species diversity, community assembly, and feedbacks between insects and ecosystem dynamics will improve prediction of species' (including insects') responses to various environmental changes and resulting changes in biotic control of ecosystem structure, function, and services.

Resolution of the debate over regulation of ecosystem processes by herbivorous insects may not be possible, given the need for large-scale manipulation of insect populations and long-term, multidisciplinary comparison of ecosystem processes necessary to test the hypothesis, within the context of rapid anthropogenic changes in the global environment. However, our perspective of insect effects determines our management approaches. Research clearly is needed to evaluate long-term effects of outbreaks on primary production, soil conditions, and sustainability of ecosystem functions and services. In particular, identifying factors underlying compensatory growth would clarify necessary conditions and contributions to long-term primary production in natural ecosystems. Effects of population changes on mass balances of energy and nutrient fluxes clearly have the capacity to mitigate or exacerbate effects of acid rain, carbon flux, and other processes affecting global change. Understanding the extent to which outbreaks modify local climate and disturbance dynamics also would improve prediction of consequences of global changes. A number of long-term ecosystem research platforms are available for this research, including Long Term Ecological Research (LTER) and National Environmental Observatory Network (NEON) sites in North America and FLUXNET sites distributed globally.

Predicting and alleviating effects of anthropogenic changes require understanding of insect roles and how these roles affect ecosystem responses to anthropogenic changes. Anthropogenic changes will continue to trigger insect outbreaks, whether as destructive events or as regulatory responses. Land use, in particular, affects patch structure and interactions among demes, greatly altering the spatial and temporal patterns of insect abundances. Ruderal plant species, valued for crop production but also adapted for rapid colonization of new habitats, are increasingly likely to dominate fragmented landscapes. The rapid growth and poor competitive ability of these species in crowded ecosystems make them targets for their associated insects. Maintenance of such ecosystems will require constant human intervention. However, ecological approaches can be used to prevent or mitigate outbreaks (Branson et al., 2006, see Chapter 17). Protection or restoration of natural ecosystems will require attention to interactions necessary to maintain key species, including pollinators, seed dispersers, and decomposers.

Accomplishment of these objectives requires broadening of research approaches to address the direct and indirect effects of insects on ecosystem structure and function. This, in turn, requires changes in research approaches and integration of population and ecosystem models. Testing of ecosystem-level hypotheses involves different approaches than does testing of population- and community-level hypotheses. At least three considerations are particularly important.

First, advances in insect ecology require attention to quantitative, as well as qualitative, methodology to test key hypotheses. Experimental design must address statistical independence of samples. Whereas individuals within populations can serve as replicates for population and community properties, data must be pooled at the site (ecosystem) level for comparison of ecosystem variables. Ecosystem studies often have provided inconclusive data, because a single site representing each of several ecosystem types or experimental treatments provides no error degrees of freedom for statistical analysis. Multiple samples collected within each site are not statistically independent (Cadogan and Scharbach, 2003; Hurlbert, 1984; D.H. Johnson, 2006), for example, may be autocorrelated (Fig. 7.15, Grilli and Gorla, 1997). Furthermore, treatment effects are subject to confounding effects of geographic gradients between treatment plots that may affect results (Hero et al., 2013). Therefore, Hurlbert (1984) referred to such sample designs as **pseudoreplication**. Valid experimental designs must incorporate multiple, geographically interspersed, replicate sites randomly assigned to treatments (Fig. 18.1, see also Cadogan and Scharbach, 2003; Hurlbert, 1984; Rambo et al., 2014; Shiels et al., 2010). A larger number of replicate sites provides a greater range of inference than do multiple samples within sites (that must be pooled for statistical analysis), whereas multiple samples within sites provide better representation of site conditions, requiring a trade-off in sampling effort within and between sites (D.H. Johnson, 2006).

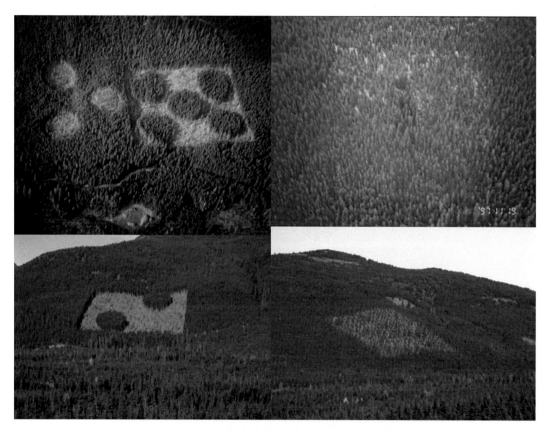

FIGURE 18.1 **Experimental treatments in the Demonstration of Ecosystem Management Options (DEMO) plots in the Butte block near Randall, WA, USA (six blocks total in Washington and Oregon).** Treatments (clockwise from upper left) are 75% aggregated and 40% aggregated retention, 40% dispersed retention; 15% dispersed retention; 15% aggregated retention and uncut control. Each experimental unit is 13 ha. Circular gaps or aggregates are each 1 ha in area. The total number of trees retained was the same in 40% aggregated or dispersed units and in 15% aggregated or dispersed units. Treatments were randomly assigned to experimental units in each of six forest blocks in western Oregon and Washington. Photos courtesy of the US Forest Service (top two photos by Jon Nakae and Jim White; bottom two by Tom Savage). *Source: From Schowalter, T.D., Zhang, Y.L., Progar, R.A., 2005. Canopy arthropod response to density and distribution of green trees retained after partial harvest. Ecol. Appl. 15, 1594–1603.*

Second, research to evaluate insect responses to, or effects on, ecosystem conditions and services should address a greater range of ecosystem variables than has been common in past studies of insect ecology. Insects respond to multiple factors simultaneously, not just one or a few factors subject to experimental manipulation, and their responses reflect trade-offs that might not be reflected in studies that control only one or a few of these factors. A greater breadth of parameters can be addressed through multidisciplinary research, with experts on different aspects of ecosystems contributing to a common goal (Fig. 18.2). Involvement of insect ecologists in established multidisciplinary projects, such as the International Long Term Ecological Research (ILTER) sites in many countries (including the US LTER network), can facilitate integration of insect ecology and ecosystem ecology. In turn, insect ecologists

FIGURE 18.2   **Interdisciplinary research on insect effects on log decomposition at the H.J. Andrews Experimental Forest Long Term Ecological Research Site in western Oregon, USA.** (A) Logs tented to exclude wood-boring insects during the first year of decomposition; (B) logs inoculated with different initial heterotroph communities (bark vs. wood borer, mold vs. decay fungi; ribbon color indicates inoculation treatment; plastic shelters reduced wood moisture relative to unsheltered logs). Data loggers at each replicate site measured ambient temperature and relative humidity and vertical and horizontal temperature and moisture profiles in logs. Sticky screens were used to measure insect colonization, emergence traps were used to measure insect emigration, PVC chambers were used to measure CO_2 flux, and funnels under logs were used to measure water and nutrient flux out of logs. Scheduled destructive sampling of logs provided data on changes in wood density, excavation by insects, and nutrient content.

can contribute to such programs by clarifying how particular species respond to, and shape, ecosystem conditions, including vegetation structure, soil properties, biogeochemical cycling processes, etc., as described in Chapters 12–14, how insects affect the balance of nutrient fluxes within and between ecosystems (eg, from aquatic to terrestrial ecosystems or across landscapes as populations move or expand, as described in Chapter 7), and how species diversity within guilds or functional groups affects the reliability of community organization and processes (see Chapters 9 and 15), as well as ecosystem services (see Chapter 16) through time (see Chapter 10).

Third, spatial and temporal scales of research and perspectives must be broadened. Most ecosystem studies address processes at relatively small spatial and temporal scales (and entomological studies at even smaller scales). However, population dynamics and capacity to influence ecosystem and global properties span landscape and watershed scales, at least, and have the capacity to affect regional and global climate (Chapters 12–14). Feedbacks often may be delayed, or operate over long time periods, especially in ecosystems with substantial buffering capacity, requiring long-term institutional and financial commitments for adequate study. Linkage of population and ecosystem variables using remote sensing and GIS techniques will become an increasingly important aspect of insect ecology. Nevertheless, ecosystems with large biomass or high complexity require simplified field mesocosms or modeling approaches to test some hypotheses.

Such studies will be particularly difficult in forests, where extreme 3-dimensional structure has limited most replicated plot manipulations to understory or early successional stages. True replication requires larger scales for experimental units to be representative of forest conditions (see Fig. 18.1). However, some studies have been conducted on large, replicated plots in which canopy trimming or tree removal patterns were manipulated to address effects of disturbances (Haddad, 2000; Haddad et al., 2003; North et al., 2007; Rambo et al., 2014; Richardson et al., 2010; Schowalter et al., 2005, 2014; Shiels et al., 2010). Another option is repetition or combination of studies using the same design and methods, a **metaanalysis** or **metareplication approach** (eg, Gruner et al., 2008; Johnson, 2006; Kratz et al., 1995; Powers et al., 2009). Such combinations increase the scope of inference for results of such experiments and/or facilitate identification of environmental factors that affect results (Gruner et al., 2008; Kratz et al., 1995; see Chapter 15).

The complexity of ecosystem interactions and information linkages has limited the incorporation of detail, such as insect population dynamics, in ecosystem models. Species modeling for ecosystem description and prediction is necessarily simplified, relative to that for population models. However, population models have largely ignored feedbacks between population and ecosystem processes. Hierarchical structure in ecosystem models facilitates integration of more detailed insect population (and other) submodels, and their linkages and feedbacks with other levels, as data become available (Figs. 11.16 and 11.17).

Several ecosystem components should be given special attention. Subterranean and forest canopy subsystems represent two ecological frontiers that may have substantial effects on carbon flux, as well as other ecosystem and global processes (Chapters 12 and 14). Logistical difficulties in gaining nondestructive or nonintrusive access to these two subsystems have limited the data available for evaluating insect effects on canopy–atmosphere and canopy–rhizosphere–soil interactions that control climate and energy and matter fluxes. Improved canopy access methods, such as construction cranes (Fig. 18.3) for ecological use (Schowalter and Ganio, 1998; D. Shaw, 1998, 2004), and rhizotron technology (Sackville Hamilton et al., 1991; Sword, 1998) offer opportunities for scientific advances in the structure and function of these subsystems.

Fourth, insects are among the most useful indicators of changing conditions (Dufrêne and Legendre, 1997; Maleque et al., 2009, see Chapter 16). Because of their short life spans, rapid

FIGURE 18.3   Canopy cranes are a new tool for experimental access to forest canopies. For example, the gondola of the Wind River Canopy Crane (75-m tall tower, 84-m long jib) can access 700,000 m^3 of 60-m tall canopy, as well as the canopy–atmosphere interface, over a 2.3 ha area in a 500-year-old *Pseudotsuga/Tsuga* forest in southwestern Washington, USA. *Source: Photo by J.F. Franklin, Reprinted from Shaw, D.C., 2004. Vertical organization of canopy biota. In: Lowman, M.D., Rinker, H.B. (Eds.), Forest Canopies, second ed. Elsevier/Academic Press, Amsterdam, pp. 73–101.*

reproductive rates, and sensitivity to biochemical changes in their resources, insects provide early warning of changes that are not visible yet in the condition or abundance of larger, longer-lived plants or vertebrates, usually favored as bioindicators. Changes in insect herbivore abundances could indicate changes in plant biochemistry before visible chlorosis or other symptoms of stress became apparent.

The succession of insect species in decomposing carcasses (Figs. 10.7 and 10.8) has proven useful to law enforcement agencies in determining time and circumstances of death (Byrd and Castner, 2001; Goff, 2000; K.G.V. Smith, 1986; E. Watson and Carlton, 2003). For example, fly colonization rate in a corpse differs between exposed or protected locations. Research on the sequence and timing of colonization by various insect species on corpses under different environmental conditions provides critical evidence in criminal cases.

Insects also provide useful indications of changing climate. Responses to increased global temperatures include measurable shifts in species' geographic ranges toward higher elevations and latitudes, earlier activity in spring as degree-day requirements are met earlier, and measurable change in frequencies of genes conferring extreme temperature tolerance.

Finally, principles of insect ecology must be applied to improved management of insect populations and ecosystem services. Ecosystem engineering can make crop systems more or less conducive to insect population irruptions (Branson et al., 2006). Alternative cropping systems include protection of soil systems to enhance energy and matter availability, and polyculture cropping and landscape patterns of crop patches and remnant native vegetation (Fig. 17.1) to restrict herbivore dispersal among hosts or patches (Gliessman, 2007; Isaacs et al., 2009; Kogan and Jepson, 2007; Landis et al., 2000, 2005; Nicholls and Altieri, 2007; Pretty et al., 2003; Tscharntke et al., 2005, 2007; Vincent et al., 2003; H. Zhou et al., 2009). These cropping systems also enhance conditions for predators that control potentially irruptive insect species. Promotion of interactions that tend to stabilize populations of irruptive species is more effective in the long term than is reliance on pesticides or genetically engineered crops. Examples include provision or retention of hedgerows, ant-attracting plants, or other refuges within agricultural landscapes that maintain predator populations (Kruess and Tscharntke, 1994; Rickson and Rickson, 1998). Furthermore, insect effects on ecosystems, including agroecosystems, are complex. Net effects of outbreaks on multiple parameters should be considered in deciding whether or not to suppress outbreaks (see Chapter 17). Given that outbreaks often reflect simplification of ecosystem conditions and function to restore complexity and, perhaps, stability, control of native species in natural ecosystems may be counterproductive. Letting outbreaks of native species run their course could serve management purposes, at least under some conditions.

Invasive species present serious challenges. Regulatory mechanisms in ecosystems have little effect on such species, which can have devastating direct and indirect effects on resource distribution, food web interactions, and ecosystem conditions (eg, Fig. 17.8). Although some invasive species may replace some effects of native species that have been lost (Ewel and Putz, 2004), rates and seasonal patterns will be altered. The threats of invasive species to ecosystem services warrant increased attention to preventing introductions and increased efforts toward eliminating exotic species before they become established. Small populations are vulnerable to eradication through mating disruption with pheromones or sterile-mate releases (Yamanaka and Liebhold, 2009), techniques that are less disruptive to other ecosystem components than are insecticides or biological control introductions. More disruptive techniques will be necessary to control established populations. This will require early detection, rapid assessment of vulnerability, and commitment by management agencies to act at this stage.

4 CONCLUSIONS

Insects are involved in virtually all interactions and processes of terrestrial and freshwater ecosystems. Current environmental issues directly or indirectly involve insects, either in their capacity to respond to environmental changes or in their capacity to alter ecosystem conditions. Therefore, insect ecology is fundamental to our ability to understand ecosystem structure and function and to solve environmental problems.

The hierarchical ecosystem approach to insect ecology emphasizes linkages and feedbacks among individual, population, community, and ecosystem levels and clarifies the basis and consequences of insect adaptive strategies. This approach also indicates which level best addresses environmental problems. For example, if the issue involves plant susceptibility to herbivores, then individual responses to environmental cues are the appropriate focus. If the issue involves spread of exotic species or restoration of native species, then metapopulation dynamics and regulatory interactions within communities are the levels of focus. If the issue involves global mass balances of carbon fluxes, then mass balances at the ecosystem level are the appropriate focus.

Our most significant scientific advances in the coming decades will be in demonstrating the degree to which species interactions in ecosystems modify environmental conditions and persist in the face of changing global conditions. Insects are major contributors to the ways in which ecosystems modify local and global conditions. Natural selection can be viewed as a primary feedback between ecosystem conditions and individual adaptations that modify or stabilize ecosystem parameters. The degree to which insects regulate ecosystem parameters remains a key issue in global change biology and one that significantly broadens the scope and value of insect ecology.

Bibliography

Abarca, M., Lill, J.T., 2015. Warming affects hatching time and early season survival of eastern tent caterpillars. Oecologia 179, 901–912.

Abbas, M., Klein, A.-M., Ebeling, A., Oelmann, Y., Ptacnik, R., Weisser, W.W., Hillebrand, H., 2014. Plant diversity effects on pollinating and herbivorous insects can be linked to plant stoichiometry. Basic Appl. Ecol. 15, 169–178.

Abbott, K.R., Dukas, R., 2009. Honeybees consider flower danger in their waggle dance. Anim. Behav. 78, 633–635.

Abdala-Roberts, L., Mooney, K.A., Quijano-Medina, T., Campos-Navarrete, M.J., González-Moreno, A., Parra-Tabla, V., 2015. Comparison of tree genotypic diversity and species diversity effects on different guilds of insect herbivores. Oikos 124, 1527–1535.

Abrams, P.A., 2001. Describing and quantifying interspecific interactions: a commentary on recent approaches. Oikos 94, 209–218.

Abrams, E.M., Rue, D.J., 1988. The causes and consequences of deforestation among the prehistoric Maya. Hum. Ecol. 16, 377–395.

Ackerman, A.S., Toon, O.B., Stevens, D.E., Heymsfield, A.J., Ramanathan, V., Welton, E.J., 2000. Reduction of tropical cloudiness by soot. Science 288, 1042–1047.

Ackerman, I.L., Teixeira, W.G., Riha, S.J., Lehmann, J., Fernandes, E.C.M., 2007. The impact of mound-building termites on surface soil properties in a secondary forest of central Amazonia. Appl. Soil Ecol. 37, 267–276.

Acuña-Soto, R., Stahle, D.W., Cleaveland, M.K., Therrell, M.D., 2002. Megadrought and megadeath in 16th century Mexico. Emerg. Infect. Dis. 8, 360–362.

Adams, T.S., Sterner, R.W., 2000. The effect of dietary nitrogen content on trophic level ^{15}N enrichment. Limnol. Oceanogr. 45, 601–607.

Adams, J.M., Fang, W., Callaway, R.M., Cipollini, D., Newell, E., Transatlantic Acer Platanoides Invasion Network, 2009. A cross-continental test of enemy release hypothesis: leaf herbivory on *Acer platanoides* (L.) is three times lower in North America than in its native Europe. Biol. Invasions 11, 1005–1016.

Adkins, J.K., Rieske, L.K., 2013. Loss of a foundation forest species due to an exotic invader impacts terrestrial arthropod communities. Forest Ecol. Manag. 295, 126–135.

Adler, P.H., McCreadie, J.W., 1997. The hidden ecology of black flies: sibling species and ecological scale. Am. Entomol. 43, 153–161.

Adler, L.S., Karban, R., Strauss, S.Y., 2001. Direct and indirect effects of alkaloids on plant fitness via herbivory and pollination. Ecology 82, 2032–2044.

Aerts, R., 1997. Climate, leaf litter chemistry and leaf litter decomposition in terrestrial ecosystems: a triangular relationship. Oikos 79, 439–449.

Agee, J.K., 1993. Fire Ecology of Pacific Northwest Forests. Island Press, Washington, DC.

Agrawal, A.A., Fishbein, M., 2008. Phylogenetic escalation and decline of plant defense strategies. Proc. Natl. Acad. Sci. USA 105, 10057–10060.

Agrawal, A.A., Konno, K., 2009. Latex: a model for understanding mechanisms, ecology, and evolution of plant defense against herbivory. Annu. Rev. Ecol. Evol. Syst. 40, 311–331.

Agrawal, A.A., Ackerly, D.D., Adler, F., Arnold, A.E., Cáceres, C., Doak, D.F., Post, E., Hudson, P.J., Maron, J., Mooney, K.A., Power, M., Schemske, D., Stachowicz, J., Strauss, S., Turner, M.G., Werner, E., 2007. Filling key gaps in population and community ecology. Front. Ecol. Environ. 5, 145–152.

Ågren, G.I., Bosatta, E., Balesdent, J., 1996. Isotope discrimination during decomposition of organic matter: a theoretical analysis. Soil Sci. Soc. Am. J. 60, 1121–1126.

Agustí, N., Aramburu, J., Gabarra, R., 1999a. Immunological detection of of *Helicoverpa armigera* (Lepidoptera: Noctuidae) ingested by heteropteran predators: time-related decay and effect of meal size on detection period. Ann. Entomol. Soc. Am. 92, 56–62.

Agustí, N., de Vincente, M.C., Gabarra, R., 1999b. Development of sequence amplified characterized region (SCAR) markers of *Helicoverpa armigera*: a new polymerase chain reaction-based technique for predator gut analysis. Mol. Ecol. 8, 1467–1474.

Ahn, J.-H., Hong, I.-P., Bok, J.-I., Kim, B.-Y., Song, J., Weon, H.-Y., 2012. Pyrosequencing analysis of the bacterial communities in the guts of honey bees *Apis cerana* and *Apis mellifera* in Korea. J. Microbiol. 50, 735–745.

Aide, T.M., 1992. Dry season leaf production: an escape from herbivory. Biotropica 24, 532–537.

Aide, T.M., 1993. Patterns of leaf development and herbivory in a tropical understory community. Ecology 74, 455–466.

Aide, T.M., Zimmerman, J.K., 1990. Patterns of insect herbivory, growth, and survivorship in juveniles of a neotropical liana. Ecology 71, 1412–1421.

Aizen, M.A., Feinsinger, P., 1994. Habitat fragmentation, native insect pollinators, and feral honey bees in Argentine 'Chaco Serrano'. Ecol. Appl. 4, 378–392.

Aizen, M.A., Garibaldi, L.A., Cunningham, S.A., Klein, A.M., 2009. How much does agriculture depend on pollinators? lessons from long-term trends in crop production. Ann. Bot. 103, 1579–1588.

Akbari, H., Menon, S., Rosenfeld, A., 2009. Global cooling: increasing world-wide urban albedos to offset CO_2. Clim. Change 94, 275–286.

Akiyama, T., Takahashi, S., Shiyomi, M., Okubo, T., 1984. Energy flow at the producer level: the energy dynamics of grazed grassland 1. Oikos 42, 129–137.

Aktakka, E.E., Kim, H., Najafi, K., 2011. Energy scavenging from insect flight. J. Micromech. Microeng. 21, 1–11.

Insect Ecology. http://dx.doi.org/10.1016/B978-0-12-803033-2.00024-8

Albeny-Simões, D., Murrell, E.G., Elliot, S.L., Andrade, M.R., Lima, E., Juliano, S.A., Vilela, E.F., 2014. Attracted to the enemy: *Aedes aegypti* prefers oviposition sites with predator-killed conspecifics. Oecologia 175, 481–492.

Alfaro, R.I., Shepherd, R.F., 1991. Tree-ring growth of interior Douglas-fir after one year's defoliation by Douglas-fir tussock moth. Forest Sci. 37, 959–964.

Alfaro, C., Navarro-Llopis, V., Primo, J., 2009. Optimization of pheromone dispenser density for managing the rice striped stem borer, *Chilo suppressalis* (Walker), by mating disruption. Crop Prot. 28, 567–572.

Algar, A.C., Kerr, J.T., Currie, D.J., 2007. A test of Metabolic Theory as the mechanism underlying broad-scale species-richness gradients. Global Ecol. Biogeogr. 16, 170–178.

Allan, J.D., Wipfli, M.S., Caouette, J.P., Prussian, A., Rodgers, J., 2003. Influence of streamside vegetation on inputs of terrestrial invertebrates to salmonid food webs. Can. J. Fish. Aquat. Sci. 60, 309–320.

Allan, B.F., Langerhans, R.B., Ryberg, W.A., Landesman, W.J., Griffin, N.W., Katz, R.S., Oberle, B.J., Schutzenhofer, M.R., Smyth, K.N., de, A., Maurice, St., Clark, L., Crooks, K.R., Hernandez, D.E., McLean, R.G., Ostfeld, R.S., Chase, J.M., 2009. Ecological correlates of risk and incidence of West Nile virus in the United States. Oecologia 158, 699–708.

Allan, E., Weisser, W., Weigelt, A., Roscher, C., Fischer, M., Hillebrand, H., 2011. More diverse plant communities have higher functioning over time due to turnover in complementary dominant species. Proc. Natl. Acad. Sci. USA 108, 17034–17039.

Allan, E., Weisser, W.W., Fischer, M., Schulze, E.-D., Weigelt, A., Roscher, C., Baade, J., Barnard, R.L., Beßler, H., Buchmann, N., Ebeling, A., Eisenhauer, N., Engels, C., Fergus, A.J.F., Gleixner, G., Gubsch, M., Halle, S., Klein, A.M., Kertscher, I., Kuu, A., Lange, M., Le Roux, X., Meyer, S.T., Migunova, V.D., Milcu, A., Niklaus, P.A., Oelmann, Y., Pašalić, E., Petermann, J.S., Poly, F., Rottstock, T., Sabais, A.C.W., Scherber, C., Scherer-Lorenzen, M., Scheu, S., Steinbeiss, S., Schwichtenberg, G., Temperton, V., Tscharntke, T., Voigt, W., Wilcke, W., Wirth, C., Schmid, B., 2013. A comparison of the strength of biodiversity effects across multiple functions. Oecologia 173, 223–237.

Allee, W.C., 1931. Animal Aggregations: A Study in General Sociology. University of Chicago Press, Chicago, IL, pp. 431.

Allen, E.B., Allen, M.F., 1990. The mediation of competition by mycorrhizae in successional and patchy environments. In: Grace, J.B., Tilman, D. (Eds.), Perspectives on Plant Competition. Academic Press, San Diego, CA, pp. 367–389.

Allen, A.P., Brown, J.H., Gillooly, J.F., 2002. Global biodiversity, biochemical kinetics, and the energetic-equivalence rule. Science 297, 1545–1548.

Allen, C.R., Epperson, D.M., Garmestani, A.S., 2004. Red imported fire ant impacts on wildlife: a decade of research. Am. Midl. Nat. 152, 88–103.

Allison, J.D., McKenney, J.L., Miller, D.R., Gimmel, M.L., 2012. Role of ipsdienol, ipsenol, and cis-verbenol in chemical ecology of *Ips avulsus*, *Ips calligraphus*, and *Ips grandicollis* (Coleoptera: Curculionidae: Scolytinae). J. Econ. Entomol. 105, 923–929.

Allred, B.W., Smith, W.K., Twidwell, D., Haggerty, J.H., Running, S.W., Naugle, D.E., Fuhlendorf, S.D., 2015. Ecosystem services lost to oil and gas in North America. Science 348, 401–402.

Allsup, C.M., Paige, K.N., 2016. Belowground fungal associations and water interact to influence the compensatory response of *Ipomopsis aggregata*. Oecologia 180, 463–575.

Alroy, J., 2008. Dynamics of origination and extinction in the marine fossil record. Proc. Natl. Acad. Sci. USA 105, 11536–11542.

Alstad, D.N., Andow, D.A., 1995. Managing the evolution of insect resistance to transgenic plants. Science 268, 1894–1896.

Alstad, D.N., Edmunds, Jr., G.F., Weinstein, L.H., 1982. Effects of air pollutants on insect populations. Annu. Rev. Entomol. 27, 369–384.

Altieri, M.A., 2002. Agroecology: the science of natural resource management for poor farmers in marginal environments. Agr. Ecosyst. Environ. 93, 1–24.

Altizer, S.M., Oberhauser, K., Brower, L.P., 2000. Associations between host migration and the prevalence of a protozoan parasite in natural populations of adult monarch butterflies. Ecol. Entomol. 25, 125–139.

Amaranthus, M.P., Perry, D.A., 1987. Effect of soil transfer on ectomycorrhiza formation and the survival and growth of conifer seedlings on old, nonreforested clear-cuts. Can. J. Forest. Res. 17, 944–950.

Amiro, B.D., Barr, A.G., Barr, J.G., Black, T.A., Bracho, R., Brown, M., Chen, J., Clark, K.L., Davis, K.J., Desai, A.R., Dore, S., Engel, V., Fuentes, J.D., Goldstein, A.H., Goulden, M.L., Kolb, T.E., Lavigne, M.B., Law, B.E., Margolis, H.A., Martin, T., McCaughey, J.H., Misson, L., Montes-Helu, M., Noormets, A., Randerson, J.T., Starr, G., Xiao, J., 2010. Ecosystem carbon dioxide fluxes after disturbance in forests of North America. J. Geophys. Res. 115, G00K02.

Amman, G.D., McGregor, M.D., Schmitz, R.F., Oakes, R.D., 1988. Susceptibility of lodgepole pine to infestation by mountain pine beetles following partial cutting of stands. Can. J. Forest. Res. 18, 688–695.

Amoo, A.O.J., Dipeolu, O.O., Capstick, P.B., Munyinyi, D.M., Gichuru, L.N., Odhiambo, T.R., 1993. Ixodid ticks (Acari: Ixodidae) and livestock production: effect of varying acaricide treatments on ticks and productivity in east coast fever-immunized weaner and dairy cattle. J. Med. Entomol. 30, 503–512.

Andersen, A.N., 1988. Soil of the nest-mound of the seed-dispersing ant, *Aphaenogaster longiceps*, enhances seedling growth. Aust. J. Ecol. 13, 469–471.

Andersen, A.N., Lonsdale, W.M., 1990. Herbivory by insects in Australian tropical savannas: a review. J. Biogeogr. 17, 433–444.

Andersen, D.C., MacMahon, J.A., 1985. Plant succession following the Mount St. Helens volcanic eruption: facilitation by a burrowing rodent, *Thomomys talpoides*. Am. Midl. Nat. 114, 63–69.

Andersen, A.N., Majer, J.D., 2004. Ants show the way Down Under: invertebrates as bioindicators in land management. Front. Ecol. Environ. 2, 291–298.

Andersen, P.C., Brodbeck, B.V., Mizell, III, R.F., 1992. Feeding by the leafhopper, *Homalodisca coagulata*, in relation to xylem fluid chemistry and tension. J. Insect Physiol. 38, 611–622.

Anderson, J.M., 1988. Invertebrate-mediated transport processes in soils. Agr. Ecosyst. Environ. 24, 5–19.

Anderson, V.J., Briske, D.D., 1995. Herbivore-induced species replacement in grasslands: is it driven by herbivory tolerance or avoidance? Ecol. Appl. 5, 1014–1024.

Anderson, J.M., Swift, M.J., 1983. Decomposition in tropical forests. In: Sutton, S.L., Whitmore, T.C., Chadwick, A.C. (Eds.), Tropical Rain Forest: Ecology and Management. Blackwell, London, pp. 287–309.

Anderson, J.M., Proctor, J., Vallack, H.W., 1983. Ecological studies in four contrasting lowland rain forests in Gunung Mulu National Park, Sarawak. J. Ecol. 71, 503–527.

Anderson, N.H., Steedman, R.J., Dudley, T., 1984. Patterns of exploitation by stream invertebrates of wood debris (xylophagy). Verhandlungen der Internationalen Vereinigung für Theoretische und Angewandte Limnologie 22, 1847–1852.

Anderson, C., Theraulaz, G., Deneubourg, J.-L., 2002. Self-assemblages in insect societies. Insect. Soc. 49, 99–110.

Anderson, K.E., Inouye, B.D., Underwood, N., 2009. Modeling herbivore competition mediated by inducible changes in plant quality. Oikos 118, 1633–1646.

Andersson, P., Löfstedt, C., Hambäck, P.A., 2013a. Insect density–plant density relationships: a modified view of insect responses to resource concentrations. Oecologia 173, 1333–1344.

Andersson, P., Löfstedt, C., Hambäck, P.A., 2013b. How insects sense olfactory patches – the spatial scaling of olfactory information. Oikos 122, 1009–1016.

Andreae, M.O., Rosenfeld, D., Artaxo, P., Costa, A.A., Frank, G.P., Longo, K.M., Silva-Dias, M.A.F., 2004. Smoking rain clouds over the Amazon. Science 303, 1337–1342.

Andresen, E., 2002. Dung beetles in a central Amazonian rainforest and their ecological role as secondary seed dispersers. Ecol. Entomol. 27, 257–270.

Andresen, E., Laurance, S.G.W., 2007. Possible indirect effects of mammal hunting on dung beetle assemblages in Panama. Biotropica 39, 141–146.

Andreux, F., Cerri, C., Vose, P.B., Vitorello, V.A., 1990. Potential of stable isotope, ^{15}N and ^{13}C, methods for determining input and turnover in soils. In: Harrison, A.F., Ineson, P., Heal, O.W. (Eds.), Nutrient Cycling in Terrestrial Ecosystems: Field Methods, Application and Interpretation. Elsevier, London, pp. 259–275.

Andrew, N.R., Hughes, L., 2007. Potential host colonization by insect herbivores in a warmer climate: a transplant experiment. Glob. Change Biol. 13, 1539–1549.

Andrewartha, H.G., Birch, L.C., 1954. The Distribution and Abundance of Animals. University of Chicago, Chicago, IL.

Andrews, W.V., 1868. The Cynthia silk-worm. Am. Nat. 2, 311–320.

Anelli, C.M., Prischmann-Voldseth, D.A., 2009. Silk batik using beeswax and cochineal dye: an interdisciplinary approach to teaching entomology. American Entomologist 55, 95–105.

Anilkumar, K.J., Pusztai-carey, M., Moar, W.J., 2008. Fitness costs associated with Cry1Ac-resistance *Helicoverpa zea* (Lepidoptera: Noctuidae): a factor countering selection for resistance to Bt cotton? J. Econ. Entomol. 101, 1421–1431.

Anstey, M.L., Rogers, S.M., Ott, S.R., Burrows, M., Simpson, S.J., 2009. Serotonin mediates behavioral gregarization underlying swarm formation in desert locusts. Science 323, 627–630.

Appanah, S., 1990. Plant-pollinator interactions in Malaysian rain forests. In: Bawa, K., Hadley, M. (Eds.), Reproductive Ecology of Tropical Forest Plants. UNESCO/Parthenon, Paris, pp. 85–100.

Appel, H.M., Cocroft, R.B., 2014. Plants respond to leaf vibrations caused by insect herbivore chewing. Oecologia 175, 1257–1266.

Aranibar, J.N., Berry, J.A., Riley, W.J., Pataki, D.E., Law, B.E., Ehleringer, J.R., 2006. Combining meteorology, eddy fluxes, isotope measurements, and modeling to understand environmental controls of carbon isotope discrimination at the canopy scale. Glob. Change Biol. 12, 710–730.

Archer, S., Pyke, D.A., 1991. Plant-animal interactions affecting plant establishment and persistence on revegetated rangeland. J. Range Manage. 44, 558–565.

Ardón, M., Pringle, C.M., 2008. Do secondary compounds inhibit microbial- and insect-mediated leaf breakdown in a tropical rainforest stream, Puerto Rico? Oecologia 155, 311–323.

Arim, M., Marquet, P.A., 2004. Intraguild predation: a widespread interaction related to species biology. Ecol. Lett. 7, 557–564.

Arnfield, A.J., 2003. Two decades of urban climate research: a review of turbulence, exchanges of energy and water, and the urban heat island. Int. J. Climatol. 23, 1–26.

Arnone, III, J.A., Zaller, J.G., Ziegler, C., Zandt, H., Körner, C., 1995. Leaf quality and insect herbivory in model tropical plant communities after long-term exposure to elevated atmospheric CO_2. Oecologia 104, 72–78.

Arriaga-Weiss, S.L., Calme, S., Kampichler, C., 2008. Bird communities in rainforest fragments: guild responses to habitat variables in Tabasco, Mexico. Biol. Conserv. 17, 173–190.

Aschehaug, E.T., Callaway, R.M., 2015. Diversity increases indirect interactions, attenuates the intensity of competition, and promotes coexistence. Am. Nat. 186, 452–459.

Ascunce, M.S., Yang, C.-C., Oakey, J., Calcaterra, L., Wu, W.-J., Shih, C.-J., Goudet, J., Ross, K.G., Shoemaker, D., 2011. Global invasion history of the fire ant *Solenopsis invicta*. Science 331, 1066–1068.

Ashman, T.-L., Knight, T.M., Steets, J.A., Amarasekare, P., Burd, M., Campbell, D.R., Dudash, M.R., Johnston, M.O., Mazer, S.J., Mitchell, R.J., Morgan, M.T., Wilson, D.W.G., 2004. Pollen limitation of plant reproduction: ecological and evolutionary causes and consequences. Ecology 85, 2408–2421.

Aslan, C.E., Zavaleta, E.S., Tershy, B., Croll, D., 2013. Mutualism disruption threatens global plant biodiversity: a systematic review. PLoS ONE 8 (6), e66993.

Asquith, N.M., Terbourgh, J., Arnold, A.E., Riveros, C.M., 1999. The fruits the agouti ate: Hymenaea courbaril seed fate when its disperser is absent. J. Trop. Ecol. 15, 229–235.

Astrup, R., Coates, K.D., Hall, E., 2008. Finding the appropriate level of complexity for a simulation model: an example with a forest growth model. Forest Ecol. Manag. 256, 1659–1665.

Auclair, J.L., 1958. Honeydew excretion in the pea aphid Acyrthosiphum pisum (Harr.) (Homoptera: Aphididae). J. Insect Physiol. 2, 330–337.

Auclair, J.L., 1959. Feeding and excretion of the pea aphid, Acyrthosiphum pisum (Harr.), reared on different varieties of peas. Entomol. Exp. Appl. 2, 279–286.

Auclair, J.L., 1965. Feeding and nutrition of the pea aphid, Acyrthosiphum pisum (Harr.) (Homoptera: Aphididae), on chemically defined diets of various pH and nutrient levels. Ann. Entomol. Soc. Am. 58, 855–875.

Aukema, B.H., Zhu, J., Moller, J., Rasmussen, J.G., Raffa, K.F., 2010. Predisposition to bark beetle attack by root herbivores and associated pathogens: roles in forest decline, gap formation, and persistence of endemic bark beetle populations. Forest Ecol. Manag. 259, 374–382.

Ausmus, B.S., 1977. Regulation of wood decomposition rates by arthropod and annelid populations. Ecol. Bull. 25, 180–192.

Autry, A.R., Fitzgerald, J.W., 1993. Relationship between microbial activity, biomass and organosulfur formation in forest soil. Soil Biol. Biochem. 25, 33–39.

Axelrod, R., Hamilton, W.D., 1981. The evolution of cooperation. Science 211, 1390–1396.

Ayres, M.P., Wilkens, R.T., Ruel, J.J., Lombardero, M.J., Vallery, E., 2000. Nitrogen budgets of phloem-feeding bark beetles with and without symbiotic fungi. Ecology 81, 2198–2210.

Ayres, E., Dromph, K.M., Cook, R., Ostle, N., Bardgett, R.D., 2007. The influence of below-ground herbivory and defoliation of a legume on nitrogen transfer to neighbouring plants. Funct. Ecol. 21, 256–263.

Bach, C.E., 1990. Plant successional stage and insect herbivory: flea beetles on sand-dune willow. Ecology 71, 598–609.

Baer, W.S., 1931. The treatment of chronic osteomyelitis with the maggot (larvae of the blowfly). J. Bone Joint Surg. 13, 438–475.

Baik, J.-J., Kim, Y.-H., Chun, H.-Y., 2001. Dry and moist convection forced by an urban heat island. J. Appl. Meteorol. 40, 1462–1475.

Bailey, J.K., Schweitzer, J.A., Rehill, B.J., Lindroth, R.L., Martinsen, G.D., Whitham, T.G., 2004. Beavers as molecular geneticists: a genetic basis to the foraging of an ecosystem engineer. Ecology 85, 603–608.

Baker, R.R., 1972. Territorial behaviour of the nymphalid butterflies, Aglais urticae (L.) and Inachis io (L.). J. Anim. Ecol. 41, 453–469.

Baker, T.C., Kuenen, L.P.S., 1982. Pheromone source location by flying moths: a supplementary non-anemotactic mechanism. Science 216, 424–427.

Balanyá, J., Oller, J.M., Huey, R.B., Gilchrist, G.W., Serra, L., 2006. Global genetic change tracks global climate warming in Drosophila subobscura. Science 313, 1773–1775.

Baldwin, I.T., 1990. Herbivory simulations in ecological research. Trends Ecol. Evol. 5, 91–93.

Baldwin, I.T., 1998. Jasmonate-induced responses are costly but benefit plants under attack in native populations. Proc. Natl. Acad. Sci. USA 95, 8113–8118.

Baldwin, I.T., Schultz, J.C., 1983. Rapid changes in tree leaf chemistry induced by damage: evidence for communication between plants. Science 221, 277–279.

Baldwin, D.H., Spromberg, J.A., Collier, T.K., Scholz, N.L., 2009. A fish of many scales: extrapolating sublethal pesticide exposures to the productivity of wild salmon populations. Ecol. Appl. 19, 2004–2115.

Bale, J.S., Masters, G.J., Hodkinson, I.D., Awmack, C., Bezemer, T.M., Brown, V.K., Butterfield, J., Buse, A., Coulson, J.C., Farrar, J., Good, J.E.G., Harrington, R., Hartley, S., Jones, T.H., Lindroth, R.L., Press, M.C., Symrnioudis, I., Watt, A.D., Whittaker, J.B., 2002. Herbivory in global climate change research: direct effects of rising temperature on insect herbivores. Glob. Change Biol. 8, 1–16.

Ballhorn, D.J., Kautz, S., Schadler, M., 2013. Induced plant defense via volatile production is dependent on rhizobial symbiosis. Oecologia 172, 833–846.

Balvanera, P., Kremen, C., Martínez-Ramos, M., 2005. Applying community structure analysis to ecosystem function: examples from pollination and carbon storage. Ecol. Appl. 15, 360–375.

Bangert, R.K., Turek, R.J., Rehill, B., Wimp, G.M., Schweitzer, J.A., Allan, G.J., Bailey, J.K., Martinsen, G.D., Keim, P., Lindroth, R.L., Whitham, T.G., 2006. A genetic similarity rule determines arthropod community structure. Mol. Ecol. 15, 1379–1391.

Banks, C.J., Macaulay, E.D.M., 1964. The feeding, growth and reproduction of Aphis fabae Scop. on Vicia faba under experimental conditions. Ann. Appl. Biol. 53, 229–242.

Banks, C.J., Nixon, H.L., 1958. Effects of the ant, Lasius niger (L.) on the feeding and excretion of the bean aphid, Aphis fabae Scop. J. Exp. Biol. 35, 703–711.

Banks, C.J., Nixon, H.L., 1959. The feeding and excretion rates of Aphis fabae Scop. on Vicia faba L. Entomol. Exp. Appl. 2, 77–81.

Barbehenn, R., Weir, Q., Salminen, J.-P., 2008. Oxidation of ingested phenolics in the tree-feeding caterpillar Orgyia leucostigma depends on foliar chemical composition. J. Chem. Ecol. 34, 748–756.

Barbosa, P. (Ed.), 1998. Conservation Biological Control. Academic Press, San Diego, CA.

Barbosa, P., Krischik, V.A., 1987. Influence of alkaloids on feeding preference of eastern deciduous forest trees by the gypsy moth Lymantria dispar. Am. Nat. 130, 53–69.

Barbosa, P., Wagner, M.R., 1989. Introduction to Forest and Shade Tree Insects. Academic Press, San Diego, CA.

Barbosa, P., Hines, J., Kaplan, I., Martinson, H., Szczepaniec, A., Szendrei, Z., 2009. Associational resistance and associational susceptibility: having right or wrong neighbors. Annu. Rev. Ecol. Evol. S. 40, 1–20.

Bardgett, R.D., Leemans, D.K., Cook, R., Hobbs, P.J., 1997. Seasonality of the soil biota of grazed and ungrazed hill grasslands. Soil Biol. Biochem. 29, 1285–1294.

Bardgett, R.D., Wardle, D.A., Yeates, G.W., 1998. Linking above-ground and below-ground interactions: how plant responses to foliar herbivory influence soil organisms. Soil Biol. Biochem. 30, 1867–1878.

Barker, J.S., 2008. Decomposition of Douglas-fir coarse woody debris in response to differing moisture content and initial heterotrophic colonization. Forest Ecol. Manag. 255, 598–604.

Barnosky, A.D., Matzke, N., Tomiya, S., Wogan, G.O.U., Swartz, B., Quental, T.B., Marshall, C., McGuire, J.L., Lindsey, E.L., Maguire, K.C., Mersey, B., Ferrer, E.A., 2011. Has the Earth's sixth mass extinction already arrived? Nature 471, 51–57.

Barras, S.J., 1970. Antagonism between *Dendroctonus frontalis* and the fungus *Ceratocystis minor*. Ann. Entomol. Soc. Am. 63, 1187–1190.

Barz, W., Weltring, K., 1985. Biodegradation of aromatic extractives of wood. In: Higuchi, T. (Ed.), Biosynthesis and Biodegradation of Wood Components. Academic Press, New York, pp. 607–666.

Bascompte, J., 2009. Mutualistic networks. Front. Ecol. Environ. 7, 429–436.

Baskerville, G.L., Emin, P., 1969. Rapid estimation of heat accumulation from maximum and minimum temperatures. Ecology 50, 514–522.

Basset, Y., 1996. Local communities of arboreal herbivores in Papua New Guinea: predictors of insect variables. Ecology 77, 1906–1919.

Basset, Y., 2001. Communities of insect herbivores foraging on saplings versus mature trees of Pourouma bicolor (Cecropiaceae) in Panama. Oecologia 129, 253–260.

Basset, Y., Corbara, B., Barrios, H., Cuénoud, P., LePonce, M., Aberlenc, H.-P., Bail, J., Bito, D., Bridle, J.R., Castaño-Meneses, G., Cizek, L., Cornejo, A., Curletti, G., Delabie, J.H.C., DeJean, A., Didham, R.K., Dufrêne, M., Fagan, L.L., Floren, A., Frame, D.M., Hallé, F., Hardy, O.J., Hernandez, A., Kitching, R.L., Lewinsohn, T.M., Lewis, O.T., Manubor, M., Medianero, E., Missa, O., Mitchell, A.W., Mogia, M., Novotny, V., Ødegaard, F., de Oliveira, E.G., Orivel, J., Ozanne, C.M.P., Pascal, O., Pinzón, S., Rapp, M., Ribeiro, S.P., Roisin, Y., Roslin, T., Roubik, D.W., Samaniego, M., Schmidl, J., Sørensen, L.L., Tishechkin, A., Van Osselaer, C., Winchester, N.N., 2007. IBISCA-Panama, a large-scale study of arthropod beta-diversity and vertical stratification in a lowland rainforest: rationale, study sites and field protocols. Entomologie 77, 39–69.

Bastolla, U., Fortuna, M.A., Pascual-García, A., Ferrera, A., Luque, B., Bascompte, J., 2009. The architecture of mutualistic networks minimizes competition and increases biodiversity. Nature 458, 1018–1021.

Batra, L.R., 1966. Ambrosia fungi: extent of specificity to ambrosia beetles. Science 153, 193–195.

Battisti, A., Stastny, M., Buffo, E., Larsson, S., 2006. A rapid altitudinal range expansion in the pine processionary moth produced by the 2003 climatic anomaly. Glob. Change Biol. 12, 662–671.

Batzer, D.P., Wissinger, S.A., 1996. Ecology of insect communities in nontidal wetlands. Annu. Rev. Entomol. 41, 75–100.

Batzer, D.P., Jackson, C.R., Mosner, M., 2000a. Influences of riparian logging on plants and invertebrates in small, depressional wetlands of Georgia, U.S.A. Hydrobiologia 441, 123–132.

Batzer, D.P., Pusateri, C.R., Vetter, R., 2000b. Impacts of fish predation on marsh invertebrates: direct and indirect effects. Wetlands 20, 307–312.

Baum, K.A., Haynes, K.J., Dillemuth, F.P., Cronin, J.T., 2004. The matrix enhances the effectiveness of corridors and stepping stones. Ecology 85, 2671–2676.

Baumann, P., Baumann, L., Lai, C.-Y., Rouhbakhsh, D., 1995. Genetics, physiology, and evolutionary relationships of the genus *Buchnera*: intracellular symbionts of aphids. Annu. Rev. Microbiol. 49, 55–94.

Bawa, K.S., 1990. Plant-pollinator interactions in tropical rain forests. Annu. Rev. Ecol. Syst. 21, 399–422.

Baxter, C.V., Fausch, K.D., Saunders, W.C., 2005. Tangled webs: reciprocal flows of invertebrate prey link streams and riparian zones. Freshwater Biol. 50, 201–220.

Baxter, C.V., Fausch, K.D., Murakami, M., Chapman, P.L., 2007. Invading rainbow trout usurp a terrestrial prey subsidy from native charr and reduce their growth and abundance. Oecologia 153, 461–470.

Bayliss-Smith, T.P., 1990. The integrated analysis of seasonal energy deficits: problems and prospects. Eur. J. Clin. Nutr. 44 (Suppl. 1), 113–121.

Bazykin, A.D., Berezovskaya, F.S., Isaev, A.S., Khlebopros, R.G., 1997. Dynamics of forest insect density: bifurcation approach. J. Theor. Biol. 186, 267–278.

Bazzaz, F.A., 1975. Plant species diversity in old-field successional ecosystems in southern Illinois. Ecology 56, 485–488.

Bazzaz, F.A., 1990. The response of natural ecosystems to the rising global CO_2 levels. Annu. Rev. Ecol. Syst. 21, 167–196.

Beard, K.H., Eschtruth, A.K., Vogt, K.A., Vogt, D.J., Scatena, F.N., 2003. The effects of the frog *Eleuthrodactylus coqui* on invertebrates and ecosystem processes at two scales in the Luquillo Experimental Forest, Puerto Rico. J. Trop. Ecol. 19, 607–617.

Beard, K.H., Vogt, K.A., Vogt, D.J., Scatena, F.N., Covich, A.P., Sigurdardottir, R., Siccama, T.G., Crowl, T.A., 2005. Structural and functional responses of a subtropical forest to 10 years of hurricanes and droughts. Ecol. Monogr. 75, 345–361.

Beare, M.H., Coleman, D.C., Crossley, Jr., D.A., Hendrix, P.F., Odum, E.P., 1995. A hierarchical approach to evaluating the significance of soil biodiversity to biogeochemical cycling. Plant Soil 170, 5–22.

Beaver, L.M., Gvakharia, B.O., Vollintine, T.S., Hege, D.M., Stanewsky, R., Giebultowicz, J.M., 2002. Loss of circadian clock function decreases reproductive fitness in males of *Drosophila melanogaster*. Proc. Natl. Acad. Sci. USA 99, 2134–2139.

Bebi, P., Kilakowski, D., Veblen, T.T., 2003. Interactions between fire and spruce beetles in a subalpine Rocky Mountain forest landscape. Ecology 84, 362–371.

Becerra, J.X., 1994. Squirt-gun defense in *Bursera* and the chrysomelid counterploy. Ecology 75, 1991–1996.

Becerra, J.X., 1997. Insects on plants: macroevolutionary chemical trends in host use. Science 276, 253–256.

Bede, J.C., Musser, R.O., Felton, G.W., Korth, K.L., 2006. Caterpillar herbivory and salivary enzymes decrease transcript levels of *Medicago truncatula* genes encoding early enzymes in terpenoid biosynthesis. Plant Mol. Biol. 60, 519–531.

Beebe, C.W., 1905. Two Bird-Lovers in Mexico. Houghton Mifflin, Boston, MA.

Beedlow, P.A., Tingey, D.T., Phillips, D.L., Hogset, W.E., Olszyk, D.M., 2004. Rising atmospheric CO_2 and carbon sequestration in forests. Front. Ecol. Environ. 2, 315–322.

Beedlow, P.A., Lee, E.H., Tingey, D.T., Waschmann, R.S., Burdick, C.A., 2013. The importance of seasonal temperature and moisture patterns on growth of Douglas-fir in western Oregon, USA. Agr. Forest Meteorol. 169, 174–185.

Begon, M., Mortimer, M., 1981. Population Ecology: A Unified Study of Animals and Plants. Blackwell Scientific, Oxford, UK, pp. 200.

Behan, V.M., Hill, S.B., 1978. Feeding habits and spore dispersal of oribatid mites in the North American Arctic. Rev. Ecol. Biol. Sol. 15, 497–516.

Behmer, S.T., 2009. Insect herbivore nutrient regulation. Annu. Rev. Entomol. 54, 165–187.

Bell, W.J., 1990. Searching behavior patterns in insects. Annu. Rev. Entomol. 35, 447–467.

Bell, G., Gonzalez, A., 2009. Evolutionary rescue can prevent extinction following environmental change. Ecol. Lett. 12, 942–948.

Bellard, C., Bertelsmeier, C., Leadley, P., Thuiller, W., Courchamp, F., 2012. Impacts of climate change on the future of biodiversity. Ecol. Lett. 15, 365–377.

Belle-Isle, J., Kneeshaw, D., 2007. A stand and landscape comparison of the effects of a spruce budworm (*Choristoneura fumiferana* (Clem.)) outbreak to the combined effects of harvesting and thinning on forest structure. Forest Ecol. Manag. 246, 163–174.

Belnap, J., Gillette, D.A., 1998. Vulnerability of desert biological soil crusts to wind erosion: the influences of crust development, soil texture, and disturbance. J. Arid. Environ. 39, 133–142.

Belovsky, G.E., Slade, J.B., 2000. Insect herbivory accelerates nutrient cycling and increases plant production. Proc. Natl. Acad. Sci. USA 97, 14412–14417.

Belovsky, G.E., Laws, A.N., Slade, J.B., 2011. Prey change behaviour with predation threat, but demographic effects vary with prey density: experiments with grasshoppers and birds. Ecol. Lett. 14, 335–340.

Belsky, A.J., 1986. Does herbivory benefit plants? A review of the evidence. Am. Nat. 127, 870–892.

Bender, M.A., Knutson, T.R., Tuleya, R.E., Sirutis, J.J., Vecchi, G.A., Garner, S.T., Held, I.M., 2010. Modeled impact of anthropogenic warming on the frequency of intense Atlantic hurricanes. Science 327, 454–458.

Benedek, S., 1988. Aquatic life (K., Misley, Ed., J. Pasek, Trans.). In: Lake Balaton: Research, Management, . Hungarian Ministry of Environment and Water Management, pp. 25–28.

Benedict, F. 1976. Herbivory Rates and Leaf Properties in Four Forests in Puerto Rico and Florida. Ph.D. Dissertation, University of Florida, Gainesville, Florida.

Benke, A.C., Wallace, J.B., 1997. Trophic basis of production among riverine caddisflies: implications for food web analysis. Ecology 78, 1132–1145.

Bennett, A., Krebs, J., 1987. Seed dispersal by ants. Trends Ecol. Evol. 2, 291–292.

Benstead, J.P., Green, A.C., Deegan, L.A., Peterson, B.J., Slavik, K., Bowden, W.B., Hershey, A.E., 2007. Recovery of three arctic stream reaches from experimental nutrient enrichment. Freshwater Biol. 52, 1077–1089.

Berenbaum, M.R., 1987. Charge of the light brigade: phototoxicity as a defense against insects. In: Heitz, J.R., Downum, K.R. (Eds.), Light-Activated Pesticides. American Chemical Society, Washington, DC, pp. 206–216.

Berenbaum, M.R., Zangerl, A.R., 1988. Stalemates in the coevolutionary arms race: syntheses, synergisms, and sundry other sins. In: Spencer, K.C. (Ed.), Chemical Mediation of Coevolution. Academic Press, Inc, San Diego, CA, pp. 113–132.

Berenbaum, M.R., Zangerl, A.R., 2008. Facing the future of plant-insect interaction research: le retour á la "Raison d'Être". Plant Physiol. 146, 804–811.

Berlow, E.L., Navarrete, S.A., Briggs, C.J., Power, M.E., Menge, B.A., 1999. Quantifying variation in the strengths of species interactions. Ecology 80, 2206–2224.

Bernays, E.A., Woodhead, S., 1982. Plant phenols utilized as nutrients by a phytophagous insect. Science 216, 201–203.

Bernays, E.A., Bright, K.L., Gonzalez, N., Angel, J., 1994. Dietary mixing in a generalist herbivore: tests of two hypotheses. Ecology 75, 1997–2006.

Bernhard-Reversat, F., 1982. Measuring litter decomposition in a tropical forest ecosystem: comparison of some methods. Int. J. Ecol. Environ. Sci. 8, 63–71.

Berryman, A.A., 1981. Population Systems: A General Introduction. Plenum Press, New York.

Berryman, A.A., 1996. What causes population cycles of forest Lepidoptera? Trends Ecol. Evol. 11, 28–32.

Berryman, A.A., 1997. On the principles of population dynamics and theoretical models. Am. Entomol. 43, 147–151.

Berryman, A.A., Stenseth, N.C., Isaev, A.S., 1987. Natural regulation of herbivorous forest insect populations. Oecologia 71, 174–184.

Beynon, S.A., Wainwright, W.A., Cristie, M., 2015. The application of an ecosystem services framework to estimate the economic value of dung beetles to the U.K. cattle industry. Ecol. Entomol. 40, 124–135.

Bezemer, T.M., Jones, T.H., 1998. Plant-herbivore interactions in elevated atmospheric CO_2: quantitative analysis and guild effects. Oikos 82, 212–222.

Bezemer, T.M., Jones, T.H., Knight, K.J., 1998. Long-term effects of elevated CO_2 and temperature on populations of the peach potato aphid *Myzus persicae* and its parasitoid *Aphidius matricarieae*. Oecologia 116, 128–135.

Bezemer, T.M., De Deyn, G.B., Bossinga, T.M., van Dam, N.M., Harvey, J.A., Van der Putten, W.H., 2005. Soil community composition drives aboveground plant-herbivore-parasitoid interactions. Ecol. Lett. 8, 652–661.

Biesmeijer, J.C., Roberts, S.P.M., Reemer, M., Ohlemüller, R., Edwards, M., Peeters, T., Schaffers, A.P., Potts, S.G., Kleukers, R., Thomas, C.D., Settele, J., Kunin, W.E., 2006. Parallel declines in pollinators and insect-pollinated plants in Britain and The Netherlands. Science 313, 351–354.

Biondini, M.E., Mielke, Jr., P.W., Berry, K.J., 1988. Data-dependent permutation techniques for the analysis of ecological data. Vegetatio 75, 161–168.

Birk, E.M., 1979. Disappearance of overstorey and understorey litter in an open eucalypt forest. Aust. J. Ecol. 4, 207–222.

Birks, H.J.B., 1980. British trees and insects: a test of the time hypothesis over the last 13,000 years. Am. Nat. 115, 600–605.

Bisch-Knaden, S., Wehner, R., 2003. Local vectors in desert ants: context-dependent landmark learning during outbound and homebound runs. J. Comp. Physiol. A 189, 181–187.

Bishop, J.G., 2002. Early primary succession on Mount St. Helens: impact of insect herbivores on colonizing lupines. Ecology 83, 191–202.

Bishop, J.A., Armbruster, W.S., 1999. Thermoregulatory abilities of Alaskan bees: effects of size, phylogeny and ecology. Funct. Ecol. 13, 711–724.

Bjorksten, T.A., Hoffmann, A.A., 1998. Persistence of experience effects in the parasitoid *Trichogramma* nr. *brassicae*. Ecol. Entomol. 23, 110–117.

Blacquière, T., Smagghe, G., van Gestel, C.A.M., Mommaerts, V., 2012. Neonicotinoids in bees: a review on concentrations, side-effects and risk assessment. Ecotoxicology 21, 973–992.

Blair, R.B., 2001. Birds and butterflies along urban gradients in two ecoregions of the U.S. In: Lockwood, J.L., McKinney, M.L. (Eds.), Biotic Homogenization. Kluwer, Norwell, MA, pp. 33–56.

Blancas, J., Casas, A., Rangel-Landa, S., Moreno-Calles, A., Torres, I., Pérez-Negrón, E., Solís, L., Delgado-Lemus, A., Parra, F., Arellanes, Y., Caballero, J., Cortés, L., Lira, R., Dávila, P., 2010. Plant management in the Tehuacán-Cuicatlán Valley, Mexico. Econ. Bot. 64, 287–302.

Blanchette, R.A., Shaw, C.G., 1978. Associations among bacteria, yeasts, and basiodiomycetes during wood decay. Phytopathology 68, 631–637.

Blanton, C.M., 1990. Canopy arthropod sampling: a comparison of collapsible bag and fogging methods. J. Agr. Entomol. 7, 41–50.

Blatt, S.E., Janmaat, J.A., Harmsen, R., 2001. Modelling succession to include a herbivore effect. Ecol. Model. 139, 123–136.

Bloesch, U., 2008. Thicket clumps: a characteristic feature of the Kagera savanna landscape, East Africa. J. Veg. Sci. 19, 31–44.

Blossey, B., Hunt-Joshi, T.R., 2003. Belowground herbivory by insects: influence on plants and aboveground herbivores. Annu. Rev. Entomol. 48, 521–547.

Blum, M.S., 1980. Arthropods and ecomones: better fitness through ecological chemistry. In: Gilles, R. (Ed.), Animals and Environmental Fitness. Pergamon Press, Oxford, UK, pp. 207–222.

Blum, M.S., 1981. Chemical Defenses of Arthropods. Academic Press, New York.

Blum, M.S., 1992. Ingested allelochemicals in insect wonderland: a menu of remarkable functions. Am. Entomol. 38, 222–234.

Blum, M.S., Severson, R.F., Arrendale, R.F., Whitman, D.W., Escoubas, P., Adeyeye, O., Jones, C.G., 1990. A generalist herbivore in a specialist mode: metabolic, sequestrative, and defensive consequences. J. Chem. Ecol. 16, 223–244.

Blumer, P., Diemer, M., 1996. The occurrence and consequences of grasshopper herbivory in an alpine grassland, Swiss central Alps. Arctic Alpine Res. 28, 435–440.

Blüthgen, N., Gebauer, G., Fiedler, K., 2003. Disentangling a rainforest food web using stable isotopes: dietary diversity in a species-rich ant community. Oecologia 137, 426–435.

Bodbyl Roels, S.A., Kelly, J.K., 2011. Rapid evolution caused by pollinator loss in *Mimulus guttatus*. Evolution 65, 2541–2552.

Boddum, T., Skals, N., Wirén, M., Baur, R., Rauscher, S., Hillbur, Y., 2009. Optimisation of the pheromone blend of the swede midge, *Contarinia nasturii*, for monitoring. Pest Manag. Sci. 65, 851–856.

Boecklen, W.J., 1991. The conservation status of insects: mass extinction, scientific interest, and statutory protection. In: Vinson, S.B., Metcalf, R.L. (Eds.), Entomology Serving Society: Emerging Technologies and Challenges. Entomological Society of America, Lanham, MD, pp. 40–57.

Boettner, G.H., Elkinton, J.S., Boettner, C.J., 2000. Effects of a biological control introduction on three nontarget native species of saturniid moths. Conserv. Biol. 14, 1798–1806.

Boggs, C.L., Freeman, K.D., 2005. Larval food limitation in butterflies: effects on adult resource allocation and fitness. Oecologia 144, 353–361.

Boggs, C.L., Ross, C.L., 1993. The effect of adult food limitation on the life history traits in *Speyeria mormonia* (Lepidoptera: Nymphalidae). Ecology 74, 433–441.

Boman, H.G., Hultmark, D., 1987. Cell-free immunity in insects. Annu. Rev. Microbiol. 41, 103–126.

Boman, H.G., Faye, I., Gudmundsson, G.H., Lee, J.-Y., Lidholm, D.-A., 1991. Cell-free immunity in cecropia: a model system for antibacterial proteins. Eur. J. Biochem. 201, 23–31.

Bommarco, R., Marini, L., Vaissiere, B.E., 2012. Insect pollination enhances seed yield, quality, and market value in oilseed rape. Oecologia 169, 1025–1032.

Bonada, N., Prat, N., Resh, V.H., Statzner, B., 2006. Developments in aquatic insect biomonitoring: a comparative analysis of recent approaches. Annu. Rev. Entomol. 51, 495–523.

Bond, W.J., 1993. Keystone species. In: Schulze, E.D., Mooney, H.A. (Eds.), Biodiversity and Ecosystem Function. Springer-Verlag, Berlin, pp. 237–253.

Bongiovanni, R., Lowenberg-Deboer, J., 2004. Precision agriculture and sustainability. Precis. Agric. 5, 359–387.

Boorman, S.A., Levitt, P.R., 1972. Group selection on the boundary of a stable population. Proc. Natl. Acad. Sci. USA 69, 2711–2713.

Bora, S., Ceccacci, I., Delgado, C., Townsend R., 2010. World Development Report 2011 Food Security and Conflict. Agriculture and Rural Development Department, World Bank, Washington, DC.

Bordeaux, J.M., Lorenz, W.W., Johnson, D., Badgett, M.J., Glushka, J., Orlando, R., Dean, J.F.D., 2014. Noctilisin, a venom glycopeptide of *Sirex noctilio* (Hymenoptera: Siricidae), causes needle wilt and defense gene responses in pines. J. Econ. Entomol. 107, 1931–1945.

Boring, L.R., Swank, W.T., Monk, C.D., 1988. Dynamics of early successional forest structure and processes in the Coweeta Basin. In: Swank, W.T., Crossley, Jr., D.A. (Eds.), Forest Hydrology and Ecology at Coweeta. Springer-Verlag, New York, pp. 161–179.

Bormann, F.H., Likens, G.E., 1979. Pattern and Process in a Forested Ecosystem. Springer-Verlag, New York.

Bos, M.M., Veddeler, D., Bogdanski, A.K., Klein, A.-M., Tscharntke, T., Steffan-Dewenter, I., Tylianakis, J.M., 2007. Caveats to quantifying ecosystem services: fruit abortion blurs benefits from crop pollination. Ecol. Appl. 17, 1841–1849.

Bossart, J.L., Opuni-Frimpong, E., Kuudaar, S., Nkrumah, E., 2006. Richness, abundance and complementarity of fruit-feeding butterfly species in relict sacred forests and forest reserves of Ghana. Biodivers. Conserv. 15, 33–359.

Bostock, R.M., Karban, R., Thaler, J.S., Weyman, P.D., Gilchrist, D., 2001. Signal interactions in induced resistance to pathogens and insect herbivores. Eur. J. Plant Pathol. 107, 103–111.

Botkin, D.B., 1981. Causality and succession. In: West, D.C., Shugart, H.H., Botkin, D.B. (Eds.), Forest Succession: Concepts and Application. Springer-Verlag, New York, pp. 36–55.

Boucot, A.J., 1990. Evolutionary Paleobiology of Behavior and Coevolution. Elsevier, Amsterdam.

Boucot, A.J., Poinar, Jr., G.O., 2010. Fossil Behavior Compendium. Taylor and Francis Group LLC, Boca Raton, FL.

Bouget, C., Brin, A., Tellez, D., Archaux, F., 2015. Intraspecific variations in dispersal ability of saproxylic beetles in fragmented forest patches. Oecologia 177, 911–920.

Boulton, A.J., Peterson, C.G., Grimm, N.B., Fisher, S.G., 1992. Stability of an aquatic macroinvertebrate community in a multiyear hydrologic disturbance regime. Ecology 73, 2192–2207.

Bowers, M.D., Puttick, G.M., 1988. Response of generalist and specialist insects to qualitative allelochemical variation. J. Chem. Ecol. 14, 319–334.

Boyce, M.S., 1984. Restitution of r- and K-selection as a model of density-dependent natural selection. Annu. Rev. Ecol. Syst. 15, 427–447.

Boyd, R.S., 2002. Does elevated body Ni concentration protect insects against pathogens? A test using *Melanotrichus boydi* (Heteroptera: Miridae). Am. Midl. Nat. 147, 225–236.

Boyd, R.S., 2004. Ecology of metal hyperaccumulation. New Phytol. 162, 563–567.

Boyd, R.S., 2007. The defense hypothesis of elemental hyperaccumulation: status, challenges and new directions. Plant Soil 293, 153–176.

Boyd, R.S., 2009. High-nickel insects and nickel hyperaccumulator plants: a review. Insect Sci. 16, 19–31.

Boyd, R.S., 2013. Exploring tradeoffs in hyperaccumulator ecology and evolution. New Phytol. 199, 871–872.

Boyd, R.S., Martens, S.N., 1994. Nickel hyperaccumulated by *Thlaspi montanum* var. *montanum* is acutely toxic to an insect herbivore. Oikos 70, 21–25.

Boyd, R.S., Moar, W.J., 1999. The defensive function of Ni in plants: response of the polyphagous herbivore *Spodoptera exigua* (Lepidoptera: Noctuidae) to hyperaccumulator and accumulator species of Streptanthus (Brassicaceae). Oecologia 118, 218–224.

Boyd, R.S., Wall, M.A., 2001. Respoonses of generalist predators fed high-Ni *Melanotrichus boydi* (Heteroptera: Miridae): elemental defense against the third trophic level. Am. Midl. Nat. 146, 186–198.

Boyero, L., Pearson, R.G., Dudgeon, D., Graça, M.A.S., Gessner, M.O., Albariño, R.J., Ferreira, V., Yule, C.M., Boulton, A.J., Arunachalam, M., Callisto, M., Chauvet, E., Ramírez, A., Chará, J., Moretti, M.S., Gonçalves, Jr., J.F., Helson, J.E., Chará-Serna, A.M., Encalada, A.C., Davies, J.N., Lamothe, S., Cornejo, A., Li, A.O.Y., Buria, L.M., Villanueva, V.D., Zúñiga, M.C., Pringle, C.M., 2011. Global distribution of a key trophic guild contrasts with common latitudinal diversity patterns. Ecology 92, 1839–1848.

Boyero, L., Pearson, R.G., Dudgeon, D., Ferreira, V., Graça, M.A.S., Gessner, M.O., Boulton, A.J., Chauvet, E., Yule, C.M., Albariño, R.J., Ramírez, A., Helson, J.E., Callisto, M., Arunachalam, M., Chará, J., Figueroa, R., Mathooko, J.M., Gonçalves, Jr., J.F., Moretti, M.S., Chará-Serna, A.M., Davies, J.N., Encalada, A., Lamothe, S., Buria, L.M., Castela, J., Cornejo, A., Li, A.O.Y., M'Erimba, C., Villanueva, V.D., Zúñiga, M.C., Swan, C.M., Barmuta, L.A., 2012. Global patterns of stream detritivore distribution: implications for biodiversity loss in changing climates. Global Ecol. Biogeog. 21, 134–141.

Bozer, S.F., Traugott, M.S., Stamp, N.E., 1996. Combined effects of allelochemical-fed and scarce prey of the generalist insect predator *Podisus maculiventris*. Ecol. Entomol. 21, 328–334.

Braaker, S., Ghazoul, J., Obrist, M.K., Moretti, M., 2014. Habitat connectivity shapes urban arthropod communities: the key role of green roofs. Ecology 95, 1010–1021.

Bradbear, N., 2009. Bees and Their Role in Forest Livelihoods: A Guide to the Services Provided by Bees and the Sustainable Harvesting, Processing and Marketing of Their Products. FAO, Rome.

Bradley, C.A., Altizer, S., 2005. Parasites hinder monarch butterfly flight: implications for disease spread in migratory hosts. Ecol. Lett. 8, 290–300.

Bradshaw, W.E., Holzapfel, C.M., 2001. Genetic shift in photoperiodic response correlated with global warming. Proc. Natl. Acad. Sci. USA 98, 14509–14511.

Bradshaw, J.W.S., Howse, P.E., 1984. Sociochemicals of ants. In: Bell, W.J., Cardé, R.T. (Eds.), Chemical Ecology of Insects. Chapman and Hall, London, UK, pp. 429–473.

Braithwaite, R.W., Estbergs, J.A., 1985. Fire patterns and woody litter vegetation trends in the Alligator Rivers region of northern Australia. In: Tothill, J.C., Mott, J.J. (Eds.), Ecology and Management of the World's Savannahs. Australian Academy of Science, Canberra, ACT, pp. 359–364.

Branson, D.H., 2010. Density-dependent effects of an early season insect herbivore on a later developing insect herbivore. Environ. Entomol. 39, 346–350.

Branson, D.H., Haferkamp, M.A., 2014. Insect herbivory and vertebrate grazing impact food limitation and grasshopper populations during a severe outbreak. Ecol. Entomol. 39, 371–381.

Branson, D.H., Joern, A., Sword, G.A., 2006. Sustainable management of insect herbivores in grassland ecosystems: new perspectives in grasshopper control. BioScience 56, 743–755.

Braschler, B., Marini, L., Thommen, G.H., Baur, B., 2009. Effects of small-scale grassland fragmentation and frequent mowing on population density and species diversity of orthopterans: a long-term study. Ecol. Entomol. 34, 321–329.

Brauman, A., Kane, M.D., Labat, M., Breznak, J.A., 1992. Genesis of acetate and methane by gut bacteria of nutritionally diverse termites. Science 257, 1384–1387.

Bravo, S.P., 2008. Seed dispersal and ingestion of insect-infested seeds by black howler monkeys in flooded forests of the Parana River, Argentina. Biotropica 40, 471–476.

Bray, R.S., 1996. Armies of Pestilence: The Impact of Disease on History. Barnes and Noble, New York.

Bray, D.P., Brandi, K.K., Brazil, R.P., Oliveira, A.G., Hamilton, J.G.C., 2009. Synthetic sex pheromone attracts the Leishmaniasis vector *Lutzomyia longipalpis* (Diptera: Psychodidae) to traps in the field. J. Med. Entomol. 46, 428–434.

Brazzel, J.R., Newsom, L.D., 1959. Diapause in *Anthomomus grandis* Boh. J. Econ. Entomol. 52, 603–611.

Bredenhand, E., Samways, M.J., 2009. Impact of a dam on benthic macroinvertebrates in a small river in a biodiversity hotspot: Cape Floristic Region, South Africa. J. Insect Conserv. 13, 297–307.

Brener, A.G.F., Silva, J.F., 1995. Leaf-cutting ants and forest groves in a tropical parkland savanna of Venezuela: facilitated succession? J. Trop. Ecol. 11, 651–669.

Brenes-Arguedas, T., Coley, P.D., Kursar, T.A., 2008. Divergence and diversity in the defensive ecology of *Inga* at two Neotropical sites. J. Ecol. 96, 127–135.

Breshears, D.D., Cobb, N.S., Rich, P.M., Price, K.P., Allen, C.D., Balice, R.G., Romme, W.H., Kastens, J.H., Floyd, M.L., Belnap, J., Anderson, J.J., Meyers, O.B., Meyer, C.W., 2005. Regional vegetation die-off in response to global-change-type drought. Proc. Natl. Acad. Sci. USA 102, 15144–15148.

Brévault, T., Heuberger, S., Zhang, M., Ellers-Kirk, C., Ni, X., Masson, L., Li, X., Tabashnik, B.E., Carrière, Y., 2013. Potential shortfall of pyramided transgenic cotton for insect resistance management. Proc. Natl. Acad. Sci. USA 110, 5806–5811.

Brey, P.T., Lee, W.-J., Yamakawa, M., Koizumi, Y., Perrot, S., François, M., Ashida, M., 1993. Role of the interument in insect immunity: epicuticular abrasion and induction of cecropin synthesis in cuticular epithelial cells. Proc. Natl. Acad. Sci. USA 90, 6275–6279.

Breznak, J.A., Brune, A., 1994. Role of microorganisms in the digestion of lignocellulose by termites. Annu. Rev. Entomol. 39, 453–487.

Briand, F., Cohen, J.E., 1984. Community food webs have scale-invariant structure. Nature 307, 264–267.

Briant, G., Gond, V., Laurance, S.G.W., 2010. Habitat fragmentation and the desiccation of forest canopies: a case study from eastern Amazonia. Biol. Conserv. 143, 2763–2769.

Bridges, J.R., 1983. Mycangial fungi of *Dendroctonus frontalis* (Coleoptera: Scolytidae) and their relationship to beetle population trends. Environ. Entomol. 12, 858–861.

Bridges, J.R., Moser, J.C., 1983. Role of two phoretic mites in transmission of bluestain fungus, *Ceratocystis minor*. Ecol. Entomol. 8, 9–12.

Bridges, J.R., Moser, J.C., 1986. Relationship of phoretic mites (Acari: Tarsonemidae) to the bluestaining fungus, *Ceratocystis minor*, in trees infested by southern pine beetle (Coleoptera: Scolytidae). Environ. Entomol. 15, 951–953.

Bridges, J.R., Perry, T.J., 1985. Effects of mycangial fungi on gallery construction and distribution of bluestain in southern pine beetle-infested pine bolts. J. Entomol. Sci. 20, 271–275.

Bridges, J.R., Nettleton, W.A., Connor, M.D., 1985. Southern pine beetle (Coleoptera: Scolytidae) infestations without the bluestain fungus, *Ceratocystis minor*. J. Econ. Entomol. 78, 325–327.

Bristow, C.M., 1991. Why are so few aphids ant-tended? In: Huxley, C.R., Cutler, D.F. (Eds.), Ant-Plant Interactions. Oxford University Press, Oxford, UK, pp. 104–119.

Brittain, C., Williams, N., Kremen, C., Klein, A.-M., 2013. Synergistic effects of non-*Apis* bees and honey bees for pollination services. Proc. Roy. Soc. B-Biol. Sci. 280, 20122767.

Broadway, R.M., 1995. Are insects resistant to plant-proteinase inhibitors? J. Insect Physiol. 41, 107–116.

Broadway, R.M., 1997. Dietary regulation of serine proteinases that are resistant to serin proteinase inhibitors. J. Insect Physiol. 43, 855–874.

Broderick, N.A., Raffa, K.F., Goodman, R.M., Handelsman, J., 2004. Census of the bacterial community of the gypsy moth larval midgut by using culturing and culture-independent methods. Appl. Environ. Microb. 70, 293–300.

Brody, A.K., Palmer, T.M., Fox-Dobbs, K., Doak, D.F., 2010. Termites, vertebrate herbivores, and the fruiting success of *Acacia drepanolobium*. Ecology 91, 399–407.

Brokaw, N.V.L., 1985. Treefalls, regrowth, and community structure in tropical forests. In: Pickett, S.T.A., White, P.S. (Eds.), The Ecology of Natural Disturbance and Patch Dynamics. Academic Press, Orlando, FL, pp. 53–69.

Bronstein, J.L., 1998. The contribution of ant-plant protection studies to our understanding of mutualism. Biotropica 30, 150–161.

Brookes, M.H., Stark, R.W., Campbell, R.W. (Eds.), 1978. The Douglas-fir Tussock Moth: a Synthesis. USDA Forest Service Tech. Bull. 1585. USDA, Washington, DC.

Brosi, B.J., Briggs, H.M., 2013. Single pollinator species losses reduce floral fidelity and plant reproductive function. Proc. Natl. Acad. Sci. USA 110, 13044–13048.

Brosi, B.J., Daily, G.C., Chamberlain, C.P., Mills, M., 2009. Detecting changes in habitat-scale bee foraging in tropical fragmented landscape using stable isotopes. Forest Ecol. Manag. 258, 1846–1855.

Brouqui, P., 2011. Arthropod-borne diseases associated with political and social disorder. Annu. Rev. Entomol. 56, 357–374.

Brower, A.V.Z., 1996. Parallel race formation and the evolution of mimicry in *Heliconius* butterflies: a phylogenetic hypothesis from mitochondrial DNA sequences. Evolution 50, 195–221.

Brower, L.P., Brower, J.V.Z., Cranston, F.P., 1965. Courtship behavior of the queen butterfly, *Danaus gilippus berenice* (Cramer). Zoologica 50, 1–39.

Brower, L.P., Ryerson, W.N., Coppinger, L.L., Glazier, S.C., 1968. Ecological chemistry and the palatability spectrum. Science 161, 1349–1351.

Brown, V.K., 1984. Secondary succession: insect-plant relationships. BioScience 34, 710–716.

Brown, V.K., 1986. Life cycle strategies and plant succession. In: Taylor, F., Karban, R. (Eds.), The Evolution of Insect Life Cycles. Springer-Verlag, New York, pp. 105–124.

Brown, V.C., 1995. Insect herbivores and gaseous air pollutants—current knowledge and predictions. In: Harrington, R., Stork, N.E. (Eds.), Insects in a Changing Environment, Academic Press, London, UK, pp. 219–249.

Brown, B.J., Allen, T.F.H., 1989. The importance of scale in evaluating herbivory impacts. Oikos 54, 189–194.

Brown, B.J., Ewel, J.J., 1987. Herbivory in complex and simple tropical successional ecosystems. Ecology 68, 108–116.

Brown, V.K., Gange, A.C., 1989. Differential effects of above- and below-ground insect herbivory during early plant succession. Oikos 54, 67–76.

Brown, V.K., Gange, A.C., 1991. Effects of root herbivory on vegetation dynamics. In: Atkinson, D. (Ed.), Plant Root Growth: an Ecological Perspective. Blackwell Scientific, Oxford, UK, pp. 453–470.

Brown, J.H., Heske, E.J., 1990. Control of a desert grassland transition by a keystone rodent guild. Science 250, 1705–1707.

Brown, V.K., Hyman, P.S., 1986. Successional communities of plants and phytophagous Coleoptera. J. Ecol. 74, 963–975.

Brown, S., Lugo, A.E., 1982. Storage and production of organic matter in tropical forests and their role in the global carbon cycle. Biotropica 14, 161–187.

Brown, V.K., Southwood, T.R.E., 1983. Trophic diversity, niche breadth and generation times of exopterygote insects in a secondary succession. Oecologia 56, 220–225.

Brown, J.H., Reichman, O.J., Davidson, D.W., 1979. Granivory in desert ecosystems. Annu. Rev. Ecol. Syst. 10, 201–227.

Brown, M.V., Nebeker, T.E., Honea, C.R., 1987. Thinning increases loblolly pine vigor and resistance to bark beetles. South. J. Appl. For. 11, 28–31.

Brown, V.K., Gange, A.C., Evans, I.M., Storr, A.L., 1987. The effect of insect herbivory on the growth and reproduction of two annual *Vicia* species at different stages in plant succession. J. Ecol. 75, 1173–1189.

Brown, V.K., Jepson, M., Gibson, C.W.D., 1988. Insect herbivory: effects on early old field succession demonstrated by chemical exclusion methods. Oikos 52, 293–302.

Brown, S.C., Smith, K., Batzer, D., 1997. Macroinvertebrate responses to wetland restoration in northern New York. Environ. Entomol. 26, 1016–1024.

Brown, J.H., Whitham, T.G., Ernest, S.K.M., Gehring, C.A., 2001. Complex species interactions and the dynamics of ecological systems: long-term experiments. Science 293, 643–650.

Brown, J.H., Gillooly, J.F., Allen, A.P., Savage, V.M., West, G.B., 2004. Toward a metabolic theory of ecology. Ecology 85, 1771–1789.

Brown, M., Black, T.A., Nesic, Z., Foord, V.N., Spittlehouse, D.L., Fredeen, A.L., Grant, N.J., Burton, P.J., Trofymow, J.A., 2010. Impact of mountain pine beetle on the net ecosystem production of lodgepole pine stands in British Columbia. Agr. Forest Meteorol. 150, 254–264.

Brown, G., Scherber, C., Ramos, Jr., P., Ebrahim, E.K., 2012. The effects of harvester ant (*Messor ebeninus* Forel) nests on vegetation and soil properties in a desert dwarf shrub community in north-eastern Arabia. Flora 207, 503–511.

Brück, E., Elbert, A., Fischer, R., Krueger, S., Kühnhold, J., Klueken, A.M., Nauen, R., Niebes, J.-F., Reckmann, U., Schnorbach, H.-J., Steffens, R., van Waetermeulen, X., 2009. Movento®, an innovative ambimobile insecticide for sucking insect pest control in agriculture: biological profile and field performance. Crop Prot. 28, 838–844.

Bruinsma, M., Dicke, M., 2008. Herbivore-induced indirect defense: from induction mechanisms to community ecology. In: Schaller, A. (Ed.), Induced Plant Resistance to Herbivory. Springer, Dordrecht, The Netherlands, pp. 31–60.

Brust, M.L., Hoback, W.W., 2009. Hypoxia tolerance in adult and larval *Cicindela* tiger beetles varies by life history but not habitat association. Ann. Entomol. Soc. Am. 102, 462–466.

Brusven, M.A., Hornig, C.E., 1984. Effects of suspended and deposited volcanic ash on survival and behavior of stream insects. J. Kansas Entomol. Soc. 57, 55–62.

Bryan, B.A., Kandulu, J.M., 2009. Cost-effective alternatives for mitigating *Cryptosporidium* risk in drinking water and enhancing ecosystem services. Water. Resour. Res. 45, W08437.

Bryant, J., Chapin, F., Klein, D., 1983. Carbon/nutrient balance of boreal plants in relation to vertebrate herbivory. Oikos 40, 357–368.

Brys, R., Jacquemyn, H., Hermy, M., 2008. Pollination efficiency and reproductive patterns in relation to local plant density, population size, and floral display in the rewarding *Listera ovata* (Orchidaceae). Bot. J. Linn. Soc. 157, 713–721.

Bubliy, O.A., Kristensen, T.N., Kellermann, V., Loeschcke, V., 2012. Plastic responses to four environmental stresses and cross-resistance in a laboratory population of *Drosophila melanogaster*. Funct. Ecol. 26, 245–253.

Buchanan, A.L., Underwood, N., 2013. Attracting pollinators and avoiding herbivores: insects influence plant traits within and across years. Oecologia 173, 473–482.

Bucher, R., Menzel, F., Entling, M.H., 2015. Risk of spider predation alters food web structure and reduces local herbivory in the field. Oecologia 178, 571–577.

Buddle, C.M., Langor, D.W., Pohl, G.R., Spence, J.R., 2006. Arthropod responses to harvesting and wildfire: implications for emulation of natural disturbance in forest management. Biol. Conserv. 128, 346–357.

Buhl, J., Sumpter, D.J.T., Couzin, I.D., Hale, J.J., Despland, E., Miller, E.R., Simpson, S.J., 2006. From disorder to order in marching locusts. Science 312, 1402–1406.

Bujang, N.S., Harrison, N.A., Su, N.-Y., 2014. A phylogenetic study of endo-beta-1,4-glucanase in higher termites. Insect. Soc. 61, 29–40.

Bullock, S.H., 1991. Herbivory and the demography of the chaparral shrub *Ceanothus greggii* (Rhamnaceae). Madroño 38, 63–72.

Burd, M., Howard, J.J., 2005. Central-place foraging continues beyond the nest entrance; the underground performance of leaf-cutting ants. Anim. Behav. 70, 737–744.

Burdon, F.J., Harding, J.S., 2008. The linkage between riparian predators and aquatic insects across a stream-resource spectrum. Freshwater Biol. 53, 330–346.

Burger, J.M.S., Kolss, M., Pont, J., Kawecki, T.J., 2008. Learning ability and longevity: a symmetrical evolutionary trade-off in *Drosophila*. Evolution 62, 1294–1304.

Burkle, L., Irwin, R., 2009. The importance of interannual variation and bottom-up nitrogen enrichment for plant-pollinator networks. Oikos 118, 1816–1829.

Burkle, L., Irwin, R., 2010. Beyond biomass: measuring the effects of community-level nitrogen enrichment on floral traits, pollinator visitation and plant reproduction. J. Ecol. 98, 705–717.

Burkle, L.A., Marlin, J.C., Knight, T.M., 2013. Plant-pollinator interactions over 120 years: loss of species, co-occurrence, and function. Science 339, 1611–1615.

Burney, D.A., Flannery, T.F., 2005. Fifty millennia of catastrophic extinctions after human contact. Trends Ecol. Evol. 20, 395–401.

Burt, A., 2003. Site-specific selfish genes as tools for the control and genetic engineering of natural populations. Proc. Roy. Soc. B-Biol. Sci. 270, 921–928.

Buteler, M., Weaver, D.K., Peterson, R.K.D., 2009. Oviposition behavior of the wheat stem sawfly when encountering plants infested with cryptic conspecifics. Environ. Entomol. 38, 1707–1715.

Butler, C.D., Trumble, J.T., 2008. Effects of pollutants on bottom-up and top-down processes in insect-plant interactions. Environ. Pollut. 156, 1–10.

Butler, C.D., Beckage, N.E., Trumble, J.T., 2009. Effects of terrestrial pollutants on insect parasitoids. Environ. Toxicol. Chem. 28, 1111–1119.

Byers, J.A., Wood, D.L., 1981. Antibiotic-induced inhibition of pheromone synthesis in a bark beetle. Science 213, 763–764.

Byrd, J.H., Castner, J.L. (Eds.), 2001. The Utility of Arthropods in Legal Investigations. CRC Press, Boca Raton, FL.

Cadogan, B.L., Scharbach, R.D., 2003. Design and evaluation of an aerial spray trial with true replicates to test the efficacy of *Bacillus thuringiensis* insecticide in a boreal forest. J. Econ. Entomol. 96, 388–395.

Cairns, D.M., Lafon, C.L., Waldron, J.D., Tchakerian, M., Coulson, R.N., Klepzig, K.D., Birt, A.G., Xi, W., 2008. Simulating the reciprocal interaction of forest landscape structure and southern pine beetle herbivory using LANDIS. Landscape Ecol. 23, 403–415.

Calder, I.R., 2001. Canopy processes: iimplications for transpiration, interception and splash induced erosion, ultimately for forst management and water resources. Plant Ecol. 153, 203–214.

Callaway, R.M., DeLuca, R.H., Belliveau, W.M., 1999. Biological-control herbivores may increase competitive ability of the noxious weed *Centaurea maculosa*. Ecology 80, 1196–1201.

Cambefort, Y., 1991. From saprophagy to coprophagy. In: Hanski, I., Cambefort, Y. (Eds.), Dung Beetle Ecology. Princeton University Press, Princeton, NJ, pp. 22–35.

Camilo, G.R., Willig, M.R., 1995. Dynamics of a food chain model from an arthropod-dominated lotic community. Ecol. Model. 79, 121–129.

Campbell, B.C., Shea, P.J., 1990. A simple staining technique for assessing feeding damage by *Leptoglossus occidentalis* Heidemann (Hemipera: Coreidae) on cones. Can. Entomol. 122, 963–968.

Canham, C.D., Thompson, J., Zimmerman, J.K., Uriarte, M., 2010. Variation in susceptibility to hurricane damage as a function of storm intensity in Puerto Rican tree species. Biotropica 42, 87–94.

Cantrell, R.S., Cosner, C., Lou, Y., 2010. Evolution of dispersal in heterogeneous landscapes. In: Cantrell, S., Cosner, C., Ruan, S. (Eds.), Spatial Ecology. Chapman and Hall/CRC/Taylor & Francis, Boca Raton, FL, pp. 213–229.

Capinera, J.L., 1987. Population ecology of rangeland grasshoppers. In: Capinera, J.L. (Ed.), Integrated Pest Management on Rangeland: a Shortgrass Prairie Perspective. Westview Press, Boulder, CO, pp. 162–182.

Cappuccino, N., 1992. The nature of population stability in *Eurosta solidaginis*, a nonoutbreaking herbivore of goldenrod. Ecology 73, 1792–1801.

Cárcamo, H.A., Abe, T.A., Prescott, C.E., Holl, F.B., Chanway, C.P., 2000. Influence of millipedes on liter decomposition, N mineralization, and microbial communities in a coastal forest in British Columbia, Canada. Can. J. Forest. Res. 30, 817–826.

Cardé, R.T., 1996. Odour plumes and odour-mediated flight in insects. In: Olfaction in Mosquito-host Interactions. Ciba Foundation Symposium 200. John Wiley & Sons, Chichester, UK, pp. 54–70.

Cardé, R.T., Baker, T.C., 1984. Sexual communication with pheromones. In: Bell, W.J., Cardé, R.T. (Eds.), Chemical Ecology of Insects. Chapman and Hall, London, pp. 355–383.

Cardinale, B.J., Palmer, M.A., Collins, S.L., 2002. Species diversity enhances ecosystem functioning through interspecific facilitation. Nature 415, 426–429.

Cardinale, B.J., Weis, J.J., Forbes, A.E., Tilmon, K.J., Ives, A.R., 2006. Biodiversity as both a cause and consequence of resource availability: a study of reciprocal causality in a predator-prey system. J. Anim. Ecol. 75, 497–505.

Carlson, J.C., Dyer, L.A., Omlin, F.X., Beier, J.C., 2009. Diversity cascades and malaria vectors. J. Med. Entomol. 46, 460–464.

Carlton, R.G., Goldman, C.R., 1984. Effects of a massive swarm of ants on ammonium concentrations in a subalpine lake. Hydrobiologia 111, 113–117.

Carmona, D., Lajeunesse, M.J., Johnson, M.T.J., 2011. Plant traits that predict resistance to herbivores. Funct. Ecol. 25, 358–367.

Carpenter, D., Cappuccino, N., 2005. Herbivory, time since introduction and invasiveness of exotic plants. J. Ecol. 93, 315–321.

Carpenter, S.R., Kitchell, J.F., 1984. Plankton community structure and limnetic primary production. Am. Nat. 124, 159–172.

Carpenter, S.R., Kitchell, J.F., 1987. The temporal scale of variance in lake productivity. Am. Nat. 129, 417–433.

Carpenter, S.R., Kitchell, J.F., 1988. Consumer control of lake productivity. BioScience 38, 764–769.

Carpenter, S.R., Kitchell, J.F., Hodgson, J.R., 1985. Cascading trophic interactions and lake productivity. BioScience 35, 634–639.

Carrière, Y., Masaki, S., Roff, D.A., 1997. The coadaptation of female morphology and offspring size: a comparative analysis in crickets. Oecologia 110, 197–204.

Carrière, Y., Dennehy, T.J., Pedersen, B., Haller, S., Ellers-Kirk, C., Antilla, L., Liu, Y.-B., Willott, E., Tabashnik, B.E., 2001a. Large-scale management of insect resistance to transgenic cotton in Arizona: can transgenic insecticidal crops be sustained? J. Econ. Entomol. 94, 315–325.

Carrière, Y., Ellers-Kirk, C., Patin, A.L., Sims, M.A., Meyer, S., Liu, Y.-B., Dennehy, T.J., Tabashnik, B.E., 2001b. Overwintering cost associated with resistance to transgenic cotton in the pink bollworm (Lepidoptera: Gelechiidae). J. Econ. Entomol. 94, 935–941.

Carrière, Y., Ellers-Kirk, C., Sisterson, M., Antilla, L., Whitlow, M., Dennehy, T.J., Tabashnik, B.E., 2003. Long-term regional suppression of pink bollworm by *Bacillus thuringiensis* cotton. Proc. Natl. Acad. Sci. USA 100, 1519–1523.

Carroll, G., 1988. Fungal endophytes in stems and leaves: from latent pathogen to mutualistic symbiont. Ecology 69, 2–9.

Carroll, C.R., Hoffman, C.A., 1980. Chemical feeding deterrent mobilized in response to insect herbivory and counter adaptation by *Epilachna trededimnotata*. Science 209, 414–416.

Carson, R., 1962. Silent Spring. Houghton-Mifflin, New York, NY.

Carson, W.P., Root, R.B., 2000. Herbivory and plant species coexistence: community regulation by an outbreaking phytophagous insect. Ecol. Monogr. 70, 73–99.

Carter, V., Hurd, H., 2010. Choosing anti-*Plasmodium* molecules for genetically modifying mosquitoes: focus on peptides. Trends Parasitol. 26, 582–590.

Carter, G.A., Knapp, A.K., 2001. Leaf optical properties in higher plants: linking spectral characteristics to stress and chlorophyll concentration. Am. J. Bot. 88, 677–684.

Carter, J.L., Fend, S.V., Kennelly, S.S., 1996. The relationships among three habitat scales and stream benthic invertebrate community structure. Freshwater Biol. 35, 109–124.

Cartron, J.-L.E., Molles, Jr., M.C., Schuetz, J.F., Crawford, C.S., Dahm, C.N., 2003. Ground arthropods as potential indicators of flooding regime in the riparian forest of the middle Rio Grande, New Mexico. Environ. Entomol. 32, 1075–1084.

Carvalheiro, L.G., Veldtman, R., Shenkute, A.G., Tesfay, G.B., Walter, C., Pirk, W., Donaldson, J.S., Nicolson, S.W., 2011. Natural and within-farmland biodiversity enhances crop productivity. Ecol. Lett. 14, 251–259.

Carvalheiro, L.G., Seymour, C.L., Nicolson, S.W., Veldtman, R., 2012. Creating patches of native flowers facilitates crop pollination in large agricultural fields: mango as a case study. J. Appl. Ecol. 49, 1373–1383.

Carvell, C., Meek, W.R., Pywell, R.F., Goulson, D., Nowakowski, M., 2007. Comparing the efficacy of agri-environment schemes to enhance bumble bee abundance and diversity on arable field margins. J. Appl. Ecol. 44, 29–40.

Casey, T.M., 1988. Thermoregulation and heat exchange. Adv. Insect Physiol. 20, 119–146.

Cates, R.G., 1980. Feeding patterns of monophagous, oligophagous, and polyphagous insect herbivores: the effect of resource abundance and plant chemistry. Oecologia 46, 22–31.

Cattaneo, M.G., Yafuso, C., Schmidt, C., Huang, C.-Y., Rahman, M., Olson, C., Ellers-Kirk, C., Orr, B.J., Marsh, S.E., Antilla, L., Dutilleul, P., Carrière, Y., 2006. Farm-scale evaluation of the impacts of transgenic cotton on biodiversity, pesticide use, and yield. Proc. Natl. Acad. Sci. USA 103, 7571–7576.

Cattânio, J.H., Davidson, E.A., Nepstad, D.C., Verchot, L.V., Ackerman, I.L., 2002. Unexpected results of a pilot throughfall exclusion experiment on soil emissions of CO_2, CH_4, N_2O and NO in eastern Amazonia. Biol. Fert. Soils 36, 102–108.

Caughlin, T.T., Ferguson, J.M., Lichstein, J.L., Zuidema, P.A., Bunyavejchewin, B., Levey, D.J., 2015. Loss of animal seed dispersal increases extinction risk in a tropical tree species due to pervasive negative density dependence across life stages. Proc. Roy. Soc. B-Biol. Sci. 282, 20142095.

Cavalieri, L.F., Koçak, H., 1994. Chaos in biological control systems. J. Theor. Biol. 169, 179–187.

Cavalieri, L.F., Koçak, H., 1995a. Intermittent transition between order and chaos in an insect pest population. J. Theor. Biol. 175, 231–234.

Cavalieri, L.F., Koçak, H., 1995b. Chaos: a potential problem in the biological control of insect pests. Math. Biosci. 127, 1–17.

Cave, R.D., 2001. Jewel scarabs. Natl. Geogr. 199 (2), 52–61.

Cayuela, L., Hódar, J.A., Zamora, R., 2011. Is insecticide spraying a viable and cost-efficient management practice to control pine processionary moth in Mediterranean woodlands? Forest Ecol. Manag. 261, 1732–1737.

Cease, A.J., Elser, J.J., Ford, C.F., Hao, S., Kang, L., Harrison, J.F., 2012. Heavy livestock grazing promotes locust outbreaks by lowering plant nitrogen content. Science 335, 467–469.

Cease, A.J., Elser, J.J., Fenichel, E.P., Hadrich, J.C., Harrison, J.F., Robinson, B.E., 2015. Living with locusts: connecting soil nitrogen, locust outbreaks, livelihoods, and livestock markets. BioScience 65, 551–558.

Cebrián, J., Duarte, C.M., 1994. The dependence of herbivory on growth rate in natural plant communities. Funct. Ecol. 8, 518–525.

Céréghino, R., Leroy, C., Dejean, A., Corbara, B., 2010. Ants mediate the structure of phytotelm communities in an ant-garden brome-liad. Ecology 91, 1549–1556.

Cerritos, R., Cano-Santana, Z., 2008. Harvesting grasshoppers *Sphenarium purpurascens* in Mexico for human consumption: a compari-son with insecticidal control for managing pest outbreaks. Crop Prot. 27, 473–480.

Cerritos Flores, R., Ponce-Reyes, R., Rojas-García, F., 2015. Exploiting a pest insect species *Spenarium purpurascens* for human con-sumption: ecological, social and economic repercussions. J. Insects Food Feed 1, 75–84.

Chabot, B.F., Hicks, D.J., 1982. The ecology of leaf life spans. Annu. Rev. Ecol. Syst. 13, 229–259.

Chalk, R., Townson, H., Ham, P.J., 1995. *Brugia pahangi*: the effects of cecropins on microfilariae *in vitro* and in *Aedes aegypti*. Exp. Parasitol. 80, 401–406.

Chamberlain, K., Guerrieri, E., Pennacchio, F., Pettersson, J., Pickett, J.A., Poppy, G.M., Powell, W., Wadhams, L.J., Woodcock, C.M., 2001. Can aphid-induced plant signals be transmitted aerially and through the rhizosphere? Biochem. Syst. Ecol. 29, 1063–1074.

Chambers, J.Q., Asner, G.P., Morton, D.C., Anderson, L.O., Saatchi, S.S., Espríto-Santo, F.D.B., Palace, M., Souza, Jr., C., 2007. Regional ecosystem structure and function: ecological insights from remote sensing of tropical forests. Trends Ecol. Evol. 22, 414–423.

Changnon, S.A., Shealy, R.T., Scott, R.W., 1991. Precipitation changes in fall, winter, and spring caused by St. Louis. J. Appl. Meteorol. 30, 126–134.

Chapelle, G., Peck, L.S., 1999. Polar gigantism dictated by oxygen availability. Nature 399, 114–115.

Chapin, III, F.S., Bloom, A.J., Field, C.B., Waring, R.H., 1987. Plant responses to multiple environmental factors. BioScience 37, 49–57.

Chapman, R.F., 1982. The Insects: Structure and Function, third ed. Harvard University Press, Cambridge, MA.

Chapman, R.F., 2003. Contact chemoreception in feeding by phytophagous insects. Annu. Rev. Entomol. 48, 455–484.

Chapman, R.F., Bernays, E.A., 1989. Insect behavior at the leaf surface and learning as aspects of host plant selection. Experientia 45, 215–222.

Chapman, C.A., Chapman, L.J., 1996. Frugivory and the fate of dispersed and non-dispersed seeds of six African tree species. J. Trop. Ecol. 12, 491–504.

Chapman, S.K., Hart, S.C., Cobb, N.S., Whitham, T.G., Koch, G.W., 2003. Insect herbivory increases litter quality and decomposition: an extension of the acceleration hypothesis. Ecology 84, 2867–2876.

Chapman, L.J., Schneider, K.R., Apodaca, C., Chapman, C.A., 2004. Respiratory ecology of macroinvertebrates in a swamp-river system of East Africa. Biotropica 36, 572–585.

Chapman, S.K., Schweitzer, J.A., Whitham, T.G., 2006. Herbivory differentially alters plant litter dynamics of evergreen and decidu-ous trees. Oikos 114, 566–574.

Chase, J.M., 1996. Abiotic controls of trophic cascades in a simple grassland food chain. Oikos 77, 495–506.

Chase, J.M., 2007. Drought mediates the importance of stochiastic community assembly. Proc. Natl. Acad. Sci. USA 104, 17430–17434.

Chase, T.N., Pielke, R.A., Kittel, T.G.F., Nemani, R., Running, S.W., 1996. Sensitivity of a general circulation model to global changes in leaf area index. J. Geophys. Res. 101, 7393–7408.

Chase, J.M., Leibold, M.A., Downing, A.L., Shurin, J.B., 2000. The effects of productivity, herbivory, and plant species turnover in grassland food webs. Ecology 81, 2485–2497.

Chávez-Moreno, C.K., Tecante, A., Casas, A., 2009. The *Opuntia* (Cactaceae) and *Dactylopius* (Hemiptera: Dactylopiidae) in Mexico: a historical perspective of use, interaction and distribution. Biodivers. Conserv. 18, 3337–3355.

Chazdon, R., Fetcher, N., 1984. Photosynthetic light environments in a lowland tropical rainforest in Costa Rica. J. Ecol. 72, 553–564.

Chen, J., Henderson, G., Laine, R.A., 1988. Isolation and identification of 2-phenoxyethanol from a ballpoint pen ink as a trail-follow-ing substance of *Coptotermes formosanus* Shiraki and *Reticulitermes* sp. J. Entomol. Sci. 33, 97–105.

Chen, J., Franklin, J.F., Spies, T.A., 1995. Growing-season microclimatic gradients from clearcut edges into old-growth Douglas-fir forests. Ecol. Appl. 5, 74–86.

Chen, J., Henderson, G., Grimm, C.C., Lloyd, S.W., Laine, R.A., 1998. Termites fumigate their nests with naphthalene. Nature 392, 558–559.

Chen, Y., Giles, K.L., Payton, M.E., Greenstone, M.H., 2000. Identifying key cereal aphid predators by molecular gut analysis. Mol. Ecol. 9, 1887–1898.

Chen, I.-C., Hill, J.K., Ohlemüller, R., Roy, D.B., Thomas, C.D., 2011. Rapid range shifts of species associated with high levels of climate warming. Science 333, 1024–1026.

Chilcutt, C.F., Tabashnik, B.E., 2004. Contamination of refuges by *Bacillus thuringiensis* toxin genes from transgenic maize. Proc. Natl. Acad. Sci. USA 101, 7526–7529.

Chittka, L., Menzel, R., 1992. The evolutionary adaptation of flower colours and the insect pollinators' colour vision. J. Comp. Physiol. A. 171, 171–181.

Chittka, L., Raine, N.E., 2006. Recognition of flowers by pollinators. Curr. Opin. Plant Biol. 9, 428–435.

Chittka, L., Gumbert, A., Kunze, J., 1997. Foraging dynamics of bumble bees: correlates of movements within and between plant species. Behav. Ecol. 8, 239–249.

Cho, J.-Y., Mizutani, M., Smimizu, B.-I., Kinoshita, T., Ogura, M., Tokoro, K., Lin, M.-L., Sakata, K., 2007. Chemical profiling and gene expression profiling during the manufacturing process of Taiwan oolong tee "Oriental Beauty". Biosci. Biotechnol. Biochem. 71, 1476–1486.

Christensen, K.M., Whitham, T.G., 1991. Indirect herbivore mediation of avian seed dispersal in pinyon pine. Ecology 72, 534–542.

Christensen, Jr., N.L., Gregory, S.V., Hagenstein, P.R., Heberlein, T.A., Hendee, J.C., Olson, J.T., Peek, J.M., Perry, D.A., Schowalter, T.D., Sullivan, K., Tilman, G.D., Vogt, K.A., 2000. Environmental Issues in Pacific Northwest Forest Management. National Academy Press, Washington, DC.

Christensen, K.L., Gallacher, A.P., Martin, L., Tong, D., Elgar, M.A., 2010. Nutrient compensatory foraging in a free-living social insect. Naturwissenschaften 97, 941–944.

Christenson, L.M., Lovett, G.M., Mitchell, M.J., Groffman, P.M., 2002. The fate of nitrogen in gypsy moth frass deposited to an oak forest floor. Oecologia 131, 444–452.

Chung, H., Muraoka, H., Nakamura, M., Han, S., Muller, O., Son, Y., 2013. Experimental warming studies on tree species and forest ecosystems: a literature review. J. Plant Res. 126, 447–460.

Cipollini, Jr., D.F., 1997. Wind-induced mechanical stimulation increases pest resistance in common bean. Oecologia 111, 84–90.

Clark, W.C., 1979. Spatial structure relationship in a forest insect system: simulation models and analysis. Mitteilungen der Schweizerischen Entomologischen Gesellschaft 52, 235–257.

Clark, D.B., Clark, D.A., 1985. Seedling dynamics of a tropical tree: impacts of herbivory and meristem damage. Ecology 66, 1884–1892.

Clark, L.R., Geier, P.W., Hughes, R.D., Morris, R.F., 1967. The Ecology of Insect Populations in Theory and Practice. Methuen, London.

Clark, J.S., Fastie, C., Hurtt, G., 1998. Reid's paradox of rapid plant migration: dispersal theory and interpretation of paleoecological records. BioScience 48, 13–24.

Clark, K.L., Skowronski, N., Hom, J., 2010. Invasive insects impact forest carbon dynamics. Glob. Change Biol. 16, 88–101.

Clarke, C.M., Kitching, R.L., 1995. Swimming ants and pitcher plants: a unique ant-plant interaction from Borneo. J. Trop. Ecol. 11, 589–602.

Classen, A.T., Hart, S.C., Whitham, T.G., Cobb, N.S., Koch, G.W., 2005. Insect infestations linked to changes in microclimate: important climate change implications. Soil Sci. Soc. Am. J. 69, 2049–2057.

Classen, A.T., Chapman, S.K., Whitham, T.G., Hart, S.C., Koch, G.W., 2007a. Genetic-based plant resistance and susceptibility traits to herbivory influence needle and root litter nutrient dynamics. J. Ecol. 95, 1181–1194.

Classen, A.T., Overby, S.T., Hart, S.C., Koch, G.W., Whitham, T.G., 2007b. Season mediates herbivore effects on litter and soil microbial abundance and activity in a semi-arid woodland. Plant Soil 295, 217–227.

Claudianos, C., Ranson, H., Johnson, R.M., Biswas, S., Schuler, M.A., Berenbaum, M.R., Feyereisen, R., Oakeshott, J.G., 2006. A deficit of detoxification enzymes: pesticide sensitivity and environmental response in the honey bee. Insect Mol. Biol. 15, 615–636.

Clausen, L.W., 1954. Insect Fact and Folklore. Macmillan Co, New York.

Clay, K., 1990. Fungal endophytes of grasses. Annu. Rev. Ecol. Syst. 21, 275–297.

Clay, K., Hardy, T.N., Hammond, A.M., 1985. Fungal endophytes of grasses and their effects on an insect herbivore. Oecologia 66, 1–5.

Clay, K., Marks, S., Cheplick, G.P., 1993. Effects of insect herbivory and fungal endophyte infection on competitive interactions among grasses. Ecology 74, 1767–1777.

Clay, N.A., Lucas, J., Kaspari, M., Kay, A.D., 2013. Manna from heaven: refuse from an arboreal ant links aboveground and belowground processes in a lowland tropical forest. Ecosphere 4 (11), 141.

Clay, N.A., Yanoviak, S.P., Kaspari, M., 2014. Short-term sodium inputs attract microbi-detritivores and their predators. Soil Biol. Biochem. 75, 248–253.

Clay, N.A., Donoso, D.A., Kaspari, M., 2015. Urine as an important source of sodium increases decomposition in an inland but not coastal tropical forest. Oecologia 177, 571–579.

Cleary, D.F.R., Grill, A., 2004. Butterfly response to severe ENSO-induced forest fires in Borneo. Ecol. Entomol. 29, 666–676.

Clements, F.E., 1916. Plant Succession: An Analysis of the Development of Vegetation. Carnegie Institute of Washington Publication 242, Washington, DC.

Cloe, III, W.W., Garman, G.C., 1996. The energetic importance of terrestrial arthropod inputs to three warm-water streams. Freshwater Biol. 36, 105–114.

Cobb, R.C., 2010. Species shift drives decomposition rates following invasion by hemlock woolly adelgid. Oikos 119, 1291–1298.

Cobb, R.C., Orwig, D.A., Currie, S., 2006. Decomposition of green foliage in eastern hemlock forests of southern New England impacted by hemlock woolly adelgid infestations. Can. J. Forest. Res. 36, 1331–1341.

Cociancich, S., Bulet, P., Hetru, C., Hoffmann, J.A., 1994. The inducible antibacterial peptides of insects. Parasitol. Today 10, 132–138.

Cockerell, F.E., Sgrò, C.M., McKechnie, S.W., 2014. Latitudinal clines in heat tolerance, protein synthesis rate and transcript level of a candidate gene in *Drosophila melanogaster*. J. Insect Physiol. 60, 136–144.

Codella, Jr., S.G., Raffa, K.F., 1993. Defense strategies of folivorous sawflies. In: Wagner, M.R., Raffa, K.F. (Eds.), Sawfly Life History Adaptations to Woody Plants. Academic Press, San Diego, CA, pp. 261–294.

Coe, M., 1977. The role of termites in the removal of elephant dung in the Tsavo (East) National Park Kenya. East Afr. Wildlife J. 15, 49–55.

Coe, M., Coe, C., 1987. Large herbivores, acacia trees and bruchid beetles. S. Afr. J. Sci. 83, 624–635.

Cohen, J.E., Palka, Z.J., 1990. A stochastic theory of community food webs. V. Intervality and triangulation in the trophic-niche overlap graph. Am. Nat. 135, 435–463.

Cohen, J.E., Briand, F., Newman, C.M., 1990. Community Food Webs: Data and Theory. Springer-Verlag, Berlin, pp. 308.

Colbert, J.J., Campbell, R.W, 1978. The integrated model. In: The Douglas-fir Tussock Moth: a Synthesis Brookes, M.H., Stark, R.W., Campbell, R.W. (Eds.), USDA Forest Service, Tech. Bull. 1585, USDA Forest Service, Washington, DC, pp. 216–230.

Cole, L., Buckland, S.M., Bardgett, R.D., 2008. Influence of disturbance and nitrogen addition on plant and soil animal diversity in grassland. Soil Biol. Biochem. 40, 505–514.

Colegrave, N., 1997. Can a patchy population structure affect the evolution of competition strategies? Evolution 51, 483–492.

Coleman, D.C., Crossley, Jr., D.A., Hendrix, P.F., 2004. Fundamentals of Soil Ecology, second ed. Elsevier, Amsterdam.

Coleman, T.W., Clarke, S.R., Meeker, J.R., Rieske, L.K., 2008. Forest composition following overstory mortality from southern pine beetle and associated treatments. Can. J. Forest. Res. 38, 1406–1418.

Coley, P.D., 1980. Effects of leaf age and plant life history patterns on herbivory. Nature 284, 545–546.

Coley, P.D., 1982. Rates of herbivory on different tropical trees. In: Leigh, Jr., E.G., Rand, A.S., Windsor, D.M. (Eds.), The Ecology of a Tropical Forest: Seasonal Rhythms and Long-term Changes. Smithsonian Institution Press, Washington, DC, pp. 123–132.

Coley, P.D., 1983. Herbivory and defensive characteristics of tree species in a lowland tropical forest. Ecol. Monogr. 53, 209–233.

Coley, P.D., 1986. Costs and benefits of defense by tannins in a neotropical tree. Oecologia 70, 238–241.

Coley, P.D., Aide, T.M., 1991. Comparison of herbivory and plant defenses in temperate and tropical broad-leaved forests. In: Price, P.W., Lewinsohn, T.M., Fernandes, G.W., Benson, W.W. (Eds.), Plant-animal Interactions: Evolutionary Ecology in Tropical and Temperate Regions. John Wiley & Sons, Inc, New York, pp. 25–49.

Coley, P.D., Barone, J.A., 1996. Herbivory and plant defenses in tropical forests. Annu. Rev. Ecol. Syst. 27, 305–335.

Coley, P.D., Bryant, J.P., Chapin, III, F.S., 1985. Resource availability and plant antiherbivore defense. Science 230, 895–899.

Collinge, S.K., 2000. Effects of grassland fragmentation on insect species loss, colonization, and movement patters. Ecology 81, 2211–2226.

Collins, N.M., 1981. The role of termites in the decomposition of wood and leaf litter in the southern Guinea savanna of Nigeria. Oecologia 51, 389–399.

Collins, N.M., 1983. Termite populations and their role in litter removal in Malaysian rain forests. In: Sutton, S.L., Whitmore, T.C., Chadwick, A.C. (Eds.), Tropical Rain Forest: Ecology and Management. Blackwell, London, UK, pp. 311–325.

Collins, N.C., Mitchell, R., Wiegert, R.G., 1976. Functional analysis of a thermal spring ecosystem, with an evaluation of the role of consumers. Ecology 57, 1221–1232.

Combes, S.A., Dudley, R., 2009. Turbulence-driven instabilities limit insect flight performance. Proc. Natl. Acad. Sci. USA 106, 9105–9108.

Commins, S.P., James, H.R., Kelly, E.A., Pochan, S.L., Workman, L.J., Perzanowski, M.S., Kocan, K.M., Fahy, J.V., Nganga, L.W., Ronmark, E., Cooper, P.J., Platts-Mills, T.A.E., 2011. The relevance of tick bites to the production of IgE antibodies to the mammalian oligosaccharide galactose-α-1,3-galactose. J. Allergy Clin. Immunol. 127, 1286–1293.

Comstock, A.B., 1911. The Handbook of Nature Study. Comstock Publishing, Ithaca, NY.

Connell, J.H., 1978. Diversity in tropical rain forests and coral reefs. Science 199, 1302–1310.

Connell, J.H., 1980. Diversity and the coevolution of competitors, or the ghost of competition past. Oikos 35, 131–138.

Connell, J.H., 1983. On the prevalence and relative importance of interspecific competition: evidence from field experiments. Am. Nat. 122, 661–696.

Connell, J.H., Slatyer, R.O., 1977. Mechanisms of succession in natural communities and their role in community stability and organization. Am. Nat. 111, 1119–1144.

Conner, W.E., Boada, R., Schroeder, F.C., González, A., Meinwald, J., Eisner, T., 2000. Chemical defense: bestowal of a nuptial alkaloidal garment by a male moth on its mate. Proc. Natl. Acad. Sci. USA 97, 14406–14411.

Contarini, M., Onufrieva, K.S., Thorpe, K.W., Raffa, K.F., Tobin, P.C., 2009. Mate-finding failure as an important cause of Allee effects along the leading edge of an invading insect population. Entomol. Exp. Appl. 133, 307–314.

Cook, D.M., Doran-Peterson, J., 2010. Mining diversity of the natural biorefinery housed within Tipula abdominalis larvae for use in an industrial biorefinery for production of lignocellulosic ethanol. Insect Sci. 13, 303–312.

Cook, R.J., Barron, J.C., Papendick, R.I., Williams, III, G.J., 1981. Impact on agriculture of the Mount St. Helens eruptions. Science 211, 16–22.

Cook, D.C., Thomas, M.B., Cunningham, S.A., Anderson, D.L., De Barro, P.J., 2007. Predicting the economic impact of an invasive species on an ecosystem service. Ecol. Appl. 17, 1832–1840.

Cook, B.I., Anchukaitis, K.J., Kaplan, J.O., Puma, M.J., Kelley, M., Gueyffier, D., 2012. Pre-Columbian deforestation as an amplifier of drought in Mesoamerica. Geophys. Res. Lett. 39, L16706.

Cooper, R.A., Molan, P.C., 1999. The use of honey as an antiseptic in managing Pseudomonas infection. J. Wound Care 8, 161–164.

Cooper, S.M., Owen-Smith, N., 1986. Effects of plant spinescence on large mammalian herbivores. Oecologia 68, 446–455.

Cooper, R.A., Molan, P.C., Harding, K.G., 1999. Antibacterial activity of honey against strains of *Staphylococcus aureus* from infected wounds. J. Roy. Soc. Med. 92, 283–285.

Cooper, D., Cory, J.S., Theilmann, D.A., Myers, J.H., 2003. Nucleopolydedroviruses of forest and western ten caterpillars: cross-infectivity and evidence of activation of latent virus in high-density field populations. Ecol. Entomol. 28, 41–50.

Corbet, P.S., 1962. A Biology of Dragonflies. H.F. & G. Witherby, London, pp. 247.

Corbet, S.A., 1997. Role of pollinators in species preservation, conservation, ecosystem stability and genetic diversity. In: Pollination: from Theory to Practise, Richards, K.W. (Ed.). Proc. Seventh International Symposium on Pollination. Acta Horticulturae #437, pp. 219–229.

Cornelissen, T., 2011. Climate change and its effects on terrestrial insects and herbivory patterns. Neotrop. Entomol. 40, 155–163.

Correa, S.B., Winemiller, K.O., López-Fernández, H., Galetti, M., 2007. Evolutionary perspectives on seed consumption and dispersal by fishes. BioScience 57, 748–756.

Cory, J.S., Myers, J.H., 2003. The ecology and evolution of insect baculoviruses. Annu. Rev. Ecol. Evol. S. 34, 239–272.

Costa, M.H., Foley, J.A., 2000. Combined effects of deforestation and doubled atmospheric CO_2 concentrations on the climate of Amazonia. J. Climate 13, 18–34.

Costa, Z.J., Vonesh, J.R., 2013. Prey subsidy or predator cue? Direct and indirect effects of caged predators on aquatic consumers and resources. Oecologia 173, 1481–1490.

Costa, A., Boone, C.K., Murphy, R.J., Sharpee, W.C., Raffa, K.F., Reeve, J.D., 2013. Dispersal and edge behaviour of bark beetles and predators inhabiting red pine plantations. Agr. Forest Entomol. 15, 1–11.

Costantino, R.F., Desharnais, R.A., Cushing, J.M., Dennis, B., 1997. Chaotic dynamics in an insect population. Science 275, 389–391.

Costanza, R., d'Arge, R., de Groot, R., Farger, S., Grasso, M., Hannon, B., Limburg, K., Naeem, S., O'Neill, R.V., Paruelo, J., Raskin, R.G., Sutton, P., van den Belt, M., 1997. The value of the world's ecosystem services and natural capital. Nature 387, 253–260.

Côté, M., Ferron, J., Gagnon, R., 2005. Invertebrate predation of postdispersal seeds and juvenile seedlings of black spruce (*Picea mariana*) in the boreal forest of eastern Canada. Can. J. Forest. Res. 35, 674–681.

Cotrufo, M.F., Briones, M.J.I., Ineson, P., 1998. Elevated CO_2 affects field decomposition rate and palatability of tree leaf litter: importance of changes in substrate quality. Soil Biol. Biochem. 12, 1565–1571.

Cottonie, K., 2005. Integrating environmental and spatial processes in ecological community dynamics. Ecol. Lett. 8, 1175–1182.

Cottonie, K., de Meester, L., 2004. Metacommunity structure: synergy of biotic interactions as selective agents and dispersal as fuel. Ecology 85, 114–119.

Couceiro, S.R.M., Hamada, N., Ferreira, R.L.M., Forsberg, B.R., da Silva, J.O., 2007. Domestic sewage and oil spills in streams: effects on edaphic invertebrates in flooded forest, Manaus, Amazonas, Brazil. Water Air Soil Poll. 180, 249–259.

Coulson, R.N., 1979. Population dynamics of bark beetles. Annu. Rev. Entomol. 24, 417–447.

Coulson, R.N., Crossley, Jr., D.A., 1987. What is insect ecology? A commentary. Bull. Entomol. Soc. Am. 33, 64–68.

Coulson, R.N., Pope, D.N., Gagne, J.A., Fargo, W.S., Pulley, P.E., Edson, L.J., Wagner, T.L., 1980. Impact of foraging by *Monochamus titillator* (Col.: Cerambycidae) on within-tree populations of *Dendroctonus frontalis* (Col.: Scolytidae). Entomophaga 25, 155–170.

Coulson, R.N., Hennier, P.B., Flamm, R.O., Rykiel, E.J., Hu, L.C., Payne, T.L., 1983. The role of lightning in the epidemiology of the southern pine beetle. Z. Angew. Entomol. 96, 182–193.

Coulson, R.N., Flamm, R.O., Pulley, P.E., Payne, T.L., Rykiel, E.J., Wagner, T.L., 1986. Response of the southern pine bark beetle guild (Coleoptera: Scolytidae) to host disturbance. Environ. Entomol. 15, 850–858.

Coulson, R.N., Fitzgerald, J.W., McFadden, B.A., Pulley, P.E., Lovelady, C.N., Giardino, J.R., 1996. Functional heterogeneity of forest landscapes: how host defenses influence epidemiology of the southern pine beetle. In: Dynamics of Forest Herbivory: Quest for Pattern and Principle, Mattson, W.J., Niemela, P., Rousi, M. (Eds.), USDA Forest Serv. Gen. Tech. Rep. NC-183. USDA Forest Serv., North Central Forest Exp. Stn., St. Paul, MN, pp. 272–286.

Coupland, R.T., Van Dyne, G.M., 1979. Systems synthesis. In: Coupland, R.T. (Ed.), Grassland Ecosystems of the World: Analysis of Grasslands and their Uses. Cambridge University Press, Cambridge, UK, pp. 97–106.

Courtney, S.P., 1985. Apparency in coevolving relationships. Oikos 44, 91–98.

Courtney, S.P., 1986. The ecology of pierid butterflies: dynamics and interactions. Adv. Ecol. Res. 15, 51–131.

Coûteaux, M.M., Mousseau, M., Célérier, M.L., Bottner, P., 1991. Increased atmospheric CO_2 and litter quality: decomposition of sweet chestnut leaf litter with animal food webs of different complexities. Oikos 61, 54–64.

Covich, A.P., McDowell, W.H., 1996. The stream community. In: Reagan, D.P., Waide, R.B. (Eds.), The Food Web of a Tropical Rain Forest. University of Chicago Press, Chicago, IL, pp. 433–459.

Covich, A.P., Crowl, T.A., Scatena, F.N., 2003. Effects of extreme low flows on freshwater shrimps in a perennial tropical stream. Freshwater Biol. 48, 1199–1206.

Covich, A.P., Crowl, T.A., Hein, C.L., Townsend, M.J., McDowell, W.H., 2009. Predator-prey interactions in river networks: comparing shrimp spatial refugia in two drainage basins. Freshwater Biol. 54, 450–465.

Cowles, H.C., 1911. The causes of vegetative cycles. Bot. Gaz. 51, 161–183.

Cowling, R.M., Pierce, S.M., Stock, W.D., Cocks, M., 1994. Why are there so many myrmecochorous species in the Cape fynbos? In: Arianoutsou, M., Graves, R.H. (Eds.), Plant-Animal Interactions in Mediterranean-Type Ecosystems. Kluwer, Dordrecht, The Netherlands, pp. 159–168.

Cox-Foster, D.L., Conlan, S., Holmes, E.C., Palacios, G., Evans, J.D., Moran, N.A., Quan, P.-L., Briese, T., Hornig, M., Geiser, D.M., Martinson, V., vanEngelsdorp, D., Kalkstein, A.L., Drysdale, A., Hui, J., Zhai, J., Cui, L., Hutchison, S.K., Simons, J.F., Egholm, M., Pettis, J.S., Lipkin, W.I., 2007. A metagenic survey of microbes in honey bee colony collapse disorder. Science 318, 283–287.

Coxson, D.S., Nadkarni, N.M., 1995. Ecological roles of epiphytes in nutrient cycles of forest canopies. In: Lowman, M.D., Nadkarni, N.M. (Eds.), Forest Canopies. Academic Press, San Diego, CA, pp. 495–543.

Crane, E., 1999. The World History of Beekeeping and Honey Hunting. Routledge, New York.

Crawford, C.S., 1978. Seasonal water balance in *Orthoporus ornatus*, a desert millipede. Ecology 59, 996–1004.

Crawford, C.S., 1986. The role of invertebrates in desert ecosystems. In: Whitford, W.G. (Ed.), Pattern and Process in Desert Ecosystems. University of New Mexico Press, Albuquerque, NM, pp. 73–91.

Crawley, M.J., 1983. Herbivory: The Dynamics of Animal-plant Interactions. University of California Press, Berkeley.

Crawley, M.J., 1989. Insect herbivores and plant population dynamics. Annu. Rev. Entomol. 34, 531–564.

Creelman, R.A., Mullet, J.E., 1997. Biosynthesis and action of jasmonates in plants. Annu. Rev. Plant Phys. 48, 355–381.

Cregg, B.M., Dix, M.E., 2001. Tree moisture stress and insect damage in urban areas in relation to heat island effects. J. Arboric. 27, 8–17.

Croft, B.A., 1990. Arthropod Biological Control Agents and Pesticides. Wiley, New York.

Croft, B.A., Gutierrez, A.P., 1991. Systems analysis role in modeling and decision-making. In: Vinson, S.B., Metcalf, R.L. (Eds.), Entomology Serving Society: Emerging Technologies and Challenges. Entomological Society of America, Lanham, MD, pp. 298–319.

Croft, B.A., Slone, D.H., 1997. Equilibrium densities of European red mite (Acari: Tetranychidae) after exposure to three levels of predaceous mite diversity on apple. Environ. Entomol. 26, 391–399.

Cromack, Jr., K., Monk, C.D., 1975. Litter production, decomposition, and nutrient cycling in a mixed-hardwood watershed and a white pine watershed. In: Howell, F.G., Gentry, J.B., Smith, M.H. (Eds.), Mineral Cycling in Southeastern Ecosystems. US Energy Research and Development Administration. Technical Information Center, Washington, DC, pp. 609–624.

Cromack, Jr., K., Todd, R.L., Monk, C.D., 1975. Patterns of basidiomycete nutrient accumulation in conifer and deciduous forest litter. Soil Biol. Biochem. 7, 265–268.

Cromack, Jr., K., Sollins, P., Todd, R.L., Crossley, Jr., D.A., Fender, W.M., Fogel, R., Todd, A.W., 1977. Soil microorganism-arthropod interactions: fungi as major calcium and sodium sources. In: Mattson, W.J. (Ed.), The Role of Arthropods in Forest Ecosystems. Springer-Verlag, New York, pp. 78–84.

Cromartie, Jr., W.J., 1975. The effect of stand size and vegetational background on the colonization of cruciferous plants by herbivorous insects. J. Appl. Ecol. 12, 517–533.

Cronin, J.T., 2003. Matrix heterogeneity and host-parasitoid interactions in space. Ecology 84, 1506–1516.

Cronin, J.T., Strong, D.R., 1999. Dispersal-dependent oviposition and the aggregation of parasitism. Am. Nat. 154, 23–36.

Cronin, J.T., Reeve, J.D., Wilkens, R., Turchin, P., 2000. The pattern and range of movement of a checkered beetle predator relative to its bark beetle prey. Oikos 90, 127–138.

Cronin, J.T., Abrahamson, W.G., Craig, T.P., 2001. Temporal variation in herbivore host-plant preference and performance: constraints on host-plant adaptation. Oikos 93, 312–320.

Cronin, J.T., Haynes, K.J., Dillemuth, F., 2004. Spider effects on planthopper mortality, dispersal, and spatial population dynamics. Ecology 85, 2134–2143.

Cronin, J.K., Bundock, P.C., Henry, R.J., Nevo, E., 2007. Adaptive climatic molecular evolution in wild barley at the *Isa* defense locus. Proc. Natl. Acad. Sci. USA 104, 2773–2778.

Crossley, Jr., D.A., 1966. Radioisotope measurement of food consumption by a leaf beetle species, *Chrysomela knabi* Brown. Ecology 47, 1–8.

Crossley, Jr., D.A., 1977. The roles of terrestrial saprophagous arthropods in forest soils: current status of concepts. In: Mattson, W.J. (Ed.), The Role of Arthropods in Forest Ecosystems. Springer-Verlag, New York, pp. 49–56.

Crossley, Jr., D.A., Hoglund, M.P., 1962. A litter-bag method for the study of microarthropods inhabiting leaf litter. Ecology 43, 571–573.

Crossley, Jr., D.A., Howden, H.F., 1961. Insect-vegetation relationships in an area contaminated by radioactive wastes. Ecology 42, 302–317.

Crossley, D.A., Jr., Witkamp, M., 1964. Effects of pesticide on biota and breakdown of forest litter. In: Proceedings of Eighth International Congress of Soil Science, Bucharest, Romania. Publishing House of the Academy of the Socialist Republic of Romania, Bucharest, pp. 887–892.

Crossley, Jr., D.A., Blood, E.R., Hendrix, P.F., Seastedt, T.R., 1995. Turnover of cobalt-60 by earthworms (*Eisenia foetida*) (Lumbricidae, Oligochaeta). Appl. Soil Ecol. 2, 71–75.

Crutsinger, G.M., Collins, M.D., Fordyce, J.A., Gompert, Z., Nice, C.C., Sanders, N.J., 2006. Plant genotypic diversity predicts predicts community structure and governs an ecosystem process. Science 313, 966–968.

Crutsinger, G.M., Habenicht, M.N., Classen, A.T., Schweitzer, J.A., Sanders, N.J., 2008. Galling by *Rhopalomyia solidaginsis* alters *Solidago altissima* architecture and litter nutrient dynamics in an old-field ecosystem. Plant Soil 303, 95–103.

Cuffney, T.F., Wallace, J.B., Lugthart, G.J., 1990. Experimental evidence quantifying the role of benthic invertebrates in organic matter dynamics in headwater streams. Freshwater Biol. 23, 281–299.

Culler, L.E., Lamp, W.O., 2009. Selective predation by larval *Agabus* (Coleoptera: Dytiscidae) on mosquitoes: support for conservation-based mosquito suppression in constructed wetlands. Freshwater Biol. 54, 2003–2014.

Culler, L.E., McPeek, M.A., Ayres, M.P., 2014. Predation risk shapes thermal physiology of a predaceous damselfly. Oecologia 176, 653–660.

Culver, D.C., Beattie, A.J., 1980. The fate of *Viola* seeds dispersed by ants. Am. J. Bot. 67, 710–714.

Culver, D.C., Beattie, A.J., 1983. Effects of ant mounds on soil chemistry and vegetation patterns in a Colorado montane meadow. Ecology 64, 485–492.

Cummins, K.W., 1973. Trophic relations of aquatic insects. Annu. Rev. Entomol. 18, 183–206.

Cunningham, J.P., West, S.A., Wright, D.J., 1998. Learning in the nectar foraging behaviour of *Helicoverpa armigera*. Ecol. Entomol. 23, 363–369.

Cunningham, J.P., Zalucki, M.P., West, S.A., 1999. Learning in Helicoverpa armigera (Lepidoptera: Noctuidae): a new look at the behaviour and control of a phytophagous pest. Bull. Entomol. Res. 89, 201–207.

Cunningham, S.A., Pullen, K.R., Colloff, M.J., 2009. Whole-tree sap flow is substantially diminished by leaf herbivory. Oecologia 158, 633–640.

Currano, E.D., Wilf, P., Wing, S.L., Labandeira, C.C., Lovelock, E.C., Royer, D.L., 2008. Sharply increased insect herbivory during the Paleocene-Eocene Thermal Maximum. Proc. Natl. Acad. Sci. USA 105, 1960–1964.

Currie, D.J., 1991. Energy and large-scale patterns of animal- and plant-species richness. Am. Nat. 137, 27–49.

Currie, C.R., 2001. A community of ants, fungi, and bacteria: a multilateral approach to studying symbiosis. Annu. Rev. Microbiol. 55, 357–380.

Currie, C.R., Mueller, U.G., Malloch, D., 1999a. The agricultural pathology of ant fungus gardens. Proc. Natl. Acad. Sci. USA 96, 7998–8002.

Currie, C.R., Scott, J.A., Summerbell, R.C., Malloch, D., 1999b. Fungus-growing ants use antibiotic-producing bacteria to control garden parasites. Nature 398, 701–704.

Curry, J.P., 1994. Grassland Invertebrates. Chapman & Hall, London.

Cushing, J.M., Costantino, R.F., Dennis, B., Desharnais, R.A., Henson, S.M., 2003. Chaos in Ecology: Experimental Nonlinear Dynamics. Academic/Elsevier, San Diego, CA.

Cushman, J.H., Addicott, J.F., 1991. Conditional interactions in ant-plant-herbivore mutualisms. In: Huxley, C.R., Cutler, D.F. (Eds.), Ant-Plant Interactions. Oxford University Press, Oxford, UK, pp. 92–103.

Cusumano, A., Weldegergis, B.T., Colazza, S., Dicke, M., Fatouros, N.E., 2015. Attraction of egg-killing parasitoids toward induced plant volatiles in a multi-herbivore context. Oecologia 179, 163–174.

Daily, G.C. (Ed.), 1997. Nature's Services: Societal Dependence on Natural Ecosystems. Island Press, Washington, DC.

Dainat, B., Evans, J.D., Chen, Y.P., Gauthier, L., Neumann, P., 2012. Predictive markers of honey bee colony collapse. PLoS ONE 7 (2), e32151.

Dale, A.G., Frank, S.D., 2014. The effects of urban warming on herbivore abundance and street tree condition. PLoS ONE 9 (7), e102996.

Daly, K.C., Durtschi, M.L., Smith, B.H., 2001. Olfactory-based discrimination learning in the moth, *Manduca sexta*. J. Insect Physiol. 47, 375–384.

Dambacher, J.M., Li, H.W., Wolff, J.O., Rossignol, P.A., 1999. Parsimonious interpretation of the impact of vegetation, food, and predation on snowshoe hare. Oikos 84, 530–532.

Dambacher, J.M., Li, H.W., Rossignol, P.A., 2002. Relevance of community structure in assessing indeterminacy of ecological predictions. Ecology 83, 1372–1385.

Dambacher, J.M., Luh, H.-K., Li, H.W., Rossignol, P.A., 2003. Qualitative stability and ambiguity in model ecosystems. Am. Nat. 161, 876–888.

Dams, L.R., 1978. Bees and honey-hunting scenes in the Mesolithic rock art of eastern Spain. Bee World 59, 45–53.

Dangerfield, J.M., 1994. Ingestion of leaf litter by millipedes: the accuracy of laboratory estimates for predicting litter turnover in the field. Pedobiologia 38, 262–265.

Dangerfield, J.M., McCarthy, T.S., Ellery, W.N., 1998. The mound-building termite *Macrotermes michaelseni* as an ecosystem engineer. J. Trop. Ecol. 14, 507–520.

Darlington, P.J., 1943. Carabidae of mountains and islands: data on the evolution of isolated faunas, and on atrophy of wings. Ecol. Monogr. 13, 37–61.

Darwin, C., 1859. The Origin of Species by Means of Natural Selection or the Preservation of Favored Races in the Struggle for Life. Murray, London.

Dasgupta, P., Levin, S., Lubchenco, J., 2000. Economic pathways to ecological sustainability. BioScience 50, 339–345.

Dauber, J., Wolters, V., 2000. Microbial activity and functional diversity in the mounds of three different ant species. Soil Biol. Biochem. 32, 93–99.

Dauber, J., Biesmeijer, J.C., Gabriel, D., Kunin, W.E., Lamborn, E., Meyer, B., Nielsen, A., Potts, S.G., Roberts, S.P.M., Söber, V., Settele, J., Steffan-Dewenter, I., Stout, J.C., Teder, T., Tscheulin, T., Vivarelli, D., Petanidou, T., 2010. Effects of patch size and density on flower visitation and seed set of wild plants: a pan-European approach. J. Ecol. 98, 188–196.

Daufresne, T., Loreau, M., 2001. Ecological stoichiometry, primary producer-decomposer interactions, and ecosystem persistence. Ecology 82, 3069–3082.

Davidson, D.W., 1993. The effects of herbivory and granivory on terrestrial plant succession. Oikos 68, 23–35.

Davidson, D.W., Fisher, B.L., 1991. Symbiosis of ants with *Cecropia* as a function of light regime. In: Huxley, C.R., Cutler, D.F. (Eds.), Ant-Plant Interactions. Oxford University Press, Oxford, UK, pp. 289–309.

Davidson, D.W., Inouye, R.S., Brown, J.H., 1984. Granivory in a desert ecosystem: experimental evidence for indirect facilitation of ants by rodents. Ecology 65, 1780–1786.

Davies, P.E., Nelson, M., 1994. Relationship between riparian buffer widths and the effect of logging in stream habitat, invertebrate community response and fish abundance. Aust. J. Mar. Fresh. Res. 45, 1289–1305.

Davis, A.L.V., 1996. Seasonal dung beetle activity and dung dispersal in selected South African habitats: implications for pasture improvement in Australia. Agr. Ecosyst. Environ. 58, 157–169.

Davis, R.L., 2005. Olfactory memory formation in *Drosophila*: from molecular to systems neuroscience. Annu. Rev. Neurosci. 28, 275–302.

Davis, T.S., Hofstetter, R.W., 2009. Effects of gallery density and species ratio on the fitness and fecundity of two sympatric bark beetles (Coleoptera: Curculionidae). Environ. Entomol. 38, 639–650.

Davis, J.M., Rosemond, A.D., Small, G.E., 2011. Increasing donor ecosystem productivity decreases terrestrial consumer reliance on a stream resource subsidy. Oecologia 167, 821–834.

Davis, J.M., Coogan, L.E., Papaj, D.R., 2015. Big maggots dig deeper: size-dependent larval dispersal in flies. Oecologia 179, 55–62.

Davis, S.E., Malfi, R.L., Roulston, T.H., 2015. Species differences in bumblebee immune response predict developmental success of a parasitoid fly. Oecologia 178, 1017–1032.

Davison, E.A., 1987. Respiration and energy flow in two Australian species of desert harvester ants, *Chelaner rothsteini* and *Chelaner whitei*. J. Arid. Environ. 12, 61–82.

Day, M.F., Irzykiewicz, H., 1953. Feeding behavior of the aphids *Myzus persicae* and *Brevicoryne brassicae*, studied with radiophosphorus. Aust. J. Biol. Sci. 6, 98–108.

Day, M.F., McKinnon, A., 1951. A study of some aspects of the feeding of the jassid *Orosius*. Aust. J. Sci. Res. Ser. B 4, 125–135.

Day, J.F., Edman, J.D., Kunz, S.E., Wikel, S.K., 2000. Direct injury: phobias, psychoses, annoyance, allergies, toxins, venoms and myiasis. In: Eldridge, B.F., Edman, J.D. (Eds.), Medical Entomology. Kluwer Academic Publishers, Dordrecht, The Netherlands, pp. 99–149.

de Carvalho, E.L., Kogan, M., 1991. Order Strepsiptera. Stehr, F. (Ed.), Immature Insects, vol. 2, Kendall/Hunt, Dubuque, IA, pp. 659–673.

de Castro, M.C., Monte-Mór, R.L., Sawyer, D.O., Singer, B.H., 2006. Malaria risk on the Amazon frontier. Proc. Natl. Acad. Sci. USA 103, 2452–2457.

de Jesús-Crespo, R., Ramírez, A., 2011. Effects of urbanization on stream physicochemistry and macroinvertebrate assemblages in a tropical urban watershed in Puerto Rico. J. N. Am. Benthol. Soc. 30, 739–750.

de la Cruz, M., Dirzo, R., 1987. A survey of the standing levels of herbivory in seedlings from a Mexican rain forest. Biotropica 19, 98–106.

de la Giroday, H.-M.C., Carroll, A.L., Aukema, B.H., 2012. Breach of the northern Rocky Mountain geoclimatic barrier: initiation of range expansion by the mountain pine beetle. J. Biogeogr. 39, 1112–1123.

de Mazancourt, C., Loreau, M., Abbadie, L., 1998. Grazing optimization and nutrient cycling: when do herbivores enhance plant production? Ecology 79, 2242–2252.

de Mazancourt, C., Johnson, E., Barradough, T.G., 2008. Biodiversity inhibits species' evolutionary responses to changing environments. Ecol. Lett. 11, 380–388.

de Oliveira, C.F., Long, E.Y., Finke, D.L., 2014. A negative effect of a pathogen on its vector? A plant pathogen increases the vulnerability of its vector to attack by natural enemies. Oecologia 174, 1169–1177.

de Oliveira, E.F., Pallini, A., Janssen, A., 2016. Herbivores with similar feeding modes interact through the induction of different plant responses. Oecologia 180, 1–10.

de Ruiter, P.C., Neutel, A.M., Moore, J.C., 1995. Energetics, patterns of interaction strengths, and stability in real ecosystems. Science 269, 1257–1260.

de Souza-Stevaux, M.C., Negrelle, R.R.B., Citadini-Zanette, V., 1994. Seed dispersal by the fish *Pterodoras granulosus* in the Paraná River Basin, Brazil. J. Trop. Ecol. 10, 621–626.

de Vries, F.T., Thébault, E., Liiri, M., Birkhofer, K., Tsiafouli, M.A., Bjørnlund, L., Jørgensen, H.B., Brady, M.V., Christensen, S., de Ruiter, P.C., d'Hertefeldt, T., Frouz, J., Hedlund, K., Hemerik, L., Hol, W.H.G., Hotes, S., Mortimer, S.R., Setälä, H., Sgardelis, S.P., Uteseny, K., van der Putten, W.H., Wolters, V., Bardgett, R.D., 2013. Soil food web properties explain ecosystem services across European land use systems. Proc. Natl. Acad. Sci. USA 110, 14296–14301.

Dean, A.M., 1983. A simple model of mutualism. Am. Nat. 121, 409–417.

DeBano, S.J., 2006. Effects of livestock grazing on aboveground insect communities in semi-arid grasslands of southeastern Arizona. Biodivers. Conserv. 15, 2547–2564.

Dedej, S., Delaplane, K.S., 2004. Nectar-robbing carpenter bees reduce seed-setting capability of honey bees (Hymenoptera: Apidae) in rabbiteye blueberry, *Vaccinium ashei*, 'Climax'. Environ. Entomol. 33, 100–106.

Deevy, E.S., 1947. Life tables for natural populations of animals. Q. Rev. Biol. 22, 283–314.

DeFoliart, G.R., 1999. Insects as food: why the Western attitude is important. Annu. Rev. Entomol. 44, 21–50.

Degen, B., Roubik, D.W., 2004. Effects of animal pollination on pollen dispersal, selfing, and effective population size of tropical trees: a simulation study. Biotropica 36, 165–179.

Delcomyn, F., 2004. Insect walking and robotics. Annu. Rev. Entomol. 49, 51–70.

Delfosse, E.S., 2005. Risk and ethics in biological control. Biol. Control 35, 319–329.

Delphin, F., 1965. The histology and possible functions of neurosecretory cells in the ventral ganglia of *Schistocerca gregaria* Forskal (Orthoptera: Acrididae). Trans. Roy. Ent. Soc. London 117, 167–214.

DeMers, M.N., 1993. Roadside ditches as corridors for range expansion of the western harvester ant (*Pogonomyrmex occidentalis* Cresson). Landscape Ecol. 8, 93–102.

Demir, E., Dickson, B.J., 2005. *Fruitless* splicing specifies male courtship behavior in *Drosophila*. Cell 121, 785–794.

Denlinger, D.L., 2002. Regulation of diapause. Annu. Rev. Entomol. 47, 93–122.

Dennis, B., Desharnais, R.A., Cushing, J.M., Henson, S.M., Costantino, R.F., 2001. Estimating chaos and complex dynamics in an insect population. Ecol. Monogr. 71, 277–303.

Denno, R.F., McClure, M.S., Ott, J.R., 1995. Interspecific interactions in phytophagous insects: competition reexamined and resurrected. Annu. Rev. Entomol. 40, 297–331.

Denno, R.F., Gratton, C., Peterson, M.A., Langellotto, G.A., Finke, D.L., Huberty, A.F., 2002. Bottom-up forces mediate natural-enemy impact in a phytophagous insect community. Ecology 83, 1443–1458.

Denno, R.F., Gratton, C., Döbel, H., Finke, D.L., 2003. Predation risk affects relative strength of top-down and bottom-up impacts on insect herbivores. Ecology 84, 1032–1044.

Denno, R.F., Mitter, M.S., Langellotto, G.A., Gratton, C., Finke, D.L., 2004. Interactions between a hunting spider and a web-builder: consequences of intraguild predation and cannibalism for prey suppression. Ecol. Entomol. 29, 566–577.

Denslow, J.S., 1985. Disturbance-mediated coexistence of species. In: Pickett, S.T.A., White, P.S. (Eds.), Ecology of Natural Disturbance and Patch Dynamics. Academic Press, Orlando, FL, pp. 307–323.

Denys, C., Tscharntke, T., 2002. Plant-insect communities and predator-prey ratios in field margin strips, adjacent crop fields, and fallows. Oecologia 130, 315–324.

DeRouen, S.M., Foil, L.D., MacKay, A.J., Fanke, D.E., Sanson, D.W., Wyatt, W.E., 2003. Effect of horn fly (*Haematobia irritans*) control on growth and reproduction of beef heifers. J. Econ. Entomol. 96, 1612–1616.

Detling, J.K., 1987. Grass response to herbivory. In: Capinera, J.L. (Ed.), Integrated Pest Management on Rangeland: a Shortgrass Prairie Perspective. Westview Press, Boulder, CO, pp. 56–68.

Detling, J.K., 1988. Grasslands and savannas: regulation of energy flow and nutrient cycling by herbivores. In: Pomeroy, L.R., Alberts, J.A. (Eds.), Ecosystems: Analysis and Synthesis. Springer-Verlag, New York, pp. 131–148.

Detling, J.K., Dyer, M.I., Procter-Gregg, C., Winn, D.T., 1980. Plant-herbivore interactions: examination of potential effects of bison saliva on regrowth of *Bouteloua gracilis* (H.B.K.) Lag. Oecologia 45, 26–31.

Deutsch, C.A., Tewksbury, J.J., Huey, R.B., Sheldon, K.S., Ghalambor, C.K., Haak, D.C., Martin, P.R., 2008. Impacts of climate warming on terrestrial ectotherms across latitude. Proc. Natl. Acad. Sci. USA 105, 6668–6672.

Dévai, G., Moldován, J., 1983. An attempt to trace eutrophication in a shallow lake (Balaton, Hungary) using chironomids. Hydrobiologia 103, 169–175.

DeWalt, S.J., Chave, J., 2004. Structure and biomass of four lowland Neotropical forests. Biotropica 36, 7–19.

Dhanker, R., Kumar, R., Hwang, J.-S., 2013. How effective are *Mesocyclops aspericornis* (Copepoda: Cyclopoida) in controlling mosquito immatures in the environment with an application of phytochemicals? Hydrobiologia 716, 147–162.

Dial, R., Roughgarden, J., 1995. Experimental removal of insectivores from rain forest canopy: direct and indirect effects. Ecology 76, 1821–1834.

Dial, R., Roughgarden, J., 1996. Natural history observations of *Anolisomyia rufianalis* (Diptera: Sarcophagidae) infesting *Anolis* lizards in a rain forest canopy. Environ. Entomol. 25, 1325–1328.

Diamond, J.M., 1990. Biological effects of ghosts. Nature 345, 769–770.

Diamond, J., 1999. Guns, Germs and Steel: the Fates of Human Societies. W.W. Norton, New York, NY.

Diamond, J.M., May, R.M., 1981. Island biogeography and the design of natural reserves. In: May, R.M. (Ed.), Theoretical Ecology: Principles and Applications. Blackwell, Oxford, UK, pp. 228–252.

Diamond, S.E., Cayton, H., Wepprich, T., Jenkins, C.N., Dunn, R.R., Haddad, N.M., Ries, L., 2014. Unexpected phenological responses of butterflies to the interaction of urbanization and geographic temperature. Ecology 95, 2613–2621.

Diaz, J.H., 2005. The evolving global epidemiology of syndromic classification, management, and prevention of caterpillar envenoming. Am. J. Trop. Med. Hyg. 72, 347–357.

Didham, R.K., Ghazoul, J., Stork, N.E., Davis, A.J., 1996. Insects in fragmented forests: a functional approach. Trends Ecol. Evol. 11, 255–260.

Dik, A.J., van Pelt, J.A., 1992. Interaction between phyllosphere yeasts, aphid honeydew and fungicide effectiveness in wheat under field conditions. Plant Pathol. 41, 661–675.

Di Prisco, G., Cavaliere, V., Annoscia, D., Varricchio, P., Caprio, E., Nazzic, F., Gargiulo, G., Pennacchio, F., 2013. Neonicotinoid clothianidin adversely affects insect immunity and promotes replication of a viral pathogen in honey bees. Proc. Natl. Acad. Sci. USA 110, 18466–18471.

Dirzo, R., 1984. Herbivory: a phytocentric overview. In: Dirzo, R., Sarukhán, J. (Eds.), Perspectives on Plant Population Ecology. Sinauer Assoc., Inc, Sunderland, MA, pp. 141–165.

Dively, G.P., Embrey, M.S., Kamel, A., Hawthorne, D.J., Pettis, J.S., 2015. Assessment of chronic sublethal effects of imidacloprid on honey bee colony health. PLoS ONE 10 (3), e0118748.

Dixon, A.F.G., 1985. Aphid Ecology. Blackie & Son Ltd, Glasgow, UK.

Djogbénou, L., Pasteur, N., Akogbéto, M., Weill, M., Chandre, F., 2011. Insecticide resistance in the *Anopheles gambiae* complex in Benin: a nationwide survey. Med. Vet. Entomol. 25, 256–267.

Dobson, A., Cattadori, I., Holt, R.D., Ostfeld, R.S., Keesing, F., Krichbaum, K., Rohr, J.R., Perkins, S.E., Hudson, P.J., 2006. Sacred cows and sympathetic squirrels: the importance of biological diversity to human health. PLoS Med. 3 (6), e231.

Dodds, K.J., Graber, C., Stephen, F.M., 2001. Facultative intraguild predation by larval Cerambycidae (Coleoptera) on bark beetle larvae (Coleoptera: Scolytidae). Environ. Entomol. 30, 17–22.

Dodds, W.K., Collins, S.M., Hamilton, S.K., Tank, J.L., Johnson, S., Webster, J.R., Simon, K.S., Whiles, M.R., Rantala, H.M., McDowell, W.H., Peterson, S.D., Rus, T., Crenshaw, C.L., Thomas, S.A., Kristensen, P.B., Cheever, B.M., Flecker, A.S., Griffiths, N.A., Crowl, T., Rosi-Marshall, E.J., El-Sabaawi, R., Marti, E., 2014. You are not always what we think you eat: selective assimilation across multiple whole-stream isotopic tracer studies. Ecology 95, 2757–2767.

Doi, H., 2009. Spatial patterns of authochthonous and allochthonous resources in aquatic food webs. Popul. Ecol. 51, 57–64.

Dolch, R., Tscharntke, T., 2000. Defoliation of alders (*Alnus glutinosa*) affects herbivory by leaf beetles on undamaged neighbors. Oecologia 125, 504–511.

Domisch, T., Finér, L., Ohashi, M., Risch, A.C., Sundström, L., Niemelä, P., Jurgensen, M.F., 2006. Contribution of red wood ant mounds to forest floor CO_2 efflux in boreal coniferous forests. Soil Biol. Biochem. 38, 2425–2433.

Donald, P.F., Evans, A.D., 2006. Habitat connectivity and matrix restoration: the wider implications of agri-environmental schemes. J. Appl. Ecol. 43, 209–218.

Dong, S.-L., Mao, L., Henderson, G., 2008. Physical contact between soldier and worker is essential in soldier self-regulation of *Coptotermes formosanus* (Isoptera, Rhinotermitidae). Insect. Soc. 56, 28–34.

Donkin, R.A., 1977. An ethnogeographical study of cochineal and the Opuntia cactus. Trans. Am. Philos. Soc. 67, 1–84.

Doube, B., Marshall, T., 2014. Dung Down Under: Dung Beetles for Australia. Dung Beetle Solutions Australia, Bridgewater, SA.

Dowding, 1984. The evolution of insect-fungus relationships in the primary invasion of forest timber. In: Invertebrate Microbial Interactions, Anderson, J.M., Rayner, A.D.M., Walton, D.W.H. (Eds.). British Mycological Society Symposium 6. Cambridge University Press, Cambridge, UK, pp. 135–153.

Downes, J.A., 1970. The feeding and mating behaviour of the specialized Empididae (Diptera); observations on four species of *Rhamphomyia* in the high Arctic and a general discussion. Can. Entomol. 102, 769–791.

Downing, A.L., Leibold, M.A., 2002. Ecosystem consequences of species richness and composition in pond food webs. Nature 416, 837–841.

Downing, K.B., Williams, W.R., 1978. Douglas-fir tussock moth: did it affect private recreational businesses in northeastern Oregon? J. Forest. 76, 29–30.

Doyle, T.W., 1981. The role of disturbance in the gap dynamics of a montane rain forest: an application of a tropical forest succession model. In: West, D.C., Shugart, H.H., Botkin, D.B. (Eds.), Forest Succession: Concepts and Application. Springer-Verlag, New York, pp. 56–73.

Doyle, R.D., Grodowitz, M., Smart, R.M., Owens, C., 2002. Impact of herbivory by *Hydrellia pakistanae* (Diptera: Ephydridae) on growth and photosynthetic potential of *Hydrilla verticillata*. Biol. Control 24, 221–229.

Dreisig, H., 1988. Foraging rate of ants collecting honeydew or extrafloral nectar and some possible constraints. Ecol. Entomol. 13, 143–154.

Drew, A.P., Boley, J.D., Zhao, Y., Johnston, M.H., Wadsworth, F.H., 2009. Sixty-two years of change in subtropical wet forest structure and composition at El Verde, Puerto Rico. Interciencia 34, 34–40.

Dreyer, J., Townsend, P.A., Hook, III, J.C., Hoekman, D., Vander Zanden, M.J., Gratton, C., 2015. Quantifying aquatic insect deposition from lake to land. Ecology 96, 499–509.

Drukker, B., Bruin, J., Sabelis, M.W., 2000. Anthocorid predators learn to associate herbivore-induced plant volatiles with presence or absence of prey. Physiol. Entomol. 25, 260–265.

Drury, W.H., Nisbet, I.C.T., 1973. Succession. J. Arnold Arboretum 54, 331–368.

Duckworth, F.S., Sandberg, J.S., 1954. The effect of cities upon horizontal and vertical temperature gradients. Bull. Am. Meteorol. Soc. 35, 198–207.

Dudley, R., 1998. Atmospheric oxygen, giant Paleozoic insects and the evolution of aerial locomotor performance. J. Exp. Biol. 201, 1043–1050.

Dudt, J.F., Shure, D.J., 1994. The influence of light and nutrients on foliar phenolics and insect herbivory. Ecology 75, 86–98.

Duffy, J.E., 2009. Why biodiversity is important to the functioning of real-world ecosystems. Front. Ecol. Environ. 7, 437–444.

Dufrêne, M., Legendre, P., 1997. Species assemblages and indicator species: the need for a flexible asymmetrical approach. Ecol. Monogr. 67, 345–366.

Dukas, R., 2008. Evolutionary biology of insect learning. Annu. Rev. Entomol. 53, 145–160.

Dukas, R., Real, L.A., 1993. Cognition in bees: from stimulus reception to behavioral change. In: Papaj, D.R., Lewis, A.C. (Eds.), Insect Learning: Ecological and Evolutionaray Perspectives. Chapman & Hall, New York, pp. 343–373.

Duman, J.G., Walters, K.R., Sformo, T., Carrasco, M.A., Nickell, P.K., Lin, X., Barnes, B.M., 2010. Antifreeze and ice-nucleator proteins. In: Denlinger, D.L., Lee, R.E. (Eds.), Low Temperature Biology of Insects. Cambridge University Press, New York, pp. 59–90.

Duncan, D.H., Nicotra, A.B., Wood, J.T., Cunningham, S.A., 2004. Plant isolation reduces outcross pollen receipt in a partially self-compatible herb. J. Ecol. 92, 977–985.

Dungan, R.J., Turnbull, M.H., Kelly, D., 2007. The carbon costs for host trees of a phloem-feeding herbivore. J. Ecol. 95, 603–613.

Dunn, P.E., Bohnert, T.J., Russell, V., 1994. Regulation of antibacterial protein synthesis following infection and during metamorphosis of *Manduca sexta*. Ann. NY Acad. Sci. 712, 117–130.

Dunn, R.R., Agosti, D., Andersen, A.N., Arnan, X., Bruhl, C.A., Cerdá, X., Ellison, A.M., et al. 2009. Climatic drivers of hemispheric asymmetry in global patterns of ant species richness. Ecol. Lett. 12, 324–333.

Dwyer, G., Dushoff, J., Yee, S.H., 2004. The combined effects of pathogens and predators on insect outbreaks. Nature 430, 341–345.

Dyer, L.A., 1995. Tasty generalists and nasty specialists? Antipredator mechanisms in tropical lepidopteran larvae. Ecology 76, 1483–1496.

Dyer, F.C., 2002. The biology of the dance language. Annu. Rev. Entomol. 47, 917–949.

Dyer, L.A., Letourneau, D.K., 1999a. Relative strengths of top-down and bottom-up forces in a tropical forest community. Oecologia 119, 265–274.

Dyer, L.A., Letourneau, D.K., 1999b. Trophic cascades in a complex terrestrial community. Proc. Natl. Acad. Sci. USA 96, 5072–5076.

Dyer, M.I., Acra, M.A., Wang, G.M., Coleman, D.C., Freckman, D.W., McNaughton, S.J., Strain, B.R., 1991. Source-sink carbon relations in two *Panicum coloratum* ecotypes in response to herbivory. Ecology 72, 1472–1483.

Dyer, M.I., Turner, C.L., Seastedt, T.R., 1993. Herbivory and its consequences. Ecol. Appl. 3, 10–16.

Dyer, M.I., Moon, A.M., Brown, M.R., Crossley, Jr., D.A., 1995. Grasshopper crop and midgut extract effects on plants: an example of reward feedback. Proc. Natl. Acad. Sci. USA 92, 5475–5478.

Dyer, L.A., Dodson, G.D., Beihoffer, J., Letourneau, D.K., 2001. Trade-offs in antiherbivore defenses in *Piper cenocladum*: ant mutualists versus plant secondary metabolites. J. Chem. Ecol. 27, 581–592.

Dyer, L.A., Singer, M.S., Lill, J.T., Stireman, J.O., Gentry, G.L., Marquis, R.J., Ricklefs, R.E., Greeney, H.F., Wagner, D.L., Morais, H.C., Diniz, I.R., Kursar, T.A., Coley, P.D., 2007. Host specificity of Lepidoptera in tropical and temperate forests. Nature 448, 696–700.

Dyer, L.A., Walla, T.R., Greeney, H.F., Stireman, III, J.O., Hazen, R.F., 2010. Diversity of interactions: a metric for studies of biodiversity. Biotropica 42, 281–289.

Eberling, H., Olesen, J.M., 1999. The structure of a high latitude plant-flower visitor system: the dominance of flies. Ecography 22, 314–323.

Edelaar, P., Siepielski, A.M., Clobert, J., 2008. Matching habitat choice causes directed gene flow: a neglected dimension in evolution and ecology. Evolution 62, 2462–2472.

Edman, J.D., 2000. Arthropod transmission of vertebrate parasites. In: Eldridge, B.F., Edman, J.D. (Eds.), Medical Entomology. Kluwer Academic Publishers, Dordrecht, The Netherlands, pp. 151–163.

Edmonds, R.L., Eglitis, A., 1989. The role of the Douglas-fir beetle and wood borers in the decomposition of and nutrient release from Douglas-fir logs. Can. J. Forest. Res. 19, 853–859.

Edmunds, Jr., G.F., Alstad, D.N., 1978. Coevolution in insect herbivores and conifers. Science 199, 941–945.

Edmunds, Jr., G.F., Alstad, D.N., 1985. Malathion-induced sex ratio changes in black pineleaf scale (Hemiptera: Diaspididae). Ann. Entomol. Soc. Am. 78, 403–405.

Edney, E.B., 1977. Water Balance in Land Arthropods. Springer-Verlag, Berlin.

Edson, K.M., Vinson, S.B., Stoltz, D.B., Summers, M.D., 1981. Virus in a parasitoid wasp: suppression of the cellular immune response in the parasitoid's host. Science 211, 582–583.

Edwards, E.P., 1982. Hummingbirds feeding on an excretion produced by scale insects. Condor 84, 122.

Edwards, N.T., 1982. The use of soda-lime for measuring respiration rates in terrestrial systems. Pedobiologia 23, 321–330.

Edwards, P.J., 1989. Insect herbivory and plant defence theory. In: Grubb, P.J., Whittaker, J.B. (Eds.), Toward a More Exact Ecology. Blackwell Scientific, Oxford, UK, pp. 275–297.

Edwards, C.A., Heath, G.W., 1963. The role of soil animals in breakdown of leaf material. In: Doeksen, J., van der Drift, J. (Eds.), Soil Organisms. North-Holland, Amsterdam, pp. 76–84.

Edwards, C.A., Lofty, J.R., 1978. The influence of arthropods and earthworms upon root growth of direct drilled cereals. J. Appl. Ecol. 15, 789–795.

Edwards, J.S., Sugg, P., 1990. Arthropod fallout as a resource in the recolonization of Mt. St. Helens. Ecology 74, 954–958.

Effiom, E.O., Birkhofer, K., Smith, H.G., Olsson, O., 2014. Changes of community composition at multiple trophic levels due to hunting in Nigerian tropical forests. Ecography 37, 367–377.

Effler, R.S., Goyer, R.A., Lenhard, G.J., 2006. Baldcypress and water tupelo responses to insect defoliation and nutrient augmentation in Maurepas Swamp, Louisiana, USA. Forest Ecol. Manag. 236, 295–304.

Eggert, S.L., Wallace, J.B., 2003. Reduced detrital resources limit *Pycnopsyche gentilis* (Trichoptera: Limnephilidae) production and growth. J. N. Am. Benthol. Soc. 22, 388–400.

Eggert, S.L., Wallace, J.B., 2007. Wood biofilm as a food resource for stream detritivores. Limnol. Oceanogr. 52, 1239–1245.

Eggert, A.-K., Reinking, M., Müller, J.K., 1998. Parental care improves offspring survival and growth in burying beetles. Anim. Behav. 55, 97–107.

Egler, F.E., 1954. Vegetation science concepts. I. initial floristic composition, a factor in old-field vegetation development. Vegetation 4, 412–417.

Ehigiator, O.A., Anyata, B.U., 2011. Effects of land clearing techniques and tillage systems on runoff and soil erosion in a tropical rain forest in Nigeria. J. Environ. Manage. 92, 2875–2880.

Ehigie, L.O., Okonji, R.E., Ehigie, F.A., 2013. Biochemical properties of thiaminase, a toxic enzyme in the gut of grasshoppers (*Zonocerus variegatus* Linn). Cameroon J. Exp. Biol. 9, 9–16.

Ehrlén, J., 1996. Spatiotemporal variation in predispersal seed predation intensity. Oecologia 108, 708–713.

Eisen, L., Eisen, R.J., 2011. Using geographic information systems and decision support systems for the prediction, prevention, and control of vector-borne diseases. Annu. Rev. Entomol. 56, 41–61.

Eisen, L., Bolling, B.G., Blair, C.D., Beaty, B.J., Moore, C.G., 2008. Mosquito species richness, composition, and abundance along habitat-climate-elevation gradients in the northern Colorado Front Range. J. Med. Entomol. 45, 800–811.

Eisner, T., Silberglied, R.E., Aneshansley, D., Carrel, J.E., Howland, H.C., 1969. Ultraviolet video-viewing: the television camera as an insect eye. Science 166, 1172–1174.

Elderd, B.D., 2006. Disturbance-mediated trophic interactions and plant performance. Oecologia 147, 261–271.

Elderd, B.D., Doak, D.F., 2006. Comparing the direct and community-mediated effects of disturbance on plant population dynamics: flooding, herbivory and *Mimulus guttatus*. J. Ecol. 94, 656–669.

Elderd, B.D., Dushoff, J., Dwyer, G., 2008. Host-pathogen interactions, insect outbreaks, and natural selection for disease resistance. Am. Nat. 172, 829–842.

Eldridge, D.J., 1993. Effect of ants on sandy soils in semi-arid eastern Australia: local distribution of nest entrances and their effect on infiltration of water. Aust. J. Soil Res. 31, 509–518.

Eldridge, D.J., 1994. Nests of ants and termites influence infiltration in a semi-arid woodland. Pedobiologia 38, 481–492.

Eldridge, D.J., Whitford, W.G., Duval, B.D., 2009. Animal disturbances promote shrub maintenance in a desertified grassland. J. Ecol. 97, 1302–1310.

Elkins, N.Z., Sabol, G.V., Ward, T.J., Whitford, W.G., 1986. The influence of subterranean termites on the hydrological characteristics of a Chihuahuan Desert ecosystem. Oecologia 68, 521–528.

Elkinton, J.S., Liebhold, A.M., 1990. Population dynamics of gypsy moth in North America. Annu. Rev. Entomol. 35, 571–596.

Elkinton, J.S., Schal, C., Ono, T., Cardé, R.T., 1987. Pheromone puff trajectory and upwind flight of male gypsy moths in a forest. Physiol. Entomol. 12, 399–406.

Elliot, S.L., Blanford, S., Thomas, M.B., 2002. Host–pathogen interactions in a varying environment: temperature, behavioural fever and fitness. Proc. Roy. Soc. B-Biol. Sci. 269, 1599–1607.

Elliot, S.L., Blanford, S., Horton, C.M., Thomas, M.B., 2003. Fever and phenotype: transgenerational effect of disease on desert locust phase state. Ecol. Lett. 6, 830–836.

Elliot, S.L., Horton, C.M., Blanford, S., Thomas, M.B., 2005. Impacts of fever on locust life-history traits: costs or benefits? Biol. Lett. UK 1, 181–184.

Elliott, E.T., Coleman, D.C., Ingham, R.E., Trofymow, J.A., 1984. Carbon and energy flow through microflora and microfauna in the soil subsystem of terrestrial ecosystems. In: Klug, M.J., Reddy, C.A. (Eds.), Current Perspectives in Microbial Ecology. American Society for Microbiology, Washington, DC, pp. 424–433.

Elliott, N.C., Simmons, G.A., Sapio, F.J., 1987. Honeydew and wildflowers as food for the parasites *Glypta fumiferanae* (Hymenoptera: Ichneumonidae) and *Apanteles fumiferanae* (Hymenoptera: Braconidae). J. Kansas Entomol. Soc. 60, 25–29.

Elser, J.J., Urabe, J., 1999. The stoichiometry of consumer-driven nutrient recycling: theory, observations, and consequences. Ecology 80, 735–751.

Elser, J.J., Dobberfuhl, D.R., MacKay, N.A., Schampel, J.H., 1996. Organism size, life history, and N:P stoichiometry: toward a unified view of cellular and ecosystem processes. BioScience 46, 674–684.

Elton, C., 1939. Animal Ecology. Macmillan, New York, pp. 209.

Elvidge, C.D., Milesi, C., Dietz, J.B., Tuttle, B.T., Sutton, P.C., Nemani, R., Vogelmann, J.E., 2004. US constructed area approaches the size of Ohio. EOS, Transactions, American Geophysical Union 85, 233.

Endo, W., Peres, C.A., Salas, E., Mori, S., Sanchez-Vega, J.-L., Shepard, G.H., Pacheco, V., Yu, D.W., 2010. Game vertebrate densities in hunted and nonhunted forest sites in Manu National Park. Peru. Biotropica 42, 251–261.

Engelberg, J., Boyarsky, L.L., 1979. The noncybernetic nature of ecosystems. Am. Nat. 114, 317–324.

Engelmann, M.D., 1961. The role of soil arthropods in the energetics of an old field community. Ecol. Monogr. 31, 221–238.

Engelmann, M.D., 1968. The role of soil arthropods in community energetics. Am. Zool. 8, 61–69.

Epstein, W.L., Kligman, A.M., 1958. Treatment of warts with cantharidin. Am. Med. Assoc. Arch. Dermatol. 77, 508–511.

Erbilgin, N., Dahlsten, D.L., Chen, P.-Y., 2004. Intraguild interactions between generalist predators and an introduced parasitoid of *Glycaspis brimblecombei* (Homoptera: Psylloidea). Biol. Control 31, 329–337.

Ernsting, G., van der Werf, D.C., 1988. Hunger, partial consumption of prey and prey size preference in a carabid beetle. Ecol. Entomol. 13, 155–164.

Erwin, T.L., 1995. Measuring arthropod diversity in the tropical forest canopy. In: Lowman, M.D., Nadkarni, N.M. (Eds.), Forest Canopies. Academic Press, San Diego, CA, pp. 109–127.

Eshleman, K.N., Morgan, II., R.P., Webb, J.R., Deviney, F.A., Galloway, J.N., 1998. Temporal patterns of nitrogen leakage from mid-Appalachian forested watersheds: role of insect defoliation. Water. Resour. Res. 34, 2005–2116.

Esper, J., Büntgen, U., Frank, D.C., Nievergelt, D., Liebhold, A., 2007. 1200 years of regular outbreaks in alpine insects. Proc. Roy. Soc. B-Biol. Sci. 274, 671–679.

Estay, S.A., Lima, M., Bozinovic, F., 2014. The role of temperature variability on insect performance and population dynamics in a warming world. Oikos 123, 131–140.

Euw, J.V., Fishelson, L., Parsons, J.A., Reichstein, T., Rothschild, M., 1967. Cardenolides (heart poisons) in a grasshopper feeding on milkweeds. Nature 214, 35–39.

Evans, W.G., 1966. Perception of infrared radiation from forest fires by *Melanophila acuminata* de Geer (Buprestidae, Coleoptera). Ecology 47, 1061–1065.

Evans, E.W., 1988. Community dynamics of prairie grasshoppers subjected to periodic fire: predictable trajectories or random walks in time? Oikos 52, 283–292.

Evans, E.W., 2004. Habitat displacement of North American ladybirds by an introduced species. Ecology 85, 637–647.

Ewel, J.J., 1986. Designing agricultural ecosystems for the humid tropics. Annu. Rev. Ecol. Syst. 17, 245–271.

Ewel, J.J., Putz, F.E., 2004. A place for alien species in ecosystem management. Front. Ecol. Environ. 2, 354–360.

Ewel, J.J., Mazzarino, M.J., Berish, C.W., 1991. Tropical soil fertility changes under monocultures and successional communities of different structure. Ecol. Appl. 1, 289–302.

Ewers, R.M., Didham, R.K., 2006. Continuous response functions for quantifying the strength of edge effects. J. Appl. Ecol. 43, 527–536.

Faeth, S.H., Connor, E.F., Simberloff, D., 1981. Early leaf abscission: a neglected source of mortality for folivores. Am. Nat. 117, 409–415.

Fagan, W.F., 1997. Omnivory as a stabilizing feature of natural communities. Am. Nat. 150, 554–567.

Fagan, W.F., Grant, E.H.C., Lynch, H.J., Unmack, P.J., 2010. Riverine landscapes: ecology for alternative geometry. In: Cantrell, S., Cosner, C., Ruan, S. (Eds.), Spatial Ecology. Chapman and Hall/CRC Press/Taylor & Francis Group, Boca Raton, FL, pp. 85–100.

Fagundes, M., Neves, F.S., Fernandes, G.W., 2005. Direct and indirect interactions involving ants, insect herbivores, parasitoids, and the host plant *Baccharis dracunculifolia* (Asteraceae). Ecol. Entomol. 30, 28–35.

Fajer, E.D., Bowers, M.D., Bazzaz, F.A., 1989. The effects of enriched carbon dioxide atmospheres on plant-insect herbivore interactions. Science 243, 1198–1200.

Falkowska, L., Reindl, A.R., Szumiło, E., Kwaśniak, J., Staniszewska, M., Bełdowska, M., Lewandowska, A., Krause, I., 2013. Mercury and chlorinated pesticides on the highest level of the food web as exemplified by herring from the southern Baltic and African penguins from the zoo. Water Air Soil Poll. 224, 1549.

Fang, J., 2010. A world without mosquitoes. Nature 466, 432–434.

Fares, Y., Sharpe, P.J.H., Magnusen, C.E., 1980. Pheromone dispersion in forests. J. Theor. Biol. 84, 335–359.

Fargo, W.S., Wagner, T.L., Coulson, R.N., Cover, J.D., McAudle, T., Schowalter, T.D., 1982. Probability functions for components of the *Dendroctonus frontalis*-host tree population system and their potential use with population models. Res. Popul. Ecol. 24, 123–131.

Farji-Brener, A.G., Chinchilla, F., Umaña, M.N., Ocasio-Torres, M.E., Chauta-Mellizo, A., Acosta-Rojas, D., Marinaro, S., De Torres Curth, M., Amador-Vargas, S., 2015. Branching angles reflect a trade-off between reducing trail maintenance costs or travel distances in leaf-cutting ants. Ecology 96, 510–517.

Farmer, E.E., Ryan, C.A., 1990. Interplant communication: airborne methyl jasmonate induces synthesis of proteinase inhibitors in plant leaves. Proc. Natl. Acad. Sci. USA 87, 7713–7716.

Farnier, K., Dyer, A.G., Steinbauer, M.J., 2014. Related but not alike: not all Hemiptera are attracted to yellow. Front. Ecol. Evol. 2, 67.

Farnsworth, N.R., 1966. Biological and phytochemical screening of plants. J. Pharm. Sci. 55, 225–276.

Farwig, N., Böhning-Gaese, K., Bleher, B., 2006. Enhanced seed dispersal of *Prunus africana* in fragmented and disturbed forests? Oecologia 147, 238–252.

Farwig, N., Brandl, R., Siemann, S., Wiener, F., Müller, J., 2014. Decomposition rate of carrion is dependent on composition not abundance of the assemblages of insect scavengers. Oecologia 175, 1291–1300.

Fastie, C.L., 1995. Causes and ecosystem consequences of multiple pathways of primary succession at Glacier Bay, Alaska. Ecology 76, 1899–1916.

Feeley, K.J., Terborgh, J.W., 2005. The effects of herbivore density on soil nutrients and tree growth in tropical forest fragments. Ecology 86, 116–124.

Feener, Jr., D.H., 1981. Competition between ant species: outcome controlled by parasitic flies. Science 214, 815–817.

Feener, Jr., D.H., Brown, B.V., 1992. Reduced foraging of *Solenopsis geminata* (Hymenoptera: Formicidae) in the presence of parasitic *Pseudacteon* spp. (Diptera: Phoridae). Ann. Entomol. Soc. Am. 85, 80–84.

Feeny, P.P., 1969. Inhibitory effect of oak leaf tannins on the hydrolysis of proteins by trypsin. Phytochemistry 8, 2119–2126.

Feeny, P.P., 1970. Seasonal changes in oak leaf tannins and nutrients as a cause of spring feeding by winter moth caterpillars. Ecology 51, 565–581.

Fei, H., Henderson, G., 1998. Effect of moisture and two nitrogen sources on nest site choice by alates and dealates of *Coptotermes formosanus* in the laboratory (Isoptera: Rhinotermitidae). Sociobiology 34, 581–589.

Feinsinger, P., 1983. Coevolution and pollination. In: Futuyma, D.J., Slatkin, M. (Eds.), Coevolution. Sinauer Associates Inc, Sunderland, MA, pp. 282–310.

Felland, C.M., Pitre, H.N., Luttrell, R.G., Hamer, J.L., 1990. Resistance to pyrethroid insecticides in soybean looper (Lepidoptera: Noctuidae) in Mississippi. J. Econ. Entomol. 83, 35–40.

Fellowes, M.D.E., Kraaijeveld, A.R., Godfray, H.C.J., 1999. Cross-resistance following artificial selection for increased defense against parasitoids in *Drosophila melanogaster*. Evolution 53, 966–972.

Felton, G.W., Eichenseer, H., 1999. Herbivore saliva and its effect on plant defense against herbivores and pathogens. In: Agrawal, A.A., Tuzun, S., Bent, E. (Eds.), Induced Plant Defenses Against Pathogens and Herbivores. American Phytopathological Society, St. Paul, MN, pp. 19–36.

Fernandez, D.S., Fetcher, N., 1991. Changes in light availability following Hurricane Hugo in a subtropical montane forest in Puerto Rico. Biotropica 23, 393–399.

Fernández-Arhex, V., Buteler, M., Armadio, M.E., Enriquez, E., Pietrantuono, A.L., Stadler, T., Becker, G., Bruzzone, O., 2013. The effects of volcanic ash from Puyehue-Caulle Range eruption on the survival of *Dichroplus vittigerum* (Orthoptera: Acrididae). Fla. Entomol. 19, 286–288.

Ferrar, P., 1975. Disintegration of dung pads in north Queensland before the introduction of exotic dung beetles. Aust. J. Exp. Agric. Animal Husbandry 15, 325–329.

Ferraz, G., Nichols, J.D., Hines, J.E., Stouffer, P.C., 2007. A large-scale deforestation experiment: effects of patch area and isolation on Amazon birds. Science 315, 238–241.

Ferrenberg, S., Mitton, J.B., 2014. Smooth bark surfaces can defend trees against insect attack: resurrecting a 'slippery' hypothesis. Funct. Ecol. 28, 837–845.

Ferrenberg, S., Kane, J.M., Mitton, J.B., 2014. Resin duct characteristics associated with tree resistance to bark beetles across lodgepole and limber pines. Oecologia 174, 1283–1292.

Ferro, M.L., Parys, K.A., Gimmel, M.L., 2010. Dragonflies and Damselfies of Louisiana. Louisiana State Arthropod Museum, Baton Rouge, LA.

Fetcher, N., Agosta, S.J., Moore, J.C., Stratford, J.A., Steele, M.A., 2015. The food web of a severely contaminated site following reclamation with warm season grasses. Restor. Ecol. 23, 421–429.

Fettig, D.J., McKelvey, S.R., Cluck, D.R., Smith, S.L., Otrosina, W.J., 2010. Effects of prescribed fire and season of burn on direct and indirect levels of tree mortality in Ponderosa and Jeffrey Pine Forests in California, USA. Forest Ecol. Manag. 260, 207–218.

Fewell, J.H., Harrison, J.F., Lighton, J.R.B., Breed, M.D., 1996. Foraging energetics of the ant, *Paraponera clavata*. Oecologia 105, 419–427.

Feyereisen, R., 1999. Insect P450 enzymes. Annu. Rev. Entomol. 44, 507–533.

Ffrench-Constant, R.H., 2013. The molecular genetics of insecticide resistance. Genetics 184, 807–815.

Fielden, L.J., Duncan, F.D., Rechav, Y., Crewe, R.M., 1994. Respiratory gas exchange in the tick *Amblyomma hebraeum* (Acari: Ixodidae). J. Med. Entomol. 31, 30–35.

Fielding, D.J., Brusven, M.A., 1993. Grasshopper (Orthoptera: Acrididae) community composition and ecological disturbances on southern Idaho rangeland. Environ. Entomol. 22, 71–81.

Fielding, D.J., Brusven, M.A., 1995. Ecological correlates between rangeland grasshopper (Orthoptera: Acrididae) and plant communities of southern Idaho. Environ. Entomol. 24, 1432–1441.

Filip, V., Dirzo, R., Maass, J.M., Sarukhán, J., 1995. Within- and among-year variation in the levels of herbivory on the foliage of trees from a Mexican tropical deciduous forest. Biotropica 27, 78–86.

Finch, V.C., Trewartha, G.T., 1949. Elements of Geography: Physical and Cultural. McGraw-Hill, New York, pp. 711.

Fine, P.V.A., Mesones, I., Coley, P.D., 2004. Herbivores promote habitat specialization by trees in Amazonian forests. Science 305, 663–665.

Finke, D.L., 2012. Contrasting the consumptive and non-consumptive cascading effects of natural enemies on vector-borne pathogens. Entomol. Exp. Appl. 144, 45–55.

Finke, D.L., Denno, R.F., 2002. Intraguild predation diminished in complex-structured vegetation: implications for prey suppression. Ecology 83, 643–652.

Finke, D.L., Denno, R.F., 2006. Spatial refuge from intraguild predation: implications for prey suppression and trophic cascades. Oecologia 149, 265–275.

Finkel, M., 2012. Tibet's golden worm. Natl. Geogr. 214 (8), .

Fischer, R.C., Richter, A., Wanek, W., Mayer, V., 2002. Plants feed ants: food bodies for myrmecophytic *Piper* and their significance for the interaction with *Pheidole bicornis* ants. Oecologia 133, 186–192.

Fischer, R.C., Wanek, W., Richter, A., Mayer, V., 2003. Do ants feed plants? A ^{15}N labelling study of nitrogen fluxes from ants to plants in the mutualism of *Pheidole* and *Piper*. J. Ecol. 91, 126–134.

Fischer, D.G., Hart, S.C., Rehill, B.J., Lindroth, R.L., Keim, P., Whitham, T.G., 2006. Do high-tannin leaves require more roots? Oecologia 149, 668–675.

Fishman, L., Willis, J.H., 2008. Pollen limitation and natural selection on floral characters in the yellow monkeyflower, *Mimulus guttatus*. New Phytol. 177, 802–810.

Fisher, S.G., 1986. Structure and dynamics of desert streams. In: Whitford, W.G. (Ed.), Pattern and Process in Desert Ecosystems. University of New Mexico Press, Albuquerque, pp. 119–139.

Fitzgerald, T.D., 1995. The Tent Caterpillars. Cornell University Press, Ithaca, NY, pp. 303.

Fitzpatrick, G., Lanan, M.C., Bronstein, J.L., 2014. Thermal tolerance affects mutualist attendance in an ant–plant protection mutualism. Oecologia 176, 129–138.

Flaherty, L., Quiring, D., 2008. Plant module size and dose of gall induction stimulus influence gall induction and galler performance. Oikos 117, 1601–1608.

Flamm, R.O., Pulley, P.E., Coulson, R.N., 1993. Colonization of disturbed trees by the southern pine beetle guild (Coleoptera: Scolytidae). Environ. Entomol. 22, 62–70.

Fleishman, E., Ray, C., Sjögren-Gulve, P., Boggs, C.L., Murphy, D.D., 2002. Assessing the roles of patch quality, area, and isolation in predicting metapopulation dynamics. Conserv. Biol. 16, 706–716.

Flinn, P.W., Hagstrum, D.W., Muir, W.E., Sudayappa, K., 1992. Spatial model for simulating changes in temperature and insect population dynamics in stored grain. Environ. Entomol. 21, 1351–1356.

Florence, L.Z., Johnson, P.C., Coster, J.E., 1982. Behavioral and genetic diversity during dispersal: analysis of a polymorphic esterase locus in southern pine beetle, *Dendroctonus frontalis*. Environ. Entomol. 11, 1014–1018.

Fogel, R., Cromack, Jr., K., 1977. Effect of habitat and substrate quality on Douglas-fir litter decomposition in western Oregon. Can. J. Bot. 55, 1632–1640.

Foley, P., 1997. Extinction models for local populations. In: Hanski, I.A., Gilpin, M.E. (Eds.), Metapopulation Biology: Ecology, Genetics, and Evolution. Academic Press, San Diego, CA, pp. 215–246.

Foley, J.A., Kutzbach, J.E., Coe, M.T., Levis, S., 1994. Feedbacks between climate and boreal forests during the Holocene epoch. Nature 371, 52–54.

Foley, J.A., Costa, M.H., Delire, C., Ramankutty, N., Snyder, P., 2003a. Green surprise? How terrestrial ecosystems could affect earth's climate. Front. Ecol. Environ. 1, 38–44.

Foley, J.A., Coe, M.T., Scheffer, M., Wang, G., 2003b. Regime shifts in the Sahara and Sahel: interactions between ecological and climatic systems in northern Africa. Ecosystems 6, 524–539.

Foley, J.A., DeFries, R., Asner, G.P., Barford, C., Bonan, G., Carpenter, S.R., Chapin, F.S., Coe, M.T., Daily, G.C., Gibbs, H.K., Helkowski, J.H., Holloway, T., Howard, E.A., Kucharik, C.J., Monfreda, C., Patz, J.A., Prentice, I.C., Ramankutty, N., Snyder, P.K., 2005. Global consequences of land use. Science 309, 570–574.

Folgarait, P.J., Gilbert, L.E., 1999. Phorid parasitoids affect foraging activity of *Solenopsis richteri* under different availability of food in Argentina. Ecol. Entomol. 24, 163–173.

Folz, H.C., Wilder, S.M., Parsons, M.H., Rypstra, A.L., 2006. Effects of predation risk on vertical habitat use and foraging of *Pardosa milvina*. Ethology 112, 1152–1158.

Fonseca, C.R., 1994. Herbivory and the long-lived leaves of an Amazonian ant-tree. J. Ecol. 82, 833–842.

Fonseca, D.M., Hart, D.D., 2001. Colonization history masks habitat preferences in local distributions of stream insects. Ecology 82, 2897–2910.

Fonseca, C.R., Fleck, T., Fernandes, G.W., 2006. Processes driving ontogenetic succession of galls in a canopy tree. Biotropica 38, 514–521.

Fonte, S.J., Schowalter, T.D., 2004. Decomposition of greenfall vs. senescent foliage in a tropical forest ecosystem in Puerto Rico. Biotropica 36, 474–482.

Fonte, S.J., Schowalter, T.D., 2005. The influence of a neotropical herbivore (*Lamponius portoricensis*) on nutrient cycling and soil processes. Oecologia 146, 423–431.

Forbush, E.H., Fernald, C.H., 1896. The Gypsy Moth. Massachusetts Board of Agriculture, Boston, MA.

Ford, C.R., Vose, J.M., 2007. *Tsuga canadensis* (L.) Carr. mortality will impact hydrologic processes in southern Appalachian forest ecosystems. Ecol. Appl. 17, 1156–1167.

Fore, T., 1970. Joe Cox and the beetle bug. Oregonizer 93, 1–3.

Forkner, R.E., Marquis, R.J., Lill, J.T., 2004. Feeny revisited: condensed tannins as anti-herbivore defences in leaf-chewing herbivore communities of *Quercus*. Ecol. Entomol. 29, 174–187.

Fornoff, F., Gross, E.M., 2013. Induced defense mechanisms in an aquatic angiosperm to insect herbivory. Oecologia 175, 173–185.

Forschler, B.T., Henderson, G., 1995. Subterranenan termite behavioral reaction to water and survival of inundation: implications for field populations. Environ. Entomol. 24, 1592–1597.

Forsman, A., Karlsson, M., Wennersten, L., Johansson, J., Karpestam, E., 2011. Rapid evolution of fire melanism in replicated populations of pygmy grasshoppers. Evolution 65, 2530–2540.

Forup, M.L., Henson, K.S.E., Craze, P.G., Memmott, J., 2008. The restoration of ecological networks: plant-pollinator networks on ancient and restored heathlands. J. Appl. Ecol. 45, 742–752.

Foster, W.J., Twitchett, R.J., 2014. Functional diversity of marine ecosystems after the Late Permian mass extinction event. Nat. Geosci.doi: 10.1038/NGEO2079.

Fowler, S.V., Lawton, J.H., 1985. Rapidly induced defenses and talking trees: the devil's advocate position. Am. Nat. 126, 181–195.

Fox, L.R., 1975a. Cannibalism in natural populations. Annu. Rev. Ecol. Syst. 6, 87–106.

Fox, L.R., 1975b. Some demographic consequences of food shortage for the predator, *Notonecta hoffmanni*. Ecology 56, 868–880.

Fox, R., 2013. The decline of moths in Great Britain: a review of possible causes. Insect Conserv. Diver. 6, 5–19.

Fox, L.R., Macauley, B.J., 1977. Insect grazing on Eucalyptus in response to variation in leaf tannins and nitrogen. Oecologia 29, 145–162.

Fox, L.R., Morrow, P.A., 1981. Specialization: species property or local phenomenon? Science 211, 887–893.

Fox, L.R., Morrow, P.A., 1983. Estimates of damage by herbivorous insects on *Eucalyptus* trees. Aust. J. Ecol. 8, 139–147.

Fox, L.R., Morrow, P.A., 1992. Eucalypt responses to fertilization and reduced herbivory. Oecologia 89, 214–222.

Fox, L.R., Ribeiro, S.P., Brown, V.K., Masters, G.J., Clarke, I.P., 1999. Direct and indirect effects of climate change on St. John's wort, *Hypericum perforatum* L. (Hypericaceae). Oecologia 120, 113–122.

Fox-Dobbs, K., Doak, D.F., Brody, A.K., Palmer, T.M., 2010. Termites create spatial structure and govern ecosystem function by affecting N_2 fixation in an east African savanna. Ecology 91, 1296–1307.

Frady, C., Johnson, S., Li, J., 2007. Stream macroinvertebrate community responses as legacies of forest harvest at the H.J. Andrews Experimental Forest, Oregon. Forest Sci. 53, 281–293.

Fraenkel, G.S., 1953. The nutritional value of green plants for insects. In: Transactions of the Ninth International Congress of Entomology. W. Junk, The Hague, The Netherlands, pp. 90–100.

Fraenkel, G., Blewett, M., 1946. Linoleic acid, vitamin E and other fat-soluble substances in the diet of certain insects, *Ephestria kuehniella*, *E. elutella*, *E. cautella*, and *Plodia interpunctella* (Lepidoptera). J. Exp. Biol. 22, 172–190.

Frank, S.A., 1995. Mutual policing and repression of competition in the evolution of cooperative groups. Nature 377, 520–522.

Frank, D.A., McNaughton, S.J., 1991. Stability increases with diversity in plant communities: empirical evidence from the 1988 Yellowstone drought. Oikos 62, 360–362.

Frankie, G.W., Ehler, L.E., 1978. Ecology of insects in urban environments. Annu. Rev. Entomol. 23, 367–387.

Franklin, J., 2010. Mapping Species Distributions: Spatial Inference and Prediction. Cambridge University Press, Cambridge, UK.

Franklin, J.F., Forman, R.T.T., 1987. Creating landscape patterns by forest cutting: ecological consequences and principles. Landscape Ecol. 1, 5–18.

Franklin, J.F., Swanson, F.J., Harmon, M.E., Perry, D.A., Spies, T.A., Dale, V.H., McKee, A., Ferrell, W.K., Means, J.E., Gregory, S.V., Lattin, J.D., Schowalter, T.D., Larsen, D., 1992. Effects of global climatic change on forests in northwestern North America. In: Peters, R.L., Lovejoy, T.E. (Eds.), Global Warming and Biological Diversity. Yale University Press, New Haven, CT, pp. 244–257.

Frantsevich, L., 2004. Righting kinematics in beetles (Insecta: Coleoptera). Arthropod Struct. Dev. 33, 221–235.

French, J.R.J., Roeper, R.A., 1972. Interactions of the ambrosia beetle, *Xyleborus dispar*, with its symbiotic fungus *Ambrosiella hartigii* (Fungi Imperfecti). Can. Entomol. 104, 1635–1641.

Freymann, B.P., Buitenwerf, R., Desouza, O., Olff, H., 2008. The importance of termites (Isoptera) for the recycling of herbivore dung in tropical ecosystems: a review. Eur. J. Entomol. 105, 165–173.

Friberg, M., Posledovich, D., Wiklund, C., 2015. Decoupling of female host plant preference and offspring performance in relative specialist and generalist butterflies. Oecologia 178, 1181–1192.

Fried, J.H., Levey, D.J., Hogsette, J.A., 2005. Habitat corridors function as both drift fences and movement conduits for dispersing flies. Oecologia 143, 645–651.

Fritz, R.S., 1983. Ant protection of a host plant's defoliator: consequences of an ant-membracid mutualism. Ecology 64, 789–797.

Frost, C.J., Hunter, M.D., 2004. Insect canopy herbivory and frass deposition affect soil nutrient dynamics and export in oak meso-cosms. Ecology 85, 3335–3347.

Frost, C.J., Hunter, M.D., 2007. Recycling of nitrogen in herbivore feces: plant recovery, herbivore assimilation, soil retention, and leaching losses. Oecologia 151, 42–53.

Frost, C.J., Hunter, M.D., 2008a. Herbivore-induced shifts in carbon and nitrogen allocation in red oak seedlings. New Phytol. 178, 835–845.

Frost, C.J., Hunter, M.D., 2008b. Insect herbivores and their frass affect *Quercus rubra* leaf quality and initial stages of subsequent decomposition. Oikos 117, 13–22.

Frouz, J., Kindlmann, P., 2006. The role of sink to source re-colonisation in the population dynamics of insects living in unstable habitats: an example of terrestrial chironomids. Oikos 93, 50–58.

Frouz, J., Lobinske, R.J., Ali, A., 2004. Influence of Chironomidae (Diptera) faecal pellet accumulation on lake sediment quality and larval abundance of pestiferous midge *Glyptotendipes paripes*. Hydrobiologia 518, 169–177.

Fry, S.N., Wehner, R., 2002. Honey bees store landmarks in an egocentric frame of reference. J. Comp. Physiol. A 187, 1009–1016.

Fu, X.-H., Ohba, N., Wang, Y.-Y., Lei, C.-L., 2005. Bioluminescent courtship behavior of aquatic firefly *Luciola substriata* (Gorh.) (Coleoptera: Lampyridae) in China. Acta Entomol. Sinica 48, 227–231, (in Chinese with English summary).

Fukami, T., Naeem, S., Wardle, D.A., 2001. On similarity among local communities in biodiversity experiments. Oikos 95, 340–348.

Fukushima, J., Kainoh, Y., Honda, H., Takabayashi, J., 2002. Learning of herbivore-induced and nonspecific plant volatiles by a parasitoid, *Cotesia kariyai*. J. Chem. Ecol. 28, 579–586.

Furniss, R.L., Carolin, V.M., 1977. Western Forest Insects. USDA Forest Service Misc. Publ. 1339. USDA Forest Service, Washington, DC.

Futuyma, D.J., Wasserman, S.S., 1980. Resource concentration and herbivory in oak forests. Science 210, 920–922.

Gagic, V., Bartomeus, I., Jonsson, T., Taylor, A., Winqvist, C., Fischer, C., Slade, E.M., Steffan-Dewenter, I., Emmerson, M., Potts, S.G., Tscharntke, T., Weisser, W., Bommarco, R., 2015. Functional identity and diversity of animals predict ecosystem functioning better than species-based indices. Proc. Roy. Soc. B-Biol. Sci. 282, 20142620.

Galatowitsch, M.L., Mumme, R.L., 2004. Escape behavior of Neotropical homopterans in response to a flush-pursuit predator. Biotropica 36, 586–595.

Galeas, M.L., Klamper, E.M., Bennett, L.E., Freeman, J.L., Kondratieff, B.C., Quinn, C.F., Pilon-Smits, E.A.H., 2008. Selenium hyperaccumulation reduces plant arthropod loads in the field. New Phytol. 177, 715–724.

Galizia, C.G., Kunze, J., Gumbert, A., Borg-Karlson, A.-K., Sachse, S., Markl, C., Menzel, R., 2005. Relationship of visual and olfactory signal parameters in a food-deceptive flower mimicry system. Behav. Ecol. 16, 159–168.

Gallai, N., Salles, J.-M., Settele, J., Vaissière, B.E., 2009. Economic valuation of the vulnerability of world agriculture confronted with pollinator decline. Ecol. Econ. 68, 810–821.

Gandar, M.V., 1982. The dynamics and trophic ecology of grasshoppers (Acridoidea) in a South African savanna. Oecologia 54, 370–378.

Gandhi, K.J.K., Spence, J.R., Langor, D.W., Morgantini, L.E., Cryer, K.J., 2004. Harvest retention patches are insufficient as stand analogues of fire residuals for litter-dwelling beetles in northern coniferous forests. Can. J. Forest. Res. 34, 1319–1331.

Gandhi, K.J.K., Gilmore, D.W., Katovich, S.A., Mattson, W.J., Spence, S.R., Seybold, S.J., 2007. Physical effects of weather events on the abundance and diversity of insects in North American forests. Environ. Rev. 15, 113–152.

Gange, A.C., Brown, V.K., 1989. Insect herbivory affects size variability in plant populations. Oikos 56, 351–356.

Gange, A.C., Bower, E., Brown, V.K., 1999. Positive effects of an arbuscular mycorrhizal fungus on aphid life history traits. Oecologia 120, 123–131.

Ganio, L.M., Torgersen, C.E., Gresswell, R.E., 2005. A geostatistical approach for describing spatial pattern in stream networks. Front. Ecol. Environ. 3, 138–144.

Gannes, L.Z., O'Brien, D.M., del Rio, C.M., 1997. Stable isotopes in animal ecology: assumptions, caveats, and a call for more laboratory experiments. Ecology 78, 1271–1276.

Gara, R.I., Geiszler, D.R., Littke, W.R., 1984. Primary attraction of the mountain pine beetle to lodgepole pine in Oregon. Ann. Entomol. Soc. Am. 77, 333–334.

Gardiner, T., Hill, J., Chesmore, D., 2005. Review of the methods frequently used to estimate the bundance of Orthoptera in grassland ecosystems. J. Insect Conserv. 9, 151–173.

Gardiner, M.M., O'Neal, M.E., Landis, D.A., 2011. Intraguild predation and native lady beetle decline. PLoS ONE 6 (9), e23576.

Gardner, K.T., Thompson, D.C., 1998. Influence of avian predation on a grasshopper (Orthoptera: Acrididae) assemblage that feeds on threadleaf snakeweed. Environ. Entomol. 27, 110–116.

Garibaldi, L.A., Steffan-Dewenter, I., Winfree, R., Aizen, M.A., Bommarco, R., Cunningham, S.A., et al. 2013. Wild pollinators enhance fruit set of crops regardless of honey bee abundance. Science 339, 1608–1611.

Garibaldi, L.A., Carvalheiro, L.G., Leonhardt, S.D., Aizen, M.A., Blaauw, B.R., Isaacs, R., Kuhlmann, M., Kleijn, D., Klein, A.M., Kremen, C., Morandin, L., Scheper, J., Winfree, R., 2014. From research to action: enhancing crop yield through wild pollinators. Front. Ecol. Environ. 12, 439–447.

Garrett, K.A., Mundt, C.C., 1999. Epidemiology in mixed host populations. Phytopathology 89, 984–990.

Garrettson, M., Stetzel, J.F., Halpern, B.S., Hearn, D.J., Lucey, B.T., McKone, M.J., 1998. Diveristy and abundance of understorey plants on active and abandoned nests of leaf-cutting ants (*Atta cephalotes*) in a Costa Rican rain forest. J. Trop. Ecol. 14, 17–26.

Garrison, R.W., Willig, M.R., 1996. Arboreal invertebrates. In: Reagan, D.P., Waide, R.B. (Eds.), The Food Web of a Tropical Rain Forest. University of Chicago Press, Chicago, IL, pp. 183–245.

Garzon-Lopez, C.X., Ballesteros-Mejia, L., Ordoñez, A., Bohlman, S.A., Olff, H., Jansen, P.A., 2015. Indirect interactions among tropical tree species through shared rodent seed predators: a novel mechanism of tree species coexistence. Ecol. Lett. 18, 752–760.

Gascoigne, J., Berec, L., Gregory, S., Courchamp, F., 2009. Dangerously few liaisons: a review of mate-finding Allee effects. Popul. Ecol. 51, 355–372.

Gassmann, A.J., Stock, S.P., Sisterson, M., Carriére, Y., Tabashnik, B.E., 2008. Synergism between entomopathogenic nematodes and Bt crops: integrating biological control and resistance management. J. Appl. Ecol. 45, 957–966.

Gassmann, A.J., Fabrick, J.A., Sisterson, M., Hannon, E.R., Stock, S.P., Carrière, Y., Tabashnik, B.E., 2009. Effects of pink bollworm resistance to *Bacillus thuringiensis* on phenoloxidase activity and susceptibility to entomopathogenic nematodes. J. Econ. Entomol. 102, 1224–1232.

Gaston, K.J., 2000. Global patterns in biodiversity. Nature 405, 220–227.

Gaston, L.K., Shorey, H.H., Saario, C.A., 1967. Insect population control by the use of sex pheromones to inhibit orientation between the sexes. Nature 213, 1155.

Gaston, L.K., Kaae, R.S., Shorey, H.H., Sellers, D., 1977. Controlling the pink bollworm by disrupting sex pheromone communication between adult moths. Science 196, 904–905.

Gaston, K.J., Duffy, J.P., Gaston, S., Bennie, J., Davies, T.W., 2014. Human alteration of natural light cycles: causes and ecological consequences. Oecologia 176, 917–931.

Gate, I.M., McNeill, S., Ashmore, M.R., 1995. Effects of air pollution on the searching behaviour of an insect parasitoid. Water Air Soil Poll. 85, 1425–1430.

Gatto, P., Zocca, A., Battisti, A., Barrento, M.J., Branco, M., Paiva, M.R., 2009. Economic assessment of managing processionary moth in pine forests: a case study in Portugal. J. Environ. Manage. 90, 683–691.

Gautam, B.K., Henderson, G., 2015. Point of no return from water loss in Coptotermes formosanus (Isoptera: Rhinotermitidae). J. Econ. Entomol. 108, 1972–1977.

Gear, A.J., Huntley, B., 1991. Rapid changes in the range limits of Scots pine 4000 years ago. Science 251, 544–547.

Gehring, C.A., Whitham, T.G., 1991. Herbivore-driven mycorrhizal mutualism in insect-susceptible pinyon pine. Nature 353, 556–557.

Gehring, C.A., Whitham, T.G., 1995. Duration of herbivore removal and environmental stress affect the ectomycorrhizae of pinyon pine. Ecology 76, 2118–2123.

Geib, S.M., Filley, T.R., Hatcher, P.G., Hoover, K., Carlson, J.E., Jimenez-Gasco, M., Nakagawa-Izumi, A., Sleighter, R.L., Tien, M., 2008. Lignin degradation in wood-feeding insects. Proc. Natl. Acad. Sci. USA 105, 12932–12937.

Genersch, E., 2010. Honey bee pathology: current threats to honey bees and beekeeping. Appl. Microbiol. Biot. 87, 87–97.

Genung, M.A., Crutsinger, G.M., Bailey, J.K., Schweitzer, J.A., Sanders, N.J., 2012. Aphid and ladybird beetle abundance depend on the interaction of spatial effects and genotypic diversity. Oecologia 168, 167–174.

Gering, J.C., Crist, T.O., Veech, J.A., 2003. Additive partitioning of species diversity across multiple spatial scales: implications for regional conservation of biodiversity. Conserv. Biol. 17, 488–499.

Gering, J.C., DeRennaux, K.A., Crist, T.O., 2007. Scale-dependence of effective specialization: its analysis and implications for estimation of global insect species richness. Divers. Distrib. 13, 115–125.

Gersich, F.M., Brusven, M.A., 1982. Volcanic ash accumulation and ash-voiding mechanisms of aquatic insects. J. Kansas Entomol. Soc. 55, 290–296.

Ghazoul, J., McLeish, M., 2001. Reproductive ecology of tropical forest trees in logged and fragmented habitats in Thailand and Costa Rica. Plant Ecol. 153, 335–345.

Gibbs, A.G., 2002a. Lipid melting and cuticular permeability: new insights into an old problem. J. Insect Physiol. 48, 391–400.

Gibbs, A.G., 2002b. Water balance in desert *Drosophila*: lessons from non-charismatic microfauna. Comp. Biochem. Phys. A 133, 781–789.

Gibbs, A.G., Markow, T.A., 2001. Effects of age on water balance in Drosophila species. Physiol. Biochem. Zool. 74, 520–530.

Gibbs, A.G., Matzkin, L.M., 2001. Evolution of water balance in the genus *Drosphila*:. J. Exp. Biol. 204, 2331–2338.

Gibbs, A.G., Fukuzato, F., Matzkin, L.M., 2003. Evolution of water conservation mechanisms in *Drosophila*. J. Exp. Biol. 206, 1183–1192.

Gibson, R.W., Pickett, J.A., 1983. Wild potato repels aphids by release of aphid alarm pheromone. Nature 302, 608–609.

Gibson, D.J., Freeman, C.C., Hulbert, L.C., 1990. Effects of small mammal and invertebrate herbivory on plant species richness and abundance in tallgrass prairie. Oecologia 84, 169–175.

Gibson, R.H., Nelson, I.L., Hopkins, G.W., Hamlett, B.J., Memmott, J., 2006. Pollinator webs, plant communities and the conservation of rare plants: arable weeds as a case study. J. Appl. Ecol. 43, 246–257.

Giebultowicz, J.M., 2000. Molecular mechanism and cellular distribution of insect circadian clocks. Annu. Rev. Entomol. 45, 769–793.

Giebultowicz, J.M., Denlinger, D.L., 1986. Role of the brain and ring gland in regulation of pupal diapause in the flesh fly *Sarcophaga crassipalpis*. J. Insect Physiol. 32, 161–166.

Gilbert, M., Grégoire, J.-C., 2003. Site condition and predation influence a bark beetle's success: a spatially realistic approach. Agr. Forest Entomol. 5, 87–96.

Gillette, N.D., Erbilgin, N., Webster, J.N., Pederson, L., Mori, S.R., Stein, J.D., Owen, D.R., Bischel, K.M., Wood, D.L., 2009. Aerially applied verbenone-releasing laminated flakes protect *Pinus contorta* stands from attack by *Dendroctonus ponderosae* in California and Idaho. Forest Ecol. Manag. 257, 1405–1412.

Girling, R.D., Higby, B.S., Cardé, R.T., 2013. The plume also rises: trajectories of pheromone plumes issuing from point sources in an orchard canopy at night. J. Chem. Ecol. 39, 1150–1160.

Gist, C.S., Crossley, Jr., D.A., 1975. The litter arthropod community in a southern Appalachian hardwood forest: numbers, biomass and mineral element content. Am. Midl. Nat. 93, 107–122.

Gleason, H.A., 1917. The structure and development of the plant association. B. Torrey Bot. Club. 44, 463–481.

Gleason, H.A., 1926. The individualistic concept of the plant association. B. Torrey Bot. Club. 53, 7–26.

Gleason, H.A., 1927. Further views on the succession-concept. Ecology 8, 299–326.

Gleason, K.L., Lawrimore, J.H., Levinson, D.H., Karl, T.R., 2008. A revised U.S. climate extremes index. J. Climate 21, 2124–2137.

Glenn-Lewin, D.C., Peet, R.K., Veblen, T.T., 1992. Plant Succession: Theory and Prediction. Chapman and Hall, New York, pp. 352.

Gliessman, S.R., 2007. Agroecology: the Ecology of Sustainable Food Systems, second ed. CRC Press, Boca Raton, FL.

Glynn, C., Herms, D.A., Egawa, M., Hansen, R., Mattson, W.J., 2003. Effects of nutrient availability on biomass allocation as well as constitutive and rapid induced herbivore resistance in poplar. Oikos 101, 385–397.

Glynn, C., Herms, D.A., Orians, C.M., Hansen, R., Mattson, W.J., 2007. Testing the growth-differentiation balance hypothesis: dynamic responses of willows to nutrient availability. New Phytol. 176, 623–634.

Godfray, H.C.J., 1994. Parasitoids: Behavioral and Evolutionary Ecology. Princeton University Press, Princeton, NJ.

Godinho, D.P., Janssen, A., Dias, T., Cruz, C., Magalhães, S., 2016. Down-regulation of plant defence in a resident spider mite species and its effect upon con- and heterospecifics. Oecologia 180, 161–167.

Goff, M.L., 2000. A Fly for the Prosecution: How Insect Evidence Helps Solve Crimes. Harvard University Press, Cambridge, MA.

Goh, B.S., 1979. Stability of models of mutualism. Am. Nat. 113, 261–275.

Golden, D.M., Crist, T.O., 1999. Experimental effects of habitat fragmentation on old-field canopy insects: community, guild and species responses. Oecologia 118, 371–380.

Goldwasser, L., Roughgarden, J., 1997. Sampling effects and the estimation of food-web properties. Ecology 78, 41–54.

Golley, F.B., 1968. Secondary productivity in terrestrial communities. Am. Zool. 8, 53–59.

Golley, F.B., 1977. Insects as regulators of forest nutrient cycling. Trop. Ecol. 18, 116–123.

Golley, F.B., 1993. A History of the Ecosystem Concept in Ecology. Yale University Press, New Haven, CT, pp. 254.

Gong, Z.-F., Xia, S.-Z., Liu, L., Feng, C.-H., Guo, A.-K., 1998. Operant visual learning and memory in *Drosphila* mutants *dunce, amnesiac* and *radish*. J. Insect Physiol. 44, 1149–1158.

Gonzalez, M., Loreau, M., 2009. The causes and consequences of compensatory dynamics in ecological communities. Annu. Rev. Ecol. Evol. S. 40, 393–414.

González, G., Seastedt, T.R., 2001. Soil fauna and plant litter decomposition in tropical and subalpine forests. Ecology 82, 955–964.

Goodale, U.M., Berlyn, G.P., Gregoire, T.G., Tennakoon, K.U., Ashton, M.S., 2014. Differences in survival and growth among tropical rain forest pioneer tree seedlings in relation to canopy openness and herbivory. Biotropica 46, 183–193.

Goodisman, M.A.D., Hahn, D.A., 2004. Colony genetic structure of the ant *Camponotus ocreatus* (Hymenoptera: Formicidae). Sociobiology 44, 21–33.

Gorb, E., Gorb, S., 2003. Seed Dispersal by Ants in a Deciduous Forest Ecosystem: Mechanisms, Strategies, Adaptations. Kluwer Academic Publishers, Dordrecht, The Netherlands.

Gordon, D.M., Kulig, A.W., 1996. Founding, foraging, and fighting: colony size and the spatial distribution of harvester ant nests. Ecology 77, 2393–2409.

Goto, S.G., Denlinger, D.L., 2002. Short-day and long-day expression patterns of genes involved in the flesh fly clock mechanism: *period, timeless, cycle* and *cryptochrome*. J. Insect Physiol. 48, 803–816.

Gottdenker, N.L., Streicker, D.G., Faust, C.L., Carroll, C.R., 2014. Anthropogenic land use change and infectious diseases: a review of the evidence. EcoHealth 11, 619–632.

Gottrup, F., Leaper, D., 2004. Wound healing: historical aspects. Eur. Wound Manage. Assoc. J. 4 (2), 21–26.

Gouinguené, S.P., Turlings, T.C.J., 2002. The effects of abiotic factors on induced volatile emissions in corn plants. Plant Physiol. 129, 1296–1307.

Gould, J.L., 1985. How bees remember flower shapes. Science 227, 1492–1494.

Gould, J.L., 1986. The locale map of honey bees: do insects have cognitive maps? Science 232, 861–863.

Gould, F., 1998. Sustainability of transgenic insecticidal cultivars. Annu. Rev. Entomol. 43, 701–726.

Gould, J.L., Towne, W.F., 1988. Honey bee learning. Adv. Insect Physiol. 20, 55–86.

Gould, S.J., Vrba, E.S., 1982. Exaptation—a missing term in the science of form. Paleobiology 8, 4–15.

Gould, J.L., Kirschvink, J.L., Deffeyes, K.S., 1978. Bees have magnetic remanence. Science 201, 1026–1028.

Goulson, D., Lye, G.C., Darvill, B., 2008. Decline and conservation of bumble bees. Annu. Rev. Entomol. 53, 191–208.

Goverde, M., van der Heijden, M.G.A., Wiemken, A., Sanders, I.R., Erhardt, A., 2000. Arbuscular mycorrhizal fungi influence life history traits of a lepidopteran herbivore. Oecologia 125, 362–369.

Goyer, R.A., Wagner, M.R., Schowalter, T.D., 1998. Current and proposed technologies for bark beetle management. J. Forest. 96 (12), 29–33.

Goyer, R.A., Lenhard, G.J., Strom, B.L., 2004. The influence of silhouette color and orientation on arrival and emergence of *Ips* pine engravers and their predators in loblolly pine. Forest Ecol. Manag. 191, 147–155.

Graça, M.A.S., Poquet, J.M., 2014. Do climate and soil influence phenotypic variability in leaf litter, microbial decomposition and shredder consumption? Oecologia 174, 1021–1032.

Grace, J.R., 1986. The influence of gypsy moth on the composition and nutrient content of litter fall in a Pennsylvania oak forest. Forest Sci. 32, 855–870.

Grace, J.K., Su, N.-Y., 2001. Evidence supporting the use of termite baiting systems for long-term structural protection. Sociobiology 37, 301–310.

Grant, G.G., Miller, W.E., 1995. Larval images on lepidopteran wings – an unrecognized defense mechanism? Am. Entomol. 41, 44–48.

Gratton, C., Denno, R.F., 2006. Arthropod food web restoration following removal of an invasive wetland plant. Ecol. Appl. 16, 622–631.

Gratton, C., Donaldson, J., Vander Zanden, M.J., 2008. Ecosystem linkages between lakes and the surrounding terrestrial landscape in Iceland. Ecosystems 11, 764–774.

Greathouse, E.A., Pringle, C.M., McDowell, W.H., Holmquist, J.G., 2006. Indirect upstream effects of dams: consequences of migratory consumer extirpation in Puerto Rico. Ecol. Appl. 16, 339–352.

Greenbank, D.O., 1957. The role of climate and dispersal in the initiation of outbreaks of the spruce budworm in New Brunswick. I. The role of climate. Can. J. Zoolog. 34, 453–476.

Greenbank, D.O., 1963. The development of the outbreak. In: The Dynamics of Epidemic Spruce Budworm Populations, Morris, R.F. (Ed.). Mem. Entomol. Soc. Can. 31: 19–23.

Greenberg, B., 1973. Flies through history. In: Greenberg, B. (Ed.), Flies and Disease. Princeton University Press, Princeton, NJ.

Greenfield, A.B., 2005. A Perfect Red: Empire, Espionage, and the Quest for the Color of Desire. Harper Collins Publications, New York.

Gressitt, J.L., Sedlacek, J., Szent-Ivany, J.J.H., 1965. Flora and fauna on backs of large Papuan moss-forest weevils. Science 150, 1833–1835.

Gressitt, J.L., Samuelson, G.A., Vitt, D.H., 1968. Moss growing on living Papuan moss-forest weevils. Nature 217, 765–767.

Gribko, L.S., Liebhold, A.M., Hohn, M.E., 1995. Model to predict gypsy moth (Lepidoptera: Lymantriidae) defoliation using kriging and logistic regression. Environ. Entomol. 24, 529–537.

Grier, C.C., Vogt, D.J., 1990. Effects of aphid honeydew on soil nitrogen availability and net primary production in an *Alnus rubra* plantation in western Washington. Oikos 57, 114–118.

Gries, R., Schaefer, P.W., Gotoh, T., Takács, S., Gries, G., 2009. Spacing of traps baited with species-specific *Lymantria* pheromones to prevent interference by antagonistic components. Can. Entomol. 141, 145–152.

Griffin, J.M., Turner, M.G., Simard, M., 2011. Nitrogen cycling following mountain pine beetle disturbance in lodgepole pine forests of Greater Yellowstone. Forest Ecol. Manag. 261, 1077–1089.

Griffiths, H.M., Louzada, J., Bardgett, R.D., Beiroz, W., França, F., Tregidgo, D., Barlow, J., 2015. Biodiversity and environmental context predict dung beetle-mediated seed dispersal in a tropical forest field experiment. Ecology 96, 1607–1619.

Grilli, M.P., Gorla, D.E., 1997. The spatio-temporal pattern of *Delphacodes kuscheli* (Homoptera: Delphacidae) abundance in central Argentina. B. Entomol. Res. 87, 45–53.

Grimbacher, P.S., Stork, N.E., 2009. How do beetle assemblages respond to cyclonic disturbance of a fragmented tropical rainforest landscape? Oecologia 161, 591–599.

Grime, J.P., 1977. Evidence for the existence of three primary strategies in plants and its relevance to ecological and evolutionary theory. Am. Nat. 111, 1169–1194.

Grime, J.P., 1997. Biodiversity and ecosystem function: the debate deepens. Science 277, 1260–1261.

Grime, J.P., Cornelissen, J.H.C., Thompson, K., Hodgson, J.G., 1996. Evidence of a causal connection between anti-herbivore defense and the decomposition rate of leaves. Oikos 77, 489–494.

Grimm, N.B., Grove, J.M., Pickett, S.T.A., Redman, C.L., 2000. Integrated approaches to long-term studies of urban ecological systems. BioScience 50, 571–584.

Groot, A.T., Horovitz, J.L., Hamilton, J., Santangelo, R.G., Schal, C., Gould, F., 2006. Experimental evidence for interspecific directional selection on moth pheromone communication. Proc. Natl. Acad. Sci. USA 103, 5858–5863.

Gross, I., Georgel, P., Kappler, C., Reichhart, J.-M., Hoffmann, J.A., 1996. *Drosophila* immunity: a comparative analysis of the Rel proteins dorsal and Dif in the induction of the genes encoding diptericin and cecropin. Nucleic Acids Res. 24, 1238–1245.

Grove, S.J., 2002. Saproxylic insect ecology and the sustainable management of forests. Annu. Rev. Ecol. Syst. 33, 1–23.

Gruner, D.S., Smith, J.E., Seabloom, E.W., Sandin, S.A., Ngai, J.T., Hillebrand, H., et al. 2008. A cross-system synthesis of consumer and nutrient resource control on producer biomass. Ecol. Lett. 11, 740–755.

Gudger, E.W., 1925. Stitching wounds with the mandibles of ants and beetles. J. Amer. Med. Assoc. 84, 1861–1864.

Gudger, E.W., 1925. Stitching wounds with the mandibles of ants and beetles. J. Am. Med. Assoc. 84, 1861–1864.

Guillot, F.S., Ring, D.R., Lax, A.R., Morgan, A., Brown, K., Riegel, C., Boykin, D., 2010. Area-wide management of the Formosan subterranean termite, *Coptotermes formosanus* Shiraki (Isoptera: Rhinotermitidae), in the New Orleans French Quarter. Sociobiology 55, 311–338.

Gumbert, A., Kunze, J., Chittka, L., 1999. Floral colour diversity in plant communities, bee colour space and a null model. Proc. Roy. Soc. Lond. B Bio 266, 1711–1716.

Gunn, D.L., Brown, H.D., 1979. The development of aircraft attack on locust swarms in Africa since 1945 and the start of operational research on control systems. Philos. T. Roy. Soc. B 287, 251–263.

Gunn, D.L., Perry, F.C., Seymour, W.G., Telford, T.M., Wright, E.N., Yeo, D., 1948. Behaviour of the desert locust (*Schistocerca gregaria* Forskål) in Kenya in relation to aircraft spraying. Anti-locust Bull. 3.

Guo, Q., 1998. Microhabitat differentiation in Chihuahuan Desert plant communities. Plant Ecol. 139, 71–80.

Guo, H., Hu, Q., Jiang, T., 2008. Annual and seasonal streamflow responses to climate and land-cover changes in the Poyang Lake Basin, China. J. Hydrol. 355, 106–122.

Gustafsson, P., Greenberg, L.A., Bergman, E., 2014. Woody debris and terrestrial invertebrates – effects on prey resources for brown trout (*Salmo trutta*) in a boreal stream. Environ. Biol. Fish. 97, 529–542.

Gutierrez, A.P., 1986. Analysis of the interactions of host plant resistance, phytophagous and entomophagous species. In: Boethel, D.J., Eikenbary, R.D. (Eds.), Interactions of Plant Resistance and Parasitoids and Predators of Insects. Ellis Horwood Ltd, Chichester, UK, pp. 198–215.

Gutierrez, A.P., 1996. Applied Population Ecology: a Supply-demand Approach. John Wiley & Sons, Inc, New York, NY.

Gutiérrez, D., Wilson, R.J., 2014. Climate conditions and resource availability drive return elevational migrations in a single-brooded insect. Oecologia 175, 861–873.

Gutiérrez-Ibáñez, C., Villagra, C.A., Niemeyer, H.M., 2007. Pre-pupation behaviour of the aphid parasitoid *Aphidius ervi* (Haliday) and its consequences for pre-imaginal learning. Naturwissenschaften 94, 595–600.

Gutschick, V.P., 1999. Biotic and abiotic consequences of differences in leaf structure. New Phytol. 143, 4–18.

Gutschick, V.P., BassiriRad, H., 2010. Biological extreme events: a research framework. EOS Trans. Am. Geophys. Union 91, 85–86.

Haase, J., Castagneyrol, B., Cornelissen, J.H.C., Ghazoul, J., Kattge, J., Koricheva, J., Scherer-Lorenzen, M., Morath, S., Jactel, H., 2015. Contrasting effects of tree diversity on young tree growth and resistance to insect herbivores across three biodiversity experiments. Oikos 124, 1674–1685.

Habermann, E., 1972. Bee and wasp venoms. Science 177, 314–322.

Hackstein, J.H.P., Stumm, C.K., 1994. Methane production in terrestrial arthropods. Proc. Natl. Acad. Sci. USA 91, 5441–5445.

Haddad, N.M., 1999. Corridor and distance effects on interpatch movements: a landscape experiment with butterflies. Ecol. Appl. 9, 612–622.

Haddad, N., 2000. Corridor length and patch colonization by a butterfly, *Junonia coenia*. Conserv. Biol. 14, 738–745.

Haddad, N.M., Baum, K.A., 1999. An experimental test of corridor effects on butterfly densities. Ecol. Appl. 9, 623–633.

Haddad, N.M., Browne, D.R., Cunningham, A., Danielson, B.J., Levey, D.J., Sargent, S., Spira, T., 2003. Corridor use by diverse taxa. Ecology 84, 609–615.

Haddad, N.M., Holyoak, M., Mata, T.M., Davies, K.F., Melbourne, B.A., Preston, K., 2008. Species' traits predict the effects of disturbance and productivity on diversity. Ecol. Lett. 11, 348–356.

Hadley, N.F., 1994. Water Relations of Terrestrial Arthropods. Academic Press, San Diego, CA.

Hadley, K.S., Veblen, T.T., 1993. Stand response to western spruce budworm and Douglas-fir bark beetle outbreaks, Colorado Front Range. Can. J. Forest. Res. 23, 479–491.

Hagen, K.S., Franz, J.M., 1973. A history of biological control. In: Smith, R.F., Mittler, T.E., Smith, C.N. (Eds.), History of Entomology. Annual Reviews Inc, Palo Alto, CA, pp. 433–476.

Haggerty, S.M., Batzer, D.P., Jackson, C.R., 2002. Macroinvertebrate assemblages in perennial headwater streams of the coastal mountain range of Washington, U.S.A. Hydrobiologia 479, 143–154.

Haggerty, S.M., Batzer, D.P., Jackson, C.R., 2004. Macroinvertebrate response to logging in coastal headwater streams in Washington, USA. Can. J. Fish. Aquat. Sci. 61, 529–537.

Hagler, J.R., Jackson, C.G., 2001. Methods for marking insects: current techniques and future prospects. Annu. Rev. Entomol. 46, 511–543.

Haglund, B.M., 1980. Proline and valine—cues which stimulate grasshopper herbivory during drought stress? Nature 288, 697–698.

Hahn, D.A., Denlinger, D.L., 2011. Energetics of insect diapause. Annu. Rev. Entomol. 56, 103–121.

Haimi, J., Huhta, V., 1990. Effects of earthworms on decomposition processes in raw humus forest soil: a microcosm study. Biol. Fert. Soils 10, 178–183.

Hain, F.P., 1980. Sampling and predicting population trends. In: The Southern Pine Beetle, Thatcher, R.C., Searcy, J.L., Coster, J.E., Hertel, G.D. (Eds.). USDA Forest Service Tech. Bull. 1631. USDA Forest Service, Washington, DC, pp. 107–135.

Haines, B.L., 1978. Element and energy flows through colonies of the leaf-cutting ant, *Atta columbica* in Panama. Biotropica 10, 270–277.

Hairston, N.G., Smith, F.E., Slobodkin, L.B., 1960. Community structure, population control, and competition. Am. Nat. 94, 421–425.

Hajek, A.E., St. Leger, R.J., 1994. Interactions between fungal pathogens and insect hosts. Annu. Rev. Entomol. 39, 293–322.

Halaj, J., Ross, D.W., Moldenke, A.R., 1997. Negative effects of ant foraging on spiders in Douglas-fir canopies. Oecologia 109, 313–322.

Haldane, J.B.S., 1932. The Causes of Evolution. Harper and Brothers, New York.

Hale, B.K., Herms, D.A., Hansen, R.C., Clausen, T.P., Arnold, D., 2005. Effects of drought stress and nutrient availability on dry matter allocation, phenolic glycosiders, and rapid induced resistance of poplar to two lymantriid defoliators. J. Chem. Ecol. 31, 2601–2620.

Hall, S.R., 2009. Stoichiometrically-explicit food webs: feedbacks between resource supply, elemental costraints, and species diversity. Annu. Rev. Ecol. Evol. S. 40, 503–528.

Hall, C.M., 2013. Glow-worm tourism in Australia and New Zealand: comodifying and conserving charismatic micro-fauna. In: Lemelin, R.H. (Ed.), The Management of Insects in Recreation and Tourism. Cambridge University Press, Cambridge, UK, pp. 217–232.

Hall, M.C., Stiling, P., Moon, D.C., Drake, B.G., Hunter, M.D., 2005. Effects of elevated CO_2 on foliar quality and herbivore damage in a scrub oak ecosystem. J. Chem. Ecol. 31, 267–286.

Hallmann, C.A., Foppen, R.P.B., van Turnhout, C.A.M., de Kroon, H., Jongejans, E., 2014. Declines in insectivorous birds are associated with high neonicotinoid concentrations. Nature 511, 341–342.

Halloy, J., Sempo, G., Caprari, G., Rivault, C., Asadpour, M., Tâche, F., Saïd, I., Durier, V., Canonge, S., Amé, J.M., Detrain, C., Correll, N., Martinoli, A., Mondada, F., Siegwart, R., Deneubourg, J.L., 2007. Social integration of robots into groups of cockroaches to control self-organized choices. Science 318, 1155–1158.

Halvorson, H.M., White, G., Scott, J.T., Evans-White, M.A., 2016. Dietary and taxonomic controls on incorporation of microbial carbon and phosphorus by detritivorous caddisflies. Oecologia 180, 567–579.

Hambäck, P.A., Pettersson, J., Ericson, L., 2003. Are associational refuges species-specific? Funct. Ecol. 17, 87–93.

Hamilton, W.D., 1964. The genetic evolution of social behavior. I. and II. J. Theor. Biol. 7, 1–52.

Hamilton, G.R., Baskett, T.F., 2000. History of anesthesia: in the arms of Morpheus: the development of morphine for postoperative pain relief. Can. J. Anaesth. 47, 367–374.

Hammer, T.J., Bowers, M.D., 2015. Gut microbes may facilitate insect herbivory of chemically defended plants. Oecologia 179, 1–14.

Hammock, B.G., Krigbaum, N.Y., Johnson, M.L., 2012. Incorporating invertebrate predators into theory regarding the timing of invertebrate drift. Aquat. Ecol. 46, 153–163.

Hamsten, C., Starkhammar, M., Tran, T.A.T., Johansson, M., Bengtsson, U., Ahlén, G., Sällberg, M., Grönlund, H., van Hage, M., 2013. Identification of galactose-α-1,3-galactose in the gastrointestinal tract of the tick *Ixodes ricinus*; possible relationship with red meat allergy. Allergy 68, 549–552.

Hance, T., van Baaren, J., Vernon, P., Boivin, G., 2007. Impact of extreme temperatures on parasitoids in a climate change perspective. Annu. Rev. Entomol. 52, 107–126.

Hanewinkel, M., Breidenbach, J., Neeff, T., Kublin, E., 2008. Seventy-seven years of natural disturbances in a mountain forest area – the influence of storm, snow, and insect damage analysed with a long-term time series. Can. J. Forest. Res. 38, 2249–2261.

Hansen, D.M., Müller, C.B., 2009. Invasive ants disrupt gecko pollination and seed dispersal of the endangered plant *Roussea simplex* in Mauritius. Biotropica 41, 202–208.

Hansen Jesse, L.C., Obrycki, J.J., 2000. Field deposition of Bt transgenic corn pollen: lethal effects on the monarch butterfly. Oecologia 125, 241–248.

Hanski, I., 1989. Metapopulation dynamics: does it help to have more of the same? Trends Ecol. Evol. 4, 113–114.

Hanski, I., 1997. Metapopulation dynamics: from concepts and observations to predictive models. In: Hanski, I.A., Gilpin, M.E. (Eds.), Metapopulation Biology: Ecology, Genetics and Evolution. Academic Press, San Diego, CA, pp. 69–91.

Hanski, I.A., Gilpin, M.E. (Eds.), 1997. Metapopulation Biology: Ecology, Genetics and Evolution. Academic Press, San Diego, CA.

Hanski, I., Simberloff, D., 1997. The metapopulation approach, its history, conceptual domain, and application to conservation. In: Hanski, I.A., Gilpin, M.E. (Eds.), Metapopulation Biology: Ecology, Genetics and Evolution. Academic Press, San Diego, CA, pp. 5–26.

Hanski, I., Singer, M.C., 2001. Extinction-colonization dynamics and host-plant choice in butterfly metapopulations. Am. Nat. 158, 341–353.

Hanson, B., Lindblom, S.D., Loeffler, M.L., Pilon-Smits, E.A.H., 2004. Selenium protects plants from phloem-feeding aphids due to both deterrence and toxicity. New Phytol. 162, 655–662.

Hanula, J.L., Horn, S., 2011. Removing an exotic shrub from riparian forests increases butterfly abundance and diversity. Forest Ecol. Manag. 262, 674–680.

Hanula, J.L., Wade, D.D., 2003. Influence of long-term dormant-season burning and fire exclusion on ground-dwelling arthropod populations in longleaf pine flatwoods ecosystems. Forest Ecol. Manag. 175, 163–184.

Harborne, J.B., 1994. Introduction to Ecological Biochemistry, fourth ed. Academic Press, London.

Hardee, D.D., Harris, F.A., 2003. Eradicating the boll weevil (Coleoptera: Curculionidae): a clash between a highly successful insect, good scientific achievement, and differing agricultural philosophies. Am. Entomol. 49, 82–97.

Hardie, J., 2001. Photoperiodism and seasonality in aphids. In: Denlinger, D.L., Giebultowicz, J.M., Saunders, D.S. (Eds.), Insect Timing: Circadian Rhythmicity to Seasonality. Elsevier, Amsterdam, pp. 85–94.

Harding, J.S., Benfield, E.F., Bolstad, P.V., Helfman, G.S., Jones, III, E.B.D., 1998. Stream biodiversity: the ghost of land use past. Proc. Natl. Acad. Sci. USA 95, 14843–14847.

Hare, J.D., 2011. Ecological role of volatiles produced by plants in response to damage by herbivorous insects. Annu. Rev. Entomol. 56, 161–180.

Hare, J.F., Alloway, T.M., 2001. Prudent *Protomognathus* and despotic *Leptothorax duloticus*: differential costs of ant slavery. Proc. Natl. Acad. Sci. USA 98, 12093–12096.

Hargrove, W.W., 1988. A photographic technique for tracking herbivory on individual leaves through time. Ecol. Entomol. 13, 359–363.

Harmon, M.E., Franklin, J.F., Swanson, F.J., Sollins, P., Gregory, S.V., Lattin, J.D., Anderson, N.H., Cline, S.P., Aumen, N.G., Sedell, J.R., Lienkaemper, G.W., Cromack, Jr., K., Cummins, K.W., 1986. Ecology of coarse woody debris in temperate ecosystems. Adv. Ecol. Res. 15, 133–302.

Harmon, M.E., Sexton, J., Caldwell, B.A., Carpenter, S.E., 1994. Fungal sporocarp mediated losses of Ca, Fe, K, Mg, Mn, N, P, and Zn from conifer logs in the early stages of decomposition. Can. J. Forest. Res. 24, 1883–1893.

Harrel, R.C., 1985. Effects of a crude oil spill on water quality and macrobenthos of a southeast Texas stream. Hydrobiologia 124, 223–228.

Harris, T.W., 1841. Treatise on some of the Insects Injurious to Vegetation. William White, Boston, MA.

Harris, L.D., 1984. The Fragmented Forest. University of Chicago Press, Chicago, IL.

Harrison, S., 1994. Resources and dispersal as factors limiting a population of the tussock moth (*Orgyia vetusta*), a flightless defoliator. Oecologia 99, 27–34.

Harrison, S., Cappuccino, N., 1995. Using density-manipulation experiments to study population regulation. In: Cappuccino, N., Price, P.W. (Eds.), Population Dynamics: New Approaches and Synthesis. Academic Press, San Diego, CA, pp. 131–147.

Harrison, S., Karban, R., 1986. Effects of an early-season folivorous moth on the success of a later-season species, mediated by a change in the quality of the shared host, *Lupinus arboreus* Sims. Oecologia 69, 354–359.

Harrison, S., Taylor, A.D., 1997. Empirical evidence for metapopulation dynamics. In: Hanski, I.A., Gilpin, M.E. (Eds.), Metapopulation Biology: Ecology, Genetics and Evolution. Academic Press, San Diego, CA, pp. 27–42.

Hart, D.D., 1992. Community organization in streams: the importance of species interactions, physical factors, and chance. Oecologia 91, 220–228.

Hartley, S.E., 1998. The chemical composition of plant galls: are levels of nutrients and secondary plant compounds controlled by the gall-former? Oecologia 113, 492–501.

Hartley, S.E., Gange, A.C., 2009. Impacts of plant symbiotic fungi on insect herbivores: mutualism in a multitrophic context. Annu. Rev. Entomol. 54, 323–342.

Harvey, B.J., Donato, D.C., Turner, M.G., 2014a. Recent mountain pine beetle outbreaks, wildfire severity, and postfire tree regeneration in the US Northern Rockies. Proc. Natl. Acad. Sci. USA 111, 15120–15125.

Harvey, B.J., Donato, D.C., Romme, W.H., Turner, M.G., 2014b. Fire severity and tree regeneration following bark beetle outbreaks: the role of outbreak stage and burning conditions. Ecol. Appl. 24, 1608–1625.

Hassall, M., Hawthorne, A., Maudsley, M., White, P., Cardwell, C., 1992. Effects of headland management on invertebrate communities in cereal fields. Agr. Ecosyst. Environ. 40, 155–178.

Hassell, M.P., Varley, G.C., 1969. New inductive population model for insect parasites and its bearing on biological control. Nature 223, 1133–1136.

Hassell, M.P., Comins, H.N., May, R.M., 1991. Spatial structure and chaos in insect population dynamics. Nature 353, 255–258.

Hatcher, P.E., Ayres, P.G., Paul, N.D., 1995. The effect of natural and simulated insect herbivory, and leaf age, on the process of infection of *Rumex crispus* L. and *R. obtusifolius* L. by *Uromyces rumicis* (Schum.) Wint. New Phytol. 130, 239–249.

Hatcher, P.E., Moore, J., Taylor, J.E., Tinney, G.W., Paul, N.D., 2004. Phytohormones and plant-herbivore-pathogen interactions: integrating the molecular with the ecological. Ecology 85, 59–69.

Hättenschwiler, S., Gasser, P., 2005. Soil animals alter plant litter diversity effects on decomposition. Proc. Natl. Acad. Sci. USA 102, 1519–1524.

Haukioja, E., 1990. Induction of defenses in trees. Annu. Rev. Entomol. 36, 25–42.

Havens, K., 1992. Scale and structure in natural food webs. Science 257, 1107–1109.

Hawkins, C.P., MacMahon, J.A., 1989. Guilds: the multiple meanings of a concept. Annu. Rev. Entomol. 34, 423–451.

Hawkins, C.P., Norris, R.H., Hogue, J.N., Feminella, J.W., 2000. Development and evaluation of predictive models for measuring the biological integrity of streams. Ecol. Appl. 10, 1456–1477.

Hawkins, T.W., Brazel, A.J., Stefanov, W.L., Bigler, W., Saffell, E.M., 2004. The role of rural variability in urban heat island determination for Phoenix, Arizona. J. Appl. Meteorol. 43, 476–486.

Hawkins, B.A., Albuquerque, F.S., Araújo, M.B., Beck, J., Bini, L.M., Cabrero-Sañudo, F.J., Castro-Parga, I., Diniz-Filho, J.A.F., Ferrer-Castán, D., Field, R., Gómez, J.F., Hortal, J., Kerr, J.T., Kitching, I.J., León-Cortés, J.L., Lobo, J.M., Montoya, D., Moreno, J.C., Olalla-Tárraga, M.Á., Pausas, J.G., Qian, H., Rahbek, C., Rodríguez, M.Á., Sanders, N.J., Williams, P., 2007. A global evaluation of metabolic theory as an explanation for terrestrial species richness gradients. Ecology 88, 1877–1888.

Hawlena, D., Schmitz, O.J., 2010. Herbivore physiological response to predation risk and implications for ecosystem nutrient dynamics. Proc. Natl. Acad. Sci. USA 107, 15503–15507.

Hawlena, D., Strickland, M.S., Bradford, M.A., Schmitz, O.J., 2012. Fear of predation slows plant-litter decomposition. Science 336, 1434–1438.

Hawthorne, D.J., Dively, G.P., 2011. Killing them with kindness? In-hive medications may inhibit xenobiotic efflux transporters and endanger honey bees. PLoS ONE 6 (11), e26796.

Hayes, J.L., Strom, B.L., Roton, L.M., Ingram, Jr., L.L., 1994. Repellent properties of the host compound 4-allylanisole to the southern pine beetle. J. Chem. Ecol. 20, 1595–1615.

Haynes, K.J., Cronin, J.T., 2003. Matrix composition affects the spatial ecology of a prairie planthopper. Ecology 84, 2856–2866.

Hazell, S.P., Groutides, C., Neve, B.P., Blackburn, T.M., Bale, J.S., 2010. A comparison of low temperature tolerance traits between closely related aphids from the tropics, temperate zone, and Arctic. J. Insect Physiol. 56, 115–122.

Hazlett, D.L., 1998. Vascular Plant Species of the Pawnee National Grasslands. USDA Forest Service Gen. Tech. Rpt. RM-GTR-17, USDA Forest Service, Rocky Mountain Exp. Stn., Ft. Collins, CO.

He, F., Alfaro, R.I., 1997. White pine weevil (Coleoptera: Curculionidae) attack on white spruce: spatial and temporal patterns. Environ. Entomol. 26, 888–895.

He, F., Legendre, P., 2002. Species diversity patterns derived from species-area models. Ecology 83, 1185–1198.

Heads, P.A., 1986. The costs of reduced feeding due to predator avoidance: potential effects on growth and fitness in *Ischnura elegans* larvae (Odonata: Zygoptera). Ecol. Entomol. 11, 369–377.

Heartsill-Scalley, T., Scatena, F.N., Estrada, C., McDowell, W.H., Lugo, A.E., 2007. Disturbance and long-term patterns of rainfall and throughfall nutrient fluxes in a subtropical wet forest in Puerto Rico. J. Hydrol. 333, 472–485.

Hedrick, P.W., Gilpin, M.E., 1997. Genetic effective size of a metapopulation. In: Hanski, I.A., Gilpin, M.E. (Eds.), Metapopulation Biology: Ecology, Genetics and Evolution. Academic Press, San Diego, CA, pp. 165–181.

Heiermann, J., Schütz, S., 2008. The effect of the tree species ratio of European beech (*Fagus sylvatica* L.) and Norway spruce (*Picea abies* (L.) Karst.) on polyphagous and monophagous pest species—*Lymantria monacha* L. and *Calliteara pudibunda* L. (Lepidoptera: Lymantriidae) as an example. Forest Ecol. Manag. 255, 1161–1166.

Heil, H., Karban, R., 2010. Explaining evolution of plant communication by airborne signals. Trends Ecol. Evol. 25, 137–144.

Heinrich, B., 1974. Thermoregulation in endothermic insects. Science 185, 747–756.

Heinrich, B., 1979. Bumblebee Economics. Harvard University Press, Cambridge, MA.

Heinrich, B., 1981. Insect Thermoregulation. Wiley, New York, NY.

Heinrich, B., 1993. The Hot-Blooded Insects: Strategies and Mechanisms of Thermoregulation. Harvard University Press, Cambridge, MA.

Heinrich, B., Raven, P., 1972. Energetics and pollination ecology. Science 176, 597–602.

Heithaus, E.R., 1979. Flower-feeding specialization in wild bee and wasp communities in seasonal neotropical habitats. Oecologia 42, 179–194.

Heithaus, E.R., 1981. Seed predation by rodents on three ant-dispersed plants. Ecology 62, 136–145.

Helbing, F., Blaeser, T.P., Löffler, F., Fartmann, T., 2014. Response of Orthoptera communities to succession in alluvial pine woodlands. J. Insect Conserv. 18, 215–224.

Heliövaara, K., 1986. Occurrence of *Petrova resinella* (Lepidoptera: Tortricidae) in a gradient of industrial air pollutants. Silva Fenn. 20, 83–90.

Heliövaara, K., Väisänen, R., 1986. Industrial air pollution and the pine bark bug, *Aradus cinnamomeus* Panz. (Het., Aradidae). Z. Angew. Entomol. 101, 469–478.

Heliövaara, K., Väisänen, R., 1993. Insects and Pollution. CRC Press, Boca Raton, FL.

Helson, J.E., Capson, T.L., Johns, T., Aiello, A., Windsor, D.M., 2009. Ecological and evolutionary bioprospecting: using aposematic insects as guides to rainforest plants active against disease. Front. Ecol. Environ. 7, 130–134.

Hemmann, D.J., Allison, J.D., Haynes, K.F., 2008. Trade-off between sensitivity and specificity in the cabbage looper response to sex pheromone. J. Chem. Ecol. 34, 1476–1486.

Hendrix, P.F., Crossley, Jr., D.A., Blair, J.M., Coleman, D.C., 1990. Soil biota as components of sustainable agroecosystems. In: Edwards, C.A., Lal, R., Madden, P., Miller, R.H., House, G. (Eds.), Sustainable Agricultural Systems. Soil and Water Conservation Society, Ankeny, IA, pp. 637–654.

Heneghan, L., Coleman, D.C., Zou, X., Crossley, Jr., D.A., Haines, B.L., 1999. Soil microarthropod contributions to decomposition dynamics: tropical-temperate comparisons of a single substrate. Ecology 80, 1873–1882.

Henne, D.C., Johnson, S.J., 2007. Zombie fire ant workers: behavior controlled by decapitating fly parasitoids. Insect. Soc. 54, 150–153.

Henne, D.C., Johnson, S.J., Cronin, J.T., 2007. Population spread of the introduced red imported fire ant parasitoid, *Pseudacteon tricuspis* Borgmeier (Diptera: Phoridae), in Louisiana. Biol. Control 42, 97–104.

Hennig, E.I., Ghazoul, J., 2012. Pollinating animals in the urban environment. Urban Ecosyst. 15, 149–166.

Henry, M., Béguin, M., Requier, F., Rollin, O., Odoux, J.-F., Aupinel, P., Aptel, J., Tchamitchian, S., Decourtye, A., 2012. A common pesticide decreases foraging success and survival in honey bees. Science 336, 348–350.

Herbert, D.A., Fownes, J.H., Vitousek, P.M., 1999. Hurricane damage to a Hawaiian forest: nutrient supply rate affects resistance and resilience. Ecology 80, 908–920.

Hermann, S.L., Thaler, J.S., 2014. Prey perception of predation risk: volatile chemical cues mediate non-consumptive effects of a predator on a herbivorous insect. Oecologia 176, 669–676.

Herms, D.A., Mattson, W.J., 1992. The dilemma of plants: to grow or defend. Q. Rev. Biol. 67, 283–335.

Herms, D.A., McCullough, D.G., 2014. Emerald ash borer invasion of North America: history, biology, ecology, impacts, and management. Annu. Rev. Entomol. 59, 13–30.

Hero, J.M., Castley, J.G., Butler, S.A., Lollback, G.W., 2013. Biomass estimation within an Australian eucalypt forest: meso-scale spatial arrangement and the influence of sampling intensity. For. Ecol. Manage. 310, 547–554.

Herrera, C.M., 1989. Vertebrate frugivores and their interaction with invertebrate fruit predators: supporting evidence from a Costa Rican dry forest. Oikos 54, 185–188.

Herrick, J.E., Lal, R., 1996. Dung decomposition and pedoturbation in a seasonally dry tropical pasture. Biol. Fert. Soils 23, 177–181.

Herz, H., Beyschlag, W., Hölldobler, B., 2007. Assessing herbivory rates of leaf-cutting ant (*Atta columbica*) colonies through short-term refuse deposition counts. Biotropica 39, 476–481.

Herzog, D.C., Reagan, T.E., Sheppard, D.C., Hyde, K.M., Nilakhe, S.S., Hussein, M.Y.B., McMahan, M.L., Thomas, R.C., Newsom, L.D., 1976. *Solenopsis invicta* Buren: influence on Louisiana pasture soil chemistry. Environ. Entomol. 5, 160–162.

Heske, E.J., Brown, J.H., Guo, Q.F., 1993. Effect of kangaroo rat exclusion on vegetation structure and plant species diversity in the Chihuahuan Desert. Oecologia 95, 520–524.

Heuberger, S., Ellers-Kirk, C., Yafuso, C., Gassmann, A.J., Tabashnik, B.E., Dennehy, T.J., Carriere, Y., 2008a. Effects of refuge contamination by transgenes on Bt resistance in pink bollworm (Lepidoptera: Gelichiidae). J. Econ. Entomol. 101, 504–514.

Heuberger, S., Yafuso, C., DeGrandi-Hoffman, G., Tabashnik, B.E., Carriere, Y., Dennehy, T.J., 2008b. Outcrossed cottonseed and adventitious Bt plants in Arizona refuges. Environ. Biosafety Res. 7, 87–96.

Heuberger, S., Ellers-Kirk, C., Tabashnik, B.E., Carriére, Y., 2010. Pollen- and seed-mediated transgene flow in commercial cotton seed production fields. PLoS ONE 5 (11), e14128.

Heystek, A., Pauw, A., 2014. Does competition for pollinators contribute to structuring Erica communities? J. Veg. Sci. 25, 648–656.

Hicks, W.T., Harmon, M.E., 2002. Diffusion and seasonal dynamics of O_2 in woody debris from the Pacific Northwest, USA. Plant Soil 243, 67–79.

Hik, D.S., Jefferies, R.L., 1990. Increases in the net above-ground primary production of a salt-marsh forage grass: a test of the predictions of the herbivore-optimization model. J. Ecol. 78, 180–195.

Hill, J.K., Griffiths, H.M., Thomas, C.D., 2011. Climate change and evolutionary adaptations at species' range margins. Annu. Rev. Entomol. 56, 144–159.

Hillstrom, M., Meehan, T.D., Kelly, K., Lindroth, R.L., 2010. Soil carbon and nitrogen mineralization following deposition of insect frass and greenfall from forests under elevated CO_2 and O_3. Plant Soil 336, 75–85.

Himler, A.G., Adachi-Hagimori, T., Bergen, J.E., Kozuch, A., Kelly, S.E., Tabashnik, B.E., Chiel, E., Duckworth, V.E., Dennehy, T.J., Zchori-Fein, E., Hunter, M.S., 2011. Rapid spread of a bacterial symbiont in an invasive whitefly is driven by fitness benefits and female bias. Science 332, 254–256.

Hinton, H.E., 1960a. A fly larva that tolerates dehydration and temperatures of −270° to +102°C. Nature 188, 336–337.

Hinton, H.E., 1960b. Cryptobiosis in the larva of *Polypedilum vanderplanki* Hint. (Chironomidae). J. Insect Physiol. 5, 286–315.

Hirai, H., Procunier, W.S., Ochoa, J.O., Uemoto, K., 1994. A cytogenetic analysis of the *Simulium ochraceum* species complex (Diptera: Simuliidae) in Central America. Genome 37, 36–53.

Hirao, T., Murakami, M., Iwamoto, J., Takafumi, H., Oguma, H., 2008. Scale-dependent effects of windthrow disturbance on forest arthropod communities. Ecol. Res. 23, 189–196.

Hirschel, G., Körner, C., Arnone, III, J.A., 1997. Will rising atmospheric CO_2 affect leaf litter quality and in situ decomposition rates in native plant communities? Oecologia 110, 387–392.

Hitchner, E.M., Kuhar, T.P., Dickens, J.C., Youngman, R.R., Schultz, P.B., Pfeiffer, D.G., 2008. Host plant choice experiments of Colorado potato beetle (Coleoptera: Chrysomelidae) in Virginia. J. Econ. Entomol. 101, 859–865.

Hobbie, S.E., 2008. Nitrogen effects on decomposition: a five-year experiment in eight temperate sites. Ecology 89, 2633–2644.

Hochberg, M.E., 1989. The potential role of pathogens in biological control. Nature 337, 262–265.

Hochwender, C.G., Fritz, R.S., 2004. Plant genetic differences influence herbivore community structure: evidence from a hybrid willow system. Oecologia 138, 547–557.

Hodges, J.D., Barras, S.J., Mauldin, J.K., 1968. Free and protein-bound amino acids in inner bark of loblolly pine. Forest Sci. 14, 330–333.

Hodkinson, I.D., Coulson, S.J., Harrison, J., Moores, J., Webb, N.R., 2001. What a wonderful web they weave: spiders, nutrient capture and early ecosystem development in the high Arctic—some counter-intuitive ideas on community assembly. Oikos 95, 349–352.

Hoehn, P., Tscharntke, T., Tylianakis, J.M., Steffan-Dewenter, I., 2008. Functional group diversity of bee pollinators increases crop yield. Proc. Roy. Soc. B-Biol. Sci. 275, 2283–2291.

Hoekman, D., Terhorst, C., Bauer, A., Braun, S., Gignac, P., Hopkins, R., Joshi, S., Laskis, K., Sanscrainte, N., Travis, J., Miller, T.E., 2007. Oviposition decreased in response to enriched water: a field study of the pitcherplant mosquito, *Wyeomyia smithii*. Ecol. Entomol. 32, 92–96.

Hofstetter, R.W., Moser, J.C., 2014. The role of mites in insect-fungus associations. Annu. Rev. Entomol. 59, 537–557.

Hohn, M.E., Liebhold, A.M., Gribko, L.S., 1993. Geostatistical model for forecasting spatial dynamics of defoliation caused by the gypsy moth (Lepidoptera: Lymantriidae). Environ. Entomol. 22, 1066–1075.

Hokkanen, H.M.T., 1991. Trap cropping in pest management. Annu. Rev. Entomol. 36, 119–138.

Hokkanen, H.M.T., van Lenteren, J.C., Menzler-Hokkanen, I., 2007. Ecological risks of biological control agents: impacts on IPM. In: Kogan, M., Jepson, P. (Eds.), Perspectives in Ecological Theory and Integrated Pest Management. Cambridge University Press, Cambridge, UK, pp. 246–268.

Holdo, R.M., McDowell, L.R., 2004. Termite mounds as nutrient-rich food patches for elephants. Biotropica 36, 231–239.

Hölker, F., Vanni, M.J., Kuiper, J.J., Meile, C., Grossart, H.-P., Stief, P., Adrian, R., Lorke, A., Dellwig, O., Brand, A., Hupfer, M., Mooij, W.M., Nützmann, G., Lewandowski, J., 2015. Tube-dwelling invertebrates: tiny ecosystem engineers have large effects in lake ecosystems. Ecol. Monogr. 85, 333–351.

Holland, J.N., 1995. Effects of above-ground herbivory on soil microbial biomass in conventional and no-tillage agroecosystems. Appl. Soil Ecol. 2, 275–279.

Holland, J.N., DeAngelis, D.L., 2009. Consumer-resource theory predicts dynamic transitions between outcomes of interspecific interactions. Ecol. Lett. 12, 1357–1366.

Holland, J.N., Cheng, W., Crossley, Jr., D.A., 1996. Herbivore-induced changes in plant carbon allocation: assessment of below-ground C fluxes using carbon-14. Oecologia 107, 87–94.

Holland, J.N., Chamberlain, S.A., Miller, T.E.X., 2011. Consequences of ants and extrafloral nectar for a pollinating seed-consuming mutualism: ant satiation, floral distraction or plant defense? Oikos 120, 381–388.

Hölldobler, B., 1995. The chemistry of social regulation: multicomponent signals in ant societies. Proc. Natl. Acad. Sci. USA 92, 19–22.

Hollenhorst, S.J., Brock, S.M., Freimund, W.A., Twery, M.J., 1993. Predicting the effects of gypsy moth on near-view aesthetic preferences and recreation appeal. Forest Sci. 39, 28–40.

Holling, C.S., 1959. Some characteristics of simple types of predation and parasitism. Can. Entomol. 91, 385–398.

Holling, C.S., 1965. The functional response of predators to prey density and its role in mimicry and population regulation. Mem. Entomol. Soc. Can. 45, 1–60.

Holling, C.S., 1966. The functional response of invertebrate predators to prey density. Mem. Entomol. Soc. Can. 48, 1–86.

Holling, C.S., 1973. Resilience and stability of ecological systems. Annu. Rev. Ecol. Syst. 4, 1–23.

Holling, C.S., 1992. Cross-scale morphology, geometry, and dynamics of ecosystems. Ecol. Monogr. 62, 447–502.

Hollinger, D.Y., 1986. Herbivory and the cycling of nitrogen and phosphorus in isolated California oak trees. Oecologia 70, 291–297.

Holopainen, J.K., Rikala, R., Kainulainen, P., Oksanen, J., 1995. Resource partitioning to growth, storage and defence in nitrogen-fertilized Scots pine and susceptibility of the seedlings to the tarnished plant bug Lygus rugulipennis. New Phytol. 131, 521–532.

Holt, R.D., Barfield, M., 2010. Metapopulation perspectives on the evolution of species' niches. In: Cantrell, S., Cosner, C., Ruan, S. (Eds.), Spatial Ecology. Chapman and Hall/CRC/Taylor & Francis, Boca Raton, FL, pp. 189–211.

Holt, J.R., Wilson, P., Brigham, C.A., 2014. A test of density-dependent pollination within three populations of endangered Pentachaeta lyonii. J. Pollination Ecol. 12, 95–100.

Holter, P., 1979. Effect of dung-beetles (Aphodius spp.) and earthworms on the disappearance of cattle dung. Oikos 32, 393–402.

Holzinger, F., Wink, M., 1996. Mediation of cardiac glycoside insensitivity in the monarch butterfly (Danaus plexippus): role of an amino acide substitution in the ouabain bindaing site of Na+, K+ -ATPase. J. Chem. Ecol. 22, 1921–1937.

Holzschuh, A., Steffan-Dewenter, I., Kleijn, D., Tscharntke, T., 2007. Diversity of flower-visiting bees in cereal fields: effects of farming system, landscape composition and regional context. J. Appl. Ecol. 44, 41–49.

Honek, A., Martinkova, Z., Saska, P., Koprdova, S., 2009. Role of post-dispersal seed and seedling predation in establishment of dandelion (Taraxacum agg.) plants. Agr. Ecosyst. Environ. 134, 126–135.

Honkanen, T., Haukioja, E., Suomela, J., 1994. Effects of simulated defoliation and debudding on needle and shoot growth in Scots pine (Pinus sylvestris): implications of plant source/sink relationships for plant-herbivore studies. Funct. Ecol. 8, 631–639.

Hoogendoorn, M., Heimpel, G.E., 2001. PCR-based gut content analysis of insect predators: using ribosomal ITS-1 fragments from prey to estimate predation frequency. Mol. Ecol. 10, 2059–2067.

Hooker, J.D., 1847. The Botany of the Antarctic Voyage of H.M. Discovery Ships Erebus and Terror, in the Years, 1839–1843. Vol. 1. Flora Antarctica. Reeve, Brothers, London.

Hooker, J.D., 1853. The Botany of the Antarctic Voyage of H.M. Discovery Ships Erebus and Terror, in the Years 1839–1843. Vol. 2. Flora Novae-Zelandiae. Reeve, Brothers, London.

Hooker, J.D., 1860. The Botany of the Antarctic Voyage of H.M. Discovery Ships Erebus and Terror, in the Years 1839–1843. Vol. 3. Flora Tasmaniae. Reeve, Brothers, London.

Hooper, D.U., Vitousek, P.M., 1997. The effects of plant composition and diversity on ecosystem processes. Science 277, 1302–1305.

Hooper, D.U., Chapin, III, F.S., Ewel, J.J., Hector, A., Inchausti, P., Lavorel, S., Lawton, J.H., Lodge, D.M., Loreau, M., Naeem, S., Schmid, B., Setälä, H., Symstad, A.J., Vandermeer, J., Wardle, D.A., 2005. Effects of biodiversity on ecosystem functioning: a concensus of current knowledge. Ecol. Monogr. 75, 3–35.

Hopkins, R.J., van Dam, N.M., van Loon, J.J.A., 2009. Role of glucosinolates in insect-plant relationships and multitrophic interactions. Annu. Rev. Entomol. 54, 57–83.

Horenstein, M.B., Linhares, A.X., 2011. Seasonal composition and temporal succession of necrophagous and predator beetles on pig carrion in central Argentina. Med. Vet. Entomol. 25, 395–401.

Horn, H.S., 1981. Some causes of variety in patterns of secondary succession. In: West, D.C., Shugart, H.H., Botkin, D.B. (Eds.), Forest Succession: Concepts and Application. Springer-Verlag, New York, pp. 24–35.

Horn, M.H., 1997. Evidence for dispersal of fig seeds by the fruit-eating characid fish Brycon guatemalensis Regan in a Costa Rican tropical rain forest. Oecologia 109, 259–264.

Horridge, A., 2003. Visual resolution of the orientation cue by the honeybee (Apis mellifera). J. Insect Physiol. 49, 1145–1152.

Horváth, G., Kriska, G., Malik, P., Robertson, B., 2009. Polarized light pollution: a new kind of ecological photopollution. Front. Ecol. Environ. 7, 317–325.

Horvitz, C.C., Schemske, D.W., 1986. Ant-nest soil and seedling growth in a neotropical ant-dispersed herb. Oecologia 70, 318–320.

Horvitz, C.C., Tuljapurkar, S., Pascarella, J.B., 2005. Plant-animal interactions in random environments: habitat-stage elasticity, seed predators and hurricanes. Ecology 86, 3312–3322.

Horwath, W.R., Paul, E.A., Harris, D., Norton, J., Jagger, L., Horton, K.A., 1996. Defining a realistic control for the chloroform fumigation-incubation method using microscopic counting and ^{14}C-substrates. Can. J. Soil Sci. 76, 459–467.

Hoskinaon, I.S., Webb, N.R., Coulson, S.J., 2002. Primary community assembly on land – the missing stages: why are the heterotrophic organisms always there first? J. Ecol. 90, 569–577.

Hossain, F., Jeyachandran, I., Pielke, R., 2009. Have large dams altered extreme precipitation patterns? EOS Trans. Am. Geophys. Union 90, 453–454.

Houlahan, J.E., Currie, J.D., Cottanie, K., Cumming, G.S., Earnest, S.K.M., Findlay, C.S., Fuhlendorf, S.D., Gaedke, U., Legendre, P., Magnuson, J.J., McArdle, B.H., Muldavin, E.H., Noble, D., Russell, R., Stevens, R.D., Willis, T.J., Woiwod, I.P., Wondzell, S.M., 2007. Compensatory dynamics are rare in natural ecological communities. Proc. Natl. Acad. Sci. USA 104, 3273–3277.

Howard, L.O., 1896. The Mexican Cotton Boll Weevil. USDA Bureau of Entomology Circular 14. Government Printing Office, Washington, DC.

Howarth, R., Chan, F., Conley, D.J., Garnier, J., Doney, S.C., Marino, R., Billen, G., 2011. Coupled biogeochemical cycles: eutrophication and hypoxia in temperate estuaries and coastal marine ecosystems. Front. Ecol. Environ. 9, 18–26.

Howden, H.F., Vogt, G.B., 1951. Insect communities of standing dead pine (*Punus virginiana* Mill.). Ann. Entomol. Soc. Am. 44, 581–595.

Howe, H.F., Brown, J.S., 2001. The ghost of granivory past. Ecol. Lett. 4, 371–378.

Howe, H.F., Smallwood, J., 1982. Ecology of seed dispersal. Annu. Rev. Ecol. Syst. 13, 201–228.

Hrabar, H., du Toit, J.T., 2014. Interactions between megaherbivores and microherbivores: elephant browsing reduces host plant quality for caterpillars. Ecosphere 5, http://dx.doi.org/10.1890/ES13-00173.1.

Hsiang, S.M., Meng, K.C., Cane, M.A., 2011. Civil conflicts are associated with the global climate. Nature 476, 438–441.

Hsu, J.-C., Feng, H.-T., Wu, W.-J., 2004. Resistance and synergistic effects of insecticides in *Bactrocera dorsalis* (Diptera: Tephritidae) in Taiwan. J. Econ. Entomol. 97, 1682–1688.

Hsu, J.-C., Haymer, D.S., Wu, W.-J., Feng, H.-T., 2006. Mutations in the acetylcholinesterase gene of *Bactrocera dorsalis* associated with resistance organophosphorus insecticides. Insect Biochem. Mol. 38, 396–402.

Hsu, J.-C., Wu, W.-J., Haymer, D.S., Liao, H.-Y., Feng, H.-T., 2008. Alterations of the acetylcholinesterase enzyme in the oriental fruit fly *Bactrocera dorsalis* are correlated with resistance to the organophosphate insecticide fenitrothion. Insect Biochem. Mol. 38, 146–154.

Huang, C.-Y., Wang, C.-P., Hou, P.-C., 2007. Toads (*Bufo bankorensis*) influence litter chemistry but not litter invertebrates and litter decomposition rates in a subtropical forest of Taiwan. J. Trop. Ecol. 23, 161–168.

Huang, F., Leonard, B.R., Andow, D.A., 2007. Sugarcane borer (Lepidoptera: Crambidae) resistance to transgenic *Bacillus thuringiensis* maize. J. Econ. Entomol. 100, 164–171.

Huberty, A.F., Denno, R.F., 2006. Trade-off in investment between dispersal and ingestion capability in phytophagous insects and its ecological implications. Oecologia 149, 226–234.

Hudgins, J.W., Christiansen, E., Franceschi, V.R., 2003. Methyl jasmonate induces changes mimicking anatomical defenses in diverse members of the Pinaceae. Tree Physiol. 23, 361–371.

Hudgins, J.W., Christiansen, E., Franceschi, V.R., 2004. Induction of anatomically based defensive responses in stems of diverse conifers by methyl jasmonate: a phylogenetic perspective. Tree Physiol. 24, 251–264.

Hudson, J.R., Hanula, J.L., Horn, S., 2013. Removing Chinese privet from riparian forests still benefits pollinators five years later. Biol. Conserv. 167, 355–362.

Hudson, J.R., Hanula, J.L., Horn, S., 2014. Impacts of removing Chinese privet from riparian forests on plant communities and tree growth five years later. Forest Ecol. Manag. 324, 101–108.

Huettl, R.F., Mueller-Dombois, D. (Eds.), 1993. Forest Declines in the Atlantic and Pacific Regions. Springer-Verlag, Berlin.

Huffaker, C.B., Messenger, P.S. (Eds.), 1976. Theory and Practice of Biological Control. Academic, New York, NY.

Hughes, L., 1990. The relocation of ant nest entrances: potential consequences for ant-dispersed seeds. Aust. J. Ecol. 16, 207–214.

Hughes, L., Bazzaz, F.A., 1997. Effect of elevated CO_2 on interactions between the western flower thrips, *Frankliniella occidentalis* (Thysanoptera: Thripidae) and the common milkweed, *Asclepias syriaca*. Oecologia 109, 286–290.

Hughes, R.D., Tyndale-Biscoe, M., Walker, J., 1978. Effects of introduced dung beetles (Coleoptera: Scarabaeidae) on the breeding and abundance of the Australian bushfly, *Musca vetustissima* Walker (Diptera: Muscidae). B. Entomol. Res. 68, 361–372.

Huhta, V., Haimi, J., Setälä, H., 1991. Role of the fauna in soil processes: techniques using simulated forest floor. Agr. Ecosyst. Environ. 34, 223–229.

Hulme, P.E., 1994. Seedling herbivory in grassland: relative impact of vertebrate and invertebrate herbivores. J. Ecol. 82, 873–880.

Hulme, M.A., Shields, J.K., 1970. Biological control of decay fungi in wood by competition for non-structural carbohydrates. Nature 227, 300–301.

Hulme, P.E., Pyšek, P., Nentwig, W., Vilà, M., 2009. Will threat of biological invasions unite the European Union? Science 324, 40–41.

Hultmark, D., Engström, A., Bennich, H., Kapur, R., Boman, H.G., 1982. Insect immunity: isolation and structure of cecropin D and four minor antibacterial components from cecropia pupae. Eur. J. Biochem. 127, 207–217.

Hung, C.F., Kao, C.H., Liu, C.C., Lin, J.G., Sun, C.N., 1990. Detoxifying enzymes of selected insect species with chewing and sucking habits. J. Econ. Entomol. 83, 361–365.

646 BIBLIOGRAPHY

Hunt, H.W., Coleman, D.C., Ingham, E.R., Ingham, R.E., Elliott, E.T., Moore, J.C., Rose, S.L., Reid, C.P.P., Morley, C.R., 1987. The detrital food web in a shortgrass prairie. Biol. Fert. Soils 3, 57–68.

Hunter, M.D., 1987. Opposing effects of spring defoliation on late season oak caterpillars. Ecol. Entomol. 12, 373–382.

Hunter, M.D., 1992. A variable insect-plant interaction: the relationship between tree budburst phenology and population levels of insect herbivores among trees. Ecol. Entomol. 17, 91–95.

Hunter, J.M., 1993. Macroterme geophagy and pregnancy clays in southern Africa. Journal of Cultural Geography 14, 69–92.

Hunter, M.D., 2001a. Out of sight, out of mind: the impacts of root-feeding insects in natural and managed systems. Agr. Forest Entomol. 3, 3–9.

Hunter, M.D., 2001b. Insect population dynamics meets ecosystem ecology: effects of herbivory on soil nutrient dynamics. Agr. Forest Entomol. 3, 77–84.

Hunter, M.D., 2009. Trophic promiscuity, intraguild predation and the problem of omnivores. Agr. Forest Entomol. 11, 125–131.

Hunter, A.F., Arssen, L.W., 1988. Plants helping plants. BioScience 38, 34–40.

Hunter, A.F., Elkinton, J.S., 2000. Effects of synchrony with host plant on populations of a spring-feeding lepidopteran. Ecology 81, 1248–1261.

Hunter, M.D., Forkner, R.E., 1999. Hurricane damage influences foliar polyphenolics and subsequent herbivory on surviving trees. Ecology 80, 2676–2682.

Hunter, W.D., Hinds, W.E., 1904. The Mexican Cotton Boll Weevil. USDA Division of Entomology Bulletin 45. Government Printing Office, Washington, DC.

Hunter, M.R., Hunter, M.D., 2008. Designing for conservation of insects in the built environment. Insect Conserv. Diver. 1, 189–196.

Hunter, M.D., Price, P.W., 1992. Playing chutes and ladders: heterogeneity and the relative roles of bottom-up and top-down forces in natural communities. Ecology 73, 724–732.

Hunter, M.D., Schultz, J.C., 1993. Induced plant defenses breached? Phytochemical induction protects an herbivore from disease. Oecologia 94, 195–203.

Hunter, M.D., Schultz, J.C., 1995. Fertilization mitigates chemical induction and herbivore responses within damaged oak trees. Ecology 76, 1226–1232.

Hunter, M.D., Adl, S., Pringle, C.M., Coleman, D.C., 2003. Relative effects of macroinvertebrates and habitat on the chemistry of litter during decomposition. Pedobiologia 47, 101–115.

Huntly, N., 1991. Herbivores and the dynamics of communities and ecosystems. Annu. Rev. Ecol. Syst. 22, 477–503.

Huntly, P.M., Van Noort, S., Hamer, M., 2005. Giving increased value to invertebrates through ecotourism. S. Afr. J. Wildl. Res. 35, 53–62.

Hurlbert, S.H., 1984. Pseudoreplication and the design of ecological field experiments. Ecol. Monogr. 54, 187–211.

Huryn, A.D., Wallace, J.B., 1987. Local geomorphology as a determinant of macrofaunal production in a mountain stream. Ecology 68, 1932–1942.

Huryn, A.D., Wallace, J.B., 2000. Life history and production of stream insects. Annu. Rev. Entomol. 45, 83–110.

Husseneder, C., Grace, J.K., 2001a. Similarity is relative: hierarchy of genetic similarities in the Formosan subterranean termite (Isoptera: Rhinotermitidae) in Hawaii. Environ. Entomol. 30, 262–266.

Husseneder, C., Grace, J.K., 2001b. Evaluation of DNA fingerprinting, aggression tests, and morphometry as tools for colony delineation of the Formosan subterranean termite. J. Insect Behav. 14, 173–186.

Husseneder, C., Brandl, R., Epplen, C., Epplen, J.T., Kaib, M., 1998. Variation between and within colonies in the termite: morphology, genomic DNA, and behaviour. Mol. Ecol. 7, 983–990.

Husseneder, C., Brandl, R., Epplen, C., Epplen, J.T., Kaib, M., 1999. Within-colony relatedness in a termite species: genetic roads to eusociality? Behaviour 136, 1045–1063.

Husseneder, C., Vargo, E.L., Grace, J.K., 2003. Molecular genetic methods: new approaches to termite biology. In: Goodell, B., Nicholas, D.D., Schultz, T.P. (Eds.), Wood Deterioration and Preservation: Advances in Our Changing World. American Chemical Society, Washington, DC, pp. 358–369.

Huston, M., 1979. A general hypothesis of species diversity. Am. Nat. 113, 81–101.

Hutchinson, G.E., 1959. Homage to Santa Rosalia or why are there so many kinds of animals? Am. Nat. 93, 145–159.

Huxley, C.R., Cutler, D.F., 1991. Ant-Plant Interactions. Oxford University Press, Oxford, UK.

Hvenegaard, G.T., Delamere, T.A., Lemelin, R.H., Brager, K., Auger, A., 2013. Insect festivals: celebrating and fostering human-insect encounters. In: Lemelin, R.H. (Ed.), The Management of Insects in Recreation, Tourism. Cambridge University Press, Cambridge, UK, pp. 198–216.

Ingham, E.R., 1985. Review of the effects of 12 selected biocides on target and non-target soil organisms. Crop Prot. 4, 3–32.

Ingham, R.E., Trofymow, J.A., Ingham, E.R., Coleman, D.C., 1985. Interactions of bacteria, fungi, and their nematode grazers: effects on nutrient cycling and plant growth. Ecol. Monogr. 55, 119–140.

Ingham, E.R., Cambardella, C., Coleman, D.C., 1986. Manipulation of bacteria, fungi and protozoa by biocides in lodgepole pine forest soil microcosms: effects on organism interactions and nitrogen mineralization. Can. J. Soil Sci. 66, 261–272.

Inkley, M.D., Wissinger, S.A., Baros, B.L., 2008. Effects of drying regime on microbial colonization and shredder preference in seasonal woodland wetlands. Freshwater Biol. 53, 435–445.

Inouye, D.W., 1982. The consequences of herbivory: a mixed blessing for *Jurinea mollis* (Asteracea). Oikos 39, 269–272.

Inouye, B., Stinchcombe, J.R., 2001. Relationships between ecological interaction modifications and diffuse coevolution: similarities, differences, and causal links. Oikos 95, 353–360.

Inouye, R.S., Byers, G.S., Brown, J.H., 1980. Effects of predation and competition on survivorship, fecundity, and community structure of desert annuals. Ecology 61, 1344–1351.

Irby, W.S., Apperson, C.S., 1988. Hosts of mosquitoes in the coastal plain of North Carolina. J. Med. Entomol. 25, 85–93.

Irvine, J., Law, B.E., Kurpius, M.R., 2005. Coupling of canopy gas exchange with root and rhizosphere respiration in a semi-arid forest. Biogeochemistry 73, 271–282.

Irvine, F., Cwynar, L.C., Vermaire, J.C., Rees, A.B.H., 2012. Midge-inferred temperature reconstructions and vegetation change over the last ~15,000 years from Trout Lake, northern Yukon Territory, eastern Beringia. J. Paleolimnol. 48, 133–146.

Irwin, R.E., Warren, P.S., Carper, A.L., Adler, L.S., 2014. Plant–animal interactions in suburban environments: implications for floral evolution. Oecologia 174, 803–815.

Isaacs, R., Byrne, D.N., 1998. Aerial distribution, flight behaviour and eggload: their inter-relationship during dispersal by the sweet-potato whitefly. J. Anim. Ecol. 67, 741–750.

Isaacs, R., Tuell, J., Fiedler, A., Gardiner, M., Landis, D., 2009. Maximizing arthropod-mediated ecosystem services in agricultural landscapes: the role of native plants. Front. Ecol. Environ. 7, 196–203.

Isaev, A.S., Khlebopros, R.G., 1979. Inertial and noninertial factors regulating forest insect population density. In: Norton, G.A., Holling, C.S. (Eds.), Pest Management. Proceedings of International Conference, Oct. 25–29, 1976, Pergamon Press, Oxford, UK, pp. 317–339.

Ishii, Y., Shimada, M., 2010. The effect of learning and search images on predator-prey interactions. Popul. Ecol. 52, 27–35.

Ishimoto, H., Sakai, T., Kitamoto, T., 2009. Ecdysone signaling regulates the formation of long-term courtship memory in adult *Drosphila melanogaster*. Proc. Natl. Acad. Sci. USA 106, 6381–6386.

Istock, C.A., 1973. Population characteristics of a species ensemble of water-boatmen (Corixidae). Ecology 54, 535–544.

Istock, C.A., 1977. Logistic interaction of natural populations of two species of waterboatmen. Am. Nat. 111, 279–287.

Istock, C.A., 1981. Natural selection and life history variation: theory plus lessons from a mosquito. In: Denno, R.F., Dingle, H. (Eds.), Insect Life History Patterns: Habitat and Geographic Variation. Springer-Verlag, New York, pp. 113–127.

Ives, A.R., Andow, D.A., 2002. Evolution of resistance to Bt crops: directional selection in structured environments. Ecol. Lett. 5, 792–801.

Iwasaki, T., 1990. Predatory behavior of the praying mantis, *Tenodera aridifolia*. I. effect of prey size on prey recognition. J. Ethol. 8, 75–79.

Iwasaki, T., 1991. Predatory behavior of the praying mantis, *Tenodera aridifolia*. II. Combined effect of prey size and predator size on the prey recognition. J. Ethol. 9, 77–81.

Izaguirre, M.M., Mazza, C.A., Astigueta, M.S., Ciarla, A.M., Ballaré, C.L., 2013. No time for candy: passionfruit (*Passiflora edulis*) plants down-regulate damage-induced extra floral nectar production in response to light signals of competition. Oecologia 173, 213–221.

Jabiol, J., Cornut, J., Danger, M., Jouffroy, M., Elger, A., Chauvet, E., 2014. Litter identity mediates predator impacts on the functioning of an aquatic detritus-based food web. Oecologia 176, 225–235.

Jabłoński, P., 1999. A rare predator exploits prey escape behavior: the role of tail-fanning and plumage contrast in foraging of the painted redstart (*Myioborus pictus*). Behav. Ecol. 10, 7–14.

Jackson, R.V., Kollmann, J., Grubb, R.J., Bee, J.N., 1999. Insect herbivory on five European tall-shrub species related to leaf ontogeny and quality, and the need to distinguish pre-expansion and expanding leaves. Oikos 87, 561–570.

Jackson, D.M., Johnson, A.W., Stephenson, M.G., 2002. Survival and development of *Heliothis virescen*s (Lepidoptera: Noctuidae) larvae on isogenic tobacco lines with different levels of alkaloids. J. Econ. Entomol. 95, 1294–1302.

Jactel, H., Brockerhoff, E.G., 2007. Tree diversity reduces herbivory by forest insects. Ecol. Lett. 10, 835–848.

Jakobsson, A., Ågren, J., 2014. Distance to semi-natural grassland influences seed production of insect-pollinated herbs. Oecologia 175, 199–208.

Jamali, H., Livesley, S.J., Dawes, T.Z., Hutley, L.B., Arndt, S.K., 2011. Termite mound emissions of CH_4 and CO_2 are primarily determined by seasonal changes in termite biomass and behavior. Oecologia 167, 525–534.

James, A.A., 2002. Engineering mosquito resistance to malaria parasites: the avian malaria model. Insect Biochem. Mol. 32, 1317–1323.

James, A.B.W., Dewson, Z.S., Death, R.G., 2008. Do stream macroinvertebrates use instream refugia in response to severe short-term flow reduction in New Zealand streams? Freshwater Biol. 53, 1316–1334.

Jandér, K.C., Herre, E.A., 2010. Host sanctions and pollinator cheating in the fig tree-fig wasp mutualism. Proc. Roy. Soc. B-Biol. Sci. 277, 1481–1488.

Janssen, R.H.H., Meinders, M.B.J., van Nes, E.H., Scheffer, M., 2008. Microscale vegetation-soil feedback boosts hysteresis in a regional vegetation-climate system. Glob. Change Biol. 14, 1104–1112.

Janssens, L., Stoks, R., 2014. Reinforcing effects of non.pathogenic bacteria and predation risk: from physiology to life history. Oecologia 176, 323–332.

Janzen, D.H., 1966. Coevolution of mutualism between ants and acacias in Central America. Evolution 20, 249–275.

Janzen, D.H., 1977. What are dandelions and aphids? Am. Nat. 111, 586–589.

Janzen, D.H., 1981. Patterns of herbivory in a tropical deciduous forest. Biotropica 13, 271–282.

Janzen, D.H., 1985. On ecological fitting. Oikos 45, 308–310.

Janzen, D.H., 2000. Costa Rica's Area de Conservacion Guanacaste: a long march to survival through non-damaging biodevelopment. Biodiversity 1, 7–20.

Janzen, D.H., 2004. Setting up tropical biodiversity for conservation through non-damaging use: participation by parataxonomists. J. Appl. Ecol. 41, 181–187.

Janzen, D.H., Martin, P.S., 1982. Neotropical anachronisms: the fruits the gomphotheres ate. Science 215, 19–27.

Janzen, D.H., Hallwachs, W., Burns, J.M., 2010. A tropical horde of counterfeit predator eyes. Proc. Natl. Acad. Sci. USA 107, 11659–11665.

Jarosz, J., 1995. Haemolymph immune proteins protect the insect body cavity from invading bacteria. Comp. Biochem. Physiol. 111C, 213–220.

Jeffries, J.M., Marquis, R.J., Forkner, R.E., 2006. Forest age influences oak insect community structure, richness, and density. Ecol. Appl. 16, 901–912.

Jenkins, M.J., Herbertson, E., Page, W., Jorgensen, C.A., 2008. Bark beetles, fuels, fires and implications for forest management in the Intermountain West. Forest Ecol. Manag. 254, 16–34.

Jentsch, A., Kreyling, J., Beierkuhnlein, C., 2007. A new generation of climate-change experiments: events, not trends. Front. Ecol. Environ. 5, 365–374.

Jentsch, A., Kreyling, J., Elmer, M., Gellesch, E., Glaser, B., Grant, K., Hein, R., Lara, M., Mirzae, H., Nadler, S.E., Nagy, L., Otieno, D., Pritsch, K., Rascher, U., Schädler, M., Schloter, M., Singh, B.K., Stadler, J., Walter, J., Wellstein, C., Jens Wöllecke, Beierkuhnlein, C., 2011. Climate extremes initiate ecosystem-regulating functions while maintaining productivity. J. Ecol. 99, 689–702.

Jepsen, J.U., Hagen, S.B., Ims, R.A., Yaccoz, N.G., 2008. Climate change and outbreaks of the geometrids Operophthera brumata and Epirrita autumnata in subarctic birch forest: evidence of a recent outbreak range expansion. J. Anim. Ecol. 77, 257–264.

Jepson, P.C., Thacker, J.R.M., 1990. Analysis of the spatial component of pesticide side-effects on non-target invertebrate populations and its relevance to hazard analysis. Funct. Ecol. 4, 349–355.

Jetz, W., Carbone, C., Fulford, J., Brown, J.H., 2004. The scaling of animal space use. Science 306, 266–268.

Jhee, E.M., Dandridge, K.L., Christy, Jr., A.M., Pollard, A.J., 1999. Selective herbivory on low-zinc phenotypes of the hyperaccumulator Thlaspi caerulescens (Brassicaceae). Chemoecology 9, 93–95.

Joel, D.M., Juniper, B.E., Dafni, A., 1985. Ultraviolet patterns in the traps of carnivorous plants. New Phytol. 101, 585–593.

Joern, A., Behmer, S.T., 1998. Impact of diet quality on demographic attributes in adult grasshoppers and the nitrogen limitation hypothesis. Ecol. Entomol. 23, 174–184.

Johansson, F., 1993. Effects of prey type, prey density and predator presence on behaviour and predation risk in a larval damselfly. Oikos 68, 481–489.

Johnsen, S., Lohmann, K.J., 2005. The physics and neurobiology of magnetoreception. Nat. Rev. Neurosci. 6, 703–712.

Johnson, M.D., 2000. Effects of shade-tree species and crop structure on the winter arthropod and bird communities in a Jamaican shade coffee plantation. Biotropica 32, 133–145.

Johnson, D.M., 2004. Life history and demography of Cephaloleia fenestrate (Hispinae: Chrysomelidae: Coleoptera). Biotropica 36, 352–361.

Johnson, C.N., 2009. Ecological consequences of late Quaternary extinctions of megafauna. Proc. Roy. Soc. B-Biol. Sci. 276, 2509–2519.

Johnson, D.H., 2006. The many faces of replication. Crop Prot. 46, 2486–2491.

Johnson, M.T.J., Agrawal, A.A., 2005. Plant genotype and environment interact to shape a diverse arthropod community on evening primrose (Oenothera biennis). Ecology 86, 874–885.

Johnson, S.D., Bond, W.J., 1994. Red flowers and butterfly pollination in the fynbos of South Africa. In: Arianoutsou, M., Graves, R.H. (Eds.), Plant-Animal Interactions in Mediterranean-Type Ecosystems. Kluwer, Dordrecht, The Netherlands, pp. 137–148.

Johnson, M.P., Simberloff, D.S., 1974. Environmental determinants of island species numbers in the British Isles. J. Biogeogr. 1, 149–154.

Johnson, K.A., Whitford, W.G., 1975. Foraging ecology and relative importance of subterranean termites in Chihuahuan Desert ecosystems. Environ. Entomol. 4, 66–70.

Johnson, D.M., Bjørnstad, O.N., Liebhold, A.M., 2004. Landscape geometry and traveling waves in the larch budmoth. Ecol. Lett. 7, 967–974.

Johnson, S.D., Collin, C.L., Wissman, H.J., Halvarsson, E., Ågren, J., 2004. Factors contributing to variation in seed production among remnant populations of the endangered daisy Gerbera aurantiaca. Biotropica 36, 148–155.

Johnson, D.M., Liebhold, A.M., Bjørnstad, O.N., McManus, M.L., 2005. Circumpolar variation in periodicity and synchrony among gypsy moth populations. J. Anim. Ecol. 74, 882–892.

Johnson, D.M., Liebhold, A.M., Tobin, P.C., Bjørnstad, O.N., 2006. Allee effects and pulsed invasion by the gypsy moth. Nature 444, 361–363.

Johnson, M.L., Armitage, S., Scholz, B.C.G., Merritt, D.J., Cribb, B.W., Zalucki, M.P., 2006. Predator presence moves Helicoverpa armigera larvae to distraction. J. Insect Behav. 20, 1–18.

Johnson, M.T.J., Lajeunesse, M.J., Agrawal, A.A., 2006. Additive and interactive effects of plant genotypic diversity on arthropod communities and plant fitness. Ecol. Lett. 9, 24–34.

Johnson, M.D., Levy, N.J., Kellermann, J.L., Robinson, D.E., 2009. Effects of shade and bird exclusion on arthropods and leaf damage on coffee farms in Jamaica's Blue Mountains. Agroforest. Syst. 76, 139–148.

Johnson, R.M., Evans, J.D., Robinson, G.E., Berenbaum, M.R., 2009. Changes in transcript abundance relating to colony collapse disorder in honey bees (Apis mellifera). Proc. Natl. Acad. Sci. USA 106, 14790–14795.

Johnson, M.D., Kellermann, J.L., Stercho, A.M., 2010. Pest reduction services by birds in shade and sun coffee in Jamaica. Anim. Conserv. 13, 140–147.

Jolivet, P., 1996. Ants and Plants: An Example of Coevolution. Backhuys Publishers, Leiden, The Netherlands.

Jonas, J.L., Joern, A., 2007. Grasshopper (Orthoptera: Acrididae) communities respond to fire, bison grazing and weather in North American tallgrass prairie: a long-term study. Oecologia 153, 699–711.

Jøndrup, P.M., Barnes, J.D., Port, G.R., 2002. The effect of ozone fumigation and different *Brassica rapa* lines on the feeding behaviour of *Pieris brassicae* larvae. Entomol. Exp. Appl. 104, 143–151.

Jones, C.R., 1917. Grasshopper Control. Colorado Agricultural Experiment Station Bulletin 233, Ft. Collins, CO.

Jones, C.G., 1984. Microorganisms as mediators of plant resource exploitation by insect herbivores. In: Price, P.W., Gaud, W.S., Slobodchikoff, C.N. (Eds.), A New Ecology: Novel Approaches to Interactive Systems. John Wiley, New York, pp. 53–100.

Jones, J.A., 1989. Environmental influences on soil chemistry in central semiarid Tanzania. Soil Sci. Soc. Am. J. 53, 1748–1758.

Jones, J.A., 1990. Termites, soil fertility and carbon cycling in dry tropical Africa: a hypothesis. J. Trop. Ecol. 6, 291–305.

Jones, C.G., Lawton, J.H., Shachak, M., 1994. Organisms as ecosystem engineers. Oikos 69, 373–386.

Jones, C.G., Ostfeld, R.S., Richard, M.P., Schauber, E.M., Wolff, J.O., 1998. Chain reactions linking acorns to gypsy moth outbreaks and Lyme disease risk. Science 279, 1023–1026.

Jones, M.E., Paine, T.D., Fenn, M.E., Poth, M.A., 2004. Influence of ozone and nitrogen deposition on bark beetle activity under drought conditions. Forest Ecol. Manag. 200, 67–76.

Jonkman, J.C.M., 1978. Nests of the leaf-cutting ant *Atta vollenweideri* as accelerators of succession in pastures. Z. Angew. Entomol. 86, 25–34.

Joo, E., van Langenhove, H., Simpraga, M., Steppe, K., Amelynck, C., Schoon, N., Muller, J.-F., Dewulf, J., 2010. Variation in biogenic volatile organic compound emission pattern of *Fagus sylvatica* L. due to aphid infection. Atmos. Environ. 44, 227–234.

Joseph, G.S., Seymour, C.L., Cumming, G.S., Cumming, D.H.M., Mahlangu, Z., 2014. Termite mounds increase functional diversity of woody plants in African savannas. Ecosystems 17, 808–819.

Jouquet, P., Dauber, J., Lagerlöf, J., Lavelle, P., Lapage, M., 2006. Soil invertebrates as ecosystem engineers: effects on soil and feedback loops. Appl. Soil Ecol. 32, 153–164.

Jouzel, J., Masson-Delmotte, V., Cattani, O., Dreyfus, G., Falourd, S., Hoffmann, G., Minster, B., Nouet, J., Barnola, J.M., Chappellaz, J., Fischer, H., Gallet, J.C., Johnsen, S., Leuenberger, M., Loulergue, L., Luethi, D., Oerter, H., Parrenin, F., Raisbeck, G., Raynaud, D., Schilt, A., Schwander, J., Selmo, E., Souchez, R., Spahni, R., Stauffer, B., Steffensen, J.P., Stenni, B., Stocker, T.F., Tison, J.L., Werner, M., Wolff, E.W., 2007. Orbital and millennial Antarctic climate variability over the ast 800,000 years. Science 317, 793–796.

Juang, J.-Y., Katul, G.G., Porporato, A., Stoy, P.C., Sequeira, M.S., Detto, M., Kim, H.-S., Oren, R., 2007. Eco-hydrological controls on summertime convective rainfall triggers. Glob. Change Biol. 13, 887–896.

Jung, C., Croft, B.A., 2001. Aerial dispersal of phytoseiid mites (Acari: Phytoseiidae): estimating falling speed and dispersal distance of adult females. Oikos 94, 182–190.

Juniper, B.E., Robins, R.J., Joel, D.M., 1989. The Carnivorous Plants. Academic Press, London.

Jurgensen, M.F., Finér, L., Domisch, T., Kilpeläinen, J., Punttila, P., Ohashi, M., Niemelä, P., Sundström, L., Neuvonen, S., Risch, A.C., 2008. Organic mound-building ants: their impact on soil properties in temperate and boreal forests. J. Appl. Entomol. 132, 266–275.

Kaae, R.S., Shorey, H.H., Gaston, L.K., 1973. Pheromone concentration as a mechanism for reproductive isolation between two lepidopterous species. Science 179, 487–488.

Käärik, A.A., 1974. Decomposition of wood. In: Dickinson, C.H., Pugh, G.J.F. (Eds.), Biology of Plant Litter Decomposition. Academic Press, London, pp. 129–174.

Kaczmarek, M., Wasilewski, A., 1977. Dynamics of numbers of the leaf-eating insects and its effect on foliage production in the "Grabowy" Reserve in the Kampinos National Park. Ekol. Pol.-Pol. J. Ecol. 25, 653–673.

Kaib, M., Husseneder, C., Epplen, C., Epplen, J.T., Brandl, R., 1996. Kin-biased foraging in a termite. Proc. Roy. Soc. Lond. B Bio. 263, 1527–1532.

Kainulainen, P., Holopainen, J.K., Hyttinen, H., Oksanen, J., 1994. Effect of ozone on the biochemistry and aphid infestation of Scots pine. Phytochemistry 35, 39–42.

Kaiser, J., 2005. Acadia: The Complete Guide: Mount Desert Island and Acadia National Park. Destination Press.

Kaiser, M.N., Sutherst, R.W., Bourne, A.S., 1982. Relationship between ticks and zebu cattle in southern Uganda. Trop. Anim. Health Pro. 14, 63–74.

Kalnay, E., Cai, M., 2003. Impact of urbanization and land-use on climate. Nature 423, 528–531.

Kamil, A.C., Krebs, J.R., Pulliam, H.R. (Eds.), 1987. Foraging Behavior. Plenum Press, New York, 676 p.

Kang, S., Kim, S., Oh, S., Lee, D., 2004. Predicting spatial and temporal patterns of soil temperature based on topography, surface cover and air temperature. Forest Ecol. Manag. 136, 173–184.

Kaplan, I., Halitschke, R., Kessler, A., Sardanelli, S., Denno, R.F., 2008. Constitutive and induced defenses to herbivory in above- and belowground plant tissues. Ecology 89, 392–406.

Kappes, H., Catalano, C., Topp, W., 2007. Coarse woody debris ameliorates chemical and biotic soil parameters of acidified broad-leaved forests. Appl. Soil Ecol. 36, 190–198.

Karban, R., 2001. Communication between sagebrush and wild tobacco in the field. Biochem. Syst. Ecol. 29, 995–1005.

Karban, R., Agrawal, A.A., 2002. Herbivore offense. Annu. Rev. Ecol. Syst. 33, 641–664.

Karban, R., Baldwin, I.T., 1997. Induced Responses to Herbivory. University of Chicago Press, Chicago, IL.

Karban, R., Maron, J., 2002. The fitness consequences of interspecific eavesdropping between plants. Ecology 83, 1209–1213.

Karban, R., Niiho, C., 1995. Induced resistance and susceptibility to herbivory: plant memory and altered plant development. Ecology 76, 1220–1225.

Karban, R., Baldwin, I.T., Baxter, K.J., Laue, G., Felton, G.W., 2000. Communication between plants: induced resistance in wild tobacco plants following clipping of neighboring sagebrush. Oecologia 125, 66–71.

Kareiva, P., 1983. Influence of vegetation texture on herbivore populations: resource concentration and herbivore movement. In: Denno, R.F., McClure, M.S. (Eds.), Variable Plants and Herbivores in Natural and Managed Systems. Academic Press, New York, pp. 259–289.

Karley, A.J., Parker, W.E., Pitchford, J.W., Douglas, A.E., 2004. The mid-season crash in aphid populations: why and how does it occur? Ecol. Entomol. 29, 383–388.

Karlsen, S.R., Jepsen, J.U., Odland, A., Ims, R.A., Elvebakk, A., 2013. Outbreaks by canopy-feeding geometrid moth cause state-dependent shifts in understorey plant communities. Oecologia 173, 859–870.

Karpestam, E., Merilaita, S., Forsman, A., 2012. Reduced predation risk for melanistic pygmy grasshoppers in post-fire environments. Ecol. Evol. 2, 2204–2212.

Kasina, J.M., Mburu, J., Kraemer, M., Holm-Mueller, K., 2009. Economic benefit of crop pollination by bees: a case of Kakamega small-holder farming in western Kenya. J. Econ. Entomol. 102, 467–473.

Kaspari, M., Clay, N.A., Donoso, D.A., Yanoviak, S.P., 2009. Sodium fertilization increases termites and enhances decomposition in an Amazonian forest. Ecology 95, 795–800.

Kaur, T., Singh, B., Kaur, A., Kaur, S., 2015. Endophyte-mediated interactions between cauliflower, the herbivore *Spodoptera litura*, and the ectoparasitoid *Bracon hebetor*. Oecologia 179, 487–494.

Kaushal, S.S., Pace, M.L., Groffman, P.M., Band, L.E., Belt, K.T., Mayer, P.M., Welty, C., 2010. Land use and climate variability amplify contaminant pulses. EOS Trans. Am. Geophys. Union 91, 221–222.

Kawaguchi, Y., Nakano, S., 2001. Contribution of terrestrial invertebrates to the annual resource budget for salmonids in forest and grassland reaches of a headwater stream. Freshwater Biol. 46, 303–316.

Kawecki, T.J., 2010. Evolutionary ecology of learning: insights from fruit flies. Popul. Ecol. 52, 15–25.

Kaye, J.P., Groffman, P.M., Grimm, N.B., Baker, L.A., Pouyat, R.V., 2006. A distinct urban biogeochemistry? Trends Ecol. Evol. 21, 192–199.

Kazemi-Dinan, A., Sauer, J., Stein, R.J., Krämer, U., Müller, C., 2015. Is there a trade-off between glucosinolate-based organic and inorganic defences in a metal hyperaccumulator in the field? Oecologia 178, 369–378.

Keeling, C.D., Whorf, T.P., Wahlen, M., van der Pilcht, J., 1995. Interannual extremes in the rate of rise of atmospheric carbon dioxide since 1980. Science 375, 666–670.

Keesing, F., Holt, R.D., Ostfeld, R.S., 2006. Effects of species diversity on disease risk. Ecol. Lett. 9, 485–498.

Keeton, W.S., Franklin, J.F., 2005. Do remnant old-growth trees accelerate rates of succession in mature Douglas-fir forests? Ecol. Monogr. 75, 103–118.

Kelber, A., 1996. Colour learning in the hawkmoth *Macroglossum stellatarum*. J. Exp. Biol. 199, 1127–1131.

Kellermann, J.L., Johnson, M.D., Stercho, A.M., Hackett, S.C., 2008. Ecological and economic services provided by birds on Jamaican Blue Mountain coffee farms. Conserv. Biol. 22, 1177–1185.

Kelly, D.J., Bothwell, M.L., Schindler, D.W., 2003. Effects of solar ultraviolet radiation on stream benthic communities: an intersite comparison. Ecology 84, 2724–2740.

Kempton, R.A., 1979. The structure of species abundance and measurement of diversity. Biometrics 35, 307–321.

Kennedy, J.S., 1975. Insect dispersal. In: Pimentel, D. (Ed.), Insects, Science, and Society. Academic Press, New York, pp. 103–119.

Kennedy, C.M., Lonsdorf, E., Neel, M.C., Williams, N.M., Ricketts, T.H., Winfree, R., Bommarco, R., Brittain, C., Burley, A.L., Cariveau, D., Carvalheiro, L.G., Chacoff, N.P., Cunningham, S.A., Danforth, B.N., Dudenhöffer, J.-H., Elle, E., Gaines, H.R., Garibaldi, L.A., Gratton, C., Holzschuh, A., Isaacs, R., Javorek, S.K., Jha, S., Klein, A.M., Krewenka, K., Mandelik, Y., Mayfield, M.M., Morandin, L., Neame, L.A., Otieno, M., Park, M., Potts, S.G., Rundlöf, M., Saez, A., Steffan-Dewenter, I., Taki, H., Viana, B.F., Westphal, C., Wilson, J.K., Greenleaf, S.S., Kremen, C., 2013. A global quantitative synthesis of local and landscape effects on wild bee pollinators in agroecosystems. Ecol. Lett. 16, 584–599.

Kerridge, A., Lappin-Scott, H., Stevens, J.R., 2005. Antibacterial properties of larval secretions of the blowfly, *Lucilia sericata*. Med. Vet. Entomol. 19, 333–337.

Kersch-Becker, M.F., Thaler, J.S., 2014. Virus strains differentially induce plant susceptibility to aphid vectors and chewing herbivores. Oecologia 174, 883–892.

Kessler, A., Baldwin, I.T., 2001. Defensive function of herbivore-induced plant volatile emissions in nature. Science 291, 2141–2144.

Kessler, A., Baldwin, I.T., 2002. Plant responses to herbivory: the emerging molecular analysis. Annu. Rev. Plant Biol. 53, 299–328.

Kessler, A., Halitschke, R., Baldwin, I.T., 2004. Silencing the jasmonate cascade: induced plant defenses and insect populations. Science 305, 665–668.

Kessler, A., Halitschke, R., Diezel, C., Baldwin, I.T., 2006. Priming of plant defense responses in nature by airborne signaling between *Artemisia tridentata* and *Nicotiana attenuata*. Oecologia 148, 280–292.

Kettlewell, H.B.D., 1956. Further selection experiments on industrial melanism in the Lepidoptera. Heredity 10, 287–301.

Key, K.H.L., Day, M.F., 1954. A temperature-controlled physiological colour response in the grasshopper *Kosciuscola tristis* Sjost. (Orthoptera: Acrididae). Aust. J. Zool. 2, 309–339.

Khalil, M.A.K., Rasmussen, R.A., French, J.R.J., Holt, J.A., 1990. The influence of termites on atmospheric trace gases: CH_4, CO_2, $CHCl_3$, N_2O, CO, H_2, and light hydrocarbons. J. Geophys. Res. 95, 3619–3634.

Kharboutli, M.S., Mack, T.P., 1993. Tolerance of the striped earwig (Dermaptera: Labiduridae) to hot and dry conditions. Environ. Entomol. 22, 663–668.

Khrimian, A., Shirali, S., Vermillion, K.E., Siegler, M.A., Guzman, F., Chauhan, K., Aldrich, J.R., Weber, D.C., 2014. Determination of the stereochemistry of the aggregation pheromone of harlequin bug, *Murgantia histrionica*. J. Chem. Ecol. 40, 1260–1268.

Kiffney, P.M., Richardson, J.S., Bull, J.P., 2003. Responses of periphyton and insects to experimental manipulation of riparian buffer width along forest streams. J. Appl. Ecol. 40, 1060–1076.

Kiilsgaard, C.W., Greene, S.E., Stafford, S.G., 1987. Nutrient concentration in litterfall from some western conifers with special reference to calcium. Plant Soil 102, 223–227.

Kikkert, D.A., Crowl, T.A., Covich, A.P., 2009. Upstream migration of amphidromous shrimps in the Luquillo Experimental Forest, Puerto Rico: temporal patterns and environmental cues. J. N. Am. Benthol. Soc. 28, 233–246.

Kim, Y., Kim, N., 1997. Cold hardiness in *Spodoptera exigua* (Lepidoptera: Noctuidae). Environ. Entomol. 26, 1117–1123.

Kim, T.N., Underwood, N., 2015. Plant neighborhood effects on herbivory: damage is both density and frequency dependent. Ecology 96, 1431–1437.

Kimberling, D.N., Karr, J.R., Fore, L.S., 2001. Measuring human disturbance using terrestrial invertebrates in the shrub-steppe of eastern Washington (USA). Ecol. Indic. 1, 63–81.

Kimmins, J.P., 1972. Relative contributions of leaching, litterfall, and defoliation by *Neodiprion sertifer* (Hymenoptera) to the removal of cesium-134 from red pine. Oikos 23, 226–234.

Kimmins, J.P., Blanco, J.A., Seely, B., Welham, C., Scoullar, K., 2008. Complexity in modelling forest ecosystems: how much is enough? Forest Ecol. Manag. 256, 1646–1658.

King, T.J., 1977a. The plant ecology of ant-hills in calcareous grasslands. I. Patterns of species in relation to ant-hills in southern England. J. Ecol. 65, 235–256.

King, T.J., 1977b. The plant ecology of ant-hills in calcareous grasslands. II. Succession on the mounds. J. Ecol. 65, 257–278.

King, E.J., Spink, W.T., 1969. Foraging galleries of the Formosan subterranean termite, Coptotermes formosanus, in Louisiana. Ann. Entomol. Soc. Am. 62, 536–542.

King, A.A., Costantino, R.F., Cushing, J.M., Henson, S.M., Desharnais, R.A., Dennis, B., 2004. Anatomy of a chaotic attractor: subtle model-predicted patterns revealed in population data. Proc. Natl. Acad. Sci. USA 101, 408–413.

Kinn, D.N., 1980. Mutualism between *Dendrolaelaps neodisetus* and *Dendroctonus frontalis*. Environ. Entomol. 9, 756–758.

Kinney, K.K., Lindroth, R.L., Jung, S.M., Nordheim, E.V., 1997. Effects of CO_2 and NO_3^- availability on deciduous trees: phytochemistry and insect performance. Ecology 78, 215–230.

Kinyamario, J.I., Imbamba, S.K., 1992. Savanna at Nairobi National Park, Nairobi. In: Long, S.P., Jones, M.B., Roberts, M.J. (Eds.), Primary Productivity of Grass Ecosystems of the Tropics and Sub-tropics. Chapman and Hall, London, pp. 25–69.

Kishtawal, C.M., Niyogi, D., Tewari, M., Pielke, Sr., R.A., Shepherd, J.M., 2010. Urbanization signature in the observed heavy rainfall climatology over India. Int. J. Climatol. 30, 1908–1916.

Kistner, E.J., Belovsky, G.E., 2014. Host dynamics determine responses to disease: additive vs. compensatory mortality in a grasshopper–pathogen system. Ecology 95, 2579–2588.

Kitchell, J.F., O'Neill, R.V., Webb, D., Gallepp, G.W., Bartell, S.M., Koonce, J.F., Ausmus, B.S., 1979. Consumer regulation of nutrient cycling. BioScience 29, 28–34.

Kitching, R.L., Bergelson, J.M., Lowman, M.D., McIntyre, S., Carruthers, G., 1993. The biodiversity of arthropods from Australian rainforest canopies: general introduction, methods, sites and ordinal results. Aust. J. Ecol. 18, 181–191.

Kizlinski, M.L., Orwig, D.A., Cobb, R.C., Foster, D.R., 2002. Direct and indirect ecosystem consequences of an invasive pest on forests dominated by eastern hemlock. J. Biogeogr. 29, 1489–1503.

Klapwijk, M.J., Gröbler, B.C., Ward, K., Wheeler, D., Lewis, O.T., 2010. Influence of experimental warming and shading on host-parasitoid synchrony. Glob. Change Biol. 16, 102–112.

Klein, B.C., 1989. Effects of forest fragmentation on dung and carrion beetle communities in central Amazonia. Ecology 70, 1715–1725.

Klein, A.-M., Steffan-Dewenter, I., Tscharntke, T., 2003. Fruit set of highland coffee increases with the diversity of pollinating bees. Proc. Roy. Soc. Lond. B Bio 270, 955–961.

Klein, A.-M., Vaissière, B.E., Cane, J.H., Steffan-Dewenter, I., Cunningham, S.A., Kremen, C., Tscharntke, T., 2007. Importance of pollinators in changing landscapes for world crops. Proc. Roy. Soc. B-Biol. Sci. 274, 303–313.

Klemola, T., Andersson, T., Ruohomäki, K., 2014. Delayed density-dependent parasitism of eggs and pupae as a contributor to the cyclic population dynamics of the autumnal moth. Oecologia 175, 1211–1225.

Klepzig, K.D., Smalley, E.B., Raffa, K.F., 1996. Combined chemical defenses against an insect-fungal complex. J. Chem. Ecol. 22, 1367–1388.

Klock, G.O., Wickman, B.E., 1978. Ecosystem effects. In: Brookes, M.H., Stark, R.W., Campbell, R.W. (Eds.), The Douglas-fir Tussock Moth: A Synthesis . USDA Forest Service, Tech. Bull. 1585, USDA Forest Service, Washington, DC, pp. 90–95.

Klok, C.J., Chown, S.L., Gaston, K.J., 2003. The geographical range structure of the holly leaf-miner. III. Cold hardiness physiology. Funct. Ecol. 17, 858–868.

Kluth, S., Kruess, A., Tscharntke, T., 2002. Insects as vectors of plant pathogens: mutualistic and antagonistic interactions. Oecologia 133, 193–199.

Knapp, A.K., Seastedt, T.R., 1986. Detritus accumulation limits productivity of tallgrass prairie. BioScience 36, 662–668.

Kneitel, J.M., Miller, T.E., 2003. Dispersal rates affect species composition in metacommunities of *Sarracenia purpurea* inquilines. Am. Nat. 162, 165–171.

Knight, T.M., Holt, R.D., 2005. Fire generates spatial gradients in herbivory: an example from a Florida sandhill ecosystem. Ecology 86, 587–593.

Knight, T.M., McCoy, M.W., Chase, J.M., McCoy, K.A., Holt, R.D., 2005a. Trophic cascades across ecosystems. Nature 437, 880–883.

Knight, T.M., Steets, J.A., Vamosi, J.A., Mazer, S.J., Burd, M., Campbell, D.R., Dudash, M.R., Johnston, M.O., Mitchell, R.J., Ashman, T.-L., 2005b. Pollen limitation of plant reproduction: pattern and process. Annu. Rev. Ecol. Evol. S. 36, 467–497.

Knops, J.M.H., Tilman, D., Haddad, N.M., Naeem, S., Mitchell, C.E., Haarstad, J., Ritchie, M.E., Howe, K.M., Reich, P.B., Siemann, E., Groth, J., 1999. Effects of plant species richness on invasion dynamics, disease outbreaks, insect abundances and diversity. Ecol. Lett. 2, 286–293.

Knorr, M., Frey, S.D., Curtis, P.S., 2005. Nitrogen additions and litter decomposition: a meta-analysis. Ecology 85, 3252–3257.

Koch, G.W., Sillett, S.C., Jennings, G.M., Davis, S.D., 2004. The limits to tree height. Nature 428, 851–854.

Koenig, W.D., Liebhold, A.M., 2005. Effects of periodical cicada emergences on abundances and synchrony of avian populations. Ecology 86, 1873–1882.

Koetsier, P., Bryan, C.F., 1989. Winter and spring macroinvertebrate drift in an outpocketing of the lower Mississippi River, Louisiana (USA). Hydrobiologia 185, 205–209.

Kogan, M., 1975. Plant resistance in pest management. In: Metcalf, R.L., Luckmann, W.H. (Eds.), Introduction to Insect Pest Management. Wiley, New York, pp. 103–146.

Kogan, M., 1981. Dynamics of insect adaptations to soybean: impact of integrated pest management. Environ. Entomol. 10, 363–371.

Kogan, M., 1998. Integrated pest management: historical perspectives and contemporary developments. Annu. Rev. Entomol. 43, 243–270.

Kogan, M., Jepson, P. (Eds.), 2007. Perspectives in Ecological Theory and Integrated Pest Management. Cambridge University Press, Cambridge, UK.

Kogan, M., Paxton, J., 1983. Natural inducers of plant resistance to insects. In: Hedin, P.A. (Ed.), Plant Resistance to Insects . ACS Symposium Series 208, American Chemical Society, Washington, DC, pp. 153–170.

Kogan, M., Turnipseed, S.G., 1987. Ecology and management of soybean arthropods. Annu. Rev. Entomol. 32, 507–538.

Kohlmann, B., 1991. Dung beetles in subtropical North America. In: Hanski, I., Cambefort, Y. (Eds.), Dung Beetle Ecology. Princeton University Press, Princeton, NJ, pp. 116–132.

Kohn, J.R., Casper, B.B., 1992. Pollen-mediated gene flow in *Cucurbita foetidissima* (Cucurbitaceae). Am. J. Bot. 79, 57–62.

Kolb, T.E., Holmberg, K.M., Wagner, M.R., Stone, J.E., 1998. Regulation of ponderosa pine foliar physiology and insect resistance mechanisms by basal area treatments. Tree Physiol. 18, 375–381.

Kolb, T.E., Dodds, K.A., Clancy, K.M., 1999. Effect of western spruce budworm defoliation on the physiology and growth of potted Douglas-fir seedlings. Forest Sci. 45, 280–291.

Kollberg, I., Bylund, H., Huitu, O., Björkman, C., 2014. Regulation of forest defoliating insects through small mammal predation: reconsidering the mechanisms. Oecologia 176, 975–983.

Konishi, M., Itô, Y., 1973. Early entomology in East Asia. In: Smith, R.F., Mittler, T.E., Smith, C.N. (Eds.), History of Entomology. Annual Reviews Inc, Palo Alto, CA, pp. 1–20.

Koptur, S., Rodriguez, M.C., Oberbauer, S.F., Weekley, C., Herndon, A., 2002. Herbivore-free time? Damage to new leaves of woody plants after Hurricane Andrew. Biotropica 34, 547–554.

Koren, I., Kaufman, Y.J., Remer, L.A., Martins, J.V., 2004. Measurement of the effect of Amazon smoke on inhibition of cloud formation. Science 303, 1342–1345.

Koricheva, J., 2002. Meta-analysis of sources of variation in fitness costs of plant antiherbivore defenses. Ecology 83, 176–190.

Koricheva, J., Larsson, S., Haukioja, E., 1998. Insect performance on experimentally stressed woody plants: a meta-analysis. Annu. Rev. Entomol. 43, 195–216.

Koricheva, J., Mulder, C.P.H., Schmid, B., Joshi, J., Huss-Danell, K., 2000. Numerical responses of different trophic groups of invertebrates to manipulations of plant diversity in grasslands. Oecologia 125, 271–282.

Körner, C., 1993. Scaling from species to vegetation: the usefulness of functional groups. In: Schulze, E.D., Mooney, H.A. (Eds.), Biodiversity and Ecosystem Function. Springer-Verlag, Berlin, pp. 117–140.

Kos, M., Bukovinszky, T., Mulder, P.P.J., Bezemer, T.M., 2015. Disentangling above- and belowground neighbor effects on the growth, chemistry, and arthropod community on a focal plant. Ecology 96, 164–175.

Kovács-Hostyánski, A., Haenke, S., Batáry, P., Jauker, B., Báldi, A., Tscharntke, T., Holzschuh, A., 2013. Contrasting effects of mass-flowering crops on bee pollination of hedge plants at different spatial and temporal scales. Ecol. Appl. 23, 1938–1946.

Koyama, J., Kakinohana, H., Miyatake, T., 2004. Eradication of the melon fly, *Bactrocera cucurbitae*, in Japan: importance of behavior, ecology, genetics, and evolution. Annu. Rev. Entomol. 49, 331–349.

Koyama, Y., Yao, I., Akimoto, S.-I., 2004. Aphid galls accumulate high concentrations of amino acids: a support for the nutrition hypothesis for gall formation. Entomol. Exp. Appl. 113, 35–44.

Kozár, F., 1991. Recent changes in the distribution of insects and the global warming. In: Proceedings of Fourth European Congress of Entomology, Gödöllö, Hungary, pp. 406–413.

Kozár, F., 1992a. Organization of arthropod communities in agroecosystems. Acta Phytopathol. Hun. 27, 365–373.

Kozár, F., 1992b. Resource partitioning of host plants by insects on a geographic scale. In: Menken, S.B.J., Visser, J.H., Harrewijn, P. (Eds.), Proceedings of Eighth International Symposium on Insect-plant Relationships. Kluwer Academic Publ., Dordrecht, The Netherlands, pp. 46–48.

Krafft, C.C., Handel, S.N., 1991. The role of carnivory in the growth and reproduction of *Drosera filiformis* and *D. rotundifolia*. Bull. Torrey Bot. Club. 118, 12–19.

Kramer, A.M., Dennis, B., Liebhold, A.M., Drake, J.M., 2009. The evidence for Allee effects. Popul. Ecol. 51, 341–354.

Krams, I.A., Krama, T., Moore, F.R., Rantal, M.J., Mänd, R., Mierauskas, P., Mänd, M., 2015. Resource availability as a proxy for terminal investment in a beetle. Oecologia 178, 339–345.

Kranthi, K.R., Naidu, S., Dhawad, C.S., Tatwawadi, A., Mate, K., Patil, E., Bharose, A.A., Behare, G.T., Wadaskar, R.M., Kranthi, S., 2005. Temporal and intra-plant variability in Cry1Ac expression in Bt cotton and its influence on the survival of the cotton bollworm, *Helicoverpa armigera* (Hübner) (Noctuidae: Lepidoptera). Curr. Sci. India 89, 291–298.

Krantz, G.W., 1978. A Manual of Acarology. Oregon State University Book Stores, Corvallis, OR.

Krantz, G.W., Mellott, J.L., 1972. Studies on phoretic specificity in *Macrocheles mycotrupetes* and *M. peltotrupetes* Krantz and Mellott (Acari: Macrochelidae), associates of geotrupine Scarabaeidae. Acarologia 14, 317–344.

Kratz, T.K.,Magnuson, J.J., Bayley, P., Benson, B.J., Berish, C.W., Bledsoe, C.S.,Blood, E.R., Bowser, C.J., Carpenter, S.R., Cunningham, G.L., Dahlgren, R.A., Frost, T.M., Halfpenny, J.C., Hansen, J.D., Heisey, D., Inouye, R.S., Kaufman, D.W., McKee, A., Yarie, J., 1995. Temporal and spatial variability as neglected ecosystem properties: lessons learned from 12 North American ecosytems. In: Rapport, D.J., Gaudet, C.L., Calow, P. (Eds.), Evaluating and Monitoring the Health of Large-scale Ecosystems. NATO ASI Series, vol. 128. Springer-Verlag, Berlin, pp. 359–383.

Kraus, J.M., Vonesh, J.R., 2012. Fluxes of terrestrial and aquatic carbon by emergent mosquitoes: a test of controls and implications for cross-ecosystem linkages. Oecologia 170, 1111–1122.

Krawchuk, M.A., Taylor, P.D., 2003. Changing importance of habitat structure across multiple spatial scales for three species of insects. Oikos 103, 153–161.

Krebs, C.J., 2008. Ecology: The Experimental Analysis of Distribution and Abundance, sixth ed. Pearson/Benjamin Cummings, San Francisco, CA.

Kremen, C., 2005. Managing ecosystem services: what do we need to know about their ecology? Ecol. Lett. 8, 468–479.

Kremen, C., Williams, N.M., Thorp, R.W., 2002. Crop pollination from native bees as risk from agricultural intensification. Proc. Natl. Acad. Sci. USA 99, 16812–16816.

Kremen, C., Williams, N.M., Bugg, R.L., Fay, J.P., Thorp, R.W., 2004. The area requirements of an ecosystem service: crop pollination by native bee communities in California. Ecol. Lett. 7, 1109–1119.

Kremen, C., Williams, N.M., Aizen, M.A., Gemmill-Herren, B., LeBuhn, G., Minckley, R., Packer, L., Potts, S.G., Roulston, T., Steffan-Dewenter, I., Vázquez, D.P., Winfree, R., Adams, L., Crone, E.E., Greenleaf, S.S., Keitt, T.H., Klein, A.-M., Regetz, J., Ricketts, T.H., 2007. Pollination and other ecosystem services produced by mobile organisms: a conceptual framework for the effects of land-use change. Ecol. Lett. 10, 299–314.

Kriska, G., Csabai, Z., Boda, P., Malik, P., Horváth, G., 2006. Why do red and dark-coloured cars lure aquatic insects? The attraction of water insects to car paintwork explained by reflection-polarization signals. Proc. Roy. Soc. B-Biol. Sci. 273, 1667–1671.

Kruess, A., Tscharntke, T., 1994. Habitat fragmentation, species loss, and biological control. Science 264, 1581–1584.

Kruess, A., Tscharntke, T., 2000. Species richness and parasitism in a fragmented landscape: experiments and field studies with insects on *Vicia sepium*. Oecologia 122, 129–137.

Krug, E.C., 2007. Coastal change and hypoxia in the northern Gulf of Mexico. Part 1. Hydrol. Earth Syst. Sc. 11, 180–190.

Kruger, M., Van Rensburg, J.B.J., van den Berg, J., 2008. Perspective on the development of stem borer resistance to Bt maize and refuge compliance at the Vaalharts irrigation scheme in South Africa. Crop Prot. 28, 684–689.

Kudo, G., Ida, T.Y., Tani, T., 2008. Linkages between phenology, pollination, photosynthesis, and reproduction in deciduous forest understory plants. Ecology 89, 321–331.

Kurokawa, H., Nakashizuka, T., 2008. Leaf herbivory and decomposability in a Malaysian tropical rain forest. Ecology 89, 2645–2656.

Kursar, T.A., Capson, T.L., Coley, P.D., Corley, D.G., Gupta, L.B., Harrison, L.A., Ortega-Barría, E., Windsor, D.M., 1999. Ecologically guided bioprospecting in Panama. Pharm. Biol. 37, 114–126.

Kurz, W.A., Dymond, C.C., Stinson, G., Rampley, G.J., Neilson, E.T., Carroll, A.L., Ebata, T., Safranyik, L., 2008. Mountain pine beetle and forest carbon feedback to climate change. Nature 452, 987–990.

Kytö, M., Niemelä, P., Larsson, S., 1996. Insects on trees: population and individual response to fertilization. Oikos 75, 148–159.

Laakso, J., Setälä, H., 1999. Sensitivity of primary production to changes in the architecture of belowground food webs. Oikos 87, 57–64.

Labandeira, C.C., 1998. Early history of arthropod and vascular plant associations. Annu. Rev. Earth Pl. Sc. 26, 329–377.

Labandeira, C.C., 2002. The history of associations between plants and animals. In: Herrera, C.M., Pellmyr, O. (Eds.), Plant-Animal Interactions: an Evolutionary Approach. Blackwell, London, pp. 26–74.

Labandeira, C.C., Sepkoski, Jr., J.J., 1993. Insect diversity in the fossil record. Science 261, 310–315.

Labandeira, C.C., LePage, B.A., Johnson, A.H., 2001. A *Dendroctonus* bark engraving (Coleoptera: Scolytidae) from a Middle Eocene *Larix* (Coniferales: Pinaceae): early or delayed colonization. Am. J. Bot. 88, 2026–2039.

Labandeira, C.C., Johnson, K.R., Wilf, P., 2002. Impact of the terminal Cretaceous event on plant-insect associations. Proc. Natl. Acad. Sci. USA 99, 2061–2066.

LaDeau, S.L., Kilpatrick, A.M., Marra, P.P., 2007. West Nile virus emergence and large-scale declines of North American bird populations. Nature 447, 710–713.

LaDeau, S.L., Marra, P.P., Kilpatrick, A.M., Calder, C.A., 2008. West Nile virus revisited: consequences for North American ecology. BioScience 58, 937–946.

Lafferty, K.D., 2009. The ecology of climate change and infectious diseases. Ecology 90, 888–900.

Lagrue, C., Besson, A.A., Lecerf, A., 2015. Interspecific differences in antipredator strategies determine the strength of nonconsumptive predator effects on stream detritivores. Oikos 124, 1589–1596.

Lai, L.-C., Huang, R.-N., Wu, W.-J., 2008. Venom alkaloids of monogyne and polygyne forms of the red imported fire ant, *Solenopsis invicta*, in Taiwan. Insect. Soc. 55, 443–449.

Lake, P.S., Bond, N., Reich, P., 2007. Linking ecological theory with stream restoration. Freshwater Biol. 52, 597–615.

Lambeets, K., Vendegehuchte, M.L., Maelfait, J.-P., Bonte, D., 2008. Understanding the impact of flooding on trait-displacements and shifts in assemblage structure of predatory arthropods on river banks. J. Anim. Ecol. 77, 1162–1174.

Lamotte, M., Bourliére, F., 1983. Energy flow and nutrient cycling in tropical savannas. In: Bourliére, F. (Ed.), Ecosystems of the World: Tropical Savannas. Elsevier, Amsterdam, pp. 583–603.

Lampert, E.C., Dyer, L.A., Bowers, M.D., 2010. Caterpillar chemical defense and parasitoid success: *Cotesia congregata* parasitism of *Ceratomia catalpae*. J. Chem. Ecol. 36, 992–998.

Lanciotti, R.S., Roehrig, J.T., Deubel, V., Smith, J., Parker, M., Steele, K., Crise, B., Volpe, K.E., Crabtree, M.B., Scherret, J.H., Hall, R.A., MacKenzie, J.S., Cropp, C.B., Panigrahy, B., Ostlund, E., Schmitt, B., Malkinson, M., Banet, C., Weissman, J., Komar, N., Savage, H.M., Stone, W., McNamara, T., Gubler, D.J., 1999. Origin of the West Nile virus responsible for an outbreak of encephalitis in the northeastern United States. Science 286, 2333–2337.

Landis, D.A., Wratten, S.D., Gurr, G.M., 2000. Habitat management to conserve natural enemies of arthropod pests in agriculture. Annu. Rev. Entomol. 45, 175–201.

Landis, D.A., Menalled, F.D., Costamagna, A.C., Wilkinson, T.K., 2005. Manipulating plant resources to enhance beneficial arthropods in agricultural landscapes. Weed Sci. 53, 902–908.

Landis, D.A., Gardiner, M.M., van der Werf, W., Swinton, S.M., 2008. Increasing corn for biofuel production reduces biocontrol services in agricultural landscapes. Proc. Natl. Acad. Sci. USA 105, 20552–20557.

Landosky, J.M., Karowe, D.N., 2014. Will chemical defenses become more effective against specialist herbivores under elevated CO_2? Glob. Change Biol. 20, 3159–3176.

Landsberg, J., 1989. A comparison of methods for assessing defoliation, tested on eucalypt trees. Aust. J. Ecol. 14, 423–440.

Landsberg, J., Ohmart, C., 1989. Levels of insect defoliation in forests: patterns and concepts. Trends Ecol. Evol. 4, 96–100.

Larsen, T.H., Williams, N.M., Kremen, C., 2005. Extinction order and altered community structure rapidly disrupt ecosystem functioning. Ecol. Lett. 8, 538–547.

Larsson, S., Bjorkman, C., Kidd, N.A.C., 1993. Outbreaks in diprionid sawflies: why some species and not others? In: Wagner, M.R., Raffa, K.F. (Eds.), Sawfly Life History Adaptations to Woody Plants. Academic Press, San Diego, CA, pp. 453–483.

Laskowski, R., Niklińska, M., Maryański, M., 1995. The dynamics of chemical elements in forest litter. Ecology 76, 1393–1406.

Lavelle, P., Pashanasi, B., 1989. Soil macrofauna and land management in Peruvian Amazonia (Yuriaguas, Loreto). Pedobiologia 33, 283–291.

Lavigne, R., Kumar, R., Scott, J.A., 1991. Additions to the Pawnee National Grasslands insect checklist. Entomol. News 102, 150–164.

Law, J.M., Regnier, F.E., 1971. Pheromones. Annual Rev. Biochem. 40, 533–548.

Lawrence, Jr., W.T., 1996. Plants: the food base. In: Reagan, D.P., Waide, R.B. (Eds.), The Food Web of a Tropical Rain Forest. University of Chicago Press, Chicago, IL, pp. 17–51.

Lawrence, R.K., Mattson, W.J., Haack, R.A., 1997. White spruce and the spruce budworm: defining the phenological window of susceptibility. Can. Entomol. 129, 291–318.

Lawton, J.H., 1982. Vacant niches and unsaturated communities: a comparison of bracken herbivores at sites on two continents. J. Anim. Ecol. 51, 573–595.

Lawton, J.H., 1983. Plant architecture and the diversity of phytophagous insects. Annu. Rev. Entomol. 28, 23–39.

Lawton, J.H., 1995. Response of insects to environmental change. In: Harrington, R., Stork, N.E. (Eds.), Insects in a Changing Environment. Academic Press, London, pp. 5–26.

Lawton, J.H., Brown, V.K., 1993. Redundancy in ecosystems. In: Schulze, E.D., Mooney, M.A. (Eds.), Biodiversity and Ecosystem Function. Springer-Verlag, Berlin, pp. 255–270.

Lawton, J.H., Strong, D.R., 1981. Community patterns and competition in folivorous insects. Am. Nat. 118, 317–338.

Le Goff, G., Boundy, S., Daborn, P.J., Yen, J.L., Sofer, L., Lind, R., Sabourault, C., Madi-Ravazzi, L., Ffrench-Constant, R.H., 2003. Microarray analysis of cytochrome P450 mediated insecticide resistance in *Drosophila*. Insect Biochem. Molec. 33, 701–708.

Le Lann, C., Visser, B., Mériaux, M., Moiroux, J., van Baaren, J., van Alphen, J.J.M., Ellers, J., 2014. Rising temperature reduces divergence in resource use strategies in coexisting parasitoid species. Oecologia 174, 967–977.

le Mellec, A., Habermann, M., Michalzik, B., 2009. Canopy herbivory altering C to N ratios and soil input patterns of different organic matter fractions in a Scots pine forest. Plant Soil 325, 255–262.

Leal, L.C., Neto, M.C.L., de Oliveira, A.F.M., Andersen, A.N., Leal, I.R., 2014. Myrmecochores can target high-quality disperser ants: variation in elaiosome traits and ant preferences for myrmecochorous Euphorbiaceae in Brazilian Caatinga. Oecologia 174, 493–500.

Leather, S.R. (Ed.), 2005. Insect Sampling in Forest Ecosystems. Blackwell Science, Malden, MA.

Leather, S.R., 2015. Influential entomology: a short review of the scientific, societal, economic and educational services provided by entomology. Ecol. Entomol. 40, 36–44.

LeCraw, R.M., Kratina, P., Srivastava, D.S., 2014. Food web complexity and stability across habitat connectivity gradients. Oecologia 176, 903–915.

Lee, K.E., Butler, J.H.A., 1977. Termites, soil organic matter decomposition and nutrient cycling. Ecological Bulletin (Stockholm) 25, 544–548.

Lee, K.P., Behmer, S.T., Simpson, S.J., Raubenheimer, D., 2002a. A geometric analysis of nutrient regulation in the generalist caterpillar *Spodoptera littoralis* (Boisduval). J. Insect Physiol. 48, 655–665.

Lee, K.Y., Horodyski, F.M., Valaitis, A.P., Denlinger, D.L., 2002b. Molecular characterization of the insect immune protein hemolin and its high induction during embryonic diapause in the gypsy moth, *Lymantria dispar*. Insect Biochem. Molec. 32, 1457–1467.

Lee, K.P., Raubenheimer, D., Behmer, S.T., Simpson, S.J., 2003. A correlation between macronutrient balancing and insect host-plant range: evidence from the specialist caterpillar *Spodoptera exempta* (Walker). J. Insect Physiol. 49, 1161–1171.

Lee Jr., R.E., 2010. A primer on insect cold-tolerance. In: Denlinger, D.L., Lee, R.E. (Eds.), Low Temperature Biology of Insects. Cambridge University Press, New York, pp. 35–58.

Lee Jr., R.E., Denlinger, D.L., 2010. Rapid cold-hardening: ecological significance and underpinning mechanisms. In: Denlinger, D.L., Lee, R.E. (Eds.), Low Temperature Biology of Insects. Cambridge University Press, New York, pp. 3–34.

Lee, R.E., Elnitsky, M.A., Rinehart, J.P., Hayward, S.A.L., Sandro, L.H., Denlinger, D.L., 2006. Rapid cold-hardening increases the freezing tolerance of the Antarctic midge *Belgica antarctica*. J. Exp. Biol. 209, 399–406.

Lehmann, P., Lyytinen, A., Piiroinen, S., Lindstrom, L., 2014. Northward range expansion requires synchronization of both overwintering behaviour and physiology with photoperiod in the invasive Colorado potato beetle (*Leptinotarsa decemlineata*). Oecologia 176, 57–68.

Leibold, M.A., Mikkelson, G.M., 2002. Coherence, species turnover, and boundary clumping: elements of meta-community structure. Oikos 97, 237–250.

Leibold, M.A., Holyoak, M., Mouquet, N., Amarasekare, P., Chase, J.M., Hoopes, M.F., Holt, R.D., Shurin, J.B., Law, R., Tilman, D., Loreau, M., Gonzalez, A., 2004. The metacommunity concept: a framework for multi-scale community ecology. Ecol. Lett. 7, 601–613.

Leigh, E.G., Smythe, N., 1978. Leaf production, leaf consumption and the regulation of folivory on Barro Colorado Island. In: Leigh, E.G., Rand, A.S., Windsor, D.M. (Eds.), The Ecology of Arboreal Folivores. Smithsonian Institution Press, Washington, DC, pp. 33–50.

Leigh, E.G., Windsor, D.M., 1982. Forest production and regulation of primary consumers on Barro Colorado Island. In: Leigh, E.G., Rand, A.S., Windsor, D.M. (Eds.), The Ecology of Arboreal Folivores. Smithsonian Institution Press, Washington, DC, pp. 109–123.

Leisnham, P.T., Jamieson, I.G., 2002. Metapopulation dynamics of a flightless alpine insect *Hemideina maori* in a naturally fragmented habitat. Ecol. Entomol. 27, 574–580.

Lelieveld, J., Butler, T.M., Crowley, J.N., Dillon, T.J., Fischer, H., Ganzeveld, L., Harder, H., Lawrence, M.G., Martinez, M., Taraborrelli, D., Williams, J., 2008. Atmospheric oxidation capacity sustained by a tropical forest. Nature 452, 737–740.

Lenoir, L., Persson, T., Bengtsson, J., 2001. Wood ant nests as potential hot spots for carbon and nitrogen mineralization. Biol. Fert. Soils 34, 235–240.

Lent, D.D., Graham, P., Collett, T.S., 2009. A motor component to the memories of habitual foraging routes in wood ants? Curr. Biol. 19, 115–121.

Lentz, D.L., Hockaday, B., 2009. Tikal timbers and temples: ancient Maya forestry and the end of time. J. Archaeol. 36, 1342–1353.

Leonard, D.E., 1970. Intrinsic factors causing qualitative changes in populations of *Porthetria dispar* (Lepidoptera: Lymantriidae). Can. Entomol. 102, 239–249.

Lerdau, M., Guenther, A., Monson, R., 1997. Plant production and emission of volatile organic compounds. BioScience 47, 373–383.

Leroux, S.J., Loreau, M., 2008. Subsidy hypothesis and strength of trophic cascades across ecosystems. Ecol. Lett. 11, 1147–1156.

Lesica, P., Kannowski, P.B., 1998. Ants create hummocks and alter structure and vegetation of a Montana fen. Am. Midl. Nat. 139, 58–68.

Letourneau, D.K., Dyer, L.A., 1998. Density patterns of Piper ant-plants and associated arthropods: top-predator trophic cascades in a terrestrial system? Biotropica 30, 162–169.

Letourneau, D.K., Jedlicka, J.A., Bothwell, S.G., Moreno, C.R., 2009. Effects of natural enemy biodiversity on the suppression of arthropod herbivores in terrestrial ecosystems. Annu. Rev. Ecol. Evol. S. 40, 573–592.

Leuschner, W.A., 1980. Impacts of the southern pine beetle. In: (Thatcher, R.C., Searcy, J.L., Coster J.E., Hertel, G.D. (Eds.), The Southern Pine Beetle. USDA Forest Service Tech. Bull. 1631. USDA Forest Service, Washington, DC, pp. 137–151.

LeVan, K.E., Hung, K.-L.J., McCann, K.R., Ludka, J.T., Holway, D.A., 2014. Floral visitation by the Argentine ant reduces pollinator visitation and seed set in the coast barrel cactus, *Ferocactus viridescens*. Oecologia 174, 163–171.

Levins, R., 1970. Extinction. Lectures on Mathematics in the Life Sciences 2: 77–107.

Lewinsohn, T.M., Price, P.W., 1996. Diversity of herbivorous insects and ecosystem processes. In: Solbrig, O.T., Medina, E., Silva, J.F. (Eds.), Biodiversity and Savanna Ecosystem Processes. Springer-Verlag, Berlin, pp. 143–157.

Lewinsohn, T.M., Roslin, T., 2008. Four ways toward tropical herbivore megadiversity. Ecol. Lett. 11, 398–416.

Lewis, A.C., 1979. Feeding preference for diseased and wilted sunflower in the grasshopper, *Melanoplus differentialis*. Entomol. Exp. Appl. 26, 202–207.

Lewis, A.C., 1986. Memory constraints and flower choice in *Pieris rapae*. Science 232, 863–865.

Lewis, A.C., 1993. Learning and the evolution of resources: pollinators and flower morphology. In: Papaj, D.R., Lewis, A.C. (Eds.), Insect Learning: Ecological and Evolutionary Perspectives. Chapman & Hall, New York, pp. 219–242.

Lewis, T., 1998. The effect of deforestation on ground surface temperatures. Global Planet. Change 18, 1–13.

Lewis, O.T., Gripenberg, S., 2008. Insect seed predators and environmental change. J. Appl. Ecol. 45, 1593–1599.

Lewis, W.J., Tumlinson, J.H., 1988. Host detection by chemically mediated associative learning in a parasitic wasp. Nature 331, 257–259.

Li, Y.-H., Liu, S.-S., 2004. Learning in phytophagous insects. Acta Entomol. Sinica 47, 106–116, (in Chinese with English summary).

Li, Y., Dicke, M., Harvey, J.A., Gols, R., 2014. Intra-specific variation in wild Brassica oleracea for aphid-induced plant responses and consequences for caterpillar–parasitoid interactions. Oecologia 174, 853–862.

Liang, D., Zhang, A., Kopanic, Jr., R.J., Roelofs, W.L., Schal, C., 1998. Field and laboratory evaluation of the female sex pheromone for detection, monitoring and and management of brownbanded cockroaches (Dictyoptera: Blattelidae). J. Econ. Entomol. 91, 480–485.

Liebhold, A.M., Elkinton, J.S., 1989. Characterizing spatial patterns of gypsy moth regional defoliation. Forest Sci. 35, 557–568.

Liebhold, A.M., Halverson, J.A., Elmes, G.A., 1992. Gypsy moth invasion of North America: a quantitative analysis. J. Biogeogr. 19, 513–520.

Liebhold, A.M., Rossi, R.E., Kemp, W.P., 1993. Geostatistics and geographic information systems in applied insect ecology. Annu. Rev. Entomol. 38, 303–327.

Liebhold, A., Koenig, W.D., Bjørnstad, O.N., 2004. Spatial synchrony in population dynamics. Annu. Rev. Ecol. Evol. S. 35, 467–490.

Liebhold, A., Johnson, D.M., Bjørnstad, O.N., 2006. Geographic variation in density-dependent dynamics impacts the synchronizing effect of dispersal and regional stochaisticity. Popul. Ecol. 48, 131–138.

Lill, J.T., Marquis, R.J., 2001. The effects of leaf quality on herbivore performance and attack from natural enemies. Oecologia 126, 418–428.

Lincoln, D.E., Fajer, E.D., Johnson, R.H., 1993. Plant-insect herbivore interactions in elevated CO_2 environments. Trends Ecol. Evol. 8, 64–68.

Lindbladh, M., Fraver, S., Edvardsson, J., Felton, A., 2013. Past forest composition, structures and processes – how paleoecology can contribute to forest conservation. Biol. Conserv. 168, 116–127.

Lindeman, R.L., 1942. The trophic-dynamic aspect of ecology. Ecology 23, 399–418.

Lindig-Cisneros, R., Benrey, B., Espinosa-García, F.J., 1997. Phytoalexins, resistance traits, and domestication status in *Phaseolus coccineus* and *Phaseolus lunatus*. J. Chem. Ecol. 23, 1997–2011.

Lindo, Z., Winchester, N.N., 2007. Oribatid mite communities and foliar litter decomposition in canopy suspended soils and forest floor habitats of western redcedar forests, Vancouver Island, Canada. Soil Biol. Biochem. 39, 2957–2966.

Lindroth, R.L., Barman, M.A., Weisbrod, A.V., 1991. Nutrient deficiencies and the gypsy moth, *Lymantria dispar*: effects of larval performance and detoxification enzyme activity. J. Insect Physiol. 37, 45–52.

Lindroth, R.L., Jung, S.M., Feuker, A.M., 1993a. Detoxification activity in the gypsy moth: effects of host CO_2 and NO_3^- availability. J. Chem. Ecol. 19, 357–367.

Lindroth, R.L., Kinney, K.K., Platz, C.L., 1993b. Responses of deciduous trees to elevated atmospheric CO: productivity, phytochemistry, and insect performance. Ecology 74, 763–777.

Lindroth, R.L., Reich, P.B., Tjoelker, M.G., Volin, J.C., Oleksyn, J., 1993c. Light environment alters response to ozone stress in seedlings of *Acer saccharum* Marsh, and hybrid *Populus* L. III. Consequences for performance of gypsy moth. New Phytol. 124, 647–651.

Linley, J.R., 1966. The ovarian cycle of *Culicoides barbosai* Wirth & Blanton and *C. furens* (Poey) (Ceratopogonidae). B. Entomol. Res. 57, 1–17.

Little, D., Gouhier-Darimont, C., Bruessow, F., Reymond, P., 2007. Oviposition by pierid butterflies triggers defense responses in Arabidopsis. Plant Physiol. 143, 784–800.

Liu, C.-Q., Huang, S.-Q., 2013. Floral divergence, pollinator partitioning and the spatiotemporal pattern of plant–pollinator interactions in three sympatric *Adenophora* species. Oecologia 173, 1411–1423.

Liu, Z.G., Zou, X.M., 2002. Exotic earthworms accelerate plant litter decomposition in a Puerto Rican pasture and a wet forest. Ecol. Appl. 12, 1406–1417.

Liu, Y., Fang, S., Chesson, P., He, F., 2015. The effect of soil-borne pathogens depends on the abundance of host tree species. Nature Communicationsdoi: 10.1038/ncomms10017.

Llewellyn, M., 1972. The effects of the lime aphid, *Eucallipterus tiliae* L. (Aphididae) on the growth of the lime, *Tilia x vulgaris* Hayne. J. Appl. Ecol. 9, 261–282.

Llewellyn, D., Tyson, C., Constable, G., Duggan, B., Beale, S., Steel, P., 2007. Containment of regulated genetically modified cotton in the field. Agr. Ecosyst. Environ. 121, 419–429.

Lloyd, J.E., 1983. Bioluminescence and communication in insects. Annu. Rev. Entomol. 28, 131–160.

Lo, N., Tokuda, G., Watanabe, H., 2011. Evolution and function of endogenous termite celulases. In: Bignell, D.E., Roisin, Y., Lo, N. (Eds.), Biology of Termites: A Modern Synthesis. Springer, New York, pp. 51–67.

Loaiza, V., Jonas, J.L., Joern, A., 2011. Grasshoppers (Orthoptera: Acrididae) select vegetation patches in local-scale responses to foliar nitrogen but not phosphorus in native grassland. Insect Sci. 18, 533–540.

Lockwood, J.A., 2001. Voices from the past: what we can learn from the Rocky Mountain locust. American Entomologist 47, 208–215.

Lockwood, J.A., 2008. Six legged soldiers: using insects as weapons of war. Oxford University Press, New York.

Lockwood, J.A., 2013. The philosophical and psychological dimensions of insects: tourism, horror and the negative sublime. In: Lemelin, R.H. (Ed.), The Management of Insects in Recreation and Tourism. Cambridge University Press, Cambridge, UK., pp. 37–55.

Lockwood, J.A., DeBrey, L.D., 1990. A solution for the sudden and unexplained extinction of the Rocky Mountain grasshopper (Orthoptera: Acrididae). Environ. Entomol. 19, 1194–1205.

Lodge, D.J., McDowell, W.H., Macy, J., Ward, S.K., Leisso, R., Claudio-Campos, K., Kuhnert, K., 2008. Distribution and role of mat-forming saprobic basidiomycetes in a tropical forest. In: Boddy, L., Frankland, J.C. (Eds.), Ecology of Saprobic Basidiomycetes. Elsevier/Academic, Amsterdam, pp. 195–208.

Logan, J.A., Allen, J.C., 1992. Nonlinear dynamics and chaos in insect populations. Annu. Rev. Entomol. 37, 455–477.

Logan, J.A., Régnière, J., Powell, J.A., 2003. Assessing the impacts of global warming on forest pest dynamics. Front. Ecol. Environ. 1, 130–137.

Logan, J.D., Wolesensky, W., Joern, A., 2006. Temperature-dependent phenology and predation in arthropod systems. Ecol. Model. 196, 471–482.

Logan, J.G., Seal, N.J., Cook, J.I., Stanczyk, N.M., Birkett, M.A., Clark, S.J., Gezan, S.A., Wadhams, L.J., Pickett, J.A., Mordue (Luntz), A.J., 2009. Identification of human-derived volatile chemicals that interfere with attraction of the Scottish biting midge and their potential use as repellents. J. Med. Entomol. 46, 208–219.

LoGiudice, K., Ostfeld, R.S., Schmidt, K.A., Keesing, F., 2003. The ecology of infectious disease: effects of host diversity and community composition on Lyme disease risk. Proc. Natl. Acad. Sci. USA 100, 567–571.

Long, E.Y., Finke, D.L., 2014. Contribution of predator identity to the suppression of herbivores by a diverse predator assemblage. Environ. Entomol. 43, 569–576.

Long, E.Y., Finke, D.L., 2015. Predators indirectly reduce the prevalence of an insect-vectored plant pathogen independent of predator diversity. Oecologia 177, 1067–1074.

Loomis, W.E., 1932. Growth-differentiation balance vs. carbohydrate-nitrogen ratio. Proc. Am. Soc. Hortic. Sci. 29, 240–245.

Loranger, H., Weisser, W.W., Ebeling, A., 2014. Invertebrate herbivory increases along an experimental gradient of grassland plant diversity. Oecologia 174, 183–193.

Lorch, P.D., Sword, G.A., Gwynne, D.T., Anderson, G.L., 2005. Radiotelemetry reveals differences in individual movement patterns between outbreak and non-outbreak Mormon cricket populations. Ecol. Entomol. 30, 548–555.

Loreau, M., 1995. Consumers as maximizers of matter and energy flow in ecosystems. Am. Nat. 145, 22–42.

Loreau, M., de Mazancourt, C., 2008. Species synchrony and its drivers: neutral and nonneutral community dynamics in fluctuating environments. Am. Nat. 172, E48–E66.

Lorenz, E.N., 1993. The Essence of Chaos. University of Washington Press, Seattle.

Lorio, Jr., P.L., 1993. Environmental stress and whole-tree physiology. In: Schowalter, T.D., Filip, G.M. (Eds.), Beetle-pathogen Interactions in Conifer Forests. Academic Press, London, pp. 81–101.

Losey, J.E., Vaughan, M., 2006. The economic value of ecological services provided by insects. BioScience 56, 311–323.

Losey, J.E., Rayor, L.S., Carter, M.E., 1999. Transgenic pollen harms monarch larvae. Nature 399, 214.

Lotka, A.J., 1925. Elements of Physical Biology. Williams and Wilkins, Baltimore, MD.

Louda, S.M., Rodman, J.E., 1996. Insect herbivory as a major factor in the shade distribution of a native crucifer (*Cardamine cordifolia* A. Gray, bittercress). J. Ecol. 84, 229–237.

Louda, S.M., Keeler, K.H., Holt, R.D., 1990a. Herbivore influences on plant performance and competitive interactions. In: Grace, J.B., Tilman, D. (Eds.), Perspectives on Plant Competition. Academic Press, San Diego, CA, pp. 413–444.

Louda, S.M., Potvin, M.A., Collinge, S.K., 1990b. Predispersal seed predation, postdispersal seed predation and competition in the recruitment of seedlings of a native thistle in sandhills prairie. Am. Midl. Nat. 124, 105–113.

Louda, S.M., Pemberton, R.W., Johnson, M.T., Follett, P.A., 2003. Non-target effects - the Achilles' heel of biocontrol? Retrospective analyses to assess risk associated with biocontrol introductions. Annu. Rev. Entomol. 48, 365–396.

Lounibos, L.P., O'Meara, G.F., Nishimura, N., Escher, R.L., 2003. Interactions with native mosquito larvae regulate the production of *Aedes albopictus* from bromeliads in Florida. Ecol. Entomol. 28, 551–558.

Lovelock, J., 1988. The Ages of Gaia. W.W. Norton, New York.

Lovelock, C.E., Posada, J., Winter, K., 1999. Effects of elevated CO2 and defoliation on compensatory growth and photosynthesis in seedlings in a tropical tree, *Copaifera aromatica*. Biotropica 31, 279–287.

Lovett, G.M., Ruesink, A.E., 1995. Carbon and nitrogen mineralization from decomposing gypsy moth frass. Oecologia 104, 133–138.

Lovett, G., Tobiessen, P., 1993. Carbon and nitrogen assimilation in red oaks (*Quercus rubra* L.) subject to defoliation and nitrogen stress. Tree Physiol. 12, 259–269.

Lovett, G.M., Nolan, S.S., Driscoll, C.T., Fahey, T.J., 1996. Factors regulating throughfall flux in a New Hampshire forested landscape. Can. J. Forest. Res. 26, 2134–2144.

Lovett, G.M., Christenson, L.M., Groffman, P.M., Jones, C.G., Hart, J.E., Mitchell, M.J., 2002. Insect defoliation and nitrogen cycling in forests. BioScience 52, 335–341.

Lowenstein, D.M., Matteson, K.C., Minor, E.S., 2015. Diversity of wild bees supports pollination services in an urbanized landscape. Oecologia 179, 811–821.

Lowman, M.D., 1982. The effects of different rates and methods of leaf area removal on coachwood (*Ceratopetalum apetalum*). Aust. J. Bot. 30, 477–483.

Lowman, M.D., 1984. An assessment of techniques for measuring herbivory: is rainforest defoliation more intense than we thought? Biotropica 16, 264–268.

Lowman, M.D., 1985. Spatial and temporal variability in herbivory of Australian rain forest canopies. Aust. J. Ecol. 10, 7–14.

Lowman, M.D., 1992. Leaf growth dynamics and herbivory in five species of Australian rain forest canopy trees. J. Ecol. 80, 433–447.

Lowman, M.D., 1995. Herbivory as a canopy process in rain forest trees. In: Lowman, M.D., Nadkarni, N.M. (Eds.), Forest Canopies. Academic Press, San Diego, CA, pp. 431–455.

Lowman, M.D., Box, J.H., 1983. Variation in leaf toughness and phenolic content among 5 species of Australia rain forest trees. Aust. J. Ecol. 8, 17–25.

Lowman, M.D., Heatwole, H.H., 1992. Spatial and temporal variability in defoliation of Australian eucalypts. Ecology 73, 129–142.

Lowman, M.D., Moffett, M., Rinker, H.B., 1993. A technique for taxonomic and ecological sampling in rain forest canopies. Selbyana 14, 75–79.

Lowman, M.D., Schowalter, T.D., Franklin, J.F., 2012. Methods in Forest Canopy Research. University of California Press, Berkeley, CA.

Lowrance, R., Stinner, B.R., House, G.J. (Eds.), 1984. Agricultural Ecosystems: Unifying Concepts. Wiley, New York.

Lubchenco, J., 1978. Plant species diversity in a marine intertidal community: importance of herbivore food preference and algal competitive abilities. Am. Nat. 112, 23–39.

Lubchenco, J., Karl, T.R., 2012. Predicting and managing extreme weather events. Phys. Today 65 (3), 31–37.

Luck, R.F., Dahlsten, D., 1975. Natural decline of a pine needle scale (*Chionaspis pinifoliae* [Fitch]), outbreak at South Lake Tahoe, California following cessation of adult mosquito control with malathion. Ecology 56, 893–904.

Lugo, A.E., 2004. The outcome of alien tree invasions in Puerto Rico. Front. Ecol. Environ. 2, 265–273.

Lugo, A.E., Gucinski, H., 2000. Function, effects, and management of forest roads. Forest Ecol. Manag. 133, 249–262.

Luh, H.-K., Croft, B.A., 1999. Classification of generalist or specialist life styles of predaceous phytoseiid mites using a computer genetic algorithm, information theory, and life history traits. Environ. Entomol. 28, 915–923.

Lunderstädt, J., 1981. The role of food as a density-determining factor for phytophagous insects with reference to the relationship between Norway spruce (*Picea abies* Karst) and *Gilpinia hercyniae* Htg. (Hymenoptera, Diprionidae). Forest Ecol. Manag. 3, 335–353.

Lundheim, R., Zachariassen, K.E., 1993. Water balance of over-wintering beetles in relation to strategies for cold tolerance. J. Comp. Physiol. B 163, 1–4.

Lundy, M.E., Parrella, M.P., 2015. Crickets are not a free lunch: protein capture from scalable organic side-streams via high-density populations of Acheta domesticus. PLoS ONE 10 (4), e0118785.

Lüscher, M., 1961. Air-conditioned termite nests. Sci. Am. 205, 138–145.

Lynch, H.J., Renkin, R.A., Crabtree, R.L., Moorcroft, P.R., 2006. The influence of previous mountain pine beetle (*Dendroctonus ponderosae*) activity on the 1988 Yellowstone fires. Ecosystems 9, 1318–1327.

Lytle, D.A., Smith, R.L., 2004. Exaptation and flash flood escape in the giant water bugs. J. Insect Behav. 17, 169–179.

Lyytikäinen-Saarenmaa, P., 1999. Growth responses of Scots pine (Pinaceae) to artificial and sawfly (Hymenoptera: Diprionidae) defoliation. Can. Entomol. 131, 455–463.

Ma, S.-C., 1958. The population dynamics of the oriental migratory locust (*Locusta migratoria manilensis* Mayen) in China. Acta Entomol. Sinica 8, 1–40, (in Chinese with English summary).

Ma, K.Y., Chirarattananon, P., Fuller, S.B., Wood, R.J., 2013. Controlled flight of a biologically inspired, insect-scale robot. Science 340, 603–607.

Macadam, C.R., Stockan, J.A., 2015. More than just fish food: ecosystem services provided by freshwater insects. Ecol. Entomol. 40, 113–123.

MacArthur, R.H., Wilson, E.O., 1967. The Theory of Island Biogeography. Princeton University Press, Princeton, NJ.

Macauley, B.J., 1975. Biodegradation of litter in *Eucalyptus pauciflora* communities. I: Techniques for comparing the effects of fungi and insects. Soil Biol. Biochem. 7, 341–344.

MacDonald, N.W., Burton, A.J., Liechty, H.O., Witter, J.A., Pregitzer, K.S., Mroz, G.D., Richter, D.D., 1992. Atmospheric pollutants: ion leaching in forest ecosystems along a Great Lakes air pollution gradient. J. Environ. Qual. 21, 614–623.

MacEvilly, C., 2000. Bugs in the system. Nutrition Bulletin 25, 267–268.

MacFadden, B.J., Jones, D.S., 1985. Magnetic butterflies: a case study of the monarch (Lepidoptera, Danaidae). In: Kirschvink, J.L., Jones, D.S., MacFadden, B.J. (Eds.), Magnetite Biomineralization and Magnetoreception in Organisms. Plenum Press, New York, pp. 407–415.

Mack, R.N., Simberloff, D., Lonsdale, W.M., Evans, H., Clout, M., Bazzazz, F.A., 2000. Biotic invasions; causes, epidemiology, global consequences and control. Ecol. Appl. 10, 689–710.

MacKenzie, D.I., Kendall, W.L., 2002. How should detection probability be incorporated into estimates of relative abundance? Ecology 83, 2387–2393.

MacLean, D.A., 2004. Predicting forest insect disturbance regimes for use in emulating natural disturbance. In: Perera, A.H., Buse, L.J., Weber, M.G. (Eds.), Emulating Natural Forest Landscape Disturbances: Concepts and Applications. Columbia University Press, New York, pp. 69–82.

MacMahon, J.A., 1981. Successional processes: comparisons among biomes with special reference to probable roles of and influences on animals. In: West, D.C., Shugart, H.H., Botkin, D.B. (Eds.), Forest Succession: Concepts and Application. Springer-Verlag, New York, pp. 277–304.

MacMahon, J.A., Mull, J.F., Crist, T.O., 2000. Harvester ants (*Pogonomyrmex* spp.): their community and ecosystem influences. Annu. Rev. Ecol. Syst. 31, 265–291.

Maddrell, S.H.P., 1962. A diuretic hormone in *Rhodnius prolixus* Stal. Nature 194, 605–606.

Madigosky, S.R., 2004. Tropical microclimatic considerations. In: Lowman, M.D., Rinker, H.B. (Eds.), Forest Canopies. second ed. Elsevier/Academic Press, San Diego, CA, pp. 24–48.

Madre, F., Vergnes, A., Machon, N., Clergeau, P., 2014. Green roofs as habitats for wild plant species in urban landscapes: first insights from a large-scale sampling. Landscape Urban Plan. 122, 100–107.

Madritch, M.D., Donaldson, J.R., Lindroth, R.L., 2007. Canopy herbivory can mediate the influence of plant genotype on soil processes through frass deposition. Soil Biol. Biochem. 39, 1192–1201.

Maeno, K., Tanaka, S., 2008. Phase-specific developmental and reproductive strategies in the desert locust. Bull. Entomol. Res. 98, 527–534.

Maeno, K., Tanaka, S., Harano, K.-I., 2011. Tactile stimuli perceived by the antennae cause the isolated females to produce gregarious offspring in the desert locust, *Schistocerca gregaria*. J. Insect Physiol. 57, 74–82.

Mafra-Neto, A., Cardé, R.T., 1995. Influence of plume structure and pheromone concentration on upwind flight by *Cadra cautella* males. Physiol. Entomol. 20, 117–133.

Magurran, A.E., 2004. Measuring Biological Diversity. Blackwell Scientific, Malden, MA.

Mahaney, W.C., Zippin, J., Milner, M.W., Sanmugadas, K., Hancock, R.G.V., Aufreiter, S., Campbell, S., Huffman, M.A., Wink, M., Malloch, D., Kalm, V., 1999. Chemistry, mineralogy and microbiology of termite mound soil eaten by the chimpanzees of the Mahale Mountains, western Tanzania. J. Trop. Ecol. 15, 565–588.

Mahon, R.J., Olsen, K.M., 2009. Limited survival of a Cry2Ab-resistant strain of *Helicoverpa armigera* (Lepidoptera: Noctuidae) on Bollgard II. J. Econ. Entomol. 102, 708–716.

Mahunka, S., (Ed.), 1981. The Fauna of the Hortobágy National Park, vol. 1. Akadémiai Kiadó, Budapest, Hungary.

Mahunka, S., (Ed.), 1983. The Fauna of the Hortobágy National Park, vol. 2. Akadémiai Kiadó, Budapest, Hungary.

Mahunka, S., (Ed.), 1986. The Fauna of the Kiskunság National Park, vol. 4. Akadémiai Kiadó, Budapest, Hungary.

Mahunka, S., (Ed.), 1987. The Fauna of the Kiskunság National Park, vol. 5. Akadémiai Kiadó, Budapest, Hungary.

Mahunka, S., (Ed.), 1991. Bátorliget Nature Reserve After Forty Years, vols. 1 and 2. Hungarian Natural History Museum, Budapest, Hungary.

Maino, J.L., Kearney, M.R., 2015. Ontogenetic and interspecific scaling of consumption in insects. Oikos 124, 1564–1570.

Majer, J.D., Recher, H.F., 1988. Invertebrate communities on Western Australian eucalypts – a comparison of branch clipping and chemical knockdown procedures. Aust. J. Ecol. 13, 269–278.

Malausa, T., Bethenod, M.-T., Bontemps, A., Bourguet, D., Cornuet, J.-M., Ponsard, S., 2005. Assortative mating in sympatric host races of the European corn borer. Science 308, 258–260.

Malcolm, S.B., 1992. Prey defense and predator foraging. In: Crawley, M.J. (Ed.), Natural Enemies: The Population Biology of Predators, Parasites and Diseases. Blackwell Scientific, London, pp. 458–475.

Maleque, M.A., Maeto, K., Ishii, H.T., 2009. Arthropods as bioindicators of sustainable forest management, with a focus on plantation forests. Appl. Entomol. Zool. 44, 1–11.

Maloney, P.E., Rizzo, D.M., 2002. Pathogens and insects in a pristine forest ecosystem: the Sierra San Pedro Martir, Baja, Mexico. Can. J. Forest. Res. 32, 448–457.

Malthus, T.R., 1789. An Essay on the Principle of Population as it Affects the Future Improvement of Society. Johnson, London.

Mankowski, M.E., Schowalter, T.D., Morrell, J.J., Lyons, B., 1998. Feeding habits and gut fauna of *Zootermopsis angusticollis* (Isoptera: Termopsidae) in response to wood species and fungal associates. Environ. Entomol. 27, 1315–1322.

Manley, G.V., 1971. A seed-cacheing carabid (Coleoptera). Ann. Entomol. Soc. Am. 64, 1474–1475.

Mann, M.E., Zhang, Z., Hughes, M.K., Bradley, R.S., Miller, S.K., Rutherford, S., Ni, F., 2008. Proxy-based reconstructions of hemispheric and global surface temperature variations over the past two millennia. Proc. Natl. Acad. Sci. USA 105, 13252–13257.

Manville, R.H., 1942. Notes on the mammals of Mount Desert Island, Maine. J. Mammal. 23, 391–398.

Mao, L., Henderson, G., 2010. Group size effect on worker juvenile hormone titers and soldier differentiation in Formosan subterranean termite. J. Insect Physiol. 56, 725–730.

Mao, W., Rupasinghe, S., Zangerl, A.R., Schuler, M.A., Berenbaum, M.R., 2006. Remarkable substrate-specificity of CYP6AB3 in *Depressaria pastinacella*, a highly specialized herbivore. Insect Mol. Biol. 15, 169–179.

Mao, Y.-B., Cai, W.-J., Wang, J.-W., Hong, G.-J., Tao, X.-Y., Wang, L.-J., Huang, Y.-P., Chen, X.-Y., 2007. Silencing a cotton bollworm P450 monooxygenase gene by plant-mediated RNAi impairs larval tolerance for gossypol. Nat. Biotechnol. 25, 1307–1313.

Mapaure, I., Moe, S.R., 2009. Changes in the structure and composition of miombo woodlands mediated by elephants (Loxodonta africana) and fire over a 26-year period in north-western Zimbabwe. Afr. J. Ecol. 47, 175–183.

Maraun, M., Scheu, S., 1996. Changes in microbial biomass, respiration and nutrient status of beech (*Fagus silvatica*) leaf litter processed by millipedes (*Glomeris marginata*). Oecologia 107, 131–140.

March, J.G., Benstead, J.P., Pringle, C.M., Ruebel, M.W., 2001. Linking shrimp assemblages with rates of detrital processing along an elevational gradient in a tropical stream. Can. J. Fish. Aquat. Sci. 58, 470–478.

Marcot, B.G., Holthausen, R.S., Raphael, M.G., Rowland, M.M., Wisdom, M.J., 2001. Using Bayesian belief networks to evaluate fish and wildlife population viability under land management alternatives from an environmental impact statement. Forest Ecol. Manag. 153, 29–42.

Marino, P.C., Landis, D.A., 1996. Effect of landscape structure on parasitoid diversity and parasitism in agroecosystems. Ecol. Appl. 6, 276–284.

Mark, S., Olesen, J.M., 1996. Importance of elaiosome size to removal of ant-dispersed seeds. Oecologia 107, 95–101.

Marks, S., Lincoln, D.E., 1996. Antiherbivore defense mutualism under elevated carbon dioxide levels: a fungal endophyte and grass. Environ. Entomol. 25, 618–623.

Marquis, R.J., 1984. Leaf herbivores decrease fitness of a tropical plant. Science 226, 537–539.

Marquis, R.J., Whelan, C.J., 1994. Insectivorous birds increase growth of white oak through consumption of leaf-chewing insects. Ecology 75, 2007–2014.

Marra, P.P., Griffing, S., Caffrey, C., Kilpatrick, A.M., McLean, R., Brand, C., Saito, E., Dupuis, A.P., Kramer, L., Novak, R., 2004. West Nile virus and wildlife. BioScience 54, 393–402.

Marron, M.T., Markow, T.A., Kain, K.J., Gibbs, A.G., 2003. Effects of starvation and desiccation on energy metabolism in desert and mesic *Drosophila*. J. Insect Physiol. 49, 261–270.

Marschner, H., 1995. The Mineral Nutrition of Higher Plants, second ed. Academic Press, San Diego, CA.

Marske, K.A., Ivie, M.A., Hilton, G.M., 2007. Effects of volcanic ash on the forest canopy insects of Montserrat, West Indies. Environ. Entomol. 36, 817–825.

Martin-R, M., Cox, J.R., Ibarra-F, F., Alston, D.G., Banner, R.E., Malecheck, J.C., 1999. Spittlebug and buffelgrass responses to summer fires in Mexico. J. Range Manage. 52, 621–625.

Martinat, P.J., Barbosa, P., 1987. Relationship between host-plant acceptability and suitability in newly eclosed first-instar gypsy moths, *Lymantria dispar* (L.) (Lepidoptera: Lymantriidae). Ann. Entomol. Soc. Am. 80, 141–147.

Martinez, N.D., 1992. Constant connectance in community food webs. Am. Nat. 139, 1208–1218.

Martínez, A.J., López-Portillo, J., Eben, A., Golubov, J., 2009. Cerambycid girdling and water stress modify mesquite architecture and reproduction. Popul. Ecol. 51, 533–541.

Martínez-Mota, R., Serio-Silva, J.C., Rico-Gray, V., 2004. The role of canopy ants in removing *Ficus perforata* seeds from howler monkey (*Alouatta palliata mexicana*) feces at Los Tuxtlas, México. Biotropica 36, 429–432.

Martius, C., Wassmann, R., Thein, U., Bandeira, A., Rennenberg, H., Junk, W., Seiler, W., 1993. Methane emission from wood-feeding termites in Amazonia. Chemosphere 26, 623–632.

Martius, C., Fearnside, P.M., Bandeira, A.G., Wassmann, R., 1996. Deforestation and methane release from termites in Amazonia. Chemosphere 33, 517–536.

Marvier, M., McCreedy, C., Regetz, J., Kareiva, P., 2007. A meta-analysis of effects of Bt cotton and maize on non-target invertebrates. Science 316, 1475–1477.

Mascanzoni, D., Wallin, H., 1986. The harmonic radar: a new method of tracing insects in the field. Ecol. Entomol. 11, 387–390.

Mascari, T.M., Foil, L.D., 2009. Evaluation of rhodamine B as an orally delivered biomarker for rodents and a feed-through transtadial biomarker for phlebotamine sand flies (Diptera: Psychodidae). J. Med. Entomol. 46, 1131–1137.

Mascari, T.M., Foil, L.D., 2010. Laboratory evaluation of novaluron as a rodent feed-through insecticide against sand fly larvae (Diptera: Psychodidae). J. Med. Entomol. 47, 205–209.

Mascari, T.M., Mitchell, M.A., Rowton, E.D., Foil, L.D., 2007a. Laboratory evaluation of diflubenzuron as a feed-through for control of immature sand flies (Diptera: Psychodidae). J. Med. Entomol. 44, 171–174.

Mascari, T.M., Mitchell, M.A., Rowton, E.D., Foil, L.D., 2007b. Laboratory evaluation of novaluron as a feed-through for control of immature sand flies (Diptera: Psychodidae). J. Med. Entomol. 44, 714–717.

Maschinski, J., Whitham, T.G., 1989. The continuum of plant responses to herbivory: the influence of plant association, nutrient availability, and timing. Am. Nat. 134, 1–19.

Masciocchi, M., Pereira, A.J., Lantschner, M.V., Corley, J.C., 2013. Of volcanoes and insects: the impact of the Puyehue–Cordon Caulle ash fall on populations of invasive social wasps, *Vespula* spp. Ecol. Res. 28, 199–205.

Mason, R.R., 1996. Dynamic behavior of Douglas-fir tussock moth populations in the Pacific Northwest. Forest Sci. 42, 182–191.

Mason, R.R., Luck, R.F., 1978. Population growth and regulation. In: Brookes, M.H., Stark, R.W., Campbell, R.W. (Eds.), The Douglas-fir Tussock Moth: A Synthesis. USDA Forest Service Tech. Bull. 1585. USDA Forest Service, Washington, DC, pp. 41–47.

Mason, C.J., Couture, J.J., Raffa, K.F., 2014. Plant.associated bacteria degrade defense chemicals and reduce their adverse effects on an insect defoliator. Oecologia 175, 901–910.

Massad, T.J., 2013. Ontogenetic differences of herbivory on woody and herbaceous plants: a meta-analysis demonstrating unique effects of herbivory on the young and the old, the slow and the fast. Oecologia 172, 1–10.

Massad, T.J., Dyer, L.A., Vega, C.G., 2012. Costs of defense and a test of the carbon-nutrient balance and growth-differentiation balance hypotheses for two co-occurring classes of plant defense. PLoS ONE 7 (10), e47554.

Massad, T.J., Balch, J.K., Davidson, E.A., Brando, P.M., Mews, C.L., Porto, P., Quintino, R.M., Vieira, S.A., Marimon, Jr., B.H., Trumbore, S.E., 2013. Interactions between repeated fire, nutrients, and insect herbivores affect the recovery of diversity in the southern Amazon. Oecologia 172, 219–229.

Massonnet, B., Simon, J.-C., Weisser, W.G., 2002. Metapopulation structure of the specialized herbivore *Macrosiphoniella tanacetaria* (Homoptera, Aphididae). Mol. Ecol. 11, 2511–2521.

Masters, G.J., Brown, V.K., Gange, A.C., 1993. Plant mediated interactions between above- and below-ground insect herbivores. Oikos 66, 148–151.

Mathias, D., Jacky, L., Bradshaw, W.E., Holzapfel, C.M., 2007. Quantitative trait loci associated with photopheriodic response and stage of diapause in the pitcher-plant mosquito, *Wyeomyia smithii*. Genetics 176, 391–402.

Matis, J.H., Kiffe, T.R., Otis, G.W., 1994. Use of birth-death-migration processes for describing the spread of insect populations. Environ. Entomol. 23, 18–28.

Matsuda, H., Abrams, P.A., Hori, M., 1993. The effect of adaptive anti-predator behavior on exploitative competition and mutualism between predators. Oikos 68, 549–559.

Matteson, K.C., Grace, J.B., Minor, E.S., 2013. Direct and indirect effects of land use on floral resources and flower-visiting insects across an urban landscape. Oikos 122, 682–694.

Matthaei, C.D., Townsend, C.R., 2000. Long term effects of local disturbance history on mobile stream invertebrates. Oecologia 125, 119–126.

Matthews, R.W., Matthews, J.R., 2010. Insect Behavior, second ed. Springer, Dordrecht, The Netherlands.

Mattson, W.J., 1980. Herbivory in relation to plant nitrogen content. Annu. Rev. Ecol. Syst. 11, 119–161.

Mattson, W.J., Addy, N.D., 1975. Phytophagous insects as regulators of forest primary production. Science 190, 515–522.

Mattson, W.J., Haack, R.A., 1987. The role of drought in outbreaks of plant-eating insects. BioScience 37, 110–118.

Mauck, K.E., de Moreas, C.M., Mescher, M.C., 2010. Deceptive chemical signals induced by a plant virus attract insect vectors to inferior hosts. Proc. Natl. Acad. Sci. USA 107, 3600–3605.

Mauricio, R., Bowers, M.D., Bazzaz, F.A., 1993. Pattern of leaf damage affects fitness of the annual plant *Raphanus sativus* (Brassicaceae). Ecology 74, 2066–2071.

May, R.M., 1973. Qualitative stability in model ecosystems. Ecology 54, 638–641.

May, R.M., 1981. Models for two interacting populations. In: May, R.M. (Ed.), Theoretical Ecology: Principles and Applications. Blackwell Scientific, Oxford, UK, pp. 78–104.

May, R.M., 1983. The structure of food webs. Nature 301, 566–568.

May, R.M., 1988. How many species are there on Earth? Science 241, 1441–1449.

Mayer, B., Feger, K.H., Giesemann, A., Jäger, H.J., 1995. Interpretation of sulfur cycling in two catchments in the Black Forest (Germany) using stable sulfur and oxygen isotope data. Biogeochemistry 30, 31–58.

Mazza, C.A., Izaguirre, M.M., Zavala, J., Scopel, A.L., Balaré, C.L., 2002. Insect perception of ultraviolet-B radiation. Ecol. Lett. 5, 722–726.

Mbata, K.J., Chidumayo, E.N., Lwatula, C.M., 2002. Traditional regulation of edible caterpillar exploitation in the Kopa area of Mpika district in northern Zambia. J. Insect Conserv. 6, 115–130.

McArthur, C., Banks, P.B., Boonstra, R., Forbey, J.S., 2014. The dilemma of foraging herbivores: dealing with food and fear. Oecologia 176, 677–689.

McBrayer, J.F., 1975. Exploitation of deciduous leaf litter by *Apheloria montana* (Diplopoda: Eurydesmidae). Pedobiologia 13, 90–98.

McCann, K.S., Rasmussen, J.B., Umbanhowar, J., 2005. The dynamics of spatially coupled food webs. Ecol. Lett. 8, 513–523.

McCarthy, E.K., White, J.W., 2016. Density-dependent prey mortality is determined by the spatial scale of predator foraging. Oecologia 180, 305–311.

McCloud, E.S., Baldwin, I.T., 1997. Herbivory and caterpillar regurgitants amplify the wound-induced increases in jasmonic acid but not nicotine in *Nicotiana sylvestris*. Planta 203, 430–435.

McClure, M.S., 1990. Role of wind, birds, deer, and humans in the dispersal of hemlock woolly adelgid (Homoptera: Adelgidae). Environ. Entomol. 19, 36–43.

McClure, M.S., 1991. Density-dependent feedback and population cycles in *Adelges tsugae* (Homoptera: Adelgidae) on *Tsuga canadensis*. Environ. Entomol. 20, 258–264.

McCoy, E.D., Frank, J.H., 2010. How should the risk associated with the introduction of biological control agents be estimated? Agr. Forest Entomol. 12, 1–8.

McCreadie, J.W., Colbo, M.H., 1993. Larval and pupal microhabitat selection by *Simulium truncatum*, *S. rostratum* and *S. verecundum* AA (Diptera: Simuliidae). Can. J. Zoolog. 71, 358–367.

McCullough, D.G., Wagner, M.R., 1993. Defusing host defenses: ovipositional adaptations of sawflies to plant resins. In: Wagner, M.R., Raffa, K.F. (Eds.), Sawfly Life History Adaptations to Woody Plants. Academic Press, San Diego, CA, pp. 157–172.

McCullough, D.G., Werner, R.A., Neumann, D., 1998. Fire and insects in northern and boreal forest ecosystems of North America. Annu. Rev. Entomol. 43, 107–127.

McDowell, W.H., Gines-Sanchez, C., Asbury, C.E., Perez, P.C.R., 1990. Influence of sea salt aerosols and long range transport on precipitation chemistry at El Verde, Puerto Rico. Atmos. Environ. 24, 2813–2821.

McElfresh, J.S., Millar, J.G., 2001. Geographic variation in the pheromone system of the saturniid moth *Hemileuca eglanterina*. Ecology 82, 3505–3518.

McElfresh, J.S., Hammond, A.M., Millar, J.G., 2001. Sex pheromone components of the buck moth, *Hemileuca maia*. J. Chem. Ecol. 27, 1409–1422.

McEvoy, P.B., Cox, C., Coombs, E., 1991. Successful biological control of ragwort, *Senecio jacobaea*, by introduced insects in Oregon. Ecol. Appl. 1, 430–442.

McEvoy, P.B., Rudd, N.T., Cox, C.S., Huso, M., 1993. Disturbance, competition, and herbivory effects on ragwort *Senecio jacobaea* populations. Ecol. Monogr. 63, 55–75.

McIntire, E.J.B., Rompre, G., Severns, P.M., 2013. Biased correlated random walk and foray loop: which movement hypothesis drives a butterfly metapopulation? Oecologia 172, 293–305.

McIntosh, R.P., 1981. Succession and ecological theory. In: West, D.C., Shugart, H.H., Botkin, D.B. (Eds.), Forest Succession: Concepts and Application. Springer-Verlag, New York, pp. 10–23.

McIntyre, N.E., Wiens, J.A., 1999. How does habitat patch size affect animal movement? An experiment with darkling beetles. Ecology 80, 2261–2270.

McKeever, D.C., 1933. Maggots in treatment of osteomyelitis: a simple inexpensive method. J. Bone Joint Surg. 15, 85–93.

McKinney, M.L., 2002. Urbanization, biodiversity and conservation. BioScience 52, 883–890.

McMahan, E.A., 1982. Bait-and-capture strategy of a termite-eating assassin bug. Insect. Soc. 29, 346–351.

McMahan, E.A., 1983. Adaptations, feeding preferences, and biometrics of a termite-baiting assassin bug (Hemiptera: Reduviidae). Ann. Entomol. Soc. Am. 76, 483–486.

McNab, B.K., 1963. Bioenergetics and the determination of home range size. Am. Nat. 97, 133–140.

McNaughton, S.J., 1977. Diversity and stability of ecological communities: a comment on the role of empiricism in ecology. Am. Nat. 111, 515–525.

McNaughton, S.J., 1979. Grazing as an optimization process: grass-ungulate relationships in the Serengeti. Am. Nat. 113, 691–703.

McNaughton, S.J., 1985. Ecology of a grazing system: the Serengeti. Ecol. Monogr. 55, 259–294.

McNaughton, S.J., 1986. On plants and herbivores. Am. Nat. 128, 765–770.

McNaughton, S.J., 1993a. Grasses and grazers, science and management. Ecol. Appl. 3, 17–20.

McNaughton, S.J., 1993b. Biodiversity and function of grazing ecosystems. In: Schulze, E.D., Mooney, H.A. (Eds.), Biodiversity and Ecosystem Function. Springer-Verlag, Berlin, pp. 361–383.

McNeill, S., Lawton, J.H., 1970. Annual production and respiration in animal populations. Nature 225, 472–474.

McNutt, D.W., Halpern, S.L., Barrows, K., Underwood, N., 2012. Intraspecific competition facilitates the evolution of tolerance to insect damage in the perennial plant *Solanum carolinense*. Oecologia 170, 1033–1044.

Meentemeyer, V., 1978. Macroclimate and lignin control of litter decomposition rates. Ecology 59, 465–472.

Mehdiabadi, N.J., Gilbert, L.E., 2002. Colony-level impacts of parasitoid flies on fire ants. Proc. Roy. Soc. Lond. B Bio 269, 1695–1699.

Meher-Homji, V.M., 1991. Probable impact of deforestation on hydrological processes. Climatic Change 19, 163–173.

Mehner, T., Ihlau, J., Dörner, H., Hupfer, M., Hölker, F., 2005. Can feeding of fish on terrestrial insects subsidize the nutrient pool of lakes? Limnol. Oceanogr. 50, 2022–2031.

Meindl, G.A., Ashman, T.-L., 2013. The effects of aluminum and nickel in nectar on the foraging behavior of bumblebees. Environ. Pollut. 177, 78–81.

Meindl, G.A., Bain, D.J., Ashman, T.-L., 2013. Edaphic factors and plant–insect interactions: direct and indirect effects of serpentine soil on florivores and pollinators. Oecologia 173, 1355–1366.

Meineke, E.K., Dunn, R.R., Sexton, J.O., Frank, S.D., 2013. Urban warming drives insect pest abundance on street trees. PLoS ONE 8 (3), e59687.

Meinwald, J., Eisner, T., 1995. The chemistry of phyletic dominance. Proc. Natl. Acad. Sci. USA 92, 14–18.

Meisel, J.E., 2006. Thermal ecology of the Neotropical army ant, *Eciton burchellii*. Ecol. Appl. 16, 913–922.

Meisner, M., Harmon, J.P., Ives, A.R., 2007. Presence of an unsuitable host diminishes the competitive superiority of an insect parasitoid: a distraction effect. Popul. Ecol. 49, 347–355.

Melander, A.L., 1923. Tolerance of San Jose Scale to Sprays. State College of Washington Agricultural Experiment Station Bulletin 174. Pullman, WA. pp. 52.

Meller, V.H., Davis, R.L., 1996. Biochemistry of insect learning: lessons from bees and flies. Insect Biochem. Molec. 26, 327–335.

Menéndez, R., 2007. How are insects responding to global warming? Tijdschrift voor Entomologie 150, 355–365.

Menninger, H.L., Palmer, M.A., Craig, L.S., Richardson, D.C., 2008. Periodical cicada detritus impacts stream ecosystem metabolism. Ecosystems 11, 1306–1317.

Menzel, R., Geiger, K., Joerges, J., Müller, U., Chittka, L., 1998. Bees travel novel homeward routes by integrating separately acquired vector memories. Anim. Behav. 55, 139–152.

Merlin, C., Gegear, R.J., Reppert, S.M., 2009. Antennal circadian clocks coordinate sun compass orientation in migratory monarch butterflies. Science 325, 1700–1704.

Merriam, C.H., 1890. Results of a biological survey of the San Francisco Mountain region and desert of the Little Colorado, Arizona. North Amer. Fauna 3, 1–136.

Merritt, R.W., Benbow, M.E., Small, P.L.C., 2005. Unraveling an emerging disease associated with disturbed aquatic environments: the case of Buruli ulcer. Front. Ecol. Environ. 3, 323–331.

Mertl, A.L., Wilkie, K.T.R., Traniello, J.F.A., 2009. Impact of flooding on the species richness, density and composition of Amazonian litter-nesting ants. Biotropica 41, 633–641.

Mery, F., Kawecki, T.J., 2005. The cost of long-term memory in *Drosophila*. Science 308, 1148.

Merzendorfer, H., 2013. Chitin synthesis inhibitors: old molecules and new developments. Insect Sci. 20, 121–138.

Meyerson, L.A., Reaser, J.K., 2003. Bioinvasions, bioterrorism, and biosecurity. Front. Ecol. Environ. 1, 307–314.

Michalson, E.L., 1975. Economic impact of mountain pine beetle on outdoor recreation. South. J. Agric. Eco. 7 (2), 43–50.

Michaud, J.P., Grant, A.K., 2009. The nature of resistance to *Dectes texanus* (Col.,Cerambycidae) in wild sunflower, *Helianthus annuus*. J. Appl. Entomol. 133, 518–523.

Michener, C.D., 1969. Comparative and social behavior of bees. Annu. Rev. Entomol. 14, 299–334.

Michener, C.D., 2007. Bees of the World. Johns Hopkins University Press, Baltimore, MD.

Milcu, A., Bonkowski, M., Collin, C.M., Crawley, M.J., 2015. Aphid honeydew-induced changes in soil biota can cascade up to tree crown architecture. Pedobiologia 58, 119–127.

Miles, P.W., 1972. The saliva of Hemiptera. Adv. Insect Physiol. 9, 183–255.

Millennium Ecosystem Assessment, 2005. Ecosystems and Human Well-being: Biodiversity Synthesis. World Resources Institute, Washington, DC.

Miller, T.E., terHorst, C.P., 2012. Testing successional hypotheses of stability, heterogeneity, and diversity in pitcher-plant inquiline communities. Oecologia 170, 243–251.

Miller, K.K., Wagner, M.R., 1984. Factors influencing pupal distribution of the pandora moth (Lepidoptera: Saturniidae) and their relationship to prescribed burning. Environ. Entomol. 13, 430–431.

Miller, S.W., Wooster, D., Li, J., 2007. Resistance and resilience of macroinvertebrates to irrigation water withdrawals. Freshwater Biol. 52, 2494–2510.

Miller, T.E., Moran, E.R., terHorst, C.P., 2014. Rethinking niche evolution: experiments with natural communities of protozoa in pitcher plants. Am. Nat. 184, 277–283.

Millspaugh, C.F., 1892. American Medicinal Plants: An Illustrated and Descriptive Guide to the American Plants Used as Homeopathic Remedies: Their History, Preparation, Chemistry and Physiological Effects. J.C. Yorston, Philadelphia, PA.

Minchin, P.R., 1987. An evaluation of the relative robustness of techniques for ecological ordination. Vegetatio 69, 89–107.

Misenendino, M.L., Archangelsky, M., Brand, C., Epele, L.B., 2012. Environmental changes and macroinvertebrate responses in Patagonian streams (Argentina) to ashfall from the Chaitén Volcano (May 2008). Sci. Total Environ. 424, 202–212.

Misra, R., 1968. Energy transfer along terrestrial food chain. Trop. Ecol. 9, 105–118.

Misson, L., Baldocchi, D.D., Black, T.A., Blanken, P.D., Brunet, Y., Curiel Yuste, J., Dorsey, J.R., Falk, M., Granier, A., Irvine, M.R., Jarosz, N., Lamaud, E., Launiainen, S., Law, B.E., Longdoz, B., Loustau, D., McKay, M., Paw U, K.T., Vesala, U.T., Vickers, D., Wilson, K.B., Goldstein, A.H., 2007. Partitioning forest carbon fluxes with overstory and understory eddy-covariance measurements: a synthesis based on FLUXNET data. Agr. Forest Meteorol. 144, 14–31.

Mitchell, R., 1970. An analysis of dispersal in mites. Am. Nat. 104, 425–431.

Mitchell, R., 1975. The evolution of oviposition tactics in the bean weevil, *Callosobruchus maculatus* (F.). Ecology 56, 696–702.

Mitchell, R.G., Martin, R.E., 1980. Fire and insects in pine culture of the Pacific Northwest. Proc. Conf. Fire Forest Meteorol. 6, 182–190.

Mitchell, R.G., Preisler, H., 1992. Analysis of spatial patterns of lodgepole pine attacked by outbreak populations of mountain pine beetle. Forest Sci. 29, 204–211.

Mitchell, J.D., Hewitt, P.H., van der Linde, T.C., de, K., 1993. Critical thermal limits and temperature tolerance in the harvester termite, *Hodotermes mossambicus* (Hagen). J. Insect Physiol. 39, 523–528.

Mitchell, R.J., Flanagan, R.J., Brown, B.J., Waser, N.M., Karron, J.D., 2009. New frontiers in competition for pollination. Ann. Bot. 103, 1403–1413.

Mittelbach, G.G., Steiner, C.F., Scheiner, S.M., Gross, K.L., Reynolds, H.L., Waide, R.B., Willig, M.R., Dodson, S.I., Gough, L., 2001. What is the observed relationship between species richness and productivity? Ecology 82, 2381–2396.

Mittler, T.E., 1958. The excretion of honeydew by *Tuberolachnus salignus* (Gmelin) (Homoptera: Aphididae). Proc. R. Entomol. Soc. A 33, 49–55.

Mittler, T.E., 1970. Uptake rates of plant sap and synthetic diet by the aphid *Myzus persicae*. Ann. Entomol. Soc. Am. 63, 1701–1705.

Mittler, T.E., Sylvester, E.S., 1961. A comparison of the injury to alfalfa by the aphids, *Therioaphis maculata* and *Microsiphum pisi*. J. Econ. Entomol. 54, 615–622.

Mizutani, M., Hijii, N., 2001. Mensuration of frass drop for evaluating arthropod biomass in canopies: a comparison among *Cryptomeria japonica*, *Larix kaempferi*, and deciduous broad-leaved trees. Forest Ecol. Manag. 154, 327–335.

Mlot, M.J., Tovey, C.A., Hu, D.L., 2011. Fire ants self-assemble into waterproof rafts to survive floods. Nature 108, 7669–7673.

Mock, K.E., Bentz, B.J., O'Neill, E.M., Chong, J.P., Orwin, J., Pfrender, M.E., 2007. Landscape-scale genetic variation in a forest outbreak species, the mountain pine beetle (*Dendroctonus ponderosae*). Mol. Ecol. 16, 553–568.

Mogren, C.L., Trumble, J.T., 2010. The impacts of metals and metalloids on insect behavior. Entomol. Exp. Appl. 135, 1–17.

Mogren, C.L., Walton, W.E., Parker, D.R., Trumble, J.T., 2013. Trophic transfer of arsenic from an aquatic insect to terrestrial insect predators. PLoS ONE 8 (6), e67817.

Mohapatra, S., Deepa, M., Lekha, S., Nethravathi, B., Radhika, B., Gourishanker, S., 2012. Residue dynamics of spirotetramat and imidacloprid in/on mango and soil. Bull. Environ. Contam. Toxicol. 89, 862–867.

Molano-Flores, B., 2009. Insect herbivory of ovules and seeds in native and restored prairies. Restor. Ecol. 17, 187–191.

Moldenke, A.R., 1976. California pollination ecology and vegetation types. Phytologia 34, 305–361.

Moldenke, A.R., 1979. Pollination ecology as an assay for ecosystemic organization: convergent evolution in Chile and California. Phytologia 42, 415–454.

Moll, E.J., McKenzie, B., 1994. Modes of dispersal of seeds in the Cape fynbos. In: Arianoutsou, M., Graves, R.H. (Eds.), Plant-Animal Interactions in Mediterranean-type Ecosystems. Kluwer, Dordrecht, The Netherlands, pp. 151–157.

Møller, A.P., Mousseau, T.A., 2009. Reduced abundance of insects and spiders linked to radiation at Chernobyl 20 years after the accident. Biol. Lett-UK 5, 356–359.

Møller, A.P., Barnier, F., Mousseau, T.A., 2012. Ecosystems effects 25 years after Chernobyl: pollinators, fruit set and recruitment. Oecologia 170, 1155–1165.

Momose, K., Nagamitsu, T., Inoue, T., 1998a. Thrips cross-pollination of *Popowia pisocarpa* (Annonaceae) in a lowland dipterocarp forest in Sarawak. Biotropica 30, 444–448.

Momose, K., Yumoto, T., Nagamitsu, T., Kato, M., Nagamasu, H., Sakai, S., Harrison, R.D., Itioka, T., Hamid, A.A., Inoue, T., 1998b. Pollination biology in a lowland dipterocarp forest in Sarawak, Malaysia. I. Characteristics of the plant-pollinator community in a lowland dipterocarp forest. Am. J. Bot. 85, 1477–1501.

Mondor, E.B., Tremblay, M.N., Awmack, C.S., Lindroth, R.L., 2004. Divergent pheromone-mediated insect behaviour under global atmospheric change. Glob. Change Biol. 10, 1820–1824.

Mondor, E.B., Tremblay, M.N., Awmack, C.S., Lindroth, R.L., 2005. Altered genotypic and phenotypic frequencies of aphid populations under enriched CO_2 and O_3 atmospheres. Glob. Change Biol. 11, 1990–1996.

Monk, C.D., Day, Jr., F.P., 1988. Biomass, primary production, and selected nutrient budgets for an undisturbed watershed. In: Swank, W.T., Crossley, Jr., D.A. (Eds.), Forest Hydrology and Ecology at Coweeta. Elsevier, New York, pp. 151–159.

Monteith, J.L., 1973. Principles of Environmental Physics. American Elsevier, New York.

Mooney, K.A., 2006. The disruption of an ant-aphid mutualism increases the effects of birds on pine herbivores. Ecology 87, 1805–1815.

Mooney, K.A., 2007. Tritrophic effects of birds and ants on a canopy food web, tree growth, and phytochemistry. Ecology 88, 2005–2014.

Moore, J.W., 2006. Animal ecosystem engineers in streams. BioScience 56, 237–246.

Moore, R., Francis, B.J., 1991. Factors influencing herbivory by insects on oak trees in pure stands and paired mixtures. J. Appl. Ecol. 28, 305–317.

Moore, J.C., Hunt, H.W., 1988. Resource compartmentation and the stability of real ecosystems. Nature 333, 261–263.

Moore, J.R., Maguire, D.A., 2005. Natural sway frequencies and damping ratios of trees: influence of crown structure. Trees 19, 363–373.

Moore, L.V., Myers, J.H., Eng, R., 1988. Western tent caterpillars prefer the sunny side of the tree, but why? Oikos 51, 321–326.

Moore, J.C., Walter, D.E., Hunt, H.W., 1988. Arthropod regulation of micro- and mesobiota in below-ground detrital food webs. Annu. Rev. Entomol. 33, 419–439.

Moore, R., Warrington, S., Whittaker, J.B., 1991. Herbivory by insects on oak trees in pure stands compared with paired mixtures. J. Appl. Ecol. 28, 290–304.

Moore, A.J., Beazley, W.D., Bibby, M.C., Devine, D.A., 1996. Antimicrobial activity of cecropins. J. Antimicrob. Chemother. 37, 1077–1089.

Moore, D.J.P., Trahan, N.A., Wilkes, P., Quaife, T., Stephens, B.B., Elder, K., Desai, A.R., Negron, J., Monson, R.K., 2013. Persistent reduced ecosystem respiration after insect disturbance in high elevation forests. Ecol. Lett. 16, 731–737.

Mooring, M.S., Samuel, W.M., 1998. Tick-removal grooming by elk (*Cervus elaphus*): testing the principles of the programmed-grooming hypothesis. Can. J. Zoolog. 76, 740–750.

Mooring, M.S., Samuel, W.M., 1999. Premature loss of winter hair in free-ranging moose (*Alces alces*) infested with winter ticks (*Dermacentor albipictus*) is correlated with grooming rate. Can. J. Zoolog. 77, 148–156.

Mopper, S., 1996. Adaptive genetic structure in phytophagous insect populations. Trends Ecol. Evol. 11, 235–238.

Mopper, S., Strauss, S.Y., 1998. Genetic Structure and Local Adaptation in Natural Insect Populations: Effects of Ecology, Life History, and Behavior. Chapman and Hall, New York, pp. 449.

Mopper, S., Wang, Y., Criner, C., Hasenstein, K., 2004. *Iris hexagona* hormonal responses to salinity stress, leafminer herbivory, and phenology. Ecology 85, 38–47.

Moran, P.A.P., 1953. The statistical analysis of the Canadian lynx cycle. II. Synchronization and meteorology. Aust. J. Zool. 1, 291–298.

Moran, V.C., Southwood, T.R.E., 1982. The guild composition of arthropod communities in trees. J. Anim. Ecol. 51, 289–306.

Moran, N.A., Whitham, T.G., 1990. Differential colonization of resistant and susceptible host plants: *Pemphigus* and *Populus*. Ecology 71, 1059–1067.

Morandin, L.A., Winston, M.L., 2005. Wild bee abundance and seed production in conventional, organic, and genetically modified canola. Ecol. Appl. 15, 871–881.

Moretti, M., Legg, C., 2009. Combining plant and animal traits to assess community functional responses to disturbance. Ecography 32, 299–309.

Morgan, F.D., 1968. Bionomics of Siricidae. Annu. Rev. Entomol. 13, 239–256.

Mori, A.S., Ota, A.T. Fujii, S., Seino, T., Kabeya, D., Okamoto, T., Ito, M.T., Kaneko, N., Hasegawa, M., 2015. Concordance and discordance between taxonomic and functional homogenization: responses of soil mite assemblages to forest conversion. Oecologia 179 527–535.

Morishima, I., Horiba, T., Iketani, M., Nishioka, E., Yamano, Y., 1995. Parallel induction of cecropin and lysozyme in larvae of the silkworm, *Bombyx mori*. Dev. Comp. Immunol. 19, 357–363.

Morón-Ríos, A., Dirzo, R., Jaramillo, V.J., 1997a. Defoliation and below-ground herbivory in the grass *Muhlenbergia quadridentata*: effects on plant performance and on the root-feeder *Phyllophaga* sp. (Coleoptera: Melolonthidae). Oecologia 110, 237–242.

Morón-Ríos, A., Jaramillo, V.J., Dirzo, R., 1997b. Species composition of root-feeding microarthropods in a subalpine grassland associated with pine forest in Mexico. Can. Entomol. 129, 71–80.

Morris, R.F., 1969. Approaches to the study of population dynamics. In: Waters, W.E. (Ed.) Forest Insect Population Dynamics, USDA Forest Service Research Paper NE-125, USDA Forest Service, Northeast Forest Exp. Stn., Hamden, CT, pp. 9–28.

Morrow, P.A., LaMarche, Jr., V.C., 1978. Tree ring evidence for chronic insect suppression of productivity in subalpine *Eucalyptus*. Science 201, 1244–1246.

Moser, J.C., 1963. Contents and structure of *Atta texana* nest in summer. Ann. Entomol. Soc. Am. 56, 286–291.

Moser, J.C., 1985. Use of sporothecae by phoretic *Tarsonemus* mites to transport ascospores of coniferous bluestain fungi. Trans. Brit. Mycol. Soc. 84, 750–753.

Moser, J.C., 2006. Complete excavation and mapping of a Texas leafcutting ant nest. Ann. Entomol. Soc. Am. 99, 891–897.

Moser, S.E., Obrycki, J.J., 2009. Competition and intraguild predation among three species of coccinellids (Coleoptera: Coccinellidae). Ann. Entomol. Soc. Am. 102, 419–425.

Mottern, J.L., Heinz, K.M., Ode, P.J., 2004. Evaluating biological control of fire ants using phorid flies: effects on competitive interactions. Biol. Control 30, 566–583.

Muff, P., Kropf, C., Frick, H., Nentwig, W., Schmidt-Entling, M.H., 2009. Co-existence of divergent communities at natural boundaries: spider (Arachnida: Araneae) diversity across an alpine timberline. Insect Conserv. Diver. 2, 36–44.

Muiruri, E.W., Milligan, H.T., Morath, S., Koricheva, J., 2015. Moose browsing alters tree diversity effects on birch growth and insect herbivory. Funct. Ecol. 29, 724–735.

Müller, C.B., Brodeur, J., 2002. Intraguild predation in biological control and conservation biology. Biol. Control 25, 216–223.

Müller, M., Job, H., 2009. Managing natural disturbance in protected areas: tourists' attitude towards the bark beetle in a German national park. Biol. Conserv. 142, 375–383.

Mumme, R.L., 2002. Scare tactics in a neotropical warbler: white tail feathers enhance flush-pursuit foraging performance in the slate-throated redstart (*Myioborus miniatus*). Auk 119, 1024–1035.

Murdock, J.N., Gido, K.B., Dodds, W.K., Bertrand, K.N., Whiles, M.R., 2010. Consumer return chronology alters recovery trajectory of stream ecosystem structure and function following drought. Ecology 91, 1048–1062.

Murlis, J., Elkinton, J.S., Cardé, R.T., 1992. Odor plumes and how insects use them. Annu. Rev. Entomol. 37, 505–532.

Murphy, S.M., Loewy, K.J., 2015. Trade-offs in host choice of an herbivorous insect based on parasitism and larval performance. Oecologia 179, 741–751.

Murrell, E.G., Ives, A.R., Juliano, S.A., 2014. Intrinsic and extrinsic drivers of succession: effects of habitat age and season on an aquatic insect community. Ecol. Entomol. 39, 316–324.

Murugan, K., Hwang, J.-S., Kovendan, K., Kumar, K.P., Vasugi, C., Kumar, A.N., 2011. Use of plant products and copepods for control of the dengue vector, *Aedes aegypti*. Hydrobiologia 666, 331–338.

Musser, R.O., Cipollini, D.F., Hum-Musser, S.M., Williams, S.A., Brown, J.K., Felton, G.W., 2005. Evidence that the caterpillar salivary enzyme glucose oxidase provides herbivore offense in solanaceous plants. Arch. Insect Biochem. 58, 128–137.

Musser, R.O., Farmer, E., Peiffer, M., Williams, S.A., Felton, G.W., 2006. Ablation of caterpillar labial salivary glands: technique for determining the role of saliva in insect-plant interactions. J. Chem. Ecol. 32, 981–992.

Mustajärvi, K., Siikamäki, P., Rytkönen, S., Lammi, A., 2001. Consequences of plant population size and density for plant-pollinator interactions and plant performance. J. Ecol. 89, 80–87.

Mustaparta, H., 1984. Olfaction. In: Bell, W.J., Carde, R.T. (Eds.), Chemical Ecology of Insects. Chapman and Hall, London, pp. 37–70.

Mustaparta, H., 2002. Encoding of plant odour information in insects: peripheral and central mechanisms. Entomol. Exp. Appl. 104, 1–13.

Mutunga, J.M., Anderson, T.D., Craft, D.T., Gross, A.D., Swale, D.R., Tong, F., Wong, D.M., Carlier, P.R., Bloomquist, J.R., 2015. Carbamate and pyrethroid resistance in the akron strain of *Anopheles gambiae*. Pestic. Biochem. Phys. 121, 116–121.

Myers, J.H., 1988. Can a general hypothesis explain population cycles of forest Lepidoptera? Adv. Ecol. Res. 18, 179–242.

Myers, N., 1996. Environmental services of biodiversity. Proc. Natl. Acad. Sci. USA 93, 2764–2769.

Myers, J.H., Cory, J.S., 2013. Population cycles in forest Lepidoptera revisited. Annu. Rev. Ecol. Evol. S. 44, 565–592.

Myers, J.H., Savoie, A., van Randen, E., 1998. Eradication and pest management. Annu. Rev. Entomol. 43, 471–491.

Myles, T.G., 1999. Review of secondary reproduction in termites (Insecta: Isoptera) with comments on its role in termite ecology and social evolution. Sociobiology 33, 1–91.

Myles, T.G., 2002. Alarm, aggregation, and defense of *Reticulitermes flavipes* in response to a naturally occurring isolate of *Metarhizium anisopliae*. Sociobiology 40, 243–255.

Mysterud, A., Qviller, L., Meisingset, E.L., Viljugrein, H., 2016. Parasite load and seasonal migration in red deer. Oecologia 180, 401–407.

Nachappa, P., Margolis, D.C., Nechols, J.R., Whitfield, A.E., Rotenberg, D., 2013. Tomato spotted wilt virus benefits a non-vector arthropod, *Tetranychus Urticae*, by modulating different plant responses in tomato. PLoS ONE 8 (9), e75909.

Naeem, S., 1998. Species redundancy and ecosystem reliability. Conserv. Biol. 12, 39–45.

Naeem, S., Li, S., 1997. Biodiversity enhances ecosystem reliability. Nature 390, 507–509.

Nagoshi, R.N., Meagher, R.L., Flanders, K., Gore, J., Jackson, R., Lopez, J., Armstrong, J.S., Buntin, G.D., Sansone, C., Leonard, B.R., 2008. Using haplotypes to monitor the migration of fall armyworm (Lepidoptera: Noctuidae) corn-strain populations from Texas and Florida. J. Econ. Entomol. 101, 742–749.

Nagoshi, R.N., Fleischer, S., Meagher, R.L., 2009. Texas is the overwintering source of fall armyworm in central Pennsylvania: implications for migration into the northeastern United States. Environ. Entomol. 38, 1546–1554.

Nagoshi, R.N., Meagher, R.L., Jenkins, D.A., 2010. Puerto Rico fall armyworm has only limited interactions with those from Brazil or Texas but could have substantial exchanges with Florida populations. J. Econ. Entomol. 103, 360–367.

Nakadai, T., Koizumi, H., Usami, Y., Satoh, M., Oikawa, T., 1993. Examination of the method for measuring soil respiration in cultivated land: effect of carbon dioxide concentration on soil respiration. Ecol. Res. 8, 65–71.

Nakano, S., Murakami, M., 2001. Reciprocal subsidies: dynamic interdependence between terrestrial and aquatic food webs. Proc. Natl. Acad. Sci. USA 98, 166–170.

Nakano, S., Miyasaka, H., Kuhara, N., 1999. Terrestrial-aquatic linkages: riparian arthropod inputs alter trophic cascades in a stream food web. Ecology 80, 2435–2441.

Nam, Y., Choi, W.I., Won, D.-S., Kim, J.-K., 2013. Density related plasticity in stand-level spatial distribution of the ambrosia beetle, *Platypus koryoensis* (Coleoptera: Curculionidae). Popul. Ecol. 55, 3–10.

Namba, T., Ma, Y.H., Inagaki, K., 1988. Insect-derived crude drugs in the Chinese Song Dynasty. J. Ethnopharmacol. 24, 247–285.

Nansen, C., Macedo, T., Swanson, R., Weaver, D.K., 2009. Use of spatial structure analysis hyperspectral data cubes for detection of insect-induced stress in wheat plants. Int. J. Remote Sens. 30, 2447–2464.

Nansen, C., Sidumo, A.J., Capareda, S., 2010. Variogram analysis of hyperspectral data to characterize the impact of biotic and abiotic stress of maize plants and to estimate biofuel potential. Appl. Spectrosc. 64, 627–636.

Nardi, J.B., Mackie, R.I., Dawson, J.O., 2002. Could microbial symbionts of arthropod guts contribute significantly to nitrogen fixation in terrestrial ecosystems? J. Insect Physiol. 48, 751–763.

Narr, C.F., Frost, P.C., 2015. Does infection tilt the scales? Disease effects on the mass balance of an invertebrate nutrient recycler. Oecologia 179, 969–979.

Nathan, R., Perry, G., Cronin, J.T., Strand, A.E., Cain, M.L., 2003. Methods for estimating long-distance dispersal. Oikos 103, 261–273.

Nault, L.R., Ammar, E.D., 1989. Leafhopper and planthopper transmission of plant viruses. Annu. Rev. Entomol. 34, 503–529.

Nealis, V.G., Noseworthy, M.K., Turnquist, R., Waring, V.R., 2009. Balancing risks of disturbance from mountain pine beetle and western spruce budworm. Can. J. Forest. Res. 39, 839–848.

Nebeker, T.E., Hodges, J.D., Blanche, C.A., 1993. Host response to bark beetle and pathogen colonization. In: Schowalter, T.D., Filip, G.M. (Eds.), Beetle-pathogen Interactions in Conifer Forests. Academic Press, London, pp. 157–173.

Negrón, J.F., Fettig, C.J., 2014. Mountain pine beetle, a major disturbance agent in US western coniferous forests: a synthesis of the state of knowledge. Forest Sci. 60, 409–413.

Ness, J.H., 2006. A mutualism's indirect costs: the most aggressive plant bodyguards also deter pollinators. Oikos 113, 506–514.

Nessimian, J.L., Venticinque, E.M., Zuanon, J., de Marco, Jr., P., Gordo, M., Fidelis, L., D'arc Batista, J., Juen, L., 2008. Land use, habitat integrity, and aquatic insect assemblages in central Amazonian streams. Hydrobiologia 614, 117–131.

Netherer, S., Schopf, A., 2010. Potential effects of climate change on insect herbivores in European forests – general aspects and the pine processionary moth as specific example. Forest Ecol. Manag. 259, 831–838.

Neville, P.J., O'Dowd, D.J., Yen, A.L., 2008. Issues and implications for research on disturbed oceanic islands illustrated through an ant survey of the Cocos (Keeling) Islands. J. Insect Conserv. 12, 313–323.

Newman, R.M., 1990. Herbivory and detritivory on freshwater macrophytes by invertebrates: a review. J. N. Am. Benthol. Soc. 10, 89–114.

Nicholls, C.I., Altieri, M.A., 2007. Agroecology: contributions towards a renewed ecological foundation for pest management. In: Kogan, M., Jepson, P. (Eds.), Perspectives in Ecological Theory and Integrated Pest Management. Cambridge University Press, Cambridge, UK, pp. 431–468.

Nichols, E., Gardner, T.A., Peres, C.A., Spector, S., The Scarabaeinae Research Network., 2009. Co-declining mammals and dung beetles: an impending ecological cascade. Oikos 118, 481–487.

Nicholson, A.J., 1933. The balance of animal populations. J. Anim. Ecol. 2 (suppl.), 131–178.

Nicholson, A.J., 1954a. Compensatory reactions of populations to stress, and their evolutionary significance. Aust. J. Zool. 2, 1–8.

Nicholson, A.J., 1954b. An outline of the dynamics of animal populations. Aust. J. Zool. 2, 9–65.

Nicholson, A.J., 1958. Dynamics of insect populations. Annu. Rev. Entomol. 3, 107–136.

Nicholson, A.J., Bailey, V.A., 1935. The balance of animal populations. Part I. Proc. Zool. Soc. London, 551–598.

Nielsen, B.O., 1978. Above ground food resources and herbivory in a beech ecosystem. Oikos 31, 273–279.

Niemelä, J., Spence, J.R., 1994. Distribution of forest dwelling carabids: spatial scale and concept of communities. Ecography 17, 166–175.

Niemelä, J., Langor, D., Spence, J.R., 1992. Effects of clear-cut harvesting on boreal ground beetle assemblages in western Canada. Conserv. Biol. 7, 551–561.

Niesenbaum, R.A., 1992. The effects of light environment on herbivory and growth in the dioecious shrub *Lindera benzoin* (Lauraceae). Am. Midl. Nat. 128, 270–275.

Nishimune, T., Watanabe, Y., Okazaki, H., Akai, H., 2000. Thiamin is decomposed due to *Anaphe* spp. entomophagy in seasonal ataxia patients in Nigeria. J. Nutr. 130, 1625–1628.

Nitschke, N., Wiesner, K., Hilke, I., Eisenhauer, N., Oelmann, Y., Weisser, W.W., 2015. Increase of fast nutrient cycling in grassland microcosms through insect herbivory depends on plant functional composition and species diversity. Oikos 124, 161–173.

Niwa, C.G., Daterman, G.E., Sartwell, C., Sower, L.L., 1988. Control of *Rhyacionia zozana* (Lepidoptera: Tortricidae) by mating disruption with synthetic sex pheromone. Environ. Entomol. 17, 593–595.

Nooten, S.S., Hughes, L., 2014. Potential impacts of climate change on patterns of insect herbivory on understorey plant species: a transplant experiment. Austral Ecol. 39, 668–676.

Norden, N., Angarita, H.A., Bongers, F., Martínez-Ramos, M., Granzow-de la Cerda, I., van Breugel, M., Lebrija-Trejos, E., Meave, J.A., Vandermeer, J., Williamson, G.B., Finegan, B., Mesquita, R., Chazdon, R.L., 2015. Successional dynamics in Neotropical forests are as uncertain as they are predictable. Proc. Natl. Acad. Sci. USA 112, 8013–8018.

Norghauer, J.M., Newbery, D.M., 2014. Herbivores differentially limit the seedling growth and sapling recruitment of two dominant rain forest trees. Oecologia 174, 459–469.

Norman, E.M., Clayton, D., 1986. Reproductive biology of two Florida pawpaws: *Asimina obovata* and *A. pygmaea* (Annonaceae). Bull. Torrey Bot. Club. 113, 16–22.

Norman, E.M., Rice, K., Cochran, S., 1992. Reproductive biology of *Asimina parviflora* (Annonaceae). Bull. Torrey Bot. Club. 119, 1–5.

Norris, L.C., Maina, B.J., Lee, Y., Collier, T.C., Fofana, A., Cornel, A.J., Lanzaroa, G.C., 2015. Adaptive introgression in an African malaria mosquito coincident with the increased usage of insecticide-treated bed nets. Proc. Natl. Acad. Sci. USA 112, 815–820.

North, M., Innes, J., Zald, H., 2007. Comparison of thinning and prescribed fire restoration treatments to Sierran mixed-conifer historic conditions. Can. J. Forest. Res. 37, 331–342.

Norton, R.A., Behan-Pelletier, V.M., 1991. Calcium carbonate and calcium oxalate as cuticular hardening agents in oribatid mites (Acari: Oribatida). Can. J. Zoolog. 69, 1504–1511.

Norval, R.A.I., Sutherst, R.W., Kurki, J., Gibson, J.D., Kerr, J.D., 1988. The effect of the brown ear-tick *Rhipicephalus appendiculatus* on the growth of sanga and European breed cattle. Vet. Parasitol. 30, 149–164.

Nothnagle, P.J., Schultz, J.C., 1987. What is a forest pest? In: Barbosa, P., Schultz, J.C. (Eds.), Insect Outbreaks. Academic Press, San Diego, CA, pp. 59–80.

Novotný, V., Basset, Y., 2000. Rare species in communities of tropical insect herbivores: pondering the mystery of singletons. Oikos 89, 564–572.

Novotný, V., Basset, Y., Miller, S.E., Weiblen, G.D., Bremer, B., Cizek, L., Drozd, P., 2002. Low host specificity of herbivorous insects in a tropical forest. Nature 416, 841–844.

Novotný, V., Drozd, P., Miller, S.E., Kulfan, M., Janda, M., Basset, Y., Weiblen, G.D., 2006. Why are there so many species of herbivorous insects in tropical rainforests? Science 313, 1115–1118.

Nowak, D.J., Hirabayashi, S., Bodine, A., Hoehn, R., 2013. Modeled $PM_{2.5}$ removal by trees in ten U.S. cities and associated health effects. Environ. Pollut. 178, 395–402.

Nowlin, W.H., González, M.J., Vanni, M.J., Stevens, M.H.H., Fields, M.W., Valenti, J.J., 2007. Allochthonous subsidy of periodical cicadas affects the dynamics and stability of pond communities. Ecology 88, 2174–2186.

O'Callaghan, M., Glare, T.R., Burgess, E.P.J., Malone, L.A., 2005. Effects of plants genetically modified for insect resistance on nontarget organisms. Annu. Rev. Entomol. 50, 271–292.

O'Connell, A.M., Menagé, P., 1983. Decomposition of litter from three major plant species of jarrah (*Eucalyptus marginata* Donn ex Sm.) forest in relation to site fire history and soil type. Aust. J. Ecol. 8, 277–286.

O'Connor, R.S., Hails, R.S., Thomas, J.A., 2014. Accounting for habitat when considering climate: has the niche of the Adonis blue butterfly changed in the UK? Oecologia 174, 1463–1472.

O'Dowd, D.J., Hay, M.E., 1980. Mutualism between harvester ants and a desert ephemeral: seed escape from rodents. Ecology 61, 531–540.

O'Dowd, D.J., Willson, M.F., 1991. Associations between mites and leaf domatia. Trends Ecol. Evol. 6, 179–182.

O'Neill, R.V., 2001. Is it time to bury the ecosystem concept? (with full military honors, of course!). Ecology 82, 3275–3284.

O'Neill, R.V., DeAngelis, D.L., Waide, J.B., Allen, T.F.H., 1986. A Hierarchical Concept of Ecosystems. Princeton University Press, Princeton, NJ.

Oberrath, R., Böhning-Gaese, K., 2002. Phenological adaptation of ant-dispersed plants to seasonal variation in ant activity. Ecology 83, 1412–1420.

Odum, E.P., 1953. Fundamentals of Ecology. W.B. Saunders, Philadelphia, PA.

Odum, H.T., 1957. Trophic structure and productivity of Silver Springs, Florida. Ecol. Monogr. 27, 55–112.

Odum, E.P., 1969. The strategy of ecosystem development. Science 164, 262–270.

Odum, H.T., 1970. Summary: an emerging view of the ecological system at El Verde. In: Odum, H.T., Pigeon, R.F. (Eds.), A Tropical Rain Forest. U.S. Atomic Energy Commission, Washington, DC, pp. I191–I289.

Odum, E.P., 1971. Fundamentals of Ecology, third ed. W.B. Saunders, Philadelphia, PA.

Odum, H.T., 1996. Environmental Accounting: Emergy and Environmental Decision Making. John Wiley, New York.

Odum, H.T., Pinkerton, R.C., 1955. Time's speed regulator: the optimum efficiency for maximum power output in physical and biological systems. Am. Sci. 43, 331–343.

Odum, H.T., Ruíz-Reyes, J., 1970. Holes in leaves and the grazing control mechanism. In: Odum, H.T., Pigeon, R.F. (Eds.), A Tropical Rain Forest. U.S. Atomic Energy Commission, Oak Ridge, Tennessee, pp. I-69–I-80.

Odum, E.P., Smalley, A.E., 1959. Comparison of population energy flow of a herbivorous and a deposit-feeding invertebrate in a salt marsh ecosystem. Proc. Natl. Acad. Sci. USA 45, 617–622.

Oertli, B., 1993. Leaf litter processing and energy flow through macroinvertebrates in a woodland pond (Switzerland). Oecologia 96, 466–477.

Oesterheld, M., McNaughton, S.J., 1988. Intraspecific variation in the response of *Themeda triandra* to defoliation: the effect of time of recovery and growth rates on compensatory growth. Oecologia 77, 181–186.

Oesterheld, M., McNaughton, S.J., 1991. Effect of stress and time for recovery on the amount of compensatory growth after grazing. Oecologia 85, 305–313.

Oesterheld, M., Sala, O.E., McNaughton, S.J., 1992. Effect of animal husbandry on herbivore-carrying capacity at a regional scale. Nature 356, 234–236.

Ohashi, K., Thomson, J.D., 2009. Trapline foraging by pollinators: its ontogeny, economics and possible consequences for plants. Ann. Bot. 103, 1365–1378.

Ohashi, K., Thomson, J.D., D'Souza, D., 2007. Trapline foraging by bumble bees. IV. Optimization of route geometry in the absence of competition. Behav. Ecol., 18.

Ohashi, K., Leslie, A., Thomson, J.D., 2008. Trapline foraging by bumble bees. V. Effects of experience and priority on competitive performance. Behav. Ecol. 19, 936–948.

Ohgushi, T., 1995. Adaptive behavior produces stability of herbivorous lady beetle populations. In: Cappuccino, N., Price, P.W. (Eds.), Population Dynamics: New Approaches and Synthesis. Academic Press, San Diego, CA, pp. 303–319.

Ohgushi, T., 2005. Indirect interaction webs: herbivore-induced effects through trait change in plants. Annu. Rev. Ecol. Evol. S. 36, 81–105.

Ohgushi, T., 2008. Herbivore-induced indirect interaction webs on terrestrial plants: the importance of non-trophic, indirect, and facultative interactions. Entomol. Exp. Appl. 128, 217–229.

Ohgushi, T., Sawada, H., 1985. Population equilibrium with respect to available food resource and its behavioural basis in an herbivorous lady beetle *Henosepilachna niponica*. J. Anim. Ecol. 54, 781–796.

Ohkawara, K., Higashi, S., Ohara, M., 1996. Effects of ants, ground beetles and the seed-fall patterns on myrmecochory of *Erythronium japonicum* Decne. (Liliaceae). Oecologia 106, 500–506.

Ohm, J.R., Miller, T.E.X., 2014. Balancing anti-herbivore benefits and anti-pollinator costs of defensive mutualists. Ecology 95, 2924–2935.

Ohmart, C.P., Stewart, L.G., Thomas, J.R., 1983. Phytophagous insect communities in the canopies of three *Eucalyptus* forest types in south-eastern Australia. Aust. J. Ecol. 8, 395–403.

Ohmart, C.P., Stewart, L.G., Thomas, J.R., 1985. Effects of nitrogen concentrations of *Eucalyptus blakelyi* foliage on the fecundity of *Paropsis atomaria* (Coleoptera: Chrysomelidae). Oecologia 68, 41–44.

Økland, B., Liebhold, A.M., Bjørnstad, A., Erbilgin, N., Krokene, P., 2005. Are bark beetle outbreaks less synchronous than forest Lepidoptera outbreaks? Oecologia 146, 365–372.

Økland, B., Skarpaas, O., Kausrud, K., 2009. Threshold facilitations of interacting species. Popul. Ecol. 51, 513–523.

Okoniewski, J.C., Stone, W.B., Hynes, K.P., 2006. Continuing organochlorine insecticide mortality in wild birds in New York, 2000–2004. Bull. Environ. Contam. Tox. 77, 726–731.

Oksanen, L., 1983. Trophic exploitation and arctic phytomass patterns. Am. Nat. 122, 45–52.

Oliveira, P.S., Brandão, C.R.F., 1991. The ant community associated with extrafloral nectaries in the Brazilian cerrados. In: Huxley, C.R., Cutler, D.F. (Eds.), Ant-Plant Interactions. Oxford University Press, Oxford, UK, pp. 198–212.

Oliveira, P.S., Galetti, M., Pedroni, F., Morellato, L.P.C., 1995. Seed cleaning by *Mycocepurus goeldii* ants (Attini) facilitates germination in *Hymenaea courbaril* (Caesalpiniaceae). Biotropica 27, 518–522.

Ollerton, J., Erenler, H., Edwards, M., Crockett, R., 2011. Extinctions of aculeate pollinators in Britain and the role of large-scale agricultural changes. Science 346, 1360–1362.

Olson, J.S., 1963. Energy storage and the balance of producers and decomposers in ecological systems. Ecology 44, 322–331.

Onstad, D.W., Crowder, D.W., Isard, S.A., Levine, E., Spencer, J.L., O'Neal, M.E., Ratcliffe, S.T., Gray, M.E., Bledsoe, L.W., Di Fonzo, C.D., Eisley, J.B., Edwards, C.R., 2003. Does landscape diversity slow the spread of rotation-resistant western corn rootworm (Coleoptera: Chrysomelidae)? Environ. Entomol. 32, 992–1001.

Opitz, S.E.W., Müller, C., 2009. Plant chemistry and insect sequestration. Chemoecology 19, 117–154.

Orr, M.R., Dahlsten, D.L., Benson, W.W., 2003. Ecological interactions among ants in the genus *Linepithema*, their phorid parasitoids, and ant competitors. Ecol. Entomol. 28, 203–210.

Orwig, D.A., 2002. Ecosystem to regional impacts of introduced pests and pathogens: historical context, questions and issues. J. Biogeogr. 29, 1471–1474.

Orwig, D.A., Foster, D.R., Mausel, D.L., 2002. Landscape patterns of hemlock decline in New England due to the introduced hemlock woolly adelgid. J. Biogeogr. 29, 1475–1487.

Ostfeld, R.S., Keesing, F., 2000. Biodiversity and disease risk: the case of Lyme disease. Conserv. Biol. 14, 722–728.

Ostfeld, R.S., Manson, R.H., Canham, C.D., 1997. Effects of rodents on survival of tree seeds and seedlings invading old fields. Ecology 78, 1531–1542.

Ostrom, P.H., Colunga-Garcia, M., Gage, S.H., 1997. Establishing pathways of energy flow for insect predators using stable isotope ratios: field and laboratory evidence. Oecologia 109, 108–113.

Otte, D., Joern, A., 1975. Insect territoriality and its evolution: population studies of desert grasshoppers on creosote bushes. J. Anim. Ecol. 44, 29–54.

Ottea, J., Leonard, R., 2006. Insecticide/acaricide resistance and management strategies. In: All, J.N., Treacy, M.F. (Eds.), Use and Management of Insecticides, Acaricides, and Transgenic Crops. Entomological Society of America, Lanham, MD, pp. 82–92.

Owen, D.F., 1978. Why do aphids synthesize melezitose? Oikos 31, 264–267.

Owen, D.F., Wiegert, R.G., 1976. Do consumers maximize plant fitness? Oikos 27, 488–492.

Ozaki, M., Wada-Katsumata, A., Fujikawa, K., Iwasaki, M., Yokohari, F., Satoji, Y., Nisimura, T., Yamaoka, R., 2005. Ant nestmate and non-nestmate discrimation by a chemosensory sensillum. Science 309, 311–314.

Packard, A.S., Jr., 1877. Report on the Rocky Mountain Locust and Other Insects Now Injuring or Likely to Injure Field and Garden Crops in the Western States and Territories. Department of the Interior, Washington, DC.

Paige, K.N., Whitham, T.G., 1987. Overcompensation in response to mammalian herbivory: the advantage of being eaten. Am. Nat. 129, 407–416.

Paine, R.T., 1966. Food web complexity and species diversity. Am. Nat. 100, 65–75.

Paine, R.T., 1969a. The *Pisaster-Tegula* interaction: prey patches, predator food preference, and intertidal community structure. Ecology 50, 950–961.

Paine, R.T., 1969b. A note on trophic complexity and community stability. Am. Nat. 103, 91–93.

Paine, T.D., Baker, F.A., 1993. Abiotic and biotic predisposition. In: Schowalter, T.D., Filip, G.M. (Eds.), Beetle-Pathogen Interactions in Conifer Forests. Academic Press, London, pp. 61–79.

Painter, R.H., 1936. The food of insects and its relation to resistance of plants to insect attack. Am. Nat. 70, 547–566.

Painter, R.H., 1951. Insect resistance in crop plants. The Macmillan Co, New York, pp. 520.

Painter, E.L., Belsky, A.J., 1993. Application of herbivore optimization theory to rangelands of the western United States. Ecol. Appl. 3, 2–9.

Palmer, T.M., 2003. Spatial habitat heterogeneity influences competition and coexistence in an African acacia ant guild. Ecology 84, 2843–2855.

Palmisano, S., Fox, L.R., 1997. Effects of mammal and insect herbivory on population dynamics of a native Californian thistle, *Cirsium occidentale*. Oecologia 111, 413–421.

Pamilo, P., Gertsch, P., Thorén, P., Seppä, P., 1997. Molecular population genetics of social insects. Annu. Rev. Ecol. Syst. 28, 1–25.

Pan, H., Chen, G., Li, F., Wu, Q., Wang, S., Xie, W., Liu, B., Xu, B., Zhang, Y., 2013. Tomato spotted wilt virus infection reduces the fitness of a nonvector herbivore on pepper. J. Econ. Entomol. 106, 924–928.

Pangesti, N., Weldegergis, B.T., Langendorf, B., van Loon, J.J.A., Dicke, M., Pineda, A., 2015. Rhizobacterial colonization of roots modulates plant volatile emission and enhances the attraction of a parasitoid wasp to host-infested plants. Oecologia 178, 1169–1180.

Paoletti, M.G., Taylor, R.A.J., Stinner, B.R., Stinner, D.H., Benzing, D.H., 1991. Diversity of soil fauna in the canopy and forest floor of a Venezuelan cloud forest. J. Trop. Ecol. 7, 373–383.

Papaj, D.R., Prokopy, R.J., 1989. Ecological and evolutionary aspects of learning in phytophagous insects. Annu. Rev. Entomol. 34, 315–350.

Papaj, D.R., Mallory, H.S., Heinz, C.A., 2007. Extreme weather change and the dynamics of oviposition behavior in the pipevine swallowtail, *Battus philenor*. Oecologia 152, 365–375.

Paquin, P., Coderre, D., 1997. Deforestation and fire impact on edaphic insect larvae and other macroarthropods. Environ. Entomol. 26, 21–30.

Paradise, C.J., 2004. Relationship of water and leaf litter variability to insects inhabiting treeholes. J. N. Am. Benthol. Soc. 23, 793–805.

Paris, R., Lohs, C., Wu, Y., Wang, J., Aksoy, S., 2008. The obligate mutualist *Wigglesworthia glossinidia* influences reproduction, digestion, and immunity processes of its host, the tsetse fly. Appl. Environ. Microb. 74, 5965–5974.

Park, T., 1948. Experimental studies of interspecies competition. I. Competition between populations of the flour beetles, *Tribolium confusum* Duval and *Tribolium castaneum* Herbst. Ecol. Monogr. 18, 265–308.

Park, T., 1954. Experimental studies of interspecies competition. II. Temperature, humidity and competition in two species of *Tribolium*. Physiol. Zool. 27, 177–238.

Parker, G.G., 1983. Throughfall and stemflow in the forest nutrient cycle. Adv. Ecol. Res. 13, 57–133.

Parker, M.A., 1985. Size dependent herbivore attack and the demography of an arid grassland shrub. Ecology 66, 850–860.

Parker, G.G., 1995. Structure and microclimate of forest canopies. In: Lowman, M.D., Nadkarni, N.M. (Eds.). Forest Canopies. Academic Press, San Diego, CA, pp. 73–106.

Parker, J., 2003. World honey prices bolstered by smaller 2002 production, U.S. antidumping tariffs and fears about contaminated Chinese honey. Am. Bee J. 143, 523–525.

Parker, L.W., Fowler, H.G., Ettershank, G., Whitford, W.G., 1982. The effects of subterranean termite removal on desert soil nitrogen and ephemeral flora. J. Arid. Environ. 5, 53–59.

Parker, T.J., Clancy, K.M., Mathiasen, R.L., 2006. Interactions among fire, insects and pathogens in coniferous forests of the interior western United States and Canada. Agr. Forest Entomol. 8, 167–189.

Parkinson, K.J., 1981. An improved method for measuring soil respiration in the field. J. Appl. Ecol. 18, 221–228.

Parmenter, R.R., Yadav, E.P., Parmenter, C.A., Ettestad, P., Gage, K.L., 1999. Incidence of plague associated with increased winter-spring precipitation in New Mexico. Am. J. Trop. Med. Hyg. 61, 814–821.

Parmesan, C., 2006. Ecological and evolutionary responses to recent climate change. Annu. Rev. Ecol. Evol. S. 37, 637–669.

Parmesan, C., 2007. Influences of species, latitudes and methodologies on estimates of phenological response to global warming. Glob. Change Biol. 13, 1860–1872.

Parmesan, C., Yohe, G., 2003. A globally coherent fingerprint of climate change impacts across natural systems. Nature 421, 37–42.

Parmesan, C., Ryrholm, N., Stefanescu, C., Hill, J.K., Thomas, C.D., Descimon, H., Huntley, B., Kaila, L., Kullberg, J., Tammaru, T., Tennent, W.J., Thomas, J.A., Warren, M., 1999. Poleward shifts in geographical ranges of butterfly species associated with regional warming. Nature 399, 579–583.

Parrella, M., 2013. A history of the entomology of fly fishing. Amer. Entomol. 59, 15–27.

Parry, D., Goyer, R.A., 2004. Variation in the suitability of host tree species for geographically discrete populations of forest tent caterpillar. Environ. Entomol. 33, 1477–1487.

Parry, D., Spence, J.R., Volney, W.J.A., 1997. Responses of natural enemies to experimentally increased populations of the forest tent caterpillar, *Malacosoma disstria*. Ecol. Entomol. 22, 97–108.

Parsons, K.A., de la Cruz, A.A., 1980. Energy flow and grazing behavior of conocephaline grasshoppers in a *Juncus roemerianus* marsh. Ecology 61, 1045–1050.

Parsons, T.J., Bradshaw, Jr., H.D., Gordon, M.P., 1989. Systematic accumulation of specific mRNAs in response to wounding in poplar trees. Proc. Natl. Acad. Sci. USA 86, 7895–7899.

Parsons, G.L., Cassis, G., Moldenke, A.R., Lattin, J.D., Anderson, N.H., Miller, J.C., Hammond, P., Schowalter, T.D., 1991. Invertebrates of the H.J. Andrews Experimental Forest, Western Cascade Range, Oregon. V: An Annotated List of Insects and Other Arthropods. Gen. Tech. Rpt. PNW-GTR-290. USDA Forest Service, Pacific Northwest Research Station, Portland, OR.

Parton, W.J., Scurlock, J.M.O., Ojima, D.S., Gilmanov, T.G., Scholes, R.J., Schimel, D.S., Kirchner, T., Menaut, J.-C., Seastedt, T., Moya, E.G., Kamnalrut, A., Kinyamario, J.I., 1993. Observations and modeling of biomass and soil organic matter dynamics for the grassland biome worldwide. Global Biogeochem. Cy. 7, 785–809.

Paschold, A., Halitschke, R., Baldwin, I.T., 2007. Co(i)-ordinating defenses: NaCOI1 mediates herbivore-induced resistance in *Nicotiana attenuata* and reveals the role of herbivore movement in avoiding defenses. Plant J. 51, 79–91.

Pascual, M., Ahumada, J.A., Chaves, L.F., Rodo, X., Bouma, M., 2006. Malaria resurgence in the East African highlands: temperature trends revisited. Proc. Natl. Acad. Sci. USA 103, 5829–5834.

Passos, L., Oliveira, P.S., 2003. Interactions between ants, fruits and seeds in a restinga forest in south-eastern Brazil. J. Trop. Ecol. 19, 261–270.

Pataki, D.E., Alig, R.J., Fung, A.S., Golubiewski, N.E., Kennedy, C.A., McPherson, E.G., Nowak, D.J., Pouyat, R.V., Lankao, P.R., 2006. Urban ecosystems and the North American carbon cycle. Glob. Change Biol. 12, 2092–2102.

Patnaik, S., Ramakrishnan, P.S., 1989. Comparative study of energy flow through village ecosystems of two co-existing communities (the Khasis and the Nepalis) of Meghalaya in north-east India. Agr. Syst. 30, 245–267.

Patrick, C.J., Fernandez, D.H., 2013. The β-richness of two detritivore caddisflies affects fine organic matter export. Oecologia 172, 1105–1115.

Patten, D.T., 1993. Herbivore optimization and overcompensation: does native herbivory on western rangelands support these theories? Ecol. Appl. 3, 35–36.

Patten, B.C., 1995. Network integration of ecological extremal principles: exergy, emergy, power, ascendency, and indirect effects. Ecol. Model. 79, 75–84.

Patten, B.C., Odum, E.P., 1981. The cybernetic nature of ecosystems. Am. Nat. 118, 886–895.

Pauw, A., Hawkins, J.A., 2011. Reconstruction of historical pollination rates reveals linked declines of pollinators and plants. Oikos 120, 344–349.

Payne, J.A., 1965. A summer carrion study of the baby pig *Sus scrofa* Linnaeus. Ecology 46, 592–602.

Peakall, R., Beattie, A.J., 1996. Ecological and genetic consequences of pollination by sexual deception in the orchid *Caladenia tentactulata*. Evolution 50, 2207–2220.

Peakall, R., Beattie, A.J., James, S.H., 1987. Pseudocopulation of an orchid by male ants: a test of two hypotheses accounting for the rarity of ant pollination. Oecologia 73, 522–524.

Pearl, R., 1928. The Rate of Living. Knopf, New York, pp. 185.

Pearl, R., Reed, L.J., 1920. On the rate of growth of the population of the United States since 1790 and its mathematical representation. Proc. Natl. Acad. Sci. USA 6, 275–288.

Pearse, I.S., Hughes, K., Shiojiri, K., Ishizaki, S., Karban, R., 2013. Interplant volatile signaling in willows: revisiting the original talking trees. Oecologia 172, 869–875.

Pearse, I.S., Funk, K.A., Kraft, T.S., Koenig, W.D., 2015. Lagged effects of early-season herbivores on valley oak fecundity. Oecologia 178, 361–368.

Pearson, T.R.H., Burslem, D.F.R.P., Goeriz, R.E., Dalling, J.W., 2003. Regeneration niche partitioning in neotropical pioneers: effects of gap size, seasonal drought and herbivory on growth and survival. Oecologia 137, 456–465.

Péch, G., 1993. Fire hazard in budworm-killed balsam fir stands on Cape Breton Highlands. Forest. Chron. 69, 178–186.

Peckarsky, B.L., Cowan, C.A., Penton, M.A., Anderson, C., 1993. Sublethal consequences of stream-dwelling predatory stoneflies on mayfly growth and fecundity. Ecology 74, 1836–1846.

Pecsenye, K., Rácz, R., Bereczki, J., Bátori, E., Varga, Z., 2014. Loss of genetic variation in declining populations of *Aricia artaxerxes* in Northern Hungary. J. Insect Conserv. 18, 233–243.

Pedigo, L.P., Hutchins, S.H., Higley, L.G., 1986. Economic injury levels in theory and practice. Annu. Rev. Entomol. 31, 341–368.

Peet, R.K., Christensen, N.L., 1980. Succession; a population process. Vegetation 43, 131–140.

Pegram, R.G., Perry, B.D., Musisi, F.L., Mwanaumo, B., 1986. Ecology and phenology of ticks in Zambia: seasonal dynamics on cattle. Exp. Appl. Acarol. 2, 25–45.

Pemberton, R.W., 1999. Insects and other arthropods used as drugs by Korean traditional medicine. J. Ethnopharmacol. 65, 207–216.

Pener, M.P., Simpson, S.J., 2009. Locust phase polyphenism: an update. Adv. Insect Physiol. 36, 272.

Peralta, G., Frost, C.M., Rand, T.A., Didham, R.K., Tylianakis, J.M., 2014. Complementarity and redundancy of interactions enhance attack rates and spatial stability in host–parasitoid food webs. Ecology 95, 1888–1896.

Perea, R., Lopez, D., San Miguel, A., Gil, L., 2012. Incorporating insect infestation into rodent seed dispersal: better if the larva is still inside. Oecologia 170, 723–733.

Pereira-Peixoto, M.H., Pufal, G., Martins, C.F., Klein, A.-M., 2014. Spillover of trap-nesting bees and wasps in an urban-rural interface. J. Insect Conserv. 18, 815–826.

Pérez-Lachaud, G., Batchelor, T.P., Hardy, I.C.W., 2004. Wasp eat wasp: facultative hyperparasitism and intra-guild predation by bethylid wasps. Biol. Control 30, 149–155.

Perlman, F., Press, E., Googins, J.A., Malley, A., Poarea, H., 1976. Tussockosis: reactions to Douglas fir tussock moth. Ann. Allergy 36, 302–307.

Perry, D.A., Oren, R., Hart, S.C., 2008. Forest Ecosystems, second ed. Johns Hopkins University Press, Baltimore, MD.

Petanidou, T., Vokou, D., 1990. Pollination and pollen energetics in Mediterranean ecosystems. Am. J. Bot. 77, 986–992.

Petelle, M., 1980. Aphids and melezitose: a test of Owen's 1978 hypothesis. Oikos 35, 127–128.

Petermann, J.S., Farjalla, V.F., Jocque, M., Kratina, P., MacDonald, A.A.M., Marino, N.A.C., de Omena, P.M., Piccoli, G.C.P., Richardson, B.A., Richardson, M.J., Romero, G.Q., Videla, M., Srivastava, D.S., 2015. Dominant predators mediate the impact of habitat size on trophic structure in bromeliad invertebrate communities. Ecology 96, 428–439.

Peters, D.P.C., Laney, C.M., Lugo, A.E., Collins, S.L., Driscoll, C.T., Groffman, P.M., Grove, J.M., Knapp, A.K., Kratz, T.K., Ohman, M.D., Waide, R.B., Yao, J., 2013. Long-term Trends in Ecological Systems: A Basis for Understanding Responses to Global Change. USDA-ARS Tech. Bull. 1931, USDA, Washington, DC.

Peterson, R.K.D., 1995. Insects, disease, and military history. American Entomologist 41, 147–160.

Peterson, R.O., 1999. Wolf-moose interaction on Isle Royale: the end of natural regulation? Ecol. Appl. 9, 10–16.

Peterson, R.K.D., 2009. The real enemy: scrub typhus and the invasion of Sansapor. Am. Entomol. 55, 91–94.

Peterson, J.R., Merrell, D.J., 1983. Rare male mating disadvantage in *Drosophila melanogaster*. Evolution 37, 1306–1316.

Peterson, L.R., Trivett, V., Baker, A.J.M., Agular, C., Pollard, A.J., 2003. Spread of metals through in invertebrate food chain as influenced by a plant that hyperaccumulates nickel. Chemoecology 13, 103–108.

Petrusewicz, K. (Ed.), 1967. Secondary Productivity of Terrestrial Ecosystems: Principles and Methods. Państwowe Wydawnictwo Naukowe, Warszawa, Poland.

Pettis, J.S., Lichtenberg, E.M., Andree, M., Stitzinger, J., Rose, R., vanEngelsdorp, D., 2013. Crop pollination exposes honey bees to pesticides which alters their susceptibility to the gut pathogen *Nosema ceranae*. PLoS ONE 8 (7), e70182.

Phillipson, J., 1981. Bioenergetic options and phylogeny. In: Townsend, C.R., Calow, P. (Eds.), Physiological Ecology: An Evolutionary Approach to Resource Use. Blackwell Scientific, Oxford, UK, pp. 20–45.

Pianka, E.R., 1974. Evolutionary Ecology. Harper & Row, New York, NY.

Pianka, E.R., 1981. Competition and niche theory. In: May, R.M. (Ed.), Theoretical Ecology: Principles and Applications. Blackwell Scientific, Oxford, UK, pp. 167–196.

Pickett, S.T.A., White, P.S., 1985. Patch dynamics: a synthesis. In: Pickett, S.T.A., White, P.S. (Eds.), The Ecology of Natural Disturbance and Patch Dynamics. Academic Press, Orlando, FL, pp. 371–384.

Pielke, R.A., Vidale, P.L., 1995. The boreal forest and the polar front. J. Geophys. Res. 100, 25755–25758.

Piesman, J., Eisen, L., 2008. Prevention of tick-borne diseases. Annu. Rev. Entomol. 53, 323–343.

Piessens, K., Adriaens, D., Jacquemyn, H., Honnay, O., 2009. Synergistic effects of an extreme weather event and habitat fragmentation on a specialised insect herbivore. Oecologia 159, 117–126.

Pimentel, D., Acquay, H., Biltonen, M., Rice, P., Silva, M., Nelson, J., Lipner, V., Giordano, S., Horowitz, A., D'Amore, M., 1992. Environmental and economic costs of pesticide use. BioScience 42, 750–760.

Pimentel, D., Lach, L., Zuniga, R., Morrison, D., 2000. Environmental and economic costs of nonindigenous species in the United States. BioScience 50, 53–65.

Pimm, S.L., 1980. Properties of food webs. Ecology 61, 219–225.

Pimm, S.L., 1982. Food Webs. Chapman and Hall, London, UK.

Pimm, S.L., Kitching, R.L., 1987. The determinants of food chain length. Oikos 50, 302–307.

Pimm, S.L., Lawton, J.H., 1977. Number of trophic levels in ecological communities. Nature 268, 329–331.

Pimm, S.L., Lawton, J.H., 1980. Are food webs divided into compartments? J. Anim. Ecol. 49, 879–898.

Pimm, S.L., Rice, J.C., 1987. The dynamics of multispecies, multi-life-stage models of aquatic food webs. Theor. Popul. Biol. 32, 303–325.

Pimm, S.L., Lawton, J.H., Cohen, J.E., 1991. Food web patterns and their consequences. Nature 350, 669–674.

Pinder, L.C.V., Morley, D.J., 1995. Chironomidae as indicators of water quality—with a comparison of the chironomid faunas of a series of contrasting Cumbrian tarns. In: Harrington, R., Stork, N.E. (Eds.), Insects in a Changing Environment. Academic Press, London, pp. 271–293.

Pinto, S.M., Pearson, D.E., Maron, J.L., 2014. Seed dispersal is more limiting to native grassland diversity than competition or seed predation. J. Ecol. 102, 1258–1265.

Plapp, F.W., 1976. Biochemical genetics of insecticide resistance. Annu. Rev. Entomol. 21, 179–197.

Platt, W.J., Connell, J.H., 2003. Natural disturbances and directional replacement of species. Ecol. Monogr. 73, 507–522.

Pleasants, J.M., Oberhauser, K.S., 2013. Milkweed loss in agricultural fields because of herbicide use: effect on the monarch butterfly population. Insect Conserv. Diver. 6, 135–144.

Plotkin, M., Hod, I., Zaban, A., Boden, S.A., Bagnall, D.M., Galushko, D., Bergman, D.J., 2010. Solar energy harvesting in the epicuticle of the oriental hornet (*Vespa orientalis*). Naturwissenschaften 97, 1067–1076.

Pohlman, C.L., Turton, S.M., Goosem, M., 2007. Edge effects of linear canopy openings on tropical rain forest understory microclimate. Biotropica 39, 62–71.

Poinar, Jr., G.O., 1993. Insects in amber. Annu. Rev. Entomol. 38, 145–159.

Poinar, Jr., G.O., 2001. Dominican amber. In: Briggs, D.E., Crowther, P.R. (Eds.), Palaeobiology II. Blackwell, Malden, MA.

Poinar, Jr., G., 2002. First fossil record of nematode parasitism of ants: a 40 million year tale. Parasitology 125, 457–459.

Poinar, Jr., G., 2005. *Plasmodium dominicana* n. sp. (Plasmodiidae: Haemospororida) from Tertiary Dominican amber. Syst. Parasitol. 61, 47–52.

Poinar, Jr., G., 2009. Early Cretaceous protest flagellates (Parabasilia: Hypermastigia: Oxymonada) of cockroaches (Insecta: Blattaria) in Burmese amber. Cretaceous Res. 30, 1066–1072.

Poinar, Jr., G., 2015. A new genus of fleas with associated microorganisms in Dominican amber. J. Med. Entomol. 52, 1234–1240.

Poinar, Jr., G., Poinar, R., 1994. The Quest for Life in Amber. Addison–Wesley, New York.

Poinar, Jr., G., Poinar, R., 1999. The Amber Forest: A Reconstruction of a Vanished World. Princeton University Press, Princeton, NJ.

Poinar, Jr., G., Poinar, R., 2004a. *Palaeoleishmania proterus* n. gen., n. sp., (Trypanosomatidae: Kinetoplastida) from Cretaceous Burmese amber. Protist 155, 305–310.

Poinar, Jr., G., Poinar, R., 2004b. Evidence of vector-borne disease of early Cretaceous reptiles. Vector-borne Zoonot. 4, 281–284.

Poinar, Jr., G., Poinar, R., 2005. Fossil evidence of insect pathogens. J. Invertebr. Pathol. 89, 243–250.

Poinar, Jr., G., Poinar, R., 2007. What Bugged the Dinosaurs? Insects, Disease, and Death in the Cretaceous. Princeton University Press, Princeton, NJ.

Poinar, Jr., G., Telford, Jr., S.R., 2005. *Paleohaemoproteus burmacis* gen. n., sp. n. (Haemospororida: Plasmodiidae) from an early Cretaceous biting midge (Diptera: Ceratopogonidae). Parasitology 131, 79–84.

Poinar, Jr., G.O., Marshall, C.J., Buckley, R., 2007. One hundred million years of chemical warfare by insects. J. Chem. Ecol. 33, 1663–1669.

Poland, T.M., Borden, J.H., Stock, A.J., Chong, L.J., 1998. Green leaf volatiles disrupt responses by the spruce beetle, *Dendroctonus rufipennis*, and the western pine beetle, *Dendroctonus brevicomis* (Coleoptera: Scolytidae) to attractant-baited traps. Journal of the Entomological Society of British Columbia 95, 17–24.

Polis, G.A., 1991a. Desert communities: an overview of patterns and processes. In: Polis, G.A. (Ed.), The Ecology of Desert Communities. University of Arizona Press, Tucson, AZ, pp. 1–26.

Polis, G.A., 1991b. Food webs in desert communities: complexity via diversity and omnivory. In: Polis, G.A. (Ed.), The Ecology of Desert Communities. University of Arizona Press, Tucson, AZ, pp. 383–429.

Polis, G.A., Strong, D.R., 1996. Food web complexity and community dynamics. Am. Nat. 147, 813–846.

Polis, G.A., Myers, C., Quinlan, M., 1986. Burrowing biology and spatial distribution of desert scorpions. J. Arid. Environ. 10, 137–145.

Polis, G.A., Anderson, W.B., Holt, R.D., 1997a. Toward an integration of landscape and food web ecology: the dynamics of spatially subsidized food webs. Annu. Rev. Ecol. Syst. 28, 289–316.

Polis, G.A., Hurd, S.D., Jackson, C.T., Sanchez-Piñero, F., 1997b. El Niño effects on the dynamics and control of an island ecosystem in the Gulf of California. Ecology 78, 1884–1897.

Polis, G.A., Hurd, S.D., Jackson, C.T., Sanchez-Piñero, F., 1998. Multifactor population limitation: variable spatial and temporal control of spiders on Gulf of California islands. Ecology 79, 490–502.

Pollard, A.J., Baker, A.J.M., 1997. Deterrence of herbivory by zinc hyperaccumulation in *Thlaspi caerulescens* (Brassicaceae). New Phytol. 135, 655–658.

Ponsard, S., Arditi, R., 2000. What can stable isotopes ($\delta^{15}N$ and $\delta^{13}C$) tell us about the food web of soil macro-invertebrates? Ecology 81, 852–864.

Ponyi, J.E., Tátrai, I., Frankó, A., 1983. Quantitative studies on Chironomidae and Oligochaeta in the benthos of Lake Balaton. Arch. Hydrobiol. 97, 196–207.

Pope, D.N., Coulson, R.N., Fargo, W.S., Gagne, J.A., Kelly, C.W., 1980. The allocation process and between-tree survival probabilities in *Dendroctonus frontalis* infestations. Res. Popul. Ecol. 22, 197–210.

Porder, S., Asner, G.P., Vitousek, P.M., 2005. Ground-based and remotely sensed nutrient availability across a tropical landscape. Proc. Natl. Acad. Sci. USA 102, 10909–10912.

Porter, E.E., Redak, R.A., 1996. Short-term recovery of grasshopper communities (Orthoptera: Acrididae) of a California native grassland after prescribed burning. Environ. Entomol. 25, 987–992.

Porter, S.D., Savignano, D.A., 1990. Invasion of polygyne fire ants decimates native ants and disrupts arthropod community. Ecology 71, 2095–2106.

Post, D.A., Jones, J.A., 2001. Hydrologic regimes of forested, mountainous, headwater basins in New Hampshire, North Carolina, Oregon, and Puerto Rico. Adv. Water Resour. 24, 1195–1210.

Potts, S.G., Biesmeijer, J.C., Kremen, C., Neumann, P., Schweiger, O., Kunin, W.E., 2010. Global pollinator declines: trends, impacts and drivers. Trends Ecol. Evol. 25, 345–353.

Poveda, K., Steffan-Dewenter, I., Scheu, S., Tscharntke, T., 2007. Plant-mediated interactions between below- and aboveground processes: decomposition, herbivory, parasitism, and pollination. In: Ohgushi, T., Craig, T.P., Price, P.W. (Eds.), Ecological Communities: Plant Mediation in Indirect Interaction Webs. Cambridge University Press, Cambridge, UK, pp. 147–163.

Póvoa, M.M., Conn, J.E., Schlichting, C.D., Amaral, J.C.O.F., Segura, M.N.O., da Silva, A.N.M., dos Santos, C.C.B., Lacerda, R.N.L., de Souza, R.T.L., Galiza, D., Rosa, E.P.S., Wirtz, R.A., 2003. Malaria vectors, epidemiology, and the re-emergence of *Anopheles darlingi* in Belém, Pará, Brazil. J. Med. Entomol. 40, 379–386.

Powell, A.H., Powell, G.V.N., 1987. Population dynamics of male euglossine bees in Amazonian forest fragments. Biotropica 19, 176–179.

Power, M.E., 1992. Top-down and bottom-up forces in food webs: do plants have primacy? Ecology 73, 733–746.

Power, M.E., Tilman, D., Estes, J.A., Menge, B.A., Bond, W.J., Mills, L.S., Daily, G., Castilla, J.C., Lubchenco, J., Paine, R.T., 1996. Challenges in the quest for keystones. BioScience 46, 609–620.

Powers, J.S., Montgomery, R.A., Adair, E.C., Brearley, F.Q., DeWalt, S.J., Castanho, C.T., Chave, J., Deinert, E., Ganzhorn, J.U., Gilbert, M.E., González-Iturbe, J.A., Bunyavejchewin, S., Grau, H.R., Harms, K.E., Hiremath, A., Iriarte-Vivar, S., Manzane, E., de Oliveira, A.A., Poorter, L., Ramanamanjato, J.-B., Salk, C., Varela, A., Weiblen, G.D., Lerdau, M.T., 2009. Decomposition in tropical forests: a pan-tropical study of the effects of litter type, litter placement and mesofaunal exclusion across a precipitation gradient. J. Ecol. 97, 801–811.

Prange, H.D., Pinshow, B., 1994. Thermoregulation of an unusual grasshopper in a desert environment: the importance of food source and body size. J. Therm. Biol. 19, 75–78.

Prather, C., 2014. Divergent responses of leaf herbivory to simulated hurricane effects in a rainforest understory. Forest Ecol. Manag. 332, 87–92.

Prather, C., Pelini, S.L., Laws, A., Rivest, E., Woltz, M., Bloch, C.P., Del Toro, I., Ho, C.-K., Kominoski, J., Newbold, T.A.S., Parsons, S., Joern, A., 2013. Invertebrates, ecosystem services and climate change. Biol. Rev. 88, 327–348.

Pray, C.L., Nowlin, W.H., Vanni, M.J., 2009. Deposition and decomposition of periodical cicadas (Homoptera: Cicadidae: Magicicada) in woodland aquatic ecosystems. J. N. Am. Benthol. Soc. 28, 181–195.

Preisser, E.L., Strong, D.R., 2004. Climate affects predator control of an herbivore outbreak. Am. Nat. 163, 754–762.

Preisser, E.L., Bolnick, D.I., Benard, M.F., 2005. Scared to death? The effects of intimidation and consumption in predatory-prey interactions. Ecology 86, 501–509.

Presley, S.J., Higgins, C.L., Willig, M.R., 2010. A comprehensive framework for the evaluation of metacommunity structure. Oikos 119, 908–917.

Pretty, J.N., Morison, J.I.L., Hine, R.E., 2003. Reducing food poverty by increasing agricultural sustainability in developing countries. Agr. Ecosyst. Environ. 95, 217–234.

Price, P.W., 1980. Evolutionary Biology of Parasites. Monographs in Population Biology 15. Princeton University Press, Princeton, NJ.

Price, P.W., 1986. Ecological aspects of host plant resistance and biological control: interactions among three trophic levels. In: Boethel, D.J., Eikenbary, R.D. (Eds.), Interactions of Plant Resistance and Parasitoids and Predators of Insects. Ellis Horwood Ltd, Chichester, UK, pp. 11–30.

Price, P.W., 1991. The plant vigor hypothesis and herbivore attack. Oikos 62, 244–251.

Price, P.W., 1997. Insect Ecology, third ed. John Wiley & Sons, New York.

Price, P.W., Bouton, C.E., Gross, P., McPheron, B.A., Thompson, J.N., Weis, A.E., 1980. Interactions among three trophic levels: influence of plants on interactions between insect herbivores and natural enemies. Annu. Rev. Ecol. Syst. 11, 41–65.

Price, P.W., Fernandes, G.W., Waring, G.W., 1987. Adaptive nature of insect galls. Environ. Entomol. 16, 15–24.

Price, P.W., Denno, R.F., Eubanks, M.D., Finke, D.L., Kaplan, I., 2011. Insect Ecology: Behavior, Populations and Communities. Cambridge University Press, Cambridge, UK.

Pridgeon, J.W., Zhao, L., Becnel, J.J., Strickman, D.A., Clark, G.G., Linthicum, K.J., 2008. Topically applied *AaeIAP1* double-stranded RNA kills female adults of *Aedes aegypti*. J. Med. Entomol. 45 (3), 414–420.

Pringle, C.M., 1997. Exploring how disturbance is transmitted upstream: going against the flow. J. N. Am. Benthol. Soc. 16, 425–438.

Pringle, R.M., Fox-Dobbs, K., 2008. Coupling of canopy and understory food webs by ground-dwelling predators. Ecol. Lett. 11, 1328–1337.

Pringle, C.M., Hemphill, N., McDowell, W.H., Bednarek, A., March, J.G., 1999. Linking species and ecosystems: differing biotic assemblages cause interstream differences in organic matter. Ecology 80, 1860–1872.

Pringle, C.M., Freeman, M.C., Freeman, B.J., 2000. Regional effects of hydrologic alternations on riverine macrobiota in the New World: tropical-temperate comparisons. BioScience 50, 807–823.

Pritchard, I.M., James, R., 1984a. Leaf mines: their effect on leaf longevity. Oecologia 64, 132–140.

Pritchard, I.M., James, R., 1984b. Leaf fall as a source of leaf miner mortality. Oecologia 64, 140–142.

Procter, W., 1946. Biological Survey of the Mount Desert Region: the Insect Fauna, with Reference to the Methods of Capture, Food Plants, the Flora and other Biological Features. Wistar Insitute of Anatomy and Biology, Philadelphia, PA.

Progar, R.A., Schowalter, T.D., 2002. Canopy arthropod assemblages along a precipitation and latitudinal gradient among Douglas-fir *Pseudotsuga menziesii* forests in the Pacific Northwest of the United States. Ecography 25, 129–138.

Progar, R.A., Schowalter, T.D., Freitag, C.M., Morrell, J.J., 2000. Respiration from coarse woody debris as affected by moisture and saprotroph functional diversity in western Oregon. Oecologia 124, 426–431.

Punttila, P., Haila, Y., Niemelä, N., Pajunen, T., 1994. Ant communities in fragments of old-growth taiga and managed surroundings. Ann. Zool. Fenn. 31, 131–144.

Pyle, R.M., 1981. The Audubon Society Field Guide to North American Butterflies. Alfred A. Knopf, New York.

Quesada, M., Bollman, K., Stephenson, A.G., 1995. Leaf damage decreases pollen production and hinders pollen performance in *Cucurbita texana*. Ecology 76, 437–443.

Quesada, M., Sanchez-Azofeifa, G.A., Alvarez-Añorve, M., Stoner, K.E., Avila-Cabadilla, L., Calvo-Alvarado, J., Castillo, A., Espírito-Santo, M.M., Fagundes, M., Fernandes, G.W., Gamon, J., Lopezaraiza-Mikel, M., Lawrence, D., Morellato, L.P.C., Powers, J.S., Neves, F.de S., Rosas-Guerrero, V., Sayago, R., Sanchez-Montoya, G., 2009. Succession and management of tropical dry forests in the Americas: review and new perspectives. Forest Ecol. Manag. 258, 1014–1024.

Rabatin, S.C., Stinner, B.R., 1988. Indirect effects of interactions between VAM fungi and soil-inhabiting invertebrates on plant processes. Agr. Ecosyst. Environ. 24, 135–146.

Rabb, R.L., DeFoliart, G.K., Kennedy, G.G., 1984. An ecological approach to managing insect populations. In: Huffaker, C.B., Rabb, R.L. (Eds.), Ecol. Entomol. John Wiley & Sons, New York, NY, pp. 697–728.

Rácz, V., Bernath, I., 1993. Dominance conditions and population dynamics of Lygus (Het., Miridae) species in Hungarian maize stands (1976-1985), as functions of climatic conditions. J. Appl. Entomol. 115, 511–518.

Radeloff, V.C., Mladenoff, D.J., Boyce, M.S., 2000. The changing relation of landscape patterns and jack pine budworm populations during an outbreak. Oikos 90, 417–430.

Raffa, K.F., Phillips, T.W., Salom, S.M., 1993. Strategies and mechanisms of host colonization by bark beetles. In: Schowalter, T.D., Filip, G.M. (Eds.), Beetle-Pathogen Interactions in Conifer Forests. Academic Press, London, pp. 103–128.

Raffa, K.F., Aukema, B.H., Bentz, B.J., Carroll, A.L., Hicke, J.A., Turner, M.G., Romme, W.H., 2008. Cross-scale drivers of natural disturbances prone to anthropogenic amplification: the dynamics of bark beetle eruptions. BioScience 58, 501–517.

Ragsdale, D.W., Landis, D.A., Brodeur, J., Heimpel, G.E., Desneux, N., 2011. Ecology and management of the soybean aphid in North America. Annu. Rev. Entomol. 56, 375–399.

Raguso, R.A., Agrawal, A.A., Douglas, A.E., Jander, G., Kessler, A., Poveda, K., Thaler, J.S., 2015. The raison d'être of chemical ecology. Ecology 96, 617–630.

Raich, J.W., Bowden, R.D., Steudler, P.A., 1990. Comparison of two static chamber techniques for determining carbon dioxide efflux from forest soils. Soil Sci. Soc. Am. J. 54, 1754–1757.

Raimondo, S., Turcáni, M., Patoèka, J., Liebhold, A.M., 2004. Interspecific synchrony among foliage-feeding forest Lepidoptera species and the potential role of generalist predators as synchronizing agents. Oikos 107, 462–470.

Rainey, R.C., 1963. Meterology and the migration of desert locusts: applications of synoptic meterology in locust control. Anti-locust Memoir 7.

Rainey, R.C., Sayer, H.J., 1953. Some recent developments in the use of aircraft against flying locust swarms. Nature 172, 224–228.

Ralph, S.G., Yueh, H., Friedmann, M., Aeschliman, D., Zeznik, J.A., Nelson, C.C., Butterfield, Y.S.N., Kirkpatrick, R., Liu, J., Jones, S.J.M., Marra, M.A., Douglas, C.J., Ritland, K., Bohlmann, J., 2006. Conifer defence against insects: microarray gene expression profiling of Sitka spruce (*Picea sitchensis*) induced by mechanical wounding or feeding by spruce budworms (*Choristoneura occidentalis*) or white pine weevils (*Pissodes strobi*) reveals large-scale changes of the host transcriptome. Plant Cell Environ. 29, 1545–1570.

Rambo, T., Schowalter, T., North, M., 2014. Canopy arthropod responses to thinning and burning treatments in old-growth mixed-conifer forest in the Sierra Nevada, California. Forest Ecol. Manag. 326, 91–100.

Ramírez, A., Engman, A., Rosas, K.G., Perez-Reyes, O., Martinó-Cardona, D.M., 2012. Urban impacts on tropical island streams: some key aspects influencing ecosystem response. Urban Ecosystems 15, 315–325.

Ramirez-Romero, R., Desneux, N., Decourtye, A., Chaffiol, A., Pham-Delègue, M.H., 2008. Does Cry1Ab protein affect learning performances of the honey bee *Apis mellifera* L. (Hymenoptera, Apidae)? Ecotox. Environ. Safe. 70, 327–333.

Ramos-Elorduy, J., 2009. Anthro-entomophagy: cultures, evolution and sustainability. Entomol. Res. 39, 271–288.

Rand, E.L., Redfield, J.H., 1894. Flora of Mount Desert Island, Maine. A Preliminary Catelogue of the Plants Growing on Mount Desert and the Adjacent Islands. John Wilson and Son, Cambridge, MA.

Rankin, M.A., Burchsted, J.C.A., 1992. The cost of migration in insects. Annu. Rev. Entomol. 37, 533–559.

Ranson, H., Claudianos, C., Ortelli, F., Abgrall, C., Hemmingway, J., Sharakhova, M.V., Unger, M.F., Collins, F.H., Feyereisen, R., 2002. Evolution of supergene families associated with insecticide resistance. Science 298, 179–181.

Rasmussen, E.M., Wallace, J.M., 1983. Meterological aspects of the El Niño/southern oscillation. Science 222, 1195–1202.

Rastetter, E.B., Ryan, M.G., Shaver, G.R., Melillo, J.M., Nadelhoffer, K.J., Hobbie, J.E., Aber, J.D., 1991. A general biogeochemical model describing the responses of the C and N cycles in terrestrial ecosystems to changes in CO_2, climate and N deposition. Tree Physiol. 9, 101–126.

Rastetter, E.B., Ågren, G.I., Shaver, G.R., 1997. Responses of N-limited ecosystems to increased CO2: a balanced-nutrition, coupled-element-cycles model. Ecol. Appl. 7, 444–460.

Raubenheimer, D., Simpson, S.J., 1999. Integrating nutrition: a geometrical approach. Entomol. Exp. Appl. 91, 67–82.

Raubenheimer, D., Simpson, S.J., 2003. Nutrient balancing in grasshoppers: behavioural and physiological correlates of diet breadth. J. Exp. Biol. 206, 1669–1681.

Raubenheimer, D., Tucker, D., 1997. Associative learning by locusts: pairing of visual cues with consumption of protein and carbohydrate. Anim. Behav. 54, 1449–1459.

Raupp, M.J., Shrewsbury, P.M., Herms, D.A., 2010. Ecology of herbivorous arthropods in urban landscapes. Annu. Rev. Entomol. 55, 19–38.

Raven, J.A., 1983. Phytophages of xylem and phloem: a comparison of animal and plant sap-feeders. Adv. Ecol. Res. 13, 136–204.

Reagan, D.P., Camilo, G.R., Waide, R.B., 1996. The community food web: major properties and patterns of organization. In: Reagan, D.P., Waide, R.B. (Eds.), The Food Web of a Tropical Rain Forest. University of Chicago Press, Chicago, IL, pp. 462–488.

Reay-Jones, F.P.F., Way, M.O., Sétamou, M., Legendre, B.L., Reagan, T.E., 2003. Resistance to the Mexican rice borer (Lepidoptera: Crambidae) among Louisiana and Texas sugarcane cultivars. J. Econ. Entomol. 96, 1929–1934.

Reay-Jones, F.P.F., Wilson, L.T., Way, M.O., Reagan, T.E., Carlton, C.E., 2007. Movement of Mexican rice borer (Lepidoptera: Crambidae) through the Texas rice belt. J. Econ. Entomol. 100, 54–60.

Reeves, J.L., Lorch, P.D., Kershner, M.W., 2009. Vision is important for plant location by the phytophagous aquatic specialist *Euhrychiopsis lecontei* Dietz (Coleoptera: Curculionidae). J. Insect Behav. 22, 54–64.

Regal, R.J., 1982. Pollination by wind and animals: ecology of geographic patterns. Annu. Rev. Ecol. Syst. 13, 497–524.

Reice, S.R., 1985. Experimental disturbance and the maintenance of species diversity in a stream community. Oecologia 67, 90–97.

Reich, P.B., Tilman, D., Isbell, F., Mueller, K., Hobbie, S.E., Flynn, D.F.B., Eisenhauer, N., 2012. Impacts of biodiversity loss escalate through time as redundancy fades. Science 336, 589–592.

Reichle, D.E., 1968. Relation of body size to food intake, oxygen consumption, and trace element metabolism in forest floor arthropods. Ecology 49, 538–542.

Reichle, D.E., Crossley, Jr., D.A., 1967. Investigation on heterotrophic productivity in forest insect communities. In: Petrusewicz, K. (Ed.), Secondary Productivity of Terrestrial Ecosystems: Principles and Methods. Państwowe Wydawnictwo Naukowe, Warszawa, Poland, pp. 563–587.

Reichle, D.E., Shanks, M.H., Crossley, Jr., D.A., 1969. Calcium, potassium, and sodium content of forest floor arthropods. Ann. Entomol. Soc. Am. 62, 57–62.

Reichle, D.E., Goldstein, R.A., Van Hook, R.I., Dodson, G.J., 1973. Analysis of insect consumption in a forest canopy. Ecology 54, 1076–1084.

Reimer, L., Fondjo, E., Patchoké, S., Diallo, B., Lee, Y., Ng, A., Ndjemai, H.M., Atangana, J., Traore, S.F., Lanzaro, G., Cornel, A.J., 2008. Relationship between kdr mutation and resistance to pyrethroid and DDT insecticides in natural populations of *Anopheles gambiae*. J. Med. Entomol. 45, 260–266.

Renwick, J.A.A., 2002. The chemical world of crucivores: lures, treats and traps. Entomol. Exp. Appl. 104, 35–42.

Resasco, J., Haddad, N.M., Orrock, J.L., Shoemaker, D., Brudvig, L.A., Damschen, E.I., Tewksbury, J.J., Levey, D.J., 2014. Landscape corridors can increase invasion by an exotic species and reduce diversity of native species. Ecology 95, 2033–2039.

Resetarits, Jr., W.J., Binckley, C.A., 2013. Patch quality and context, but not patch number, drive multi-scale colonization dynamics in experimental aquatic landscapes. Oecologia 173, 933–946.

Reudler, J.H., Lindstedt, C., Pakkanen, H., Lehtinen, I., Mappes, J., 2015. Costs and benefits of plant allelochemicals in herbivore diet in a multi enemy world. Oecologia 179, 1147–1158.

Reusch, T.B.H., Ehlers, A., Hämmerli, A., Worm, B., 2005. Ecosystem recovery after climatic extremes enhanced by genotypic diversity. Proc. Natl. Acad. Sci. USA 102, 2826–2831.

Reynolds, B.C., Crossley, Jr., D.A., Hunter, M.D., 2003. Response of soil invertebrates to forest canopy inputs along a productivity gradient. Pedobiologia 47, 127–139.

Reynolds, L.V., Ayres, M.P., Siccama, T.G., Holmes, R.T., 2007. Climatic effects on caterpillar fluctuations in northern hardwood forests. Can. J. Forest. Res. 37, 481–491.

Rhoades, D.F., 1977. The antiherbivore chemistry of *Larrea*. In: Mabry, T.J., Hunziker, J.H., DiFeo, Jr., D.R. (Eds.), Creosote Bush: Biology and Chemistry of Larrea in New World Deserts. Dowden, Hutchinson & Ross, Inc, Stroudsburg, PA, pp. 135–175.

Rhoades, D.F., 1983. Responses of alder and willow to attack by tent caterpillars and webworms: evidence for pheromonal sensitivity of willows. In: Hedin, P.A. (Ed.). Plant Resistance to Insects. ACS Symposium Series 208, American Chemical Society, Washington, DC, pp. 55–68.

Ribeiro, S.P., Pimenta, H.R., Fernandes, G.W., 1994. Herbivory by chewing and sucking insects on *Tabebuia ochracea*. Biotropica 26, 302–307.

Rice, B., Westoby, M., 1986. Evidence against the hypothesis that ant-dispersed seeds reach nutrient-enriched microsites. Ecology 67, 1270–1274.

Richards, L.A., Dyer, L.A., Forister, M.L., Smilanich, A.M., Dodson, C.D., Leonard, M.D., Jeffrey, C.S., 2015. Phytochemical diversity drives plant–insect community diversity. Proc. Natl. Acad. Sci. USA 112, 10973–10978.

Richardson, B.A., Hull, G.A., 2000. Insect colonization sequences in bracts of *Heliconia caribaea* in Puerto Rico. Ecol. Entomol. 25, 460–466.

Richardson, B.A., Richardson, M.J., Scatena, F.N., McDowell, W.H., 2000a. Effects of nutrient availability and other elevational changes on bromeliad populations and their invertebrate communities in a humid tropical forest in Puerto Rico. J. Trop. Ecol. 16, 167–188.

Richardson, B.A., Rogers, C., Richardson, M.J., 2000b. Nutrients, diversity, and community structure of two phytotelm systems in a lower montane forest, Puerto Rico. Ecol. Entomol. 25, 348–356.

Richardson, S.J., Press, M.C., Parsons, A.N., Hartley, S.E., 2002. How do nutrients and warming impact on plant communities and their insect herbivores? A 9-year study from a sub-Arctic heath. J. Ecol. 90, 544–556.

Richardson, B.A., Richardson, M.J., Gonzalez, G., Shiels, A.B., Srivastava, D.S., 2010. A canopy trimming experiment in Puerto Rico: the response of litter invertebrate communities to canopy loss and debris deposition in a tropical forest subject to hurricanes. Ecosystems 11, 286–301.

Richerson, J.V., Boldt, P.E., 1995. Phytophagous insect fauna of *Flourensia cernua* (Asteraceae: Heliantheae) in trans-Pecos Texas and Arizona. Environ. Entomol. 24, 588–594.

Richter, M.R., 1990. Hunting social wasp interactions: influence of prey size, arrival order, and wasp species. Ecology 71, 1018–1030.

Ricketts, T.H., 2004. Tropical forest fragments enhance pollinator activity in nearby coffee crops. Conserv. Biol. 18, 1262–1271.

Ricketts, T.H., Regetz, J., Steffan-Dewenter, I., Cunningham, S.A., Kremen, C., Bogdanski, A., Gemmill-Herren, B., Greenleaf, S.S., Klein, A.M., Mayfield, M.M., Morandin, L.A., Ochieng, A., Viana, B.F., 2008. Landscape effects on crop pollinator services: are there general patterns? Ecol. Lett. 11, 499–515.

Rickson, F.R., 1971. Glycogen plastids in Müllerian body cells of *Cecropia peltata* – a higher green plant. Science 173, 344–347.

Rickson, F.R., 1977. Progressive loss of ant-related traits of *Cecropia peltata* on selected Caribbean islands. Am. J. Bot. 64, 585–592.

Rickson, F.R., Rickson, M.M., 1998. The cashew nut, *Anacardium occidentale* (Anacardiaceae), and its perennial association with ants: extrafloral nectary location and the potential for ant defense. Am. J. Bot. 85, 835–849.

Ridsdill-Smith, T.J., Kirk, A.A., 1985. Selecting dung beetles (Scarabaeinae) from Spain for bushfly control in south-western Australia. Entomophaga 30, 217–223.

Riehle, M.A., Moreira, C.K., Lampe, D., Lauzon, C., Jacobs-Lorena, M., 2007. Using bacteria to express and display anti-*Plasmodium* molecules in the mosquito midgut. Int. J. Parasitol. 37, 595–603.

Ries, L., Fagan, W.F., 2003. Habitat edges as a potential ecological trap for an insect predator. Ecol. Entomol. 28, 567–572.

Ries, L., Fletcher, Jr., R.J., Battin, J., Sisk, T.D., 2004. Ecological responses to habitat edges: mechanisms, models, and variability explained. Annu. Rev. Ecol. Evol. S. 35, 491–522.

Riginos, C., Karande, M.A., Rubenstein, D.I., Palmer, T.M., 2015. Disruption of a protective ant–plant mutualismby an invasive ant increases elephant damage to savanna trees. Ecology 96, 654–661.

Riley, C.V., 1878. First Annual Report of the United States Entomological Commission for the Year 1877 Relating to the Rocky Mountain Locust and the Best Methods of Preventing its Injuries and of Guarding Against its Invasions, in Pursuance of an Appropriation Made by Congress for this Purpose. US Department of Agriculture, Washington, DC.

Riley, C.V., 1880. Second Report of the United States Entomological Commission for the Years 1878 and 1879 Relating to the Rocky Mountain Locust and the Western Cricket and Treating of the Best Means of Subduing the Locust in its Permanent Breeding Grounds, With a View of Preventing its Migrations into the More Fertile Portions of the Trans-Mississippi Country, in Pursuance of Appropriations Made by Congress for this Purpose. US Department of Agriculture, Washington, DC.

Riley, C.V., 1883. Third Report of the United States Entomological Commission, Relating to the Rocky Mountain Locust, the Western Cricket, the Army-worm, Canker Worms, and the Hessian Fly, Together with Descriptions of Larvae of Injurious Forest Insects, Studies on the Embryological Development of the Locust and of Other Insects, and on the Systematic Position of the Orthoptera in Relation to Other Orders of Insects. US Department of Agriculture, Washington, DC.

Riley, C.V., 1885. Fourth Report of the United States Entomological Commission, Being a Revised Edition of Bulletin No. 3, and the Final Report on the Cotton Worm, Together with a Chapter on the Boll Worm. US Department of Agriculture, Washington, DC.

Riley, C.V., 1893. Predaceous and parasitic insects in applied entomology. Insect Life 6, 130–141.

Riley, C.V., Howard, L.O., 1890. The imported gypsy moth (*Ocneria dispar* L.). Insect Life 2, 208–211.

Riley, C.V., Vasey, G., 1870. Imported insects and native American insects. Am. Entomol. 2, 110–112.

Riley, J.R., Smith, A.D., Reynolds, D.R., Edwards, A.S., Osborne, J.L., Williams, I.H., Carreck, N.L., Poppy, G.M., 1996. Tracking bees with harmonic radar. Nature 379, 29–30.

Rinehart, J.P., Li, A., Yocum, G.D., Robich, R.M., Hayward, S.A.L., Denlinger, D.L., 2007. Up-regulation of heat shock proteins is essential for cold survival during insect diapause. Proc. Natl. Acad. Sci. USA 104, 11130–11137.

Rio, R.V.M., Wu, Y.-N., Filardo, G., Aksoy, S., 2006. Dynamics of multiple symbiont density regulation during host development: tsetse fly and its microbial flora. Proc. Roy. Soc. B-Biol. Sci. 273, 805–814.

Ris, N., Allemand, R., Fouillet, P., Fleury, F., 2004. The joint effect of temperature and host species induce complex genotype-by-environment interactions in the larval parasitoid of *Drosophila*, *Leptopilina heterotoma* (Hymenoptera: Figitidae). Oikos 106, 451–456.

Risch, S., 1980. The population dynamics of several herbivorous beetles in a tropical agroecosystem: the effect of intercropping corn, beans and squash in Costa Rica. J. Appl. Ecol. 17, 593–612.

Risch, S.J., 1981. Insect herbivore abundance in tropical monocultures and polycultures: an experimental test of two hypotheses. Ecology 62, 1325–1340.

Risch, A.C., Jurgensen, M.F., Schütz, M., Page-Dumbroese, D.S., 2005. The contribution of red wood ants to soil C and N pools and CO_2 emissions in subalpine forests. Ecology 85, 419–430.

Risley, L.S., Crossley, Jr., D.A., 1993. Contribution of herbivore-caused greenfall to litterfall nitrogen flux in several southern Appalachian forested watersheds. Am. Midl. Nat. 129, 67–74.

Rissing, S.W., 1986. Indirect effects of granivory by harvester ants: plant species composition and reproductive increase near ant nests. Oecologia 68, 231–234.

Ritchie, M.E., 2000. Nitrogen limitation and trophic vs. abiotic influences on insect herbivores in a temperate grassland. Ecology 81, 1601–1612.

Ritchie, M.E., Tilman, D., Knops, J.M.H., 1998. Herbivore effects on plant and nitrogen dynamics in oak savanna. Ecology 79, 165–177.

Ritland, D.B., Brower, L.P., 1991. The viceroy butterfly is not a batesian mimic. Nature 350, 497–498.

Ritter, Jr., H., 1964. Defense of mate and mating chamber in a wood roach. Science 143, 1459–1460.

Ritzmann, R.E., Quinn, R.D., Fischer, M.S., 2004. Convergent evolution and locomotion through complex terrain by insects, vertebrates and robots. Arthropod Struct. Dev. 33, 361–379.

Rivers, D.B., Lee, Jr., R.E., Denlinger, D.L., 2000. Cold hardiness of the fly pupal parasitoid Nasonia vitripennis is enhanced by its host Sarcophaga crassipalpis. J. Insect Physiol. 46, 99–106.

Roberds, J.H., Hain, F.P., Nunnally, L.B., 1987. Genetic structure of southern pine beetle populations. Forest Sci. 33, 52–69.

Robertson, A.I., Giddins, R., Smith, T.J., 1990. Seed predation by insects in tropical mangrove forests: extent and effects on seed viability and the growth of seedlings. Oecologia 83, 213–219.

Robertson, C., Nelson, T.A., Jelinski, D.E., Wulder, M.A., Boots, B., 2009. Spatial–temporal analysis of species range expansion: the case of the mountain pine beetle, Dendroctonus ponderosae. J. Biogeogr. 36, 1446–1458.

Robinson, M.H., 1969. The defensive behaviour of some orthopteroid insects from Panama. Trans. Roy. Ent. Soc. Lond. 121, 281–303.

Robinson, M.H., Robinson, B.C., 1974. Adaptive complexity: the thermoregulatory postures of the golden-web spider, Nephila clavipes, at low latitudes. Am. Midl. Nat. 92, 386–396.

Rodgers, H.L., Brakke, M.P., Ewel, J.J., 1995. Shoot damage effects on starch reserves of Cedrela odorata. Biotropica 27, 71–77.

Rodrigues, A., Cunha, L., Amaral, A., Medeiros, J., Garcia, P., 2008. Bioavailability of heavy metals and their effects on the midgut cells of a phytophagous insect inhabiting volcanic environments. Sci. Total Environ. 406, 116–122.

Rodriguez, J.M. (Ed.), 1972. Insect and Mite Nutrition: Significance and Implications in Ecology and Pest Management. North-Holland Publishing Co, Amsterdam.

Rodriguez-Saona, C., Chalmers, J.A., Raj, S., Thaler, J.S., 2005. Induced plant responses to multiple damagers: differential effects on an herbivore and its parasitoid. Oecologia 143, 566–577.

Rodriguez-Saona, C., Vorsa, N., Singh, A.P., Johnson-Cicalese, J., Szendrei, Z., Mescher, M.C., Frost, C.J., 2011. Tracing the history of plant traits under domestication in cranberries: potential consequences on anti-herbivore defences. J. Exp. Bot. 62, 2633–2644.

Rodstein, J., McElfresh, J.S., Barbour, J.D., Ray, A.M., Hanks, L.M., Millar, J.G., 2009. Identification and synthesis of a female-produced sex pheromone for the cerambycid beetle Prionus californicus. J. Chem. Ecol. 35, 590–600.

Roelofs, W.L., 1995. Chemistry of sex attraction. Proc. Natl. Acad. Sci. USA 92, 44–49.

Roff, D.A., 1990. The evolution of flightlessness in insects. Ecol. Monogr. 60, 389–421.

Rogers, S.R., Tarpy, D.R., Burrack, H.J., 2014. Bee species diversity enhances productivity and stability in a perennial crop. PLoS ONE 9 (5), e97307.

Rohr, J.R., Mahan, C.G., Kim, K.C., 2009. Response of arthropod biodiversity to foundation species declines: the case of the eastern hemlock. Forest Ecol. Manag. 258, 1503–1510.

Roland, J., 1993. Large-scale forest fragmentation increases the duration of tent caterpillar outbreak. Oecologia 93, 25–30.

Roland, J., Kaupp, W.J., 1995. Reduced transmission of forest tent caterpillar (Lepidoptera: Lasiocampidae) nuclear polyhedrosis virus at the forest edge. Environ. Entomol. 24, 1175–1178.

Roland, J., Taylor, P.D., 1997. Insect parasitoid species respond to forest structure at different spatial scales. Nature 386, 710–713.

Romeis, J., Bartsch, D., Bigler, F., Candolfi, M.P., Gielkens, M.M.C., Hartley, S.E., Hellmich, R.L., Huesing, J.E., Jepson, P.C., Layton, R., Quemada, H., Raybould, A., Rose, R.I., Schiemann, J., Sears, M.K., Shelton, A.M., Sweet, J., Vaituzis, Z., Wolt, J.D., 2008. Assessment of risk of insect-resistant transgenic crops to nontarget arthropods. Nat. Biotechnol. 26, 203–208.

Romme, W.H., Knight, D.H., Yavitt, J.B., 1986. Mountain pine beetle outbreaks in the Rocky Mountains: regulators of primary productivity? Am. Nat. 127, 484–494.

Romoser, W.S., Stoffolano, Jr., J.G., 1998. The Science of Entomology, fourth ed. McGraw-Hill, Boston, MA, pp. 605.

Root, R.B., 1967. The niche exploitation pattern of the blue-gray gnatcatcher. Ecol. Monogr. 37, 317–350.

Root, R.B., 1973. Organization of a plant-arthropod association in simple and diverse habitats: the fauna of collards (Brassica oleracea). Ecol. Monogr. 43, 95–124.

Rösch, V., Tscharntke, T., Scherber, C., Batáry, P., 2015. Biodiversity conservation across taxa and landscapes requires many small as well as single large habitat fragments. Oecologia 179, 209–222.

Roscher, C., Weigelt, A., Proulx, R., Marquard, E., Schumacher, J., Weisser, W.W., Schmid, B., 2011. Identifying population- and community-level mechanisms of diversity–stability relationships in experimental grasslands. J. Ecol. 99, 1460–1469.

Rosenblatt, A.E., Nifong, J.C., Heithaus, M.R., Mazzotti, F.J., Cherkiss, M.S., Jeffery, B.M., Elsey, R.M., Decker, R.A., Silliman, B.R., Guillette, Jr., L.J., Lowers, R.H., Larson, J.C., 2015. Factors affecting individual foraging specialization and temporal diet stability across the range of a large "generalist" apex predator. Oecologia 178, 5–16.

Rosenheim, J.A., 2005. Intraguild predation of Orius tristicolor by Geocoris spp. and the paradox of irruptive spider mite dynamics in California cotton. Biol. Control 32, 172–179.

Rosenthal, G.A., Janzen, D.H. (Eds.), 1979. Herbivores: Their Interactions with Secondary Plant Metabolites. Academic Press, New York.

Rosenzweig, M.L., Abramsky, Z., 1993. How are diversity and productivity related? In: Ricklefs, R.E., Schluter, D. (Eds.), Species Diversity in Ecological Communities: Historical and Geographic Perspectives. University of Chicago Press, Chicago, IL, pp. 52–65.

Rosi-Marshall, E.J., Tank, J.L., Royer, T.V., Whiles, M.R., Evans-White, M., Chambers, C., Griffiths, N.A., Pokelsek, J., Stephen, M.L., 2007. Toxins in transgenic crop byproducts may affect headwater stream ecosystems. Proc. Natl. Acad. Sci. USA 104, 16204–16208.

Rotem, K., Agrawal, A.A., Kott, L., 2003. Parental effects in *Pieris rapae* in response to variation in food quality: adaptive plasticity across generations? Ecol. Entomol. 28, 211–218.

Roth, S.K., Lindroth, R.L., 1994. Effects of CO_2-mediated changes in paper birch and white pine chemistry on gypsy moth performance. Oecologia 98, 133–138.

Roth, J.P., MacQueen, A., Bay, D.E., 1988. Predation by the introduced phoretic mite, *Macrocheles peregrinus* (Acari: Macrochelidae), on the buffalo fly, *Haematobia irritans exigua* (Diptera: Muscidae), in Australia. Environ. Entomol. 17, 603–607.

Roubik, D.W., 1989. Ecology and Natural History of Tropical Bees. Cambridge University Press, Cambridge, UK.

Roubik, D.W., 1993. Tropical pollinators in the canopy and understory: field data and theory for stratum "preferences". J. Insect Behav. 6, 659–673.

Roubik, D.W., 2002. The value of bees to the coffee harvest. Nature 417, 708.

Rousseaux, M.C., Julkunen-Tiitto, R., Searles, P.S., Scopel, A.L., Aphalo, P.J., Ballaré, C.L., 2004. Solar UV-B radiation affects leaf quality and insect herbivory in the southern beech tree *Nothofagus antarctica*. Oecologia 138, 505–512.

Roussel, J.S., Clower, D.F., 1957. Resistance to the chlorinated hydrocarbon insecticides in the boll weevil. J. Econ. Entomol. 50, 463–468.

Roy, B.A., Alexander, H.M., Davidson, J., Campbell, F.T., Burdon, J.J., Sniezko, R., Brasier, C., 2014. Increasing forest loss worldwide from invasive pests requires new trade regulations. Front. Ecol. Environ. 12, 457–465.

Royama, T., 1984. Population dynamics of the spruce budworm *Choristoneura fumiferana*. Ecol. Monogr. 54, 429–462.

Royama, T., 1992. Analytical Population Dynamics. Chapman & Hall, London.

Ruangpanit, N., 1985. Percent crown cover related to water and soil losses in mountainous forest in Thailand. In: El-Swaify, S.A., Moldenhauer, W.C., Lo, A. (Eds.), Soil Erosion and Conservation. Soil Conservation Society of America, Ankeny, IA, pp. 462–471.

Rubenstein, D.I., 1992. The greenhouse effect and changes in animal behavior: effects on social structure and life-history strategies. In: Peters, R.L., Lovejoy, T.E. (Eds.), Global Warming and Biological Diversity. Yale University Press, New Haven, CT, pp. 180–192.

Ruberson, J.R., Kring, T.J., Elkassabany, N., 1998. Overwintering and the diapause syndrome of predatory Heteroptera. In: Coll, M., Ruberson, J.R. (Eds.), Predatory Heteroptera: Their Ecology and Use In Biological Control. Proceedings of Thomas Say Publication in Entomology, Entomologial Society of America, Lanham, MD, pp. 49–69.

Rudd, W.G., Gandour, R.W., 1985. Diffusion model for insect dispersal. J. Econ. Entomol. 78, 295–301.

Rudinsky, J.A., Ryker, L.C., 1976. Olfactory and auditory signals mediating behavioral patterns of bark beetles. In: Coll. Internat., Centre National de la Recherche Scientifique, No. 265, Paris, France, pp. 195–207.

Ruel, J., Whitham, T.G., 2002. Fast-growing juvenile pinyons suffer greater herbivory when mature. Ecology 83, 2691–2699.

Rundlöf, M., Persson, A.S., Smith, H.G., Bommarco, R., 2014. Late-season mass-flowering red clover increases bumble bee queen and male densities. Biol. Conserv. 172, 138–145.

Rundlöf, M., Andersson, G.K.S., Bommarco, R., Fries, I., Hederström, V., Herbertsson, L., Jonsson, O., Klatt, B.K., Pedersen, T.R., Yourstone, J., Smith, H.G., 2015. Seed coating with a neonicotinoid insecticide negatively affects wild bees. Nature 521, 77–80.

Running, S.W., Gower, S.T., 1991. FOREST-BGC, a general model of forest ecosystem processes for regional applications. II. Dynamic carbon allocation and nitrogen budgets. Tree Physiol. 9, 147–160.

Russell, T.L., Kay, B.H., 2008. Biologically based insecticides for the control of immature Australian mosquitoes: a review. Aust. J. Entomol. 47, 232–242.

Russell, J.A., Moran, N.A., 2006. Costs and benefits of symbiont infection in aphids: variation among symbionts and across temperatures. Proc. Roy. Soc. B-Biol. Sci. 273, 603–610.

Rykiel, E.J., Saunders, M.C., Wagner, T.L., Loh, D.K., Turnbow, R.H., Hu, L.C., Pulley, P.E., Coulson, R.N., 1984. Computer-aided decision making and information accessing in pest management systems, with emphasis on the southern pine beetle (Coleoptera: Scolytidae). J. Econ. Entomol. 77, 1073–1082.

Rykken, J.J., Capen, D.E., Mahabir, S.P., 1997. Ground beetles as indicators of land type diversity in the Green Mountains of Vermont. Conserv. Biol. 11, 522–530.

Rykken, J.J., Moldenke, A.R., Olson, D.H., 2007a. Headwater riparian forest-floor invertebrate communities associated with alternative forest management practices. Ecol. Appl. 17, 1168–1183.

Rykken, J.J., Chan, S.S., Moldenke, A.R., 2007b. Headwater riparian microclimate patterns under alternative forest management treatments. Forest Sci. 53, 270–280.

Ryszkowski, L., 1979. Consumers. In: Coupland, R.T. (Ed.), Grassland Ecosystems of the World: Analysis of Grasslands and their Uses. Cambridge University Press, Cambridge, UK, pp. 309–318.

Sabo, J.L., Sponseller, R., Dixon, M., Gade, K., Harms, T., Heffernan, J., Jani, A., Katz, G., Soykan, C., Watts, J., Welter, J., 2005. Riparian zones increase regional species richness by harboring different, not more, species. Ecology 86, 56–62.

Sackville Hamilton, C.A.G., Cherrett, J.M., Ford, J.B., Sagar, G.R., Whitbread, R., 1991. A modular rhizotron for studying soil organisms: construction and establishment. In: Atkinson, D. (Ed.), Plant Root Growth: an Ecological Perspective. Blackwell Scientific, London, pp. 49–59.

Sakai, S., Momose, K., Yumoto, T., Kato, M., Inoue, T., 1999. Beetle pollination of *Shorea parvifolia* (section *Mutica*, Dipterocarpaceae) in a general flowering period in Sarawak, Malaysia. Am. J. Bot. 86, 62–69.

Salati, E., 1987. The forest and the hydrologic cycle. In: Dickinson, R.E. (Ed.), The Geophysiology of Amazonia: Vegetation and Climate Interactions. John Wiley & Sons, New York, pp. 273–296.

Salick, J., Herrera, R., Jordan, C.F., 1983. Termitaria: nutrient patchiness in nutrient-deficient rain forests. Biotropica 15, 1–7.

Sallabanks, R., Courtney, S.P., 1992. Frugivory, seed predation, and insect-vertebrate interactions. Annu. Rev. Entomol. 37, 377–400.

Salo, J., Kallioloa, R., Häkkinen, I., Mäkinen, Y., Niemelä, P., Puhakka, M., Coley, P.D., 1986. River dyanamics and the diversity of Amazon lowland forest. Nature 322, 254–258.

Salt, D.T., Fenwick, P., Whittaker, J.B., 1996. Interspecific herbivore interactions in a high CO_2 environment: root and shoot aphids feeding on *Cardamine*. Oikos 77, 326–330.

Saltzmann, K.D., Giovanini, M.P., Zheng, C., Williams, C.E., 2008. Virulent Hessian fly larvae manipulate the free amino acid content of host wheat plants. J. Chem. Ecol. 34, 1401–1410.

Samways, M.J., 1995. Southern hemisphere insects: their variety and the environmental pressures upon them. In: Harrington, R., Stork, N.E. (Eds.), Insects in a Changing Environment. Academic Press, London, pp. 297–320.

Samways, M.J., Caldwell, P.M., Osborn, R., 1996. Ground-living invertebrate assemblages in native, planted and invasive vegetation in South Africa. Agr. Ecosyst. Environ. 59, 19–32.

San Martin y Gomez, G., Van Dyck, H., 2012. Ecotypic differentiation between urban and rural populations of the grasshopper Chorthippus brunneus relative to climate and habitat fragmentation. Oecologia 169, 125–133.

Sanchez-Bayo, F., Goka, K., 2014. Pesticide residues and bees—a risk assessment. PLoS ONE 9 (4), e94482.

Sánchez-Zapata, J.A., Donázar, J.A., Delgado, A., Forero, M.G., Ceballos, O., Hiraldo, F., 2007. Desert locust outbreaks in the Sahel: resource competition, predation and ecological effects of pest control. J. Appl. Ecol. 44, 323–329.

Sanders, D., Platner, C., 2007. Intraguild interactions between spiders and ants and top-down control in a grassland food web. Oecologia 150, 611–624.

Sanders, N.J., Gotelli, N.J., Heller, N.E., Gordon, D.M., 2003. Community disassembly by an invasive species. Proc. Natl. Acad. Sci. USA 100, 2474–2477.

Sanders, N.J., Weltzin, J.F., Crutsinger, G.M., Fitzpatrick, M.C., Nuñez, M.A., Oswalt, C.M., Lane, K.E., 2007. Insects mediate the effects of propagule supply and resource availability on a plant invasion. Ecology 88, 2383–2391.

Sanders, D., Schaefer, M., Platner, C., Griffiths, G.J.K., 2011. Intraguild interactions among generalist predator functional groups drive impact on herbivore and decomposer prey. Oikos 120, 418–426.

Sanderson, M.G., 1996. Biomass of termites and their emissions of methane and carbon dioxide: a global database. Global Biogeochem. Cy. 10, 543–557.

Sandlin, E.A., Willig, M.R., 1993. Effects of age, sex, prior experience, and intraspecific food variation on diet composition of a tropical folivore (Phasmatodea: Phasmatidae). Environ. Entomol. 22, 625–633.

Sanson, D.W., DeRosa, A.A., Oremus, G.R., Foil, L.D., 2003. Effect of horn fly and internal parasite control on growth of beef heifers. Vet. Parasitol. 117, 291–300.

Santos, P.F., Whitford, W.G., 1981. The effects of microarthropods on litter decomposition in a Chihuahuan Desert ecosystem. Ecology 62, 654–663.

Santos, P.F., Phillips, J., Whitford, W.G., 1981. The role of mites and nematodes in early stages of buried litter decomposition in a desert. Ecology 62, 664–669.

Šantrůčková, H., Bird, M.I., Frouz, J., Šustr, V., Tajovský, K., 2000. Natural abundance of ^{13}C in leaf litter as related to feeding activity of soil invertebrates and microbial mineralization. Soil Biol. Biochem. 32, 1793–1797.

Sarmiento, J.L., Le Quéré, C., 1996. Oceanic carbon dioxide uptake in a model of century-scale global warming. Science 274, 1346–1350.

Sartwell, C., Stevens, R.E., 1975. Mountain pine beetle in ponderosa pine: prospects for silvicultural control in second-growth stands. J. Forest. 73, 136–140.

Sattler, T., Obrist, M.K., Duelli, P., Moretti, M., 2011. Urban arthropod communities: added value or just a blend of surrounding biodiversity? Landscape Urban Plan. 103, 347–361.

Savage, H.M., Niebylski, M.L., Smith, G.C., Mitchell, C.J., Craig, Jr., G.B., 1993. Host-feeding patterns of *Aedes albopictus* (Diptera: Culicidae) at a temperate North American site. J. Med. Entomol. 30, 27–34.

Savely, Jr., H.E., 1939. Ecological relations of certain animals in dead pine and oak logs. Ecol. Monogr. 9, 321–385.

Savva, Y., Szlavecz, K., Pouyat, R.V., Groffman, P.M., Heisler, G., 2010. Effects of land use and vegetation cover on soil temperature in an urban ecosystem. J. Soil Sci. Soc. Am. 74, 469–480.

Scatena, F.N., Moya, S., Estrada, C., Chinea, J.D., 1996. The first five years in the reorganization of aboveground biomass and nutrient use following Hurricane Hugo in the Bisley Experimental Watersheds, Luquillo Experimental Forest, Puerto Rico. Biotropica 28, 424–440.

Schäfer, M.L., Lundström, J.O., 2006. Different responses of two floodwater mosquito species, *Aedes vexans* and *Ochlerotatus sticticus* (Diptera: Culicidae), to larval habitat drying. J. Vector Ecol. 31, 123–128.

Schäffer, M., Winkelmann, C., Hellmann, C., Benndorf, J., 2013. Reduced drift activity of two benthic invertebrate species is mediated by infochemicals of benthic fish. Aquat. Ecol. 47, 99–107.

Schäpers, A., Nylin, S., Carlsson, M.A., Janz, N., 2016. Specialist and generalist oviposition strategies in butterflies: maternal care or precocious young? Oecologia 180, 335–343.

Scheiner, S.M., Willig, M.R., 2008. A general theory of ecology. Theor. Ecol. 1, 21–28.

Schell, S.P., Lockwood, J.A., 1997. Spatial characteristics of rangeland grasshopper (Orthoptera: Acrididae) population dynamics in Wyoming: implications for pest management. Environ. Entomol. 26, 1056–1065.

Schellhorn, N.A., Gagic, V., Bommarco, R., 2015. Time will tell: resource continuity bolsters ecosystem services. Trends Ecol. Evol. 30, 524–530.

Scheper, J., Bommarco, R., Holzschuh, A., Potts, S.G., Riedinger, V., Roberts, S.P.M., Rundlof, M., Smith, H.G., Steffan-Dewenter, I., Wickens, J.B., Wickens, V.J., Kleijn, D., 2015. Local and landscape-level floral resources explain effects of wildflower strips on wild bees across four European countries. J. Appl. Ecol. 52, 1165–1175.

Scheu, S., Falca, M., 2000. The soil food web of two beech forests (*Fagus sylvatica*) of contrasting humus type: stable isotope analysis of a macro- and a mesofauna-dominated community. Oecologia 123, 285–296.

Schiff, H., 1991. Modulation of spike frequencies by varying the ambient magnetic field and magnetite candidates in bees (*Apis mellifera*). Comp. Biochem. Phys. A 100, 975–985.

Schirmel, J., Fartmann, T., 2014. Coastal heathland succession influences butterfly community composition and threatens endangered butterfly species. J. Insect Conserv. 18, 111–120.

Schlesinger, W.H., Reynolds, J.F., Cunningham, G.L., Huenneke, L.F., Jarrell, W.M., Virginia, R.A., Whitford, W.G., 1990. Biological feedbacks in global desertification. Science 247, 1043–1048.

Schmelz, E.A., Brebeno, R.J., Ohnmeiss, T.E., Bowers, W.S., 2002. Interactions between *Spinacia oleracea* and *Bradysia impatiens*: a role for phytoecdysteroids. Arch. Insect Biochem. 51, 204–221.

Schmelz, E.A., Carroll, M.J., LeClere, S., Phipps, S.M., Meredith, J., Chourey, P.S., Alborn, H.T., Teal, P.E.A., 2006. Fragments of ATP synthase mediate plant perception of insect attack. Proc. Natl. Acad. Sci. USA 103, 8894–8899.

Schmelz, E.A., LeClere, S., Carroll, M.J., Alborn, H.T., Teal, P.E.A., 2007. Cowpea chloroplastic ATP synthase is the source of multiple plant defense elicitors during insect herbivory. Plant Physiol. 144, 793–805.

Schmidt, J.O., 1982. Biochemistry of insect venoms. Annu. Rev. Entomol. 27, 339–368.

Schmidt, J.O., 2014. Evolutionary responses of solitary and social Hymenoptera to predation by primates and overwhelmingly powerful vertebrate predators. J. Hum. Evol. 71, 12–19.

Schmidt, D.D., Voelckel, C., Hartl, M., Schmidt, S., Baldwin, I.T., 2005. Specificity in ecological interactions. Attack from the same lepidopteran herbivore results in species-specific transcriptional responses in two solenaceous host plants. Plant Physiol. 138, 1763–1773.

Schmidt, J.M., Crist, T.O., Wrinn, K., Rypstra, A.L., 2014. Predator interference alters foraging behavior of a generalist predatory arthropod. Oecologia 175, 501–508.

Schmidt, J.P., Moore, R., Alber, M., 2014. Integrating ecosystem services and local government finances into land use planning: a case study from coastal Georgia. Landscape Urban Plan. 122, 56–67.

Schmitz, O.J., 2007. Predator diversity and trophic interactions. Ecology 88, 2415–2426.

Schmitz, H., Trenner, S., 2003. Electrophysiological characterization of the multipolar thermoreceptors in the "fire-beetle" *Merimna atrata* and comparison with the infrared sensilla of *Melanophila acuminata* (both Coleoptera, Buprestidae). J. Comp. Physiol. A 189, 715–722.

Schmitz, H., Bleckmann, H., Murtz, M., 1997. Infrared detection in a beetle. Nature 386, 773–774.

Schneider, J.C., 1999. Dispersal of a highly vagile insect in a heterogeneous environment. Ecology 80, 2740–2749.

Schnierla, T.C., 1953. Modifiability in insect behavior. In: Roeder, K.D. (Ed.), Insect Physiology. John Wiley & Sons, New York, pp. 723–747.

Schoeller, E.N., Husseneder, C., Allison, J.D., 2012. Molecular evidence of facultative intraguild predation by *Monochamus titillator* larvae (Coleoptera: Cerambycidae) on members of the southern pine beetle guild. Naturwissenschaften 99, 913–924.

Schoener, T.W., 1982. The controversy over interspecific competition. Am. Sci. 70, 586–595.

Scholte, E.-J., Ng'habi, K., Kihonda, J., Takken, W., Paaijmans, K., Abdulla, S., Killeen, G.F., Knols, B.G.J., 2005. An entomopathogenic fungus for control of adult African malaria mosquitoes. Science 308, 1641–1642.

Schöning, C., Njagi, W.M., Franks, N.R., 2005. Temporal and spatial patterns in the emigrations of the army ant *Dorylus (Anomma) molestus* in the montane forest of Mt. Kenya. Ecol. Entomol. 30, 532–540.

Schöpf, R., Mignat, C., Hedden, P., 1982. As to the food quality of spruce needles for forest damaging insects: 18: Resorption of secondary plant metabolites by the sawfly, *Gilpinia hercyniae* Htg. (Hym., Diprionidae). Z. Angew. Entomol. 93, 244–257.

Schowalter, T.D., 1981. Insect herbivore relationship to the state of the host plant: biotic regulation of ecosystem nutrient cycling through ecological succession. Oikos 37, 126–130.

Schowalter, T.D., 1985. Adaptations of insects to disturbance. In: Pickett, S.T.A., White, P.S. (Eds.), The Ecology of Natural Disturbance and Patch Dynamics. Academic Press, New York, pp. 235–252.

Schowalter, T.D., 1989. Canopy arthropod community structure and herbivory in old-growth and regenerating forests in western Oregon. Can. J. Forest. Res. 19, 318–322.

Schowalter, T.D., 1993. Cone and seed insect phenology in a Douglas-fir seed orchard during three years in western Oregon. J. Econ. Entomol. 87, 758–765.

Schowalter, T.D., 1995. Canopy arthropod communities in relation to forest age and alternative harvest practices in western Oregon. Forest Ecol. Manag. 78, 115–125.

Schowalter, T.D., 2000. Insects as regulators of ecosystem development. In: Coleman, D.C., Hendrix, P. (Eds.), Invertebrates as Webmasters in Ecosystems. CAB International, Wallingford, UK, pp. 99–114.

Schowalter, T.D., 2008. Insect herbivore responses to management practices in conifer forests in North America. J. Sust. For. 26, 204–222.

Schowalter, T.D., 2012. Insect responses to major landscape-level disturbance. Annu. Rev. Entomol. 57, 1–20.

Schowalter, T.D., 2013. Insects and Sustainability of Ecosystem Services. CRC Press/Taylor and Francis Group, Boca Raton, FL.

Schowalter, T.D., Crossley, Jr., D.A., 1982. Bioelimination of ^{51}Cr and ^{85}Sr by cockroaches, *Gromphadorhina portentosa* (Orthoptera: Blaberidae), as affected by mites, *Gromphadorholaelaps schaeferi* (Parasitiformes: Laelapidae). Ann. Entomol. Soc. Am. 75, 158–160.

Schowalter, T.D., Crossley, Jr., D.A., 1983. Forest canopy arthropods as sodium, potassium, magnesium and calcium pools in forests. Forest Ecol. Manag. 7, 143–148.

Schowalter, T.D., Crossley, Jr., D.A., 1988. Canopy arthropods and their response to forest disturbance. In: Swank, W.T., Crossley, Jr., D.A. (Eds.), Forest Hydrology and Ecology at Coweeta. Springer-Verlag, New York, NY, pp. 207–218.

Schowalter, T.D., Ganio, L.M., 1998. Vertical and seasonal variation in canopy arthropod communities in an old-growth conifer forest in southwestern Washington, USA. B. Entomol. Res. 88, 633–640.

Schowalter, T.D., Ganio, L.M., 1999. Invertebrate communities in a tropical rain forest canopy in Puerto Rico following Hurricane Hugo. Ecol. Entomol. 24, 1–11.

Schowalter, T.D., Ganio, L.M., 2003. Diel, seasonal and disturbance-induced variation in invertebrate assemblages. In: Basset, Y., Novotny, V., Miller, S.E., Kitching, R.L. (Eds.), Arthropods of Tropical Forests. Cambridge University Press, Cambridge, UK, pp. 315–328.

Schowalter, T.D., Lowman, M.D., 1999. Forest herbivory by insects. In: Walker, L.R. (Ed.), Ecosystems of the World: Ecosystems of Disturbed Ground. Elsevier, Amsterdam, pp. 269–285.

Schowalter, T.D., Sabin, T.E., 1991. Litter microarthropod responses to canopy herbivory, season and decomposition in litterbags in a regenerating conifer ecosystem in western Oregon. Biol. Fert. Soils 11, 93–96.

Schowalter, T.D., Turchin, P., 1993. Southern pine beetle infestation development: interaction between pine and hardwood basal areas. Forest Sci. 39, 201–210.

Schowalter, T.D., Whitford, W.G., 1979. Territorial behavior of *Bootettix argentatus* Bruner (Orthoptera: Acrididae). Am. Midl. Nat. 102, 182–184.

Schowalter, T.D., Whitford, W.G., Turner, R.B., 1977. Bioenergetics of the range caterpillar, *Hemileuca oliviae* (Ckll.). Oecologia 28, 153–161.

Schowalter, T.D., Coulson, R.N., Crossley, Jr., D.A., 1981a. Role of southern pine beetle and fire in maintenance of structure and function of the southeastern coniferous forest. Environ. Entomol. 10, 821–825.

Schowalter, T.D., Pope, D.N., Coulson, R.N., Fargo, W.S., 1981b. Patterns of southern pine beetle (*Dendroctonus frontalis* Zimm.) infestation enlargement. Forest Sci. 27, 837–849.

Schowalter, T.D., Webb, J.W., Crossley, Jr., D.A., 1981c. Community structure and nutrient content of canopy arthropods in clearcut and uncut forest ecosystems. Ecology 62, 1010–1019.

Schowalter, T.D., Coulson, R.N., Turnbow, R.H., Fargo, W.S., 1982. Accuracy and precision of procedures for estimating populations of the southern pine beetle (Coleoptera: Scolytidae) by using host tree correlates. J. Econ. Entomol. 75, 1009–1016.

Schowalter, T.D., Sabin, T.E., Stafford, S.G., Sexton, J.M., 1991. Phytophage effects on primary production, nutrient turnover, and litter decomposition of young Douglas-fir in western Oregon. Forest Ecol. Manag. 42, 229–243.

Schowalter, T.D., Zhang, Y.L., Sabin, T.E., 1998. Decomposition and nutrient dynamics of oak *Quercus* spp. logs after five years of decomposition. Ecography 21, 3–10.

Schowalter, T.D., Lightfoot, D.C., Whitford, W.G., 1999. Diversity of arthropod responses to host-plant water stress in a desert ecosystem in southern New Mexico. Am. Midl. Nat. 142, 281–290.

Schowalter, T.D., Zhang, Y.L., Rykken, J.J., 2003. Litter invertebrate responses to variable density thinning in western Washington forest. Ecol. Appl. 13, 1204–1211.

Schowalter, T.D., Zhang, Y.L., Progar, R.A., 2005. Canopy arthropod response to density and distribution of green trees retained after partial harvest. Ecol. Appl. 15, 1594–1603.

Schowalter, T.D., Fonte, S.J., Geagan, J., Wang, J., 2011. Effects of manipulated herbivore inputs on nutrient flux and decomposition in a tropical rainforest in Puerto Rico. Oecologia 167, 1141–1149.

Schowalter, T.D., Willig, M.R., Pressley, S.J., 2014. Canopy arthropod responses to experimental canopy opening and debris deposition in a tropical rainforest subject to hurricanes. Forest Ecol. Manag. 332, 93–102.

Schroll, H., 1994. Energy-flow and ecological sustainability in Danish agriculture. Agr. Ecosyst. Environ. 51, 301–310.

Schultz, J.C., 1983. Habitat selection and foraging tactics of caterpillars in heterogeneous trees. In: Denno, R.F., McClure, M.S. (Eds.), Variable Plants and Herbivores in Natural and Managed Systems. Academic Press, New York, pp. 61–90.

Schultz, J.C., 1988. Many factors influence the evolution of herbivore diets, but plant chemistry is central. Ecology 69, 896–897.

Schultz, J.C., Appel, H.M., 2004. Cross-kingdom cross-talk: hormones shared by plants and their insect herbivores. Ecology 85, 70–77.

Schulze, E.D., Mooney, H.A. (Eds.), 1993. Biodiversity and Ecosystem Function. Springer-Verlag, Berlin.

Schupp, E.W., 1988. Seed and early seedling predation in the forest understory and in treefall gaps. Oikos 51, 71–78.

Schupp, E.W., Feener, Jr., D.H., 1991. Phylogeny, lifeform, and habitat dependence of ant-defended plants in a Panamanian forest. In: Huxley, C.R., Cutler, D.F. (Eds.), Ant-Plant Interactions. Oxford University Press, Oxford, UK, pp. 175–197.

Schütz, S., Weissbecker, B., Hummel, H.E., Apel, K.-H., Schmitz, H., Bleckmann, H., 1999. Insect antenna as a smoke detector. Nature 398, 298–299.

Schütz, M., Kretz, C., Dekoninck, L., Iravani, M., Risch, A.C., 2008. Impact of *Formica exsecta* Nyl. on seed bank and vegetation patterns in a subalpine grassland ecosystem. J. Appl. Entomol. 132, 295–305.

Schuur, E.A.G., Chadwick, O.A., Matson, P.A., 2001. Carbon cycling and soil carbon storage in mesic to wet Hawaiian montane forests. Ecology 82, 3182–3196.

Schwachtje, J., Baldwin, I.T., 2008. Why does herbivore attack reconfigure primary metabolism? Plant Physiol. 146, 845–851.

Schwachtje, J., Minchin, P.E.H., Jahnke, S., van Dongen, J.T., Schittko, U., Baldwin, I.T., 2006. SNF1-related kinases allow plants to tolerate herbivory by allocating carbon to roots. Proc. Natl. Acad. Sci. USA 103, 12935–12940.

Schweitzer, J.A., Bailey, J.K., Rehill, B.J., Martinsen, G.D., Hart, S.C., Lindroth, R.L., Keim, P., Whitham, T.G., 2004. Genetically based trait in a dominant tree affects ecosystem processes. Ecol. Lett. 7, 127–134.

Schweitzer, J.A., Bailey, J.K., Hart, S.C., Wimp, G.M., Chapman, S.K., Whitham, T.G., 2005. The interaction of plant genotype and herbivory decelerate leaf litter decomposition and alter nutrient dynamics. Oikos 110, 133–145.

Schwenk, W.S., Strong, A.M., Sillett, T.S., 2010. Effects of bird predation on arthropod abundance and tree growth across an elevational gradient. J. Avian Biol. 41, 367–377.

Scoble, M.J., 1992. The Lepidoptera: Form, Function and Diversity. Oxford University Press, Oxford, UK.

Scott, A.C., Taylor, T.N., 1983. Plant/animal interactions during the Upper Carboniferous. Bot. Rev. 49, 259–307.

Scriber, J.M., Slansky, Jr., F., 1981. The nutritional ecology of immature insects. Annu. Rev. Entomol. 26, 183–211.

Sears, M.K., Hellmich, R.L., Stanley-Horn, D.E., Oberhauser, K.S., Pleasants, J.M., Mattila, H.R., Siegfried, B.D., Dively, G.P., 2001. Impact of *Bt* corn pollen on monarch butterfly populations: a risk assessment. Proc. Natl. Acad. Sci. USA 98, 11937–11942.

Seastedt, T.R., 1984. The role of microarthropods in decomposition and mineralization processes. Annu. Rev. Entomol. 29, 25–46.

Seastedt, T.R., 1985. Maximization of primary and secondary productivity by grazers. Am. Nat. 126, 559–564.

Seastedt, T.R., 2000. Soil fauna and controls of carbon dynamics: comparisons of rangelands and forests across latitudinal gradients. In: Coleman, D.C., Hendrix, P. (Eds.), Invertebrates as Webmasters in Ecosystems. CAB International, Wallingford, UK, pp. 99–114.

Seastedt, T.R., Crossley, Jr., D.A., 1981a. Microarthropod response following cable logging and clear-cutting in the southern Appalachians. Ecology 62, 126–135.

Seastedt, T.R., Crossley, Jr., D.A., 1981b. Sodium dynamics in forest ecosystems and the animal starvation hypothesis. Am. Nat. 117, 1029–1034.

Seastedt, T.R., Crossley, Jr., D.A., 1983. Nutrients in forest litter treated with naphthalene and simulated throughfall: a field microcosm study. Soil Biol. Biochem. 15, 159–165.

Seastedt, T.R., Crossley, Jr., D.A., 1984. The influence of arthropods on ecosystems. BioScience 34, 157–161.

Seastedt, T.R., Tate, C.M., 1981. Decomposition rates and nutrient contents of arthropod remains in forest litter. Ecology 62, 13–19.

Seastedt, T.R., Crossley, Jr., D.A., Hargrove, W.W., 1983. The effects of low-level consumption by canopy arthropods on the growth and nutrient dynamics of black locust and red maple trees in the southern Appalachians. Ecology 64, 1040–1048.

Seastedt, T.R., Ramundo, R.A., Hayes, D.C., 1988. Maximization of densities of soil animals by foliage herbivory: empirical evidence, graphical and conceptual models. Oikos 51, 243–248.

Seastedt, T.R., Reddy, M.V., Cline, S.P., 1989. Microarthropods in decomposing wood from temperate coniferous and deciduous forests. Pedobiologia 33, 69–78.

Seibold, S., Bässler, C., Brandl, R., Gossner, M.M., Thorn, S., Ulyshen, M.D., Müller, J., 2015. Experimental studies of dead-wood biodiversity – a review identifying global gaps in knowledge. Biol. Conserv. 191, 139–149.

Şekercioğlu, C.H., Ehrlich, P.R., Daily, G.C., Aygen, D., Goehring, D., Sandi, R.F., 2002. Disappearance of insectivorous birds from tropical forest fragments. Proc. Natl. Acad. Sci. USA 99, 263–267.

Selikhovkin, A., 2005. Main disturbance factors in north-west Russian forests: structure and databases. Scand. J. For. Res. 20, 27–32.

Senthil-Nathan, S., Kalaivani, K., Choi, M.-Y., Paik, C.-H., 2009. Effects of jasmonic acid-induced resistance in rice on the plant brownhopper, *Nilaparvata lugens* Stål (Homoptera: Delphacidae). Pestic. Biochem. Phys. 95, 77–84.

Setälä, H., Huhta, V., 1991. Soil fauna increase *Betula pendula* growth: laboratory experiments with coniferous forest floor. Ecology 72, 665–671.

Setälä, H., Marshall, V.G., Trofymow, J.A., 1996. Influence of body size of soil fauna on litter decomposition and ^{15}N uptake by poplar in a pot trial. Soil Biol. Biochem. 28, 1661–1675.

Seto, K.C., Güneralp, B., Hutyra, L.R., 2012. Global forecasts of urban expansion to 2030 and direct impacts on biodiversity and carbon pools. Proc. Natl. Acad. Sci. USA 109, 16083–16088.

Sexton, J.M., Schowalter, T.D., 1991. Physical barriers to reduce *Lepesoma lecontei* (Coleoptera: Curculionidae) damage to conelets in a Douglas-fir seed orchard in western Oregon. J. Econ. Entomol. 84, 212–214.

Seymour, A.S., Gutiérrez, D., Jordano, D., 2003. Dispersal of the lycaenid *Plebejus argus* in response to patches of its mutualist ant *Lasius niger*. Oikos 103, 162–174.

Shannon, G., Thaker, M., Vanak, A.T., Page, B.R., Grant, R., Slotow, R., 2011. Relative impacts of elephant and fire on large trees in a savanna ecosystem. Ecosystems 14, 1372–1381.

Sharkey, M.J., 2001. The all taxa biological inventory of the Great Smoky Mountains National Park. Fla. Entomol. 84, 556–564.

Shaw, D.C., 1998. Distribution of larval colonies of *Lophocampa argentata* Packard, the silver spotted tiger moth (Lepidoptera: Arctiidae), in an old growth Douglas-fir/western hemlock forest canopy, Cascade Mountains, Washington State USA. Can. Field Nat. 112, 250–253.

Shaw, D.C., 2004. Vertical organization of canopy biota. In: Lowman, M.D., Rinker, H.B. (Eds.), Forest Canopies. second ed. Elsevier/Academic Press, Amsterdam, pp. 73–101.

Shaw, III, C.G., Eav, B.B., 1993. Modeling interactions. In: Schowalter, T.D., Filip, G.M. (Eds.), Beetle-Pathogen Interactions in Conifer Forests. Academic Press, London, pp. 199–208.

Shaw, P.B., Richman, D.B., Owens, J.C., Huddleston, E.W., 1987. Ecology of the range caterpillar, *Hemileuca oliviae* Cockerell. In: Capinera, J.L. (Ed.), Integrated Pest Management on Rangeland: A Shortgrass Prairie Perspective. Westview Press, Boulder, CO, pp. 234–247.

Shea, S.R., McCormick, M., Portlock, C.C., 1979. The effect of fires on regeneration of leguminous species in the northern jarrah (*Eucalyptus marginata* Sm) forest of Western Australia. Aust. J. Ecol. 4, 195–205.

Sheldon, K.S., Tewksbury, J.J., 2014. The impact of seasonality in temperature on thermal tolerance and elevational range size. Ecology 95, 2134–2143.

Shelford, V.E., 1907. Preliminary note on the distribution of the tiger beetles (*Cicindela*) and its relation to plant succession. Biol. Bull. 14, 9–14.

Shelford, V.E., 1918. Physiological problems in the life-histories of animals with particular reference to their seasonal appearance. Am. Nat. 52, 129–154.

Shelton, T.G., Grace, J.K., 2003. Effects of exposure duration on transfer of nonrepellent termiticides among workers of *Coptotermes formosanus* Shiraki (Isoptera: Rhinotermitidae). J. Econ. Entomol. 96, 456–460.

Shepherd, W.P., Sullivan, B.T., 2013. Southern pine beetle, *Dendroctonus frontalis*, antennal and behavioral responses to nonhost leaf and bark volatiles. J. Chem. Ecol. 39, 481–493.

Shepherd, W.P., Huber, D.P.W., Seybold, S.J., Fettig, C.J., 2008. Antennal responses of the western pine beetle, *Dendroctonus brevicomis* (Coleoptera: Curculionidae), to stem volatiles of its primary host, *Pinus ponderosa*, and nine sympatric nonhost angiosperms and conifers. Chemoecology 17, 209–221.

Sheppard, S., Picard, P., 2006. Visual-quality impact of forest pest activity at the landscape level: a synthesis of published knowledge and research needs. Landscape Urban Plan. 77, 321–342.

Sheppard, P.M., Turner, J.R.G., Brown, K.S., Benson, W.W., Singer, M.C., 1985. Genetics and the evolution of Muellerian mimicry in *Heliconius* butterflies. Philos. Trans. Roy. Soc. B 308, 433–613.

Sherman, R.A., Pechter, E.A., 1988. Maggot therapy: a review of the therapeutic applications of fly larvae in human medicine, especially for treating osteomyelitis. Med. Vet. Entomol. 2, 225–230.

Sherman, R.A., Hall, M.J.R., Thomas, S., 2000. Medical maggots: an ancient remedy for some contemporary afflictions. Annu. Rev. Entomol. 45, 55–81.

Sherman, R.A., Stevens, H., Ng, D., Iversen, E., 2007. Treating wounds in small animals with maggot debridement therapy: a survey of practitioners. Vet. J. 173, 138–143.

Sherratt, T.N., Jepson, P.C., 1993. A metapopulation approach to modelling the long-term impact of pesticides on invertebrates. J. Appl. Ecol. 30, 696–705.

Shettleworth, S.J., 1984. Learning and behavioural ecology. In: Krebs, J.R., Davies, N.B. (Eds.), Behavioural Ecology: An Evolutionary Approach. Blackwell Scientific, Oxford, UK, pp. 170–194.

Shi, Y.-W., Liu, X.-S., Wang, H.-Y., Zhang, R.-J., 2009. Seasonality of insect succession on exposed rabbit carrion in Guangzhou, China. Insect Sci. 16, 425–439.

Shiels, A.B., Zimmerman, J.K., García-Montiel, D.C., Jonckheere, I., Holm, J., Horton, D., Brokaw, N., 2010. Plant responses to simulated hurricane impacts in a subtropical wet forest, Puerto Rico. J. Ecol. 98, 659–673.

Sholes, O.D.V., 2008. Effects of associational resistance and host density on woodland insect herbivores. J. Anim. Ecol. 77, 16–23.

Shonle, I., Bergelson, J., 2000. Evolutionary ecology of the tropane alkaloids of *Datura stramonium* L. (Solanaceae). Evolution 54, 778–788.

Showalter, A.M., Heuberger, S., Tabashnik, B.E., Carrière, Y., 2009. A primer for using transgenic insecticidal cotton in developing countries. J. Insect Sci. 9, 22.

Showler, A.T., 2009. Roles of host plants in boll weevil range expansion beyond tropical Mesoamerica. Am. Entomol. 55, 234–242.

Shrestha, U.B., Bawa, K.S., 2013. Trade, harvest, and conservation of caterpillar fungus (*Ophiocordyceps sinensis*) in the Himalayas. Biol. Conserv. 159, 514–520.

Shugart, H.H., West, D.C., Emanuel, W.R., 1981. Patterns and dynamics of forests: an application of simulation models. In: West, D.C., Shugart, H.H., Botkin, D.B. (Eds.), Forest Succession: Concepts and Application. Springer-Verlag, New York, pp. 74–94.

Shure, D.J., Phillips, D.L., 1991. Patch size of forest openings and arthropod populations. Oecologia 86, 325–334.

Shure, D.J., Wilson, L.A., 1993. Patch-size effects on plant phenolics in successional openings of the southern Appalachians. Ecology 74, 55–67.

Shurin, J.B., Borer, E.T., Seabloom, E.W., Anderson, K., Blanchette, C.A., Broitman, B., Cooper, S.D., Halpern, B.S., 2002. A cross-ecosystem comparison of the strength of trophic cascades. Ecol. Lett. 5, 785–791.

Shuster, S.M., Lonsdorf, E.V., Wimp, G.M., Bailey, J.K., Whitham, T.G., 2006. Community heritability measures the evolutionary consequences of indirect genetic effects on community structure. Evolution 60, 991–1003.

Siderhurst, L.A., Griscom, H.P., Hudy, M., Bortolot, Z.J., 2010. Changes in light levels and stream temperatures with loss of eastern hemlock (*Tsuga canadensis*) at a southern Appalachian stream: implications for brook trout. Forest Ecol. Manag. 260, 1677–1688.

Siemann, E., 1998. Experimental tests of effects of plant productivity and diversity on grassland arthropod diversity. Ecology 79, 2057–2070.

Siepel, H., de Ruiter-Dijkman, E.M., 1993. Feeding guilds of oribatid mites based on their carbohydrase activities. Soil Biol. Biochem. 25, 1491–1497.

Siepielski, A.M., Hung, K.-L., Bein, E.E.B., McPeek, M.A., 2010. Experimental evidence for neutral community dynamics governing an insect assemblage. Ecology 91, 847–857.

Sikes, D.S., Slowik, J., 2010. Terrestrial arthropods of pre- and post-eruption Kasatochi Island, Alaska, 2008-2009: a shift from a plant-based to a necromass-based food web. Arct. Antarct. Alp. Res. 42, 297–305.

Sillén-Tullberg, B., 1985. Higher survival of an aposematic than of a cryptic form of a distasteful bug. Oecologia 67, 411–415.

Silva, S.I., MacKay, W.P., Whitford, W.G., 1985. The relative contributions of termites and microarthropods to fluff grass litter disappearance in the Chihuahuan Desert. Oecologia 67, 31–34.

Silvey, C.J., Hayward, M.W., Gibb, H., 2015. Effects of reconstruction of a pre-European vertebrate assemblage on ground-dwelling arachnids in arid Australia. Oecologia 178, 497–509.

Simard, M., Payette, S., 2005. Reduction of black spruce seed bank by spruce budworm infestation compromises postfire stand regeneration. Can. J. Forest. Res. 35, 1686–1696.

Simard, M., Romme, W.H., Griffin, J.M., Turner, M.G., 2011. Do mountain pine beetle outbreaks change the probability of active crown fire in lodgepole pine forests? Ecol. Monogr. 81, 3–24.

Simberloff, D.S., 1969. Experimental zoology of islands: a model for insular colonization. Ecology 50, 296–314.

Simberloff, D.S., 1974. Equilibrium theory of island biogeography and ecology. Annu. Rev. Ecol. Syst. 5, 161–182.

Simberloff, D.S., 1978. Colonization of islands by insects: immigration, extinction, and diversity. In: Mound, L.A., Waloff, N. (Eds.), Diversity of Insect Faunas. Symposium of the Royal Entomological Society, London No. 9. Royal Entomological Society, London, UK, pp. 139–153.

Simberloff, D.S., Dayan, T., 1991. The guild concept and the structure of ecological communities. Annu. Rev. Ecol. Syst. 22, 115–143.

Simberloff, D.S., Wilson, E.O., 1969. Experimental zoogeography of islands: the colonization of empty islands. Ecology 50, 278–295.

Similä, M., Kouki, J., Martikainen, P., Uotila, A., 2002. Conservation of beetles in boreal pine forests: the effects of forest age and naturalness on species assemblages. Biol. Conserv. 106, 19–27.

Simone-Finstrom, M.D., Spivak, M., 2010. Increased resin collection after parasite challenge: a case of self-medication in honey bees? PLoS ONE 7 (3), e34601.

Simpson, S.J., Raubenheimer, D., 1993. A multi-level analysis of feeding behaviour: the geometry of nutritional decisions. Philos. T. Roy. Soc. B 342, 381–402.

Simpson, S.J., Raubenheimer, D., Behmer, S.T., Whitworth, A., Wright, G.A., 2002. A comparison of nutritional regulation in solitarious and gregarious phase nymphs of the desert locust Schistocerca gregaria. J. Exp. Biol. 205, 121–129.

Sims, P.L., Singh, J.S., 1978. The structure and function of ten western North American grasslands. III. Net primary production, turnover and efficiencies of energy capture and water use. J. Ecol. 66, 573–597.

Sinclair, A.R.E., 1975. The resource limitation of trophic levels in tropical grassland ecosystems. J. Anim. Ecol. 44, 497–520.

Sinclair, B.J., Worland, M.R., Wharton, D.A., 1999. Ice nucleation and freezing tolerance in New Zealand alpine and lowland weta, Hemideina spp. (Orthoptera; Stenopelmatidae). Physiol. Entomol. 24, 56–63.

Sinclair, B.J., Vernon, P., Klok, C.J., Chown, S.L., 2003. Insects at low temperatures: an ecological perspective. Trends Ecol. Evol. 18, 257–262.

Singer, M.C., Parmesan, C., 2010. Phenological asynchrony between herbivorous insects and their hosts: signal of climate change or pre-existing adaptive strategy? Philos. Trans. Roy. Soc. B 365, 3161–3176.

Singh, K.P., Jayasomu, R.S., 2002. Bombyx mori – a review of its potential as a medicinal insect. Pharm. Biol. 40, 28–32.

Singh, G., Prakash, S., 2009. Efficacy of Bacillus sphaericus against larvae of malaria and filarial vectors: an analysis of early resistance detection. Parasitol. Res. 104, 763–766.

Sivinski, J., 1982. Prey attraction by luminous larvae of the fungus gnat Orfelia fultoni. Ecol. Entomol. 7, 443–446.

Skarmoutsos, G., Millar, C., 1982. Adelges aphids and fungi causing premature defoliation of larch. Eur. J. Forest Pathol. 12, 73–78.

Skellam, J.G., 1951. Random dispersal in theoretical populations. Biometrika 38, 196–218.

Skovmand, O., Ouedraogo, T.D.A., Sanogo, E., Samuelsen, H., Toé, L.P., Baldet, T., 2009. Impact of slow-release Bacillus sphaericus granules on mosquito populations followed in a tropical urban environment. J. Med. Entomol. 46, 67–76.

Slansky, Jr., F., 1978. Utilization of energy and nitrogen by larvae of the imported cabbageworm, Pieris rapae, as affected by parasitism by Apanteles glomeratus. Environ. Entomol. 7, 179–185.

Slavik, K., Peterson, B.J., Deegan, L.A., Bowden, W.B., Hershey, A.E., Hobby, J.B., 2004. Long-term responses of the Kuparuk River ecosystem to phosphorus fertilization. Ecology 85, 939–954.

Smalley, A.E., 1960. Energy flow of a salt marsh grasshopper population. Ecology 41, 672–677.

Smedley, S.R., Eisner, T., 1995. Sodium uptake by puddling in a moth. Science 270, 1816–1818.

Smith, R.C., 1954. An analysis of 100 years of grasshopper populations in Kansas (1854 to 1954). Trans. Kansas Acad. Sci. 57, 397–433.

Smith, J.M., 1964. Group selection and kin selection. Nature 201, 1145–1147.

Smith, W.H., 1981. Air Pollution and Forests: Interactions Between Air Contaminants and Forest Ecosystems. Springer-Verlag, New York.

Smith, K.G.V., 1986. A Manual of Forensic Entomology. Cornell University Press, Ithaca, NY.

Smith, J.M.B., 1989. An example of ant-assisted plant invasion. Aust. J. Ecol. 14, 247–250.

Smith, C.M., 2005. Plant Resistance to Arthropods: Molecular and Conventional Approaches. Springer, Dordrecht, The Netherlands.

Smith, R.H., 2007. History of the Boll Weevil in Alabama. Alabama Agricultural Experiment Station Bulletin 670, Auburn, AL.

Smith, B.H., Breed, M.D., 1995. The chemical basis for nestmate recognition and mate discrimination in social insects. In: Cardé, R.T., Bell, W.J. (Eds.), Chemical Ecology of Insects 2. Chapman & Hall, New York, pp. 287–317.

Smith, J.P., Schowalter, T.D., 2001. Aphid-induced reduction of shoot and root growth in Douglas-fir seedlings. Ecol. Entomol. 26, 411–416.

Smith, F.R., Yeaton, R.I., 1998. Disturbance by the mound-buiding termite, *Trinervitermes trinervoides*, and vegetation patch dynamics in a semi-arid, southern African grassland. Plant Ecol. 137, 41–53.

Smith, S., Reagan, T.E., Flynn, J.L., Willis, G.H., 1983. Azinphosmethyl and fenvalerate runoff loss from a sugarcane-insect IPM system. J. Environ. Qual. 12, 534–537.

Smith, M.T., Tabin, P.C., Bancroft, J., Li, G., Gao, R., 2004. Dispersal and spatiotemporal dynamics of Asian longhorned beetle (Coleoptera: Cerambycidae) in China. Environ. Entomol. 33, 435–442.

Smith, K.F., Sax, D.F., Gaines, S.D., Guernier, V., Guégan, J.-F., 2007. Globalization of human infectious disease. Ecology 88, 1903–1910.

Smith, A.A., Hölldobler, B., Liebig, J., 2009. Cuticular hydrocarbons reliably identify cheaters and allow enforcement of altruism in social insects. Curr. Biol. 19, 78–81.

Šobotník, J., Bourguignon, T., Hanus, R., Demianová, Z., Pytelková, J., Mareš, M., Foltynová, P., Preisler, J., Cvačka, J., Krasulová, J., Roisin, Y., 2012. Explosive backpacks in old termite workers. Science 337, 436.

Soderlund, D.M., Bloomquist, J.R., 1990. Molecular mechanisms of insecticide resistance. In: Roush, R.T., Tabashnik, B.E. (Eds.), Pesticide Resistance in Arthropods. Chapman & Hall, New York, pp. 58–96.

Sollins, P., Cline, S.P., Verhoeven, R., Sachs, D., Spycher, G., 1987. Patterns of log decay in old-growth Douglas-fir forests. Can. J. Forest. Res. 17, 1585–1595.

Solomon, A.M., West, D.C., Solomon, J.A., 1981. Simulating the role of climate change and species immigration in forest succession. In: West, D.C., Shugart, H.H., Botkin, D.B. (Eds.), Forest Succession: Concepts and Application. Springer-Verlag, New York, pp. 154–177.

Somanathan, H., Borges, R.M., Chakravarthy, V.S., 2004. Does neighbourhood floral display matter? Fruit set in carpenter bee-pollinated *Heterophragma quadriloculare* and beetle-pollinated *Lasiosiphon eriocephalus*. Biotropica 36, 139–147.

Somlyódy, L., van Straten, G. (Eds.), 1986. Modeling and Managing Shallow Lake Eutrophication, with Application to Lake Balaton. Springer-Verlag, Berlin.

Song, M.Y., Brown, J.J., 2006. Influence of fluctuating salinity on insecticide tolerance of two euryhaline arthropods. J. Econ. Entomol. 99, 745–751.

Song, J., Wu, D., Shao, P., Hui, D., Wan, S., 2015. Exosystem carbon exchange in response to locust outbreaks in a temperate steppe. Oecologia 178, 579–590.

Sopow, S.L., Shorthouse, J.D., Strong, W., Quiring, D.T., 2003. Evidence for long-distance, chemical gall induction by an insect. Ecol. Lett. 6, 102–105.

Sorensen, B.C., Smith, E.H., Smith, J., Carton, Y., 2008. Charles V. Riley, France, and *Phylloxera*. Am. Entomol. 54, 134–149.

Soulé, M.E., Simberloff, D., 1986. What do genetics and ecology tell us about the design of nature refuges? Biol. Conserv. 35, 19–40.

Sousa, W.P., 1979. Disturbance in marine intertidal boulder bields: the nonequilibrium maintenance of species diversity. Ecology 60, 1225–1239.

Sousa, W.P., 1985. Disturbance and patch dynamics on rocky intertidal shores. In: Pickett, S.T.A., White, P.S. (Eds.), Ecology of Natural Disturbance and Patch Dynamics. Academic Press, New York, pp. 101–124.

Sousa, W.P., Quek, S.P., Mitchell, B.J., 2003. Regeneration of *Rhizophora mangle* in a Caribbean mangrove forest: interacting effects of canopy disturbance and a stem-boring beetle. Oecologia 137, 436–445.

Southwick, L.M., Willis, G.H., Reagan, T.E., Rodriguez, L.M., 1995. Residues in runoff and on leaves of azinphosmethyl and esfenvalerate applied to sugarcane. Environ. Entomol. 24, 1013–1017.

Southwood, T.R.E., 1975. The dynamics of insect populations. In: Pimentel, D. (Ed.), Insects, Science, and Society. Academic Press, San Diego, CA, pp. 151–199.

Southwood, T.R.E., 1977. The relevance of population dynamics theory to pest status. In: Cherrett, J.M., Sagar, G.R. (Eds.), Origins of Pest, Parasite, Disease and Weed Problems. Symposium of the British Ecological Society 18. British Ecological Society, London, pp. 35–54.

Southwood, T.R.E., 1978. Ecological Methods with Particular Reference to the Study of Insect Populations. Methuen, Inc, London.

Souza, L., Zelikova, T.J., Sanders, N.J., 2016. Bottom–up and top–down effects on plant communities: nutrients limit productivity, but insects determine diversity and composition. Oikos 125, 566–575.

Spaethe, J., Tautz, J., Chittka, L., 2001. Visual constraints in foraging bumblebees: flower size and color affect search time and flight behavior. Proc. Natl. Acad. Sci. USA 98, 3898–3903.

Spain, A.V., Le Feuvre, R.P., 1987. Breakdown of four litters of contrasting quality in a tropical Australian rain forest. J. Appl. Ecol. 24, 279–288.

Spain, A.V., Le Feuvre, R.P., 1997. Stable C and N isotope values of selected components of a tropical Australian sugarcane ecosystem. Biol. Fert. Soils 24, 118–122.

Speer, J.H., Swetnam, T.W., Wickman, B.E., Youngblood, A., 2001. Changes in Pandora moth outbreak dynamics during the past 622 years. Ecology 82, 679–697.

Speer, J.H., Clay, K., Bishop, G., Creech, M., 2010. The effect of periodical cicadas on growth of five tree species in Midwestern deciduous forests. Am. Midl. Nat. 164, 173–186.

Spehn, E.M., Hector, A., Joshi, J., Scherer-Lorenzen, M., Schmid, B., Bazeley-White, E., Beierkuhnlein, C., Caldeira, M.C., Diemer, M., Dimitrakopoulos, P.G., Finn, J.A., Freitas, H., Giller, P.S., Good, J., Harris, R., Högberg, P., Huss-Danell, K., Jumpponen, A., Koricheva, J., Leadley, P.W., Loreau, M., Minns, A., Mulder, C.P.H., O'Donovan, G., Otway, S.J., Palmborg, C., Pereira, J.S., Pfisterer, A.B., Prinz, A., Read, D.J., Schultze, E.-D., Siamantziouras, A.-S.D., Terry, A.C., Troumbis, A.Y., Woodward, F.I., Yachi, S., Lawton, J.H., 2005. Ecosystem effects of biodiversity manipulations in European grasslands. Ecol. Monogr. 75, 37–63.

Speight, M.R., Hunter, M.D., Watt, A.D., 2008. Ecology of Insects: Concepts and Applications. Wiley-Blackwell.

Spencer, H.J., Port, G.R., 1988. Effects of roadside conditions on plants and insects. II. Soil conditions. J. Appl. Ecol. 25, 709–715.

Spencer, H.J., Scott, N.E., Port, G.R., Davison, A.W., 1988. Effects of roadside conditions on plants and insects. I. atmospheric conditions. J. Appl. Ecol. 25, 699–707.

Springett, B.P., 1968. Aspects of the relationship between burying beetles, *Necrophorus* spp. and the mite, *Poecilochirus necrophori* Vitz. J. Anim. Ecol. 37, 417–424.

Srinivasan, M.V., Zhang, S., Alwein, M., Tautz, J., 2000. Honeybee navigation: nature and calibration of the "odometer". Science 287, 851–853.

St. Pierre, M.J., Hendrix, S.D., 2003. Movement patterns of *Rhyssomatus lineaticollis* Say (Coleoptera: Curculionidae) within and among *Asclepias syriaca* (Asclepiadaceae) patches in a fragmented landscape. Ecol. Entomol. 28, 579–586.

St. Pierre, M.J., Hendrix, S.D., Lewis, C.K., 2005. Dispersal ability and host-plant characteristics influence spatial population structure of monophagous beetles. Ecol. Entomol. 30, 105–115.

Stach, S., Benard, J., Giurfa, M., 2004. Local-feature assembling in visual pattern recognition and generalization in honeybees. Nature 429, 758–761.

Stachurski, A., Zimka, J.R., 1984. The budget of nitrogen dissolved in rainfall during its passing through the crown canopy in forest ecosystems. Ekol. Pol.-Pol. J. Ecol. 32, 191–218.

Städler, E., 1984. Perceptual mechanisms. In: Bell, W.J., Cardé, R.T. (Eds.), Chemical Ecology of Insects. Chapman and Hall, London, UK, pp. 3–35.

Stadler, B., Müller, T., 1996. Aphid honeydew and its effect on the phyllosphere microflora of *Picea abies* (L.) Karst. Oecologia 108, 771–776.

Stadler, B., Michalzik, B., Müller, T., 1998. Linking aphid ecology with nutrient fluxes in a coniferous forest. Ecology 79, 1514–1525.

Stadler, B., Solinger, S., Michalzik, B., 2001. Insect herbivores and the nutrient flow from the canopy to the soil in coniferous and deciduous forests. Oecologia 126, 104–113.

Stadler, B., Müller, T., Orwig, D., 2006. The ecology of energy and nutrient fluxes in hemlock forests invaded by hemlock woolly adelgid. Ecology 87, 1792–1804.

Stamp, N.E., 1992. Relative susceptibility to predation of two species of caterpillars on plantain. Oecologia 92, 124–129.

Stamp, N., 2004. Can the growth-differentiation balance hypothesis be tested rigorously? Oikos 107, 439–448.

Stamp, N.E., Bowers, M.D., 1990. Variation in food quality and temperature constrain foraging of gregarious caterpillars. Ecology 71, 1031–1039.

Stamp, N.E., Yang, Y., Osier, T.L., 1997. Response of an insect predator to prey fed multiple allelochemicals under representative thermal regimes. Ecology 78, 203–214.

Stanko-Golden, K.M., Swank, W.T., Fitzgerald, J.W., 1994. Factors affecting sulfate adsorption, organic sulfur formation, and mobilization in forest and grassland spodosols. Biol. Fert. Soils 17, 289–296.

Stanton, N., 1975. Herbivore pressure on 2 types of forests. Biotropica 7, 8–11.

Stanton, M.L., 1983. Spatial patterns in the plant community and their effects upon insect search. In: Ahmad, S. (Ed.), Herbivorous Insects: Host-seeking Behavior and Mechanisms. Academic Press, New York, pp. 125–157.

Stapp, P., Antolin, M.F., Ball, M., 2004. Patterns of extinction in prairie dog metapopulations: plague outbreaks follow El Niño events. Front. Ecol. Environ. 2, 235–240.

Stark, J.D., Banks, J.E., 2003. Population-level effects of pesticides and other toxicants on arthropods. Annu. Rev. Entomol. 48, 505–519.

Starzyk, J.R., Witkowski, Z., 1981. Changes in the parameters describing the cambio- and xylophagous insect communities during the secondary succession of the oak-hornbeam association in the Niepołomice Forest near Kraków. Z. Angew. Entomol. 91, 525–533.

Stastny, M., Agrawal, A.A., 2014. Love thy neighbor? reciprocal impacts between plant community structure and insect herbivory in co-occurring Asteraceae. Ecology 95, 2904–2914.

Steelman, C.D., 1976. Effects of external and internal arthropod parasites on domestic livestock production. Annu. Rev. Entomol. 21, 155–178.

Steffan, S.A., Chikaraishi, Y., Horton, D.R., Ohkouchi, N., Singleton, M.E., Miliczky, E., Hogg, D.B., Jones, V.P., 2013. Trophic hierarchies illuminated via amino acid isotopic analysis. PLoS ONE 8 (9), e76152.

Steffan-Dewenter, I., Tscharntke, T., 1999. Effects of habitat isolation on pollinator communities and seed set. Oecologia 121, 432–440.

Steffan-Dewenter, I., Münzenberg, U., Tscharntke, T., 2001. Pollination, seed set and seed predation on a landscape scale. Proc. Roy. Soc. Lond. B Bio 268, 1685–1690.

Steffan-Dewenter, I., Münzenberg, U., Bürger, C., Thies, C., Tscharntke, T., 2002. Scale-dependent effects of landscape context on three pollinator guilds. Ecology 83, 1421–1432.

Steidle, J.L.M., 1998. Learning pays off: influence of experience on host finding and parasitism in *Lariophagus distinguendus*. Ecol. Entomol. 23, 451–456.

Steiger, D.M., Johnson, P., Hilbert, D.W., Ritchie, S., Jones, D., Laurance, S.G.W., 2012. Effects of landscape disturbance on mosquito community composition in tropical Australia. J. Vector Ecol. 37, 69–76.

Stephen, F.M., Berisford, C.W., Dahlsten, D.L., Fenn, P., Moser, J.C., 1993. Invertebrate and microbial associates. In: Schowalter, T.D., Filip, G.M. (Eds.), Beetle-Pathogen Interaction in Conifer Forests. Academic Press, London, pp. 129–153.

Stephens, D.W., Krebs, J.R., 1986. Foraging Theory. Princeton University Press, Princeton, NJ.

Stermitz, F.R., Tawara, J.N., Boeckl, M., Pomeroy, M., Foderaro, T.A., Todd, F.G., 1994. Piperidine alkaloid content of *Picea* (spruce) and *Pinus* (pine). Phytochemistry 35, 951–953.

Stern, V.M., Smith, R.F., van den Bosch, R., Hagen, K.S., 1959. The integration of chemical and biological control of the spotted alfalfa aphid. Part 1. The integrated control concept. Hilgardia 29, 81–101.

Sterner, R.W., Elser, J.J., 2002. Ecological Stoichiometry: the Biology of Elements from Molecules to the Biosphere. Princeton University Press, Princeton, NJ.

Stevens, M.T., Waller, D.M., Lindroth, R.L., 2007. Resistance and tolerance in *Populus tremuloides*: genetic variation, costs, and environmental dependency. Evol. Ecol. 21, 829–847.

Stewart, M.M., Woolbright, L.L., 1996. Amphibians. In: Reagan, D.P., Waide, R.B. (Eds.), The Food Web of a Tropical Rain Forest. University of Chicago Press, Chicago, IL, pp. 273–320.

Sthultz, C.M., Gehring, C.A., Whitham, T.G., 2009. Deadly combination of genes and drought: increased mortality of herbivore-resistant trees in a foundation species. Glob. Change Biol. 15, 1949–1961.

Sticher, L., Mauch-Mani, B., Métraux, M.P., 1997. Systematic acquired resistance. Annu. Rev. Phytopathol. 35, 235–270.

Stige, L.C., Chan, K.-S., Zhang, Z., Frank, D., Stenseth, N.C., 2007. Thousand-year-long Chinese time series reveals climatic forcing of decadal locust dynamics. Proc. Natl. Acad. Sci. USA 104, 16188–16193.

Stiles, J.H., Jones, R.H., 1998. Distribution of the red imported fire ant, *Solenopsis invicta*, in road and powerline habitats. Landscape Ecol. 13, 335–346.

Stiling, P.D., 1996. Ecology: Theories and Applications. Prentice-Hall, Upper Saddle River, NJ.

Stiling, P., Moon, D.C., 2005. Quality or quantity: the direct and indirect effects of host plants on herbivores and their natural enemies. Oecologia 142, 413–420.

Stiling, P.D., Simberloff, D., Brodbeck, B.V., 1991. Variation in rates of leaf abscission between plants may affect the distribution patterns of sessile insects. Oecologia 88, 367–370.

Stireman, III, J.O., Dyer, L.A., Janzen, D.H., Singer, M.S., Lill, J.T., Marquis, R.J., Ricklefs, R.E., Gentry, G.L., Hallwachs, W., Coley, P.D., Barone, J.A., Greeney, H.F., Connahs, H., Barbosa, P., Morais, H.C., Diniz, I.R., 2005. Climatic unpredictability and parasitism of caterpillars: implications of global warming. Proc. Natl. Acad. Sci. USA 102, 17384–17387.

Støen, O.-G., Okullo, P., Eid, T., Moe, S.R., 2013. Termites facilitate and ungulates limit savanna tree regeneration. Oecologia 172, 1085–1093.

Stoepler, T.M., Edge, A., Steel, A., O'Quinn, R.L., Fishbein, M., 2012. Differential pollinator effectiveness and importance in a milkweed (*Asclepias*, Apocynaceae) hybrid zone. Am. J. Bot. 99, 448–458.

Stone, M.K., Wallace, J.B., 1998. Long-term recovery of a mountain stream from clear-cut logging: the effects of forest succession on benthic invertebrate community structure. Freshwater Biol. 39, 151–169.

Stone, J.E., Kolb, T.E., Covington, W.W., 1999. Effects of restoration thinning on presettlement *Pinus ponderosa* in northern Arizona. Restor. Ecol. 7, 172–182.

Storey, A.W., Pinder, L.C.V., 1985. Mesh-size and efficiency of sampling larval Chironomidae. Hydrobiologia 124, 193–197.

Stork, N.E., 1987. Guild structure of arthropods from Bornean rain forest trees. Ecol. Entomol. 12, 69–80.

Stouffer, P.C., Strong, C., Naka, L.N., 2009. Twenty years of understorey bird extinctions from Amazonian rain forest fragments: consistent trends and landscape-mediated dynamics. Divers. Distrib. 15, 88–97.

Stouffer, P.C., Johnson, E.I., Bierregaard, Jr., R.O., Lovejoy, T.E., 2011. Understory bird communities in Amazonian rainforest fragments: species turnover through 25 years post-isolation in recovering landscapes. PLoS ONE 6 (6), e20543.

Stout, R.J., 1989. Effects of condensed tannins on leaf processing in mid-latitude and tropical streams: a theoretical approach. Can. J. Fish. Aquat. Sci. 46, 1097–1106.

Stout, M.J., Bostock, R.M., 1999. Specificity of induced responses to arthropods and pathogens. In: Agrawal, A.A., Tuzun, S., Bent, E. (Eds.), Induced Plant Defenses Against Pathogens and Herbivores: Biochemistry, Ecology, and Agriculture. American Phytopathological Society, St. Paul, MN, pp. 183–209.

Stout, J., Vandermeer, J., 1975. Comparison of species richness for stream-inhabiting insects in tropical and mid-latitude streams. Am. Nat. 109, 263–280.

Stout, M.J., Zehnder, G.W., Baur, M.E., 2002. Potential for use of elicitors of plant resistance in arthropod management programs. Arch. Insect Biochem. 51, 222–235.

Stout, M.J., Thaler, J.S., Thomma, B.P.H.J., 2006. Plant-mediated interactions between pathogenic microorganisms and herbivorous insects. Annu. Rev. Entomol. 51, 663–689.

Strand, T., Lamb, B., Thistle, H., Allwine, E., Peterson, H., 2009. A simple model for simulation of insect pheromone dispersion within forest canopies. Ecol. Model. 220, 640–656.

Stratford, J.A., Stouffer, P.C., 1999. Local extinctions of terrestrial insectivorous birds in a fragmented landscape near Manaus, Brazil. Conserv. Biol. 13, 1416–1423.

Straub, C.S., Snyder, W.E., 2008. Increasing enemy biodiversity strengthens herbivore suppression on two plant species. Ecology 89, 1605–1615.

Strauss, S.Y., Murch, P., 2004. Towards an understanding of the mechanisms of tolerance: compensating for herbivore damage by enhancing a mutualism. Ecol. Entomol. 29, 234–239.

Strauss, S.Y., Irwin, R.E., Lambrix, V.M., 2004. Optimal defence theory and flower petal colour predict variation in the secondary chemistry of wild radish. J. Ecol. 92, 132–141.

Streams, F.A., 1994. Effect of prey size on attack components of the functional response by Notonecta undulata. Oecologia 98, 57–63.

Strickland, T.C., Fitzgerald, J.W., 1986. Organosulphur recalcitrance in soil and litter from a hardwood forest. Soil Biol. Biochem. 18, 661–662.

Strom, B.L., Roton, L.M., Goyer, R.A., Meeker, J.R., 1999. Visual and semiochemical disruption of host finding in the southern pine beetle. Ecol. Appl. 9, 1028–1038.

Strong, D.R., 1992. Are trophic cascades all wet? Differentiation and donor-control in speciose ecosystems. Ecology 73, 747–754.

Strong, D.R., Lawton, J.H., Southwood, T.R.E., 1984. Insects on Plants: Community Patterns and Mechanisms. Harvard University Press, Cambridge, MA.

Strong, D.R., Maron, J.L., Connors, P.G., Whipple, A., Harrison, S., Jeffries, R.L., 1995. High mortality, fluctuation in numbers, and heavy subterranean insect herbivory in bush lupine, Lupinus arboreus. Oecologia 104, 85–92.

Strong, W.B., Croft, B.A., Slone, D.H., 1997. Spatial aggregation and refugia of the mites Tetranychus urticae and Neoseiulus fallacis (Acari: Tetranychidae, Phytoseiidae) on hop. Environ. Entomol. 26, 859–865.

Stubbington, R., Wood, P.J., Boulton, A.J., 2009. Low flow controls on benthic and hyporheic macroinvertebrate assemblages during supra-seasonal drought. Hydrol. Process. 23, 2252–2263.

Sturgeon, K.B., Mitton, J.B., 1986. Allozyme and morphological differentiation of mountain pine beetles Dendroctonus ponderosae Hopkins (Coleoptera: Scolytidae) associated with host trees. Evolution 40, 290–302.

Suarez, M.E., Thorne, B.L., 2000. Rate, amount, and distribution pattern of alimentary fluid transfer via trophallaxis in three species of termites (Isoptera: Rhinotermitidae). Ann. Entomol. Soc. Am. 93, 145–155.

Suarez, A.V., Bolger, D.T., Case, T.J., 1998. Effects of fragmentation and invasion on native ant communities in coastal southern California. Ecology 79, 2041–2056.

Suarez, A.V., Holway, D.A., Case, T.J., 2001. Patterns of spread in biological invasions dominated by long-distance jump dispersal: insights from Argentine ants. Proc. Natl. Acad. Sci. USA 98, 1095–1100.

Sullivan, L.L., Johnson, B.L., Brudvig, L.A., Haddad, N.M., 2011. Can dispersal mode predict corridor effects on plant parasites? Ecology 92, 1559–1564.

Summerville, K.S., Crist, T.O., 2001. Effects of experimental habitat fragmentation on patch use by butterflies and skippers (Lepidoptera). Ecology 82, 1360–1370.

Summerville, K.S., Crist, T.O., 2004. Contrasting effects of habitat quantity and quality on moth communities in fragmented landscapes. Ecography 27, 3–12.

Summerville, K.S., Veech, J.A., Crist, T.O., 2002. Does variation in patch use among butterfly species contribute to nestedness at fine spatial scales? Oikos 97, 195–204.

Summerville, K.S., Courard-Hauri, D., Dupont, M.M., 2009. The legacy of timber harvest: do patterns of species dominance suggest recovery of lepidopteran communities in managed hardwood stands? Forest Ecol. Manag. 259, 8–13.

Sun, Y., Zhao, J., Sheng, Y., Xiao, Y.-F., Zhang, Y.-J., Bai, L.-X., Tan, Y., Xiao, L.-B., Xu, G.-C., 2016. Identification of heat shock cognate protein 70 gene (Alhsc70) of Apolygus lucorum and its expression in response to different temperature and pesticide stresses. Insect Sci. 23, 37–49.

Supp, S.R., Ernest, S.K.M., 2014. Species-level and community-level responses to disturbance: a cross-community analysis. Ecology 95, 1717–1723.

Swank, W.T., Waide, J.B., Crossley, Jr., D.A., Todd, R.L., 1981. Insect defoliation enhances nitrate export from forest ecosystems. Oecologia 51, 297–299.

Swenson, N.G., Mahler, D.L., Ferro, M., Ritchie, A., 2007. The energetic determination, spatial dispersion and density dependence of Mymeleon ant lion pits in Las Cruces, Cost Rica. Biotropica 39, 774–777.

Swetnam, T.W., Lynch, A.M., 1989. A tree-ring reconstruction of western spruce budworm history in the southern Rocky Mountains. Forest Sci. 35, 962–986.

Swetnam, T.W., Lynch, A.M., 1993. Multicentury, regional-scale patterns of western spruce budworm outbreaks. Ecol. Monogr. 63, 399–424.

Swift, M.J., 1977. The ecology of wood decomposition. Sci. Prog. 64, 175–199.

Swift, M.J., Heal, O.W., Anderson, J.M., 1979. Decomposition in Terrestrial Ecosystems. Blackwell Scientific, Oxford, UK.

Sword, M.A., 1998. Seasonal development of loblolly pine lateral roots in response to stand density and fertilization. Plant Soil 200, 21–25.

Sword, G.A., Lorch, P.D., Gwynne, D.T., 2005. Migratory bands give crickets protection. Nature 433, 703.

Symondson, W.O.C., Sunderland, K.D., Greenstone, M.H., 2002. Can generalist predators be effective biocontrol agents? Annu. Rev. Entomol. 47, 561–594.

Szujko-Lacza, J., 1982. The Flora of the Hortobágy National Park. Akadémiai Kiadó, Budapest, Hungary.

Szujko-Lacza, J., Kovacs, D. (Eds.), 1993. The Flora of the Kiskunság National Park. Akadémiai Kiadó, Budapest, Hungary.

Tabashnik, B.E., 1994. Evolution of resistance to *Bacillus thuringiensis*. Annu. Rev. Entomol. 39, 47–79.

Tabashnik, B.E., Groeters, F.R., Finson, N., Liu, Y.B., Johnson, M.W., Heckel, D.G., Luo, K., Adang, M.L., 1996. Resistance to *Bacillus thuringiensis* in *Plutella xylostella*: the moth heard round the world. In: Molecular Genetics and Evolution of Pesticide Resistance, American Chemical Society Symposium Series 645. Washington, DC, pp. 130–140.

Tabashnik, B.E., Liu, Y.B., Finson, N., Masson, L., Heckel, D.G., 1997. One gene in diamondback moth confers resistance to four *Bacillus thuringiensis* toxins. Proc. Natl. Acad. Sci. USA 94, 1640–1644.

Tabashnik, B.E., Gassmann, A.J., Crowder, D.W., Carriére, Y., 2008. Insect resistance to Bt crops: evidence vs. theory. Nat. Biotechnol. 26, 199–202.

Tack, A.J.M., Ovsskainen, O., Harrison, P.J., Roslin, T., 2009. Competition as a structuring force in leaf miner communities. Oikos 118, 809–818.

Tahvanainen, J.O., Root, R.B., 1972. The influence of vegetational diversity on the population ecology of a specialized herbivore, *Phyllotreta cruciferae* (Coleoptera: Chrysomelidae). Oecologia 10, 321–346.

Taki, H., Kevan, P.G., Ascher, J.S., 2007. Landscape effects of forest loss in a pollination system. Landscape Ecol. 22, 1575–1587.

Tallamy, D.W., Halaweish, F.T., 1993. Effects of age, reproductive activity, sex, and prior exposure on sensitivity to cucurbitacins in southern corn rootworm (Coleoptera: Chrysomelidae). Environ. Entomol. 22, 922–925.

Tallamy, D.W., Stull, J., Ehresman, N.P., Gorski, P.M., Mason, C.E., 1997. Cucurbitacins as feeding and oviposition deterrents to insects. Environ. Entomol. 26, 678–683.

Tallamy, D.W., Whittington, D.P., Defurio, F., Fontaine, D.A., Gorski, P.M., Gothro, P.W., 1998. Sequestered cucurbitacins and pathogenicity of *Metarhizium anisopliae* (Moniliales: Moniliaceae) on spotted cucumber beetle eggs and larvae (Coleoptera: Chrysomelidae). Environ. Entomol. 27, 366–372.

Tamasi, E., Stokes, A., Lasserre, B., Danjon, F., Berthier, S., Fourcaud, T., Chiatante, D., 2005. Influence of wind loading on root system development and architecture in oak *(Quercus robur* L.) seedlings. Trees 19, 374–384.

Tanaka, S., Nishide, Y., 2012. Do desert locust hoppers develop gregarious characteristics by watching a video? J. Insect Physiol. 58, 1060–1071.

Tanaka, S., Suzuki, Y., 1998. Physiological trade-offs between reproduction, flight capability and longevity in a wing-dimorphic cricket, *Modicogryllus confirmatus*. J. Insect Physiol. 44, 121–129.

Tanaka, Y., Yoshimura, J., Simon, C., Cooley, J.R., Tainaka, K., 2009. Allee effect in the selection of prime-numbered cycles in periodical cicadas. Proc. Natl. Acad. Sci. USA 106, 8975–8979.

Tansley, A.G., 1935. The use and abuse of vegetational concepts and terms. Ecology 16, 284–307.

Tantawi, T.I., El-Kady, E.M., Greenberg, B., El-Ghaffar, H.A., 1996. Arthropod succession on exposed rabbit carrion in Alexandria, Egypt. J. Med. Entomol. 33, 566–580.

Tayasu, I., Abe, T., Eggleton, P., Bignell, D.E., 1997. Nitrogen and carbon isotope ratios in termites: an indicator of trophic habit along the gradient from wood-feeding to soil-feeding. Ecol. Entomol. 22, 343–351.

Taylor, S.L., MacLean, D.A., 2009. Legacy of insect defoliators: increased wind-related mortality two decades after a spruce budworm outbeak. Forest Sci. 55, 256–267.

Taylor, B.J., Schalk, D.R., Jeanne, R.L., 2010. Yellowjackets use nest-based cues to differentially exploit higher-quality resources. Naturwissenschaften 97, 1041–1046.

Teal, J.M., 1957. Community metabolism in a temperate cold spring. Ecol. Monogr. 27, 283–302.

Teal, J.M., 1962. Energy flow in the salt marsh ecosystem of Georgia. Ecology 43, 614–624.

Teets, N.M., Denlinger, D.L., 2014. Surviving in a frozen desert: environmental stress physiology of terrestrial Antarctic arthropods. J. Exp. Biol. 217, 84–93.

Teets, N.M., Peyton, J.T., Colinet, H., Renault, D., Kelley, J.L., Kawarasaki, Y., Lee, Jr., R.E., Denlinger, D.L., 2012a. Gene expression changes governing extreme dehydration tolerance in an Antarctic insect. Proc. Natl. Acad. Sci. USA 109, 20744–20749.

Teets, N.M., Peyton, J.T., Ragland, G.J., Colinet, H., Renault, D., Hahn, D.A., Denlinger, D.L., 2012b. Combined transcriptomic and metabolomics approach uncovers molecular mechanisms of cold tolerance in a temperate flesh fly. Physiol. Genomics 44, 764–777.

Teets, N.M., Yi, S.-X., Lee, Jr., R.E., Denlinger, D.L., 2013. Calcium signaling mediates cold sensing in insect tissues. Proc. Natl. Acad. Sci. USA 110, 9154–9159.

Tegelaar, K., Glinwood, R., Pettersson, J., Leimar, O., 2013. Transgenerational effects and the cost of ant tending in aphids. Oecologia 173, 779–790.

Temple, S.A., 1977. Plant-animal mutualism: coevolution with dodo leads to near extinction of plant. Science 197, 885–886.

Terborgh, J., 1973. On the notion of favorableness in plant ecology. Am. Nat. 107, 481–501.

Terborgh, J., 1985. The vertical component of plant species diversity in temperate and tropical forests. Am. Nat. 126, 760–776.

Thaler, J.S., 1999a. Jasmonic acid mediated interactions between plants, herbivores, parasitoids, and pathogens: a review of field experiments with tomato. In: Agrawal, A.A., Tuzun, S., Bent, E. (Eds.), Induced Plant Defenses Against Pathogens and Herbivores: Biochemistry, Ecology, and Agriculture. American Phytopathological Society, St. Paul, MN, pp. 319–334.

Thaler, J.S., 1999b. Jasmonate-inducible plant defenses cause increased parasitism of herbivores. Nature 399, 686–687.

Thaler, J.S., Stout, M.J., Karban, R., Duffey, S.S., 2001. Jasmonate-mediated induced plant resistance affects a community of herbivores. Ecol. Entomol. 26, 312–324.

Thaler, J.S., Farag, M.A., Pare, P.W., Dicke, M., 2002. Jasmonate-deficient plants have reduced direct and indirect defences against herbivores. Ecol. Lett. 5, 764–774.

Thébault, E., Fontaine, C., 2010. Stability of ecological communities and the architecture of mutualistic and trophic networks. Science 329, 853–856.

Theis, N., Raguso, R.A., 2005. The effect of pollination on floral fragrance in thistles. J. Chem. Ecol. 31, 2581–2600.

Therrell, M.D., Stahle, D.W., Acuña-Soto, R., 2004. Aztec drought and the "curse of one rabbit". Bull. Am. Meteorol. Soc. 85, 1263–1272.

Therrien, J., Mason, C.J., Cale, J.A., Adams, A., Aukema, B.H., Currie, C.R., Raffa, K.F., Erbilgin, N., 2015. Bacteria influence mountain pine beetle brood development through interactions with symbiotic and antagonistic fungi: implications for climate-driven host range expansion. Oecologia 179, 467–485.

Thézé, J., Bézier, A., Periquet, G., Drezen, J.-M., Herniou, E.A., 2011. Paleozoic origin of insect large dsDNA viruses. Proc. Natl. Acad. Sci. 108, 15931–15935.

Thies, C., Tscharntke, T., 1999. Landscape structure and biological control in agroecosystems. Science 285, 893–895.

Thies, C., Steffan-Dewenter, I., Tscharntke, T., 2003. Effects of landscape context on herbivory and parasitism at different spatial scales. Oikos 101, 18–25.

Thistle, H.W., Peterson, H., Allwine, G., Lamb, B., Strand, T., Holsten, E.H., Shea, P.J., 2004. Surrogate pheromone plumes in three forest trunk spaces: composite statistics and case studies. Forest Sci. 50, 610–625.

Thomas, C.D., Hanski, I., 1997. Butterfly populations. In: Hanski, I.A., Gilpin, M.E. (Eds.), Metapopulation Biology: Ecology, Genetics and Evolution. Academic Press, San Diego, CA, pp. 359–386.

Thomas, M.D., Wratten, S.D., Sotherton, N.W., 1992. Creation of 'island' habitats in farmland to manipulate populations of beneficial arthropods: predator densities and species composition. J. Appl. Ecol. 29, 524–531.

Thomas, J.A., Telfer, M.G., Roy, D.B., Preston, C.D., Greenwood, J.J.D., Asher, J., Fox, R., Clarke, R.T., Lawton, J.H., 2004. Comparative losses of British butterflies, birds, and plants and the global extinction crisis. Science 303, 1879–1881.

Thompson, C.G., 1978. Nuclear polyhedrosis epizootiology. In: Brookes, M.H., Stark, R.W., Campbell, R.W. (Eds.), The Douglas-fir Tussock Moth: A Synthesis . USDA Forest Service Tech. Bull. 1585. USDA Washington, DC, pp. 136–140.

Thompson, D.C., Gardner, K.T., 1996. Importance of grasshopper defoliation period on southwestern blue grama-dominated rangeland. J. Range Manage. 49, 494–498.

Thomson, L.J., Hoffmann, A.A., 2010. Natural enemy responses and pest control: importance of local vegetation. Biol. Control 52, 160–166.

Thorne, B.L., 1997. Evolution of eusociality in termites. Annu. Rev. Ecol. Syst. 28, 27–54.

Thorne, B.L., Traniello, J.F.A., 2003. Comparative social biology of basal taxa of ants and termites. Annu. Rev. Entomol. 48, 283–306.

Thorne, C.M., Otvos, I.S., Conder, N., Levin, D.B., 2007. Development and evaluation of methods to detects nucleopolyhedroviruses in larvae of the Douglas-fir tussock moth, Orgyia pseudotsugata (McDunnough). Appl. Environ. Microb. 73, 1101–1106.

Thornhill, R., 1976. Sexual selection and nuptial feeding behavior in Bittacus apicalis (Insecta: Mecoptera). Am. Nat. 110, 529–548.

Thorsen, M.J., Seddon, P.J., Dickinson, K.J.M., 2011. Faunal influences on New Zealand seed dispersal characteristics. Evol. Ecol. 25, 1397–1426.

Throop, H.L., Lerdau, M.T., 2004. Effects of nitrogen deposition on insect herbivory: implications for community and ecosystem processes. Ecosystems 7, 109–133.

Throop, H.L., Holland, E.A., Parton, W.J., Ojima, D.S., Keough, C.A., 2004. Effects of nitrogen deposition and insect herbivory on patterns of ecosystem-level carbon and nitrogen dynamics: results from the CENTURY model. Glob. Change Biol. 10, 1092–1105.

Tian, G., Brussaard, L., Kang, B.T., 1995. Breakdown of plant residues with contrasting chemical compositions under humid tropical conditions: effects of earthworms and millipedes. Soil Biol. Biochem. 27, 277–280.

Tian, H., Stige, L.C., Cazelles, B., Kausrud, K.L., Svarverud, R., Stenseth, N.C., Zhang, Z., 2011. Reconstruction of a 1,910-y-long locust series reveals consistent associations with climate fluctuations in China. Proc. Natl. Acad. Sci. USA 108, 14521–14526.

Tilman, D., 1978. Cherries, ants, and tent caterpillars: timing of nectar production in relation to susceptibility of caterpillars to ant predation. Ecology 59, 686–692.

Tilman, D., Downing, J.A., 1994. Biodiversity and stability in grasslands. Nature 367, 363–365.

Tilman, D., Pacala, S., 1993. The maintenance of species richness in plant communities. In: Ricklefs, R.E., Schluter, D. (Eds.), Species Diversity in Ecological Communities: Historical and Geographic Perspectives. University of Chicago Press, Chicago, IL, pp. 13–25.

Tilman, D., Knops, J., Wedin, D., Reich, P., Ritchie, M., Siemann, E., 1997. The influence of functional diversity and composition on ecosystem processes. Science 277, 1300–1302.

Tinbergen, L., 1960. The natural control of insects in pinewoods. I. Factors influencing the intensity of predation by songbirds. Archives Neerlandaises de Zoologie 13, 265–343.

Tisdale, R.A., Wagner, M.R., 1990. Effects of photoperiod, temperature, and humidity on oviposition and egg development of Neodiprion fulviceps (Hymenoptera: Diprionidae) on cut branches of ponderosa pine. Environ. Entomol. 19, 456–458.

Tobin, P.C., Liebhold, A.M., Roberts, E.A., 2007. Comparison of methods for estimating the spread of a non-indigenous species. J. Biogeogr. 34, 305–312.

Torchin, M.E., Mitchell, C.E., 2004. Parasites, patholgens, and invasions by plants and animals. Front. Ecol. Environ. 4, 183–190.

Torell, L.A., Davis, J.H., Huddleston, E.W., Thompson, D.C., 1989. Economic injury levels for interseasonal control of rangeland insects. J. Econ. Entomol. 82, 1289–1294.

Torres, J.A., 1988. Tropical cyclone effects on insect colonization and abundance in Puerto Rico. Acta Cientifica 2, 40–44.

Torres, J.A., 1992. Lepidoptera outbreaks in response to successional changes after the passage of Hurricane Hugo in Puerto Rico. J. Trop. Ecol. 8, 285–298.

Towne, E.G., 2000. Prairie vegetation and soil nutrient responses to ungulate carcasses. Oecologia 122, 232–239.

Towne, W.F., Gould, J.L., 1985. Magnetic field sensitivity in honeybees. In: Kirschvink, J.L., Jones, D.S., MacFadden, B.J. (Eds.), Magnetite Biomineralization and Magnetoreception in Organisms. Plenum Press, New York, pp. 385–406.

Townsend, C.R., Hughes, R.N., 1981. Maximizing net energy returns from foraging. In: Townsend, C.R., Calow, P. (Eds.), Physiological Ecology: An Evolutionary Approach to Resource Use. Blackwell Scientific, Oxford, UK, pp. 86–108.

Traniello, J.F.A., Robson, S.K., 1995. Trail and territorial communication in social insects. In: Cardé, R.T., Bell, W.J. (Eds.), Chemical Ecology of Insects 2. Chapman & Hall, New York, pp. 241–286.

Traugott, M.S., Stamp, N.E., 1996. Effects of chlorogenic acid- and tomatine-fed prey on behavior of an insect predator. J. Insect Behav. 9, 461–476.

Trenberth, K.E., 1999. Atmospheric moisture recycling: role of advection and local evaporation. J. Climate 12, 1368–1381.

Treseder, K.K., 2008. Nitrogen additions and microbial biomass: a meta-analysis of ecosystem studies. Ecol. Lett. 11, 1111–1120.

Treuhaft, R.N., Law, B.E., Asner, G.P., 2004. Forest attributes from radar interferometric structure and its fusion with optical remote sensing. BioScience 54, 561–571.

Trivers, R.L., 1971. The evolution of reciprocal altruism. Q. Rev. Biol. 46, 35–57.

Trlica, M.J., Rittenhouse, L.R., 1993. Grazing and plant performance. Ecol. Appl. 3, 21–23.

Trouvelot, L., 1867. The American silk worm. Am. Nat. 1, 30-38, 85-94, 145-149.

Trumble, J.T., Jensen, P.D., 2004. Ovipositional response, developmental effecs and toxicity of hexavalent chromium to *Magaselia scalaris*, a terrestrial detritivore. Arch. Environ. Con. Tox. 46, 372–376.

Trumble, J., Sorensen, M., 2008. Selenium and the elemental defense hypothesis. New Phytol. 177, 569–572.

Trumble, J.T., Kolodny-Hirsch, D.M., Ting, I.P., 1993. Plant compensation for arthropod herbivory. Annu. Rev. Entomol. 38, 93–119.

Tscharntke, T., 1992. Fragmentation of *Phragmites* habitats, minimum viable population size, habitat suitability, and local extinction of moths, midges, flies, aphids, and birds. Conserv. Biol. 6, 530–536.

Tscharntke, T., Thiessen, S., Dolch, R., Boland, W., 2001. Herbivory, induced resistance, and interplant signal transfer in *Alnus glutinosa*. Biochem. Syst. Ecol. 29, 1025–1047.

Tscharntke, T., Klein, A.M., Kruess, A., Steffan-Dewenter, I., Thies, C., 2005. Landscape perspectives on agricultural intensification and biodiversity – ecosystem service management. Ecol. Lett. 8, 857–874.

Tscharntke, T., Bommarco, R., Clough, Y., Crist, T.O., Kleijn, T., Rand, T.A., Tylianakis, J.M., van Nouhoys, S., Vidal, S., 2007. Conservation biological control and enemy diversity on a landscape scale. Biol. Control 43, 294–309.

Tschinkel, W.R., 1999. Sociometry and sociogenesis of colonies of the harvester ant, *Pogonomyrmex badius*: distribution of workers, brood and seeds within the nest in relation to colony size and season. Ecol. Entomol. 24, 222–237.

Tschinkel, W.R., 2004. The nest architecture of the Florida harvester ant, *Pogonomyrmex badius*. J. Insect Sci. 4, 21.

Tschinkel, W.R., 2005. The nest architecture of the ant, *Camponotus socius*. J. Insect Sci. 5, 9.

Tullis, K., Goff, M.L., 1987. Arthropod succession in exposed carrion in a tropical rainforest on O'ahu Island, Hawai'i. J. Med. Entomol. 24, 332–339.

Tumlinson, J.H., Teal, P.E.A., 1987. Relationship of structure and function to biochemistry in insect pheromone systems. In: Prestwich, G.D., Blomquist, G.J. (Eds.), Pheromone Biochemistry. Academic Press, Orlando, FL, pp. 3–26.

Tumlinson, J.H., Hardee, D.D., Gueldner, R.C., Thompson, A.C., Hedin, P.A., Minyard, J.P., 1969. Sex pheromones produced by male boll weevils: isolation, identification and synthesis. Science 166, 1010–1012.

Tuomi, J., Niemela, P., Haukioja, E., Siren, S., Neuvonen, S., 1984. Nutrient stress: an explanation for plant anti-herbivore responses to defoliation. Oecologia 61, 208–210.

Turcek, F.J., 1971. On vertebrate secondary production of forests. In: Duvigneaud, P. (Ed.), Productivity of Forest Ecosystems. UNESCO, Paris, pp. 379–385.

Turchin, P., 1988. The effect of host-plant density on the numbers of Mexican bean beetles, *Epilachna varivestis*. Am. Midl. Nat. 119, 15–20.

Turchin, P., 1990. Rarity of density dependence or population regulation with lags? Nature 344, 660–663.

Turchin, P., 1998. Quantitative Analysis of Movement. Sinauer Associates, Sunderland, MA.

Turchin, P., Tayler, A.D., Reeve, J.D., 1999. Dynamical role of predators in population cycles of a forest insect. Science 285, 1068–1971.

Turgeon, J.J., Roques, A., de Groot, P., 1994. Insect fauna of coniferous seed cones: diversity, host plant interactions, and management. Annu. Rev. Entomol. 39, 179–212.

Turlings, T.C.J., Tumlinson, J.H., Lewis, W.J., 1990. Exploitation of herbivore-induced plant odors by host-seeking parasitic wasps. Science 250, 1251–1253.

Turlings, T.C.J., McCall, P.J., Alborn, H.T., Tumlinson, J.H., 1993. An elicitor in caterpillar oral secretions that induces corn seedlings to emit chemical signals attractive to parasitic wasps. J. Chem. Ecol. 19, 411–425.

Turlings, T.C.J., Loughrin, J.H., McCall, P.J., Röse, U.S.R., Lewis, W.J., Tumlinson, J.H., 1995. How caterpillar-damaged plants protect themselves by attracting parasiting wasps. Proc. Natl. Acad. Sci. USA 92, 4169–4174.

Turnbow, R.H., Coulson, R.N., Hu, L., Billings, R.F., 1982. Procedural Guide for Using the Interactive Version of the TAMBEETLE Model of Southern Pine Beetle Population and Spot Dynamics. Texas Agricultural Experiment Station, Miscellaneous Publication MP-1518, Texas A&M University, College Station, TX.

Turner, G.B., 1970. The ecological efficiency of consumer populations. Ecology 51, 741–742.

Turner, M.G., 1989. Landscape ecology: the effect of pattern on process. Annu. Rev. Ecol. Syst. 20, 171–197.

Turner, II, B.L., Sabloff, J.A., 2012. Classic Period collapse of the Central Maya Lowlands: insights about human–environment relationships for sustainability. Proc. Natl. Acad. Sci. USA 109, 13908–13914.

Turner, D.P., Ritts, W.D., Cohen, W.B., Maeirsperger, T.K., Gower, S.T., Kirschbaum, A.A., Running, S.W., Zhao, M., Wofsy, S.C., Dunn, A.L., Law, B.E., Campbell, J.L., Oechel, W.C., Kwon, H.J., Meyers, T.P., Small, E.E., Kurc, S.A., Gamon, J.A., 2005. Site-level evaluation of satellite-based terrestrial gross primary production and net primary production monitoring. Glob. Change Biol. 11, 666–684.

Turney, S., Gonzalez, A., Millien, V., 2014. The negative relationship between mammal host diversity and Lyme disease incidence strengthens through time. Ecology 95, 3244–3250.

Turrell, M.J., Dohm, D.J., Sardelis, M.R., O'Guinn, M.L., Andreadis, T.G., Blow, J.A., 2005. An update on the potential of North American mosquitoes (Diptera: Culicidae) to transmit West Nile virus. J. Med. Entomol. 42, 57–62.

Twardochleb, L.A., Olden, J.D., Larson, E.R., 2013. A global meta-analysis of the ecological impacts of nonnative crayfish. Freshw. Sci. 32, 1367–1382.

Tyler, C.M., 1995. Factors contributing to postfire seedling establishment in chaparral: direct and indirect effects of fire. J. Ecol. 83, 1009–1020.

Tylianakis, J.M., Tscharntke, T., Lewis, O.T., 2007. Habitat modification alters the structure of tropical host-parasitoid food webs. Nature 445, 202–205.

Tylianakis, J.M., Didham, R.K., Bascompte, J., Wardle, D.A., 2008. Global change and species interactions in terrestrial ecosystems. Ecol. Lett. 11, 1351–1363.

Tyndale-Biscoe, M., 1994. Dung burial by native and introduced dung beetles (Scarabaeidae). Aust. J. Agr. Res. 45, 1799–1808.

Tyndale-Biscoe, M., Vogt, W.G., 1996. Population status of the bush fly, *Musca vetustissima* (Diptera: Muscidae), and native dung beetles (Coleoptera: Scarabaeinae) in south-eastern Australia in relation to establishment of exotic dung beetles. Bull. Entomol. Res. 86, 183–192.

Tzean, S.S., Hsieh, L.S., Wu, W.J., 1997. Atlas of Entomopathogenic Fungi from Taiwan. Council of Agriculture, Taipei, Taiwan, R.O.C.

Uesugi, A., Poelman, E.H., Kessler, A., 2013. A test of genotypic variation in specificity of herbivore-induced responses in *Solidago altissima* L. (Asteraceae). Oecologia 173, 1387–1396.

Ulanowicz, R.E., 1995. Utricularia's secret: the advantage of positive feedback in oligotrophic environments. Ecol. Model. 79, 49–57.

Ulyshen, M.D., 2014. Interacting effects of insects and flooding on wood decomposition. PLoS ONE 9 (7), e101867.

Ulyshen, M.D., Hanula, J.L., 2010. Patterns of saproxylic beetle succession in loblolly pine. Agr. Forest Entomol. 12, 187–194.

Ulyshen, M.D., Horn, S., Hanula, J.L., 2010. Response of beetles (Coleoptera) at three heights to the experimental removal of an invasive shrub, Chinese privet (*Ligustrum sinense*), from floodplain forests. Biol. Invasions 12, 1573–1579.

Ulyshen, M.D., Wagner, T.L., Mulrooney, J.E., 2014. Contrasting effects of insect exclusion on wood loss in a temperate forest. Ecosphere 5 (4), 47.

Unlu, I., Farajollahi, A., Strickman, D., Fonseca, D.M., 2013. Crouching tiger, hidden trouble: urban sources of *Aedes albopictus* (Diptera: Culicidae) refractory to source-reduction. PLoS ONE 8 (10), e77999.

Unlu, I., Faraji, A., Indelicato, N., Fonsecaa, D.M., 2014. The hidden world of Asian tiger mosquitoes: immature Aedes albopictus (Skuse) dominate in rainwater corrugated extension spouts. Trans. Roy. Soc. Trop. Med. H 108 (11), doi: 10.1093/trstmh/tru139.

Urbas, P., Araújo, Jr., M.V., Leal, I.R., Wirth, R., 2007. Cutting more from cut forests: edge effects on foraging and herbivory of leaf-cutting ants in Brazil. Biotropica 39, 489–495.

Ushimaru, A., Kobayashi, A., Dohzono, I., 2014. Does urbanization promote floral diversification? Implications from changes in herkogamy with pollinator availability in an urban-rural area. Am. Nat. 184, 258–267.

Uvarov, B.P., 1954. Present trends in locust research. Report of the Sixth Commonwealth Entomological Conference. Commonwealth Institute of Entomology, London.

Vacas, S., Alfaro, C., Navarro-Llopis, V., Zarzo, M., Primo, J., 2009. Study on the optimal pheromone release rate for attraction of *Chilo suppressalis* (Lepidoptera: Pyralidae). J. Econ. Entomol. 102, 1094–1100.

Valkama, E., Koricheva, J., Oksanen, E., 2007. Effects of elevated O_3, alone and in combination with elevated CO_2, on tree leaf chemistry and insect herbivore performance: a meta-analysis. Glob. Change Biol. 13, 184–201.

Vamosi, J.C., Knight, T.M., Steets, J.A., Mazer, S.J., Burd, M., Ashman, T.-L., 2006. Pollination decays in biodiversity hotspots. Proc. Natl. Acad. Sci. USA 103, 956–961.

Van Baaren, J., Le Lann, C., Pichenot, J., Pierre, J.S., Krespi, L., Outreman, Y., 2009. How could host discrimination abilities influence the structure of a parasitoid community? Bull. Entomol. Res. 99, 299–306.

Van Bael, S.A., Aiello, A., Valderrama, A., Medianero, E., Samaniego, M., Wright, S.J., 2004. General herbivore outbreak following an El Niño-related drought in a lowland Panamanian forest. J. Trop. Ecol. 20, 625–633.

Van Bael, S.A., Valencia, M.C., Rojas, E.I., Gómez, N., Windsor, D.M., Herre, E.A., 2009. Effects of foliar endophytic fungi on the preference and performance of the leaf beetle *Chelymorpha alternans* in Panama. Biotropica 41, 221–225.

van Biervliet, O., Wiśniewski, K., Daniels, J., Vonesh, J.R., 2009. Effects of tea plantations on stream invertebrates in a global biodiversity hotspot in Africa. Biotropica 41, 469–475.

Van Cleve, K., Martin, S., 1991. Long-term Ecological Research in the United States, sixth ed. University of Washington, Seattle, WA.

van Dam, N.M., 2009. Belowground herbivory and plant defenses. Annu. Rev. Ecol. Evol. S. 40, 373–391.

van den Bosch, R., Stern, V.M., 1962. The integration of chemical and biological control of arthropod pests. Annu. Rev. Entomol. 7, 367–386.

van den Bosch, R., Messenger, P.S., Gutierrez, A.P., 1982. An Introduction to Biological Control. Plenum Press, New York.

van der Maarel, E., Titlyanova, A., 1989. Above-ground and below-ground biomass relations in steppes under different grazing conditions. Oikos 56, 364–370.

van der Zee, B., Behmer, S.T., Simpson, S.J., 2002. Food mixing strategies in the desert locust: effects of phase, distance between foods, and food nutrient content. Entomol. Exp. Appl. 103, 227–237.

Van Driesche, R.G., Bellows, T., 1996. Biological Control. Chapman & Hall, New York, NY.

Van Hook, Jr., R.I., Nielsen, M.G., Shugart, H.H., 1980. Energy and nitrogen relations for a *Macrosiphum liriodendri* (Homoptera: Aphididae) population in an east Tennessee *Liriodendron tulipifera* stand. Ecology 61, 960–975.

van Huis, A., 2013. Potential of insects as food and feed in assuring food security. Annu. Rev. Entomol. 58, 563–583.

Van Langevelde, R., van de Vijver, C.A.D.M., Kumar, L., van de Koppel, J., de Ridder, N., van Andel, J., Skidmore, A.K., Hearne, J.W., Stroosnijder, L., Bond, W.J., Prins, H.H.T., Rietkerk, M., 2003. Effects of fire and herbivory on the stability of savanna ecosystems. Ecology 84, 337–350.

van Lenteren, J.C., Bale, J., Bigler, F., Hokkanen, H.M.T., Loomans, A.J.M., 2006. Assessing risks of releasing exotic biological control agents of arthropod pests. Annu. Rev. Entomol. 51, 609–634.

van Panhuis, W.G., Choisy, M., Xiong, X., Chok, N.S., Akarasewi, P., Iamsirithaworn, S., Lam, S.K., Chong, C.K., Lam, F.C., Phommasak, B., Vongphrachanh, P., Bouaphanh, K., Rekol, H., Hien, N.T., Thai, P.Q., Duong, T.N., Chuang, J.-H., Liu, Y.-L., Ng, L.-C., Shi, Y., Tayag, E.A., Roque, Jr., V.G., Suy, L.L.L., Jarman, R.G., Gibbons, R.V., Velasco, J.M.S., Yoon, I.-K., Burke, D.S., Cummings, D.A.T., 2015. Region-wide synchrony and traveling waves of dengue across eight countries in Southeast Asia. Proc. Natl. Acad. Sci. USA 112, 13069–13074.

Van Pelt, A.F., 2002. Insect species described from Big Bend National Park. Insecta Mundi 16, 125–145.

van Straalen, N., 1998. Evaluation of bioindicator systems derived from soil arthropod communities. Appl. Soil Ecol. 9, 429–437.

van Straalen, N.M., Verhoef, H.A., 1997. The development of a bioindicator system for soil acidity based on arthropod pH preferences. J. Appl. Ecol. 34, 217–232.

Van Zandt, P.A., Agrawal, A.A., 2004. Community-wide impacts of herbivore-induced plant responses in milkweed (*Asclepias syriaca*). Ecology 85, 2616–2629.

Vanbergen, A.J., The Insect Pollinators Initiative, 2013. Threats to an ecosystem service: pressures on pollinators. Front. Ecol. Environ. 11, 251–259.

Vandermeer, J., Granzow de la Cerda, I., Perfecto, I., Boucher, D., Ruiz, J., Kaufmann, A., 2004. Multiple basins of attraction in a tropical forest: evidence for nonequilibrium community structure. Ecology 85, 575–579.

Vanni, M.J., Layne, G.D., 1997. Nutrient recycling and herbivory as mechanisms in the "top-down" effect of fish on algae in lakes. Ecology 78, 21–40.

Vannote, R.L., Minshaw, G.W., Cummins, K.W., Sedell, J.R., Cushing, C.E., 1980. The river continuum concept. Can. J. Fish. Aquat. Sci. 37, 130–137.

Vargas, R.I., Piñero, J.C., Mau, R.F.L., Stark, J.D., Hertlein, M., Mafra-Neto, A., Coler, R., Getchell, A., 2009. Attraction and mortality of oriental fruit flies to SPLAT-MAT-methyl eugenol with spinosad. Entomol. Exp. Appl. 131, 286–293.

Vargo, E.L., Husseneder, C., Grace, J.K., 2003. Colony and population genetic structure of the Formosan subterranean termite, *Coptotermes formosanus*, in Japan. Mol. Ecol. 12, 2599–2608.

Várkonyi, G., Kuussaari, M., Lappalainen, H., 2003. Use of forest corridors by boreal *Xestia* moths. Oecologia 137, 466–474.

Varley, G.C., Gradwell, G.R., 1970. Recent advances in insect population dynamics. Annu. Rev. Entomol. 15, 1–24.

Varley, G.C., Gradwell, G.R., Hassell, M.P., 1973. Insect Population Ecology: An Analytical Approach. Blackwell Scientific, Oxford, UK.

Vasconcelos, H.L., Vieira-Neto, E.H.M., Mundim, F.M., 2006. Roads alter the colonization dynamics of a keystone herbivore in Neotropical savannas. Biotropica 38, 661–665.

Veblen, T.T., Hadley, K.S., Nel, E.M., Kitzberger, T., Reid, M., Villalba, R., 1994. Disturbance regime and disturbance interactions in a Rocky Mountain subalpine forest. J. Ecol. 82, 125–135.

Vega, F.E., Kaya, H.K., 2012. Insect Pathology, second ed. Elsevier/Academic Press, Amsterdam.

Veltman, K., 2013. Butterfly conservatories, butterfly ranches and insectariums: generating income while promoting social and environmental justice. In: Lemelin, R.H. (Ed.), The Management of Insects in Recreation and Tourism. Cambridge University Press, Cambridge, UK, pp. 189–197.

Venarsky, M.P., Huntsman, B.M., Huryn, A.D., Benstead, J.P., Kuhajda, B.R., 2014. Quantitative food web analysis supports the energy limitation hypothesis in cave stream ecosystems. Oecologia 176, 859–869.

Via, S., 1990. Ecological genetics and host adaptation in herbivorous insects: the experimental study of evolution in natural and agricultural systems. Annu. Rev. Entomol. 35, 421–446.

Via, S., 1991a. The genetic structure of host plant adaptation in a spatial patchwork: demographic variability among reciprocally transplanted pea aphid clones. Evolution 45, 827–852.

Via, S., 1991b. Specialized host plant performance of pea aphid clones is not altered by experience. Ecology 72, 1420–1427.

Vickerman, D.B., Trumble, J.T., 2003. Biotransfer of selenium: effects on an insect predator, *Podisus maculiventris*. Ecotoxicology 12, 497–504.

Vickerman, D.B., Young, J.K., Trumble, J.T., 2002. Effect of selenium-treated alfalfa on development, survival, feeding, and oviposition preferences of *Spodoptera exigua* (Lepidoptera: Noctuidae). Environ. Entomol. 31, 953–959.

Vigueras G., A.L., Portillo, L., 2001. Uses of Opuntia species and the potential impact of *Cactoblastis cactorum* (Lepidoptera: Pyralidae) in Mexico. Fla. Entomol. 84 493–498.

Vincent, C., Hallman, G., Panneton, B., Fleurat-Lessard, F., 2003. Management of agricultural insects with physical control methods. Annu. Rev. Entomol. 48, 261–281.

Vinson, M.R., Hawkins, C.P., 1998. Biodiversity of stream insects: variation at local, basin, and regional scales. Annu. Rev. Entomol. 43, 271–293.

Visser, J.H., 1986. Host odor perception in phytophagous insects. Annu. Rev. Entomol. 31, 121–144.

Visser, M.E., Both, C., 2005. Shifts in phenology due to global climate change: the need for a yardstick. Proc. Roy. Soc. B-Biol. Sci. 272, 2561–2569.

Visser, M.E., Holleman, L.J.M., 2001. Warmer springs disrupt the synchrony of oak and winter moth phenology. Proc. Roy. Soc. B-Biol. Sci. 268, 289–294.

Vitousek, P., 1982. Nutrient cycling and nutrient use efficiency. Am. Nat. 119, 553–572.

Vitousek, P.M., Hooper, D.U., 1993. Biological diversity and terrestrial ecosystem biogeochemistry. In: Schulze, E.D., Mooney, H.A. (Eds.), Biodiversity and Ecosystem Function. Springer-Verlag, Berlin, pp. 3–14.

Vitousek, P.M., Mooney, H.A., Lubchenco, J., Melillo, J.M., 1997. Human domination of Earth's ecosystems. Science 277, 494–499.

Vittor, A.Y., Gilman, R.H., Tielsch, J., Glass, G., Shields, T., Lozano, W.S., Pinedo-Cancino, V., Patz, J.A., 2006. The effect of deforestation on the human-biting rate of *Anopheles darlingi*, the primary vector of falciparum malaria in the Peruvian Amazon. Am. J. Trop. Med. Hyg. 74, 3–11.

Vittor, A.Y., Pan, W., Gilman, R.H., Tielsch, J., Glass, G., Shields, T., Sánchez-Lozano, W., Pinedo, V.V., Salas-Cobos, E., Flores, S., Patz, J.A., 2009. Linking deforestation to malaria in the Amazon: characterization of the breeding habitat of the principal malaria vector, Anopheles darlingi. Am. J. Trop. Med. Hyg. 81, 5–12.

Voelckel, C., Schittko, U., Baldwin, I.T., 2001. Herbivore-induced ethylene burst reduces fitness costs of jasmonate- and oral secretion-induced defenses in *Nicotiana attenuata*. Oecologia 127, 274–280.

Voelz, N.J., Poff, N.L., Ward, J.V., 1994. Differential effects of a brief thermal disturbance on caddisflies (Trichoptera) in a regulated river. Am. Midl. Nat. 132, 173–182.

Vogel, A., Scherer-Lorenzen, M., Weigelt, A., 2012. Grassland resistance and resilience after drought depends on management intensity and species richness. PLoS ONE 7 (5), e36992.

Volney, W.J.A., Milstead, J.E., Lewis, V.R., 1983. Effect of food quality, larval density and photoperiod on the feeding rate of the California oakworm (Lepidoptera: Diopidae). Environ. Entomol. 12, 792–798.

Volterra, V., 1926. Fluctuations in the abundance of a species considered mathematically. Nature 118, 558–560.

von Frisch, K., 1967. The Dance Language and Orientation of Bees. Harvard University Press, Cambridge, MA.

Vos, V.C.A., van Ruijven, J., Berg, M.P., Peeters, E.T.H.M., Berendse, F., 2013. Leaf litter quality drives litter mixing effects through complementary resource use among detritivores. Oecologia 173, 269–280.

Vossbrinck, C.R., Coleman, D.C., Woolley, T.A., 1979. Abiotic and biotic factors in litter decomposition in a semiarid grassland. Ecology 60, 265–271.

Wäckers, F.L., Bonifay, C., Lewis, W.J., 2002. Conditioning of appetitive behavior in the Hymenopteran parasitoid *Microplitis croceipes*. Entomol. Exp. Appl. 103, 135–138.

Wagner, R., 1991. The influence of the diel activity pattern of the larvae of *Sericostoma personatum* (Kirby & Spence) (Trichoptera) on organic matter distribution in stream sediments: a laboratory study. Hydrobiologia 224, 65–70.

Wagner, D., 1997. The influence of ant nests on *Acacia* seed production, herbivory and soil nutrients. J. Ecol. 85, 83–93.

Wagner, D.L., 2009. Ode to *Alabama*: the meteoric fall of a once extraordinarily abundant moth. Am. Entomol. 55, 170–173.

Wagner, D., Jones, J.B., 2004. The contribution of harvester ant nests, *Pogonomyrmex rugosus* (Hymenoptera, Formicidae), to soil nutrient stocks and microbial biomass in the Mojave Desert. Environ. Entomol. 33, 599–607.

Wagner, T.L., Feldman, R.M., Gagne, J.A., Cover, J.D., Coulson, R.N., Schoolfield, R.M., 1981. Factors affecting gallery construction, oviposition, and reemergence of *Dendroctonus frontalis* in the laboratory. Ann. Entomol. Soc. Am. 74, 255–273.

Wagner, D., Brown, M.J.F., Gordon, D.M., 1997. Harvest ant nests, soil biota and soil chemistry. Oecologia 112, 232–236.

Waide, R.B., Willig, M.R., 2012. Conceptual overview: disturbance, gradients, and ecological response. In: Brokaw, N., Crowl, T.A., Lugo, A.E., McDowell, W.H., Scaten, F.N., Waide, R.B., Willig, M.R. (Eds.), A Caribbean Forest Tapestry: The Multidimensional Nature of Disturbance and Response. Oxford University Press, New York, pp. 42–71.

Waide, R.B., Willig, M.R., Mittelbach, G., Steiner, C., Gough, L., Dodson, S.I., Judy, G.P., Parmenter, R., 1999. The relationship between primary productivity and species richness. Annu. Rev. Ecol. Syst. 30, 257–300.

Walker, L.R., Willig, M.R., 1999. An introduction to terrestrial disturbances. In: Walker, L.R. (Ed.), Ecosystems of the World 16: Ecosystems of Disturbed Ground. Elsevier, Amsterdam, pp. 1–16.

Wallace, A.R., 1876. The Geographical Distribution of Animals, with a Study of the Relations of Living and Extinct Faunas as Elucidating the Past Changes of the Earth's Surface. Macmillan, London.

Wallace, A.R., 1911. Island Life, or the Phenomena and Causes of Insular Faunas and Floras Including a Revision and Attempted Solution of the Problem of Geological Climates. Macmillan, London.

Wallace, J.B., Hutchens, Jr., J.J., 2000. Effects of invertebrates in lotic ecosystem processes. In: Coleman, D.C., Hendrix, P. (Eds.), Invertebrates as Webmasters in Ecosystems. CAB International, Wallingford, UK, pp. 99–114.

Wallace, J., McJannet, D., 2010. Processes controlling transpiration in the rainforests of north Queensland, Australia. J. Hydrol. 384, 107–117.

Wallace, J.B., O'Hop, J., 1985. Life on a fast pad: waterlily leaf beetle impact on water lilies. Ecology 66, 1534–1544.

Wallace, J.B., Webster, J.R., 1996. The role of macroinvertebrates in stream ecosystem function. Annu. Rev. Entomol. 41, 115–139.

Wallace, J.B., Cuffney, T.F., Webster, J.R., Lugthart, G.J., Chung, K., Goldwitz, G.S., 1991. Export of fine organic particles from headwater streams: effects of season, extreme discharges, and invertebrate manipulation. Limnol. Oceanogr. 36, 670–682.

Wallace, J.B., Webster, J.R., Lowe, R.L., 1992. High-gradient streams of the Appalachians. In: Hackney, C.T., Adams, S.M., Martin, W.A. (Eds.), Biodiversity of Southeastern United States: Aquatic Communities. John Wiley, New York, pp. 133–191.

Wallace, J.B., Whiles, M.R., Eggert, S., Cuffney, T.F., Lugthart, G.J., Chung, K., 1995. Long-term dynamics of coarse particulate organic matter in three Appalachian Mountain streams. J. N. Am. Benthol. Soc. 14, 217–232.

Wallace, J.B., Eggert, S.L., Meyer, J.L., Webster, J.B., 1997. Multiple trophic levels of a forest stream linked to terrestrial litter inputs. Science 277, 102–104.

Wallace, J.B., Eggert, S.L., Meyer, J.L., Websters, J.L., 2015. Stream invertebrate productivity linked to forest subsidies: 37 stream-years of reference and experimental data. Ecology 96, 1213–1228.

Wallin, K.F., Raffa, K.F., 2001. Effects of folivory on subcortical plant defenses: can defense theories predict interguild processes? Ecology 82, 1387–1400.

Wallner, W.E., 1996. Invasive pests ('biological pollutants') and US forests: whose problem, who pays? EPPO Bull. 26, 167–180.

Waloff, N., Thompson, P., 1980. Census data of populations of some leafhoppers (Auchenorrhyncha, Homoptera) of acid grassland. J. Anim. Ecol. 49, 395–416.

Walter, D.E., O'Dowd, D.J., 1995. Life on the forest phyllophane: hairs, little houses, and myriad mites. In: Lowman, M.D., Nadkarni, N.M. (Eds.), Forest Canopies. Academic Press, San Diego, CA, pp. 325–351.

Walter, J.K., Bilby, R.E., Fransen, B.R., 2006. Effects of Pacific salmon spawning and carcass availability on the caddisfly *Ecclisomyia conspersa* (Trichoptera: Limnephilidae). Freshwater Biol. 51, 1211–1218.

Wang, Y., Mopper, S., Hasenstein, K., 2001. Effects of salinity on endogenous ABA, IAA, JA, and SA in *Iris hexagona*. J. Chem. Ecol. 27, 327–342.

Wang, J., Wu, Y., Yang, G., Aksoy, S., 2009. Interactions between mutualist *Wigglesworthia* and tsetse peptidoglycan recognition protein (PGRP-LB) influence trypanosome transmission. Proc. Natl. Acad. Sci. USA 106, 12133–12138.

Wang, M., Biere, A., van der Putten, W.H., Bezemer, T.M., 2014. Sequential effects of root and foliar herbivory on aboveground and belowground induced plant defense responses and insect performance. Oecologia 175, 187–198.

Ward, J.V., 1992. Aquatic Insect Ecology. 1. Biology and Habitat. John Wiley & Sons, New York.

Ward, J.V., Stanford, J.A., 1982. Thermal responses in the evolutionary ecology of aquatic insects. Annu. Rev. Entomol. 27, 97–117.

Waring, G.L., Cobb, N.S., 1992. The impact of plant stress on herbivore population dynamics. Bernays, E.A. (Ed.), Plant-Insect Interactions, vol. 4, CRC Press, Boca Raton, FL, pp. 167–226.

Waring, R.H., Pitman, G.B., 1983. Physiological stress in lodgepole pine as a precursor for mountain pine beetle attack. Z. Angew. Entomol. 96, 265–270.

Waring, G.L., Price, P.W., 1990. Plant water stress and gall formation (Cecidomyiidae: *Asphondylia* spp.) on creosote bush (*Larrea tridentata*). Ecol. Entomol. 15, 87–95.

Waring, R.H., Running, S.W., 1998. Forest Ecosystems: Analysis at Multiple Scales. Academic Press, San Diego, CA.

Warren, M.S., Hill, J.K., Thomas, J.A., Asher, J., Fox, R., Huntley, B., Roy, D.B., Telfer, M.G., Jeffcoate, S., Harding, P., Jeffcoate, G., Willis, S.G., Greatorex-Davies, J.N., Moss, D., Thomas, C.D., 2001. Rapid responses of British butterflies to opposing forces of climate and habitat change. Nature 414, 65–69.

Wason, W.L., Hunter, M.D., 2014. Genetic variation in plant volatile emission does not result in differential attraction of natural enemies in the field. Oecologia 174, 479–491.

Wassenaar, L.I., Hobson, K.A., 1998. Natal origins of migratory monarch butterflies at wintering colonies in Mexico: new isotopic evidence. Proc. Natl. Acad. Sci. USA 95, 15436–15439.

Waterman, P.G., 2007. The current status of chemical systematics. Phytochemistry 68, 2896–2903.

Watson, E.J., Carlton, C.E., 2003. Spring succession of necrophilous insects on wildlife carcasses in Louisiana. J. Med. Entomol. 40, 338–347.

Watson, M.A., Nixon, H.L., 1953. Studies on the feeding of *Myzus persicae* (Sulz.) on radioactive plants. Ann. Appl. Biol. 40, 537–545.

Watson, F., Vertessy, R., McMahon, T., Rhodes, B., Watson, I., 2001. Improved methods to assess water yield changes from paired-catchment studies: application to the Maroondah catchments. Forest Ecol. Manag. 143, 189–204.

Watt, A.D., McFarlane, A.M., 2002. Will climate change have a different impact on different trophic levels? Phenological development of winter moth *Opherophtera brumata* and its host plants. Ecol. Entomol. 27, 254–256.

Watt, A.D., Whittaker, J.B., Docherty, M., Brooks, G., Lindsay, E., Salt, D.T., 1995. The impact of elevated atmospheric CO_2 on insect herbivores. In: Harrington, R., Stork, N.E. (Eds.), Insects in a Changing Environment. Academic Press, London, pp. 197–217.

Watts, D.M., Burke, D.S., Harrison, B.A., Whitmire, R.E., Nisalak, A., 1987. Effect of temperature on the vector efficiency of *Aedes aegypti* for dengue 2 virus. Am. J. Trop. Med. Hyg. 36, 143–152.

Wauer, R.H., 1980. Naturalist's Big Bend: An Introduction to the Trees and Shrubs, Wildflowers, Cacti, Mammals, Birds, Reptiles and Amphibians, Fish, and Insects. Texas A&M University Press, College Station.

Webb, W.L., 1978. Effects of defoliation and tree energetics. In: Brookes, M.H., Stark, R.W., Campbell, R.W. (Eds.), The Douglas-fir Tussock Moth: a Synthesis. USDA Forest Service Tech. Bull. 1585, USDA Forest Service, Washington, DC, pp. 77–81.

Webb, W.L., Karchesy, J.J., 1977. Starch content of Douglas-fir defoliated by the tussock moth. Can. J. Forest. Res. 7, 186–188.

Webb, M.R., Pullin, A.S., 1998. Effects of submergence by winter floods on diapausing caterpillars of a wetland butterfly, *Lycaena dispar batavus*. Ecol. Entomol. 23, 96–99.

Webb, W.L., Lauenroth, W.K., Szarek, S.R., Kinerson, R.S., 1983. Primary production and abiotic controls in forests, grasslands, and desert ecosystems in the United States. Ecology 64, 134–151.

Webb, J.R., Cosby, B.J., Diviney, Jr., F.A., Eshleman, K.N., Galloway, J.N., 1995. Change in the acid-base status of an Appalachian Mountain catchment following forest defoliation by the gypsy moth. Water Air Soil Poll. 85, 535–540.

Webb, B., Harrison, R.R., Willis, M.A., 2004. Sensorimotor control of navigation in arthropod and artificial systems. Arthropod Struct. Dev. 33, 301–329.

Weber, N.A., 1966. The fungus-growing ants. Science 121, 587–604.

Weber, D.C., Lundgren, J.G., 2009. Detection of predation using qPCR: effect of prey quantity, elapsed time, chaser diet, and sample preservation on detectable quantity of prey DNA. J. Insect Sci. 9, 41.

Weber, D.C., Leskey, T.C., Walsh, G.C., Khrimian, A., 2014a. Synergy of aggregation pheromone with methyl (E,E,Z)-2,4,6-decatrienoate in attraction of *Halyomorpha halys* (Hemiptera: Pentatomidae). J. Econ. Entomol. 107, 1061–1068.

Weber, D.C., Walsh, G.C., DiMeglio, A.S., Athanas, M.M., Leskey, T.C., Khrimian, A., 2014b. Attractiveness of harlequin bug, *Murgantia histrionica*, aggregation pheromone: field response to isomers, ratios, and dose. J. Chem. Ecol. 40, 1251–1259.

Webster, J.R., Waide, J.B., Patten, B.C., 1975. Nutrient recycling and the stability of ecosystems. In: Howell, F.G., Gentry, J.B., Smith, M.H. (Eds.), Mineral Cycling in Southeastern Ecosystems. CONF-740513 . USDOE Energy Research and Development Administration, Washington, DC, pp. 1–27.

Weckerle, C.S., Yang, Y., Huber, F.K., Li, Q., 2010. People, money, and protected areas: the collection of the caterpillar mushroom *Ophiocordyceps sinensis* in the Baima Xueshan Nature Reserve, Southwest China. Biodivers. Conserv. 19, 2685–2698.

Wedin, D.A., Tieszen, L.L., Dewey, B., Pastor, J., 1995. Carbon isotope dynamics during grass decomposition and soil organic matter formation. Ecology 76, 1383–1392.

Wegener, A.L., 1924. Entstehung der Kontinente und Ozeane. English Trans, third ed. Methuen, London.

Wehner, R., 2003. Desert ant navigation: how miniature brains solve complex tasks. J. Comp. Physiol. A 189, 579–588.

Wehner, R., Gallizzi, K., Frei, C., Vesely, M., 2002. Calibration processes in desert ant navigation: vector courses and systematic search. J. Comp. Physiol. A 188, 683–693.

Wei, C.A., Rafalko, S.L., Dyer, F.C., 2002. Deciding to learn: modulation of learning flight in honeybees, *Apis mellifera*. J. Comp. Physiol. A 188, 725–737.

Weinstock, G.M., Robinson, G.E., Members of the Honeybee Genome Sequencing Consortium, 2006. Insights into social insects from the genome of the honeybee *Apis mellifera*. Nature 443, 931–949.

Weisser, W.W., Siemann, E. (Eds.), 2004. Insects and Ecosystem Function. Springer-Verlag, Berlin.

Welker, J.M., Menke, J.W., 1990. The influence of simulated browsing on tissue water relations, growth and survival of *Quercus douglasii* (Hook and Arn.) seedlings under slow and rapid rates of soil drought. Funct. Ecol. 4, 807–817.

Wellington, W.G., 1980. Dispersal and population change. In: Berryman, A.A., Safranyik, L. (Eds.), Dispersal of Forest Insects: Evaluation, Theory and Management Implications. Proceedings of International Union of Forest Research Organizations Conference . Washington State University Cooperative Extension Service, Pullman, WA, pp. 11–24.

Wellington, W.G., Cameron, P.J., Thompson, W.A., Vertinsky, I.B., Landsberg, A.S., 1975. A stochastic model for assessing the effects of external and internal heterogeneity of an insect population. Res. Popul. Ecol. 17, 1–28.

Wells, J.D., Greenberg, B., 1994. Effect of the red imported fire ant (Hymenoptera: Formicidae) and carcass type on the daily occurrence of postfeeding carrion-fly larvae (Diptera: Calliphoridae, Sarcophagidae). J. Med. Entomol. 31, 171–174.

Weslien, J., Djupström, L.B., Schroeder, M., Widenfalk, O., 2011. Long-term priority effects among insects and fungi colonizing decaying wood. J. Anim. Ecol. 80, 1155–1162.

West, D.C., Shugart, H.H., Botkin, D.B. (Eds.), 1981. Forest Succession: Concepts and Application. Springer-Verlag, New York.

Westoby, M., French, K., Hughes, L., Rice, B., Rodgerson, L., 1991. Why do more plant species use ants for dispersal on infertile compared with fertile soils? Aust. J. Ecol. 16, 445–455.

Weygoldt, P., 1969. The Biology of Pseudoscorpions. Harvard University Press, Cambridge, MA.

Wheeler, G.S., Tokoro, M., Scheffrahn, R.H., Su, N.Y., 1996. Comparative respiration and methane production rates in Nearctic termites. J. Insect Physiol. 42, 799–806.

Wheelwright, N.T., Orians, G.H., 1982. Seed dispersal by animals: contrasts with pollen dispersal, problems of terminology, and constraints on coevolution. Am. Nat. 119, 402–413.

Whiles, M.R., Wallace, J.B., 1997. Leaf litter decomposition and macroinvertebrate communities in headwater streams draining pine and hardwood catchments. Hydrobiologia 353, 107–119.

Whiles, M.R., Wallace, J.B., Chung, K., 1993. The influence of *Lepidostoma* (Trichoptera: Lepidostomatidae) on recovery of leaf-litter processing in disturbed headwater streams. Am. Midl. Nat. 130, 356–363.

Whiles, M.R., Callaham, Jr., M.A., Meyer, C.K., Brock, B.L., Charlton, R.E., 2001. Emergence of periodical cicadas (*Magicicada cassini*) from a Kansas riparian forest: densities, biomass and nitrogen flux. Am. Midl. Nat. 145, 176–187.

Whitaker, I.S., Twine, C., Whitaker, M.J., Welck, M., Brown, C.S., Shandall, A., 2007. Larval therapy from antiquity to the present day: mechanisms of action, clinical applications and future potential. Postgrad. Med. J. 83, 409–413.

White, T.C.R., 1969. An index to measure weather-induced stress of trees associated with outbreaks of psyllids in Australia. Ecology 50, 905–909.

White, T.C.R., 1976. Weather, food and plagues of locusts. Oecologia 22, 119–134.

White, R.E., 1983. Peterson Field Guide to the Beetles of North America. Houghton Mifflin, Boston, MA.

White, T.C.R., 1984. The abundance of invertebrate herbivores in relation to the availability of nitrogen in stressed food plants. Oecologia 63, 90–105.

White, P.S., 1988. Prickle distribution in *Aralia spinosa* (Araliaceae). Am. J. Bot. 75, 282–285.

White, T.C.R., 2009. Plant vigour versus plant stress: a false dichotomy. Oikos 118, 807–808.

White, P.S., Pickett, S.T.A., 1985. Natural disturbance and patch dynamics: an introduction. In: Pickett, S.T.A., White, P.S. (Eds.), Ecology of Natural Disturbance and Patch Dynamics. Academic Press, New York, pp. 3–13.

White, G.M., Boshier, D.H., Powell, W., 2002. Increased pollen flow counteracts fragmentation in a tropical dry forest: an example from *Swietenia humilis* Zuccarini. Proc. Natl. Acad. Sci. USA 99, 2038–2042.

Whitehouse, N.J., Langdon, P.G., Bustin, R., Galsworthy, S., 2008. Fossil insects and ecosystem dynamics in wetlands: implications for biodiversity and conservation. Biodivers. Conserv. 17, 2055–2078.

Whitford, W.G., 1978. Foraging by seed-harvesting ants. In: Brian, M.V. (Ed.), Production Ecology of Ants and Termites. Cambridge University Press, Cambridge, UK., pp. 107–110.

Whitford, W.G., 1986. Decomposition and nutrient cycling in deserts. In: Whitford, W.G. (Ed.), Pattern and Process in Desert Ecosystems. University of New Mexico Press, Albuquerque, NM, pp. 93–117.

Whitford, W.G., 1992. Effects of climate change on soil biotic communities and soil processes. In: Peters, R.L., Lovejoy, T.E. (Eds.), Global Warming and Biological Diversity. Yale University Press, New Haven, CT, pp. 124–136.

Whitford, W.G., 2000. Arthropods as keystone webmasters in desert ecosystems. In: Coleman, D.C., Hendrix, P. (Eds.), Invertebrates as Webmasters in Ecosystems. CAB International, Wallingford, UK, pp. 99–114.

Whitford, W.G., 2002. Ecology of Desert Systems. Elsevier/Academic Press, San Diego, CA.

Whitford, W.G., Eldridge, D.J., 2013. Effects of ants and termites on soil and geomorphological processes. In: Schroder, J.F. (Ed.), A Treatise on Geomorphology 14. Methods in Geomorphology. Elsevier/Academic Press, San Diego, CA, pp. 281–292.

Whitford, W.G., Jackson, E., 2007. Seed harvester ants (*Pogonomyrmex rugosus*) as "pulse" predators. J. Arid. Environ. 70, 549–552.

Whitford, W.G., Johnson, P., Ramirez, J., 1976. Comparative ecology of the harvester ants *Pogonomyrmex barbatus* (F. Smith) and *Pogonomyrmex rugosus* (Emery). Insect. Soc. 23, 117–132.

Whitford, W.G., Meentemeyer, V., Seastedt, T.R., Cromack, Jr., K., Crossley, Jr., D.A., Santos, P., Todd, R.L., Waide, J.B., 1981. Exceptions to the AET model: deserts and clear-cut forest. Ecology 62, 275–277.

Whitford, W.G., Steinberger, Y., Ettershank, G., 1982. Contributions of subterranean termites to the "economy" of Chihuahuan Desert ecosystems. Oecologia 55, 298–302.

Whitham, T.G., 1983. Host manipulation of parasites: within-plant variation as a defense against rapidly evolving pests. In: Denno, R.F., McClure, M.S. (Eds.), Variable Plants and Herbivores in Natural and Managed Systems. Academic Press, New York, pp. 15–41.

Whitham, T.G., Young, W.P., Martinsen, G.D., Gehring, C.A., Schweitzer, J.A., Shuster, S.M., Wimp, G.M., Fischer, D.G., Bailey, J.K., Lindroth, R.L., Woolbright, S., Kuske, C.R., 2003. Community and ecosystem genetics: a consequence of the extended phenotype. Ecology 84, 559–573.

Whitham, T.G., Bailey, J.K., Schweitzer, J.A., Shuster, S.M., Bangert, R.K., LeRoy, C.J., Lonsdorf, E.V., Allan, G.J., DiFazio, S.P., Potts, B.M., Fischer, D.G., Gehring, C.A., Lindroth, R.L., Marks, J.C., Hart, S.C., Wimp, G.M., Wooley, S.C., 2006. A framework for community and ecosystem genetics: from genes to ecosystems. Nat. Rev. Genet. 7, 510–523.

Whitmee, S., Haines, A., Beyrer, C., Boltz, F., Capon, A.G., de Souza Dias, B.F., Ezeh, A., Frumkin, H., Gong, P., Head, P., Horton, R., Mace, G.M., Marten, R., Myers, S.S., Nishtar, S. Osofsky, S.A.,Pattanayak, S.K., Pongsiri, M.J., Romanelli, C., Soucat, A., Vega, J., Yach, D., 2015. Safeguarding Human Health in the Anthropocene Epoch: Report of The Rockefeller Foundation–Lancet Commission on Planetary Health. Lancet doi.org/10.1016/S0140-6736(15)60901-1.

Whittaker, R.H., 1953. A consideration of climax theory: the climax as a population and pattern. Ecol. Monogr. 23, 41–78.

Whittaker, R.H., 1970. Communities and Ecosystems. Macmillan, London.

Whittaker, R.H., Feeny, P.P., 1971. Allelochemics: chemical interactions between species. Science 171, 757–770.

Wickler, W., 1968. Mimicry in Plants and Animals. R.D. Martin (Trans.). Weidenfeld and Nicolson, Ltd., London.

Wickman, B.E., 1964. Attack habits of *Melanophila consputa* on fire-killed pines. Pan-Pac. Entomol. 40, 183–186.

Wickman, B.E., 1980. Increased growth of white fir after a Douglas-fir tussock moth outbreak. J. Forest. 78, 31–33.

Wickman, B.E., 1992. Forest Health in the Blue Mountains: the Influence of Insects and Diseases. USDA Forest Serv. Gen. Tech. Rpt. PNW-GTR-295. USDA Forest Serv., Pacific Northwest Res. Stn., Portland, OR.

Wiegert, R.G., 1964. Population energetics of meadow spittlebugs (*Philaenus spumarius* L.) as affected by migration and habitat. Ecol. Monogr. 34, 217–241.

Wiegert, R.G., 1968. Thermodynamic considerations in animal nutrition. Am. Zool. 8, 71–81.

Wiegert, R.G., Evans, F.C., 1967. Investigations of secondary productivity in grasslands. In: Petrusewicz, K. (Ed.), Secondary Productivity of Terrestrial Ecosystems: Principles and Methods. Panstwowe Wydawnictwo Naukowe, Warszawa, Poland, pp. 499–518.

Wiegert, R.G., Petersen, C.E., 1983. Energy transfer in insects. Annu. Rev. Entomol. 28, 455–486.

Wiens, J.J., Graham, C.H., Moen, D.S., Smith, S.A., Reeder, T.W., 2006. Evolutionary and ecological causes of the latitudinal diversity gradient in hylid frogs: treefrog trees unearth the roots of high tropical diversity. Am. Nat. 168, 579–596.

Wier, A., Dolan, M., Grimaldi, D., Guerrero, R., Wagensberg, J., Margulis, L., 2002. Spirochete and protist symbionts of a termite (*Mastotermes electrodominicus*) in Miocene amber. Proc. Natl. Acad. Sci. USA 99, 1410–1413.

Wiescher, P.T., Pearce-Duvet, J.M.C., Feener, D.H., 2012. Assembling an ant community: species functional traits reflect environmental filtering. Oecologia 169, 1063–1074.

Wiggins, G.B., MacKay, R.J., Smith, I.M., 1980. Evolutionary and ecological strategies of animals in annual temporary pools. Arch. Hydrobiol. Supplement 58, 97–206.

Wikars, L.-O., Schimmel, J., 2001. Immediate effects of fire-severity on soil invertebrates in cut and uncut pine forests. Forest Ecol. Manag. 141, 189–200.

Wiklund, C., Friberg, M., 2009. The evolutionary ecology of generalization: among-year variation in host plant use and offspring survival in a butterfly. Ecology 90, 3406–3417.

Wilby, A., Orwin, K.H., 2013. Herbivore species richness, composition and community structure mediate predator richness effects and top-down control of herbivore biomass. Oecologia 172, 1167–1177.

Wilf, P., Labandeira, C.C., 1999. Response of plant-insect associations to Paleocene-Eocene warming. Science 284, 2153–2156.

Wilf, P., Labandeira, C.C., Johnson, K.R., Coley, P.D., Cutter, A.D., 2001. Insect herbivory, plant defense, and early Cenozoic climate change. Proc. Natl. Acad. Sci. USA 98, 6221–6226.

Willers, J.L., Jenkins, J.N., Ladner, W.L., Gerard, P.D., Boykin, D.L., Hood, K.B., McKibben, P.L., Samson, S.A., Bethel, M.M., 2005. Site specific approaches to cotton insect control. Sampling and remote sensing analysis techniques. Precis. Agric. 6, 431–452.

Williams, N.M., Kremen, C., 2007. Resource distributions among habitats determine solitary bee offspring production in a mosaic landscape. Ecol. Appl. 17, 910–921.

Williams, D.W., Liebhold, A.M., 1995. Forest defoliators and climatic change: potential changes in spatial distribution of outbreaks of western spruce budworm (Lepidoptera: Tortricidae) and gypsy moth (Lepidoptera: Lymantriidae). Environ. Entomol. 24, 1–9.

Williams, D.W., Liebhold, A.M., 2002. Climate change and the outbreak ranges of two North American bark beetles. Agr. Forest Entomol. 4, 87–99.

Williams, K.S., Simon, C., 1995. The ecology, behavior, and evolution of periodical cicadas. Annu. Rev. Entomol. 40, 269–295.

Williams, D.D., Nesterovitch, A.I., Tavares, A.F., Muzzatti, E.G., 2001. Morphological deformities occurring in Belarusian chironomids (Diptera: Chironomidae) subsequent to the Chernobyl nuclear disaster. Freshwater Biol. 46, 503–512.

Williams, J.L., Ellers-Kirk, C., Orth, R.G., Gassmann, A.J., Head, G., Tabashnik, B.E., Carriére, Y., 2011. Fitness cost of resistance to Bt cotton linked with increased gossypol content in pink bollworm larvae. PLoS ONE 6 (6), e21863.

Williams, A.P., Schwartz, R.E., Iacobellis, S., Seager, R., Cook, B.I., Still, C.J., Husak, D., Michaelsen, J., 2015. Urbanization causes increased cloud base height and decreased fog in coastal Southern California, Geophysical. Res. Lett. 42, 1527–1536.

Williamson, M., 1972. The Analysis of Biological Populations. Edward Arnold, London, pp. 180.

Williamson, S.C., Detling, J.K., Dodd, J.L., Dyer, M.I., 1989. Experimental evaluation of the grazing optimization hypothesis. J. Range Manage. 42, 149–152.

Willig, M.R., Camilo, G.R., 1991. The effect of Hurricane Hugo on six invertebrate species in the Luquillo Experimental Forest of Puerto Rico. Biotropica 23, 455–461.

Willig, M.R., Lyons, S.K., 1998. An analytical model of latitudinal gradient in species richness with an empirical test for marsupials and bats in the New World. Oikos 81, 93–98.

Willig, M.R., McGinley, M.A., 1999. Animal responses to natural disturbance and roles as patch generating phenomena. In: Walker, L.R. (Ed.), Ecosystems of the World: Ecosystems of Disturbed Ground. Elsevier Science, Amsterdam, The Netherlands, pp. 667–689.

Willig, M.R., Walker, L.R., 1999. Disturbance in terrestrial ecosystems: salient themes, synthesis, and future directions. In: Walker, L.R. (Ed.), Ecosystems of the World: Ecosystems of Disturbed Ground. Elsevier Science, Amsterdam, The Netherlands, pp. 747–767.

Willig, M.R., Presley, S.J., Bloch, C.P., 2011. Long-term dynamics of topical walking sticks in response to multiple large-scale and intense disturbances. Oecologia 165, 357–368.

Willmer, P.G., Hughes, J.P., Woodford, J.A.T., Gordon, S.C., 1996. The effects of crop microclimate and associated physiological contraints on the seasonal and diurnal distribution patterns of raspberry beetle (*Byturus tomentosus*) on the host plant *Rubus idaeus*. Ecol. Entomol. 21, 87–97.

Wilmers, C.C., Post, E., Peterson, R.O., Vucetich, J.A., 2006. Predator disease out-break modulates top-down, bottom-up and climatic effects on herbivore population dynamics. Ecol. Lett. 9, 383–389.

Wilson, E.O., 1969. The species equilibrium. In: Woodwell, G.M., Smith, H.H. (Eds.), Diversity and Stability in Ecological Systems. Brookhaven Symposium in Biology 22. Brookhaven National Laboratory, Upton, NY, pp. 38–47.

Wilson, E.O., 1973. Group selection and its significance for ecology. BioScience 23, 631–638.

Wilson, E.O., 1975. Sociobiology: the New Synthesis. Belknap Press of Harvard University Press, Cambridge, MA, pp. 697.

Wilson, D.S., 1976. Evolution on the level of communities. Science 192, 1358–1360.

Wilson, E.O., 1986. The organization of flood evacuation in the ant genus *Pheidole* (Hymenoptera: Formicidae). Insect. Soc. 33, 458–469.

Wilson, E.O., 1992. The Diversity of Life. Harvard University Press, Cambridge, MA.

Wilson, D.S., 1997. Biological communities as functionally organized units. Ecology 78, 2018–2024.

Wilson, E.O., Eisner, T., 1957. Quantitative studies of liquid food transmission in ants. Insect. Soc. 4, 157–166.

Wilson, D., Faeth, S.H., 2001. Do fungal endophytes result in selection for leafminer ovipositional preference? Ecology 82, 1097–1111.

Wilson, E.O., Simberloff, D.S., 1969. Experimental zoogeography of islands: defaunation and monitoring techniques. Ecology 50, 267–278.

Wilson, M.V., Hammond, P.C., Schultz, C.B., 1997. The interdependence of native plants and Fender's blue butterfly. In: Kaye, T.N., Liston, A., Love, R.M., Luoma, D.L., Meinke, R.J., Wilson, M.V. (Eds.), Conservation and Management of Native Plants and Fungi. Native Plant Society of Oregon, Corvallis, OR, pp. 83–87.

Wimp, G.M., Wooley, S., Bangert, R.K., Young, W.P., Martinsen, G.D., Keim, P., Rehill, B., Lindroth, R.L., Whitham, T.G., 2007. Plant genetics predicts intra-annual variation in phytochemistry and arthropod community structure. Mol. Ecol. 16, 5057–5069.

Winchester, N.N., 1997. Canopy arthropods of coastal Sitka spruce trees on Vancouver Island, British Columbia, Canada. In: Stork, N.E., Adis, J., Didham, R.K. (Eds.), Canopy Arthropods. Chapman & Hall, London, pp. 151–168.

Winchester, N.N., Behan-Pelletier, V., Ring, R.A., 1999. Arboreal specificity, diversity and abundance of canopy-dwelling oribatid mites (Acari: Oribatida). Pedobiologia 43, 391–400.

Windsor, D.M., 1990. Climate and Moisture Variability in a Tropical Forest: Long-term Records from Barro Colorado Island, Panamá. Smithsonian Institution Press, Washington, DC., pp. 145.

Winfree, R., Williams, N.M., Dushoff, J., Kremen, C., 2007. Native bees provide insurance against ongoing honey bee losses. Ecol. Lett. 10, 1105–1113.

Winfree, R., Aguilar, R., Vázquez, D.P., LeBuhn, G., Aizen, M.A., 2009. A meta-analysis of bees' responses to anthropogenic disturbance. Ecology 90, 2068–2076.

Winfree, R., Bartomeus, I., Cariveau, D.P., 2011. Native pollinators in anthropogenic habitats. Annu. Rev. Ecol. Evol. S. 42, 1–22.

Wint, G.R.W., 1983. Leaf damage in tropical rain forest canopies. In: Sutton, S.L., Whitmore, T.C., Chadwick, A.C. (Eds.), Tropical Rain Forest: Ecology and Management. Blackwell Scientific, Oxford, UK, pp. 229–240.

Winter, K., Smith, J.A.C., 1996. An introduction to crassulacean acid metabolism: biochemical principles and ecological diversity. In: Winter, K., Smith, J.A.C. (Eds.), Crassulacean Acid Metabolism: Biochemistry, Ecophysiology, and Evolution. Springer-Verlag, New York, pp. 1–13.

Winter, C., Lehmann, S., Diekmann, M., 2008. Determinants of reproductive success: a comparative study of five endangered river corridor plants in fragmented habitats. Biol. Conserv. 141, 1095–1104.

Winter, M.-B., Ammer, C., Baier, R., Donato, D.C., Seibold, S., Müller, J., 2015. Multi-taxon alpha diversity following bark beetle disturbance: evaluating multi-decade persistence of a diverse early-seral phase. Forest Ecol. Manag. 338, 32–45.

Wipfli, M.S., 1997. Terrestrial invertebrates as salmonid prey and nitrogen sources in streams: contrasting old-growth and young-growth riparian forests in southeastern Alaska, U.S.A. Can. J. Fish. Aquat. Sci. 54, 1259–1269.

Wipfli, M.S., Musslewhite, J., 2004. Density of red alder (*Alnus rubra*) in headwaters influences invertebrate and detritus subsidies to downstream fish habitats in Alaska. Hydrobiologia 520, 153–163.

Wipfli, M.S., Richardson, J.S., Naiman, R.J., 2007. Ecological linkages between headwaters and downstream ecosystems: transport of organic matter, invertebrates, and wood down headwater channels. J. Am. Water Resour. As. 43, 72–85.

Wisdom, C.S., Crawford, C.S., Aldon, E.F., 1989. Influence of insect herbivory on photosynthetic area and reproduction in *Gutierrezia* species. J. Ecol. 77, 685–692.

Wise, M.J., 2009. Competition among herbivores of *Solanum carolinense* as a constraint on the evolution of host-plant resistance. Evol. Ecol. 23, 347–361.

Wise, D.H., Schaefer, M., 1994. Decomposition of leaf litter in a mull beech forest: comparison between canopy and herbaceous species. Pedobiologia 38, 269–288.

Wise, D.H., Moldenhauer, D.M., Halaj, J., 2006. Using stable isotopes to reveal shifts in prey consumption by generalist predators. Ecol. Appl. 16, 865–876.

Wissinger, S.A., McGrady, J., 1993. Intra-guild predation and competition between larval dragonflies: direct and indirect effects of shared prey. Ecology 74, 207–218.

Witcosky, J.J., Schowalter, T.D., Hansen, E.M., 1986. The influence of time of precommercial thinning on the colonization of Douglas-fir by three species of root-colonizing insects. Can. J. Forest. Res. 16, 745–749.

Witkamp, M., 1971. Soils as components of ecosystems. Annu. Rev. Ecol. Syst. 2, 85–110.

Witkamp, M., Olson, J.S., 1963. Breakdown of confined and unconfined oak litter. Oikos 14, 138–147.

Witkamp, M., Crossley, Jr., D.A., 1966. The role of microarthropods and microflora in the breakdown of white oak litter. Pedobiologia 6, 293–303.

Witmer, M.C., Cheke, A.S., 1991. The dodo and the tambalocoque tree: an obligate mutualism reconsidered. Oikos 61, 133–137.

Wold, E.N., Marquis, R.J., 1997. Induced defense in white oak: effects on herbivores and consequences for the plant. Ecology 78, 1356–1369.

Wolkovich, E.M., Allesina, S., Cottingham, K.L., Moore, J.C., Sandin, S.A., de Mazancourt, C., 2014. Linking the green and brown worlds: the prevalence and effect of multichannel feeding in food webs. Ecology 95, 3376–3386.

Wood, T.K., 1976. Alarm behavior of brooding female *Umbonia crassicornis* (Homoptera: Membracidae). Ann. Entomol. Soc. Am. 69, 340–344.

Wood, D.M., Andersen, M.C., 1990. The effect of predispersal seed predators on colonization of *Aster ledophyllus* on Mount St. Helens, Washington. Am. Midl. Nat. 123, 193–201.

Wood, E.M., Yasutake, W.T., 1956. Tissue damage in salmonids caused by Halisidota argentata Packard. J. Parasitol. 42, 544–546.

Wood, T.E., Lawrence, D., Clark, D.A., Chazdon, R.L., 2009. Rain forest nutrient cycling and productivity in response to large-scale litter manipulation. Ecology 90, 109–121.

Woodcock, B.A., Edwards, M., Redhead, J., Meek, W.R., Nuttall, P., Falk, S., Nowakowski, M., Pywell, R.F., 2013. Crop flower visitation by honeybees, bumblebees and solitary bees: behavioural differences and diversity responses to landscape. Agr. Ecosyst. Environ. 171, 1–8.

Woods, P.V., Raison, R.J., 1982. An appraisal of techniques for the study of litter decomposition in eucalypt forests. Aust. J. Ecol. 7, 215–225.

Woods, P.V., Raison, R.J., 1983. Decomposition of litter in sub-alpine forests of *Eucalyptus delegatensis*, *E. pauciflora* and *E. dives*. Aust. J. Ecol. 8, 287–299.

Woodwell, F.I., 1993. How many species are required for a functional ecosystem? In: Schulze, E.D., Mooney, H.A. (Eds.), Biodiversity and Ecosystem Function. Springer-Verlag, Berlin, pp. 271–291.

Woolhouse, H.W., 1981. Aspects of the carbon and energy requirements of photosynthesis considered in relation to environmental constraints. In: Townsend, C.R., Calow, P. (Eds.), Physiological Ecology: an Evolutionary Approach to Resource Use. Blackwell Scientific, Oxford, UK, pp. 51–85.

Wotton, R.S., Malmqvist, B., Muotka, T., Larsson, K., 1998. Fecal pellets from a dense aggregation of suspension-feeders in a stream: an example of ecosystem engineering. Limnol. Oceanogr. 43, 719–725.

Wragg, P.D., Johnson, S.D., 2011. Transition from wind pollination to insect pollination in sedges: experimental evidence and functional traits. New Phytol. 191, 1128–1140.

Wright, L.C., Berryman, A.A., Wickman, B.E., 1986. Abundance of the fir engraver, *Scolytus ventralis*, and the Douglas-fir beetle, *Dendroctonus pseudotsugae*, following tree defoliation by the Douglas-fir tussock moth, *Orgyia pseudotsugata*. Can. Entomol. 116, 293–305.

Wright, S., Hall, R.W., Peacock, J.W., 1989. Effect of simulated insect damage on growth and survival of northern red oak (*Quercus rubra* L.) seedlings. Environ. Entomol. 18, 235–239.

Wynne-Edwards, V.C., 1963. Intergroup selection in the evolution of social systems. Nature 200, 623–626.

Wynne-Edwards, V.C., 1965. Self-regulating systems in populations of animals. Science 147, 1543–1548.

Wystrach, A., Beugnon, G., 2009. Ants learn geometry and features. Curr. Biol. 19, 61–66.

Xiao, J., Zhuang, Q., Baldocchi, D.D., Law, B.E., Richardson, A.D., Chen, J., Oren, R., Starr, G., Noormets, A., Ma, S., Verma, S.B., Wharton, S., Wofsy, S.C., Bolstad, P.V., Burns, S.P., Cook, D.R., Curtis, P.S., Drake, B.G., Falk, M., Fischer, M.L., Foster, D.R., Gu, L., Hadley, J.L., Hollinger, D.Y., Katul, G.G., Litvak, M., Martin, T.A., Matamala, R., McNulty, S., Meyers, T.P., Monson, R.K., Munger, J.W., Oechel, W.C., Paw U, K.T., Schmid, H.P., Scott, R.L., Sun, G., Suyker, A.E., Torn, M.S., 2008. Estimation of net ecosystem carbon exchange for the coterminous United States by combining MODIS and AmeriFlux data. Agr. Forest Meteorol. 148, 1827–1847.

Xu, Z.-B., Zou, X.-P., Zhang, N., Feng, Q.-L., Zheng, S.-C., 2015. Detoxification of insecticides, allechemicals and heavy metals by glutathione S-transferase SlGSTE1 in the gut of *Spodoptera litura*. Insect Sci. 22, 503–511.

Xue, Y., Liou, K.N., Kashahara, A., 1990. Investigation of biophysical feedback on the African climate using a two-dimensional model. J. Climate 3, 337–352.

Yachi, S., Loreau, M., 1999. Biodiversity and ecosystem productivity in a fluctuating environment: The insurance hypothesis. Proc. Natl. Acad. Sci. USA 96, 1463–1468.

Yachi, S., Loreau, M., 2007. Does complementary resource use enhance ecosystem functioning? A model of light competition in plant communities. Ecol. Lett. 10, 54–62.

Yamada, D., Imura, O., Shi, K., Shibuya, T., 2007. Effect of tunneler dung beetles on cattle dung decomposition, soil nutrients and herbage growth. Grassland Sci. 53, 121–129.

Yamanaka, T., Liebhold, A.M., 2009. Spatially implicit approaches to understand the manipulation of mating success for insect invasion management. Popul. Ecol. 51, 427–444.

Yang, L.H., 2004. Periodical cicadas as resource pulses in North American forests. Science 306, 1565–1567.

Yang, G.-H., 2005. Harm of introducing the western honeybee Apis mellifera L. to the Chinese honeybee Apis cerana F. and its ecological impact. Acta Entomol. Sinica 48, 401–406, (in Chinese with English summary).

Yang, L.H., 2008. Pulses of dead periodical cicadas increase herbivory of American bellflowers. Ecology 89, 1497–1502.

Yang, L.H., 2013. Resource pulses of dead periodical cicadas increase the growth of American bellflower rosettes under competitive and non-competitive conditions. Arthropod-Plant Inte. 7, 93–98.

Yang, S., Bishop, J.G., Webster, M.S., 2008. Colonization genetics of an animal-dispersed plant (*Vaccinium membranaceum*) at Mount St. Helens, Washington. Mol. Ecol. 17, 731–740.

Yang, C.-C., Yu, Y.-C., Valles, S.M., Oi, D.H., Chen, Y.-C., Shoemaker, D., Wu, W.-J., Shih, C.-J., 2010. Loss of microbial (pathogen) infections associated with recent invasions of the red imported fire ant *Solenopsis invicta*. Biol. Invasions 12, 3307–3318.

Yang, M.-X., Wang, Z.-W., Li, H., Zhang, Z.-Y., Tan, K., Radloff, S.E., Hepburn, H.R., 2010. Thermoregulation in mixed-species colonies of honeybees (*Apis cerana and Apis mellifera*). J. Insect Physiol. 56, 706–709.

Yao, F.-L., You, M.-S., Vasseur, L., Yang, G., Zheng, Y.-K., 2012. Polycultural manipulation for better regulation of planthopper populations in irrigated rice-based ecosystems. Crop Prot. 34, 104–111.

Yee, D.A., Willig, M.R., 2007. Colonisation of *Heliconia caribaea* by aquatic invertebrates: resource and microsite characteristics. Ecol. Entomol. 32, 603–612.

Yen, A.L., 2009. Entomophagy and insect conservation: some thoughts for digestion. J. Insect Conserv. 13, 667–670.

Yoder, J.A., Theriot, G.C., Rivers, D.B., 1996. Venom from *Nasonia vitripennis* alters water loss from the flesh fly, *Sarcophaga bullata*. Entomol. Exp. Appl. 81, 235–238.

Yodzis, P., 1980. The connectance of real ecosystems. Nature 284, 544–545.

Yokoyama, K., Kai, H., Tsuchiyama, H., 1991. Paracoprid dung beetles and gaseous loss of nitrogen from cow dung. Soil Biol. Biochem. 23, 643–647.

Yorks, T.E., Leopold, D.J., Raynal, D.J., 2003. Effects of *Tsuga canadensis* mortality on soil water chemistry and understory vegetation: possible consequences of an invasive insect herbivore. Can. J. Forest. Res. 33, 1525–1537.

Yoshimura, M., Akama, A., 2014. Radioactive contamination of aquatic insects in a stream impacted by the Fukushima nuclear power plant accident. Hydrobiologia 722, 19–30.

Yu, G., Shen, H., Liu, J., 2009. Impacts of climate change on historical locust outbreaks in China. J. Geophys. Res. 114, D18104.

Yuan, W., Liu, S., Zhou, G., Zhou, G., Tieszen, L.L., Baldocchi, D., Bernhofer, C., Gholz, H., Goldstein, A.H., Goulden, M.L., Hollinger, D.Y., Hu, Y., Law, B.E., Stoy, P.C., Vesala, T., Wofsy, S.C., 2007. Deriving a light use efficiency model from eddy covariance flux data for predicting daily gross primary production across biomes. Agr. Forest Meteorol. 143, 189–207.

Zabel, J., Tscharntke, T., 1998. Does fragmentation of *Urtica* habitats affect phytophagous and predatory insects differentially? Oecologia 116, 419–425.

Zaidi, R.H., Jaal, Z., Hawkes, N.J., Hemingway, J., Symondson, W.O.C., 1999. Can multiple-copy sequences of prey DNA be detected amongst the gut contents of invertebrate predators? Mol. Ecol. 8, 2081–2087.

Zalucki, M.P., Brower, L.P., Alonso-M., A., 2001. Detrimental effects of latex and cardiac glycosides on survival and growth of first-instar monarch butterfly larvae *Danaus plexippus* feeding on the sandhill milkweed *Asclepias humistrata*. Ecol. Entomol. 26, 212–224.

Zangerl, A.R., McKenna, D., Wraight, C.L., Carroll, M., Ficarello, P., Warner, R., Berenbaum, M.R., 2001. Effects of exposure to event 176 *Bacillus thuringiensis* corn pollen on monarch and black swallowtail caterpillars under field conditions. Proc. Natl. Acad. Sci. USA 98, 11908–11912.

Zaragoza, S.R., Whitford, W.G., Steinberger, Y., 2007. Effects of temporally persistent ant nests on soil protozoan communities and the abundance of morphological types of amoeba. Appl. Soil Ecol. 37, 81–87.

Zavala, J.A., Casteel, C.L., DeLucia, E.H., Berenbaum, M.R., 2008. Anthropogenic increase in carbon dioxide compromises plant defense against invasive insects. Proc. Natl. Acad. Sci. USA 105, 5129–5133.

Zavala, J.A., Nabity, P.D., DeLucia, E.H., 2013. An emerging understanding of mechanisms governing insect herbivory under elevated CO_2. Annu. Rev. Entomol. 58, 79–97.

Zehnder, C.B., Hunter, M.D., 2008. Effects of nitrogen deposition on the interaction between an aphid and its host plant. Ecol. Entomol. 33, 24–30.

Zehnder, C.B., Hunter, M.D., 2009. More is not necessarily better: the impact of limiting and excessive nutrients on herbivore population growth rates. Ecol. Entomol. 34, 535–543.

Zeng, R.S., Wen, Z., Niu, G., Schuler, M.A., Berenbaum, M.R., 2009. Enhanced toxicity and induction of cytochrome P450s suggest a cost of "eavesdropping" in a multitrophic interaction. J. Chem. Ecol. 35, 526–532.

Zenk, M.H., Juenger, M., 2007. Evolution and current status of the phytochemistry of nitrogenous compounds. Phytochemistry 68, 2757–2772.

Zera, A.J., Denno, R.F., 1997. Physiology and ecology of dispersal polymorphism in insects. Annu. Rev. Entomol. 42, 207–230.

Zera, A.J., Zhao, Z., 2006. Intermediary metabolism and life-history trade-offs: differential metabolism of amino acids underlies the dispersal-reproduction trade-off in a wing-polymorphic cricket. Am. Nat. 167, 889–900.

Zeringue, Jr., H.J., 1987. Changes in cotton leaf chemistry induced by volatile elicitors. Phytochemistry 26, 1357–1360.

Zerm, M., Adis, J., 2003. Exceptional anoxia resistance in larval tiger beetle, *Phaeoxantha klugii* (Coleoptera: Cicindelidae). Physiol. Entomol. 28, 150–153.

Zettler, J.A., Taylor, M.D., Allen, C.R., Spira, T.P., 2004. Consequences of forest clear-cuts for native and nonindigenous ants. Ann. Entomol. Soc. Am. 97, 513–518.

Zhang, Q.-H., Schlyter, F., 2004. Olfactory recognition and behavioural avoidance of angiosperm nonhost volatiles by conifer-inhabiting bark beetles. Agr. Forest Entomol. 6, 1–19.

Zhang, D.D., Brecke, P., Lee, H.F., He, Y.Q., Zhang, J., 2007. Global climate change, war and population decline in recent human history. Proc. Natl. Acad. Sci. USA 104, 19214–19219.

Zhang, J., Hu, Y., Xiao, X., Chen, P., Han, S., Song, G., Yu, G., 2009. Satellite-based estimation of evapotranspiration of an old-growth temperate mixed forest. Agr. Forest Meteorol. 149, 976–984.

Zhang, G., Han, X., Elser, J.J., 2011. Rapid top-down regulation of plant C:N:P stoichiometry by grasshoppers in an Inner Mongolia grassland ecosystem. Oecologia 166, 253–264.

Zhang, W., Rudolf, V.H.W., Ma, C.-S., 2015. Stage-specific heat effects: timing and duration of heat waves alter demographic rates of a global insect pest. Oecologia 179, 947–957.

Zheng, X., Eltahir, E.A.B., 1998. The role of vegetation in the dynamics of West African monsoons. J. Climate 11, 2078–2096.

Zhong, H., Schowalter, T.D., 1989. Conifer bole utilization by wood-boring beetles in western Oregon. Can. J. Forest. Res. 19, 943–947.

Zhong, H., Yan, Z., Jones, F., Brock, C., 2003. Ecological analysis of mosquito light trap collections from west central Florida. Environ. Entomol. 32, 807–815.

Zhou, J., Lau, W.K.-M., Masuoka, P.M., Andre, R.G., Chamberlin, J., Lawyer, P., Laughlin, L.W., 2002. El Niño helps spread Bartonellosis epidemics in Peru. EOS Trans. Am. Geophys. Union 83, 160–161.

Zhou, H.-B., Chen, J.-L., Cheng, D.-F., Liu, Y., Sun, J.-R., 2009. Effects of wheat-pea intercropping on the population dynamics of *Sitobion avenae* (Homoptera: Aphididae) and its main natural enemies. Acta Entomol. Sinica 52, 775–782, (in Chinese with English summary).

Zimmerman, P.R., Greenberg, J.P., Wandiga, S.O., Crutzen, P.J., 1982. Termites: a potentially large source of atmospheric methane, carbon dioxide, and molecular hydrogen. Science 218, 563–565.

Zimmerman, J.K., Willig, M.R., Walker, L.R., Silver, W.L., 1996. Introduction: disturbance and Caribbean ecosystems. Biotropica 28, 414–423.

Zimmerman, K., Lockwood, J.A., Latchininsky, A.V., 2004. A spatial, Markovian model of rangeland grasshopper (Orthoptera: Acrididae) population dynamics: do long-term benefits justify suppression of infestations? Environ. Entomol. 33, 257–266.

Zimmerman, J.K., Comita, L.S., Thompson, J., Uriarte, M., Brokaw, N., 2010. Patch dynamics and community metastability of a subtropical forest: compound effects of natural disturbance and human land use. Landscape Ecol. 25, 1099–1111.

Zjhra, M.L., 2008. Facilitating sympatric species coexistence via pollinator partitioning in endemic tropical trees of Madagascar. Plant Syst. Evol. 271, 157–176.

Zlotin, R.I., Khodashova, K.S., 1980. The Role of Animals in Biological Cycling of Forest-steppe Ecosystems. N.R. French (Trans.). Dowden, Hutchinson & Ross, Stroudsburg, PA.

Zorrilla-Miras, P., Palomo, I., Gómez-Baggethun, E., Martín-López, B., Lomas, P.L., Montes, C., 2014. Effects of land-use change on wetland ecosystem services: a case study in the Doñana marshes (SW Spain). Landscape Urban Plan. 122, 160–174.

Żywiec, M., Holeksa, J., Ledwon, M., Seget, P., 2013. Reproductive success of individuals with different fruit production patterns. What does it mean for the predator satiation hypothesis? Oecologia 172, 461–467.

Author Index

Taxonomic Index (Arthropods only)

Subject Index

A

Abies balsamea, 67
Abiotic conditions, 5, 6, 13, 22, 31, 34, 36, 38
Abiotic factors, 5, 31, 221, 342
Above-ground net primary production (ANPP)
  peak foliar standing crop (FSC), relation with, 376
Acclimation, 38
Acetate, 114, 493
Acetic acid, 73
Acetylcholine, 73
Acid precipitation, 12, 190, 199
Acokanthera ouabaio, 66
Adaptation, 1, 13, 22, 43, 90, 150, 187, 231, 575, 598
  rate, 230
Adaptive management, 570
Adaptive physiological responses, 37
Adelphoparasitism, 263
ADP, 387
Aflatoxin B, 68
Aflatoxins, 65, 69
AFLP. *See* Amplified fragment length polymorphism
      (AFLP)
Ageratum houstoniatum, 67
Age structure, 147, 169
Aggregation, 119, 122, 144, 241, 373, 585
Aggression, 153, 154, 516
Aggression tests, 154
Agricultural chemicals, 5
Agricultural intensification, 458, 461, 567, 590
Agricultural landscapes, 225, 590
Agricultural pests, 231
Agriculture, 32, 446
Airborne visible and infrared imaging spectrometry
      (AVIRIS), 367
Airflow, 22, 26, 97, 391, 440
Air pollution, 1
Air temperatures, 37
Alarm, 69, 130
Albedo, 13, 29, 370, 391, 395, 439
Alfalfa, 150
Algae, 36
Algal biomass, 36
Alkaloid-deprived males, 118
Alkaloids, 69, 70, 73, 77, 91, 500
Allee effect, 195

Alleles, 91, 148, 156
Allelopathy, 284
Allocthonous inputs, organic matter, 27
Allylanisole, 100
Alpine climate, 39
Altruism, 513–515
Altruist genes, 515
Altruistic behaviors, 513
Always defect strategy, 515
Alyssum pintodasilvae, 73
Amazonian myrmecophytic canopy tree, 414
Ambient temperature, 38
Amino acids, 63, 69, 88, 94, 382, 387
Ammonia, 109, 371, 386, 403
Ammonifying bacteria, 387
Amplified fragment length polymorphism
      (AFLP), 148
Anadromous species, 291
Anaerobiosis, 186
Anal pedicel, 159
Anatomical modification, 268
ANCOVA model, 136
Anhydrobiosis, 44
Animal activity, importance to successional
      transitions, 344
Animal dung, 17
Antarctic ecosystems, 12
Ant-attracting plants, 607
Anthropogenic activities, 1, 29, 230, 561
  disrupt ecosystem, 537
Anthropogenic alteration, 36
Anthropogenic changes, 12, 22
  in environmental conditions, 10
Anthropogenic chemicals, 150
Anthropogenic conditions, 39
Anthropogenic conversion, 31
Anthropogenic transport, 238
Antibiotic proteins, 37
Antibiotics, 74, 264, 445
Antimicrobial defenses, 74–75
Antioxidant, 89
Ant nest, 498, 501
Ant-plant interaction 206, 209, 215, 220, 246, 249, 261,
      274, 278-9
Ant-removal plots, 255

Printed in the United States
By Bookmasters